UNITEXT for Physics

UNITEXT for Physics series publishes textbooks in physics and astronomy, characterized by a didactic style and comprehensiveness. The books are addressed to upper-undergraduate and graduate students, but also to scientists and researchers as important resources for their education, knowledge, and teaching.

More information about this series at https://link.springer.com/bookseries/13351

Stefan Weinzierl

Feynman Integrals

A Comprehensive Treatment for Students and Researchers

Stefan Weinzierl
Institut für Physik
Universität Mainz
Mainz, Rheinland-Pfalz, Germany

ISSN 2198-7882 ISSN 2198-7890 (electronic)
UNITEXT for Physics
ISBN 978-3-030-99560-7 ISBN 978-3-030-99558-4 (eBook)
https://doi.org/10.1007/978-3-030-99558-4

This Springer imprint is published by the registered company Springer Nature Switzerland AG
The registered company address is: Gewerbestrasse 11, 6330 Cham, Switzerland

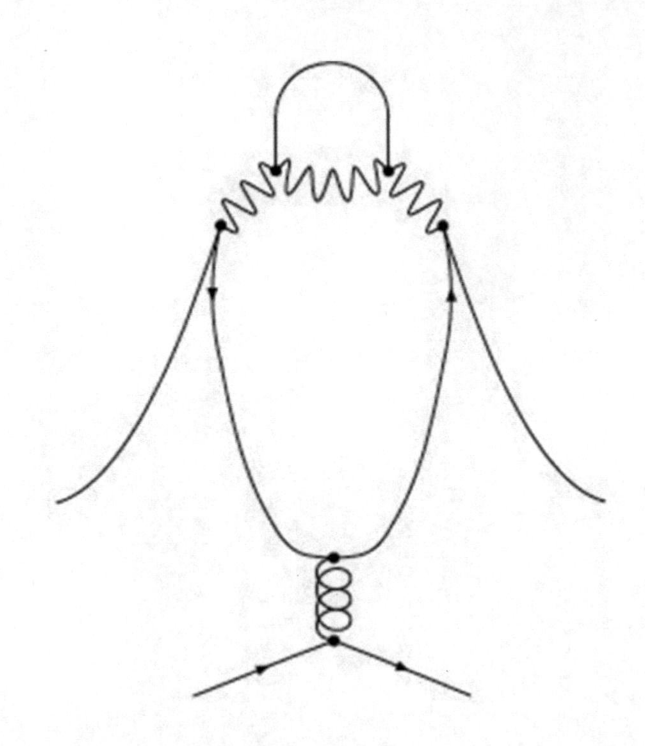

Contents

Symbols

a_j	Feynman parameter
α_j	Schwinger parameter
$\mathcal{B}$	Baikov polynomial
c	Speed of light
D	Dimension of space-time
D_{int}	(integer) expansion point within dimensional regularisation
ε	Dimensional regularisation parameter
$\mathcal{F}$	Second Symanzik polynomial
$\mathcal{G}$	Lee-Pomeransky polynomial
γ_{E}	Euler-Mascheroni constant
Grp	Category of groups
$\hbar$	Reduced Planck constant
Hom	Morphisms of a category
HS	Category of (pure) Hodge structures
$\mathcal{K}$	Kirchhoff polynomial
l	Loop number
MHS	Category of mixed Hodge structures
MixMot	Category of mixed motives
$\mathbf{Mod}_R$	Category of finitely generated R-modules
n	Number of edges
N_B	Number of kinematic variables
N_{cohom}	Dimension of the twisted cohomology group
n_{ext}	Number of external edges
n_{int}	Number of internal edges
N_L	Number of letters
N_{master}	Number of master integrals
N_V	Number of Baikov variables
Obj	Objects of a category
$\mathbf{Proj}_R$	Category of finitely generated projective R-modules
PureMot	Category of pure motives
r	Number of vertices

r_{int}	Number of internal vertices
Set	Nategory of sets
$\mathbf{SmProj}_{\mathbb{Q}}$	Category of smooth projective varieties over $\mathbb{Q}$
$\mathcal{U}$	First Symanzik polynomial
u_j	Lee-Pomeransky variable
$\mathbf{Var}_{\mathbb{Q}}$	Category of algebraic varieties defined over $\mathbb{Q}$
$\mathbf{Vect}_{\mathbb{F}}$	Category of finite-dimensional $\mathbb{F}$-vector spaces
ζ_n	Zeta value
z_j	Baikov variable

Chapter 1
Introduction

Feynman integrals are indispensable for precision calculations in quantum field theory. They occur as soon as one goes beyond the leading order in perturbative quantum field theory. Feynman integrals are also fascinating from a mathematical point of view. They can be used to teach and illustrate a large part of modern mathematics with concrete non-trivial examples.

In recent years there were some exiting developments in the field of Feynman integrals, enlarging significantly our knowledge and understanding of Feynman integrals. Although there are some excellent older books on the subject [1–6], a modern introduction to the theory of Feynman integrals, which includes the recent developments, will be helpful.

This book is intended for two types of readers: First, there is the physicist interested in precision calculations in quantum field theory, where she/he encounters Feynman integrals. Her/his primary motivation is to be able to calculate these integrals. For this audience the book provides current state-of-the-art techniques, covering all aspects of the computations, from the starting definition of a Feynman integral to the final step of getting a number out.

Secondly, there is the mathematician, interested in the mathematical aspects underlying Feynman integrals. These are rich and the book provides wherever possible the connection to mathematics.

Of course, these two topics are not independent but interwoven, which makes the theory of Feynman integrals so enthralling. The book is written in this spirit, showing the deep connection between physics and mathematics in the field of Feynman integrals.

This book is intended for students at the master level in physics or mathematics. I tried to keep the essential requirements to a minimum. As minimum requirements I assume that all readers are familiar with special relativity on the physics side and the theory of complex functions (i.e., Cauchy's residue theorem) and differential

S. Weinzierl, *Feynman Integrals*, UNITEXT for Physics,
https://doi.org/10.1007/978-3-030-99558-4_1

forms on the mathematics side. Students of physics or mathematics with an interest in mathematical physics should have covered these topics during their bachelor studies. Of course, a knowledge of quantum field theory or algebraic geometry is extremely helpful for the topic of this book. However, as most readers might have followed one of these courses, but not both, I arranged the material covered in this book in such a way that no prior knowledge of quantum field theory nor algebraic geometry is assumed. The relevant topics are introduced as they are needed. In this way the book can complement a course in quantum field theory or algebraic geometry. Of course, the book cannot substitute a course in quantum field theory nor a course in algebraic geometry and readers are encouraged to familiarise themselves with quantum field theory and algebraic geometry beyond the topics required and introduced in this book.

There are always some readers, who are impatient: Reading Chaps. 2, 6 and 7 should bring them to the point that they can perform state-of-the-art Feynman integral calculations with the method of differential equations. One might also be lucky that a particular Feynman integral is already known in the literature. The database Loopedia [7] is a good place to check out first.

For all others, who would like to take the recommended long and scenic route, we start in Chap. 2 with introducing the central objects of this book: Feynman integrals. We do this by requiring only a basic knowledge of special relativity, avoiding quantum field theory as a prerequisite. The chapter also introduces the most popular integral representations for Feynman integrals.

The Feynman parameter representation and the Schwinger parameter representation involve two graph polynomials. These graph polynomials have many interesting properties and we discuss them in detail in Chap. 3.

As we deliberately did not build upon quantum field theory in Chap. 2, we should nevertheless discuss how Feynman integrals arise in quantum field theory. This is done in Chap. 4.

In many applications within perturbation theory the next-to-leading order correction requires one-loop integrals. The one-loop integrals are therefore of particular importance. We devote a special chapter to them (Chap. 5).

For all other Feynman integrals the most commonly method to compute these integrals (at the time of writing this book) is the method of differential equations. We introduce this technique in Chap. 6. The most important result of this chapter is the fact, that the computation of Feynman integrals can be reduced to the problem of finding a suitable transformation for the associated differential equation. Methods to find such a transformation are discussed in Chap. 7.

In Chap. 8 we discuss an important class of functions, which appear in Feynman integral computations: These are the multiple polylogarithms.

Apart from an iterated integral representation (which we use extensively in Chaps. 6 and 7) Feynman integrals may also be represented as nested sums. We discuss this aspect in Chap. 9. In this chapter we also show the relation of Feynman integrals to Gelfand-Kapranov-Zelevinsky hypergeometric systems.

Chapter 10 is devoted to sector decomposition. On the one hand sector decomposition (or in a more mathematical language: blow-ups) allow us to device an algorithm, which computes numerically the coefficients of the Laurent expansion in the dimensional regularisation parameter of a Feynman integral. On the other hand (and on the more formal side), we may use this algorithm to prove that these coefficients are numerical periods for rational input parameters.

In Chap. 8 we discussed the algebraic properties of the multiple polylogarithms. Chapter 11 continues this theme and explores the coalgebra side: Coproducts, Hopf algebras, coactions, symbols and single-valued projections are discussed in this chapter.

With the methods of Chaps. 6 and 7 we may transform the differential equation of a Feynman integral, which evaluates to multiple polylogarithms, to a dlog-form. We may ask if there is any relation between the arguments of the dlog's and the original kinematic variables. This will lead us to cluster algebras, which we introduce in Chap. 12.

In Chap. 13 we discuss integrals, which do not evaluate to multiple polylogarithms. We focus on the next-more-complicated case: These are Feynman integrals related to an elliptic curve. We introduce elliptic curves, elliptic functions, modular transformations, modular forms and the moduli space of a genus one curve with marked points.

Chapter 14 is the most mathematical chapter of this book: We introduce motives and mixed Hodge structures and their relation to Feynman integrals. This chapter continues a thread, which started on the one hand in Chap. 10 with (numerical) periods and on the other hand in Chap. 11 with coactions. In Chap. 14 we bring these concepts together. We will see that each coefficient of the Laurent expansion in the dimensional regularisation parameter of a Feynman integral corresponds to a motivic period.

At the end of the day physics is about numbers: We would like to get for specified input parameters (i.e., for specified kinematic variables) a number for a Feynman integral (more precisely for the coefficients of the Laurent expansion in the dimensional regularisation parameter). In Chap. 15 we discuss numerical evaluation routines. These methods can be used to obtain numerical values to a high numerical precision (up to a few hundred or thousand digits). This in turn opens the possibility to use the heuristic PSLQ algorithm to simplify analytic expressions. The main application is the simplification of boundary constants. In this chapter we also introduce the PSLQ algorithm.

In the last chapter of this book (Chap. 16) we carry out a full project: We show in detail how the two-loop penguin diagram on the title page is computed from the starting Feynman diagram to the final numerical result. The purpose of this chapter is to show how the methods and algorithms introduced in this book are used in practice.

This book is supplemented with several appendices:

In Appendix A we review spinors in four space-time dimensions. These are useful for the methods discussed in Chap. 5. Chapter 5 is devoted to one-loop integrals. There are only a finite number of one-loop integrals which we need to know. We list all relevant one-loop integrals for massless theories in Appendix B.

Appendix C is a supplement to Chap. 9: We summarise the definitions and main properties of a few transcendental functions: Hypergeometric functions, Appell functions, Lauricella functions and Horn functions are reviewed in this appendix.

Appendix D is devoted to Lie groups and Lie algebras. Of course Lie groups and Lie algebras are omnipresent in particle physics. In the context of Feynman integrals it is useful to know the classification of simple Lie algebras and their relation to Dynkin diagrams (which we will need in Chap. 12). The appendix gives a concise discussion of the classification of simple Lie algebras.

Appendices E and F supplement Chap. 13: Appendix E introduces Dirichlet characters, while Appendix F discusses the moduli space $\mathcal{M}_{g,n}$ of a smooth algebraic curve of genus g with n marked points.

Appendix G is a concise introduction to the main concepts of algebraic geometry. We give the definitions of sheaves and schemes. These are avoided (as much as possible) in the main text of the book, but one is confronted with these terms as soon as one consults the mathematical literature.

Appendices H and I are supplements to Chaps. 6 and 7. Appendix H reviews standard algorithms in polynomial rings for computing a Gröbner basis, a Nullstellensatz certificate and an annihilator. These are used in Chaps. 6 and 7. Appendix I introduces finite field methods, which can be used to speed-up the integration-by-parts reduction discussed in Sect. 6.1.

There are many exercises included in the main text of this book. The solutions to the exercises are given in Appendix J.

This book grew out of lectures on Feynman integrals given at the Johannes Gutenberg University Mainz in the summer term 2021 (covering the first half of the book) and of lectures given at the Higgs Centre School of Theoretical Physics 2021 (covering some of the more advanced topics). I am grateful to the students attending these lectures for their feedback and to Luigi del Debbio, Einan Gardi and Roman Zwicky for organising the Higgs Centre School. My particular thanks go to Alexander Aycock, Christian Bogner, Ina Hönemann, Philipp Kreer, Sascha Kromin, Hildegard Müller, Farroukh Peykar Negar Khiabani, Robert Runkel, Juan Pablo Vesga and Xing Wang for valuable suggestions on the manuscript. I am also grateful to John Gracey for helpful comments.

In writing this book I recycled some existing material. In particular, Chap. 3 is a revision of a review article [8] on graph polynomials written together with Christian Bogner. Other chapters have their origin in shorter contributions to summer schools and conference proceedings [9–16].

I would like to thank my collaborators Luise Adams, Marco Besier, Isabella Bierenbaum, Christian Bogner, Ekta Chaubey, Andre van Hameren, Philipp Kreer, Dirk Kreimer, Sven Moch, Stefan Müller-Stach, Armin Schweitzer, Duco van Straten, Kirsten Tempest, Peter Uwer, Jens Vollinga, Moritz Walden, Pascal Wasser and Raphael Zayadeh, for their shared work and research interest related to various topics in relation with Feynman integrals.

Finally, I would like to thank a few colleagues, from whom I learned through discussions, conversations or lectures about different aspects of Feynman integrals: My thanks go to David Broadhurst, Johannes Brödel, Francis Brown, Lance Dixon, Claude Duhr, Johannes Henn, David Kosower, Erik Panzer and Lorenzo Tancredi. I also would like to thank Jacob Bourjaily and Henriette Elvang for useful information on book projects.

The figures in this book have been produced with the help of the programs `Axodraw` [17], `TikZ` [18] and `ROOT` [19].

Chapter 2
Basics

In this chapter we introduce the central object of this book: Feynman integrals. We first review special relativity in Sect. 2.1 and the basic concepts of graphs in Sect. 2.2. With these preparations and the Feynman rules for a scalar theory we define Feynman integrals in Sect. 2.3. Before we embark on calculating the first Feynman integrals we need to introduce two fundamental concepts: Wick rotation and regularisation. We do this in Sect. 2.4. We conclude this chapter with a section containing an overview of various integral representations for Feynman integrals (Sect. 2.5). This includes the Feynman parameter representation, the Schwinger parameter representation, the Baikov representation, the Lee-Pomeransky representation and the Mellin-Barnes representation.

2.1 Special Relativity

Let us denote by D the number of space-time dimensions. In our real world D equals 4 (one time dimension and three spatial dimensions), but it is extremely helpful to keep this number arbitrary. We will always assume that space-time consists of one time dimension and $(D-1)$ spatial dimensions.

The momentum of a particle is a D-dimensional vector, whose first component gives the energy E (divided by the speed of light c) and the remaining $(D-1)$ components give the components of the spatial momentum, which we label with superscripts:

$$p = \left(\frac{E}{c}, p^1, \dots, p^{D-1}\right). \tag{2.1}$$

It is common practice in high-energy physics to work in natural units, where

$$c = \hbar = 1. \tag{2.2}$$

S. Weinzierl, *Feynman Integrals*, UNITEXT for Physics,
https://doi.org/10.1007/978-3-030-99558-4_2

We use this convention throughout this book from now on. Let us set $p^0 = E$. We then have

$$p = \left(p^0, p^1, \dots, p^{D-1}\right). \tag{2.3}$$

We write

$$p^\mu \text{ with } 0 \le \mu \le D-1 \tag{2.4}$$

for a component of p. The index μ is called a **Lorentz index**.

We denote by $g_{\mu\nu}$ the components of the metric tensor. The indices μ and ν take integer values between 0 and $(D-1)$. We are primarily concerned with the **Minkowski metric**. Our convention for the Minkowski metric is

$$g_{\mu\nu} = \begin{cases} 1, \; \mu = \nu = 0, \\ -1, \; \mu = \nu \in \{1, \dots, D-1\}, \\ 0, \; \text{otherwise.} \end{cases} \tag{2.5}$$

This convention is the standard convention in high-energy physics phenomenology. Some authors (mostly in the field of formal high-energy physics theory) use a convention, where the roles of $(+1)$ and (-1) are interchanged. Working consistently within one or the other convention does not change physics. The transition from one convention to the other is rather easy and given by a minus sign. In this book we use the convention as given by Eq. (2.5).

The **Minkowski scalar product** of two momentum vectors p_a and p_b is

$$p_a \cdot p_b = \sum_{\mu=0}^{D-1} \sum_{\nu=0}^{D-1} p_a^\mu \, g_{\mu\nu} \, p_b^\nu. \tag{2.6}$$

Einstein's summation convention is the convention to drop the summation symbol for any Lorentz index, which occurs twice, once as an upper index and once as a lower index. The summation is then implicitly assumed. With Einstein's summation convention we may write

$$p_a \cdot p_b = p_a^\mu \, g_{\mu\nu} \, p_b^\nu. \tag{2.7}$$

The Minkowski scalar product $p_a \cdot p_b$ is an example of a **Lorentz invariant**: A Lorentz invariant is a quantity, whose value is not changed under Lorentz transformations.

With the Minkowski scalar product at hand, we may in particular take the scalar product of a momentum vector with itself. Let us write this out explicitly:

$$p^2 \;=\; p \cdot p \;=\; \left(p^0\right)^2 - \left(p^1\right)^2 - \left(p^2\right)^2 - \dots - \left(p^{D-1}\right)^2. \tag{2.8}$$

Please note that on the left-hand side p^2 denotes the Minkowski scalar product of p with itself, while on the right-hand side p^2 (appearing in the term $(p^2)^2$) denotes the third component (the second spatial component) of p. As the meaning should be clear from the context, we follow common practice and do not disambiguate the notation.

Apart from the momentum, we also associate a mass m to a particle. We represent a particle propagating in space-time by a line:

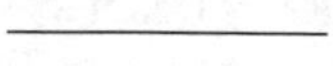

(2.9)

If we would like to indicate the direction of the momentum flow, we optionally put an arrow:

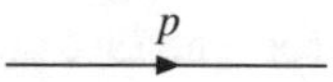

p

(2.10)

The line in Eq. (2.9) is our first building block for a Feynman graph. A Feynman graph is a graphical notation for a mathematical formula and a line for a particle with momentum p and mass m stands for

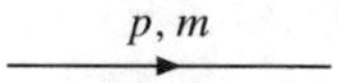

$$\overset{p,\,m}{\longrightarrow} = \frac{1}{-p^2 + m^2}. \tag{2.11}$$

On the right-hand side the Minkowski scalar product of p with itself appears. Note that the right-hand side is independent of the orientation of p.

If one follows standard conventions in quantum field theory the propagator of a scalar particle with momentum p and mass m is given by

$$\frac{i}{p^2 - m^2}. \tag{2.12}$$

This differs by a factor $(-i)$ from Eq. (2.11). This is just a prefactor and easily adjusted. Throughout this book we use the convention that the mathematical expression corresponding to a line is given by Eq. (2.11).

2.2 Graphs

Let us now turn to graphs. An **unoriented graph** consists of edges and vertices, where an edge connects two vertices. A graph may be connected or disconnected. We will mainly consider connected graphs. An **oriented graph** is a graph, where for every edge an orientation is chosen. An oriented graph is also called a **quiver**. As any edge connects two vertices, say v_a and v_b, an orientation is equivalent to declaring one of the two vertices the source for this edge (for example v_a) and the other vertex the sink for this edge (for example v_b). An orientated edge is usually drawn with an arrow line:

$$\text{source} \bullet\!\!\longrightarrow\!\!\bullet \text{sink} \tag{2.13}$$

The **valency of a vertex** is the number of edges attached to it. Vertices of valency 0, 1 and 2 are special. A vertex of valency 0 is necessarily disconnected from the rest of graph and therefore not relevant for connected graphs. A vertex of valency 1 has exactly one edge attached to it. This edge is called an **external edge**. All other edges are called **internal edges**. In the physics community it is common practice not to draw a vertex of valency 1, but just the external edge. A vertex of valency 2 is also called a **dot**. In physics the use of the word "vertex" sometimes implies a vertex of valency 3 or greater. This derives from the fact that in a particle picture a vertex of valency 3 or greater corresponds to a genuine interaction among particles.

As an example for a graph let us look at Fig. 2.1. Figure 2.1 shows a disconnected graph with five edges and six vertices. There, vertex v_6 has valency 0 and is disconnected from the rest of the graph. Vertices v_1 and v_5 have valency 1. The edges attached to them (e_1 and e_5) are the external edges of the graph. Vertex v_3 has valency 2 and is an example of a dot. Vertices v_2 and v_4 have valency 3 and are genuine interaction vertices.

The internal edges of the graph are e_2, e_3 and e_4.

A Feynman graph is a graph with additional information. In the basic version we associate to each edge an orientation (thus our graph becomes an oriented graph), a D-dimensional vector p (the momentum) and a number m (the mass). The physics picture is that an oriented edge represents the propagation of a particle with momentum p and mass m. The momentum flow is in the direction of the orientation of the edge. Note that the choice of the orientation of an edge does not matter if a change in the orientation is accompanied by reversing the momentum $p \to -p$. Consider again an edge e, which connects the vertices v_a and v_b. We have

$$\overset{v_a}{\bullet}\!\!\xrightarrow{\;p,m\;}\!\!\overset{v_b}{\bullet} \;=\; \overset{v_a}{\bullet}\!\!\xleftarrow{\;-p,m\;}\!\!\overset{v_b}{\bullet} \tag{2.14}$$

Let us now consider a graph G with n edges and r vertices. Assume that the graph has k connected components. The **loop number** l is defined by

$$l = n - r + k. \tag{2.15}$$

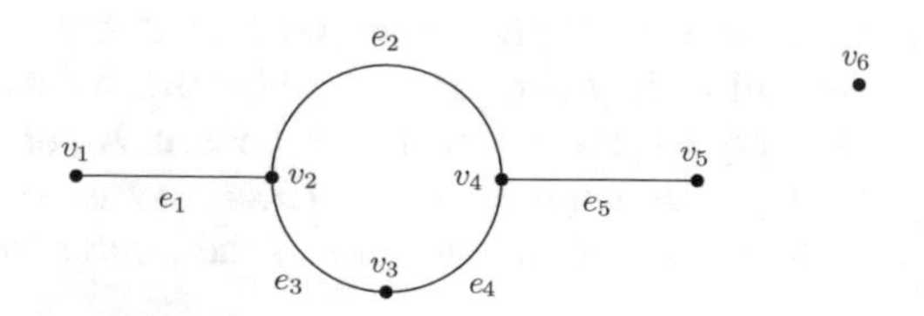

Fig. 2.1 A (disconnected) graph with five edges and six vertices. The vertex v_6 is disconnected from the rest of the graph

If the graph is connected we have $l = n - r + 1$. The loop number l is also called the **first Betti number** of the graph or the **cyclomatic number**. In the physics context it has the following interpretation: If we fix all momenta of the external lines and if we impose momentum conservation at each vertex, then the loop number is equal to the number of independent momentum vectors not constrained by momentum conservation.

A connected graph of loop number 0 is called a **tree**. A graph of loop number 0, connected or not, is called a **forest**. If the forest has k connected components, it is called a k-forest. A tree is a 1-forest. Feynman graphs which are trees pose no conceptual problem. Our focus in this book is on connected Feynman graphs, which are not trees, e.g. Feynman graphs with loop number $l > 0$.

Unless stated otherwise we consider in this book **from now on only connected graphs**.

Let us fix some notation: Consider a graph G with n_{ext} external edges, n_{int} internal edges and loop number l. As we now always assume that the graph is connected, we know from Eq. (2.15) that the graph G must have

$$r = n_{\text{ext}} + n_{\text{int}} + 1 - l \tag{2.16}$$

vertices. Out of these r vertices exactly n_{ext} vertices are vertices of valency 1, this leaves

$$r_{\text{int}} = n_{\text{int}} + 1 - l \tag{2.17}$$

vertices of valency > 1. For each edge we choose an orientation. We label the edges such that $e_1, \dots, e_{n_{\text{int}}}$ are the internal edges and $e_{n_{\text{int}}+1}, \dots, e_{n_{\text{int}}+n_{\text{ext}}}$ are the external edges. For any edge e_j (internal or external) we denote the momentum flowing through this edge (with respect to the chosen orientation) by q_j. For external momenta we use a second notation: We label the momentum flowing through the external edge $e_{n_{\text{int}}+j}$ by p_j. Thus we have

$$p_j = q_{n_{\text{int}}+j}. \tag{2.18}$$

We will soon see that this redundant notation is useful and simplifies the notation in some formulae. Consider a vertex v_a. We denote by

$$\begin{aligned} E^{\text{source}}(v_a) &: \text{set of edges, which have vertex } v_a \text{as source,} \\ E^{\text{sink}}(v_a) &: \text{set of edges, which have vertex } v_a \text{as sink.} \end{aligned} \tag{2.19}$$

At each vertex v_a of valency > 1 we impose momentum conservation:

$$\sum_{e_j \in E^{\text{source}}(v_a)} q_j = \sum_{e_j \in E^{\text{sink}}(v_a)} q_j. \tag{2.20}$$

Furthermore, we denote by

$$\begin{aligned} E^{\text{in}} &: \text{set of edges, which have a vertex of valency 1as source,} \\ E^{\text{out}} &: \text{set of edges, which have a vertex of valency 1as sink.} \end{aligned} \tag{2.21}$$

The edges in E^{in} and E^{out} are necessarily external edges.

Exercise 1 *Consider a connected graph G with the notation as above. Show that momentum conservation at each vertex of valency > 1 implies momentum conservation of the external momenta:*

$$\sum_{e_j \in E^{\text{in}}} q_j = \sum_{e_j \in E^{\text{out}}} q_j. \tag{2.22}$$

If we choose an orientation such that all external edges have a vertex of valency 1 as sink (e.g. $E^{\text{in}} = \emptyset$) this translates to

$$\sum_{j=1}^{n_{\text{ext}}} p_j = 0. \tag{2.23}$$

Let us now investigate how many independent momenta we have. Our graph has $n = n_{\text{ext}} + n_{\text{int}}$ edges and thus we start from n momenta q_i $(1 \leq i \leq n)$. Clearly, in a space of dimension D there can only be D linear independent momenta. This is not the effect we want to study here. Therefore we assume that the dimension D of space-time is large enough ($D \geq n_{\text{ext}} - 1 + l$ will be sufficient). We have seen in the exercise above that the external momenta satisfy momentum conservation. Thus they are not independent and there is at least one linear relation among them. We will assume that the external momenta are generic, e.g. there are besides momentum conservation no further linear relations among the external momenta. Thus we have $(n_{\text{ext}} - 1)$ linear independent external momenta. (A non-generic or special configuration of external momenta is for example given by the four momenta $p_1, p_2, p_3 = 2p_1, p_4 = -3p_1 - p_2$.) Each vertex of valency > 1 gives us through momentum conservation at this vertex a relation among the n momenta q_j. Assuming that the external momenta are known quantities, this leaves

$$n - (n_{\text{ext}} - 1) - r_{\text{int}} = l \tag{2.24}$$

momenta undetermined. We label these momenta by $k_1, \dots, k_l$ and call them the **independent loop momenta**. We may then express any other momentum q_j as linear combination of the l independent loop momenta and the $(n_{\text{ext}} - 1)$ independent external momenta with coefficients $\{-1, 0, 1\}$:

$$q_j = \sum_{r=1}^{l} \lambda_{jr} k_r + \sum_{r=1}^{n_{\text{ext}}-1} \sigma_{jr} p_r, \qquad \lambda_{jr}, \sigma_{jr} \in \{-1, 0, 1\}. \tag{2.25}$$

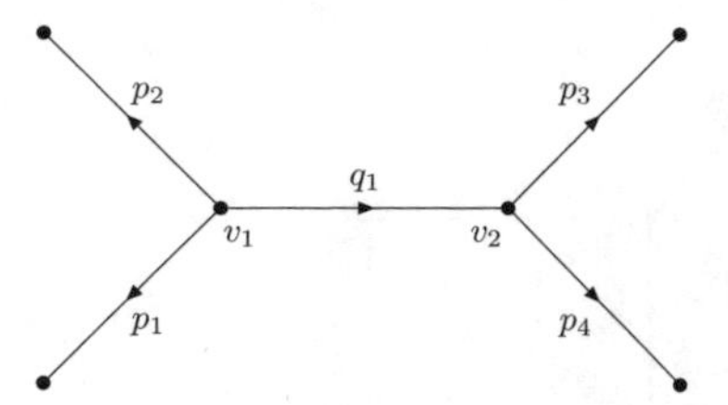

Fig. 2.2 A tree graph with five edges and six vertices. There are four external edges (labelled with momenta $p_1, \dots, p_4$) and one internal edge (labelled with momentum q_1). The orientation of the external edges is chosen such that all external momenta are outgoing

Let us look at some examples: We start with a tree graph, shown in Fig. 2.2.

This graph has five edges ($n = 5$) and six vertices ($r = 6$). The graph is connected, hence $k = 1$. Equation (2.15) gives then the loop number as

$$l = 5 - 6 + 1 \; = \; 0, \tag{2.26}$$

confirming that it is a tree graph. Two vertices have valency > 1, in Fig. 2.2 these two vertices are labelled v_1 and v_2. Momentum conservation at these two vertices yields

$$\begin{aligned} v_1 : \quad & p_1 + p_2 + q_1 = 0, \\ v_2 : \quad & p_3 + p_4 = q_1. \end{aligned} \tag{2.27}$$

From the first equation we have $q_1 = -p_1 - p_2$ and combining the two equations we obtain momentum conservation for the external momenta:

$$p_1 + p_2 + p_3 + p_4 = 0. \tag{2.28}$$

In particular, $p_4 = -p_1 - p_2 - p_3$ and all momenta can be expressed as a linear combination of p_1, p_2, p_3 with coefficients $\{-1, 0, 1\}$. Since we considered a tree graph there are no independent loop momenta in this example.

In the next example we look at a loop graph. The graph is shown in Fig. 2.3.

This graph has eleven edges ($n = 11$) and ten vertices ($r = 10$). Again, the graph is connected, hence $k = 1$. The loop number is therefore

$$l = 11 - 10 + 1 \; = \; 2. \tag{2.29}$$

Six vertices have valency > 1, in Fig. 2.3 these vertices are labelled by $v_1, \dots, v_6$. Momentum conservation at these vertices gives

$$\begin{aligned} v_1 : \quad & p_1 + q_1 = q_3, \\ v_2 : \quad & p_2 + q_2 = q_1, \end{aligned}$$

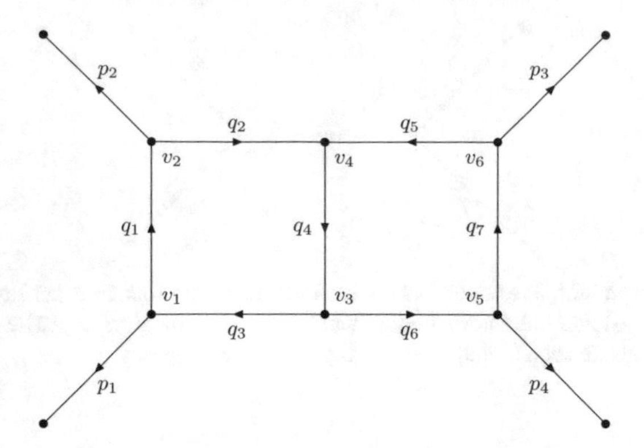

Fig. 2.3 A two-loop graph with eleven edges and ten vertices. There are four external edges (labelled with momenta $p_1, \dots, p_4$) and seven internal edges (labelled with momenta $q_1, \dots, q_7$). Six vertices have valency > 1. These are labelled by $v_1, \dots, v_6$. The orientation of the external edges is chosen such that all external momenta are outgoing

$$\begin{aligned} v_3: &\quad q_3 + q_6 = q_4, \\ v_4: &\quad q_4 = q_2 + q_5, \\ v_5: &\quad p_4 + q_7 = q_6, \\ v_6: &\quad p_3 + q_5 = q_7. \end{aligned} \tag{2.30}$$

Let us take p_1, p_2, p_3 as the independent external momenta and

$$k_1 = q_3, \; k_2 = q_6 \tag{2.31}$$

as the independent loop momenta. All other momenta may then be expressed as a linear combination of k_1, k_2, p_1, p_2, p_3 with coefficients $\{-1, 0, 1\}$. This is nothing else than solving the linear system in Eq. (2.30) for the momenta $q_1, q_2, q_4, q_5, q_7, p_4$. Explicitly we have

$$\begin{aligned} q_1 &= k_1 - p_1, \\ q_2 &= k_1 - p_1 - p_2, \\ q_4 &= k_1 + k_2, \\ q_5 &= k_2 + p_1 + p_2, \\ q_7 &= k_2 + p_1 + p_2 + p_3 \end{aligned} \tag{2.32}$$

and $p_4 = -p_1 - p_2 - p_3$.

Exercise 2 *We stepped from a tree example immediately to a two-loop example. As an exercise consider the one-loop graph shown in Fig. 2.4. Write down the equations expressing momentum conservation at each vertex of valency* > 1. *Use* p_1, p_2, p_3 *as independent external momenta and* $k_1 = q_4$ *as the independent loop momentum. Express all other momenta as linear combinations of these.*

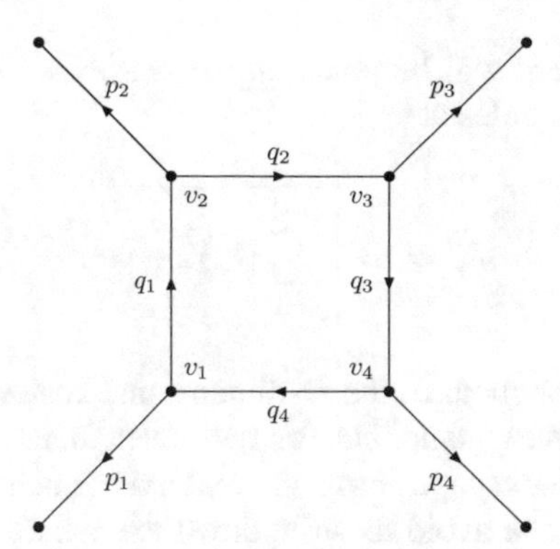

Fig. 2.4 A one-loop graph with eight edges and eight vertices. There are four external edges (labelled with momenta $p_1, \dots, p_4$) and four internal edges (labelled with momenta $q_1, \dots, q_4$). Four vertices have valency > 1. These are labelled by $v_1, \dots, v_4$. The orientation of the external edges is chosen such that all external momenta are outgoing

Let us now return to a general graph with n_{ext} external momenta (satisfying momentum conservation) and l independent loop momenta. We may consider the external momenta as input data, but what shall we do with the l independent loop momenta? As they are independent of the external momenta there is no reason to prefer a particular configuration over any other configuration. Quantum field theory instructs us to integrate over the independent loop momenta. Thus we include for every independent loop momentum k_r $(1 \le r \le l)$ a D-dimensional integration

$$\int \frac{d^D k_r}{i \pi^{\frac{D}{2}}}. \tag{2.33}$$

i denotes the imaginary unit. This is the measure which we will use in this book. It is normalised conveniently. If one follows standard conventions in quantum field theory, the measure is given by

$$\int \frac{d^D k_r}{(2\pi)^D}. \tag{2.34}$$

The difference is a simple prefactor and one easily converts from one convention to the other convention.

We should also specify the integration contour. Our naive expectation is that we integrate each of the D components of k_r along the real axis from $-\infty$ to $+\infty$. However, we have to be more careful. An internal edge with momentum k_r and mass m contributes a factor

$$\xrightarrow{\quad k_r,\, m \quad} \;=\; \frac{1}{-k_r^2 + m^2} \tag{2.35}$$

to the integrand and there will be poles on the real axis. For example, for the k_r^0-integration we will have poles at

$$k_r^0 = \pm \sqrt{\sum_{i=1}^{D-1} \left(k_r^i\right)^2 + m^2}. \tag{2.36}$$

(k_r^i denotes the i-th component of the D-dimensional vector k_r.) We can go around these poles in the complex plane, but for two poles there a four possibilities: For each poles we may either escape above the real axis or below the real axis into the complex plane. In order to avoid to write down the square root more often, let us define

$$E_{\mathbf{k}_r} = \sqrt{\sum_{i=1}^{D-1} \left(k_r^i\right)^2 + m^2}. \tag{2.37}$$

Quantum field theory and causality in particular dictates us that the correct integration contour is the following:

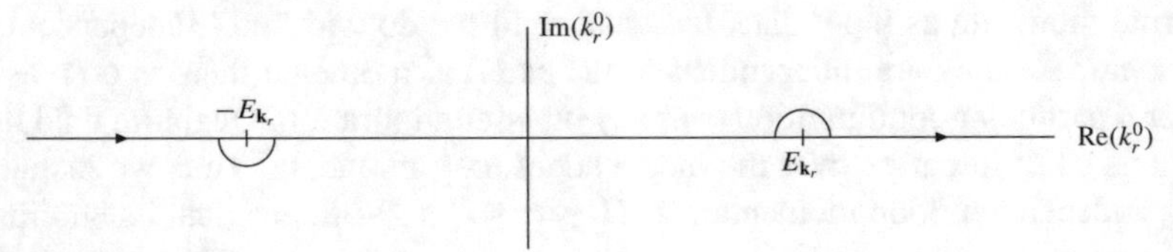

Thus we escape for the pole at $k_r^0 = -E_{\mathbf{k}_r}$ into the complex lower half-plane and for the pole at $k_r^0 = E_{\mathbf{k}_r}$ into the complex upper half-plane. Alternatively, we may keep the contour along the real axis and move the pole at $k_r^0 = -E_{\mathbf{k}_r}$ an infinitesimal amount above the real axis and the pole as $k_r^0 = E_{\mathbf{k}_r}$ an infinitesimal amount below the real axis. This can be done by adding an infinitesimal small imaginary part to the Feynman rule for an internal edge:

$$\xrightarrow{\quad q, m \quad} \quad = \frac{1}{-q^2 + m^2 - i\delta} \tag{2.38}$$

with δ an infinitesimal small positive number. This is **Feynman's $i\delta$-prescription**. In this book we usually don't write the $i\delta$ explicitly and only include it where it is needed. This is common practice in the field.

We started defining the basic version of a Feynman graph as a graph, where we associate to every edge a momentum and a mass. Let us add some additional information. The motivation is as follows: Consider a vertex of valency 2 inside a Feynman graph. By momentum conservation, the momenta flowing through the two edges attached to this vertex are (with an appropriate choice of orientation of the two edges) identical. Let us assume that the masses associated with these two edges are identical as well. The two edges would then contribute two identical factors

$$\xrightarrow{q,\,m} \bullet \xrightarrow{q,\,m} \quad = \frac{1}{\left(-q^2+m^2\right)^2}. \tag{2.39}$$

We get the same effect if we associate to each edge in addition to the momentum and the mass a further number ν, corresponding to the power to which the propagator occurs:

$$\xrightarrow{q,\,m,\,\nu} \quad = \frac{1}{\left(-q^2+m^2\right)^\nu} \tag{2.40}$$

This convention will prove useful in later chapters of the book.

2.3 Feynman Rules

Feynman rules translate a Feynman graph into a mathematical formula. In the previous paragraph we already discussed the essential ingredients. Let's wrap them up and finalise them.

We consider a (connected) Feynman graph G with n_{ext} external edges, n_{int} internal edges and l loops. This graph has

$$r = n_{\text{ext}} + n_{\text{int}} + 1 - l \tag{2.41}$$

vertices. To each external edge we associate an external momentum. We label the external momenta by $p_1, \dots, p_{n_{\text{ext}}}$. To each internal edge e_j we associate a triple (q_j, m_j, ν_j), where q_j is the momentum flowing through this edge, m_j the mass and ν_j the power to which the propagator occurs. Momentum conservation at each vertex of valency > 1 allows us to express any q_j as a linear combination of $(n_{\text{ext}} - 1)$ linear independent external momenta and l independent loop momenta. We denote the latter by $k_1, \dots, k_l$. Thus

$$q_j = \sum_{r=1}^{l} \lambda_{jr} k_r + \sum_{r=1}^{n_{\text{ext}}-1} \sigma_{jr} p_r, \qquad \lambda_{jr}, \sigma_{jr} \in \{-1, 0, 1\}. \tag{2.42}$$

The Feynman integral I corresponding to this Feynman graph is obtained as follows:

1. For each internal edge e_j include a factor

$$\frac{1}{\left(-q_j^2+m_j^2\right)^{\nu_j}}. \tag{2.43}$$

2. For each independent loop momentum k_r include an integration

$$\int \frac{d^D k_r}{i\pi^{\frac{D}{2}}}. \tag{2.44}$$

3. Multiply by the prefactor

$$e^{l\varepsilon\gamma_E} \left(\mu^2\right)^{\nu - \frac{lD}{2}} \text{ where } \varepsilon = \frac{[D] - D}{2}, \quad \nu = \sum_{j=1}^{n_{\text{int}}} \nu_j. \tag{2.45}$$

The prefactor in rule 3 requires some explanation. First of all, it is just a prefactor. All complications in computing Feynman integrals come from the integrations in rule 2. The sole purpose of the prefactor is to make the final result as simple as possible. We will see examples later on.

Let us discuss the ingredients of the prefactor. ε is called the dimensional regularisation parameter. For a start we may define ε as follows: If D denotes the dimension of space-time, we define $[D]$ to be the closest integer to D. The dimensional regularisation parameter ε is then defined to be

$$\varepsilon = \frac{[D] - D}{2}. \tag{2.46}$$

This may sound weird at first sight. Up to now we implicitly assumed that the dimension D of space-time is a positive integer and the closest integer to an integer is the integer itself: If D is an integer we have $[D] = D$ and therefore $\varepsilon = 0$. Later on in this book we will work with the assumption that D is an arbitrary complex number, not necessarily an integer. This is called dimensional regularisation. Very often we will write

$$D = 4 - 2\varepsilon, \tag{2.47}$$

where we assume that $|\varepsilon|$ is a small number. In this case

$$[D] = 4 \tag{2.48}$$

and Eq. (2.47) agrees with the definition in Eq. (2.46). Rearranging Eq. (2.46) gives

$$D = [D] - 2\varepsilon, \tag{2.49}$$

and once we specify $D_{\text{int}} = [D]$ as the integer dimension of the physical space-time we are interested in, we may give up the restriction $|\text{Re}(\varepsilon)| < 1/2$ and consider ε as an arbitrary complex parameter. In other words, as we will be using perturbation theory in ε, the quantity D_{int} denotes the (integer) expansion point in the complex D-plane and ε the expansion parameter. We will also denote the integer dimension of the physical space-time by D_{int}. Equation (2.49) reads then

$$D = D_{\text{int}} - 2\varepsilon. \tag{2.50}$$

The symbol γ_{E} denotes **Euler's constant** (also called the Euler-Mascheroni constant). This constant is defined by

$$\gamma_{\text{E}} = \lim_{n\to\infty} \left(-\ln n + \sum_{j=1}^{n} \frac{1}{j} \right). \tag{2.51}$$

The numerical value is

$$\gamma_{\text{E}} = 0.57721566490153286\ldots \tag{2.52}$$

We will later see that without the prefactor $e^{l\varepsilon\gamma_{\text{E}}}$ Euler's constant will appear in the final result for a Feynman integral. The dependence of the result on γ_{E} is rather simple (later we will see explicitly that it is given by $e^{-l\varepsilon\gamma_{\text{E}}}$) and it is convenient to factor this out. Thus the prefactor $e^{l\varepsilon\gamma_{\text{E}}}$ removes this factor and with this prefactor included Euler's constant will not clutter the final result of a Feynman integral.

In physics we like to work with dimensionless quantities. The momentum squared p^2 is not dimensionless, it has mass dimension 2 (recall that we work in natural units where $c = 1$). We have n_{int} internal edges, each bringing a factor $1/(-q_j^2 + m_j^2)^{\nu_j}$. The total mass dimension of these factors is

$$\dim \left(\prod_{j=1}^{n_{\text{int}}} \frac{1}{\left(-q_j^2 + m_j^2\right)^{\nu_j}} \right) = -2\nu, \qquad \nu = \sum_{j=1}^{n_{\text{int}}} \nu_j. \tag{2.53}$$

The integral measure has mass dimension

$$\dim \left(\prod_{r=1}^{l} \frac{d^D k_r}{i\pi^{\frac{D}{2}}} \right) = lD. \tag{2.54}$$

In order to enforce that our Feynman integral is dimensionless, we introduce an arbitrary parameter μ with mass dimension

$$\dim(\mu) = 1 \tag{2.55}$$

and multiply by $(\mu^2)^{\nu - lD/2}$.

In summary, the Feynman integral corresponding to a Feynman graph G with n_{ext} external edges, n_{int} internal edges and l loops is given in D space-time dimensions by

$$I = e^{l\varepsilon\gamma_{\text{E}}} \left(\mu^2\right)^{\nu - \frac{lD}{2}} \int \prod_{r=1}^{l} \frac{d^D k_r}{i\pi^{\frac{D}{2}}} \prod_{j=1}^{n_{\text{int}}} \frac{1}{\left(-q_j^2 + m_j^2\right)^{\nu_j}}, \tag{2.56}$$

where each internal edge e_j of the graph is associated with a triple (q_j, m_j, ν_j), specifying the momentum q_j flowing through this edge, the mass m_j and the power ν_j to which the propagator occurs. The external momenta are labelled by $p_1, \dots, p_{n_{\text{ext}}}$. Furthermore

$$\varepsilon = \frac{D_{\text{int}} - D}{2}, \quad \nu = \sum_{j=1}^{n_{\text{int}}} \nu_j, \quad q_j = \sum_{r=1}^{l} \lambda_{jr} k_r + \sum_{r=1}^{n_{\text{ext}}-1} \sigma_{jr} p_r. \tag{2.57}$$

The coefficients λ_{jr} and σ_{jr} can be obtained from momentum conservation at each vertex of valency > 1. The integration contour is given by Feynman's $i\delta$-prescription.

Equation (2.56) defines the central object of this book. We would like to compute integrals of this type. We will learn about methods how to approach this task and the underlying mathematics in the sequel of this book.

We already mentioned that the Feynman integral defined in Eq. (2.56) differs by a prefactor from standard conventions in quantum field theory. Let us summarise the (small) differences: If we come from quantum field theory, each edge corresponds to a single power of a propagator. Thus $\nu_j = 1$. A scalar propagator is given by

$$\frac{i}{q_j^2 - m_j^2}, \tag{2.58}$$

the integral measure is given by

$$\frac{d^D k_r}{(2\pi)^D} \tag{2.59}$$

and the prefactor $e^{l\varepsilon\gamma_{\text{E}}}(\mu^2)^{\nu - lD/2}$ is replaced by one. If one later chooses the **modified minimal subtraction scheme** for renormalisation, Euler's constant is removed in the same way as the factor $e^{l\varepsilon\gamma_{\text{E}}}$ removes γ_{E} from the final result. The ε-dependent part of $(\mu^2)^{\nu - lD/2}$ is $(\mu^2)^{l\varepsilon}$. The latter is introduced to keep the action dimensionless (in natural units $\hbar = 1$). The parameter μ is known as the **renormalisation scale**.

2.4 Fundamental Concepts

2.4.1 Wick Rotation

Minkowski space comes with the Minkowski metric, given by $g_{\mu\nu} = \text{diag}(1, -1, -1, -1, \dots)$. It will be simpler to work with the standard Euclidean metric $d_{\mu\nu}^{\text{eucl}} = \text{diag}(1, 1, 1, 1, \dots)$. The Wick rotation [20] allows us effectively to go from the Minkowski metric to the standard Euclidean metric.

We explain the basic idea for the one-loop case, where the integrand depends on the integration variables only through k^2. The simplest example is given by the one-loop tadpole integral

$$T_1 = e^{\varepsilon\gamma_E} \left(\mu^2\right)^{1-\frac{D}{2}} \int \frac{d^D k}{i\pi^{\frac{D}{2}}} \frac{1}{\left(-k^2 + m^2\right)}. \tag{2.60}$$

Remember, that k^2 written out in components in D-dimensional Minkowski space reads

$$k^2 = \left(k^0\right)^2 - \left(k^1\right)^2 - \left(k^2\right)^2 - \left(k^3\right)^2 - \dots - \left(k^{D-1}\right)^2. \tag{2.61}$$

Furthermore, when integrating over k^0, we encounter poles which are avoided by Feynman's $i\delta$-prescription

$$\frac{1}{-k^2 + m^2 - i\delta}. \tag{2.62}$$

In the complex k^0-plane we consider the integration contour shown in Fig. 2.5. Since the contour does not enclose any poles, the integral along the complete contour is zero:

$$\oint dk^0 f(k^0) = 0. \tag{2.63}$$

If the quarter-circles at infinity give a vanishing contribution (it can be shown that this is the case) we obtain

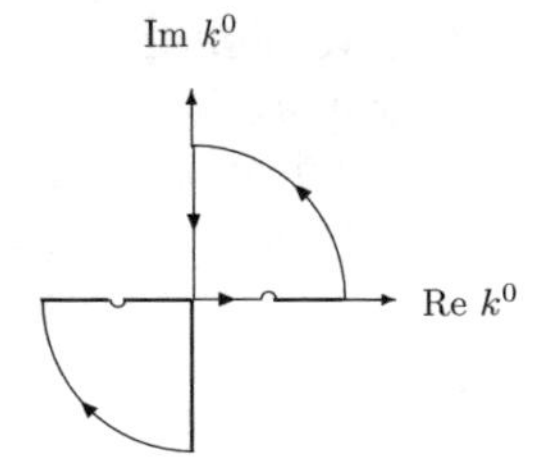

Fig. 2.5 Integration contour for the Wick rotation. The little circles along the real axis exclude the poles

$$\int_{-\infty}^{\infty} dk^0 f(k^0) = - \int_{i\infty}^{-i\infty} dk^0 f(k^0). \tag{2.64}$$

We now make the following change of variables:

$$\begin{aligned} k^0 &= i K^0, \\ k^j &= K^j, \quad \text{for } 1 \leq j \leq D-1. \end{aligned} \tag{2.65}$$

As a consequence we have

$$\begin{aligned} k^2 &= -K^2, \\ d^D k &= i d^D K, \end{aligned} \tag{2.66}$$

where K^2 is now given with Euclidean signature:

$$K^2 = \left(K^0\right)^2 + \left(K^1\right)^2 + \left(K^2\right)^2 + \left(K^3\right)^2 + \dots + \left(K^{D-1}\right)^2. \tag{2.67}$$

In this book we use lower case letters for vectors in Minkowski space and upper case letters for vectors in Euclidean space. Combining the exchange of the integration contour with the change of variables we obtain for the integration of a function $f(k^2)$ in D dimensions

$$\int \frac{d^D k}{i \pi^{\frac{D}{2}}} f(-k^2) = \int \frac{d^D K}{\pi^{\frac{D}{2}}} f(K^2), \tag{2.68}$$

whenever there are no poles inside the contour of Fig. 2.5 and the arcs at infinity give a vanishing contribution. The integral on the right-hand side is now in D-dimensional Euclidean space. This equation justifies our convention to introduce in the definition of the Feynman integral a factor i in the denominator of the measure and a minus sign for each propagator. These conventions are just such that after Wick rotation we have simple formulae.

2.4.2 Dimensional Regularisation

Before we start with an actual calculation of a Feynman integral, we should mention one complication: Loop integrals are often divergent! Let us first look at the simple example of a one-loop tadpole integral with a double propagator and vanishing internal mass in four space-time dimensions:

$$T_2 = \int \frac{d^4 k}{i \pi^2} \frac{1}{(-k^2)^2} = \int \frac{d^4 K}{\pi^2} \frac{1}{(K^2)^2} = \int_0^{\infty} \frac{dK^2}{K^2} = \int_0^{\infty} \frac{dx}{x}. \tag{2.69}$$

Here, we first performed a Wick rotation to Euclidean space. We then used spherical coordinates in four dimensions and integrated over the angles. We are left with the radial integration, where we used as variable the norm squared. This integral diverges at

- $x \to \infty$, which is called an **ultraviolet** (UV) divergence and at
- $x \to 0$, which is called an **infrared** (IR) divergence.

Any quantity, which is given by a divergent integral, is of course an ill-defined quantity. Therefore the first step is to make these integrals well-defined by introducing a regulator. There are several possibilities how this can be done. One possibility is cut-off regularisation with an ultraviolet regulator Λ and an infrared regulator λ:

$$\int\limits_0^\infty \frac{dx}{x} \to \int\limits_\lambda^\Lambda \frac{dx}{x} = \ln \Lambda - \ln \lambda. \tag{2.70}$$

For infrared divergences mass regularisation can be used:

$$\int \frac{d^4K}{\pi^2} \frac{1}{(K^2)^2} \to \int \frac{d^4K}{\pi^2} \frac{1}{(K^2+m^2)^2} = \int\limits_0^\infty dK^2 \frac{K^2}{(K^2+m^2)^2}. \tag{2.71}$$

This cures the problem at $K^2 \to 0$, but not at $K^2 \to \infty$. A third possibility is lattice regularisation, where in position space space-time is approximated by a finite lattice with finite lattice spacing. Ultraviolet divergences are regulated by the finite lattice spacing, infrared divergences are regulated by the finite lattice. However, within perturbative quantum field theory the method of **dimensional regularisation** [21–23] has almost become a standard, as the calculations in this regularisation scheme turn out to be the simplest. Within dimensional regularisation one replaces the four-dimensional integral over the loop momentum by a D-dimensional integral, where D is now an additional parameter, which can be a non-integer or even a complex number. We consider the result of the integration as a function of D and we are interested in the behaviour of this function as D approaches 4.

At first sight the concept of dimensional regularisation may sound strange, as it is hard to imagine a space of non-integer dimension, but the concept is closely related to the following situation: Consider a function $f(z)$, which is defined for any positive integer $n \in \mathbb{N}$ by

$$f(n) = n! = 1 \cdot 2 \cdot 3 \cdot \ldots \cdot n. \tag{2.72}$$

We then would like to define $f(z)$ for any value $z \in \mathbb{C}$ (except for a countable set of isolated points, where $f(z)$ is allowed to have poles). Of course, the answer is well known in mathematics and given by Euler's gamma function

$$f(z) = \Gamma(z+1). \tag{2.73}$$

We will soon see that dimensional regularisation is based on the analytic properties of Euler's gamma function.

Let us start with general properties of dimensional regularisation: The D-dimensional integration still fulfils the standard laws for integration, like linearity, translation invariance and scaling behaviour [24, 25]. If f and g are two functions, and if a and b are two constants, linearity states that

$$\int \frac{d^D K}{\pi^{\frac{D}{2}}} \left[af(K) + bg(K) \right] = a \int \frac{d^D K}{\pi^{\frac{D}{2}}} f(K) + b \int \frac{d^D K}{\pi^{\frac{D}{2}}} g(K). \quad (2.74)$$

Translation invariance requires that

$$\int \frac{d^D K}{\pi^{\frac{D}{2}}} f(K+P) = \int \frac{d^D K}{\pi^{\frac{D}{2}}} f(K) \quad (2.75)$$

for any vector P. The scaling law states that

$$\int \frac{d^D K}{\pi^{\frac{D}{2}}} f(\lambda K) = \lambda^{-D} \int \frac{d^D K}{\pi^{\frac{D}{2}}} f(K). \quad (2.76)$$

The integral measure is normalised such that it agrees with the result for the integration of a Gaussian function for all integer values D:

$$\int \frac{d^D K}{\pi^{\frac{D}{2}}} \exp\left(-K^2\right) = 1. \quad (2.77)$$

In Eq. (2.50) we introduced D_{int} as an integer, giving the dimension of space-time we are interested in. We further introduced the dimensional regularisation parameter ε such that

$$D - D_{\mathrm{int}} = -2\varepsilon. \quad (2.78)$$

Let us take D_{int} and $D - D_{\mathrm{int}} = -2\varepsilon$ as two independent quantities. We will assume that we can always decompose any vector into a D_{int}-dimensional part and a $(D - D_{\mathrm{int}})$-dimensional part

$$K_{(D)} = K_{(D_{\mathrm{int}})} + K_{(D-D_{\mathrm{int}})}, \quad (2.79)$$

and that the D_{int}-dimensional and $(D - D_{\mathrm{int}})$-dimensional subspaces are orthogonal to each other:

$$K_{(D_{\mathrm{int}})} \cdot K_{(D-D_{\mathrm{int}})} = 0. \quad (2.80)$$

If we substitute $k^0 = iK^0$ and $k^j = K^j$ for $1 \leq j \leq (D-1)$ this also implies the Minkowski version

$$k_{(D_{\text{int}})} \cdot k_{(D-D_{\text{int}})} = 0. \tag{2.81}$$

If D is an integer greater than D_{int}, Eqs. (2.79) and (2.80) are obvious. We postulate that these relations are true for any value of D. One can think of the underlying vector space as a space of sufficiently high dimension (possibly infinite), where the integral measure mimics the one in D dimensions.

Digression Constructing vector spaces associated to dimensional regularisation *Let us digress and discuss how Eqs. (2.79) and (2.80) can actually be realised. This will give us some insight how to interpret a space-time of dimension* $D = 3.99$.

We start with a simpler but related question: Suppose we know the natural numbers $\mathbb{N}$ *together addition and multiplication. Subtraction and division are not yet known to us. How do we construct the integer numbers* $\mathbb{Z}$*, the rational numbers* $\mathbb{Q}$*, the real numbers* $\mathbb{R}$ *and the complex numbers* $\mathbb{C}$*?*

The way to do it is well-known to mathematicians: Let's consider the first step, constructing the integer numbers $\mathbb{Z}$ *from the natural numbers* $\mathbb{N}$*. Consider pairs* (a, b) *with* $a, b \in \mathbb{N}$ *together with an equivalence relation. Two pairs* (a_1, b_1) *and* (a_2, b_2) *are equivalent if there is a* $n \in \mathbb{N}$ *such that*

$$a_1 + b_2 + n = a_2 + b_1 + n. \tag{2.82}$$

We denote the equivalence classes by $[(a, b)]$*. One defines an addition on the set of equivalence classes by*

$$[(a_1, b_1)] + [(a_2, b_2)] = [(a_1 + a_2, b_1 + b_2)]. \tag{2.83}$$

One can show that this definition does not depend on the representatives of the equivalence classes. The set of equivalence classes together with the addition defined by Eq. (2.83) forms a group, isomorphic to $\mathbb{Z}$*. The neutral element is the equivalence class* $[(1, 1)]$*, the inverse of* $[(a, b)]$ *is* $[(b, a)]$*. Note that we never used a minus sign in this construction.*

The construction of the rational numbers $\mathbb{Q}$ *from the integer numbers* $\mathbb{Z}$ *proceeds in a similar way: One considers pairs* (p, q) *with* $p, q \in \mathbb{Z}$ *together with an equivalence relation. Two pairs* (p_1, q_1) *and* (p_2, q_2) *are equivalent if there is a* $n \in \mathbb{Z}$ *such that*

$$p_1 \cdot q_2 \cdot n = p_2 \cdot q_1 \cdot n. \tag{2.84}$$

Let us denote the equivalence classes by $[(p, q)]$*. One defines a multiplication on the set of equivalence classes by*

$$[(p_1, q_1)] \cdot [(p_2, q_2)] = [(p_1 \cdot p_2, q_1 \cdot q_2)]. \tag{2.85}$$

Again, one can show this does not depend on the representatives and that the set of equivalence classes together with the multiplication defines a group isomorphic to $\mathbb{Q}$*. The neutral element is* $[(1, 1)]$*, the inverse of* $[(p, q)]$ *is* $[(q, p)]$*.*

This construction is quite general. We started from the natural numbers $\mathbb{N}$*, an Abelian semi-group with respect to addition, and constructed the integer numbers* $\mathbb{Z}$*, an Abelian group with respect to addition. We then used the integer numbers* $\mathbb{Z}$*, which form an Abelian semi-group with respect to multiplication and constructed the rational numbers* $\mathbb{Q}$*, an Abelian group with respect to multiplication. The mathematical framework, which associates to each Abelian semi-group an Abelian group, is the domain of K-theory. If A is an Abelian semi-group with composition* $\circ$*, the* **Grothendieck group** $K(A)$ *of A is constructed in the same way as in the examples above: We consider pairs* (a, b) *with* $a, b \in A$ *together with the equivalence relation*

$$(a_1, b_1) \sim (a_2, b_2) \Leftrightarrow \exists\ p \in A\ :\ a_1 \circ b_2 \circ p\ =\ a_2 \circ b_1 \circ p. \tag{2.86}$$

Then by definition $K(A) = A \times A/\sim$*. The Grothendieck group* $K(A)$ *is an Abelian group. Elements of* $K(A)$ *are denoted* $[(a, b)]$*.*

Let us now consider a set of vector space $\mathcal{V} = \{V_1, V_2, V_3, \dots\}$*, where the vector space* V_j *has dimension* $\dim V_j = j$*. On* $\mathcal{V}$ *we have two operations, the direct sum and the tensor product, such that*

$$V_i \oplus V_j \in \mathcal{V} \text{ and } V_i \otimes V_j \in \mathcal{V} \tag{2.87}$$

are again elements of $\mathcal{V}$*. It is easy to see that with respect to each of these operations* $\mathcal{V}$ *is an Abelian semi-group. The dimensions of the resulting vector spaces are:*

$$\dim\left(V_i \oplus V_j\right) = i + j,\ \dim\left(V_i \otimes V_j\right) = i \cdot j. \tag{2.88}$$

We may therefore proceed as above and construct the Grothendieck K-groups. As an example we consider the K-group with respect to the direct sum: One considers pairs (V_a, V_b) *with* $V_a, V_b \in \mathcal{V}$ *together with an equivalence relation. Two pairs* (V_{a_1}, V_{b_1}) *and* (V_{a_2}, V_{b_2}) *are equivalent if there is a* $V_n \in \mathcal{V}$ *such that*

$$V_{a_1} \oplus V_{b_2} \oplus V_n = V_{a_2} \oplus V_{b_1} \oplus V_n, \tag{2.89}$$

where the equal sign refers to an isomorphism between vector spaces. The equivalence classes are denoted by $[(V_a, V_b)]$*. On the set of equivalence classes one defines the operation* $\oplus$ *by*

$$\left[\left(V_{a_1}, V_{b_1}\right)\right] \oplus \left[\left(V_{a_2}, V_{b_2}\right)\right] = \left[\left(V_{a_1} \oplus V_{a_2}, V_{b_1} \oplus V_{b_2}\right)\right]. \tag{2.90}$$

Again, one can show that this is independent of the chosen representative. For example we have

$$[(V_5, V_1)] = [(V_{42}, V_{38})]. \tag{2.91}$$

Since we are considering equivalence classes of pairs of vector spaces, the dimensions of the vector spaces representing an equivalence class have no particular meaning. However, the rank defined by

$$\text{rank}\ [(V_i, V_j)] = \dim V_i - \dim V_j \qquad (2.92)$$

is an integer number and independent of the representative. The rank of an equivalence class corresponds to our variable D, and we managed to construct equivalence classes, where the rank is a negative integer. We then repeat the argumentation where we replace the direct sum with the tensor product. This gives us equivalence classes, where the rank is a rational number.

It remains to define the integration on these equivalence classes. It is no problem to define the integration on a representative of an equivalence class. The subtle point is that we have to show that this definition is independent of the chosen representative. Integration is a linear functional on a space of functions, and the space of functions we are interested is rather special. The functions we want to integrate depend only on a few components

$$k^0, k^1, \dots, k^{(D-1)} \qquad (2.93)$$

explicitly and on the rest of the components only through the combination

$$\left(k^0\right)^2 + \left(k^1\right)^2 + \dots + \left(k^{d-1}\right)^2, \qquad (2.94)$$

where d is the sum of dimensions of the vector spaces making up the representative. In this case it is possible to define an integration, which is independent of the chosen representative and has the desired properties [26].

Thus we managed to give a meaning to integration in spaces of dimension $D \in \mathbb{Q}$, where D corresponds to the rank of the equivalence class. In the last step one extends the integration to $D \in \mathbb{R}$ and $D \in \mathbb{C}$ in the same way as the real and complex numbers are constructed: Each real number is the limit of a sequence of rational numbers and each complex number can be represented by a pair of real numbers.

In practice we will always arrange things such that every function we integrate over D dimensions is rotational invariant, e.g. is a function of k^2. In this case the integration over the $(D - 1)$ angles is trivial and can be expressed in a closed form as a function of D. Let us assume that we have an integral, original in four space-time dimensions, which has a UV-divergence, but no IR-divergences. Let us further assume that this integral would diverge logarithmically, if we would use a cut-off regularisation instead of dimensional regularisation, e.g. the integral behaves for large x in four space-time dimensions as

$$\text{logarithmically divergent}: \int\limits_1^{\Lambda} \frac{dx}{x} = \ln \Lambda. \qquad (2.95)$$

It turns out that this integral will be convergent if the real part of D is smaller than 4. Therefore we may compute this integral under the assumption that $\text{Re}(D) < 4$ and we will obtain as a result a function of D. This function can be analytically continued to the whole complex plane. We are mainly interested in what happens close to the point $D = 4$. For an ultraviolet divergent one-loop integral we will find that the analytically continued result will exhibit a pole at $D = 4$. It should be mentioned that there are also integrals which in the ultraviolet region diverge stronger. An integral diverges for large x linearly, respectively quadratically if it behaves as a function of the cut-off as

$$\begin{aligned} \text{linearly divergent}: & \int\limits_0^\Lambda dx = \Lambda, \\ \text{quadratically divergent}: & \; 2\int\limits_0^\Lambda x\, dx = \Lambda^2. \end{aligned} \tag{2.96}$$

For example, if the integral diverges quadratically for $D = 4$ we can repeat the argumentation above with the replacement $\text{Re}(D) < 2$.

The terminology also applies to infrared divergences. An integral is said to be logarithmically divergent at $x = 0$ if it behaves as

$$\text{logarithmically divergent}: \int\limits_\lambda^1 \frac{dx}{x} = -\ln \lambda. \tag{2.97}$$

The integrand is said to have a simple pole at $x = 0$. Integrands with a double pole correspond to linearly divergent integrals, integrands with a pole of order three to quadratically divergent integrals. This is most easily seen by the substitution $\lambda = 1/\Lambda'$, for example

$$\int\limits_\lambda^\infty \frac{dx}{x^2} = \frac{1}{\lambda} = \Lambda'. \tag{2.98}$$

Let us now consider a logarithmic IR-divergent integral, which has no UV-divergence. This integral will be convergent if $\text{Re}(D) > 4$. Again, we can compute the integral in this domain and continue the result to $D = 4$. Logarithmic IR-divergent one-loop integral may have a divergence in the radial integration as well as in the angular integration. The former is called a **soft divergence**, the latter a **collinear divergence**. Each divergence will give a pole at $D = 4$, and if both divergences are present this will lead in total to a double pole at $D = 4$.

We will use dimensional regularisation to regulate both the ultraviolet and infrared divergences. The attentive reader may ask how this goes together, as we argued above that UV-divergences require $\text{Re}(D) < 4$ or even $\text{Re}(D) < 2$, whereas IR-divergences are regulated by $\text{Re}(D) > 4$. Suppose for the moment that we use dimensional regularisation just for the UV-divergences and that we use a second regulator for the

IR-divergences. For the IR-divergences we could keep all external momenta off-shell, or introduce small masses for all massless particles or even raise the original propagators to some power ν. The exact implementation of this regulator is not important, as long as the IR-divergences are screened by this procedure. We then perform the loop integration in the domain where the integral is UV-convergent. We obtain a result, which we can analytically continue to the whole complex D-plane, in particular to $\text{Re}(D) > 4$. There we can remove the additional regulator and the IR-divergences are now regulated by dimensional regularisation. Then the infrared divergences will also show up as poles at $D = 4$.

In summary, within dimensional regularisation the initial divergences show up as poles in the complex D-plane at the point $D = 4$. What shall we do with these poles? The answer has to come from physics and we distinguish again the case of UV-divergences and IR-divergences. The UV-divergences are removed through renormalisation. On the level of Feynman diagrams we can associate to any UV-divergent part a counter-term, which has exactly the same pole term at $D = 4$, but with an opposite sign, such that in the sum the two pole terms cancel.

As far as infrared-divergences are concerned we first note that any detector has a finite resolution. Therefore two particles which are sufficiently close to each other in phase space will be detected as one particle. The Kinoshita-Lee-Nauenberg theorem [27, 28] guarantees that all infrared divergences cancel, when summed over all degenerate physical states. To make this cancellation happen in practice requires usually quite some work, as the different contributions live on phase spaces of different dimensions. We will discuss the cancellation of ultraviolet and infrared divergences in Chap. 4.

A Feynman integral I in D space-time dimensions will therefore have a Laurent expansion around $D = D_{\text{int}}$:

$$I = \sum_{j=j_{\min}}^{\infty} \varepsilon^j \, I^{(j)}, \qquad \varepsilon = \frac{D_{\text{int}} - D}{2}, \tag{2.99}$$

where $I^{(j)}$ denotes the coefficient of ε^j. For precision calculations we are interested in the first few terms $I^{(j_{\min})}$, $I^{(j_{\min}+1)}$, ... of this Laurent series. The exact number of required terms depends on the order of perturbation theory we are calculating. Let us stress that we would like to get the $I^{(j)}$'s, not necessarily I itself. There are situations where a closed form expression for I is readily obtained, but the Laurent expansion in ε is not immediate.

Let us now start to get our hands dirty. We compute the first Feynman integrals. For the moment we focus on one-loop integrals, which only depend on the loop momentum squared (and no scalar product $k \cdot p$ with some external momentum). At first sight it seems that there aren't too many Feynman integrals of this type. The one-loop tadpole integral already exhausts these specifications:

$$T_\nu = e^{\varepsilon \gamma_E} \left(\mu^2\right)^{\nu - \frac{D}{2}} \int \frac{d^D k}{i \pi^{\frac{D}{2}}} \frac{1}{\left(-k^2 + m^2\right)^{\nu}}. \tag{2.100}$$

However, we will later see that we can always arrange the integrand of any Feynman integral in such a way that it only depends on k^2. Thus this covers an important case and doing this loop-by-loop allows us to perform all loop momenta integrations. However, there is no free lunch: Re-organising the integrand such that it depends only on k^2 introduces additional integrations (typically over Schwinger or Feynman parameters) and we merely shifted the complications from the loop momentum integration to the Schwinger or Feynman parameter integration.

After Wick rotation we have

$$T_\nu = e^{\varepsilon \gamma_E} \left(\mu^2\right)^{\nu - \frac{D}{2}} \int \frac{d^D K}{\pi^{\frac{D}{2}}} \frac{1}{\left(K^2 + m^2\right)^\nu}. \tag{2.101}$$

As the integrand only depends on K^2, it is natural to introduce spherical coordinates. In D dimensions they are given by

$$\begin{aligned} K^0 &= K \cos\theta_1, \\ K^1 &= K \sin\theta_1 \cos\theta_2, \\ &\ldots \\ K^{D-2} &= K \sin\theta_1 \ldots \sin\theta_{D-2} \cos\theta_{D-1}, \\ K^{D-1} &= K \sin\theta_1 \ldots \sin\theta_{D-2} \sin\theta_{D-1}. \end{aligned} \tag{2.102}$$

In D dimensions we have one radial variable K, $(D-2)$ polar angles θ_j (with $1 \le j \le D-2$) and one azimuthal angle θ_{D-1}. The measure becomes

$$d^D K = K^{D-1} dK \, d\Omega_D, \qquad d\Omega_D = \prod_{i=1}^{D-1} \sin^{D-1-i}\theta_i \, d\theta_i. \tag{2.103}$$

Integration over the angles yields

$$\int d\Omega_D = \int\limits_0^\pi d\theta_1 \sin^{D-2}\theta_1 \ldots \int\limits_0^\pi d\theta_{D-2} \sin\theta_{D-2} \int\limits_0^{2\pi} d\theta_{D-1} = \frac{2\pi^{\frac{D}{2}}}{\Gamma\left(\frac{D}{2}\right)}. \tag{2.104}$$

$\Gamma(z)$ denotes Euler's gamma function.

Digression Euler's gamma function and Euler's beta function
It is now the appropriate place to introduce two special functions, Euler's gamma function and Euler's beta function, which are used within dimensional regularisation to continue the results from integer D towards non-integer values. The **gamma function** *is defined for* $Re(z) > 0$ *by*

$$\Gamma(z) = \int\limits_0^\infty e^{-t} t^{z-1} dt. \tag{2.105}$$

It fulfils the functional equation

$$\Gamma(z+1) = z\,\Gamma(z). \tag{2.106}$$

For positive integers n it takes the values

$$\Gamma(n+1) = n! \;=\; 1 \cdot 2 \cdot 3 \cdot \ldots \cdot n. \tag{2.107}$$

The gamma function $\Gamma(z)$ has simple poles located on the negative real axis at $z = 0, -1, -2, \ldots$. Quite often we will need the expansion around these poles. This can be obtained from the expansion around $z = 1$ and the functional equation. The expansion around $z = 1$ reads

$$\Gamma(1+\varepsilon) = \exp\left(-\gamma_{\mathrm{E}}\varepsilon + \sum_{n=2}^{\infty} \frac{(-1)^n}{n} \zeta_n \varepsilon^n\right), \tag{2.108}$$

where γ_{E} is Euler's constant and ζ_n is given by

$$\zeta_n = \sum_{j=1}^{\infty} \frac{1}{j^n}. \tag{2.109}$$

ζ_n is called a **zeta value**. *As an example we obtain for the Laurent expansion around $z = 0$*

$$\Gamma(\varepsilon) = \frac{1}{\varepsilon} - \gamma_{\mathrm{E}} + O(\varepsilon). \tag{2.110}$$

It will be useful to know the residues of $\Gamma(z+a)$ and $\Gamma(-z+a)$ at the poles

$$\begin{aligned} \operatorname{res}\left(\Gamma(z+a), z = -a-n\right) &= \frac{(-1)^n}{n!}, \qquad n \in \mathbb{N}_0, \\ \operatorname{res}\left(\Gamma(-z+a), z = a+n\right) &= -\frac{(-1)^n}{n!}. \end{aligned} \tag{2.111}$$

For integers n we have the reflection identity

$$\frac{\Gamma(z-n)}{\Gamma(z)} = (-1)^n \frac{\Gamma(1-z)}{\Gamma(1-z+n)}. \tag{2.112}$$

Furthermore

$$\Gamma(z)\Gamma(1-z) = \frac{\pi}{\sin \pi z}, \tag{2.113}$$

from which we may deduce the value at $z = 1/2$:

$$\Gamma\left(\frac{1}{2}\right) = \sqrt{\pi}. \tag{2.114}$$

There is a duplication formula for the gamma function:

$$\prod_{j=0}^{k-1} \Gamma\left(z+\frac{j}{k}\right) = (2\pi)^{\frac{k-1}{2}} \, k^{\frac{1}{2}-kz} \, \Gamma(kz), \qquad k \in \mathbb{N}. \tag{2.115}$$

In particular we have for $k = 2$

$$\Gamma(z)\,\Gamma\left(z+\frac{1}{2}\right) = 2^{1-2z} \sqrt{\pi}\,\Gamma(2z). \tag{2.116}$$

Euler's beta function *is defined for* $Re(z_1) > 0$ *and* $Re(z_2) > 0$ *by*

$$B(z_1, z_2) = \int\limits_0^1 t^{z_1-1}(1-t)^{z_2-1} dt, \tag{2.117}$$

or equivalently by

$$B(z_1, z_2) = \int\limits_0^\infty \frac{t^{z_1-1}}{(1+t)^{z_1+z_2}} dt. \tag{2.118}$$

The beta function can be expressed in terms of gamma functions:

$$B(z_1, z_2) = \frac{\Gamma(z_1)\,\Gamma(z_2)}{\Gamma(z_1+z_2)}. \tag{2.119}$$

Note that the integration on the left-hand side of Eq. (2.104) is defined for any natural number D, whereas the result on the right-hand side is an analytic function of D, which can be continued to any complex value. Performing the angular integrations for our tadpole integral we obtain

$$T_\nu = \frac{e^{\varepsilon \gamma_E} \left(\mu^2\right)^{\nu-\frac{D}{2}}}{\Gamma\left(\frac{D}{2}\right)} \int\limits_0^\infty dK^2 \frac{\left(K^2\right)^{\frac{D}{2}-1}}{\left(K^2+m^2\right)^\nu}. \tag{2.120}$$

Please note that in Eq. (2.120) we may now allow non-integer values for D without any problems. Let us proceed and let us substitute $t = K^2/m^2$. We obtain

$$T_\nu = \frac{e^{\varepsilon \gamma_E}}{\Gamma\left(\frac{D}{2}\right)} \left(\frac{m^2}{\mu^2}\right)^{\frac{D}{2}-\nu} \int\limits_0^\infty dt \frac{t^{\frac{D}{2}-1}}{(t+1)^\nu}. \tag{2.121}$$

The remaining integral is just Euler's beta function

$$\int_0^\infty dt \frac{t^{\frac{D}{2}-1}}{(1+t)^\nu} = \frac{\Gamma\left(\frac{D}{2}\right)\Gamma\left(\nu - \frac{D}{2}\right)}{\Gamma(\nu)}. \tag{2.122}$$

Thus we computed our first Feynman integral (recall $D = D_{\text{int}} - 2\varepsilon$):

Tadpole integral:

$$T_\nu\left(D, \frac{m^2}{\mu^2}\right) = \frac{e^{\varepsilon\gamma_E}\Gamma\left(\nu - \frac{D}{2}\right)}{\Gamma(\nu)}\left(\frac{m^2}{\mu^2}\right)^{\frac{D}{2}-\nu}. \tag{2.123}$$

We are interested in the Laurent expansion of Feynman integrals. With $D = 4 - 2\varepsilon$, $\nu = 1$ and $L = \ln(m^2/\mu^2)$ we obtain

$$\begin{aligned} T_1(4-2\varepsilon) &= \frac{m^2}{\mu^2} e^{\varepsilon\gamma_E}\Gamma(-1+\varepsilon)\, e^{-\varepsilon L} \\ &= \frac{m^2}{\mu^2}\left[-\frac{1}{\varepsilon} + (L-1) + \left(-\frac{1}{2}L^2 - \frac{1}{2}\zeta_2 + L - 1\right)\varepsilon\right] + O\left(\varepsilon^2\right). \end{aligned} \tag{2.124}$$

The pole at $\varepsilon = 0$ originates from the ultraviolet divergence of tadpole integral with $\nu = 1$ in four space-time dimensions.

The result in Eq. (2.123) is valid for any D, so we may as well expand it around two space-time dimensions. Just for fun, let's do it:

$$\begin{aligned} \varepsilon\, T_1(2-2\varepsilon) &= e^{\varepsilon\gamma_E}\Gamma(1+\varepsilon)\, e^{-\varepsilon L} \\ &= 1 - L\varepsilon + \left(\frac{1}{2}L^2 + \frac{1}{2}\zeta_2\right)\varepsilon^2 + O\left(\varepsilon^3\right). \end{aligned} \tag{2.125}$$

Let us define a **weight**. We declare that $L = \ln(m^2/\mu^2)$ has weight one and that L^n and ζ_n have weight n. A rational number has weight 0. The weight of a product is the sum of the weights of its factors. We then spot a difference between $\varepsilon\, T_1(2-2\varepsilon)$ and $\varepsilon\, (\mu^2/m^2) T_1(4-2\varepsilon)$. In $\varepsilon\, T_1(2-2\varepsilon)$ we see that the j-th term in the ε-expansion only involves terms of weight j (we have only given the first three terms of the ε-expansion, but this statement holds to any order), whereas in $\varepsilon\, (\mu^2/m^2) T_1(4-2\varepsilon)$ the j-th term in the ε-expansion involves terms of weight j and terms of lower weight. In this sense, $\varepsilon\, T_1(2-2\varepsilon)$ has a simpler ε-expansion. We call $\varepsilon\, T_1(2-2\varepsilon)$ to be of **uniform weight**. We will discuss this issue in more detail in Chap. 6.

From Eq. (2.123) we also deduce

$$T_\nu(2-2\varepsilon) = \nu T_{\nu+1}(4-2\varepsilon). \tag{2.126}$$

This is an example of a **dimensional shift relation**, relating integrals in $D = 2 - 2\varepsilon$ and $D = 4 - 2\varepsilon$ space-time dimensions. Also dimensional shift relations will be discussed in more detail in Chap. 6.

Exercise 3 *Prove*

$$T_\nu(D) = \nu T_{\nu+1}(D+2). \tag{2.127}$$

Let us set temporarily

$$J_1 = \varepsilon\, T_1(2-2\varepsilon) \;=\; e^{\varepsilon\gamma_E}\Gamma(1+\varepsilon)\, e^{-\varepsilon L}. \tag{2.128}$$

It is not too difficult to show that

$$T_\nu(D) = \frac{\Gamma\left(\nu - \frac{D_{\text{int}}}{2} + \varepsilon\right)}{\Gamma(\nu)\,\Gamma(1+\varepsilon)} \left(\frac{m^2}{\mu^2}\right)^{\left(\frac{D_{\text{int}}}{2}-\nu\right)} J_1. \tag{2.129}$$

For $\nu \in \mathbb{N}$ and D_{int} even, the prefactor is always a rational function in ε and m^2, for example for $D = 4 - 2\varepsilon$ and $\nu = 1$ we have

$$\frac{\Gamma\left(\nu - \frac{D_{\text{int}}}{2} + \varepsilon\right)}{\Gamma(\nu)\,\Gamma(1+\varepsilon)} \left(\frac{m^2}{\mu^2}\right)^{\left(\frac{D_{\text{int}}}{2}-\nu\right)} = -\frac{1}{\varepsilon(1-\varepsilon)}\frac{m^2}{\mu^2}. \tag{2.130}$$

Equation (2.129) expresses any integral $T_\nu(D)$ as a coefficient times J_1. For the tadpole integrals this is a trivial statement, as the coefficient is just the ratio $T_\nu(D)/J_1$. Later on we will see, that this generalises as follows: We may express any member of a family of Feynman integrals as a linear combination of Feynman integrals from a finite set. The Feynman integrals from this finite set are called **master integrals**. A master integral, which is of uniform weight, is called a **canonical master integral**. Thus J_1 is a canonical master integral.

Let us close this section with some results on related integrals. The first result is a generalisation of the tadpole integrals and is helpful whenever we iteratively integrate out loop momenta (after having arranged that the integrand only depends on k^2). We consider

$$\tilde{T} = e^{\varepsilon\gamma_E}\left(\mu^2\right)^{\nu-\frac{D}{2}-a} \int \frac{d^Dk}{i\pi^{\frac{D}{2}}} \frac{\left(-k^2\right)^a}{\left(-Uk^2+V\right)^\nu}, \tag{2.131}$$

where U, V and a do not depend on k. For $a = 0$, $U = 1$ and $V = m^2$ we recover the tadpole integral. Following the same steps as for the tadpole integral one finds

$$\tilde{T} = e^{\varepsilon\gamma_E}\left(\mu^2\right)^{\nu-\frac{D}{2}-a} \frac{\Gamma\left(\frac{D}{2}+a\right)}{\Gamma\left(\frac{D}{2}\right)} \frac{\Gamma\left(\nu-\frac{D}{2}-a\right)}{\Gamma(\nu)} \frac{U^{-\frac{D}{2}-a}}{V^{\nu-\frac{D}{2}-a}} \tag{2.132}$$

and the

one-loop master formula:

$$\int \frac{d^D k}{i\pi^{\frac{D}{2}}} \frac{\left(-k^2\right)^a}{\left(-Uk^2+V\right)^\nu} = \frac{\Gamma\left(\frac{D}{2}+a\right)}{\Gamma\left(\frac{D}{2}\right)} \frac{\Gamma\left(\nu-\frac{D}{2}-a\right)}{\Gamma(\nu)} \frac{U^{-\frac{D}{2}-a}}{V^{\nu-\frac{D}{2}-a}}. \quad (2.133)$$

In the definition of $\tilde{T}$ we allowed additional powers $(-k^2)^a$ of the loop momentum in the numerator. Note that the dependency of the result on a, apart from a factor $\Gamma(D/2+a)/\Gamma(D/2)$, occurs only in the combination $D/2+a$. Therefore adding additional powers $(-k^2)^a$ to the numerator is almost equivalent to consider the integral without this factor in dimensions $(D+2a)$.

Exercise 4 *Derive Eq. (2.132).*

There is one more generalisation: Sometimes it is convenient to decompose k^2 into a D_{int}-dimensional piece and a remainder:

$$k_{(D)}^2 = k_{(D_{\text{int}})}^2 + k_{(-2\varepsilon)}^2. \quad (2.134)$$

If D is an integer greater than D_{int} we have

$$k_{(D_{\text{int}})}^2 = \left(k^0\right)^2 - \left(k^1\right)^2 - \dots - \left(k^{D_{\text{int}}-1}\right)^2,$$
$$k_{(-2\varepsilon)}^2 = -\left(k^{D_{\text{int}}}\right)^2 - \dots - \left(k^{D-1}\right)^2. \quad (2.135)$$

We also need loop integrals where additional powers of $(-k_{(-2\varepsilon)}^2)$ appear in the numerator. These are related to integrals in higher dimensions as follows:

ε-components in the numerator:

$$\int \frac{d^D k}{i\pi^{\frac{D}{2}}} \left(-k_{(-2\varepsilon)}^2\right)^r f\left(k_{(D_{\text{int}})}, k_{(-2\varepsilon)}^2\right) = \frac{\Gamma(r-\varepsilon)}{\Gamma(-\varepsilon)} \int \frac{d^{D+2r} k}{i\pi^{\frac{D+2r}{2}}} f\left(k_{(D_{\text{int}})}, k_{(-2\varepsilon)}^2\right). \quad (2.136)$$

Here, $f(k_{(D_{\text{int}})}, k_{(-2\varepsilon)}^2)$ is a function which depends on $k^{D_{\text{int}}}, k^{D_{\text{int}}+1}, \dots, k^{D-1}$ only through $k_{(-2\varepsilon)}^2$. The dependency on $k^0, k^1, \dots, k^{D_{\text{int}}-1}$ is not constrained.

Exercise 5 *Derive Eq. (2.136).*

Hint: Split the D-dimensional integration into a D_{int}-dimensional part and a (-2ε)-dimensional part. Equation (2.136) can be derived by just considering the (-2ε)-dimensional part.

Finally it is worth noting that

$$\int \frac{d^D k}{i\pi^{\frac{D}{2}}} \left(-k^2\right)^a = \begin{cases} (-1)^{\frac{D}{2}} \Gamma\left(1-\frac{D}{2}\right), & \text{if } \frac{D}{2}+a=0, \\ 0, & \text{otherwise.} \end{cases} \quad (2.137)$$

Exercise 6 *Derive Eq. (2.137).*

Hint: Consider the mass dimension of the integral to prove the statement for $D/2 + a \neq 0$ and the normalisation of the integral measure in Eq. (2.77) to prove the statement for $D/2 + a = 0$.

2.5 Representations Of Feynman Integrals

There are several integral representations for Feynman integrals. We introduce the various integral representations in this section.

As before, we consider a Feynman graph G with n_{ext} external edges, n_{int} internal edges and l loops. To each external edge we associate an external momentum, labelled by $p_1, \dots, p_{n_{\text{ext}}}$. To each internal edge e_j we associate a triple (q_j, m_j, ν_j), where q_j is the momentum flowing through this edge, m_j the mass and ν_j the power to which the propagator occurs. Momentum conservation at each vertex of valency > 1 allows us to express any q_j as a linear combination of $(n_{\text{ext}} - 1)$ linear independent external momenta and l independent loop momenta. We denote the latter by $k_1, \dots, k_l$.

2.5.1 The Momentum Representation Of Feynman Integrals

The momentum representation of Feynman integrals is the one we started with in Eq. (2.56):

$$I = e^{l\varepsilon\gamma_{\text{E}}} \left(\mu^2\right)^{\nu - \frac{lD}{2}} \int \prod_{r=1}^{l} \frac{d^D k_r}{i\pi^{\frac{D}{2}}} \prod_{j=1}^{n_{\text{int}}} \frac{1}{\left(-q_j^2 + m_j^2\right)^{\nu_j}}, \tag{2.138}$$

where

$$\varepsilon = \frac{D_{\text{int}} - D}{2}, \qquad \nu = \sum_{j=1}^{n_{\text{int}}} \nu_j, \qquad q_j = \sum_{r=1}^{l} \lambda_{jr} k_r + \sum_{r=1}^{n_{\text{ext}}-1} \sigma_{jr} p_r. \tag{2.139}$$

The coefficients λ_{jr} and σ_{jr} can be obtained from momentum conservation at each vertex of valency > 1. The integration contour is given by Feynman's $i\delta$-prescription.

Let us discuss the variables the Feynman integral depends on. First of all, the Feynman integrals depends on the dimension of space-time $D \in \mathbb{C}$ and through the prefactor $e^{l\varepsilon\gamma_{\text{E}}}$ on D_{int}. Secondly, the Feynman integral depends also on the n_{int}-tuple $(\nu_1, \dots, \nu_{n_{\text{int}}})$. In principle we may allow $\nu_j \in \mathbb{C}$, but very often we will limit us to the case $\nu_j \in \mathbb{Z}$. Thirdly, the Feynman integral depends on kinematic variables. The Feynman integral in Eq. (2.138) is a scalar integral, thus the dependence on the $(n_{\text{ext}} -$

1) linear independent external momenta is only through the Lorentz invariants

$$p_i \cdot p_j. \tag{2.140}$$

The Feynman integral in Eq. (2.138) is dimensionless, therefore the dependence on the Lorentz invariants, the internal masses and the scale μ is only through the dimensionless ratios

$$\frac{-p_i \cdot p_j}{\mu^2}, \ \frac{m_i^2}{\mu^2}. \tag{2.141}$$

We call these the **kinematic variables**. We will denote the kinematic variables by $x_1, x_2, \dots$. Let us count how many kinematic variables there can be. We may have $n_{\text{ext}}(n_{\text{ext}} - 1)/2$ kinematic variables of the type

$$\frac{-p_i \cdot p_j}{\mu^2}, \quad 1 \leq i \leq j \leq (n_{\text{ext}} - 1), \tag{2.142}$$

and n_{int} kinematic variables of the type

$$\frac{m_i^2}{\mu^2}. \tag{2.143}$$

However, if we rescale all kinematic variables by a factor λ we have

$$I(\lambda x_1, \lambda x_2, \dots) = \lambda^{\frac{lD}{2} - \nu} I(x_1, x_2, \dots). \tag{2.144}$$

This is most easily seen by substituting $\mu^2 \to \mu^2/\lambda$ in Eq. (2.138). Thus, we may set one kinematic variable to 1 and recover the full dependence on all kinematic variables from the scaling relation in Eq. (2.144).

In total we may have up to

$$\frac{n_{\text{ext}}(n_{\text{ext}} - 1)}{2} + n_{\text{int}} - 1 \tag{2.145}$$

kinematic variables. In typical applications some of them may be zero (for example some internal masses might be zero) or identical (for example some internal masses might be identical). We denote the number of independent kinematic variables by N_B and the independent kinematic variables by $x_1, \dots, x_{N_B}$:

Notation:

number of independent kinematic variables: N_B
independent kinematic variables: $x_1, x_2, \dots, x_{N_B}$

The x_j's are dimensionless quantities of the form as in Eq. (2.141).

Exercise 7 *Consider again the one-loop box graph shown in Fig. 2.4. Assume first that all internal masses are non-zero and pairwise distinct and that the external momenta are as generic as possible. How many kinematic variables are there?*

Secondly, assume that all internal masses are zero and that the external momenta satisfy $p_1^2 = p_2^2 = p_3^2 = p_4^2 = 0$. *How many kinematic variables are there now?*

We allow the x_j's to be complex numbers. We will often encounter the situation, where a kinematic variable is given by a real number plus an infinitesimal small imaginary part. The infinitesimal small imaginary part is inherited from Feynman's $i\delta$-prescription. The special case, where all kinematic variables are real and non-negative is called the **Euclidean region**. We have defined the kinematic variables involving Lorentz invariants of the external momenta with a minus sign as in Eq. (2.142). We may define **Euclidean external momenta** in the same way as we defined the Euclidean loop momentum in Eq. (2.65):

$$\begin{aligned} p^0 &= i P^0, \\ p^j &= P^j, \qquad \text{for } 1 \leq j \leq D-1. \end{aligned} \tag{2.146}$$

Then

$$\frac{-p_i \cdot p_j}{\mu^2} = \frac{P_i \cdot P_j}{\mu^2}, \tag{2.147}$$

where the scalar product on the right-hand side is calculated with Euclidean signature

$$(+, +, +, +, \dots). \tag{2.148}$$

Thus, if everything is expressed in Euclidean variables, no minus sign appears.

Let us add one word on our notation: We denote the Feynman integral in Eq. (2.138) by I. If we want to emphasise that this integral corresponds to the graph G we write

$$I_G, \tag{2.149}$$

if we want to give the dependence on all variables we write

$$I_{\nu_1 \dots \nu_{n_{\text{int}}}} \left(D, x_1, \dots, x_{N_B}, D_{\text{int}} \right). \tag{2.150}$$

In situations, where the dependence on some specific variables is relevant, while the dependence on the other variables is not, we may write the former and suppress the latter. In particular we will almost always suppress the dependence on D_{int} and write

$$I_{\nu_1 \dots \nu_{n_{\text{int}}}} \left(D, x_1, \dots, x_{N_B} \right). \tag{2.151}$$

Examples of even shorter notations are $I_{\nu_1 \dots \nu_{n_{\text{int}}}}$ or $I(D)$. Thus we will use the notation which is most appropriate within a given context.

2.5.2 *The Schwinger Parameter Representation*

The Schwinger parameter representation is the first representation, where we trade the (lD) momentum integrations for some auxiliary integrations. In the case of the Schwinger parameter representation we will treat the momentum integration for an integration over n_{int} Schwinger parameters. We start from the following identity, also called **Schwinger's trick**: Let $A > 0$ and $\text{Re}(\nu) > 0$. We have

$$\frac{1}{A^\nu} = \frac{1}{\Gamma(\nu)} \int_0^\infty d\alpha \, \alpha^{\nu-1} \, e^{-\alpha A}. \tag{2.152}$$

Equation (2.152) follows immediately from the definition of Euler's gamma function. We apply Eq. (2.152) for $A = Q^2 + m^2 = -q^2 + m^2$ (the quantity A is positive after Wick rotation and for external Euclidean kinematics):

$$\frac{1}{(-q_j^2 + m_j^2)^{\nu_j}} = \frac{1}{\Gamma(\nu_j)} \int_{\alpha_j \geq 0} d\alpha_j \, \alpha_j^{\nu_j - 1} \exp\left(-\alpha_j(-q_j^2 + m_j^2)\right) \tag{2.153}$$

The variable α_j is called a **Schwinger parameter**. Doing this for all internal edges gives

$$I = \frac{e^{l\varepsilon\gamma_{\text{E}}} \left(\mu^2\right)^{\nu - \frac{lD}{2}}}{\prod\limits_{j=1}^{n_{\text{int}}} \Gamma(\nu_j)} \int_{\alpha_j \geq 0} d^{n_{\text{int}}}\alpha \left(\prod_{j=1}^{n_{\text{int}}} \alpha_j^{\nu_j - 1} \right) \int \prod_{r=1}^{l} \frac{d^D k_r}{i\pi^{\frac{D}{2}}} \exp\left(-\sum_{j=1}^{n_{\text{int}}} \alpha_j \left(-q_j^2 + m_j^2 \right) \right). \tag{2.154}$$

Using

$$q_j = \sum_{r=1}^{l} \lambda_{jr} k_r + \sum_{r=1}^{n_{\text{ext}}-1} \sigma_{jr} p_r \tag{2.155}$$

we express the argument of the exponential function as

$$\sum_{j=1}^{n_{\text{int}}} \alpha_j(-q_j^2 + m_j^2) = -\sum_{r=1}^{l} \sum_{s=1}^{l} k_r M_{rs} k_s + \sum_{r=1}^{l} 2k_r \cdot v_r + J. \tag{2.156}$$

where M is a $l \times l$ matrix with scalar entries, v is a l-vector with D-dimensional momentum vectors as entries and J is scalar. Let us define

$$\mathcal{U} = \det(M), \ \ \mathcal{F} = \det(M) \left(J + v^T M^{-1} v \right) / \mu^2. \tag{2.157}$$

The functions $\mathcal{U}$ and $\mathcal{F}$ are called **graph polynomials** (they are polynomials in the α_j's) and are discussed in detail in Chap. 3. The polynomials $\mathcal{U}$ and $\mathcal{F}$ have the following properties:

- They are homogeneous in the Schwinger parameters, $\mathcal{U}$ is of degree l, $\mathcal{F}$ is of degree $l+1$.
- $\mathcal{U}$ is linear in each Schwinger parameter. If all internal masses are zero, then also $\mathcal{F}$ is linear in each Schwinger parameter.
- In expanded form each monomial of $\mathcal{U}$ has coefficient $+1$.

We call $\mathcal{U}$ the **first Symanzik polynomial** and $\mathcal{F}$ the **second Symanzik polynomial**.

From linear algebra we know that for a real symmetric positive definite $(n \times n)$-matrix A we have

$$\int\limits_{-\infty}^{\infty} dy_1 ... dy_n \ \exp\left(-\mathbf{y}^T A \mathbf{y} + 2 \mathbf{w}^T \mathbf{y} + c \right) = \pi^{n/2} \left(\det A \right)^{-\frac{1}{2}} \exp\left(\mathbf{w}^T A^{-1} \mathbf{w} + c \right), \tag{2.158}$$

and due to Eq. (2.77) this extends to dimensional regularisation. We may therefore perform the loop momentum integration and obtain the

Schwinger parameter representation:

$$I = \frac{e^{l \varepsilon \gamma_E}}{\prod\limits_{j=1}^{n_{\text{int}}} \Gamma(\nu_j)} \int\limits_{\alpha_j \geq 0} d^{n_{\text{int}}} \alpha \left(\prod_{j=1}^{n_{\text{int}}} \alpha_j^{\nu_j - 1} \right) \left[\mathcal{U}(\alpha) \right]^{-\frac{D}{2}} \exp\left(-\frac{\mathcal{F}(\alpha)}{\mathcal{U}(\alpha)} \right). \tag{2.159}$$

Thus we went from a $(l \cdot D)$-fold momentum integration to a n_{int}-fold Schwinger parameter integration. Note that the number of space-time dimensions D enters only the exponent of $\mathcal{U}^{-D/2}$ (and the prefactor $e^{l\varepsilon\gamma}$ through $\varepsilon = (D_{\text{int}} - D)/2$).

We now encountered for the first time the sum on the left-hand side of Eq. (2.156) and the two graph polynomials $\mathcal{U}$ and $\mathcal{F}$. As these ingredients will occur quite frequently throughout this book, it is worth working them out in a non-trivial example. We consider the two-loop graph shown in Fig. 2.3 for the case

$$p_1^2 = 0, \quad p_2^2 = 0, \quad p_3^2 = 0, \quad p_4^2 = 0,$$
$$m_1 = m_2 = m_3 = m_4 = m_5 = m_6 = m_7 = 0. \tag{2.160}$$

We define

$$s = (p_1 + p_2)^2 = (p_3 + p_4)^2 , \quad t = (p_2 + p_3)^2 = (p_1 + p_4)^2 . \quad (2.161)$$

We have

$$\sum_{j=1}^{7} \alpha_j \left(-q_j^2\right) = -(\alpha_1 + \alpha_2 + \alpha_3 + \alpha_4) k_1^2 - 2\alpha_4 k_1 \cdot k_2 - (\alpha_4 + \alpha_5 + \alpha_6 + \alpha_7) k_2^2 \quad (2.162)$$
$$+2\left[\alpha_1 p_1 + \alpha_2 (p_1 + p_2)\right] \cdot k_1 + 2\left[\alpha_5 (p_3 + p_4) + \alpha_7 p_4\right] \cdot k_2 - (\alpha_2 + \alpha_5) s.$$

In comparing with Eq. (2.156) we find

$$M = \begin{pmatrix} \alpha_1 + \alpha_2 + \alpha_3 + \alpha_4 & \alpha_4 \\ \alpha_4 & \alpha_4 + \alpha_5 + \alpha_6 + \alpha_7 \end{pmatrix},$$
$$v = \begin{pmatrix} \alpha_1 p_1 + \alpha_2 (p_1 + p_2) \\ \alpha_5 (p_3 + p_4) + \alpha_7 p_4 \end{pmatrix},$$
$$J = (\alpha_2 + \alpha_5)(-s) . \quad (2.163)$$

Plugging this into Eq. (2.157) we obtain the graph polynomials as

$$\mathcal{U} = (\alpha_1 + \alpha_2 + \alpha_3)(\alpha_5 + \alpha_6 + \alpha_7) + \alpha_4 (\alpha_1 + \alpha_2 + \alpha_3 + \alpha_5 + \alpha_6 + \alpha_7),$$
$$\mathcal{F} = \left[\alpha_2\alpha_3 (\alpha_4 + \alpha_5 + \alpha_6 + \alpha_7) + \alpha_5\alpha_6 (\alpha_1 + \alpha_2 + \alpha_3 + \alpha_4) + \alpha_2\alpha_4\alpha_6 + \alpha_3\alpha_4\alpha_5\right] \left(\frac{-s}{\mu^2}\right)$$
$$+\alpha_1\alpha_4\alpha_7 \left(\frac{-t}{\mu^2}\right). \quad (2.164)$$

We see in this example that $\mathcal{U}$ is of degree 2 and $\mathcal{F}$ is of degree 3 in the Schwinger parameters. Each polynomial is linear in each Schwinger parameter. Furthermore, when we write $\mathcal{U}$ in expanded form

$$\mathcal{U} = \alpha_1\alpha_5 + \alpha_1\alpha_6 + \alpha_1\alpha_7 + \alpha_2\alpha_5 + \alpha_2\alpha_6 + \alpha_2\alpha_7 + \alpha_3\alpha_5 + \alpha_3\alpha_6 + \alpha_3\alpha_7$$
$$+\alpha_1\alpha_4 + \alpha_2\alpha_4 + \alpha_3\alpha_4 + \alpha_4\alpha_5 + \alpha_4\alpha_6 + \alpha_4\alpha_7, \quad (2.165)$$

each term has coefficient $+1$.

Exercise 8 *Determine with the method above the graph polynomials $\mathcal{U}$ and $\mathcal{F}$ for the graph shown in Fig. 2.6 for the case where all internal masses are zero.*

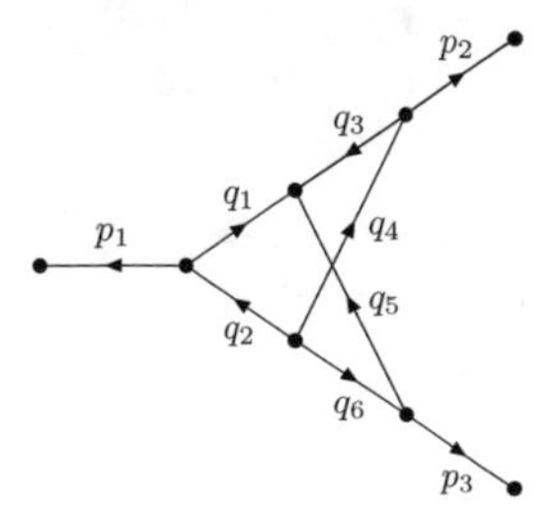

Fig. 2.6 A two-loop non-planar vertex graph

2.5.3 The Feynman Parameter Representation

Probably the most popular parameter representation is the Feynman parameter representation. It is effectively a $(n_{\mathrm{int}} - 1)$-fold integral representation. We obtain the Feynman parameter representation from the Schwinger parameter representation as follows: We first note that the sum of the Schwinger parameters is non-negative:

$$\sum_{j=1}^{n} \alpha_j \geq 0. \tag{2.166}$$

We then insert a 1 in the form of

$$1 = \int_{-\infty}^{\infty} dt \, \delta\left(t - \sum_{j=1}^{n} \alpha_j\right) = \int_{0}^{\infty} dt \, \delta\left(t - \sum_{j=1}^{n} \alpha_j\right), \tag{2.167}$$

where in the last step we used the fact that the sum of the Schwinger parameters is non-negative. $\delta(x)$ denotes Dirac's delta distribution. Changing variables according to $a_j = \alpha_j / t$ gives us the identity

$$\int_{\alpha_j \geq 0} d^n\alpha \; f(\alpha_1, \ldots, \alpha_n) = \int_{a_j \geq 0} d^n a \; \delta\left(1 - \sum_{j=1}^{n} a_j\right) \int_{0}^{\infty} dt \; t^{n-1} \; f(ta_1, \ldots, ta_n) . \tag{2.168}$$

We apply this identity to the Schwinger parameter representation and use the fact that $\mathcal{U}$ and $\mathcal{F}$ are homogeneous of degree l and $(l + 1)$, respectively. We obtain

$$\begin{aligned} I = & \frac{e^{l\varepsilon\gamma_{\mathrm{E}}}}{\prod\limits_{j=1}^{n_{\mathrm{int}}} \Gamma(\nu_j)} \int_{a_j \geq 0} d^{n_{\mathrm{int}}} a \; \delta\left(1 - \sum_{j=1}^{n_{\mathrm{int}}} a_j\right) \left(\prod_{j=1}^{n_{\mathrm{int}}} a_j^{\nu_j - 1}\right) \left[\mathcal{U}(a)\right]^{-\frac{D}{2}} \\ & \times \int_{0}^{\infty} dt \; t^{\nu - \frac{lD}{2} - 1} \exp\left(-\frac{\mathcal{F}(a)}{\mathcal{U}(a)} t\right) \\ = & \frac{e^{l\varepsilon\gamma_{\mathrm{E}}}}{\prod\limits_{j=1}^{n_{\mathrm{int}}} \Gamma(\nu_j)} \int_{a_j \geq 0} d^{n_{\mathrm{int}}} a \; \delta\left(1 - \sum_{j=1}^{n_{\mathrm{int}}} a_j\right) \left(\prod_{j=1}^{n_{\mathrm{int}}} a_j^{\nu_j - 1}\right) \frac{\left[\mathcal{U}(a)\right]^{\nu - \frac{(l+1)D}{2}}}{\left[\mathcal{F}(a)\right]^{\nu - \frac{lD}{2}}} \int_{0}^{\infty} dt \; t^{\nu - \frac{lD}{2} - 1} e^{-t}. \end{aligned} \tag{2.169}$$

In the step towards the last line we substituted $t \to t\mathcal{U}(a)/\mathcal{F}(a)$. The final integral over t gives $\Gamma(\nu - lD/2)$. We thus arrive at the

Feynman parameter representation:

$$I = \frac{e^{l\varepsilon\gamma_E}\Gamma\left(\nu - \frac{lD}{2}\right)}{\prod\limits_{j=1}^{n_{\text{int}}}\Gamma(\nu_j)} \int\limits_{a_j \geq 0} d^{n_{\text{int}}}a\, \delta\left(1 - \sum_{j=1}^{n_{\text{int}}} a_j\right) \left(\prod_{j=1}^{n_{\text{int}}} a_j^{\nu_j - 1}\right) \frac{[\mathcal{U}(a)]^{\nu - \frac{(l+1)D}{2}}}{[\mathcal{F}(a)]^{\nu - \frac{lD}{2}}}. \quad (2.170)$$

The polynomials $\mathcal{U}$ and $\mathcal{F}$ are as before, with α_j substituted by a_j. The variable a_j is called a **Feynman parameter**.

We have derived the Feynman parameter representation from the Schwinger parameter representation. We may go directly from the momentum representation to the Feynman parameter representation with the help of **Feynman's trick**: For $A_j > 0$ and $\text{Re}(\nu_j) > 0$ we have

$$\prod_{j=1}^{n} \frac{1}{A_j^{\nu_j}} = \frac{\Gamma(\nu)}{\prod\limits_{j=1}^{n}\Gamma(\nu_j)} \int\limits_{a_j \geq 0} d^n a\; \delta(1 - \sum_{j=1}^{n} a_j) \left(\prod_{j=1}^{n} a_j^{\nu_j - 1}\right) \frac{1}{\left(\sum\limits_{j=1}^{n} a_j A_j\right)^{\nu}},$$

$$\nu = \sum_{j=1}^{n} \nu_j. \quad (2.171)$$

Exercise 9 *Prove Eq. (2.171).*

We use this formula with $A_j = -q_j^2 + m_j^2$. We may then use translational invariance (i.e. Eq. (2.75)) for each D-dimensional momentum integral and shift each loop momentum k_r to complete the square, such that the integrand depends only on k_r^2. Then all D-dimensional momentum integrals can be performed with the help of Eq. (2.133) and we recover Eq. (2.170).

Let us look at an example.

We consider the one-loop bubble diagram shown in Fig. 2.7. The momentum representation is

$$I_{\nu_1\nu_2} = e^{\varepsilon\gamma_E}\left(\mu^2\right)^{\nu - \frac{D}{2}} \int \frac{d^D k}{i\pi^{\frac{D}{2}}} \frac{1}{\left(-q_1^2 + m_1^2\right)^{\nu_1}\left(-q_2^2 + m_2^2\right)^{\nu_2}}, \quad (2.172)$$

with

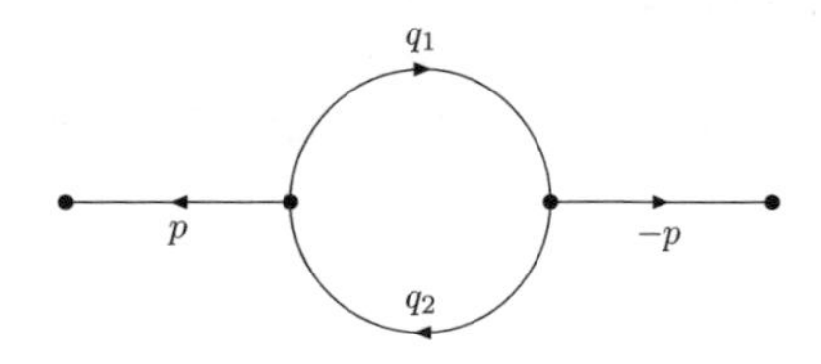

Fig. 2.7 The one-loop bubble diagram

$$q_1 = k - p, \; q_2 = k. \tag{2.173}$$

Feynman parametrisation gives us

$$I_{\nu_1\nu_2} = \frac{e^{\varepsilon\gamma_E}\Gamma(\nu)}{\Gamma(\nu_1)\Gamma(\nu_2)}\left(\mu^2\right)^{\nu-\frac{D}{2}} \int d^2a\, \delta(1-a_1-a_2)\, a_1^{\nu_1-1} a_2^{\nu_2-1} \times \int \frac{d^Dk}{i\pi^{\frac{D}{2}}} \frac{1}{\left[a_1\left(-q_1^2+m_1^2\right)+a_2\left(-q_2^2+m_2^2\right)\right]^{\nu}}. \tag{2.174}$$

Completing the square we obtain

$$a_1\left(-q_1^2+m_1^2\right)+a_2\left(-q_2^2+m_2^2\right) = -\left(a_1+a_2\right)\left(k-\frac{a_1}{a_1+a_2}p\right)^2 + \frac{a_1a_2}{a_1+a_2}\left(-p^2\right) + a_1m_1^2 + a_2m_2^2. \tag{2.175}$$

Let us set

$$\mathcal{U} = a_1 + a_2,$$
$$\mathcal{F} = a_1a_2\left(\frac{-p^2}{\mu^2}\right) + \left(a_1+a_2\right)\left[a_1\left(\frac{m_1^2}{\mu^2}\right) + a_2\left(\frac{m_2^2}{\mu^2}\right)\right]. \tag{2.176}$$

These are the two graph polynomials. With the substitution $k \to k + a_1/(a_1+a_2)p$ we obtain

$$I_{\nu_1\nu_2} = \frac{e^{\varepsilon\gamma_E}\Gamma(\nu)}{\Gamma(\nu_1)\Gamma(\nu_2)}\left(\mu^2\right)^{\nu-\frac{D}{2}} \int d^2a\, \delta(1-a_1-a_2)\, a_1^{\nu_1-1} a_2^{\nu_2-1} \int \frac{d^Dk}{i\pi^{\frac{D}{2}}} \frac{1}{\left[-\mathcal{U}k^2 + \frac{\mathcal{F}}{\mathcal{U}}\mu^2\right]^{\nu}}. \tag{2.177}$$

This is now in the form of Eq. (2.133) and using the one-loop master formula yields

$$I_{\nu_1\nu_2} = \frac{e^{\varepsilon\gamma_E}\Gamma\left(\nu-\frac{D}{2}\right)}{\Gamma(\nu_1)\Gamma(\nu_2)} \int d^2a\, \delta(1-a_1-a_2)\, a_1^{\nu_1-1} a_2^{\nu_2-1} \frac{\mathcal{U}^{\nu-D}}{\mathcal{F}^{\nu-\frac{D}{2}}}. \tag{2.178}$$

This is again the Feynman parameter representation of Eq. (2.170), which we recovered "by foot". Of course, it is just sufficient to determine the two graph polynomials and use Eq. (2.170) directly. We have seen one method to determine the two graph polynomials in Sect. 2.5.2. We will learn more (efficient) methods to determine the graph polynomials in Chap. 3.

Let us now specialise to the case, where the internal masses vanish: $m_1 = m_2 = 0$. In this case the second graph polynomial simplifies to

$$\mathcal{F} = a_1a_2\left(\frac{-p^2}{\mu^2}\right) \tag{2.179}$$

and the one-loop bubble integral to

$$I_{\nu_1\nu_2} = \frac{e^{\varepsilon\gamma_E}\Gamma\left(\nu - \frac{D}{2}\right)}{\Gamma(\nu_1)\Gamma(\nu_2)} \left(\frac{-p^2}{\mu^2}\right)^{\frac{D}{2}-\nu} \int\limits_0^1 da\, a^{\frac{D}{2}-\nu_2-1} (1-a)^{\frac{D}{2}-\nu_1-1}. \tag{2.180}$$

The integral over a is just Euler's beta function and we obtain

$$I_{\nu_1\nu_2} = e^{\varepsilon\gamma_E} \frac{\Gamma\left(\nu - \frac{D}{2}\right)\Gamma\left(\frac{D}{2} - \nu_1\right)\Gamma\left(\frac{D}{2} - \nu_2\right)}{\Gamma(\nu_1)\Gamma(\nu_2)\Gamma(D-\nu)} \left(\frac{-p^2}{\mu^2}\right)^{\frac{D}{2}-\nu}. \tag{2.181}$$

If we further specialise to $\nu_1 = \nu_2 = 1$, $D = 4 - 2\varepsilon$ and set $L = \ln(-p^2/\mu^2)$ we find

$$I_{11} = \frac{1}{\varepsilon} + 2 - L + O(\varepsilon). \tag{2.182}$$

Thus we calculated our second Feynman integral, the massless one-loop two-point function.

Exercise 10 *Calculate with the help of the Feynman parameter representation the one-loop triangle integral*

$$I_{\nu_1\nu_2\nu_3} = e^{\varepsilon\gamma_E} \left(\mu^2\right)^{\nu-\frac{D}{2}} \int \frac{d^Dk}{i\pi^{\frac{D}{2}}} \frac{1}{\left(-q_1^2\right)^{\nu_1}\left(-q_2^2\right)^{\nu_2}\left(-q_3^2\right)^{\nu_3}}, \tag{2.183}$$

shown in Fig. 2.8 for the case where all internal masses are zero ($m_1 = m_2 = m_3 = 0$) and for the kinematic configuration $p_1^2 = p_2^2 = 0$, $p_3^2 \neq 0$.

The Feynman parameter representation of Eq. (2.170) treats every internal edge equal. This is called the **democratic approach**. However, this is not the only possibility. We may also use a **hierarchical approach**. The following exercise shows, that there are situations where this is useful.

Exercise 11 *Consider again the one-loop box graph in Fig. 2.4, this time for the kinematic configuration*

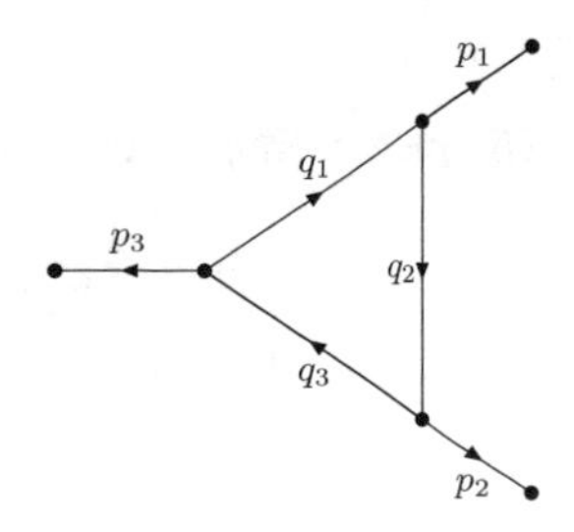

Fig. 2.8 The one-loop triangle diagram

$$p_2^2 = p_4^2 = 0, \; m_1 = m_2 = m_3 = m_4 = 0. \tag{2.184}$$

Write down the Feynman parameter representation as in Eq. (2.170). Obtain a second integral representation by first combining propagators 1 *and* 2 *with a pair of Feynman parameters, then combining propagators* 3 *and* 4 *with a second pair of Feynman parameters and finally the two results with a third pair of Feynman parameters.*

2.5.3.1 Projective Integrals

The Feynman parameter representation is actual a projective integral. It pays off to rewrite the Feynman parameter representation in terms of differential forms on projective space. Below we will state the Cheng-Wu theorem, which follows directly from the fact the Feynman parameter representation is a projective integral.

We start with introducing the essential facts about projective space.

Digression Projective space
Let $\mathbb{F}$ be a field. Relevant to us are the cases where the field $\mathbb{F}$ is either the field of real numbers $\mathbb{R}$ or the field of complex numbers $\mathbb{C}$.

The projective space $\mathrm{P}^n(\mathbb{F})$ is the set of lines through the origin in $\mathbb{F}^{n+1}$. Equivalently, it is the set of points in $\mathbb{F}^{n+1}\backslash\{0\}$ modulo the equivalence relation

$$(x_0, x_1, ..., x_n) \sim (y_0, y_1, ..., y_n) \Leftrightarrow \exists\, \lambda \neq 0 : (x_0, x_1, ..., x_n) = (\lambda y_0, \lambda y_1, ..., \lambda y_n)\,. \tag{2.185}$$

Points in $\mathrm{P}^n(\mathbb{F})$ will be denoted by

$$[z_0 : z_1 : ... : z_n]\,. \tag{2.186}$$

The coordinates in Eq. (2.186) are called **homogeneous coordinates**. *Affine coordinate patches are defined as follows: We consider the open subsets*

$$U_j = \left\{ [z_0 : z_1 : ... : z_n] \mid z_j \neq 0 \right\}, \qquad 0 \leq j \leq n. \tag{2.187}$$

We have the homeomorphisms

$$\begin{aligned} \varphi_j &: U_j \to \mathbb{F}^n, \\ &[z_0 : z_1 : ... : z_n] \to \left(\frac{z_0}{z_j}, ..., \frac{z_{j-1}}{z_j}, \frac{z_{j+1}}{z_j}, ..., \frac{z_n}{z_j} \right). \end{aligned} \tag{2.188}$$

The inverse mapping is given by

$$\begin{aligned} \varphi_j^{-1} &: \mathbb{F}^n \to U_j, \\ &\left(z_0, ..., z_{j-1}, z_{j+1}, ..., z_n\right) \to \left[z_0 : ... : z_{j-1} : 1 : z_{j+1} : ... : z_n\right]. \end{aligned} \tag{2.189}$$

The pair (U_j, φ_j) *defines a chart for* $\mathrm{P}^n(\mathbb{F})$*, and the collection of all* (U_j, φ_j) *for* $0 \leq j \leq n$ *provides an atlas for* $\mathrm{P}^n(\mathbb{F})$*.*

For $\mathbb{F} = \mathbb{C}$ *or* $\mathbb{F} = \mathbb{R}$ *we will also use the notation*

$$\mathbb{CP}^n = \mathrm{P}^n(\mathbb{C}), \quad \mathbb{RP}^n = \mathrm{P}^n(\mathbb{R}), \tag{2.190}$$

and we will speak about the complex projective space and the real projective space, respectively.

The **positive real projective space** $\mathbb{RP}^n_{>0}$ *is the set of all points of* $\mathbb{RP}^n$*, which can be represented by*

$$[x_0 : x_1 : ... : x_n] \text{ with } x_j > 0, \qquad 0 \leq j \leq n. \tag{2.191}$$

Thus we have $[1 : 2 : 3] \in \mathbb{RP}^2_{>0}$ *and* $[(-4) : (-5) : (-6)] \in \mathbb{RP}^2_{>0}$ *(we may choose* $\lambda = -1$ *in Eq. (2.185)), but* $[7 : (-8) : 9] \notin \mathbb{RP}^2_{>0}$*.*

The **non-negative real projective space** $\mathbb{RP}^n_{\geq 0}$ *is the set of all points of* $\mathbb{RP}^n$*, which can be represented by*

$$[x_0 : x_1 : ... : x_n] \text{ with } x_j \geq 0, \qquad 0 \leq j \leq n. \tag{2.192}$$

In the literature the notation is sometimes used in a sloppy way, e.g. the word positive real projective space is also used where the non-negative real projective space is meant. However, the symbols $\mathbb{RP}^n_{>0}$ *and* $\mathbb{RP}^n_{\geq 0}$ *clearly indicate what is meant.*

We return to the Feynman parameter representation. We denote the integrand of the Feynman parameter representation by

$$f(a) = \frac{e^{l\varepsilon\gamma_E}\Gamma\left(\nu - \frac{lD}{2}\right)}{\prod\limits_{j=1}^{n_{\mathrm{int}}}\Gamma(\nu_j)} \left(\prod_{j=1}^{n_{\mathrm{int}}} a_j^{\nu_j - 1}\right) \frac{[\mathcal{U}(a)]^{\nu - \frac{(l+1)D}{2}}}{[\mathcal{F}(a)]^{\nu - \frac{lD}{2}}}. \tag{2.193}$$

If we rescale all Feynman parameters by λ we have

$$f\left(\lambda a_1, \ldots, \lambda a_{n_{\mathrm{int}}}\right) = \lambda^{-n_{\mathrm{int}}} f\left(a_1, \ldots, a_{n_{\mathrm{int}}}\right). \tag{2.194}$$

This follows easily from the homogeneity of $\mathcal{U}$ and $\mathcal{F}$ in the Feynman parameters. We also have

$$\left(n_{\mathrm{int}} + \sum_{j=1}^{n_{\mathrm{int}}} a_j \frac{\partial}{\partial a_j}\right) f(a) = 0. \tag{2.195}$$

Exercise 12 *Prove Eq. (2.195).*

Let us further introduce the differential $(n_{\mathrm{int}} - 1)$-form

$$\omega = \sum_{j=1}^{n_{\rm int}} (-1)^{n_{\rm int}-j} \; a_j \; da_1 \wedge ... \wedge \widehat{da_j} \wedge ... \wedge da_{n_{\rm int}}, \tag{2.196}$$

where the hat indicates that the corresponding term is omitted. Let us denote by Δ the $(n_{\rm int} - 1)$-dimensional standard simplex:

$$\Delta = \left\{ \left(a_1, \dots, a_{n_{\rm int}}\right) \in \mathbb{R}^{n_{\rm int}} \Big| \sum_{j=1}^{n_{\rm int}} a_j = 1, a_j \geq 0 \right\}. \tag{2.197}$$

With these definitions we may write the Feynman parameter representation as

$$I = \int\limits_{\Delta} f \omega. \tag{2.198}$$

Exercise 13 *Show explicitly that Eq. (2.198) is equivalent to Eq. (2.170).*

The differential $(n_{\rm int} - 1)$-form ω has the property that when we integrate it along a line through the origin, the result vanishes. In general, integrating a $(n_{\rm int} - 1)$-form along a curve gives a $(n_{\rm int} - 2)$-form. If the curve is defined by the vector field X, integrating along an infinitesimal interval gives a $(n_{\rm int} - 2)$-form proportional to the **interior product** $\iota_X \omega$. A line through the origin is defined by the vector field

$$X = \lambda_1 e_1 + \cdots + \lambda_{n_{\rm int}} e_{n_{\rm int}}, \tag{2.199}$$

where $\lambda_1, \dots, \lambda_{n_{\rm int}}$ are constants and $(\lambda_1, \dots, \lambda_{n_{\rm int}}) \neq (0, \dots, 0)$. $e_1, \dots, e_{n_{\rm int}}$ are basis vectors of the tangent space. The statement above, that the integration of ω along a line through the origin vanishes, is equivalent to

$$\iota_X \omega = 0. \tag{2.200}$$

Exercise 14 *Prove Eq. (2.200).*

From Eq. (2.195) it follows that $f\omega$ is closed:

$$d\left(f\omega\right) = 0. \tag{2.201}$$

Exercise 15 *Prove Eq. (2.201).*

Let us temporarily assume that $f\omega$ is non-singular for all points $a \in \Delta$. (This assumption is not as innocent as it may seem. The case, where $f\omega$ is singular for some points $a \in \Delta$ will stalk us in the sequel of the book. For the moment, we are certainly safe in the Euclidean region, if all internal propagators have a non-zero mass (this avoids infrared singularities) and $D < 0$ (this avoids ultraviolet singularities)). We now have all ingredients to claim that the Feynman parameter representation is a projective integral over the non-negative real projective space $\mathbb{RP}^{n_{\rm int}-1}_{\geq 0}$:

Projective Feynman parameter integral representation:

$$I = \int\limits_{\mathbb{RP}^{n_{\text{int}}-1}_{\geq 0}} f\omega, \tag{2.202}$$

where f is defined in Eq. (2.193), ω is defined in Eq. (2.196) and $\mathbb{RP}^{n_{\text{int}}-1}_{\geq 0}$ denotes the non-negative real projective space of dimension $(n_{\text{int}} - 1)$.

In particular we may integrate over any hyper-surface covering the solid angle $a_j \geq 0$.

The proof is based on Stoke's theorem's: Since $d(f\omega)$ is closed, integration over any n_{int}-dimensional domain Σ in $\mathbb{R}^{n_{\text{int}}}$ gives zero. Thus

$$\int\limits_{\Sigma} d\left(f\omega\right) = \int\limits_{\partial\Sigma} f\omega. \tag{2.203}$$

We may choose Σ as a domain bounded by the standard simplex Δ, the hyper-surface $\tilde{\Delta}$ we are interested in and covering the solid angle $a_j \geq 0$, and additional $(n_{\text{int}} - 1)$-dimensional domains in the coordinate sub-spaces $a_j = 0$.

An example for $n_{\text{int}} = 2$ is shown in Fig. 2.9. Due to Eq. (2.200) the integration over the $(n_{\text{int}} - 1)$-dimensional domains in the coordinate sub-spaces $a_j = 0$ vanishes. and the result follows.

Let us now relax the assumption, that $f\omega$ is regular for all points $a \in \Delta$. We may allow integrable singularities on the boundary $\partial\Delta$. The Feynman integral exists as an improper integral. In the proof above we replace the domain Σ by Σ_δ, which avoids all singular points by an angle δ, such that the domains connecting Δ and $\tilde{\Delta}$ (these

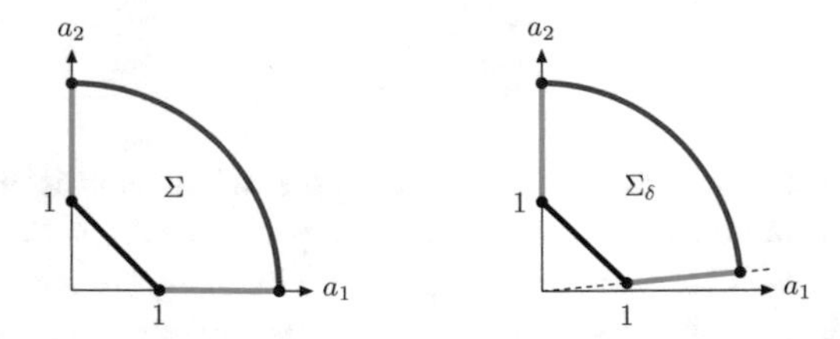

Fig. 2.9 The left picture shows an example for $n_{\text{int}} = 2$. The domain Σ is bounded by the one-dimensional standard simplex Δ (shown in black), a hyper-surface $\tilde{\Delta}$, covering the solid angle $a_1, a_2 \geq 0$ (shown in red) and two one-dimensional domains along the coordinate sub-space $a_1 = 0$ and $a_2 = 0$ (shown in green). Integration over the domains along the coordinate sub-spaces vanishes, and therefore the integration over Δ gives the same result as the integration over $\tilde{\Delta}$. If the integrand has an integrable singularity for $[a_1 : a_2] = [1 : 0]$ we first consider the domain Σ_δ, where we avoid the a_1-axis by an angle δ. The domains shown in green are still along the radial direction. In the end we take the limit $\delta \to 0$

domains are shown in green in Fig. 2.9) have always a direction along a line through the origin. For Σ_δ we may use the proof as above, and in particular the integration over the domains connecting Δ and $\tilde{\Delta}$ give a vanishing contribution. Taking then the limit $\delta \to 0$ shows that the two improper integrals over Δ and $\tilde{\Delta}$ are equal.

A consequence of the fact, that we may choose any hyper-surface covering the solid angle $a_j \geq 0$ is the **Cheng-Wu theorem** [29]. To state this theorem, let S be a non-empty subset of $\{1, \dots, n_{\text{int}}\}$.

Theorem 1 *(Cheng-Wu theorem): We may replace the argument*

$$1 - \sum_{j=1}^{n_{\text{int}}} a_j \tag{2.204}$$

of the delta distribution in Eq. (2.170) by

$$1 - \sum_{j \in S} a_j. \tag{2.205}$$

The Feynman integral is then given by

$$I = \frac{e^{l\varepsilon\gamma_E} \Gamma\left(\nu - \frac{lD}{2}\right)}{\prod\limits_{j=1}^{n_{\text{int}}} \Gamma(\nu_j)} \int\limits_{a_j \geq 0} d^{n_{\text{int}}}a \; \delta\left(1 - \sum_{j \in S} a_j\right) \left(\prod_{j=1}^{n_{\text{int}}} a_j^{\nu_j - 1}\right) \frac{[\mathcal{U}(a)]^{\nu - \frac{(l+1)D}{2}}}{[\mathcal{F}(a)]^{\nu - \frac{lD}{2}}}. \tag{2.206}$$

In particular one may choose $S = \{j_0\}$, which sets a_{j_0} to one. The integration is then over all other Feynman parameters from zero to infinity.

Proof The proof uses again Stoke's theorem, where the n_{int}-dimensional domain Σ is now bounded by the standard simplex Δ, the hyper-plane defined by

$$\sum_{j \in S} a_j = 1, \tag{2.207}$$

coordinate sub-spaces $a_j = 0$ for $j \in S$ and hyper-surfaces at infinity. The hyper-surfaces at infinity can be taken as $a_j = \Lambda$ for $j \notin S$ with $\Lambda \to \infty$.

An example for $n_{\text{int}} = 2$ is shown in Fig. 2.10. The integration over the $(n_{\text{int}} - 1)$-dimensional domains in the coordinate sub-spaces gives a vanishing contribution for the same reasons as before. The integration over the hyper-planes at infinity corresponds to an infinitesimal small solid angle and gives therefore a vanishing contribution. Hence, the claim follows.

Exercise 16 *An alternative proof of the Cheng-Wu theorem: Prove the Cheng-Wu theorem directly from the Schwinger parameter representation by inserting*

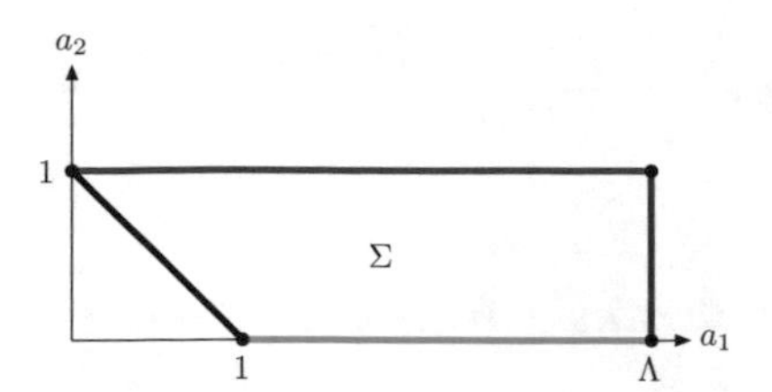

Fig. 2.10 Illustration for the Cheng-Wu theorem for the case $n_{\text{int}} = 2$. The domain Σ is bounded by the one-dimensional standard simplex Δ (shown in black), the hyper-plane $a_2 = 1$ (shown in red), a one-dimensional domain along the coordinate sub-space $a_2 = 0$ (shown in green) and a domain at infinity (shown in blue)

$$1 = \int\limits_{-\infty}^{\infty} dt \, \delta\left(t - \sum_{j\in S} \alpha_j\right) = \int\limits_{0}^{\infty} dt \, \delta\left(t - \sum_{j\in S} \alpha_j\right), \tag{2.208}$$

where in the last step we used again the fact that the sum of the Schwinger parameters is non-negative.

Let us give an example for the application of the Cheng-Wu theorem: We consider again the one-loop bubble integral with vanishing internal masses. We choose

$$S = \{2\}. \tag{2.209}$$

Setting $x = -p^2/\mu^2$, the Feynman integral is then given by

$$I_{\nu_1\nu_2} = \frac{e^{\varepsilon\gamma_E}\Gamma\left(\nu - \frac{D}{2}\right)}{\Gamma\left(\nu_1\right)\Gamma\left(\nu_2\right)} \int\limits_{a_j\geq 0} d^2a \, \delta\left(1 - a_2\right) a_1^{\nu_1-1} a_2^{\nu_2-1} \frac{\mathcal{U}^{\nu-D}}{\mathcal{F}^{\nu-\frac{D}{2}}},$$
$$\mathcal{U} = a_1 + a_2, \qquad \mathcal{F} = a_1 a_2 x. \tag{2.210}$$

The integration over a_2 is trivial due to the delta distribution, leaving us with

$$\begin{aligned} I_{\nu_1\nu_2} &= \frac{e^{\varepsilon\gamma_E}\Gamma\left(\nu - \frac{D}{2}\right)}{\Gamma\left(\nu_1\right)\Gamma\left(\nu_2\right)} \int\limits_{0}^{\infty} da_1 \, a_1^{\nu_1-1} \frac{\left(a_1+1\right)^{\nu-D}}{\left(a_1 x\right)^{\nu-\frac{D}{2}}} \\ &= \frac{e^{\varepsilon\gamma_E}\Gamma\left(\nu - \frac{D}{2}\right)}{\Gamma\left(\nu_1\right)\Gamma\left(\nu_2\right)} x^{\frac{D}{2}-\nu} \int\limits_{0}^{\infty} da_1 \, \frac{a_1^{\frac{D}{2}-\nu_2-1}}{\left(1+a_1\right)^{D-\nu}}. \end{aligned} \tag{2.211}$$

The integral over a_1 gives Euler's beta function $B(D/2 - \nu_2, D/2 - \nu_1)$ and we recover the result of Eq. (2.181):

$$I_{\nu_1\nu_2} = e^{\varepsilon\gamma_E} \frac{\Gamma\left(\nu - \frac{D}{2}\right)\Gamma\left(\frac{D}{2} - \nu_1\right)\Gamma\left(\frac{D}{2} - \nu_2\right)}{\Gamma(\nu_1)\Gamma(\nu_2)\Gamma(D-\nu)} x^{\frac{D}{2}-\nu}. \tag{2.212}$$

2.5.4 The Lee-Pomeransky Representation

The Lee-Pomeransky representation is as the Schwinger parameter representation a n_{int}-fold integral representation. It has the advantage that only one polynomial $\mathcal{G}$ enters, given by the sum of the two graph polynomials $\mathcal{G} = \mathcal{U} + \mathcal{F}$. The Lee-Pomeransky representation reads [30]:

Lee-Pomeransky representation:

$$I = \frac{e^{l\varepsilon\gamma_E}\Gamma\left(\frac{D}{2}\right)}{\Gamma\left(\frac{(l+1)D}{2} - \nu\right)\prod\limits_{j=1}^{n_{\text{int}}}\Gamma(\nu_j)} \int\limits_{u_j \geq 0} d^{n_{\text{int}}}u \left(\prod_{j=1}^{n_{\text{int}}} u_j^{\nu_j - 1}\right) [\mathcal{G}(u)]^{-\frac{D}{2}},$$

$$\text{with} \quad \mathcal{G}(u) = \mathcal{U}(u) + \mathcal{F}(u). \tag{2.213}$$

In order to derive the Lee-Pomeransky representation it is simplest to work backwards: We start with the Lee-Pomeransky representation and show that the Lee-Pomeransky representation is equivalent to the Feynman parameter representation. In order to do this, we use the same trick as before and we insert a one in the form of

$$1 = \int\limits_0^\infty dt\, \delta\left(t - \sum_{j=1}^{n_{\text{int}}} u_j\right) \tag{2.214}$$

into the Lee-Pomeransky representation. We then change variables according to $a_j = u_j/t$ and exploit again that $\mathcal{U}$ and $\mathcal{F}$ are homogeneous of degree l and $(l+1)$, respectively. We arrive at

$$I = \frac{e^{l\varepsilon\gamma_E}\Gamma\left(\frac{D}{2}\right)}{\Gamma\left(\frac{(l+1))D}{2} - \nu\right)\prod\limits_{j=1}^{n_{\text{int}}}\Gamma(\nu_j)} \int\limits_{a_j \geq 0} d^{n_{\text{int}}}a\, \delta\left(1 - \sum_{j=1}^{n_{\text{int}}} a_j\right) \left(\prod_{j=1}^{n_{\text{int}}} a_j^{\nu_j - 1}\right)$$
$$\times \int\limits_0^\infty dt\, t^{\nu - \frac{lD}{2} - 1} \left[\mathcal{U}(a) + \mathcal{F}(a)\, t\right]^{-\frac{D}{2}}. \tag{2.215}$$

We then substitute $t \to t\mathcal{U}(a)/\mathcal{F}(a)$. This gives us

$$I = \frac{e^{l\varepsilon\gamma_E}\Gamma\left(\frac{D}{2}\right)}{\Gamma\left(\frac{(l+1))D}{2}-\nu\right)\prod\limits_{j=1}^{n_{\mathrm{int}}}\Gamma(\nu_j)} \int\limits_{a_j\geq 0} d^{n_{\mathrm{int}}}a\;\delta\left(1-\sum_{j=1}^{n_{\mathrm{int}}}a_j\right)\left(\prod_{j=1}^{n_{\mathrm{int}}}a_j^{\nu_j-1}\right)\frac{[\mathcal{U}(a)]^{\nu-\frac{(l+1)D}{2}}}{[\mathcal{F}(a)]^{\nu-\frac{lD}{2}}}$$
$$\times \int\limits_0^\infty dt\; t^{\nu-\frac{lD}{2}-1}\;(1+t)^{-\frac{D}{2}}. \tag{2.216}$$

We recognise the integral over t as the second integral representation of Euler's beta function:

$$\int\limits_0^\infty dt\; t^{\nu-\frac{lD}{2}-1}\;(1+t)^{-\frac{D}{2}} = \frac{\Gamma\left(\nu-\frac{lD}{2}\right)\Gamma\left(\frac{(l+1)D}{2}-\nu\right)}{\Gamma\left(\frac{D}{2}\right)}. \tag{2.217}$$

With this result we recover the Feynman parameter representation. We call the variables u_j **Lee-Pomeransky variables.**

Let us also consider for the Lee-Pomeransky representation an example. Again, we choose the one-loop two-point integral with vanishing internal masses. The Lee-Pomeransky polynomial is in this case

$$\mathcal{G} = u_1 + u_2 + u_1 u_2 x, \tag{2.218}$$

where we set $x = -p^2/\mu^2$. The Lee-Pomeransky representation is then given by

$$I_{\nu_1\nu_2} = \frac{e^{\varepsilon\gamma_E}\Gamma\left(\frac{D}{2}\right)}{\Gamma(D-\nu)\Gamma(\nu_1)\Gamma(\nu_2)}\int\limits_0^\infty du_1 \int\limits_0^\infty du_2\; u_1^{\nu_1-1}u_2^{\nu_2-1}\left[u_1+u_2+u_1u_2x\right]^{-\frac{D}{2}}. \tag{2.219}$$

2.5.5 The Baikov Representation

Up to now we always considered arbitrary Feynman graphs with n_{ext} external edges, n_{int} internal edges and l loops. In particular we never assumed that there is an additional relation between n_{ext}, n_{int} and l.

The Baikov representation [31] applies to a subset of Feynman graphs, where the number of internal edges n_{int} equals the number of independent scalar products involving the loop momenta.

Let $p_1, p_2, ..., p_{n_{\mathrm{ext}}}$ denote the external momenta and denote by

$$e = \dim\left\langle p_1, p_2, ..., p_{n_{\mathrm{ext}}}\right\rangle \tag{2.220}$$

the dimension of the span of the external momenta. For generic external momenta and $D \geq n_{\mathrm{ext}} - 1$ we have $e = n_{\mathrm{ext}} - 1$. Lorentz invariants involving the loop momenta are of the form

$$
\begin{aligned}
&-k_i^2, \; 1 \le i \le l, \\
&-k_i \cdot k_j, \; 1 \le i < j \le l, \\
&-k_i \cdot p_j, \; 1 \le i \le l, \quad 1 \le j \le e.
\end{aligned} \tag{2.221}
$$

In total we have

$$N_V = \frac{1}{2} l (l+1) + el \tag{2.222}$$

linear independent scalar products involving the loop momenta. We denote this number by N_V and the linear independent scalar products involving the loop momenta by

$$\sigma = \left(\sigma_1, ..., \sigma_{N_V}\right) = \left(-k_1 \cdot k_1, -k_1 \cdot k_2, ..., -k_l \cdot k_l, -k_1 \cdot p_1, ..., -k_l \cdot p_e\right). \tag{2.223}$$

A Feynman graph G has a Baikov representation if

$$N_V = n_{\text{int}} \tag{2.224}$$

and if we may express any internal inverse propagator as a linear combination of the linear independent scalar products involving the loop momenta and terms independent of the loop momenta. The second condition says, that there is an invertible $N_V \times N_V$-dimensional matrix C and a loop-momentum independent N_V-dimensional vector f such that

$$-q_s^2 + m_s^2 = C_{st} \sigma_t + f_s \tag{2.225}$$

for all $1 \le s \le n_{\text{int}}$. At first sight it might seem that the Baikov representation applies due to the conditions in Eqs. (2.224) and (2.225) only to a very special subset of Feynman graphs. However, we will soon see that for a given graph G, which not necessarily satisfies the conditions Eqs. (2.224) and (2.225), we can always find a graph $\tilde{G}$ which does, and obtain the induced Baikov representation of the graph G from the Baikov representation of the graph $\tilde{G}$.

In order to arrive at the Baikov representation we change the integration variables to the **Baikov variables** z_j:

$$z_j = -q_j^2 + m_j^2. \tag{2.226}$$

The Baikov variables are nothing else than the inverse propagators. From Eq. (2.225) we have the inverse relation

$$\sigma_t = \left(C^{-1}\right)_{ts} \left(z_s - f_s\right). \tag{2.227}$$

The Baikov representation of I is given by

$$I = e^{l\varepsilon\gamma_E} \left(\mu^2\right)^{\nu-\frac{lD}{2}} \frac{\pi^{-\frac{1}{2}(N_V-l)}}{\prod\limits_{j=1}^{l} \Gamma\left(\frac{D-e+1-j}{2}\right)} \frac{\left[\det G\left(p_1, ..., p_e\right)\right]^{\frac{-D+e+1}{2}}}{\det C}$$
$$\times \int\limits_C d^{N_V}z \; \left[\det G\left(k_1, ..., k_l, p_1, ..., p_e\right)\right]^{\frac{D-l-e-1}{2}} \prod_{s=1}^{N_V} z_s^{-\nu_s}, \qquad (2.228)$$

where the Gram determinants are defined by

$$\det G\left(q_1, ..., q_n\right) = \det\left(-q_i \cdot q_j\right) \;=\; \det\left(Q_i \cdot Q_j\right), \qquad (2.229)$$

e.g.

$$\det G\left(q_1, q_2\right) = \begin{vmatrix} -q_1^2 & -q_1 \cdot q_2 \\ -q_1 \cdot q_2 & -q_2^2 \end{vmatrix} \;=\; q_1^2 q_2^2 - \left(q_1 \cdot q_2\right)^2. \qquad (2.230)$$

Note that exchanging two vectors leaves the Gram determinant invariant

$$\det G\left(\dots, q_i, \dots, q_j, \dots\right) = \det G\left(\dots, q_j, \dots, q_i, \dots\right), \qquad (2.231)$$

as this corresponds to the exchange of two rows and two columns.

The determinant $\det G(k_1, ..., k_l, p_1, ..., p_e)$ expressed in the variables z_s's through Eq. (2.227) is called the **Baikov polynomial**:

$$\mathcal{B}\left(z_1, ..., z_{N_V}\right) = \det G\left(k_1, ..., k_l, p_1, ..., p_e\right). \qquad (2.232)$$

The domain of integration C is given by

$$C = C_1 \cap C_2 \cap \dots \cap C_l \qquad (2.233)$$

with

$$C_j = \left\{ \frac{\det G\left(k_j, k_{j+1}, ..., k_l, p_1, ..., p_e\right)}{\det G\left(k_{j+1}, ..., k_l, p_1, ..., p_e\right)} \geq 0 \right\}. \qquad (2.234)$$

Putting everything together we arrive at the

Baikov representation (for a graph G satisfying Eqs. (2.224) and (2.225)):

$$I = \frac{e^{l\varepsilon\gamma_E} \left(\mu^2\right)^{\nu - \frac{lD}{2}} \left[\det G\left(p_1, ..., p_e\right)\right]^{\frac{-D+e+1}{2}}}{\pi^{\frac{1}{2}(N_V - l)} \left(\det C\right) \prod\limits_{j=1}^{l} \Gamma\left(\frac{D-e+1-j}{2}\right)} \int\limits_C d^{N_V} z \left[\mathcal{B}(z)\right]^{\frac{D-l-e-1}{2}} \prod_{s=1}^{N_V} z_s^{-\nu_s}. \quad (2.235)$$

The Baikov representation is very useful if we would like to calculate cuts of Feynman integrals, i.e. the residue when one or several propagators go on-shell.

Before we go into the details of the derivation of the Baikov representation, let us first discuss how to get around the restrictions imposed by Eqs. (2.224) and (2.225). The most common situation is that the number of internal propagators n_{int} is smaller than the number N_V of linear independent scalar products involving the loop momenta:

$$n_{\text{int}} < N_V. \quad (2.236)$$

This does not occur at one-loop, but it frequently occurs beyond one-loop. As an example let us look at the two-loop double box graph in Fig. 2.3. In this example we have seven propagators ($n_{\text{int}} = 7$), but nine linear independent scalar products involving the loop momenta:

$$-k_1^2,\ -k_2^2,\ -k_1 \cdot k_2,\ -k_1 \cdot p_1,\ -k_1 \cdot p_2,\ -k_1 \cdot p_3,\ -k_2 \cdot p_1,\ -k_2 \cdot p_2,\ -k_2 \cdot p_3. \quad (2.237)$$

We may express seven scalar products

$$-k_1^2,\ -k_2^2,\ -k_1 \cdot k_2,\ -k_1 \cdot p_1,\ -k_1 \cdot p_2,\ -k_2 \cdot (p_1 + p_2),\ -k_2 \cdot p_3 \quad (2.238)$$

in terms of inverse propagators, but not

$$-k_1 \cdot p_3,\ -k_2 \cdot p_1. \quad (2.239)$$

The scalar products in Eq. (2.239) are called **irreducible scalar products**. The solution is to consider the original graph G as a subgraph of a larger graph $\tilde{G}$.

An example of an auxiliary graph $\tilde{G}$ for the double box graph G is shown in Fig. 2.11. This graph has nine internal edges and six external edges. However, the dimension of the span of the external momenta is still

$$e = \dim \langle p_1, p_2, p_3, p_4, p_1 + p_3, p_2 + p_4 \rangle = 3, \quad (2.240)$$

and hence the number of linear independent scalar products involving the loop momenta remains $N_V = 9$. For the graph $\tilde{G}$ we may express any internal inverse propagator as a linear combination of the linear independent scalar products involving the loop momenta and terms independent of the loop momenta. The Baikov

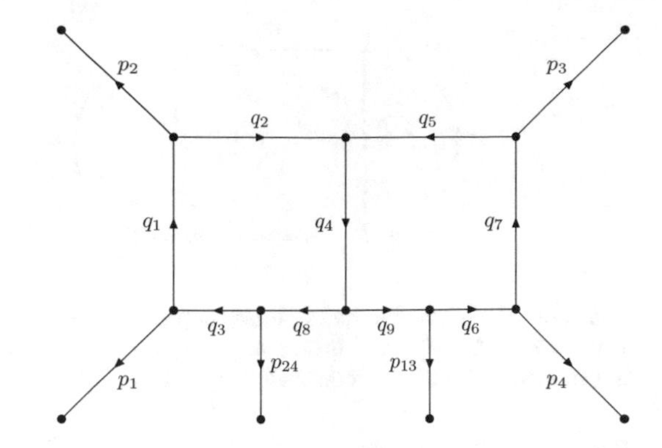

Fig. 2.11 An auxiliary graph $\tilde{G}$ with nine internal propagators. For this graph we have as many internal propagators as linear independent scalar products involving the loop momenta. We use the notation $p_{ij} = p_i + p_j$

representation of the graph G is then given by the Baikov representation of the graph $\tilde{G}$ with $\nu_8 = \nu_9 = 0$. Setting $\nu_8 = \nu_9 = 0$ ensures that the two additional propagators corresponding to the edges e_8 and e_9 are absent.

It is always possible to find a graph $\tilde{G}$. To see this, let us first introduce for a graph G the associated **chain graph** G^{chain} as follows: We group the internal propagators into chains. Two propagators belong to the same chain, if their momenta differ only by a linear combination of the external momenta. Obviously, each internal line can only belong to one chain. To each graph G we associate a new graph G^{chain} called the chain graph by deleting all external lines and by choosing one propagator for each chain as a representative.

Up to three loops, all chain graphs are (sub-) graphs of the three chain graphs shown in Fig. 2.12. A non-trivial example is given by the three-loop ladder graph shown in the left figure of Fig. 2.13.

In this example we note that propagators 1 and $6'$ belong to the same chain, as the same loop momentum is flowing through both propagators. Hence, the associated chain graph is the one shown in the right figure of Fig. 2.13. This chain graph is a subgraph of the three-loop Mercedes-Benz graph shown in the right figure of Fig. 2.12.

With the chain graph G^{chain} at hand, we re-insert external edges to the chains, such that all linear independent scalar products involving the loop momenta can

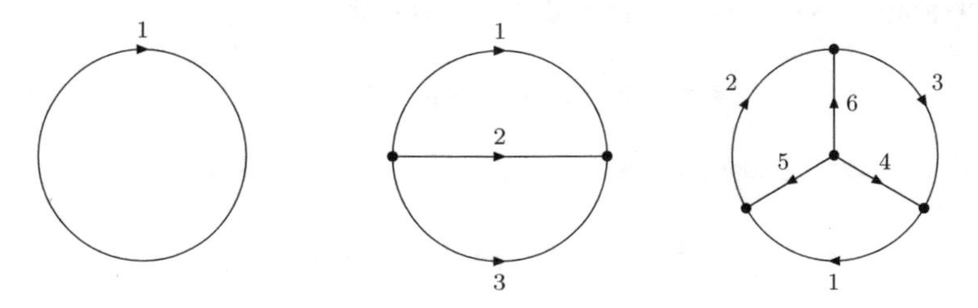

Fig. 2.12 The basic chain graphs up to three loops. Up to this loop order, all other chain graphs are subgraphs of these three graphs

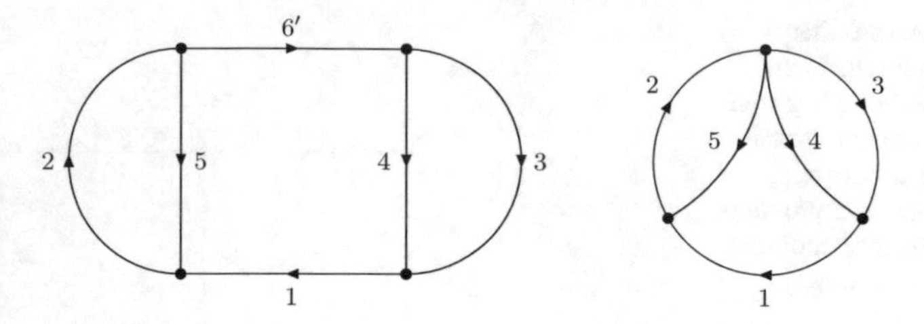

Fig. 2.13 The left figure shows a three-loop graph. Propagators 1 and 6′ belong to the same chain. The underlying chain graph is the five-propagator graph shown in the right figure. This chain graph is a subgraph of the three-loop chain graph of Fig. 2.12

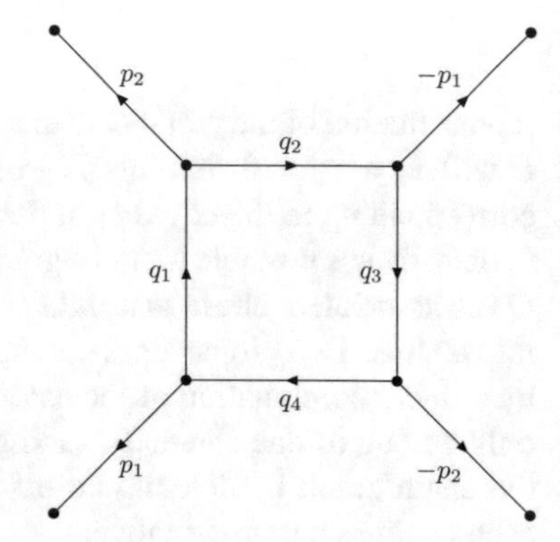

Fig. 2.14 A one-loop box graph with external momenta $p_1, p_2, -p_1, -p_2$. For this graph we have $e = \dim\langle p_1, p_2, -p_1, -p_2\rangle = 2$

be expressed in terms of inverse propagators and terms independent of the loop momenta. In practice one starts with the original propagators and then adds additional propagators such that the irreducible scalar products may be expressed in terms of the inverse propagators. In a final step one adjusts the external momenta appropriately.

Having discussed the case $n_{\text{int}} < N_V$, let us now turn to the other case:

$$n_{\text{int}} > N_V. \tag{2.241}$$

In practice, this case is rather special. We discuss this case for completeness. An example is shown in Fig. 2.14.

This graph has four internal propagators, but with

$$e = \dim\langle p_1, p_2, -p_1, -p_2\rangle = 2 \tag{2.242}$$

we only have three linear independent scalar products involving the loop momenta:

$$-k^2, \quad -k \cdot p_1 \quad -k \cdot p_2. \tag{2.243}$$

At the same time, the four inverse propagators of the graph in Fig. 2.14 are not independent. We have

$$
\begin{aligned}
& -q_3^2+m_3^2 = \\
& \quad -\left[-q_1^2+m_1^2\right]+\left[-q_2^2+m_2^2\right]+\left[-q_4^2+m_4^2\right]+2p_1\cdot p_2+m_1^2-m_2^2+m_3^2-m_4^2.
\end{aligned} \tag{2.244}
$$

We may therefore consider as graph $\tilde{G}$ a one-loop triangle graph with internal edges $\{e_1, e_2, e_4\}$. With

$$
z_1 = -q_1^2+m_1^2, \quad z_2 = -q_2^2+m_2^2, \quad z_4 = -q_4^2+m_4^2 \tag{2.245}
$$

we obtain the induced Baikov representation of the graph G from the Baikov representation of the graph $\tilde{G}$ by replacing

$$
\frac{1}{\left(-q_3^2+m_3^2\right)^{\nu_3}} = \frac{1}{\left(-z_1+z_2+z_4+f\right)^{\nu_3}}, \qquad f = 2p_1\cdot p_2+m_1^2-m_2^2+m_3^2-m_4^2. \tag{2.246}
$$

Digression Details on the derivation of the Baikov representation
The basic idea to derive the Baikov representation is to split the loop momentum variables into a set, on which the integrands depends non-trivially and a set, on which the integrand depends trivially. For the former set we perform a change of variables to the Baikov variables, while for the latter set we integrate over these variables.

Let us start with the one-loop case. We decompose the loop momentum

$$
k = k_\parallel + k_\perp \tag{2.247}
$$

into a component living in the parallel space, defined by

$$
k_\parallel \in \langle p_1, p_2, \dots, p_{n_{\text{ext}}} \rangle \tag{2.248}
$$

and a component living in the complement, called the orthogonal space. The dependence of the integrand on $k_\perp$ is only through

$$
k^2 = k_\parallel^2 + k_\perp^2. \tag{2.249}
$$

For the measure we have

$$
d^D k = d^e k_\parallel \, d^{D-e} k_\perp. \tag{2.250}
$$

The relation with the Euclidean measure is

$$
d^D k = i d^D K, \qquad d^e k_\parallel = i d^e K_\parallel, \qquad d^{D-e} k_\perp = d^{D-e} K_\perp. \tag{2.251}
$$

The energy component is part of the parallel space. From the standard formulae

$$
d^D K = \left(K^2\right)^{\frac{D-2}{2}} dK^2 \, \frac{1}{2} d\Omega_D \text{ and } \int d\Omega_D = \frac{2\pi^{\frac{D}{2}}}{\Gamma\left(\frac{D}{2}\right)} \tag{2.252}
$$

we obtain

$$d^{D-e}K_{\perp} = \frac{\pi^{\frac{D-e}{2}}}{\Gamma\left(\frac{D-e}{2}\right)} \left(K_{\perp}^2\right)^{\frac{D-e-2}{2}} dK_{\perp}^2. \tag{2.253}$$

This allows us to perform all angular integrations in the orthogonal space.

The linear independent scalar products involving the loop momenta are

$$\sigma = \left(-k^2, -k \cdot p_1, \cdots - k \cdot p_e\right) = \left(K^2, K \cdot P_1, \ldots K \cdot P_e\right). \tag{2.254}$$

We have

$$J = \frac{\partial\left(\sigma_2, ..., \sigma_{e+1}\right)}{\partial\left(K^0, ..., K^{e-1}\right)} = \begin{pmatrix} P_1^0 & P_1^1 & ... & P_1^{e-1} \\ P_2^0 & P_2^1 & ... & P_2^{e-1} \\ & & ... & \\ P_e^0 & P_e^1 & ... & P_e^{e-1} \end{pmatrix}. \tag{2.255}$$

We have $J J^T = (P_i \cdot P_j)$ *and therefore*

$$\det J = \sqrt{\det G^{\mathrm{eucl}}\left(P_1, \ldots, P_e\right)} = \sqrt{\det G\left(p_1, \ldots, p_e\right)}. \tag{2.256}$$

Here we defined the Euclidean Gram determinant by

$$\det G^{\mathrm{eucl}}\left(Q_1, \ldots, Q_n\right) = \det(Q_i \cdot Q_j) = \det(-q_i \cdot q_j) = \det G\left(q_1, \ldots, q_n\right). \tag{2.257}$$

In addition, we may trade the integration over $dK_{\perp}^2$ *for an integration over* $\sigma_1 = -k^2 = K^2$ *with Jacobian*

$$\frac{\partial \sigma_1}{\partial K_{\perp}^2} = \frac{\partial\left(K_{\parallel}^2 + K_{\perp}^2\right)}{\partial K_{\perp}^2} = 1. \tag{2.258}$$

Our final change of variables is to change from the variables $\sigma = (\sigma_1, \ldots, \sigma_{e+1})$ *to the Baikov variables* $(z_1, \ldots, z_{N_V})$*, where* $N_V = e + 1$*. From Eq. (2.225)*

$$z_s = C_{st}\sigma_t + f_s \tag{2.259}$$

we obtain the Jacobian

$$\frac{\partial z_s}{\partial \sigma_t} = \det C. \tag{2.260}$$

It remains to express $-k_{\perp}^2 = K_{\perp}^2$ *in terms of the Baikov variables. This can be done by first noting that*

$$-k_{\perp}^2 = K_{\perp}^2 = \frac{\det G^{\mathrm{eucl}}\left(K, P_1, ..., P_e\right)}{\det G^{\mathrm{eucl}}\left(P_1, ..., P_e\right)} = \frac{\det G\left(k, p_1, ..., p_e\right)}{\det G\left(p_1, ..., p_e\right)}, \tag{2.261}$$

and then replacing the scalar products involving the loop momenta with the Baikov variables with the help of Eq. (2.227).

Exercise 17 *Prove Eq. (2.261).*

Putting everything together we obtain for the measure

$$\frac{d^D k}{i\pi^{\frac{D}{2}}} = \frac{1}{\pi^{\frac{e}{2}} (\det C) \Gamma\left(\frac{D-e}{2}\right)} \left[\det G\left(p_1, ..., p_e\right)\right]^{\frac{-D+e+1}{2}} \left[\det G\left(k, p_1, ..., p_e\right)\right]^{\frac{D-e-2}{2}} d^{N_V} z. \quad (2.262)$$

The domain of integration follows from the requirement

$$-k_\perp^2 = K_\perp^2 = \frac{\det G\left(k, p_1, ..., p_e\right)}{\det G\left(p_1, ..., p_e\right)} \geq 0. \quad (2.263)$$

For a multi-loop integral ($l > 1$) we may apply the argument above l-times, starting with loop momentum k_1 and ending with loop momentum k_l, keeping in mind that the parallel space for loop momentum k_j is spanned by the external momenta and all loop momenta not yet integrated out:

$$\text{parallel space for} k_j : \left\langle k_{j+1}, \dots, k_l, p_1, p_2, \dots, p_{n_{\text{ext}}} \right\rangle. \quad (2.264)$$

Doing so, one derives Eq. (2.235).

Let us now look at a simple example for the Baikov representation. We consider the one-loop tadpole integral

$$T_\nu\left(D, x\right) = e^{\varepsilon \gamma_E} \left(\mu^2\right)^{\nu - \frac{D}{2}} \int \frac{d^D k}{i\pi^{\frac{D}{2}}} \frac{1}{\left(-k^2 + m^2\right)^\nu}, \qquad x = \frac{m^2}{\mu^2}, \quad (2.265)$$

which we already computed in Eq. (2.123). This integral does not depend on any external momenta, therefore

$$e = 0, \text{ and } N_V = 1. \quad (2.266)$$

There is one Baikov variable $z_1 = -k^2 + m^2$. The Gram determinant is given by

$$\det G\left(k\right) = -k^2 = z_1 - m^2. \quad (2.267)$$

Thus, we obtain the Baikov polynomial as

$$\mathcal{B}\left(z_1\right) = z_1 - m^2. \quad (2.268)$$

The requirement $\det G(k) \geq 0$ defines the integration region $z_1 \in [m^2, \infty[$. We arrive at the Baikov representation of the tadpole integral:

$$T_\nu(D,x) = \frac{e^{\gamma_E \varepsilon}\left(\mu^2\right)^{\nu-\frac{D}{2}}}{\Gamma\left(\frac{D}{2}\right)} \int\limits_C dz_1 \left[\mathcal{B}(z_1)\right]^{\frac{D}{2}-1} \frac{1}{z_1^\nu}, \tag{2.269}$$

where the contour is given by $C = [m^2, \infty[$.

Exercise 18 *Perform the integration in Eq. (2.269).*

We close this section by introducing a variant of the Baikov representation. In the Baikov representation in Eq. (2.235) we treated any scalar product of a loop momentum with any other momentum on equal footing. This is referred to as the **democratic approach**. This approach allows us to express any irreducible scalar product in terms of the Baikov variables. This is useful, when irreducible scalar products appear in the numerator of the loop momentum representation of the Feynman integral and we would like to convert the Feynman integral to the Baikov representation. Up to now we didn't discuss this case. We will treat this case in Chap. 4. The price to pay for being able to express any irreducible scalar product in terms of Baikov variables is the number of Baikov variables:

$$N_V = \frac{1}{2} l (l+1) + el. \tag{2.270}$$

However, if we are only interested in scalar Feynman integrals (without any irreducible scalar product in the numerator)—and we will see in Chap. 4 that it is enough to focus on these integrals—we might have an additional interest in keeping the number of integration variables as low as possible. There is a variant of the Baikov representation, known as the **loop-by-loop approach** [32], which achieves this. This is best explained by an example.

Consider the two-loop graph shown in Fig. 2.15. With

$$e = \dim \langle p_1, p_2, p_3, p_4 \rangle = 3 \tag{2.271}$$

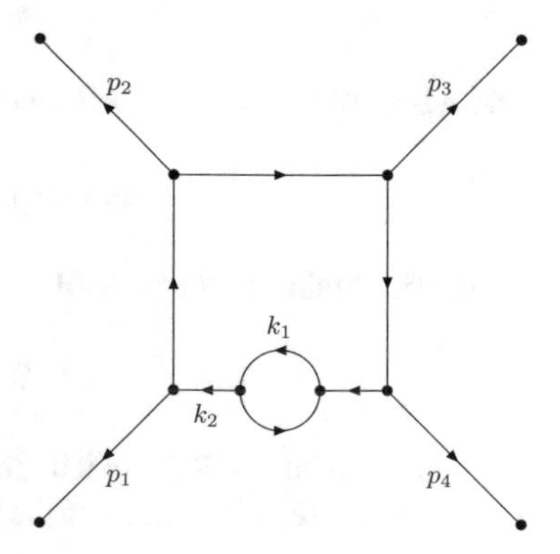

Fig. 2.15 A two-loop graph, where a one-loop bubble is inserted into a one-loop box

we have

$$N_V = 9 \tag{2.272}$$

independent scalar products involving the loop momenta:

$$-k_1^2,\ -k_1 \cdot k_2,\ -k_2^2,\ -k_1 \cdot p_1,\ -k_1 \cdot p_2,\ -k_1 \cdot p_3,\ -k_2 \cdot p_1,\ -k_2 \cdot p_2,\ -k_2 \cdot p_3. \tag{2.273}$$

However, the external momenta relative to the loop with k_1 are just k_2 (and $-k_2$) and not the full set k_2, p_1, p_2, p_3. Thus we may decompose k_1 in

$$k_1 = k_{1,\parallel} + k_{1,\perp}, \tag{2.274}$$

where $k_{1,\parallel}$ is spanned by

$$\text{loop-by-loop parallel space for } k_1 :\ \langle k_2 \rangle \tag{2.275}$$

instead of

$$\text{democratic parallel space for } k_1 :\ \langle k_2, p_1, p_2, p_3 \rangle \tag{2.276}$$

We then use Eq. (2.262) for the integration over k_1 with the loop-by-loop parallel space for k_1. This introduces only two Baikov variables. For the integration over k_2 the parallel space is spanned by

$$\text{parallel space for } k_2 :\ \langle p_1, p_2, p_3 \rangle\,, \tag{2.277}$$

giving four additional Baikov variables. We therefore obtain the loop-by-loop Baikov representation with only six integration variables.

Exercise 19 *Derive the Baikov representation of the graph shown in Fig. 2.16 within the democratic approach and within the loop-by-loop approach. Assume that all internal masses are non-zero and equal.*

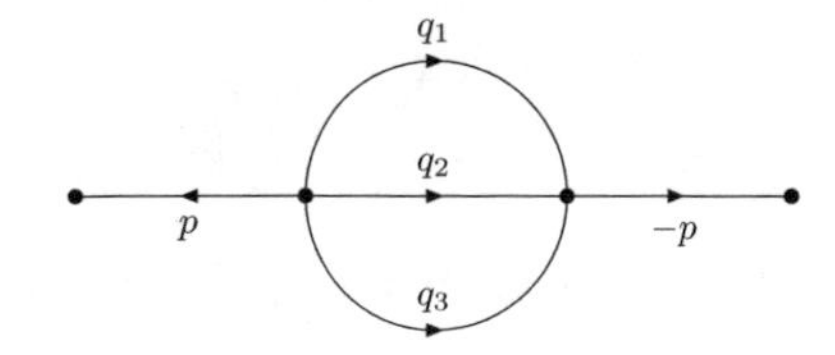

Fig. 2.16 The two-loop sunrise diagram (also known as sunset diagram or London transport diagram)

2.5.6 The Mellin-Barnes Representation

Up to now we presented for all representations of Feynman integrals (momentum representation, Schwinger parameter representation, Feynman parameter representation, Lee-Pomeransky representation and Baikov representation) a closed integral formula for the Feynman integral. We won't do this for the Mellin-Barnes representation. The Mellin-Barnes representation [33, 34] can be applied to any representation discussed so far and allows to trivialise the integrations of the representation we start with at the expense of introducing new integrations over Mellin-Barnes variables. We explain the Mellin-Barnes representation by starting from the Feynman parameter representation.

The Feynman parameter representation in Eq. (2.170) depends on two graph polynomials $\mathcal{U}$ and $\mathcal{F}$. Assume for the moment that the two graph polynomials $\mathcal{U}$ and $\mathcal{F}$ are absent from the Feynman parameter integral. In this case we have an integral of the form

$$\int\limits_{a_j \geq 0} d^n a \; \delta\left(1 - \sum_{j=1}^{n} a_j\right) \left(\prod_{j=1}^{n} a_j^{\nu_j - 1}\right) = \frac{\prod\limits_{j=1}^{n} \Gamma(\nu_j)}{\Gamma(\nu_1 + ... + \nu_n)}. \tag{2.278}$$

The Mellin-Barnes transformation allows us to reduce the Feynman parameter integration to this case. The Mellin-Barnes transformation reads

$$\begin{aligned}
\left(A_1 + A_2 + ... + A_n\right)^{-c} &= \frac{1}{\Gamma(c)} \frac{1}{(2\pi i)^{n-1}} \int\limits_{-i\infty}^{i\infty} d\sigma_1 ... \int\limits_{-i\infty}^{i\infty} d\sigma_{n-1} \\
&\times \Gamma(-\sigma_1) ... \Gamma(-\sigma_{n-1}) \Gamma(\sigma_1 + ... + \sigma_{n-1} + c) \; A_1^{\sigma_1} ... A_{n-1}^{\sigma_{n-1}} A_n^{-\sigma_1 - ... - \sigma_{n-1} - c}.
\end{aligned} \tag{2.279}$$

Each contour is such that the poles of $\Gamma(-\sigma)$ are to the right and the poles of $\Gamma(\sigma + c)$ are to the left. This transformation can be used to convert the sum of monomials of the polynomials $\mathcal{U}$ and $\mathcal{F}$ into a product, such that all Feynman parameter integrals are of the form of Eq. (2.278). As this transformation converts a sum into a product it is the "inverse" of Feynman parametrisation, which converts a product into a sum.

Equation (2.279) is derived from the theory of Mellin transformations: Let $h(x)$ be a function which is bounded by a power law for $x \to 0$ and $x \to \infty$, e.g.

$$\begin{aligned}
|h(x)| &\leq K x^{-c_0} \quad \text{for} \quad x \to 0, \\
|h(x)| &\leq K' x^{c_1} \quad \text{for} \quad x \to \infty.
\end{aligned} \tag{2.280}$$

Then the Mellin transform is defined for $c_0 < \text{Re}\, \sigma < c_1$ by

$$h_{\mathcal{M}}(\sigma) = \int\limits_{0}^{\infty} dx \; h(x) \; x^{\sigma - 1}. \tag{2.281}$$

The inverse Mellin transform is given by

$$h(x) = \frac{1}{2\pi i} \int\limits_{\gamma-i\infty}^{\gamma+i\infty} d\sigma \, h_{\mathcal{M}}(\sigma) \, x^{-\sigma}. \tag{2.282}$$

The integration contour is parallel to the imaginary axis and $c_0 < \text{Re}\,\gamma < c_1$. As an example for the Mellin transform we consider the function

$$h(x) = \frac{x^c}{(1+x)^c} \tag{2.283}$$

with Mellin transform $h_{\mathcal{M}}(\sigma) = \Gamma(-\sigma)\Gamma(\sigma+c)/\Gamma(c)$. For $\text{Re}(-c) < \text{Re}\,\gamma < 0$ we have

$$\frac{x^c}{(1+x)^c} = \frac{1}{2\pi i} \int\limits_{\gamma-i\infty}^{\gamma+i\infty} d\sigma \, \frac{\Gamma(-\sigma)\Gamma(\sigma+c)}{\Gamma(c)} \, x^{-\sigma}. \tag{2.284}$$

From Eq. (2.284) one obtains with $x = B/A$ the Mellin-Barnes formula

$$(A+B)^{-c} = \frac{1}{2\pi i} \int\limits_{\gamma-i\infty}^{\gamma+i\infty} d\sigma \, \frac{\Gamma(-\sigma)\Gamma(\sigma+c)}{\Gamma(c)} \, A^{\sigma} B^{-\sigma-c}. \tag{2.285}$$

Equation (2.279) is then obtained by repeated use of Eq. (2.285).

With the help of Eqs. (2.278) and (2.279) we may exchange a Feynman parameter integral against a multiple contour integral. A typical single contour integral is of the form

$$I = \frac{1}{2\pi i} \int\limits_{\gamma-i\infty}^{\gamma+i\infty} d\sigma \, \frac{\Gamma(\sigma+a_1)...\Gamma(\sigma+a_m)}{\Gamma(\sigma+c_1)...\Gamma(\sigma+c_p)} \frac{\Gamma(-\sigma+b_1)...\Gamma(-\sigma+b_n)}{\Gamma(-\sigma+d_1)...\Gamma(-\sigma+d_q)} \, x^{-\sigma}. \tag{2.286}$$

If $\max\left(\text{Re}(-a_1), ..., \text{Re}(-a_m)\right) < \min\left(\text{Re}(b_1), ..., \text{Re}(b_n)\right)$ the contour can be chosen as a straight line parallel to the imaginary axis with

$$\max\left(\text{Re}(-a_1), ..., \text{Re}(-a_m)\right) \quad < \quad \text{Re}\,\gamma \quad < \quad \min\left(\text{Re}(b_1), ..., \text{Re}(b_n)\right), \tag{2.287}$$

otherwise the contour is indented, such that the residues of $\Gamma(\sigma+a_1), ..., \Gamma(\sigma+a_m)$ are to the left of the contour, whereas the residues of $\Gamma(-\sigma+b_1), ..., \Gamma(-\sigma+b_n)$ are to the right of the contour, as shown in Fig. 2.17.

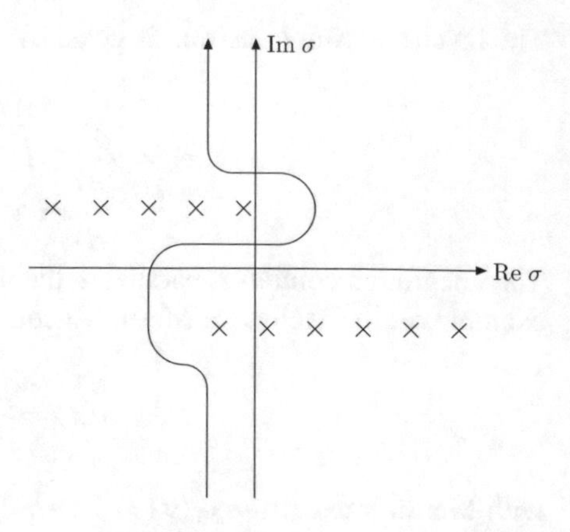

Fig. 2.17 The integration contour for Mellin-Barnes integrals: Residues of $\Gamma(\sigma + a_1)$, ..., $\Gamma(\sigma + a_m)$ are to the left of the contour, residues of $\Gamma(-\sigma + b_1)$, ..., $\Gamma(-\sigma + b_n)$ are to the right of the contour

The integral Eq. (2.286) is most conveniently evaluated with the help of the residue theorem by closing the contour to the left or to the right. To sum up all residues which lie inside the contour it is useful to know the residues of the gamma function:

$$\text{res}\ (\Gamma(\sigma + a), \sigma = -a - n) = \frac{(-1)^n}{n!}, \quad \text{res}\ (\Gamma(-\sigma + a), \sigma = a + n) = -\frac{(-1)^n}{n!}. \tag{2.288}$$

Remember that in the case where we close the contour to the right, there is an extra minus sign from the negative winding number.

In general there are multiple contour integrals, and as a consequence one obtains multiple sums. In particular simple cases the contour integrals can be performed in closed form with the help of two lemmas of Barnes. **Barnes' first lemma** states that

$$\frac{1}{2\pi i} \int\limits_{-i\infty}^{i\infty} d\sigma\ \Gamma(a + \sigma)\Gamma(b + \sigma)\Gamma(c - \sigma)\Gamma(d - \sigma) = \frac{\Gamma(a + c)\Gamma(a + d)\Gamma(b + c)\Gamma(b + d)}{\Gamma(a + b + c + d)}, \tag{2.289}$$

if none of the poles of $\Gamma(a + \sigma)\Gamma(b + \sigma)$ coincides with the ones from $\Gamma(c - \sigma)\Gamma(d - \sigma)$. **Barnes' second lemma** reads

$$\begin{aligned} & \frac{1}{2\pi i} \int\limits_{-i\infty}^{i\infty} d\sigma\ \frac{\Gamma(a + \sigma)\Gamma(b + \sigma)\Gamma(c + \sigma)\Gamma(d - \sigma)\Gamma(e - \sigma)}{\Gamma(a + b + c + d + e + \sigma)} \\ & = \frac{\Gamma(a + d)\Gamma(b + d)\Gamma(c + d)\Gamma(a + e)\Gamma(b + e)\Gamma(c + e)}{\Gamma(a + b + d + e)\Gamma(a + c + d + e)\Gamma(b + c + d + e)}. \end{aligned} \tag{2.290}$$

Chapter 3
Graph Polynomials

The Schwinger parameter representation and the Feynman parameter representation involve two graph polynomials $\mathcal{U}$ and $\mathcal{F}$, the Lee-Pomeransky representation involves the sum of these two polynomials $\mathcal{G} = \mathcal{U} + \mathcal{F}$. These polynomials encode the essential information of the integrands. In this chapter we study these polynomials in more detail. Previously, we defined the polynomials $\mathcal{U}$ and $\mathcal{F}$ in Eq. (2.157). In this chapter we will also learn alternative methods to compute these polynomials.

3.1 Spanning Trees and Spanning Forests

In this section we delve deeper into concepts of graph theory. We define spanning trees and spanning forests. These concepts lead to a second method for the computation of the graph polynomials. We consider a connected graph G with n edges and r vertices. We label the edges by $\{e_1, \dots, e_n\}$ and the vertices by $\{v_1, \dots, v_r\}$. Vertices of valency 1 are called external vertices, all other vertices are internal vertices. There is exactly one edge connected to an external vertex. Such an edge is called an external edge. Edges, which are not external edges are called internal edges. We label the edges of the graph G such that the first n_{int} edges are the internal edges and the remaining n_{ext} edges are the external edges:

$$\begin{aligned} \text{internal edges} &: \{e_1, e_2, \dots, e_{n_{\text{int}}}\}, \\ \text{external edges} &: \{e_{n_{\text{int}}+1}, e_{n_{\text{int}}+2}, \dots, e_{n_{\text{int}}+n_{\text{ext}}}\}. \end{aligned} \tag{3.1}$$

We have $n = n_{\text{int}} + n_{\text{ext}}$. In a similar way we label the vertices such that the first r_{int} vertices are the internal vertices and the remaining n_{ext} vertices are the external vertices (there are exactly n_{ext} external vertices):

S. Weinzierl, *Feynman Integrals*, UNITEXT for Physics,
https://doi.org/10.1007/978-3-030-99558-4_3

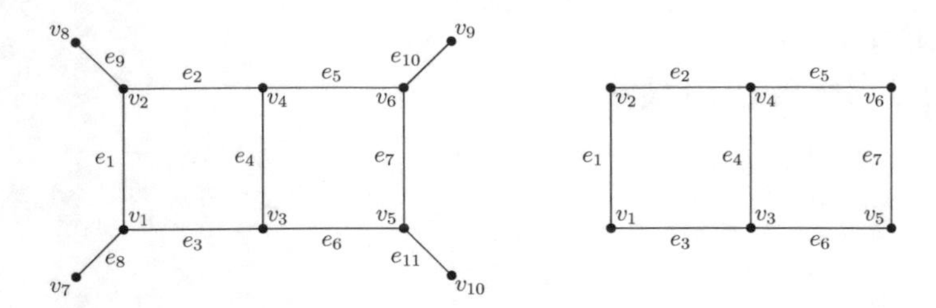

Fig. 3.1 An example for a Feynman graph G (left) and the associated internal graph G_{int} (right). The latter is obtained by deleting all external vertices and all external edges

$$\begin{aligned} \text{internal vertices} &: \{v_1, v_2, \dots, v_{r_{\text{int}}}\}, \\ \text{external vertices} &: \{v_{r_{\text{int}}+1}, v_{r_{\text{int}}+2}, \dots, v_{r_{\text{int}}+n_{\text{ext}}}\}. \end{aligned} \tag{3.2}$$

The distinction between internal edges and external edges (and internal vertices and external vertices) is necessary for the application towards Feynman integrals: The Feynman rules for the internal and external objects differ (and the variables of the graph polynomials are in one-to-one correspondence with the internal edges of the graph). However, pure mathematicians might prefer to work just with a graph (consisting of edges and vertices), without any particular distinction between internal and external vertices and edges. In order to reconcile the two approaches, let us introduce the **internal graph** G_{int} associated to G as the sub-graph of G obtained by deleting the external vertices and the external edges from G. The internal graph G_{int} has n_{int} edges $\{e_1, e_2, \dots, e_{n_{\text{int}}}\}$ and r_{int} vertices $\{v_1, v_2, \dots, v_{r_{\text{int}}}\}$. An example is shown in Fig. 3.1.

As before we denote by l the first Betti number of the graph (or in physics jargon: the number of loops). We have the relation

$$l = n - r + 1 \;=\; n_{\text{int}} - r_{\text{int}} + 1. \tag{3.3}$$

If we would allow for disconnected graphs, the corresponding formula for the first Betti number would be $n - r + k$, where k is the number of connected components.

Definition 1 *(spanning tree):* A **spanning tree** for the graph G is a sub-graph T of G satisfying the following requirements:

- T contains all the vertices of G,
- the first Betti number of T is zero,
- T is connected.

If T is a spanning tree for G, then it can be obtained from G by deleting l edges. In general a given graph G has several spanning trees. We will later obtain a formula which counts the number of spanning trees for a given graph G.

Definition 2 *(spanning forest):* A **spanning forest** for the graph G is a sub-graph F of G satisfying just the first two requirements:

- F contains all the vertices of G,
- the first Betti number of F is zero.

It is not required that a spanning forest is connected. If F has k connected components, we say that F is a k-forest. A spanning tree is a spanning 1-forest. If F is a spanning k-forest for G, then it can be obtained from G by deleting $l + k - 1$ edges.

For the application towards Feynman graphs we need a refinement of this definition:

Definition 3 *(spanning forest with respect to an edge set):* A spanning forest for the graph G with respect to an edge set E is a sub-graph F of G satisfying:

- F contains all the vertices of G,
- the first Betti number of F is zero.
- F contains all edges $\{e_1, \dots, e_n\}\backslash E$.

The third requirement states that we may only delete edges from the set E, but not from the complement $\{e_1, \dots, e_n\}\backslash E$. For $E = \{e_1, \dots, e_n\}$ the two definitions agree: A spanning forest with respect to the edge set $\{e_1, \dots, e_n\}$ is a spanning forest in the sense of Definition 2. The typical application towards Feynman graphs is the case, where E is the set of internal edges

$$E = \left\{e_1, e_2, \dots, e_{n_{\text{int}}}\right\}. \tag{3.4}$$

In this case the third condition ensures that no external edges are deleted. As this is the most common case, we will from now on always assume that a k-forest of a Feynman graph G is a k-forest of G with respect to the internal edges, unless stated otherwise.

Figure 3.2 shows an example for a spanning tree and a spanning 2-forest for the graph of Fig. 2.3.

We denote by $\mathcal{T}$ the set of spanning forests of G (by default with respect to the internal edges, unless stated otherwise) and by $\mathcal{T}_k$ the set of spanning k-forests

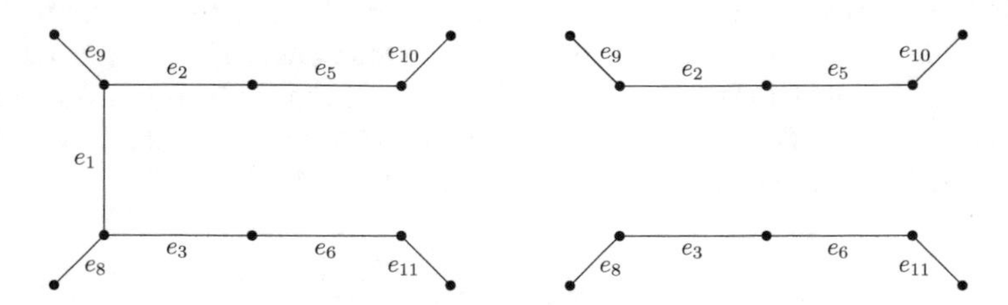

Fig. 3.2 The left picture shows a spanning tree for the graph of Fig. 2.3, the right picture shows a spanning 2-forest for the same graph. The spanning tree is obtained by deleting edges 4 and 7, the spanning 2-forest is obtained by deleting edges 1, 4 and 7

of G (again by default with respect to the internal edges unless stated otherwise). Obviously, we can write $\mathcal{T}$ as the disjoint union

$$\mathcal{T} = \bigcup_{k=1}^{r_{\text{int}}} \mathcal{T}_k. \tag{3.5}$$

$\mathcal{T}_1$ is the set of spanning trees. For an element of $\mathcal{T}_k$ we write

$$(T_1, T_2, \ldots, T_k) \in \mathcal{T}_k. \tag{3.6}$$

The T_i are the connected components of the k-forest. They are necessarily trees. We denote by P_{T_i} the set of external momenta attached to T_i. For the example of the 2-forest in the right picture of Fig. 3.2 we have (compare with the momentum labelling in Fig. 2.3)

$$P_{T_1} = \{p_2, p_3\}, \quad P_{T_2} = \{p_1, p_4\}. \tag{3.7}$$

The spanning trees and the spanning 2-forests (with respect to the internal edges) of a graph G are closely related to the graph polynomials $\mathcal{U}$ and $\mathcal{F}$ of the graph:

Graph polynomials from spanning trees and the spanning 2-forests:

$$\mathcal{U}(a) = \sum_{T \in \mathcal{T}_1} \prod_{e_i \notin T} a_i, \tag{3.8}$$

$$\mathcal{F}(a) = \sum_{(T_1,T_2) \in \mathcal{T}_2} \left(\prod_{e_i \notin (T_1,T_2)} a_i \right) \left(\sum_{p_j \in P_{T_1}} \sum_{p_k \in P_{T_2}} \frac{p_j \cdot p_k}{\mu^2} \right) + \mathcal{U}(a) \sum_{i=1}^{n_{\text{int}}} a_i \frac{m_i^2}{\mu^2}.$$

The sum is over all spanning trees for $\mathcal{U}$, and over all spanning 2-forests (with respect to the internal edges) in the first term of the formula for $\mathcal{F}$.

Equation (3.8) provides a second method for the computation of the graph polynomials $\mathcal{U}$ and $\mathcal{F}$. Let us first look at the formula for $\mathcal{U}$. For each spanning tree T we take the edges e_i, which have been removed from the graph G to obtain T. The product of the corresponding parameters a_i gives a monomial. The first formula says, that $\mathcal{U}$ is the sum of all the monomials obtained from all spanning trees. The formula for $\mathcal{F}$ has two parts: One part is related to the external momenta and the other part involves the masses. The latter is rather simple and we write

$$\mathcal{F}(a) = \mathcal{F}_0(a) + \mathcal{U}(a) \sum_{i=1}^{n_{\text{int}}} a_i \frac{m_i^2}{\mu^2}. \tag{3.9}$$

We focus on the polynomial $\mathcal{F}_0$. Here the 2-forests with respect to the internal edges are relevant. For each 2-forest (T_1, T_2) we consider again the edges e_i, which have been removed from the graph G to obtain (T_1, T_2). The product of the corresponding parameters a_i defines again a monomial, which in addition is multiplied by a quantity which depends on the external momenta. We define the square of the sum of momenta through the deleted lines of (T_1, T_2) by

$$s_{(T_1,T_2)} = \left(\sum_{e_j \notin (T_1,T_2)} q_j \right)^2 . \tag{3.10}$$

Here we assumed for simplicity that the orientation of the momenta of the deleted internal lines are chosen such that all deleted momenta flow from T_1 to T_2 (or alternatively that all deleted momenta flow from T_2 to T_1, but not mixed). From momentum conservation it follows that the sum of the momenta flowing through the deleted edges out of T_1 is equal to the negative of the sum of the external momenta of T_1. With the same reasoning the sum of the momenta flowing through the deleted edges into T_2 is equal to the sum of the external momenta of T_2. Therefore we can equally write

$$s_{(T_1,T_2)} = - \left(\sum_{p_i \in P_{T_1}} p_i \right) \cdot \left(\sum_{p_j \in P_{T_2}} p_j \right) \tag{3.11}$$

and $\mathcal{F}_0$ is given by

$$\mathcal{F}_0(a) = \sum_{(T_1,T_2) \in \mathcal{T}_2} \left(\prod_{e_i \notin (T_1,T_2)} a_i \right) \left(\frac{-s_{(T_1,T_2)}}{\mu^2} \right). \tag{3.12}$$

Since we have to remove l edges from G to obtain a spanning tree and $(l+1)$ edges to obtain a spanning 2-forest, it follows that $\mathcal{U}(a)$ and $\mathcal{F}(a)$ are homogeneous in the parameters a of degree l and $(l+1)$, respectively. From the fact, that an internal edge can be removed at most once, it follows that $\mathcal{U}$ and $\mathcal{F}_0$ are linear in each parameter a_j. Finally it is obvious from Eq. (3.8) that each monomial in the expanded form of $\mathcal{U}$ has coefficient $+1$.

Let us look at an example. Figure 3.3 shows the graph of a two-loop two-point integral. We take again all internal masses to be zero. The set of all spanning trees for this graph is shown in Fig. 3.4. There are eight spanning trees. Figure 3.5 shows the set of spanning 2-forests with respect to the internal edges for this graph. There are ten spanning 2-forests. The last example in each row of Fig. 3.5 does not contribute to the graph polynomial $\mathcal{F}$, since the momentum sum flowing through all deleted edges is zero. Therefore we have in this case $s_{(T_1,T_2)} = 0$. In all other cases we have $s_{(T_1,T_2)} = p^2$. We arrive therefore at the graph polynomials

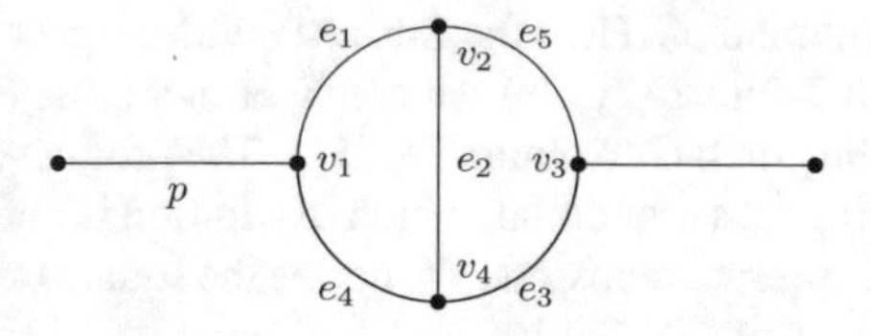

Fig. 3.3 A two-loop two-point graph

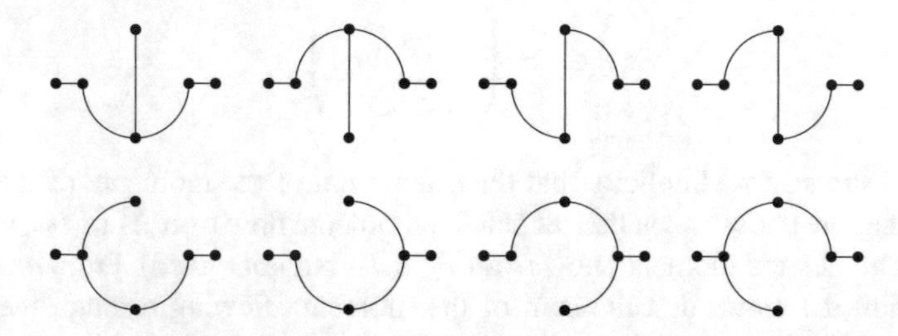

Fig. 3.4 The set of spanning trees for the two-loop two-point graph of Fig. 3.3

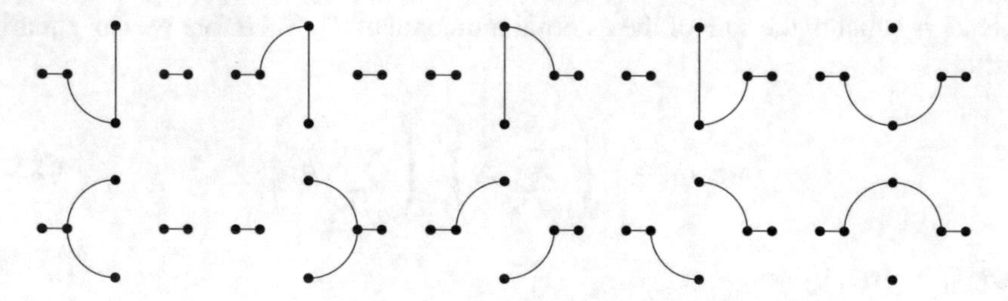

Fig. 3.5 The set of spanning 2-forests for the two-loop two-point graph of Fig. 3.3

$$\mathcal{U} = (a_1 + a_4)(a_3 + a_5) + (a_1 + a_3 + a_4 + a_5)a_2,$$
$$\mathcal{F} = [(a_1 + a_5)(a_3 + a_4)a_2 + a_1a_4(a_3 + a_5) + a_3a_5(a_1 + a_4)] \left(\frac{-p^2}{\mu^2}\right). \tag{3.13}$$

Exercise 20 *Re-compute the first graph polynomial $\mathcal{U}$ for the graph shown in Fig. 2.6 from the set of spanning trees.*

3.2 The Matrix-Tree Theorem

In this section we introduce the Laplacian of a graph. The Laplacian is a matrix constructed from the topology of the graph. The determinant of a minor of this matrix where the i-th row and column have been deleted gives us the Kirchhoff polynomial of the graph, which in turn upon a simple substitution leads to the first

Symanzik polynomial. We then show how this construction generalises for the second Symanzik polynomial. This provides a third method for the computation of the two graph polynomials. This method is very well suited for computer algebra systems, as it involves just the computation of a determinant of a matrix. The matrix is easily constructed from the data defining the graph.

We begin with the **Kirchhoff polynomial** of a graph G. We associate a parameter a_j to any edge e_j (internal or external). The Kirchhoff polynomial is defined by

$$\mathcal{K}(a_1, \dots, a_n) = \sum_{T \in \mathcal{T}_1} \prod_{e_j \in T} a_j. \tag{3.14}$$

In physics we associate parameters a_j only to internal edges. We set

$$\mathcal{K}_{\text{int}}(G) = \mathcal{K}(G_{\text{int}}) = \sum_{T \in \mathcal{T}_1} \prod_{e_j \in (T \cap E)} a_j, \tag{3.15}$$

where E is the set of internal edges. The definition is very similar to the expression for the first Symanzik polynomial in Eq. (3.8). Again we have a sum over all spanning trees, but this time we take for each spanning tree the monomial of the Feynman parameters corresponding to the edges which have not been removed. The Kirchhoff polynomial is therefore homogeneous of degree $(n_{\text{int}} - l)$ in the parameters a. There is a simple relation between the Kirchhoff polynomial $\mathcal{K}_{\text{int}}$ and the first Symanzik polynomial $\mathcal{U}$:

$$\begin{aligned} \mathcal{U}(a_1, \dots, a_{n_{\text{int}}}) &= a_1 \dots a_{n_{\text{int}}} \; \mathcal{K}_{\text{int}}\left(\frac{1}{a_1}, \dots, \frac{1}{a_{n_{\text{int}}}}\right), \\ \mathcal{K}_{\text{int}}(a_1, \dots, a_{n_{\text{int}}}) &= a_1 \dots a_{n_{\text{int}}} \; \mathcal{U}\left(\frac{1}{a_1}, \dots, \frac{1}{a_{n_{\text{int}}}}\right). \end{aligned} \tag{3.16}$$

These equations are immediately evident from the fact that $\mathcal{U}$ and $\mathcal{K}_{\text{int}}$ are homogeneous polynomials which are linear in each variable together with the fact that a monomial corresponding to a specific spanning tree in one polynomial contains exactly those Feynman parameters which are not in the corresponding monomial in the other polynomial.

We now define the **Laplacian of a graph** G.

Definition 4 *(Laplacian of a graph):* Let G be a graph with n edges and r vertices. To each edge e_j one associates a parameter a_j. The Laplacian of the graph G is a $r \times r$-matrix L, whose entries are given by

$$\mathrm{L_{ij}} = \begin{cases} \sum \mathrm{a_k} \text{ if } \mathrm{i} = \mathrm{j} \text{ and edge } \mathrm{e_k} \text{ is attached to } \mathrm{v_i} \text{ and is not a self-loop,} \\ -\sum \mathrm{a_k} \text{ if } \mathrm{i} \neq \mathrm{j} \text{ and edge } \mathrm{e_k} \text{ connects } \mathrm{v_i} \text{ and } \mathrm{v_j}. \end{cases} \tag{3.17}$$

The graph may contain multiple edges and self-loops.

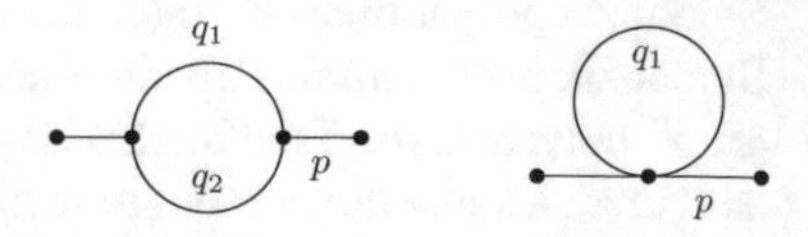

Fig. 3.6 The left picture shows a graph with a double edge, the right picture shows a graph with a self-loop

We speak of a **multiple edge**, if two vertices are connected by more than one edge. We speak of a **self-loop** if an edge starts and ends at the same vertex. In the physics literature a self-loop is known as a tadpole. Figure 3.6 shows a simple example for a double edge and a self-loop. If the vertices v_i and v_j are connected by two edges e_{k_1} and e_{k_2}, then the Laplacian depends only on the sum $a_{k_1} + a_{k_2}$. If an edge e_k is a self-loop attached to a vertex v_i, then it does not contribute to the Laplacian.

For the application towards Feynman graphs we need a refinement of this definition. (This can already be anticipated from the fact, that we associate parameters a_j to all internal edges, but not to the external edges.)

Definition 5 *(Laplacian of a graph with respect to internal vertices and edges):* Let G be a graph with n_{int} internal edges and r_{int} internal vertices. To each internal edge e_j one associates a parameter a_j. Denote by G_{int} the internal graph of G. The Laplacian of the graph G with respect to internal vertices and edges is the (ordinary) Laplacian of the graph G_{int}:

$$L_{\text{int}}(G) = L(G_{\text{int}}). \tag{3.18}$$

Phrased differently, the Laplacian of the graph G with respect to internal vertices and edges is a $r_{\text{int}} \times r_{\text{int}}$-matrix L_{int}, whose entries are given by

$$(\mathrm{L_{int}})_{\mathrm{ij}} = \begin{cases} \sum \mathrm{a_k} \ \text{if}\ \mathrm{i} = \mathrm{j} \ \text{and edge}\ \mathrm{e_k} \ \text{is attached to}\ \mathrm{v_i} \ \text{and is not a self-loop}, \\ -\sum \mathrm{a_k} \ \text{if}\ \mathrm{i} \neq \mathrm{j} \ \text{and edge}\ \mathrm{e_k} \ \text{connects}\ \mathrm{v_i} \ \text{and}\ \mathrm{v_j}. \end{cases} \tag{3.19}$$

In Eq. (3.19) only internal vertices and edges are considered.

Unless stated otherwise, we will from now on always assume that the Laplacian of a Feynman graph G refers to the Laplacian of the graph G with respect to internal vertices and edges.

Let us consider an example: The Laplacian of the two-loop two-point graph of Fig. 3.3 is given by

$$L_{\text{int}} = \begin{pmatrix} a_1 + a_4 & -a_1 & 0 & -a_4 \\ -a_1 & a_1 + a_2 + a_5 & -a_5 & -a_2 \\ 0 & -a_5 & a_3 + a_5 & -a_3 \\ -a_4 & -a_2 & -a_3 & a_2 + a_3 + a_4 \end{pmatrix}. \tag{3.20}$$

In the sequel we will need minors of the matrices L and L_{int}. It is convenient to introduce the following notation: For a $r \times r$ matrix A we denote by

$A[i_1, \ldots, i_k; j_1, \ldots, j_k]$ the $(r-k) \times (r-k)$ matrix, which is obtained from A by deleting the rows i_1, ..., i_k and the columns j_1, ..., j_k. For $A[i_1, \ldots, i_k; i_1, \ldots, i_k]$ we will simply write $A[i_1, \ldots, i_k]$.

Let v_i be an arbitrary vertex of G. The **matrix-tree theorem** states [35]

$$\mathcal{K} = \det\ L[i], \tag{3.21}$$

i.e. the Kirchhoff polynomial is given by the determinant of the minor of the Laplacian, where the i-th row and column have been removed. One can choose for i any number between 1 and r. For an arbitrary internal vertex v_i of G we have

$$\mathcal{K}_{\text{int}} = \det\ L_{\text{int}}[i]. \tag{3.22}$$

Choosing for example $i = 4$ in Eq. (3.20) one finds for the Kirchhoff polynomial of the two-loop two-point graph of Fig. 3.3

$$\begin{aligned} \mathcal{K}_{\text{int}} &= \begin{vmatrix} a_1 + a_4 & -a_1 & 0 \\ -a_1 & a_1 + a_2 + a_5 & -a_5 \\ 0 & -a_5 & a_3 + a_5 \end{vmatrix} \\ &= a_1 a_5 (a_3 + a_4) + (a_1 + a_5) a_3 a_4 + (a_1 a_5 + a_1 a_3 + a_4 a_5 + a_3 a_5)\, a_2. \end{aligned} \tag{3.23}$$

Using Eq. (3.16) one recovers the first Symanzik polynomial of this graph as given in Eq. (3.13).

Exercise 21 *Re-compute the first graph polynomial $\mathcal{U}$ for the graph shown in Fig. 2.6 from the Laplacian of the graph.*

The matrix-tree theorem allows to determine the number of spanning trees of a given graph G. Setting $a_1 = \cdots = a_n = 1$, each monomial in $\mathcal{K}$, $\mathcal{K}_{\text{int}}$ and $\mathcal{U}$ reduces to 1. There is exactly one monomial for each spanning tree, therefore one obtains

$$|\mathcal{T}_1| = \mathcal{K}(1, \ldots, 1) \ = \ \mathcal{K}_{\text{int}}(1, \ldots, 1) \ = \ \mathcal{U}(1, \ldots, 1). \tag{3.24}$$

The matrix-tree theorem as in Eq. (3.21) relates the determinant of the minor of the Laplacian, where the i-th row and the i-th column have been deleted to a sum over the spanning trees of the graph. There are two generalisations we can think of:

1. We delete more than one row and column.
2. We delete different rows and columns, i.e. we delete row i and column j with $i \neq j$.

The all-minors matrix-tree theorem relates the determinant of the corresponding minor to a specific sum over spanning forests [36–38]. We first state the version for the Laplacian L and then specialise to L_{int}. To state this theorem we need some notation: We consider a graph with r vertices (internal and external). Let $I = (i_1, \ldots, i_k)$ with

$1 \leq i_1 < \cdots < i_k \leq r$ denote the rows, which we delete from the Laplacian L, and let $J = (j_1, \ldots, j_k)$ with $1 \leq j_1 < \cdots < j_k \leq r$ denote the columns to be deleted from the Laplacian L. We set $|I| = i_1 + \cdots + i_k$ and $|J| = j_1 + \cdots + j_k$. We denote by $\mathcal{T}_k^{I,J}$ the spanning k-forests (in sense of Definition 2), such that each tree of an element of $\mathcal{T}_k^{I,J}$ contains exactly one vertex v_{i_α} and exactly one vertex v_{j_β}. The set $\mathcal{T}_k^{I,J}$ is a sub-set of all spanning k-forests. We now consider an element F of $\mathcal{T}_k^{I,J}$. Since the element F is a k-forest, it consists therefore of k trees and we can write it as

$$F = (T_1, \ldots, T_k) \in \mathcal{T}_k^{I,J}. \tag{3.25}$$

We can label the trees such that $v_{i_1} \in T_1$, ..., $v_{i_k} \in T_k$. By assumption, each tree T_α contains also exactly one vertex from the set $\{v_{j_1}, \ldots, v_{j_k}\}$, although not necessarily in the order $v_{j_\alpha} \in T_\alpha$. In general it will be in a different order, which we can specify by a permutation $\pi_F \in S_k$:

$$v_{j_\alpha} \in T_{\pi_F(\alpha)}. \tag{3.26}$$

The **all-minors matrix-tree theorem** reads then

$$\det\ L[I, J] = (-1)^{|I|+|J|} \sum_{F \in \mathcal{T}_k^{I,J}} \text{sign}(\pi_F) \prod_{e_j \in F} a_j. \tag{3.27}$$

Let us now specialise to L_{int}. We consider a graph with r_{int} internal vertices. As before we denote the deleted rows by $I = (i_1, \ldots, i_k)$, now with $1 \leq i_1 < \cdots < i_k \leq r_{\text{int}}$. The deleted columns are denoted by $J = (j_1, \ldots, j_k)$ with $1 \leq j_1 < \cdots < j_k \leq r_{\text{int}}$. As before we set $|I| = i_1 + \cdots + i_k$ and $|J| = j_1 + \cdots + j_k$. The set $\mathcal{T}_k^{I,J}$ now denotes the spanning k-forests in sense of Definition 3, such that each tree of an element of $\mathcal{T}_k^{I,J}$ contains exactly one vertex v_{i_α} and exactly one vertex v_{j_β}. As before we consider

$$F = (T_1, \ldots, T_k) \in \mathcal{T}_k^{I,J} \tag{3.28}$$

and define $\pi_F \in S_k$ by $v_{j_\alpha} \in T_{\pi_F(\alpha)}$. The all-minors matrix-tree theorem for the internal graph reads

$$\det\ L_{\text{int}}[I, J] = (-1)^{|I|+|J|} \sum_{F \in \mathcal{T}_k^{I,J}} \text{sign}(\pi_F) \prod_{e_j \in F} a_j. \tag{3.29}$$

In the special case $I = J$ this reduces to

$$\det\ L_{\text{int}}[I] = \sum_{F \in \mathcal{T}_k^{I,I}} \prod_{e_j \in F} a_j. \tag{3.30}$$

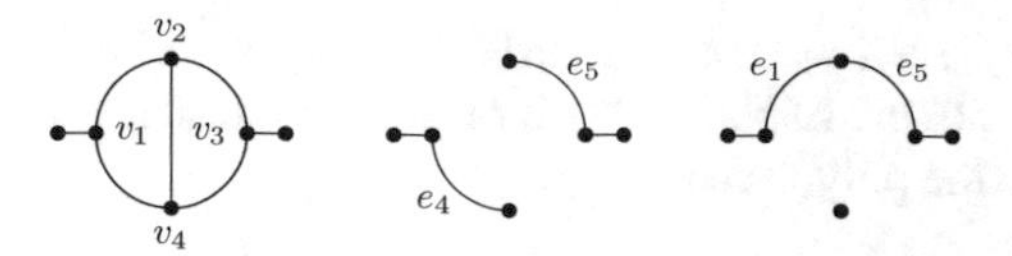

Fig. 3.7 The left picture shows the labelling of the vertices for the two-loop two-point function. The middle and the right picture show the two 2-forests contributing to $\mathcal{T}_2^{I,J}$ with $I = (2, 4)$ and $J = (3, 4)$

If we specialise further to $I = J = (i)$, the sum equals the sum over all spanning trees (since each spanning 1-forest of $\mathcal{T}_1^{(i),(i)}$ necessarily contains the vertex v_i). We recover the classical matrix-tree theorem:

$$\det\ L_{\text{int}}[i] = \sum_{T \in \mathcal{T}_1} \prod_{e_j \in T} a_j. \tag{3.31}$$

Let us illustrate the all-minors matrix-tree theorem with an example. We consider again the two-loop two-point graph with the labelling of the vertices as shown in Fig. 3.7. Taking as an example

$$I = (2, 4) \text{ and } J = (3, 4) \tag{3.32}$$

we find for the determinant of $L_{\text{int}}[I; J]$:

$$\det\ L_{\text{int}}[2, 4; 3, 4] = \begin{vmatrix} a_1 + a_4 & -a_1 \\ 0 & -a_5 \end{vmatrix} = -a_1 a_5 - a_4 a_5. \tag{3.33}$$

On the other hand there are exactly two 2-forests, such that in each 2-forest the vertices v_2 and v_3 are contained in one tree, while the vertex v_4 is contained in the other tree. These two 2-forests are shown in Fig. 3.7. The monomials corresponding to these two 2-trees are $a_1 a_5$ and $a_4 a_5$, respectively. The permutation π_F is in both cases the identity and with $|I| = 6$, $|J| = 7$ we have an overall minus sign

$$(-1)^{|I|+|J|} = -1. \tag{3.34}$$

Therefore, the right hand side of Eq. (3.27) equals $-a_1 a_5 - a_4 a_5$, showing the agreement with the result of Eq. (3.33).

Equation (3.21) together with Eq. (3.16) allows to determine the first Symanzik polynomial $\mathcal{U}$ from the Laplacian of the graph.

We may ask if also the polynomial $\mathcal{F}_0$ can be obtained in a similar way. We consider again a graph G with n_{int} internal edges $(e_1, \dots, e_{n_{\text{int}}})$, r_{int} internal vertices $(v_1, \dots, v_{r_{\text{int}}})$, n_{ext} external edges $(e_{n_{\text{int}}+1}, \dots, e_{n_{\text{int}}+n_{\text{ext}}})$ and n_{ext} external vertices $(v_{r_{\text{int}}+1}, \dots, e_{r_{\text{int}}+n_{\text{ext}}})$. As before we associate the parameters a_i to the edges e_i $(1 \leq i \leq n_{\text{int}})$ and new parameters b_j to the edges $e_{n_{\text{int}}+j}$ $(1 \leq j \leq n_{\text{ext}})$. The Laplacian of G is a $(r_{\text{int}} + n_{\text{ext}}) \times (r_{\text{int}} + n_{\text{ext}})$ matrix. We are now considering the Laplacian as in Definition 4, not the Laplacian with respect to internal vertices and edges.

Let us consider the polynomial

$$\mathcal{W}(a_1, \dots, a_{n_{\text{int}}}, b_1, \dots, b_{n_{\text{ext}}}) = \det\ L\,(G)\,[r_{\text{int}} + 1, \dots, r_{\text{int}} + n_{\text{ext}}]. \tag{3.35}$$

$\mathcal{W}$ is a polynomial of degree $r_{\text{int}} = n_{\text{int}} - l + 1$ in the variables a_i and b_j. We can expand $\mathcal{W}$ in polynomials homogeneous in the variables b_j:

$$\mathcal{W} = \mathcal{W}^{(0)} + \mathcal{W}^{(1)} + \mathcal{W}^{(2)} + \dots + \mathcal{W}^{(m)}, \tag{3.36}$$

where $\mathcal{W}^{(k)}$ is homogeneous of degree k in the variables b_j. We further write

$$\mathcal{W}^{(k)} = \sum_{(j_1,\dots,j_k)} \mathcal{W}^{(k)}_{(j_1,\dots,j_k)}(a_1, \dots, a_{n_{\text{int}}})\, b_{j_1} \dots b_{j_k}. \tag{3.37}$$

The sum is over all indices with $1 \leq j_1 < \dots < j_j \leq n_{\text{ext}}$. The $\mathcal{W}^{(k)}_{(j_1,\dots,j_k)}$ are homogeneous polynomials of degree $r_{\text{int}} - k$ in the variables a_i. For $\mathcal{W}^{(0)}$ and $\mathcal{W}^{(1)}$ one finds

$$\mathcal{W}^{(0)} = 0, \quad \mathcal{W}^{(1)} = \mathcal{K}_{\text{int}}\left(a_1, \dots, a_{n_{\text{int}}}\right) \sum_{j=1}^{n_{\text{ext}}} b_j, \tag{3.38}$$

therefore

$$\mathcal{U} = a_1 \dots a_{n_{\text{int}}}\ \mathcal{W}^{(1)}_{(j)}\left(\frac{1}{a_1}, \dots, \frac{1}{a_{n_{\text{int}}}}\right), \tag{3.39}$$

for any $j \in \{1, \dots, n_{\text{ext}}\}$. $\mathcal{F}_0$ is related to $\mathcal{W}^{(2)}$:

$$\mathcal{F}_0 = a_1 \dots a_{n_{\text{int}}} \sum_{(j,k)} \left(\frac{p_j \cdot p_k}{\mu^2}\right) \cdot \mathcal{W}^{(2)}_{(j,k)}\left(\frac{1}{a_1}, \dots, \frac{1}{a_{n_{\text{int}}}}\right). \tag{3.40}$$

The proof of Eqs. (3.38)–(3.40) follows from the all-minors matrix-tree theorem. The all-minors matrix-tree theorem states

$$\mathcal{W}\left(a_1, \dots, a_{n_{\text{int}}}, b_1, \dots, b_{n_{\text{ext}}}\right) = \sum_{F \in \mathcal{T}^{I,I}_{n_{\text{ext}}}(G)} \prod_{e_j \in F} c_j, \tag{3.41}$$

with $I = (r_{\text{int}} + 1, \dots, r_{\text{int}} + n_{\text{ext}})$ and $c_j = a_j$ if e_j is an internal edge or $c_j = b_{j-n_{\text{int}}}$ if e_j is an external edge. The sum is over all n_{ext}-forests of G (in the sense of Definition 2), such that each tree in an n_{ext}-forest contains exactly one of the external vertices $v_{r_{\text{int}}+1}$, ..., $v_{r_{\text{int}}+n_{\text{ext}}}$. Each n_{ext}-forest has n_{ext} connected components. The polynomial $\mathcal{W}^{(0)}$ by definition does not contain any variable b_j. $\mathcal{W}^{(0)}$ would therefore correspond to forests where all edges connecting the external vertices $e_{r_{\text{int}}+1}, \dots, e_{r_{\text{int}}+n_{\text{ext}}}$ have been cut. The external vertices appear therefore as isolated vertices in the forest. For $l > 0$, such a forest must necessarily have more than n_{ext} connected components. This is a contradiction with the requirement of having exactly n_{ext} connected components and therefore $\mathcal{W}^{(0)} = 0$. Next, we consider $\mathcal{W}^{(1)}$. Each term is linear in the variables b_j. Therefore $(n_{\text{ext}} - 1)$ vertices of the external vertices $v_{r_{\text{int}}+1}, \dots, v_{r_{\text{int}}+n_{\text{ext}}}$ appear as isolated vertices in the n_{ext}-forest. The remaining added vertex is connected to a spanning tree of G_{int}. Summing over all possibilities one sees that $\mathcal{W}^{(1)}$ is given by the product of $(b_1 + \dots + b_{n_{\text{ext}}})$ with the Kirchhoff polynomial of G_{int}. Finally we consider $\mathcal{W}^{(2)}$. Here, $(n_{\text{ext}} - 2)$ of the added vertices appear as isolated vertices. The remaining two are connected to a spanning 2-forest of the graph G_{int}, one to each tree of the 2-forest. Summing over all possibilities one obtains Eq. (3.40).

Let us summarise the results on the Laplacian:

Graph polynomials from the Laplacian of the graph:

$$\mathcal{U} = a_1 \dots a_{n_{\text{int}}} \, \mathcal{W}^{(1)}_{(j)}\left(\frac{1}{a_1}, \dots, \frac{1}{a_{n_{\text{int}}}}\right), \qquad \text{for any } j \in \{1, \dots, n_{\text{ext}}\},$$

$$\mathcal{F}_0 = a_1 \dots a_{n_{\text{int}}} \sum_{(j,k)} \left(\frac{p_j \cdot p_k}{\mu^2}\right) \cdot \mathcal{W}^{(2)}_{(j,k)}\left(\frac{1}{a_1}, \dots, \frac{1}{a_{n_{\text{int}}}}\right). \tag{3.42}$$

The quantities $\mathcal{W}^{(1)}_{(j)}$ and $\mathcal{W}^{(2)}_{(j,k)}$ are obtained from the Laplacian of the graph by Eqs. (3.35)–(3.37). The graph polynomial $\mathcal{F}$ is obtained from $\mathcal{F}_0$ and $\mathcal{U}$ by Eq. (3.9).

Equation (3.42) together with Eq. (3.9) allow the computation of the first and second Symanzik polynomial from the Laplacian of the graph G. This provides a third method for the computation of the graph polynomials $\mathcal{U}$ and $\mathcal{F}$.

As an example we consider the double-box graph of Fig. 2.3. Figure 3.8 shows the labelling of the vertices and the Feynman parameters for the graph G. The Laplacian of G (in the sense of Definition 2) is a 10×10-matrix. We are interested in the minor, where—with the labelling of Fig. 3.8—we delete the rows and columns 7, 8, 9 and 10. The determinant of this minor reads

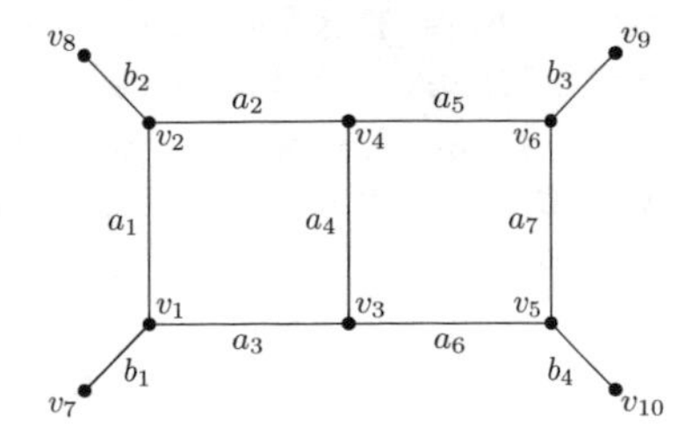

Fig. 3.8 The labelling of the vertices and the Feynman parameters for the "double box"-graph

$$\mathcal{W} = \det L[7,8,9,10] =$$
$$= \begin{vmatrix} a_1+a_3+b_1 & -a_1 & -a_3 & 0 & 0 & 0 \\ -a_1 & a_1+a_2+b_2 & 0 & -a_2 & 0 & 0 \\ -a_3 & 0 & a_3+a_4+a_6 & -a_4 & -a_6 & 0 \\ 0 & -a_2 & -a_4 & a_2+a_4+a_5 & 0 & -a_5 \\ 0 & 0 & -a_6 & 0 & a_6+a_7+b_4 & -a_7 \\ 0 & 0 & 0 & -a_5 & -a_7 & a_5+a_7+b_3 \end{vmatrix}$$
$$= \mathcal{W}^{(1)} + \mathcal{W}^{(2)} + \mathcal{W}^{(3)} + \mathcal{W}^{(4)}. \tag{3.43}$$

For the polynomials $\mathcal{W}^{(1)}$ and $\mathcal{W}^{(2)}$ one finds

$$\begin{aligned} \mathcal{W}^{(1)} = & (b_1+b_2+b_3+b_4) \\ & (a_1a_2a_3a_5a_6 + a_1a_2a_3a_5a_7 + a_1a_2a_3a_6a_7 + a_1a_2a_4a_5a_6 + a_1a_2a_4a_5a_7 \\ & + a_1a_2a_4a_6a_7 + a_1a_2a_5a_6a_7 + a_1a_3a_4a_5a_6 + a_1a_3a_4a_5a_7 + a_1a_3a_4a_6a_7 \\ & + a_1a_3a_5a_6a_7 + a_2a_3a_4a_5a_6 + a_2a_3a_4a_5a_7 + a_2a_3a_4a_6a_7 + a_2a_3a_5a_6a_7), \end{aligned}$$

$$\begin{aligned} \mathcal{W}^{(2)} = & (b_1+b_4)(b_2+b_3)a_2a_3a_5a_6 \\ & +(b_1+b_2)(b_4+b_3)\,(a_1a_2a_3a_7 + a_1a_2a_4a_7 + a_1a_2a_6a_7 + a_1a_3a_4a_7 + a_1a_3a_5a_7 + a_1a_4a_5a_6 \\ & + a_1a_4a_5a_7 + a_1a_4a_6a_7 + a_1a_5a_6a_7 + a_2a_3a_4a_7) \\ & + b_1(b_2+b_3+b_4)a_2\,(a_3a_5a_7 + a_4a_5a_6 + a_4a_5a_7 + a_4a_6a_7 + a_5a_6a_7) \\ & + b_2(b_1+b_3+b_4)a_3\,(a_2a_6a_7 + a_4a_5a_6 + a_4a_5a_7 + a_4a_6a_7 + a_5a_6a_7) \\ & + b_3(b_1+b_2+b_4)a_6\,(a_1a_2a_3 + a_1a_2a_4 + a_1a_3a_4 + a_1a_3a_5 + a_2a_3a_4) \\ & + b_4(b_1+b_2+b_3)a_5\,(a_1a_2a_3 + a_1a_2a_4 + a_1a_2a_6 + a_1a_3a_4 + a_2a_3a_4). \end{aligned} \tag{3.44}$$

With the help of Eqs. (3.39) and (3.40) and using the kinematic specifications of Eq. (2.160) we recover $\mathcal{U}$ and $\mathcal{F}$ of Eq. (2.164).

We would like to make a few remarks: The polynomial $\mathcal{W}$ is obtained from the determinant of the matrix $L = L(G)\,[r_{\text{int}}+1, \ldots, r_{\text{int}}+n_{\text{ext}}]$. This matrix was constructed from the Laplacian of the graph G, taking external vertices and external edges into account. Then one deletes the rows and columns corresponding to the external vertices. There are two alternative ways to arrive at the same matrix L:

The first alternative consists in merging the external vertices $v_{r_{\text{int}}+1}, v_{r_{\text{int}}+2}, \ldots, v_{r_{\text{int}}+n_{\text{ext}}}$ into a single new vertex v_∞, which connects to all external lines. This defines a new graph $\hat{G}$, which by construction no longer has any external lines. As before we associate variables b_1, ..., $b_{n_{\text{ext}}}$ to the edges connected to v_∞. Figure 3.9 shows an example for the graph $\hat{G}$ associated to a one-loop graph with three external legs. The Laplacian of $\hat{G}$ is a $(r_{\text{int}}+1) \times (r_{\text{int}}+1)$-matrix. It is easy to see that

$$L = L\left(\hat{G}\right)[r_{\text{int}}+1]. \tag{3.45}$$

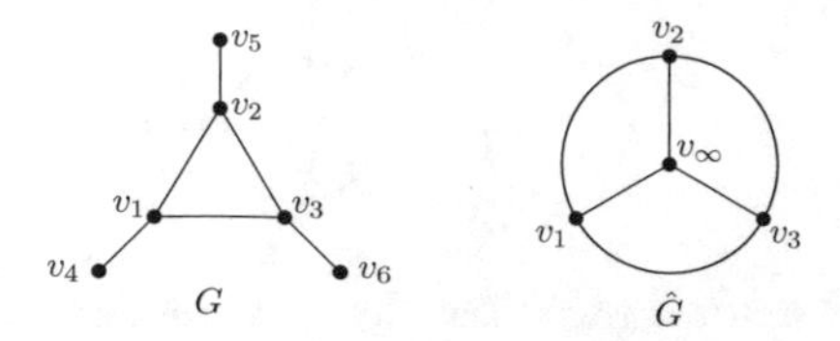

Fig. 3.9 The left picture shows a one-loop graph with three external edges. The right picture shows the graph $\hat{G}$ associated to G, where all external vertices have been joined in one additional vertex v_∞

From Eq. (3.21) we see that the polynomial $\mathcal{W}$ is nothing else than the Kirchhoff polynomial of the graph $\hat{G}$:

$$\mathcal{W}(G) = \mathcal{K}\left(\hat{G}\right) = \det L\left(\hat{G}\right)[j], \tag{3.46}$$

where j is any number between 1 and $r_{\text{int}} + 1$.

For the second alternative one starts from the Laplacian with respect to internal vertices and edges of the original graph G. Let π be a permutation of $(1, \dots, r_{\text{int}})$. We consider the diagonal matrix $\text{diag}\left(b_{\pi(1)}, \dots, b_{\pi(r_{\text{int}})}\right)$. We can choose the permutation π such that

$$\pi(i) = j, \quad \text{if the external edge } e_{n_{\text{int}}+j} \text{ is attached to vertex } v_i. \tag{3.47}$$

We then have

$$L = L_{\text{int}}(G) + \text{diag}\left(b_{\pi(1)}, \dots, b_{\pi(r_{\text{int}})}\right)\Big|_{b_{n_{\text{ext}}+1}=\dots=b_{r_{\text{int}}}=0}. \tag{3.48}$$

3.3 Deletion and Contraction Properties

In this section we study two operations on a graph: the deletion of an edge and the contraction of an edge. This leads to a recursive algorithm for the calculation of the graph polynomials $\mathcal{U}$ and $\mathcal{F}$. In addition we discuss the multivariate Tutte polynomial and Dodgson's identity.

In graph theory an edge is called a **bridge**, if the deletion of the edge increases the number of connected components. In the physics literature the term "one-particle-reducible" is used for a connected graph containing at least one bridge as an internal edge. The contrary is called "one-particle-irreducible", i.e. a connected graph containing no internal bridges. All external edges are necessarily bridges. Figure 3.10 shows an example. The edge e_3 is a bridge, while the edges e_1, e_2, e_4 and e_5 are not bridges. Note that all edges of a tree graph are bridges. An edge which is neither a bridge nor a self-loop is called a **regular edge**. All regular edges are internal edges.

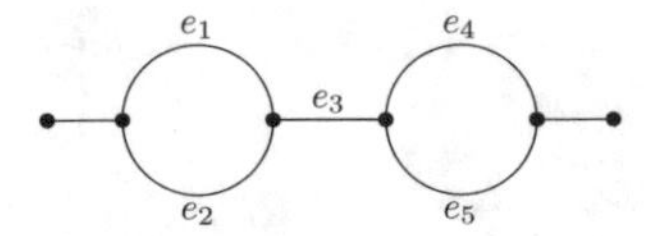

Fig. 3.10 A one-particle-reducible graph: The edge e_3 is called a bridge. Deleting e_3 results in two connected components

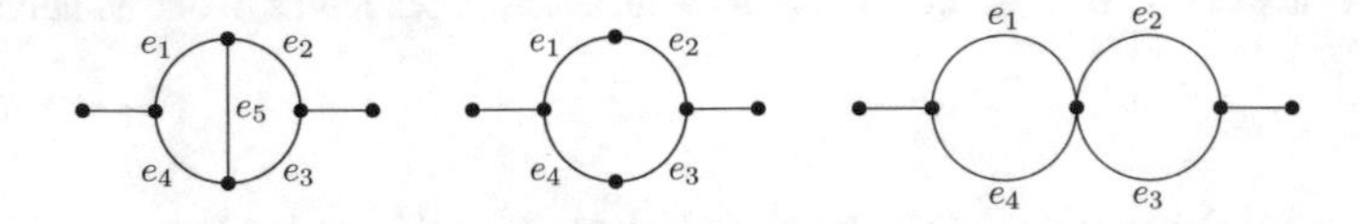

Fig. 3.11 The left picture shows the graph G of the two-loop two-point function. The middle picture shows the graph $G - e_5$, where edge e_5 has been deleted. The right picture shows the graph G/e_5, where the two vertices connected to e_5 have been joined and the edge e_5 has been removed

For a graph G and a regular edge e we define

$$\begin{aligned} G/e \quad & \text{to be the graph obtained from } G \text{ by contracting the regular edge } e, \\ G - e \quad & \text{to be the graph obtained from } G \text{ by deleting the regular edge } e. \end{aligned} \tag{3.49}$$

Figure 3.11 shows an example. If the graph G has loop number l it follows that $G - e$ has loop number $(l - 1)$, while G/e has loop number l. This follows easily from the formula $l = n - r + 1$ for a connected graph: $G - e$ has one edge less, but the same number of vertices, while G/e has one edge and one vertex less.

Let us now study the behaviour of the Laplacian under these operations. Under deletion the Laplacian behaves as

$$L_{\text{int}}(G - e_k) = L_{\text{int}}(G)|_{a_k=0}, \tag{3.50}$$

i.e. the Laplacian of the graph $G - e_k$ is obtained from the Laplacian of the graph G by setting the variable a_k to zero. The behaviour of the Laplacian under contraction is slightly more complicated: As before we consider a graph with r_{int} internal vertices. Assume that edge e_k connects the vertices v_a and $v_{r_{\text{int}}}$. The Laplacian $L_{\text{int}}(G/e_k)$ is then a $(r_{\text{int}} - 1) \times (r_{\text{int}} - 1)$-matrix with entries

$$L_{\text{int}}(G/e_k)_{ij} = \begin{cases} L_{\text{int}}(G)_{aa} + L_{\text{int}}(G)_{r_{\text{int}} r_{\text{int}}} + L_{\text{int}}(G)_{a r_{\text{int}}} + L_{\text{int}}(G)_{r_{\text{int}} a}, & \text{if } i = j = a, \\ L_{\text{int}}(G)_{aj} + L_{\text{int}}(G)_{r_{\text{int}} j}, & \text{if } i = a,\ j \neq a, \\ L_{\text{int}}(G)_{ia} + L_{\text{int}}(G)_{i r_{\text{int}}}, & \text{if } j = a,\ i \neq a, \\ L_{\text{int}}(G)_{ij}, & \text{otherwise.} \end{cases} \tag{3.51}$$

Therefore the Laplacian of $L_{\text{int}}(G/e_k)$ is identical to the minor $L_{\text{int}}(G)[r_{\text{int}}]$ except for the row and column a. The choice that the edge e_k is connected to the last internal vertex $v_{r_{\text{int}}}$ was merely made to keep the notation simple. If the edge connects the vertices v_a and v_b with $a < b$ one deletes from $L_{\text{int}}(G)$ row and column b and modifies row and column a analogously to the formula above with b substituted for r_{int}. In particular we have [39]

$$L_{\text{int}}\left(G/e_k\right)[a] = L_{\text{int}}\left(G\right)[a, b]. \tag{3.52}$$

The deletion/contraction operations can be used for a recursive definition of the graph polynomials. For any regular edge e_k we have

$$\begin{aligned} \mathcal{U}(G) &= \mathcal{U}(G/e_k) + a_k \mathcal{U}(G - e_k), \\ \mathcal{F}_0(G) &= \mathcal{F}_0(G/e_k) + a_k \mathcal{F}_0(G - e_k). \end{aligned} \tag{3.53}$$

The recursion terminates when all edges are either bridges or self-loops. This is then a graph, which can be obtained from a tree graph by attaching self-loops to some vertices. These graphs are called **terminal forms**. If a terminal form has r_{int} internal vertices and l (self-) loops, then there are $(r_{\text{int}} - 1)$ "tree-like" propagators, where the momenta flowing through these propagators are linear combinations of the external momenta p_i alone and independent of the independent loop momenta k_j. The momenta of the remaining l propagators are on the other hand independent of the external momenta and can be taken as the independent loop momenta k_j, $j = 1, \ldots, l$. Let us agree that we label the $(r_{\text{int}} - 1)$ "tree-like" internal edges from 1 to $r_{\text{int}} - 1$, and the remaining l internal edges from r_{int} to n_{int} (with $n_{\text{int}} = r_{\text{int}} + l - 1$). We further denote the momentum squared flowing through edge j by q_j^2. For a terminal form we have

$$\mathcal{U} = a_{r_{\text{int}}} \ldots a_{n_{\text{int}}}, \quad \mathcal{F}_0 = a_{r_{\text{int}}} \ldots a_{n_{\text{int}}} \sum_{j=1}^{r_{\text{int}}-1} a_j \left(\frac{-q_j^2}{\mu^2} \right). \tag{3.54}$$

In the special case that the terminal form is a tree graph, this reduces to

$$\mathcal{U} = 1, \quad \mathcal{F}_0 = \sum_{j=1}^{r_{\text{int}}-1} a_j \left(\frac{-q_j^2}{\mu^2} \right). \tag{3.55}$$

The Kirchhoff polynomial has for any regular edge the recursion relation

$$\begin{aligned} \mathcal{K}(G) &= a_k \mathcal{K}(G/e_k) + \mathcal{K}(G - e_k), \\ \mathcal{K}_{\text{int}}(G) &= a_k \mathcal{K}_{\text{int}}(G/e_k) + \mathcal{K}_{\text{int}}(G - e_k), \end{aligned} \tag{3.56}$$

Note that the factor a_k appears here in combination with the contracted graph G/e_k. The recursion ends again on terminal forms. For these graphs we have with the conventions as above

$$\mathcal{K}_{\text{int}} = a_1 \ldots a_{(r_{\text{int}}-1)}, \tag{3.57}$$

and a similar formula holds for the terminal forms of $\mathcal{K}$. The recursion relations Eqs. (3.53) and (3.56) are proven with the help of the formulae, which express the polynomials $\mathcal{U}$, $\mathcal{K}$ and $\mathcal{K}_{\text{int}}$ in terms of spanning trees. For $\mathcal{F}_0$ one uses the corresponding formula, which expresses this polynomial in terms of spanning 2-forests. As an example consider the polynomial $\mathcal{U}$ and the set of all spanning trees. This set can be divided into two sub-sets: the first sub-set is given by the spanning trees, which contain the edge e_k, while the second subset is given by those which do not. The spanning trees in the first sub-set are in one-to-one correspondence with the spanning trees of G/e_k, the relation is given by contracting and decontracting the edge e_k. The second subset is identical to the set of spanning trees of $G - e_k$. The graphs G and $G - e_k$ differ by the edge e_k which has been deleted, therefore the explicit factor a_k in front of $\mathcal{U}(G - e_k)$.

We summarise the results on the deletion and contraction properties:

Graph polynomials from recursion: For any regular edge e_k we have

$$\begin{aligned} \mathcal{U}(G) &= \mathcal{U}(G/e_k) + a_k \mathcal{U}(G - e_k), \\ \mathcal{F}_0(G) &= \mathcal{F}_0(G/e_k) + a_k \mathcal{F}_0(G - e_k). \end{aligned} \tag{3.58}$$

The recursion terminates when all edges are either bridges or self-loops, in which case the graph polynomials are given by

$$\mathcal{U} = a_{r_{\text{int}}} \dots a_{n_{\text{int}}}, \quad \mathcal{F}_0 = a_{r_{\text{int}}} \dots a_{n_{\text{int}}} \sum_{j=1}^{r_{\text{int}}-1} a_j \left(\frac{-q_j^2}{\mu^2} \right), \tag{3.59}$$

where we associate the parameters $a_1, \dots, a_{r_{\text{int}}-1}$ to the internal bridges and the parameters $a_{r_{\text{int}}}, \dots, a_{n_{\text{int}}}$ to the self-loops. The graph polynomial $\mathcal{F}$ is obtained from $\mathcal{F}_0$ and $\mathcal{U}$ by Eq. (3.9).

Equations (3.58) and (3.59) together with Eq. (3.9) provide a fourth method for the computation of the graph polynomials $\mathcal{U}$ and $\mathcal{F}$.

Exercise 22 *Consider a massless theory. Show that in this case the Lee-Pomeransky polynomial $\mathcal{G}$ satisfies for any regular edge e_k the recursion*

$$\mathcal{G}(G) = \mathcal{G}(G/e_k) + a_k \mathcal{G}(G - e_k). \tag{3.60}$$

We now look at a generalisation of the Kirchhoff polynomial satisfying a recursion relation similar to Eq. (3.56). For a graph G – not necessarily connected – we denote by $\mathcal{S}$ the set of all **spanning sub-graphs** of G, i.e. sub-graphs H of G, which contain all vertices of G. It is not required that a spanning sub-graph is a forest or a tree. We denote by $k(H)$ the number of connected components of H. As before we associate to each edge e_i a variable a_i. We will need one further formal variable q. We recall that the loop number of a graph G with n internal edges and r vertices is given by

$$l = n - r + k, \tag{3.61}$$

where k is the number of connected components of G. We can extend the definition of the deletion and contraction properties to edges which are not regular. It is straightforward to define the operation of deleting a bridge or a self-loop (just delete the edge). It is also straightforward to define the operation of contracting a bridge (just contract the bridge). Only the operation of contracting a self-loop needs a dedicated definition: If the edge e is a self loop, we define the contracted graph G/e to be identical to $G - e$. The multivariate Tutte polynomial is defined by [40]

$$\mathcal{Z}(q, a_1, \dots, a_n) = \sum_{H \in \mathcal{S}} q^{k(H)} \prod_{e_i \in H} a_i. \tag{3.62}$$

It is a polynomial in q and a_1, ..., a_n. The multivariate Tutte polynomial generalises the standard Tutte polynomial [41–46], which is a polynomial in two variables. For the multivariate Tutte polynomial we have the recursion relation

$$\mathcal{Z}(G) = a_k \mathcal{Z}(G/e_k) + \mathcal{Z}(G - e_k), \tag{3.63}$$

where e_k is any edge, not necessarily regular. The terminal forms are graphs which consists solely of vertices without any edges. For a graph with r vertices and no edges one has

$$\mathcal{Z} = q^r. \tag{3.64}$$

The multivariate Tutte polynomial starts as a polynomial in q with q^k if G is a graph with k connected components. If the graph G is not connected we write $G = (G_1, \dots, G_k)$, where G_1 to G_k are the connected components. For a disconnected graph the multivariate Tutte polynomial factorises:

$$\mathcal{Z}(G) = \mathcal{Z}(G_1) \dots \mathcal{Z}(G_k). \tag{3.65}$$

Some examples for the multivariate Tutte polynomial are

$$\begin{aligned}
\mathcal{Z}\left(\bullet\!\!-\!\!\bullet\right) &= qa + q^2, \\
\mathcal{Z}\left(\circlearrowleft\right) &= q\,(a+1), \\
\mathcal{Z}\left(\bullet\!\!\bigcirc\!\!\bullet\right) &= q\,(a_1 a_2 + a_1 + a_2) + q^2.
\end{aligned} \tag{3.66}$$

If G is a connected graph we recover the Kirchhoff polynomial $\mathcal{K}(G)$ from the Tutte polynomial $\mathcal{Z}(G)$ by first taking the coefficient of the linear term in q and then retaining only those terms with the lowest degree of homogeneity in the variables a_i. Expressed in a formula we have

$$\mathcal{K}(a_1,\ldots,a_n) = \lim_{\lambda\to 0}\lim_{q\to 0} \lambda^{1-r} q^{-1} \mathcal{Z}(q,\lambda a_1,\ldots,\lambda a_n). \tag{3.67}$$

To prove this formula one first notices that the definition in Eq. (3.62) of the multivariate Tutte polynomial is equivalent to

$$\mathcal{Z}(q,a_1,\ldots,a_n) = q^r \sum_{H\in\mathcal{S}} q^{l(H)} \prod_{e_i\in H} \frac{a_i}{q}. \tag{3.68}$$

One then obtains

$$\lambda^{1-r} q^{-1} \mathcal{Z}(q,\lambda a_1,\ldots,\lambda a_n) = \sum_{H\in\mathcal{S}} q^{k(H)-1} \lambda^{l(H)-k(H)+1} \prod_{e_i\in H} a_i. \tag{3.69}$$

The limits $q \to 0$ and $\lambda \to 0$ select $k(H) = 1$ and $l(H) = 0$, hence the sum over the spanning sub-graphs reduces to a sum over spanning trees and one recovers the Kirchhoff polynomial.

At the end of this section we want to discuss Dodgson's identity[1]. Dodgson's identity states that for any $n \times n$ matrix A and integers i, j with $1 \le i, j \le n$ and $i \neq j$ one has [47, 48]

$$\det(A)\det(A[i,j]) = \det(A[i])\det(A[j]) - \det(A[i;j])\det(A[j;i]). \tag{3.70}$$

We remind the reader that $A[i, j]$ denotes a $(n-2)\times(n-2)$ matrix obtained from A by deleting the rows and columns i and j. On the other hand $A[i; j]$ denotes a $(n-1)\times(n-1)$ matrix which is obtained from A by deleting the i-th row and the j-th column. The identity in Eq. (3.70) has an interesting application towards graph polynomials: Let e_a and e_b be two regular edges of a graph G, which share one common vertex. Assume that the edge e_a connects the vertices v_i and v_k, while the edge e_b connects the vertices v_j and v_k. The condition $i \neq j$ ensures that after contraction of one edge the other edge is still regular. (If we would allow $i = j$ we have a multiple edge and the contraction of one edge leads to a self-loop for the other edge.) For the Kirchhoff polynomial of the graph $G - e_a - e_b$ we have

$$\mathcal{K}(G - e_a - e_b) = \det L(G - e_a - e_b)[k]. \tag{3.71}$$

Let us now consider the Kirchhoff polynomials of the graphs $G/e_a - e_b$ and $G/e_b - e_a$. One finds

$$\begin{aligned} \mathcal{K}(G/e_a - e_b) &= \det L(G - e_a - e_b)[i,k], \\ \mathcal{K}(G/e_b - e_a) &= \det L(G - e_a - e_b)[j,k]. \end{aligned} \tag{3.72}$$

[1] Dodgson's even more famous literary work contains the novel "Alice in wonderland" which he wrote using the pseudonym Lewis Carroll.

Here we made use of the fact that the operations of contraction and deletion commute (i.e. $G/e_a - e_b = (G - e_b)/e_a$) as well as of the fact that the variable a_a occurs in the Laplacian of G only in rows and columns i and k, therefore $L(G - e_b)[i, k] = L(G - e_a - e_b)[i, k]$. Finally we consider the Kirchhoff polynomial of the graph $G/e_a/e_b$, for which one finds

$$\mathcal{K}(G/e_a/e_b) = \det L(G - e_a - e_b)[i, j, k]. \tag{3.73}$$

The Laplacian of any graph is a symmetric matrix, therefore

$$\det L(G - e_a - e_b)[i, k; j, k] = \det L(G - e_a - e_b)[j, k; i, k]. \tag{3.74}$$

We can now apply Dodgson's identity to the matrix $L(G - e_a - e_b)[k]$. Using the fact that $L(G - e_a - e_b)[i, k; j, k] = L(G)[i, k; j, k]$ one finds [49]

$$\mathcal{K}(G/e_a - e_b)\,\mathcal{K}(G/e_b - e_a) - \mathcal{K}(G - e_a - e_b)\,\mathcal{K}(G/e_a/e_b) = (\det L(G)[i, k; j, k])^2. \tag{3.75}$$

The version for the internal graph reads

$$\mathcal{K}_{\text{int}}(G/e_a - e_b)\,\mathcal{K}_{\text{int}}(G/e_b - e_a) - \mathcal{K}_{\text{int}}(G - e_a - e_b)\,\mathcal{K}_{\text{int}}(G/e_a/e_b) = (\det L_{\text{int}}(G)[i, k; j, k])^2. \tag{3.76}$$

This equation shows that the expression on the left-hand side factorises into a square. The expression on the right-hand side can be re-expressed using the all-minors matrix-tree theorem as a sum over 2-forests, such that the vertex v_k is contained in one tree of the 2-forest, while the vertices v_i and v_j are both contained in the other tree.

Expressed in terms of the first Symanzik polynomial we have

$$\mathcal{U}(G/e_a - e_b)\,\mathcal{U}(G/e_b - e_a) - \mathcal{U}(G - e_a - e_b)\,\mathcal{U}(G/e_a/e_b) = \left(\frac{\Delta_1}{a_a a_b}\right)^2. \tag{3.77}$$

The expression Δ_1 is given by

$$\Delta_1 = \sum_{F \in \mathcal{T}_2^{(i,k),(j,k)}} \prod_{e_t \notin F} a_t. \tag{3.78}$$

The sum is over all 2-forests $F = (T_1, T_2)$ of G such that $v_i, v_j \in T_1$ and $v_k \in T_2$. Note that each term of Δ_1 contains a_a and a_b. The factorisation property of Eq. (3.77) plays a crucial role in the algorithms of [50–52].

A factorisation formula similar to Eq. (3.77) can be derived for an expression containing both the first Symanzik polynomial $\mathcal{U}$ and the polynomial $\mathcal{F}_0$. As before we assume that e_a and e_b are two regular edges of a graph G, which share one common vertex. The derivation uses the results of Sect. 3.2 and starts from Eq. (3.75) for the graph $\hat{G}$ associated to G. Equation (3.46) relates then the Kirchhoff polynomial of $\hat{G}$ to the $\mathcal{W}$-polynomial of G. The $\mathcal{W}$-polynomial is then expanded in powers of b. The lowest order terms reproduce Eq. (3.77). The next order yields

$$\begin{aligned} &\mathcal{U}(G/e_a - e_b)\,\mathcal{F}_0(G/e_b - e_a) - \mathcal{U}(G - e_a - e_b)\,\mathcal{F}_0(G/e_a/e_b) \\ &+\mathcal{F}_0(G/e_a - e_b)\,\mathcal{U}(G/e_b - e_a) - \mathcal{F}_0(G - e_a - e_b)\,\mathcal{U}(G/e_a/e_b) = 2\left(\frac{\Delta_1}{a_a a_b}\right)\left(\frac{\Delta_2}{a_a a_b}\right). \end{aligned} \tag{3.79}$$

The quantity Δ_2 appearing on the right-hand side is obtained from the all-minors matrix-tree theorem. We can express this quantity in terms of spanning three-forests of G as follows: Let us denote by $\mathcal{T}_3^{((i,j),\cdot,k)}$ the set of spanning three-forests (T_1, T_2, T_3) of G such that $v_i, v_j \in T_1$ and $v_k \in T_3$. Similar we denote by $\mathcal{T}_3^{(i,j,k)}$ the set of spanning three-forests (T_1, T_2, T_3) of G such that $v_i \in T_1$, $v_j \in T_2$ and $v_k \in T_3$. Then

$$\begin{aligned} \Delta_2 = &\sum_{(T_1,T_2,T_3)\in\mathcal{T}_3^{(i,j,k)}} \;\sum_{v_c\in T_1, v_d\in T_2} \left(\frac{p_c \cdot p_d}{\mu^2}\right) \prod_{e_t\notin(T_1,T_2,T_3)} a_t \\ &- \sum_{(T_1,T_2,T_3)\in\mathcal{T}_3^{((i,j),\cdot,k)}} \;\sum_{v_c, v_d\in T_2} \left(\frac{p_c \cdot p_d}{\mu^2}\right) \prod_{e_t\notin(T_1,T_2,T_3)} a_t. \end{aligned} \tag{3.80}$$

In this formula we used the convention that the momentum p_j equals zero if no external leg is attached to vertex v_j. Expanding the $\mathcal{W}$-polynomial to order b^4 we have terms of order b^2 squared as well as terms which are products of order b with order b^3. We are interested in an expression which arises from terms of order b^2 squared alone. In this case we obtain a factorisation formula only for special kinematic configurations. If for all external momenta one has

$$\left(p_{i_1} \cdot p_{i_2}\right) \cdot \left(p_{i_3} \cdot p_{i_4}\right) = \left(p_{i_1} \cdot p_{i_3}\right) \cdot \left(p_{i_2} \cdot p_{i_4}\right), \qquad i_1, i_2, i_3, i_4 \in \{1, \dots, m\} \tag{3.81}$$

then

$$\mathcal{F}_0(G/e_a - e_b)\,\mathcal{F}_0(G/e_b - e_a) - \mathcal{F}_0(G - e_a - e_b)\,\mathcal{F}_0(G/e_a/e_b) = \left(\frac{\Delta_2}{a_a a_b}\right)^2. \tag{3.82}$$

Equation (3.81) is satisfied for example if all external momenta are collinear. A second example is given by a three-point function. In the kinematic configuration where

$$\left(p_1^2\right)^2 + \left(p_2^2\right)^2 + \left(p_3^2\right)^2 - 2p_1^2p_2^2 - 2p_1^2p_3^2 - 2p_2^2p_3^2 = 0, \tag{3.83}$$

Equation (3.81) is satisfied.

3.4 Duality

We have seen that the Kirchhoff polynomial $\mathcal{K}_{\text{int}}$ and the first Symanzik polynomial $\mathcal{U}$ of a graph G with n_{int} internal edges are related by the Eq. (3.16):

$$\mathcal{U}\left(a_1, \ldots, a_{n_{\text{int}}}\right) = a_1 \ldots a_{n_{\text{int}}} \ \mathcal{K}_{\text{int}}\left(\frac{1}{a_1}, \ldots, \frac{1}{a_{n_{\text{int}}}}\right),$$

$$\mathcal{K}_{\text{int}}\left(a_1, \ldots, a_{n_{\text{int}}}\right) = a_1 \ldots a_{n_{\text{int}}} \ \mathcal{U}\left(\frac{1}{a_1}, \ldots, \frac{1}{a_{n_{\text{int}}}}\right).$$

Let G be a graph with n edges (internal and external). In this section we will ask if one can find a graph G^* with n edges such that $\mathcal{K}(G^*) = \mathcal{U}(G)$ and $\mathcal{K}(G) = \mathcal{U}(G^*)$. Such a graph G^* will be called a dual graph of G. In this section we will show that for a planar graph one can always construct a dual graph. The dual graph of G need not be unique, there might be several topologically distinct graphs G^* fulfilling the above mentioned relation. In other words two topologically distinct graphs G_1 and G_2 both of them with n edges can have the same Kirchhoff polynomial.

In this section we associate parameters a_j to all edges (internal and external). This is no restriction, as the following exercise shows:

Exercise 23 *Let G be a graph with n_{int} edges and n_{ext} edges and set $n = n_{\text{int}} + n_{\text{ext}}$. Label the edges as*

$$\begin{aligned} &\textit{internal edges}: \{e_1, e_2, \ldots, e_{n_{\text{int}}}\}, \\ &\textit{external edges}: \{e_{n_{\text{int}}+1}, e_{n_{\text{int}}+2}, \ldots, e_{n_{\text{int}}+n_{\text{ext}}}\}. \end{aligned} \tag{3.84}$$

Let G_{int} be the internal graph of G. Define $\mathcal{U}$, $\mathcal{K}$ and $\mathcal{K}_{\text{int}}$ as before. Define $\tilde{\mathcal{U}}$ by

$$\tilde{\mathcal{U}}\left(a_1, \ldots, a_n\right) = a_1 \ldots a_n \ \mathcal{K}\left(\frac{1}{a_1}, \ldots, \frac{1}{a_n}\right). \tag{3.85}$$

Show

$$\begin{aligned} \tilde{\mathcal{U}}\left(a_1, \ldots, a_n\right) &= \mathcal{U}\left(a_1, \ldots, a_{n_{\text{int}}}\right), \\ \mathcal{K}\left(a_1, \ldots, a_n\right) &= a_{n_{\text{int}}+1} \ldots a_n \mathcal{K}\left(a_1, \ldots, a_{n_{\text{int}}}\right). \end{aligned} \tag{3.86}$$

The exercise also shows that $\tilde{\mathcal{U}} = \mathcal{U}$ and in particular $\tilde{\mathcal{U}}$ is independent of $a_{n_{\text{int}}+1}$, $\ldots$, a_n, so we can simply use $\mathcal{U}$ to denote both polynomials.

A graph is called planar if it can be embedded in a plane without crossings of edges. We would like to note that the "crossed double-box"-graph shown in Fig. 3.12 is a planar graph. The right picture of Fig. 3.12 shows how this graph can be drawn in the plane without any crossing of edges.

Figure 3.13 shows two examples of non-planar graphs. The first graph is the complete graph with five vertices K_5. The second example is denoted $K_{3,3}$. A theorem

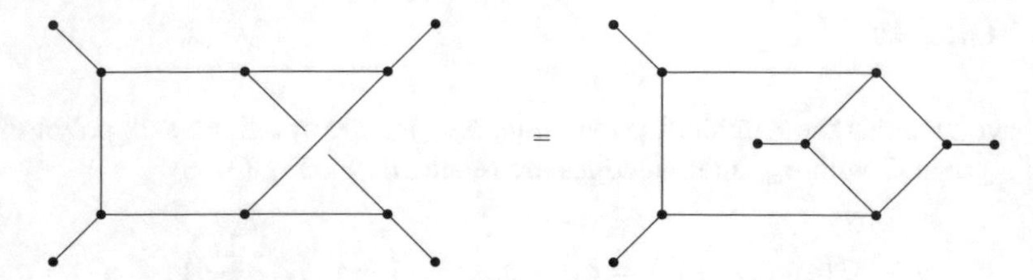

Fig. 3.12 The "crossed double-box"-graph can be drawn as a planar graph

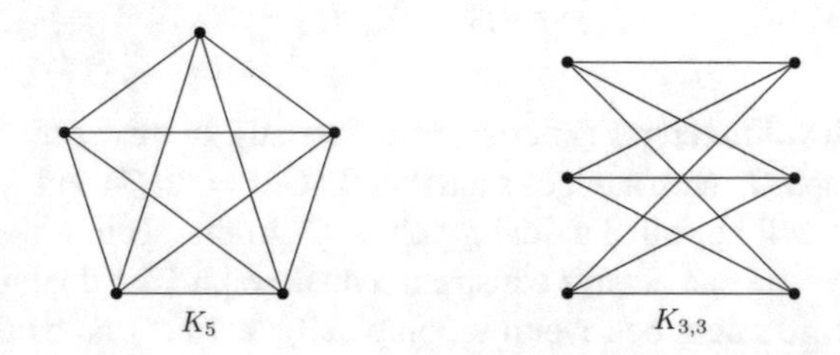

Fig. 3.13 The 'smallest' non-planar graphs

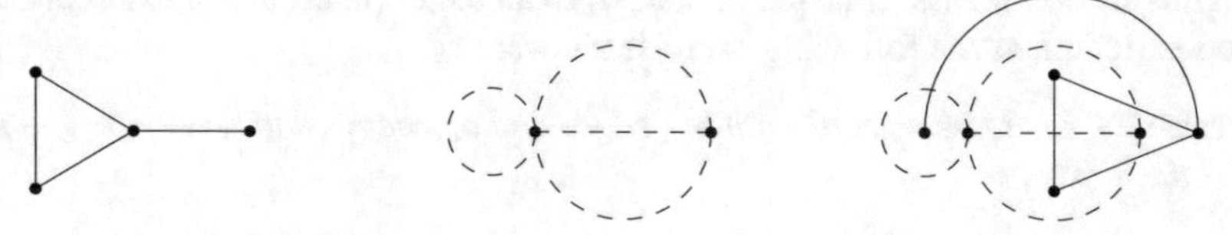

Fig. 3.14 The first two pictures show a graph G and its dual graph G^*. The right picture shows the construction of G^* from G (or vice versa)

states that a graph G is planar if and only if none of the graphs obtained from G by a (possibly empty) sequence of contractions of edges contains K_5 or $K_{3,3}$ as a sub-graph [53–55].

Exercise 24 *Determine the number of loops for K_5 and $K_{3,3}$.*

Each planar graph G has a dual graph $G^\star$ which can be obtained as follows:

- Draw the graph G in a plane, such that no edges intersect. In this way, the graph divides the plane into open subsets, called faces.
- Draw a vertex inside each face. These are the vertices of $G^\star$.
- For each edge e_i of G draw a new edge e_i^* connecting the two vertices of the faces, which are separated by e_i. The new edges e_i^* are the edges of $G^\star$.

An example for this construction is shown in Fig. 3.14. We note from the construction of the dual graph, that for each external edge in G there is a self-loop in $G^\star$ and that for each self-loop in G there is an external edge in $G^\star$.

If we now associate the variable a_i to the edge e_i of G as well as to the edge e_i^* of G^* we have

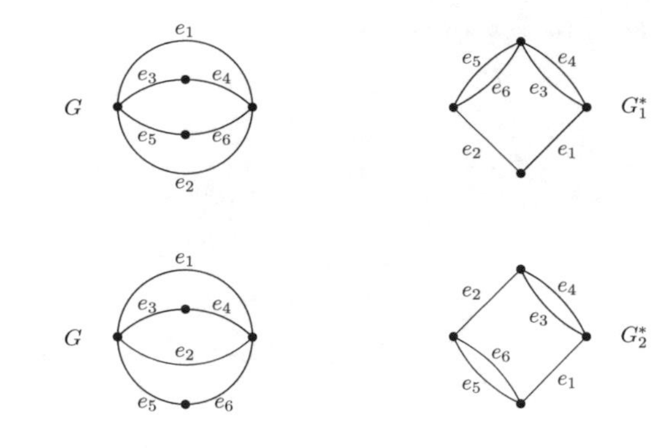

Fig. 3.15 An example showing that different embeddings of a planar graph G into the plane yield different dual graphs G_1^* and G_2^*

$$\mathcal{K}(G^*) = \mathcal{U}(G), \quad \mathcal{K}(G) = \mathcal{U}(G^*). \tag{3.87}$$

It is important to note that the above construction of the dual graph $G^\star$ depends on the way, how G is drawn in the plane. A given graph G can have several topologically distinct dual graphs. These dual graphs have the same Kirchhoff polynomial. An example is shown in Fig. 3.15. For this example one finds

$$\begin{aligned}
\mathcal{K}(G) &= \mathcal{U}(G_1^*) = \mathcal{U}(G_2^*) = (a_1 + a_2)(a_3 + a_4)(a_5 + a_6) + a_3 a_4 (a_5 + a_6) + (a_3 + a_4) a_5 a_6, \\
\mathcal{U}(G) &= \mathcal{K}(G_1^*) = \mathcal{K}(G_2^*) = a_1 a_2 (a_3 + a_4 + a_5 + a_6) + (a_1 + a_2)(a_3 + a_4)(a_5 + a_6).
\end{aligned} \tag{3.88}$$

3.5 Matroids

In this section we introduce the basic terminology of matroid theory. We are in particular interested in cycle matroids. A cycle matroid can be constructed from a graph and it contains the information which is needed for the construction of the Kirchhoff polynomial and therefore as well for the construction of the first Symanzik polynomial $\mathcal{U}$. In terms of matroids we want to discuss the fact, that two different graphs can have the same Kirchhoff polynomial. We have already encountered an example in Fig. 3.15. We review a theorem on matroids which determines the classes of graphs whose Kirchhoff polynomials are the same. For a detailed introduction to matroid theory we refer to [56, 57].

We introduce cycle matroids by an example and consider the graph G of Fig. 3.16. The graph G has three vertices $V = \{v_1, v_2, v_3\}$ and four edges $E = \{e_1, e_2, e_3, e_4\}$. The graph has five spanning trees given by the sets of edges $\{e_1, e_3\}, \{e_1, e_4\}, \{e_2, e_3\}$, $\{e_2, e_4\}$, $\{e_3, e_4\}$, respectively. We obtain the Kirchhoff polynomial $\mathcal{K} = a_1 a_3 + a_1 a_4 + a_2 a_3 + a_2 a_4 + a_3 a_4$. The **unoriented incidence matrix** of a graph G with r vertices and n edges is a $r \times n$-matrix $B_{\text{incidence}} = (b_{ij})$, defined by

Fig. 3.16 A graph G

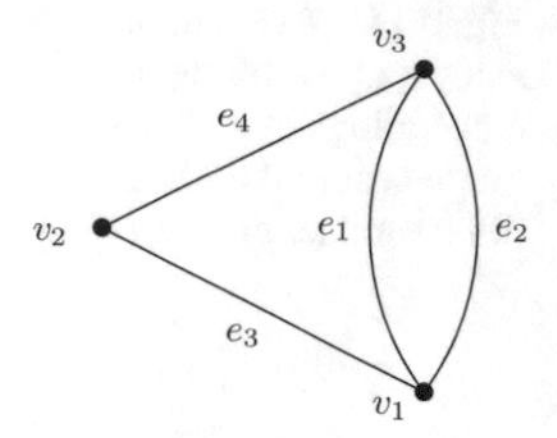

$$b_{ij} = \begin{cases} 1, & \text{if } e_j \text{ is incident to } v_i \text{ and } e_j \text{ is not a self-loop,} \\ 0, & \text{else.} \end{cases} \tag{3.89}$$

There is also the definition of the **oriented incidence matrix** of an oriented graph G with r vertices and n edges. This is again $r \times n$-matrix $B_{\text{oriented incidence}} = (b_{ij})$, whose entries are

$$b_{ij} = \begin{cases} +1, & \text{if } v_i \text{ is the source of } e_j \text{ and } e_j \text{ is not a self-loop,} \\ -1, & \text{if } v_i \text{ is the sink of } e_j \text{ and } e_j \text{ is not a self-loop,} \\ 0, & \text{else.} \end{cases} \tag{3.90}$$

The entries in each column of $B_{\text{oriented incidence}}$ sum up to zero, as every edge has exactly one source and one sink. A self-loop corresponds to a zero column in $B_{\text{oriented incidence}}$ and $B_{\text{incidence}}$. Given $B_{\text{oriented incidence}}$ for an oriented graph G, we obtain the unoriented incidence matrix $B_{\text{incidence}}$ for G as

$$B_{\text{incidence}} = B_{\text{oriented incidence}} \mod 2. \tag{3.91}$$

For the graph G of Fig. 3.16 the unoriented incidence matrix reads

$$\begin{matrix} e_1 & e_2 & e_3 & e_4 \end{matrix}$$

$$\begin{pmatrix} 1 & 1 & 1 & 0 \\ 0 & 0 & 1 & 1 \\ 1 & 1 & 0 & 1 \end{pmatrix},$$

where we indicated that each column vector corresponds to one edge of the graph. Let us focus on the set of these four column vectors. We want to consider all subsets of these vectors which are linearly independent over $\mathbb{Z}_2$. Obviously the set of all four vectors and the set of any three of the given vectors are linearly dependent over $\mathbb{Z}_2$. Furthermore the first two columns corresponding to $\{e_1, e_2\}$ are equal and therefore linearly dependent. Hence the linearly independent subsets are all sets with only one vector and all sets with two vectors, except for the just mentioned one consisting of the first and second column. For each set of linearly independent vectors let us now write the set of the corresponding edges. The set of all these sets shall be denoted $\mathcal{I}$. We obtain

$$\mathcal{I} = \{\emptyset, \{e_1\}, \{e_2\}, \{e_3\}, \{e_4\},\\ \{e_1, e_3\}, \{e_1, e_4\}, \{e_2, e_3\}, \{e_2, e_4\}, \{e_3, e_4\}\}. \quad (3.92)$$

The empty set is said to be independent and is included here by definition. Let us make the important observation, that the sets in $\mathcal{I}$ which have two elements, i.e. the maximal number of elements, are exactly the sets of edges of the spanning trees of the graph given above.

The pair $(E, \mathcal{I})$ consisting of the set of edges E and the set of linearly independent sets $\mathcal{I}$ is an example of a matroid. **Matroids** are defined as ordered pairs $(E, \mathcal{I})$ where E is a finite set, the ground set, and where $\mathcal{I}$ is a collection of subsets of E, called the independent sets, fulfilling the following conditions:

1. $\emptyset \in \mathcal{I}$.
2. If $I \in I$ and $I' \subseteq I$, then $I' \in \mathcal{I}$.
3. If I_1 and I_2 are in $\mathcal{I}$ and $|I_1| < |I_2|$, then there is an element e of $I_2 - I_1$ such that $I_1 \cup e \in \mathcal{I}$.

All subsets of E which do not belong to $\mathcal{I}$ are called dependent. The definition goes back to Whitney who wanted to describe the properties of linearly independent sets in an abstract way. In a similar way as a topology on a space is given by the distinction between open and closed sets, a matroid is given by deciding, which of the subsets of a ground set E shall be called independent and which dependent. A matroid can be defined on any kind of ground set, but if we choose E to be a set of vectors, we can see that the conditions for the independent sets match with the well-known linear independence of vectors.

Let us go through the three conditions. The first condition simply says that the empty set shall be called independent. The second condition states that a subset of an independent set is again an independent set. This is fulfilled for sets of linearly independent vectors as we already have seen in the above example. The third condition is called the independence augmentation axiom and may be clarified by an example. Consider the sets

$$I_1 = \left\{ \begin{pmatrix} 1 \\ 0 \\ 0 \\ 0 \end{pmatrix}, \begin{pmatrix} 0 \\ 1 \\ 0 \\ 0 \end{pmatrix} \right\}, \quad I_2 = \left\{ \begin{pmatrix} 1 \\ 0 \\ 0 \\ 0 \end{pmatrix}, \begin{pmatrix} 0 \\ 1 \\ 1 \\ 1 \end{pmatrix}, \begin{pmatrix} 1 \\ 0 \\ 0 \\ 1 \end{pmatrix} \right\}.$$

Both sets are sets of linearly independent vectors. The set I_2 has one element more than I_1. $I_2 - I_1$ is the set of vectors in I_2 which do not belong to I_1. The set $I_2 - I_1$ contains for example $e = (1, 0, 0, 1)^T$ and if we include this vector in I_1 then we obtain again a linearly independent set. The third condition states that such an e can be found for any two independent sets with different numbers of elements.

The most important origins of examples of matroids are linear algebra and graph theory. The **cycle matroid** (or **polygon matroid**) of a graph G is the matroid whose ground set E is given by the edges of G and whose independent sets $\mathcal{I}$ are given by the linearly independent subsets over $\mathbb{Z}_2$ of column vectors in the incidence matrix of G. We can convince ourselves, that $\mathcal{I}$ fulfils the conditions laid out above.

Let us consider the **bases** or **maximal independent sets** of a matroid $(E, \mathcal{I})$. These are the sets in $\mathcal{I}$ which are not proper subsets of any other sets in $\mathcal{I}$. The set of these bases of a matroid shall be denoted $\mathcal{B}$ and it can be defined by the following conditions:

1. $\mathcal{B}$ is non-empty.
2. If B_1 and B_2 are members of $\mathcal{B}$ and $x \in B_1 - B_2$, then there is an element y of $B_2 - B_1$ such that $(B_1 - \{x\}) \cup y \in \mathcal{B}$.

One can show, that all sets in $\mathcal{B}$ have the same number of elements. Furthermore $\mathcal{I}$ is uniquely determined by $\mathcal{B}$: it contains the empty set and all subsets of members of $\mathcal{B}$.

Let $M = (E, \mathcal{I})$ be the cycle matroid of a connected graph G and let $\mathcal{B}(M)$ be the set of bases. Then one can show that $\mathcal{B}(M)$ consists of the sets of edges of the spanning trees in G. In other words, T is a spanning tree of G if and only if its set of edges are a basis in $\mathcal{B}(M)$. We can therefore relate the Kirchhoff polynomial to the bases of the cycle matroid:

$$\mathcal{K} = \sum_{B_j \in \mathcal{B}(M)} \prod_{e_i \in B_j} a_i . \tag{3.93}$$

The Kirchhoff polynomial of G is called a basis generating polynomial of the matroid M associated to G. The Kirchhoff polynomial allows us to read off the set of bases $\mathcal{B}(M)$. Therefore two graphs without any self-loops have the same Kirchhoff polynomial if and only if they have the same cycle matroid associated to them.

Let us cure a certain ambiguity which is still left when we consider cycle matroids and Kirchhoff polynomials and which comes from the freedom in giving names to the edges of the graph. In the graph of Fig. 3.16 we could obviously choose different names for the edges and their edge variables, for example the edge e_2 could instead be named e_3 and vice versa. As a consequence we would obtain a different cycle matroid where compared to the above one, e_3 takes the place of e_2 and vice versa. Similarly, we would obtain a different Kirchhoff polynomial, where the variables a_2 and a_3 are exchanged. Of course we are not interested in any such different cases which simply result from a change of variable names. Therefore it makes sense to consider classes of isomorphic matroids and Kirchhoff polynomials.

Let M_1 and M_2 be two matroids and let $E(M_1)$ and $E(M_2)$ be their ground sets, respectively. The matroids M_1 and M_2 are called isomorphic if there is a bijection ψ from $E(M_1)$ to $E(M_2)$ such that $\psi(I)$ is an independent set in M_2 if and only if I is an independent set in M_1. The mentioned interchange of e_2 and e_3 in the above example would be such a bijection: $\psi(e_2) = e_3$, $\psi(e_3) = e_2$, $\psi(e_i) = e_i$ for $i = 1, 4$. The new matroid obtained this way is isomorphic to the above one and its independent sets are given by interchanging e_2 and e_3 in $\mathcal{I}$ of Eq. (3.92). In the same sense we want to say that two Kirchhoff polynomials are isomorphic if they are equal up to bijections on their sets of variables.

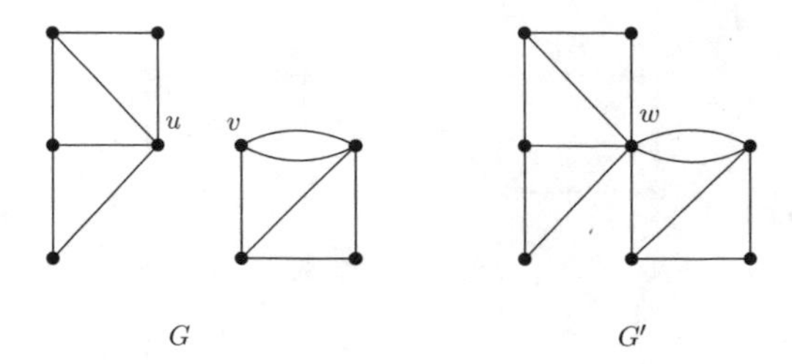

Fig. 3.17 Vertex identification and cleaving

Now let us come to the question when the Kirchhoff polynomials of two different graphs are isomorphic, which means that after an appropriate change of variable names they are equal. From the above discussion and Eq. (3.93) it is now clear, that a sufficient condition is that the cycle matroids of the graphs are isomorphic. The question when two graphs have isomorphic cycle matroids was answered in the following theorem of Whitney [58] (also see [59]) which was one of the foundational results of matroid theory:

Isomorphic cycle matroids:
Let G and H be graphs having no isolated vertices. Then the cycle matroids $M(G)$ and $M(H)$ are isomorphic if and only if G is obtained from H after a sequence of the following three transformations:

1. **Vertex identification**: Let u and v be vertices of distinct components of a graph G. Then a new graph G' is obtained from the identification of u and v as a new vertex w in G' (see the transition from G to G' in Fig. 3.17).
2. **Vertex cleaving**: Vertex cleaving is the reverse operation of vertex identification, such that from cleaving at vertex w in G' we obtain u and v in distinct components of G (see the transition from G' to G in Fig. 3.17).
3. **Twisting**: Let G be a graph which is obtained from two disjoint graphs G_1 and G_2 by identifying the vertices u_1 of G_1 and u_2 of G_2 as a vertex u of G and by identifying the vertices v_1 of G_1 and v_2 of G_2 as a vertex v of G. Then the graph G' is called the twisting of G about $\{u, v\}$ if it is obtained from G_1 and G_2 by identifying instead u_1 with v_2 and v_1 with u_2 (see Fig. 3.18).

Proofs can be obtained from [56, 58–60]. Whitney's theorem does not exclude self-loops. If self-loops are allowed, isomorphic cycle matroids are a sufficient condition for isomorphic Kirchhoff polynomials, but not a necessary condition, as the next exercise shows:

Exercise 25 *Consider the two graphs G_1 and G_2 shown in Fig. 3.19, which differ by a self-loop. For each of the two graphs, give the Kirchhoff polynomial $\mathcal{K}$ and the first graph polynomial $\mathcal{U}$. Show that the cycle matroids are not isomorphic.*

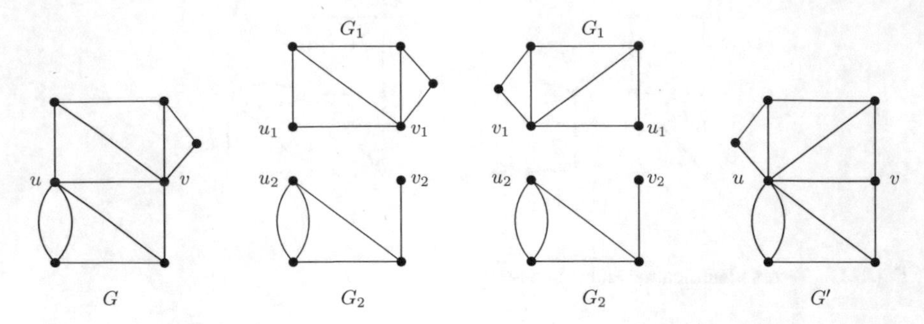

Fig. 3.18 Twisting about u and v

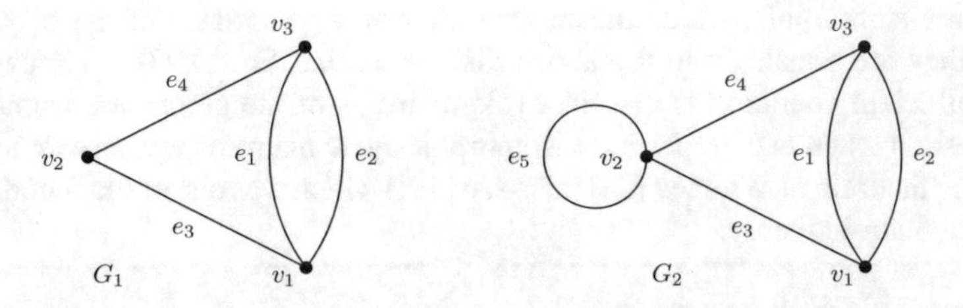

Fig. 3.19 Two graphs G_1 and G_2, which differ by the self-loop formed by e_5

As a consequence of Whitney's theorem, the Kirchhoff polynomials $\mathcal{K}(G)$ and $\mathcal{K}(H)$ of the connected graphs G and H, both without any self-loops, are isomorphic if and only if G is obtained from H by a sequence of the above three transformations. For the transformations of vertex identification and vertex cleaving this is obvious from a well-known observation: If two distinct components G_1 and G_2 are obtained from G after vertex cleaving, then $\mathcal{K}(G)$ is the product of $\mathcal{K}(G_1)$ and $\mathcal{K}(G_2)$. Therefore any other graph G' obtained from G_1 and G_2 after vertex identification has the same Kirchhoff polynomial $\mathcal{K}(G') = \mathcal{K}(G_1) \cdot \mathcal{K}(G_2) = \mathcal{K}(G)$. The non-trivial part of the statement on the Kirchhoff polynomials of G and H is the relevance of the operation of twisting. In the initial example of Fig. 3.15 the two graphs G_1^* and G_2^* can be obtained from each other by twisting.

Let us now discuss the implications for the two graph polynomials $\mathcal{U}$ and $\mathcal{F}_0$. Consider two connected graphs G and H, possibly with external edges and self-loops. Suppose that G can be obtained from H by sequence of vertex identifications, vertex cleavings and twisting. Then the first graph polynomials $\mathcal{U}(G)$ and $\mathcal{U}(H)$ are isomorphic. Denote by $\hat{G}$ the graph obtained from G by merging all external vertices into a single new vertex v_∞, which connects to all external lines, as discussed at the end of Sect. 3.2. Similarly, denote by $\hat{H}$ the graph obtained from H through the same operation. If $\hat{G}$ can be obtained from $\hat{H}$ by sequence of vertex identifications, vertex cleavings and twisting, then the graph polynomials $\mathcal{F}_0(G)$ and $\mathcal{F}_0(H)$ are isomorphic. This follows directly from Eq. (3.46).

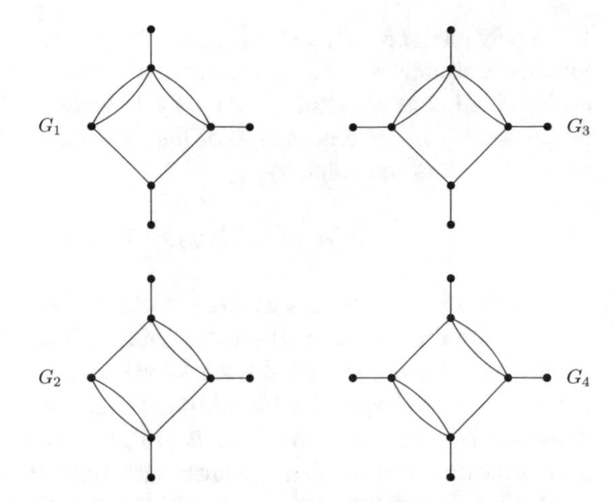

Fig. 3.20 The graphs G_1 and G_2 have three external edges, the graphs G_3 and G_4 have four external edges

Exercise 26 *Consider first the two graphs G_1 and G_2 shown in Fig. 3.20, both with three external legs. Assume that all internal masses vanish. Show that*

$$\mathcal{U}(G_1) = \mathcal{U}(G_2), \quad \mathcal{F}(G_1) = \mathcal{F}(G_2). \tag{3.94}$$

Consider then the graphs G_3 and G_4 with four external legs. Show that

$$\mathcal{U}(G_3) = \mathcal{U}(G_4), \tag{3.95}$$

but

$$\mathcal{F}(G_3) \neq \mathcal{F}(G_4). \tag{3.96}$$

There is an alternative definition of a matroid: Instead of specifying the set $\mathcal{I}$ of independent sets a matroid can be defined by a **rank function**, which associates a non-negative integer to every sub-set of the ground set. The rank function has to satisfy for all $S, S' \subseteq E$ the following three conditions:

1. $\mathrm{rk}(S) \leq |S|$.
2. $S' \subset S$ implies $\mathrm{rk}(S') \leq \mathrm{rk}(S)$.
3. $\mathrm{rk}(S \cup S') + \mathrm{rk}(S \cap S') \leq \mathrm{rk}(S) + \mathrm{rk}(S')$.

The independent sets are exactly those for which $\mathrm{rk}(S) = |S|$ holds. For the cycle matroid of a graph G we can associate to a subset S of E the spanning sub-graph H of G obtained by taking all the vertices of G, but just the edges which are in S. In this case the rank of S equals the number of vertices of H minus the number of connected components of H. The multivariate Tutte polynomial for a matroid is defined by

$$\tilde{Z}(q, a_1, \dots, a_n) = \sum_{S \subseteq E} q^{-\mathrm{rk}(S)} \prod_{e_i \in S} a_i. \tag{3.97}$$

It is a polynomial in $1/q$ and $a_1, \ldots, a_n$. Since a matroid can be defined by giving the rank for each subset S of E, it is clear that the multivariate Tutte polynomial encodes all information of a matroid. For the cycle matroid of a graph G with r vertices the multivariate Tutte polynomial $\tilde{\mathcal{Z}}$ of the matroid is related to the multivariate Tutte polynomial $\mathcal{Z}$ of the graph by

$$\tilde{\mathcal{Z}}(q, a_1, \ldots, a_n) = q^{-r} \mathcal{Z}(q, a_1, \ldots, a_n). \tag{3.98}$$

For a matroid there are as well the notions of duality, deletion and contraction. Let us start with the definition of the dual of a matroid. We consider a matroid M with the ground set E and the set of bases $\mathcal{B}(M) = \{B_1, , B_2, \ldots, B_n\}$. The dual matroid $M^\star$ of M is the matroid with ground set E and whose set of bases is given by $\mathcal{B}(M^\star) = \{E - B_1, E - B_2, \ldots, E - B_n\}$. It should be noted that in contrast to graphs the dual matroid can be constructed for any arbitrary matroid.

Deletion and contraction for matroids are defined as follows: Let us consider a matroid $M = (E, \mathcal{I})$ with ground set E and $\mathcal{I}$ being the set of independent sets. Let us divide the ground set E into two disjoint sets X and Y:

$$E = X \cup Y, \quad X \cap Y = \emptyset. \tag{3.99}$$

We denote by $\mathcal{I}_Y$ the elements of $\mathcal{I}$, which are sub-sets of Y. The matroid $M - X$ is then defined as the matroid with ground set $E - X = Y$ and whose set of independent sets is given by $\mathcal{I}_Y$. We say that the matroid $M - X$ is obtained from the matroid M by deleting X. The contracted matroid M/X is defined as follows:

$$M/X = \left(M^* - X\right)^*, \tag{3.100}$$

i.e. one deletes from the dual M^* the set X and takes then the dual. The contracted matroid M/X has the ground set $E - X$. With these definitions of deletion and contraction we can now state the recursion relation for the multivariate Tutte polynomial of a matroid: We have to distinguish two cases. If the one-element set $\{e\}$ is of rank zero (corresponding to a self-loop in a graph) we have

$$\tilde{\mathcal{Z}}(M) = \tilde{\mathcal{Z}}(M - \{e\}) + a_e \tilde{\mathcal{Z}}(M/\{e\}). \tag{3.101}$$

Otherwise we have

$$\tilde{\mathcal{Z}}(M) = \tilde{\mathcal{Z}}(M - \{e\}) + \frac{a_e}{q} \tilde{\mathcal{Z}}(M/\{e\}). \tag{3.102}$$

The recursion terminates for a matroid with an empty ground set, in this case we have $\tilde{\mathcal{Z}} = 1$. The fact that one has a different recursion relation for the case where the one-element set $\{e\}$ is of rank zero is easily understood from the definition of $\tilde{\mathcal{Z}}$ and the relation to graphs: For a cycle matroid $\tilde{\mathcal{Z}}$ differs from $\mathcal{Z}$ by the extra factor q^{-r}, where r is the number of vertices of the graph. If e is a self-loop of G, the contracted graph G/e equals $G - e$ and in particular it has the same number of vertices as G. In all other cases the contracted graph G/e has one vertex less than G, which is reflected by the factor $1/q$ in Eq. (3.102).

Chapter 4
Quantum Field Theory

We introduced Feynman integrals in Chap. 2, building only on the knowledge of special relativity and graphs. We did not discuss how Feynman integrals arise in perturbative quantum field theory. This is of course the main application for Feynman integrals. In this chapter we fill this gap and give a brief outline in Sect. 4.1, how Feynman integrals arise in the perturbative expansion for scattering amplitudes in quantum field theory. This is also covered in depth in many books on quantum field theory and readers not yet familiar with quantum field theory are invited to consult one of these textbooks [61–64].

We have seen quite early on with the example of Eq. (2.69) that Feynman integrals in four space-time dimensions are often divergent. In order to have a well-defined expression, we introduced dimensional regularisation. This regulates all divergences. The original divergences show up as poles in the dimensional regularisation parameter ε. While this procedure allows us to work with well-defined expressions, it does not tell us anything what we shall do with these poles. The answer comes again from quantum field theory. In the end we would like to have finite results, where we can take the limit $\varepsilon \to 0$. The way divergences cancel is explained in Sect. 4.2.

The definition of a Feynman integral in Eq. (2.56) corresponds to a "scalar" integral. From the Feynman rules for most quantum field theories (like Yang-Mills theory, QED, QCD or more generally any quantum field theory, which is not a scalar theory) we get Feynman integrals, which are "tensor" integrals. In Sect. 4.3 we show that tensor integrals can be expressed in terms of scalar integrals. It is therefore sufficient to focus our attention on scalar integrals.

Quantum field theories with spin $1/2$-fermions involve the Dirac matrices and the weak interactions involve γ_5. As we use dimensional regularisation as our regularisation scheme, the Dirac algebra has to be continued from four space-time dimensions to D space-time dimensions. For the most part of the Dirac algebra this is straightforward, but the treatment of γ_5 is a little bit more subtle. We discuss this in Sect. 4.4.

S. Weinzierl, *Feynman Integrals*, UNITEXT for Physics,
https://doi.org/10.1007/978-3-030-99558-4_4

4.1 Basics of Perturbative Quantum Field Theory

Elementary particle physics is described by quantum field theory. To begin with let us start with a single field $\phi(x)$. Important concepts in quantum field theory are the Lagrangian, the action and the generating functional. If $\phi(x)$ is a scalar field, a typical Lagrangian is

$$\mathcal{L} = \frac{1}{2}\left(\partial_\mu \phi(x)\right)\left(\partial^\mu \phi(x)\right) - \frac{1}{2} m^2 \phi(x)^2 + \frac{1}{4}\lambda \phi(x)^4. \tag{4.1}$$

The quantity m is interpreted as the mass of the particle described by the field $\phi(x)$, the quantity λ describes the strength of the interactions among the particles. Integrating the Lagrangian over Minkowski space yields the action:

$$S[\phi] = \int d^D x \; \mathcal{L}(\phi). \tag{4.2}$$

The action is a functional of the field ϕ. In order to arrive at the generating functional we introduce an auxiliary field $J(x)$, called the source field, and integrate over all field configurations $\phi(x)$:

$$Z[J] = \mathcal{N} \int \mathcal{D}\phi \; e^{i\left(S[\phi] + \int d^D x J(x)\phi(x)\right)}. \tag{4.3}$$

The integral over all field configurations is an infinite-dimensional integral. It is called a path integral. The prefactor $\mathcal{N}$ is chosen such that $Z[0] = 1$. The n_{ext}-point Green function is given by

$$\langle 0 | T(\phi(x_1) ... \phi(x_{n_{\mathrm{ext}}})) | 0 \rangle = \frac{\int \mathcal{D}\phi \; \phi(x_1) ... \phi(x_{n_{\mathrm{ext}}}) e^{i S(\phi)}}{\int \mathcal{D}\phi \; e^{i S(\phi)}}. \tag{4.4}$$

With the help of functional derivatives this can be expressed as

$$\langle 0 | T(\phi(x_1) ... \phi(x_{n_{\mathrm{ext}}})) | 0 \rangle = (-i)^{n_{\mathrm{ext}}} \left. \frac{\delta^{n_{\mathrm{ext}}} Z[J]}{\delta J(x_1) ... \delta J(x_{n_{\mathrm{ext}}})} \right|_{J=0}. \tag{4.5}$$

We are in particular interested in connected Green functions. These are obtained from a functional $W[J]$, which is related to $Z[J]$ by

$$Z[J] = e^{i W[J]}. \tag{4.6}$$

The connected Green functions are then given by

$$G_{n_{\mathrm{ext}}}\left(x_1, ..., x_{n_{\mathrm{ext}}}\right) = (-i)^{n_{\mathrm{ext}}-1} \left. \frac{\delta^{n_{\mathrm{ext}}} W[J]}{\delta J(x_1) ... \delta J(x_{n_{\mathrm{ext}}})} \right|_{J=0}. \tag{4.7}$$

It is convenient to go from position space to momentum space by a Fourier transformation. We define the Green functions in momentum space by

$$G_{n_{\text{ext}}}\left(x_1, ..., x_{n_{\text{ext}}}\right) = \tag{4.8}$$
$$\int \frac{d^D p_1}{(2\pi)^D} ... \frac{d^D p_{n_{\text{ext}}}}{(2\pi)^D} e^{-i\sum p_j x_j} (2\pi)^D \delta^D\left(p_1 + \cdots + p_{n_{\text{ext}}}\right) \tilde{G}_{n_{\text{ext}}}\left(p_1, ..., p_{n_{\text{ext}}}\right).$$

Note that the Fourier transform $\tilde{G}_{n_{\text{ext}}}$ is defined by explicitly factoring out the δ-function $\delta(p_1 + \cdots + p_{n_{\text{ext}}})$ and a factor $(2\pi)^D$. We denote the two-point function in momentum space by $\tilde{G}_2(p)$. In this case we have to specify only one momentum, since the momentum flowing into the Green function on one side has to be equal to the momentum flowing out of the Green function on the other side due to the presence of the δ-function in Eq. (4.8). We now are in a position to define the scattering amplitude: In momentum space the **scattering amplitude** with n_{ext} external particles is given by the connected n_{ext}-point Green function multiplied by the inverse two-point function for each external particle:

$$i\mathcal{A}_{n_{\text{ext}}}\left(p_1, ..., p_{n_{\text{ext}}}\right) = \tilde{G}_2\left(p_1\right)^{-1} ... \tilde{G}_2\left(p_{n_{\text{ext}}}\right)^{-1} \tilde{G}_{n_{\text{ext}}}\left(p_1, ..., p_{n_{\text{ext}}}\right). \tag{4.9}$$

The multiplication with the inverse two-point function for each external particle amputates the external propagators. This is the reason, why we distinguish in a graph external and internal edges.

The scattering amplitude enters directly the calculation of a physical observable. Let us first consider the scattering process of two incoming elementary spinless particles with no further internal degrees of freedom (like colour) and momenta p_a' and p_b' and $(n_{\text{ext}} - 2)$ outgoing particles with momenta p_1 to $p_{n_{\text{ext}}-2}$. Let us further assume that we are interested in an observable $O\left(p_1, ..., p_{n_{\text{ext}}-2}\right)$ which depends on the momenta of the outgoing particles. In general the observable depends on the experimental set-up and can be an arbitrary complicated function of the momenta. In the simplest case this function is just a constant equal to one, corresponding to the situation where we count every event with $(n_{\text{ext}} - 2)$ particles in the final state. In more realistic situations one takes for example into account that it is not possible to detect particles close to the beam pipe. The function O would then be zero if all final state particles are in this region of phase space. Furthermore any experiment has a finite resolution. Therefore it will not be possible to detect particles which are very soft nor will it be possible to distinguish particles which are very close in angle. We will therefore sum over the number of final state particles. In order to obtain finite results within perturbation theory we have to require that in the case where one or more particles become unresolved the value of the observable O has a continuous limit agreeing with the value of the observable for a configuration where the unresolved particles have been merged into "hard" (or resolved) pseudo-particles. Observables having this property are called **infrared-safe observables**. The expectation value for an infrared-safe observable O is given by

$$\langle O \rangle = \frac{1}{2(p'_a + p'_b)^2} \sum_{n_{\text{ext}}} \int d\phi_{n_{\text{ext}}-2} O \left(p_1, ..., p_{n_{\text{ext}}-2}\right) \left|\mathcal{A}_{n_{\text{ext}}}\right|^2, \tag{4.10}$$

where $1/2/(p'_a + p'_b)^2$ is a normalisation factor taking into account the incoming flux. The phase space measure is given by

$$d\phi_n = \frac{1}{n!} \prod_{i=1}^{n} \frac{d^{D-1} p_i}{(2\pi)^3 2E_i} (2\pi)^D \delta^D \left(p'_a + p'_b - \sum_{i=1}^{n} p_i \right). \tag{4.11}$$

The quantity E_i is the energy of particle i, given by

$$E_i = \sqrt{\vec{p}_i^{\,2} + m_i^2}. \tag{4.12}$$

We see that the expectation value of O is given by the phase space integral over the observable, weighted by the norm squared of the scattering amplitude. As the integrand can be a rather complicated function, the phase space integral is usually performed numerically by Monte Carlo integration.

Let us now look towards a more realistic theory relevant to LHC physics. LHC is the abbreviation for the Large Hadron Collider at CERN, Geneva. As an example for a more realistic theory we consider quantum chromodynamics (QCD) consisting of quarks and gluons. Quarks and gluons are collectively called partons. There are a few modifications to Eq. (4.10). The master formula reads now

$$\langle O \rangle = \sum_{f_a, f_b} \int dx_a f_{f_a}(x_a) \int dx_b f_{f_b}(x_b) \tag{4.13}$$

$$\frac{1}{2\hat{s} n_s(a) n_s(b) n_c(a) n_c(b)} \sum_{n_{\text{ext}}} \int d\phi_{n_{\text{ext}}-2} O \left(p_1, ..., p_{n_{\text{ext}}-2}\right) \sum_{\text{spins,colour}} \left|\mathcal{A}_{n_{\text{ext}}}\right|^2.$$

The partons have internal degrees of freedom, given by the spin and the colour of the partons. In squaring the amplitude we sum over these degrees of freedom. For the particles in the initial state we would like to average over these degrees of freedom. This is done by dividing by the factors $n_s(i)$ and $n_c(i)$, giving the number of spin degrees of freedom (2 for quarks and gluons) and the number of colour degrees of freedom (3 for quarks, 8 for gluons). The second modification is due to the fact that the particles brought into collision are not partons, but composite particles like protons. At high energies the elementary constituents of the protons interact and we have to include a function $f_{f_a}(x_a)$ giving us the probability of finding a parton of flavour f_a with momentum fraction x_a of the original proton momentum inside the proton. If the momenta of the incoming protons are P'_a and P'_b, then the momenta of the two incoming partons are given by

$$p'_a = x_a P'_a, \quad p'_b = x_b P'_b. \tag{4.14}$$

$\hat{s}$ is the centre-of-mass energy squared of the two partons entering the hard interaction. Neglecting particle masses we have

$$\hat{s} = \left(p'_a + p'_b\right)^2 = x_a x_b \left(P'_a + P'_b\right)^2. \quad (4.15)$$

In addition there is a small change in Eq. (4.11). The quantity $(n!)$ is replaced by $(\prod n_j!)$, where n_j is the number of times a parton of type j occurs in the final state.

It is very convenient to calculate the amplitude with the convention that all particles are out-going. To this aim we set

$$p_{n_{\text{ext}}-1} = -p'_a, \quad p_{n_{\text{ext}}} = -p'_b \quad (4.16)$$

and calculate the amplitude for the momentum configuration

$$\left\{p_1, \ldots, p_{n_{\text{ext}}-2}, p_{n_{\text{ext}}-1}, p_{n_{\text{ext}}}\right\}. \quad (4.17)$$

Momentum conservation reads

$$p_1 + \cdots + p_{n_{\text{ext}}-2} + p_{n_{\text{ext}}-1} + p_{n_{\text{ext}}} = 0. \quad (4.18)$$

Note that the momenta $p_{n_{\text{ext}}-1}$ and $p_{n_{\text{ext}}}$ have negative energy components.

We have seen through Eqs. (4.10) and (4.13) that the scattering amplitudes $\mathcal{A}_{n_{\text{ext}}}$ with n_{ext} external particles enter the theory predictions for an observable O. Thus we need to compute the scattering amplitudes. Unfortunately, it is usually not possible to calculate the scattering amplitudes exactly. However, we may calculate scattering amplitudes within perturbation theory, if all couplings describing the strengths of interactions among the particles are small. Let us assume for simplicity that there is only one coupling, which we denote by g. We expand the scattering amplitude in powers of g:

$$\mathcal{A}_{n_{\text{ext}}} = \mathcal{A}^{(0)}_{n_{\text{ext}}} + \mathcal{A}^{(1)}_{n_{\text{ext}}} + \mathcal{A}^{(2)}_{n_{\text{ext}}} + \mathcal{A}^{(3)}_{n_{\text{ext}}} + \cdots, \quad (4.19)$$

where $\mathcal{A}^{(l)}_{n_{\text{ext}}}$ contains $(n_{\text{ext}} - 2 + 2l)$ factors of g. Equation (4.19) gives the perturbative expansion of the scattering amplitude. In this expansion, $\mathcal{A}^{(l)}_{n_{\text{ext}}}$ is an amplitude with n_{ext} external particles and l loops. The recipe for the computation of $\mathcal{A}^{(l)}_{n_{\text{ext}}}$ based on Feynman diagrams is as follows:

Algorithm 1 *Calculation of scattering amplitudes from Feynman diagrams.*

1. *Draw all Feynman diagrams for the given number of external particles* n_{ext} *and the given number of loops* l.
2. *Translate each graph into a mathematical formula with the help of the Feynman rules.*
3. *The quantity* $i\mathcal{A}^{(l)}_{n_{\text{ext}}}$ *is then given as the sum of all these terms.*

Tree-level amplitudes are amplitudes with no loops and are denoted by $\mathcal{A}^{(0)}_{n_{\mathrm{ext}}}$. They give the leading contribution to the full amplitude. The computation of tree-level amplitudes involves only basic mathematical operations: Addition, multiplication, contraction of indices, etc. The above algorithm allows therefore in principle for any n_{ext} the computation of the corresponding tree-level amplitude. (In practice, the number of contributing Feynman diagrams is a limiting factor for tree-level amplitudes with a large number of external particles. For a review of methods to tackle this problem see [65].) The situation is different for loop amplitudes $\mathcal{A}^{(l)}_{n_{\mathrm{ext}}}$ (with $l \geq 1$). Here, the Feynman rules involve an integration over each internal momentum not constrained by momentum conservation. That's where Feynman integrals enter the game.

In the algorithm above a Feynman diagram is translated into a mathematical expression with the help of the Feynman rules. The starting point for a physical theory (or a model) of particle physics is usually the Lagrangian. In Eq. (4.1) we specified a scalar ϕ^4-theory by giving the Lagrangian. For a more realistic theory let's look at quantum chromodynamics (QCD). For QCD the Lagrange density reads in Lorenz gauge:

$$\begin{aligned} \mathcal{L}_{\mathrm{QCD}} &= -\frac{1}{4} F^a_{\mu\nu}(x) F^{a\mu\nu}(x) - \frac{1}{2\xi}\left(\partial^\mu A^a_\mu(x)\right)^2 - \bar{c}^a(x)\partial^\mu D^{ab}_\mu c^b(x), \\ &+ \sum_{\text{quarks } q} \bar{\psi}_q(x)\left(i\gamma^\mu D_\mu - m_q\right)\psi_q(x), \end{aligned} \tag{4.20}$$

with

$$F^a_{\mu\nu}(x) = \partial_\mu A^a_\nu(x) - \partial_\nu A^a_\mu(x) + g f^{abc} A^b_\mu(x) A^c_\nu. \tag{4.21}$$

The gluon field is denoted by $A^a_\mu(x)$, the Faddeev-Popov ghost fields are denoted by $c^a(x)$ and the quark fields are denoted by $\psi_q(x)$. The sum is over all quark flavours. The masses of the quarks are denoted by m_q. The quark fields also carry a colour index j and a Dirac index α, which we didn't denote explicitly. The variable g gives the strength of the strong coupling. QCD is a $SU(3)$-gauge theory. Indices referring to the fundamental representation of $SU(3)$ are chosen from the middle of the alphabet $i, j, k, \dots$ and range from 1 to 3, while indices referring to the adjoint representation of $SU(3)$ are chosen from the beginning of the alphabet $a, b, c, \dots$ and range from 1 to 8. The generators of the group $SU(3)$ are denoted by T^a and satisfy

$$\left[T^a, T^b\right] = i f^{abc} T^c. \tag{4.22}$$

The standard normalisation is

$$\mathrm{Tr}\left(T^a T^b\right) = T_R \delta^{ab} = \frac{1}{2}\delta^{ab}. \tag{4.23}$$

For later use we set

$$N_c = 3, \quad T_R = \frac{1}{2}, \quad C_A = N_c = 3, \quad C_F = \frac{N_c^2 - 1}{2N_c} = \frac{4}{3}. \tag{4.24}$$

The quantity $F^a_{\mu\nu}$ is called the field strength, the quantity $D_\mu = D_{\mu,jk}$ denotes the covariant derivative in the fundamental representation of $SU(3)$, D^{ab}_μ denotes the covariant derivative in the adjoint representation of $SU(3)$:

$$\begin{aligned} D_{\mu,jk} &= \delta_{jk}\partial_\mu - igT^a_{jk}A^a_\mu, \\ D^{ab}_\mu &= \delta^{ab}\partial_\mu - gf^{abc}A^c_\mu. \end{aligned} \tag{4.25}$$

The variable ξ is called the gauge-fixing parameter. Gauge-invariant quantities like scattering amplitudes are independent of this parameter.

In order to derive the Feynman rules from the Lagrangian one proceeds as follows: We first order the terms in the Lagrangian according to the number of fields they involve. From the terms bilinear in the fields one obtains the propagators, while the terms with three or more fields give rise to vertices. Note that a "normal" Lagrangian does not contain terms with just one or zero fields. Furthermore we always assume within perturbation theory that all fields fall off rapidly enough at infinity. Therefore we can use partial integration and ignore boundary terms. Using partial integration we may re-write the Lagrangian of ϕ^4-theory of Eq. (4.1) as

$$\mathcal{L} = \frac{1}{2}\phi(x)\left[-\Box - m^2\right]\phi(x) + \frac{1}{4}\lambda\phi(x)^4, \tag{4.26}$$

where we denoted the d'Alembert operator by $\Box = \partial_\mu\partial^\mu$. As a second example we consider the gluonic part of the QCD Lagrange density:

$$\begin{aligned} \mathcal{L}_{\mathrm{QCD}} = & \frac{1}{2}A^a_\mu(x)\left[\partial_\rho\partial^\rho g^{\mu\nu}\delta^{ab} - \left(1-\frac{1}{\xi}\right)\partial^\mu\partial^\nu\delta^{ab}\right]A^b_\nu(x) \\ & -gf^{abc}\left(\partial_\mu A^a_\nu(x)\right)A^{b\mu}(x)A^{c\nu}(x) - \frac{1}{4}g^2 f^{eab}f^{ecd}A^a_\mu(x)A^b_\nu(x)A^{c\mu}(x)A^{d\nu}(x) \\ & +\mathcal{L}_{\mathrm{quarks}} + \mathcal{L}_{\mathrm{FP}}. \end{aligned} \tag{4.27}$$

Let us first consider the terms bilinear in the fields, giving rise to the propagators. A generic term for real boson fields ϕ_i has the form

$$\mathcal{L}_{\mathrm{bilinear}}(x) = \frac{1}{2}\phi_i(x)P_{ij}(x)\phi_j(x), \tag{4.28}$$

where P is a real symmetric operator that may contain derivatives and must have an inverse. Define the inverse of P by

$$\sum_j P_{ij}(x)P^{-1}_{jk}(x-y) = \delta_{ik}\delta^D(x-y), \tag{4.29}$$

and its Fourier transform by

$$P^{-1}_{ij}(x) = \int \frac{d^Dq}{(2\pi)^D} e^{-iq\cdot x}\tilde{P}^{-1}_{ij}(q). \tag{4.30}$$

Then the **propagator** is given by

$$\Delta_F(q)_{ij} = i\tilde{P}_{ij}^{-1}(q). \tag{4.31}$$

Let's see how this works out for the scalar propagator in ϕ^4-theory and for the gluon propagator in QCD. We start with ϕ^4-theory: From Eq. (4.26) we deduce

$$P(x) = -\Box - m^2. \tag{4.32}$$

It is not too difficult to show that

$$\tilde{P}^{-1}(q) = \frac{1}{q^2 - m^2}. \tag{4.33}$$

The propagator of a scalar particle is therefore given by

$$\Delta_F(k) = \quad \text{———} \quad = \frac{i}{q^2 - m^2} \tag{4.34}$$

and drawn as a line.

Let us now look into a more involved example. We consider the gluon propagator in QCD. The first line of Eq. (4.27) gives the terms bilinear in the gluon fields. This defines an operator

$$P^{\mu\nu\, ab}(x) = \partial_\rho \partial^\rho g^{\mu\nu} \delta^{ab} - \left(1 - \frac{1}{\xi}\right) \partial^\mu \partial^\nu \delta^{ab}. \tag{4.35}$$

For the propagator we are interested in the inverse of this operator

$$P^{\mu\sigma\, ac}(x) \left(P^{-1}\right)^{cb}_{\sigma\nu}(x-y) = g^\mu{}_\nu \delta^{ab} \delta^4(x-y). \tag{4.36}$$

Working in momentum space we are more specifically interested in the Fourier transform of the inverse of this operator:

$$\left(P^{-1}\right)^{ab}_{\mu\nu}(x) = \int \frac{d^D q}{(2\pi)^D} e^{-iq\cdot x} \left(\tilde{P}^{-1}\right)^{ab}_{\mu\nu}(q). \tag{4.37}$$

The Feynman rule for the propagator is then given by $(\tilde{P}^{-1})^{ab}_{\mu\nu}(q)$ times the imaginary unit. For the gluon propagator one finds the Feynman rule

$$\mu, a \quad \text{(gluon line)} \quad \nu, b \quad = \frac{i}{q^2}\left(-g_{\mu\nu} + (1-\xi)\frac{q_\mu q_\nu}{q^2}\right)\delta^{ab}. \tag{4.38}$$

Exercise 27 *Derive Eq. (4.38) from Eqs. (4.36) and (4.37).*
Hint: It is simpler to work directly in momentum space, using the Fourier representation of $\delta^D(x-y)$.

Let us now consider a generic interaction term with $n \geq 3$ fields. We may write this term as

$$\mathcal{L}_{\text{int}}(x) = O_{i_1 \dots i_n}(\partial_1, \dots, \partial_n)\, \phi_{i_1}(x) \dots \phi_{i_n}(x), \tag{4.39}$$

with the notation that ∂_j acts only on the j-th field $\phi_{i_j}(x)$. For each field we have the Fourier transform

$$\phi_i(x) = \int \frac{d^D q}{(2\pi)^D}\, e^{-iqx}\, \tilde{\phi}_i(q), \quad \tilde{\phi}_i(q) = \int d^D x\, e^{iqx}\, \phi_i(x), \tag{4.40}$$

where q denotes an in-coming momentum. We thus have

$$\begin{aligned} \mathcal{L}_{\text{int}}(x) = \int \frac{d^D q_1}{(2\pi)^D} \dots \frac{d^D q_n}{(2\pi)^D} e^{-i(q_1+\dots+q_n)x} \\ O_{i_1 \dots i_n}(-iq_1, \dots, -iq_n)\, \tilde{\phi}_i(q_1) \dots \tilde{\phi}_i(q_n)\,. \end{aligned} \tag{4.41}$$

Changing to outgoing momenta we replace q_j by $-q_j$. The **vertex** is then given by

$$V = i \sum_{\text{permutations}} (-1)^{P_F}\, O_{i_1 \dots i_n}(iq_1, \dots, iq_n)\,, \tag{4.42}$$

where the momenta are taken to flow outward. The summation is over all permutations of indices and momenta of identical particles. In the case of identical fermions there is in addition a minus sign for every odd permutation of the fermions, indicated by $(-1)^{P_F}$.

Let us also work out some examples here. We start again with scalar ϕ^4-theory. There is only one interaction term, containing four field $\phi(x)$:

$$\mathcal{L}_{\text{int}}(x) = \frac{\lambda}{4!}\phi(x)\phi(x)\phi(x)\phi(x). \tag{4.43}$$

Thus $O = \lambda/4!$ and the Feynman rule for the vertex is given by

$$= i\lambda. \tag{4.44}$$

The factor $1/4!$ is cancelled by summing over the 4! permutations of the four identical particles.

It is instructive to consider also a more involved example. We derive the Feynman rule for the three-gluon vertex in QCD. The relevant term in the Lagrangian is the first term in the second line of Eq. (4.27):

$$\mathcal{L}_{ggg} = -g f^{abc} \left(\partial_\mu A_\nu^a(x)\right) A^{b\mu}(x) A^{c\nu}(x). \tag{4.45}$$

This term contains three gluon fields and will give rise to the three-gluon vertex. We may rewrite this term as

$$\mathcal{L}_{ggg} = -g f^{abc} g^{\mu\rho} \partial_1^\nu\, A_\mu^a(x) A_\nu^b(x) A_\rho^c(x). \tag{4.46}$$

Thus

$$O^{abc,\mu\nu\rho}(\partial_1, \partial_2, \partial_3) = -gf^{abc}g^{\mu\rho}\partial_1^\nu, \quad O^{abc,\mu\nu\rho}(iq_1, iq_2, iq_3) = -gf^{abc}g^{\mu\rho}iq_1^\nu. \tag{4.47}$$

The Feynman rule for the vertex is given by the sum over all permutations of identical particles of the function $O^{abc,\mu\nu\rho}(iq_1, iq_2, iq_3)$ multiplied by the imaginary unit i. For the case at hand, we have three identical gluons and we have to sum over $3! = 6$ permutations. One finds

$$\begin{aligned} V_{ggg} &= i \sum_{\text{permutations}} \left(-gf^{abc}g^{\mu\rho}iq_1^\nu\right) \\ &= -gf^{abc}\left[g^{\mu\nu}\left(q_1^\rho - q_2^\rho\right) + g^{\nu\rho}\left(q_2^\mu - q_3^\mu\right) + g^{\rho\mu}\left(q_3^\nu - q_1^\nu\right)\right]. \end{aligned} \tag{4.48}$$

Note that we have momentum conservation at each vertex, for the three-gluon vertex this implies

$$q_1 + q_2 + q_3 = 0. \tag{4.49}$$

In a similar way one obtains the Feynman rules for the four-gluon vertex, the ghost-antighost-gluon vertex and the quark-antiquark-gluon vertex.

Let us summarise the Feynman rules for the propagators and the vertices of QCD: The gluon propagator (in Feynman gauge, corresponding to $\xi = 1$), the ghost propagator and the quark propagator are given by

μ, a (gluon line) ν, b $$= \frac{-ig^{\mu\nu}}{q^2}\delta^{ab},$$

a (ghost line) b $$= \frac{i}{q^2}\delta^{ab},$$

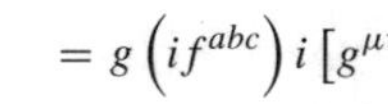

j (quark line) k $$= i\frac{\not{q} + m}{q^2 - m^2}\delta_{jk}. \tag{4.50}$$

Here we used the notation $\not{q} = q_\mu \gamma^\mu$. The Feynman rules for the vertices of QCD are

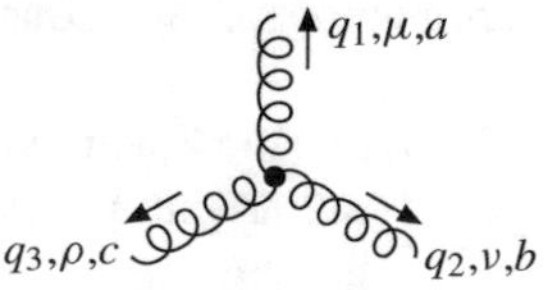

(three-gluon vertex: q_1,μ,a; q_2,ν,b; q_3,ρ,c)

$$= g\left(if^{abc}\right)i\left[g^{\mu\nu}\left(q_1^\rho - q_2^\rho\right) + g^{\nu\rho}\left(q_2^\mu - q_3^\mu\right) + g^{\rho\mu}\left(q_3^\nu - q_1^\nu\right)\right],$$

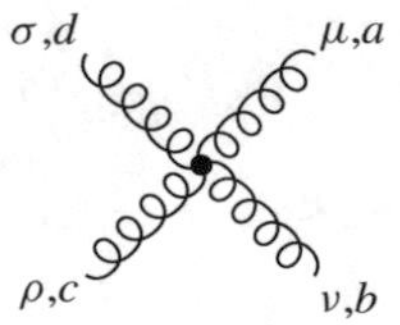

(four-gluon vertex: μ,a; ν,b; ρ,c; σ,d)

$$\begin{aligned} &= ig^2\left[\left(if^{abe}\right)\left(if^{ecd}\right)\left(g^{\mu\rho}g^{\nu\sigma} - g^{\nu\rho}g^{\mu\sigma}\right)\right. \\ &\quad + \left(if^{bce}\right)\left(if^{ead}\right)\left(g^{\nu\mu}g^{\rho\sigma} - g^{\rho\mu}g^{\nu\sigma}\right) \end{aligned}$$

$$+\left(if^{cae}\right)\left(if^{ebd}\right)\left(g^{\rho\nu}g^{\mu\sigma}-g^{\mu\nu}g^{\rho\sigma}\right)\Big],$$

q_1,a / c — μ,b

$$= ig\left(if^{abc}\right)q_1^\mu.$$

j / l — μ,a

$$= ig\gamma^\mu T^a_{jl}, \tag{4.51}$$

When translating a Feynman diagram to a mathematical expression, the Feynman rules distinguish between internal edges and external edges. Whereas an internal edge is translated to the mathematical formula for the corresponding propagator, an external edge translates to a factor describing the spin polarisation of the corresponding particle. Thus, there is a **polarisation vector** $\varepsilon^\mu(q)$ for each external spin-1 boson and a **spinor** $\bar{u}(q)$, $u(q)$, $\bar{v}(q)$ or $v(q)$ for each external spin-$1/2$ fermion. For spin-0 bosons there is no non-trivial spin polarisation to be described, hence an external edge corresponding to a spin-0 boson translates to the trivial factor 1.

In addition, there are a few additional Feynman rules:

- There is an integration

$$\int \frac{d^D q}{(2\pi)^D} \tag{4.52}$$

  for each internal momentum not constrained by momentum conservation. Such an integration is called a "loop integration" and the number of independent loop integrations in a diagram is called the loop number of the diagram.
- A factor (-1) for each closed fermion loop.
- Each diagram is multiplied by a factor $1/S$, where S is the order of the permutation group of the internal lines and vertices leaving the diagram unchanged when the external lines are fixed.

With the methods outlined above we may obtain the Feynman rules for any theory specified by a Lagrangian. As examples we considered a scalar ϕ^4-theory and QCD. The list of Feynman rules for the various propagators and interaction vertices of the full Standard Model of particle physics is rather long and not reproduced here. The

Feynman rules for the Standard Model comprise apart from the Feynman rules for QCD discussed above also the Feynman rules for the electro-weak sector and the Higgs sector. These rules can be found in many textbooks of quantum field theory, for example [62]. However, we would like to show one particular Feynman rule: The Feynman rule for the coupling of a Z-boson to a fermion-antifermion pair reads

$$\mu \;=\; \frac{ie}{2\sin\theta_W\cos\theta_W}\gamma^\mu\left(v_f - a_f\gamma_5\right), \tag{4.53}$$

where e denotes the elementary electric charge (i.e., the magnitude of the electric charge of the electron), θ_W denotes the Weinberg angle and the quantities v_f and a_f are given by

$$v_f = I_3 - 2Q\sin^2\theta_W, \quad a_f = I_3. \tag{4.54}$$

Here Q denotes the electric charge of the fermion in units of e and I_3 equals $1/2$ for up-type fermions and $-1/2$ for down-type fermions. We picked this specific Feynman rule for the following reason: The Feynman rule involves the Dirac matrix γ_5. In four space-time dimensions γ_5 is defined by

$$\gamma_5 = i\gamma^0\gamma^1\gamma^2\gamma^3. \tag{4.55}$$

γ_5 is an inherently four-dimensional object. Therefore, the treatment of γ_5 within dimensional regularisation requires some care and is discussed in Sect. 4.4.

Exercise 28 *Compute the four-gluon amplitude $\mathcal{A}_4^{(0)}$ from the four diagrams shown in Fig. 4.1. Assume that all momenta are outgoing. The result will involve scalar products $2p_i \cdot p_j$, $2p_i \cdot \varepsilon_j$ and $2\varepsilon_i \cdot \varepsilon_j$. For a $2 \to 2$ process (more precisely for a $0 \to 4$ process, since we take all momenta to be outgoing), the Mandelstam variables are defined by*

$$s = (p_1 + p_2)^2, \quad t = (p_2 + p_3)^2, \quad u = (p_1 + p_3)^2. \tag{4.56}$$

The four momenta p_1, p_2, p_3 and p_4 are on-shell, $p_i^2 = 0$ for $i = 1, ..., 4$, and satisfy momentum conservation. Derive the Mandelstam relation

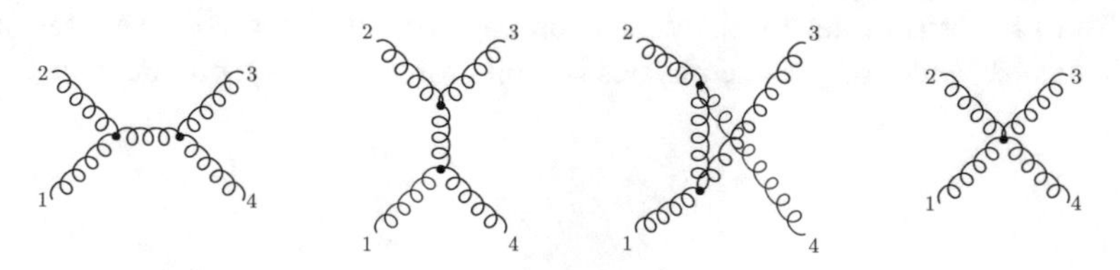

Fig. 4.1 The four Feynman diagrams contributing to the tree-level four-gluon amplitude $\mathcal{A}_4^{(0)}$

$$s + t + u = 0. \tag{4.57}$$

This relation allows to eliminate in the result for $\mathcal{A}_4^{(0)}$ *one variable, say* u*. Furthermore the polarisation vector of gluon* j *is orthogonal to the momentum of gluon* j*, i.e., we have the relation* $2p_j \cdot \varepsilon_j = 0$*. Combined with momentum conservation we may eliminate several scalar products* $2p_i \cdot \varepsilon_j$*, such that for a given* j *we only have* $2p_{j-1} \cdot \varepsilon_j$ *and* $2p_{j+1} \cdot \varepsilon_j$*, where the indices* $(j-1)$ *and* $(j+1)$ *are understood modulo 4. You might want to use a computer algebra system to carry out the calculations. The open-source computer algebra systems* `FORM` *[66] and* `GiNaC` *[67] have their roots in particle physics and were originally invented for calculations of this type.*

Let us summarise:

Feynman rules:

- For each internal edge include a propagator. The propagator is derived from the terms in the Lagrangian bilinear in the fields.
- For each external edge include a factor, describing the spin polarisation of the particle.
- For each internal vertex include a vertex factor. The vertex factor for a vertex of valency n is derived from the terms of the Lagrangian containing exactly the n fields meeting at this vertex.
- For each internal momentum not constrained by momentum conservation integrate with measure

$$\int \frac{d^D q}{(2\pi)^D} \tag{4.58}$$

- Include a factor (-1) for each closed fermion loop.
- Include a factor $1/S$, where S is the order of the permutation group of the internal lines and vertices leaving the diagram unchanged when the external lines are fixed.

Let us now look at an example how to translate a Feynman diagram with a loop into a mathematical expression. Figure 4.2 shows a Feynman diagram contributing to the one-loop correction for the process $e^+e^- \to qg\bar{q}$. At high energies we can ignore the masses of the electron and the light quarks. From the Feynman rules one obtains for this diagram:

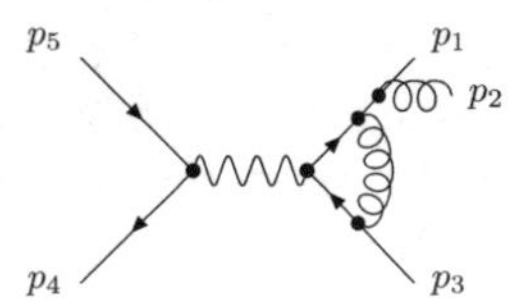

Fig. 4.2 A one-loop Feynman diagram contributing to the process $e^+e^- \to qg\bar{q}$

$$-e^2 g^3 C_F T^a_{jl} \bar{v}(p_4)\gamma^\mu u(p_5) \frac{1}{p_{123}^2}$$
$$\int \frac{d^D k_1}{(2\pi)^D} \frac{1}{k_2^2} \bar{u}(p_1) \not{\varepsilon}(p_2) \frac{\not{p}_{12}}{p_{12}^2} \gamma_\nu \frac{\not{k}_1}{k_1^2} \gamma_\mu \frac{\not{k}_3}{k_3^2} \gamma^\nu v(p_3). \tag{4.59}$$

Here, $p_{12} = p_1 + p_2$, $p_{123} = p_1 + p_2 + p_3$, $k_2 = k_1 - p_{12}$, $k_3 = k_2 - p_3$. Further $\not{\varepsilon}(p_2) = \gamma_\tau \varepsilon^\tau(p_2)$, where $\varepsilon^\tau(p_2)$ is the polarisation vector of the outgoing gluon. All external momenta are assumed to be massless: $p_i^2 = 0$ for $i = 1, \dots, 5$. We can reorganise this formula into a part, which depends on the loop integration and a part, which does not. The loop integral to be calculated reads:

$$\int \frac{d^D k_1}{(2\pi)^D} \frac{k_1^\rho k_3^\sigma}{k_1^2 k_2^2 k_3^2}, \tag{4.60}$$

while the remaining factor, which is independent of the loop integration is given by

$$-e^2 g^3 C_F T^a_{jl} \bar{v}(p_4)\gamma^\mu u(p_5) \frac{1}{p_{123}^2 p_{12}^2} \bar{u}(p_1) \not{\varepsilon}(p_2) \not{p}_{12} \gamma_\nu \gamma_\rho \gamma_\mu \gamma_\sigma \gamma^\nu v(p_3). \tag{4.61}$$

The loop integral in Eq. (4.60) contains in the denominator three propagator factors and in the numerator two factors of the loop momentum. We call a loop integral, in which the loop momentum occurs also in the numerator a "**tensor integral**". A loop integral, in which the numerator is independent of the loop momentum is called a "**scalar integral**". The scalar integral associated to Eq. (4.60) reads

$$\int \frac{d^D k_1}{(2\pi)^D} \frac{1}{k_1^2 k_2^2 k_3^2}. \tag{4.62}$$

It is always possible to reduce tensor integrals to scalar integrals and we will discuss a method to achieve that in Sect. 4.3.

4.2 How to Obtain Finite Results

We have already seen in Eq. (2.124) that the result of a regularised Feynman integral may contain poles in the regularisation parameter ε. These poles reflect the original ultraviolet and infrared singularities of the unregularised integral. What shall we do with these poles? The answer has to come from physics and we distinguish again the case of UV-divergences and IR-divergences. The UV-divergences are removed through **renormalisation**. Ultraviolet divergences are absorbed into a redefinition of the parameters. As an example we consider the renormalisation of the coupling in QCD:

$$\underbrace{g}_{\text{divergent}} = \underbrace{Z_g}_{\text{divergent}} \underbrace{g_r}_{\text{finite}} . \tag{4.63}$$

The renormalisation constant Z_g absorbs the divergent part. However Z_g is not unique: One may always shift a finite piece from g_r to Z_g or vice versa. Different choices for Z_g correspond to different **renormalisation schemes**. Two different renormalisation schemes are always connected by a finite renormalisation. Note that different renormalisation schemes give numerically different answers. Therefore one always has to specify the renormalisation scheme. Some popular renormalisation schemes are the **on-shell scheme**, where the renormalisation constants are defined by conditions at a scale where the particles are on-shell. A second widely used scheme is **modified minimal subtraction** ($\overline{\text{MS}}$-scheme). In this scheme one always absorbs the combination

$$\Delta = \frac{1}{\varepsilon} - \gamma_{\text{E}} + \ln 4\pi \tag{4.64}$$

into the renormalisation constants. One proceeds similar with all other quantities appearing in the original Lagrangian. For example:

$$A_\mu^a = \sqrt{Z_3} A_{\mu,r}^a, \quad \psi_q = \sqrt{Z_2} \psi_{q,r}, \quad g = Z_g g_r, \quad m = Z_m m_r, \quad \xi = Z_\xi \xi_r. \tag{4.65}$$

The fact that square roots appear for the field renormalisation is just convention. Let us look a little bit closer into the coupling renormalisation within dimensional regularisation and the $\overline{\text{MS}}$-renormalisation scheme. Within dimensional regularisation the renormalised coupling g_r is a dimensionfull quantity. We define a dimensionless quantity g_R by

$$g_r = g_R \mu^\varepsilon, \tag{4.66}$$

where μ is an arbitrary mass scale, called the **renormalisation scale**. From a one-loop calculation one obtains

$$Z_g = 1 - \frac{1}{2} \beta_0 \frac{g_R^2}{(4\pi)^2} \Delta + O(g_R^4), \qquad \beta_0 = \frac{11}{3} N_c - \frac{2}{3} N_f. \tag{4.67}$$

N_c is the number of colours and N_f the number of light quarks. The quantity g_R will depend on the arbitrary scale μ. To derive this dependence one first notes that the unrenormalised coupling constant g is of course independent of μ:

$$\frac{d}{d\mu} g = 0 \tag{4.68}$$

Substituting $g = Z_g \mu^\varepsilon g_R$ into this equation one obtains

$$\mu \frac{d}{d\mu} g_R = -\varepsilon g_R - \left(Z_g^{-1} \mu \frac{d}{d\mu} Z_g \right) g_R. \tag{4.69}$$

From Eq. (4.67) one obtains

$$Z_g^{-1} \mu \frac{d}{d\mu} Z_g = \beta_0 \frac{g_R^2}{(4\pi)^2} (\varepsilon \Delta) + O(g_R^4). \tag{4.70}$$

Instead of g_R one often uses the quantity $\alpha_s = g_R^2/(4\pi)$, Going to $D = 4$ (in this limit we have $\varepsilon\Delta = 1 + O(\varepsilon)$) one arrives at

$$\mu^2 \frac{d}{d\mu^2} \frac{\alpha_s}{4\pi} = -\beta_0 \left(\frac{\alpha_s}{4\pi} \right)^2 + O\left(\left(\frac{\alpha_s}{4\pi} \right)^3 \right) + O(\varepsilon). \tag{4.71}$$

This differential equation gives the dependence of α_s on the renormalisation scale μ. At leading order the solution is given by

$$\frac{\alpha_s(\mu)}{4\pi} = \frac{1}{\beta_0 \ln\left(\frac{\mu^2}{\Lambda^2}\right)}, \tag{4.72}$$

where Λ is an integration constant. The quantity Λ is called the QCD scale parameter. For QCD β_0 is positive and $\alpha_s(\mu)$ decreases with larger μ. This property is called asymptotic freedom: The coupling becomes smaller at high energies. In QED β_0 has the opposite sign and the fine-structure constant $\alpha(\mu)$ increases with larger μ. The electromagnetic coupling becomes weaker when we go to smaller energies.

Let us now look at the infrared divergences: We first note that any detector has a finite resolution. Therefore two particles which are sufficiently close to each other in phase space will be detected as one particle. Now let us look again at Eqs. (4.10) and (4.19). The next-to-leading order term will receive contributions from the interference term of the one-loop amplitude $\mathcal{A}_{n_{\text{ext}}}^{(1)}$ with the leading-order amplitude $\mathcal{A}_{n_{\text{ext}}}^{(0)}$, both with $(n_{\text{ext}} - 2)$ final state particles. This contribution is of order $g^{2n_{\text{ext}}-2}$. Of the same order is the square of the leading-order amplitude $\mathcal{A}_{n_{\text{ext}}+1}^{(0)}$ with $(n_{\text{ext}} - 1)$ final state particles. This contribution we have to take into account whenever our detector resolves only $(n_{\text{ext}} - 2)$ final-state particles. It turns out that the phase space integration over the regions where one or more particles become unresolved is also divergent, and, when performed in D dimensions, leads to poles with the opposite sign as the one encountered in the loop amplitudes. Therefore the sum of the two contributions is finite. The **Kinoshita-Lee-Nauenberg theorem** [27, 28] guarantees that all infrared divergences cancel, when summed over all degenerate physical states. As an example we consider the NLO corrections to $\gamma^* \to 2$ jets, where we treat the quarks as massless. The interference term of the one-loop amplitude with the Born amplitude is given for one flavour by

$$2 \text{ Re } \mathcal{A}_3^{(0)*} \mathcal{A}_3^{(1)} = \frac{\alpha_s}{\pi} C_F \left(-\frac{1}{\varepsilon^2} - \frac{3}{2\varepsilon} - 4 + \frac{7}{12}\pi^2 \right) S_\varepsilon \left| \mathcal{A}_3^{(0)} \right|^2 + O(\varepsilon). \tag{4.73}$$

$S_\varepsilon = (4\pi)^\varepsilon e^{-\varepsilon\gamma_E}$ is the typical phase-space volume factor in $D = 4 - 2\varepsilon$ dimensions. For simplicity we have set the renormalisation scale μ equal to the centre-of-mass energy squared s. The square of the Born amplitude is given by

$$\left|\mathcal{A}_3^{(0)}\right|^2 = 16\pi N_c \alpha (1-\varepsilon)\, s. \tag{4.74}$$

This is independent of the final state momenta and the integration over the phase space can be written as

$$\begin{aligned}&\int d\phi_2 \left(2\,\mathrm{Re}\ \mathcal{A}_3^{(0)\,*}\,\mathcal{A}_3^{(1)}\right) \\ &= \frac{\alpha_s}{\pi} C_F \left(-\frac{1}{\varepsilon^2} - \frac{3}{2\varepsilon} - 4 + \frac{7}{12}\pi^2\right) S_\varepsilon \int d\phi_2 \left|\mathcal{A}_3^{(0)}\right|^2 + O(\varepsilon)\,.\end{aligned} \tag{4.75}$$

The real corrections are given by the leading order matrix element for $\gamma^* \to qg\bar{q}$ and read

$$\left|\mathcal{A}_4^{(0)}\right|^2 = 128\pi^2 \alpha\alpha_s C_F N_c (1-\varepsilon) \left[\frac{2}{x_1 x_2} - \frac{2}{x_1} - \frac{2}{x_2} + (1-\varepsilon)\frac{x_2}{x_1} + (1-\varepsilon)\frac{x_1}{x_2} - 2\varepsilon\right], \tag{4.76}$$

where $x_1 = s_{12}/s_{123}$, $x_2 = s_{23}/s_{123}$ and $s_{123} = s$ is again the centre-of-mass energy squared. The quantities s_{ij} and s_{ijk} are defined by

$$s_{ij} = \left(p_i + p_j\right)^2, \ s_{ijk} = \left(p_i + p_j + p_k\right)^2. \tag{4.77}$$

For massless particles these quantities are equal to

$$s_{ij} = 2p_i \cdot p_j, \ s_{ijk} = 2p_i \cdot p_j + 2p_i \cdot p_k + 2p_j \cdot p_k. \tag{4.78}$$

For this particular simple example we can write the three-particle phase space in D dimensions as

$$\begin{aligned} d\phi_3 &= d\phi_2 d\phi_{\mathrm{unres}}, \\ d\phi_{\mathrm{unres}} &= \frac{(4\pi)^{\varepsilon-2}}{\Gamma(1-\varepsilon)} s_{123}^{1-\varepsilon} d^3x\, \delta(1 - x_1 - x_2 - x_3)\,(x_1 x_2 x_3)^{-\varepsilon}\,. \end{aligned} \tag{4.79}$$

Integration over the phase space ϕ_{unres} yields

$$\int d\phi_3 \left|\mathcal{A}_4^{(0)}\right|^2 = \frac{\alpha_s}{\pi} C_F \left(\frac{1}{\varepsilon^2} + \frac{3}{2\varepsilon} + \frac{19}{4} - \frac{7}{12}\pi^2\right) S_\varepsilon \int d\phi_2 \left|\mathcal{A}_3^{(0)}\right|^2 + O(\varepsilon)\,. \tag{4.80}$$

We see that in the sum the poles cancel and we obtain the finite result

$$\int d\phi_2 \left(2 \text{ Re } \mathcal{A}_3^{(0)*} \mathcal{A}_3^{(1)}\right) + \int d\phi_3 \left|\mathcal{A}_4^{(0)}\right|^2 = \frac{3}{4} C_F \frac{\alpha_s}{\pi} \int d\phi_2 \left|\mathcal{A}_3^{(0)}\right|^2 + O(\varepsilon). \tag{4.81}$$

In this example we have seen the cancellation of the infrared (soft and collinear) singularities between the virtual and the real corrections according to the Kinoshita-Lee-Nauenberg theorem. In this example we integrated over the phase space of all final state particles. In practise one is often interested in differential distributions. In these cases the cancellation is technically more complicated, as the different contributions live on phase spaces of different dimensions and one integrates only over restricted regions of phase space. Methods to overcome this obstacle are known under the name "phase-space slicing" and "subtraction method" [68–75].

The Kinoshita-Lee-Nauenberg theorem is related to the finite experimental resolution in detecting final state particles. In addition we have to discuss initial state particles. Let us go back to Eq. (4.13). The differential cross section we can write schematically

$$d\sigma_{H_1 H_2} = \sum_{a,b} \int dx_a f_{H_1 \to a}(x_a) \int dx_b f_{H_2 \to b}(x_b) d\sigma_{ab}(x_a, x_b), \tag{4.82}$$

where $f_{H \to a}(x)$ is the parton distribution function, giving us the probability to find a parton of type a in a hadron of type H carrying a fraction x to $x + dx$ of the hadron's momentum. $d\sigma_{ab}(x_a, x_b)$ is the differential cross section for the scattering of partons a and b. Now let us look at the parton distribution function $f_{a \to b}$ of a parton inside another parton. At leading order this function is trivially given by $\delta_{ab}\delta(1 - x)$, but already at the next order a parton can radiate off another parton and thus loose some of its momentum and/or convert to another flavour. One finds in D dimensions

$$f_{a \to b}(x, \varepsilon) = \delta_{ab}\delta(1 - x) - \frac{1}{\varepsilon} \frac{\alpha_s}{4\pi} P^0_{a \to b}(x) + O(\alpha_s^2), \tag{4.83}$$

where $P^0_{a \to b}$ is the lowest order Altarelli-Parisi splitting function. To calculate a cross section $d\sigma_{H_1 H_2}$ at NLO involving parton densities one first calculates the cross section $d\hat{\sigma}_{ab}$ where the hadrons H_1 and H_2 are replaced by partons a and b to NLO:

$$d\hat{\sigma}_{ab} = d\hat{\sigma}^0_{ab} + \frac{\alpha_s}{4\pi} d\hat{\sigma}^1_{ab} + O(\alpha_s^2) \tag{4.84}$$

The hard scattering part $d\sigma_{ab}$ is then obtained by inserting the perturbative expansions for $d\hat{\sigma}_{ab}$ and $f_{a \to b}$ into the factorisation formula.

$$\begin{aligned} & d\hat{\sigma}^0_{ab} + \frac{\alpha_s}{4\pi} d\hat{\sigma}^1_{ab} \\ & = d\sigma^0_{ab} + \frac{\alpha_s}{4\pi} d\sigma^1_{ab} - \frac{1}{\varepsilon} \frac{\alpha_s}{4\pi} \sum_c \int dx_1 P^0_{a \to c} d\sigma^0_{cb} - \frac{1}{\varepsilon} \frac{\alpha_s}{4\pi} \sum_d \int dx_2 P^0_{b \to d} d\sigma^0_{ad}. \end{aligned}$$

One therefore obtains for the LO- and the NLO-terms of the hard scattering part

$$\begin{aligned} d\sigma^0_{ab} &= d\hat{\sigma}^0_{ab} \\ d\sigma^1_{ab} &= d\hat{\sigma}^1_{ab} + \frac{1}{\varepsilon}\sum_c \int dx_1 P^0_{a\to c} d\hat{\sigma}^0_{cb} + \frac{1}{\varepsilon}\sum_d \int dx_2 P^0_{b\to d} d\hat{\sigma}^0_{ad}. \end{aligned} \tag{4.85}$$

The last two terms remove the collinear initial state singularities in $d\hat{\sigma}^1_{ab}$.

4.3 Tensor Reduction

In Sect. 4.1 we listed the Feynman rules for a theory like QCD and worked out one example in Eq. (4.59) how a one-loop Feynman diagram translates into a mathematical formula involving a Feynman integral. The attentive reader will have noticed, that the Feynman integral in Eq. (4.60) does not fit directly the definition of Feynman integrals in Eq. (2.56). There are two issues here: One issue is rather trivial and concerns prefactors: If we follow standard conventions within quantum field theory, the propagator of a scalar particle is given by (see Eq. (4.34))

$$\frac{i}{q^2 - m^2}, \tag{4.86}$$

and the integral measure for every internal momentum not constrained by momentum conservation is given by (see Eq. (4.58))

$$\int \frac{d^D q}{(2\pi)^D}. \tag{4.87}$$

On the other hand, we used in Chap. 2 the convention, that an edge corresponds to (see Eq. (2.43))

$$\frac{1}{-q^2 + m^2}, \tag{4.88}$$

and the integral measure is given by (see Eq. (2.44))

$$\int \frac{d^D q}{i\pi^{\frac{D}{2}}}. \tag{4.89}$$

In addition, we included in Chap. 2 a factor (see Eq. (2.45))

$$e^{l\varepsilon\gamma_E} \left(\mu^2\right)^{\nu - \frac{lD}{2}} \tag{4.90}$$

for each Feynman integral. This issue is rather trivial and only concerns prefactors: If we are able to compute a Feynman integral within one convention for prefactors, we also are able to compute this Feynman integral within any other convention for prefactors. The only point to remember is, that we should not forget to adjust the prefactors appropriately in the final result.

The second issue is more serious: The integral in Eq. (4.60) is a tensor integral, while in Chap. 2 we only discussed scalar integrals. Thus we have to show that any tensor integral can always be reduced to a linear combination of scalar integrals. This can be done if we allow in the linear combination of scalar integrals shifted space-time dimensions ($D \rightarrow D+2$) and raised propagators ($\nu_j \rightarrow \nu_j + 1$) [76, 77]. We denote a **tensor integral** by giving the numerator in square brackets:

$$I_{\nu_1 \dots \nu_{n_{\mathrm{int}}}}\left(D, x_1, \dots, x_{N_B}\right)\left[k_{i_1}^{\mu_1} \dots k_{i_t}^{\mu_t}\right] = e^{l \varepsilon \gamma_{\mathrm{E}}} \left(\mu^2\right)^{\nu - \frac{lD}{2}} \int \prod_{r=1}^{l} \frac{d^D k_r}{i \pi^{\frac{D}{2}}} \frac{k_{i_1}^{\mu_1} \dots k_{i_t}^{\mu_t}}{\prod\limits_{j=1}^{n_{\mathrm{int}}} \left(-q_j^2 + m_j^2\right)^{\nu_j}},$$

$$k_{i_1}, \dots, k_{i_t} \in \{k_1, \dots, k_l\}. \tag{4.91}$$

In this notation, a **scalar integral** is an integral, where the numerator equals one:

$$\begin{aligned} I_{\nu_1 \dots \nu_{n_{\mathrm{int}}}}\left(D, x_1, \dots, x_{N_B}\right)[1] &= I_{\nu_1 \dots \nu_{n_{\mathrm{int}}}}\left(D, x_1, \dots, x_{N_B}\right) \\ &= e^{l \varepsilon \gamma_{\mathrm{E}}} \left(\mu^2\right)^{\nu - \frac{lD}{2}} \int \prod_{r=1}^{l} \frac{d^D k_r}{i \pi^{\frac{D}{2}}} \frac{1}{\prod\limits_{j=1}^{n_{\mathrm{int}}} \left(-q_j^2 + m_j^2\right)^{\nu_j}}. \end{aligned} \tag{4.92}$$

Let us introduce two operators $\mathbf{D}^+$ and $\mathbf{D}^-$, which raise, respectively lower, the number of space-time dimensions by two units:

$$\mathbf{D}^{\pm} I_{\nu_1 \dots \nu_{n_{\mathrm{int}}}}\left(D, x_1, \dots, x_{N_B}\right) = I_{\nu_1 \dots \nu_{n_{\mathrm{int}}}}\left(D \pm 2, x_1, \dots, x_{N_B}\right). \tag{4.93}$$

The operators $\mathbf{D}^{\pm}$ are called **dimensional-shift operators**. In addition, we introduce **raising operators** $\mathbf{j}^+$ (with $j \in \{1, \dots, n_{\mathrm{int}}\}$), which raise the power of the propagator j by one unit:

$$\mathbf{j}^+ I_{\nu_1 \dots \nu_j \dots \nu_{n_{\mathrm{int}}}}\left(D, x_1, \dots, x_{N_B}\right) = \nu_j \cdot I_{\nu_1 \dots (\nu_j + 1) \dots \nu_{n_{\mathrm{int}}}}\left(D, x_1, \dots, x_{N_B}\right). \tag{4.94}$$

Note that we defined $\mathbf{j}^+$ such that it raises the index $\nu_j \rightarrow \nu_j + 1$ and multiplies the integral with a factor ν_j. With this definition we have for example

$$\left(\mathbf{j}^+\right)^2 I_{\nu_1 \dots \nu_j \dots \nu_{n_{\mathrm{int}}}}\left(D, x_1, \dots, x_{N_B}\right) = \nu_j \left(\nu_j + 1\right) \cdot I_{\nu_1 \dots (\nu_j + 2) \dots \nu_{n_{\mathrm{int}}}}\left(D, x_1, \dots, x_{N_B}\right). \tag{4.95}$$

Let us now study how the operators $\mathbf{D}^+$ and $\mathbf{j}^+$ act on the integrand of the Schwinger parameter representation of a scalar Feynman integral. We will see that they act in a rather simple way. Let us write for the graph polynomials in the Schwinger parameter representation $\mathcal{U} = \mathcal{U}(\alpha)$ and $\mathcal{F} = \mathcal{F}(\alpha)$. From

$$I_{\nu_1 \dots \nu_{n_{\mathrm{int}}}}(D) = \frac{e^{l\varepsilon\gamma_{\mathrm{E}}}}{\prod\limits_{k=1}^{n_{\mathrm{int}}} \Gamma(\nu_k)} \int\limits_{\alpha_k \geq 0} d^{n_{\mathrm{int}}}\alpha \left(\prod_{k=1}^{n_{\mathrm{int}}} \alpha_k^{\nu_k - 1} \right) \frac{1}{\mathcal{U}^{\frac{D}{2}}} e^{-\frac{\mathcal{F}}{\mathcal{U}}} \tag{4.96}$$

we find

$$\mathbf{D}^{+} I_{\nu_1 \dots \nu_{n_{\mathrm{int}}}}(D) = \frac{e^{l\varepsilon\gamma_{\mathrm{E}}}}{\prod\limits_{k=1}^{n_{\mathrm{int}}} \Gamma(\nu_k)} \int\limits_{\alpha_k \geq 0} d^{n_{\mathrm{int}}}\alpha \left(\prod_{k=1}^{n_{\mathrm{int}}} \alpha_k^{\nu_k - 1} \right) \frac{1}{\mathcal{U} \cdot \mathcal{U}^{\frac{D}{2}}} e^{-\frac{\mathcal{F}}{\mathcal{U}}},$$

$$\mathbf{j}^{+} I_{\nu_1 \dots \nu_j \dots \nu_{n_{\mathrm{int}}}}(D) = \frac{e^{l\varepsilon\gamma_{\mathrm{E}}}}{\prod\limits_{k=1}^{n_{\mathrm{int}}} \Gamma(\nu_k)} \int\limits_{\alpha_k \geq 0} d^{n_{\mathrm{int}}}\alpha \left(\prod_{k=1}^{n_{\mathrm{int}}} \alpha_k^{\nu_k - 1} \right) \frac{\alpha_j}{\mathcal{U}^{\frac{D}{2}}} e^{-\frac{\mathcal{F}}{\mathcal{U}}}. \tag{4.97}$$

We see that an additional factor of the first graph polynomial $\mathcal{U}$ in the denominator corresponds to a shift $D \to D + 2$. An additional factor of the Schwinger parameter α_j in the numerator corresponds to the application of $\mathbf{j}^{+}$ (i.e., a multiplication of the integral by ν_j and a shift $\nu_j \to \nu_j + 1$).

Exercise 29 *Let $n \in \mathbb{N}$. Show that the action of $(\mathbf{j}^{+})^n$ on the integrand of the Schwinger parameter representation is given by*

$$\left(\mathbf{j}^{+}\right)^n I_{\nu_1 \dots \nu_j \dots \nu_{n_{\mathrm{int}}}}(D) = \frac{e^{l\varepsilon\gamma_{\mathrm{E}}}}{\prod\limits_{k=1}^{n_{\mathrm{int}}} \Gamma(\nu_k)} \int\limits_{\alpha_k \geq 0} d^{n_{\mathrm{int}}}\alpha \left(\prod_{k=1}^{n_{\mathrm{int}}} \alpha_k^{\nu_k - 1} \right) \frac{\alpha_j^n}{\mathcal{U}^{\frac{D}{2}}} e^{-\frac{\mathcal{F}}{\mathcal{U}}}. \tag{4.98}$$

Applying $\mathbf{j}^{+}$ to a Feynman integral with $\nu_j = 0$ gives zero, due to explicit prefactor ν_j in Eq. (4.94):

$$\mathbf{j}^{+} I_{\nu_1 \dots \nu_{j-1} 0 \nu_{j+1} \dots \nu_{n_{\mathrm{int}}}} = 0 \cdot I_{\nu_1 \dots \nu_{j-1} 1 \nu_{j+1} \dots \nu_{n_{\mathrm{int}}}} = 0. \tag{4.99}$$

The algorithm for reducing tensor integrals to scalar integrals proceeds as follows: We start from a tensor integral in the momentum representation. For each propagator we introduce a Schwinger parameter as in Eq. (2.153). We then obtain an integral over the loop momenta and the Schwinger parameters similar to Eq. (2.154), but with the additional tensor structure in the integrand. The argument of the exponential function is as in the scalar case the quadric (see Eq. (2.156))

$$\sum_{j=1}^{n_{\mathrm{int}}} \alpha_j \left(-q_j^2 + m_j^2\right) = -\sum_{r=1}^{l} \sum_{s=1}^{l} k_r M_{rs} k_s + \sum_{r=1}^{l} 2 k_r \cdot v_r + J. \tag{4.100}$$

By a suitable change of the independent loop momenta variables $k_r \to k'_r$ we bring this quadric to the form

$$\sum_{j=1}^{n_{\text{int}}} \alpha_j(-q_j^2 + m_j^2) = -\sum_{r=1}^{l} \lambda_r k_r'^2 + J'. \tag{4.101}$$

This decouples the l momentum integrations and we can treat each momentum integration separately. We have to consider integrals of the form

$$\int \frac{d^D k}{i\pi^{D/2}} k^{\mu_1} \dots k^{\mu_t} f(k^2), \tag{4.102}$$

where $f(k^2) = e^{\lambda k^2}$. Integrals with an odd power of the loop momentum in the numerator vanish by symmetry:

$$\int \frac{d^D k}{i\pi^{D/2}} k^{\mu_1} \dots k^{\mu_{2t-1}} f(k^2) = 0, \qquad t \in \mathbb{N}. \tag{4.103}$$

Integrals with an even power of the loop momentum must be proportional to a symmetric tensor build from the metric tensor due to Lorentz symmetry. For the simplest cases we have

$$\begin{aligned} \int \frac{d^D k}{i\pi^{D/2}} k^\mu k^\nu f(k^2) &= -\frac{1}{D} g^{\mu\nu} \int \frac{d^D k}{i\pi^{D/2}} (-k^2) f(k^2), \\ \int \frac{d^D k}{i\pi^{D/2}} k^\mu k^\nu k^\rho k^\sigma f(k^2) &= \frac{1}{D(D+2)} \left(g^{\mu\nu} g^{\rho\sigma} + g^{\mu\rho} g^{\nu\sigma} + g^{\mu\sigma} g^{\nu\rho} \right) \\ &\quad \int \frac{d^D k}{i\pi^{D/2}} (-k^2)^2 f(k^2). \end{aligned} \tag{4.104}$$

The generalisation to arbitrary higher tensor structures is obvious.

Exercise 30 *Work out the corresponding formula for*

$$\int \frac{d^D k}{i\pi^{D/2}} k^{\mu_1} k^{\mu_2} k^{\mu_3} k^{\mu_4} k^{\mu_5} k^{\mu_6} f(k^2). \tag{4.105}$$

Each loop momentum integral is now of the form

$$\int \frac{d^D k}{i\pi^{D/2}} \left(-k^2\right)^a e^{\lambda k^2} = \int \frac{d^D K}{\pi^{D/2}} \left(K^2\right)^a e^{-\lambda K^2} = \frac{\Gamma\left(\frac{D}{2}+a\right)}{\Gamma\left(\frac{D}{2}\right)} \frac{1}{\lambda^{\frac{D}{2}+a}}. \tag{4.106}$$

This leaves us with the Schwinger parameter integrals. The change of variables $k_r \to k'_r$ and the integration in Eq. (4.106) may introduce additional powers of the

Schwinger parameters in the numerator and additional powers of the first graph polynomial $\mathcal{U}$ in the denominator. With the help of Eq. (4.97) we may write these integrals as scalar integrals with raised powers of the propagators and shifted space-time dimensions. This completes the algorithm for the tensor reduction.

Let us consider an example. We consider the two-loop double-box graph shown in Fig. 2.3 for the case

$$
\begin{aligned}
&p_1^2 = 0, \quad p_2^2 = 0, \quad p_3^2 = 0, \quad p_4^2 = 0, \\
&m_1 = m_2 = m_3 = m_4 = m_5 = m_6 = m_7 = 0.
\end{aligned} \tag{4.107}
$$

This example is a continuation of the example discussed in Sect. 2.5.2. Suppose we would like to reduce the tensor integral

$$
I_{1111111}(D)\left[k_1^\mu k_2^\nu\right] = e^{2\varepsilon\gamma_E}\left(\mu^2\right)^{7-D}\int \frac{d^D k_1}{i\pi^{\frac{D}{2}}}\frac{d^D k_2}{i\pi^{\frac{D}{2}}}\frac{k_1^\mu k_2^\nu}{\prod\limits_{j=1}^{7}\left(-q_j^2\right)} \tag{4.108}
$$

to scalar integrals. Introducing Schwinger parameters we have

$$
\begin{aligned}
&I_{1111111}(D)\left[k_1^\mu k_2^\nu\right] \\
&= e^{2\varepsilon\gamma_E}\left(\mu^2\right)^{7-D}\int\limits_{\alpha_j\geq 0} d^7\alpha \int \frac{d^D k_1}{i\pi^{\frac{D}{2}}}\frac{d^D k_2}{i\pi^{\frac{D}{2}}} k_1^\mu k_2^\nu e^{-\sum\limits_{j=1}^{7}\alpha_j\left(-q_j^2\right)}.
\end{aligned} \tag{4.109}
$$

We may write the argument of the exponential function as

$$
\begin{aligned}
\sum_{j=1}^{7}\alpha_j\left(-q_j^2\right) &= -\alpha_{1234}\left(k_1 - \frac{1}{\alpha_{1234}}\left(\alpha_{12}p_1 + \alpha_2 p_2 - \alpha_4 k_2\right)\right)^2 \\
&\quad -\frac{\mathcal{U}}{\alpha_{1234}}\left(k_2 - \frac{1}{\mathcal{U}}\left(-\alpha_{12}\alpha_4 p_1 - \alpha_2\alpha_4 p_2 + \alpha_{1234}\alpha_5 p_3 + \alpha_{1234}\alpha_{57} p_4\right)\right)^2 \\
&\quad +\mu^2\frac{\mathcal{F}}{\mathcal{U}}.
\end{aligned} \tag{4.110}
$$

Here we used the notation $\alpha_{i_1 i_2 \dots i_n} = \alpha_{i_1} + \alpha_{i_2} + \dots + \alpha_{i_n}$. We substitute

$$
k_1' = k_1 - \frac{1}{\alpha_{1234}}\left(\alpha_{12}p_1 + \alpha_2 p_2 - \alpha_4 k_2\right) \tag{4.111}
$$

followed by

$$
k_2' = k_2 - \frac{1}{\mathcal{U}}\left(-\alpha_{12}\alpha_4 p_1 - \alpha_2\alpha_4 p_2 + \alpha_{1234}\alpha_5 p_3 + \alpha_{1234}\alpha_{57} p_4\right). \tag{4.112}
$$

This gives

$$k_1 = k_1' - \frac{\alpha_4}{\alpha_{1234}} k_2' + \frac{1}{\mathcal{U}} \left[\alpha_{4567} \left(\alpha_{12} p_1 + \alpha_2 p_2 \right) - \alpha_4 \left(\alpha_5 p_3 + \alpha_{57} p_4 \right) \right],$$
$$k_2 = k_2' + \frac{1}{\mathcal{U}} \left[-\alpha_4 \left(\alpha_{12} p_1 + \alpha_2 p_2 \right) + \alpha_{1234} \left(\alpha_5 p_3 + \alpha_{57} p_4 \right) \right]. \qquad (4.113)$$

The Jacobian of the transformation $(k_1, k_2) \to (k_1', k_2')$ is one. With the help of Eq. (4.103) we obtain

$$I_{1111111}(D) \left[k_1^\mu k_2^\nu \right] = \qquad (4.114)$$
$$e^{2\varepsilon\gamma_E} \left(\mu^2\right)^{7-D} \int\limits_{\alpha_j \geq 0} d^7\alpha \int \frac{d^D k_1'}{i\pi^{\frac{D}{2}}} \frac{d^D k_2'}{i\pi^{\frac{D}{2}}} e^{\alpha_{1234} k_1'^2 + \frac{\mathcal{U}}{\alpha_{1234}} k_2'^2 - \mu^2 \frac{\mathcal{F}}{\mathcal{U}}} \left\{ -\frac{\alpha_4}{\alpha_{1234}} k_2'^\mu k_2'^\nu \right.$$
$$+ \frac{1}{\mathcal{U}^2} \left[\alpha_{4567} \left(\alpha_{12} p_1^\mu + \alpha_2 p_2^\mu \right) - \alpha_4 \left(\alpha_5 p_3^\mu + \alpha_{57} p_4^\mu \right) \right] \left[-\alpha_4 \left(\alpha_{12} p_1^\nu + \alpha_2 p_2^\nu \right) \right.$$
$$\left. \left. + \alpha_{1234} \left(\alpha_5 p_3^\nu + \alpha_{57} p_4^\nu \right) \right] \right\}.$$

Let us consider two terms in more detail (the others are similar). We first consider the term proportional to $k_2'^\mu k_2'^\nu$:

$$\tilde{I}_1 = -e^{2\varepsilon\gamma_E} \left(\mu^2\right)^{7-D} \int\limits_{\alpha_j \geq 0} d^7\alpha \int \frac{d^D k_1'}{i\pi^{\frac{D}{2}}} \frac{d^D k_2'}{i\pi^{\frac{D}{2}}} \frac{\alpha_4}{\alpha_{1234}} k_2'^\mu k_2'^\nu e^{\alpha_{1234} k_1'^2 + \frac{\mathcal{U}}{\alpha_{1234}} k_2'^2 - \mu^2 \frac{\mathcal{F}}{\mathcal{U}}}$$
$$= \frac{g^{\mu\nu}}{D} e^{2\varepsilon\gamma_E} \left(\mu^2\right)^{7-D} \int\limits_{\alpha_j \geq 0} d^7\alpha \, \frac{\alpha_4}{\alpha_{1234}} \int \frac{d^D k_1'}{i\pi^{\frac{D}{2}}} \frac{d^D k_2'}{i\pi^{\frac{D}{2}}} \left(-k_2'^2 \right) e^{\alpha_{1234} k_1'^2 + \frac{\mathcal{U}}{\alpha_{1234}} k_2'^2 - \mu^2 \frac{\mathcal{F}}{\mathcal{U}}}.$$

With

$$\int \frac{d^D k_1'}{i\pi^{\frac{D}{2}}} e^{\alpha_{1234} k_1'^2} = \frac{1}{\alpha_{1234}^{\frac{D}{2}}},$$
$$\int \frac{d^D k_2'}{i\pi^{\frac{D}{2}}} \left(-k_2'^2 \right) e^{\frac{\mathcal{U}}{\alpha_{1234}} k_2'^2} = \frac{\Gamma\left(\frac{D}{2}+1\right)}{\Gamma\left(\frac{D}{2}\right)} \left(\frac{\alpha_{1234}}{\mathcal{U}} \right)^{\frac{D}{2}+1} \qquad (4.115)$$

we obtain

$$\tilde{I}_1 = \frac{g^{\mu\nu}}{2} e^{2\varepsilon\gamma_E} \int\limits_{\alpha_j \geq 0} d^7\alpha \, \frac{\alpha_4}{\mathcal{U}^{\frac{D}{2}+1}} e^{-\frac{\mathcal{F}}{\mathcal{U}}} = \frac{1}{2} g^{\mu\nu} \, I_{1112111}(D+2). \qquad (4.116)$$

Note that the powers of α_{1234} have cancelled out. As second term we consider

$$\tilde{I}_2 = -e^{2\varepsilon\gamma_E} \left(\mu^2\right)^{7-D} \int\limits_{\alpha_j \geq 0} d^7\alpha \int \frac{d^D k_1'}{i\pi^{\frac{D}{2}}} \frac{d^D k_2'}{i\pi^{\frac{D}{2}}} \frac{\left(\alpha_4 \alpha_1 p_1^\mu\right)\left(\alpha_4 \alpha_1 p_1^\nu\right)}{\mathcal{U}^2} e^{\alpha_{1234} k_1'^2 + \frac{\mathcal{U}}{\alpha_{1234}} k_2'^2 - \mu^2 \frac{\mathcal{F}}{\mathcal{U}}}$$

$$
\begin{aligned}
&= -p_1^\mu p_1^\nu e^{2\varepsilon\gamma_E} \left(\mu^2\right)^{7-D} \int\limits_{\alpha_j \geq 0} d^7\alpha \, \frac{\alpha_1^2 \alpha_4^2}{\mathcal{U}^2} \int \frac{d^D k_1'}{i\pi^{\frac{D}{2}}} \frac{d^D k_2'}{i\pi^{\frac{D}{2}}} e^{\alpha_{1234} k_1'^2 + \frac{\mathcal{U}}{\alpha_{1234}} k_2'^2 - \mu^2 \frac{\mathcal{F}}{\mathcal{U}}} \\
&= -p_1^\mu p_1^\nu e^{2\varepsilon\gamma_E} \int\limits_{\alpha_j \geq 0} d^7\alpha \, \frac{\alpha_1^2 \alpha_4^2}{\mathcal{U}^{\frac{D}{2}+2}} e^{-\frac{\mathcal{F}}{\mathcal{U}}} = -4 p_1^\mu p_1^\nu I_{3113111}(D+4) .
\end{aligned}
\tag{4.117}
$$

All other terms are similar to the last one. Thus we are able to express the tensor integral $I_{1111111}(D)[k_1^\mu k_2^\nu]$ in terms of scalar integrals.

4.4 Dimensional Regularisation and Spins

We introduced dimensional regularisation to regulate divergent Feynman integrals. In the calculation of amplitudes the tensor integrals are multiplied by loop momenta independent prefactors. In theories with particles with spin, these prefactors include the polarisation factors for the external particles (i.e., spinors for spin-$1/2$ fermions and polarisation vectors for spin-1 bosons). Within dimensional regularisation we have to specify how to continue these polarisation factors from four space-time dimensions to D space-time dimensions.

There are several schemes on the market which treat this issue differently. To discuss these schemes it is best to look how they treat the momenta and the polarisation factors of observed and unobserved particles. Unobserved particles are particles circulating inside loops or emitted particles not resolved within a given detector resolution. The most commonly used schemes are the **conventional dimensional regularisation scheme** (CDR) [25], where all momenta and all polarisation factors are taken to be in D dimensions (the momenta of the observed particles can be taken to lie in a four-dimensional sub-space of the D-dimensional space) and the **'t Hooft-Veltman scheme** (HV) [21, 78], where the momenta and the polarisation factors of the unobserved particles are D-dimensional, whereas the momenta and the polarisation factors of the observed particles are four-dimensional.

Let us also mention two further schemes, dimensional reduction and the four-dimensional helicity scheme. These two schemes introduce an additional space of dimension D_s. In the modern formulation of **dimensional reduction** (DRED) [79–81] all momenta are taken to be in D dimensions, whereas all polarisation factors are taken to be in D_s dimensions. As above, the momenta of the observed particles can be taken to lie in a four-dimensional sub-space of the D-dimensional space. In the **four-dimensional helicity scheme** (FDH) [26, 82, 83] the momenta of the unobserved particles are D-dimensional, the momenta of the observed particles are four-dimensional, the polarisation factors of the unobserved particles are D_s-dimensional and the polarisation factors of the observed particles are four-dimensional. Let us summarise:

Dimensional regularisation schemes for particles with spin:

	Momenta observed	Momenta unobserved	Polarisation observed	Polarisation unobserved
CDR	D	D	D	D
HV	4	D	4	D
DRED	D	D	D_s	D_s
FDH	4	D	4	D_s

One assumes that the four-dimensional space can be embedded into the D-dimensional space, and that the D-dimensional space can be embedded into the D_s-dimensional space. Thus, there is a projection from the D_s-dimensional space to the D-dimensional space, which forgets the $(D_s - D)$-dimensional components. Likewise, there is a projection from the D-dimensional space to the four-dimensional space, which forgets the $(D - 4)$-dimensional components. This implies for example the algebraic rules

$$g^{(4)}_{\mu\rho} g^{(D)\,\rho}{}_{\nu} = g^{(4)}_{\mu\nu}, \quad g^{(D)}_{\mu\rho} g^{(D_s)\,\rho}{}_{\nu} = g^{(D)}_{\mu\nu}, \quad g^{(4)}_{\mu\rho} g^{(D_s)\,\rho}{}_{\nu} = g^{(4)}_{\mu\nu}. \tag{4.118}$$

As mentioned in Sect. 2.4.2 we may realise spaces of non-integer dimensions as equivalence classes of tuples of vector space. In this construction, the quantities D_s, D and 4 corresponds to the rank of the tuples of vector spaces.

In dimensional reduction and in the four-dimensional helicity scheme the space of dimension D_s is only used for the polarisations. The final result will be an analytic function of D_s. As the polarisations only enter in the numerator, the limit $D_s \to 4$ will not lead to additional poles. Thus, we may take the limit $D_s \to 4$ at the end of the calculation without any problems. This also explains the name "four-dimensional helicity scheme".

It is possible to relate results obtained in one scheme to another scheme, using simple and universal transition formulae [84–86].

4.4.1 The Dirac Algebra Within Dimensional Regularisation

In four space-time dimensions the Dirac matrices are 4×4-matrices satisfying the anti-commutation relation

$$\left\{ \gamma^{\mu}_{(4)}, \gamma^{\nu}_{(4)} \right\} = 2\, g^{\mu\nu}_{(4)} \cdot \mathbf{1}, \tag{4.119}$$

where$\mathbf{1}$ denotes the unit matrix in spinor space. The hermitian properties are

$$\left(\gamma^{0}_{(4)}\right)^{\dagger} = \gamma^{0}_{(4)}, \quad \left(\gamma^{i}_{(4)}\right)^{\dagger} = -\gamma^{i}_{(4)}, \quad 1 \le i \le 3. \tag{4.120}$$

The matrix γ_5 is defined by

$$\gamma_5 = i\gamma^0_{(4)}\gamma^1_{(4)}\gamma^2_{(4)}\gamma^3_{(4)} = \frac{i}{24}\varepsilon_{\mu\nu\rho\sigma}\gamma^\mu_{(4)}\gamma^\nu_{(4)}\gamma^\rho_{(4)}\gamma^\sigma_{(4)}, \qquad (4.121)$$

where $\varepsilon_{\mu\nu\rho\sigma}$ denotes the totally anti-symmetric tensor with $\varepsilon_{0123} = 1$. The matrix γ_5 satisfies

$$\left\{\gamma^\mu_{(4)}, \gamma_5\right\} = 0, \quad \gamma_5^2 = \mathbf{1} \qquad (4.122)$$

and

$$\gamma_5^\dagger = \gamma_5. \qquad (4.123)$$

In evaluating traces of Dirac matrices we have the following rules:

1. Traces of an even number of Dirac matrices are evaluated with the rules

$$\mathrm{Tr}\left(\gamma^\mu_{(4)}\gamma^\nu_{(4)}\right) = 4g^{\mu\nu}_{(4)},$$

$$\mathrm{Tr}\left(\gamma^{\mu_1}_{(4)}\gamma^{\mu_2}_{(4)}...\gamma^{\mu_{2n}}_{(4)}\right) = \sum_{j=2}^{2n}(-1)^j\, g^{\mu_1\mu_j}_{(4)}\mathrm{Tr}\left(\gamma^{\mu_2}_{(4)}...\gamma^{\mu_{j-1}}_{(4)}\gamma^{\mu_{j+1}}_{(4)}...\gamma^{\mu_{2n}}_{(4)}\right). \qquad (4.124)$$

2. Traces of an odd number of Dirac matrices vanish:

$$\mathrm{Tr}\left(\gamma^{\mu_1}_{(4)}\gamma^{\mu_2}_{(4)}...\gamma^{\mu_{2n-1}}_{(4)}\right) = 0 \qquad (4.125)$$

3. For traces involving γ_5 we have

$$\mathrm{Tr}\left(\gamma_5\right) = 0,$$

$$\mathrm{Tr}\left(\gamma^\mu_{(4)}\gamma^\nu_{(4)}\gamma_5\right) = 0,$$

$$\mathrm{Tr}\left(\gamma^\mu_{(4)}\gamma^\nu_{(4)}\gamma^\rho_{(4)}\gamma^\sigma_{(4)}\gamma_5\right) = 4i\varepsilon^{\mu\nu\rho\sigma}. \qquad (4.126)$$

In particular, we have a non-zero value for $\mathrm{Tr}(\gamma^\mu_{(4)}\gamma^\nu_{(4)}\gamma^\rho_{(4)}\gamma^\sigma_{(4)}\gamma_5)$.

Exercise 31 *Prove Eqs. (4.124)–(4.126).*

Let us now consider the Dirac algebra in D dimensions. The generalisation of Eq. (4.119) reads

$$\left\{\gamma^\mu_{(D)}, \gamma^\nu_{(D)}\right\} = 2\, g^{\mu\nu}_{(D)} \cdot \mathbf{1}. \qquad (4.127)$$

This is unproblematic. **1** denotes again the unit matrix in spinor space. If D is a positive even integer, the standard representation of the Dirac matrices is given by matrices of size $2^{\frac{D}{2}} \times 2^{\frac{D}{2}}$. It is common practice to use for simplicity the convention that the trace of the unit matrix in spinor space equals

$$\mathrm{Tr}\,(\mathbf{1}) = 4, \tag{4.128}$$

and not the more natural choice from a mathematical point of view

$$\mathrm{Tr}\,(\mathbf{1}) = 2^{\frac{D}{2}}. \tag{4.129}$$

As a trace of Dirac matrices is always associated with a closed fermion loop, we may convert easily between these two conventions. We will follow standard conventions and use the convention of Eq. (4.128) from now on.

However, there is no way of continuing the definition of γ_5 to D dimensions, maintaining the cyclicity of the trace and the relations

$$\begin{aligned} \left\{\gamma^\mu_{(D)}, \gamma_5\right\} &= 0, && 0 \le \mu \le D-1, \\ \mathrm{Tr}\left(\gamma^\mu_{(D)}\gamma^\nu_{(D)}\gamma^\rho_{(D)}\gamma^\sigma_{(D)}\gamma_5\right) &= 4i\varepsilon^{\mu\nu\rho\sigma}, && \mu,\nu,\rho,\sigma \in \{0,1,2,3\}. \end{aligned} \tag{4.130}$$

In order to see this, consider

$$g^{(D)}_{\alpha\beta}\,\varepsilon_{\mu\nu\rho\sigma}\,\mathrm{Tr}\left(\gamma^\alpha_{(D)}\gamma^\mu_{(D)}\gamma^\nu_{(D)}\gamma^\rho_{(D)}\gamma^\sigma_{(D)}\gamma^\beta_{(D)}\gamma_5\right). \tag{4.131}$$

We first note that from Eq. (4.127) we have

$$\gamma^\mu_{(D)}\gamma^{(D)}_\mu = D\cdot\mathbf{1}. \tag{4.132}$$

We then evaluate the expression in Eq. (4.131) in two ways: We first use the cyclicity of the trace, anti-commute γ_5 with $\gamma^\alpha_{(D)}$ and use Eq. (4.132):

$$\begin{aligned} & g^{(D)}_{\alpha\beta}\,\varepsilon_{\mu\nu\rho\sigma}\,\mathrm{Tr}\left(\gamma^\alpha_{(D)}\gamma^\mu_{(D)}\gamma^\nu_{(D)}\gamma^\rho_{(D)}\gamma^\sigma_{(D)}\gamma^\beta_{(D)}\gamma_5\right) \\ &= g^{(D)}_{\alpha\beta}\,\varepsilon_{\mu\nu\rho\sigma}\,\mathrm{Tr}\left(\gamma^\mu_{(D)}\gamma^\nu_{(D)}\gamma^\rho_{(D)}\gamma^\sigma_{(D)}\gamma^\beta_{(D)}\gamma_5\gamma^\alpha_{(D)}\right) \\ &= -g^{(D)}_{\alpha\beta}\,\varepsilon_{\mu\nu\rho\sigma}\,\mathrm{Tr}\left(\gamma^\mu_{(D)}\gamma^\nu_{(D)}\gamma^\rho_{(D)}\gamma^\sigma_{(D)}\gamma^\beta_{(D)}\gamma^\alpha_{(D)}\gamma_5\right) \\ &= -D\varepsilon_{\mu\nu\rho\sigma}\,\mathrm{Tr}\left(\gamma^\mu_{(D)}\gamma^\nu_{(D)}\gamma^\rho_{(D)}\gamma^\sigma_{(D)}\gamma_5\right). \end{aligned} \tag{4.133}$$

On the other hand, we may anti-commute $\gamma^\alpha_{(D)}$ through the string $\gamma^\mu_{(D)}\gamma^\nu_{(D)}\gamma^\rho_{(D)}\gamma^\sigma_{(D)}$, followed by the application of Eq. (4.132):

$$\begin{aligned} & g^{(D)}_{\alpha\beta}\,\varepsilon_{\mu\nu\rho\sigma}\,\mathrm{Tr}\left(\gamma^\alpha_{(D)}\gamma^\mu_{(D)}\gamma^\nu_{(D)}\gamma^\rho_{(D)}\gamma^\sigma_{(D)}\gamma^\beta_{(D)}\gamma_5\right) \\ &= 2\varepsilon_{\mu\nu\rho\sigma}\left[\mathrm{Tr}\left(\gamma^\nu_{(D)}\gamma^\rho_{(D)}\gamma^\sigma_{(D)}\gamma^\mu_{(D)}\gamma_5\right) - \mathrm{Tr}\left(\gamma^\mu_{(D)}\gamma^\rho_{(D)}\gamma^\sigma_{(D)}\gamma^\nu_{(D)}\gamma_5\right) + \mathrm{Tr}\left(\gamma^\mu_{(D)}\gamma^\nu_{(D)}\gamma^\sigma_{(D)}\gamma^\rho_{(D)}\gamma_5\right)\right. \end{aligned}$$

$$
\begin{aligned}
& -\mathrm{Tr}\left(\gamma_{(D)}^{\mu}\gamma_{(D)}^{\nu}\gamma_{(D)}^{\rho}\gamma_{(D)}^{\sigma}\gamma_5\right)\Big] + D\varepsilon_{\mu\nu\rho\sigma}\,\mathrm{Tr}\left(\gamma_{(D)}^{\mu}\gamma_{(D)}^{\nu}\gamma_{(D)}^{\rho}\gamma_{(D)}^{\sigma}\gamma_5\right) \\
& = (D-8)\,\varepsilon_{\mu\nu\rho\sigma}\,\mathrm{Tr}\left(\gamma_{(D)}^{\mu}\gamma_{(D)}^{\nu}\gamma_{(D)}^{\rho}\gamma_{(D)}^{\sigma}\gamma_5\right). \qquad (4.134)
\end{aligned}
$$

Combining Eqs. (4.133) and (4.134) we arrive at

$$
2\,(D-4)\,\varepsilon_{\mu\nu\rho\sigma}\,\mathrm{Tr}\left(\gamma_{(D)}^{\mu}\gamma_{(D)}^{\nu}\gamma_{(D)}^{\rho}\gamma_{(D)}^{\sigma}\gamma_5\right) = 0. \qquad (4.135)
$$

At $D = 4$ this equation permits the usual non-zero trace of γ_5 with four other Dirac matrices. However, for $D \neq 4$ we conclude that the trace equals zero, and there is no smooth limit $D \to 4$ which reproduces the non-zero trace at $D = 4$.

Thus we cannot have simultaneously the cyclicity of the trace, the anti-commutation relation as in Eq. (4.130) and a non-zero value for the trace as in Eq. (4.130). On physical grounds we insist on keeping the non-zero value for the trace. It enters the theoretical description of certain decays of particles. If the trace would be zero, the predicted decay rate would be zero as well, in contradiction with experiment. We also would like to maintain the cyclicity of the trace. Therefore we have to give up the simple anti-commutation relation of γ_5 in D dimensions.

The 't Hooft-Veltman prescription [21] defines γ_5 as the product of the first four Dirac matrices in D dimensions:

$$
\gamma_5 = i\gamma_{(D)}^{0}\gamma_{(D)}^{1}\gamma_{(D)}^{2}\gamma_{(D)}^{3}. \qquad (4.136)
$$

With this definition, γ_5 anti-commutes with the first four Dirac matrices (as in four space-time dimensions), but commutes with the remaining ones:

$$
\begin{aligned}
\left\{\gamma_{(D)}^{\mu}, \gamma_5\right\} &= 0, \;\; \text{if} \;\; \mu \in \{0, 1, 2, 3\}, \\
\left[\gamma_{(D)}^{\mu}, \gamma_5\right] &= 0, \;\; \text{otherwise}.
\end{aligned} \qquad (4.137)
$$

Exercise 32 *Show that with the definitions and conventions as above the rules for the traces of Dirac matrices carry over to D dimensions. In detail, show:*

1. *Traces of an even number of Dirac matrices are evaluated with the rules*

$$
\begin{aligned}
\mathrm{Tr}\left(\gamma_{(D)}^{\mu}\gamma_{(D)}^{\nu}\right) &= 4g_{(D)}^{\mu\nu}, \\
\mathrm{Tr}\left(\gamma_{(D)}^{\mu_1}\gamma_{(D)}^{\mu_2}\ldots\gamma_{(D)}^{\mu_{2n}}\right) &= \sum_{j=2}^{2n}(-1)^j\, g_{(D)}^{\mu_1\mu_j}\,\mathrm{Tr}\left(\gamma_{(D)}^{\mu_2}\ldots\gamma_{(D)}^{\mu_{j-1}}\gamma_{(D)}^{\mu_{j+1}}\ldots\gamma_{(D)}^{\mu_{2n}}\right).
\end{aligned} \qquad (4.138)
$$

2. *Traces of an odd number of Dirac matrices vanish:*

$$
\mathrm{Tr}\left(\gamma_{(D)}^{\mu_1}\gamma_{(D)}^{\mu_2}\ldots\gamma_{(D)}^{\mu_{2n-1}}\right) = 0 \qquad (4.139)
$$

3. *For traces involving γ_5 we have*

$$\mathrm{Tr}\left(\gamma_5\right) = 0,$$
$$\mathrm{Tr}\left(\gamma^{\mu}_{(D)}\gamma^{\nu}_{(D)}\gamma_5\right) = 0,$$
$$\mathrm{Tr}\left(\gamma^{\mu}_{(D)}\gamma^{\nu}_{(D)}\gamma^{\rho}_{(D)}\gamma^{\sigma}_{(D)}\gamma_5\right) = \begin{cases} 4i\varepsilon^{\mu\nu\rho\sigma}, & \mu,\nu,\rho,\sigma \in \{0,1,2,3\}, \\ 0, & \textit{otherwise}. \end{cases} \tag{4.140}$$

Let us now look at the implications of the 't Hooft-Veltman prescription for γ_5. We first show that with this definition of γ_5 we correctly obtain the triangle anomaly within dimensional regularisation. On the other hand, the 't Hooft-Veltman prescription for γ_5 treats the first four indices differently from the remaining indices, as can be seen in the anti-commutation/commutation relations in Eq. (4.137). As such, the regularisation scheme breaks a symmetry of the original unregularised theory. This is unavoidable and not a problem. However, we have to include finite renormalisations, which restore Ward identities reflecting the original symmetry. We discuss this in the context of the non-singlet axial vector current.

From now on we always take the Dirac algebra in D dimensions and we drop the subscript (D).

4.4.1.1 The Singlet Axial-Vector Current and the Triangle Anomaly

The triangle anomaly for one axial-vector coupling and two vector couplings originates from the two diagrams shown in Fig. 4.3. For massless fermions we obtain for the sum of the two graphs (ignoring coupling factors)

$$A_{\alpha\beta\mu} = \int \frac{d^D k}{(2\pi)^D} \frac{N_{\alpha\beta\mu}}{q_0^2 q_1^2 q_2^2}, \tag{4.141}$$

where

$$N_{\alpha\beta\mu} = \mathrm{Tr}\left(\not{q}_1\gamma_\beta\not{q}_0\gamma_\alpha\not{q}_2\gamma_\mu\gamma_5\right) - \mathrm{Tr}\left(\not{q}_2\gamma_\alpha\not{q}_0\gamma_\beta\not{q}_1\gamma_\mu\gamma_5\right) \tag{4.142}$$

and $q_0 = k$, $q_1 = k - p_2$ and $q_2 = k + p_1$. It is convenient to calculate the graphs for the kinematic configuration where p_1^2, p_2^2 and $(p_1 + p_2)^2$ are non-zero. In that case there will be no infrared divergences, which are not relevant to the discussion of the anomaly. Contracting $A_{\alpha\beta\mu}$ with $(p_1 + p_2)^\mu$ gives the anomaly:

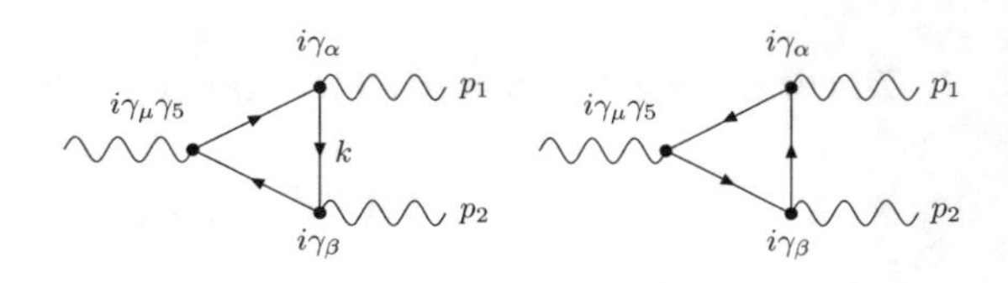

Fig. 4.3 The triangle graphs for the anomaly

$$A^{AVV} = (p_1 + p_2)^\mu A_{\alpha\beta\mu}. \qquad (4.143)$$

Let us now calculate the anomaly. In the first trace of $(p_1 + p_2)^\mu N_{\alpha\beta\mu}$ we use

$$\left(\not{p}_1 + \not{p}_2\right) \gamma_5 = \left(\not{q}_2 - \not{q}_1\right) \gamma_5 = \not{q}_2 \gamma_5 + \gamma_5 \not{q}_1 - 2\not{k}_{(-2\varepsilon)} \gamma_5. \qquad (4.144)$$

For the second trace we use

$$\left(\not{p}_1 + \not{p}_2\right) \gamma_5 = \left(\not{q}_2 - \not{q}_1\right) \gamma_5 = -\not{q}_1 \gamma_5 - \gamma_5 \not{q}_2 + 2\not{k}_{(-2\varepsilon)} \gamma_5. \qquad (4.145)$$

The terms $\not{q}_1 \not{q}_1$ and $\not{q}_2 \not{q}_2$ inside the traces cancel propagators and the resulting tensor bubble integrals can be shown to vanish after integration. Therefore the only relevant term is

$$-2\left(\text{Tr}\ \not{q}_1 \gamma_\beta \not{q}_0 \gamma_\alpha \not{q}_2 \not{k}_{(-2\varepsilon)} \gamma_5 + \text{Tr}\ \not{q}_2 \gamma_\alpha \not{q}_0 \gamma_\beta \not{q}_1 \not{k}_{(-2\varepsilon)} \gamma_5\right). \qquad (4.146)$$

The traces evaluate to

$$\begin{aligned}
\text{Tr}\ \not{q}_1 \gamma_\beta \not{q}_0 \gamma_\alpha \not{q}_2 \not{k}_{(-2\varepsilon)} \gamma_5 &= k^2_{(-2\varepsilon)} \cdot 4i \varepsilon_{\alpha\lambda\beta\kappa} p_1^\lambda p_2^\kappa + \cdots, \\
\text{Tr}\ \not{q}_2 \gamma_\alpha \not{q}_0 \gamma_\beta \not{q}_1 \not{k}_{(-2\varepsilon)} \gamma_5 &= k^2_{(-2\varepsilon)} \cdot 4i \varepsilon_{\alpha\lambda\beta\kappa} p_1^\lambda p_2^\kappa + \cdots,
\end{aligned} \qquad (4.147)$$

where the dots stand for terms, which vanish after integration.

Exercise 33 *Derive Eq. (4.147).*

We then obtain for the anomaly

$$A^{AVV} = 16 \varepsilon_{\alpha\lambda\beta\kappa} p_1^\lambda p_2^\kappa \int \frac{d^D k}{(2\pi)^D i} \frac{k^2_{(-2\varepsilon)}}{k_0^2 k_1^2 k_2^2} = \frac{1}{(4\pi)^2} 8 \varepsilon_{\alpha\beta\lambda\kappa} p_1^\lambda p_2^\kappa + O(\varepsilon), \qquad (4.148)$$

which is the well-known result for the anomaly in the 't Hooft-Veltman scheme [21].

Exercise 34 *Show*

$$\int \frac{d^D k}{(2\pi)^D i} \frac{k^2_{(-2\varepsilon)}}{k_0^2 k_1^2 k_2^2} = -\frac{1}{2} \frac{1}{(4\pi)^2} + O(\varepsilon). \qquad (4.149)$$

4.4.1.2 The Non-Singlet Axial-Vector Current

The Ward identity for the non-singlet axial-vector current for massless fermions reads

$$(p_1 - p_2)^\mu \Gamma_{\mu 5} = S_F^{-1}(p_1) \gamma_5 + \gamma_5 S_F^{-1}(p_2), \qquad (4.150)$$

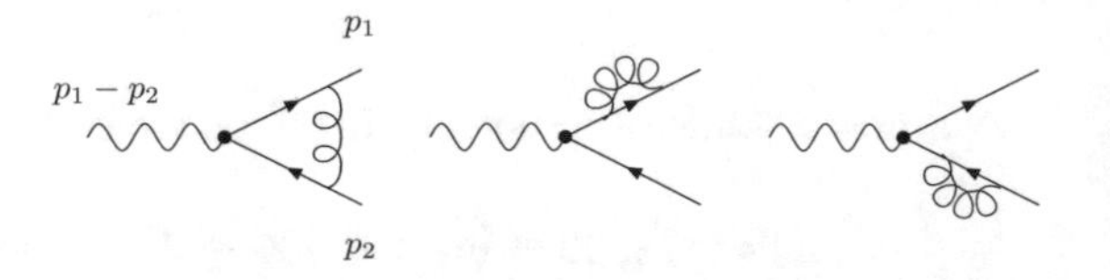

Fig. 4.4 Feynman graphs for the non-singlet axial-vector Ward identity

where $iS_F(p)$ denotes the full fermion propagator and $i\Gamma_{\mu 5}$ denotes the full $i\gamma_\mu\gamma_5$-vertex. We are now going to check the Ward identity at one-loop level. The relevant diagrams are shown in Fig. 4.4. The momentum p_1 is flowing outwards, whereas we take the momentum p_2 to be directed inwards. The one-loop contribution from the right-hand-side of Eq. (4.150) reads:

$$-\int \frac{d^D k}{(2\pi)^D i} \frac{\gamma_\nu \not{q}_2 \gamma^\nu}{q_0^2 q_2^2} \gamma_5 - \gamma_5 \int \frac{d^D k}{(2\pi)^D i} \frac{\gamma_\nu \not{q}_1 \gamma^\nu}{q_0^2 q_1^2} \gamma_5, \tag{4.151}$$

where we used the notation $q_0 = k$, $q_1 = k + p_2$ and $q_2 = k + p_1$. The contribution from the three-point diagram reads:

$$\int \frac{d^D k}{(2\pi)^D i} \frac{\gamma_\nu \not{q}_2 \gamma_\mu \gamma_5 \not{q}_1 \gamma^\nu}{q_0^2 q_1^2 q_2^2} \tag{4.152}$$

Contracting with $(p_1 - p_2)^\mu$ and rewriting $p_1 - p_2 = q_2 - q_1$ we obtain

$$\begin{aligned} \gamma_\nu \not{q}_2 (\not{p}_1 - \not{p}_2) \gamma_5 \not{q}_1 \gamma^\nu &= \frac{1}{2} \left[\gamma_\nu \not{q}_2 (\not{p}_1 - \not{p}_2) \gamma_5 \not{q}_1 \gamma^\nu - \gamma_\nu \not{q}_2 \gamma_5 (\not{p}_1 - \not{p}_2) \not{q}_1 \gamma^\nu \right] \\ &= \frac{1}{2} \left[\gamma_\nu \not{q}_2 (\not{q}_2 - \not{q}_1) \gamma_5 \not{q}_1 \gamma^\nu - \gamma_\nu \not{q}_2 \gamma_5 (\not{q}_2 - \not{q}_1) \not{q}_1 \gamma^\nu \right] \\ &= -q_1^2 \gamma_\nu \not{q}_2 \gamma^\nu \gamma_5 - q_2^2 \gamma_5 \gamma_\nu \not{q}_1 \gamma^\nu \\ &\quad + 4\varepsilon q_1^2 \not{q}_2 \gamma_5 + 4\varepsilon q_2^2 \gamma_5 \not{q}_1 - \left(k_{(-2\varepsilon)}\right)^2 \gamma_\nu \left(\gamma_5 \not{q}_1 + \not{q}_2 \gamma_5\right) \gamma^\nu . \end{aligned} \tag{4.153}$$

The two terms on the second-to-last line correspond exactly to the right-hand-side of Eq. (4.150). However, the terms on the last line spoil the Ward identity. These terms give the contribution

$$-4 \frac{1}{(4\pi)^2} (\not{p}_1 - \not{p}_2) \gamma_5 + O(\varepsilon). \tag{4.154}$$

In order to restore the Ward identity we have to perform a finite renormalisation on the non-singlet axial-vector current

$$\Gamma^r_{\mu 5} = Z_5^{ns} \Gamma^0_{\mu 5}. \tag{4.155}$$

For QCD the finite renormalisation Z_5^{ns} is given in the 't Hooft-Veltman scheme (including a factor $g^2 C_F$, where $C_F = \mathrm{Tr} T^a T^a$ is the fundamental Casimir of the gauge group) by

$$Z_5^{ns} = 1 - 4\frac{\alpha_s}{4\pi} C_F, \tag{4.156}$$

with $\alpha_s = g^2/(4\pi)$.

Chapter 5
One-Loop Integrals

The simplest, but most important loop integrals are the one-loop integrals. Within perturbative quantum field theory they enter the first quantum corrections. Contributions of this order are called **next-to-leading order (NLO)** contributions. The **leading-order (LO)** contribution is (usually) the tree-level approximation or Born approximation. As the perturbative expansion is an expansion in a small coupling, the next-to-leading order corrections are expected to give numerically the dominant corrections to the tree-level approximation.

The one-loop Feynman integrals are very well understood. They are simpler than the full class of all Feynman integrals. First of all, at one-loop there are no irreducible scalar products in the numerator. As a consequence, there is an alternative algorithm for the reduction of tensor integrals to scalar integrals. This algorithm is known as Passarino-Veltman reduction and discussed in Sect. 5.1. The Passarino-Veltman method does not shift the space-time dimension, nor does it raise the powers of the propagators.

In relativistic quantum field theory (and by using dimensional regularisation) we are usually interested in results for loop integrals in $D = 4 - 2\varepsilon$ space-time dimensions. A second important result states, that we may reduce any scalar one-loop integral to scalar one-loop integrals with no more than 5 external legs. Furthermore, we are usually only interested in the Laurent expansion up to and including the $O(\varepsilon^0)$-term. In this case, only scalar integrals with no more than 4 external legs are relevant. This is discussed in Sect. 5.2. The number of one-loop integrals is therefore finite, and they can be calculated (and have been calculated) once and for all. In Appendix B we provide the full list of scalar one-loop integrals for massless theories and give references, where results for one-loop integrals with internal masses can be found.

As mentioned above, for NLO calculations we are usually only interested in the Laurent expansion up to and including the $O(\varepsilon^0)$-term. We may ask, what transcendental functions appear in these terms. The answer at one-loop is amazingly simple: Up to and including the $O(\varepsilon^0)$-term there are just two transcendental functions. These

S. Weinzierl, *Feynman Integrals*, UNITEXT for Physics,
https://doi.org/10.1007/978-3-030-99558-4_5

are the logarithm and the dilogarithm

$$\text{Li}_1(x) = -\ln(1-x) = \sum_{n=1}^{\infty} \frac{x^n}{n}, \qquad \text{Li}_2(x) = \sum_{n=1}^{\infty} \frac{x^n}{n^2}. \quad (5.1)$$

In Sect. 5.3 we will study a Feynman integral, which leads to a dilogarithm. We will also discuss the properties of the dilogarithm.

The Passarino-Veltman method mentioned above is conceptually simple and historically important, but it also has a short-coming: Expressing a tensor integral as a linear combination of scalar integrals, the coefficients of this linear combination may contain Gram determinants in the denominator. This can lead to numerical instabilities in certain regions of phase-space. We discuss a more efficient method in Sect. 5.4.

At the end of the day our real interest are loop amplitudes, i.e. the sum of all relevant Feynman integrals. Of course, if we know how to compute all relevant Feynman integrals, we may sum up the individual results and obtain the loop amplitude. However, as the number of Feynman diagrams growths, this approach becomes inefficient and methods which directly deal with loop amplitudes are preferred. For one-loop amplitudes we have efficient methods which bypass individual Feynman diagrams. These methods exploit the fact that we know all relevant one-loop integrals, therefore only the coefficients in front of these integrals need to be determined. These methods are discussed in Sect. 5.5.

5.1 Passarino-Veltman Reduction

The Passarino-Veltman reduction method [87] reduces one-loop tensor integrals to scalar integrals and offers for one-loop integrals an alternative to the general method discussed in Sect. 4.3. The Passarino-Veltman reduction method exploits the fact, that at one-loop, any scalar product involving the loop momentum and the external momenta can be expressed as a linear combination of inverse propagators and a loop-momentum independent term. This is not true for a general Feynman integral beyond one-loop, as there might be irreducible scalar products. This is exactly the same issue as in our discussion of the Baikov representation in Sect. 2.5.5. In other words, the Passarino-Veltman reduction method exploits the fact that any one-loop graph G has a Baikov representation, which implies that we may express any scalar product involving the loop momentum and the external momenta as a linear combination of inverse propagators and a loop-momentum independent term. In particular, there is no need to consider a larger graph $\tilde{G}$.

Let us first introduce the Passarino-Veltman notation for one-loop tensor integrals. For scalar integrals with one, two or three external legs we write

$$A_0(m) = e^{\varepsilon \gamma_E} \mu^{2\varepsilon} \int \frac{d^D k}{i \pi^{D/2}} \frac{1}{(-k^2 + m^2)}, \quad (5.2)$$

$$B_0(p, m_1, m_2) = e^{\varepsilon \gamma_E} \mu^{2\varepsilon} \int \frac{d^D k}{i\pi^{D/2}} \frac{1}{(-k^2 + m_1^2)(-(k-p)^2 + m_2^2)},$$

$$C_0(p_1, p_2, m_1, m_2, m_3) =$$
$$e^{\varepsilon \gamma_E} \mu^{2\varepsilon} \int \frac{d^D k}{i\pi^{D/2}} \frac{1}{(-k^2 + m_1^2)(-(k-p_1)^2 + m_2^2)(-(k-p_1-p_2)^2 + m_3^2)},$$

with an obvious generalisation towards more external legs. Four-point functions are denoted with the letter D, five-point functions are denoted with the letter E, etc. For tensor integrals we use the notation $X^{\mu_1\mu_2\dots\mu_r}$, with $X \in \{A, B, C, \dots\}$ and the superscripts $\mu_1\mu_2\dots\mu_r$ indicate that the numerator is

$$k^{\mu_1} k^{\mu_2} \dots k^{\mu_r}. \tag{5.3}$$

To give an example:

$$\begin{aligned} B^{\mu_1}(p, m_1, m_2) &= e^{\varepsilon \gamma_E} \mu^{2\varepsilon} \int \frac{d^D k}{i\pi^{D/2}} \frac{k^{\mu_1}}{(-k^2 + m_1^2)(-(k-p)^2 + m_2^2)}, \\ B^{\mu_1\mu_2}(p, m_1, m_2) &= e^{\varepsilon \gamma_E} \mu^{2\varepsilon} \int \frac{d^D k}{i\pi^{D/2}} \frac{k^{\mu_1} k^{\mu_2}}{(-k^2 + m_1^2)(-(k-p)^2 + m_2^2)}. \end{aligned} \tag{5.4}$$

The reduction technique according to Passarino and Veltman uses the fact that due to Lorentz symmetry the result can only depend on tensor structures which can be build from the external momenta p_j^μ and the metric tensor $g^{\mu\nu}$. We therefore write the tensor integrals in the most general form in terms of form factors times external momenta and/or the metric tensor. For example

$$\begin{aligned} B^{\mu_1} &= p^{\mu_1} B_1, \\ B^{\mu_1\mu_2} &= p^{\mu_1} p^{\mu_2} B_{21} + g^{\mu_1\mu_2} B_{22}, \\ C^{\mu_1} &= p_1^{\mu_1} C_{11} + p_2^{\mu_1} C_{12}, \\ C^{\mu_1\mu_2} &= p_1^{\mu_1} p_1^{\mu_2} C_{21} + p_2^{\mu_1} p_2^{\mu_2} C_{22} + \left(p_1^{\mu_1} p_2^{\mu_2} + p_1^{\mu_2} p_2^{\mu_1}\right) C_{23} + g^{\mu_1\mu_2} C_{24}. \end{aligned} \tag{5.5}$$

One then solves for the form factors B_1, B_{21}, B_{22}, C_{11}, etc. by first contracting both sides with the external momenta and the metric tensor $g_{\mu\nu}$. On the left-hand side the resulting scalar products between the loop momentum k^μ and the external momenta are rewritten in terms of inverse propagators, as for example

$$2p \cdot k = \left[-(k-p)^2 + m_2^2\right] - \left[-k^2 + m_1^2\right] + \left(p^2 + m_1^2 - m_2^2\right). \tag{5.6}$$

The first two terms of the right-hand side above cancel propagators, whereas the last term does not involve the loop momentum any more. The remaining step is to solve for the form-factors by inverting the matrix which one obtains on the right-hand side of Eq. (5.5).

As an example we consider the two-point function: Contraction with p_{μ_1} or $p_{\mu_1} p_{\mu_2}$ and $g_{\mu_1 \mu_2}$ yields

$$p^2 B_1 = -\frac{1}{2} \left(\left(m_2^2 - m_1^2 - p^2 \right) B_0 - A_0(m_1) + A_0(m_2) \right),$$

$$\begin{pmatrix} p^2 & 1 \\ p^2 & D \end{pmatrix} \begin{pmatrix} B_{21} \\ B_{22} \end{pmatrix} = \begin{pmatrix} -\frac{1}{2}(m_2^2 - m_1^2 - p^2) B_1 - \frac{1}{2} A_0(m_2) \\ m_1^2 B_0 - A_0(m_2) \end{pmatrix}. \tag{5.7}$$

Solving for the form factors we obtain

$$\begin{aligned} B_1 &= -\frac{1}{2p^2} \left(\left(m_2^2 - m_1^2 - p^2 \right) B_0 - A_0(m_1) + A_0(m_2) \right), \\ B_{21} &= \frac{1}{(D-1)p^2} \left(-\frac{D}{2} (m_2^2 - m_1^2 - p^2) B_1 - m_1^2 B_0 - \frac{D-2}{2} A_0(m_2) \right), \\ B_{22} &= \frac{1}{2(D-1)} \left((m_2^2 - m_1^2 - p^2) B_1 + 2 m_1^2 B_0 - A_0(m_2) \right). \end{aligned} \tag{5.8}$$

Due to the matrix inversion in the last step determinants usually appear in the denominator of the final expression. For a three-point function we would encounter the Gram determinant

$$\det G\left(p_1, p_2\right) = \begin{vmatrix} -p_1^2 & -p_1 \cdot p_2 \\ -p_1 \cdot p_2 & -p_2^2 \end{vmatrix}. \tag{5.9}$$

One drawback of this algorithm is closely related to these determinants: In a phase space region where p_1 becomes collinear to p_2, the Gram determinant will tend to zero, and the form factors will take large values, with possible large cancellations among them. This makes it difficult to set up a stable numerical program for automated evaluation of tensor loop integrals. Methods to overcome this obstacle are discussed in Sect. 5.4.

Exercise 35 *Reduce the tensor integral*

$$A^{\mu_1 \mu_2 \mu_3 \mu_4}(m) = e^{\varepsilon \gamma_E} \mu^{2\varepsilon} \int \frac{d^D k}{i \pi^{D/2}} \frac{k^{\mu_1} k^{\mu_2} k^{\mu_3} k^{\mu_4}}{(-k^2 + m^2)} \tag{5.10}$$

to $A_0(m)$.

Exercise 36 *Reduce*

$$\begin{aligned} & g_{\mu_1 \mu_2} g_{\mu_3 \mu_4} C^{\mu_1 \mu_2 \mu_3 \mu_4}(p_1, p_2, 0, 0, 0) \\ & = -g_{\mu_1 \mu_2} g_{\mu_3 \mu_4} e^{\varepsilon \gamma_E} \mu^{2\varepsilon} \int \frac{d^D k}{i \pi^{D/2}} \frac{k^{\mu_1} k^{\mu_2} k^{\mu_3} k^{\mu_4}}{k^2 (k - p_1)^2 (k - p_1 - p_2)^2} \end{aligned} \tag{5.11}$$

to scalar integrals.

The Passarino-Veltman algorithm is based on the observation, that for one-loop integrals a scalar product of the loop momentum with an external momentum can be expressed as a combination of inverse propagators and a loop-momentum independent term. This property does no longer hold if one goes to two or more loops. If we consider Fig. 2.3 and Eq. (2.239), we see for example in the double-box Feynman integral that the scalar product

$$-k_2 \cdot p_1 \tag{5.12}$$

cannot be expressed in terms of inverse propagators and a loop-momentum independent term. In order to be able to express any scalar product involving the loop momenta and the external momenta as a linear combination of inverse propagators and a loop-momentum independent term we must introduce a larger graph $\tilde{G}$. We may still use the Passarino-Veltman ansatz based on Lorentz symmetry, that a tensor integral is written in terms of form factors times external momenta and/or the metric tensor. If one now tries to solve for the form factors by contracting with external momenta and the metric tensor, not all scalar products cancel propagators or give loop-momentum independent terms. Inverse propagators related to the propagators in $\tilde{G}$, which are not in the original graph G, may remain in the numerator. These inverse propagators in the numerator are called irreducible scalar products. In the notation of Eq. (2.150)

$$I_{\nu_1 \dots \nu_{n_{\mathrm{int}}}}\left(D, x_1, \dots, x_{N_B}\right) \tag{5.13}$$

for a Feynman integral associated to the graph $\tilde{G}$ these irreducible scalar products correspond to negative integer values for some ν_j's.

5.2 Reduction of Higher Point Integrals

In this section we discuss the reduction of scalar one-loop integrals with n_{ext} external legs to a set of scalar one-, two-, three-, four- and five-point functions. By an appropriate choice of the basis integrals for the five-point functions it can be arranged, that the five-point functions only contribute at order $O(\varepsilon)$ and are thus not relevant to NLO calculations, where we only need the ε-expansion of scalar one-loop integrals up to and including $O(\varepsilon^0)$. Thus, we may reduce any one-loop Feynman integral to scalar one-, two-, three- and four-point functions plus terms of order $O(\varepsilon)$ and beyond, which are not relevant to NLO calculations. This is a finite set of one-loop Feynman integrals, which can (and has been) calculated once and for all.

The reason a scalar one-loop n_{ext}-point function can be reduced to scalar one-loop Feynman integrals with no more than five external legs is the following: We assume all external momenta to lie in a four-dimensional space. Thus, even for $n_{\mathrm{ext}} > 5$, the external momenta span maximally a space of dimension 4. The scalar one-loop n-point functions with $n_{\mathrm{ext}} \geq 5$ are always ultraviolet-finite, but they may have infrared-divergences. Let us first assume that there are no IR-divergences. Then the integral is finite and can be performed in four dimensions. In a space of four dimensions we can have no more than four linearly independent vectors, therefore it comes to no surprise that in a one-loop integral with five or more propagators, one propagator can be expressed through the remaining ones. This is the basic idea for the reduction of the higher point scalar integrals. For infrared-finite integrals this fact has been known for a long time [88, 89]. For infrared-divergent integrals we have to use a regulator. With slight modifications the basic idea above can be generalised to dimensional regularisation [90–95]. Within dimensional regularisation, the external momenta and the loop momentum span maximally a space of dimension 5.

Let us now look at the details. We discuss the method for massless one-loop integrals. For one-loop integrals we have $n_{\mathrm{ext}} = n_{\mathrm{int}}$. The first step is to set up an appropriate notation. In this section we denote a scalar one-loop integral with n_{ext} legs (and massless propagators) by

$$I_{n_{\mathrm{ext}}}(D) = e^{\varepsilon\gamma_{\mathrm{E}}}\mu^{2\varepsilon} \int \frac{d^D k}{i\pi^{\frac{D}{2}}} \frac{1}{(-k^2)(-(k-p_1)^2)...(-(k-p_1-...p_{n_{\mathrm{ext}}-1})^2)}. \tag{5.14}$$

In terms of our previous notation

$$I_{n_{\mathrm{ext}}}(D) = \left(\mu^2\right)^{\frac{D_{\mathrm{int}}}{2}-n_{\mathrm{ext}}} I_{\underbrace{11...1}_{n_{\mathrm{ext}}}}(D). \tag{5.15}$$

We denote by $I^{(i)}_{n_{\mathrm{ext}}-1}(D)$ the scalar one-loop integral, where the i'th propagator has been removed:

$$I^{(i)}_{n_{\mathrm{ext}}-1}(D) = \left(\mu^2\right)^{\frac{D_{\mathrm{int}}}{2}-(n_{\mathrm{ext}}-1)} I_{\underbrace{1...1}_{i-1}0\underbrace{1...1}_{n_{\mathrm{ext}}-i}}(D). \tag{5.16}$$

Let us also introduce the sums of the external momenta:

$$p_i^{\mathrm{sum}} = \sum_{j=1}^{i} p_j, \qquad 1 \leq i \leq n_{\mathrm{ext}} - 1. \tag{5.17}$$

We associate two matrices S and G to the integral in Eq. (5.14). The entries of the $n_{\mathrm{ext}} \times n_{\mathrm{ext}}$ kinematic matrix S are given by

$$S_{ij} = \left(p_i^{\mathrm{sum}} - p_j^{\mathrm{sum}}\right)^2, \tag{5.18}$$

and the entries of the $(n_{\mathrm{ext}} - 1) \times (n_{\mathrm{ext}} - 1)$ Gram matrix are defined by

$$G_{ij} = -p_i^{\mathrm{sum}} \cdot p_j^{\mathrm{sum}}. \tag{5.19}$$

For the reduction one distinguishes three different cases: Scalar pentagons (i.e. scalar five-point functions), scalar hexagons (scalar six-point functions) and scalar integrals with more than six propagators.

Let us start with the pentagon. A five-point function in $D = 4 - 2\varepsilon$ dimensions can be expressed as a sum of four-point functions, where one propagator is removed, plus a five-point function in $6 - 2\varepsilon$ dimensions [90]. Since the $(6 - 2\varepsilon)$-dimensional pentagon is finite and comes with an extra factor of ε in front, it does not contribute at $O(\varepsilon^0)$. In detail we have:

Reduction of the massless five-point integral:

$$I_5(4-2\varepsilon) = -2\varepsilon B I_5(6-2\varepsilon) - \sum_{i=1}^{5} b_i I_4^{(i)}(4-2\varepsilon)$$
$$= -\sum_{i=1}^{5} b_i I_4^{(i)}(4-2\varepsilon) + O(\varepsilon), \quad (5.20)$$

where the coefficients B and b_i are obtained from the kinematic matrix S_{ij} as follows:

$$b_i = \sum_{j=1}^{5} \left(S^{-1}\right)_{ij}, \quad B = \sum_{i=1}^{5} b_i. \quad (5.21)$$

In Eq. (5.20) $I_5(6 - 2\varepsilon)$ denotes the $(6 - 2\varepsilon)$-dimensional pentagon and $I_4^{(i)}(4 - 2\varepsilon)$ denotes the four-point function, which is obtained from the pentagon by removing propagator i. The proof of Eq. (5.20) (as the proofs of Eqs. (5.22) and (5.24) below) uses integration-by-parts identities, which will be introduced in Chap. 6.

The six-point function can be expressed as a sum of five-point functions [91] without any correction of $O(\varepsilon)$:

Reduction of the massless six-point integral:

$$I_6(4-2\varepsilon) = -\sum_{i=1}^{6} b_i I_5^{(i)}(4-2\varepsilon). \quad (5.22)$$

The coefficients b_i are again related to the kinematic matrix S_{ij}:

$$b_i = \sum_{j=1}^{6} \left(S^{-1}\right)_{ij}. \quad (5.23)$$

For the seven-point function and beyond we can again express the n_{ext}-point function as a sum over $(n_{\text{ext}} - 1)$-point functions [94]:

Reduction of the massless n_{ext}-point integral ($n_{\text{ext}} \geq 7$):

$$I_{n_{\text{ext}}}(4-2\varepsilon) = -\sum_{i=1}^{n_{\text{ext}}} r_i I^{(i)}_{n_{\text{ext}}-1}(4-2\varepsilon), \tag{5.24}$$

where the coefficients r_i are defined below in Eq. (5.26).

In contrast to Eq. (5.22), the decomposition in Eq. (5.24) is no longer unique. A possible set of coefficients r_i can be obtained from the singular value decomposition of the Gram matrix

$$G_{ij} = \sum_{k=1}^{4} U_{ik} w_k \left(V^\dagger\right)_{kj}. \tag{5.25}$$

as follows [96]

$$r_i = \frac{V_{i5}}{W_5}, \quad 1 \leq i \leq n_{\text{ext}} - 1, \quad r_{n_{\text{ext}}} = -\sum_{j=1}^{n_{\text{ext}}-1} r_j, \quad W_5 = -\sum_{j=1}^{n_{\text{ext}}-1} G_{jj} V_{j5}. \tag{5.26}$$

Digression Singular value decomposition

Let M be a complex $m \times n$-matrix of rank r. The singular value decomposition of M is a decomposition of the form

$$M = U \Sigma V^\dagger, \tag{5.27}$$

where U is a unitary $m \times m$-matrix, V is a unitary $n \times n$-matrix, $V^\dagger$ denotes the Hermitian transpose of V and Σ is a real $m \times n$-matrix of the form

$$\Sigma = \left(\begin{array}{ccc|ccc} w_1 & & & & \vdots & \\ & \ddots & & \dots & 0 & \dots \\ & & w_r & & \vdots & \\ \hline & \vdots & & & \vdots & \\ \dots & 0 & \dots & \dots & 0 & \dots \\ & \vdots & & & \vdots & \end{array} \right), \tag{5.28}$$

with $w_k > 0$. The diagonal entries w_k are called the singular values of M. If M is real, the matrices U and V can be chosen as orthogonal matrices.

5.3 The Basic One-Loop Integrals

With the results of the previous two sections, we may reduce any one-loop Feynman integral to scalar one-, two-, three- and four-point functions plus terms of order $O(\varepsilon)$ and beyond, which are not relevant to NLO calculations. This is a finite set of one-loop Feynman integrals, which can (and has been) calculated once and for all. We have already calculated the massive one-loop one-point function in Eq. (2.124) (the massless one-loop one-point function is zero, see Eq. (2.137)) and the massless one-loop two-point function in Eq. (2.182). In the results of Eqs. (2.124) and (2.182) we already saw the appearance of a logarithm ($\ln(m^2/\mu^2)$ for the tadpole and $\ln(-p^2/\mu^2)$ for the massless bubble). It turns out that up to order $O(\varepsilon)$ there is just another transcendental function, which we have to know: This is Euler's dilogarithm. As an example for the appearance of the dilogarithm let us discuss the one-loop three-point function with no internal masses and the kinematic configuration

$$p_1^2 \neq 0, \quad p_2^2 \neq 0, \quad p_3^2 = (p_1+p_2)^2 \neq 0. \tag{5.29}$$

We consider the integral

$$I_3 = e^{\varepsilon\gamma_E}\mu^{2\varepsilon}\int \frac{d^Dk}{i\pi^{D/2}} \frac{1}{(-k^2)(-(k-p_1)^2)(-(k-p_1-p_2)^2)}. \tag{5.30}$$

The integral is finite and can be evaluated in four dimensions. In the Feynman parameter representation the Feynman integral is given by

$$I_3 = \int\limits_0^1 da_1 \int\limits_0^{a_1} da_2 \frac{1}{-a_1^2p_3^2 - a_2^2p_2^2 + a_1a_2(p_1^2-p_2^2-p_3^2) - a_1p_3^2 + a_2(p_3^2-p_1^2)} + O(\varepsilon). \tag{5.31}$$

We follow here closely the original work of 't Hooft and Veltman [97]. We make the change of variables $a_2' = a_2 - \alpha a_1$ and choose α as a root of the equation

$$-\alpha^2 p_2^2 + \alpha\left(p_1^2 - p_2^2 - p_3^2\right) - p_3^2 = 0. \tag{5.32}$$

With this choice we eliminate the quadratic term in a_1. We then perform the a_1-integration and we end up with three integrals of the form

$$\int\limits_0^1 \frac{dt}{t-t_0}\left[\ln\left(at^2+bt+c\right) - \ln\left(at_0^2+bt_0+c\right)\right]. \tag{5.33}$$

Factorising the arguments of the logarithms, these integrals are reduced to the type

$$R = \int_0^1 \frac{dt}{t - t_0} \left[\ln (t - t_1) - \ln (t_0 - t_1) \right]. \tag{5.34}$$

This integral is expressed in terms of a new function, the dilogarithm, as follows:

$$R = \text{Li}_2 \left(\frac{t_0}{t_1 - t_0} \right) - \text{Li}_2 \left(\frac{t_0 - 1}{t_1 - t_0} \right), \tag{5.35}$$

provided $-t_1$ and $1/(t_0 - t_1)$ have imaginary part of opposite sign, otherwise additional logarithms occur.

Digression The dilogarithm
The dilogarithm is defined by

$$\text{Li}_2(x) = - \int_0^1 dt \frac{\ln(1 - xt)}{t} = - \int_0^x dt \frac{\ln(1 - t)}{t}. \tag{5.36}$$

If we take the main branch of the logarithm with a cut along the negative real axis, then the dilogarithm has a cut along the positive real axis, starting at the point $x = 1$*. For* $|x| \leq 1$ *the dilogarithm has the power series expansion*

$$\text{Li}_2(x) = \sum_{n=1}^{\infty} \frac{x^n}{n^2}. \tag{5.37}$$

Some important numerical values are

$$\text{Li}_2(0) = 0, \quad \text{Li}_2(1) = \frac{\pi^2}{6}, \quad \text{Li}_2(-1) = -\frac{\pi^2}{12}, \quad \text{Li}_2 \left(\frac{1}{2} \right) = \frac{\pi^2}{12} - \frac{1}{2} (\ln 2)^2. \tag{5.38}$$

The dilogarithm with argument x *can be related to the dilogarithms with argument* $(1 - x)$ *or* $1/x$*:*

$$\begin{aligned} \text{Li}_2(x) &= -\text{Li}_2(1 - x) + \frac{1}{6}\pi^2 - \ln(x) \ln(1 - x), \\ \text{Li}_2(x) &= -\text{Li}_2 \left(\frac{1}{x} \right) - \frac{1}{6}\pi^2 - \frac{1}{2} (\ln(-x))^2. \end{aligned} \tag{5.39}$$

Another important relation is the five-term relation:

$$\text{Li}_2(xy) = \text{Li}_2(x) + \text{Li}_2(y) + \text{Li}_2 \left(\frac{xy - x}{1 - x} \right) + \text{Li}_2 \left(\frac{xy - y}{1 - y} \right) + \frac{1}{2} \ln^2 \left(\frac{1 - x}{1 - y} \right). \tag{5.40}$$

In Appendix B we provide a list of all basic one-loop integrals for massless theories up to order $O(\varepsilon)$ and references for the basic integrals with internal masses.

5.4 Spinor Techniques

The reduction methods for one-loop tensor integrals discussed in Sect. 5.1 (and in Sect. 4.3) are rather general and independent of the tensor structure into which the tensor integral is contracted. By taking into account information from this external tensor structure, more efficient reduction algorithms can be derived [98–103]. These algorithms significantly soften the problem with Gram determinants inherent in the Passarino-Veltman tensor reduction method. We will discuss as an example a method for one-loop integrals with massless propagators. The method is most conveniently explained within the FDH-scheme of dimensional regularisation. A generic one-loop tensor integral of rank r is denoted by

$$
\begin{aligned}
& I_{n_{\text{ext}}}^{\mu_1 \ldots \mu_r}(D) \\
& = e^{\varepsilon \gamma_{\mathrm{E}}} \mu^{2\varepsilon} \int \frac{d^D k}{i \pi^{\frac{D}{2}}} \frac{k^{\mu_1} \ldots k^{\mu_r}}{(-k^2)(-(k-p_1)^2) \ldots (-(k-p_1-\ldots p_{n_{\text{ext}}-1})^2)}.
\end{aligned} \tag{5.41}
$$

Let us assume that this integral is contracted into $J_{\mu_1 \ldots \mu_r}$, e.g., we are considering

$$
J_{\mu_1 \ldots \mu_r} I_{n_{\text{ext}}}^{\mu_1 \ldots \mu_r}(D). \tag{5.42}
$$

The tensor $J_{\mu_1 \ldots \mu_r}$ does not depend on the loop momentum k. In the FDH-scheme we can assume without loss of generality that the tensor structure $J_{\mu_1 \ldots \mu_r}$ is given by

$$
J_{\mu_1 \ldots \mu_r} = \langle a_1 - | \gamma_{\mu_1} | b_1 - \rangle \ldots \langle a_r - | \gamma_{\mu_r} | b_r - \rangle, \tag{5.43}
$$

where $\langle a_i - |$ and $| b_j - \rangle$ are Weyl spinors of definite helicity. Spinors are reviewed in Appendix A. Therefore we consider tensor integrals of the form

$$
\begin{aligned}
I_{n_{\text{ext}}}^r = {} & e^{\varepsilon \gamma_{\mathrm{E}}} \mu^{2\varepsilon} \langle a_1 - | \gamma_{\mu_1} | b_1 - \rangle \ldots \langle a_r - | \gamma_{\mu_r} | b_r - \rangle \\
& \int \frac{d^D k}{i \pi^{\frac{D}{2}}} \frac{k_{(4)}^{\mu_1} \ldots k_{(4)}^{\mu_r}}{(-k^2)(-(k-p_1)^2) \ldots (-(k-p_1-\ldots p_{n_{\text{ext}}-1})^2)},
\end{aligned} \tag{5.44}
$$

where $k_{(4)}^{\mu}$ denotes the projection of the D dimensional vector k^{μ} onto the four-dimensional subspace. The quantity $\langle a - | \gamma_{\mu} | b - \rangle$ is a vector in a complex vector-space of dimension 4 and can therefore be expressed as a linear combination of four basis vectors.

The first step for the construction of the reduction algorithm based on spinor methods is to associate to each n-point loop integral a pair of two **light-like** momenta l_1 and l_2, which are linear combinations of two external momenta p_i and p_j of the loop integral under consideration [101]. Obviously, this construction only makes sense for three-point integrals and beyond, as for two-point integrals there is only one independent external momentum. This is not a limitation, tensor two-point functions can be reduced with the Passarino-Veltman technique. The only Gram determinant occurring in this process is the determinant of the 1×1-matrix $G = -p^2$, where p denotes the external momentum of the two-point function. This is harmless.

For three-point functions and beyond we write

$$l_1 = \frac{1}{1-\alpha_1\alpha_2}\left(p_i - \alpha_1 p_j\right), \quad l_2 = \frac{1}{1-\alpha_1\alpha_2}\left(-\alpha_2 p_i + p_j\right), \tag{5.45}$$

where α_1 and α_2 are two constants, which can be determined from p_i and p_j.

Exercise 37 *Determine the constants α_1 and α_2 in Eq. (5.45) from the requirement that l_1 and l_2 are light-like, i.e. $l_1^2 = l_2^2 = 0$. Distinguish the cases*

(i) p_i and p_j are light-like.
(ii) p_i is light-like, p_j is not.
(iii) both p_i and p_j are not light-like.

In the second step we use l_1 and l_2 to write $\langle a - | \gamma_\mu | b - \rangle$ as a linear combination of the four basis vectors

$$\langle l_1 - | \gamma_\mu | l_1 - \rangle, \quad \langle l_2 - | \gamma_\mu | l_2 - \rangle, \quad \langle l_1 - | \gamma_\mu | l_2 - \rangle, \quad \langle l_2 - | \gamma_\mu | l_1 - \rangle. \tag{5.46}$$

The contraction of $k_{(4)}^\mu$ with the first or second basis vector leads to

$$\begin{aligned} \langle l_1 - | \gamma_\mu | l_1 - \rangle k_{(4)}^\mu = 2k_l l_1 &= \frac{1}{1-\alpha_1\alpha_2}\left(2p_i k - \alpha_1 2 p_j k\right), \\ \langle l_2 - | \gamma_\mu | l_2 - \rangle k_{(4)}^\mu = 2k_l l_2 &= \frac{1}{1-\alpha_1\alpha_2}\left(-\alpha_2 2 p_i k + 2 p_j k\right), \end{aligned} \tag{5.47}$$

and therefore reduces immediately the rank of the tensor integral. Repeating this procedure we end up with integrals, where the numerator is given by products of

$$\langle l_1 - | \not{k}_{(4)} | l_2 - \rangle \text{ and } \langle l_2 - | \not{k}_{(4)} | l_1 - \rangle, \tag{5.48}$$

plus additional reduced integrals. Therefore the tensor integral is now in a standard form. In the next step one reduces any product of factors as in Eq. (5.48). For example, if in the tensor structure both spinor types appear, we can use

$$\left\langle l_1 - | \not{k}^{(4)} | l_2 - \right\rangle \left\langle l_2 - | \not{k}^{(4)} | l_1 - \right\rangle = (2 l_1 k)\,(2 l_2 k) - (2 l_1 l_2)\left(k^{(4)}\right)^2 \tag{5.49}$$

and

$$\left(k^{(4)}\right)^2 = \left(k^{(D)}\right)^2 - \left(k^{(-2\varepsilon)}\right)^2. \tag{5.50}$$

After repeated use of Eq. (5.49) we end up with a tensor integral, where only one spinor type appears. These tensor integrals can be reduced with formulae, which depend on the number of external legs. These formulae are not reproduced here, but can be found in the literature [103]. After completion of this step, all tensor integrals are reduced to rank 1 integrals. Finally, the rank 1 integrals are reduced to scalar integrals. The relevant formulae for the last step depend again on the number of external legs and can be found in the literature [103].

Therefore the only non-zero higher-dimensional integrals which occur in the Feynman gauge result from the two-point function with a single power of $k_{(-2\varepsilon)}^2$ in the numerator ($n = 2$ and $s = 1$), the three-point function with a single power of $k_{(-2\varepsilon)}^2$ in the numerator ($n = 3$ and $s = 1$) and the four-point function with two powers of $k_{(-2\varepsilon)}^2$ in the numerator ($n = 4$ and $s = 2$). With $q_j = k - p_j^{\mathrm{sum}}$ we find for these cases:

$$e^{\varepsilon\gamma_E}\mu^{2\varepsilon}\int\frac{d^Dk}{i\pi^{\frac{D}{2}}}\frac{\left(-k_{(-2\varepsilon)}^2\right)}{\left(-q_1^2\right)\left(-q_2^2\right)} = -\frac{p^2}{6}+\mathcal{O}(\varepsilon),$$
$$e^{\varepsilon\gamma_E}\mu^{2\varepsilon}\int\frac{d^Dk}{i\pi^{\frac{D}{2}}}\frac{\left(-k_{(-2\varepsilon)}^2\right)}{\left(-q_1^2\right)\left(-q_2^2\right)\left(-q_3^2\right)} = -\frac{1}{2}+\mathcal{O}(\varepsilon),$$
$$e^{\varepsilon\gamma_E}\mu^{2\varepsilon}\int\frac{d^Dk}{i\pi^{\frac{D}{2}}}\frac{\left(-k_{(-2\varepsilon)}^2\right)^2}{\left(-q_1^2\right)\left(-q_2^2\right)\left(-q_3^2\right)\left(-q_4^2\right)} = -\frac{1}{6}+\mathcal{O}(\varepsilon). \tag{5.51}$$

Contributions of this type are called **rational terms** (as they do not involve logarithms or dilogarithms at order $\mathcal{O}(\varepsilon^0)$).

Exercise 38 *The method above does not apply to a tensor two-point function, as there is only one linear independent external momentum. However, the tensor two-point functions is easily reduced with standard methods to the scalar two-point function. In this exercise you are asked to work this out for the massless tensor two-point function. The most general massless tensor two-point function is given by*

$$I_2^{\mu_1\dots\mu_r,s} = e^{\varepsilon\gamma_E}\mu^{2\varepsilon}\int\frac{d^Dk}{i\pi^{\frac{D}{2}}}\left(-k_{(-2\varepsilon)}^2\right)^s\frac{k^{\mu_1}\dots k^{\mu_r}}{k^2(k-p)^2}. \tag{5.52}$$

Reduce this tensor integral to a scalar integral.

5.5 Amplitude Methods

It is the scattering amplitude $\mathcal{A}_{n_{\mathrm{ext}}}$ which enters the formula Eq. (4.10) for the expectation value of an observable. By Algorithm 1 the scattering amplitude is computed through the sum of all relevant Feynman diagrams. However, the number of Feynman diagrams growths rapidly with the number of external particles and this approach can become inefficient in practice. Fortunately, we have for one-loop amplitudes methods which bypass individual Feynman diagrams.

From Sects. 5.1 and 5.2 we know that we can reduce any one-loop tensor integral to a set of scalar one-, two-, three-, four- and five-point functions. By an appropriate choice of the basis integrals for the five-point functions it can be arranged, that the five-point functions only contribute at order $O(\varepsilon)$ and are thus not relevant to NLO calculations, where we only need the ε-expansion of scalar one-loop integrals up to and including $O(\varepsilon^0)$. Thus, we may reduce any one-loop Feynman integral to scalar one-, two-, three- and four-point functions plus terms of order $O(\varepsilon)$ and beyond, which are not relevant to NLO calculations. These scalar integrals are known (see Sect. 5.3), therefore only the coefficients of these integrals need to be determined.

In this section it is convenient to use the notation

$$I_n^{(i_1 \dots i_n)} = e^{\varepsilon \gamma_E} \mu^{2\varepsilon} \int \frac{d^D k}{i \pi^{\frac{D}{2}}} \frac{1}{\left(-q_{i_1}^2\right) \dots \left(-q_{i_n}^2\right)}, \tag{5.53}$$

where the superscript $(i_1 \dots i_n)$ indicate the propagators present in the one-loop Feynman integral. With this notation we may write for a one-loop amplitude in a massless theory

$$\mathcal{A}_{n_{\text{ext}}}^{(1)} = \sum_{i_1 < i_2 < i_3 < i_4} c_{i_1 i_2 i_3 i_4} I_4^{(i_1 i_2 i_3 i_4)} + \sum_{i_1 < i_2 < i_3} c_{i_1 i_2 i_3} I_3^{(i_1 i_2 i_3)} + \sum_{i_1 < i_2} c_{i_1 i_2} I_2^{(i_1 i_2)} + O(\varepsilon). \tag{5.54}$$

$I_2^{(i_1 i_2)}$, $I_3^{(i_1 i_2 i_3)}$ and $I_4^{(i_1 i_2 i_3 i_4)}$ are the scalar bubble, triangle and box integral functions. In a massive theory we would have in addition also scalar one-point functions. In a massless theory these functions are zero within dimensional regularisation. Note that there are no integral functions with more than four internal propagators. These higher-point functions can always be reduced to the set above, as we have seen in Sect. 5.2. The coefficients $c_{i_1 i_2}$, $c_{i_1 i_2 i_3}$ and $c_{i_1 i_2 i_3 i_4}$ depend on the external momenta and the dimensional regularisation parameter ε. All poles in the dimensional regularisation parameter ε are contained in the scalar integral functions. The coefficients $c_{i_1 i_2}$, $c_{i_1 i_2 i_3}$ and $c_{i_1 i_2 i_3 i_4}$ have a Taylor expansion in ε. We write

$$c_{i_1 \dots i_n} = c_{i_1 \dots i_n}^{(0)} + O(\varepsilon), \tag{5.55}$$

where $c_{i_1 \dots i_n}^{(0)}$ is the coefficient in four space-time dimensions. We therefore have

$$\mathcal{A}_{n_{\text{ext}}}^{(1)} = \sum_{i_1 < i_2 < i_3 < i_4} c_{i_1 i_2 i_3 i_4}^{(0)} I_4^{(i_1 i_2 i_3 i_4)} + \sum_{i_1 < i_2 < i_3} c_{i_1 i_2 i_3}^{(0)} I_3^{(i_1 i_2 i_3)} + \sum_{i_1 < i_2} c_{i_1 i_2}^{(0)} I_2^{(i_1 i_2)} + R + O(\varepsilon), \tag{5.56}$$

where the correction term R contains all terms up to order ε^0 originating from the $O(\varepsilon)$-terms in Eq. (5.55) hitting a pole in ε from the scalar integral functions. It can be shown that R is of order ε^0 and does not contain any logarithms. R is called the **rational term**. The set of all occurring integral functions

$$\{I_2^{(i_1 i_2)}, I_3^{(i_1 i_2 i_3)}, I_4^{(i_1 i_2 i_3 i_4)}\} \tag{5.57}$$

is rather easily obtained from pinching in all possible ways internal propagators in all occurring diagrams. We can assume that we know this set in advance. Furthermore all integral functions in this set are known (see Sect. 5.3). To compute the amplitude requires therefore only the determination of the coefficients $c^{(0)}_{i_1 i_2}$, $c^{(0)}_{i_1 i_2 i_3}$, $c^{(0)}_{i_1 i_2 i_3 i_4}$ and of the rational term R.

Below we discuss methods how this information can be obtained. Readers only interested in the most practical method may directly jump to Sect. 5.5.3, where we discuss the Ossola-Papadopoulos-Pittau (OPP) method. We follow the historical path and discuss first the unitarity-based method, followed by a discussion of generalised unitarity before finally arriving at the OPP-method. We do this, because some of the ideas like generalised cuts will reappear in the next chapter.

5.5.1 The Unitarity-Based Method

Loop amplitudes have branch cuts. We denote the discontinuity across a branch cut in a particular channel s by

$$\text{Disc}_s\, \mathcal{A}^{(l)}_{n_{\text{ext}}} = \mathcal{A}^{(l)}_{n_{\text{ext}}}(s + i\delta) - \mathcal{A}^{(l)}_{n_{\text{ext}}}(s - i\delta), \tag{5.58}$$

where $\delta > 0$ denotes an infinitesimal quantity. Please note that the discontinuity across a branch cut is a well-defined quantity, while the imaginary part of a loop amplitude depends on several phase conventions, like the ones used in the expressions for the polarisation factors for the external particles. For one-loop amplitudes, the discontinuity across a branch cut stems from the imaginary parts of the logarithm and the dilogarithm in certain regions of phase space. We have for example

$$\begin{aligned}
\text{Im}\, \ln\left(\frac{-s - i\delta}{-t - i\delta}\right) &= -\pi\left[\theta(s) - \theta(t)\right], \\
\text{Im}\, \text{Li}_2\left(1 - \frac{(-s - i\delta)}{(-t - i\delta)}\right) &= -\ln\left(1 - \frac{s}{t}\right) \text{Im}\, \ln\left(\frac{-s - i\delta}{-t - i\delta}\right).
\end{aligned} \tag{5.59}$$

The unitarity-based method [104, 105] exploits the fact, that the discontinuities of the basic integral functions are characteristic: Knowing the discontinuity we may uniquely reconstruct the integral function and the coefficient accompanying it. In general there will be discontinuities corresponding to different channels (e.g., to the different possibilities to cut a one-loop diagram into two parts). The discontinuity in one channel of a one-loop amplitude can be obtained via unitarity from a phase space integral over two tree-level amplitudes. Let us consider a process with n_{ext} particles, which we label from 1 to n_{ext}. Divide the set $\{1, \dots, n_{\mathrm{ext}}\}$ into two disjoint sets I and J

$$I \cup J = \{1, \dots, n_{\mathrm{ext}}\}, \ \ I \cap J = \emptyset, \tag{5.60}$$

and consider the channel

$$s = \left(\sum_{i \in I} p_i \right)^2 = \left(\sum_{j \in J} p_j \right)^2. \tag{5.61}$$

From the unitarity of the S-matrix (hence the name unitarity-based method)

$$S^\dagger S = 1 \tag{5.62}$$

and the Cutkosky rules [106] one arrives at

$$\mathrm{Disc}_s \, \mathcal{A}^{(1)}_{n_{\mathrm{ext}}} = \sum_{\lambda_1, \lambda_2} \int \frac{d^D k}{(2\pi)^D i} \, (2\pi i) \, \delta_+ \left(q_1^2\right) (2\pi i) \, \delta_+ \left(q_2^2\right) \mathcal{A}^{(0)}_{|I|+2} \mathcal{A}^{(0)}_{|J|+2}. \tag{5.63}$$

$\mathcal{A}^{(0)}_{|I|+2}$ and $\mathcal{A}^{(0)}_{|J|+2}$ are tree-level amplitudes appearing on the left and right side of the cut in a given channel, as shown in the first picture of Fig. 5.1. The momenta crossing the cut are q_1 and q_2. We set $q_1 = k$ (setting $q_2 = k$ would equally be possible). The sub-script "+" of $\delta_+(q^2)$ selects the solution of $q^2 = 0$ with positive energy. The factors of i follow from our convention, that $i\mathcal{A}^{(1)}_{n_{\mathrm{ext}}}$ equals the sum of all relevant Feynman diagrams. The amplitude $\mathcal{A}^{(0)}_{|I|+2}$ has $|I| + 2$ external particles, the (outgoing) momenta of these particles are

$$q_1, q_2, p_i, \ \ i \in I. \tag{5.64}$$

The amplitude $\mathcal{A}^{(0)}_{|J|+2}$ has $|J| + 2$ external particles, the (outgoing) momenta of these particles are

$$-q_1, -q_2, p_j, \ \ j \in J. \tag{5.65}$$

The sum over λ_1 and λ_2 in Eq. (5.63) is over the spins of the two particles crossing the cut. Note that in $\mathcal{A}^{(0)}_{|I|+2}$ and $\mathcal{A}^{(0)}_{|J|+2}$ all external particles are on-shell, also the ones with momenta q_1 and q_2 (respectively $(-q_1)$ and $(-q_2)$).

Let us denote by $\mathcal{A}^{(0),\mathrm{off}}_{|I|+2}$ and $\mathcal{A}^{(0),\mathrm{off}}_{|J|+2}$ off-shell continuations with respect to q_1 and q_2 of $\mathcal{A}^{(0)}_{|I|+2}$ and $\mathcal{A}^{(0)}_{|J|+2}$, respectively. An off-shell continuation is neither unique nor gauge-invariant. Two off-shell continuations of $\mathcal{A}^{(0)}_{|I|+2}$ (or $\mathcal{A}^{(0)}_{|J|+2}$) may differ by terms proportional to q_1^2 or q_2^2. However, neither the non-uniqueness nor the gauge-dependence matter for the subsequent argument. Lifting Eq. (5.63) one obtains

$$\mathcal{A}^{(1)}_{n_{\mathrm{ext}}} = \sum_{\lambda_1,\lambda_2} \int \frac{d^D k}{(2\pi)^D i} \frac{1}{q_1^2} \frac{1}{q_2^2} \mathcal{A}^{(0),\mathrm{off}}_{|I|+2} \mathcal{A}^{(0),\mathrm{off}}_{|J|+2} + \text{ cut free pieces}, \tag{5.66}$$

where "cut free pieces" denote contributions which do not develop an imaginary part in this particular channel. By evaluating the cut, one determines the coefficients of the integral functions, which have a discontinuity in this channel. Iterating over all possible cuts, one finds all coefficients. One advantage of a cut-based calculation is that one starts with tree amplitudes on both sides of the cut, which are already sums of Feynman diagrams. Therefore cancellations and simplifications, which usually occur between various diagrams, can already be performed before we start the calculation of the loop amplitude. The rational part R can be obtained by calculating higher order terms in ε within the cut-based method. At one-loop order an arbitrary scale $\mu^{2\varepsilon}$ is introduced in order to keep the coupling dimensionless. In a massless theory the factor $\mu^{2\varepsilon}$ is always accompanied by some kinematical invariant $s^{-\varepsilon}$ for dimensional reasons. If we write symbolically

$$\mathcal{A}^{(1)}_{n_{\mathrm{ext}}} = \frac{c_2}{\varepsilon^2}\left(\frac{s_2}{\mu^2}\right)^{-\varepsilon} + \frac{c_1}{\varepsilon}\left(\frac{s_1}{\mu^2}\right)^{-\varepsilon} + c_0\left(\frac{s_0}{\mu^2}\right)^{-\varepsilon} + \varepsilon\tilde{R} + O\left(\varepsilon^2\right), \tag{5.67}$$

where $\tilde{R}$ is independent of ε and free of discontinuities, the cut-free pieces $c_0(s_0/\mu^2)^{-\varepsilon}$ can be detected at order ε:

$$c_0\left(\frac{s_0}{\mu^2}\right)^{-\varepsilon} = c_0 - \varepsilon c_0 \ln\left(\frac{s_0}{\mu^2}\right) + O(\varepsilon^2). \tag{5.68}$$

5.5.2 *Generalised Unitarity*

The unitarity-based method allows us to bypass the set of all one-loop diagrams. We sew two tree amplitudes together (for which very often compact expressions are known) and perform a traditional tensor reduction on the resulting integrand.

However, we may push things even further. Apart from the two-particle cut discussed in the previous section, one can also consider triple or quadruple cuts as shown in Fig. 5.1. These more general cuts motivate the name "generalised unitarity". A particular nice result follows from quadruple cuts [107]: Let us consider the coefficient $c^{(0)}_{i_1 i_2 i_3 i_4}$ of the scalar box integral functions $I_4^{(i_1 i_2 i_3 i_4)}$ in Eq. (5.56). The quadruple cut is defined by the equations

$$q_{i_1}^2 = q_{i_2}^2 = q_{i_3}^2 = q_{i_4}^2 = 0, \tag{5.69}$$

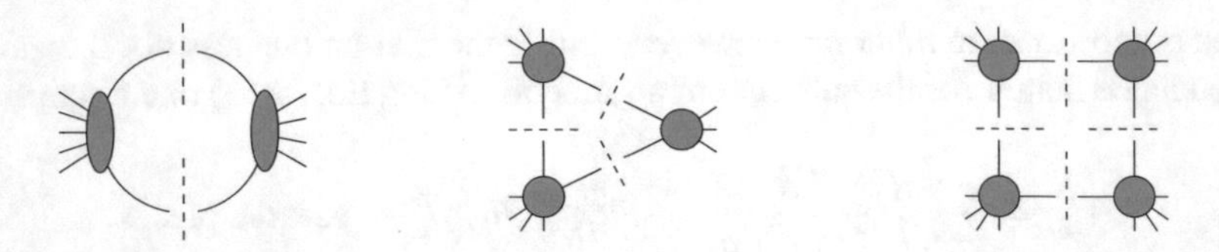

Fig. 5.1 Double, triple and quadruple cuts

where $q_j = k - p_j^{\text{sum}}$. These equations have in four space-time dimensions two solutions for k, which we denote by k^+ and k^-. We also set $q_j^\pm = k^\pm - p_j^{\text{sum}}$.

Exercise 39 *Consider a one-loop four-point function with external momenta p_1, p_2, p_3, p_4 and $p_1^2 = p_2^2 = p_3^2 = p_4^2 = 0$. The external momenta satisfy momentum conservation $p_1 + p_2 + p_3 + p_4 = 0$. For $j \in \{1, 2, 3, 4\}$ set $q_j = k - p_j^{\text{sum}}$. Solve the equations for the quadruple cut*

$$q_1^2 = q_2^2 = q_3^2 = q_4^2 = 0. \tag{5.70}$$

Hint: Start from an ansatz

$$k_\mu = c \left\langle a - |\gamma_\mu| b- \right\rangle, \tag{5.71}$$

with $c \in \mathbb{C}$ and a, b light-like.

The coefficient of the box integral function is proportional to a product of four tree amplitudes, summed over the spins of the particles crossing the cuts and averaged over the two solutions of the on-shell conditions. Let us say that the quadruple cut divides the labels $\{1, \dots, n_{\text{ext}}\}$ into four disjoint sets I_1, I_2, I_3 and I_4, corresponding to the four corners in the right picture of Fig. 5.1. Assume further that q_{i_1}, q_{i_2}, q_{i_3} and q_{i_4} are labelled such that the external momenta of the four tree amplitudes are

$$\mathcal{A}^{(0)}_{|I_1|+2}\left(q_{i_1}, \dots, -q_{i_4}\right), \quad \mathcal{A}^{(0)}_{|I_2|+2}\left(q_{i_2}, \dots, -q_{i_1}\right),$$
$$\mathcal{A}^{(0)}_{|I_3|+2}\left(q_{i_3}, \dots, -q_{i_2}\right), \quad \mathcal{A}^{(0)}_{|I_4|+2}\left(q_{i_4}, \dots, -q_{i_3}\right).$$

Then

$$c^{(0)}_{i_1 i_2 i_3 i_4} = \frac{1}{2} \frac{1}{(4\pi)^2} \sum_{\lambda_1,\lambda_2,\lambda_3,\lambda_4} \sum_{\sigma=\pm} \mathcal{A}^{(0)}_{|I_1|+2}\left(q^\sigma_{i_1}, \dots, -q^\sigma_{i_4}\right) \mathcal{A}^{(0)}_{|I_2|+2}\left(q^\sigma_{i_2}, \dots, -q^\sigma_{i_1}\right)$$
$$\times \mathcal{A}^{(0)}_{|I_3|+2}\left(q^\sigma_{i_3}, \dots, -q^\sigma_{i_2}\right) \mathcal{A}^{(0)}_{|I_4|+2}\left(q^\sigma_{i_4}, \dots, -q^\sigma_{i_3}\right). \tag{5.72}$$

The factor $1/(4\pi)^2$ comes from our convention for the integral measure in $d^D k/\pi^{D/2}$ in Eq. (5.53) instead of $d^D k/(2\pi)^D$. The sum over $\lambda_1, \dots, \lambda_4$ is over the spins of the particles crossing the cuts.

Exercise 40 *Consider the one-loop eight-point amplitude in massless ϕ^4 theory. Verify Eq. (5.72) for the box coefficient.*

Having determined all box coefficients with quadrupole cuts, one may move on to the triangle coefficients by considering triple cuts. The triple cut receives contributions from box integrals and triangle integrals. As we already have determined the coefficient of all box integrals, we may subtract out the box contributions and uniquely identify the triangle coefficients. This can then be repeated for the bubble coefficients: Having all the coefficients of the box and triangle integrals at hand, we consider double cuts. These cuts receive contributions from box integrals, triangle integrals and bubble integrals. Subtracting out the contributions from the box integrals and triangle integrals, one may extract the coefficients of the bubble integrals.

5.5.3 The OPP Method

We now discuss the method of Ossola, Papadopoulos and Pittau [108]. Let us start with a preliminary remark: It is always possible to decompose a one-loop amplitude $\mathcal{A}^{(1)}_{n_{\mathrm{ext}}}$ into cyclic-ordered primitive amplitudes $A^{(1)}_{n_{\mathrm{ext}}}$:

$$\mathcal{A}^{(1)}_{n_{\mathrm{ext}}} = \sum_{\sigma \in S_{n_{\mathrm{ext}}}/\mathbb{Z}_{n_{\mathrm{ext}}}} A^{(1)}_{n_{\mathrm{ext}}}(\sigma). \tag{5.73}$$

For non-gauge amplitudes this is a trivial statement, for gauge amplitudes the non-trivial part of this decomposition is the fact that the primitive amplitudes $A^{(1)}_{n_{\mathrm{ext}}}$ are themselves gauge-invariant. The cyclic order of the external legs is specified by the permutation $\sigma \in S_{n_{\mathrm{ext}}}/\mathbb{Z}_{n_{\mathrm{ext}}}$. Without loss of generalisation we will consider in the following the cyclic order $\sigma = (1, 2, \dots, n_{\mathrm{ext}})$. Working with cyclic-ordered primitive amplitudes has the advantage that only n_{ext} different loop propagators may appear. For simplicity let us discuss – as before – massless theories. We may write

$$A^{(1)}_{n_{\mathrm{ext}}} = e^{\varepsilon\gamma_{\mathrm{E}}}\mu^{2\varepsilon} \int \frac{d^D k}{i\pi^{\frac{D}{2}}} \frac{P(k)}{\prod\limits_{j=1}^{n_{\mathrm{ext}}} \left(-q_j^2\right)}, \tag{5.74}$$

with $q_j = k - p_j^{\mathrm{sum}}$ and $p_j^{\mathrm{sum}} = \sum_{i=1}^{j} p_i$. The numerator $P(k)$ is a polynomial in the loop momentum k. The degree of this polynomial is bounded. For example, we have in gauge theories in a renormalisable gauge

$$\deg P(k) \le n_{\mathrm{ext}}. \tag{5.75}$$

Furthermore, $P(k)$ can be computed easily by a tree-like calculation. Let us split the numerator into a four-dimensional part and a remainder

$$P(k) = P\left(k^{(4)}\right) + \tilde{P}\left(k^{(4)}, k^{(-2\varepsilon)}\right), \tag{5.76}$$

The OPP method consists in writing

$$\begin{aligned} P\left(k^{(4)}\right) = & \sum_{i_1<i_2<i_3<i_4} \left[c^{(0)}_{i_1 i_2 i_3 i_4} + \tilde{c}_{i_1 i_2 i_3 i_4}\left(k^{(4)}\right)\right] \prod_{i \notin \{i_1,i_2,i_3,i_4\}} \left[-\left(q_i^{(4)}\right)^2\right] \\ & + \sum_{i_1<i_2<i_3} \left[c^{(0)}_{i_1 i_2 i_3} + \tilde{c}_{i_1 i_2 i_3}\left(k^{(4)}\right)\right] \prod_{i \notin \{i_1,i_2,i_3\}} \left[-\left(q_i^{(4)}\right)^2\right] \\ & + \sum_{i_1<i_2} \left[c^{(0)}_{i_1 i_2} + \tilde{c}_{i_1 i_2}\left(k^{(4)}\right)\right] \prod_{i \notin \{i_1,i_2\}} \left[-\left(q_i^{(4)}\right)^2\right]. \end{aligned} \tag{5.77}$$

The terms $\tilde{c}_{i_1 i_2 i_3 i_4}$, $\tilde{c}_{i_1 i_2 i_3}$ and $\tilde{c}_{i_1 i_2}$ have the property, that they vanish after integration. Their dependence on $k^{(4)}$ is known up to some yet to be determined parameters, on which these terms depend linearly. To give an example, let's consider $\tilde{c}^{(0)}_{i_1 i_2 i_3 i_4}$. We denote by p_1', p_2', p_3', p_4' the external momenta of the box function with propagators $(-q_{i_1})^2$, $(-q_{i_2})^2$, $(-q_{i_3})^2$ and $(-q_{i_4})^2$. Then

$$\begin{aligned} \tilde{c}_{i_1 i_2 i_3 i_4}\left(k^{(4)}\right) &= \tilde{C}_{i_1 i_2 i_3 i_4} \mathrm{Tr}\left(\not{q}_{i_4} \not{p}'_1 \not{p}'_2 \not{p}'_3 \gamma_5\right) \\ &= 4i \tilde{C}_{i_1 i_2 i_3 i_4} \varepsilon_{\mu\nu\rho\sigma} q^{\mu}_{i_4} p_1'^{\nu} p_2'^{\rho} p_3'^{\sigma}. \end{aligned} \tag{5.78}$$

It is clear that a rank one box integral with this numerator will vanish after integration: We may choose $k' = q_{i_4}$. From Passarino-Veltman reduction we know that k'^{μ} will become proportional to either $p_1'^{\mu}$, $p_2'^{\mu}$ or $p_3'^{\mu}$ after integration. But this vanishes when contracted into the antisymmetric tensor. $\tilde{C}_{i_1 i_2 i_3 i_4}$ is the yet to be determined parameter.

These parameters and the constants $c^{(0)}_{i_1 i_2 i_3 i_4}$, $c^{(0)}_{i_1 i_2 i_3}$ and $c^{(0)}_{i_1 i_2}$ can be determined by evaluating the left-hand side and the right-hand side of Eq. (5.77) for various values of $k^{(4)}$. Solving this linear system yields the coefficients $c^{(0)}_{i_1 i_2 i_3 i_4}$, $c^{(0)}_{i_1 i_2 i_3}$ and $c^{(0)}_{i_1 i_2}$ of the scalar integral functions. It remains to extract the rational term R. There are two sources contributing to R and we write [109]

$$R = R_1 + R_2. \tag{5.79}$$

First of all, the factors $(-(q_i^{(4)})^2)$ in Eq. (5.77) do not cancel exactly the denominators $(-q_i^2)$. There is a mismatch

$$-\left(q_i^{(4)}\right)^2 = -q_i^2 + \left(q_i^{(-2\varepsilon)}\right)^2 = -q_i^2 + \left(k^{(-2\varepsilon)}\right)^2. \tag{5.80}$$

The terms proportional to $(k^{(-2\varepsilon)})^2$ make up the rational term R_1.

Secondly, we split in Eq. (5.76) the numerator $P(k)$ into a four-dimensional part and a remainder. The remainder $\tilde{P}(k^{(4)}, k^{(-2\varepsilon)})$ makes up the rational term R_2.

Chapter 6
Iterated Integrals

In this chapter we introduce modern methods to tackle Feynman integrals. The main tool will be the method of differential equations. This builds on integration-by-parts identities and dimensional shift relations, which we discuss first. If the system of differential equations can be brought into a particular simple form (the ε-form which is introduced in Sect. 6.3.2), a solution in terms of iterated integrals is immediate. The methods of this chapter reduce the problem of computing Feynman integrals to the problem of finding an appropriate transformation for the system of differential equations. Algorithms to construct an appropriate transformation are discussed in Chap. 7. We will see that integration-by-parts identities allow us to express any Feynman integral as a linear combination of basis integrals, which we call master integrals. The master integrals span a vector space and viewing the Feynman integrals as functions of the kinematic variables gives us a vector bundle. We discuss fibre bundles in Sect. 6.4.

Sections 6.5 and 6.6 are devoted to cuts of Feynman integrals and singularities of Feynman integrals, respectively.

As we may express any Feynman integral as a linear combination of master integrals, we may ask if this vector space is equipped with an inner product. An inner product would allow us to compute the coefficient in front of each master integral directly, bypassing the need to solve a linear system of integration-by-parts identities. This leads us to twisted cohomology, which we introduce in Sect. 6.7.

6.1 Integration-by-Parts

In this section we study more closely the family of Feynman integrals

$$I_{\nu_1 \dots \nu_{n_{\mathrm{int}}}}\left(D, x_1, \dots, x_{N_B}\right) = e^{l \varepsilon \gamma_{\mathrm{E}}} \left(\mu^2\right)^{\nu - \frac{lD}{2}} \int \prod_{r=1}^{l} \frac{d^D k_r}{i \pi^{\frac{D}{2}}} \prod_{j=1}^{n_{\mathrm{int}}} \frac{1}{\left(-q_j^2 + m_j^2\right)^{\nu_j}}. \quad (6.1)$$

S. Weinzierl, *Feynman Integrals*, UNITEXT for Physics,
https://doi.org/10.1007/978-3-030-99558-4_6

We are in particular interested in relations between members of this family, which differ by the values of the indices $(\nu_1, \dots, \nu_{n_{\text{int}}})$. Integration-by-parts identities provide these relations [110, 111]. Integration-by-parts identities are based on the fact that within dimensional regularisation the integral of a total derivative vanishes

$$\int \frac{d^D k}{i\pi^{\frac{D}{2}}} \frac{\partial}{\partial k^\mu} \left[q^\mu \cdot f(k) \right] = 0, \tag{6.2}$$

i.e. there are no boundary terms. The vector q can be any linear combination of the external momenta and the loop momentum k. Equation (6.2) is derived as follows: Let us first assume that q is a linear combination of the external momenta. Integrals within dimensional regularisation are translation invariant (see Eq. (2.75))

$$\int \frac{d^D k}{i\pi^{\frac{D}{2}}} f(k) = \int \frac{d^D k}{i\pi^{\frac{D}{2}}} f(k + \lambda q). \tag{6.3}$$

The right-hand side has to be independent of λ. This implies in particular that the $O(\lambda)$-term has to vanish. From

$$\left[\frac{d}{d\lambda} f(k + \lambda q) \right] \Bigg|_{\lambda=0} = q^\mu \frac{\partial}{\partial k^\mu} f(k) = \frac{\partial}{\partial k^\mu} \left[q^\mu \cdot f(k) \right] \tag{6.4}$$

Eq. (6.2) follows.

Equation (6.2) also holds for $q = k$. This is the task of the next exercise:

Exercise 41 *Show that Eq. (6.2) holds for $q = k$.*
Hint: Consider the scaling relation Eq. (2.76).

Let us formulate the integration-by-parts identities for l-loop integrals:

Integration-by-parts identities:
Within dimensional regularisation we have for any loop momentum k_i $(1 \leq i \leq l)$ and any vector $q_{\text{IBP}} \in \{p_1, ..., p_{N_{\text{ext}}}, k_1, ..., k_l\}$

$$e^{l\varepsilon\gamma_{\text{E}}} \left(\mu^2\right)^{\nu - \frac{lD}{2}} \int \prod_{r=1}^{l} \frac{d^D k_r}{i\pi^{\frac{D}{2}}} \frac{\partial}{\partial k_i^\mu} q_{\text{IBP}}^\mu \prod_{j=1}^{n_{\text{int}}} \frac{1}{\left(-q_j^2 + m_j^2\right)^{\nu_j}} = 0. \tag{6.5}$$

Working out the derivatives leads to relations among integrals with different sets of indices $(\nu_1, \dots, \nu_{n_{\text{int}}})$.

Let's see how this works in an example: We consider the one-loop two-point function with an equal internal mass:

$$I_{\nu_1 \nu_2}(D, x) = e^{\varepsilon\gamma_{\text{E}}} \left(m^2\right)^{\nu_{12} - \frac{D}{2}} \int \frac{d^D k}{i\pi^{\frac{D}{2}}} \frac{1}{\left(-q_1^2 + m^2\right)^{\nu_1} \left(-q_2^2 + m^2\right)^{\nu_2}}. \tag{6.6}$$

with $q_1 = k - p$, $q_2 = k$, $\nu_{12} = \nu_1 + \nu_2$ and $x = -p^2/m^2$. We have set $\mu^2 = m^2$. As vector q_{IBP} we may take $q_{\text{IBP}} \in \{p, k\}$. Let us start with $q_{\text{IBP}} = p$. We obtain

$$\begin{aligned}
0 &= e^{\varepsilon\gamma_E} \left(m^2\right)^{\nu_{12}-\frac{D}{2}} p^\mu \int \frac{d^Dk}{i\pi^{\frac{D}{2}}} \frac{\partial}{\partial k^\mu} \frac{1}{\left(-q_1^2+m^2\right)^{\nu_1}\left(-q_2^2+m^2\right)^{\nu_2}} \\
&= e^{\varepsilon\gamma_E} \left(m^2\right)^{\nu_{12}-\frac{D}{2}} \int \frac{d^Dk}{i\pi^{\frac{D}{2}}} \left[\frac{\nu_1\left(q_2^2-q_1^2-p^2\right)}{\left(-q_1^2+m^2\right)^{\nu_1+1}\left(-q_2^2+m^2\right)^{\nu_2}} \right. \\
&\quad \left. + \frac{\nu_2\left(q_2^2-q_1^2+p^2\right)}{\left(-q_1^2+m^2\right)^{\nu_1}\left(-q_2^2+m^2\right)^{\nu_2+1}} \right] \\
&= \nu_1 \left[I_{\nu_1\nu_2} - I_{(\nu_1+1)(\nu_2-1)} + x I_{(\nu_1+1)\nu_2} \right] + \nu_2 \left[I_{(\nu_1-1)(\nu_2+1)} - I_{\nu_1\nu_2} - x I_{\nu_1(\nu_2+1)} \right].
\end{aligned} \tag{6.7}$$

In deriving this result we used

$$\begin{aligned}
2p\cdot q_1 &= q_2^2 - q_1^2 - p^2, \\
2p\cdot q_2 &= q_2^2 - q_1^2 + p^2.
\end{aligned} \tag{6.8}$$

Thus we obtain a relation between integrals with different values of (ν_1, ν_2):

$$(\nu_1 - \nu_2) I_{\nu_1\nu_2} - \nu_1 I_{(\nu_1+1)(\nu_2-1)} + \nu_2 I_{(\nu_1-1)(\nu_2+1)} + \nu_1 x I_{(\nu_1+1)\nu_2} - \nu_2 x I_{\nu_1(\nu_2+1)} = 0. \tag{6.9}$$

Exercise 42 *Repeat the derivation with $q_{\text{IBP}} = k$ and show*

$$(D - \nu_1 - 2\nu_2) I_{\nu_1\nu_2} - \nu_1 I_{(\nu_1+1)(\nu_2-1)} + \nu_1 (2+x) I_{(\nu_1+1)\nu_2} + 2\nu_2 I_{\nu_1(\nu_2+1)} = 0. \tag{6.10}$$

Instead of Eqs. (6.9) and (6.10) we may consider two independent linear combinations, where either the integral $I_{(\nu_1+1)\nu_2}$ or the integral $I_{\nu_1(\nu_2+1)}$ is absent:

$$\begin{aligned}
& \nu_1 x (4+x) I_{(\nu_1+1)\nu_2} = \\
& [2(-\nu_1+\nu_2) + (\nu_1+2\nu_2-D)x] I_{\nu_1\nu_2} + \nu_1(2+x) I_{(\nu_1+1)(\nu_2-1)} - 2\nu_2 I_{(\nu_1-1)(\nu_2+1)}, \\
& \nu_2 x (4+x) I_{\nu_1(\nu_2+1)} = \\
& [2(\nu_1-\nu_2) + (2\nu_1+\nu_2-D)x] I_{\nu_1\nu_2} - 2\nu_1 I_{(\nu_1+1)(\nu_2-1)} + \nu_2(2+x) I_{(\nu_1-1)(\nu_2+1)}.
\end{aligned} \tag{6.11}$$

For $\nu_1 > 0$ and $\nu_2 > 0$ we may use either the first or the second equation to reduce the sum $\nu_1 + \nu_2$: In both equations, the sum of the indices equals $\nu_1 + \nu_2 + 1$ on the left-hand side, while on the right-hand side the sum of the indices equals for all terms $\nu_1 + \nu_2$.

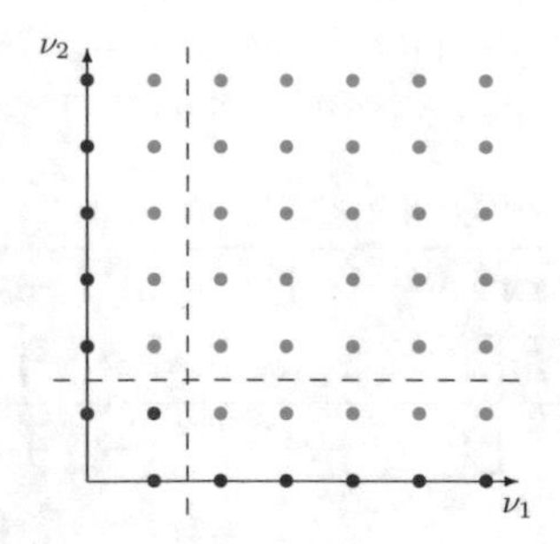

Fig. 6.1 Integration-by-parts reduction for the one-loop two-point function: For all integrals to the right of the vertical dashed line and indicated by a green dot, we may use the first equation of Eq. (6.11) to lower ν_{12}, for all integrals above the horizontal dashed line and indicated by a green dot, we may use the second equation of Eq. (6.11) to lower ν_{12}. Integrals represented by a blue dot are reduced with Eq. (6.14). The integrals with a red dot cannot be reduced to simpler integrals

If either $\nu_1 = 0$ (and $\nu_2 > 0$) or $\nu_2 = 0$ (and $\nu_1 > 0$) we have a simpler integral: The integral reduces to a tadpole integral. As the two internal masses are equal, we have

$$I_{\nu 0} = I_{0\nu}. \tag{6.12}$$

Exercise 43 *Derive the integration-by-parts identity for the integral*

$$I_{0\nu_2} = e^{\varepsilon \gamma_E} \left(m^2\right)^{\nu_2 - \frac{D}{2}} \int \frac{d^D k}{i\pi^{\frac{D}{2}}} \frac{1}{\left(-k^2 + m^2\right)^{\nu_2}}. \tag{6.13}$$

Verify the identity with the explicit result from Eq. (2.123).

In the previous exercise you were supposed to derive the identity

$$\nu_2 I_{0(\nu_2+1)} = \left(\nu_2 - \frac{D}{2}\right) I_{0\nu_2}. \tag{6.14}$$

This identity can be used to reduce for $\nu_2 > 0$ the integral $I_{0(\nu_2+1)}$. The situation is summarised in Fig. 6.1. We may reduce with integration-by-parts identities any integral $I_{\nu_1\nu_2}$ with $\nu_1 \geq 0$, $\nu_2 \geq 0$ and $\nu_1 + \nu_2 > 0$ to a linear combination of I_{11}, I_{10} and I_{01}. In the equal mass case we have the symmetry $I_{01} = I_{10}$, and therefore any integral $I_{\nu_1\nu_2}$ with $\nu_1 \geq 0$, $\nu_2 \geq 0$ and $\nu_1 + \nu_2 > 0$ can be reduced to a linear combination of I_{11} and I_{10}. We call I_{11} and I_{10} **master integrals**.

Let us now return to the general case. We consider a graph G which has a Baikov representation. This ensures that we may express any scalar product involving a loop momentum as a linear combination of inverse propagators and loop momentum independent terms. In our example above we needed this property in Eq. (6.8). We

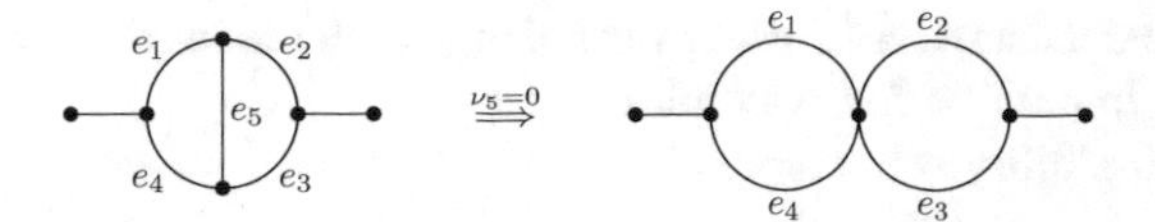

Fig. 6.2 Integrals, where one or more indices satisfy $\nu_j < 1$, belong to a sub-topology. The figure shows as an example the two-loop two-point integral: The case $\nu_5 = 0$ corresponds to the sub-topology obtained by pinching the edge e_5. Note that also the case $\nu_5 < 0$ corresponds to the sub-topology shown in the right picture, in this case with an irreducible scalar product in the numerator

consider integrals

$$I_{\nu_1 \dots \nu_{n_{\mathrm{int}}}}, \quad \nu_j \in \mathbb{Z}. \tag{6.15}$$

We call integrals, where all indices satisfy $\nu_j > 0$, integrals of the **top topology** (or of the **top sector**). Integrals, where one or more indices satisfy $\nu_j < 1$, belong to a sub-topology (or belong to a sub-sector).

This is illustrated in Fig. 6.2.

For a given set of indices $(\nu_1, \dots, \nu_{n_{\mathrm{int}}})$ we define

$$\begin{aligned} N_{\mathrm{prop}} &= \sum_{j=1}^{n_{\mathrm{int}}} \Theta\left(\nu_j - \frac{1}{2}\right), \quad N_{\mathrm{id}} = \sum_{j=1}^{n_{\mathrm{int}}} 2^{j-1} \Theta\left(\nu_j - \frac{1}{2}\right), \\ r &= \sum_{j=1}^{n_{\mathrm{int}}} \nu_j \Theta\left(\nu_j - \frac{1}{2}\right), \quad s = \sum_{j=1}^{n_{\mathrm{int}}} |\nu_j| \, \Theta\left(-\nu_j + \frac{1}{2}\right). \end{aligned} \tag{6.16}$$

$\Theta(x)$ denotes the Heaviside step function. (Adding/subtracting the constant $1/2$ avoids ambiguities in the definition of $\Theta(0)$.) N_{prop} counts the number of propagators having positive indices. N_{id} assigns a sector identity to the integral (a number between 0 and $2^{n_{\mathrm{int}}} - 1$). The variable r counts the sum of the powers of the propagators having positive indices, the variable s does the same thing for the propagators having negative indices. With the help of these variables we may now define a criteria which allows us to compare two integrals and to decide which integral is considered to be simpler. One possibility is the tuple

$$\left(N_{\mathrm{prop}}, N_{\mathrm{id}}, r, s, \dots\right), \tag{6.17}$$

together with the lexicographical order. Thus integrals with a smaller number of propagators N_{prop} are considered simpler. Within the group of integrals with the same number of propagators, integrals with a smaller sector identity are considered simpler. Within one sector, one first selects integrals with a smaller value of r as simpler, and in the case of an equal value of r, one uses as a secondary criteria the variable s. The dots stand for further variables, which are used if two non-identical integrals agree in the first four variables.

A second possibility is the tuple

$$\left(N_{\text{prop}}, N_{\text{id}}, s, r, \dots\right), \tag{6.18}$$

again with the lexicographical order.

We may now write down all possible integration-by-parts identities according to Eq. (6.5). This is a system of linear equations for the Feynman integrals $I_{\nu_1 \dots \nu_{n_{\text{int}}}}$. With the help of an ordering criteria as in Eqs. (6.17) or in (6.18) we may eliminate the more complicated integrals in favour of the simpler ones. This procedure is known as the **Laporta algorithm** [112]. At the end of the day we are able to express most of the integrals in terms of a few remaining integrals. The remaining integrals are called **master integrals**. The set of master integrals depends on the chosen ordering criteria. It should be noted that for the ordering criteria given in Eqs. (6.17) and (6.18) the set of master integrals will also depend on the way we label the internal edges. This dependence enters through the sector identity N_{id}, which depends on the labelling of the internal edges. The choice of Eq. (6.17) will lead to a basis of master integrals with irreducible scalar products (i.e. with some negative indices), avoiding positive indices larger than one. Such a basis is called an **ISP-basis**. On the other hand, the choice of Eq. (6.18) will avoid irreducible scalar products (i.e. negative indices) at the expense of allowing positive indices larger than one. Such a basis is called a **dot-basis**.

As in the example discussed above we may supplement the integration-by-parts identities with symmetry relations. (In the example of the one-loop two-point function with equal internal masses we had the symmetry $I_{\nu 0} = I_{0\nu}$.) We denote the number of master integrals obtained by taking integration-by-parts identities and symmetries into account by N_{master}. If we are only interested in the number of unreduced integrals obtained from integration-by-parts identities alone, we denote this number by N_{cohom}. In physics we are mainly interested in the master integrals, which remain after applying integration-by-parts identities and symmetries. However, in Sect. 6.7 we analyse in more detail the effects of the integration-by-parts identities alone. In this context, N_{cohom} is the relevant quantity.

We denote the indices of the master integrals by

$$\begin{aligned}
\boldsymbol{\nu}_1 &= \left(\nu_{11}, \dots, \nu_{1n_{\text{int}}}\right), \\
\boldsymbol{\nu}_2 &= \left(\nu_{21}, \dots, \nu_{2n_{\text{int}}}\right), \\
&\dots \\
\boldsymbol{\nu}_{N_{\text{master}}} &= \left(\nu_{N_{\text{master}}1}, \dots, \nu_{N_{\text{master}}n_{\text{int}}}\right).
\end{aligned} \tag{6.19}$$

We define a N_{master}-dimensional vector $\vec{I}$ by

$$\vec{I} = \left(I_{\boldsymbol{\nu}_1}, I_{\boldsymbol{\nu}_2}, \dots, I_{\boldsymbol{\nu}_{N_{\text{master}}}}\right)^T. \tag{6.20}$$

For the specific example discussed in Eq. (6.6) and below we have

$$\vec{I} = \begin{pmatrix} I_{10} \\ I_{11} \end{pmatrix} \tag{6.21}$$

Integration-by-parts identities and symmetries allow us to express a generic integral $I_{\nu_1 \dots \nu_{n_{\mathrm{int}}}}$ as a linear combination of the master integrals. Only the latter need to be computed. There are public available computer programs, which perform the task of reducing Feynman integrals to master integrals. These programs are `Fire` [113, 114], `Reduze` [115, 116] and `Kira` [117, 118]. The following exercise will help you to get acquainted with these programs:

Exercise 44 *Consider the double-box graph G shown in Fig. 2.3 and the auxiliary graph $\tilde{G}$ with nine propagators shown in Fig. 2.11. This exercise is about the family of Feynman integrals*

$$I_{\nu_1 \nu_2 \nu_3 \nu_4 \nu_5 \nu_6 \nu_7 \nu_8 \nu_9} \tag{6.22}$$

with $\nu_8, \nu_9 \leq 0$. Use the notation of the momenta as in Fig. 2.11. Assume that all external momenta are light-like ($p_1^2 = p_2^2 = p_3^2 = p_4^2 = 0$) and that all internal propagators are massless. Use one of the public available computer programs `Kira`, `Reduze` *or* `Fire` *to reduce the Feynman integral*

$$I_{1111111(-1)(-1)} \tag{6.23}$$

to master integrals. For the choice of master integrals you may use the default ordering criteria of the chosen computer program.

We note that integration-by-parts reduction is based only on linear algebra with rational functions in the kinematic variables x and the dimension of space-time D. However, the simplification of the rational functions (i.e. cancelling common factors in the numerator and in the denominator) is actually a performance bottle-neck. For this reason, many of the programs mentioned above employ finite field methods to improve the performance. Finite field methods are reviewed in Appendix I.

6.2 Dimensional Shift Relations

Given a basis of master integrals $\vec{I}$ in D space-time dimensions and a basis of master integrals $\vec{I}'$ in $(D+2)$ space-time dimensions, we may express any element of the basis $\vec{I}$ (i.e. any component of $\vec{I}$) as a linear combination of the elements of $\vec{I}'$ and vice versa. In dimensional regularisation we have $\dim \vec{I} = \dim \vec{I}'$. The relation between Feynman integrals in D and $(D+2)$ space-time dimensions (or $(D-2)$ space-time dimensions) is known as the **dimensional shift relations** [76, 77]. In this section we will derive these relations.

In Sect. 4.3 we introduced the dimensional-shift operators $\mathbf{D}^{\pm}$, which act on a Feynman integral as

$$\mathbf{D}^{\pm} I_{\nu_1 \dots \nu_{n_{\text{int}}}}(D, x) = I_{\nu_1 \dots \nu_{n_{\text{int}}}}(D \pm 2, x) \tag{6.24}$$

and the raising operators $\mathbf{j}^+$ (with $j \in \{1, \dots, n_{\text{int}}\}$), which act on a Feynman integral as

$$\mathbf{j}^+ I_{\nu_1 \dots \nu_j \dots \nu_{n_{\text{int}}}}(D, x) = \nu_j \cdot I_{\nu_1 \dots (\nu_j + 1) \dots \nu_{n_{\text{int}}}}(D, x). \tag{6.25}$$

We start from the Schwinger parameter representation of the Feynman integral

$$I_{\nu_1 \dots \nu_{n_{\text{int}}}}(D) = \frac{e^{l \varepsilon \gamma_{\text{E}}}}{\prod\limits_{k=1}^{n_{\text{int}}} \Gamma(\nu_k)} \int\limits_{\alpha_k \ge 0} d^{n_{\text{int}}} \alpha \left(\prod_{k=1}^{n_{\text{int}}} \alpha_k^{\nu_k - 1} \right) \frac{1}{\mathcal{U}^{\frac{D}{2}}} e^{-\frac{\mathcal{F}}{\mathcal{U}}}. \tag{6.26}$$

In Eq. (4.97) we have already seen that the operators $\mathbf{D}^+$ and $\mathbf{j}^+$ act on the Schwinger parameter representation as

$$\begin{aligned} \mathbf{D}^+ I_{\nu_1 \dots \nu_{n_{\text{int}}}}(D) &= \frac{e^{l \varepsilon \gamma_{\text{E}}}}{\prod\limits_{k=1}^{n_{\text{int}}} \Gamma(\nu_k)} \int\limits_{\alpha_k \ge 0} d^{n_{\text{int}}} \alpha \left(\prod_{k=1}^{n_{\text{int}}} \alpha_k^{\nu_k - 1} \right) \frac{1}{\mathcal{U} \cdot \mathcal{U}^{\frac{D}{2}}} e^{-\frac{\mathcal{F}}{\mathcal{U}}}, \\ \mathbf{j}^+ I_{\nu_1 \dots \nu_j \dots \nu_{n_{\text{int}}}}(D) &= \frac{e^{l \varepsilon \gamma_{\text{E}}}}{\prod\limits_{k=1}^{n_{\text{int}}} \Gamma(\nu_k)} \int\limits_{\alpha_k \ge 0} d^{n_{\text{int}}} \alpha \left(\prod_{k=1}^{n_{\text{int}}} \alpha_k^{\nu_k - 1} \right) \frac{\alpha_j}{\mathcal{U}^{\frac{D}{2}}} e^{-\frac{\mathcal{F}}{\mathcal{U}}}. \end{aligned} \tag{6.27}$$

In order to derive the dimensional shift relations we use Eq. (6.26) and expand the fraction with $\mathcal{U}$:

$$I_{\nu_1 \dots \nu_{n_{\text{int}}}}(D) = \frac{e^{l \varepsilon \gamma_{\text{E}}}}{\prod\limits_{k=1}^{n_{\text{int}}} \Gamma(\nu_k)} \int\limits_{\alpha_k \ge 0} d^{n_{\text{int}}} \alpha \left(\prod_{k=1}^{n_{\text{int}}} \alpha_k^{\nu_k - 1} \right) \frac{\mathcal{U}}{\mathcal{U} \cdot \mathcal{U}^{\frac{D}{2}}} e^{-\frac{\mathcal{F}}{\mathcal{U}}}. \tag{6.28}$$

The additional factor of $\mathcal{U}$ in the denominator shifts the space-time dimension of the Feynman integral by two units according to the first formula of Eq. (6.27). The additional factor of $\mathcal{U}$ in the numerator is treated as follows: Recall that the graph polynomial $\mathcal{U}(\alpha_1, \dots, \alpha_{n_{\text{int}}})$ is a homogeneous polynomial of degree l in the Schwinger parameters α. Furthermore, the graph polynomial $\mathcal{U}(\alpha_1, \dots, \alpha_{n_{\text{int}}})$ is linear in each Schwinger parameter α_j. We may interpret each occurrence of a Schwinger parameter α_j in the numerator as the result of applying the raising operator $\mathbf{j}^+$ to the Feynman integral, according to the second formula in Eq. (6.27). Thus we have

$$I_{\nu_1 \dots \nu_{n_{\text{int}}}}(D) = \mathcal{U}\left(\mathbf{1}^+, \dots, \mathbf{n}_{\text{int}}{}^+\right) \mathbf{D}^+ I_{\nu_1 \dots \nu_{n_{\text{int}}}}(D). \tag{6.29}$$

The action of $\mathcal{U}(\mathbf{1}^+, \dots, \mathbf{n_{int}}^+)$ on $I_{\nu_1 \dots \nu_{n_{int}}}(D)$ is defined in the obvious way: If $\mathcal{U}(\alpha_1, \alpha_2, \alpha_3) = \alpha_1\alpha_2 + \alpha_2\alpha_3 + \alpha_1\alpha_3$ we have

$$\begin{aligned}\mathcal{U}\left(\mathbf{1}^+, \mathbf{2}^+, \mathbf{3}^+\right) I_{111}(D) &= \left(\mathbf{1}^+\mathbf{2}^+ + \mathbf{2}^+\mathbf{3}^+ + \mathbf{1}^+\mathbf{3}^+\right) I_{111}(D) \\ &= I_{221}(D) + I_{122}(D) + I_{212}(D). \end{aligned} \tag{6.30}$$

It is also clear that the operators $\mathbf{i}^+$ and $\mathbf{j}^+$ commute

$$\left[\mathbf{i}^+, \mathbf{j}^+\right] = 0 \tag{6.31}$$

and that the operators $\mathbf{j}^+$ commute with $\mathbf{D}^+$:

$$\left[\mathbf{j}^+, \mathbf{D}^\pm\right] = 0. \tag{6.32}$$

We may bring in Eq. (6.29) the dimensional shift operator to the other side and obtain

$$\mathbf{D}^- I_{\nu_1 \dots \nu_{n_{int}}}(D) = \mathcal{U}\left(\mathbf{1}^+, \dots, \mathbf{n_{int}}^+\right) I_{\nu_1 \dots \nu_{n_{int}}}(D) \tag{6.33}$$

or

$$I_{\nu_1 \dots \nu_{n_{int}}}(D-2) = \mathcal{U}\left(\mathbf{1}^+, \dots, \mathbf{n_{int}}^+\right) I_{\nu_1 \dots \nu_{n_{int}}}(D). \tag{6.34}$$

Let's go back to Eq. (6.29), which we may also write as

$$I_{\nu_1 \dots \nu_{n_{int}}}(D) = \mathcal{U}\left(\mathbf{1}^+, \dots, \mathbf{n_{int}}^+\right) I_{\nu_1 \dots \nu_{n_{int}}}(D+2). \tag{6.35}$$

The right-hand side consists of integrals with raised propagators in $(D+2)$ space-time dimensions. Let $\vec{I} = (I_{\boldsymbol{\nu}_1}, \dots, I_{\boldsymbol{\nu}_{N_{master}}})^T$ be a basis in D space-time dimensions and $\vec{I}' = (I'_{\boldsymbol{\nu}_1}, \dots, I'_{\boldsymbol{\nu}_{N_{master}}})^T$ be a basis in $(D+2)$ space-time dimensions. On the left-hand side we may consider all master integrals from the basis $\vec{I}$ in D dimension. For each of these Feynman integrals we may use on the right-hand side integration-by-parts identities and express all integrals as linear combinations of the master integrals $\vec{I}'$ in $(D+2)$ dimensions. This allows us to express any master integral of the basis $\vec{I}$ in D dimensions as a linear combination of the master integrals $\vec{I}'$ in $(D+2)$ dimensions. Thus we find a $(N_{master} \times N_{master})$-matrix S

$$\vec{I} = S\,\vec{I}'. \tag{6.36}$$

Within dimensional regularisation the matrix S is invertible. Inverting this matrix allows us to express any master integral in $(D+2)$ dimensions as a linear combination of master integrals in D dimensions:

$$\vec{I}' = S^{-1}\,\vec{I}. \tag{6.37}$$

Exercise 45 *Consider the example of the one-loop two-point function with equal internal masses, discussed below Eq. (6.6). Let*

$$\vec{I} = \begin{pmatrix} I_{10}(D, x) \\ I_{11}(D, x) \end{pmatrix} \tag{6.38}$$

be a basis in D space-time dimensions and

$$\vec{I}' = \begin{pmatrix} I_{10}(D+2, x) \\ I_{11}(D+2, x) \end{pmatrix} \tag{6.39}$$

be a basis in $(D+2)$ *space-time dimensions. Work out the* 2×2*-matrices* S *and* S^{-1}.

6.3 Differential Equations

We now introduce one of the most important methods to compute Feynman integrals: The method of differential equations [119–122]. The idea is the following: Instead of calculating the Feynman integral directly, we first derive a differential equation of the Feynman integral under consideration with respect to the kinematic variables. In a second step we solve this differential equation and obtain in this way the answer for the sought-after Feynman integral. To be more precise, we study a system of differential equations, namely the system of differential equations for a basis of master integrals. This has the advantage that we have to consider only first-order differential equations.

Solving a differential equation requires boundary values. As boundary value we may use the master integrals, where one of the kinematic variables has a special value, for example zero or equal to another kinematic variable. The Feynman integrals for this special kinematic configuration are simpler, as they depend on one kinematic variable less. We may assume that they are already known. If not, we first solve this simpler problem first.

The power of the method of differential equations lies in the following facts: We will soon see that it is always possible to derive the system of differential equations for a basis of master integrals. There are no principle obstacles to do this, we only might be limited by the available computing resources and the fact that the used algorithms are not particular efficient. We will also see that if the system of differential equations is in a particular nice form (the ε-form), it is always possible to solve the system of differential equations in terms of iterated integrals. Here we assume—as remarked above—that the boundary values are known. Thus the only missing piece is the transformation of an original system of differential equations to the nice ε-form. In the cases where we know how to do this, this is achieved by a redefinition of the master integrals and/or a variable transformation of the kinematic variables. We call a redefinition of the master integrals a fibre transformation and a transformation of the kinematic variables a base transformation. We discuss these transformations in Sects. 6.4.3 and 6.4.4, respectively. Let us stress that this reduces the task of computing a Feynman integral to finding a suitable fibre transformation and/or base transformation for the associated system of differential equations.

6.3.1 Deriving the System of Differential Equation

Let's start to derive the system of differential equations. We consider a basis of master integrals

$$\vec{I} = \left(I_{\boldsymbol{\nu}_1}, I_{\boldsymbol{\nu}_2}, \dots, I_{\boldsymbol{\nu}_{N_{\mathrm{master}}}}\right)^T, \tag{6.40}$$

depending on N_B kinematic variables $x_1, x_2, \dots, x_{N_B}$. Let's recall from Sect. 2.5.1 that we start with $N_B + 1$ variables of the form

$$\frac{-p_i \cdot p_j}{\mu^2}, \quad \frac{m_i^2}{\mu^2}. \tag{6.41}$$

We denote these variables by $x_1, x_2, \dots, x_{N_B}, x_{N_B+1}$. Due to the scaling relation Eq. (2.144) we may set one of these variables to one, say $x_{N_B+1} = 1$. Having done so, we usually view the Feynman integrals as functions of $x_1, x_2, \dots, x_{N_B}$ (and D). For the moment, let's keep the dependence on all variables $x_1, x_2, \dots, x_{N_B}, x_{N_B+1}$, without setting one kinematic variable to one. The second Symanzik polynomial $\mathcal{F}$ is linear in the kinematic variables x_j. We may write

$$\mathcal{F}(\alpha, x) = \sum_{j=1}^{N_B+1} \mathcal{F}'_{x_j}(\alpha) \cdot x_j, \tag{6.42}$$

where $\mathcal{F}'_{x_j}$ denotes the coefficient of x_j. As $\mathcal{F}$ is linear in x_j, we have

$$\mathcal{F}'_{x_j}(\alpha) = \frac{\partial}{\partial x_j} \mathcal{F}(\alpha, x). \tag{6.43}$$

In order to derive the differential equation we start again from the Schwinger parameter representation

$$I_{\nu_1 \dots \nu_{n_{\mathrm{int}}}} = \frac{e^{l \varepsilon \gamma_{\mathrm{E}}}}{\prod\limits_{k=1}^{n_{\mathrm{int}}} \Gamma(\nu_k)} \int\limits_{\alpha_k \geq 0} d^{n_{\mathrm{int}}} \alpha \left(\prod_{k=1}^{n_{\mathrm{int}}} \alpha_k^{\nu_k - 1} \right) \frac{1}{\left[\mathcal{U}(\alpha)\right]^{\frac{D}{2}}} e^{-\frac{\mathcal{F}(\alpha, x)}{\mathcal{U}(\alpha)}}. \tag{6.44}$$

The only dependence on the kinematic variables is through the second Symanzik polynomial $\mathcal{F}(\alpha, x)$. We therefore find

$$\frac{\partial}{\partial x_j} I_{\nu_1 \dots \nu_{n_{\text{int}}}} = - \frac{e^{l \varepsilon \gamma_{\text{E}}}}{\prod\limits_{k=1}^{n_{\text{int}}} \Gamma(\nu_k)} \int\limits_{\alpha_k \geq 0} d^{n_{\text{int}}} \alpha \left(\prod_{k=1}^{n_{\text{int}}} \alpha_k^{\nu_k - 1} \right) \frac{\mathcal{F}'_{x_j}(\alpha)}{\mathcal{U}(\alpha) \cdot [\mathcal{U}(\alpha)]^{\frac{D}{2}}} e^{-\frac{\mathcal{F}(\alpha, x)}{\mathcal{U}(\alpha)}} \tag{6.45}$$

for $x_j \in \{x_1, \dots, x_{N_B+1}\}$. The additional factor of $\mathcal{U}(\alpha)$ in the denominator is again equivalent to shifting the space-time dimension of the Feynman integral by two units. The additional factor of $\mathcal{F}'_{x_j}(\alpha)$ in the numerator is a polynomial in the Schwinger parameters, equivalent to the action of the polynomial $\mathcal{F}'_{x_j}(\mathbf{1}^+, \dots, \mathbf{n}_{\text{int}}{}^+)$ in the raising operators on the Feynman integral. Thus

$$\frac{\partial}{\partial x_j} I_{\nu_1 \dots \nu_{n_{\text{int}}}} = -\mathcal{F}'_{x_j}\left(\mathbf{1}^+, \dots, \mathbf{n}_{\text{int}}{}^+\right) \mathbf{D}^+ I_{\nu_1 \dots \nu_{n_{\text{int}}}} \tag{6.46}$$

for $x_j \in \{x_1, \dots, x_{N_B+1}\}$.

If the kinematic variable x_j corresponds to an internal mass, there is a slightly simpler formula, which follows directly from the momentum representation. Let us assume

$$x_j = \frac{m_j^2}{\mu^2}. \tag{6.47}$$

Let's further assume that m_j denotes the mass of the j-th internal edge, that this mass is distinct from all other internal masses and that the kinematic configuration is defined without any reference to this mass (i.e. we exclude on-shell conditions like $p^2 = m_j^2$). In other words, x_j enters only as the mass of the the j-th internal propagator. From

$$\frac{\partial}{\partial x_j} \frac{1}{\left(-q_j^2 + m_j^2\right)^{\nu_j}} = -\nu_j \frac{\mu^2}{\left(-q_j^2 + m_j^2\right)^{\nu_j + 1}} \tag{6.48}$$

we obtain in this case

$$\frac{\partial}{\partial x_j} I_{\nu_1 \dots \nu_{n_{\text{int}}}} = -\mathbf{j}^+ I_{\nu_1 \dots \nu_{n_{\text{int}}}}. \tag{6.49}$$

We may relax the condition that the mass of the j-th internal propagator has to be distinct from all other internal masses. Let S_{x_j} be the subset of $\{1, \dots, n_{\text{int}}\}$ containing all indices of internal edges whose internal mass equals m_j. We keep the condition, that the kinematic configuration is defined without any reference to m_j^2. Then

$$\frac{\partial}{\partial x_j} I_{\nu_1 \dots \nu_{n_{\mathrm{int}}}} = - \sum_{j \in S_{x_j}} \mathbf{j}^+ I_{\nu_1 \dots \nu_{n_{\mathrm{int}}}}, \tag{6.50}$$

which follows directly from the product rule for differentiation.

Exercise 46 *Show that*

$$\sum_{j=1}^{N_B+1} x_j \frac{\partial}{\partial x_j} I_{\nu_1 \dots \nu_{n_{\mathrm{int}}}} = \left(\frac{lD}{2} - \nu \right) \cdot I_{\nu_1 \dots \nu_{n_{\mathrm{int}}}}. \tag{6.51}$$

Hint: Consider the Feynman parameter representation.

From Eq. (6.51) we may extract the derivative with respect to x_{N_B+1}, provided we know all derivatives with respect to $x_1, \dots, x_{N_B}$.

From now on we consider again the case, where we set one kinematic variable to one (i.e. $x_{N_B+1} = 1$). We view the Feynman integral $I_{\nu_1 \dots \nu_{n_{\mathrm{int}}}}(D, x_1, \dots, x_{N_B})$ as a function of D and the N_B kinematic variables $x_1, \dots, x_{N_B}$. The derivatives with respect to the kinematic variables are given by Eq. (6.46). The expression on the right-hand side of Eq. (6.46) is a linear combination of Feynman integrals in $(D+2)$ space-time dimensions. Using integration-by-parts identities we may reduce this expression to a linear combination of master integrals in $(D+2)$ space-time dimensions. Using the dimensional shift relations discussed in Sect. 6.2 we may express each master integral in $(D+2)$ space-time dimensions as a linear combination of master integrals in D space-time dimensions. Combining these two operations we may express the right-hand side of Eq. (6.46) as a linear combination of master integrals in D space-time dimensions.

Let us now specialise to the basis

$$\vec{I} = \left(I_{\boldsymbol{\nu}_1}, I_{\boldsymbol{\nu}_2}, \dots, I_{\boldsymbol{\nu}_{N_{\mathrm{master}}}} \right)^T, \tag{6.52}$$

For each $I_{\boldsymbol{\nu}_i} \in \{ I_{\boldsymbol{\nu}_1}, \dots, I_{\boldsymbol{\nu}_{N_{\mathrm{master}}}} \}$ we therefore have

$$\frac{\partial}{\partial x_j} I_{\boldsymbol{\nu}_i} = - \sum_{k=1}^{N_{\mathrm{master}}} A_{x_j, ik} \, I_{\boldsymbol{\nu}_k}, \quad 1 \le i \le N_{\mathrm{master}}, \quad 1 \le j \le N_B, \tag{6.53}$$

where the coefficients $A_{x_j,ik}$ are rational functions of D and $x_1, \dots, x_{N_B}$. The fact that the coefficients $A_{x_j,ik}$ are rational functions follows from the fact that integration-by-parts identities and dimensional shift relations involve only rational functions.

Let us make a few definitions: We use the standard notation for the total differential with respect to the kinematic variables $x_1, \dots, x_{N_B}$ (D is treated as a parameter)

$$d I_{\boldsymbol{\nu}_i} = \sum_{j=1}^{N_B} \left(\frac{\partial I_{\boldsymbol{\nu}_i}}{\partial x_j} \right) dx_j. \tag{6.54}$$

We denote by A_{x_j} the $(N_{\text{master}} \times N_{\text{master}})$-matrix with entries $A_{x_j,ik}$. We also define a matrix-valued one-form A by

$$A = \sum_{j=1}^{N_B} A_{x_j} dx_j. \tag{6.55}$$

We may then write the system of differential equations compactly as

$$(d + A)\,\vec{I} = 0. \tag{6.56}$$

This is the sought-after system of first-order differential equations for the master integrals $I_{\nu_1}, \dots, I_{\nu_{N_{\text{master}}}}$. This system is integrable, which puts a constraint on A. The integrability condition reads

$$dA + A \wedge A = 0. \tag{6.57}$$

Let us summarise:

System of differential equations:
The vector of master integrals $\vec{I} = (I_{\nu_1}, I_{\nu_2}, \dots, I_{\nu_{N_{\text{master}}}})^T$ satisfies the differential equation

$$(d + A)\,\vec{I} = 0, \tag{6.58}$$

where d denotes the total differential with respect to the kinematic variables $x_1, \dots, x_{N_B}$ and A is a matrix-valued one-form, which satisfies the integrability condition

$$dA + A \wedge A = 0. \tag{6.59}$$

If we write

$$A = \sum_{j=1}^{N_B} A_{x_j} dx_j, \tag{6.60}$$

then the A_{x_j}'s are $(N_{\text{master}} \times N_{\text{master}})$-matrices with entries, which are rational functions in $x_1, \dots, x_{N_B}$ and D. The matrix-valued one-form A is computable with the help Eq. (6.46), integration-by-parts identities and dimensional shift relations.

Example 1

Let's look at a few examples. As our first example we consider the one-loop two-point function with equal internal masses, discussed below Eq. (6.6):

$$I_{\nu_1\nu_2}(D,x) = e^{\varepsilon\gamma_E}\left(m^2\right)^{\nu_{12}-\frac{D}{2}}\int\frac{d^Dk}{i\pi^{\frac{D}{2}}}\frac{1}{\left(-q_1^2+m^2\right)^{\nu_1}\left(-q_2^2+m^2\right)^{\nu_2}}, \tag{6.61}$$

with $x = -p^2/m^2$. As a basis of master integrals we choose

$$\vec{I} = \begin{pmatrix} I_{10} \\ I_{11} \end{pmatrix}. \tag{6.62}$$

The graph polynomials read

$$\mathcal{U} = \alpha_1 + \alpha_2, \ \mathcal{F} = \alpha_1\alpha_2 x + (\alpha_1+\alpha_2)^2. \tag{6.63}$$

From Eq. (6.46) we have

$$\frac{\partial}{\partial x}I_{\nu_1\nu_2}(D,x) = -\mathbf{1}^+\mathbf{2}^+\mathbf{D}^+ I_{\nu_1\nu_2}(D,x) = -\nu_1\nu_2 I_{(\nu_1+1)(\nu_2+1)}(D+2,x). \tag{6.64}$$

Using integration-by-parts identities and dimensional shift relations we obtain

$$\begin{aligned}\frac{\partial}{\partial x}I_{10}(D,x) &= 0,\\ \frac{\partial}{\partial x}I_{11}(D,x) &= -\frac{D-2}{x(4+x)}I_{10}(D,x) - \frac{4+(4-D)x}{2x(4+x)}I_{11}(D,x).\end{aligned} \tag{6.65}$$

Hence

$$A = \begin{pmatrix} 0 & 0 \\ \frac{D-2}{x(4+x)} & \frac{4+(4-D)x}{2x(4+x)} \end{pmatrix} dx. \tag{6.66}$$

Exercise 47 *The steps from Eqs. (6.64) to (6.65) can still be carried out by hand. Fill in the missing details.*
Hint: Use Eqs. (6.11), (6.14) and the result from Exercise 45.

Example 2

As our second example we consider the one-loop four-point function shown in Fig. 2.4

$$I_{\nu_1\nu_2\nu_3\nu_4}(D, x_1, x_2) = e^{\varepsilon\gamma_E}\left(-p_4^2\right)^{\nu_{1234}-\frac{D}{2}} \int \frac{d^D k}{i\pi^{\frac{D}{2}}} \frac{1}{\left(-q_1^2\right)^{\nu_1}\left(-q_2^2\right)^{\nu_2}\left(-q_3^2\right)^{\nu_3}\left(-q_4^2\right)^{\nu_4}}, \tag{6.67}$$

with $q_1 = k - p_1, q_2 = k - p_1 - p_2, q_3 = k - p_1 - p_2 - p_3, q_4 = k$ and vanishing internal masses. We consider the kinematic configuration $p_1^2 = p_2^2 = p_3^2 = 0$ but $p_4^2 \neq 0$. The integral depends on two kinematic variables, which we take as

$$x_1 = \frac{2p_1 \cdot p_2}{p_4^2}, \quad x_2 = \frac{2p_2 \cdot p_3}{p_4^2}. \tag{6.68}$$

The graph polynomials are given by

$$\mathcal{U} = \alpha_1 + \alpha_2 + \alpha_3 + \alpha_4, \quad \mathcal{F} = \alpha_2\alpha_4 x_1 + \alpha_1\alpha_3 x_2 + \alpha_3\alpha_4. \tag{6.69}$$

There are four master integral and we choose as basis

$$\vec{I} = \begin{pmatrix} I_{0011} \\ I_{0101} \\ I_{1010} \\ I_{1111} \end{pmatrix}. \tag{6.70}$$

The computations to determine the differential equation are best done with the help of a computer algebra program and we only quote the result here. One finds

$$A = A_{x_1} dx_1 + A_{x_2} dx_2, \tag{6.71}$$

$$A_{x_1} = \begin{pmatrix} 0 & 0 & 0 & 0 \\ 0 & \frac{D-4}{2x_1} & 0 & 0 \\ 0 & 0 & 0 & 0 \\ \frac{2(D-3)}{x_1(1-x_1)(1-x_1-x_2)} & -\frac{2(D-3)}{x_1(1-x_1)(1-x_1-x_2)} & -\frac{2(D-3)}{x_1 x_2(1-x_1-x_2)} & -\frac{2x_1+(D-6)(1-x_2)}{2x_1(1-x_1-x_2)} \end{pmatrix},$$

$$A_{x_2} = \begin{pmatrix} 0 & 0 & 0 & 0 \\ 0 & 0 & 0 & 0 \\ 0 & 0 & -\frac{D-4}{2x_2} & 0 \\ \frac{2(D-3)}{x_2(1-x_2)(1-x_1-x_2)} & -\frac{2(D-3)}{x_1 x_2(1-x_1-x_2)} & -\frac{2(D-3)}{x_2(1-x_2)(1-x_1-x_2)} & -\frac{2x_2+(D-6)(1-x_1)}{2x_2(1-x_1-x_2)} \end{pmatrix}.$$

Exercise 48 *This example depends on two kinematic variables x_1 and x_2, hence the integrability condition is non-trivial. Check explicitly the integrability condition*

$$dA + A \wedge A = 0. \tag{6.72}$$

In order to derive the differential equation, we first make a choice for the master integrals. In this example the choice of master integrals is given by Eq. (6.70). However, there is no particular reason for this specific choice (except that it is the default choice of the computer program `Kira`). In general, we may choose a set of four linear independent linear combinations of these four master integrals with coefficients being functions of D and x. In the simplest case the coefficients are rational functions of D and x. However, we will soon also consider the case, where the coefficients are algebraic functions of x (e.g. expressions with square roots). The dependence on D of the coefficients will usually remain rational. In Chap. 13 we will also consider the case, where the coefficients are transcendental functions of x.

Let us now explore this freedom. Suppose we start from the basis

$$\vec{I}' = \begin{pmatrix} -\frac{1}{2}(D-3)(D-4)\, I_{0011} \\ -\frac{1}{2}(D-3)(D-4)\, I_{0101} \\ -\frac{1}{2}(D-3)(D-4)\, I_{1010} \\ \frac{1}{8}(D-4)^2\, x_1 x_2 I_{1111} \end{pmatrix}. \tag{6.73}$$

We repeat the calculation with this basis of master integrals. Again we find a differential equation

$$\left(d + A'\right)\vec{I}' = 0, \qquad A' = A'_{x_1} dx_1 + A'_{x_2} dx_2, \tag{6.74}$$

where the matrices A'_{x_1} and A'_{x_2} are now given by

$$A'_{x_1} = \frac{4-D}{2}\begin{pmatrix} 0 & 0 & 0 & 0 \\ 0 & \frac{1}{x_1} & 0 & 0 \\ 0 & 0 & 0 & 0 \\ \frac{1}{x_1-1} - \frac{1}{x_1+x_2-1} & -\frac{1}{x_1-1} + \frac{1}{x_1+x_2-1} & \frac{1}{x_1+x_2-1} & \frac{1}{x_1} - \frac{1}{x_1+x_2-1} \end{pmatrix},$$
$$A'_{x_2} = \frac{4-D}{2}\begin{pmatrix} 0 & 0 & 0 & 0 \\ 0 & 0 & 0 & 0 \\ 0 & 0 & \frac{1}{x_2} & 0 \\ \frac{1}{x_2-1} - \frac{1}{x_1+x_2-1} & \frac{1}{x_1+x_2-1} & -\frac{1}{x_2-1} + \frac{1}{x_1+x_2-1} & \frac{1}{x_2} - \frac{1}{x_1+x_2-1} \end{pmatrix}. \tag{6.75}$$

We observe that the only dependence on D is now through the prefactor $(4-D)/2$. Within dimensional regularisation we usually set $D = 4 - 2\varepsilon$. Then

$$\frac{4-D}{2} = \varepsilon. \tag{6.76}$$

Let us further introduce five one-forms

$$\omega_1 = d\ln(x_1) = \frac{dx_1}{x_1}, \qquad \omega_2 = d\ln(x_1-1) = \frac{dx_1}{x_1-1},$$
$$\omega_3 = d\ln(x_2) = \frac{dx_2}{x_2}, \qquad \omega_4 = d\ln(x_2-1) = \frac{dx_2}{x_2-1},$$
$$\omega_5 = d\ln(x_1+x_2-1) = \frac{dx_1+dx_2}{x_1+x_2-1}. \tag{6.77}$$

Differential one-forms as in Eq. (6.77) are called **dlog-forms**. We may then write A' as

$$A' = \varepsilon \begin{pmatrix} 0&0&0&0 \\ 0&1&0&0 \\ 0&0&0&0 \\ 0&0&0&1 \end{pmatrix} \omega_1 + \varepsilon \begin{pmatrix} 0&0&0&0 \\ 0&0&0&0 \\ 0&0&0&0 \\ 1&-1&0&0 \end{pmatrix} \omega_2 + \varepsilon \begin{pmatrix} 0&0&0&0 \\ 0&0&0&0 \\ 0&0&1&0 \\ 0&0&0&1 \end{pmatrix} \omega_3$$
$$+\varepsilon \begin{pmatrix} 0&0&0&0 \\ 0&0&0&0 \\ 0&0&0&0 \\ 1&0&-1&0 \end{pmatrix} \omega_4 + \varepsilon \begin{pmatrix} 0&0&0&0 \\ 0&0&0&0 \\ 0&0&0&0 \\ -1&1&1&-1 \end{pmatrix} \omega_5. \quad (6.78)$$

In the basis $\vec{I}'$ the differential equation has a particular simple form, which we will discuss in more detail in Sect. 6.3.2. In particular there is a systematic (and easy) way to solve this differential equation order-by-order in the dimensional regularisation parameter. We discuss this in Sect. 6.3.3.

We call the one-forms $\omega_1, \omega_2, \omega_3, \omega_4$ and ω_5 **letters** and the set of all independent letters the **alphabet**. In this example the alphabet contains five letters. We denote the number of letters in the alphabet by N_L.

Example 3

As third example let us consider the double-box integral discussed in Exercise 44. We use the same notation and consider

$$I_{\nu_1 \nu_2 \nu_3 \nu_4 \nu_5 \nu_6 \nu_7 \nu_8 \nu_9} \quad (6.79)$$

with $\nu_8, \nu_9 \leq 0$. We assume that all external momenta are light-like ($p_1^2 = p_2^2 = p_3^2 = p_4^2 = 0$) and that all internal propagators are massless. As usual we define the Mandelstam variables by

$$s = (p_1 + p_2)^2, \; t = (p_2 + p_3)^2. \quad (6.80)$$

We set $\mu^2 = t$ and $x = s/t$. There are eight master integral and we choose as basis

$$\vec{I} = \begin{pmatrix} I_{001110000} \\ I_{100100100} \\ I_{011011000} \\ I_{100111000} \\ I_{111100100} \\ I_{101110100} \\ I_{111111100} \\ I_{1111111(-1)0} \end{pmatrix}. \quad (6.81)$$

The calculations to determine the differential equation are again best carried out with the help of a computer program and one finds $A = A_x dx$ with

$$A_x = \begin{pmatrix} -\frac{D-3}{x} & 0 & 0 & 0 & 0 & 0 & 0 & 0 \\ 0 & 0 & 0 & 0 & 0 & 0 & 0 & 0 \\ 0 & 0 & -\frac{D-4}{x} & 0 & 0 & 0 & 0 & 0 \\ 0 & 0 & 0 & -\frac{D-4}{x} & 0 & 0 & 0 & 0 \\ 0 & \frac{(3D-8)(3D-10)}{2(D-4)x(1+x)} & 0 & \frac{3D-10}{2x(1+x)} & \frac{2x-(D-6)}{2x(1+x)} & 0 & 0 & 0 \\ -\frac{(D-3)(3D-8)(3D-10)}{(D-4)^2x^2(1+x)} & \frac{(D-3)(3D-8)(3D-10)}{(D-4)^2x(1+x)} & 0 & 0 & 0 & \frac{x-(D-4)}{x(1+x)} & 0 & 0 \\ \frac{3(D-3)(3D-8)(3D-10)}{(D-4)^2x^3(1+x)} & \frac{3(D-3)(3D-8)(3D-10)}{(D-4)^2x^2(1+x)} & 0 & \frac{3(D-3)(3D-10)}{(D-4)x^2(1+x)} & \frac{6(D-3)}{x(1+x)} & -\frac{3(D-4)}{x^2} & \frac{2}{x} & \frac{D-4}{x(1+x)} \\ A_{x,81} & A_{x,82} & A_{x,83} & A_{x,84} & A_{x,85} & A_{x,86} & A_{x,87} & A_{x,88} \end{pmatrix}.$$

The entries in the eighth row are a little bit longer and we list them separately below:

$$\begin{aligned}
A_{x,81} &= \frac{3(D-3)(3D-8)(3D-10)\left[(3D-14)x+2(D-5)\right]}{2(D-4)^3 x^3 (1+x)}, \\
A_{x,82} &= \frac{3(D-3)(3D-8)(3D-10)\left[(3D-14)x+2(2D-9)\right]}{2(D-4)^3 x^2 (1+x)}, \\
A_{x,83} &= \frac{4(D-3)^2}{(D-4)x^3}, \\
A_{x,84} &= \frac{3(D-3)(3D-10)\left[(3D-14)x+2(2D-9)\right]}{2(D-4)^2 x^2 (1+x)}, \\
A_{x,85} &= \frac{3(D-3)(3D-14)}{[(D-4)x}, \\
A_{x,86} &= -\frac{3\left[(3D-14)x+2(2D-9)\right]}{2x^2}, \\
A_{x,87} &= -\frac{D-4}{x}, \\
A_{x,88} &= -\frac{(3D-16)x+4(D-5)}{2x(1+x)}.
\end{aligned} \tag{6.82}$$

We will always order the master integrals such that master integrals which can be obtained through pinching (and possibly symmetry relations) from other master integrals appear before their parent integrals. The matrices A_{x_j} have then always a lower block triangular structure, induced by the sub-sectors. In Eq. (6.16) we introduced the sector identification number N_{id}. In the example of the double box integral we have master integrals from seven sectors:

$$\begin{aligned} N_{\text{id}} &= 28, \quad && I_{001110000}, \\ N_{\text{id}} &= 73, \quad && I_{100100100}, \\ N_{\text{id}} &= 54, \quad && I_{011011000}, \\ N_{\text{id}} &= 57, \quad && I_{100111000}, \\ N_{\text{id}} &= 79, \quad && I_{111100100}, \\ N_{\text{id}} &= 93, \quad && I_{101110100}, \\ N_{\text{id}} &= 127, \quad && I_{111111100}, I_{1111111(-1)0}. \end{aligned} \tag{6.83}$$

The first six sectors have one master integral per sector, while the seventh sector (sector 127) has two master integrals.

Example 4

As fourth and final example let us consider the two-loop sunrise integral with equal internal masses, shown in Fig. 2.16:

$$I_{\nu_1\nu_2\nu_3}(D, x) = e^{2\varepsilon\gamma_E} \left(m^2\right)^{\nu_{123}-D} \int \frac{d^D k_1}{i\pi^{\frac{D}{2}}} \frac{d^D k_2}{i\pi^{\frac{D}{2}}} \frac{1}{\left(-q_1^2+m^2\right)^{\nu_1} \left(-q_2^2+m^2\right)^{\nu_2} \left(-q_3^2+m^2\right)^{\nu_3}}, \tag{6.84}$$

with $x = -p^2/m^2$ and $q_1 = k_1, q_2 = k_2 - k_1, q_3 = -k_2 - p$. We have set $\mu^2 = m^2$. There are three master integrals and we choose the basis

$$\vec{I} = \begin{pmatrix} I_{110} \\ I_{111} \\ I_{211} \end{pmatrix}. \tag{6.85}$$

We obtain the differential equation

$$(d + A)\,\vec{I} = 0 \tag{6.86}$$

with

$$\begin{aligned} A = &\begin{pmatrix} 0 & 0 & 0 \\ 0 & -(D-3) & -3 \\ 0 & \frac{1}{6}(D-3)(3D-8) & \frac{1}{2}(3D-8) \end{pmatrix} \frac{dx}{x} \\ &+ \begin{pmatrix} 0 & 0 & 0 \\ 0 & 0 & 0 \\ -\frac{1}{4} & -\frac{1}{8}(D-3)(3D-8) & -(D-3) \end{pmatrix} \frac{dx}{x+1} \\ &+ \begin{pmatrix} 0 & 0 & 0 \\ 0 & 0 & 0 \\ \frac{1}{4} & -\frac{1}{24}(D-3)(3D-8) & -(D-3) \end{pmatrix} \frac{dx}{x+9}. \end{aligned} \tag{6.87}$$

The two-loop sunrise integral with equal non-vanishing internal masses is a Feynman integral which cannot be expressed in terms of multiple polylogarithms. We will discuss this integral in more detail in Chap. 13.

6.3.2 The ε-form of the System of Differential Equations

In the previous section we have seen that we can always systematically obtain the differential equation for a set of master integrals $\vec{I}$:

$$(d + A)\,\vec{I} = 0. \tag{6.88}$$

A is a matrix-valued one-form, which we write as

$$A = \sum_{j=1}^{N_B} A_{x_j} dx_j. \tag{6.89}$$

The $(N_{\text{master}} \times N_{\text{master}})$-matrices A_{x_j} can be computed with the help Eq. (6.46), integration-by-parts identities and dimensional shift relations. In general, the entries of A_{x_j} are rational functions of $x_1, \dots, x_{N_B}$ and D.

It will be convenient to exchange the D-dependence for a dependence on the dimensional regularisation parameter ε. We fix an even integer D_{int}, giving us the dimension of space-time we are interested in and set $D = D_{\text{int}} - 2\varepsilon$. Our usual interest is $D_{\text{int}} = 4$, hence

$$D = 4 - 2\varepsilon, \;\; \varepsilon = \frac{4 - D}{2}. \tag{6.90}$$

The entries of A_{x_j} are then rational functions of x and ε. In Eq. (6.78) we have already seen an example where the dependence on ε is rather simple: In Eq. (6.78) the only dependence on ε is through a prefactor ε. Furthermore, the dependence on the kinematic variables x_1 and x_2 has also particular features: The one-forms $\omega_1, \omega_2, \dots, \omega_5$ have only simple poles. Thirdly, the entries of the (4×4)-matrices multiplying $\varepsilon \cdot \omega_j$ are integer numbers.

We say that the set of master integrals $\vec{I}$ satisfies a differential equation in ε-form [123] if the following conditions are met:

ε-form of the differential equation:
The differential equation

$$(d + A)\,\vec{I} = 0. \tag{6.91}$$

is in ε-form, if A is of the form

$$A = \varepsilon \sum_{j=1}^{N_L} C_j \, \omega_j, \tag{6.92}$$

where

1. C_j is a $(N_{\mathrm{master}} \times N_{\mathrm{master}})$-matrix, whose entries are algebraic numbers,
2. the only dependence on ε is given by the explicit prefactor,
3. the only singularities of the differential one-forms ω_j are simple poles,
4. the non-zero boundary constants have uniform weight zero.

We call the ω_j's **letters** and N_L the **number of letters**.

The first requirement (the entries of the matrices C_j are algebraic numbers) forbids transcendental numbers like π^2 as entries.

The fourth requirement is a condition on the boundary constants. This is best explained by an example. Consider the one-loop tadpole integral, given by Eq. (2.123). For $\mu^2 = m^2$ this integral does not depend on any kinematic variable and can be considered in the framework of differential equations as a pure boundary constant. We have

$$\begin{aligned} I = T_1\,(4-2\varepsilon) &= e^{\varepsilon\gamma_{\mathrm{E}}}\Gamma\,(-1+\varepsilon) \\ &= -\frac{1}{\varepsilon} - 1 - \left(1 - \frac{1}{2}\zeta_2\right)\varepsilon + \left(\frac{1}{3}\zeta_3 - \frac{1}{2}\zeta_2 - 1\right)\varepsilon^2 + O\left(\varepsilon^3\right). \end{aligned} \tag{6.93}$$

We assign rational numbers and more generally any algebraic expression in the kinematic variables the weight zero, π the weight 1, zeta values ζ_n the weight n and the dimensional regularisation parameter ε the weight (-1). The weight of a product is the sum of the weights of its factors. We say that an expression, given as a sum of terms, is of **uniform weight** if every term has the same weight. With these assignments the expression in Eq. (6.93) is not of uniform weight. However, if we consider instead of I the integral I', defined by

$$\begin{aligned} I' = \varepsilon T_2\,(4-2\varepsilon) &= \varepsilon T_1\,(2-2\varepsilon) = e^{\varepsilon\gamma_{\mathrm{E}}}\Gamma\,(1+\varepsilon) \\ &= 1 + \frac{1}{2}\zeta_2\varepsilon^2 - \frac{1}{3}\zeta_3\varepsilon^3 + \frac{9}{16}\zeta_4\varepsilon^4 - \left(\frac{1}{5}\zeta_5 + \frac{1}{6}\zeta_2\zeta_3\right)\varepsilon^5 + O\left(\varepsilon^6\right), \end{aligned} \tag{6.94}$$

we see that I' is of uniform weight zero. A uniform weight zero implies that the coefficient of the ε^j-term in the ε-expansion has weight j.

The differential equation with A' given by Eq. (6.78) is in ε-form.

Digression Poles and residues of differential forms

Let us digress and discuss poles and residues of differential forms. Let X be a complex manifold of dimension n and ω a differential k-form. ω is **closed**, *if*

$$d\omega = 0. \tag{6.95}$$

ω is **exact**, *if there is a $(k-1)$-form η such that*

$$\omega = d\eta. \tag{6.96}$$

The k-th **de Rham cohomology group** *$H^k_{\mathrm{dR}}(X)$ is the set of equivalence classes of closed k-forms modulo exact k-forms. The group law is the addition of k-forms.*

We are in particular interested in holomorphic and meromorphic k-forms on X:

$$\omega = \sum_I \omega_I(x)\, dx_I, \qquad dx_I \;=\; dx_{i_1} \wedge dx_{i_2} \wedge \cdots \wedge dx_{i_k}. \tag{6.97}$$

ω is holomorphic if all the $\omega_I(x)$ are holomorphic functions, ω is meromorphic if all the $\omega_I(x)$ are meromorphic functions. Note that ω in Eq. (6.97) does not contain any antiholomorphic differentials $d\bar{x}_j$.

Let Y be a complex codimension one submanifold, defined locally by an equation

$$Y = \{x \in X | f(x) = 0\}, \tag{6.98}$$

where f is meromorphic and $df \neq 0$ on Y. (Don't worry about the poles of f, Y is defined by the zeros of f and as we are only interested in local properties, we are essentially saying that f is holomorphic in a neighbourhood of $f(x) = 0$.) The k-form ω has a pole of order r on the manifold Y, if r is the smallest integer such that $f^r \cdot \omega$ is holomorphic in a neighbourhood of Y. Let us further assume that ω is closed. We may write ω as

$$\omega = \frac{df}{f^r} \wedge \psi + \theta, \tag{6.99}$$

where the $(k-1)$-form ψ is holomorphic in a neighbourhood of Y, and the k-form θ has at most a pole of order $(r-1)$ on Y. We may reduce poles of order $r > 1$ to poles of order 1 and exact forms due to the identity

$$\frac{df}{f^r} \wedge \psi + \theta = d\left(-\frac{\psi}{(r-1)\, f^{r-1}}\right) + \frac{d\psi}{(r-1)\, f^{r-1}} + \theta. \tag{6.100}$$

Thus every form ω is equivalent up to an exact form to a form ω_1 with at most a simple pole on Y. For a form ω_1 with at most a simple pole on Y

$$\omega_1 = \frac{df}{f} \wedge \psi_1 + \theta_1, \tag{6.101}$$

we define the **Leray residue** *[124] of ω_1 along Y by*

$$\text{Res}_Y(\omega_1) = \psi_1|_Y, \tag{6.102}$$

where $\psi_1|_Y$ denotes the restriction of ψ_1 to Y. If ω_1 is equivalent to ω up to an exact form we set

$$\text{Res}_Y(\omega) = \text{Res}_Y(\omega_1). \tag{6.103}$$

Since we assumed ω to be closed and having a pole (of order r) along Y, the k-form ω defines a class $[\omega] \in H^k_{\text{dR}}(X\backslash Y)$. It can be shown that $\text{Res}_Y(\omega)$ is independent of the chosen representative and the decomposition in Eq. (6.99). Furthermore, $\text{Res}_Y(\omega)$ is again closed. Therefore the Leray residue defines a map

$$\text{Res}_Y : H^k_{\text{dR}}(X\backslash Y) \to H^{k-1}_{\text{dR}}(Y). \tag{6.104}$$

Multivariate (Leray) residues are defined as follows: Suppose we have two codimension one sub-varieties Y_1 and Y_2 defined by $f_1 = 0$ and $f_2 = 0$, respectively. Again we may reduce higher poles to simple poles modulo exact forms. Let us therefore consider

$$\omega = \frac{df_1}{f_1} \wedge \frac{df_2}{f_2} \wedge \psi_{12} + \frac{df_1}{f_1} \wedge \psi_1 + \frac{df_2}{f_2} \wedge \psi_2 + \theta, \tag{6.105}$$

where ψ_{12} is regular on $Y_1 \cap Y_2$, ψ_j is regular on Y_j and θ is regular on $Y_1 \cup Y_2$. One sets

$$\text{Res}_{Y_1,Y_2}(\omega) = \psi_{12}|_{Y_1 \cap Y_2}. \tag{6.106}$$

Note that the residue is anti-symmetric with respect to the order of the hypersurfaces:

$$\text{Res}_{Y_2,Y_1}(\omega) = -\text{Res}_{Y_1,Y_2}(\omega). \tag{6.107}$$

Multivariate residues for several codimension one sub-varieties $Y_1, \ldots, Y_m$ are defined analogously.

Exercise 49 *Let $X = \mathbb{C}^2$ and*

$$Y = \{x \in X | x_1 + x_2 = 0\}. \tag{6.108}$$

Compute

$$\text{Res}_Y \left(\frac{x_1 x_2^2 dx_1 \wedge dx_2}{x_1 + x_2} \right). \tag{6.109}$$

There is a second definition of the Leray residue: Let us first introduce the **tubular neighbourhood** of Y: These are all points in X, with distance to Y less or equal to a small quantity δ. Let's denote the boundary of this tubular neighbourhood by δY. As X has real dimension $2n$, and Y has real dimension $(2n-2)$, δY has real dimension $(2n-1)$. δY consists of all points in X, which are a distance δ away from Y. By construction, δY does not intersect Y. Locally, we may choose n complex coordinates $(y_1, \dots, y_{n-1}, z)$ on X such that $(y_1, \dots, y_{n-1}, 0) \in Y$. Then for each point $(y_1, \dots, y_{n-1}, 0) \in Y$ the points $(y_1, \dots, y_{n-1}, z) \in \delta Y$ are the ones with

$$|z| = \delta. \tag{6.110}$$

This is a circle in the complex z-plane with radius δ. Mathematically we say that the boundary δY of the tubular neighbourhood fibres over Y with fibre S^1. We may then integrate for each $y \in Y$ the differential k-form ω over S^1, yielding a $(k-1)$-form. This gives the second definition of the Leray residue:

$$\text{Res}_Y(\omega) = \frac{1}{2\pi i} \int\limits_{S^1} \omega|_{\delta Y}. \tag{6.111}$$

The orientation of S^1 is induced by the complex structure on X. Multivariate residues are then defined iteratively.

The two definitions of Leray's residue generalise two expressions for the ordinary residue known from complex analysis: Let $f(z)$ be meromorphic function of one complex variable with a pole of order r at z_0. We may give the residue of $f(z)$ at z_0 either as

$$\text{res}(f, z_0) = \frac{1}{(r-1)!} \left(\frac{d^{r-1}}{dz^{r-1}} \left[(z-z_0)^r f(z) \right] \right) \Bigg|_{z=z_0} \tag{6.112}$$

or as

$$\text{res}(f, z_0) = \frac{1}{2\pi i} \int\limits_{\gamma} f(z)\, dz, \tag{6.113}$$

where γ denotes a small circle around z_0, oriented counter-clockwise. Equation (6.112) corresponds to the first definition, Eq. (6.113) corresponds to the second definition.

A special case of the Leray residue is the **Grothendieck residue**, where we consider the n-fold residue of a n-form. Let X be a complex manifold of dimension

n as above and consider n meromorphic functions $f_1, f_2, \dots, f_n$, defining

$$Y_j = \left\{ x \in X | f_j(x) = 0 \right\}, \qquad 1 \leq j \leq n. \tag{6.114}$$

Assume further that the system of equations

$$f_1(x) = f_2(x) = \dots f_n(x) = 0 \tag{6.115}$$

has as solutions a finite number of isolated points $x^{(j)} = (x_1^{(j)}, ..., x_n^{(j)})$, where j labels the individual solutions. Let us further consider a function $g(x)$, regular at the solutions $x^{(j)}$. We now consider a n-form of the form

$$\omega = \frac{g}{f_1 f_2 \dots f_n} dx_1 \wedge dx_2 \wedge \dots \wedge dx_n. \tag{6.116}$$

We first continue to consider the situation locally. Let $x^{(j)}$ be one of the solutions of Eq. (6.115). We define the **local residue** or **Grothendieck residue** [125] of ω with respect to Y_1, ..., Y_n at $x^{(j)}$ by

$$\text{Res}_{Y_1,...,Y_n}\left(\omega, x^{(j)}\right) = \frac{1}{(2\pi i)^n} \oint\limits_{\Gamma_\delta} \frac{g(x)\ dx_1 \wedge ... \wedge dx_n}{f_1(x) ... f_n(x)}. \tag{6.117}$$

The integration in Eq. (6.117) is around a small n-torus

$$\Gamma_\delta = \{\ (x_1, ..., x_n) \in X \mid |f_i(x)| = \delta\ \}, \tag{6.118}$$

encircling $x^{(j)}$ with orientation

$$d \arg f_1 \wedge d \arg f_2 \wedge ... \wedge d \arg f_n \geq 0. \tag{6.119}$$

In order to evaluate a local residue it is advantageous to perform a change of variables

$$x_i' = f_i(x), \qquad i = 1, ..., n. \tag{6.120}$$

Let us denote the Jacobian of this transformation by

$$J(x) = \frac{1}{\det\left(\frac{\partial(f_1,...,f_n)}{\partial(x_1,...,x_n)}\right)}. \tag{6.121}$$

The local residue at $x^{(j)}$ is then given by

$$\text{Res}_{Y_1,\ldots,Y_n}\left(\omega, x^{(j)}\right) = \frac{1}{(2\pi i)^n} \oint\limits_{\Gamma_\delta} \frac{g\left(x\right)\, dx_1 \wedge \ldots \wedge dx_n}{f_1\left(x\right) \ldots f_n\left(x\right)} = J\left(x^{(j)}\right) g\left(x^{(j)}\right). \tag{6.122}$$

This also shows that the local residue at $x^{(j)}$ agrees with the Leray residue at $x^{(j)}$.

The **global residue** of ω with respect to f_1, ..., f_n is defined as

$$\text{Res}_{Y_1,\ldots,Y_n}\left(\omega\right) = \sum_{\text{solutions } j} \text{Res}_{Y_1,\ldots,Y_n}\left(\omega, x^{(j)}\right), \tag{6.123}$$

where the sum is over all solutions $x^{(j)}$ of Eq. (6.115).

Equation (6.123) defines the global residue of a n-form ω. As this n-form is given by the meromorphic function g and the n meromorphic functions $f_1, \ldots, f_n$ as in Eq. (6.116), it will be convenient to define the global residue of the function g by

$$\text{res}_{Y_1,\ldots,Y_n}\left(g\right) = \text{Res}_{Y_1,\ldots,Y_n}\left(\omega\right). \tag{6.124}$$

6.3.3 Solution in Terms of Iterated Integrals

In this section we solve a differential equation, which is in ε-form. We show that this can be done systematically.

We fix an even integer D_{int} and set $D = D_{\text{int}} - 2\varepsilon$. The main application will be $D_{\text{int}} = 4$, hence

$$D = 4 - 2\varepsilon, \quad \varepsilon = \frac{4 - D}{2}. \tag{6.125}$$

The Feynman integrals $I_{\nu_1 \ldots \nu_{n_{\text{int}}}}$ are then functions of ε and x. We are interested in the Laurent expansion in ε:

$$I_{\nu_1 \ldots \nu_{n_{\text{int}}}}\left(\varepsilon, x\right) = \sum_{j=j_{\min}}^{\infty} I^{(j)}_{\nu_1 \ldots \nu_{n_{\text{int}}}}\left(x\right) \cdot \varepsilon^j. \tag{6.126}$$

We would like to determine the coefficients $I^{(j)}_{\nu_1 \ldots \nu_{n_{\text{int}}}}\left(x\right)$. In applications towards perturbation theory we usually need only the first few terms of this Laurent expansion. The method discussed here is systematic and allows us to obtain as many terms of the Laurent expansion as desired.

Let $\vec{I}$ be a vector of N_{master} master integrals.

$$\vec{I} = \left(I_{\boldsymbol{\nu}_1}, I_{\boldsymbol{\nu}_2}, \ldots, I_{\boldsymbol{\nu}_{N_{\text{master}}}}\right)^T. \tag{6.127}$$

We make the following assumptions:

1. The differential equation for $\vec{I}$ is in ε-form:

$$(d + A)\,\vec{I} \;=\; 0, \;\; A \;=\; \varepsilon \sum_{j=1}^{N_L} C_j \,\omega_j. \tag{6.128}$$

2. All master integrals have a Taylor expansion in ε:

$$I_{\nu_i}\,(\varepsilon, x) = \sum_{j=0}^{\infty} I^{(j)}_{\nu_i}\,(x) \cdot \varepsilon^j. \tag{6.129}$$

3. We know suitable boundary values for all master integrals.

Assumption 2 may seem at first sight rather restrictive, as it forbids any pole terms in ε. However, it isn't. It only means that we multiplied a Feynman integral with a sufficient high power of ε, such that the Laurent expansion starts with the ε^0-term or later. We have already seen an example in Eq. (6.73): The first three master integrals I_{0011}, I_{0101} and I_{1010} are one-loop two-point function with vanishing internal masses. They have been calculated in Eq. (2.181). For example, the Laurent expansion for I_{0101} starts as

$$\begin{aligned} I_{0101} &= \frac{1}{\varepsilon} + (2 - L) + \left(\frac{1}{2}L^2 - 2L - \frac{\pi^2}{12} + 4\right)\varepsilon \\ &+ \left(-\frac{1}{6}L^3 + L^2 - 4\,L - \frac{7}{3}\zeta_3 - \frac{\pi^2}{6} + \frac{\pi^2}{12}L + 8\right)\varepsilon^2 + O\left(\varepsilon^3\right), \end{aligned} \tag{6.130}$$

with $L = \ln x_1$. This integral has a pole in ε. However, in going to the ε-form for the differential equation we defined a new master integral $I'_{0101} = \varepsilon\,(1 - 2\varepsilon)\,I_{0101}$. The Laurent expansion for I'_{0101} starts as

$$I'_{0101} = 1 - L\varepsilon + \left(\frac{1}{2}L^2 - \frac{\pi^2}{12}\right)\varepsilon^2 + \left(-\frac{1}{6}L^3 - \frac{7}{3}\zeta_3 + \frac{\pi^2}{12}L\right)\varepsilon^3 + O\left(\varepsilon^4\right). \tag{6.131}$$

This expansion starts at ε^0.

In order to present the solution of a differential equation in ε-form, we introduce **iterated integrals**. Let us start with the general definition of an iterated integrals [126]: Let X be a n-dimensional (complex) manifold and

$$\gamma : [a, b] \to X \tag{6.132}$$

a path with start point $x_a = \gamma(a)$ and end point $x_b = \gamma(b)$. Suppose further that $\omega_1, \dots, \omega_r$ are differential 1-forms on X. Let us write

$$f_j(\lambda)\, d\lambda = \gamma^* \omega_j \tag{6.133}$$

for the pull-backs to the interval $[a, b]$. If

$$\omega_j = \sum_{k=1}^{n} \omega_{jk}(x)\, dx_k, \quad \gamma(\lambda) = \begin{pmatrix} \gamma_1(\lambda) \\ \gamma_2(\lambda) \\ \vdots \\ \gamma_n(\lambda) \end{pmatrix}, \tag{6.134}$$

the pull-back $\gamma^* \omega_j$ is given by

$$\gamma^* \omega_j = \sum_{k=1}^{n} \omega_{jk}(\gamma(\lambda)) \frac{d\gamma_k(\lambda)}{d\lambda} d\lambda, \tag{6.135}$$

hence

$$f_j(\lambda) = \sum_{k=1}^{n} \omega_{jk}(\gamma(\lambda)) \frac{d\gamma_k(\lambda)}{d\lambda}. \tag{6.136}$$

Exercise 50 *Let*

$$\omega = 3dx_1 + (5 + x_1)dx_2 + x_3 dx_3 \tag{6.137}$$

and

$$\gamma : [0, 1] \to \mathbb{C}^3, \qquad \gamma(\lambda) = \begin{pmatrix} \lambda \\ \lambda^2 \\ 1 + \lambda \end{pmatrix}. \tag{6.138}$$

Compute

$$\int_\gamma \omega. \tag{6.139}$$

For $\lambda \in [a, b]$ the r-fold iterated integral of $\omega_1, \dots, \omega_r$ along the path γ is defined by

$$I_\gamma(\omega_1, \dots, \omega_r; \lambda) = \int_a^\lambda d\lambda_1 f_1(\lambda_1) \int_a^{\lambda_1} d\lambda_2 f_2(\lambda_2) \dots \int_a^{\lambda_{r-1}} d\lambda_r f_r(\lambda_r). \tag{6.140}$$

We call r the **depth** of the iterated integral. We define the 0-fold iterated integral to be

$$I_\gamma\left(;\lambda\right) = 1. \tag{6.141}$$

We then have the recursive structure

$$I_\gamma\left(\omega_1, \omega_2, \dots, \omega_r; \lambda\right) = \int\limits_a^\lambda d\lambda_1 f_1\left(\lambda_1\right) I_\gamma\left(\omega_2, \dots, \omega_r; \lambda_1\right). \tag{6.142}$$

Digression Basic properties of iterated integrals

Let $\gamma_1 : [a, b] \rightarrow X$ and $\gamma_2 : [a, b] \rightarrow X$ be two paths with $\gamma_1(b) = \gamma_2(a)$. In this case we may form a new path by gluing the endpoint of γ_1 to the starting point of γ_2. The combined path starts at $\gamma_1(a)$ and ends at $\gamma_2(b)$. In detail we define the path $\gamma_2 \circ \gamma_1 : [a, b] \rightarrow X$ to be given by

$$\left(\gamma_2 \circ \gamma_1\right)\left(\lambda\right) = \begin{cases} \gamma_1\left(2\lambda - a\right) \text{ for } a \leq \lambda \leq \frac{1}{2}\left(a+b\right), \\ \gamma_2\left(2\lambda - b\right) \text{ for } \frac{1}{2}\left(a+b\right) \leq \lambda \leq b. \end{cases}$$

For the iterated integral along the path $\gamma_2 \circ \gamma_1$ we have

$$I_{\gamma_2 \circ \gamma_1}\left(\omega_1, \dots, \omega_r; \lambda\right) = \sum_{j=0}^{r} I_{\gamma_2}\left(\omega_1, \dots, \omega_j; \lambda\right) I_{\gamma_1}\left(\omega_{j+1}, \dots, \omega_r; \lambda\right). \tag{6.143}$$

For a path $\gamma : [a, b] \rightarrow X$ we denote by $\gamma^{-1} : [a, b] \rightarrow X$ the reverse path given by

$$\gamma^{-1}\left(\lambda\right) = \gamma\left(a + b - \lambda\right). \tag{6.144}$$

For the iterated integral along the path γ^{-1} we have

$$I_{\gamma^{-1}}\left(\omega_1, \dots, \omega_r; b\right) = \left(-1\right)^r I_\gamma\left(\omega_r, \dots, \omega_1; b\right). \tag{6.145}$$

Exercise 51 *Prove Eq. (6.143) for the case $r = 2$, i.e. show*

$$I_{\gamma_2 \circ \gamma_1}\left(\omega_1, \omega_2; \lambda\right) = I_{\gamma_1}\left(\omega_1, \omega_2; \lambda\right) + I_{\gamma_2}\left(\omega_1; \lambda\right) I_{\gamma_1}\left(\omega_2; \lambda\right) + I_{\gamma_2}\left(\omega_1, \omega_2; \lambda\right). \tag{6.146}$$

Exercise 52 *Prove Eq. (6.145).*

Let us discuss the path (in-) dependence of iterated integrals. We consider two paths $\gamma_1 : [a, b] \rightarrow X$ and $\gamma_2 : [a, b] \rightarrow X$ with the same starting point and the same end point

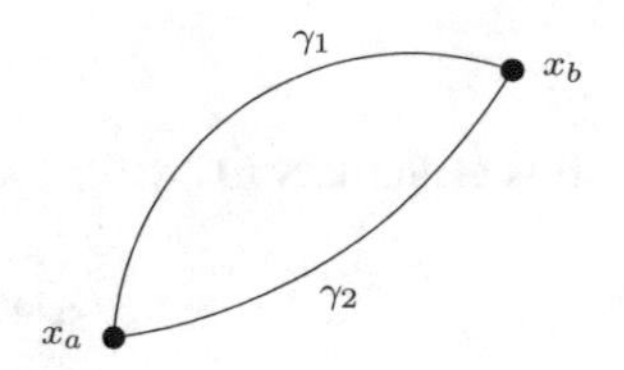

Fig. 6.3 Two paths γ_1 and γ_2 with the same starting point x_a and the same end point x_b

$$\gamma_1(a) = \gamma_2(a) = x_a, \ \gamma_1(b) = \gamma_2(b) = x_b, \tag{6.147}$$

see Fig. 6.3. We call two paths γ_1 and γ_2 **homotopic** if there is a continuous map $\phi : [a, b] \times [0, 1] \to X$ such that

$$\phi(\lambda, 0) = \gamma_1(\lambda), \ \phi(\lambda, 1) = \gamma_2(\lambda) \tag{6.148}$$

and $\phi(a, \kappa) = x_a$, $\phi(b, \kappa) = x_b$ for all $\kappa \in [0, 1]$. This defines an equivalence relation between paths with the same starting point and the same end point. We would like to investigate the question under which conditions iterated integrals depend only on the equivalence class of homotopic paths and not on a specific representative within this equivalence class. We call such iterated integrals **homotopy functionals**. In general, an individual iterated integral is not a homotopy functional. This is easily seen as follows: Consider

$$\omega_1 = dx, \ \omega_2 = dy \tag{6.149}$$

and a family of paths ($\kappa > 0$)

$$\begin{aligned} \gamma_\kappa &: [0, 1] \to \mathbb{R}^2, \\ \gamma_\kappa(\lambda) &= (\lambda, \lambda^\kappa). \end{aligned} \tag{6.150}$$

γ_κ defines a family of homotopic paths with starting point $\gamma_\kappa(0) = (0, 0)$ and end point $\gamma_\kappa(1) = (1, 1)$. Members of this family are indexed by κ. Let us consider

$$I_{\gamma_\kappa}(\omega_1, \omega_2; 1) = \int\limits_0^1 d\lambda_1 \int\limits_0^{\lambda_1} \kappa \lambda_2^{\kappa-1} d\lambda_2 = \frac{1}{\kappa + 1}. \tag{6.151}$$

The result depends on κ and $I_{\gamma_\kappa}(\omega_1, \omega_2)$ is therefore not a homotopy functional.

Let's consider an iterated integral of depth r

$$I_\gamma(\omega_1, \omega_2, \dots, \omega_r; \lambda). \tag{6.152}$$

There is a one-to-one correspondence between ordered sequences of differential one-forms $\omega_1, \omega_2, \dots, \omega_r$ and elements in the tensor algebra $(\Omega(X))^{\otimes r}$ (where $\Omega(X)$ denotes the space of differential forms on X) of the form

$$\omega_1 \otimes \omega_2 \otimes \cdots \otimes \omega_r. \tag{6.153}$$

It is customary to denote the latter as

$$[\omega_1|\omega_2|\dots|\omega_r] = \omega_1 \otimes \omega_2 \otimes \cdots \otimes \omega_r. \tag{6.154}$$

This is called the **bar construction**. In the tensor algebra we define

$$d\,[\omega_1|\omega_2|\dots|\omega_r] = \sum_{j=1}^{r} \left[\omega_1|\dots|\omega_{j-1}|d\omega_j|\omega_{j+1}|\dots|\omega_r\right] + \sum_{j=1}^{r-1} \left[\omega_1|\dots|\omega_{j-1}|\omega_j \wedge \omega_{j+1}|\omega_{j+2}|\dots|\omega_r\right]. \tag{6.155}$$

If all our ω's are closed, this reduces to

$$d\,[\omega_1|\omega_2|\dots|\omega_r] = \sum_{j=1}^{r-1} \left[\omega_1|\dots|\omega_{j-1}|\omega_j \wedge \omega_{j+1}|\omega_{j+2}|\dots|\omega_r\right]. \tag{6.156}$$

Let us now consider a linear combination of iterated integrals of depth $\leq r$ with constant coefficients:

$$I = \sum_{j=1}^{r} \sum_{i_1,\dots,i_j} c_{i_1\dots i_j} I_\gamma\left(\omega_{i_1}, \dots, \omega_{i_j}; \lambda\right) \tag{6.157}$$

and the corresponding element in the tensor algebra

$$B = \sum_{j=1}^{r} \sum_{i_1,\dots,i_j} c_{i_1\dots i_j} \left[\omega_{i_1}|\dots|\omega_{i_j}\right]. \tag{6.158}$$

I is a homotopy functional if and only if [126]

$$dB = 0. \tag{6.159}$$

This is the sought-after criteria when a linear combination of iterated integrals is path independent. We call any B satisfying Eq. (6.159) an **integrable word**.
We are interested in differential one-form ω_j, which have simple poles. It may happen that some ω has a simple pole at the starting point of the integration path $x = x_a$. In the special case where

$$\gamma^*\omega = \frac{d\lambda}{\lambda}, \tag{6.160}$$

e.g. the pull-back has just a simple pole and no regular part, we define the iterated integral by

$$I_\gamma\left(\omega, ..., \omega; \lambda\right) = \frac{1}{r!} \ln^r\left(\lambda\right). \tag{6.161}$$

In the case, where ω has a simple pole at $x = x_a$ and a regular part, we may always decompose ω as

$$\omega = L + \omega_{\text{reg}}, \tag{6.162}$$

with $\gamma^* L = d\lambda/\lambda$ and ω_{reg} having no pole at $x = x_a$. In general we define the iterated integral by Eqs. (6.142) and (6.161). We say that the iterated integral $I_\gamma(\omega_1, \omega_2, ..., \omega_r; \lambda)$ has a **trailing zero**, if the last differential one-form ω_r has a simple pole at $x = x_a$.

Let us discuss a specific class of iterated integrals: We take $X = \mathbb{C}$ with coordinate x and

$$\omega^{\text{mpl}}\left(z_j\right) = \frac{dx}{x - z_j}. \tag{6.163}$$

In this example we treat z_j as a (fixed) parameter. Let $\gamma : [0, \lambda] \to \mathbb{C}$ be the line segment from zero along the positive real axis to $y \in \mathbb{R}_+$, e.g. $\gamma(0) = 0$ and $\gamma(\lambda) = y$. Let us assume that none of the z_j's lie on the path γ. We introduce a special notation for iterated integrals build from differential one-forms as in Eq. (6.163). We set

$$G\left(z_1, \dots, z_r; y\right) = I_\gamma\left(\omega^{\text{mpl}}\left(z_1\right), \dots, \omega^{\text{mpl}}\left(z_r\right); \lambda\right). \tag{6.164}$$

The functions $G\left(z_1, \dots, z_r; y\right)$ are called **multiple polylogarithms**. We discuss these functions in detail in Chap. 8. The general definition of iterated integrals translates in the case of multiple polylogarithms to

$$\begin{aligned} G(\underbrace{0, \dots, 0}_{r-\text{times}}; y) &= \frac{1}{r!} \ln^r\left(y\right), \\ G\left(z_1, z_2 \dots, z_r; y\right) &= \int\limits_0^y \frac{dy_1}{y_1 - z_1} G\left(z_2 \dots, z_r; y_1\right). \end{aligned} \tag{6.165}$$

Let us now return to the differential equation. According to Eq. (6.129), each master integral has a Taylor expansion in ε. It is convenient to write

$$\vec{I}\left(\varepsilon, x\right) = \sum_{j=0}^{\infty} \vec{I}^{(j)}\left(x\right) \cdot \varepsilon^j. \tag{6.166}$$

We plug Eq. (6.166) into the differential equation (6.128)

$$\left(d + \varepsilon \sum_{k=1}^{N_L} C_k \, \omega_k \right) \left(\sum_{j=0}^{\infty} \vec{I}^{(j)}(x) \cdot \varepsilon^j \right) = 0, \tag{6.167}$$

and compare term-by-term in the ε-expansion. We obtain

$$\begin{aligned} d\vec{I}^{(0)}(x) &= 0, \\ d\vec{I}^{(j)}(x) &= - \sum_{k=1}^{N_L} \omega_k \, C_k \, \vec{I}^{(j-1)}(x), \qquad j \geq 1. \end{aligned} \tag{6.168}$$

This system can easily be solved: The first equation of Eq. (6.168) states that $\vec{I}^{(0)}(x)$ is a constant, which is determined by the boundary condition. Knowing $\vec{I}^{(j-1)}(x)$ we obtain $\vec{I}^{(j)}(x)$ by integration. The integration constant is again fixed by the boundary condition. The integration can be done in the class of iterated integrals. Each integration increases the depth of the iterated integrals by one. At order j we obtain iterated integrals of depth $\leq j$. The integrability condition Eq. (6.57) ensures that the result is a homotopy functional. Thus each

$$I^{(j)}_{\boldsymbol{\nu}_i}, \quad 1 \leq i \leq N_{\text{master}} \tag{6.169}$$

is a path-independent linear combination of iterated integrals. Recall that $I^{(j)}_{\boldsymbol{\nu}_i}$ denotes the ε^j-term of the i-th master integral. The individual iterated integrals appearing in $I^{(j)}_{\boldsymbol{\nu}_i}$ are in general not path independent.

Let's look at a simple example. We consider a system with $N_{\text{master}} = 1$ and $N_B = 1$, e.g. one function $I(\varepsilon, x)$ depending on ε and one variable x. Let us assume that the function $I(\varepsilon, x)$ satisfies the differential equation

$$(d + A) \, I \;=\; 0, \quad A \;=\; -\varepsilon \, \frac{dx}{x-1} \tag{6.170}$$

with the boundary condition $I(\varepsilon, 0) = 1$. Then

$$I(x) = 1 + \varepsilon G(1; x) + \varepsilon^2 G(1, 1; x) + \varepsilon^3 G(1, 1, 1; x) + \ldots \tag{6.171}$$

Let us also discuss a non-trivial example. We consider the one-loop four point function in Eq. (6.67) with the basis of master integrals as in Eq. (6.73). In this basis the differential equation is in ε-form and given by Eq. (6.74). In this example the Feynman integrals depend on two variables x_1 and x_2. Let assume that we would like to integrate the differential equation from the start point $x_a = (0, 0)$ to the point of interest $x_b = (x_1, x_2)$. We can do this by first integrating from $(0, 0)$ to $(x_1, 0)$ along the x_1-direction, and then from $(x_1, 0)$ to (x_1, x_2) along the x_2 direction. Doing so, we find

$$I'_1 = B_1^{(0)} + B_1^{(1)}\varepsilon + B_1^{(2)}\varepsilon^2 + O\left(\varepsilon^3\right), \tag{6.172}$$
$$I'_2 = B_2^{(0)} + \left[B_2^{(1)} - B_2^{(0)}G(0;x_1)\right]\varepsilon + \left[B_2^{(2)} - B_2^{(1)}G(0;x_1) + B_2^{(0)}G(0,0;x_1)\right]\varepsilon^2 + O\left(\varepsilon^3\right),$$
$$I'_3 = B_3^{(0)} + \left[B_3^{(1)} - B_3^{(0)}G(0;x_2)\right]\varepsilon + \left[B_3^{(2)} - B_3^{(1)}G(0;x_2) + B_3^{(0)}G(0,0;x_2)\right]\varepsilon^2 + O\left(\varepsilon^3\right),$$
$$I'_4 = B_4^{(0)} + \left[B_4^{(1)} - B_4^{(0)}G(0;x_1) - B_4^{(0)}G(0;x_2) + \left(B_4^{(0)} - B_3^{(0)}\right)G(1;x_1)\right.$$
$$\left. + \left(B_3^{(0)} - B_1^{(0)}\right)G(1;x_2) + \left(B_1^{(0)} - B_2^{(0)} - B_3^{(0)} + B_4^{(0)}\right)G(1-x_1;x_2)\right]\varepsilon + O\left(\varepsilon^2\right).$$

where the $B_i^{(j)}$ are integration constants. The $O\left(\varepsilon^2\right)$-term of I'_4 is already rather long, therefore we stopped for I'_4 at order $O\left(\varepsilon^1\right)$. In order to fix the integration constants $B_i^{(j)}$ we have to know the integrals at one specific kinematic point. This does not have to be the starting point $x_a = (0, 0)$ of the integration. For the case at hand it is convenient to choose the point $x_{\text{boundary}} = (1, 1)$. The first three master integrals I'_1, I'_2 and I'_3 are given at this kinematic point by (see Eq. (2.181))

$$I'_1 = I'_2 = I'_3 = e^{\varepsilon\gamma_E}\frac{\Gamma(1+\varepsilon)\Gamma(1-\varepsilon)^2}{\Gamma(1-2\varepsilon)} = 1 - \frac{\pi^2}{12}\varepsilon^2 + O\left(\varepsilon^3\right), \tag{6.173}$$

yielding

$$\begin{aligned} B_1^{(0)} &= B_2^{(0)} = B_3^{(0)} = 1, \\ B_1^{(1)} &= B_2^{(1)} = B_3^{(1)} = 0, \\ B_1^{(2)} &= B_2^{(2)} = B_3^{(2)} = -\frac{1}{2}\zeta_2, \end{aligned} \tag{6.174}$$

with $\zeta_2 = \pi^2/6$.

Exercise 53 *Show that I'_4 is given at the kinematic point $(x_1, x_2) = (1, 1)$ by*

$$I'_4 = e^{\varepsilon\gamma_E}\frac{\Gamma(1+\varepsilon)\Gamma(1-\varepsilon)^2}{\Gamma(1-2\varepsilon)}\left(1 - \sum_{k=2}^{\infty}\zeta_k\varepsilon^k\right). \tag{6.175}$$

Hint: Use the trick from Exercise 11 and the Mellin-Barnes technique.

From Eq. (6.175) we deduce

$$B_4^{(0)} = 1, \qquad B_4^{(1)} = 0, \qquad B_4^{(2)} = \frac{1}{2}\zeta_2, \tag{6.176}$$

and obtain as final result

$$
\begin{aligned}
I_1' &= 1 - \frac{1}{2}\zeta_2\varepsilon^2 + O\left(\varepsilon^3\right), \\
I_2' &= 1 - G(0;x_1)\,\varepsilon + \left[G(0,0;x_1) - \frac{1}{2}\zeta_2\right]\varepsilon^2 + O\left(\varepsilon^3\right), \\
I_3' &= 1 - G(0;x_2)\,\varepsilon + \left[G(0,0;x_2) - \frac{1}{2}\zeta_2\right]\varepsilon^2 + O\left(\varepsilon^3\right), \\
I_4' &= 1 - \left[G(0;x_1) + G(0;x_2)\right]\varepsilon + \left[G(0,0;x_1) + G(0,0;x_2) - G(1,0;x_1) - G(1,0;x_2)\right. \\
&\quad \left. + G(0;x_1)\,G(0;x_2) + \frac{1}{2}\zeta_2\right]\varepsilon^2 + O\left(\varepsilon^3\right).
\end{aligned}
\tag{6.177}
$$

Up to order $O(\varepsilon^2)$ we may express alternatively the result in terms of logarithms and dilogarithms:

$$
\begin{aligned}
I_1' &= 1 - \frac{1}{2}\zeta_2\varepsilon^2 + O\left(\varepsilon^3\right), \\
I_2' &= 1 - \ln(x_1)\,\varepsilon + \frac{1}{2}\left[\ln^2(x_1) - \zeta_2\right]\varepsilon^2 + O\left(\varepsilon^3\right), \\
I_3' &= 1 - \ln(x_2)\,\varepsilon + \frac{1}{2}\left[\ln^2(x_1) - \zeta_2\right]\varepsilon^2 + O\left(\varepsilon^3\right), \\
I_4' &= 1 - \left[\ln(x_1) + \ln(x_2)\right]\varepsilon + \left[-\text{Li}_2(x_1) - \text{Li}_2(x_2) + \frac{1}{2}\ln^2(x_1) + \frac{1}{2}\ln^2(x_2) + \ln(x_1)\ln(x_2)\right. \\
&\quad \left. - \ln(x_1)\ln(1-x_1) - \ln(x_2)\ln(1-x_2) + \frac{1}{2}\zeta_2\right]\varepsilon^2 + O\left(\varepsilon^3\right).
\end{aligned}
\tag{6.178}
$$

Exercise 54 *Show the equivalence of the $O(\varepsilon^2)$-term of I_4' between Eqs. (6.177) and (6.178).*

In practical applications our main interest is $I_4 = I_{1111}$ and not so much $I_4' = 1/2 \cdot \varepsilon^2 x_1 x_2 I_{1111}$. Inverting this relation gives us

$$
I_{1111} = \frac{2}{\varepsilon^2}\frac{1}{x_1 x_2} I_4'.
\tag{6.179}
$$

We have calculated I_4' as a function of ε, $x_1 = s/p_4^2$ and $x_2 = t/p_4^2$. From the scaling relation of Eq. (2.144) we may reinstate the full dependence on ε, s, t, p_4^2 and μ^2:

$$
I_{1111}\left(\varepsilon, s, t, p_4^2, \mu^2\right) = \frac{2}{\varepsilon^2}\left(\frac{-p_4^2}{\mu^2}\right)^{-2-\varepsilon} \frac{1}{x_1 x_2} I_4'\left(\varepsilon, x_1, x_2\right)\Bigg|_{x_1 = s/p_4^2, x_2 = t/p_4^2}.
\tag{6.180}
$$

One may check that this agrees with Eq. (B.13) given in Appendix B.

Let us make a few remarks: Firstly, instead of integrating from $(0, 0)$ to $(x_1, 0)$ and from there to (x_1, x_2), we could have used alternatively a path from $(0, 0)$ to $(0, x_2)$ and from there to (x_1, x_2). This gives the same result. In general, we are free to choose any path we like, as long as we don't cross any branch cuts. In the present example there are no problems with branch cuts for real values of x_1 and x_2 as long as

$$x_1 > 0, \quad x_2 > 0, \quad x_1 + x_2 < 1. \tag{6.181}$$

Branch cuts originate from the singularities in the differential equation. In the present example the singularities are determined by ω_1 - ω_5 (defined in Eq. (6.77)) and located at

$$x_1 = 0, \quad x_1 = 1, \quad x_2 = 0, \quad x_2 = 1, \quad x_1 + x_2 = 1. \tag{6.182}$$

We may integrate the differential equation beyond a singularity. In this case Feynman's $i\delta$-prescription tells us which branch we have to choose.

As a second remark let us notice that the multiple polylogarithms in Eq. (6.175) have trailing zeros and that the coefficients of the ε-expansion of I_2', I_3' and I_4' diverge as a power of a logarithm in the limit $x_1 \to 0$ or $x_2 \to 0$. This may happen and there is nothing wrong with it. This was the reason why we didn't use $(x_1, x_2) = (0, 0)$ as the kinematic point to fix the integration constants. Let's see what happens, if we set $x_1 = x_2 = 0$ from the start and calculate these integrals within dimensional regularisation. We denote these integrals, where we set $x_1 = x_2 = 0$ from the start by $\tilde{I}_2'$, $\tilde{I}_3'$ and $\tilde{I}_4'$. The first two are scaleless integrals and hence equal to zero within dimensional regularisation (see Eq. (2.137)). Also $\tilde{I}_4'$ is zero due to the explicit prefactor $x_1 x_2$. Thus

$$\begin{aligned} \tilde{I}_2' &= 0, \\ \tilde{I}_3' &= 0, \\ \tilde{I}_4' &= 0. \end{aligned} \tag{6.183}$$

This seems puzzling, as Eq. (6.183) is quite different from Eq. (6.177). To analyse the situation, let's consider I_2'. For $x_1 > 0$ and $\varepsilon < 0$ the integral I_2' is unambiguously given by

$$I_2' = e^{\varepsilon \gamma_E} \frac{\Gamma(1+\varepsilon)\,\Gamma(1-\varepsilon)^2}{\Gamma(1-2\varepsilon)} x_1^{-\varepsilon}. \tag{6.184}$$

If we first expand in ε and then take the limit $x_1 \to 0$, we find that the coefficients of the ε-expansion are given by Eq. (6.177) and that they diverge as a power of a logarithm in this limit. If on the other hand we first take the limit $x_1 \to 0$ and then expand in ε, we recover Eq. (6.183). (In our particular example, we already obtain zero after taking the limit $x_1 \to 0$, so for the expansion in ε there is nothing left to expand.) In other words, the expansion in ε does not commute with the limit $x_1 \to 0$. In a nutshell, we have

$$\lim_{x\to 0+}\lim_{\varepsilon\to 0-} x^{-\varepsilon} = \lim_{x\to 0+} 1 = 1,$$
$$\lim_{\varepsilon\to 0-}\lim_{x\to 0+} x^{-\varepsilon} = \lim_{\varepsilon\to 0-} 0 = 0. \tag{6.185}$$

6.4 Fibre Bundles

We have seen that with the help of integration-by-parts identities we may express any Feynman integral $I_{\nu_1\dots\nu_{n_{\mathrm{int}}}}$ as a linear combination of N_{master} master integrals $I_{\boldsymbol{\nu}_1}, I_{\boldsymbol{\nu}_2}, \dots, I_{\boldsymbol{\nu}_{N_{\mathrm{master}}}}$. We may view the master integrals as a basis of a N_{master}-dimensional vector space. The master integrals depend on the dimensional regularisation parameter ε and N_B kinematic variables $x = (x_1, \dots x_{N_B})$. Let us now focus on the dependence on x. For the moment we treat ε as a (fixed) parameter. For every value of x we have a separate vector space spanned by $I_{\boldsymbol{\nu}_1}(x), \dots, I_{\boldsymbol{\nu}_{N_{\mathrm{master}}}}(x)$ The master integrals satisfy a differential equation

$$(d + A)\,\vec{I}\,(x) = 0, \tag{6.186}$$

where $\vec{I}(x) = (I_{\boldsymbol{\nu}_1}(x), \dots, I_{\boldsymbol{\nu}_{N_{\mathrm{master}}}}(x))^T$.

In mathematical terms we are looking at a vector bundle of rank N_{master} over a base space parametrised by the coordinates x. The vector bundle is equipped with a flat connection defined locally by the matrix-valued one-form A. In Sect. 6.4.1 we introduce these terms.

We also know by now that the computation of Feynman integrals reduces to the task of finding appropriate transformations, which bring the differential equation for the master integrals into a ε-form. The mathematical reformulation gives us a clear picture what type of transformation we should consider: These are transformations, which correspond to a basis change in the fibre or transformations, which correspond to a coordinate transformation on the base manifold. These are discussed in Sect. 6.4.3 and Sect. 6.4.4, respectively.

6.4.1 Mathematical Background

Let us introduce the required terminology for fibre bundles. We give a concise summary with the essential definitions. For a more detailed introduction into this topic for a readership with a physics background we refer to the books by Nakahara [127] and Isham [128].

We start with the definition of a manifold. Let M be a topological space. The basics of topology are summarised in Appendix G.1.

An **open chart** on M is a pair (U, φ), where U is an open subset of M and φ is a homeomorphism of U onto an open subset of $\mathbb{R}^n$.

A **differentiable manifold** of dimension n is a Hausdorff space with a collection of open charts $(U_\alpha, \varphi_\alpha)_{\alpha \in A}$ such that

M1:

$$M = \bigcup_{\alpha \in A} U_\alpha . \tag{6.187}$$

M2: For each pair $\alpha, \beta \in A$ the mapping $\varphi_\beta \circ \varphi_\alpha^{-1}$ is an infinitely differentiable mapping of $\varphi_\alpha \left(U_\alpha \cap U_\beta\right)$ onto $\varphi_\beta \left(U_\alpha \cap U_\beta\right)$.

A differentiable manifold is also often denoted as a C^∞ manifold. As we will only be concerned with differentiable manifolds, we will often omit the word "differentiable" and just speak about manifolds.
The collection of open charts $(U_\alpha, \varphi_\alpha)_{\alpha \in A}$ is called an **atlas**.
If $p \in U_\alpha$ and

$$\varphi_\alpha(p) = (x_1(p), ..., x_n(p)) , \tag{6.188}$$

the set U_α is called the **coordinate neighbourhood** of p and the numbers $x_i(p)$ are called the **local coordinates** of p.
Note that in each coordinate neighbourhood M looks like an open subset of $\mathbb{R}^n$. But note that we do not require that M be $\mathbb{R}^n$ globally.
The definition of a **complex manifold** requires only small modifications:

- An open chart of a complex manifold M is a pair (U, φ), where U is an open subset of M and φ is a homeomorphism of U onto an open subset of $\mathbb{C}^n$ (which we may also view as an open subset of $\mathbb{R}^{2n}$, so the real dimension of any complex manifold is even).
- Axiom (M2) in the definition above is modified as follows: We require that for each pair $\alpha, \beta \in A$ the mapping $\varphi_\beta \circ \varphi_\alpha^{-1} : \varphi_\alpha \left(U_\alpha \cap U_\beta\right) \to \varphi_\beta \left(U_\alpha \cap U_\beta\right)$ is holomorphic.

Let us now turn to fibre bundles. A **differentiable fibre bundle** (E, M, F, π, G) (or simply fibre bundle for short) consists of the following elements:

- A differentiable manifold E called the **total space**.
- A differentiable manifold M called the **base space**.
- A differentiable manifold F called the **fibre**.
- A surjection $\pi : E \to M$ called the **projection**. The inverse image $\pi^{-1}(p) = F_p$ is called the fibre at p.
- A Lie group G called the **structure group**, which acts on F from the left.

We require that

F1: there is a set of open coverings $\{U_i\}$ of M with diffeomorphisms $\phi_i : U_i \times F \to \pi^{-1}(U_i)$ such that $\pi \phi_i(p, f) = p$. The map ϕ_i is called the **local trivialisation**, since ϕ_i^{-1} maps $\pi^{-1}(U_i)$ onto the direct product $U_i \times F$,

F2: if we write $\phi_i(p, f) = \phi_{i,p}(f)$, the map $\phi_{i,p} : F \to F_p$ is a diffeomorphism. On $U_i \cap U_j \neq \emptyset$ we require that $t_{ij}(p) = \phi_{i,p}^{-1}\phi_{j,p} : F \to F$ be an element of G, satisfying the consistency conditions $t_{ii} = \text{id}$, $t_{ij} = t_{ji}^{-1}$, $t_{ij}t_{jk} = t_{ik}$. The $\{t_{ij}\}$ are called the **transition functions**.

A **local section** of a fibre bundle $E \xrightarrow{\pi} M$ over $U \subset M$ is a smooth map $\sigma : U \to E$, which satisfies $\pi\sigma = \text{id}_M$. A **global section** of a fibre bundle $E \xrightarrow{\pi} M$ is a smooth map $\sigma : M \to E$, which satisfies $\pi\sigma = \text{id}_M$. The space of local sections and global sections is denoted by $\Gamma(U, E)$ and $\Gamma(M, E)$, respectively.

Two important special cases of fibre bundles are vector bundles and principal bundles:

A **principal bundle** is a fibre bundle, whose fibre is identical with the structure group G. A principal bundle is also often called a G-bundle over M and denoted $P(M, G)$. P denotes the total space.

A **vector bundle** is a fibre bundle, whose fibre is a vector space. The dimension r of the fibre F is called the **rank** of the vector bundle. A vector bundle of rank 1 is called a **line bundle**. A **complex vector bundle** of rank r is a vector space, where the fibre is $\mathbb{C}^r$. Examples of vector bundles are the tangent bundle TM and the cotangent bundle T^*M. For the **tangent bundle** the fibre at the point $p \in M$ is the vector space of all tangent vectors to M at p. Similar, the fibre of the **cotangent bundle** at the point $p \in M$ is the vector space of all cotangent vectors at p.

A **frame** of a rank r vector bundle over $U \subset M$ is an ordered set of local sections $\sigma_j : U \to E$ with $1 \leq j \leq r$, such that $\sigma_1(p), \sigma_2(p), \dots, \sigma_r(p)$ is a basis of F_p for any $p \in U$.

Let E and M be complex manifolds and $\pi : E \to M$ be a holomorphic surjection. E is a **holomorphic vector bundle** of rank r if the typical fibre is $\mathbb{C}^r$, the structure group is $G = \text{GL}_r(\mathbb{C})$ and

H1: the local trivialisation $\phi_i : U_i \times \mathbb{C}^r \to \pi^{-1}(U_i)$ is a biholomorphism,
H2: the transition functions $t_{ij} : U_i \cap U_j \to G = \text{GL}_r(\mathbb{C})$ are holomorphic maps.

Let us now turn to connections. We start from a principal bundle $P(M, G)$. Let u be a point in the total space P and let G_p be the fibre at $p = \pi(u)$. The **vertical subspace** $V_u P$ is a subspace of the tangent space $T_u P$, which is tangent to G_p at u. The vertical subspace $V_u P$ is isomorphic as a vector space to the Lie algebra $\mathfrak{g}$ of G. There is a right action of an element $g \in G$ on a point $u \in P$. Within a local trivialisation a point u in the total space is given by $u = (x, g')$ and the right action by g is given by $ug = (x, g'g)$. The right action by g maps the point u to another point ug on the same fibre.

A **connection** on $P(M, G)$ is a unique separation of the tangent space $T_u P$ into the vertical subspace $V_u P$ and a **horizontal subspace** $H_u P$ such that

C1: $T_u P = H_u P \oplus V_u P$
C2: a smooth vector field X on $P(M, G)$ is separated into smooth vector fields $X^H \in H_u P$ and $X^V \in V_u P$ as $X = X^H + X^V$.
C3: Let $g \in G$. The horizontal subspaces $H_u P$ and $H_{ug} P$ on the same fibre are related by a linear map R_{g*} induced by the right action of $g : H_{ug} P = R_{g*} H_u P$. Accordingly a subspace $H_u P$ at u generates all the horizontal subspaces on the same fibre.

A **connection one-form** $\omega \in \mathfrak{g} \otimes T^* P$, which takes values in the Lie algebra $\mathfrak{g}$ of G, is a projection of $T_u P$ onto the vertical component $V_u P \cong \mathfrak{g}$, compatible with axiom (C3) above. In detail we require that for $X \in T_u P$

CF1: $\omega(X) \in \mathfrak{g}$,
CF2: $\omega_{ug}(R_{g*} X) = g^{-1} \omega_u(X) g$, where $R_{g*} X$ denotes the push-forward of the tangent vector X at the point u to the point ug by the right action of g.

The horizontal subspace $H_u P$ is defined to be the kernel of ω. Condition (CF2) may be stated equivalently as $R_g^* \omega = \mathrm{Ad}_{g^{-1}} \omega$, with $R_g^* \omega_{ug}(X) = \omega_{ug}(R_{g*} X)$ and $\mathrm{Ad}_{g^{-1}} \omega(X) = g^{-1} \omega_u(X) g$ we recover (CF2).

Let $U \subset M$ be an open subset of M and $s : U \to P$ a local section. We denote by A the pull-back of ω to M:

$$A = s^* \omega. \tag{6.189}$$

Let us now consider two sections σ_1 and σ_2, with associated pull-backs A_1 and A_2. We can always relate the two sections σ_1 and σ_2 by

$$\sigma_1(x) = \sigma_2(x) g(x), \tag{6.190}$$

where $g(x)$ is a x-dependent element of the Lie group G. Then we obtain for the pull-backs A_1 and A_2 of the connection one-form the relation

$$A_2 = g A_1 g^{-1} + g d g^{-1}. \tag{6.191}$$

In the sequel of the book we will use the notation $g(x) = U(x)$. Equation (6.192) reads then

$$A_2 = U A_1 U^{-1} + U d U^{-1}. \tag{6.192}$$

Readers familiar with gauge theories in physics will recognise that Eq. (6.192) is nothing else than a gauge transformation.

Exercise 55 *Derive Eqs. (6.192) from (6.190).*
Hint: Recall that the action of the pull-back A_2 on a tangent vector is defined as the action of the original form ω on the push-forward of the tangent vector. Recall further that a tangent vector at a point x can be given as a tangent vector to a curve through x. It is sufficient to show that the actions of A_2 and $U A_1 U^{-1} + U d U^{-1}$ on

an arbitrary tangent vector give the same result. In order to prove the claim you will need in addition the defining relations for the connection one-form ω, given in (CF1) and (CF2).

Given a principal bundle $P(M, G)$ and a r-dimensional vector space F, on which G acts on the left through a r-dimensional representation ρ, we may always construct the associated vector bundle (E, M, F, π, G). The total space E is given by $(P \times F)/G$, where points (u, v) and $(u, \rho(g^{-1})v)$ are identified. The projection π in the associated vector bundle E is given by $\pi(u, v) = \pi_P(u)$, where π_P denotes the projection of the principal bundle P. The transition functions of E are given by $\rho(t_{ij})$, where t_{ij} denote the transition functions of P.

Conversely, a rank r vector bundle with fibre $\mathbb{R}^r$ or $\mathbb{C}^r$ induces a principal bundle with structure group $\mathrm{GL}_r(\mathbb{R})$ or $\mathrm{GL}_r(\mathbb{C})$, respectively. The transition functions of the induced principal bundle are the ones of the vector bundle, and this defines the principal bundle. (In general, the minimal information required to define a fibre bundle are M, F, G, a set of open coverings $\{U_i\}$ and the transition functions t_{ij}.)

We continue to work in a local chart (U, φ). The local connection one-form A defines the **covariant derivative**

$$D_A = d + A. \tag{6.193}$$

If A is a $\mathfrak{g}$-valued p-form and B is a $\mathfrak{g}$-valued q-form, the commutator of the two is defined by

$$[A, B] = A \wedge B - (-1)^{pq} B \wedge A, \tag{6.194}$$

the factor $(-1)^{pq}$ takes into account that we have to permute p differentials dx_i from A past q differentials dx_j from B. If A and B are both one-forms we have

$$[A, B] = A \wedge B + B \wedge A = \left[A_i, B_j\right] dx_i \wedge dx_j. \tag{6.195}$$

In particular

$$A \wedge A = \frac{1}{2} [A, A] = \frac{1}{2} \left[A_i, A_j\right] dx_i \wedge dx_j. \tag{6.196}$$

We define the **curvature two-form** of the fibre bundle by

$$F = D_A A = dA + A \wedge A = dA + \frac{1}{2} [A, A]. \tag{6.197}$$

The fibre bundle is **flat** (or pure gauge) if

$$F = 0. \tag{6.198}$$

6.4.2 *Fibre Bundles in Physics*

6.4.2.1 Gauge Theories

Let us now see where fibre bundles occur in physics. Our first example is a gauge theory with a real scalar particle in the fundamental representation of a gauge group G. We denote the generators of the gauge group by T^a and the coupling by g. Let us further denote by r the dimension of the fundamental representation of G. The Lagrange density reads

$$\mathcal{L} = \frac{1}{2}\left(D_{\mu,jk}\phi_k(x)\right)^\dagger \left(D^\mu{}_{jl}\phi_l(x)\right) - \frac{1}{4}F^a_{\mu\nu}(x)F^{a\mu\nu}(x) \tag{6.199}$$

where the covariant derivative $D_{\mu,jk}$ and the field strength $F^a_{\mu\nu}$ are given by

$$\begin{aligned} D_{\mu,jk} &= \delta_{jk}\partial_\mu - igT^a_{jk}A^a_\mu(x), \\ F^a_{\mu\nu}(x) &= \partial_\mu A^a_\nu(x) - \partial_\nu A^a_\mu(x) + gf^{abc}A^b_\mu(x)A^c_\nu. \end{aligned} \tag{6.200}$$

There are two fibre bundles here: a principal bundle and a vector bundle. In both cases the base space M is given by flat Minkowski space. For simplicity let's assume that we are not interested in topologically non-trivial configurations (like instantons), therefore a trivial fibre bundle and a global section are fine for us. The discussion below can easily be extended to the general case by introducing an atlas of coordinate patches and local sections glued together in the appropriate way. The gauge field A^a_μ defines a local connection one-form by

$$A = A_\mu\, dx^\mu = \frac{g}{i}T^a A^a_\mu\, dx^\mu, \qquad A_\mu = \frac{g}{i}T^a A^a_\mu. \tag{6.201}$$

Let's see what the curvature of the connection is: We have

$$\begin{aligned} dA = d\left(A_\nu dx^\nu\right) &= \partial_\mu A_\nu dx^\mu \wedge dx^\nu = \frac{1}{2}\left(\partial_\mu A_\nu - \partial_\nu A_\mu\right) dx^\mu \wedge dx^\nu, \\ A \wedge A = A_\mu dx^\mu \wedge A_\nu dx^\nu &= \frac{1}{2}\left(A_\mu A_\nu - A_\nu A_\mu\right) dx^\mu \wedge dx^\nu = \frac{1}{2}\left[A_\mu, A_\nu\right] dx^\mu \wedge dx^\nu, \end{aligned} \tag{6.202}$$

and therefore

$$F = \frac{1}{2}\left(\partial_\mu A_\nu - \partial_\nu A_\mu + \left[A_\mu, A_\nu\right]\right) dx^\mu \wedge dx^\nu = \frac{1}{2}F_{\mu\nu}dx^\mu \wedge dx^\nu. \tag{6.203}$$

With the notation

$$F_{\mu\nu} = \frac{g}{i}T^a F^a_{\mu\nu} \tag{6.204}$$

we have

$$F = \frac{1}{2} F_{\mu\nu} dx^\mu \wedge dx^\nu = \frac{1}{2} \frac{g}{i} T^a F^a_{\mu\nu} dx^\mu \wedge dx^\nu, \tag{6.205}$$

and $F^a_{\mu\nu}$ is given by Eq. (6.200). Thus we see that a gauge field is equivalent to the pull-back of the connection one-form of a principal bundle.

The real scalar field $\phi_k(x)$ (with $1 \leq k \leq r$) we may view as a section in a vector bundle with fibre $\mathbb{R}^r$.

A more realistic example is QCD, where the real scalar field is replaced by one or more spinor fields (depending on the number of quarks).

6.4.2.2 Feynman Integrals

Our main interest are Feynman integrals. In Sect. 2.5.1 we discussed the variables on which a Feynman integral $I_{\nu_1 \dots \nu_{n_{\text{int}}}}$ depends. These were the dimension of space-time D (or alternatively the dimensional regularisation parameter ε) and $N_B + 1$ dimensionless kinematic variables $x_1, \dots, x_{N_B+1}$ of the form

$$\frac{-p_i \cdot p_j}{\mu^2} \quad \text{or} \quad \frac{m_i^2}{\mu^2}. \tag{6.206}$$

Due to the scaling relation in Eq. (2.144) we may set without loss of information one kinematic variable to one, leaving N_B non-trivial kinematic variables. That's the way we usually approach the problem from the physics side: We prefer dimensionless quantities and the arbitrary scale μ can be related to the renormalisation scale introduced in Eq. (4.66).

If we approach the problem from the mathematical side we may view the situation as a complex vector bundle over a base space, which is parametrised by the kinematic variables x. As we are mainly interested in the local properties and not the global properties, we may view the base space as an open subset U of $\mathbb{CP}^{N_B}$. The open subset U is defined by the requirement that the master integrals are single-valued and non-degenerate on U. To get there let's go one step back to Eq. (2.138) and start from $N_B + 1$ dimensionfull variables $X_1, \dots, X_{N_B+1}$ of the form

$$- p_i \cdot p_j \quad \text{or} \quad m_i^2 \tag{6.207}$$

of mass dimension 2 and an arbitrary scale μ^2, again of mass dimension 2. Assume that one particular variable X_j is not equal to zero. Let us now make the choice $\mu^2 = X_j$. Again, no information is lost: μ^2 enters only as a trivial prefactor in Eq. (2.138), and once we know the integral for one particular choice of μ^2, we recover the integral for any other choice of μ^2 from a scaling relation derived from Eq. (2.138). Once we made the choice $\mu^2 = X_j$, our integral depends on D (or ε) and $X_1, \dots, X_{N_B+1}$. The dependence on D (or ε) will play no further role in the discussion below and we focus on the dependence on the variables $X_1, \dots, X_{N_B+1}$. If we scale all variables $X_1, \dots, X_{N_B+1}$ by a factor λ we now have

$$I^{\text{chart } j}_{\nu_1 \dots \nu_{n_{\text{int}}}} \left(\lambda X_1, \dots, \lambda X_{N_B+1} \right) = I^{\text{chart } j}_{\nu_1 \dots \nu_{n_{\text{int}}}} \left(X_1, \dots, X_{N_B+1} \right). \tag{6.208}$$

The superscript chart j indicates that we assumed $X_j \neq 0$ and made the choice $\mu^2 = X_j$. Thus the Feynman integral $I^{\text{chart } j}_{\nu_1 \dots \nu_{n_{\text{int}}}}$ is invariant under a simultaneous scaling of all variables $X_1, \dots, X_{N_B+1}$ and defines therefore a (in general multi-valued) function on the chart $X_j \neq 0$ of $\mathbb{CP}^{N_B}$. We have

$$I^{\text{chart } j}_{\nu_1 \dots \nu_{n_{\text{int}}}} \left(X_1, \dots, X_{j-1}, X_j, X_{j+1}, \dots, X_{N_B+1} \right) = I^{\text{chart } j}_{\nu_1 \dots \nu_{n_{\text{int}}}} \left(x_1, \dots, x_{j-1}, 1, x_{j+1}, \dots, x_{N_B+1} \right) \tag{6.209}$$

with $x_i = X_i / X_j$. Since we made the choice $\mu^2 = X_j$ we have $x_i = X_i / \mu^2$. This shows the equivalence of this approach with the previous one and with the implicit assumption that in the previous approach we always set x_{N_B+1} to one we have

$$I_{\nu_1 \dots \nu_{n_{\text{int}}}} \left(x_1, \dots, x_{N_B} \right) = I^{\text{chart } N_B+1}_{\nu_1 \dots \nu_{n_{\text{int}}}} \left(x_1, \dots, x_{N_B}, 1 \right). \tag{6.210}$$

We may take

$$\left[X_1 : X_2 : \dots : X_{N_B+1} \right] \tag{6.211}$$

as homogeneous coordinates on $\mathbb{CP}^{N_B}$. Feynman integrals are in general multivalued function of the kinematic variables. A simple toy example follows directly from Eq. (6.184)

$$I^{\text{chart } 2} \left(X_1, X_2 \right) = \left(\frac{X_1}{X_2} \right)^{-\varepsilon}, \quad I \left(x_1 \right) = I^{\text{chart } 2} \left(x_1, 1 \right) = x_1^{-\varepsilon}, \tag{6.212}$$

where we ignored an x-independent prefactor not relevant to the discussion here. $I^{\text{chart } 2}(X_1, X_2)$ clearly satisfies Eq. (6.208) and defines a multi-valued function on $\mathbb{CP}^1$. The multi-valuedness arises as follows: Consider the chart $X_2 = 1$ of $\mathbb{CP}^1$ with coordinate x_1. If we go anti-clockwise around the origin $x_1 = 0$, $I(x_1)$ changes by a multiplicative prefactor

$$e^{-2\pi i \varepsilon}, \tag{6.213}$$

which follows from

$$x_1^{-\varepsilon} = e^{-\varepsilon \ln x_1} \tag{6.214}$$

and the multi-valuedness of the logarithm. In order to get a single-valued function we restrict to an open subset, where we choose a branch of the logarithm which allows us to view $I(x_1)$ as a single-valued function. In this example we may choose the open subset as

$$U_2 = \left\{ x_1 \in \mathbb{C} \mid x_1 \notin \mathbb{R}_{\leq 0} \right\}. \tag{6.215}$$

Let us now return to the general case. Let U be an open subset of $\mathbb{CP}^{N_B}$, where the Feynman integrals are single-valued and where the master integrals are non-degenerate. We set U_j to be the intersection of U with the set of points of $\mathbb{CP}^{N_B}$ for which $X_j \neq 0$:

$$U_j = U \cap \left\{ \left[X_1 : \dots : X_{N_B+1} \right] \in \mathbb{CP}^{N_B} \mid X_j \neq 0 \right\}. \tag{6.216}$$

Equation (6.209) defines then $I^{\text{chart } j}_{\nu_1 \dots \nu_{n_{\text{int}}}}$ as a single-valued function on U_j. An analogous definition applies to $I^{\text{chart } i}_{\nu_1 \dots \nu_{n_{\text{int}}}}$ for U_i. Let us now assume $X_i \neq 0$, $X_j \neq 0$ and $U_i \cap U_j \neq 0$. We have

$$I^{\text{chart } i}_{\nu_1 \dots \nu_{n_{\text{int}}}} \left(X_1, \dots, X_i, \dots, X_j, \dots, X_{N_B+1} \right) = \left(\frac{X_i}{X_j} \right)^{\nu - \frac{lD}{2}} I^{\text{chart } j}_{\nu_1 \dots \nu_{n_{\text{int}}}} \left(X_1, \dots, X_i, \dots, X_j, \dots, X_{N_B+1} \right). \tag{6.217}$$

For the transition function we have

$$t_{ij} = \left(\frac{X_i}{X_j} \right)^{\nu - \frac{lD}{2}}. \tag{6.218}$$

Thus an individual Feynman integral $I_{\nu_1 \dots \nu_{n_{\text{int}}}}$ defines a complex line bundle over U. A set of N_{master} master integrals defines a complex vector bundle of rank N_{master} over U.

On U each master integral $I_{\boldsymbol{\nu}_1}, \dots, I_{\boldsymbol{\nu}_{N_{\text{master}}}}$ defines a local section. It may happen that for specific values of x some master integrals degenerate and become linearly dependent.

An example is given by the one-loop two-point function with two unequal internal masses. For $m_1^2 \neq m_2^2$ we have three master integrals, which may be taken as I_{11}, I_{10} and I_{01}. For $m_1^2 = m_2^2$ the two master integrals I_{10} and I_{01} degenerate and are equal to each other (see the discussion in Sect. 6.1). For $m_1^2 = m_2^2$ we only have two master integrals, which may be taken as I_{11} and I_{10}.

We say that the set of master integrals are **ramified** at a point x if the set of master integrals is linearly dependent for the value x. We excluded those points from the definition of U. Thus, the set of all master integrals defines a frame on U. This says that on U the N_{master} master integrals are linearly independent. The vector bundle is equipped with a connection. The local connection one-form is given by the matrix-valued one-form A appearing in the differential equation Eq. (6.58).

Usually our primary interest is not the global structure, but only a single chart, which we may take as U_{N_B+1}. The reason is the following: The chart U_{N_B+1} covers all cases except the ones with $X_{N_B+1} = 0$. The case $X_{N_B+1} = 0$ has one kinematic variable less and can be considered as simpler. With these remarks we take in the sequel the base space M to be U_{N_B+1}. U_{N_B+1} is homeomorphic to an open subset of $\mathbb{C}^{N_B}$.

In a mathematical language we are considering a local system on U_{N_B+1} with the corresponding Gauß-Manin connection given by $\nabla = d + A$.

Digression Local systems and the Gauß-Manin connection

Let M be a topological space. By a **local system** *one understands either:*

1. *a vector bundle $\pi : E \to M$ with parallel transport, i.e. for each homotopy class of paths in M there is a vector space isomorphism between the fibres. The vector space isomorphisms are compatible with the composition of paths.*
2. *if M is a differentiable manifold, a vector bundle $\pi : E \to M$ with a flat connection $\nabla = d + A$. The connection is flat if $dA + A \wedge A = 0$. ∇ is called the* **Gauß-Manin connection**.
3. *a locally constant sheaf of vector spaces on M. (Appendix G gives a definition of sheaves.)*

The three definitions are equivalent (of course, for definition 2 we have to assume that M is differentiable).

6.4.3 Fibre Transformations

We have learned that in Feynman integral computations we are considering a vector bundle over a base space M, which is parametrised by the kinematic variables $x_1, \dots, x_{N_B}$. The master integrals $I_{\nu_1}(x), \dots, I_{\nu_{N_{\mathrm{master}}}}(x)$ can be viewed as local sections, and for each x they define a basis of the vector space in the fibre. The master integrals $\vec{I} = (I_{\nu_1}, I_{\nu_2}, \dots, I_{\nu_{N_{\mathrm{master}}}})^T$ satisfy the differential equation

$$(d + A)\,\vec{I} = 0. \tag{6.219}$$

The matrix-valued one-form A gives the local connection one-form. The integrability condition in Eq. (6.57) says that the connection is flat:

$$dA + A \wedge A = 0. \tag{6.220}$$

Let us now turn back to more practical questions: We would like to transform a given differential equation (not necessarily in ε-form) to the ε-form. In order to achieve this goal, we first have to understand what transformations can be done. The first transformation, which we discuss in this section, is a change of the master integrals. This amounts to changing the sections and the frame of the vector bundle. We consider the transformation

$$\vec{I}'\,(\varepsilon, x) = U\,(\varepsilon, x)\,\vec{I}\,(\varepsilon, x)\,, \tag{6.221}$$

where $U(\varepsilon, x)$ is an invertible $(N_{\mathrm{master}} \times N_{\mathrm{master}})$-matrix, which may depend on ε and x. Let's work out the differential equation satisfies by the new master integrals. We have

$$\vec{I} = U^{-1}\vec{I}' \tag{6.222}$$

and

$$d\vec{I} = d\left(U^{-1}\vec{I}'\right) = U^{-1}d\vec{I}' + \left(dU^{-1}\right)\vec{I}'. \tag{6.223}$$

Thus

$$d\vec{I}' = Ud\vec{I} - \left(UdU^{-1}\right)\vec{I}' = -UA\vec{I} - \left(UdU^{-1}\right)\vec{I}' = -\left(UAU^{-1} + UdU^{-1}\right)\vec{I}'. \tag{6.224}$$

In summary we have

Fibre transformation:
Let $\vec{I} = (I_{\nu_1}, \dots, I_{\nu_{N_{\mathrm{master}}}})^T$ be a set of master integrals satisfying the differential equation

$$(d + A)\,\vec{I} = 0. \tag{6.225}$$

If we change the master integrals according to

$$\vec{I}' = U\vec{I}, \tag{6.226}$$

where U is an invertible $(N_{\mathrm{master}} \times N_{\mathrm{master}})$-matrix, which may depend on ε and x, the new master integrals satisfy the differential equation

$$\left(d + A'\right)\vec{I}' = 0, \tag{6.227}$$

where A' is related to A by

$$A' = UAU^{-1} + UdU^{-1}. \tag{6.228}$$

We have already seen in example 2 in Sect. 6.3.1 that a suitable fibre transformation may bring the differential equation into an ε-form. Equation (6.73) may also be written as

$$\vec{I}' = U\vec{I}, \tag{6.229}$$

with

$$U(\varepsilon, x_1, x_2) = \begin{pmatrix} \varepsilon(1-2\varepsilon) & 0 & 0 & 0 \\ 0 & \varepsilon(1-2\varepsilon) & 0 & 0 \\ 0 & 0 & \varepsilon(1-2\varepsilon) & 0 \\ 0 & 0 & 0 & \frac{1}{2}\varepsilon^2 x_1 x_2 \end{pmatrix}. \tag{6.230}$$

Let us also consider example 3 from Sect. 6.3.1. This is the family of the two-loop double box integral with eight master integrals, given by Eq. (6.81). We consider the transformation

$$\vec{I}' = U\vec{I}, \tag{6.231}$$

where the (8×8)-matrix U is given by

$$U(\varepsilon, x) = \tag{6.232}$$

$$\begin{pmatrix} \frac{g_1(\varepsilon)}{x} & 0 & 0 & 0 & 0 & 0 & 0 & 0 \\ 0 & g_1(\varepsilon) & 0 & 0 & 0 & 0 & 0 & 0 \\ 0 & 0 & \varepsilon^2(1-2\varepsilon)^2 & 0 & 0 & 0 & 0 & 0 \\ 0 & 0 & 0 & g_2(\varepsilon) & 0 & 0 & 0 & 0 \\ 0 & 0 & 0 & g_2(\varepsilon) & 6\varepsilon^3(1-2\varepsilon)x & 0 & 0 & 0 \\ 0 & 0 & 0 & 0 & 0 & 3\varepsilon^4(1+x) & 0 & 0 \\ 0 & 0 & 0 & 0 & 0 & 0 & \varepsilon^4 x^2 & 0 \\ -g_1(\varepsilon) & -g_1(\varepsilon)x & 0 & g_2(\varepsilon)x & 6\varepsilon^3(1-2\varepsilon)x^2 & 6\varepsilon^4 x(1+x) & 0 & 2\varepsilon^4 x^2 \end{pmatrix},$$

with

$$\begin{aligned} g_1(\varepsilon) &= -3\varepsilon(1-2\varepsilon)(1-3\varepsilon)(2-3\varepsilon), \\ g_2(\varepsilon) &= -3\varepsilon^2(1-2\varepsilon)(1-3\varepsilon). \end{aligned} \tag{6.233}$$

This transforms the differential equation into ε-form. We have

$$A' = \varepsilon\left(C_0\frac{dx}{x} + C_{-1}\frac{dx}{x+1}\right), \tag{6.234}$$

with

$$C_0 = \begin{pmatrix} 2 & 0 & 0 & 0 & 0 & 0 & 0 & 0 \\ 0 & 0 & 0 & 0 & 0 & 0 & 0 & 0 \\ 0 & 0 & 2 & 0 & 0 & 0 & 0 & 0 \\ 0 & 0 & 0 & 2 & 0 & 0 & 0 & 0 \\ 0 & 0 & 0 & 1 & 1 & 0 & 0 & 0 \\ 1 & -1 & 0 & 0 & 0 & 2 & 0 & 0 \\ 0 & 0 & 0 & 0 & 0 & 0 & 0 & -1 \\ 0 & 0 & -4 & 0 & 0 & 0 & 4 & 4 \end{pmatrix}, \quad C_{-1} = \begin{pmatrix} 0 & 0 & 0 & 0 & 0 & 0 & 0 & 0 \\ 0 & 0 & 0 & 0 & 0 & 0 & 0 & 0 \\ 0 & 0 & 0 & 0 & 0 & 0 & 0 & 0 \\ 0 & 0 & 0 & 0 & 0 & 0 & 0 & 0 \\ 0 & 2 & 0 & -1 & -1 & 0 & 0 & 0 \\ 0 & 0 & 0 & 0 & 0 & -2 & 0 & 0 \\ -2 & -2 & 0 & 0 & 2 & 4 & 0 & 1 \\ 2 & -2 & 0 & 2 & 0 & 4 & 0 & -1 \end{pmatrix}.$$

As our last example let's consider example 1 from Sect. 6.3.1. This is the family of the one-loop two-point function with equal internal masses. You might be tempted to consider this example to be the easiest example among the three examples discussed in Sect. 6.3.1, but we will soon see that this example requires an additional transformation not discussed so far. We start from the master integrals $\vec{I} = (I_{10}, I_{11})^T$ and the differential equation (see Eq. (6.66))

$$(d + A)\,\vec{I} \;=\; 0, \qquad A \;=\; \begin{pmatrix} 0 & 0 \\ \frac{1-\varepsilon}{2x} - \frac{1-\varepsilon}{2(x+4)} & \frac{1}{2x} - \frac{1-2\varepsilon}{2(x+4)} \end{pmatrix} dx. \tag{6.235}$$

Let's first see if we can transform the differential equation by a fibre transformation such that ε only appears as a prefactor. You are welcomed to play around and to convince yourself that there is no transformation rational in x and ε, which achieves that. However, if we allow the transformation to be algebraic, we may achieve this goal. Let's consider the transformation

$$\vec{I}' \;=\; U\vec{I}, \quad U \;=\; \begin{pmatrix} 2\varepsilon\,(1-\varepsilon) & 0 \\ 2\varepsilon\,(1-\varepsilon)\sqrt{\frac{x}{4+x}} & 2\varepsilon\,(1-2\varepsilon)\sqrt{\frac{x}{4+x}} \end{pmatrix}. \tag{6.236}$$

For the transformed system we find

$$\left(d + A'\right)\vec{I}' \;=\; 0, \;\; A' \;=\; \varepsilon \begin{pmatrix} 0 & 0 \\ -\frac{dx}{\sqrt{x(4+x)}} & \frac{dx}{4+x} \end{pmatrix}. \tag{6.237}$$

We have achieved that ε only appears as a prefactor, however the condition that the only singularities are simple poles is not met: The differential one-form

$$\frac{dx}{\sqrt{x\,(4+x)}} \tag{6.238}$$

has square root singularities at $x = 0$ and $x = -4$. In the next section we will learn how to transform these away. As a side remark let us note that we may force Eq. (6.238) into a dlog-form:

$$\frac{dx}{\sqrt{x\,(4+x)}} = d\ln\left(2 + x + \sqrt{x\,(4+x)}\right). \tag{6.239}$$

We see that in this case the argument of the logarithm is no longer a polynomial, but an algebraic function of x.

6.4.4 Base Transformations

In this section we consider coordinate transformation on the base manifold. Whereas the fibre transformations discussed in the previous section are like gauge transformation in gauge theories, the coordinate transformations on the base manifold discussed in this section are like coordinate transformation in general relativity.

On the base manifold M we perform a change of coordinates: We go from old coordinates $x_1, \dots, x_{N_B}$ to new coordinates $x'_1, \dots, x'_{N_B}$. Let's assume that the new coordinates are given in terms of the old coordinates as

$$x'_i = f_i(x), \qquad 1 \le i \le N_B. \tag{6.240}$$

If the matrix-valued differential one-form is written in terms of the old coordinates as

$$A = \sum_{i=1}^{N_B} A_i dx_i, \tag{6.241}$$

and in terms of the new coordinates as

$$A = \sum_{i=1}^{N_B} A'_i dx'_i, \tag{6.242}$$

then A'_i and A_j are related by

$$A'_i = \sum_j^{N_B} A_j \frac{\partial x_j}{\partial x'_i}. \tag{6.243}$$

Base transformation:
Let $\vec{I} = (I_{\mathfrak{v}_1}, \dots, I_{\mathfrak{v}_{N_{\text{master}}}})^T$ be a set of master integrals satisfying the differential equation

$$(d + A)\,\vec{I} = 0, \qquad A = \sum_{i=1}^{N_B} A_i dx_i, \tag{6.244}$$

If we change the coordinates on the base manifold M according to

$$x'_i = f_i(x), \qquad 1 \le i \le N_B, \tag{6.245}$$

and write A in terms of the new coordinates as

$$A = \sum_{i=1}^{N_B} A'_i dx'_i, \tag{6.246}$$

then A'_i and A_j are related by

$$A'_i = \sum_j^{N_B} A_j \frac{\partial x_j}{\partial x'_i}. \tag{6.247}$$

Let see how this works in an example: We continue with example 1 from Sect. 6.3.1. In the basis $\vec{I}'$ we had

$$\left(d + A'\right)\vec{I}' = 0, \; A' = \varepsilon \begin{pmatrix} 0 & 0 \\ -\frac{dx}{\sqrt{x(4+x)}} & \frac{dx}{4+x} \end{pmatrix}. \tag{6.248}$$

Let's define x' by

$$x = \frac{\left(1-x'\right)^2}{x'}. \tag{6.249}$$

The inverse relation reads

$$x' = \frac{1}{2}\left(2 + x - \sqrt{x\left(4+x\right)}\right), \tag{6.250}$$

where we made a choice for the sign of the square root. We have

$$\frac{\partial x}{\partial x'} = -\frac{\left(1-x'^2\right)}{x'^2} \tag{6.251}$$

and

$$\frac{dx}{\sqrt{x\left(4+x\right)}} = -\frac{dx'}{x'}, \; \frac{dx}{4+x} = \frac{2dx'}{x'+1} - \frac{dx'}{x'}. \tag{6.252}$$

Thus in term of the new variable x' we have

$$A' = \varepsilon \begin{pmatrix} 0 & 0 \\ \frac{dx'}{x'} & \frac{2dx'}{x'+1} - \frac{dx'}{x'} \end{pmatrix}. \tag{6.253}$$

The differential equation is now in ε-form: The dimensional regularisation parameter occurs only as a prefactor and the only singularities of A' are simple poles. For the case at hand, A' has simple poles at $x' = 0$ and $x' = -1$.

6.5 Cuts of Feynman Integrals

Up to now we focused entirely on Feynman integrals, given in the momentum representation by

$$I_{\nu_1 \dots \nu_{n_{\mathrm{int}}}}\left(D, x_1, \dots, x_{N_B}\right) = e^{l\varepsilon\gamma_{\mathrm{E}}}\left(\mu^2\right)^{\nu - \frac{lD}{2}} \int \prod_{r=1}^{l} \frac{d^D k_r}{i\pi^{\frac{D}{2}}} \prod_{j=1}^{n_{\mathrm{int}}} \frac{1}{\left(-q_j^2 + m_j^2\right)^{\nu_j}}. \tag{6.254}$$

The integration contour is along the real axes with deformations into the complex domain dictated by Feynman's $i\delta$-prescription. The Baikov representation of the Feynman integral reads

$$I_{\nu_1 \dots \nu_n}\left(D, x_1, \dots, x_{N_B}\right) = C_{\text{pre}} \int\limits_C d^{N_V} z \; \left[\mathcal{B}\left(z\right)\right]^{\frac{D-l-e-1}{2}} \prod_{s=1}^{N_V} z_s^{-\nu_s}, \tag{6.255}$$

where the prefactor C_{pre} is given by

$$C_{\text{pre}} = \frac{e^{l\varepsilon\gamma_{\text{E}}} \left(\mu^2\right)^{\nu - \frac{lD}{2}} \left[\det G\left(p_1, ..., p_e\right)\right]^{\frac{-D+e+1}{2}}}{\pi^{\frac{1}{2}(N_V - l)} \left(\det C\right) \prod\limits_{j=1}^{l} \Gamma\left(\frac{D-e+1-j}{2}\right)}. \tag{6.256}$$

$\mathcal{B}(z)$ denotes the Baikov polynomial and the Baikov variables are given by $z_j = -q_j^2 + m_j^2$. The domain of integration C is defined by Eqs. (2.233) and (2.234).

In this section we enlarge the set of integrals we are interested in and include integrals, which have the same integrands as in Eq. (6.255), but are integrated over a different domain. The new integration domains are not completely arbitrary, but should satisfy the following requirements:

1. Integration-by-parts identities still hold.
2. The variation of the integral with respect to the kinematic variables comes entirely from the integrand.
3. The symmetries among the integrals are respected.

Condition 1 says that in the language of differential forms

$$\int\limits_C d\xi \;=\; \int\limits_{\partial C} \xi \;=\; 0, \tag{6.257}$$

either because C is a cycle and therefore $\partial C = 0$ or because ξ vanishes on ∂C. Condition 2 is best explained with an example. Consider

$$\omega = \frac{x}{z - x} dz \tag{6.258}$$

and C a circle around $z = x$ of radius δ oriented anti-clockwise. Then

$$I \;=\; \int\limits_C \omega \;=\; 2\pi i x, \text{ and } \frac{d}{dx} I \;=\; 2\pi i. \tag{6.259}$$

On the other hand, we have

$$\frac{d}{dx}\omega = \frac{dz}{z - x} - x \left(\frac{d}{dz} \frac{1}{(z - x)} \right) dz, \tag{6.260}$$

and

$$\int\limits_C \frac{dz}{z - x} = 2\pi i, \quad -x \int\limits_C \left(\frac{d}{dz} \frac{1}{(z - x)} \right) dz = 0, \tag{6.261}$$

in agreement with our previous result. Although the integration contour moves with x (it is a circle of radius δ around $z = x$), in a small neighbourhood of $x = x_0$ (say of radius $\delta' \ll \delta$) we may deform the integration contour to a constant integration contour (for example for $x \in [x_0 - \delta', x_0 + \delta']$ we may deform the integration contour to a circle around $z = x_0$ of radius δ, independently of x). Hence, the derivative of the integral is calculated from the derivative of the integrand.

Also condition 3 is best explained by an example: For the one-loop two-point function with equal internal masses we have the symmetry $I_{\nu_1 \nu_2} = I_{\nu_2 \nu_1}$. Changing the integration contour may break this symmetry. This can be restored by considering a suitable symmetrised contour. We will require condition 3 in this section, but we will dispense ourselves from condition 3 in the next section.

Let us now consider a set of Feynman master integrals $\vec{I}$, satisfying the differential equation

$$(d + A)\,\vec{I} = 0. \tag{6.262}$$

We now consider a set of new integrals, where the original integration contour C is replaced by a new integration contour C', which satisfies conditions 1-3. Let us denote the new integrals by $\vec{I}'$. Conditions 1-3 are sufficient to show that the new integrals $\vec{I}'$ satisfy the same differential equation as the old integrals $\vec{I}$ [129]:

$$(d + A)\,\vec{I}' = 0. \tag{6.263}$$

The proof is rather simple: The differential equations rely only on the forms of the integrands (which are the same for $\vec{I}$ and $\vec{I}'$) and the properties given by conditions 1-3.

Let us now come to the topic of this section: **Feynman integrals with cuts**. We define a cut Feynman integral as a special case of the general situation discussed above. Let us consider the case, where we cut the internal edge e_j. We define the Feynman integral with the internal edge e_j cut to be given by the Baikov representation, where the domain of integration C is replaced by a modified domain of integration C' [32, 130, 131]. The modified domain C' consists of a small anti-clockwise circle around $z_j = 0$ in the complex z_j-plane. In all other variables the domain of integration is given by equations similar to Eqs. (2.233) and (2.234), with $\mathcal{B}$ replaced by

Fig. 6.4 A Feynman graph with a cut. The edge e_2 is cut

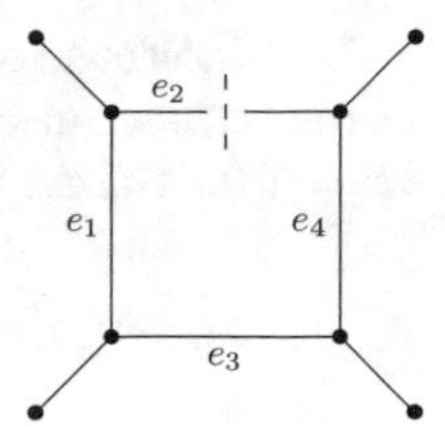

$$\mathcal{B}_j\left(z_1,\ldots,z_{j-1},z_{j+1},\ldots,z_{N_V}\right)=\mathcal{B}\left(z_1,\ldots,z_{j-1},0,z_{j+1},\ldots,z_{N_V}\right), \tag{6.264}$$

or shortly

$$\mathcal{B}_j=\mathcal{B}|_{z_j=0}\,. \tag{6.265}$$

This is just the intersection of the original integration domain C with the hyperplane $z_j=0$.

Thus we see that the cut integral is given by an integrand, which is $(2\pi i)$ times the residue at $z_j=0$ of the original integrand, integrated over the remaining variables. We draw a cut graph as shown in Fig. 6.4.

The name "cut Feynman integral" stems from the fact that if a propagator occurs only to power one, cutting this propagators corresponds in the momentum representation to the replacement

$$\frac{1}{-q_j^2+m_j^2}\rightarrow 2\pi i\;\delta\left(-q_j^2+m_j^2\right). \tag{6.266}$$

The δ-distribution forces the propagator on-shell. It is easily seen that for the case where a propagator occurs only to power one Eq. (6.266) is equivalent to taking the residue. In the Baikov variables Eq. (6.266) reads

$$\frac{1}{z_j}\rightarrow 2\pi i\;\delta\left(z_j\right). \tag{6.267}$$

We have for a small anti-clockwise circle γ_j around $z_j=0$ and $f(z_j)$ regular at $z_j=0$:

$$2\pi i\int dz_j f\left(z_j\right)\delta\left(z_j\right)\;=\;2\pi i f\left(0\right)\;=\;2\pi i\;\text{res}\left(\frac{f\left(z_j\right)}{z_j},z_j=0\right)\;=\;\oint\limits_{\gamma_j} dz_j\frac{f\left(z_j\right)}{z_j}. \tag{6.268}$$

Of course we may iterate the procedure and take multiple cuts. Of particular importance is the **maximal cut**, where we take for a Feynman integral $I_{\nu_1\ldots\nu_{n_{\text{int}}}}$ the cut for all edges e_j for which $\nu_j>0$.

Let's look at an example: We consider the one-loop two-point function with equal internal masses discussed as example 1 in Sect. 6.3.1. With $x = -p^2/m^2$ and $\mu^2 = m^2 = 1$ the Baikov polynomial is given by

$$\mathcal{B}(z_1, z_2) = -\frac{1}{4}\left[(z_1 - z_2)^2 - 2x(z_1 + z_2) + x(4 + x)\right], \tag{6.269}$$

and the Baikov representation of I_{11} is given by

$$I_{11} = \frac{e^{\varepsilon\gamma_E} x^{-\frac{D-2}{2}}}{2\sqrt{\pi}\,\Gamma\left(\frac{D-1}{2}\right)} \int\limits_C d^2z\; [\mathcal{B}(z_1, z_2)]^{\frac{D-3}{2}} \frac{1}{z_1 z_2}. \tag{6.270}$$

The cut of edge e_1 is given by

$$\mathrm{Cut}_{e_1} I_{11} = (2\pi i) \frac{e^{\varepsilon\gamma_E} x^{-\frac{D-2}{2}}}{2\sqrt{\pi}\,\Gamma\left(\frac{D-1}{2}\right)} \int\limits_{C'} dz_2\; [\mathcal{B}(0, z_2)]^{\frac{D-3}{2}} \frac{1}{z_2}. \tag{6.271}$$

We have

$$\mathcal{B}(0, z_2) = -\frac{1}{4}\left[z_2 - \left(x - 2\sqrt{-x}\right)\right]\left[z_2 - \left(x + 2\sqrt{-x}\right)\right]. \tag{6.272}$$

Let assume for the moment that $x < -4$. Then the integration domain C' is from $(x - 2\sqrt{-x})$ to $(x + 2\sqrt{-x})$ and $0 \notin [x - 2\sqrt{-x}, x + 2\sqrt{-x}]$. If $x \nless -4$ we may first pretend that $x < -4$, perform the integration and then continue analytically to the desired value of x.

In a similar way, the cut of the edge e_2 is given by

$$\mathrm{Cut}_{e_2} I_{11} = (2\pi i) \frac{e^{\varepsilon\gamma_E} x^{-\frac{D-2}{2}}}{2\sqrt{\pi}\,\Gamma\left(\frac{D-1}{2}\right)} \int\limits_{C''} dz_1\; [\mathcal{B}(z_1, 0)]^{\frac{D-3}{2}} \frac{1}{z_1}. \tag{6.273}$$

Cutting the edges e_1 and e_2 gives the maximal cut. We have

$$\begin{aligned} \mathrm{MaxCut}\; I_{11} &= \mathrm{Cut}_{e_1,e_2} I_{11} = (2\pi i)^2 \frac{e^{\varepsilon\gamma_E} x^{-\frac{D-2}{2}}}{2\sqrt{\pi}\,\Gamma\left(\frac{D-1}{2}\right)} [\mathcal{B}(0,0)]^{\frac{D-3}{2}} \\ &= (2\pi i)^2 \frac{e^{\varepsilon\gamma_E} x^{-\frac{D-2}{2}}}{2\sqrt{\pi}\,\Gamma\left(\frac{D-1}{2}\right)} \left(-\frac{1}{4}x(4+x)\right)^{\frac{D-3}{2}}. \end{aligned} \tag{6.274}$$

In $D = 2 - 2\varepsilon$ dimensions we have to leading order in the ε-expansion

$$\text{MaxCut}\ I_{11}\left(2-2\varepsilon\right) = -\frac{4\pi}{\sqrt{-x\left(4+x\right)}} + O\left(\varepsilon\right). \tag{6.275}$$

In Eq. (6.236) we found a fibre transformation, which puts the differential equation into a form, where the dimensional regularisation parameter ε appears only as a prefactor. The new master integral in the top sector was

$$I_2' = 2\varepsilon\sqrt{\frac{x}{4+x}}\left[\left(1-\varepsilon\right) I_{10}\left(4-2\varepsilon\right) + \left(1-2\varepsilon\right) I_{11}\left(4-2\varepsilon\right)\right]. \tag{6.276}$$

With the help of the dimensional shift relations (see also Exercise 45), we may rewrite this expression as

$$I_2' = -\varepsilon\sqrt{x\left(4+x\right)} I_{11}\left(2-2\varepsilon\right). \tag{6.277}$$

We see that up to a constant prefactor I_2' is $I_{11}\left(2-2\varepsilon\right)$ divided by the leading term in the ε-expansion of its maximal cut:

$$I_2' = \frac{4\pi\varepsilon}{\sqrt{-1}}\,\frac{I_{11}\left(2-2\varepsilon\right)}{\text{MaxCut}\ I_{11}\left(2\right)}. \tag{6.278}$$

The cut of the edge e_j of a Feynman integral where the corresponding propagator occurs to a higher power (i.e. $\nu_j > 1$) is obtained by first expanding the integrand of the Baikov representation as a Laurent series in the corresponding Baikov variable and by determining the residue as the coefficient of the $1/z_j$-term. For example

$$\text{Cut}_{e_1} I_{21} = \left(2\pi i\right) \frac{e^{\varepsilon\gamma_{\text{E}}} x^{-\frac{D-2}{2}}}{2\sqrt{\pi}\,\Gamma\left(\frac{D-3}{2}\right)} \int\limits_{C'} dz_2\ \left[\mathcal{B}\left(0,z_2\right)\right]^{\frac{D-5}{2}} \frac{\mathcal{B}'\left(0,z_2\right)}{z_2}, \tag{6.279}$$

with

$$\mathcal{B}'\left(0,z_2\right) = \left.\frac{\partial}{\partial z_1}\mathcal{B}\left(z_1,z_2\right)\right|_{z_1=0} = \frac{1}{2}\left(z_2+x\right). \tag{6.280}$$

The cut of the edge e_j of a Feynman integral where the corresponding propagator occurs to power $\nu_j \leq 0$ is zero, as there is no residue in this variable. For example

$$\text{Cut}_{e_1} I_{01} = \text{Cut}_{e_1} I_{(-1)1} = 0. \tag{6.281}$$

This reveals the power of considering cuts: A Feynman integral with $\nu_j \leq 0$ corresponds to a sub-topology, where the edge e_j is pinched. The corresponding cut integral is zero. Let us now consider the differential equation for a set of master integrals $\vec{I}$. Replacing the original integration contour by a contour with a cut in the edge e_j has the effect of setting all sub-topologies with edge e_j pinched to zero. The

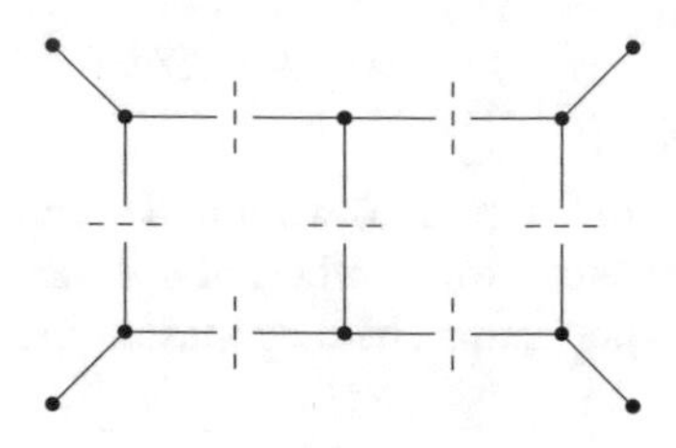

Fig. 6.5 The maximal cut of the double box graph

resulting differential equation is simpler. In particular, the maximal cut will set all sub-topologies to zero.

Exercise 56 *Work out the maximal cut of the double box integral $I_{111111100}$ shown in Fig. 6.5. Use the notation as in example 2 in Sect. 6.3.1. To work out the maximal cut it is simpler to use the loop-by-loop approach as discussed in Sect. 2.5.5.*

6.6 Singularities of Feynman Integrals

In this section we discuss singularities of Feynman integrals. There are two aspects to it. First there may be singularities, which occur for any values of the kinematic variables. These are the **ultraviolet** or **infrared divergences** of Feynman integrals. Secondly, there may be singularities, which only occur for specific values of the kinematic variables. These are called **Landau singularities**. A classic textbook on this subject is [132].

The singularities of Feynman integrals are most easily discussed within the Feynman parameter representation. We denote the kinematic variables by $(x_1, \dots, x_{N_B})$. We let U be an open subset of $\mathbb{CP}^{N_B}$ such that

$$(x_1, \dots, x_{N_B}) \in U \subset \mathbb{CP}^{N_B}. \tag{6.282}$$

The Feynman parameters $[a_1 : \dots : a_{n_{\text{int}}}]$ denote a point in $\mathbb{CP}^{n_{\text{int}}-1}$

$$[a_1 : \dots : a_{n_{\text{int}}}] \in \mathbb{CP}^{n_{\text{int}}-1}. \tag{6.283}$$

Let us first discuss ultraviolet and infrared singularities. These manifest themselves as poles in the dimensional regularisation parameter ε after integration. Let us assume that U is a subset of the Euclidean region, hence $x_j \geq 0$. This assumption will simplify the discussion below.

From the Feynman parameter integral in Eq. (2.170) we see that there are three possibilities how poles in ε can arise: First of all the gamma function $\Gamma(\nu - lD/2)$ of the prefactor can give rise to a (single) pole if the argument of this function is close to zero or to a negative integer value. This divergence is called the **overall ultraviolet divergence**.

Secondly, we consider the polynomial $\mathcal{U}$. Depending on the exponent $\nu - (l + 1)D/2$ of $\mathcal{U}$ the vanishing of the polynomial $\mathcal{U}$ in some part of the integration region can lead to poles in ε after integration. As mentioned in Sect. 2.5.2, each term of the expanded form of the polynomial $\mathcal{U}$ has coefficient $+1$, therefore $\mathcal{U}$ can only vanish if some of the Feynman parameters are equal to zero. In other words, $\mathcal{U}$ is non-zero (and positive) inside the integration region, but may vanish on the boundary of the integration region. Poles in ε resulting from the vanishing of $\mathcal{U}$ are related to **ultraviolet sub-divergences**.

Thirdly, we consider the polynomial $\mathcal{F}$. In the Euclidean region the polynomial $\mathcal{F}$ is also non-zero (and positive) inside the integration region. Therefore if all kinematic variables are within the Euclidean region the polynomial $\mathcal{F}$ can only vanish on the boundary of the integration region, similar to what has been observed for the the polynomial $\mathcal{U}$. Depending on the exponent $\nu - lD/2$ of $\mathcal{F}$ the vanishing of the polynomial $\mathcal{F}$ on the boundary of the integration region may lead to poles in ε after integration. These poles are related to **infrared divergences**.

Let us now turn to Landau singularities. We consider a Feynman integral as a function of the kinematic variables. We no longer impose any restrictions (like for example within the Euclidean region) on the kinematic variables. In particular, the kinematic variables are now allowed to lie in the physical region. Landau's equations give a necessary condition for a singularity to occur in the Feynman integral as we vary the kinematic variables.

The Feynman integral I as given in Eq. (2.170) depends through the polynomial $\mathcal{F}$ on the kinematic variables x_j. As we no longer restrict the kinematic variables x_j to the Euclidean region, the region where the polynomial $\mathcal{F}$ vanishes is no longer restricted to the boundary of the Feynman parameter integration region and we may encounter zeros of the polynomial $\mathcal{F}$ inside the integration region for specific values of the kinematic variables. The vanishing of $\mathcal{F}$ may in turn result in singularities after integration for specific values of the kinematic variables. These singularities are called Landau singularities. Necessary conditions for the occurrence of a Landau singularity are given as follows: A Landau singularity may occur if there exists a subset S of $\{1, \dots, n_{\text{int}}\}$ such that

$$\begin{aligned} a_i &= 0 \quad \text{for } i \in S \\ \text{and} \quad \frac{\partial}{\partial a_j} \mathcal{F} &= 0 \quad \text{for } j \in \{1, ..., n_{\text{int}}\} \backslash S. \end{aligned} \tag{6.284}$$

The Eq. 6.284 are called **Landau equations**.

Exercise 57 *Show that the Landau equations imply* $\mathcal{F} = 0$.

The case corresponding to $S = \emptyset$ is called the **leading Landau singularity**, and cases corresponding to $S \neq \emptyset$ are called **non-leading Landau singularities**. It is sufficient to focus on the leading Landau singularity, since a non-leading singularity is the leading Landau singularity of a sub-graph of G obtained by contracting the propagators corresponding to the Feynman parameters a_i with $i \in S$.

Let us now consider the leading Landau singularity of a graph G with n_{ext} external lines. We set

$$A = \mathbb{CP}^{n_{\mathrm{int}}-1} \backslash V\left(a_1 \cdot \ldots \cdot a_{n_{\mathrm{int}}} \mathcal{U}\right), \tag{6.285}$$

where

$$V(f) = \left\{ a \in \mathbb{CP}^{n_{\mathrm{int}}-1} \mid f(a) = 0 \right\}. \tag{6.286}$$

This takes out the regions in Feynman parameter space where we may have sub-leading Landau singularities or ultraviolet sub-divergences. Note that if we restrict the Feynman parameters to $\mathbb{RP}_{\geq 0}^{n_{\mathrm{int}}-1}$ we already know that the first graph polynomial can only vanish on the boundary, hence

$$V(\mathcal{U}) \cap \mathbb{RP}_{\geq 0}^{n_{\mathrm{int}}-1} \subset V\left(a_1 \cdot \ldots \cdot a_{n_{\mathrm{int}}}\right) \cap \mathbb{RP}_{\geq 0}^{n_{\mathrm{int}}-1}. \tag{6.287}$$

Let us consider

$$Y = \left\{ (a, x) \in A \times U \mid \frac{\partial}{\partial a_j} \mathcal{F} = 0, \; j = 1, \ldots, n_{\mathrm{int}} \right\} \tag{6.288}$$

together with the projection

$$\begin{aligned} \pi : \quad & Y \to U, \\ & (a, x) \to x. \end{aligned} \tag{6.289}$$

The **Landau discriminant** D_{Landau} is defined as the Zariski closure

$$D_{\mathrm{Landau}} = \overline{\pi(Y)} \subset U. \tag{6.290}$$

Essentially, this corresponds to all points $x \in U$, where the equations $\partial \mathcal{F} / \partial a_j = 0$ have a solution in the space A. A computer program to compute the Landau discriminant is described in [133].

Let us look at a simple example. We consider the one-loop two-point Feynman integral with equal internal masses in $D = 4 - 2\varepsilon$ space-time dimensions (example 1 in Sect. 6.3.1). This integral is given by

$$I_{11}(D, x) = e^{\varepsilon \gamma_{\mathrm{E}}} \left(m^2\right)^{\varepsilon} \int \frac{d^D k}{i \pi^{\frac{D}{2}}} \frac{1}{(-k^2 + m^2)(-(k-p)^2 + m^2)}, \tag{6.291}$$

where we set $\mu^2 = m^2$ and $x = -p^2/m^2$. The second graph polynomial is given by

$$\mathcal{F} = a_1 a_2 x + (a_1 + a_2)^2 . \tag{6.292}$$

The Landau equations for the leading Landau singularities are

$$a_2 x + 2(a_1 + a_2) = 0, \quad a_1 x + 2(a_1 + a_2) = 0. \tag{6.293}$$

These equations have a solution with $a_1 \neq 0$ and $a_2 \neq 0$ for $x \in \{-4, 0\}$, hence

$$D_{\text{Landau}} = \{-4, 0\} . \tag{6.294}$$

The Feynman parameter representation for this integral reads

$$I_{11}(4 - 2\varepsilon, x) = e^{\varepsilon \gamma_E} \Gamma(\varepsilon) \int_0^1 da_1 \; [1 + a_1 (1 - a_1) x]^{-\varepsilon} . \tag{6.295}$$

Working out this integral we find

$$I_{11}(4 - 2\varepsilon, x) = \frac{1}{\varepsilon} + 2 + \sqrt{\frac{4+x}{x}} \ln \frac{\sqrt{4+x} - \sqrt{x}}{\sqrt{4+x} + \sqrt{x}} + O(\varepsilon). \tag{6.296}$$

The $1/\varepsilon$-term corresponds to an ultraviolet divergence. As a function of x the Feynman integral has a Landau singularity at $x = -4$ (corresponding to $p^2 = 4m^2$). Note that the Feynman integral is finite at $x = 0$ and we see that Landau's equations give only a necessary condition for a Landau singularity, but not a sufficient condition. The Landau singularity at $x = -4$ is called a normal **threshold singularity**. The normal threshold manifests itself as a branch point in the complex x-plane.

Exercise 58 *Work out the Landau discriminant for the double box graph discussed in Exercise 44.*

6.7 Twisted Cohomology

We have seen that we can express any Feynman integral $I_{\nu_1 \dots \nu_{n_{\text{int}}}}$ from a family of Feynman integrals as a linear combination of master integrals $I_{\boldsymbol{\nu}_1}, \dots, I_{\boldsymbol{\nu}_{N_{\text{master}}}}$. The coefficients can be obtained by solving a linear system of equations. The equations themselves are symmetry relations and integration-by-parts identities. The system of linear equations can systematically be solved with the help of the Laporta algorithm. There is no principal problem in obtaining the coefficients, after all this is just linear algebra. However, there is a practical problem: Feynman integrals for cutting-edge precision calculations often lead to linear systems which barely can be treated with current computing resources.

We are therefore interested in alternative and more efficient methods to compute the coefficients. We have seen that the master integrals span a vector space and any Feynman integral from the family of Feynman integrals which we are investigating corresponds to a specific vector in this vector space. Writing this Feynman integral as a linear combination of master integrals is nothing else than expressing an arbitrary vector as a linear combination of basis vectors. Finding the coefficients is particular easy if the vector space is equipped with an inner product.

This leads directly to the question: Is there an inner product on the vector space of Feynman integrals? We will see in this section that the answer is almost yes, however we will not work with Feynman integrals, but with the integrands of Feynman integrals. As discussed in Sect. 6.1, the difference is that for Feynman integrals we take integration-by-parts identities and symmetry relations into account, while for the integrands of Feynman integrals we only take integration-by-parts identities into account. We denote the number of master integrals obtained by taking integration-by-parts identities and symmetries into account by N_{master}. If we are only interested in the number of unreduced integrals obtained from integration-by-parts identities alone, we denote this number by N_{cohom}. The simplest example is the one-loop two-point function with equal internal masses, where we have the symmetry $I_{\nu 0} = I_{0\nu}$. In this case $N_{\text{master}} = 2$, but $N_{\text{cohom}} = 3$. The integrands of the two tadpole integrals I_{10} and I_{01} differ, but yield the same result after integration.

In this section we focus on the integrands of Feynman integrals, therefore N_{cohom} is the relevant quantity. The mathematical setting is called twisted de Rham cohomology, which we now introduce.

Textbooks on twisted de Rham cohomology are the books by Yoshida [134] and Aomoto and Kita [135]. The application towards Feynman integrals started with [136, 137]. Many examples are provided in [137, 138]. Review articles on this subject are [139, 140].

6.7.1 *Twisted Cocycles*

We start from the n-dimensional complex space $\mathbb{C}^n$ and a **divisor** D, on which we will allow singularities. Instead of $\mathbb{C}^n$ we will later also consider other spaces like $\mathbb{CP}^n$. For our purpose it is sufficient to think of the divisor D as a union of hypersurfaces, where each hypersurface is defined by a polynomial equation. In detail, consider m polynomial equations

$$p_i(z_1, \dots, z_n) = 0, \qquad 1 \leq i \leq m, \tag{6.297}$$

with $p_i \in \mathbb{F}[z_1, \dots, z_n]$, where $\mathbb{F}$ is a field, typically $\mathbb{Q}$ or $\mathbb{Q}(x_1, \dots, x_m)$ (the field of rational functions in $x_1, \dots, x_m$ with rational coefficients). Each polynomial equation defines a hypersurface

$$D_i = \{(z_1, \dots, z_n) \in \mathbb{C}^n | p_i(z_1, \dots, z_n) = 0\}, \tag{6.298}$$

and D is the union of the m hypersurfaces:

$$D = \bigcup_{i=1}^{m} D_i. \tag{6.299}$$

We consider rational differential n-forms φ in the variables $z = (z_1, \dots, z_n)$, which are holomorphic on $\mathbb{C}^n - D$. The rational n-forms φ are of the form

$$\varphi = \frac{q}{p_1^{n_1} \dots p_m^{n_m}} \, dz_n \wedge \dots \wedge dz_1, \qquad q \in \mathbb{F}[z_1, \dots, z_n], \quad n_i \in \mathbb{N}_0. \tag{6.300}$$

Using the reversed wedge product $dz_n \wedge \dots \wedge dz_1$ instead of the standard order $dz_1 \wedge \dots \wedge dz_n$ is at this stage just a convention. $q(z_1, \dots, z_n)$ is a polynomial and the only singularities of φ in $\mathbb{C}^n$ are on D.

In cohomology theory we call the differential n-form φ a cocycle. It is closed on $\mathbb{C}^n - D$, since it is a holomorphic n-form: Obviously we have for $1 \leq j \leq n$

$$\left(\frac{\partial}{\partial \bar{z}_j} \frac{q}{p_1^{n_1} \dots p_m^{n_m}} \right) d\bar{z}_j \wedge dz_n \wedge \dots \wedge dz_1 = 0, \tag{6.301}$$

since the derivative in the bracket vanishes, but also

$$\left(\frac{\partial}{\partial z_j} \frac{q}{p_1^{n_1} \dots p_m^{n_m}} \right) dz_j \wedge dz_n \wedge \dots \wedge dz_1 = 0, \tag{6.302}$$

since the wedge product contains $dz_j \wedge dz_j$.

Let C be a n-dimensional integration cycle (i.e. an integration domain with no boundary $\partial C = 0$). We may now consider the integral

$$\langle \varphi | C \rangle = \int_C \varphi. \tag{6.303}$$

This is a pairing between a cycle and a cocycle. The quantity $\langle \varphi | C \rangle$ will not change if we add to φ the exterior derivative of a $(n-1)$-form ξ:

$$\varphi \to \varphi + d\xi. \tag{6.304}$$

Due to $\partial C = 0$ and Stokes' theorem

$$\int_C d\xi = \int_{\partial C} \xi = 0, \tag{6.305}$$

we have

$$\langle \varphi + d\xi | C \rangle = \langle \varphi | C \rangle . \tag{6.306}$$

We call two n-forms φ and φ' equivalent, if they differ by the exterior derivative of a $(n-1)$-form ξ as in Eq. (6.304):

$$\varphi' \sim \varphi \;\Leftrightarrow\; \varphi' = \varphi + d\xi. \tag{6.307}$$

The set of equivalence classes defines the (untwisted) de Rham cohomology group H^n.

Let us now introduce the **twist**: For m complex numbers $\gamma = (\gamma_1, \dots, \gamma_m)$ we set

$$u = \prod_{i=1}^{m} p_i^{\gamma_i}. \tag{6.308}$$

Since the exponents γ_i of the polynomials p_i are allowed to be complex numbers, u is in general a multi-valued function on $\mathbb{C}^n - D$. It will be convenient to define

$$\omega = d \ln u \;=\; \sum_{i=1}^{m} \gamma_i d \ln p_i \;=\; \sum_{j=1}^{n} \omega_j dz_j. \tag{6.309}$$

Let us fix a branch of u. We then consider the integral

$$\langle \varphi | C \rangle_{\omega} = \int\limits_{C} u \, \varphi. \tag{6.310}$$

C is again an integration cycle. We may allow C to have a boundary contained in the divisor D: $\partial C \subset D$. The integral remains well defined, if we assume that $\mathrm{Re}(\gamma_i)$ is sufficiently large, such that $u\varphi$ vanishes on D. It is not too difficult to see that now the integral remains invariant under

$$\varphi \to \varphi + \nabla_{\omega} \xi, \tag{6.311}$$

where we introduced the covariant derivative $\nabla_{\omega} = d + \omega$. In fact we have

$$\int\limits_{C} u \nabla_{\omega} \xi = \int\limits_{C} \left[u d\xi + u \left(d \ln u \right) \xi \right] \;=\; \int\limits_{C} d \left(u \xi \right) \;=\; \int\limits_{\partial C} u \xi \;=\; 0. \tag{6.312}$$

Introducing the twist amounts to going from the normal derivative d in Eq. (6.304) to the covariant derivative $\nabla_{\omega} = d + \omega$ in Eq. (6.311). The invariance under Eq. (6.311) motivates the definition of equivalence classes of n-forms φ: Two n-forms φ' and φ are called equivalent, if they differ by a covariant derivative

$$\varphi' \sim \varphi \quad \Leftrightarrow \quad \varphi' \;=\; \varphi + \nabla_\omega \xi \tag{6.313}$$

for some $(n-1)$-form ξ. We denote the equivalence classes by $\langle \varphi |$. Being n-forms, each φ is closed with respect to ∇_ω and the equivalence classes define the **twisted cohomology group** H^n_ω:

$$\langle \varphi | \in H^n_\omega. \tag{6.314}$$

Exercise 59 *Show that for φ as in Eq. (6.300) and ω as in Eq. (6.309) the differential n-form φ is closed with respect to ∇_ω:*

$$\nabla_\omega \varphi = 0. \tag{6.315}$$

The dual twisted cohomology group is given by

$$\left(H^n_\omega\right)^* = H^n_{-\omega}. \tag{6.316}$$

Elements of $(H^n_\omega)^*$ are denoted by $|\varphi\rangle$. We have

$$\left|\varphi'\right\rangle = |\varphi\rangle \quad \Leftrightarrow \quad \varphi' \;=\; \varphi + \nabla_{-\omega} \xi \tag{6.317}$$

for some $(n-1)$-form ξ. A representative of a dual cohomology class is of the form

$$\varphi = \frac{q}{p_1^{n_1} \dots p_m^{n_m}} \, dz_1 \wedge \dots \wedge dz_n, \qquad q \in \mathbb{F}\left[z_1, \dots, z_n\right], \quad n_i \in \mathbb{N}_0. \tag{6.318}$$

It will be convenient to use here the order $dz_1 \wedge \dots \wedge dz_n$ in the wedge product.

Digression Divisors *In the one-dimensional case the concept of a divisor originates from describing the set of zeros and the set of poles of a rational function. The concept of a divisor can be generalised to higher-dimensional algebraic varieties. Two different generalisations are in common use: Weil divisors and Cartier divisors. They agree on non-singular varieties.*

Let us start from the one-dimensional case: We consider divisors in the complex plane. Codimension one sub-varieties are zero-dimensional (i.e. points).

Let U be a connected open sub-set of $\mathbb{C}$. A divisor is a function

$$D : U \to \mathbb{Z}, \tag{6.319}$$

which takes non-zero values $D(z) \neq 0$ at most on a discrete set $\Sigma \subset U$. We write a divisor as a formal linear combination

$$D = \sum_{z \in U} D(z) \cdot z \tag{6.320}$$

Note that the symbol · in Eq. (6.320) is a convention for writing a divisor and has nothing to do with ordinary multiplication. The **degree of a divisor** *is the sum of its coefficients:*

$$\deg(D) = \sum_{z \in U} D(z). \tag{6.321}$$

A divisor is called **effective***, if all coefficients are non-negative:*

$$D(z) \geq 0, \qquad \forall z \in U. \tag{6.322}$$

In this case we also write

$$D \geq 0. \tag{6.323}$$

The divisors form an Abelian group, denoted by $\mathrm{Div}(U)$*. The zero divisor is given by* $D(z) = 0$ *for all* z*, the addition is defined by*

$$D_1 + D_2 = \sum_{z \in U} (D_1(z) + D_2(z)) \cdot z. \tag{6.324}$$

Let us now consider the field of meromorphic functions $K(U)$ *on* U*. We denote by* $K^*(U)$ *the set of meromorphic functions on* U *without the function* $f(z) = 0$*. Every* $f \in K^*(U)$ *has a Laurent series*

$$f(z) = \sum_{j=j_0}^{\infty} a_j \cdot (z - z_0)^j \tag{6.325}$$

around z_0*, starting with* j_0*. If* $f(z_0)$ *is finite and non-zero, we have* $j_0 = 0$*. If* $f(z_0) = 0$*, then* j_0 *gives the order of the zero. If* $f(z_0)$ *has a pole at* $z = z_0$*, then* $(-j_0)$ *denotes the order of the pole. A divisor* D_f *is defined by*

$$D_f(z_0) = j_0. \tag{6.326}$$

A divisor D *is called* **principal divisor***, if there is* $f \in K^*(U)$ *such that* $D = D_f$*. We have*

$$D_{fg} = D_f + D_g, \quad D_{\frac{1}{f}} = -D_f, \tag{6.327}$$

and therefore the map

$$\begin{aligned} K^*(U) &\to \mathrm{Div}(U), \\ f &\to D_f, \end{aligned} \tag{6.328}$$

is a group homomorphism. On $\mathbb{C}$ *we have the following statements:*

1. *If* $D \in \mathrm{Div}(U)$*, then there exists a* $f \in K^*(U)$ *such that* $D = D_f$*.*
2. *If* $f, g \in K^*(U)$ *with* $D_f = D_g$*, then* $h = f/g$ *is a holomorphic function with* $h(z) \neq 0$ *for all* z*.*

The first statement says, that on $\mathbb{C}$ *every divisor is a principal divisor. This is true for* $\mathbb{C}$*, but not for compact Riemann surfaces. The generalisation is given by the Riemann-Roch theorem.*

Weil divisor*: We have defined a divisor in the complex plane as a linear combination of points (i.e. codimension one sub-varieties) with integer coefficients. A Weil divisor is the generalisation of this idea of codimension one sub-varieties in higher dimensions*

We follow the book of Griffiths and Harris [125]. Let X be a complex manifold (or algebraic variety) of dimension n, not necessarily compact.

A Weil divisor is a locally finite linear combination with integral coefficients of irreducible sub-varieties of codimension one. We write

$$D = \sum n_i \cdot V_i, \qquad (6.329)$$

with $n_i \in \mathbb{Z}$ *and* V_i *irreducible sub-varieties of dimension* $(n-1)$*. Locally finite means that for any* $p \in M$ *there exists a neighbourhood of* p *meeting only a finite number of the* V_i*'s appearing in* D*.*

Cartier divisor*: We have seen that in the complex plane every divisor is a principal divisor and that the associated function* $f \in K^*(U)$ *is determined up to a multiple of a holomorphic function, everywhere non-zero. A Cartier divisor is a generalisation of this idea. For readers familiar with sheaves we give the definition of a Cartier divisor below. A definition of sheaves is given in Appendix G.*

Let X be a complex manifold (or algebraic variety) of dimension n, not necessarily compact. Further, denote by $\mathcal{O}$ *the sheaf of holomorphic functions on* X *and by* $\mathcal{M}$ *the sheaf of meromorphic functions on* X*. Let* $\mathcal{O}^*$ *denote the sheaf of holomorphic functions which are nowhere zero on* X *and let* $\mathcal{M}^*$ *denote the sheaf of meromorphic functions on* X *without the zero function. The quotient sheaf* $\mathcal{D} = \mathcal{M}^*/\mathcal{O}^*$ *is called the sheaf of divisors and a section of* $\mathcal{D}$ *is called Cartier divisor. The set of all sections* $\Gamma(X, \mathcal{D})$ *forms an Abelian group.*

6.7.2 Intersection Numbers

There is a non-degenerate bilinear pairing between a cohomology class $\langle \varphi_L |$ and a dual cohomology class $| \varphi_R \rangle$, given by the intersection number

$$\langle \varphi_L | \varphi_R \rangle_\omega . \qquad (6.330)$$

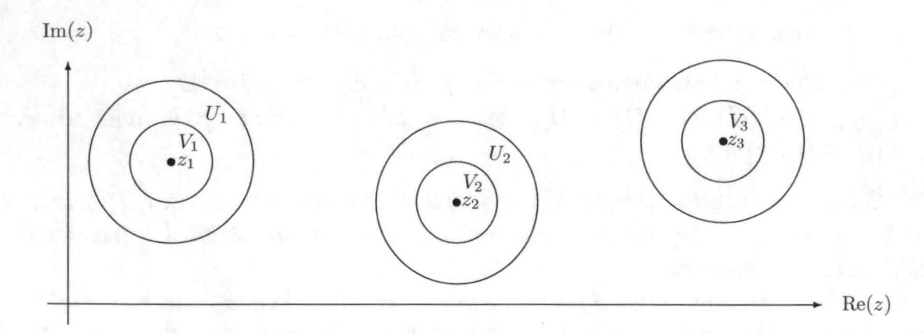

Fig. 6.6 The construction of the differential form with compact support in the one-dimensional case. The divisor is given by the union of three points: $D = \{z_1\} \cup \{z_2\} \cup \{z_3\}$. V_j and U_j are small discs around z_j with $V_j \subset U_j$

φ_L and φ_R are representatives of the cohomology classes $\langle \varphi_L |$ and $| \varphi_R \rangle$, respectively. φ_L and φ_R are differential n-forms as in Eqs. (6.300) and (6.318), respectively. It will be convenient to define $\hat{\varphi}_L$ and $\hat{\varphi}_R$ to be the functions obtained by stripping $dz_n \wedge \cdots \wedge dz_1$ and $dz_1 \wedge \cdots \wedge dz_n$ off, respectively:

$$\varphi_L = \hat{\varphi}_L \, dz_n \wedge \cdots \wedge dz_1, \;\; \varphi_R = \hat{\varphi}_R \, dz_1 \wedge \cdots \wedge dz_n. \tag{6.331}$$

The intersection number is defined by [135, 141]

$$\langle \varphi_L \, | \varphi_R \rangle_\omega = \frac{1}{(2\pi i)^n} \int \iota_\omega \left(\varphi_L\right) \wedge \varphi_R = \frac{1}{(2\pi i)^n} \int \varphi_L \wedge \iota_{-\omega} \left(\varphi_R\right), \tag{6.332}$$

where ι_ω maps φ_L to a differential form in the same cohomology class as φ_L but with compact support. Similarly, $\iota_{-\omega}$ maps φ_R to a differential form in the same cohomology class as φ_R but with compact support. That is to say, that $\iota_\omega(\varphi_L)$ and $\iota_{-\omega}(\varphi_R)$ vanish in a tubular neighbourhood of D (and at infinity). Although we started from differential forms φ_L and φ_R which are holomorphic on $\mathbb{C}^n - D$, the compactly supported versions $\iota_\omega(\varphi_L)$ and $\iota_{-\omega}(\varphi_R)$ are no longer holomorphic on $\mathbb{C}^n - D$.

Please note that the pairing $\langle \varphi | C \rangle_\omega$ between an integrand and an integration contour denotes the integral defined in Eq. (6.310), while the pairing $\langle \varphi_L | \varphi_R \rangle_\omega$ between an integrand and a dual integrand denotes the intersection number defined in Eq. (6.332).

In order to see how $\iota_\omega(\varphi_L)$ (or $\iota_{-\omega}(\varphi_R)$) is constructed, let's consider $\mathbb{CP}^1 - D$ [142]. The divisor is a set of points $D = \{z_1, \dots, z_m\}$. Let's assume that none of these points is at infinity. Around a point z_j we consider two small discs V_j and U_j, both centred at z_j and such that the radius of V_j is smaller than the radius of U_j. This is shown in Fig. 6.6. We assume that the U_j's do not overlap. We introduce non-holomorphic functions $h_j(z, \bar{z})$ equal to 1 on V_j, equal to 0 outside U_j and interpolating smoothly in the region $U_j - V_j$. As φ_L and $\iota_\omega(\varphi_L)$ are in the same cohomology class, they differ by a covariant derivative:

$$\varphi_L - \iota_\omega (\varphi_L) = \nabla_\omega \xi. \tag{6.333}$$

Let $\psi_{L,j}$ be a solution of

$$\nabla_\omega \psi_{L,j} = \varphi_L \tag{6.334}$$

on $U_j \backslash \{z_j\}$. We set

$$\xi = \sum_{j=1}^{m} h_j \psi_{L,j}. \tag{6.335}$$

By construction, $\iota_\omega(\varphi_L)$ is in the same cohomology class as φ_L. Let's verify that $\iota_\omega(\varphi_L)$ has compact support. We show that $\iota_\omega(\varphi_L)$ vanishes on V_i. We have

$$\begin{aligned}
\iota_\omega (\varphi_L) = \varphi_L - \nabla_\omega \xi \; &= \varphi_L - \sum_{j=1}^{m} \nabla_\omega \left(h_j \psi_{L,j}\right) \\
&= \varphi_L - \sum_{j=1}^{m} h_j \nabla_\omega \psi_{L,j} - \sum_{j=1}^{m} \left(dh_j\right) \psi_{L,j} \\
&= \varphi_L - \sum_{j=1}^{m} h_j \varphi_L - \sum_{j=1}^{m} \left(dh_j\right) \psi_{L,j}.
\end{aligned} \tag{6.336}$$

On V_i we have $h_i = 1$ and $h_j = 0$ for $j \neq i$. Furthermore we have on V_i that the derivative of all functions h_j vanishes: $dh_j = 0$ for all j. Thus we find on V_i

$$\iota_\omega (\varphi_L) = \varphi_L - h_i \varphi_L \; = \; \varphi_L - \varphi_L \; = \; 0. \tag{6.337}$$

Let us now turn to the intersection number for the case $\mathbb{CP}^1 - D$. We have

$$\begin{aligned}
\langle \varphi_L | \varphi_R \rangle_\omega = \frac{1}{2\pi i} \int \iota_\omega (\varphi_L) \wedge \varphi_R \; &= \; \frac{1}{2\pi i} \int \left[\varphi_L - \sum_{j=1}^{m} h_j \varphi_L - \sum_{j=1}^{m} \left(dh_j\right) \psi_{L,j} \right] \wedge \varphi_R \\
&= -\frac{1}{2\pi i} \sum_{j=1}^{m} \int \left(dh_j\right) \psi_{L,j} \wedge \varphi_R.
\end{aligned} \tag{6.338}$$

The first two terms yield $dz \wedge dz$, only the last term yields a non-vanishing wedge product $dz \wedge d\bar{z}$. As dh_j is non-zero only on $U_j - V_j$ we obtain

$$\begin{aligned}
\langle \varphi_L | \varphi_R \rangle_\omega &= -\frac{1}{2\pi i} \sum_{j=1}^{m} \int_{U_j - V_j} (dh_j)\, \psi_{L,j} \wedge \varphi_R \\
&= -\frac{1}{2\pi i} \sum_{j=1}^{m} \int_{U_j - V_j} \left[d\left(h_j \psi_{L,j} \varphi_R\right) - h_j \left(d\psi_{L,j}\right) \wedge \varphi_R - h_j \psi_{L,j} d\varphi_R \right] \\
&= -\frac{1}{2\pi i} \sum_{j=1}^{m} \int_{U_j - V_j} d\left(h_j \psi_{L,j} \varphi_R\right).
\end{aligned} \tag{6.339}$$

In the second line the last two terms yield a vanishing contribution, again due to $dz \wedge dz = 0$. We may now use Stokes' theorem and obtain

$$\langle \varphi_L | \varphi_R \rangle_\omega = \frac{1}{2\pi i} \sum_{j=1}^{m} \int_{\partial V_j} \psi_{L,j} \varphi_R = \sum_{j=1}^{m} \mathrm{Res}_{D_j} \left(\psi_{L,j} \varphi_R\right), \tag{6.340}$$

where we used that $h_j = 0$ on ∂U_j and $h_j = 1$ on ∂V_j.

Alternatively, we could have used $\iota_{-\omega}(\varphi_R)$. Let $\psi_{R,j}$ be a solution of

$$\nabla_{-\omega} \psi_{R,j} = \varphi_R \tag{6.341}$$

on $U_j \backslash \{z_j\}$. Then

$$\begin{aligned}
\langle \varphi_L | \varphi_R \rangle_\omega &= \frac{1}{2\pi i} \int \varphi_L \wedge \iota_{-\omega}(\varphi_R) \\
&= -\frac{1}{2\pi i} \sum_{j=1}^{m} \int_{\partial V_j} \varphi_L \psi_{R,j} = -\sum_{j=1}^{m} \mathrm{Res}_{D_j} \left(\varphi_L \psi_{R,j}\right).
\end{aligned} \tag{6.342}$$

Computing the intersection number through the definition in Eq. (6.332) is not the most practical way. In Sect. 6.7.3 we will learn more efficient methods.

Let's consider an example: We take $n = 1$ and we consider $p_1(z) = z$, $p_2(z) = 1 - z$. Thus we have $D_1 = \{0\}$ and $D_2 = \{1\}$. We consider

$$u(z) = z^{\gamma_1} (1 - z)^{\gamma_2}. \tag{6.343}$$

The one-form ω is then given by

$$\omega = \gamma_1 \frac{dz}{z} - \gamma_2 \frac{dz}{1 - z}. \tag{6.344}$$

Let us further consider

$$\varphi_L = \frac{dz}{z^{1+n_1}(1-z)^{1+n_2}}, \quad \varphi_R = \frac{dz}{z^{1+n_3}(1-z)^{1+n_4}}, \qquad n_1, n_2, n_3, n_4 \in \mathbb{Z}. \tag{6.345}$$

Around $z = 0$ let $\psi_{L,1}$ be a solution of

$$\nabla_\omega \psi_{L,1} = \varphi_L. \tag{6.346}$$

For $\psi_{L,1}$ we make the ansatz

$$\psi_{L,1} = \sum_{j=-n_1}^{\infty} a_j z^j. \tag{6.347}$$

Equation (6.346) becomes

$$\left(\frac{d}{dz} + \frac{\gamma_1}{z} - \frac{\gamma_2}{1-z}\right) \sum_{j=-n_1}^{\infty} a_j z^j = \frac{1}{z^{1+n_1}(1-z)^{1+n_2}}. \tag{6.348}$$

This equation can be solved recursively for the unknown coefficients a_j, starting with

$$a_{-n_1} = \frac{1}{\gamma_1 - n_1}. \tag{6.349}$$

In a similar way, we let $\psi_{L,2}$ be a solution of

$$\nabla_\omega \psi_{L,2} = \varphi_L. \tag{6.350}$$

around $z = 1$. We make the ansatz

$$\psi_{L,2} = \sum_{j=-n_2}^{\infty} b_j (1-z)^j \tag{6.351}$$

and determine the coefficients b_j recursively. The intersection number $\langle \varphi_L | \varphi_R \rangle_\omega$ is then obtained with the help of Eq. (6.340) as

$$\langle \varphi_L | \varphi_R \rangle_\omega = \mathrm{Res}_{D_1}\left(\psi_{L,1}\varphi_R\right) + \mathrm{Res}_{D_2}\left(\psi_{L,2}\varphi_R\right). \tag{6.352}$$

We obtain

$$\langle \varphi_L | \varphi_R \rangle_\omega = \frac{(\gamma_1+\gamma_2)}{\gamma_1\gamma_2} \frac{\Gamma(1-\gamma_1)\Gamma(1-\gamma_2)}{\Gamma(1-\gamma_1-\gamma_2)} \frac{\Gamma(1+\gamma_1)\Gamma(1+\gamma_2)}{\Gamma(1+\gamma_1+\gamma_2)} \frac{\Gamma(1+n_1+n_2-\gamma_1-\gamma_2)}{\Gamma(1+n_1-\gamma_1)\Gamma(1+n_2-\gamma_2)} \frac{\Gamma(1+n_3+n_4+\gamma_1+\gamma_2)}{\Gamma(1+n_3+\gamma_1)\Gamma(1+n_4+\gamma_2)}. \tag{6.353}$$

Exercise 60 *Proof Eq. (6.353) for the special case* $n_1 = n_2 = n_3 = n_4 = 0$.

Under certain assumptions it can be shown [135, 143–145] that the twisted cohomology groups H^k_ω vanish for $k \neq n$, thus H^n_ω is the only interesting twisted cohomology group. We denote the dimensions of the twisted cohomology groups by

$$\nu = \dim H^n_\omega \; = \; \dim \left(H^n_\omega\right)^*. \tag{6.354}$$

Let $\langle e_j|$ with $1 \leq j \leq \nu$ be a basis of H^n_ω and let $|h_j\rangle$ with $1 \leq j \leq \nu$ be a basis of $(H^n_\omega)^*$. We denote the $(\nu \times \nu)$-dimensional intersection matrix by C. The entries are given by

$$C_{jk} = \langle e_j | \, h_k \rangle. \tag{6.355}$$

The matrix C is invertible. Given a basis $\langle e_j|$ of H^n_ω we say that a basis $|d_j\rangle$ of $(H^n_\omega)^*$ is the **dual basis** with respect to $\langle e_j|$ if

$$\langle e_j | \, d_k \rangle = \delta_{jk}. \tag{6.356}$$

Starting from an arbitrary basis $|h_j\rangle$ of $(H^n_\omega)^*$ we may always construct the dual basis $|d_j\rangle$ of $(H^n_\omega)^*$ with respect to $\langle e_j|$. The dual basis is given by

$$|d_j\rangle = |h_k\rangle \left(C^{-1}\right)_{kj}. \tag{6.357}$$

The dimension of the twisted cohomology groups is related to the number of critical points of $f = \ln(u)$. We have

$$f = \ln(u) \; = \; \sum_{i=1}^{m} \gamma_i \ln(p_i). \tag{6.358}$$

A point z is called a **critical point** if

$$df|_z = 0. \tag{6.359}$$

A critical point z is called a **non-degenerate critical point** if the Hessian matrix is invertible, i.e.

$$\det\left(\frac{\partial^2 f}{\partial z_i \partial z_j}\right)\Bigg|_z \neq 0. \tag{6.360}$$

A critical point z is called a **proper critical point** if

$$z \notin D. \tag{6.361}$$

By the definition of ω in Eq. (6.309) we have $df = \omega$. Assuming that all critical points are proper and non-degenerate we have [30, 137]

$$\dim H^n_\omega = \left(\text{\# solutions of } \omega = 0 \text{ on } \mathbb{C}^n - D\right). \tag{6.362}$$

Usually it is not an issue to find a basis. For completeness, we give here a systematic algorithm to construct a basis for H^n_ω and $(H^n_\omega)^*$ for the case where all critical points are proper and non-degenerate. We write

$$\omega = \sum_{j=1}^{n} \omega_j dz_j, \quad \omega_j = \frac{P_j}{Q_j}, \quad P_j, Q_j \in \mathbb{F}[z_1, \dots, z_n], \quad \gcd\left(P_j, Q_j\right) = 1. \tag{6.363}$$

We consider the ideal

$$I_n = \langle P_1, \dots, P_n \rangle \subset \mathbb{F}[z_1, \dots, z_n]. \tag{6.364}$$

In the case where all critical points are proper and non-degenerate we have

$$\dim H^n_\omega = \dim\left(\mathbb{F}[z_1, \dots, z_n] / I_n\right). \tag{6.365}$$

Let $G_1, \dots, G_r$ be a Gröbner basis of I_n with respect to some term order $<$:

$$I_n = \langle G_1, \dots, G_r \rangle. \tag{6.366}$$

For a basis $\langle e_j|$ of H^n_ω we write as in Eq. (6.331)

$$e_j = \hat{e}_j \, dz_n \wedge \dots \wedge dz_1. \tag{6.367}$$

Similarly, we write for a basis $|h_j\rangle$ of $(H^n_\omega)^*$

$$h_j = \hat{h}_j \, dz_1 \wedge \dots \wedge dz_n. \tag{6.368}$$

Then $\hat{e}_j$ and $\hat{h}_j$ are given by all monomials

$$\prod_{k=1}^{n} z_k^{\nu_k}, \quad \nu_k \in \mathbb{N}_0 \tag{6.369}$$

which are not divisible by any leading term of the Gröbner basis:

$$\mathrm{lt}\left(G_j\right) \not| \prod_{k=1}^{n} z_k^{\nu_k} \qquad \forall\, 0 \le j \le r. \tag{6.370}$$

Here, lt denotes the leading term of a polynomial with respect to the chosen term order. In general, $\langle e_j|$ and $|h_j\rangle$ defined in this way will not be dual to each other.

Let's look at an example. We consider a case with two variables z_1 and z_2 (i.e. $n = 2$) and three polynomials (i.e. $m = 3$)

$$p_1 = z_1, \quad p_2 = z_2, \quad p_3 = z_2^2 - 4z_1^3 + 11z_1 - 7. \tag{6.371}$$

We set

$$u = (p_1 p_2 p_3)^{\gamma}\,. \tag{6.372}$$

The differential one-form ω reads

$$\omega = \gamma \frac{z_2^2 - 16z_1^3 + 22z_1 - 7}{p_1 p_3} dz_1 + \gamma \frac{3z_2^2 - 4z_1^3 + 11z_1 - 7}{p_2 p_3} dz_2. \tag{6.373}$$

We therefore have to consider the ideal

$$I_2 = \left\langle z_2^2 - 16z_1^3 + 22z_1 - 7, 3z_2^2 - 4z_1^3 + 11z_1 - 7\right\rangle. \tag{6.374}$$

A Gröbner basis with respect to the graded reverse lexicographic order is given by

$$I_2 = \left\langle 11z_2^2 + 22z_1 - 21, 44z_1^3 - 55z_1 + 14\right\rangle. \tag{6.375}$$

The leading terms of the elements of the Gröbner basis are $11z_2^2$ and $44z_1^3$. A basis $\langle e_j|$ of H_ω^2 with $e_j = \hat{e}_j dz_2 \wedge dz_1$ is therefore given by

$$\hat{e}_j \in \left\{1, z_1, z_2, z_1 z_2, z_1^2, z_1^2 z_2\right\}. \tag{6.376}$$

Similarly, a basis $|h_j\rangle$ of $(H_\omega^n)^*$ with $h_j = \hat{h}_j dz_1 \wedge dz_2$ is given by

$$\hat{h}_j \in \left\{1, z_1, z_2, z_1 z_2, z_1^2, z_1^2 z_2\right\}. \tag{6.377}$$

Digression Gröbner bases

Gröbner bases are useful in many situation. The most prominent application of Gröbner bases is probably the simplification of a polynomial with respect to polynomial siderelations. A good introduction to Gröbner bases is the book by Adams and Loustaunau [146].

Assume that we have a (possibly rather long) expression f, which is a polynomial in several variables $x_1, \ldots, x_k$. In addition we have several siderelations of the form

$$s_j(x_1, \dots, x_k) = 0, \quad 1 \le j \le r, \tag{6.378}$$

which are also polynomials in x_1, ..., x_k. *A standard task is now to simplify* f *with respect to the siderelations* s_j, *e.g. to rewrite* f *in the form*

$$f = a_1 s_1 + \dots + a_r s_r + g, \tag{6.379}$$

where g is "simpler" than f *The precise meaning of "simpler" requires the introduction of an order relation on the multivariate polynomials. As an example let us consider the expressions*

$$f_1 = x + 2y^3, \quad f_2 = x^2, \tag{6.380}$$

which we would like to simplify with respect to the siderelations

$$\begin{aligned} s_1 &= x^2 + 2xy, \\ s_2 &= xy + 2y^3 - 1. \end{aligned} \tag{6.381}$$

As an order relation we may choose lexicographic ordering, e.g. x is "more complicated" as y, and x^2 *is "more complicated" than x. This definition will be made more precise below. A naive approach would now take each siderelation, determine its "most complicated" element, and replace each occurrence of this element in the expression* f *by the more simpler terms of the siderelation. As an example let us consider for this approach the simplification of* f_2 *with respect to the siderelations* s_1 *and* s_2*:*

$$f_2 = x^2 = s_1 - 2xy = s_1 - 2ys_2 + 4y^4 - 2y, \tag{6.382}$$

and f_2 *would simplify to* $4y^4 - 2y$. *In addition, since* f_1 *does not contain* x^2 *nor* xy, *the naive approach would not simplify* f_1 *at all. However, this is not the complete story, since if* s_1 *and* s_2 *are siderelations, any linear combination of those is again a valid siderelation. In particular,*

$$s_3 = ys_1 - xs_2 = x \tag{6.383}$$

is a siderelation which can be deduced from s_1 *and* s_2. *This implies that* f_2 *simplifies to* 0 *with respect to the siderelations* s_1 *and* s_2. *Clearly, some systematic approach is needed. The appropriate tools are ideals in rings, and Gröbner bases for these ideals.*

We consider multivariate polynomials in the ring $R[x_1, \dots, x_k]$. *Each element can be written as a sum of monomials of the form*

$$c x_1^{m_1} \dots x_k^{m_k}. \tag{6.384}$$

We define the **lexicographic order** *of these terms by*

$$c x_1^{m_1} \dots x_k^{m_k} > c' x_1^{m'_1} \dots x_k^{m'_k}, \tag{6.385}$$

if the leftmost non-zero entry in $(m_1 - m'_1, \dots, m_k - m'_k)$ *is positive. With this ordering we can write any element* $f \in R[x_1, \dots, x_k]$ *as*

$$f = \sum_{i=0}^{n} h_i \tag{6.386}$$

where the h_i *are monomials and* $h_{i+1} > h_i$ *with respect to the lexicographic order. The term* h_n *is called the* **leading term** *and denoted* $\mathrm{lt}(f) = h_n$.

Let $B = \{b_1, \dots, b_r\} \subset R[x_1, \dots, x_k]$ *be a (finite) set of polynomials. The set*

$$\langle B \rangle = \langle b_1, \dots, b_r \rangle = \left\{ \sum_{i=1}^{r} a_i b_i \,\middle|\, a_i \in R[x_1, \dots, x_k] \right\} \tag{6.387}$$

is called the **ideal** *generated by the set* B. *The set* B *is also called a* **basis** *for this ideal. (In general, given a ring* R *and a subset* $I \subset R$, I *is called an ideal if* $a + b \in I$ *for all* $a, b, \in I$ *and* $ra \in I$ *for all* $a \in I$ *and* $r \in R$. *Note the condition for the multiplication: The multiplication has to be closed with respect to elements from* R *and not just* I.*)*
Suppose that we have an ideal I *and a finite subset* $H \subset I$. *We denote by* $\mathrm{lt}(H)$ *the set of leading terms of* H *and, correspondingly by* $\mathrm{lt}(I)$ *the set of leading terms of* I. *Now suppose that the ideal generated by* $\mathrm{lt}(H)$ *is identical with the one generated by* $\mathrm{lt}(I)$, *e.g.* $\mathrm{lt}(H)$ *is a basis for* $\langle \mathrm{lt}(I) \rangle$. *Then a mathematical theorem guarantees that* H *is also a basis for* I, *e.g.*

$$\langle \mathrm{lt}(H) \rangle = \langle \mathrm{lt}(I) \rangle \Rightarrow \langle H \rangle = I \tag{6.388}$$

However, the converse is in general not true, e.g. if H *is a basis for* I *this does not imply that* $\mathrm{lt}(H)$ *is a basis for* $\langle \mathrm{lt}(I) \rangle$. *A further theorem (due to Hilbert) states however that there exists a subset* $G \subset I$ *such that*

$$\langle G \rangle = I \quad \textit{and} \quad \langle \mathrm{lt}(G) \rangle = \langle \mathrm{lt}(I) \rangle, \tag{6.389}$$

e.g. G *is a basis for* I *and* $\mathrm{lt}(G)$ *is a basis for* $\langle \mathrm{lt}(I) \rangle$. *Such a set* G *is called a* **Gröbner basis** *for* I. *Buchberger [147] gave an algorithm to compute* G, *which nowadays is implemented in many computer algebra systems.*

The importance of Gröbner bases for simplifications stems from the following theorem: Let G *be a Gröbner basis for an ideal* $I \subset R[x_1, \dots, x_k]$ *and* $f \in R[x_1, \dots, x_k]$. *Then there is a unique polynomial* $g \in R[x_1, \dots, x_k]$ *with*

$$f - g \in I \tag{6.390}$$

and no term of g is divisible by any monomial in $\text{lt}(G)$.
In plain text: f *is an expression which we would like to simplify according to the siderelations defined by* I. *This ideal is originally given by a set of polynomials* $\{s_1, \dots, s_r\}$ *and the siderelations are supposed to be of the form* $s_i = 0$. *From this set of siderelations a Gröbner basis* $\{b_1, \dots, b_{r'}\}$ *for this ideal is calculated. This is the natural basis for simplifying the expression* f. *The result is the expression g, from which the "most complicated" terms of G have been eliminated, e.g. the terms* $\text{lt}(G)$. *The precise meaning of "most complicated" terms depends on the definition of the order relation.*
In our example, $\{s_1, s_2\}$ *is not a Gröbner basis for* $\langle s_1, s_2\rangle$, *since* $\text{lt}(s_1) = x^2$ *and* $\text{lt}(s_2) = xy$ *and*

$$\text{lt}\,(y s_1 - x s_2) = x \quad \notin \quad \langle \text{lt}(s_1), \text{lt}(s_2)\rangle. \tag{6.391}$$

A Gröbner basis for $\langle s_1, s_2\rangle$ *is given by*

$$\left\{x, 2y^3 - 1\right\}. \tag{6.392}$$

With $b_1 = x$ *and* $b_2 = 2y^3 - 1$ *as a Gröbner basis,* f_1 *and* f_2 *can be simplified as follows:*

$$\begin{aligned} f_1 &= b_1 + b_2 + 1, \\ f_2 &= x b_1 + 0, \end{aligned} \tag{6.393}$$

e.g. f_1 *simplifies to* 1 *and* f_2 *simplifies to* 0.

We are not forced to use the lexicographic order introduced above. We may use any **term order**. *A term order is a total order (this means that between two quantities exactly one of the relations* $<, =$ *or* $>$ *must be true) on the monomials* $c x_1^{m_1} \dots x_k^{m_k}$ *which satisfies*

1. $x_1^{m_1} \dots x_k^{m_k} > 1$ *for* $(m_1, \dots, m_k) \neq (0, \dots, 0)$.
2. $x_1^{m_1} \dots x_k^{m_k} > x_1^{m'_1} \dots x_k^{m'_k}$ *implies*

$$\left(x_1^{m_1} \dots x_k^{m_k}\right)\left(x_1^{m''_1} \dots x_k^{m''_k}\right) > \left(x_1^{m'_1} \dots x_k^{m'_k}\right)\left(x_1^{m''_1} \dots x_k^{m''_k}\right) \tag{6.394}$$

for any monomial $x_1^{m''_1} \dots x_k^{m''_k}$.

Apart from the lexicographic order introduced above other popular choices for a term order are the degree lexicographic order and the degree reverse lexicographic order. The **degree lexicographic order** *(or graded lexicographic order) is defined as follows: We have*

$$\left(x_1^{m_1} \dots x_k^{m_k}\right) > \left(x_1^{m'_1} \dots x_k^{m'_k}\right) \tag{6.395}$$

if either

$$\sum_{j=1}^{k} m_j > \sum_{j=1}^{k} m'_j,$$

or in the case that the total degrees are equal

$$\sum_{j=1}^{k} m_j = \sum_{j=1}^{k} m'_j \text{ and the leftmost non-zero entry in } (m_1 - m'_1, \dots, m_k - m'_k) \text{ is positive,}$$

The **degree reverse lexicographic order** *(or graded reverse lexicographic order) is defined as follows: We have*

$$\left(x_1^{m_1} \dots x_k^{m_k}\right) > \left(x_1^{m'_1} \dots x_k^{m'_k}\right) \tag{6.396}$$

if either

$$\sum_{j=1}^{k} m_j > \sum_{j=1}^{k} m'_j,$$

or in the case that the total degrees are equal

$$\sum_{j=1}^{k} m_j = \sum_{j=1}^{k} m'_j \text{ and the rightmost non-zero entry in } (m_1 - m'_1, \dots, m_k - m'_k) \text{ is negative.}$$

Exercise 61 *Consider the monomials*

$$p_1 = x_1^2 x_2 x_3, \quad p_2 = x_1 x_2^3. \tag{6.397}$$

Order the two monomials with respect to the degree lexicographic order and the degree reverse lexicographic order (assuming $x_1 > x_2 > x_3$).

The critical points of

$$f = \ln(u) = \sum_{i=1}^{m} \gamma_i \ln(p_i) \tag{6.398}$$

allow us to construct a basis $|C_j\rangle$ of the twisted homology groups H_n^ω as well: We first fix a branch of u. As before we assume that all critical points are proper and non-degenerate. For simplicity let us further assume that $\gamma_i \in \mathbb{R}_{<0}$ and $\gamma_i \notin \mathbb{Z}$. We split f into the real part and the imaginary part:

$$f(z) = \sum_{i=1}^{m} \gamma_i \ln |p_i(z)| + i \sum_{i=1}^{m} \gamma_i \arg(p_i(z)). \tag{6.399}$$

We denote the real part by $h(z)$ and the imaginary part by $\phi(z)$. Thus

$$h(z) = \sum_{i=1}^{m} \gamma_i \ln |p_i(z)|, \quad \phi(z) = \sum_{i=1}^{m} \gamma_i \arg(p_i(z)). \tag{6.400}$$

The value of ϕ at a critical points $z_{\mathrm{crit}}^{(j)}$ is called a critical phase and denoted by

$$\phi^{(j)} = \phi\left(z_{\mathrm{crit}}^{(j)}\right). \tag{6.401}$$

Since we assumed that all critical points are non-degenerate it follows that h is a Morse function. We will now consider the gradient flow equations for h. If we temporarily introduce $(2n)$ real coordinates $x_1, \dots, x_n, y_1, \dots, y_n$ such that $z_j = x_j + i y_j$, the gradient flow equations read

$$\frac{dx_j}{d\lambda} = -\frac{\partial h}{\partial x_j}, \quad \frac{dy_j}{d\lambda} = -\frac{\partial h}{\partial y_j}. \tag{6.402}$$

Changing back to complex coordinates, we may write the gradient flow equations as

$$\frac{dz_j}{d\lambda} = -2\frac{\partial h}{\partial \bar{z}_j}. \tag{6.403}$$

The gradient flow equations define curves in $\mathbb{C}^n - D$. We denote by C_j the union of curves with

$$\lim_{\lambda \to -\infty} z(\lambda) = z_{\mathrm{crit}}^{(j)}, \tag{6.404}$$

and by $\mathcal{D}_j$ the union of curves with

$$\lim_{\lambda \to \infty} z(\lambda) = z_{\mathrm{crit}}^{(j)}. \tag{6.405}$$

C_j and $\mathcal{D}_j$ are called **Lefschetz thimbles**. The curves which make up $\mathcal{D}_j$ start at a point $z \in D$, where $h(z)$ is plus infinity and approach the critical point $z_{\mathrm{crit}}^{(j)}$ for $\lambda \to \infty$. The curves which make up C_j end at points where $|z| = \infty$ and where $h(z)$ is minus infinity. Tracing back these curves one reaches the critical point $z_{\mathrm{crit}}^{(j)}$ for $\lambda \to -\infty$. Let us reformulate this slightly: The critical point $z_{\mathrm{crit}}^{(j)}$ is a saddle point of $h(z)$. The Lefschetz thimble $\mathcal{D}_j$ is the union of all trajectories, which descend by the steepest descent towards $z_{\mathrm{crit}}^{(j)}$. The Lefschetz thimble C_j is the union of all trajectories, which descend from $z_{\mathrm{crit}}^{(j)}$ by the steepest descent to minus infinity. Of interest to us are the Lefschetz thimbles C_j. Due to the Cauchy-Riemann equations the phase $\phi(z)$ is constant on a Lefschetz thimbles. If all critical phases $\phi^{(j)}$ are pairwise distinct, it follows that for $i \neq j$ the Lefschetz thimbles C_i and C_j do not intersect.

The real dimension of C_j equals the Morse index of the critical point $z_{\mathrm{crit}}^{(j)}$. For the application towards Feynman integrals we may assume that all critical points have Morse index n. Then C_j has real dimension n and defines a representative for $|C_j\rangle$.

In summary, each critical point $z_{\mathrm{crit}}^{(j)}$ defines a Lefschetz thimble C_j. The set of all Lefschetz thimbles satisfying Eq. (6.404) defines a basis $|C_j\rangle$ of H_n^{ω}.

Let us look at an example. We take $n = 1$ (one complex variable z) and $m = 1$ (one polynomial $p(z)$). We discuss

$$p(z) = z^3 + z^2 + z + 1. \tag{6.406}$$

We take $\gamma = -\frac{1}{10}$. Therefore

$$u(z) = [p(z)]^{-\frac{1}{10}}, \quad f(z) = \ln(u(z)) = -\frac{1}{10}\ln(p(z)). \tag{6.407}$$

The divisor D is given by the roots of the polynomial $p(z)$:

$$D = \{-1, i, -i\}. \tag{6.408}$$

We call the three roots “singular points” and denote them by $z_{\mathrm{sing}}^{(1)} = -1$, $z_{\mathrm{sing}}^{(2)} = i$, and $z_{\mathrm{sing}}^{(3)} = -i$. We have

$$p'(z) = 3z^2 + 2z + 1. \tag{6.409}$$

The critical points are given by the roots of $p'(z)$, hence we have two critical points $z_{\mathrm{crit}}^{(1)}$ and $z_{\mathrm{crit}}^{(2)}$. Therefore

$$\text{critical points} = \left\{z_{\mathrm{crit}}^{(1)}, z_{\mathrm{crit}}^{(2)}\right\} = \left\{-\frac{1}{3} + \frac{i}{3}\sqrt{2}, -\frac{1}{3} - \frac{i}{3}\sqrt{2},\right\}. \tag{6.410}$$

The Morse function is given by

$$\begin{aligned} h(x, y) &= -\gamma \ln|p(x+iy)| \\ &= -\frac{1}{10}\ln\left|x^6 + 3x^4y^2 + 3x^2y^4 + y^6 + 2x^5 + 4x^3y^2 + 2xy^4 + 3x^4 + 2x^2y^2 - y^4\right. \\ &\quad \left. +4x^3 - 4xy^2 + 3x^2 - y^2 + 2x + 1\right|. \end{aligned} \tag{6.411}$$

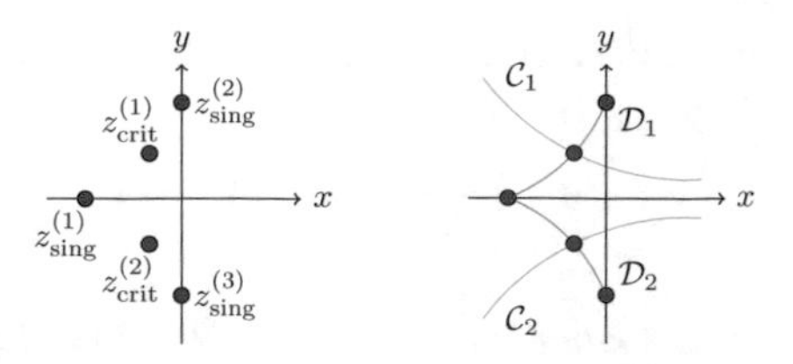

Fig. 6.7 The left picture shows the location of the singular points $z_{\text{sing}}^{(1)}$, $z_{\text{sing}}^{(2)}$ and $z_{\text{sing}}^{(3)}$ (red points) and the location of the critical points $z_{\text{crit}}^{(1)}$ and $z_{\text{crit}}^{(2)}$ (blue points). The right picture shows a sketch of the Lefschetz thimbles $\mathcal{C}_1$ and $\mathcal{C}_2$ (green) as well as the Lefschetz thimbles $\mathcal{D}_1$ and $\mathcal{D}_2$ (orange)

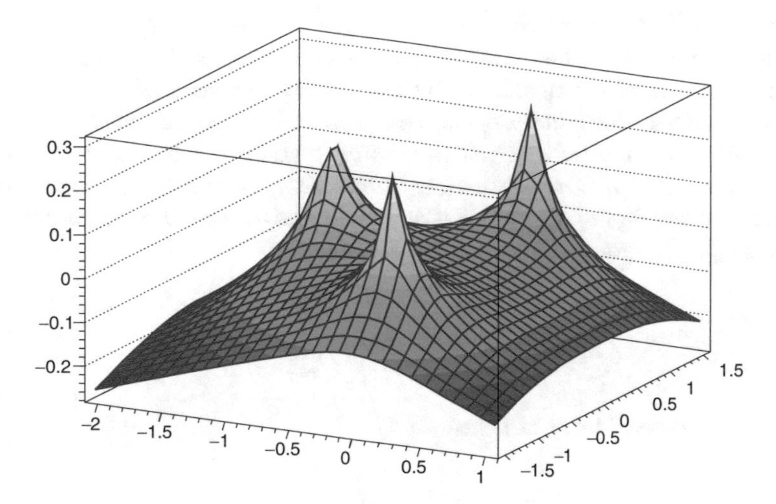

Fig. 6.8 The Morse function $h(x, y)$ plotted in the range $x \in [-2, 1]$ and $y \in [-1.5, 1.5]$. The Morse function h tends to $+\infty$ at the three singular points $z_{\text{sing}}^{(1)} = -1$, $z_{\text{sing}}^{(2)} = i$ and $z_{\text{sing}}^{(3)} = -i$

The left picture of Fig. 6.7 shows the location of the singular points and the location of the critical points. Figure 6.8 shows a plot of the Morse function h. The right picture of Fig. 6.7 shows a sketch of the Lefschetz thimbles $\mathcal{C}_1, \mathcal{C}_2, \mathcal{D}_1$ and $\mathcal{D}_2$.

Digression Morse theory

Morse theory studies the critical points of a real function [148]. There is also a complex analogue, called Picard–Lefschetz theory.

Let M be a compact manifold and $f : M \to \mathbb{R}$ a smooth real valued function on M. Coordinates on M are denoted by $x_1, \dots, x_n$.

A critical point of f is a point $x \in M$, where all partial derivatives vanish

$$\frac{\partial f}{\partial x_1} = \dots = \frac{\partial f}{\partial x_n} = 0, \tag{6.412}$$

or $df = 0$ in short. A critical point x is called non-degenerate, if

$$\det\left(\frac{\partial^2 f}{\partial x_i \partial x_j}\right) \neq 0. \tag{6.413}$$

Let $x^{(0)}$ be a non-degenerate critical point. In a neighbourhood of $x^{(0)}$ we may choose a coordinate system $(x_1', \dots, x_n')$ such that

$$f = f\left(x^{(0)}\right) - x_1'^2 - \dots - x_\lambda'^2 + x_{\lambda+1}'^2 + \dots + x_n'^2 + O\left(x'^3\right), \tag{6.414}$$

i.e. there are λ downward directions and $(n - \lambda)$ upward directions. As $x^{(0)}$ is assumed to be a non-degenerate critical point there are no flat directions. This implies that a non-degenerate critical point is isolated. The number λ of downward directions is called the **Morse index** *of the critical point. We denote the number of critical points with Morse index λ by C_λ. The function f is called a* **Morse function** *if all critical points of f are non-degenerate. Morse theory relates the number of critical points to the topology of M. The relation is provided by the Morse inequalities. The Morse inequalities for a Morse function f read*

$$\sum_{k=0}^{\lambda} (-1)^{\lambda-k} \dim H_k(M) \leq \sum_{k=0}^{\lambda} (-1)^{\lambda-k} C_k. \tag{6.415}$$

The Euler characteristic of M is defined by

$$\chi(M) = \sum_{k=0}^{n} (-1)^k \dim H_k(M). \tag{6.416}$$

The Morse inequalities imply

$$\chi(M) = \sum_{k=0}^{n} (-1)^k C_k. \tag{6.417}$$

Exercise 62 *Assume $C_{\lambda+1} = C_{\lambda-1} = 0$. Show that this implies*

$$\dim H_{\lambda+1}(M) = 0, \qquad \dim H_\lambda(M) = C_\lambda, \qquad \dim H_{\lambda-1}(M) = 0. \tag{6.418}$$

Exercise 63 *Derive Eqs. (6.417) from (6.415).*

6.7.3 Computation of Intersection Numbers

As the integral appearing in Eq. (6.332) in the definition of the intersection number of twisted cocycles is not the most practical way to compute intersection numbers, let us now turn to a more efficient method to compute intersection numbers. With a few technical assumptions, outlined in [137, 149, 150] we may compute multivariate intersection numbers in n variables $z_1, \dots z_n$ recursively by splitting the problem into the computation of an intersection number in $(n-1)$ variables $z_1, \dots, z_{n-1}$ and the computation of a (generalised) intersection number in the variable z_n. By recursion, we therefore have to compute only (generalised) intersection numbers in a single variable z_i. This reduces the multivariate problem to an univariate problem.

Let us comment on the word "generalised" intersection number: We only need to discuss the univariate case. Consider two cohomology classes $\langle \varphi_L |$ and $| \varphi_R \rangle$. Representatives φ_L and φ_R for the two cohomology classes $\langle \varphi_L |$ and $| \varphi_R \rangle$ are in the univariate case differential one-forms and of the form as in Eqs. (6.300) or (6.318). We may view the representatives φ_L and φ_R, the cohomology classes $\langle \varphi_L |$ and $| \varphi_R \rangle$, and the twist ω as scalar quantities.

Consider now a vector of ν differential one-forms $\varphi_{L,j}$ in the variable z, where j runs from 1 to ν. Similar, consider for the dual space a ν-dimensional vector $\varphi_{R,j}$ and generalise ω to a $(\nu \times \nu)$-dimensional matrix Ω. The equivalence classes $\langle \varphi_{L,j} |$ and $| \varphi_{R,j} \rangle$ are now defined by

$$\varphi'_{L,j} = \varphi_{L,j} + \partial_z \xi_j + \xi_i \Omega_{ij} \quad \text{and} \quad \varphi'_{R,j} = \varphi_{R,j} + \partial_z \xi_j - \Omega_{ji} \xi_i, \tag{6.419}$$

for some zero-forms ξ_j (i.e. functions). Readers familiar with gauge theories will certainly recognise that the generalisation is exactly the same step as going from an Abelian gauge theory (like QED) to a non-Abelian gauge theory (like QCD).

Let us now set up the notation for the recursive structure. We fix an ordered sequence $(z_{\sigma_1}, \dots, z_{\sigma_n})$, indicating that we first integrate out z_{σ_1}, then z_{σ_2}, etc. Without loss of generality we will always consider the order $(z_1, \dots, z_n)$, unless indicated otherwise.

For $i = 1, \dots, n$ we consider a fibration $E_i : \mathbb{C}^n \to B_i$ with total space $\mathbb{C}^n$, fibre $V_i = \mathbb{C}^i$ parametrised by the coordinates $(z_1, \dots, z_i)$ and base $B_i = \mathbb{C}^{n-i}$ parametrised by the coordinates $(z_{i+1}, \dots, z_n)$. The covariant derivative splits as

$$\nabla_\omega = \nabla_\omega^{(\mathrm{i}),F} + \nabla_\omega^{(\mathrm{i}),B}, \tag{6.420}$$

with

$$\nabla_\omega^{(\mathrm{i}),F} = \sum_{j=1}^{i} dz_j \left(\frac{\partial}{\partial z_j} + \omega_j \right), \quad \nabla_\omega^{(\mathrm{i}),B} = \sum_{j=i+1}^{n} dz_j \left(\frac{\partial}{\partial z_j} + \omega_j \right). \tag{6.421}$$

One sets

$$\omega^{(\mathbf{i})} = \sum_{j=1}^{i} \omega_j dz_j. \tag{6.422}$$

Clearly, for $i = n$ we have

$$\omega^{(\mathbf{n})} = \omega, \quad \nabla_{\omega}^{(\mathbf{n}),F} = \nabla_{\omega}. \tag{6.423}$$

We may now study for each i the twisted cohomology group in the fibre, defined by replacing ω with $\omega^{(\mathbf{i})}$. The additional variables $(z_{i+1}, \dots, z_n)$ are treated as parameters, that is to say we consider all polynomials as polynomials with coefficients in $\tilde{\mathbb{F}} = \mathbb{F}(z_{i+1}, \dots, z_n)$. For each i only the i-th cohomology group is of interest and for simplicity we write

$$H_{\omega}^{(\mathbf{i})} = H_{\omega^{(\mathbf{i})}}^{i}, \quad \left(H_{\omega}^{(\mathbf{i})}\right)^{*} = \left(H_{\omega^{(\mathbf{i})}}^{i}\right)^{*}. \tag{6.424}$$

We denote the dimensions of the twisted cohomology groups by

$$\nu_{\mathbf{i}} = \dim H_{\omega}^{(\mathbf{i})} = \dim \left(H_{\omega}^{(\mathbf{i})}\right)^{*}. \tag{6.425}$$

Let $\langle e_j^{(\mathbf{i})}|$ with $1 \le j \le \nu_{\mathbf{i}}$ be a basis of $H_{\omega}^{(\mathbf{i})}$ and let $|h_j^{(\mathbf{i})}\rangle$ with $1 \le j \le \nu_{\mathbf{i}}$ be a basis of $(H_{\omega}^{(\mathbf{i})})^{*}$. We denote the $(\nu_{\mathbf{i}} \times \nu_{\mathbf{i}})$-dimensional intersection matrix by $C_{\mathbf{i}}$. The entries are given by

$$(C_{\mathbf{i}})_{jk} = \left\langle e_j^{(\mathbf{i})} \middle| h_k^{(\mathbf{i})} \right\rangle. \tag{6.426}$$

The matrix $C_{\mathbf{i}}$ is invertible. We denote by $|d_j^{(\mathbf{i})}\rangle$ with $1 \le j \le \nu_{\mathbf{i}}$ the dual basis with respect to $\langle e_j^{(\mathbf{i})}|$. From Eq. (6.357) the dual basis is given by

$$\left|d_j^{(\mathbf{i})}\right\rangle = \left|h_k^{(\mathbf{i})}\right\rangle \left(C_{\mathbf{i}}^{-1}\right)_{kj}, \tag{6.427}$$

and satisfies

$$\left\langle e_j^{(\mathbf{i})} \middle| d_k^{(\mathbf{i})} \right\rangle = \delta_{jk}. \tag{6.428}$$

The essential step in the recursive approach is to expand the twisted cohomology class $\langle \varphi_L^{(\mathbf{n})}| \in H_{\omega}^{(\mathbf{n})}$ in the basis of $H_{\omega}^{(\mathbf{n-1})}$

$$\left\langle \varphi_L^{(\mathbf{n})} \right| = \sum_{j=1}^{\nu_{\mathbf{n-1}}} \left\langle \varphi_{L,j}^{(\mathbf{n})} \right| \wedge \left\langle e_j^{(\mathbf{n-1})} \right|, \tag{6.429}$$

and to expand $|\varphi_R^{(\mathbf{n})}\rangle \in (H_\omega^{(\mathbf{n})})^*$ in the dual basis of $(H_\omega^{(\mathbf{n-1})})^*$:

$$\left|\varphi_R^{(\mathbf{n})}\right\rangle = \sum_{j=1}^{\nu_{\mathbf{n-1}}} \left|d_j^{(\mathbf{n-1})}\right\rangle \wedge \left|\varphi_{R,j}^{(\mathbf{n})}\right\rangle. \tag{6.430}$$

Classes in $H_\omega^{(\mathbf{n-1})}$ are represented by rational functions in $z_1, \dots, z_n$ times $dz_{n-1} \wedge \dots \wedge dz_1$. The coefficients $\langle\varphi_{L,j}^{(\mathbf{n})}|$ and $|\varphi_{R,j}^{(\mathbf{n})}\rangle$ are one-forms proportional to dz_n and independent of $z_1, \dots, z_{n-1}$. They are given by

$$\left\langle\varphi_{L,j}^{(\mathbf{n})}\right| = \left\langle\varphi_L^{(\mathbf{n})}\middle|d_j^{(\mathbf{n-1})}\right\rangle, \quad \left|\varphi_{R,j}^{(\mathbf{n})}\right\rangle = \left\langle e_j^{(\mathbf{n-1})}\middle|\varphi_R^{(\mathbf{n})}\right\rangle. \tag{6.431}$$

The coefficients $\langle\varphi_{L,j}^{(\mathbf{n})}|$ and $|\varphi_{R,j}^{(\mathbf{n})}\rangle$ are obtained by computing only intersection numbers in $(n-1)$ variables. This is compatible with the recursive approach. It also shows that the coefficients do not depend on the variables $(z_1, \dots, z_{n-1})$, as these variables are integrated out. Let us define a $(\nu_{\mathbf{n-1}} \times \nu_{\mathbf{n-1}})$-matrix $\Omega^{(\mathbf{n})}$ by

$$\Omega_{ij}^{(\mathbf{n})} = \left\langle\left(\partial_{z_n} + \omega_n\right) e_i^{(\mathbf{n-1})}\middle| d_j^{(\mathbf{n-1})}\right\rangle = -\left\langle e_i^{(\mathbf{n-1})}\middle| \left(\partial_{z_n} - \omega_n\right) d_j^{(\mathbf{n-1})}\right\rangle. \tag{6.432}$$

The invariance of the original class $\langle\varphi_L|$ under a transformation as in Eq. (6.311) translates into the invariance of the vector of coefficients $\langle\varphi_{L,j}|$ as in Eq. (6.419).

The algorithm for computing a multivariate intersection number consists of three steps:

1. Recursive approach: The algorithm integrates out one variable at a time. This part has been outlined above. It has the advantage to reduce a multivariate problem to a univariate problem.
2. Reduction to simple poles: In general we deal in cohomology with equivalence classes. We may replace a representative of an equivalence class with higher poles with an equivalent representative with only simple poles. This is similar to integration-by-part reduction. However, let us stress that the involved systems of linear equations are usually significantly smaller compared to standard integration-by-part reduction.
3. Evaluation of the intersection number as a global residue. Having reduced our objects to simple poles we may use a mathematical theorem which states that in this case the intersection number equals a global residue. The theorem does not hold for higher poles, therefore step 2 is required. The global residue is easily computed and does not involve algebraic extensions like square roots.

Let us now fill in the technical details: We would like to compute the intersection number

$$\left\langle\varphi_L^{(\mathbf{n})}\middle|\varphi_R^{(\mathbf{n})}\right\rangle_\omega, \tag{6.433}$$

where $\varphi_L^{(\mathbf{n})}$ and $\varphi_R^{(\mathbf{n})}$ are differential n-forms in the variables $z_1, \dots, z_n$. Expanding $\langle \varphi_L^{(\mathbf{n})} |$ as in Eq. (6.429), $| \varphi_R^{(\mathbf{n})} \rangle$ as in Eq. (6.429) and using the fact that $\langle e_j^{(\mathbf{n-1})} |$ and $| d_j^{(\mathbf{n-1})} \rangle$ are dual bases of $H_\omega^{(\mathbf{n-1})}$ and $(H_\omega^{(\mathbf{n-1})})^*$, respectively, reduces the problem to

$$\left\langle \varphi_L^{(\mathbf{n})} \middle| \varphi_R^{(\mathbf{n})} \right\rangle_\omega = \sum_{j=1}^{\nu_{\mathbf{n-1}}} \left\langle \varphi_{L,j}^{(\mathbf{n})} \middle| \varphi_{R,j}^{(\mathbf{n})} \right\rangle_\Omega . \tag{6.434}$$

The right-hand side is an univariate generalised intersection number in the variable z_n.

In the next step we reduce the vector of coefficients $\langle \varphi_{L,j}^{(\mathbf{n})} |$ and $| \varphi_{R,j}^{(\mathbf{n})} \rangle$ to a form where only simple poles in the variable z_n occur. A rational function in the variable z_n

$$r(z_n) = \frac{P(z_n)}{Q(z_n)}, \qquad P, Q \in \mathbb{F}[z_n] \qquad \gcd(P, Q) = 1, \tag{6.435}$$

has only simple poles if $\deg P < \deg Q$ and if in the partial fraction decomposition each irreducible polynomial in the denominator occurs only to power 1. The condition $\deg P < \deg Q$ ensures that there are no higher poles at infinity.

It is sufficient to discuss the reduction to simple poles for a ν-dimensional vector $\hat{\varphi}_j$ $(1 \le j \le \nu)$ which transforms as

$$\hat{\varphi}_j \to \hat{\varphi}_j + \left(\delta_{jk} \partial_{z_n} + \Omega_{jk} \right) \xi_k . \tag{6.436}$$

The reduction of $\langle \varphi_{L,j}^{(\mathbf{n})} |$ is then achieved by setting $\Omega = (\Omega^{(\mathbf{n})})^T$, the reduction of $| \varphi_{R,j}^{(\mathbf{n})} \rangle$ is achieved by setting $\Omega = -\Omega^{(\mathbf{n})}$. In both case we have $\nu = \nu_{\mathbf{n-1}}$.

We first treat poles at infinity: Assume that Ω has only simple poles and that the vector $\hat{\varphi}_j$ has a pole of order $o > 1$ at infinity. A transformation as in Eq. (6.436) with the seed

$$\xi_j(z_n) = c_j z_n^{o-1}, \qquad c_j \in \mathbb{F} \tag{6.437}$$

reduces the order of the pole at infinity, provided the linear system obtained from the condition that the ν equations

$$\hat{\varphi}_j + \left(\delta_{jk} \partial_{z_n} + \Omega_{jk} \right) \xi_k \tag{6.438}$$

have only poles of order $(o - 1)$ at infinity yield a solution for the ν coefficients c_j. Furthermore, this gauge transformation does not introduce higher poles elsewhere.

The procedure is only slightly more complicated for higher poles at finite points. Assume that Ω has only simple poles. Let $q \in \mathbb{F}[z_n]$ be an irreducible polynomial appearing in the denominator of the partial fraction decomposition of the $\hat{\varphi}_j$'s at worst to the power o. A transformation as in Eq. (6.436) with the seed

$$\xi_j\left(z_n\right) = \frac{1}{q^{o-1}} \sum_{k=0}^{\deg(q)-1} c_{j,k}\, z_n^k, \qquad c_{j,k} \in \mathbb{F}. \tag{6.439}$$

reduces the order, provided the linear system obtained from the condition that in the partial fraction decomposition of

$$\hat{\varphi}_j + \left(\delta_{jk}\partial_{z_n} + \Omega_{jk}\right)\xi_k \tag{6.440}$$

terms of the form z_n^k/q^o are absent (with $0 \leq k \leq \deg(q) - 1$) yield a solution for the $(\nu \cdot \deg(q))$ coefficients $c_{j,k}$. Furthermore, this gauge transformation does not introduce higher poles elsewhere.

In the third step we relate the intersection number to a global residue. Let's assume that $\Omega^{(\mathbf{n})}$, $\langle\varphi_{L,j}^{(\mathbf{n})}|$ and $|\varphi_{R,j}^{(\mathbf{n})}\rangle$ have at most only simple poles in z_n. Define two polynomials P and Q by

$$\det\left(\Omega^{(\mathbf{n})}\right) = \frac{P}{Q}, \qquad P, Q \in \mathbb{F}\left[z_n\right], \qquad \gcd\left(P, Q\right) = 1, \tag{6.441}$$

and denote by adj $\Omega^{(\mathbf{n})}$ the adjoint matrix of $\Omega^{(\mathbf{n})}$. This matrix satisfies

$$\Omega^{(\mathbf{n})} \cdot \left(\text{adj } \Omega^{(\mathbf{n})}\right) = \left(\text{adj } \Omega^{(\mathbf{n})}\right) \cdot \Omega^{(\mathbf{n})} = \det\left(\Omega^{(\mathbf{n})}\right) \cdot \mathbf{1}. \tag{6.442}$$

Let further

$$Y = \left\{ z_n \in \mathbb{C} \mid P\left(z_n\right) = 0 \right\}. \tag{6.443}$$

Then

$$\langle\varphi_L | \varphi_R\rangle_\omega = -\text{res}_Y\left(Q\, \hat{\varphi}_{L,i} \left(\text{adj } \Omega^{(\mathbf{n})}\right)_{ij} \hat{\varphi}_{R,j}\right). \tag{6.444}$$

Digression Computation of a global residue

Consider n meromorphic functions $f_1, f_2, \dots, f_n$ of n variables $x_1, \dots, x_n$ and assume that the system of equations

$$f_1\left(x\right) = f_2\left(x\right) = \dots f_n\left(x\right) = 0 \tag{6.445}$$

has as solutions a finite number of isolated points $x^{(j)} = (x_1^{(j)}, ..., x_n^{(j)})$, where j labels the individual solutions. Denote by

$$Y_j = \left\{x \in \hat{\mathbb{C}}^n | f_j\left(x\right) = 0\right\}, \qquad 1 \leq j \leq n, \tag{6.446}$$

with $\hat{\mathbb{C}} = \mathbb{C} \cup \{\infty\}$. In Eq. (6.124) we defined the global residue

$$\text{res}_{Y_1,...,Y_n}(g) \tag{6.447}$$

of a meromorphic function g, regular at the solutions $x^{(j)}$, as a sum over the local residues at the solutions $x^{(j)}$. Let us now assume that g is a rational function. The local residues may involve algebraic extensions (i.e. roots), however the global residue does not. We may compute the global residue without the need to introduce algebraic extensions as follows:

Consider the ring $\mathrm{R} = \mathbb{C}[x_1, ..., x_n]$ and the ideal $I = \langle f_1, \dots, f_n \rangle$. The zero locus of $f_1 = \cdots = f_n = 0$ is a zero-dimensional variety. It follows that the quotient ring R/I is a finite-dimensional $\mathbb{C}$-vector space. Let $\{e_i\}$ be a basis of this vector space and let $P_1, P_2 \in \mathrm{R}/I$ be two polynomials (i.e. vectors) in this vector space. A theorem of algebraic geometry states that the global residue defines a **symmetric non-degenerate inner product** *[125]:*

$$\langle P_1, P_2 \rangle = \text{res}_{Y_1,...,Y_n}(\,P_1 \cdot P_2\,). \tag{6.448}$$

Since the inner product is non-degenerate there exists a **dual basis** *$\{d_i\}$ with the property*

$$\langle e_i, d_j \rangle = \delta_{ij}. \tag{6.449}$$

To compute the global residue of a polynomial $P(z)$ we therefore obtain the following method: We express P in the basis $\{e_i\}$ and 1 in the dual basis $\{d_i\}$:

$$P = \sum_i \alpha_i e_i, \quad 1 = \sum_i \beta_i d_i, \qquad \alpha_i, \beta_i \in \mathbb{C}. \tag{6.450}$$

We then have

$$\text{res}_{Y_1,...,Y_n}(\,P\,) = \text{res}_{Y_1,...,Y_n}(\,P \cdot 1\,) = \sum_i \sum_j \alpha_i \beta_j \langle e_i, d_j \rangle = \sum_i \alpha_i \beta_i. \tag{6.451}$$

Given a basis $\{e_i\}$ and the associated dual basis $\{d_i\}$, Eq. (6.451) allows us to compute the global residue of a polynomial P without knowing the solutions $x^{(j)}$. Equation (6.451) simplifies, if the dual basis contains a constant polynomial $d_{i_0} = c$. We then have

$$\text{res}_{Y_1,...,Y_n}(\,P\,) = \frac{\alpha_{i_0}}{c}. \tag{6.452}$$

We would like to compute the global residue of the rational function g. Equation (6.451) is not yet directly applicable to our problem, since $g(x)$ is a rational function, not a polynomial. We write $g(x) = P(x)/Q(x)$. We may assume that $\{f_1, ..., f_n, Q\}$ have no common zeros, since we assumed that g is regular on the *solutions $x^{(j)}$. Hilbert's Nullstellensatz guarantees then that there exist polynomials $p_1, \dots, p_n, \tilde{Q} \in R$, such that*

$$p_1 f_1 + \cdots + p_n f_n + \tilde{Q} Q = 1. \tag{6.453}$$

We call $\tilde{Q}$ the polynomial inverse of Q with respect to $\langle f_1, \dots, f_n \rangle$. For the global residue we have

$$\mathrm{res}_{Y_1,\dots,Y_n}\left(g \right) = \mathrm{res}_{Y_1,\dots,Y_n}\left(\frac{P}{Q} \right) = \mathrm{res}_{Y_1,\dots,Y_n}\left(P\tilde{Q} \right). \tag{6.454}$$

Note $P\tilde{Q}$ is a polynomial. We have therefore reduced the case of a rational function $g(x)$ to the polynomial case $P(x)\tilde{Q}(x)$.

The above calculations can be carried out with the help of a Gröbner basis for the ideal I [151, 152].

Let us apply these ideas to the computation of the global residue in Eq. (6.444). As we are in the univariate case, the calculation simplifies significantly. We have to compute the global residue of a rational function in z_n. Let us write

$$\frac{P_g}{Q_g} = Q\, \hat{\varphi}_{L,i} \left(\mathrm{adj}\ \Omega^{(\mathbf{n})}\right)_{ij} \hat{\varphi}_{R,j}, \qquad P_g, Q_g \in \mathbb{F}[z_n]. \tag{6.455}$$

We may assume $\gcd(P, Q_g) = \gcd(P_g, Q_g) = 1$. Let $\nu = \deg P$ and let $\tilde{Q}_g$ be the polynomial inverse of Q_g with respect to the ideal $\langle P \rangle$. Then

$$\mathrm{res}_Y \left(\frac{P_g}{Q_g} \right) = \frac{a_\nu}{c_\nu}, \tag{6.456}$$

where a_ν is the coefficient of $z^{\nu-1}$ in the reduction of $P_g\tilde{Q}_g$ modulus P and c_ν is the coefficient of z^ν of P.

6.7.4 Inner Product for Feynman Integrals

Let us now make contact with Feynman integrals [136–139, 153–157]. From section 2.5.5 we recall the Baikov representation:

$$I_{\nu_1\dots\nu_n} = C \int\limits_{\mathcal{C}} d^{N_V} z\ \left[\mathcal{B}(z)\right]^{\frac{D-l-e-1}{2}} \prod_{s=1}^{N_V} z_s^{-\nu_s}. \tag{6.457}$$

C is a prefactor (given in Eq. (2.235)) and not relevant for the further discussion. $\mathcal{B}(z)$ denotes the Baikov polynomial. It is obtained from a Gram determinant. The domain of integration is such that the Baikov polynomial vanishes on the boundary of the integration region. We note that the indices ν_s enter only the last factor. Equation (6.457) is an integral of the form as in Eq. (6.310) with

$$\varphi = \left(\prod_{s=1}^{N_V} z_s^{-\nu_s} \right) d^{N_V} z \tag{6.458}$$

and

$$u = \left[\mathcal{B}(z) \right]^{\frac{D-l-e-1}{2}}, \qquad \omega = d \ln u. \tag{6.459}$$

As the Baikov polynomial vanishes on the boundary of the integration region, the Feynman integral is invariant under

$$\varphi \to \varphi + \nabla_\omega \xi \tag{6.460}$$

for any $(N_V - 1)$-form ξ and we may group the integrands of the Feynman integrals $I_{\nu_1 \dots \nu_n}$ corresponding to different sets of indices $\nu_1, \dots, \nu_n$ into cohomology classes. The number of independent cohomology classes in $H_\omega^{N_V}$ is finite, and we may express any φ as a linear combination of a basis of $H_\omega^{N_V}$. Let $\langle e_j|$ be a basis of $H_\omega^{N_V}$ and $|d_j\rangle$ a basis of the dual cohomology group $(H_\omega^{N_V})^*$, chosen such that

$$\left\langle e_i | d_j \right\rangle_\omega = \delta_{ij}. \tag{6.461}$$

We then have

$$\langle \varphi | = \sum_j c_j \left\langle e_j \right|, \tag{6.462}$$

where the coefficients are given by the intersection numbers

$$c_j = \left\langle \varphi | d_j \right\rangle_\omega. \tag{6.463}$$

This provides an alternative to integration-by-parts reduction.

Note that the dimension of $H_\omega^{N_V}$ can be larger than the number of master integrals, as the latter takes symmetries of integrals into account, while the former operates on integrands. This is most easily explained by the simplest example, the one-loop two-point function with two equal internal masses. This system has two master integrals. A standard choice is I_{11} and I_{10}. By symmetry, the integral I_{01} is identical to I_{10}. At the level of the integrands we have $\dim H_\omega^2 = 3$. A basis for H_ω^2 is given by $\langle \varphi_{11}|$, $\langle \varphi_{10}|$, $\langle \varphi_{01}|$ with

$$\varphi_{\nu_1 \nu_2} = \frac{dz_2 \wedge dz_1}{z_1^{\nu_1} z_2^{\nu_2}}. \tag{6.464}$$

The 2-forms φ_{10} and φ_{01}

$$\varphi_{10} = -\frac{dz_1}{z_1} \wedge dz_2, \quad \varphi_{01} = -dz_1 \wedge \frac{dz_2}{z_2}. \tag{6.465}$$

are not identical (but of course one is obtained from the other up to a sign through the substitution $z_1 \leftrightarrow z_2$), only the integrals as in Eq. 6.310) give identical results.

Let us now illustrate the technique of intersection numbers by an example. We consider the double-box integral discussed as example 3 in Sect. 6.3.1. As a basis of master integrals we take the eight master integrals given in Eq. (6.81). Suppose we would like to express the integral $I_{1111111(-2)0}$ as a linear combination of the master integrals:

$$I_{1111111(-2)0} = \sum_{j=1}^{8} c_j I_{\nu_j}, \tag{6.466}$$

with I_{ν_j} denoting the master integrals in the order as they appear in Eq. (6.81), e.g. $I_{\nu_7} = I_{111111100}$ and $I_{\nu_8} = I_{1111111(-1)0}$. Let's compute the coefficients c_7 and c_8 from intersection numbers. This can be done on the maximal cut, as

$$\text{MaxCut } I_{1111111(-2)0} = c_7 \text{MaxCut } I_{\nu_7} + c_8 \text{MaxCut } I_{\nu_8}, \tag{6.467}$$

with the same coefficients c_7 and c_8 as in Eq. (6.466). For simplicity we set $\mu^2 = 1$. A Baikov representation of MaxCut $I_{1111111\nu 0}$ is

$$\text{MaxCut } I_{1111111\nu 0} = C \int\limits_{\mathcal{C}} dz_8 \, z_8^{-1-2\varepsilon} \left(t - z_8\right)^{-1-\varepsilon} \left(s + t - z_8\right)^{\varepsilon} z_8^{-\nu}. \tag{6.468}$$

This form of the Baikov representation is obtained within the loop-by-loop approach by first considering the loop with loop momentum k_2 and then the remaining loop with loop momentum k_1. The prefactor C and the integration contour $\mathcal{C}$ are not particularly relevant. The multi-valued function u is given by

$$u = C \, z_8^{-1-2\varepsilon} \left(t - z_8\right)^{-1-\varepsilon} \left(s + t - z_8\right)^{\varepsilon}, \tag{6.469}$$

the rational one-form φ_ν (recall that the index ν in $I_{1111111\nu 0}$ is an integer) by

$$\varphi_\nu = z_8^{-\nu} dz_8. \tag{6.470}$$

The one-form ω is therefore given by

$$\omega = d \ln u = \left[-\frac{1+2\varepsilon}{z_8} - \frac{1+\varepsilon}{z_8 - t} + \frac{\varepsilon}{z_8 - s - t} \right] dz_8. \tag{6.471}$$

The equation $\omega = 0$ leads to a quadratic equation for z_8, which has two solutions. We therefore have

$$\dim H^1_\omega = 2. \tag{6.472}$$

This is consistent with the fact that there are two master integrals (I_{ν_7} and I_{ν_8}) in this sector. The φ's corresponding to I_{ν_7} and I_{ν_8} give us immediately a basis of H^1_ω:

$$e_1 = 1 \cdot dz_8, \quad e_2 = z_8 \cdot dz_8. \tag{6.473}$$

We then compute the dual basis. We start from an arbitrary basis of $(H^1_\omega)^*$, which we take to be

$$h_1 = 1 \cdot dz_8, \quad h_2 = z_8 \cdot dz_8, \tag{6.474}$$

compute the intersection matrix between $(\langle e_1|, \langle e_2|)$ and $(|h_1\rangle, |h_2\rangle)$ and obtain the dual basis according to Eq. (6.357). We find

$$\begin{aligned}
d_1 = & \left[\frac{(1+\varepsilon)(2+\varepsilon)(2z_8 - s - 2t)}{(1+2\varepsilon)\, s\, (s+t)^2} - \frac{3\,(2z_8 - s - 2t)}{4\,(1+2\varepsilon)\, s t^2} - \frac{4\,(s+t)\, z_8}{(1+\varepsilon)\, s^2 t^2} + \frac{2\,(2+\varepsilon)\, z_8}{st\,(s+t)} \right. \\
& \left. + \frac{4\,(t - z_1)}{s^2 t} + \frac{9\,(3+2\varepsilon)\, z_8}{2st^2} - \frac{27\,(1+2\varepsilon)}{4t^2} - \frac{(11+30\varepsilon)}{2st} \right] dz_8, \\
d_2 = & \left[\frac{3\,(2z_8 - s - 2t)}{2\,(1+2\varepsilon)\, t^2\, (s+t)^2} - \frac{(5+2\varepsilon)(2z_8 - t)}{2st\,(s+t)^2} + \frac{2\,(2+\varepsilon)}{st\,(s+t)} - \frac{6\varepsilon z_8}{st^2\,(s+t)} - \frac{(3+7\varepsilon)\, z_8}{(1+\varepsilon)\, s^2 t^2} \right. \\
& \left. + \frac{11 z_8}{s^2 t\,(s+t)} + \frac{9\varepsilon}{st^2} - \frac{4}{s^2 t} + \frac{3}{2st^2} \right] dz_8.
\end{aligned} \tag{6.475}$$

We would like to find the coefficients c_7 and c_8 in the reduction of $I_{1111111(-2)0}$ to master integrals. The integral MaxCut $I_{1111111(-2)0}$ corresponds to

$$\varphi_{(-2)} = z_8^2 \cdot dz_8. \tag{6.476}$$

c_7 and c_8 are then given by the intersection numbers

$$\begin{aligned}
c_7 = \langle \varphi_{(-2)} | d_1 \rangle_\omega &= \frac{2\varepsilon t\,(s+t)}{1 - 2\varepsilon}, \\
c_8 = \langle \varphi_{(-2)} | d_2 \rangle_\omega &= \frac{(1 - 4\varepsilon)\, t - 3\varepsilon s}{1 - 2\varepsilon}.
\end{aligned} \tag{6.477}$$

This agrees with the result obtained from integration-by-parts reduction (compare with Exercise 44).

Chapter 7
Transformations of Differential Equations

In Chap. 6 we learned that the computation of Feynman integrals can be reduced to finding appropriate fibre transformations (see Sect. 6.4.3) and base transformations (see Sect. 6.4.4). However, up to now we didn't discuss methods how to find these fibre and base transformations.

Currently, there is no known method how to do this in full generality. In this chapter we introduce methods, which allow us to construct the required fibre or base transformation in special cases. We focus in this chapter mainly on rational and algebraic transformations. This covers many Feynman integrals, which evaluate to multiple polylogarithms. But there are also Feynman integrals, where the required transformations involve transcendental functions. We discuss an example for this case in Chap. 13.

7.1 Fibre Transformations

We denote by $\mathbf{I} = (I_{\nu_1}, \dots, I_{\nu_{N_{\text{master}}}})^T$ a set of master integrals satisfying the differential equation

$$(d + A)\,\mathbf{I} = 0. \tag{7.1}$$

A fibre transformations, given by an $(N_{\text{master}} \times N_{\text{master}})$-matrix $U(\varepsilon, x)$, redefines the set of master integrals as

$$\mathbf{I}' = U\mathbf{I}, \tag{7.2}$$

and transforms the differential equation to

$$\left(d + A'\right)\mathbf{I}' = 0, \tag{7.3}$$

S. Weinzierl, *Feynman Integrals*, UNITEXT for Physics,
https://doi.org/10.1007/978-3-030-99558-4_7

where A' is related to A by

$$A' = UAU^{-1} + UdU^{-1}. \tag{7.4}$$

The goal is to find a transformation $U(\varepsilon, x)$, such that the dependence on ε of A' is only through an explicit prefactor ε as in Eq. (6.92).

In the following subsections we discuss a variety of methods: We start with block decomposition in Subsect. 7.1.1. This allows us to reduce the original problem involving a $(N_{\text{master}} \times N_{\text{master}})$-matrix A to matrices of smaller size. In addition we derive a differential equation for the transformation we are seeking. In Subsect. 7.1.2 we reduce a multivariate problem depending on N_B kinematic variables $x_1, \dots, x_{N_B}$ to an univariate problem depending only on a single kinematic variable x. Obviously, some information is lost in this reduction, but the solution of the simpler univariate problem can be useful to find a solution for the multivariate problem. The next three subsections deal with univariate problems: In Subsect. 7.1.3 we convert a system of N_{master} first-order differential equations to a higher-order differential equation for one selected master integral. The order of this differential equation is at most N_{master}. The differential operator of this differential equation is called the Picard-Fuchs operator. We then study the factorisation properties of the Picard-Fuchs operator when the parameter D denoting the number of space-time dimensions is an (even) integer. In Subsect. 7.1.4 we study the Magnus expansion. This is particularly useful if the matrix A is linear in ε, i.e. $A = A^{(0)} + \varepsilon A^{(1)}$ with $A^{(0)}$ and $A^{(1)}$ being independent of ε. In Subsect. 7.1.5 we discuss Moser's algorithm. This algorithm is at the core of several computer programs for finding an appropriate fibre transformation. In Subsect. 7.1.6 we return from the univariate case to the (general) multivariate case. We discuss the Leinartas decomposition, which can be thought of as a generalisation of partial fraction decomposition from the univariate case to the multivariate case. Finally, Subsect. 7.1.7 is devoted to maximal cuts and constant leading singularities. This method allows us often to make an educated guess for a suitable fibre transformation.

7.1.1 Block Decomposition

We start with an elementary method [158]: We may order the set of master integrals

$$\mathbf{I} = \left(I_{\boldsymbol{\nu}_1}, \dots, I_{\boldsymbol{\nu}_{N_{\text{master}}}}\right)^T \tag{7.5}$$

such that $I_{\boldsymbol{\nu}_1}$ is the simplest integral and $I_{\boldsymbol{\nu}_{N_{\text{master}}}}$ the most complicated integral. We may do this with an order criteria as in Eqs. (6.17) or (6.18). Doing so, the matrix A has a lower block-triangular structure. To give an example, consider the situation with three sectors. Suppose that the simplest sector has one master integral, the next sector two master integrals and the most complicated sector one master integral. The matrix A has then the structure

$$A = \begin{pmatrix} A_1 & 0 & 0 & 0 \\ A_3 & A_2 & & 0 \\ & & & 0 \\ A_6 & A_5 & & A_4 \end{pmatrix}, \tag{7.6}$$

where only the coloured entries are non-zero. The blocks on the diagonal, A_1, A_2 and A_4 have the size (1×1), (2×2) and (1×1), respectively.

In order to find the fibre transformation which transforms the system to an ε-form we may split the problem into smaller tasks and first find transformations, which transform a specific block into an ε-form. For the example in Eq. (7.6) we may do this in the order A_1, A_2, A_3, A_4, A_5 and A_6. This has the advantage that for most blocks we may work with matrices of smaller size. The size of the matrices required for the individual blocks are

$$\begin{aligned} A_1 &: 1 \times 1 \\ A_2 &: 2 \times 2 \\ A_3 &: 3 \times 3 \\ A_4 &: 1 \times 1 \\ A_5 &: 3 \times 3 \\ A_6 &: 4 \times 4 \end{aligned} \tag{7.7}$$

For example, if $\mathbf{I} = (I_{\nu_1}, I_{\nu_2}, I_{\nu_3}, I_{\nu_4},)$ the (3×3)-system for the block A_5 is obtained by using $I_{\nu_2}, I_{\nu_3}, I_{\nu_4}$ and setting I_{ν_1} to zero. Only for the last block (A_6) we need a (4×4)-system. We do this bottom-up: We first put the block A_1 into an ε-form, then block A_2, etc.. In this approach we may assume as we try to find a transformation for block A_i that all blocks A_j with $j < i$ have already been put into an ε-form.

There are two types of blocks: Blocks on the diagonal (A_1, A_2 and A_4 in the example above) and off-diagonal blocks (A_3, A_5 and A_6 in the example above). Let's see how a fibre transformation acts on these blocks. It is sufficient to discuss the case, where A is of the form

$$A = \begin{pmatrix} A_1 & 0 & 0 \\ A_3 & A_2 & 0 \\ A_6 & A_5 & A_4 \end{pmatrix} \tag{7.8}$$

and to consider the transformation of the blocks A_2 (a diagonal block) and A_3 (an off-diagonal block). We assume that block A_1 is already in ε-form, and the sought-after transformation should preserve block A_1. The blocks A_4, A_5 and A_6 will be dealt with in a later step. The transformation for the blocks A_2 and A_3 is allowed to modify these blocks.

Let's start with block A_2. We consider a transformation of the form

$$U = \begin{pmatrix} 1 & 0 & 0 \\ 0 & U_2 & 0 \\ 0 & 0 & 1 \end{pmatrix}, \; U^{-1} = \begin{pmatrix} 1 & 0 & 0 \\ 0 & U_2^{-1} & 0 \\ 0 & 0 & 1 \end{pmatrix}. \tag{7.9}$$

The transformed A' is given by

$$A' = \begin{pmatrix} A_1 & 0 & 0 \\ U_2 A_3 & U_2 A_2 U_2^{-1} + U_2 d U_2^{-1} & 0 \\ A_6 & A_5 U_2^{-1} & A_4 \end{pmatrix}. \tag{7.10}$$

Suppose the block A_2 contains an unwanted term F and a remainder R:

$$A_2 = F + R. \tag{7.11}$$

The term F can be removed by a fibre transformation of the form as in Eq. (7.9) with U_2 given as a solution of the differential equation

$$dU_2^{-1} = -F U_2^{-1}. \tag{7.12}$$

We consider a simple example: Assume that we have only one kinematic variable $x_1 = x$ (e.g. $N_B = 1$) and that A_2 is of size (1×1) and given by

$$A_2 = \left(\frac{1}{x-1} + \frac{2\varepsilon}{x-1} \right) dx. \tag{7.13}$$

We would like to remove the first term $F = dx/(x-1)$ by a fibre transformation. We have to solve the differential equation

$$\frac{d}{dx} U_2^{-1} + \frac{1}{x-1} U_2^{-1} = 0. \tag{7.14}$$

A solution is easily found and given by

$$U_2^{-1} = \frac{C}{x-1}, \; U_2 = C^{-1}(x-1). \tag{7.15}$$

The integration constant C is of no particular relevance, as it corresponds to multiplying a master integral with a constant prefactor. We may set $C = 1$ and $U_2 = x - 1$ is the sought-after transformation.

Let us stay with one kinematic variable and A_2 of size (1×1). Let us now consider $F = f(x)dx$ and assume that $f(x)$ is a rational function in x. We have to solve the differential equation

$$\left[\frac{d}{dx} + f(x) \right] U_2^{-1} = 0. \tag{7.16}$$

A solution is easily found and given by

$$U_2^{-1} = \exp\left[- \int\limits_{x_0}^{x} dx' f\left(x'\right) \right], \quad U_2 = \exp\left[\int\limits_{x_0}^{x} dx' f\left(x'\right) \right]. \tag{7.17}$$

By using partial fraction decomposition we may write $f(x)$ as a sum of a polynomial, terms with simple poles and terms with higher poles. We have with a suitable choice for the integration constant

$$\begin{aligned} f(x) &= c_n x^n, & U_2(x) &= \exp\left(\frac{c_n x^{n+1}}{n+1}\right), \\ f(x) &= \frac{r_0}{x-z}, & U_2(x) &= (x-z)^{r_0}, \\ f(x) &= \frac{d_n}{(x-z)^n}, & U_2(x) &= \exp\left(-\frac{d_n}{(n-1)(x-z)^{n-1}}\right). \end{aligned} \tag{7.18}$$

A first-order differential equation as in Eq. (7.16) is said to be in **Fuchsian form**, if $f(x)$ is a rational function in x and if in the partial fraction decomposition of f polynomial terms and higher poles are absent (i.e. f has only simple poles). From Eq. (7.18) we see that simple poles with integer residues can be removed from a (1×1)-block on the diagonal by a rational fibre transformation. If the residue is not an integer the fibre transformation is algebraic.

Let us now consider block A_3. At this stage we would like to preserve the blocks A_1 and A_2. We consider a transformation of the form

$$U = \begin{pmatrix} 1 & 0 & 0 \\ U_3 & 1 & 0 \\ 0 & 0 & 1 \end{pmatrix}, \quad U^{-1} = \begin{pmatrix} 1 & 0 & 0 \\ -U_3 & 1 & 0 \\ 0 & 0 & 1 \end{pmatrix}. \tag{7.19}$$

The transformed A' is given by

$$A' = \begin{pmatrix} A_1 & 0 & 0 \\ A_3 - A_2 U_3 + U_3 A_1 - dU_3 & A_2 & 0 \\ A_6 - A_5 U_3 & A_5 & A_4 \end{pmatrix}. \tag{7.20}$$

Suppose the block A_3 contains an unwanted term F and a remainder R:

$$A_3 = F + R. \tag{7.21}$$

The term F can be removed by a fibre transformation of the form as in Eq. (7.19) with U_3 given as a solution of the differential equation

$$dU_3 + A_2 U_3 - U_3 A_1 = F. \tag{7.22}$$

Let us also consider an example here. We again consider the case of one kinematic variable x (e.g. $N_B = 1$). We further assume that A_1 and A_2 are both blocks of size (1×1). Then A_3 is also a block of size (1×1). Assume that A_1 and A_2 are already in ε-form and given by

$$A_1 = \frac{\varepsilon dx}{x-1}, \quad A_2 = \frac{2\varepsilon dx}{x-1}. \tag{7.23}$$

Assume further that F is given by

$$F = \frac{dx}{(x-1)^2}. \tag{7.24}$$

We have to solve the differential equation

$$\left[\frac{d}{dx} + \frac{\varepsilon}{x-1}\right] U_3 = \frac{1}{(x-1)^2}. \tag{7.25}$$

A solution is given by

$$U_3 = \frac{1}{(1-\varepsilon)(1-x)}. \tag{7.26}$$

7.1.2 Reduction to an Univariate Problem

In general our Feynman integrals depend on N_B kinematic variables $x_1, \dots, x_{N_B}$. If $N_B > 1$ we are considering a multivariate problem, if $N_B = 1$ we are considering an univariate problem. Clearly, an univariate problem is simpler than a multivariate problem.

To any multivariate problem we may associate an univariate problem as follows [159]: Let $\alpha = [\alpha_1 : ... : \alpha_{N_B}] \in \mathbb{CP}^{N_B - 1}$ be a point in projective space. Without loss of generality we work in the chart $\alpha_{N_B} = 1$. We consider a path $\gamma_\alpha : [0, 1] \to \mathbb{C}^n$, indexed by α and parametrised by a variable λ. Explicitly, we have

$$x_j(\lambda) = \alpha_j \lambda, \qquad 1 \le j \le N_B. \tag{7.27}$$

We then view the master integrals as functions of λ. In other words, we look at the variation of the master integrals in the direction specified by α. Consider now a set of master integrals $\mathbf{I}$ with differential equation

$$(d + A)\,\mathbf{I} = 0, \quad A = \sum_{j=1}^{N_B} A_{x_j} dx_j. \tag{7.28}$$

For the derivative with respect to λ we have

$$\left(\frac{d}{d\lambda} + B_\lambda\right)\mathbf{I} = 0, \quad B_\lambda = \sum_{j=1}^{N_B} \alpha_j A_{x_j}. \tag{7.29}$$

B is a $(N_{\text{master}} \times N_{\text{master}})$-matrix, whose entries are functions. Equation (7.29) is now an univariate problem in the variable λ. This problem depends on the additional parameters $\alpha = [\alpha_1 : \cdots : \alpha_{N_B-1} : 1]$ specifying the direction of the path in Eq. (7.27). We may now try to find with univariate methods a fibre transformation, which transforms Eq. (7.29) into an ε-form. Let's denote this transformation by V:

$$\mathbf{I}' = V\mathbf{I}. \tag{7.30}$$

V is a function of λ and the parameters α:

$$V = V\left(\alpha_1, ..., \alpha_{N_B-1}, \lambda\right). \tag{7.31}$$

We recall that we work in the chart $\alpha_{N_B} = 1$. We may now try to lift the transformation V to the original kinematic space with coordinates $x_1, \dots, x_{N_B}$. Let us set

$$U = V\left(\frac{x_1}{x_{N_B}}, ..., \frac{x_{N_B-1}}{x_{N_B}}, x_{N_B}\right). \tag{7.32}$$

U defines a transformation in terms of the original variables $x_1, ..., x_{N_B}$.

It is important to note that there is no guarantee that the transformation U puts the original system in Eq. (7.28) into an ε-form, even if the transformation V puts the system in Eq. (7.29) into an ε-form. The reason is that going from the original system in Eq. (7.28) to the simpler univariate system in Eq. (7.29) we threw away information, which we cannot recover by lifting the solution of the univariate system to the multivariate system. The information we threw away is not so easy to spot, after all we kept the dependence on all directions in the kinematic space by introducing the parameters α. The information we threw away comes from the specific paths we consider in Eq. (7.27): We only consider lines through the origin. Therefore, there might be terms in the original A, which map to zero in B_λ for the class of paths considered in Eq. (7.27). These terms are derivatives of functions being constant on lines through the origin. An example is given by

$$d \ln Z\left(x_1, ..., x_{N_B}\right), \tag{7.33}$$

where $Z(x_1, ..., x_{N_B})$ is a rational function in $(x_1, ..., x_{N_B})$ and homogeneous of degree zero in $(x_1, ..., x_{N_B})$.

Nevertheless, it can be a promising strategy to first solve the simpler univariate problem and to remove then any offending terms of the form as in Eq. (7.33) by a subsequent transformation, which usually is rather easy to find.

Exercise 64 *Let* $N_{\mathrm{master}} = 1$, $N_B = 2$ *and*

$$A = d \ln \left(\frac{x_1}{x_1 + x_2} \right). \tag{7.34}$$

Show that B_λ, *defined as in Eq. (7.29), equals zero.*

7.1.3 Picard-Fuchs Operators

In this section and the two following sections we investigate differential equations for Feynman integrals, which only depend on one kinematic variable x. Thus we consider the case $N_B = 1$. In the previous section we have seen that we may reduce a multivariate problem to an univariate problem, solve the latter first and finally lift the result to the multivariate case. Let us therefore consider a set of master integrals $\mathbf{I} = (I_{\boldsymbol{\nu}_1}, \dots, I_{\boldsymbol{\nu}_{N_{\mathrm{master}}}})^T$ satisfying the differential equation

$$\left(\frac{d}{dx} + A_x \right) \mathbf{I} = 0. \tag{7.35}$$

This is a system of N_{master} coupled first-order differential equations.

In this section we derive a single differential equation, usually of higher order, for one Feynman integral I [159, 160]. Let I be one of the master integrals $\{I_{\boldsymbol{\nu}_1}, \dots, I_{\boldsymbol{\nu}_{N_{\mathrm{master}}}}\}$. Equation (7.35) allows us to express the k-th derivative of I with respect to x as a linear combination of the original master integrals. We now determine the largest number r, such that the matrix which expresses

$$I, \quad \frac{d}{dx} I, \quad \dots, \quad \left(\frac{d}{dx} \right)^{r-1} I \tag{7.36}$$

in terms of the original set $\{I_{\boldsymbol{\nu}_1}, \dots, I_{\boldsymbol{\nu}_{N_{\mathrm{master}}}}\}$ has full rank. Obviously, we have $r \leq N_{\mathrm{master}}$. In the case $r < N_{\mathrm{master}}$ we complement the set $I, (d/dx)I, \dots, (d/dx)^{r-1}I$ by $(N_{\mathrm{master}} - r)$ elements $I_{\boldsymbol{\nu}_{\sigma_{r+1}}}, \dots, I_{\boldsymbol{\nu}_{\sigma_{N_{\mathrm{master}}}}} \in \{I_{\boldsymbol{\nu}_1}, \dots, I_{\boldsymbol{\nu}_{N_{\mathrm{master}}}}\}$ such that the transformation matrix has rank N_{master}. The elements $I_{\boldsymbol{\nu}_{\sigma_{r+1}}}, \dots, I_{\boldsymbol{\nu}_{\sigma_{N_{\mathrm{master}}}}}$ must exist, since we assumed that the set $\{I_{\boldsymbol{\nu}_1}, \dots, I_{\boldsymbol{\nu}_{N_{\mathrm{master}}}}\}$ forms a basis of master integrals. The basis

$$I, \quad \frac{d}{dx} I, \quad \dots, \quad \left(\frac{d}{dx} \right)^{r-1} I, \quad I_{\boldsymbol{\nu}_{\sigma_{r+1}}}, \quad \dots, \quad I_{\boldsymbol{\nu}_{\sigma_{N_{\mathrm{master}}}}} \tag{7.37}$$

decouples the system into a block of size r, which is closed under differentiation and a remaining sector of size $(N_{\mathrm{master}} - r)$.

We recall that r is the largest number such that I, $(d/dx)I$, ..., $(d/dx)^{r-1}I$ are independent. It follows that $(d/dx)^r I$ can be written as a linear combination of I, $(d/dx)I$, ..., $(d/dx)^{r-1}I$. This defines the **Picard-Fuchs operator** L_r for the master integral I:

$$L_r I = 0, \qquad L_r = \sum_{k=0}^{r} R_k \frac{d^k}{dx^k}, \tag{7.38}$$

where the coefficients R_k are rational functions in x and we use the normalisation $R_r = 1$, hence

$$L_r = \frac{d^r}{dx^r} + \sum_{k=0}^{r-1} R_k \frac{d^k}{dx^k}. \tag{7.39}$$

The Picard-Fuchs operator $L_r = L_r(D, x, d/dx)$ depends on D, x and d/dx. The Picard-Fuchs operator L_r is called **Fuchsian**, if R_k has maximally poles of order $r - k$ (including possible poles at infinity). The Picard-Fuchs operator is a differential operator, which annihilates the Feynman integral I. The Picard-Fuchs operator is easily obtained by a transformation to the basis given in Eq. (7.37). In this basis the upper-left $r \times r$-block of the transformed matrix A'_x has the form

$$\begin{pmatrix} 0 & -1 & \dots & 0 & 0 \\ & & \dots & & \\ 0 & 0 & \dots & 0 & -1 \\ R_0 & R_1 & \dots & R_{r-2} & R_{r-1} \end{pmatrix}, \tag{7.40}$$

and the coefficients R_k of the Picard-Fuchs operator can easily be read off.

Let us look at a few examples. We start with example 1 from Sect. 6.3.1 and work out the Picard-Fuchs operator for the one-loop two-point integral I_{11}. The transformation matrix from the basis $\mathbf{I} = (I_{10}, I_{11})^T$ to the basis $\mathbf{I}' = (I_{11}, (d/dx)I_{11})^T$ is

$$\begin{pmatrix} I_{11} \\ \frac{d}{dx} I_{11} \end{pmatrix} = \begin{pmatrix} 0 & 1 \\ -\frac{D-2}{x(4+x)} & -\frac{4+(4-D)x}{2x(4+x)} \end{pmatrix} \begin{pmatrix} I_{10} \\ I_{11} \end{pmatrix}. \tag{7.41}$$

Note that the second line of Eq. (7.41) is given as the negative of the second line of Eq. (6.66). In the basis $\mathbf{I}'$ the differential equation reads

$$\left(d + A'\right) \mathbf{I}' = 0, \tag{7.42}$$

with

$$A' = \begin{pmatrix} 0 & -1 \\ -\frac{D-4}{2x(4+x)} & \frac{12+(8-D)x}{2x(4+x)} \end{pmatrix} dx. \tag{7.43}$$

From Eq. (7.43) we may now read off the Picard-Fuchs operator for the integral I_{11}:

$$\left[\frac{d^2}{dx^2} + \frac{12 + (8 - D)\, x}{2x\,(4 + x)} \frac{d}{dx} - \frac{D - 4}{2x\,(4 + x)} \right] I_{11} = 0. \tag{7.44}$$

In this case the Picard-Fuchs operator is a second-order differential operator. In this example it is a particular simple differential operator, as it factorises for any D into two first-order differential operators:

$$\frac{d^2}{dx^2} + \frac{12+(8-D)\,x}{2x\,(4+x)}\frac{d}{dx} - \frac{D-4}{2x\,(4+x)} = \left(\frac{d}{dx} + \frac{1}{x} + \frac{1}{4+x}\right)\left(\frac{d}{dx} + \frac{1}{2x} - \frac{D-3}{2\,(4+x)}\right). \tag{7.45}$$

Let's now look at a second example: We consider the double box integral discussed as example 3 in Sect. 6.3.1. Proceeding along the same lines, we work out the Picard-Fuchs operator for the integral $I_{111111100}$. This is now a differential operator of order eight, which factorises for any D as

$$L_{1,1}\, L_{2,1}^3\, L_{3,1}\, L_{4,1}\, L_{5,2}\, I_{111111100} = 0, \tag{7.46}$$

with

$$\begin{aligned} L_{1,1} &= \frac{d}{dx} + \frac{7}{x} + \frac{3}{x+1}, \\ L_{2,1} &= \frac{d}{dx} - \frac{D-10}{x} + \frac{3}{x+1}, \\ L_{3,1} &= \frac{d}{dx} - \frac{D-8}{x} + \frac{D-2}{x+1}, \\ L_{4,1} &= \frac{d}{dx} - \frac{D-10}{2x} + \frac{D-2}{2\,(x+1)}, \\ L_{5,2} &= \frac{d^2}{dx^2} + \left(-\frac{2D-13}{x} + \frac{D-2}{2\,(x+1)}\right)\frac{d}{dx} + \frac{(D-6)\,(D-6-3x)}{x^2\,(x+1)}. \end{aligned} \tag{7.47}$$

For systems with a larger number of master integrals the calculation of the full Picard-Fuchs operators becomes soon impractical. However, the essential information can already be extracted from the Picard-Fuchs operator for the maximal cut. Let us therefore consider the maximal cut for the double box integral. We set

$$\mathbf{J} = \begin{pmatrix} J_1 \\ J_2 \end{pmatrix} = \begin{pmatrix} \text{MaxCut } I_{111111100} \\ \text{MaxCut } I_{1111111(-1)0} \end{pmatrix} \tag{7.48}$$

The differential equation for $\mathbf{J}$ reads

$$(d + A)\,\mathbf{J} = 0, \tag{7.49}$$

with A given by the lower-right (2×2)-block of Eq. (6.82):

$$A = \begin{pmatrix} \frac{2}{x} & \frac{D-4}{x(1+x)} \\ -\frac{D-4}{x} & -\frac{(3D-16)x+4(D-5)}{2x(1+x)} \end{pmatrix} dx. \tag{7.50}$$

One obtains for the Picard-Fuchs operator for $J_1 = \text{MaxCut } I_{111111100}$ the second-order differential operator $L_{5,2}$ defined in Eq. (7.47):

$$L_{5,2}\; J_1 = 0. \tag{7.51}$$

For generic D, the second-order differential operator $L_{5,2}$ does not factor into two first-order differential operators. However, if D equals an even integer, the second-order differential operator $L_{5,2}(D, x, d/dx)$ factorises, for example [159–161]

$$\begin{aligned} L_{5,2}\left(4, x, \frac{d}{dx}\right) &= \left(\frac{d}{dx} + \frac{3}{x} + \frac{1}{x+1}\right)\left(\frac{d}{dx} + \frac{2}{x}\right), \\ L_{5,2}\left(6, x, \frac{d}{dx}\right) &= \left(\frac{d}{dx} + \frac{1}{x} + \frac{2}{x+1}\right)\frac{d}{dx}. \end{aligned} \tag{7.52}$$

As our last example we look at the two-loop sunrise integral with equal internal masses, discussed as example 4 in Sect. 6.3.1. We consider the Picard-Fuchs operator for MaxCut I_{111}. This is a second-order differential operator

$$L_2 = \frac{d^2}{dx^2} + \left(\frac{D}{2x} - \frac{D-3}{x+1} - \frac{D-3}{x+9}\right)\frac{d}{dx} + (D-3)\left(-\frac{D+4}{18x} + \frac{D}{8\,(x+1)} - \frac{5D-16}{72\,(x+9)}\right) \tag{7.53}$$

which annihilates the integral MaxCut I_{111}:

$$L_2 \text{ MaxCut } I_{111} = 0. \tag{7.54}$$

This differential operator does not factorise for generic D, and remains irreducible in even integer dimensions. In two space-time dimensions L_2 reads

$$L_2\left(2, x, \frac{d}{dx}\right) = \frac{d^2}{dx^2} + \left(\frac{1}{x} + \frac{1}{x+1} + \frac{1}{x+9}\right)\frac{d}{dx} + \frac{1}{3x} - \frac{1}{4\,(x+1)} - \frac{1}{12\,(x+9)}. \tag{7.55}$$

As already mentioned, $L_2(2, x, d/dx)$ does not factorise as a differential operator, i.e. L_2 is an **irreducible differential operator**. In Chap. 13 we will see that $L_2(2, x, d/dx)$ is also the Picard-Fuchs operator of a family of elliptic curves, parametrised by the variable x. We will also see in Chap. 13 that I_{111} cannot be expressed in terms of multiple polylogarithms. The appearance of an irreducible differential operator of order greater than one in the factorisation of the Picard-Fuchs operator in even integer space-time dimensions is an indication that the Feynman integral cannot be expressed in terms of multiple polylogarithms.

Digression The Frobenius method
Let us consider the differential equation

$$L_r I = 0, \qquad L_r = \sum_{k=0}^{r} P_k(x) \frac{d^k}{dx^k}, \tag{7.56}$$

where the coefficients $P_k(x)$ are polynomials in x. This form is equivalent to Eq. (7.38). We obtain Eq. (7.56) by multiplying Eq. (7.38) with the least common multiple of all denominators. We assume that Eq. (7.56) is a Fuchsian differential equation. The zeros of $P_r(x)$ (and possibly the point $x = \infty$) are the **singular points of the differential equation**.

The Frobenius method allows us to construct r independent solutions around a point $x_0 \in \mathbb{C}$ in the form of power series. The power series converge up to the next nearest singularity. Without loss of generality we assume $x_0 = 0$. A variable transformation $x' = x - x_0$ (or $x' = 1/x$ for the point $x_0 = \infty$) will transform to the case $x_0 = 0$.

Let us introduce the **Euler operator** *θ defined by*

$$\theta = x \frac{d}{dx}. \tag{7.57}$$

We may rewrite the differential operator L_r in terms of the Euler operator:

$$L_r = \sum_{k=0}^{r} Q_k(x) \theta^k. \tag{7.58}$$

The conversion between the two forms can be done with the help of

$$\frac{d^k}{dx^k} = x^{-k} \prod_{j=0}^{k-1} (\theta - j), \; \theta^k = \sum_{j=1}^{k} S(k, j) x^j \frac{d^j}{dx^j}, \tag{7.59}$$

where $S(n, k)$ denotes the Stirling numbers of the second kind:

$$S(n, k) = \frac{1}{k!} \sum_{j=0}^{k} (-1)^j \binom{k}{j} (k - j)^n. \tag{7.60}$$

Exercise 65 *Prove the two relations in Eq. (7.59).*

After possibly multiplying by an appropriate power of x we arrive at

$$\tilde{L}_r = \sum_{k=0}^{r} \tilde{Q}_k(x) \theta^k, \tag{7.61}$$

where the coefficients $\tilde{Q}_k(x)$ are again polynomials in x and $\tilde{L}_r$ annihilates the integral I as well:

$$\tilde{L}_r I = 0. \tag{7.62}$$

Exercise 66 *Rewrite*

$$L_2 = x\,(x+1)\,(x+9)\,\frac{d^2}{dx^2} + \left(3x^2 + 20x + 9\right)\frac{d}{dx} + x + 3 \tag{7.63}$$

in Euler operators. (This is the differential operator of Eq. (7.55) multiplied with $x(x+1)(x+9)$).

Let us now discuss how the solutions are constructed [162]. We consider the **indicial equation**

$$\sum_{k=0}^{r} \tilde{Q}_k\,(0)\,\alpha^k = 0. \tag{7.64}$$

This is a polynomial equation of degree r in the variable α. The r solutions for α are called the **indicials** *or* **local exponents** *at $x_0 = 0$. We denote them by $\alpha_1, \ldots, \alpha_r$.*

Let us first assume that $\alpha_i - \alpha_j \notin \mathbb{Z}$ for $i \neq j$. Then the r independent solutions are given by the power series

$$x^{\alpha_i} \sum_{j=0}^{\infty} c_{i,j} x^j, \;\; c_{i,0} \; = \; 1, \qquad 1 \, \leq \, i \, \leq \, r. \tag{7.65}$$

The coefficients $c_{i,j}$ for $j > 0$ can be computed recursively by plugging the ansatz into the differential equation. Note that the fact that α_i is a root of the indicial equation ensures that

$$\tilde{L}_r x^{\alpha_i} = 0. \tag{7.66}$$

The condition $\alpha_i - \alpha_j \notin \mathbb{Z}$ ensures that

$$\tilde{L}_r x^{\alpha_i + j} \neq 0, \qquad j \, \in \, \mathbb{N}. \tag{7.67}$$

Let us now relax the condition $\alpha_i - \alpha_j \notin \mathbb{Z}$. We now allow that a root α_i occurs with multiplicity λ_i, but we maintain to condition $\alpha_i - \alpha_j \notin \mathbb{Z}$ for $i \neq j$. Let t denote the number of distinct roots. We have

$$\lambda_1 + \cdots + \lambda_t = r. \tag{7.68}$$

The solutions are now spanned by series of the form

$$x^{\alpha_i} \sum_{k=0}^{b} \frac{1}{(b-k)!} \ln^{(b-k)}(x) \sum_{j=0}^{\infty} c_{i,j,k} x^j, \tag{7.69}$$

where $b \in \{0, 1, \dots, \lambda_i - 1\}$ *and* $c_{i,0,k} = \delta_{k0}$.

Exercise 67 *Consider*

$$\tilde{L} = (\theta - \alpha)^{\lambda}. \tag{7.70}$$

Show that the solution space is spanned by

$$x^{\alpha},\; x^{\alpha} \ln(x),\; \dots,\; \frac{x^{\alpha} \ln^{\lambda-1}(x)}{(\lambda-1)!}. \tag{7.71}$$

In the general case we allow $\alpha_i - \alpha_j \in \mathbb{Z}$. *Suppose that the indicials* α_i *and* α_j *have multiplicities* λ_i *and* λ_j, *respectively. Assume further that* $\alpha_i - \alpha_j \in \mathbb{Z}$ *and* $\mathrm{Re}(\alpha_i) > \mathrm{Re}(\alpha_j)$. *We start with* λ_i *solutions of the form as in Eq. (7.69). These are supplemented by* λ_j *solutions starting with the power* x^{α_j}. *Up to the power* $x^{\alpha_i - 1}$ *these solutions follow the pattern of Eq. (7.69). Starting from the power* x^{α_i} *we have to allow logarithms up to the power* $(\lambda_i + \lambda_j - 1)$. *The following exercise illustrates this:*

Exercise 68 *Consider the differential operators*

$$\begin{aligned} \tilde{L}_a &= (\theta - 1)(\theta - x), \\ \tilde{L}_b &= (\theta - x)(\theta - 1). \end{aligned} \tag{7.72}$$

Construct for both operators two independent solutions around $x_0 = 0$.

Let $x_1, \dots, x_s$ *be the set of the singular points of the* r*-th order differential equation, including possibly the point at infinity. We denote the indicials at the* j*-th singular point by* $\alpha_1^{(j)}, \dots, \alpha_s^{(j)}$. *The* **Riemann** P**-symbol** *can be viewed as a* $(r \times s)$*-matrix, which collects the information on the indicials at all singular points:*

$$P \begin{pmatrix} \alpha_1^{(1)} & \alpha_1^{(2)} & \dots & \alpha_1^{(s)} \\ \vdots & \vdots & & \vdots \\ \alpha_r^{(1)} & \alpha_r^{(2)} & \dots & \alpha_r^{(s)} \end{pmatrix}. \tag{7.73}$$

The Fuchsian relation states that the sum of all indicials equals

$$\sum_{i=1}^{r} \sum_{j=1}^{s} \alpha_i^{(j)} = \frac{1}{2}(s-2)(r-1)\,r. \tag{7.74}$$

A singular point x_j, *where the indicial equation takes the form*

$$(\alpha - \alpha_0)^r = 0, \tag{7.75}$$

e.g. where there is only one indicial α_0 *with multiplicity* r *is called a* **point of maximal unipotent monodromy**. *A basis of solutions around this point is called a* **Frobenius basis**.

7.1.4 Magnus Expansion

Let us again consider the univariate problem

$$\left(\frac{d}{dx} + A_x\right) \mathbf{I} = 0. \tag{7.76}$$

A solution to this differential equation with boundary value $\mathbf{I}_0$ at $x = 0$ is given by the infinite series

$$\mathbf{I}(x) = \left[1 - \int_0^x dx_1 A_x(x_1) + \int_0^x dx_1 A_x(x_1) \int_0^{x_1} dx_2 A_x(x_2) \right.$$
$$\left. - \int_0^x dx_1 A_x(x_1) \int_0^{x_1} dx_2 A_x(x_2) \int_0^{x_2} dx_3 A_x(x_3) + \dots \right] \mathbf{I}_0. \tag{7.77}$$

The individual terms of this infinite series are iterated integrals. If we introduce the **path ordering operator** $\mathcal{P}$ by

$$\mathcal{P}\left(A(x_1) A(x_2) \dots A(x_n)\right) = A\left(x_{\sigma_1}\right) A\left(x_{\sigma_2}\right) \dots A\left(x_{\sigma_n}\right), \tag{7.78}$$

where σ is the permutation of $\{1, \dots, n\}$ such that

$$x_{\sigma_1} > x_{\sigma_2} > \dots > x_{\sigma_n}, \tag{7.79}$$

we may write Eq. (7.77) as

$$\mathbf{I}(x) = \mathcal{P} \exp\left(- \int_0^x dx_1 A_x(x_1) \right) \mathbf{I}_0. \tag{7.80}$$

Exercise 69 *Show the equivalence of Eq. (7.80) with Eq. (7.77).*

The infinite series in Eq. (7.77) is in general not yet particular useful, as there is no truncation criteria. If the differential equation is in ε-form, then there is a clear truncation criteria: We truncate to the desired order in ε and the solution coincides with the solution discussed in Sect. 6.3.3.

But let's go on with our formal investigations: If there is only one master integral ($N_{\text{master}} = 1$), $A_x(x_1)$ is a (1×1)-matrix and the path ordering can be ignored, as (1×1)-matrices always commute. In this case the solution is simply

$$I(x) = \exp\left(-\int_0^x dx_1 A_x(x_1)\right) I_0. \tag{7.81}$$

Let us return to the general case, where A_x is a $(N_{\text{master}} \times N_{\text{master}})$-matrix. Let us now insist that we write the solution for Eq. (7.76) in the form of an exponential as in Eq. (7.81) as opposed to the form of a path-ordered exponential (as in Eq. (7.80)). Thus we write

$$\mathbf{I}(x) = \exp(\Omega(x))\, \mathbf{I}_0. \tag{7.82}$$

$\Omega(x)$ is in general again given by an infinite series, called the **Magnus series** [163, 164]. That's the price we have to pay in order to write the solution in terms of an ordinary exponential. We write

$$\Omega(x) = \sum_{n=1}^{\infty} \Omega_n(x). \tag{7.83}$$

The first few terms are

$$\begin{aligned}
\Omega_1(x) &= -\int_0^x dx_1 A_x(x_1), \\
\Omega_2(x) &= \frac{1}{2}\int_0^x dx_1 \int_0^{x_1} dx_2 \left[A_x(x_1), A_x(x_2)\right], \\
\Omega_3(x) &= -\frac{1}{6}\int_0^x dx_1 \int_0^{x_1} dx_2 \int_0^{x_2} dx_3 \left(\left[A_x(x_1), \left[A_x(x_2), A_x(x_3)\right]\right] + \left[\left[A_x(x_1), A_x(x_2)\right], A_x(x_3)\right]\right).
\end{aligned} \tag{7.84}$$

The higher terms $\Omega_n(x)$ (with $n \geq 2$) correct for the fact that in general $A_x(x_1)$ and $A_x(x_2)$ don't commute. In general, $\Omega_1(x)$ is given by Eq. (7.84) and $\Omega_n(x)$ is given for $n \geq 2$ by

$$\Omega_n(x) = \sum_{j=1}^{n-1} \frac{B_j}{j!} \int_0^x dx_1 S_n^{(j)}(x_1), \tag{7.85}$$

where B_j denote the j-th **Bernoulli number**, defined by

$$\frac{x}{e^x - 1} = \sum_{j=0}^{\infty} \frac{B_j}{j!} x^j, \tag{7.86}$$

and the $S_n^{(j)}$ are defined recursively by

$$S_n^{(1)}(x) = \left[A_x(x), \Omega_{n-1}(x)\right],$$
$$S_n^{(j)}(x) = \sum_{k=1}^{n-j} \left[\Omega_k(x), S_{n-k}^{(j-1)}(x)\right], \quad 2 \le j \le n-1. \tag{7.87}$$

The Magnus expansion is useful when the Magnus series in Eq. (7.83) terminates. This is for example the case, whenever $A_x(x)$ is a diagonal matrix (in which case $\Omega_n = 0$ for $n \ge 2$) or a nilpotent matrix. Let us consider the case where A_x can be decomposed into a diagonal matrix D_x and a nilpotent matrix N_x:

$$A_x = D_x + N_x. \tag{7.88}$$

We write

$$\Omega\left[A_x\right](x) \tag{7.89}$$

for the Magnus series of $A_x(x)$. For a diagonal matrix D_x we have

$$\Omega\left[D_x\right](x) = -\int_0^x dx_1 D_x(x_1). \tag{7.90}$$

We set

$$N_x' = e^{-\Omega[D_x]} N_x e^{\Omega[D_x]}. \tag{7.91}$$

Since we assumed that N_x is nilpotent, the matrix N_x' is nilpotent as well. Therefore the Magnus series $\Omega[N_x']$ terminates. A solution of the differential equation

$$\left(\frac{d}{dx} + D_x + N_x\right) \mathbf{I} = 0 \tag{7.92}$$

with boundary value $\mathbf{I}_0$ at $x = 0$ is given by

$$\mathbf{I}(x) = e^{\Omega[D_x](x)}\, e^{\Omega[N_x'](x)}\, \mathbf{I}_0. \tag{7.93}$$

Exercise 70 *Prove Eq. (7.93).*

As an application consider the case where A_x is linear in the dimensional regularisation parameter ε. We write

$$A_x = A_x^{(0)} + \varepsilon A_x^{(1)}, \tag{7.94}$$

where $A_x^{(0)}$ and $A_x^{(1)}$ are independent of ε. Set

$$\mathbf{I}' = U\mathbf{I}, \qquad U = e^{-\Omega[A_x^{(0)}](x)}. \tag{7.95}$$

The matrix U is independent of ε as well. The differential equation

$$\left(\frac{d}{dx} + A_x^{(0)} + \varepsilon A_x^{(1)} \right) \mathbf{I} = 0 \tag{7.96}$$

transforms under Eq. (7.95) into

$$\left(\frac{d}{dx} + \varepsilon U A_x^{(1)} U^{-1} \right) \mathbf{I}' = 0 \tag{7.97}$$

The only dependence on ε is now given by the explicit prefactor in Eq. (7.97).

Exercise 71 *Show that the transformation in Eq. (7.95) transforms the differential Eq. (7.96) into Eq. (7.97).*

As an example let us consider the one-loop two-point function with equal internal masses discussed in example 1 in Sect. 6.3.1. The differential equation in $D = 4 - 2\varepsilon$ space-time dimensions is linear in ε and reads

$$\left(\frac{d}{dx} + A_x^{(0)} + \varepsilon A_x^{(1)} \right) \begin{pmatrix} I_{10} \\ I_{11} \end{pmatrix} = 0 \tag{7.98}$$

with

$$A_x^{(0)} = \begin{pmatrix} 0 & 0 \\ \frac{2}{x(4+x)} & \frac{2}{x(4+x)} \end{pmatrix}, \; A_x^{(1)} = \begin{pmatrix} 0 & 0 \\ -\frac{2}{x(4+x)} & \frac{1}{4+x} \end{pmatrix}. \tag{7.99}$$

We may write $A_x^{(0)}$ as a sum of a diagonal matrix and a nilpotent matrix:

$$A_x^{(0)} = D_x^{(0)} + N_x^{(0)} = \begin{pmatrix} 0 & 0 \\ 0 & \frac{2}{x(4+x)} \end{pmatrix} + \begin{pmatrix} 0 & 0 \\ \frac{2}{x(4+x)} & 0 \end{pmatrix}. \tag{7.100}$$

The matrix $A_x^{(0)}$ has a singular point at $x = 0$. For the iterated integrals we take as lower boundary x_0. In the end we take the limit $x_0 \to 0$ and discard any logarithmic divergent terms $\ln^k(x_0)$. With this prescription we obtain

$$e^{-\Omega[D_x^{(0)}](x)} = \begin{pmatrix} 1 & 0 \\ 0 & \sqrt{\frac{x}{4+x}} \end{pmatrix}, \tag{7.101}$$

and for $N_x^{(0)\prime} = e^{-\Omega[D_x^{(0)}](x)} N_x^{(0)} e^{\Omega[D_x^{(0)}](x)}$

$$e^{-\Omega[N_x^{(0)\prime}](x)} = \begin{pmatrix} 1 & 0 \\ \sqrt{\frac{x}{4+x}} & 1 \end{pmatrix}. \tag{7.102}$$

Thus

$$U = e^{-\Omega[A_x^{(0)}](x)} = e^{-\Omega[N_x^{(0)\prime}](x)} e^{-\Omega[D_x^{(0)}](x)} = \begin{pmatrix} 1 & 0 \\ \sqrt{\frac{x}{4+x}} & \sqrt{\frac{x}{4+x}} \end{pmatrix} \quad (7.103)$$

equals up to a prefactor $2\varepsilon(1-\varepsilon)$ the transformation matrix given in Eq. (6.236). (The prefactor $2\varepsilon(1-\varepsilon)$ comes from the requirement that the non-zero boundary constants have uniform weight zero.)

7.1.5 Moser's Algorithm

We stay with the univariate problem

$$\left(\frac{d}{dx} + A_x\right)\mathbf{I} = 0. \quad (7.104)$$

Suppose that there exists a rational fibre transformation, which brings the differential equation into ε-form. Moser's algorithm [165–167] allows us to systematically construct such a transformation. This algorithm has been implemented in several computer programs [168–170].

In this section we will always take the complex numbers as ground field. This is an algebraically closed field. In particular, any polynomial $p(x) \in \mathbb{C}[x]$ factorises into linear factors.

The entries of A_x are rational functions of the kinematic variable x (and the dimensional regularisation parameter ε). Let us denote by $S = \{x_1, x_2, \dots\}$ the set of points, where A_x is singular, including possibly the point at infinity and by S' the set of singular points excluding the point at infinity (i.e. the set of finite singular points).

Using partial fraction decomposition in x we may write A_x as

$$A_x = \sum_{j=0}^{o_\infty - 2} M_{\infty, j+2}(\varepsilon)\, x^j + \sum_{x_i \in S'} \sum_{j=1}^{o_{x_i}} M_{x_i, j}(\varepsilon) \frac{1}{(x-x_i)^j}. \quad (7.105)$$

The entries of the $(N_{\text{master}} \times N_{\text{master}})$-matrices $M_{\infty,j}(\varepsilon)$ and $M_{x_i,j}(\varepsilon)$ are rational functions in ε. o_{x_i} denotes the order of the pole at x_i.

We say that the differential equation in Eq. (7.104) is in **Fuchsian form**, if A_x has only simple poles. In this case, A_x can be written as

$$A_x = \sum_{x_i \in S'} M_{x_i, 1}(\varepsilon) \frac{1}{(x-x_i)}. \quad (7.106)$$

We call $M_{x_i,1}(\varepsilon)$ the matrix residue at $x = x_i$.

Exercise 72 *Assume that A_x is in Fuchsian form (i.e. of the form as in Eq. (7.106)). Show that the matrix residue at $x = \infty$ is given by*

$$M_{\infty,1}(\varepsilon) = -\sum_{x_i \in S'} M_{x_i,1}(\varepsilon). \tag{7.107}$$

Moser's algorithm proceeds in three steps: In the first step one reduces A_x to a Fuchsian form. In the second step we treat ε as an infinitesimal quantity and transform the eigenvalues of the matrices $M_{x_i,1}(\varepsilon)$ into the interval $[-\frac{1}{2}, \frac{1}{2}[$. If all eigenvalues are proportional to ε, the algorithm succeeds and one may factor out in a third step ε as a prefactor.

Digression Jordan normal form and generalised eigenvectors
We review a view basic facts from linear algebra. A quadratic matrix $A \in \mathrm{M}(n \times n, \mathbb{C})$ may or may not be diagonalisable. However, the matrix can always be put into the Jordan normal form, e.g. there exists an invertible matrix $Q \in \mathrm{GL}(n, \mathbb{C})$ such that

$$A = Q J Q^{-1} \tag{7.108}$$

and J is in the Jordan normal form. The Jordan normal form consists of Jordan block matrices on the diagonal

$$J = \begin{pmatrix} J_1 & & \\ & \ddots & \\ & & J_r \end{pmatrix}, \tag{7.109}$$

and the Jordan block matrices J_i's are of the form

$$J_i = \begin{pmatrix} \lambda_i & 1 & & \\ & \lambda_i & \ddots & \\ & & \ddots & 1 \\ & & & \lambda_i \end{pmatrix}. \tag{7.110}$$

Note that the same eigenvalue λ_i may occur in different Jordan blocks. The Jordan normal form is unique up to permutations of the Jordan blocks. The number of times the eigenvalue λ_i appears on the diagonal in Eq. (7.109) is called the **algebraic multiplicity of the eigenvalue** *λ_i. The number of Jordan blocks corresponding to the eigenvalue λ_i is called the* **geometric multiplicity of the eigenvalue** *λ_i. A right eigenvector $\mathbf{v}_R$ to the matrix A for the eigenvalue λ_i satisfies*

$$(A - \lambda_i \mathbf{1}) \cdot \mathbf{v}_R = 0. \tag{7.111}$$

A right **generalised eigenvector** $\mathbf{v}_R$ *to the matrix A for the eigenvalue* λ_i *satisfies*

$$(A - \lambda_i \mathbf{1})^r \cdot \mathbf{v}_R = 0. \tag{7.112}$$

To each Jordan block of size $(r \times r)$ *there corresponds a set of r (right) generalised eigenvectors. A left eigenvector* $\mathbf{v}_L^T$ *to the matrix A for the eigenvalue* λ_i *satisfies*

$$\mathbf{v}_L^T \cdot (A - \lambda_i \mathbf{1}) \;=\; 0, \; or \; \left(A^T - \lambda_i \mathbf{1}\right) \cdot \mathbf{v}_L \;=\; 0. \tag{7.113}$$

Similar, a left generalised eigenvector $\mathbf{v}_L^T$ *to the matrix A for the eigenvalue* λ_i *satisfies*

$$\mathbf{v}_L^T \cdot (A - \lambda_i \mathbf{1})^r \;=\; 0, \; or \; \left(A^T - \lambda_i \mathbf{1}\right)^r \cdot \mathbf{v}_L \;=\; 0. \tag{7.114}$$

For non-diagonalisable matrices we have to distinguish between left and right eigenvectors as the following example shows: Consider

$$A = \begin{pmatrix} \lambda & 1 & 0 \\ 0 & \lambda & 1 \\ 0 & 0 & \lambda \end{pmatrix}. \tag{7.115}$$

The right eigenvectors are spanned by

$$\mathbf{v}_R = \begin{pmatrix} 1 \\ 0 \\ 0 \end{pmatrix}, \tag{7.116}$$

while the left eigenvectors are spanned by

$$\mathbf{v}_L^T = \begin{pmatrix} 0 & 0 & 1 \end{pmatrix}. \tag{7.117}$$

Let us now look at the technical details of Moser's algorithm. In the first step we reduce A_x to a Fuchsian form. The strategy is to remove successively for each singular point $x_i \in S$ the highest pole until only simple poles are left. A necessary condition for the existence of a fibre transformation which removes the highest pole at $x_i \in S$ is that the matrix $M_{x_i,o_{x_i}}(\varepsilon)$ is nilpotent [165]. In the applications towards Feynman integrals this is usually the case and no counter-examples are known. Very often higher poles in Feynman integral calculations can be removed by a suitable ansatz. This is usually the most efficient way. There is also a systematic algorithm based on balance transformations, with projectors constructed from generalised eigenvectors [166, 171, 172]. The balance transformations are discussed below. This algorithm removes all higher poles and introduces at worst a spurious singularity with a simple pole at a regular point. The spurious singularity is then removed in the second step.

Let us now assume that A_x is in Fuchsian form:

$$A_x = \sum_{x_i \in S'} M_{x_i,1}(\varepsilon) \frac{1}{(x - x_i)}. \tag{7.118}$$

We treat the dimensional regularisation parameter ε as an infinitesimal quantity. We now look at the eigenvalues and the (left and right) eigenvectors of the matrices $M_{x_1,1}$ and $M_{x_2,1}$, where $x_1, x_2 \in S$. Let $x_1 \in S$ be a singular point such that $M_{x_1,1}$ has an eigenvalue $\lambda_1 \geq \frac{1}{2}$. Let $\mathbf{v}_{R,x_1}$ be a right eigenvector of $M_{x_1,1}$ to the eigenvalue λ_1. Similar, let $x_2 \in S$ (with $x_2 \neq x_1$) be a singular point such that $M_{x_2,1}$ has an eigenvalue $\lambda_2 < -\frac{1}{2}$. Let $\mathbf{v}_{L,x_2}^T$ be a left eigenvector of $M_{x_2,1}$ to the eigenvalue λ_2. Assume further that

$$\mathbf{v}_{L,x_2}^T \cdot \mathbf{v}_{R,x_1} \neq 0. \tag{7.119}$$

We define a $(N_{\text{master}} \times N_{\text{master}})$-matrix P by

$$P = \frac{\mathbf{v}_{R,x_1} \, \mathbf{v}_{L,x_2}^T}{\left(\mathbf{v}_{L,x_2}^T \cdot \mathbf{v}_{R,x_1}\right)}. \tag{7.120}$$

We denote the $(N_{\text{master}} \times N_{\text{master}})$-unit matrix by $\mathbf{1}$.

Exercise 73 *Show that P and $\mathbf{1} - P$ are projectors, i.e.*

$$P^2 = P, \; (\mathbf{1} - P)^2 = \mathbf{1} - P. \tag{7.121}$$

Show further

$$\left[(\mathbf{1} - P) + \frac{x - x_2}{x - x_1} P\right] \left[(\mathbf{1} - P) + \frac{x - x_1}{x - x_2} P\right] = \mathbf{1}. \tag{7.122}$$

We define a fibre transformation, called a balance transformation, by

$$U^{\text{balance}}(x_1, x_2, P) = \begin{cases} (\mathbf{1} - P) + \frac{x - x_1}{x - x_2} P, & x_1, x_2 \neq \infty, \\ (\mathbf{1} - P) - \frac{1}{x - x_2} P, & x_1 = \infty, \\ (\mathbf{1} - P) - (x - x_1) P, & x_2 = \infty. \end{cases} \tag{7.123}$$

The inverse transformation is then given by

$$\left[U^{\text{balance}}(x_1, x_2, P)\right]^{-1} = \begin{cases} (\mathbf{1} - P) + \frac{x - x_2}{x - x_1} P, & x_1, x_2 \neq \infty, \\ (\mathbf{1} - P) - (x - x_2) P, & x_1 = \infty, \\ (\mathbf{1} - P) - \frac{1}{x - x_1} P, & x_2 = \infty. \end{cases} \tag{7.124}$$

The balance transformation lowers the eigenvalue at x_1 by one unit and raises the eigenvalue at x_2 by one unit. By a sequence of balance transformations we may try to make all eigenvalues proportional to ε or reduce them to zero. This may fail for several reasons. One reason can be that some eigenvalue is of the form "half-integer plus ε". As we only shift the eigenvalues by units of one, we can never make this eigenvalue proportional to ε. Another reason can be that there are unbalanced eigenvalues, but the scalar product between the corresponding eigenvectors is zero:

$$\mathbf{v}_{L,x_2}^T \cdot \mathbf{v}_{R,x_1} = 0. \tag{7.125}$$

However, if we succeed we may construct a rational fibre transformation which puts the differential equation into ε-form.

In the third step we factor out ε. At this stage we may assume that A_x is in Fuchsian form as in Eq. (7.118) and that all eigenvalues of the matrices $M_{x_i,1}(\varepsilon)$ are proportional to ε or are zero. We seek an x-independent fibre transformation U, which transforms the differential equation into the ε-form. Let $V(\varepsilon)$ be such a transformation. This transformation must fulfil

$$V(\varepsilon) \frac{M_{x_i,1}(\varepsilon)}{\varepsilon} V(\varepsilon)^{-1} = N_{x_i}, \quad \forall\, x_i \in S', \tag{7.126}$$

where the N_{x_i}'s are x- and ε-independent $(N_{\mathrm{master}} \times N_{\mathrm{master}})$-matrices. We don't know the matrix $V(\varepsilon)$ nor do we know matrices N_{x_i}. However, the right-hand side of Eq. (7.126) is independent of ε. This implies that

$$V(\varepsilon) \frac{M_{x_i,1}(\varepsilon)}{\varepsilon} V(\varepsilon)^{-1} = V(\varepsilon') \frac{M_{x_i,1}(\varepsilon')}{\varepsilon'} V(\varepsilon')^{-1} \tag{7.127}$$

for any ε' and all $x_i \in S'$. Let us now set

$$U(\varepsilon, \varepsilon') = V(\varepsilon')^{-1} V(\varepsilon). \tag{7.128}$$

We may re-write Eq. (7.127) as

$$U(\varepsilon, \varepsilon') \frac{M_{x_i,1}(\varepsilon)}{\varepsilon} = \frac{M_{x_i,1}(\varepsilon')}{\varepsilon'} U(\varepsilon, \varepsilon'), \quad \forall\, x_i \in S', \tag{7.129}$$

Eq. (7.129) yield $N_{\mathrm{master}}^2 \cdot |S'|$ linear equations for N_{master}^2 unknowns (i.e. the entries of $U(\varepsilon, \varepsilon')$). This system can be solved with standard tools from linear algebra. This yields the sought-after transformation $U(\varepsilon, \varepsilon')$, which factors out (for any choice of ε') the dimensional regularisation parameter ε. We may then set ε' to any suitable value.

Let us look at an example. We consider again the maximal cut of the double box integral (see Eqs. (7.48)–(7.50)). With

$$\mathbf{J} = \begin{pmatrix} J_1 \\ J_2 \end{pmatrix} = \begin{pmatrix} \mathrm{MaxCut}\ I_{111111100} \\ \mathrm{MaxCut}\ I_{1111111(-1)0} \end{pmatrix} \tag{7.130}$$

the matrix A of the differential equation $(d + A)\mathbf{J} = 0$ is given by

$$A = \begin{pmatrix} 2 & -2\varepsilon \\ 2\varepsilon & 2+4\varepsilon \end{pmatrix} \frac{dx}{x} + \begin{pmatrix} 0 & 2\varepsilon \\ 0 & -\varepsilon \end{pmatrix} \frac{dx}{x+1}. \tag{7.131}$$

The singular points are $S = \{0, -1, \infty\}$. The residues are

$$M_{0,1} = \begin{pmatrix} 2 & -2\varepsilon \\ 2\varepsilon & 2+4\varepsilon \end{pmatrix}, \quad M_{-1,1} = \begin{pmatrix} 0 & 2\varepsilon \\ 0 & -\varepsilon \end{pmatrix}, \quad M_{\infty,1} = \begin{pmatrix} -2 & 0 \\ -2\varepsilon & -2-3\varepsilon \end{pmatrix}. \tag{7.132}$$

We may read off $M_{0,1}$ and $M_{-1,1}$ directly from Eq. (7.131). The residue at infinity is given by $M_{\infty,1} = -M_{0,1} - M_{-1,1}$ (see Exercise 72). The matrix $M_{0,1}$ has only the eigenvalue $2+2\varepsilon$ with multiplicity 2. The corresponding right eigenspace is one-dimensional and spanned by $(1, -1)^T$. The matrix $M_{-1,1}$ has the eigenvalues 0 and $-\varepsilon$. These are already as they should be. The matrix $M_{\infty,1}$ has the eigenvalues -2 and $-2-3\varepsilon$. The left eigenspace for the eigenvalue -2 is spanned by $(1, 0)$, the left eigenspace for the eigenvalue $-2-3\varepsilon$ is spanned by $(2, 3)$. We may balance the eigenvalue $2+2\varepsilon$ at $x = 0$ against one of the eigenvalues at $x = \infty$. Let us pick the eigenvalue -2 at $x = \infty$. Thus we choose

$$\mathbf{v}_{R,0} = \begin{pmatrix} 1 \\ -1 \end{pmatrix}, \quad \mathbf{v}_{L,\infty} = \begin{pmatrix} 1 \\ 0 \end{pmatrix}. \tag{7.133}$$

The projector P reads then

$$P = \begin{pmatrix} 1 & 0 \\ -1 & 0 \end{pmatrix} \tag{7.134}$$

and the balance transformation reads

$$U_1 = U^{\text{balance}}(0, \infty, P) = \begin{pmatrix} -x & 0 \\ x+1 & 1 \end{pmatrix}. \tag{7.135}$$

This gives

$$\mathbf{J}' = U_1 \mathbf{J}, \qquad (d + A')\,\mathbf{J}' = 0, \qquad A' = M'_{0,1} \frac{dx}{x} + M'_{-1,1} \frac{dx}{x+1} \tag{7.136}$$

with

$$M'_{0,1} = \begin{pmatrix} 1+2\varepsilon & 0 \\ 1+\varepsilon & 2+2\varepsilon \end{pmatrix}, \quad M'_{-1,1} = \begin{pmatrix} 0 & 2\varepsilon \\ 0 & -\varepsilon \end{pmatrix}, \quad M'_{\infty,1} = \begin{pmatrix} -1-2\varepsilon & -2\varepsilon \\ -1-\varepsilon & -2-\varepsilon \end{pmatrix}.$$

The situation has improved: $M'_{0,1}$ has now the eigenvalues $1+2\varepsilon$ and $2+2\varepsilon$, the matrix $M'_{\infty,1}$ has now the eigenvalues -1 and $-2-3\varepsilon$. We may iterated this procedure. With

$$U = U_4 U_3 U_2 U_1, \qquad \mathbf{J}'' = U\mathbf{J}, \qquad (d + A'')\,\mathbf{J}'' = 0, \tag{7.137}$$

and

$$U_2 = \begin{pmatrix} -\frac{x-2\varepsilon}{1+2\varepsilon} & \frac{2\varepsilon(1+x)}{(1+\varepsilon)(1+2\varepsilon)} \\ \frac{(1+\varepsilon)(1+x)}{1+2\varepsilon} & \frac{1-2x\varepsilon}{1+2\varepsilon} \end{pmatrix},$$

$$U_3 = \begin{pmatrix} \frac{1-2\varepsilon x}{1+2\varepsilon} & -\frac{2\varepsilon(1+x)}{(1+\varepsilon)(1+2\varepsilon)} \\ -\frac{(1+\varepsilon)(1+x)}{1+2\varepsilon} & -\frac{x-2\varepsilon}{1+2\varepsilon} \end{pmatrix},$$

$$U_4 = \begin{pmatrix} 1 & 0 \\ -(1+x) & -x \end{pmatrix} \tag{7.138}$$

we arrive at

$$A'' = \varepsilon \begin{pmatrix} 0 & -2 \\ 2 & 4 \end{pmatrix} \frac{dx}{x} + \varepsilon \begin{pmatrix} 0 & 2 \\ 0 & -1 \end{pmatrix} \frac{dx}{x+1}. \tag{7.139}$$

A'' is now already in ε-form and there is nothing to be done in the third step.

Let us now look at the limitations of Moser's algorithm: We first consider the one-loop two-point function with equal internal masses (example 1 in Sect. 6.3.1). We already know from Sect. 6.4.3 that in the transformation to the ε-form square roots appear (see Eq. (6.236)). We expect that there is no rational fibre transformation, which brings the differential equation into the ε-form. Let's see where Moser's algorithm fails: The differential equation in Eq. (6.235) is already in Fuchsian form, so we may start directly with step 2 of Moser's algorithm. We write the quantity A appearing in Eq. (6.235) as

$$A = M_{0,1} \frac{dx}{x} + M_{-4,1} \frac{dx}{x+4} \tag{7.140}$$

with

$$M_{0,1} = \begin{pmatrix} 0 & 0 \\ \frac{1-\varepsilon}{2} & \frac{1}{2} \end{pmatrix}, \quad M_{-4,1} = \begin{pmatrix} 0 & 0 \\ -\frac{1-\varepsilon}{2} & -\frac{1}{2}+\varepsilon \end{pmatrix}, \quad M_{\infty,1} = \begin{pmatrix} 0 & 0 \\ 0 & -\varepsilon \end{pmatrix}. \tag{7.141}$$

The matrix $M_{0,1}$ has the eigenvalues 0 and $\frac{1}{2}$, the matrix $M_{-4,1}$ has the eigenvalues 0 and $-\frac{1}{2}+\varepsilon$. As we can only shift eigenvalues by units of 1 with Moser's algorithm, there is no way to balance them such that they become zero or proportional to ε.

A a second counter example we look at the two-loop sunrise integral with equal internal masses (example 4 in Sect. 6.3.1). We look at the maximal cut in $D = 2 - 2\varepsilon$ dimensions. With

$$\mathbf{J} = \begin{pmatrix} I_{111}(2-2\varepsilon) \\ I_{211}(2-2\varepsilon) \end{pmatrix}, \quad (d+A)\,\mathbf{J} = 0, \quad M_{0,1}\frac{dx}{x} + M_{-1,1}\frac{dx}{x+1} + M_{-9,1}\frac{dx}{x+9} \tag{7.142}$$

we have

$$M_{0,1} = \begin{pmatrix} 1+2\varepsilon & -3 \\ \frac{1}{3}(1+2\varepsilon)(1+3\varepsilon) & -1-3\varepsilon \end{pmatrix}, \quad M_{-1,1} = \begin{pmatrix} 0 & 0 \\ -\frac{1}{4}(1+2\varepsilon)(1+3\varepsilon) & 1+2\varepsilon \end{pmatrix},$$
$$M_{-9,1} = \begin{pmatrix} 0 & 0 \\ -\frac{1}{12}(1+2\varepsilon)(1+3\varepsilon) & 1+2\varepsilon \end{pmatrix}, \quad M_{\infty,1} = \begin{pmatrix} -1-2\varepsilon & 3 \\ 0 & -1-\varepsilon \end{pmatrix}. \tag{7.143}$$

We do not expect that this differential equation can be put into an ε-form with a rational transformation. We will discuss this example in more detail in the context of elliptic curves in Chap. 13. Let's see where Moser's algorithm fails: We find for the eigenvalues

$$\begin{aligned} \text{Eigenvalues}\left(M_{0,1}\right) &= \{0, -\varepsilon\}, \\ \text{Eigenvalues}\left(M_{-1,1}\right) &= \{0, 1+2\varepsilon\}, \\ \text{Eigenvalues}\left(M_{-9,1}\right) &= \{0, 1+2\varepsilon\}, \\ \text{Eigenvalues}\left(M_{\infty,1}\right) &= \{-1-\varepsilon, -1-2\varepsilon\}. \end{aligned} \tag{7.144}$$

For example, we may balance the eigenvalue $1+2\varepsilon$ of $M_{-1,1}$ against one of the eigenvalues of $M_{\infty,1}$. In the next step we would like to balance the eigenvalue $1+2\varepsilon$ of $M_{-9,1}$ against the other eigenvalue of $M_{\infty,1}$. However, we discover that in this case the corresponding eigenspaces are orthogonal, i.e.

$$\mathbf{v}_{L,\infty}^T \cdot \mathbf{v}_{R,-9} = 0, \tag{7.145}$$

which prohibits the definition of the projector in Eq. (7.120). This observation is independent of the choices we made for the first balance transformation.

7.1.6 Leinartas Decomposition

Up to now we showed in Sect. 7.1.2 how to reduce a multivariate problem to a univariate problem. In Sects. 7.1.3–7.1.5 we treated the univariate case. Of course, once the univariate case is solved, we have to lift the result to the multivariate case.

In this section we start to treat the multivariate case directly. The first challenge we have to face is how to represent a rational function. In the univariate case we may use partial fractioning: Let $p(x), q(x) \in \mathbb{C}[x]$ and assume that $q(x)$ factorises as

$$q(x) = c \prod_{j=1}^{r} \left(x - x_j\right)^{o_j}. \tag{7.146}$$

We set further $o_\infty = 2 + \deg p - \deg q$. The quantity o_∞ denotes the order of the pole at infinity. Using partial fraction decomposition we may write the rational function $p(x)/q(x)$ as

$$\frac{p(x)}{q(x)} = \sum_{j=0}^{o_\infty - 2} a_j x^j + \sum_{j=1}^{r} \sum_{k=1}^{o_j} \frac{b_{j,k}}{\left(x - x_j\right)^k}. \tag{7.147}$$

If $\deg p < \deg q$ the polynomial part is absent.

Extending partial fractioning iteratively to the multivariate case may introduce spurious poles and may lead to infinite loops. Consider as an example the rational function

$$f(x_1, x_2) = \frac{1}{(x_1 + x_2)(x_1 - x_2)}. \tag{7.148}$$

Partial fractioning with respect to x_1 leads to

$$f(x_1, x_2) = \frac{1}{2x_2(x_1 - x_2)} - \frac{1}{2x_2(x_1 + x_2)}. \tag{7.149}$$

This introduces the spurious singularity $x_2 = 0$. A subsequent partial fraction decomposition with respect to x_2 leads to

$$f(x_1, x_2) = \frac{1}{2x_1(x_1 + x_2)} + \frac{1}{2x_1(x_1 - x_2)}. \tag{7.150}$$

This step introduces the spurious singularity $x_1 = 0$ and spoils the partial fractioning with respect to x_1.

In the multivariate case we may use the Leinartas decomposition [173]. Let $\mathbb{F}$ be a field and $\overline{\mathbb{F}}$ the algebraic closure. We start with polynomials $p, q \in \mathbb{F}[x_1, \dots, x_n]$ in n variables $x = (x_1, \dots, x_n)$. A rational function is a quotient of two polynomials:

$$f(x) = \frac{p(x)}{q(x)}. \tag{7.151}$$

Let's assume that we know the factorisation of the denominator polynomial into irreducible polynomials:

$$q(x) = \prod_{j=1}^{r} \left(q_j(x)\right)^{o_j}. \tag{7.152}$$

For each irreducible polynomial $q_j(x)$ we denote the corresponding algebraic variety by

$$V_j = \left\{ x \in \overline{\mathbb{F}}^n \mid q_j(x) = 0 \right\}. \tag{7.153}$$

Let S be a subset of $\{1, \dots, r\}$. We say that the polynomials $q_j(x)$, $j \in S$ have no common zero, if

$$\bigcap_{j \in S} V_j = \emptyset. \tag{7.154}$$

In this case **Hilbert's Nullstellensatz** guarantees that there are polynomials $h_j(x) \in \mathbb{F}[x]$, $j \in S$ such that

$$\sum_{j \in S} h_j(x) q_j(x) = 1. \tag{7.155}$$

Equation (7.155) is called a **Nullstellensatz certificate**. An algorithm to compute the polynomials $h_j(x)$ is reviewed in Appendix H.2. If the denominator of our rational function contains a set of irreducible polynomials, which do not share a common zero, we may insert the left-hand side of Eq. (7.155) in the numerator and cancel in each term the common factor $q_j(x)$ in the numerator and the denominator.

To give an example consider

$$f_1(x) = \frac{1}{x_1 x_2 (x_1 + x_2 - 1)}. \tag{7.156}$$

The polynomials $q_1 = x_1$, $q_2 = x_2$ and $q_3 = x_1 + x_2 - 1$ do not share a common zero and we have

$$q_1 + q_2 - q_3 = 1. \tag{7.157}$$

Thus

$$f_1(x) = \frac{1}{x_2 (x_1 + x_2 - 1)} + \frac{1}{x_1 (x_1 + x_2 - 1)} - \frac{1}{x_1 x_2}. \tag{7.158}$$

The polynomials in the denominators of the individual terms in Eq. (7.158) have common zeros, hence a further reduction with the help of Hilbert's Nullstellensatz is not possible.

We say that a set of polynomials $q_1(x), \dots, q_r(x)$ is **algebraically independent** if there exists no non-zero polynomial a in r variables with coefficients in $\mathbb{F}$ such that

$$a(q_1, \dots, q_r) = 0 \tag{7.159}$$

in $\mathbb{F}[x]$, otherwise the set of polynomials $q_1(x), \dots, q_r(x)$ is called **algebraically dependent** and the polynomial a is called an **annihilator**. If $q_1, \dots, q_r$ are algebraically dependent, then also $q_1^{b_1}, \dots, q_r^{b_r}$ with $b_j \in \mathbb{N}$ are algebraically dependent. A set of r polynomials $q_1, \dots, q_r$ is always algebraically dependent if $r > n$, where n denotes the number of variables $x_1, \dots, x_n$.

In Appendix H.3 we review an algorithm which allows us to decide if a given set of polynomials is algebraically dependent, and in the case it is, computes an annihilating polynomial.

We may then reduce the denominators further, until the polynomials in the denominator are algebraically independent. This is done as follows: Let us introduce a multi-index notation:

$$x^b = \prod_{j=1}^{n} x_j^{b_j}, \quad q^\nu = \prod_{j=1}^{r} q_j^{\nu_j}. \tag{7.160}$$

Let's assume that $q_1, \dots, q_r$ are algebraically dependent and that the denominator of the rational function is given by

$$q = \prod_{j=1}^{r} q_j^{o_j}. \tag{7.161}$$

We set $Q_j = q_j^{o_j}$. Then also $Q_1, \dots, Q_r$ are algebraically dependent and with the notation above we write the annihilating polynomial as

$$a(Q_1, \dots, Q_r) = \sum_{\nu \in I} c_\nu Q^\nu. \tag{7.162}$$

Let $\nu^{(0)}$ be an r-tuple $(\nu_1^{(0)}, \dots, \nu_r^{(0)})$ with the smallest degree

$$\deg \nu^{(0)} = \sum_{j=1}^{r} \nu_j^{(0)}. \tag{7.163}$$

Then

$$1 = \sum_{\nu \in I \backslash \{\nu^{(0)}\}} \frac{c_\nu}{c_{\nu^{(0)}}} Q^{\nu - \nu^{(0)}}. \tag{7.164}$$

As $\nu^{(0)}$ is an r-tuple with smallest norm, in each term at least one Q_j in $Q^{\nu - \nu^{(0)}}$ has a positive exponent and removes the corresponding Q_j from the denominator. As a result, each term will have fewer polynomials in the denominator and repeating this procedure we arrive at denominators, which are algebraically independent.

Let look at an example:

$$f_1(x) = \frac{1}{x_1 x_2 (x_1 + x_2)}. \tag{7.165}$$

The polynomials $q_1 = x_1$, $q_2 = x_2$ and $q_3 = x_1 + x_2$ share a common zero ($x_1 = x_2 = 0$), so a decomposition with Hilbert's Nullstellensatz is not possible. However, they are algebraically dependent. An annihilating polynomial is given by

$$a(q_1, q_2, q_3) = q_1 + q_2 - q_3. \tag{7.166}$$

The three terms (q_1, q_2 and $(-q_3)$) all are of degree one. Let's pick the last one. We have

$$1 = \frac{q_1}{q_3} + \frac{q_2}{q_3} \tag{7.167}$$

and

$$f_1(x) = \frac{1}{x_2 (x_1 + x_2)^2} + \frac{1}{x_1 (x_1 + x_2)^2}. \tag{7.168}$$

Note that the decomposition is not unique, we may picked q_1 or q_2 as the term $c_{\nu^{(0)}} q^{\nu^{(0)}}$. Note also that the decomposition may increase the power of the remaining polynomials in the denominator.

Putting Hilbert's Nullstellensatz decomposition and the decomposition based on algebraic dependence together, we arrive at the Leinartas decomposition:

Leinartas decomposition: A rational function

$$f = \frac{p}{q}, \quad q = \prod_{j=1}^{r} q_j^{o_j}, \qquad p, q \in \mathbb{F}[x_1, \dots, x_n] \tag{7.169}$$

may be written as

$$f = \sum_S \frac{p_s}{\prod\limits_{j \in S} q_j^{b_j}}, \tag{7.170}$$

where the sum is over subsets S of $\{1, \dots, r\}$ such that the polynomials q_j, $j \in S$ have a common zero and are algebraically independent.

Applications towards Feynman integrals have been considered in [174–176].

7.1.7 *Maximal Cuts and Constant Leading Singularities*

The study of the maximal cuts is one of the most efficient ways of finding an appropriate fibre transformation, in particular if the Feynman integrals evaluate to multiple polylogarithms. Suppose somebody gives us a transformation matrix U

$$\mathbf{I}' = U\mathbf{I}. \tag{7.171}$$

Then it is easy to check if this fibre transformation transforms the differential equation to an ε-form. We simply calculate

$$A' = UAU^{-1} + UdU^{-1} \tag{7.172}$$

and check if A' is in ε-form. The problem is only to come up initially with the concrete form of the transformation matrix U. This is a situation where a heuristic method may work well: Guessing a suitable U may outperform any systematic algorithm to construct the matrix U.

For the technique discussed below we will focus on the diagonal blocks (e.g. the blocks A_1, A_2 and A_4 in Eq. (7.6)). The study of the maximal cut allows us to obtain the transformation matrix for this diagonal block up to an ε-dependent prefactor (i.e. the unknown prefactor may depend on ε, but not on the kinematic variables x).

A diagonal block corresponds to the maximal cut of a particular sector [129]. Let us denote the number of master integrals for this sector by N_{sector} and the integrands of the master integrals by $\varphi_1, \dots, \varphi_{N_{\text{sector}}}$. The number N_{sector} equals the dimension of the diagonal block. As before, we denote by N_{prop} the number of propagators having positive indices. Let's consider a Baikov representation for these integrals. The number of Baikov variables is denoted by N_V. For the maximal cut we take a N_{prop}-fold residue. This leaves us with

$$N_V - N_{\text{prop}} \tag{7.173}$$

integrations for the maximal cut integrals. We now choose N_{sector} independent integration domains for the remaining integrations. We denote these integration domains each combined with the N_{prop}-fold residue integration domain by $\mathcal{C}_1, \dots, \mathcal{C}_{N_{\text{sector}}}$. Thus, $\mathcal{C}_j$ defines an N_V-dimensional integration domain. The integration domains are independent, if the $N_{\text{sector}} \times N_{\text{sector}}$-matrix with entries

$$\langle \varphi_i | \mathcal{C}_j \rangle = \int\limits_{\mathcal{C}_j} \varphi_i \tag{7.174}$$

has full rank. We are interested in choosing the integration domains $\mathcal{C}_j$ as simple as possible. Particular simple integration domains are products of circles around singular points. These correspond to additional residue calculations.

Having fixed N_{sector} independent integration domains, we then look for N_{sector} integrands $\varphi'_1, \dots, \varphi'_{N_{\text{sector}}}$ such that the first term in the Laurent expansion in the dimensional regularisation parameter ε of

$$\langle \varphi'_i | \mathcal{C}_j \rangle = \int\limits_{\mathcal{C}_j} \varphi'_i \tag{7.175}$$

is a constant (i.e. independent of the kinematic variables x) of weight zero for all j. More precisely, let $j_{\min}$ be defined by

$$j_{\min} = \min_{j} \left(\text{ldegree} \left(\langle \varphi_i' | \mathcal{C}_j \rangle , \varepsilon \right) \right), \tag{7.176}$$

where ldegree denotes the low degree of a Laurent series. Note that $j_{\min} = j_{\min}(i)$ depends on i, two integrands φ_{i_1}' and φ_{i_2}' may have $j_{\min}(i_1) \neq j_{\min}(i_2)$. We require that for all j the term of order $\varepsilon^{j_{\min}}$ is a constant of weight zero:

$$\text{coeff} \left(\langle \varphi_i' | \mathcal{C}_j \rangle , \varepsilon^{j_{\min}} \right) \cdot \varepsilon^{j_{\min}} = \text{constant of weight zero}, \tag{7.177}$$

where coeff (f, ε^j) denotes the coefficient of ε^j in the Laurent expansion of f around $\varepsilon = 0$. The weight counting is as follows: We define the weight of rational numbers to be zero. The transcendental constant π has weight one, the dimensional regularisation parameter ε has weight (-1). The weight of a product is the sum of the weights of its factors.

Let us denote by $\mathcal{C}_{\text{MaxCut}}$ the integration domain for the original maximal cut. If φ' satisfies Eq. (7.177), we say that

$$\text{MaxCut } I = \int\limits_{\mathcal{C}_{\text{MaxCut}}} \varphi' \tag{7.178}$$

has **constant leading singularities**.

There is no principal obstruction for restricting us to a diagonal block. In theory at least we could consider the full system of N_{master} master integrals. Let $\mathcal{C}$ denote the original integration domain for the N_{master} master integrals and let $\mathcal{C}_1, \dots, \mathcal{C}_{N_{\text{master}}}$ denote a set of N_{master} independent integration domains. If φ' satisfies the condition of Eq. (7.177) for $\mathcal{C}_1, \dots, \mathcal{C}_{N_{\text{master}}}$, we say that

$$I = \int\limits_{\mathcal{C}} \varphi' \tag{7.179}$$

has constant leading singularities [177, 178].

Integrals with constant leading singularities are a guess for a basis of master integrals, which puts the differential equation into an ε-form. In practice we will be using the requirement of Eq. (7.177). We mention that the requirement of Eq. (7.177) is not a necessary requirement for transforming the differential equation into an ε-form. We will see in chapter 13 an explicit example, which does not satisfy Eq. (7.177) but nevertheless puts the differential equation into an ε-form.

Let us now look at an example. We consider the two-loop double box integral (example 3 in Sect. 6.3.1). This is a system with eight master integrals. Suppose we already found suitable master integrals, which puts the sub-system of the first six master integrals into an ε-form. A possible choice can be read off from Eq. (6.232):

$$
\begin{aligned}
I'_{\nu_1} &= \frac{g_1(\varepsilon)}{x} I_{001110000}, \\
I'_{\nu_2} &= g_1(\varepsilon)\, I_{100100100}, \\
I'_{\nu_3} &= \varepsilon^2 (1-2\varepsilon)^2\, I_{011011000}, \\
I'_{\nu_4} &= g_2(\varepsilon)\, I_{100111000} \\
I'_{\nu_5} &= 6\varepsilon^3 (1-2\varepsilon)\, I_{111100100} + g_2(\varepsilon)\, I_{100111000}, \\
I'_{\nu_6} &= 3\varepsilon^4 (1+x)\, I_{101110100},
\end{aligned}
\tag{7.180}
$$

where $g_1(\varepsilon)$ and $g_2(\varepsilon)$ have been defined in Eq. (6.233). Thus we are left with finding a fibre transformation, which transforms the last sector, consisting of the two master integrals $I_{111111100}$ and $I_{1111111(-1)0}$ into an ε-form. We consider the maximal cut of this sector for the integrals $I_{1111111\nu 0}$ (see Eq. (6.468)). With $\mu^2 = t$ we have

$$
\text{MaxCut } I_{1111111\nu 0} = \\
(2\pi i)^7 \frac{2^{4\varepsilon}(s+t)^{\varepsilon} t^{3+\nu+3\varepsilon}}{4\pi^3 \left(\Gamma\left(\frac{1}{2}-\varepsilon\right)\right)^2 s^{2+2\varepsilon}} \int\limits_{\mathcal{C}_{\text{MaxCut}}} dz_8\, z_8^{-1-2\varepsilon} (t-z_8)^{-1-\varepsilon} (s+t-z_8)^{\varepsilon} z_8^{-\nu}. \tag{7.181}
$$

We now choose two independent integration domains:

$$
\begin{aligned}
\mathcal{C}_1 &: \text{small circle around } z_8 = 0 \text{ for the } z_8\text{-integration}, \\
\mathcal{C}_2 &: \text{small circle around } z_8 = t \text{ for the } z_8\text{-integration}.
\end{aligned}
\tag{7.182}
$$

We set

$$
\varphi_\nu = \frac{2^{4\varepsilon}(s+t)^{\varepsilon} t^{3+\nu+3\varepsilon}}{4\pi^3 \left(\Gamma\left(\frac{1}{2}-\varepsilon\right)\right)^2 s^{2+2\varepsilon}} z_8^{-1-2\varepsilon} (t-z_8)^{-1-\varepsilon} (s+t-z_8)^{\varepsilon} z_8^{-\nu} d^8 z. \tag{7.183}
$$

With $x = s/t$ we have

$$
\langle \varphi_0 | \mathcal{C}_1 \rangle = \frac{64\pi^4}{x^2} + \mathcal{O}(\varepsilon), \quad \langle \varphi_0 | \mathcal{C}_2 \rangle = -\frac{64\pi^4}{x^2} + \mathcal{O}(\varepsilon). \tag{7.184}
$$

The integral

$$
\text{MaxCut } I_{111111100} = \langle \varphi_0 | \mathcal{C}_{\text{MaxCut}} \rangle \tag{7.185}
$$

does not have constant leading singularities, but it is easy to fix this issue: We multiply the integrand by x^2. If in addition we multiply by ε^4, the leading singularities are constants of weight zero. Strictly speaking we can only infer from the first term of the ε-expansion of $\langle \varphi_0 | \mathcal{C}_j \rangle$ that we should multiply by an ε-dependent prefactor, whose ε-expansion starts at ε^4. In this example we can verify a posteriori that ε^4 is the correct ε-dependent prefactor. We now set

$$\varphi_0' = \varepsilon^4 x^2 \varphi_0. \qquad (7.186)$$

Then

$$\langle \varphi_0' | \mathcal{C}_1 \rangle = 64\pi^4 \varepsilon^4 + \mathcal{O}\left(\varepsilon^5\right), \ \langle \varphi_0' | \mathcal{C}_2 \rangle = -64\pi^4 \varepsilon^4 + \mathcal{O}\left(\varepsilon^5\right). \qquad (7.187)$$

Thus

$$\text{MaxCut}\left(\varepsilon^4 x^2 I_{111111100}\right) = \langle \varphi_0' | \mathcal{C}_{\text{MaxCut}} \rangle \qquad (7.188)$$

has constant leading singularities.

As this sector has two master integrals, we need a second master integral. We consider φ_{-1} and compute the leading singularities. We obtain

$$\langle \varphi_{-1} | \mathcal{C}_1 \rangle = 0 + \mathcal{O}(\varepsilon), \ \langle \varphi_{-1} | \mathcal{C}_2 \rangle = -\frac{64\pi^4}{x^2} + \mathcal{O}(\varepsilon). \qquad (7.189)$$

It follows that

$$\text{MaxCut}\left(2\varepsilon^4 x^2 I_{1111111(-1)0}\right) = \langle 2\varepsilon^4 x^2 \varphi_{-1} | \mathcal{C}_{\text{MaxCut}} \rangle \qquad (7.190)$$

has constant leading singularities. Including a prefactor of 2 or not is irrelevant at this stage. We included it to be consistent with Eq. (6.232).

It is easily verified, that the two master integrals

$$\varepsilon^4 x^2 I_{111111100} \text{ and } 2\varepsilon^4 x^2 I_{1111111(-1)0} \qquad (7.191)$$

put the 2×2-diagonal block for this sector into an ε-form. It remains to treat the off-diagonal block with entries $A_{i,j}$, $i \in \{7, 8\}$, $j \in \{1, 2, 3, 4, 5, 6\}$. This is most easily done with the methods of Sect. 7.1.1 (see Eqs. (7.19)-(7.22)). One finds

$$\begin{aligned} I_{\nu_7}' &= \varepsilon^4 x^2 I_{111111100}, \\ I_{\nu_8}' &= 2\varepsilon^4 x^2 I_{1111111(-1)0} + x\left[2I_{\nu_6}' + I_{\nu_5}' + I_{\nu_4}' - I_{\nu_2}' - I_{\nu_1}'\right]. \end{aligned} \qquad (7.192)$$

7.2 Base Transformations

Let us now discuss base transformations. We assume that through an appropriate fibre transformation we transformed the differential equation

$$(d + A)\, \mathbf{I} = 0 \qquad (7.193)$$

into the form

$$A = \varepsilon \sum_{j=1}^{N_L} C_j \, \omega_j, \tag{7.194}$$

where the C_j's are $(N_{\text{master}} \times N_{\text{master}})$-matrices, whose entries are algebraic numbers and the ω_j's are dlog-forms

$$\omega_j = d \ln f_j, \tag{7.195}$$

with the f_j's being **algebraic** functions of the kinematic variables x. Equation (6.237) is an example. In this example the differential one-form

$$d \ln \left(2 + x + \sqrt{x \, (4 + x)} \right) \tag{7.196}$$

appears. We would like to find a base transformation, which transforms all f_j's to **rational** functions of the new kinematic variables x'. If we achieve this, we may express the Feynman integrals **I** in terms of multiple polylogarithms.

For the example in Eq. (7.196) we have already seen in Sect. 6.4.4 that the substitution

$$x = \frac{\left(1 - x'\right)^2}{x'} \tag{7.197}$$

rationalises the argument of the logarithm

$$\ln \left(2 + x + \sqrt{x \, (4 + x)} \right) = \ln \left(\frac{2}{x'} \right). \tag{7.198}$$

We look for a systematic way to find such a transformation.

7.2.1 Mathematical Set Up

Assume that we have n kinematic variables $x_1, \dots, x_n$. (In this section we write for simplicity $n = N_B$.) Consider a polynomial

$$f \left(x_1, \dots, x_n \right) \in \mathbb{C} \left[x_1, \dots, x_n \right]. \tag{7.199}$$

We are interested in

$$\sqrt{f \left(x_1, \dots, x_n \right)} \tag{7.200}$$

and we seek a change of variables from $x_1, \dots, x_n$ to $x_1', \dots, x_n'$ such that Eq. (7.200) becomes a rational function in the new variables $x_1', \dots, x_n'$.

We first introduce a few concept from algebraic geometry. An **affine hypersurface** V is the zero set $V(f)$ of a polynomial $f \in \mathbb{C}[x_1, \ldots, x_n]$ in n variables, embedded in $\mathbb{C}^n$:

$$V(f) \subset \mathbb{C}^n. \tag{7.201}$$

The **degree d of the hypersurface** is the degree of the defining polynomial f.

Besides affine hypersurfaces we will also deal with projective hypersurfaces. These are defined by homogeneous polynomials. A polynomial $F \in \mathbb{C}[x_0, \ldots, x_n]$ in $(n+1)$ variables $x_0, \ldots, x_n$ is called **homogeneous of degree** d if all its terms have the same degree d. In particular, a degree-d homogeneous polynomial satisfies

$$F(\lambda x_0, \ldots, \lambda x_n) = \lambda^d F(x_0, \ldots, x_n), \qquad \lambda \in \mathbb{C}. \tag{7.202}$$

Note that if a point $(x_0, \ldots, x_n) \in \mathbb{C}^{n+1}$ is a zero of a homogeneous polynomial F, then every point $(\lambda x_0, \ldots, \lambda x_n)$ is a zero of F. Thus, the zero set of F is a union of complex lines through the origin in $\mathbb{C}^{n+1}$. A **projective hypersurface** is the set of zeros of a homogeneous polynomial $F \in \mathbb{C}[x_0, \ldots, x_n]$, embedded in $\mathbb{CP}^n$:

$$V(F) \subset \mathbb{CP}^n. \tag{7.203}$$

The **projective closure** of an affine hypersurface $V(f) \subset \mathbb{C}^n$ is the projective hypersurface $\overline{V} = V(F) \subset \mathbb{CP}^n$, where F is the **homogenisation** of f. We can homogenise a degree-d polynomial f in n variables $x_1, \ldots, x_n$ to turn it into a degree-d homogeneous polynomial F in $n+1$ variables $x_0, x_1, \ldots, x_n$ in the following way: decompose f into the sum of its homogeneous components of various degrees, $f = g_0 + \cdots + g_d$, where g_i has degree i. Note that some g_j's may be zero, but $g_d \neq 0$. We have $g_j \in \mathbb{C}[x_1, \ldots, x_n]$. The homogenisation F of f is defined by

$$F = x_0^d g_0 + x_0^{d-1} g_1 + \cdots + x_0 g_{d-1} + g_d \in \mathbb{C}[x_0, x_1, \ldots, x_n]. \tag{7.204}$$

We call x_0 the homogenising variable. Note that the restriction of F to the plane $x_0 = 1$ gives the original polynomial f.

As an example, consider the affine parabola $V(y - x^2) \subset \mathbb{C}^2$. The homogenisation of $f = y - x^2 \in \mathbb{C}[x, y]$ is

$$F = zy - x^2 \in \mathbb{C}[x, y, z]. \tag{7.205}$$

The projective closure $\overline{V} = V(F) \subset \mathbb{CP}^2$ consists of all points in $\mathbb{CP}^2$, which correspond to lines in $\mathbb{C}^3$ that connect the points on the original parabola in the plane $z = 1$ with the origin plus the line $x = z = 0$, i.e., the y-axis. The latter line corresponds to the "infinitely distant point" $[0 : 1 : 0] \in \mathbb{CP}^2$.

If $V(f)$ is a hypersurface, affine or projective, a point $p \in V$ is said to be of **multiplicity** $o \in \mathbb{N}$ if all partial derivatives of order $< o$ vanish at p

Fig. 7.1 The nodal cubic defined by $y^2 - x^3 - x^2 = 0$

$$\frac{\partial^{i_1+\ldots+i_n} f}{\partial x_1^{i_1} \cdots \partial x_n^{i_n}}(p) = 0 \text{ with } i_1 + \cdots + i_n < o \tag{7.206}$$

and if there exists at least one non-vanishing o-th partial derivative

$$\frac{\partial^{i_1+\cdots+i_n} f}{\partial x_1^{i_1} \cdots \partial x_n^{i_n}}(p) \neq 0 \text{ with } i_1 + \cdots + i_n = o. \tag{7.207}$$

We write $\mathrm{mult}_p(V) = o$. Points of multiplicity 1 are called **regular points**, points of multiplicity $o > 1$ are called **singular points** of V.

As an example of an (affine) variety with a singular point consider the nodal cubic $V(f) \in \mathbb{C}^2$ defined by $f = y^2 - x^3 - x^2$.

The curve is shown in Fig. 7.1. The point $p = (0, 0)$ is a singular point of multiplicity $o = 2$: One easily verifies

$$f(p) = \frac{\partial f}{\partial x}(p) = \frac{\partial f}{\partial y}(p) = 0, \quad \frac{\partial^2 f}{\partial x^2}(p) \neq 0. \tag{7.208}$$

If d denotes the degree of a given hypersurface, we will be particularly interested in the points of multiplicity $d - 1$.

To a square root we associate a hypersurface as follows: Consider a square root $\sqrt{p/q}$ of a rational function, where $p, q \in \mathbb{C}[x_1, \ldots, x_n]$ are polynomials. We introduce a new variable r and set $r = \sqrt{p/q}$. After squaring and clearing the denominator we obtain $qr^2 = p$. Thus we define

$$f = q \cdot r^2 - p \in \mathbb{C}[r, x_1, \ldots, x_n] \tag{7.209}$$

and call $V(f)$ the **associated hypersurface**. Note that we can also associate a hypersurface to more general algebraic functions such as roots of degree greater than 2 or nested roots. For example, $V(r^3 - x^3 - x^2)$ is associated to $\sqrt[3]{x^3 + x^2}$ and

$$V((r^2 - x^2)^2 - x^4 - y^3) \text{ is associated to } \sqrt{x^2 + \sqrt{x^4 + y^3}}. \tag{7.210}$$

7.2.2 Rationalisation Algorithms

Let us start with an example: We consider the square root $\sqrt{1-x^2}$ and we look for an appropriate transformation $\varphi_x : x' \mapsto \varphi_x(x')$ that turns

$$\sqrt{1-(\varphi_x(x'))^2} \qquad (7.211)$$

into a rational function of x'. One easily checks that the parametrisation

$$\varphi_x(x') = \frac{1-x'^2}{1+x'^2} \qquad (7.212)$$

solves the problem, leading to

$$\sqrt{1-(\varphi_x(x'))^2} = \frac{2x'}{1+x'^2}. \qquad (7.213)$$

There is a systematic way to construct φ_x. We start with the associated hypersurface: We introduce a new variable y, set $y = \sqrt{1-x^2}$ and arrive after squaring at the defining equation for the associated hypersurface

$$x^2 + y^2 - 1 = 0. \qquad (7.214)$$

This equation describes the unit circle. We see that asking for a rational change of variables $\varphi_x(x')$ which rationalises the square root $y = \sqrt{1-x^2}$ is the same as asking for rational functions $(\varphi_x(x'), \varphi_y(x'))$ which parametrise the unit circle. If one can find such rational functions, one would call the circle a **rational algebraic hypersurface**. For the square root $\sqrt{1-x^2}$, the solution to the rationalisation problem is known since 1500 BC [179]:

Consider a fixed point P on the circle and a variable point Q moving on a line not passing through P (see Fig. 7.2). Then look at the second point of intersection R of the line PQ with the circle. We observe that, if Q traces its line, then R traces the

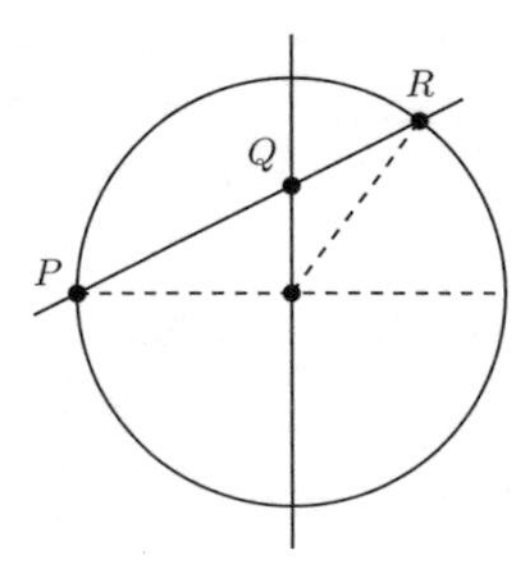

Fig. 7.2 Parametrising the circle by a 1-parameter family of lines

circle. If we take the point P to be $(-1, 0)$ and assume Q to move along the y-axis, i.e., $Q = (0, x')$, then the defining equation of the line PQ is given by $y = x'(1 + x)$ from which we find the parametrisation

$$R(x') = (\varphi_x(x'), \varphi_y(x')) = \left(\frac{1 - x'^2}{1 + x'^2}, \frac{2x'}{1 + x'^2}\right) \tag{7.215}$$

of the unit circle by a short calculation: simply determine the intersection points of the line $PQ : y = x'(1 + x)$ and the circle $x^2 + y^2 = 1$. The first point of intersection is P, the second one yields $R(x')$. Note that, to calculate the expression for $R(x')$, one solely needs rational operations (addition, subtraction, multiplication, division) on polynomial expressions with coefficients in $\mathbb{Q}$. This is precisely the reason why the above method returns a rational function of x'.

We ensure rational coefficients by choosing P to be a point with all coordinate entries lying in $\mathbb{Q}$. In principle, nothing prevents us from taking $P \notin \mathbb{Q}^2$, e.g., choosing $P = \left(-\frac{1}{\sqrt{2}}, -\frac{1}{\sqrt{2}}\right)$ as the starting point of our construction. Still, the method would return a rational function. However, the coefficients of this rational function would no longer be rational, but rather contain factors of $\sqrt{2}$.

This construction generalises to hypersurfaces of degree d, whenever the hypersurface possesses a point of multiplicity $(d - 1)$: A generic line through this point will intersect the hypersurface at one other point and provide a rational parametrisation of the hypersurface. If we consider affine hypersurfaces, it is not necessary that the affine hypersurface possesses a point of multiplicity $(d - 1)$, it suffices if the projective closure possesses a point of multiplicity $(d - 1)$. In other words, the point of multiplicity $(d - 1)$ may be a point at infinity. This leads to the following algorithm:

Algorithm 2 *Rationalisation of an irreducible degree-d hypersurface V defined by f whose projective closure $\overline{V}$ has at least one point of multiplicity $d - 1$.*

1. *Choose a point p_0 with $\text{mult}_{p_0}(\overline{V}) = d - 1$.*
2. *If p_0 is not at infinity, go on with step 3. and finish with step 4. If on the other hand p_0 is at infinity, consider another affine chart V' of the projective closure $\overline{V}$ in which p_0 is not at infinity, continue with steps 3., 4., and finish with step 5.*
3. *With $p_0 = (a_0, \ldots, a_n)$, compute*

$$g(r, x_1, \ldots, x_n) = f(r + a_0, x_1 + a_1, \ldots, x_n + a_n) \tag{7.216}$$

and write

$$g(r, x_1, \ldots, x_n) = g_d(r, x_1, \ldots, x_n) + g_{d-1}(r, x_1, \ldots, x_n), \tag{7.217}$$

where g_d and g_{d-1} are homogeneous components of degree d and $d-1$.

4. *Return*

$$\begin{aligned} \varphi_r(x'_0, \ldots, x'_n) &= -x'_0 \frac{g_{d-1}(x'_0, x'_1, \ldots, x'_n)}{g_d(x'_0, x'_1, \ldots, x'_n)} + a_0, \\ \varphi_{x_1}(x'_0, \ldots, x'_n) &= -x'_1 \frac{g_{d-1}(x'_0, x'_1, \ldots, x'_n)}{g_d(x'_0, x'_1, \ldots, x'_n)} + a_1, \\ &\vdots \\ \varphi_{x_n}(x'_0, \ldots, x'_n) &= -x'_n \frac{g_{d-1}(x'_0, x'_1, \ldots, x'_n)}{g_d(x'_0, x'_1, \ldots, x'_n)} + a_n, \end{aligned} \tag{7.218}$$

where one sets for a single $i \in \{0, \ldots, n\}$ the corresponding variable $x'_i = 1$.

5. *Change coordinates to switch from V' to the original affine chart V.*

As an example let us consider our original problem $\sqrt{x(4+x)}$. The associated hypersurface $V(f)$ is defined by

$$f(r, x) = r^2 - x(4+x). \tag{7.219}$$

f is of degree 2, thus we a need a point p_0 of multiplicity 1, i.e. a regular point. It is easily checked that $p_0 = (r, x) = (0, 0)$ is a regular point, as

$$\frac{\partial f}{\partial x}(p) \neq 0. \tag{7.220}$$

We then have

$$g_2(r, x) = r^2 - x^2, \quad g_1(r, x) = -4x. \tag{7.221}$$

The rationalisation algorithm with $x'_0 = 1$ gives

$$\varphi_r = \frac{4x'_1}{(1 - x_1'^2)}, \quad \varphi_x = \frac{4x_1'^2}{(1 - x_1'^2)}. \tag{7.222}$$

This is already a valid rationalisation:

$$x = \frac{4x_1'^2}{(1 - x_1'^2)}, \quad \sqrt{x(4+x)} = \frac{4x'_1}{(1 - x_1'^2)}. \tag{7.223}$$

We recover the rationalisation of Eq. (6.249) through the substitution

$$x_1' = \frac{1-x'}{1+x'}, \quad x' = \frac{1-x_1'}{1+x_1'}. \tag{7.224}$$

A slightly more general example is the square root of a quadratic polynomial in one variable

$$\sqrt{(x-a)(x-b)}, \qquad \text{with } a \neq b \tag{7.225}$$

and where we treat a and b as constants. The associated hypersurface is now defined by

$$f = r^2 - (x-a)(x-b). \tag{7.226}$$

This is again a hypersurface of degree two and a point of multiplicity 1 is given by $p_0 = (r, x) = (0, a)$. We find

$$g_2(r, x) = r^2 - x^2, \quad g_1(r, x) = (b-a)x. \tag{7.227}$$

Thus

$$x = -(b-a)\frac{x_1'^2}{\left(1-x_1'^2\right)} + a = \frac{a - bx_1'^2}{1-x_1'^2} \tag{7.228}$$

is a rationalisation of the square root in Eq. (7.225). Alternatively, we may additionally perform the substitution of Eq. (7.224). This gives another rationalisation

$$x = \frac{(a-b)}{4x'}\left(1 + 2\frac{(a+b)}{(a-b)}x' + x'^2\right). \tag{7.229}$$

Exercise 74 *Consider the square roots*

$$r_1 = \sqrt{x(4+x)} \text{ and } r_2 = \sqrt{x(36+x)}. \tag{7.230}$$

Find a transformation, which simultaneously rationalises r_1 and r_2.

The above algorithm relies on the existence of a point of multiplicity $(d-1)$. The following theorem allows us under certain conditions to obtain the rationalisation of a degree d hypersurface from the rationalisation of a hypersurface of lower degree. To state the theorem, we first have to introduce the concept of **k-homogenisation**. This is a generalisation of the homogenisation introduced in Eq. (7.204). Let $f \in \mathbb{C}[x_1, \dots, x_n]$ be a polynomial of degree d and write $f = g_0 + \dots + g_d$, where g_i is homogeneous of degree i. Let k be a positive integer with $k \geq d$. The k-homogenisation of f is the degree-k homogeneous polynomial

$$F = x_0^k g_0 + x_0^{k-1} g_1 + \dots + x_0^{k-(d-1)} g_{d-1} + x_0^{k-d} g_d \in \mathbb{C}[x_0, x_1, \dots, x_n]. \tag{7.231}$$

In other words, the k-homogenisation of f is x_0^{k-d} times the usual homogenisation of f. To give an example, the 4-homogenisation of the polynomial $f(x_1, x_2) = x_1 x_2$ is given by $F(x_0, x_1, x_2) = x_0^2 x_1 x_2$.

Theorem 2 *(F-decomposition theorem): Let $V = V(r^2 - f_{\frac{d}{2}}^2 + 4 f_{\frac{d}{2}+1} f_{\frac{d}{2}-1})$ be the hypersurface associated to*

$$\sqrt{f_{\frac{d}{2}}^2 - 4 f_{\frac{d}{2}+1} f_{\frac{d}{2}-1}}, \tag{7.232}$$

where each $f_k \in \mathbb{C}[x_1, \ldots, x_n]$ is a polynomial of degree $\deg(f_k) \leq k$. Then V has a rational parametrisation if $W = V(F_{\frac{d}{2}+1} + F_{\frac{d}{2}} + F_{\frac{d}{2}-1})$ has a rational parametrisation with F_k being the k-homogenization of f_k using the same homogenising variable for each of the three homogenisations. The rational parametrisation of V is obtained from the rational parametrisation $(\varphi_{x_0}^W, \varphi_{x_1}^W, \ldots, \varphi_{x_n}^W)$ of W by

$$\begin{aligned}
\varphi_r^V &= 2 \cdot \varphi_{x_0}^W \cdot f_{\frac{d}{2}+1}\left(\varphi_{x_1}^W / \varphi_{x_0}^W, \ldots, \varphi_{x_n}^W / \varphi_{x_0}^W\right) + f_{\frac{d}{2}}\left(\varphi_{x_1}^W / \varphi_{x_0}^W, \ldots, \varphi_{x_n}^W / \varphi_{x_0}^W\right), \\
\varphi_{x_1}^V &= \frac{\varphi_{x_1}^W}{\varphi_{x_0}^W}, \\
&\vdots \\
\varphi_{x_n}^V &= \frac{\varphi_{x_n}^W}{\varphi_{x_0}^W}.
\end{aligned} \tag{7.233}$$

The proof of this theorem can be found in [180]. The following example illustrates many of the facets discussed so far. Consider the square root

$$\sqrt{x^4 + y^3}. \tag{7.234}$$

The associated affine hypersurface $V(f)$ is defined by $f = r^2 - x^4 - y^3$. Because V has degree 4, we need to find a point p with $\mathrm{mult}_p(V) = 3$ to apply the rationalisation algorithm. Computing the partial derivatives of the homogenisation F of f, however, we see that V does not have a point of multiplicity 3 — not even at infinity.

We, therefore, use the F-decomposition: as a first step, we rewrite the square root as

$$\sqrt{x^4 + y^3} = \sqrt{f_2^2 - 4 f_3 f_1} \tag{7.235}$$

with

$$f_1(x, y) = -\frac{1}{4}, \quad f_2(x, y) = x^2, \quad f_3(x, y) = y^3, \tag{7.236}$$

and k-homogenisations

$$F_1(x,y,z) = -\frac{1}{4}z, \quad F_2(x,y,z) = x^2, \quad F_3(x,y,z) = y^3. \quad (7.237)$$

According to the theorem, V has a rational parametrisation if the hypersurface

$$W = V(F_1 + F_2 + F_3) = V\left(-\frac{z}{4} + x^2 + y^3\right) \quad (7.238)$$

has a rational parametrisation. W is an affine hypersurface of degree 3. We apply algorithm 2 to W. Because $\deg(W) = 3$, we need to find a point of multiplicity 2. Looking at the partial derivatives of $F_1 + F_2 + F_3$, we see that W does not have such a point. There is, however, a point of multiplicity 2 at infinity. We see this by considering the projective closure

$$\overline{W} = V\left(v^2 F_1 + v F_2 + F_3\right). \quad (7.239)$$

This projective hypersurface has a single point of multiplicity 2, namely

$$p_0 = [x_0 : y_0 : z_0 : v_0] = [0 : 0 : 1 : 0]. \quad (7.240)$$

Viewed from the affine chart W, p_0 is at infinity, because v_0 is zero. Therefore, we have to consider a different affine chart W' of $\overline{W}$ in which p_0 is not at infinity. In this particular example, we only have one choice, namely to consider the chart where $z = 1$. Switching from $\overline{W}$ to W' corresponds to the map

$$[x : y : z : v] \mapsto (x', y', v') = (x/z, y/z, v/z). \quad (7.241)$$

Under this mapping, $p_0 \in \overline{W}$ is send to $p_0' = (0,0,0) \in W'$. The affine hypersurface W' is given by

$$W' = V\left(-\frac{1}{4}(v')^2 + v'(x')^2 + (y')^3\right). \quad (7.242)$$

Set

$$\begin{aligned} g(x', y', v') &= -\frac{1}{4}(v' + 0)^2 + (v' + 0)(x' + 0)^2 + (y' + 0)^3 \\ &= g_3(x', y', v') + g_2(x', y', v'), \end{aligned} \quad (7.243)$$

where

$$g_3(x', y', v') = v'(x')^2 + (y')^3 \text{ and } g_2(x', y', v') = -\frac{1}{4}(v')^2. \quad (7.244)$$

A rational parametrisation of W' is then given by

$$\phi_{x'}(t_1, t_2) = -\frac{g_2(1, t_1, t_2)}{g_3(1, t_1, t_2)} = \frac{t_2^2}{4(t_1^3 + t_2)},$$

$$\phi_{y'}(t_1, t_2) = -t_1 \frac{g_2(1, t_1, t_2)}{g_3(1, t_1, t_2)} = \frac{t_1 t_2^2}{4(t_1^3 + t_2)},$$

$$\phi_{v'}(t_1, t_2) = -t_2 \frac{g_3(1, t_1, t_2)}{g_4(1, t_1, t_2)} = \frac{t_2^3}{4(t_1^3 + t_2)}. \tag{7.245}$$

We then translate the rational parametrisation for W' to a rational parametrisation for W. To do this, we solve

$$\phi_{x'} = \frac{\phi_x}{\phi_z}, \quad \phi_{y'} = \frac{\phi_y}{\phi_z}, \quad \text{and} \quad \phi_{v'} = \frac{\phi_v}{\phi_z} \tag{7.246}$$

for ϕ_x, ϕ_y, and ϕ_z while putting $\phi_v = 1$. In this way, we obtain a rational parametrisation of W as

$$\phi_x^W(t_1, t_2) = \frac{1}{t_2}, \quad \phi_y^W(t_1, t_2) = \frac{t_1}{t_2}, \quad \phi_z^W(t_1, t_2) = \frac{4(t_1^3 + t_2)}{t_2^3}. \tag{7.247}$$

In the last step we use the F-decomposition theorem to obtain the change of variables that rationalises $\sqrt{x^4 + y^3}$:

$$\phi_x^V(t_1, t_2) = \frac{\phi_x^W(t_1, t_2)}{\phi_z^W(t_1, t_2)} = \frac{t_2^2}{4(t_1^3 + t_2)}, \quad \phi_y^V(t_1, t_2) = \frac{\phi_y^W(t_1, t_2)}{\phi_z^W(t_1, t_2)} = \frac{t_1 t_2^2}{4(t_1^3 + t_2)}. \tag{7.248}$$

We may verify that Eq. (7.248) rationalises the original square root:

$$\sqrt{\left(\phi_x^V(t_1, t_2)\right)^4 + \left(\phi_y^V(t_1, t_2)\right)^3} = \frac{t_2^3(2t_1^3 + t_2)}{16(t_1^3 + t_2)^2}. \tag{7.249}$$

An implementation of these algorithms has been given in [181].

7.2.3 Theorems on Rationalisations

It is useful to know theorems, which allow us to decide if a given hypersurface has a rational parametrisation or not. Proofs of the theorems stated below can be found in [182]. We start with a simple theorem:

Theorem 3 *Let* $r = \sqrt{p/q}$ *be the square root of a ratio of two polynomials* $p, q \in \mathbb{C}[x_1, \ldots, x_n]$ *and* q *non-zero. Write* $p \cdot q = f h^2$, *where* f *is square free. Then* r *is rationalisable if and only if* $\sqrt{f}$ *is.*

First of all, this theorem reduces the rationalisation of a square root of a rational function to the problem of the rationalisation of a square root of a polynomial. Secondly, it states that for the rationalisation of a square root of a polynomial only the square free part is relevant. Square factors are not relevant.

Theorem 4 *Let $f \in \mathbb{C}[x_1, \dots, x_n]$ be a non-constant square free polynomial of degree d and denote by $F \in \mathbb{C}[x_0, x_1, \dots, x_n]$ the homogenisation of f. We have:*

1. *If d is even, $\sqrt{f}$ is rationalisable if and only if $\sqrt{F}$ is.*
2. *If d is odd, $\sqrt{f}$ is rationalisable if and only if $\sqrt{x_0 F}$ is.*

We may unify the two cases of even degree and odd degree as follows: Define $h = \lceil d/2 \rceil$ by the ceiling function of $d/2$. For example, for $d = 4$ we have $h = 2$ and for $d = 3$ we have $h = 2$. Denote by $\tilde{F}$ the $(2h)$-homogenisation of f. If d is even, this is the usual d-homogenisation of f, if d is odd it is the $(d+1)$-homogenisation of f. The theorem above states that $\sqrt{f}$ is rationalisable if and only if $\sqrt{\tilde{F}}$ is.

Let us now look at square roots in one variable. We have:

Theorem 5 *Let $f \in \mathbb{C}[x]$ be a square free polynomial of degree d in one variable. Then $\sqrt{f}$ is rationalisable if and only if $d \leq 2$.*

For $f \in \mathbb{C}[x]$ as above let V be the affine curve in $\mathbb{C}^2$ defined by $y^2 - f(x) = 0$ and $\overline{V}$ its projective closure in $\mathbb{CP}^2$. The above theorem is equivalent to the statement that $\sqrt{f}$ is rationalisable if and only if $\overline{V}$ has geometric genus zero.

Similar theorems for cases with more variables are more difficult to state and to obtain. Already for the case of square roots in two variables we first need to introduce a few technicalities: Let $\mathbb{F}$ be a field. In chapter 2 we introduced the projective space $\mathrm{P}^n(\mathbb{F})$ as the set of points in $\mathbb{F}^{n+1}\backslash\{0\}$ modulo the equivalence relation

$$(x_0, x_1, ..., x_n) \sim (y_0, y_1, ..., y_n) \Leftrightarrow \exists\, \lambda \neq 0 : (x_0, x_1, ..., x_n) = (\lambda y_0, \lambda y_1, ..., \lambda y_n). \quad (7.250)$$

The **weighted projective space** $\mathrm{P}^n_{w_0,\dots,w_n}(\mathbb{F})$ with weights $(w_0, w_1, \dots, w_n)$ is the set of points in $\mathbb{F}^{n+1}\backslash\{0\}$ modulo the equivalence relation

$$(x_0, x_1, ..., x_n) \sim (y_0, y_1, ..., y_n) \Leftrightarrow \exists\, \lambda \neq 0 : (x_0, x_1, ..., x_n) = \left(\lambda^{w_0} y_0, \lambda^{w_1} y_1, ..., \lambda^{w_n} y_n\right). \quad (7.251)$$

We are mainly concerned with the case $\mathbb{F} = \mathbb{C}$. We write

$$\mathbb{CP}^n_{w_0,\dots,w_n} = \mathrm{P}^n_{w_0,\dots,w_n}(\mathbb{C}). \quad (7.252)$$

Let $f \in \mathbb{C}[x_1, \dots, x_n]$ be a non-constant square free polynomial of degree d and set $h = \lceil d/2 \rceil$ as above. We consider the weighted projective space $\mathbb{CP}^{n+1}_{1,1,\dots,1,h}$ with homogeneous coordinates $(x_0, x_1, \dots, x_n, r)$. The coordinates $x_0, x_1, \dots, x_n$ have weight one, while the coordinate r has weight h. x_0 is the homogenising coordinate, r names the square root. Denote by

$$F(x_0, x_1, \dots, x_n) \in \mathbb{C}[x_0, x_1, \dots, x_n] \quad (7.253)$$

the $(2h)$-homogenisation of f. We associate to $\sqrt{f}$ a hypersurface $\overline{W}$ in the weighted projective space $\mathbb{CP}^{n+1}_{1,1,\dots,1,h}$. The hypersurface $\overline{W}$ is defined by

$$r^2 - F(x_0, x_1, \dots, x_n) = 0. \tag{7.254}$$

(It is worth thinking about the differences in the definition of $\overline{V}$ and $\overline{W}$: In defining $\overline{V}$ we start from $\sqrt{f}$ and first consider the affine hypersurface $r^2 - f(x_1, \dots, x_n) \in \mathbb{C}^{n+1}$. We then take the d-homogenisation. This gives a projective hypersurface in $\mathbb{CP}^{n+1}$. In defining $\overline{W}$ we again start from $\sqrt{f}$ but first consider the $(2h)$-homogenisation $F(x_0, x_1, \dots, x_n) \in \mathbb{C}[x_0, x_1, \dots, x_n]$. In the second step we add the variable r naming the root and consider the hypersurface defined by $r^2 - F(x_0, \dots, x_n)$. This is a hypersurface in the weighted projective space $\mathbb{CP}^{n+1}_{1,1,\dots,1,h}$.) One can show that the hypersurface $\overline{W}$ is birationally equivalent to the hypersurface $\overline{V}$ defined previously. Some theorems can be formulated more elegantly by referring to $\overline{W}$ instead of $\overline{V}$.

We also need the concept of simple singularities. We denote by $\mathbb{C}[[x_1, \dots, x_n]]$ the **ring of formal power series** in $x_1, \dots, x_n$. Let $f_1, f_2 \in \mathbb{C}[x_1, \dots, x_n]$ be two polynomials and assume that both $V(f_1)$ and $V(f_2)$ have a singular point at the origin. We say that these two singularities are **of the same type** if the two quotient rings $\mathbb{C}[[x_1, \dots, x_n]]/\langle f_1 \rangle$ and $\mathbb{C}[[x_1, \dots, x_n]]/\langle f_2 \rangle$ are isomorphic. We call $\mathbb{C}[[x_1, \dots, x_n]]/\langle f_i \rangle$ the **associated quotient ring** of $V(f_i)$.

Let $f \in \mathbb{C}[x_1, \dots, x_n]$ be a polynomial and assume that affine hypersurface $V(f)$ has a singular point at the origin of $\mathbb{C}^n$. We say that the origin is a **simple singularity** (or **ADE singularity** or **Du Val singularity**), if the associated quotient ring is isomorphic to a quotient ring $\mathbb{C}[[x_1, \dots, x_n]]/\langle g \rangle$, where g is a polynomial from the following list:

$$\begin{array}{lll} A_k: & x_1^2 + x_2^{k+1} + X, & k \geq 1, \\ D_k: & \left(x_1^2 + x_2^{k-2}\right) x_2 + X, & k \geq 4, \\ E_6: & x_1^3 + x_2^4 + X, & \\ E_7: & x_1 \left(x_1^2 + x_2^3\right) + X, & \\ E_8: & x_1^3 + x_2^5 + X, & \end{array} \tag{7.255}$$

with

$$X = x_3^2 + \dots + x_n^2. \tag{7.256}$$

The type of a singularity is invariant under linear coordinate transformations, hence we may always translate any singular point to the origin.

With these preparations we may now state the theorem on the rationalisation of square roots in two variables:

Theorem 6 *Let $f \in \mathbb{C}[x_1, x_2]$ be a square free polynomial of degree d in two variables and assume that the hypersurface $\overline{W} \in \mathbb{CP}^{n+1}_{1,1,\dots,1,h}$ has at most ADE-singularities. Then $\sqrt{f}$ is rationalisable if and only if $d \leq 4$.*

We close this section with a theorem on multiple square roots:

Theorem 7 *Let* $f_1, \ldots, f_r \in \mathbb{C}[x_1, \ldots, x_n]$*. If the set of roots* $\{\sqrt{f_1}, \ldots, \sqrt{f_r}\}$ *is simultaneously rationalisable, then for every non-empty subset* $J \subseteq \{1, \ldots, r\}$ *the square root*

$$\sqrt{\prod_{j \in J} f_j} \tag{7.257}$$

is rationalisable.

The main application of this theorem is to prove that a certain set of square root cannot be rationalised simultaneously by showing that a specific product as in Eq. (7.257) is not rationalisable.

Chapter 8
Multiple Polylogarithms

In Chap. 6 we already encountered multiple polylogarithms. Multiple polylogarithms are an important class of functions in the context of Feynman integrals. In this chapter we study them in more detail.

There are two frequently used notations for multiple polylogarithms, either $G(z_1, \dots, z_r; y)$ (which we already introduced in Sect. 6.3.3) or $\mathrm{Li}_{m_1 \dots m_k}(x_1, \dots, x_k)$. The former is directly related to the iterated integral representation, while the latter is related to the nested sum representation. This reveals already the fact, that multiple polylogarithms may either be defined in terms of iterated integrals or nested sums. We will study both cases.

Each of the two representations gives rise to a product: A shuffle product in the case of the iterated integral representation and a quasi-shuffle product in the case of the nested sum representation. This turns the vector space spanned by the multiple polylogarithms into an algebra. This is actually a Hopf algebra. We will discuss the coalgebra properties in more detail in Chap. 11.

Multiple polylogarithms are generalisations of the logarithms and it comes to no surprise that these functions have branch cuts. We will study the monodromy around a branch point.

In the last two sections of this chapter we study fibration bases and linearly reducible Feynman integrals. The linearly reducible Feynman integrals have the property that they evaluate to multiple polylogarithms. They can be computed efficiently from the Feynman parameter representation.

Multiple polylogarithms surfaced in the work of Kummer [183–185], Poincaré [186] and Lappo-Danilevsky [187]. Modern references are Goncharov [188, 189] and Borwein et al. [190]. An introductory survey article can be found in [191].

S. Weinzierl, *Feynman Integrals*, UNITEXT for Physics,
https://doi.org/10.1007/978-3-030-99558-4_8

8.1 The Integral Representation

In Sect. 6.3.3 we introduced multiple polylogarithms as a special case of iterated integrals. Let us recall the definition: If all z's are equal to zero, we define $G(z_1, \dots, z_r; y)$ by

$$G(\underbrace{0, \dots, 0}_{r-\text{times}}; y) = \frac{1}{r!} \ln^r(y). \tag{8.1}$$

This definition includes as a trivial case

$$G(; y) = 1. \tag{8.2}$$

If at least one variable z is not equal to zero we define recursively

$$G(z_1, z_2 \dots, z_r; y) = \int\limits_0^y \frac{dy_1}{y_1 - z_1} G(z_2 \dots, z_r; y_1). \tag{8.3}$$

We have for example

$$G(0; y) = \ln(y), \quad G(z; y) = \ln\left(\frac{z-y}{z}\right). \tag{8.4}$$

If one further defines $g(z; y) = 1/(y - z)$, then one has

$$\frac{d}{dy} G(z_1, \dots, z_r; y) = g(z_1; y) G(z_2, \dots, z_r; y) \tag{8.5}$$

and if at least one variable z is not equal to zero

$$G(z_1, z_2, \dots, z_r; y) = \int\limits_0^y dt \; g(z_1; t) G(z_2, \dots, z_r; t). \tag{8.6}$$

The function $G(z_1, \dots, z_r; y)$ is said to have a **trailing zero**, if $z_r = 0$. We will soon see that with the help of the shuffle product we can always remove trailing zeros. Let us therefore focus on multiple polylogarithms $G(z_1, \dots, z_r; y)$ without trailing zeros (i.e. $z_r \neq 0$). For $z_r \neq 0$ the recursive definition translates to

$$G(z_1, \dots, z_r; y) = \int\limits_0^y \frac{dt_1}{t_1 - z_1} \int\limits_0^{t_1} \frac{dt_2}{t_2 - z_2} \dots \int\limits_0^{t_{r-1}} \frac{dt_r}{t_r - z_r}. \tag{8.7}$$

The number r is referred to as the **depth** of the integral representation. In the case of multiple polylogarithms the number r is also referred to as the **weight** of the multiple polylogarithm. The differential of $G(z_1, \dots, z_r; y)$ is

$$dG(z_1, \dots, z_r; y) = \sum_{j=1}^{r} G(z_1, \dots, \hat{z}_j, \dots, z_r; y) \left[d \ln \left(z_{j-1} - z_j \right) - d \ln \left(z_{j+1} - z_j \right) \right], \tag{8.8}$$

where we set $z_0 = y$ and $z_{r+1} = 0$. A hat indicates that the corresponding variable is omitted. In addition one uses the convention that for $z_{j+1} = z_j$ the one-form $d \ln(z_{j+1} - z_j)$ equals zero. The proof of Eq. (8.8) is based on the identity

$$\frac{\partial}{\partial z} \frac{1}{t - z} = - \frac{\partial}{\partial t} \frac{1}{t - z} \tag{8.9}$$

and partial integration.

For $z_r \neq 0$ we also have the scaling relation

$$G(z_1, \dots, z_r; y) = G(x z_1, \dots, x z_r; x y), \qquad z_r \in \mathbb{C} \backslash \{0\}, \quad x \in \mathbb{C} \backslash \{0\}. \tag{8.10}$$

This allows us to scale the variable y to one:

$$G(z_1, \dots, z_r; y) = G\left(\frac{z_1}{y}, \dots, \frac{z_r}{y}; 1 \right), \qquad z_r \in \mathbb{C} \backslash \{0\}, \quad y \in \mathbb{C} \backslash \{0\}. \tag{8.11}$$

Note that the scaling relation does not hold for multiple polylogarithms with trailing zeros. We have for example

$$G(0; xy) = \ln(xy) = \ln(x) + \ln(y) \neq \ln(y) = G(0; y). \tag{8.12}$$

In order to relate the integral representation of the multiple polylogarithms to the sum representation of the multiple polylogarithms it is convenient to introduce the following short-hand notation:

$$G_{m_1 \dots m_k}(z_1, \dots, z_k; y) = G(\underbrace{0, \dots, 0}_{m_1 - 1}, z_1, \dots, z_{k-1}, \underbrace{0, \dots, 0}_{m_k - 1}, z_k; y) \tag{8.13}$$

Here, all z_j for $j = 1, \dots, k$ are assumed to be non-zero. For example,

$$G_{12}(z_1, z_2; y) = G(z_1, 0, z_2; y). \tag{8.14}$$

The multiply polylogarithm $G_{m_1 \dots m_k}(z_1, \dots, z_k; y)$ has weight $m_1 + \dots + m_k$.

8.2 The Sum Representation

The multiple polylogarithms have also a sum representation. The standard notation for the sum representation is $\text{Li}_{m_1 \ldots m_k}(x_1, \ldots, x_k)$. The sum representation is defined by

$$\text{Li}_{m_1 \ldots m_k}(x_1, \ldots, x_k) = \sum_{n_1 > n_2 > \ldots > n_k > 0}^{\infty} \frac{x_1^{n_1}}{n_1{}^{m_1}} \cdots \frac{x_k^{n_k}}{n_k{}^{m_k}}. \tag{8.15}$$

The sum converges for

$$\left|x_1 x_2 \ldots x_j\right| \leq 1 \text{ for all } j \in \{1, \ldots, k\} \text{ and } (m_1, x_1) \neq (1, 1). \tag{8.16}$$

In the following we will always assume that the arguments x_j are such that Eq. (8.16) is satisfied. The number k in the definition of the sum representation is referred to as the **depth** of the sum representation of the multiple polylogarithm. The number $m_1 + \cdots + m_k$ is referred to the **weight** of the multiple polylogarithm. Note that for the sum representation of multiple polylogarithms the weight and the depth of the sum representation are in general not equal. Equation (8.15) is a nested sum, which we may also write as

$$\text{Li}_{m_1 \ldots m_k}(x_1, \ldots, x_k) = \sum_{n_1=1}^{\infty} \frac{x_1^{n_1}}{n_1^{m_1}} \sum_{n_2=1}^{n_1-1} \frac{x_2^{n_2}}{n_2^{m_2}} \cdots \sum_{n_{k-1}=1}^{n_{k-2}-1} \frac{x_{k-1}^{n_{k-1}}}{n_{k-1}^{m_{k-1}}} \sum_{n_k=1}^{n_{k-1}-1} \frac{x_k^{n_k}}{n_k^{m_k}}, \tag{8.17}$$

with the convention that

$$\sum_{n=a}^{b} f(n) = 0, \text{ for } b < a. \tag{8.18}$$

The relation between the sum representation in Eq. (8.15) and the integral representation in Eq. (8.13) is given by

$$\text{Li}_{m_1 \ldots m_k}(x_1, \ldots, x_k) = (-1)^k G_{m_1 \ldots m_k}\left(\frac{1}{x_1}, \frac{1}{x_1 x_2}, \ldots, \frac{1}{x_1 \ldots x_k}; 1\right), \tag{8.19}$$

and

$$G_{m_1 \ldots m_k}(z_1 \ldots, z_k; y) = (-1)^k \, \text{Li}_{m_1 \ldots m_k}\left(\frac{y}{z_1}, \frac{z_1}{z_2}, \ldots, \frac{z_{k-1}}{z_k}\right). \tag{8.20}$$

Exercise 75 *Prove Eq. (8.19).*

The multiple polylogarithms include several special cases. The **classical polylogarithms** are defined by

$$\mathrm{Li}_m(x) = \sum_{n=1}^{\infty} \frac{x^n}{n^m} \tag{8.21}$$

and are the special case of depth one. The most prominent examples are

$$\mathrm{Li}_1(x) = \sum_{i_1=1}^{\infty} \frac{x^{i_1}}{i_1} = -\ln(1-x), \;\; \mathrm{Li}_2(x) = \sum_{i_1=1}^{\infty} \frac{x^{i_1}}{i_1^2}. \tag{8.22}$$

Nielsen's generalised polylogarithms $S_{n,p}(x)$ are defined by Nielsen [192]

$$S_{n,p}(x) = \mathrm{Li}_{(n+1)1...1}(x, \underbrace{1, \dots, 1}_{p-1}), \tag{8.23}$$

Multiple polylogarithms with $x_2 = x_3 = \dots = x_k = 1$ are a subset of the **harmonic polylogarithms** $H_{m_1,\dots,m_k}(x)$ [193, 194]

$$H_{m_1 \dots m_k}(x) = \mathrm{Li}_{m_1 \dots m_k}(x, \underbrace{1, \dots, 1}_{k-1}). \tag{8.24}$$

If one restricts in the integral representation $G(z_1, \dots, z_r; y)$ the possible values of z_j's to zero and the n-th roots of unity, one arrives at the n-th **cyclotomic harmonic polylogarithms** [195]. The harmonic polylogarithms in Eq. (8.24) are just the first cyclotomic harmonic polylogarithms, corresponding to $z_j \in \{0, 1\}$. The word "harmonic polylogarithms" is used as a synonym for the second cyclotomic harmonic polylogarithms, i.e. multiple polylogarithms with $z_j \in \{-1, 0, 1\}$.

The values of the multiple polylogarithms at $x_1 = \dots = x_k = 1$ are known as **multiple ζ-values**:

$$\begin{aligned} \zeta_{m_1 m_2 \dots m_k} &= \mathrm{Li}_{m_1 m_2 \dots m_k}(1, 1, \dots, 1) \\ &= \sum_{n_1 > n_2 > \dots > n_k > 0}^{\infty} \frac{1}{n_1^{m_1}} \frac{1}{n_2^{m_2}} \cdots \frac{1}{n_k^{m_k}}, \qquad m_1 \neq 1. \end{aligned} \tag{8.25}$$

Digression The Clausen and Glaisher functions

As an excursion let us turn to the Clausen and Glaisher functions. These are related to linear combinations of classical polylogarithms.

The **Clausen function** *is defined by*

$$\mathrm{Cl}_n(\theta) = \begin{cases} \mathrm{Im}\ \mathrm{Li}_n\left(e^{i\theta}\right) = \frac{1}{2i}\left[Li_n\left(e^{i\theta}\right) - Li_n\left(e^{-i\theta}\right)\right], & n \text{ even}, \\ \mathrm{Re}\ \mathrm{Li}_n\left(e^{i\theta}\right) = \frac{1}{2}\left[Li_n\left(e^{i\theta}\right) + Li_n\left(e^{-i\theta}\right)\right], & n \text{ odd}, \end{cases} \tag{8.26}$$

the **Glaisher function** *is defined by*

$$\mathrm{Gl}_n(\theta) = \begin{cases} \mathrm{Re}\,\mathrm{Li}_n\left(e^{i\theta}\right) = \frac{1}{2}\left[Li_n\left(e^{i\theta}\right) + Li_n\left(e^{-i\theta}\right)\right], & n \text{ even}, \\ \mathrm{Im}\,\mathrm{Li}_n\left(e^{i\theta}\right) = \frac{1}{2i}\left[Li_n\left(e^{i\theta}\right) - Li_n\left(e^{-i\theta}\right)\right], & n \text{ odd}. \end{cases} \tag{8.27}$$

From the definition it is clear that these functions are periodic with period 2π*:*

$$\mathrm{Cl}_n(\theta + 2\pi) = \mathrm{Cl}_n(\theta), \;\; \mathrm{Gl}_n(\theta + 2\pi) = \mathrm{Gl}_n(\theta). \tag{8.28}$$

It is therefore sufficient to restrict the argument to

$$0 \leq \mathrm{Re}\,(\theta) < 2\pi. \tag{8.29}$$

Note that $\mathrm{Cl}_n(\theta)$ *and* $\mathrm{Gl}_n(\theta)$ *are not necessarily continuous as* $\theta \to 2\pi$*, as* $\mathrm{Li}_n(x)$ *has a branch cut along the positive real axis, starting at* $x = 1$*.*

For $l \in \mathbb{N}_0$ *we have*

$$\mathrm{Li}_{2l}\left(e^{i\theta}\right) = \mathrm{Gl}_{2l}(\theta) + i\mathrm{Cl}_{2l}(\theta), \;\; \mathrm{Li}_{2l+1}\left(e^{i\theta}\right) = \mathrm{Cl}_{2l+1}(\theta) + i\mathrm{Gl}_{2l+1}(\theta). \tag{8.30}$$

It is worth knowing the special value

$$\mathrm{Cl}_2\left(\frac{\pi}{2}\right) = G, \tag{8.31}$$

where **Catalan's constant** G *is given by*

$$G = \sum_{n=0}^{\infty} \frac{(-1)^n}{(2n+1)^2}. \tag{8.32}$$

It is also worth noting that the Glaisher functions $\mathrm{Gl}_n(\theta)$ *are (for* $0 \leq \mathrm{Re}\,(\theta) < 2\pi$*) polynomials in* θ *of degree n. In detail we have for* $0 \leq \mathrm{Re}\,(\theta) < 2\pi$

$$\mathrm{Gl}_n(\theta) = \begin{cases} -\frac{1}{2}\frac{(2\pi i)^n}{n!} B_n\left(\frac{\theta}{2\pi}\right), & n \text{ even}, \\ -\frac{1}{2i}\frac{(2\pi i)^n}{n!} B_n\left(\frac{\theta}{2\pi}\right), & n \text{ odd}, \end{cases} \tag{8.33}$$

where $B_n(x)$ *is the n'th* **Bernoulli polynomial** *defined by*

$$\frac{te^{xt}}{e^t - 1} = \sum_{n=0}^{\infty} \frac{B_n(x)}{n!} t^n. \tag{8.34}$$

We return to the multiple polylogarithms. The differential of $\mathrm{Li}_{m_1 \dots m_k}(x_1, \dots, x_k)$ with respect to the variables $x_1, \dots, x_k$ is

$$d\mathrm{Li}_{m_1 \dots m_k}(x_1, \dots, x_k) = \sum_{j=1}^{k} \mathrm{Li}_{m_1 \dots m_{j-1}(m_j - 1)m_{j+1} \dots m_k}(x_1, \dots, x_k) \cdot d\ln(x_j). \tag{8.35}$$

This follows easily from

$$d\left(\frac{x_j^{n_j}}{n_j^{m_j}}\right) = n_j \frac{x_j^{n_j-1}}{n_j^{m_j}} dx_j = \frac{x_j^{n_j}}{n_j^{m_j-1}} \frac{dx_j}{x_j} = \frac{x_j^{n_j}}{n_j^{m_j-1}} d\ln(x_j). \tag{8.36}$$

If an index m_j equals one, we obtain in the differential an index with the value zero. These multiple polylogarithms can be reduced. We have

$$\mathrm{Li}_0(x_1) = \frac{x_1}{1-x_1} \tag{8.37}$$

and

$$\begin{aligned}
&\mathrm{Li}_{m_1 \dots m_{i-1} 0 m_{i+1} \dots m_k}(x_1, \dots, x_{i-1}, x_i, x_{i+1}, \dots, x_k) = \\
&\quad \mathrm{Li}_0(x_i) \mathrm{Li}_{m_1 \dots m_{i-1} m_{i+1} \dots m_k}(x_1, \dots, x_{i-1}, x_{i+1}, \dots, x_k) \\
&\quad - \sum_{j=i+1}^{k} \mathrm{Li}_{m_1 \dots m_{i-1} m_{i+1} \dots m_j 0 m_{j+1} \dots m_k}(x_1, \dots, x_{i-1}, x_{i+1}, \dots, x_j, x_i, x_{j+1}, \dots, x_k) \\
&\quad - \sum_{j=i+1}^{k} \mathrm{Li}_{m_1 \dots m_{i-1} m_{i+1} \dots m_j \dots m_k}(x_1, \dots, x_{i-1}, x_{i+1}, \dots, x_i \cdot x_j, \dots, x_k) \\
&\quad - \left[1 + \mathrm{Li}_0(x_i)\right] \mathrm{Li}_{m_1 \dots m_{i-1} m_{i+1} \dots m_k}(x_1, \dots, x_{i-1} \cdot x_i, x_{i+1}, \dots, x_k).
\end{aligned} \tag{8.38}$$

If $i = 1$ the last term is absent. Equation (8.38) allows us to shift recursively the zero index to the last position. If the zero index is in the last position, the sums from $(i+1)$ to k are empty and the recursion terminates. We will prove Eq. (8.38) in Exercise 82 in Chap. 9, once we learned about the quasi-shuffle product and Z-sums.

8.3 The Shuffle Product

In this section we introduce the shuffle product for multiple polylogarithms. The shuffle product is associated with the iterated integral representation. It is not specific to multiple polylogarithms, but holds for any iterated integral.

We start with the definition of a shuffle algebra. Consider a finite set of objects, which we will call **letters**. We denote the letters by $l_1, l_2, \dots$, and the set of all letters the **alphabet** $A = \{l_1, l_2, \dots\}$. A **word** is an ordered sequence of letters:

$$w = l_1 l_2 ... l_k. \tag{8.39}$$

The word of length zero is denoted by e. Let $\mathbb{F}$ be a field and consider the vector space of words over $\mathbb{F}$. We may turn this vector space into an algebra by supplying a product for words. We say that a permutation σ is a shuffle of $(1, 2, ..., k)$ and of $(k+1, ..., r)$, if in

$$(\sigma(1), \sigma(2), \dots, \sigma(r)) \tag{8.40}$$

the relative order of 1, 2, ..., k and of $k+1$, ..., r is preserved. Thus (1, 3, 2) is a shuffle of (1, 2) and (3), while (2, 1, 3) is not. The **shuffle product** of two words is defined by

$$l_1 l_2 \dots l_k \sqcup\!\sqcup\, l_{k+1} \dots l_r = \sum_{\text{shuffles } \sigma} l_{\sigma(1)} l_{\sigma(2)} \dots l_{\sigma(r)}, \tag{8.41}$$

where the sum runs over all permutations σ which are shuffles of $(1, \dots, k)$ and $(k+1, \dots, r)$, i.e. which preserve the relative order of 1, 2, ..., k and of $k+1$, ..., r. This product turns the vector space of words into a shuffle algebra $\mathcal{A}$.

The name "shuffle algebra" is related to the analogy of shuffling cards: If a deck of cards is split into two parts and then shuffled, the relative order within the two individual parts is conserved. A shuffle algebra is also known under the name "mould symmetral" [196].

The empty word e is the unit in this algebra:

$$e \sqcup\!\sqcup\, w = w \sqcup\!\sqcup\, e = w. \tag{8.42}$$

A recursive definition of the shuffle product is given by

$$l_1 l_2 \dots l_k \sqcup\!\sqcup\, l_{k+1} \dots l_r = l_1 \left(l_2 \dots l_k \sqcup\!\sqcup\, l_{k+1} \dots l_r \right) + l_{k+1} \left(l_1 l_2 \dots l_k \sqcup\!\sqcup\, l_{k+2} \dots l_r \right), \tag{8.43}$$

where concatenation of letters is extended on the vector space of words by linearity:

$$l \left(c_1 w_1 + c_2 w_2 \right) = c_1 l w_1 + c_2 l w_2, \qquad c_1, c_2 \in \mathbb{F}, \quad l \in A, \quad w_1, w_2 \in \mathcal{A}. \tag{8.44}$$

Of course, concatenation of words would also define a product on the vector space of words, but this is not the product we are interested in. The shuffle product is commutative

$$w_1 \sqcup\!\sqcup\, w_2 = w_2 \sqcup\!\sqcup\, w_1, \tag{8.45}$$

while the concatenation product is non-commutative. A few examples are

$$\begin{aligned} l_1 l_2 \sqcup\!\sqcup\, l_3 &= l_1 l_2 l_3 + l_1 l_3 l_2 + l_3 l_1 l_2, \\ l_1 l_2 \sqcup\!\sqcup\, l_2 &= 2 l_1 l_2 l_2 + l_2 l_1 l_2, \\ l_1 l_2 \sqcup\!\sqcup\, l_3 l_4 &= l_1 l_2 l_3 l_4 + l_1 l_3 l_2 l_4 + l_3 l_1 l_2 l_4 + l_1 l_3 l_4 l_2 + l_3 l_1 l_4 l_2 + l_3 l_4 l_1 l_2. \end{aligned} \tag{8.46}$$

The shuffle algebra (with the shuffle product as product) is generated by the Lyndon words [197]. If one introduces a lexicographic ordering on the letters of the alphabet A, a **Lyndon word** is defined by the property $w < v$ for any sub-words u and v such that $w = uv$. To give an example, consider the alphabet $A = \{l_1, l_2\}$ with $l_1 < l_2$. The words

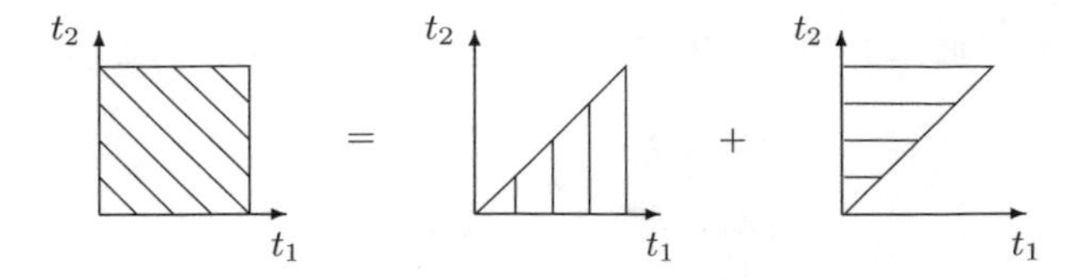

Fig. 8.1 Shuffle algebra from the integral representation: The shuffle product follows from replacing the integral over the square by an integral over the lower triangle and an integral over the upper triangle

$$w_1 = l_1 l_1 l_2, \quad w_2 = l_1 l_1 l_2 l_1 l_2 l_2 \tag{8.47}$$

are Lyndon words, while $w_3 = l_1 l_2 l_1$ is not. The word w_3 may be written as $w_3 = uv$ with $u = l_1 l_2$, $v = l_1$ and $v < w_3$.

Exercise 76 *Consider the alphabet $A = \{l_1, l_2\}$ with $l_1 < l_2$. Write down all Lyndon words of depth ≤ 3.*

Let us now make the connection to multiple polylogarithms $G(z_1, \ldots, z_r; y)$. We take the z_j's as letters. The alphabet A is given by the distinct z_j's. A multiple polylogarithm $G(z_1, \ldots, z_r; y)$ is therefore specified by a word $w = z_1 z_2 \ldots z_r$ (i.e. an ordered sequence) and a value y. The non-trivial statement is the shuffle product for multiple polylogarithms:

$$G(z_1, z_2, ..., z_k; y) \cdot G(z_{k+1}, ..., z_r; y) = \sum_{\text{shuffles } \sigma} G(z_{\sigma(1)}, z_{\sigma(2)}, ..., z_{\sigma(r)}; y), \tag{8.48}$$

where the sum runs over all permutations σ which are shuffles of $(1, \ldots, k)$ and $(k+1, \ldots, r)$, i.e. which preserve the relative order of $1, 2, ..., k$ and of $k+1, ..., r$. An simple example for the shuffle product of two multiple polylogarithms is given by

$$G(z_1; y) \cdot G(z_2; y) = G(z_1, z_2; y) + G(z_2, z_1; y). \tag{8.49}$$

The proof that the integral representation of the multiple polylogarithms fulfils the shuffle product formula in Eq. (8.48) is sketched for the example in Eq. (8.49) in Fig. 8.1 and can easily be extended to multiple polylogarithms of higher depth by recursively replacing the two outermost integrations by integrations over the upper and lower triangle.

It is clear that the proof does not depend on the specific form of the integrands of the iterated integral, only the iterated structure is relevant. This implies that the shuffle product is not specific to the iterated integral representation of multiple polylogarithms, but holds for any iterated integral of the form as in Eq. (6.140).

A non-trivial example for the shuffle product of two multiple polylogarithms is given by

$$G(z_1, z_2; y) \cdot G(z_3; y) = G(z_1, z_2, z_3; y) + G(z_1, z_3, z_2; y) + G(z_3, z_1, z_2; y). \tag{8.50}$$

For fixed y we may view the multiple polylogarithm $G(z_1, \dots, z_r; y)$ as a function

$$\begin{aligned} G \ : \ \mathbb{C}^r &\to \mathbb{C}, \\ (z_1, \dots, z_r) &\to G(z_1, \dots, z_r; y). \end{aligned} \tag{8.51}$$

By linearity this extends to a map from the vector space of words $\mathcal{A}$ to the complex numbers $\mathbb{C}$

$$G : \mathcal{A} \to \mathbb{C}, \tag{8.52}$$

e.g. for $w_1 = z_1 \dots z_k$, $w_2 = z_{k+1} \dots z_r$ and $c_1, c_2 \in \mathbb{C}$ the map is given by

$$\begin{aligned} G(c_1 w_1 + c_2 w_2) &= c_1 G(w_1) + c_2 G(w_2) \\ &= c_1 G(z_1, \dots, z_k; y) + c_2 G(z_{k+1}, \dots, z_r; y). \end{aligned} \tag{8.53}$$

Equation (8.48) says that this map is an algebra homomorphism, i.e.

$$G(w_1 \sqcup\!\sqcup\, w_2) = G(w_1) \cdot G(w_2). \tag{8.54}$$

We may use the shuffle product to remove trailing zeros: We say that a multiple polylogarithm of the form

$$G(z_1, \dots, z_j, \underbrace{0, \dots, 0}_{r-j}; y) \tag{8.55}$$

with $z_j \neq 0$ has $(r - j)$ trailing zeroes. Multiple polylogarithms with trailing zeroes do not have a Taylor expansion in y around $y = 0$, but logarithmic singularities at $y = 0$. In removing the trailing zeroes, one explicitly separates these logarithmic terms, such that the rest has a regular expansion around $y = 0$. The starting point is the shuffle relation

$$\begin{aligned} & G(0; y) G(z_1, \dots, z_j, \underbrace{0, \dots, 0}_{r-j-1}; y) = \\ & (r - j) G(z_1, \dots, z_j, \underbrace{0, \dots, 0}_{r-j}; y) + \sum_{(s_1 \dots s_j) = (z_1 \dots z_{j-1}) \sqcup\!\sqcup (0)} G(s_1, \dots, s_j, z_j, \underbrace{0, \dots, 0}_{r-j-1}; y). \end{aligned} \tag{8.56}$$

Solving this equation for $G(z_1, \dots, z_j, 0, \dots, 0; y)$ yields

$$G(z_1, \dots, z_j, \underbrace{0, \dots, 0}_{r-j}; y) = \tag{8.57}$$
$$\frac{1}{r-j} \left[G(0; y) G(z_1, \dots, z_j, \underbrace{0, \dots, 0}_{r-j-1}; y) \right.$$
$$\left. - \sum_{(s_1 \dots s_j) = (z_1 \dots z_{j-1}) ⧢ (0)} G(s_1, \dots, s_j, z_j, \underbrace{0, \dots, 0}_{r-j-1}; y) \right].$$

In the first term, one logarithm has been explicitly factored out:

$$G(0; y) = \ln y. \tag{8.58}$$

All remaining terms have at most $(r - j - 1)$ trailing zeroes. Using recursion, we may therefore eliminate all trailing zeroes. Let's consider an example: Let's assume $z_1 \neq 0$. We have

$$\begin{aligned} G(z_1, 0; y) &= G(0; y) G(z_1; y) - G(0, z_1; y) \\ &= \ln(y) G(z_1; y) - G(0, z_1; y). \end{aligned} \tag{8.59}$$

Both $G(z_1; y)$ and $G(0, z_1; y)$ are free of trailing zeros.

Exercise 77 *Express the product*

$$G_2(z; y) \cdot G_3(z; y) \tag{8.60}$$

as a linear combination of multiple polylogarithms.

8.4 The Quasi-Shuffle Product

In the previous section we have seen that the iterated integral representation induces the shuffle product for multiple polylogarithms. In this section we work out the analogy based on the nested sum representation for multiple polylogarithms. We will see that the nested sum representation induces a quasi-shuffle product. Again, it is not specific to multiple polylogarithms, but holds for any nested sum.

We start by considering a generalisation of shuffle algebras. Assume that on the alphabet A of letters we have an additional operation

$$\begin{aligned} \circ \ : \ & A \times A \to A, \\ & (l_1, l_2) \to l_1 \circ l_2, \end{aligned} \tag{8.61}$$

which is commutative and associative. Then we can define a new product $ш_q$ of words recursively through

$$l_1 l_2 ... l_k \; ш_q \; l_{k+1} ... l_r = l_1 \left(l_2 ... l_k \; ш_q \; l_{k+1} ... l_r \right) + l_{k+1} \left(l_1 l_2 ... l_k \; ш_q \; l_{k+2} ... l_r \right) \\ + (l_1 \circ l_{k+1}) \left(l_2 ... l_k \; ш_q \; l_{k+2} ... l_r \right) \quad (8.62)$$

together with

$$e \; ш_q \; w = w \; ш_q \; e = w. \quad (8.63)$$

This product is a generalisation of the shuffle product and differs from the recursive definition of the shuffle product in Eq. (8.43) through the extra term in the last line. This modified product is known under the names quasi-shuffle product [198], mixable shuffle product [199], stuffle product [190] or mould symmetrel [196]. This product turns the vector space of words into a quasi-shuffle algebra $\mathcal{A}_q$.

We have for example

$$l_1 l_2 \; ш_q \; l_3 = l_1 l_2 l_3 + l_1 l_3 l_2 + l_3 l_1 l_2 + l_1 l_{23} + l_{13} l_2, \quad (8.64)$$

with $l_{13} = l_1 \circ l_3$ and $l_{23} = l_2 \circ l_3$.

The quasi-shuffle algebra (with the quasi-shuffle product as product) is generated as an algebra by the Lyndon words [197]. This is not too surprising: We already know that the shuffle algebra is generated by the Lyndon words. Furthermore, the quasi-shuffle product differs from the shuffle product only by terms of lower depth.

Let us now make the connection to multiple polylogarithms $\text{Li}_{m_1 ... m_k}(x_1, \dots, x_k)$. As letters we now take pairs $l_j = (m_j, x_j)$. A multiple polylogarithms $\text{Li}_{m_1 ... m_k}(x_1, \dots, x_k)$ is uniquely specified by a word $w = l_1 l_2 \dots l_k$ in these letters. We define the additional operation in Eq. (8.61) by

$$(m_1, x_1) \circ (m_2, x_2) = (m_1 + m_2; x_1 x_2), \quad (8.65)$$

i.e. the first entries are added, while the second entries are multiplied. We may view the multiple polylogarithm $\text{Li}_{m_1 ... m_k}(x_1, \dots, x_k)$ as a function

$$\begin{aligned} \text{Li} \; : \; & \mathbb{N}^k \times \mathbb{C}^k \; \to \mathbb{C}, \\ & (m_1, \dots, m_k, x_1, \dots, x_k) \; \to \; \text{Li}_{m_1 ... m_k}(x_1, \dots, x_k). \end{aligned} \quad (8.66)$$

Again, we may extend this by linearity to a map from the vector space of words $\mathcal{A}_q$ to the complex numbers $\mathbb{C}$

$$\text{Li} : \mathcal{A}_q \; \to \; \mathbb{C}, \quad (8.67)$$

e.g. for $w_1 = l_1 \dots l_k$, $w_2 = l_{k+1} \dots l_r$, $l_j = (m_j, z_j)$ and $c_1, c_2 \in \mathbb{C}$ we have

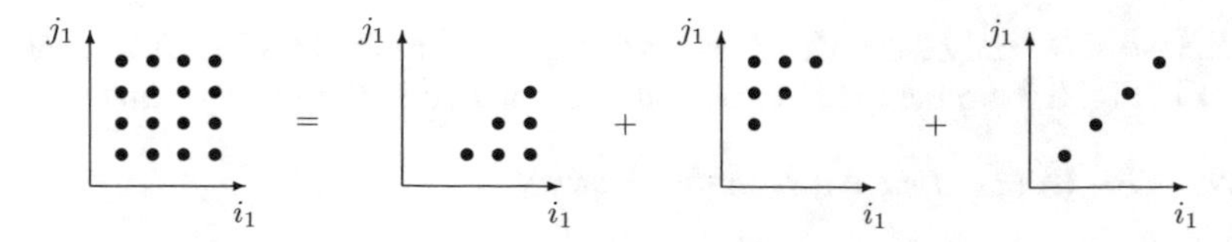

Fig. 8.2 Quasi-shuffle algebra from the sum representation: The quasi-shuffle product follows from replacing the sum over the square by a sum over the lower triangle, a sum over the upper triangle, and a sum over the diagonal

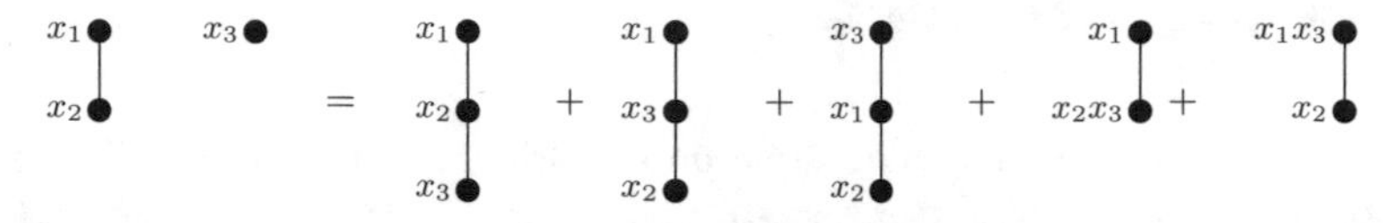

Fig. 8.3 Pictorial representation of the quasi-shuffle multiplication law. The first three terms on the right-hand side correspond to the ordinary shuffle product, whereas the two last terms are the additional "stuffle"-terms

$$\begin{aligned} \mathrm{Li}\,(c_1 w_1 + c_2 w_2) &= c_1 \mathrm{Li}\,(w_1) + c_2 \mathrm{Li}\,(w_2) \\ &= c_1 \mathrm{Li}_{m_1 \dots m_k}(x_1, \dots, x_k) + c_2 \mathrm{Li}_{m_{k+1} \dots m_r}(x_{k+1}, \dots, x_r). \end{aligned} \tag{8.68}$$

This map is again an algebra homomorphism, i.e.

$$\mathrm{Li}\left(w_1 ⧢_q w_2\right) = \mathrm{Li}\,(w_1) \cdot \mathrm{Li}\,(w_2)\,. \tag{8.69}$$

A simple example for the quasi-shuffle product is given by

$$\mathrm{Li}_{m_1}(x_1)\mathrm{Li}_{m_2}(x_2) = \mathrm{Li}_{m_1,m_2}(x_1, x_2) + \mathrm{Li}_{m_2,m_1}(x_2, x_1) + \mathrm{Li}_{m_1+m_2}(x_1 x_2). \tag{8.70}$$

The proof that the sum representation of the multiple polylogarithms fulfils the quasi-shuffle product formula in Eq. (8.69) is sketched for the example in Eq. (8.70) in Fig. 8.2 and can easily be extended to multiple polylogarithms of higher depth by recursively replacing the two outermost summations by summations over the upper triangle, the lower triangle, and the diagonal.

Let us provide one further example for the quasi-shuffle product. Working out the recursive definition of the quasi-shuffle product we obtain

$$\begin{aligned} &\mathrm{Li}_{m_1 m_2}(x_1, x_2) \cdot \mathrm{Li}_{m_3}(x_3) = \\ &= \mathrm{Li}_{m_1 m_2 m_3}(x_1, x_2, x_3) + \mathrm{Li}_{m_1 m_3 m_2}(x_1, x_3, x_2) + \mathrm{Li}_{m_3 m_1 m_2}(x_3, x_1, x_2) \\ &\quad + \mathrm{Li}_{m_1 (m_2+m_3)}(x_1, x_2 x_3) + \mathrm{Li}_{(m_1+m_3) m_2}(x_1 x_3, x_2) \end{aligned} \tag{8.71}$$

This is shown pictorially in Fig. 8.3. The first three terms correspond to the ordinary shuffle product, whereas the two last terms are the additional "stuffle"-terms. In

Fig. 8.3 we show only the x-variables, which are multiplied in the stuffle-terms. Not shown in Fig. 8.3 are the indices m_j, which are added in the stuffle-terms.

Exercise 78 *Work out the quasi-shuffle product*

$$\text{Li}_{m_1 m_2}(x_1, x_2) \cdot \text{Li}_{m_3 m_4}(x_3, x_4). \tag{8.72}$$

8.5 Double-Shuffle Relations

We recall that we may denote multiple polylogarithms either as $G(z_1, \dots, z_r; y)$ (the notation for the integral representation) or as $\text{Li}_{m_1 \dots m_k}(x_1, \dots, x_k)$ (the notation for the sum representation). The conversion between these two notations is given by Eqs. (8.19) and (8.20).

We have seen that shuffle product associated with the integral representation gives relations among the multiple polylogarithms. In the notation of Sect. 8.3 we have for fixed y and $w_1 = z_1 \dots z_k$, $w_2 = z_{k+1} \dots z_r$

$$G(w_1 \sqcup\!\sqcup w_2) = G(w_1) \cdot G(w_2). \tag{8.73}$$

At the same time the quasi-shuffle product associated with the sum representation provides another set of relations among the multiple polylogarithms. In the notation of Sect. 8.4 we have for $l_j = (m_j, x_j)$ and $w_1 = l_1 \dots l_k$, $w_2 = l_{k+1} \dots l_r$,

$$\text{Li}\left(w_1 \sqcup\!\sqcup_q w_2\right) = \text{Li}(w_1) \cdot \text{Li}(w_2). \tag{8.74}$$

The union of the relations given by Eqs. (8.73) and (8.74) are called the **double-shuffle relations**.

Multiple zeta values are special values of multiple polylogarithms and as such also have a sum representation and an integral representation:

$$\begin{aligned} \zeta_{m_1 \dots m_k} &= \text{Li}_{m_1 \dots m_k}(1, \dots, 1), \qquad\qquad m_1 \neq 1. \\ &= (-1)^k G_{m_1 \dots m_k}(1, \dots, 1; 1). \end{aligned} \tag{8.75}$$

Hence we have double-shuffle relations for multiple zeta values. Using these, we may for example derive

$$\zeta_{31} = \frac{1}{4}\zeta_4. \tag{8.76}$$

This follows easily from the shuffle relation

$$\zeta_2^2 = \left[-G(0, 1; 1)\right]^2 = 2G(0, 1, 0, 1; 1) + 4G(0, 0, 1, 1; 1) = 2\zeta_{22} + 4\zeta_{31} \tag{8.77}$$

and the quasi-shuffle relation

$$\zeta_2^2 = [\text{Li}_2(1)]^2 \;=\; 2\text{Li}_{22}(1,1) + \text{Li}_4(1) \;=\; 2\zeta_{22} + \zeta_4. \tag{8.78}$$

As a second example consider the relation

$$\zeta_{21} = \zeta_3. \tag{8.79}$$

This relation is due to Euler. We may derive this relation from the double-shuffle relations in a way similar to what we did above, but we have to be careful since the Riemann zeta function $\zeta(s)$ diverges at $s = 1$ (i.e. ζ_1 does not exist). To do it properly, we are going to use regularised (quasi-) shuffle relations. Let

$$L = -\ln\lambda \;=\; \text{Li}_1(1-\lambda) \;=\; -G(1; 1-\lambda). \tag{8.80}$$

L is well-defined for $\lambda > 0$, but diverges logarithmically for $\lambda \to 0$. From the quasi-shuffle product we have

$$L \cdot \zeta_2 = \text{Li}_1(1-\lambda) \cdot \text{Li}_2(1) \;=\; \text{Li}_{12}(1-\lambda, 1) + \text{Li}_{21}(1, 1-\lambda) + \text{Li}_3(1-\lambda). \tag{8.81}$$

For the shuffle product we consider

$$\begin{aligned} -L \cdot G(0,1;1-\lambda) &= G(1;1-\lambda) \cdot G(0,1;1-\lambda) \\ &= G(1,0,1;1-\lambda) + 2G(0,1,1;1-\lambda). \end{aligned} \tag{8.82}$$

Expressed in the Li-notation Eq. (8.82) reads

$$L \cdot \text{Li}_2(1-\lambda) = \text{Li}_{12}(1-\lambda, 1) + 2\text{Li}_{21}(1-\lambda, 1). \tag{8.83}$$

We now subtract Eq. (8.81) from Eq. (8.83). This yields

$$L \cdot [\text{Li}_2(1-\lambda) - \zeta_2] = 2\text{Li}_{21}(1-\lambda, 1) - \text{Li}_{21}(1, 1-\lambda) - \text{Li}_3(1-\lambda). \tag{8.84}$$

It is easy to see that $\text{Li}_2(1-\lambda) - \zeta_2 = O(\lambda)$ and therefore

$$\lim_{\lambda\to 0} \{-\ln\lambda \cdot [\text{Li}_2(1-\lambda) - \zeta_2]\} = 0. \tag{8.85}$$

On the right-hand side all terms are finite and we have

$$\lim_{\lambda\to 0} [2\text{Li}_{21}(1-\lambda, 1) - \text{Li}_{21}(1, 1-\lambda) - \text{Li}_3(1-\lambda)] = \zeta_{21} - \zeta_3, \tag{8.86}$$

yielding $\zeta_{21} = \zeta_3$. The relations in Eq. (8.81) or Eq. (8.82) are examples of regularised (quasi-) shuffle relations.

Exercise 79 *Use the (regularised) double-shuffle relations to show*

$$\zeta_2^2 = \frac{5}{2}\zeta_4. \tag{8.87}$$

From Eq. (5.38) we know that

$$\zeta_2 = \text{Li}_2(1) = \frac{\pi^2}{6}. \tag{8.88}$$

The above exercise shows that

$$\zeta_4 = \frac{\pi^4}{90}. \tag{8.89}$$

In general, the even zeta values are powers of π:

$$\zeta_n = -\frac{B_n}{2n!}(2\pi i)^n, \qquad n = 2, 4, 6, 8, \ldots, \tag{8.90}$$

where B_n are the Bernoulli numbers defined in Eq. (7.86).

A database of relations among multiple zeta values can be found at [200, 201].

8.6 Monodromy

In order to motivate the study of monodromies, let us first consider the logarithm $\ln(x)$ for a complex variable x. The logarithm is singular for $x = 0$, therefore we consider the logarithm on the punctured complex plane $\mathbb{C}\backslash\{0\}$. It is well-known that the logarithm is a multi-valued function on $\mathbb{C}\backslash\{0\}$. This is easily seen as follows: By the definition of the logarithm, $y = \ln(x)$ is a number, which fulfils

$$e^y = x. \tag{8.91}$$

Now let y be a number, which fulfils Eq. (8.91). Then

$$y + 2\pi i n, \ \ n \in \mathbb{Z} \tag{8.92}$$

fulfils Eq. (8.91) as well. We may turn the logarithm into a single-valued function by viewing the logarithm as a function on a covering space of $\mathbb{C}\backslash\{0\}$, or by restricting the logarithm to an open subset of $\mathbb{C}\backslash\{0\}$, for example $\mathbb{C}\backslash\mathbb{R}_{\leq 0}$. The restriction of the logarithm to an open subset U of $\mathbb{C}\backslash\{0\}$ such that $\ln(x)$ is single-valued on U is called a **branch of the logarithm**.

Let us now patch together branches of the logarithm, such that we may analytically continue $\ln(x)$ counter clockwise around $x_0 = 0$. After analytically continuing $\ln(x)$ around a small loop counter clockwise around $x_0 = 0$ we do not recover $\ln(x)$ but obtain $\ln(x) + 2\pi i$. This is called the **monodromy**.

In order to prepare for the discussion of the monodromy of the multiple polylogarithms, let us be more explicit. We consider the analytic continuation of $f(x) = \ln(x)$ around a small counter clockwise loop around $x_0 = 0$. We parametrise the loop by

$$x_\varepsilon(t) = x_0 + \varepsilon e^{2\pi i t}, \tag{8.93}$$

with $t \in [0, 1]$. We have

$$\ln(x_\varepsilon(1)) - \ln(x_\varepsilon(0)) = 2\pi i. \tag{8.94}$$

We denote by $\mathcal{M}_{x_0} f(x)$ the analytic continuation of $f(x)$ around x_0. Thus

$$\mathcal{M}_0 \ln(x) = \ln(x) + 2\pi i. \tag{8.95}$$

Let us now turn to the classical polylogarithms. We first consider

$$\text{Li}_1(x) = -\ln(1-x) \quad = \int\limits_0^x \frac{dt}{1-t}. \tag{8.96}$$

$\text{Li}_1(x)$ has a branch cut along the positive real axis starting at $x = 1$. Here we find

$$\mathcal{M}_1 \text{Li}_1(x) = \text{Li}_1(x) - 2\pi i. \tag{8.97}$$

The classical polylogarithms are given by

$$\text{Li}_n(x) = \sum_{j=1}^{\infty} \frac{x^j}{j^n} \quad = \int\limits_0^x \frac{dt}{t} \text{Li}_{n-1}(t). \tag{8.98}$$

$\text{Li}_n(x)$ is analytic at $x_0 = 0$, therefore

$$\mathcal{M}_0 \text{Li}_n(x) = \text{Li}_n(x). \tag{8.99}$$

For the monodromy around $x_1 = 1$ one finds

$$\mathcal{M}_1 \text{Li}_n(x) = \text{Li}_n(x) - 2\pi i \frac{\ln^{n-1}(x)}{(n-1)!}. \tag{8.100}$$

Equation (8.100) is proven by induction [202]: We may write

$$\text{Li}_n(x) = \int\limits_0^{1-\varepsilon} \frac{dt}{t} \text{Li}_{n-1}(t) + \int\limits_{1-\varepsilon}^{x} \frac{dt}{t} \text{Li}_{n-1}(t), \tag{8.101}$$

by splitting the integration path into a piece from 0 to $1-\varepsilon$, followed by second piece from $1-\varepsilon$ to x. The path does not encircle the point $x = 1$. Let's work out $\mathcal{M}_1 \mathrm{Li}_n(x)$. As $\mathrm{Li}_n(x)$ is given by an integral from 0 to x, we obtain $\mathcal{M}_1 \mathrm{Li}_n(x)$ by using an integration path which encircles the point $x = 1$. We may deform this path into a path, which we split into three pieces: A first piece from 0 to $1-\varepsilon$ as above, followed by second piece given by a small circle counter clockwise around $x = 1$ and finally a third piece from $1-\varepsilon$ to x. For the integrand of the third piece we have to use the formula after analytically continuing around a small loop around $x = 1$. This formula is given by the induction hypothesis. We obtain

$$\mathcal{M}_1 \mathrm{Li}_n(x) - \mathrm{Li}_n(x) = \lim_{\varepsilon\to 0}\left[\oint \frac{dx'}{x'} \mathrm{Li}_{n-1}(x') - \frac{2\pi i}{(n-2)!}\int\limits_{1-\varepsilon}^{x} \frac{dt}{t}\ln^{n-2}(t)\right]. \tag{8.102}$$

The first integral is around

$$x'(t) = 1 - \varepsilon e^{2\pi i t}, \qquad t \in [0,1], \tag{8.103}$$

and corresponds to the second piece (a small circle counter clockwise around $x = 1$) mentioned above. This integral vanishes for $\varepsilon \to 0$. For $n > 2$ this follows from the fact that $\mathrm{Li}_{n-1}(x)$ is bounded in a neighbourhood of $x = 1$. For $n = 2$ we have to consider

$$\lim_{\varepsilon\to 0}\oint \frac{dx'}{x'} \mathrm{Li}_1(x') = \lim_{\varepsilon\to 0}\left\{2\pi i\varepsilon\int\limits_0^1 dt \frac{e^{2\pi i t}}{1-\varepsilon e^{2\pi i t}}\left[\ln(\varepsilon) + 2\pi i t\right]\right\} = 0. \tag{8.104}$$

The second integral in Eq. (8.102) is along the path $1-\varepsilon$ to x and corresponds to the difference of the integrands

$$\mathcal{M}_1 \mathrm{Li}_{n-1}(x) - \mathrm{Li}_{n-1}(x). \tag{8.105}$$

We may use the induction hypothesis for this difference. With

$$\int\limits_{1-\varepsilon}^{x} \frac{dt}{t}\ln^{n-2}(t) = \left.\frac{\ln^{n-1}(t)}{n-1}\right|_{1-\varepsilon}^{x} \tag{8.106}$$

the claim follows:

$$\mathcal{M}_1 \mathrm{Li}_n(x) = \mathrm{Li}_n(x) - 2\pi i \frac{\ln^{n-1}(x)}{(n-1)!}. \tag{8.107}$$

Exercise 80 *Let*

$$f_0(x) = \frac{1}{r!}\ln^r(x), \quad f_1(x) = \frac{(-1)^r}{r!}\ln^r(1-x). \tag{8.108}$$

Determine

$$\mathcal{M}_0 f_0(x), \quad \mathcal{M}_0 f_1(x), \quad \mathcal{M}_1 f_0(x), \quad \mathcal{M}_1 f_1(x). \tag{8.109}$$

The monodromy of the multiple polylogarithms can be worked out along the same lines. This is most conveniently done by using the integral representation $G(z_1, \dots, z_r; y)$. We consider the case where we analytically continue y around a point z. We assume $(z_1, \dots, z_r) \neq (0, \dots, 0)$, as the case $(z_1, \dots, z_r) = (0, \dots, 0)$ follows from Exercise 80. We then have

$$\mathcal{M}_z G(z_1, \dots, z_r; y) - G(z_1, \dots, z_r; y) = \tag{8.110}$$
$$\lim_{\varepsilon \to 0} \left\{ \oint \frac{dy'}{y'-z_1} G(z_2, \dots, z_r; y') + \int\limits_{z+\varepsilon}^{y} \frac{dy'}{y'-z_1} \left[\mathcal{M}_z G(z_2, \dots, z_r; y') - G(z_2, \dots, z_r; y') \right] \right\}.$$

The first integral is around

$$y'(t) = z + \varepsilon e^{2\pi i t}, \qquad t \in [0, 1]. \tag{8.111}$$

For $z \neq z_1$ the first integral does not contribute, for the same reasons as above: For $z \neq z_1$ we obtain from the Jacobian an explicit prefactor ε, which in the limit $\varepsilon \to 0$ kills any logarithmic singularity which might arise from $G(z_2, \dots, z_r; y)$.

For $z = z_1$ the first integral equals

$$\oint \frac{dy'}{y' - z_1} G(z_2, \dots, z_r; y') = 2\pi i \int\limits_0^1 dt G\left(z_2, \dots, z_r; z_1 + \varepsilon^{2\pi i t}\right). \tag{8.112}$$

Equation (8.110) allows us to compute the monodromy of the multiple polylogarithms.

Exercise 81 *Compute the monodromy of* $G(1, 1; y)$ *around* $y = 1$.

8.7 The Drinfeld Associator

In this section we study the Knizhnik-Zamolodchikov Equation [203] and the Drinfeld associator [204]. Both topics are related to the first cyclotomic harmonic polylogarithms (i.e. multiple polylogarithms, where all z_j's are from the set $z_j \in \{0, 1\}$). It is common practice to use the notation

$$H(z_1, \dots, z_r; x) = (-1)^{n_1} G(z_1, \dots, z_r; x), \qquad z_j \in \{0, 1\}, \tag{8.113}$$

and n_1 is the number of times the value 1 occurs in the sequence $z_1, \dots, z_r$. We call the functions $H(z_1, \dots, z_r; x)$ harmonic polylogarithms for short. We also use for harmonic polylogarithms without trailing zeros the notation

$$H_{m_1 \ldots m_k}(x) = (-1)^k \, G_{m_1 \ldots m_k}(1, \ldots, 1; x). \tag{8.114}$$

If we write out $H_{m_1 \ldots m_k}(x)$ in the long form $H(z_1, \ldots, z_r; x)$, the sequence of z_j's so obtained contains exactly k times the letter 1. From Eq. (8.24) we have

$$H_{m_1 \ldots m_k}(x) = \text{Li}_{m_1 \ldots m_k}(x, \underbrace{1, ..., 1}_{k-1}). \tag{8.115}$$

We denote by $A = \{0, 1\}$ the alphabet corresponding to Eq. (8.113). As in Sect. 8.3 and Sect. 8.4 it is also convenient to denote alternatively harmonic polylogarithms with words. A word $w = l_1 l_2 ... l_r$ with letters from the alphabet $A = \{0, 1\}$ defines a harmonic polylogarithm as follows:

$$H(w; x) = H(l_1, \ldots, l_r; x). \tag{8.116}$$

In the discussion above we introduced the harmonic polylogarithms as a special cases of the multiple polylogarithms. We may also define them from scratch. We introduce two differential one-forms

$$\omega_0(x) = \frac{dx}{x}, \quad \omega_1(x) = \frac{dx}{1-x}. \tag{8.117}$$

A word $w = l_1 l_2 \ldots l_r$ defines a harmonic polylogarithm as follows: For the empty word e we set

$$H(e; x) = 1. \tag{8.118}$$

For a word consisting only of zeros ($l_1 = l_2 = \cdots = l_r = 0$) we set

$$H\left(0^r; x\right) = \frac{1}{r!} \ln^r(x). \tag{8.119}$$

For all other words we define

$$H(lw; x) = \int_0^x \omega_l(t) \, H(w; t). \tag{8.120}$$

Note that we have for all words w not of the form 0^r

$$\lim_{x \to 0} H(w; x) = 0. \tag{8.121}$$

Digression Harmonic polylogarithms up to weight 3
It is convenient to know explicit expressions of harmonic polylogarithms of low weight. We list here the explicit expressions for the harmonic polylogarithms up to weight 3. The tables can be found in [205]. At weight 1 we have

$$\begin{aligned} H(0;x) &= \ln(x), \\ H(1;x) &= -\ln(1-x). \end{aligned} \tag{8.122}$$

At weight 2 we have

$$\begin{aligned} H(0,0;x) &= \frac{1}{2}\ln^2(x), \\ H(0,1;x) &= \text{Li}_2(x), \\ H(1,0;x) &= -\ln(x)\ln(1-x) - \text{Li}_2(x), \\ H(1,1;x) &= \frac{1}{2}\ln^2(1-x). \end{aligned} \tag{8.123}$$

At weight 3 we have

$$\begin{aligned} H(0,0,0;x) &= \frac{1}{6}\ln^3(x), \\ H(0,0,1;x) &= \text{Li}_3(x), \\ H(0,1,0;x) &= \ln(x)\,\text{Li}_2(x) - 2\text{Li}_3(x), \\ H(0,1,1;x) &= \zeta_3 + \ln(1-x)\,\zeta_2 - \ln(1-x)\,\text{Li}_2(x) \\ &\quad -\frac{1}{2}\ln(x)\ln^2(1-x) - \text{Li}_3(1-x), \\ H(1,0,0;x) &= -\frac{1}{2}\ln^2(x)\ln(1-x) - \ln(x)\,\text{Li}_2(x) + \text{Li}_3(x), \\ H(1,0,1;x) &= -2\zeta_3 - 2\ln(1-x)\,\zeta_2 + \ln(1-x)\,\text{Li}_2(x) + \ln(x)\ln^2(1-x) \\ &\quad +2\text{Li}_3(1-x), \\ H(1,1,0;x) &= \zeta_3 + \ln(1-x)\,\zeta_2 - \text{Li}_3(1-x), \\ H(1,1,1;x) &= -\frac{1}{6}\ln^3(1-x). \end{aligned} \tag{8.124}$$

Let e_0 and e_1 be two non-commutative variables. (We may think of e_0 and e_1 as two generators of a Lie algebra or as two $(N \times N)$-matrices.) Strings of e_0 and e_1 are denoted by e_w, for example

$$e_{0101} = e_0 e_1 e_0 e_1. \tag{8.125}$$

With these preparations, we may now state the Knizhnik-Zamolodchikov equation.

The Knizhnik-Zamolodchikov equation:

$$\frac{d}{dx} L(x) = \left(\frac{e_0}{x} + \frac{e_1}{1-x} \right) L(x). \tag{8.126}$$

This equation is solved by

$$L(x) = \sum_w H(w; x)\, e_w, \tag{8.127}$$

where the sum runs over all words, which can be formed from the alphabet $A = \{0, 1\}$. The sum includes the empty word.

Proof: We have

$$L(x) = 1 + \sum_w H(0, w; x)\, e_{0w} + \sum_w H(1, w; x)\, e_{1w} \tag{8.128}$$

and therefore

$$\frac{d}{dx} L(x) = \sum_w \frac{H(w; x)}{x} e_{0w} + \sum_w \frac{H(w; x)}{1-x} e_{1w} = \left(\frac{e_0}{x} + \frac{e_1}{1-x} \right) \sum_w H(w; x)\, e_w. \tag{8.129}$$

Digression Relation to physics

The Knizhnik-Zamolodchikov equation is not too far away from physics. To see this, consider again the two-loop double box Feynman integral discussed as example 3 in Sect. 6.3.1. In Sect. 6.4.3 we have shown that a fibre transformation puts the differential equation into ε-form. Setting $x' = -x$ we obtain from Eq. (6.234):

$$\frac{d}{dx'} \mathbf{I}'(x) = \left(-\frac{\varepsilon C_0}{x'} + \frac{\varepsilon C_{-1}}{1-x'} \right) \mathbf{I}'(x), \tag{8.130}$$

where the (8×8)-matrices C_0 and C_{-1} have been given immediately after Eq. (6.234). Setting $e_0 = -\varepsilon C_0$ and $e_1 = \varepsilon C_{-1}$ gives the relation to Eq. (8.126). Note that with the choice $e_0 = -\varepsilon C_0$ and $e_1 = \varepsilon C_{-1}$ the solution $L(x)$ of Eq. (8.126) is a (8×8)-matrix, while $\mathbf{I}'$ is a vector of dimension 8. In order to obtain a vector-valued solution from the Knizhnik-Zamolodchikov equation, we may always multiply Eq. (8.126) by a constant vector $\mathbf{I}'_0$ from the right.

Let us return to the general case. $L(x)$ has for $x \to 0^+$ the asymptotic value

$$L(x) = e^{e_0 \ln x} + O\left(\sqrt{x}\right). \tag{8.131}$$

The exponential term $e^{e_0 \ln x}$ is related to the words consisting of zeros only:

$$e^{e_0 \ln x} = \sum_{n=0}^{\infty} \frac{\ln^n(x)}{n!} e_0^n = \sum_{n=0}^{\infty} H\left(0^n; x\right) e_{0^n}. \tag{8.132}$$

We further have

$$H\left(1^n; x\right) = (-1)^n \frac{\ln^n(1-x)}{n!} \tag{8.133}$$

and therefore

$$\sum_{n=0}^{\infty} H\left(1^n; x\right) e_{1^n} = e^{-e_1 \ln(1-x)}. \tag{8.134}$$

For the first terms of $L(x)$ we have

$$\begin{aligned} L(x) = {} & 1 + H(0; x) e_0 + H(1; x) e_1 \\ & + H(0, 0; x) e_0^2 + H(0, 1; x) e_0 e_1 + H(1, 0; x) e_1 e_0 + H(1, 1; x) e_1^2 + \ldots \end{aligned} \tag{8.135}$$

We define the regularised boundary values by

$$C_0 = \lim_{x \to 0} e^{-e_0 \ln x} L(x), \quad C_1 = \lim_{x \to 1} e^{e_1 \ln(1-x)} L(x). \tag{8.136}$$

The first few terms of C_0 and C_1 are

$$\begin{aligned} C_0 &= \lim_{x \to 0} \left\{1 + H(1; x) e_1 + H(1, 0; x) [e_1, e_0] + H(1, 1; x) e_1^2 \right. \\ & \quad + H(1, 0, 0; x) (e_0 [e_0, e_1] + [e_1, e_0] e_0) + H(1, 0, 1; x) [e_1, e_0] e_1 \\ & \quad \left. + H(1, 1, 0; x) \left[e_1^2, e_0\right] + H(1, 1, 1; x) e_1^3 + \ldots \right\} \\ &= 1, \\ C_1 &= \lim_{x \to 1} \left\{1 + H(0; x) e_0 + H(0, 0; x) e_0^2 + H(0, 1; x) [e_0, e_1] \right. \\ & \quad + H(0, 0, 0; x) e_0^3 + H(0, 0, 1; x) \left[e_0^2, e_1\right] + H(0, 1, 0; x) [e_0, e_1] e_0 \\ & \quad \left. + H(0, 1, 1; x) ([e_0, e_1] e_1 + e_1 [e_1, e_0]) + \ldots \right\} \\ &= 1 + \zeta_2 [e_0, e_1] + \zeta_3 [e_0 - e_1, [e_0, e_1]] + \ldots \end{aligned} \tag{8.137}$$

$C_0 = 1$ follows from Eq. (8.121). C_1 and C_0 are related by the **Drinfeld associator**

$$C_1 = \Phi(e_0, e_1) C_0. \tag{8.138}$$

The associator is given by

$$\Phi(e_0, e_1) = \sum_{w} \zeta(w) e_w, \tag{8.139}$$

where $\zeta(w)$ denotes a multiple zeta value in the expanded notation. We set

$$\zeta(e) = 1, \quad \zeta(0) = 0, \quad \zeta(1) = 0. \tag{8.140}$$

The relation with the standard notation $\zeta_{n_1 \dots n_k}$ is given for $n_1 \geq 2$ by

$$\zeta\left(0^{n_1-1}, 1, \dots, 0^{n_k-1}, 1\right) = \zeta_{n_1 \dots n_k}. \tag{8.141}$$

Furthermore we have the shuffle relation

$$\zeta(w_1)\,\zeta(w_2) = \zeta(w_1 \sqcup\!\sqcup\, w_2). \tag{8.142}$$

Up to weight 3 we therefore have

$$\begin{aligned}
\Phi(e_0, e_1) &= 1 + \zeta(0,0)\,e_0^2 + \zeta(0,1)\,e_0 e_1 + \zeta(1,0)\,e_1 e_0 + \zeta(1,1)\,e_1 e_1 \\
&\quad + \zeta(0,0,0)\,e_0^3 + \zeta(0,0,1)\,e_0^2 e_1 + \zeta(0,1,0)\,e_0 e_1 e_0 + \zeta(0,1,1)\,e_0 e_1^2 \\
&\quad + \zeta(1,0,0)\,e_1 e_0^2 + \zeta(1,0,1)\,e_1 e_0 e_1 + \zeta(1,1,0)\,e_1^2 e_0 + \zeta(1,1,1)\,e_1^3 + \dots \\
&= 1 + \zeta_2\,[e_0, e_1] + \zeta_3\,[e_0 - e_1, [e_0, e_1]] + \dots
\end{aligned} \tag{8.143}$$

Here we used

$$\begin{aligned}
\zeta(0,0) = \frac{1}{2}\zeta(0)^2 &= 0, \\
\zeta(1,1) = \frac{1}{2}\zeta(1)^2 &= 0, \\
\zeta(1,0) = \zeta(0)\zeta(1) - \zeta(0,1) &= -\zeta(0,1),
\end{aligned} \tag{8.144}$$

and similar relations at weight 3.

With the help of the Drinfeld associator we may give the monodromy of $L(x)$ (and hence the monodromy of the harmonic polylogarithms) in a compact form as

$$\begin{aligned}
\mathcal{M}_0 L(x) &= L(x)\, e^{2\pi i e_0}, \\
\mathcal{M}_1 L(x) &= L(x)\, \Phi^{-1} e^{-2\pi i e_1}\, \Phi.
\end{aligned} \tag{8.145}$$

With

$$\Phi^{-1} = 1 - \zeta_2\,[e_0, e_1] - \zeta_3\,[e_0 - e_1, [e_0, e_1]] + \dots \tag{8.146}$$

we find

$$\begin{aligned}
\mathcal{M}_1 L(x) &= 1 + H(0; x)\,e_0 + [H(1; x) - 2\pi i]\,e_1 + H(0,0; x)\,e_0^2 \\
&\quad + [H(0,1; x) - 2\pi i H(0; x)]\,e_0 e_1 \\
&\quad + H(1,0; x)\,e_1 e_0 + \left[H(1,1; x) - 2\pi i H(1; x) + \frac{1}{2}(2\pi i)^2\right] e_1^2 + \dots
\end{aligned} \tag{8.147}$$

Taking the coefficient of a particular word in e_0 and e_1 on the left-hand side and on the right-hand side, we obtain the monodromy of the harmonic polylogarithms. For example, taking the coefficient of e_1^2 we find

$$\mathcal{M}_1 H(1,1;x) = H(1,1;x) - 2\pi i H(1;x) + \frac{1}{2}(2\pi i)^2, \qquad (8.148)$$

in agreement with Exercise 81.

8.8 Fibration Bases

Let's look at

$$\begin{aligned} f_1(x) &= G(1,0;x), \\ f_2(x) &= -G(0,1;x) + \ln(x)\ln(1-x) \end{aligned} \qquad (8.149)$$

and

$$\begin{aligned} g_1(x,y) &= G(0,0;1-x) - G(0;1-x)\,G(0;1-y) + G(0,0;1-y) \\ &\quad - G(0,1;x) - G(0,1;y), \\ g_2(x,y) &= G(0,1-x;xy-x) + G(0,1-y;xy-y) - G(0,1;xy). \end{aligned} \qquad (8.150)$$

A priori it is not obvious that $f_1(x) = f_2(x)$ and $g_1(x,y) = g_2(x,y)$. An attentive reader might notice, that a proof of $f_1(x) = f_2(x)$ can be reduced to a shuffle relation, while a proof of $g_1(x,y) = g_2(x,y)$ can be reduced to the five-term relation for the dilogarithm of Eq. (5.40). Here we are interested in the more general situation: Given two expressions in multiple polylogarithms, can be prove or disprove that they are equal?

In order to prove $f_1(x) = f_2(x)$ we may use the fact that two functions of a variable x are equal, if their derivatives with respect to x are equal and the two functions agree at one point. Thus instead of showing $f_1(x) = f_2(x)$ we may show

$$f_1'(x) = f_2'(x) \text{ and } f_1(0) = f_2(0). \qquad (8.151)$$

This is simpler, as the derivatives are of lower weight

$$f_1'(x) = \frac{\ln(x)}{x-1}, \quad f_2'(x) = -\frac{\ln(1-x)}{x} + \frac{\ln(1-x)}{x} + \frac{\ln(x)}{x-1} = \frac{\ln(x)}{x-1} \qquad (8.152)$$

and the equation $f_1(0) = f_2(0)$ has one variable less. For the case at hand we have

$$f_1(0) = f_2(0) = 0, \qquad (8.153)$$

where we used

$$\lim_{x \to 0} (x \cdot \ln(x)) = 0. \tag{8.154}$$

If $f_1(x)$ and $f_2(x)$ are of higher weight we may iterate this process.

Once we have the derivatives, we may integrate back:

$$f_j(x) = f_j(0) + \int_0^x d\tilde{x} f_j'(\tilde{x}), \qquad j \in \{1, 2\}. \tag{8.155}$$

This puts $f_j(x)$ into a standardised form:

$$f_1(x) = G(1, 0; x), \quad f_2(x) = G(1, 0; x). \tag{8.156}$$

A comparison of $f_1(x)$ and $f_2(x)$ is now straightforward.

We may tackle the proof of $g_1(x, y) = g_2(x, y)$ in a similar way: We show for example

$$\frac{\partial}{\partial y} g_1(x, y) = \frac{\partial}{\partial y} g_2(x, y) \tag{8.157}$$

and

$$g_1(x, 0) = g_2(x, 0). \tag{8.158}$$

Note that we have to show $g_1(x, 0) = g_2(x, 0)$ for all points on the line $y = 0$. For the case at hand we find

$$g_1(x, y) = g_2(x, y) = \frac{1}{y-1}[G(1; y) - G(1; x)] - \frac{1}{y} G(1; y) \tag{8.159}$$

and

$$g_1(x, 0) = g_2(x, 0) = -G(0, 1; x) + G(1, 1; x). \tag{8.160}$$

Integrating back gives

$$g_j(x, y) = g_j(x, 0) + \int_0^y d\tilde{y} \left(\frac{\partial}{\partial \tilde{y}} g_j(x, \tilde{y}) \right), \qquad j \in \{1, 2\}. \tag{8.161}$$

Thus

$$g_j(x, y) = -G(0, 1; x) + G(1, 1; x) - G(1; x) G(1; y) + G(1, 1; y) - G(0, 1; y) \tag{8.162}$$

and a comparison of $g_1(x, y)$ and $g_2(x, y)$ is now straightforward.

Let us now formulate this in generality [206]:

Fibration basis:
We consider n variables $x_1, \dots, x_n$ and let $A = \{l_1, l_2, \dots\}$ be a set of rational functions in the variables $x_1, \dots, x_n$. We call A an alphabet and denote words by $w = l_1 l_2 \dots l_r$. We further denote a multiple polylogarithm by $G(w; z) = G(l_1, \dots, l_r; z)$. Consider now $G(w; 1)$. We may write $G(w; 1)$ as

$$G(w; 1) = \sum_j c_j \, G(w_{1,j}; x_1) \dots G(w_{n-1,j}; x_{n-1}) \, G(w_{n,j}; x_n), \quad (8.163)$$

where $w_{i,j}$ is a word from an alphabet A_i. The letters in the alphabet A_i are algebraic functions in the variables $x_1, \dots, x_{i-1}$. The important point here is that the letters in A_i no longer depend on the variables $x_i, \dots, x_n$. The c_j are constants with respect to $x_1, \dots, x_n$.
A multiple polylogarithm written as in the right-hand side of Eq. (8.163) is said to be expressed in the fibration basis with respect to the order $[x_1, \dots, x_n]$. The expression on the right-hand side of Eq. (8.163) depends on the order of the variables $x_1, \dots, x_n$.

Please note that although we start with the alphabet A with letters which are rational functions of the variables $x_1, \dots, x_n$, the letters of the alphabets A_i are in general algebraic functions of the variables $x_1, \dots, x_{i-1}$. This can be seen from the following simple example, where the algebraic function $\sqrt{x_1}$ appears. We assume $x_1, x_2 > 0$ and $x_2^2 > x_1$.

$$\begin{aligned} G\left(\frac{x_2^2}{x_1}; 1\right) &= \ln\left(\frac{x_2^2 - x_1}{x_2^2}\right) \\ &= G\left(\sqrt{x_1}; x_2\right) + G\left(-\sqrt{x_1}; x_2\right) - 2G(0; x_2) + G(0; x_1) - i\pi. \quad (8.164) \end{aligned}$$

This example also shows a second important point: For $x_1, x_2 > 0$ and $x_2^2 > x_1$ the function $G(x_2^2/x_1; 1)$ gives a real number and there are no singularities on the integration path from 0 to 1. On the other hand we have $\sqrt{x_1} < x_2$ and there is for the function $G(\sqrt{x_1}; x_2)$ a singularity on the integration path, resulting in a branch cut on the real x_2-axis starting at $x_2 = \sqrt{x_1}$. We have to specify how this singularity is avoided. A standard mathematical convention is that a function with a branch cut starting at a finite point and extending to infinity is taken to be continuous as the cut is approached coming around the finite endpoint of the cut in a counter clockwise direction [207]. For the case at hand this amounts to deforming the integration contour into the lower complex plane. Therefore, the function $G(\sqrt{x_1}; x_2)$ has an imaginary part for $x_2 > \sqrt{x_1}$. This imaginary part is compensated by the term $(-i\pi)$ and the full result is real.

8.9 Linearly Reducible Feynman Integrals

In this section we study an algorithm, which allows us for a special class of Feynman integrals to perform all integrations in the Feynman parameter representation [50, 208]. The class of Feynman integrals we would like to consider must satisfy two conditions:

1. The integrand of the Feynman parameter representation is integrable in an integer dimension D_{int}.
2. The integrand is linearly reducible for at least one ordering σ of the Feynman parameters $a_{\sigma_1}, \dots, a_{\sigma_{n_{\text{int}}}}$.

The first condition allows us to expand the integrand in the dimensional regularisation parameter ε:

$$[\mathcal{U}(a)]^{\nu-\frac{(l+1)D}{2}} = [\mathcal{U}(a)]^{\nu-\frac{(l+1)D_{\text{int}}}{2}} \sum_{j=0}^{\infty} \frac{(l+1)^j \varepsilon^j}{j!} [\ln(\mathcal{U})]^j,$$
$$[\mathcal{F}(a)]^{\frac{lD}{2}-\nu} = [\mathcal{F}(a)]^{\frac{lD_{\text{int}}}{2}-\nu} \sum_{j=0}^{\infty} \frac{(-l)^j \varepsilon^j}{j!} [\ln(\mathcal{F})]^j. \tag{8.165}$$

The second condition will be explained below.

Let's consider the order $a_1, \dots, a_{n_{\text{int}}}$, corresponding to the case where we first integrate over $a_{n_{\text{int}}}$, then $a_{(n_{\text{int}}-1)}$ until a_1. It is convenient to use the Cheng-Wu theorem with the delta distribution $\delta(1-a_{n_{\text{int}}})$. The Feynman integral we are interested in is

$$I = \frac{e^{l\varepsilon\gamma_E}\Gamma\left(\nu-\frac{lD}{2}\right)}{\prod\limits_{j=1}^{n_{\text{int}}}\Gamma(\nu_j)} \int\limits_0^\infty da_1 \dots \int\limits_0^\infty da_{n_{\text{int}}}\, \delta\left(1-a_{n_{\text{int}}}\right) \cdot R \cdot G. \tag{8.166}$$

where

$$R = \left(\prod_{j=1}^{n_{\text{int}}} a_j^{\nu_j-1}\right) \frac{[\mathcal{U}(a)]^{\nu-\frac{(l+1)D_{\text{int}}}{2}}}{[\mathcal{F}(a)]^{\nu-\frac{lD_{\text{int}}}{2}}},$$
$$G = \left[\sum_{j_1=0}^{\infty} \frac{(l+1)^{j_1} \varepsilon^{j_1}}{j_1!} [\ln(\mathcal{U})]^{j_1}\right] \left[\sum_{j_2=0}^{\infty} \frac{(-l)^{j_2} \varepsilon^{j_2}}{j_2!} [\ln(\mathcal{F})]^{j_2}\right]. \tag{8.167}$$

R is a rational function in the Feynman parameters, while the function G contains logarithms. The function G has an ε-expansion and we may consider each term in the ε-expansion separately.

Let's consider the integration order $a_1, \dots, a_{n_{\text{int}}}$, i.e. we integrate over $a_{n_{\text{int}}}$ first and a_1 last. The integrand for the integration over a_j is then a function of $a_1, \dots, a_j$.

At this stage the variables $a_{j+1}, \dots, a_{n_{\mathrm{int}}}$ have already been integrated out. We say that the integrand is **linearly reducible at stage** j, if the integrand at stage j can be written as a sum of terms, where each term is a product of a rational function and a multiple polylogarithm subject to the conditions that the denominator of the rational function factorises into linear factors with respect to a_j and the multiple polylogarithm can be cast into a form, where a_j appears only as upper integration limit and nowhere else. We say that the integrand is **linearly reducible for the order** $1, \dots, n_{\mathrm{int}}$, if it is linearly reducible at all stages j.

Note that also the condition on the multiple polylogarithm is non-trivial: If the letters of the multiple polylogarithm depend algebraically on a_j, but not rationally, Eq. (8.163) does not apply.

The two conditions are tailored such that all integrations may be performed within the class of multiple polylogarithms. The essential integration is

$$\int_0^\infty \frac{da_j}{a_j - l_1} G\left(l_2, \dots, l_r; a_j\right) = G\left(l_1, l_2, \dots, l_r; a_j\right)\Big|_0^\infty, \tag{8.168}$$

where $l_1, l_2, \dots, l_r$ may depend on $a_1, \dots, a_{j-1}$, but not on a_j. The condition on the rational function ensures that we may use partial fractioning:

$$R = P + \sum_i \sum_n \frac{c_{i,n}}{\left(a_j - l_i\right)^n}, \tag{8.169}$$

where P is a polynomial in a_j. Any terms, which are not simple poles can be reduced by partial integration:

$$\begin{aligned}
\int_0^\Lambda da_j\, a_j^n G\left(w; a_j\right) &= \frac{1}{n+1}\left[a_j^{n+1} G\left(w; a_j\right)\Big|_0^\Lambda - \int_0^\Lambda da_j\, a_j^{n+1} \frac{\partial}{\partial a_j} G\left(w; a_j\right)\right], \\
\int_0^\Lambda da_j \frac{G\left(w; a_j\right)}{\left(a_j - l\right)^{n+1}} &= -\frac{1}{n}\left[\frac{G\left(w; a_j\right)}{\left(a_j - l\right)^n}\Bigg|_0^\Lambda - \int_0^\Lambda \frac{da_j}{\left(a_j - l\right)^n} \frac{\partial}{\partial a_j} G\left(w; a_j\right)\right].
\end{aligned} \tag{8.170}$$

Let's look at an example: We consider the graph shown in Fig. 3.3 with vanishing internal masses. We are interested in I_{11111} in $D = 4$ space-time dimensions. This integral is finite and we may calculate it without regularisation. We choose the integration order a_3, a_4, a_5, a_1, a_2. We have

$$I_{11111}\left(4, \frac{-p^2}{\mu^2}\right) = \int_{a_j \geq 0} d^5a \frac{\delta\left(1 - a_2\right)}{\mathcal{U}\mathcal{F}} \tag{8.171}$$

with

$$\mathcal{U} = (a_1 + a_4)(a_3 + a_5) + (a_1 + a_3 + a_4 + a_5)a_2,$$
$$\mathcal{F} = [(a_1 + a_5)(a_3 + a_4)a_2 + a_1a_4(a_3 + a_5) + a_3a_5(a_1 + a_4)] \left(\frac{-p^2}{\mu^2}\right). \quad (8.172)$$

We set $\mu^2 = -p^2$. Thus we have

$$I_{11111}(4, 1) = \int\limits_0^\infty da_3 \int\limits_0^\infty da_4 \int\limits_0^\infty da_5 \int\limits_0^\infty da_1 \frac{1}{\mathcal{U}_1\mathcal{F}_1}$$
$$\mathcal{U}_1 = (a_1 + a_4)(a_3 + a_5) + a_1 + a_3 + a_4 + a_5,$$
$$\mathcal{F}_1 = (a_1 + a_5)(a_3 + a_4) + a_1a_4(a_3 + a_5) + a_3a_5(a_1 + a_4). \quad (8.173)$$

Partial fractioning with respect to a_1 and integration in a_1 yields

$$\int\limits_0^\infty da_1 \frac{1}{\mathcal{U}_1\mathcal{F}_1} =$$
$$\frac{1}{[a_3 + a_4 + (a_3 + a_5)\,a_4]^2} [\ln(a_3 + a_4 + a_5 + a_3a_4 + a_4a_5)$$
$$+ \ln(a_3 + a_4 + a_3a_4 + a_3a_5 + a_4a_5) - \ln(1 + a_3 + a_5)$$
$$- \ln(a_3 + a_4 + a_3a_4) - \ln(a_5)]. \quad (8.174)$$

We then continue with the integration in a_5, followed by the integration in a_4 and the final integration in a_3. The final result is

$$I_{11111}(4, 1) = 6\zeta_3. \quad (8.175)$$

It might seem that the integration algorithm for linearly reducible Feynman integrals applies only to a very narrow set of Feynman integrals (the ones satisfying the two conditions mentioned at the beginning of this section). However, this set is larger than one might naively expect. Let's focus on the first non-trivial integration (in the example above this corresponds to the integration in a_1). Consider for $j \neq 2$ the partial fraction decomposition in a_j of $1/(\mathcal{U}(G)\mathcal{F}_0(G))$ (the fact that we set one Feynman parameter to one does not affect the argument). For $\mathcal{U}(G)$ and $\mathcal{F}_0(G)$ we may use the recursion formulae from Eq. (3.58). We thus have

$$\frac{1}{(\mathcal{U}(G/e_j)+\mathcal{U}(G-e_j)a_j)(\mathcal{F}_0(G/e_j)+\mathcal{F}_0(G-e_j)a_j)} =$$
$$\frac{1}{\mathcal{U}(G-e_j)\mathcal{F}_0(G/e_j)-\mathcal{U}(G/e_j)\mathcal{F}_0(G-e_j)} \left[\frac{\mathcal{U}(G-e_j)}{\mathcal{U}(G/e_j)+\mathcal{U}(G-e_j)a_j}\right.$$
$$\left.- \frac{\mathcal{F}_0(G-e_j)}{\mathcal{F}_0(G/e_j)+\mathcal{F}_0(G-e_j)a_j}\right]. \quad (8.176)$$

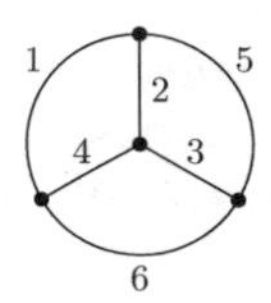

Fig. 8.4 The graph $\tilde{G}$ obtained from the two-loop two-point function by closing the two external edges

$\mathcal{U}(G/e_j), \mathcal{U}(G-e_j), \mathcal{F}_0(G/e_j)$ and $\mathcal{F}_0(G-e_j)$ are linear in the remaining Feynman parameters. We then expect that

$$\mathcal{U}(G-e_j)\mathcal{F}_0(G/e_j) - \mathcal{U}(G/e_j)\mathcal{F}_0(G-e_j) \tag{8.177}$$

is quadratic in the remaining Feynman parameters. However, in the example above we saw that this combination factorises as

$$[a_3 + a_4 + (a_3 + a_5)\, a_4]^2 \tag{8.178}$$

and each factor is again linear in each of the remaining integration variables. This is no accident: Consider the graph $\tilde{G}$ obtained from G by closing the two external edges. This gives a three-loop vacuum graph with six edges, as shown in Fig. 8.4. It is not too difficult to show that for $\mu^2 = -p^2$ and vanishing internal masses

$$\mathcal{U}\left(\tilde{G}\right) = \mathcal{F}_0(G) + a_6\mathcal{U}(G). \tag{8.179}$$

Hence

$$\mathcal{U}(G) = \mathcal{U}\left(\tilde{G} - e_6\right), \quad \mathcal{F}_0(G) = \mathcal{U}\left(\tilde{G}/e_6\right) \tag{8.180}$$

and

$$\begin{aligned} &\mathcal{U}(G-e_j)\mathcal{F}_0(G/e_j) - \mathcal{U}(G/e_j)\mathcal{F}_0(G-e_j) = \\ &\mathcal{U}\left(\tilde{G} - e_j - e_6\right)\mathcal{U}\left(\tilde{G}/e_j/e_6\right) - \mathcal{U}\left(\tilde{G}/e_j - e_6\right)\mathcal{U}\left(\tilde{G}/e_6 - e_j\right). \end{aligned} \tag{8.181}$$

For $j \neq 2$ the edges e_j and e_6 share one common vertex and the factorisation of the expression follows then from Dodgson's identity Eq. (3.77):

$$\mathcal{U}(G-e_j)\mathcal{F}_0(G/e_j) - \mathcal{U}(G/e_j)\mathcal{F}_0(G-e_j) = \left(\frac{\Delta_1}{a_j a_6}\right)^2, \tag{8.182}$$

where Δ_1 is defined in Eq. (3.78).

The computer program `HyperInt` implements the algorithm for linearly reducible Feynman integrals [208]. Extensions of the algorithm are discussed in Hidding and Moriello [209], Bourjaily [210], Bourjaily et al. [211].

Chapter 9
Nested Sums

In Chaps. 6 and 7 we developed the method of differential equations for the computation of Feynman integrals. An essential part of this method was the reduction to master integrals. Only the master integrals need to be calculated, all other Feynman integrals may expressed as a linear combination of the master integrals.

However, the reduction to master integrals through integration-by-parts identities is very often the most CPU-time consuming part of an calculation and we are interested in alternatives. Turning the argument around that any Feynman integral from a family of Feynman integrals can be written as a linear combination of master integrals, shows that all Feynman integrals in this family will involve the same final functions (namely the union of all functions appearing in the master integrals). We now look for algorithms for the computation of Feynman integrals, which can be applied to all members of a family of Feynman integrals and give the result directly, bypassing the need for a reduction to master integrals.

Furthermore, there are Feynman integrals which do not depend on any kinematic variable (e.g. $N_B = 0$). A non-trivial example is given by the massless two-loop two-point function discussed as an example in Sect. 8.9. These Feynman integrals cannot be treated directly with the method of differential equations.

Often it is possible to express a Feynman integral with arbitrary values of the space-time dimension D and the powers of the propagators $\nu_1, \dots, \nu_{n_{\mathrm{int}}}$ in terms of generalisations of hypergeometric functions. In order to arrive at such a representation one may use the Mellin-Barnes representation, close the integration contour and sum up the residues as discussed in Sect. 2.5.6. In this way one obtains the sum representation of a transcendental function.

For a particular member of the family of Feynman integrals we then specialise the indices $\nu_1, \dots, \nu_{n_{\mathrm{int}}}$ to the desired integers and D to $D_{\mathrm{int}} - 2\varepsilon$. We then have to compute the Laurent expansion in ε. We will discuss algorithms for this task in Sect. 9.1.

There are some well-known generalisations of hypergeometric functions: Appell functions, Lauricella functions, Horn functions and $\mathcal{A}$-hypergeometric functions

S. Weinzierl, *Feynman Integrals*, UNITEXT for Physics,
https://doi.org/10.1007/978-3-030-99558-4_9

(also known as GKZ hypergeometric functions), etc. We collect useful information on the first three classes of functions in appendix C. In Sect. 9.2 we discuss $\mathcal{A}$-hypergeometric functions. $\mathcal{A}$-hypergeometric functions are defined as solutions of a system of partial differential equations, known as a GKZ system. We may view any Feynman integral as a special case of an $\mathcal{A}$-hypergeometric function. At the same time, a GKZ system is holonomic, i.e. the solution space is finite dimensional.

9.1 Expansion of Special Transcendental Functions

Let us start from two examples to motivate the content of this section. The first example is the one-loop triangle graph shown in Fig. 2.8. We are now interested in the case where all internal masses are zero and for the kinematic configuration $p_2^2 = 0$, but $p_1^2 \neq 0$ and $p_3^2 \neq 0$. The integral depends then on one kinematic variable, which we take as $x = p_1^2/p_3^2$. From the Feynman parameter representation we obtain with $\mu^2 = -p_3^2$, $\nu_{ij} = \nu_i + \nu_j$ and $\nu_{ijk} = \nu_i + \nu_j + \nu_k$

$$\begin{aligned} I_{\nu_1\nu_2\nu_3} &= e^{\varepsilon\gamma_E} \int \frac{d^D k}{i\pi^{\frac{D}{2}}} \frac{1}{(-q_1^2)^{\nu_1}} \frac{1}{(-q_2^2)^{\nu_2}} \frac{1}{(-q_3^2)^{\nu_3}} \\ &= e^{\varepsilon\gamma_E} \frac{\Gamma(\nu_{123} - \frac{D}{2})}{\Gamma(\nu_1)\Gamma(\nu_2)\Gamma(\nu_3)} \int\limits_0^1 da\, a^{\nu_2 - 1} (1-a)^{\nu_3 - 1} \\ &\quad \times \int\limits_0^1 db\, b^{\frac{D}{2} - \nu_{23} - 1} (1-b)^{\frac{D}{2} - \nu_1 - 1} \left[1 - a(1-x)\right]^{\frac{D}{2} - \nu_{123}} \\ &= e^{\varepsilon\gamma_E} \frac{\Gamma(\frac{D}{2} - \nu_1)\Gamma(\frac{D}{2} - \nu_{23})}{\Gamma(\nu_1)\Gamma(\nu_2)\Gamma(D - \nu_{123})} \sum_{n=0}^{\infty} \frac{\Gamma(n+\nu_2)\Gamma(n - \frac{D}{2} + \nu_{123})}{\Gamma(n+1)\Gamma(n+\nu_{23})} (1-x)^n . \end{aligned} \quad (9.1)$$

As a second example we consider the two-loop two-point graph already discussed in Sect. 8.9 and shown in Fig. 3.3 with vanishing internal masses and $\mu^2 = -p^2$. Using the Mellin-Barnes representation one arrives at the following representation: To present the result after all residues have been taken in a compact form, we introduce two functions $F_\pm$ with ten arguments each:

$$\begin{aligned} F_\pm(a_1, a_2, a_3, a_4; b_1, b_2, b_3; c_1, c_2, c_3) &= \sum_{n=0}^{\infty} \sum_{j=0}^{\infty} \frac{(-1)^{n+j}}{n!j!} \\ \frac{\Gamma(\mp n - j - a_1)\Gamma(\pm n + j + a_2)\Gamma(\pm n + j + a_3)}{\Gamma(\pm n + j + a_4)} & \frac{\Gamma(\mp n \mp b_1)\Gamma(n + b_2)}{\Gamma(\mp n \mp b_3)} \frac{\Gamma(-j - c_1)\Gamma(j + c_2)}{\Gamma(-j - c_3)}, \end{aligned} \quad (9.2)$$

together with two operators $\mathcal{L}_d$ and $\mathcal{R}_d$ acting on the arguments as follows:

$$\begin{aligned} & \mathcal{L}_d F_\pm(a_1, a_2, a_3, a_4; b_1, b_2, b_3; c_1, c_2, c_3) = \\ & \quad F_\pm(a_1 + d, a_2 + d, a_3 + d, a_4 + d; b_1 + 2d, b_2 + d, b_3 + d; c_1, c_2, c_3), \end{aligned}$$

$$\begin{aligned}
&\mathcal{R}_d F_\pm(a_1, a_2, a_3, a_4; b_1, b_2, b_3; c_1, c_2, c_3) = \\
&\quad F_\pm(a_1 + d, a_2 + d, a_3 + d, a_4 + d; b_1, b_2, b_3; c_1 + 2d, c_2 + d, c_3 + d). \quad (9.3)
\end{aligned}$$

Then [212]

$$\begin{aligned}
I_{\nu_1\nu_2\nu_3\nu_4\nu_5} &= c\left(\mathbf{1} + \mathcal{L}_{\frac{D}{2}-\nu_{23}} + \mathcal{R}_{\frac{D}{2}-\nu_{25}} + \mathcal{L}_{\frac{D}{2}-\nu_{23}}\mathcal{R}_{\frac{D}{2}-\nu_{25}}\right) \quad (9.4)\\
&F_+\left(\frac{D}{2} - \nu_{14}, \nu_{235} - \frac{D}{2}, \nu_2, D - \nu_{14}; \nu_{23} - \frac{D}{2}, \frac{D}{2} - \nu_4, -\nu_4; \nu_{25} - \frac{D}{2}, \frac{D}{2} - \nu_1, -\nu_1\right)\\
&+c\left(\mathbf{1} + \mathcal{R}_{\frac{D}{2}-\nu_{25}}\right)\\
&F_-\left(-\nu_1, \frac{D}{2} - \nu_{14}, D - \nu_{1234}, \frac{D}{2} - \nu_1; \nu_{12345} - D, \nu_{124} - \frac{D}{2}, \frac{D}{2}; \nu_{25} - \frac{D}{2}, \frac{D}{2} - \nu_1, -\nu_1\right).
\end{aligned}$$

Here, **1** denotes the identity operator with a trivial action on the arguments of the functions $F_\pm$ and the prefactor c is given by

$$c = \frac{e^{2\varepsilon\gamma_E}}{\Gamma(\nu_2)\Gamma(\nu_3)\Gamma(\nu_5)\Gamma(D - \nu_{235})}. \quad (9.5)$$

In both example we obtained for arbitrary indices $\nu_1, \nu_2, \dots$ (multiple) sums. The task is then to expand all terms in the dimensional regularisation parameter ε and to re-express the resulting multiple sums in terms of known functions. If this can be done we bypass integration-by-parts reduction. It depends on the form of the multiple sums if this can be done systematically. The following types of multiple sums occur often and can be evaluated systematically if all $a_n, a_n', b_n, b_n', c_n$ and c_n' are of the form

$$p + q\varepsilon \quad \text{with } p \in \mathbb{Z} \text{ and } q \in \mathbb{C} \quad (9.6)$$

(the typical case is $q \in \mathbb{Z}$ as well) [213]:

Type A:

$$\sum_{i=0}^{\infty} \frac{\Gamma(i + a_1)}{\Gamma(i + a_1')} \cdots \frac{\Gamma(i + a_k)}{\Gamma(i + a_k')} x^i \quad (9.7)$$

Up to prefactors the hypergeometric functions ${}_{J+1}F_J$ fall into this class.

Type B:

$$\sum_{i=0}^{\infty}\sum_{j=0}^{\infty} \frac{\Gamma(i + a_1)}{\Gamma(i + a_1')} \cdots \frac{\Gamma(i + a_k)}{\Gamma(i + a_k')} \frac{\Gamma(j + b_1)}{\Gamma(j + b_1')} \cdots \frac{\Gamma(j + b_l)}{\Gamma(j + b_l')} \frac{\Gamma(i + j + c_1)}{\Gamma(i + j + c_1')} \cdots \frac{\Gamma(i + j + c_m)}{\Gamma(i + j + c_m')} x^i y^j \quad (9.8)$$

An example for a function of this type is given by the first Appell function F_1.

Type C:

$$\sum_{i=0}^{\infty}\sum_{j=0}^{\infty}\binom{i+j}{j}\frac{\Gamma(i+a_1)}{\Gamma(i+a_1')}\cdots\frac{\Gamma(i+a_k)}{\Gamma(i+a_k')}\frac{\Gamma(i+j+c_1)}{\Gamma(i+j+c_1')}\cdots\frac{\Gamma(i+j+c_m)}{\Gamma(i+j+c_m')}x^i y^j \tag{9.9}$$

Here, an example is given by the Kampé de Fériet function S_1.

Type D:

$$\sum_{i=0}^{\infty}\sum_{j=0}^{\infty}\binom{i+j}{j}\frac{\Gamma(i+a_1)}{\Gamma(i+a_1')}\cdots\frac{\Gamma(i+a_k)}{\Gamma(i+a_k')}\frac{\Gamma(j+b_1)}{\Gamma(j+b_1')}\cdots\frac{\Gamma(j+b_l)}{\Gamma(j+b_l')}\frac{\Gamma(i+j+c_1)}{\Gamma(i+j+c_1')}\cdots\frac{\Gamma(i+j+c_m)}{\Gamma(i+j+c_m')}x^i y^j \tag{9.10}$$

An example for a function of this type is the second Appell function F_2.

Note that in these examples there are always as many gamma functions in the numerator as in the denominator. The task is now to expand these functions systematically into a Laurent series in ε. We start with the formula for the expansion of the gamma function $\Gamma(n+\varepsilon)$ with $n \in \mathbb{N}$:

$$\Gamma(n+\varepsilon) = \Gamma(1+\varepsilon)\Gamma(n)\left[1+\varepsilon Z_1(n-1)+\varepsilon^2 Z_{11}(n-1)+\varepsilon^3 Z_{111}(n-1)+\cdots+\varepsilon^{n-1}Z_{11\ldots1}(n-1)\right], \tag{9.11}$$

where $Z_{m_1\ldots m_k}(n)$ denotes a **Euler-Zagier sum** defined by

$$Z_{m_1\ldots m_k}(n) = \sum_{n\geq i_1>i_2>\ldots>i_k>0}\frac{1}{{i_1}^{m_1}}\cdots\frac{1}{{i_k}^{m_k}}. \tag{9.12}$$

This motivates the following definition of a special form of nested sums, called **Z-sums**:

$$Z_{m_1\ldots m_k}(x_1,\ldots,x_k;n) = \sum_{n\geq i_1>i_2>\ldots>i_k>0}\frac{x_1^{i_1}}{{i_1}^{m_1}}\cdots\frac{x_k^{i_k}}{{i_k}^{m_k}}. \tag{9.13}$$

k is called the **depth** of the Z-sum and $w = m_1+\cdots+m_k$ is called the **weight**. If the sums go to infinity ($n=\infty$) the Z-sums are multiple polylogarithms:

$$Z_{m_1\ldots m_k}(x_1,\ldots,x_k;\infty) = \mathrm{Li}_{m_1\ldots m_k}(x_1,\ldots,x_k). \tag{9.14}$$

For $x_1=\cdots=x_k=1$ the definition reduces to the Euler-Zagier sums:

$$Z_{m_1\ldots m_k}(1,\ldots,1;n) = Z_{m_1\ldots m_k}(n). \tag{9.15}$$

For $n = \infty$ and $x_1 = \cdots = x_k = 1$ the sum is a multiple ζ-value:

$$Z_{m_1 \dots m_k}(1, \dots, 1; \infty) = \zeta_{m_1 \dots m_k}. \tag{9.16}$$

The usefulness of the Z-sums lies in the fact, that they interpolate between multiple polylogarithms and Euler-Zagier sums. For fixed n the Z-sums form a quasi-shuffle algebra in the same way as multiple polylogarithms do. The letters are pairs $l_j = (m_j, x_j)$. On the alphabet of letter we have an additional operation "$\circ$" defined by

$$(m_1, x_1) \circ (m_2, x_2) = (m_1 + m_2; x_1 x_2). \tag{9.17}$$

This is exactly the same operation as in Eq. (8.65). On the vector space of words $\mathcal{A}_q$ we have a map

$$Z_n : \mathcal{A}_q \rightarrow \mathbb{C}, \tag{9.18}$$

which sends the word $w = l_1 \dots l_r$ with $l_j = (m_j, x_j)$ to $Z_{m_1 \dots m_r}(x_1, \dots, x_r, n)$. This map is an algebra homomorphism, i.e.

$$Z_n \left(w_1 ⧢_q w_2\right) = Z_n (w_1) \cdot Z_n (w_2), \tag{9.19}$$

where $⧢_q$ denotes the quasi-shuffle product introduced in Sect. 8.4. Thus we have for example

$$Z_{m_1}(x_1; n) Z_{m_2}(x_2; n) = Z_{m_1 m_2}(x_1, x_2; n) + Z_{m_2 m_1}(x_2, x_1; n) + Z_{m_1+m_2}(x_1 x_2; n). \tag{9.20}$$

Exercise 82 *Prove Eq. (8.38) from Chap. 8.*

In addition to Z-sums, it is sometimes useful to introduce as well S-sums. A **S-sum** is defined by

$$S_{m_1 \dots m_k}(x_1, \dots, x_k; n) = \sum_{n \geq i_1 \geq i_2 \geq \dots \geq i_k \geq 1} \frac{x_1^{i_1}}{{i_1}^{m_1}} \dots \frac{x_k^{i_k}}{{i_k}^{m_k}}. \tag{9.21}$$

The S-sums reduce for $x_1 = \cdots = x_k = 1$ to **harmonic sums** [214]:

$$S_{m_1 \dots m_k}(1, \dots, 1; n) = S_{m_1 \dots m_k}(n). \tag{9.22}$$

The S-sums are closely related to the Z-sums, the difference being the upper summation boundary for the nested sums: $(i - 1)$ for Z-sums, i for S-sums. The introduction of S-sums is redundant, since S-sums can be expressed in terms of Z-sums and vice versa. It is however convenient to introduce both Z-sums and S-sums, since some properties are more naturally expressed in terms of Z-sums while others are more naturally expressed in terms of S-sums. An algorithm for the conversion from Z-sums to S-sums and vice versa can be found in [213].

The quasi-shuffle product of Eq. (9.19) is the essential ingredient to expand functions of type A as in Eq. (9.7) in a small parameter. As a simple example let us consider the function

$$\sum_{i=0}^{\infty} \frac{\Gamma(i+a_1+t_1\varepsilon)\Gamma(i+a_2+t_2\varepsilon)}{\Gamma(i+1)\Gamma(i+a_3+t_3\varepsilon)} x^i. \tag{9.23}$$

Here a_1, a_2 and a_3 are assumed to be integers. Up to prefactors the expression in Eq. (9.23) is a hypergeometric function ${}_2F_1$. We are interested in the Laurent expansion of the function above in the small parameter ε.

Using $\Gamma(x+1) = x\Gamma(x)$, partial fractioning and an adjustment of the summation index one can transform Eq. (9.23) into terms of the form

$$\sum_{i=1}^{\infty} \frac{\Gamma(i+t_1\varepsilon)\Gamma(i+t_2\varepsilon)}{\Gamma(i)\Gamma(i+t_3\varepsilon)} \frac{x^i}{i^m}, \tag{9.24}$$

where m is an integer. Now using Eq. (9.11) one obtains

$$\Gamma(1+\varepsilon) \sum_{i=1}^{\infty} \frac{(1+\varepsilon t_1 Z_1(i-1)+\ldots)(1+\varepsilon t_2 Z_1(i-1)+\ldots)}{(1+\varepsilon t_3 Z_1(i-1)+\ldots)} \frac{x^i}{i^m}. \tag{9.25}$$

Inverting the power series in the denominator and truncating in ε one obtains in each order in ε terms of the form

$$\sum_{i=1}^{\infty} \frac{x^i}{i^{m_0}} Z_{m_1 \ldots m_k}(i-1)\; Z_{m'_1 \ldots m'_l}(i-1)\; Z_{m''_1 \ldots m''_n}(i-1). \tag{9.26}$$

Using the quasi-shuffle product for Z-sums the three Euler-Zagier sums can be reduced to single Euler-Zagier sums and one finally arrives at terms of the form

$$\sum_{i=1}^{\infty} \frac{x^i}{i^{m_0}} Z_{m_1 \ldots m_k}(i-1), \tag{9.27}$$

which are special cases of multiple polylogarithms, called harmonic polylogarithms $H_{m_0 m_1 \ldots m_k}(x)$. This completes the algorithm for the expansion in ε for sums of the form as in Eq. (9.23), and more generally any sum of type A as in Eq. (9.7).

Let us now consider expressions of the form

$$\frac{x_0^n}{n^{m_0}} Z_{m_1 \ldots m_k}(x_1, \ldots, x_k; n), \tag{9.28}$$

e.g. Z-sums multiplied by a letter. Then the following convolution product

$$\sum_{i=1}^{n-1} \frac{x^i}{i^{m_0}} Z_{m_1...}(x_1, \ldots; i-1) \frac{y^{n-i}}{(n-i)^{m_0'}} Z_{m_1'...}(x_1', \ldots; n-i-1) \quad (9.29)$$

can again be expressed by partial fractioning and relabellings of summation indices in terms of expressions of the form (9.28). An example is

$$\sum_{i=1}^{n-1} \frac{x^i}{i} Z_1(i-1) \frac{y^{n-i}}{(n-i)} Z_1(n-i-1) = \quad (9.30)$$
$$\frac{x^n}{n} \left[Z_{111}\left(\frac{y}{x}, \frac{x}{y}, \frac{y}{x}; n-1\right) + Z_{111}\left(\frac{y}{x}, 1, \frac{x}{y}; n-1\right) + Z_{111}\left(1, \frac{y}{x}, 1; n-1\right)\right] + (x \leftrightarrow y).$$

Combing this algorithm with the previous algorithms allows to expand any sum of type B as in Eq. (9.8).

In addition there is for terms of the form as in Eq. (9.28) a conjugation, e.g. sums of the form

$$-\sum_{i=1}^{n} \binom{n}{i} (-1)^i \frac{x^i}{i^{m_0}} S_{m_1...}(x_1, \ldots; i) \quad (9.31)$$

can also be reduced to terms of the form (9.28). Although one can easily convert between the notations for S-sums and Z-sums, expressions involving a conjugation tend to be shorter when expressed in terms of S-sums. The name conjugation stems from the following fact: To any function $f(n)$ of an integer variable n one can define a conjugated function $C * f(n)$ as the following sum

$$C * f(n) = \sum_{i=1}^{n} \binom{n}{i} (-1)^i f(i). \quad (9.32)$$

Then conjugation satisfies the following two properties:

$$C * 1 = 1,$$
$$C * C * f(n) = f(n). \quad (9.33)$$

An example for a sum involving a conjugation is

$$-\sum_{i=1}^{n} \binom{n}{i} (-1)^i \frac{x^i}{i} S_1(i) = S_{11}\left(1-x, \frac{1}{1-x}; n\right) - S_{11}(1-x, 1; n). \quad (9.34)$$

Conjugation in combination with the previous algorithms allow us to expand all functions of type C as in Eq. (9.9).

Finally there is the combination of conjugation and convolution, e.g. sums of the form

$$-\sum_{i=1}^{n-1} \binom{n}{i} (-1)^i \; \frac{x^i}{i^{m_0}} S_{m_1 \dots}(x_1, \dots; i) \; \frac{y^{n-i}}{(n-i)^{m_0'}} S_{m_1' \dots}(x_1', \dots; n-i) \tag{9.35}$$

can also be reduced to terms of the form (9.28). An example is given by

$$\begin{aligned} -\sum_{i=1}^{n-1} \binom{n}{i} (-1)^i \; S_1(x; i) \; S_1(y; n-i) = \\ \frac{1}{n} \left\{ S_1(y; n) + (1-x)^n \left[S_1\left(\frac{x}{x-1}; n\right) - S_1\left(\frac{x-y}{x-1}; n\right) \right] \right\} \\ + \frac{(-1)^n}{n} \left\{ S_1(x; n) + (1-y)^n \left[S_1\left(\frac{y}{y-1}; n\right) - S_1\left(\frac{y-x}{y-1}; n\right) \right] \right\}. \end{aligned} \tag{9.36}$$

This allows us to expand functions of type D as in Eq. (9.10). There are computer packages implementing the algorithms for the expansion of sums of type A-D [215–218].

Exercise 83 *Consider I_{111} from Eq. (9.1) with $\mu^2 = -p_3^2$ and $x = p_1^2/p_3^2$ in $D = 4 - 2\varepsilon$ space-time dimensions:*

$$I_{111} = e^{\varepsilon \gamma_E} \frac{\Gamma(-\varepsilon)\Gamma(1-\varepsilon)}{\Gamma(1-2\varepsilon)} \sum_{n=0}^{\infty} \frac{\Gamma(n+1+\varepsilon)}{\Gamma(n+2)} (1-x)^n . \tag{9.37}$$

Expand the sum in ε and give the first two terms of the ε-expansion for the full expression.

Up to now we assumed through Eq. (9.6) that the gamma functions are expanded around an integer value. The extension to rational numbers is straightforward for sums of type A and B if the gamma functions always occur in ratios of the form

$$\frac{\Gamma(n + a - \frac{p}{q} + b\varepsilon)}{\Gamma(n + c - \frac{p}{q} + d\varepsilon)}, \tag{9.38}$$

where the same rational number $p/q \in \mathbb{Q}$ occurs in the numerator and in the denominator [219]. The generalisation of Eq. (9.11) reads

$$\begin{aligned} \Gamma\left(n + 1 - \frac{p}{q} + \varepsilon\right) = \frac{\Gamma\left(1 - \frac{p}{q} + \varepsilon\right) \Gamma\left(n + 1 - \frac{p}{q}\right)}{\Gamma\left(1 - \frac{p}{q}\right)} \\ \times \exp\left(-\frac{1}{q} \sum_{l=0}^{q-1} \left(r_q^l\right)^p \sum_{k=1}^{\infty} \varepsilon^k \frac{(-q)^k}{k} Z_k(r_q^l; q \cdot n) \right) \end{aligned} \tag{9.39}$$

and introduces the q-th roots of unity

$$r_q^p = \exp\left(\frac{2\pi i p}{q}\right). \tag{9.40}$$

With the help of the q-th roots of unity we may express any Z-sum $Z_{m_1...}(x_1, \dots; n)$ as a combination of Z-sums $Z_{m_1'...}(x_1', \dots; q \cdot n)$, where the summation goes now up to $q \cdot n$.

9.2 GKZ Hypergeometric Functions

In this section we show that a Feynman integral can be viewed as a special case of a Gelfand-Kapranov-Zelevinsky (GKZ) hypergeometric function. GKZ hypergeometric functions are solutions to a system of partial differential equations, called a GKZ system. In order to present GKZ hypergeometric functions we may either use a geometric language or an algebraic language. In both cases we need some preparation. In preparation for the geometric setting we introduce polytopes in Sect. 9.2.1 and in preparation for the algebraic setting we introduce D-modules in Sect. 9.2.2. GKZ systems and GKZ hypergeometric functions are then introduced in Sect. 9.2.3. The application towards Feynman integrals is discussed in Sect. 9.2.5. More information on polytopes can be found in the books by Zieger [220] and by De Loera, Rambau and Santos [221]. Further information on D-modules can be found in the books by Björk [222], by Coutinho [223] and by Saito, Sturmfels and Takayama [224]. An introduction into the topic of GKZ hypergeometric structures can be found in the lecture notes by Stienstra [225] and by Cattani [226]. Additional background on GKZ hypergeometric functions can be found in the book by Gelfand, Kapranov and Zelevinsky [227].

In this section it is convenient to use **multi-index notation**: For $x = (x_1, \dots, x_n)$ and $\alpha = (\alpha_1, \dots, \alpha_n)$ we set

$$x^\alpha = x_1^{\alpha_1} \dots x_n^{\alpha_n}. \tag{9.41}$$

We denote by $(\mathbb{R}^n)^*$ the dual space of $\mathbb{R}^n$, i.e. the space of linear functionals $\varphi : \mathbb{R}^n \to \mathbb{R}$. We may view $a \in \mathbb{R}^n$ as a column vector (or as a ket vector) and $b \in (\mathbb{R}^n)^*$ as a row vector (or as a bra vector). The row vector b defines a linear functional by

$$\varphi_b : a \to b \cdot a. \tag{9.42}$$

9.2.1 Polytopes

Let $A = \{a_1, \dots, a_n\} \subset \mathbb{R}^d$ be a non-empty finite set of points in $\mathbb{R}^d$. The convex hull of these points defines a **polytope**, denoted by $\text{conv}(A)$. In a formula

$$\text{conv}(A) = \left\{ \alpha_1 a_1 + \dots + \alpha_n a_n | \alpha_j \geq 0, \sum_{j=1}^{n} \alpha_j = 1 \right\}. \tag{9.43}$$

A **cone** is defined by

$$\text{cone}(A) = \left\{ \alpha_1 a_1 + \dots + \alpha_n a_n | \alpha_j \geq 0 \right\}. \tag{9.44}$$

We define the cone of an empty set to be the set containing the origin, i.e. $\text{cone}(\{\}) = \{0\}$. Every cone contains the origin. Given two sets $P, Q \subseteq \mathbb{R}^d$ their **Minkowski sum** is defined to be

$$P + Q = \{a + b | a \in P, b \in Q\}. \tag{9.45}$$

Figure 9.1 shows an example for a polytope and a cone. The figure shows also the Minkowski sum of the polytope and the cone.

A **polyhedron** P is the Minkowski sum of a polytope $\text{conv}(A)$ and a cone $\text{cone}(B)$:

$$P = \text{conv}(A) + \text{cone}(B). \tag{9.46}$$

A hyperplane in $\mathbb{R}^d$ divides the space into two halfspaces. Let $a \in \mathbb{R}^d$, $b \in (\mathbb{R}^d)^*$ and $c \in \mathbb{R}$. A hyperplane is given by

$$\left\{ x \in \mathbb{R}^d \mid b \cdot a = c \right\}. \tag{9.47}$$

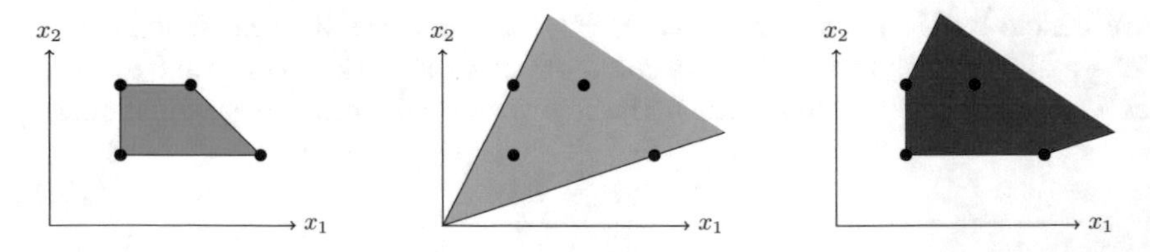

Fig. 9.1 The left picture shows the polytope in $\mathbb{R}^2$ defined by the points $a_1 = (1, 1)^T$, $a_2 = (1, 2)^T$, $a_3 = (2, 2)^T$, $a_4 = (3, 1)^T$. The middle picture shows the cone defined by these points. The right picture shows the Minkowski sum of the polytope and the cone

The two closed halfspace are

$$\left\{ x \in \mathbb{R}^d \mid b \cdot a \geq c \right\} \text{ and } \left\{ x \in \mathbb{R}^d \mid b \cdot a \leq c \right\}. \tag{9.48}$$

This can be used to give alternative definitions: We may define a polyhedron as an intersection of finitely many closed halfspaces and a polytope as a bounded polyhedron. A cone is the intersection of finitely many closed halfspaces, where all hyperplanes defining the halfspaces contain the origin. The **lineality space of a cone** is the largest linear subspace contained in the cone. A cone is called **pointed**, if its lineality space is $\{0\}$. Let us illustrate the last two definitions by an example: The cone in $\mathbb{R}^2$ generated by

$$\begin{pmatrix} 1 \\ 0 \end{pmatrix}, \begin{pmatrix} 0 \\ 1 \end{pmatrix} \tag{9.49}$$

is pointed. It corresponds to the first quadrant and $\{0\}$ is the largest linear subspace. On the other hand, the cone in $\mathbb{R}^2$ generated by

$$\begin{pmatrix} 1 \\ 0 \end{pmatrix}, \begin{pmatrix} 0 \\ 1 \end{pmatrix}, \begin{pmatrix} -1 \\ -1 \end{pmatrix} \tag{9.50}$$

is not pointed. Any point of $\mathbb{R}^2$ belongs to the cone and hence the largest linear subspace is $\mathbb{R}^2$.

Consider now a hyperplane such that all points of a polytope / cone / polyhedron lie in one closed halfspace (this includes the points on the hypersurface). A **face** is the intersection of a polytope / cone / polyhedron with such a hyperplane. Faces of dimension zero are called **vertices** and faces of dimension $\dim(P) - 1$ are called **facets**. A face F with $\dim(F) < \dim(P)$ is called a proper face.

The dimension of a polytope $P = \text{conv}(A)$ is the dimension of its convex hull. Let P be a k-dimensional polytope. We define the normalised volume $\text{vol}_0(P)$ of P as

$$\text{vol}_0(P) = k!\text{vol}(P), \tag{9.51}$$

where $\text{vol}(P)$ denotes the standard (Euclidean) volume.

Exercise 84 *Consider the k-dimensional standard simplex in $\mathbb{R}^{k+1}$. This is the polytope with vertices given by the $(k+1)$ standard unit vectors $e_j \in \mathbb{R}^{k+1}$. Show that the standard simplex has Euclidean volume $1/k!$ and therefore the normalised volume 1.*

Let σ be a k-dimensional simplex in $\mathbb{R}^{k+1}$, defined by $(k+1)$ points $A = \{a_1, \dots, a_{k+1}\} \in \mathbb{R}^{k+1}$. By abuse of notation, we also denote by A the $(k+1) \times (k+1)$-matrix, whose columns are given by $a_1, \dots, a_{k+1}$. Then

$$\text{vol}_0(\sigma) = |\det A|. \tag{9.52}$$

A **triangulation** of a polytope $P = \text{conv}(A) \subset \mathbb{R}^d$ is a set of simplices $\{\sigma_1, \dots, \sigma_r\}$ with vertices from the set $A = \{a_1, \dots, a_n\}$ such that all faces of a simplex are contained in this set, the union of all simplices is the full polytope and the intersection of two distinct simplices is a proper face of the two, possibly empty.

A triangulation $\{\sigma_1, \dots, \sigma_r\}$ of a polytope $P = \text{conv}(A) = \text{conv}(a_1, \dots, a_n) \subset \mathbb{R}^d$ is called **regular**, if there exists a height vector $h \in \mathbb{R}^n$, such that for every simplex σ_i of this triangulation there exists a vector $r_i \in \mathbb{R}^d$ satisfying

$$\begin{aligned} r_i \cdot a_j &= h_j \qquad a_j \in \sigma_i, \\ r_i \cdot a_j &< h_j \qquad a_j \notin \sigma_i. \end{aligned} \tag{9.53}$$

These equations say that if we consider points $\tilde{a}_1, \dots, \tilde{a}_n \in \mathbb{R}^{d+1}$ obtained from $a_1, \dots, a_n \in \mathbb{R}^d$ by adjoining the height as last coordinate

$$\tilde{a}_j = \begin{pmatrix} a_{1j} \\ \vdots \\ a_{dj} \\ h_j \end{pmatrix}, \tag{9.54}$$

the resulting geometrical object obtained from lifting the triangulation to $\mathbb{R}^{d+1}$ is convex. As an example consider the six points in $\mathbb{R}^1$:

$$a_1 = (1), \quad a_2 = (2), \quad a_3 = (3), \quad a_4 = (4), \quad a_5 = (5), \quad a_6 = (6), \tag{9.55}$$

and the triangulation

$$\{\sigma_{12}, \sigma_{23}, \sigma_{34}, \sigma_{45}, \sigma_{56}, \sigma_1, \sigma_2, \sigma_3, \sigma_4, \sigma_5, \sigma_6\}, \tag{9.56}$$

where σ_{ij} denotes the 1-dimensional simplex defined by a_i and a_j, and σ_j denotes the 0-dimensional simplex defined by a_j. The triangulation in Eq. (9.56) is regular. A height vector is given by

$$h = (3, 1, 0, 0, 1, 3)^T, \tag{9.57}$$

see Fig. 9.2.

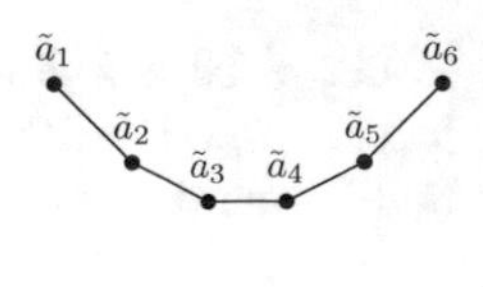

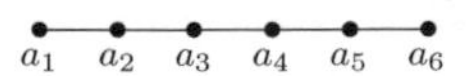

Fig. 9.2 The triangulation of $(a_1, a_2, a_3, a_4, a_5, a_6)$ in $\mathbb{R}^1$ and the lift $(\tilde{a}_1, \tilde{a}_2, \tilde{a}_3, \tilde{a}_4, \tilde{a}_5, \tilde{a}_6)$ to $\mathbb{R}^2$. The lift is convex

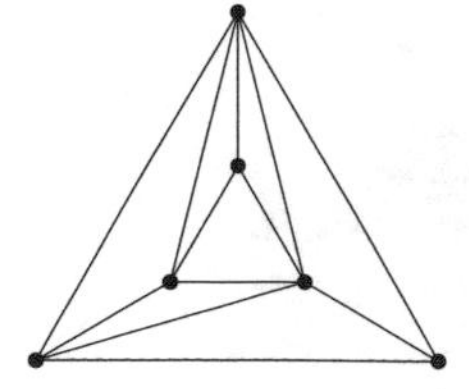

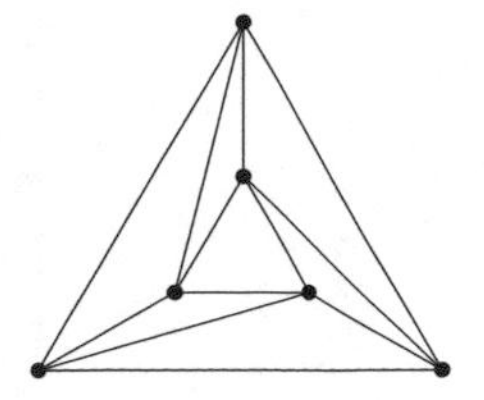

Fig. 9.3 The left picture shows a regular triangulation of the polytope P, the right picture shows a non-regular triangulation of the polytope P

As an example for a non-regular triangulations consider the six points

$$A = \left\{ \begin{pmatrix} 4 \\ 0 \\ 0 \end{pmatrix}, \begin{pmatrix} 0 \\ 4 \\ 0 \end{pmatrix}, \begin{pmatrix} 0 \\ 0 \\ 4 \end{pmatrix}, \begin{pmatrix} 2 \\ 1 \\ 1 \end{pmatrix}, \begin{pmatrix} 1 \\ 2 \\ 1 \end{pmatrix}, \begin{pmatrix} 1 \\ 1 \\ 2 \end{pmatrix} \right\} \tag{9.58}$$

in $\mathbb{R}^3$. These points lie in a plane with normal vector $n = (1, 1, 1)$. The polytope $P = \text{conv}(A)$ is two-dimensional.

Figure 9.3 shows a regular triangulation of P (left picture) and a non-regular triangulation of P (right picture). The right picture of Fig. 9.3 is the standard example for a non-regular triangulation.

Exercise 85 *Show that the left picture of Fig. 9.3 defines a regular triangulation.*

Exercise 86 *Show that the right picture of Fig. 9.3 defines a non-regular triangulation.*

A regular triangulation $\{\sigma_1, \dots, \sigma_r\}$ where all simplices have normalised volume one ($\text{vol}_0(\sigma_j) = 1$) is called a **unimodular triangulation**.

A **fan** in $\mathbb{R}^d$ is a finite set of cones

$$\mathcal{F} = \{C_1, C_2, \dots, C_r\} \tag{9.59}$$

such that

1. Every face of a cone in $\mathcal{F}$ is also a cone in $\mathcal{F}$.
2. The intersection of any two cones in $\mathcal{F}$ is a face of both

The fan is called **complete**, if the union of all cones in $\mathcal{F}$ equals $\mathbb{R}^d$:

$$C_1 \cup C_2 \cup \dots \cup C_r = \mathbb{R}^d. \tag{9.60}$$

A fan is called **rational**, if all the cones in the fan are given by inequalities with rational coefficients.

Consider now a polytope $P \in \mathbb{R}^d$ and let F be a face of P. The **normal cone** $N_F(P)$ of a face F is a cone in the dual space $(\mathbb{R}^d)^*$ defined by

$$N_F(P) = \left\{ b \in (\mathbb{R}^d)^* \mid F \subseteq \left\{ x \in P \mid b \cdot x = \max_{y \in P} (b \cdot y) \right\} \right\}. \tag{9.61}$$

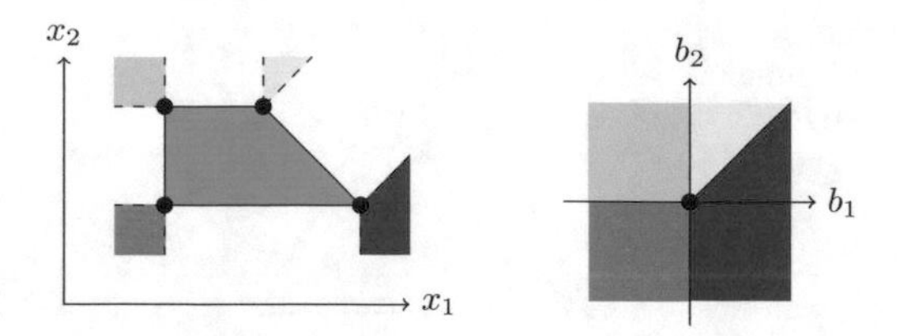

Fig. 9.4 The left picture shows the polytope in $\mathbb{R}^2$ defined by the points $a_1 = (1, 1)^T$, $a_2 = (1, 2)^T$, $a_3 = (2, 2)^T$, $a_4 = (3, 1)^T$. The right picture shows the normal fan of the polytope

The collection of the normal cones $N_F(P)$ for the faces F of a polytope P forms a complete fan in $(\mathbb{R}^d)^*$. This fan is called the **normal fan** of P and denoted by $N(P)$.

An example is shown in Fig. 9.4.

As before consider a non-empty set of n points $A = \{a_1, \dots, a_n\} \subset \mathbb{R}^d$. The convex hull of these points defines a polytope $P = \text{conv}(A)$, see Eq. (9.43). The polytope P is also called the **primary polytope** of A. For any regular triangulation $T = \{\sigma_1, \sigma_2, \dots\}$ of P we define a point $q_T \in \mathbb{R}^n$ with coordinates $q_{1T}, \dots, q_{nT}$ by

$$q_{jT} = \sum_{\sigma \in T, a_j \in \text{vertices}(\sigma)} \text{vol}_0\,(\sigma)\,. \tag{9.62}$$

Let S be the set $\{q_{T_1}, q_{T_2}, \dots\} \subset \mathbb{R}^n$ obtained from all regular triangulations T_i of P. The **secondary polytope** $\Sigma(A)$ of A is defined to be

$$\Sigma(A) = \text{conv}\,(S) \subset \mathbb{R}^n. \tag{9.63}$$

The normal fan of the secondary polytope $\Sigma(A)$ is called the **secondary fan** $N(\Sigma(A))$.

Given a multivariate polynomial f in d variables $x_1, \dots, x_d$, written in multi-index notation (e.g. $x^{a_j} = x_1^{a_{j1}} \dots x_d^{a_{jd}}$) as

$$f(x) = \sum_{j=1}^{n} c_{a_j} x^{a_j}, \qquad c_{a_j} \neq 0, \tag{9.64}$$

the **Newton polytope** Δ_f of f is defined by

$$\Delta_f = \text{conv}\,(\{a_1, \dots, a_n\})\,. \tag{9.65}$$

We denote by $\mathcal{V}_f$ the hypersurface defined by f:

$$\mathcal{V}_f = \left\{ x \in \mathbb{C}^d \mid f(x) = 0 \right\}. \tag{9.66}$$

The **amoeba** $\mathcal{A}_f$ of $\mathcal{V}_f$ is defined as

$$\mathcal{A}_f = \{ (\ln|x_1|, \ldots, \ln|x_d|) \in \mathbb{R}^d \mid x \in \mathcal{V}_f \}, \tag{9.67}$$

the **coamoeba** $\mathcal{A}'_f$ of $\mathcal{V}_f$ is defined as

$$\mathcal{A}'_f = \{ (\arg(x_1), \ldots, \arg(x_d)) \in \mathbb{R}^d \mid x \in \mathcal{V}_f \}. \tag{9.68}$$

9.2.2 D-modules

Let $\mathbb{F}$ be a field of characteristic zero and $n \in \mathbb{N}$. The ring of differential operators $\partial_1, \ldots, \partial_n$ (where $\partial_j = \partial/\partial x_j$) with coefficients in the polynomial ring $\mathbb{F}[x_1, \ldots, x_n]$ is called the **Weyl algebra** A_n in n variables. The Weyl algebra is generated by

$$x_1, \ldots, x_n, \partial_1, \ldots, \partial_n \tag{9.69}$$

subject to the commutation relation

$$[\partial_i, x_i] = 1, \qquad 1 \leq i \leq n. \tag{9.70}$$

All other commutation relations are trivial. With the multi-index notation

$$x^\alpha = x_1^{\alpha_1} \ldots x_n^{\alpha_n}, \quad \partial^\alpha = \partial_1^{\alpha_1} \ldots \partial_n^{\alpha_n} \tag{9.71}$$

any $P \in A_n$ can be written uniquely as

$$P = \sum_{\alpha,\beta} c_{\alpha\beta}\, x^\alpha\, \partial^\beta, \qquad c_{\alpha\beta} \in \mathbb{F}. \tag{9.72}$$

We call P written in the form as in Eq. (9.72) a **normal ordered expression**. Let $\xi_1, \ldots, \xi_n$ be commutative variables. It is often useful to replace $\partial_j \to \xi_j$. From Eq. (9.72) we see that if we view A_n and $\mathcal{F}[x_1, \ldots, x_n, \xi_1, \ldots, \xi_n]$ as vector spaces over $\mathbb{F}$ there is a vector space isomorphism between

$$P(x, \partial) = \sum_{\alpha,\beta} c_{\alpha\beta}\, x^\alpha\, \partial^\beta \text{ and } P(x, \xi) = \sum_{\alpha,\beta} c_{\alpha\beta}\, x^\alpha\, \xi^\beta. \tag{9.73}$$

Note that this is not an algebra homomorphism.

Let D be a ring of differential operators. A ***D*-module** is a left module over the ring D. If we take $D = A_n$, examples for D-modules are the polynomial ring $\mathbb{F}[x_1, \ldots, x_n]$ or the field of rational functions $\mathbb{F}(x_1, \ldots, x_n)$.

Let us now consider a system of linear partial differential equations, given by differential operators $P_1, \dots, P_r \in A_n$:

$$P_j f(x_1, \dots, x_n) = 0. \tag{9.74}$$

These define an ideal $I = \langle P_1, \dots, P_r \rangle$ in the Weyl algebra A_n, as

$$Q P_j f(x_1, \dots, x_n) = 0, \qquad \text{for any } Q \in A_n. \tag{9.75}$$

A vector $(v, w) = (v_1, \dots, v_n, w_1, \dots, w_n) \in \mathbb{Z}^{2n}$ is called a **weight vector** for the Weyl algebra A_n if

$$v_j + w_j \geq 0, \qquad 1 \leq j \leq n. \tag{9.76}$$

The order of a monomial $x^\alpha \partial^\beta$ with respect to the weight vector (v, w) is

$$\mathrm{ord}_{(v,w)}\left(x^\alpha \partial^\beta\right) = v_1\alpha_1 + \dots + v_n\alpha_n + w_1\beta_1 + \dots + w_n\beta_n \tag{9.77}$$

and we define the order of a differential operator $P \in A_n$ as the maximum of the orders of the individual monomials when P is written in normal form. A weight vector (v, w) induces a filtration

$$F_k^{(v,w)}(A_n) = \left\{ P \in A_n \mid \mathrm{ord}_{(v,w)}(P) \leq k \right\}. \tag{9.78}$$

The **associated graded ring** is

$$\mathrm{gr}^{(v,w)}(A_n) = \bigoplus_{k \in \mathbb{Z}} \mathrm{gr}_k^{(v,w)}(A_n), \qquad \mathrm{gr}_k^{(v,w)}(A_n) = F_k^{(v,w)}(A_n) / F_{k-1}^{(v,w)}(A_n). \tag{9.79}$$

We may think of $\mathrm{gr}^{(v,w)}(A_n)$ as the algebra generated by $x_1, \dots, x_n$ and for $1 \leq j \leq n$ either ∂_j or ξ_j according to

$$\begin{aligned} v_j + w_j &= 0 : \partial_j, \\ v_j + w_j &> 0 : \xi_j. \end{aligned} \tag{9.80}$$

The non-trivial commutation relations are the ones given in Eq. (9.70). If $v_j + w_j > 0$ the commutator $[\partial_j, x_j]$ is of lower weight and we may replace ∂_j with the commutative variable ξ. Only for $v_j + w_j = 0$ we have to keep track of order.

Let $P \in A_n$ and assume that P is written in normal form as in Eq. (9.72). We set $o = \mathrm{ord}_{(v,w)}(P)$. The **initial form** $\mathrm{in}_{(v,w)}(P)$ of P with respect to the weight vector (v, w) is defined to be

$$\mathrm{in}_{(v,w)}(P) = \sum_{\substack{\alpha,\beta \\ \alpha v + \beta w = o}} c_{\alpha\beta}\, x^\alpha \prod_{v_j + w_j > 0} \xi_j^{\beta_j} \prod_{v_j + w_j = 0} \partial_j^{\beta_j}. \tag{9.81}$$

In other words: We take only the terms of order o and make the replacement $\partial_j \to \xi_j$, whenever $v_j + w_j > 0$.

Consider an ideal $I = \langle P_1, \dots, P_r \rangle \subset A_n$. Then

$$\langle \, \mathrm{in}_{(v,w)} \, (P) \mid P \in I \, \rangle \tag{9.82}$$

is an ideal in the associated graded ring $\mathrm{gr}^{(v,w)}(A_n)$, called the **initial ideal** of I with respect to the weight vector (v, w).

Let us now consider the weight vector

$$(\mathbf{0}, \mathbf{1}) = (\underbrace{0, \dots, 0}_{n}, \underbrace{1, \dots, 1}_{n}). \tag{9.83}$$

For this weight vector the associated graded ring $\mathrm{gr}^{(\mathbf{0},\mathbf{1})}(A_n)$ of the Weyl algebra A_n is the commutative ring $\mathbb{F}[x_1, \dots, x_n, \xi_1, \dots, \xi_n]$. Let I be an ideal in the Weyl algebra A_n. We define the **characteristic ideal** of I to be the initial ideal

$$\mathrm{in}_{(\mathbf{0},\mathbf{1})} \, (I) \tag{9.84}$$

with respect to the weight vector $(\mathbf{0}, \mathbf{1})$. The **characteristic variety** $\mathrm{ch}(I)$ is the zero set of $\mathrm{in}_{(\mathbf{0},\mathbf{1})} \, (I)$ in the affine space of dimension $2n$ with coordinates $x_1, \dots, x_n, \xi_1, \dots, \xi_n$.

We say that the ideal I is **holonomic** if the characteristic ideal $\mathrm{in}_{(\mathbf{0},\mathbf{1})}(I)$ of I has dimension n. Let us denote by $\mathbb{F}(x) = \mathbb{F}(x_1, \dots, x_n)$ the field of rational functions in $x = (x_1, \dots, x_n)$. The **holonomic rank** of I is the dimension of the vector space $\mathbb{F}(x)[\xi]/(\mathbb{F}(x)[\xi] \cdot \mathrm{in}_{(\mathbf{0},\mathbf{1})}(I))$:

$$\mathrm{rank} \, (I) = \dim \left(\, \mathbb{F}(x)[\xi] \, / \, \left(\mathbb{F}(x)[\xi] \cdot \mathrm{in}_{(\mathbf{0},\mathbf{1})}(I)\right) \, \right). \tag{9.85}$$

An important theorem states that if I is holonomic, than $\mathrm{rank}(I)$ is finite.

Let us now consider the weight vector $(-w, w)$ with $w \in \mathbb{Z}^n$. Consider an ideal $I = \langle P_1, \dots, P_r \rangle \subset A_n$. The ideal

$$\mathrm{in}_{(-w,w)} \, (I) = \langle \, \mathrm{in}_{(-w,w)} \, (P) \mid P \in I \, \rangle \tag{9.86}$$

is called a **Gröbner deformation** of I. We define the **Euler operators** θ_j by

$$\theta_j = x_j \frac{\partial}{\partial x_j}. \tag{9.87}$$

Let us denote by $\hat{A}_n = \mathbb{F}(x) \cdot A_n$, i.e. the ring of differential operators $\partial_1, \dots, \partial_n$ with coefficients being rational functions of $x_1, \dots, x_n$. To spell out the difference between A_n and $\hat{A}_n$: In the Weyl algebra the coefficients are polynomials in $x_1, \dots, x_n$, in $\hat{A}_n$ the coefficients are allowed to be rational functions in $x_1, \dots, x_n$. Given an ideal $I \subset$

A_n we denote in a similar way by $\hat{I}$ the ideal $\mathbb{F}(x) \cdot I$. We further denote by $\mathbb{F}[\theta] = \mathbb{F}[\theta_1, \dots, \theta_n]$ the (commutative) ring of the Euler operators. We may now define the **indicial ideal** $\text{ind}_w(I)$ of $I \subset A_n$ relative to the weight $w \in \mathbb{Z}^n$: It is given by

$$\text{ind}_w(I) = \widehat{\text{in}}_{(-w,w)}(I) \cap \mathbb{F}[\theta]. \quad (9.88)$$

9.2.3 GKZ Systems

A **GKZ hypergeometric system** [228, 229] or $\mathcal{A}$-**hypergeometric system** is defined by a vector $c \in \mathbb{C}^{k+1}$ and an n-element subset

$$\mathcal{A} = \{a_1, \dots, a_n\} \subset \mathbb{Z}^{k+1}, \quad (9.89)$$

which satisfies the two conditions

1. $\mathcal{A}$ generates $\mathbb{Z}^{k+1}$ as an Abelian group. (9.90)
2. There exists a group homomorphism $h : \mathbb{Z}^{k+1} \to \mathbb{Z}$ such that $h(a) = 1$ for all $a \in \mathcal{A}$.

We must have $n > k$, otherwise the elements of $\mathcal{A}$ cannot generate $\mathbb{Z}^{k+1}$. The second condition is equivalent to the statement that all elements $a \in \mathcal{A}$ lie in a hyperplane. We may think of $\mathcal{A}$ as a $(k+1) \times n$-matrix with integer entries and where the columns are given by the a_j's.

We denote by $\mathbb{L} \subset \mathbb{Z}^n$ the lattice of relations in $\mathcal{A}$:

$$\mathbb{L} = \left\{ (l_1, \dots, l_n) \in \mathbb{Z}^n \mid l_1 a_1 + \dots + l_n a_n = 0 \right\}. \quad (9.91)$$

GKZ hypergeometric system:
The GKZ hypergeometric system associated with $\mathcal{A}$ and c is the following system of differential equations for a function ϕ of n variables $x_1, \dots, x_n$: For every $(l_1, \dots, l_n) \in \mathbb{L}$ one differential equation

$$\left[\prod_{l_j > 0} \left(\frac{\partial}{\partial x_j} \right)^{l_j} - \prod_{l_j < 0} \left(\frac{\partial}{\partial x_j} \right)^{-l_j} \right] \phi = 0, \quad (9.92)$$

and $(k+1)$ differential equations

$$\left[\sum_{j=1}^{n} a_j x_j \frac{\partial}{\partial x_j} - c \right] \phi = 0. \quad (9.93)$$

The differential equations in Eq. (9.92) are called the toric differential equations, the ones in Eq. (9.93) are called the homogeneity equations. We may write the system of differential equations in a slightly modified way: Let $l \in \mathbb{L}$ and write

$$l = u - v, \text{ with } u, v \in \mathbb{N}_0^n. \tag{9.94}$$

Thus u contains all positive entries of l, while v is the negative of the negative entries of l. The condition $l_1 a_1 + \cdots + l_n a_n = 0$ is equivalent to

$$\mathcal{A}u = \mathcal{A}v. \tag{9.95}$$

Then Eq. (9.92) is equivalent to

$$\left(\partial^u - \partial^v\right) \phi = 0 \tag{9.96}$$

for any pair $u, v \in \mathbb{N}_0^n$ with $\mathcal{A}u = \mathcal{A}v$.

Let us denote the vector of Euler operators by

$$\boldsymbol{\theta} = (\theta_1, \dots, \theta_n)^T = \left(x_1 \frac{\partial}{\partial x_1}, \dots, x_n \frac{\partial}{\partial x_n}\right)^T. \tag{9.97}$$

Then Eq. (9.93) is equivalent to

$$(\mathcal{A}\boldsymbol{\theta} - c)\,\phi = 0. \tag{9.98}$$

Let $\gamma \in \mathbb{C}^n$. We define the **Γ -series** associated with $\mathbb{L}$ and γ by

$$\phi_{\mathbb{L},\gamma}(x_1, \dots, x_n) = \sum_{(l_1,\dots,l_n)\in\mathbb{L}} \prod_{j=1}^{n} \frac{x_j^{l_j+\gamma_j}}{\Gamma\left(l_j + \gamma_j + 1\right)}. \tag{9.99}$$

The Γ-series is in its domain of convergence a solution of the GKZ system associated to $\mathcal{A}$ and

$$c = \sum_{j=1}^{n} a_j \gamma_j. \tag{9.100}$$

Recall that $c \in \mathbb{C}^{k+1}$, $\gamma \in \mathbb{C}^n$ and $\mathcal{A} \in M(k+1, n, \mathbb{Z})$. Alternatively we may write Eq. (9.100) as

$$c = \mathcal{A}\gamma. \tag{9.101}$$

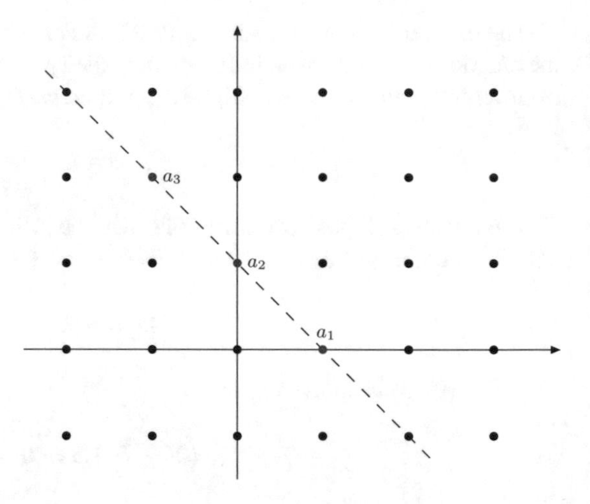

Fig. 9.5 The set $\mathcal{A} = \{a_1, a_2, a_3\} \in \mathbb{Z}^2$. The vectors a_1 and a_2 generate $\mathbb{Z}^2$. Viewed as points, a_1, a_2 and a_3 lie in the hyperplane indicated by the dashed line

Let's look at a simple example: We take $k = 2$ and $n = 3$. For the set $\mathcal{A}$ we choose

$$\mathcal{A} = \left\{ \begin{pmatrix} 1 \\ 0 \end{pmatrix}, \begin{pmatrix} 0 \\ 1 \end{pmatrix}, \begin{pmatrix} -1 \\ 2 \end{pmatrix} \right\}. \tag{9.102}$$

This set spans $\mathbb{Z}^2$ (the first two elements already span $\mathbb{Z}^2$) and the three points lie in a hyperplane, as shown in Fig. 9.5. The lattice of relations is generated by $(1, -2, 1)$, e.g.

$$\mathbb{L} = \left\{ (l, -2l, l) \in \mathbb{Z}^3 | l \in \mathbb{Z} \right\}. \tag{9.103}$$

The Γ-series associated to $\mathcal{A}$ and γ is

$$\phi_{\mathbb{L},\gamma}(x_1, x_2, x_3) = \sum_{l \in \mathbb{Z}} \frac{x_1^{l+\gamma_1}}{\Gamma(l + \gamma_1 + 1)} \frac{x_2^{-2l+\gamma_2}}{\Gamma(-2l + \gamma_2 + 1)} \frac{x_3^{l+\gamma_3}}{\Gamma(l + \gamma_3 + 1)}. \tag{9.104}$$

In general, we are interested in the space of local solutions of a GKZ hypergeometric system. Let us denote by $\Delta_{\mathcal{A}}$ the polytope in $\mathbb{R}^{k+1}$ defined by $\mathcal{A} = (a_1, \dots, a_n)$. If c is generic and $\Delta_{\mathcal{A}}$ admits a unimodular triangulation, than the dimension of the solution space is is given by

$$\text{vol}_0(\Delta_{\mathcal{A}}). \tag{9.105}$$

We may describe the solution space either in a geometrical language, using polytopes, or in an algebraic language, using D-modules. Let's start with the geometrical picture:

The key player in the geometrical picture is the secondary fan $N(\Sigma(\mathcal{A}))$ of $\mathcal{A}$. Let C be a maximal cone of the secondary fan $N(\Sigma(\mathcal{A}))$. Such a maximal cone corresponds to a regular triangulation of the primary polytope $\Delta_{\mathcal{A}}$. Let us denote by T_C the list of subsets of $\{1, \dots, n\}$ such that each subset denotes the indices of the vertices of the maximal simplices in this triangulation. For example, if $\{a_1, a_4, a_5, a_7\}$ are the vertices of a maximal simplex in the regular triangulation, we would include in T_C the subset $\{1, 4, 5, 7\}$. Let us now consider vectors $\gamma \in \mathbb{C}^n$ such that

$$\begin{aligned} c &= \mathcal{A}\gamma, \\ \exists\, J &\in T_C \quad \text{such that} \quad \gamma_j \in \mathbb{Z}_{\leq 0} \quad \text{for} \quad j \notin J. \end{aligned} \tag{9.106}$$

The first condition is the same as in Eq. (9.101). We call γ and γ' equivalent, if they only differ by $l \in \mathbb{L}$:

$$\gamma \sim \gamma' \Leftrightarrow \gamma - \gamma' \in \mathbb{L}. \tag{9.107}$$

Given a maximal cone C of the secondary fan and a vector $c \in \mathbb{C}^{k+1}$ we say that the vector c is C **-resonant** if the number of equivalence classes of solutions of Eq. (9.106) is less than $\text{vol}_0(\Delta_{\mathcal{A}})$. If c is not C-resonant, we say that c is generic.

We may now describe the solution space: We start in a geometric language. Let's assume that the primary polytope $\Delta_{\mathcal{A}}$ admits a unimodular triangulation. Let C be a maximal cone of the secondary fan $N(\Sigma(\mathcal{A}))$ and assume that c is generic. Then there are $\text{vol}_0(\Delta_{\mathcal{A}})$ inequivalent solutions of Eq. (9.106) and the corresponding Γ-series are linearly independent and span the solution space of the GKZ hypergeometric system.

Alternatively, we may give a description in algebraic terms: We first note that the differential equations of Eq. (9.92) and Eq. (9.92) generate an ideal I in the Weyl algebra A_n. For the solution of the GKZ system one constructs a basis of logarithmic series. A logarithmic series is a series of the form (in multi-index notation)

$$\sum_{\alpha\in A}\sum_{\beta\in B} c_{\alpha\beta} x^{\alpha} \left(\ln(x)\right)^{\beta}, \qquad \left(\ln(x)\right)^{\beta} = \left(\ln(x_1)\right)^{\beta_1} \dots \left(\ln(x_n)\right)^{\beta_n}, \tag{9.108}$$

where $A \subset \mathbb{C}^n$ is a discrete set and $B \subset \{0, \dots, b_{\text{max}}\}^n$ for some $b_{\text{max}} \in \mathbb{N}_0$. Given a weight vector $w \in \mathbb{Z}^n$ we may define a partial order on the terms of a logarithmic series by

$$x^{\alpha} \left(\ln(x)\right)^{\beta} > x^{\alpha'} \left(\ln(x)\right)^{\beta'} \Leftrightarrow \text{Re}\left(w \cdot \alpha\right) > \text{Re}\left(w \cdot \alpha'\right). \tag{9.109}$$

This partial order can be refined to a total order by using the lexicographic order to break ties. This total order is denoted by $>_w$. The initial term of a logarithmic series is the minimal term with respect to the total order $>_w$. For a generic weight vector w a basis of solutions of the GKZ system consists of logarithmic series, such that for each logarithmic series the exponent α of the initial term

$$x^{\alpha} \left(\ln(x)\right)^{\beta} \tag{9.110}$$

is a root of the indicial ideal $\text{ind}_w(I)$, i.e. $\alpha \in V(\text{ind}_w(I))$. Moreover, for a given root α the number of logarithmic series in the basis with initial term as in Eq. (9.110) equals the multiplicity of the root α in $V(\text{ind}_w(I))$. The initial terms of these logarithmic series differ in β. Starting from the initial term, it is possible to construct the full logarithmic series. For an algorithm we refer to the book by Saito, Sturmfels and Takayama [224].

9.2.4 Euler-Mellin Integrals

Euler-Mellin integrals are examples of $\mathcal{A}$-hypergeometric functions [230]. In order to define Euler-Mellin integrals we start with a function $p(z, x) = p(z_1, \dots, z_k, x_1, \dots, x_n)$ of the form

$$p\left(z_1, \dots, z_k, x_1, \dots, x_n\right) = \sum_{j=1}^{n} x_j \, z_1^{a_{1j}} \dots z_k^{a_{kj}}, \qquad a_{ij} \in \mathbb{Z}. \tag{9.111}$$

We call a function of this type a **Laurent polynomial**. Note that the exponents a_{ij} are allowed to be negative integers. If for all exponents a_{ij} we have $a_{ij} \in \mathbb{N}_0$, then the function $p(z, x)$ is a polynomial in z and x. Note further that $p(z, x)$ is linear in each x_j and that there are as many x_j's as there are monomials in the sum in Eq. (9.111).

Let us now consider a GKZ hypergeometric system with

$$\mathcal{A} = \begin{pmatrix} 1 & 1 & \dots & 1 \\ a_{11} & a_{12} & \dots & a_{1n} \\ \dots & & & \dots \\ a_{k1} & a_{k2} & \dots & a_{kn} \end{pmatrix}, \qquad c = \begin{pmatrix} -\nu_0 \\ -\nu_1 \\ \dots \\ -\nu_k \end{pmatrix}. \tag{9.112}$$

$\mathcal{A}$ is a $(k + 1) \times n$-matrix, c is a vector in $\mathbb{C}^{k+1}$. $\mathcal{A}$ defines a Laurent polynomial $p(x, z)$ through Eq. (9.111). We then consider integrals of the form

$$\int\limits_{C} d^k z \left(\prod_{j=1}^{k} z_j^{\nu_j - 1} \right) \left[p\left(z, x\right) \right]^{-\nu_0}, \tag{9.113}$$

where C is a cycle. An integral of the form as in Eq. (9.113) is called an **Euler-Mellin integral**.

We may extent the definition of Euler-Mellin integrals to integrals involving m Laurent polynomials $p_1(z, x), \dots, p_m(z, x)$ as follows: We start from a GKZ hypergeometric system of the form

$$\mathcal{A} = \begin{pmatrix} 1 & \dots & 1 & 0 & \dots & 0 & \dots & 0 & \dots & 0 \\ 0 & \dots & 0 & 1 & \dots & 1 & & 0 & \dots & 0 \\ \dots & & \dots & \dots & & \dots & & \dots & & \dots \\ 0 & \dots & 0 & 0 & \dots & 0 & & 1 & \dots & 1 \\ a_{111} & \dots & a_{11n_1} & a_{211} & \dots & a_{21n_2} & \dots & a_{m11} & \dots & a_{m1n_m} \\ \dots & & \dots & \dots & & \dots & & \dots & & \dots \\ a_{1k1} & \dots & a_{1kn_1} & a_{2k1} & \dots & a_{2kn_2} & \dots & a_{mk1} & \dots & a_{mkn_m} \end{pmatrix}, \quad c = \begin{pmatrix} -\mu_1 \\ -\mu_2 \\ \dots \\ -\mu_m \\ -\nu_1 \\ \dots \\ -\nu_k \end{pmatrix}. \tag{9.114}$$

We set $n = n_1 + \dots + n_m$. $\mathcal{A}$ is a $(k+m) \times n$-matrix, c is a vector in $\mathbb{C}^{k+m}$. $\mathcal{A}$ defines m Laurent polynomials $p_1(z, x), \dots, p_m(z, x)$ as follows: We set

$$p_i\,(z, x) = \sum_{j=1}^{n_i} x_{n_1 + \dots + n_{i-1} + j}\, z_1^{a_{i1j}} \dots z_k^{a_{ikj}}. \tag{9.115}$$

$p_1(z, x)$ depends on the first n_1 variables $x_1, \dots, x_{n_1}$, $p_2(z, x)$ on the next n_2 variables $x_{n_1+1}, \dots, x_{n_1+n_2}$ etc. The associated Euler-Mellin integral is

$$\int\limits_C d^k z \left(\prod_{j=1}^{k} z_j^{\nu_j - 1} \right) \left(\prod_{i=1}^{m} [p_i\,(z, x)]^{-\mu_i} \right). \tag{9.116}$$

9.2.5 *Feynman Integrals as GKZ Hypergeometric Functions*

Let us now consider a Feynman integral $I_{\nu_1 \dots \nu_{n_{\mathrm{int}}}}(D, x_1, \dots, x_{N_B})$. The Lee-Pomeransky representation reads

$$I_{\nu_1 \dots \nu_{n_{\mathrm{int}}}}\left(D, x_1, \dots, x_{N_B}\right) = C \int\limits_{z_j \geq 0} d^{n_{\mathrm{int}}} z \left(\prod_{j=1}^{n_{\mathrm{int}}} z_j^{\nu_j - 1} \right) [\mathcal{G}(z, x)]^{-\frac{D}{2}}, \tag{9.117}$$

with the prefactor (irrelevant for the discussion here)

$$C = \frac{e^{l \varepsilon \gamma_{\mathrm{E}}} \Gamma\left(\frac{D}{2}\right)}{\Gamma\left(\frac{(l+1))D}{2} - \nu\right) \prod\limits_{j=1}^{n_{\mathrm{int}}} \Gamma(\nu_j)} \tag{9.118}$$

and the Lee-Pomeransky polynomial

$$\mathcal{G}(z, x) = \mathcal{U}(z) + \mathcal{F}(z, x). \tag{9.119}$$

The Lee-Pomeransky representation of the Feynman integral is close to an Euler-Mellin integral as in Eq. (9.113), but not quite: First of all, the number of monomials in $\mathcal{G}(z, x)$ will in general not match the number of kinematic variables $x_1, \ldots, x_{N_B}$. This issue is easily fixed: We consider a generalised Lee-Pomeransky polynomial $G(z, x') = G(z_1, \ldots, z_{n_{\text{int}}}, x'_1, \ldots, x'_n)$ with as many variables x'_j as there are monomials in the original Lee-Pomeransky polynomial $\mathcal{G}(z, x)$. The original Lee-Pomeransky polynomial $\mathcal{G}(z, x)$ is then recovered as the special case, where the additional variables take special values. As an example consider the one-loop two-point function with equal internal masses. With $x = -p^2/m^2$ the original Lee-Pomeransky polynomial reads

$$\mathcal{G}(z_1, z_2, x) = z_1 + z_2 + (2 + x)\, z_1 z_2 + z_1^2 + z_2^2. \tag{9.120}$$

The generalised Lee-Pomeransky polynomial reads

$$G\left(z_1, z_2, x'_1, x'_2, x'_3, x'_4, x'_5\right) = x'_1 z_1 + x'_2 z_2 + x'_3 z_1 z_2 + x'_4 z_1^2 + x'_5 z_2^2. \tag{9.121}$$

We recover the original Lee-Pomeransky polynomial as

$$\mathcal{G}(z_1, z_2, x) = G\left(z_1, z_2, 1, 1, 2 + x, 1, 1\right). \tag{9.122}$$

The exponents of the monomials in $G(z, x')$ define a $(n_{\text{int}} + 1) \times n$-matrix $\mathcal{A}$ through Eq. (9.114). For the example from Eq. (9.121) we obtain

$$\mathcal{A} = \begin{pmatrix} 1 & 1 & 1 & 1 & 1 \\ 1 & 0 & 1 & 2 & 0 \\ 0 & 1 & 1 & 0 & 2 \end{pmatrix}. \tag{9.123}$$

Secondly, the matrix $\mathcal{A}$ should satisfy the two conditions of Eq. (9.90). The second condition does not pose any problem: As the first row of $\mathcal{A}$ contains only 1's, all points lie in a hyperplane (with first coordinate equal to one) and the $(n_{\text{int}} + 1)$-dimensional row vector

$$(1, \underbrace{0, \ldots, 0}_{n_{\text{int}}}) \tag{9.124}$$

defines the group homomorphism $h : \mathbb{Z}^{k+1} \to \mathbb{Z}$. It may happen that $\mathcal{A}$ does not satisfy the first condition (that the columns of $\mathcal{A}$ generate $\mathbb{Z}^{k+1}$ as an Abelian group. In this case we add additional monomials to the generalised Lee-Pomeransky polynomial $G(z, x')$ until $\mathcal{A}$ does satisfy the first condition. These additional monomials come with new variables x'_j and we take the limit $x'_j \to 0$ in the end.

Theorem 8 (Feynman integrals and) $\mathcal{A}$ *-hypergeometric functions Any Feynman integral*

$$I_{\nu_1 \dots \nu_{n_{\text{int}}}} (D, x_1, \dots, x_{N_B}) \tag{9.125}$$

depending on the kinematic variables $x_1, \dots, x_{N_B}$ *is a special case of a* $\mathcal{A}$*-hypergeometric function in more variables* $x'_1, \dots, x'_n$*, where the variables* x'_j *take special values.*

The relation between Feynman integrals and $\mathcal{A}$-hypergeometric functions has been considered in [231–234]. The above theorem is due to de la Cruz [233].

Exercise 87 *In this exercise we are going to prove Theorem 8. Let* $G(z, x') = G(z_1, \dots, z_{n_{\text{int}}}, x'_1, \dots, x'_n)$ *be a generalised Lee-Pomeransky polynomial such that the associated* $(n_{\text{int}} + 1) \times n$*-matrix* $\mathcal{A}$ *satisfies Eq. (9.90). Consider the integral*

$$I = C \int\limits_{z_j \geq 0} d^{n_{\text{int}}} z \left(\prod_{j=1}^{n_{\text{int}}} z_j^{\nu_j - 1} \right) \left[G\left(z, x'\right) \right]^{-\frac{D}{2}}. \tag{9.126}$$

Show that I *satisfies the differential equations in Eq. (9.92) and Eq. (9.93) with* $c = (-D/2, -\nu_1, \dots, -\nu_{n_{\text{int}}})^T$.

The number of variables $x'_1, \dots, x'_n$ for the $\mathcal{A}$-hypergeometric function can be quite large, as the following example illustrates: Consider the two-loop double box integral where all internal masses vanish and the external momenta are light-like: $p_1^2 = p_2^2 = p_3^2 = p_4^2$. This integral depends only on one kinematic variable, which can be taken as $x = s/t$. The original Lee-Pomeransky polynomial reads

$$\begin{aligned} \mathcal{G}(z_1, \dots, z_7, x) = \\ & z_1 z_5 + z_1 z_6 + z_1 z_7 + z_2 z_5 + z_2 z_6 + z_2 z_7 + z_3 z_5 + z_3 z_6 + z_3 z_7 + z_1 z_4 + z_2 z_4 + z_3 z_4 + z_4 z_5 + z_4 z_6 \\ & + z_4 z_7 + x z_2 z_3 z_4 + x z_2 z_3 z_5 + x z_2 z_3 z_6 + x z_2 z_3 z_7 + x z_5 z_6 z_1 + x z_5 z_6 z_2 + x z_5 z_6 z_3 + x z_5 z_6 z_4 \\ & + x z_2 z_4 z_6 + x z_3 z_4 z_5 + z_1 z_4 z_7. \end{aligned} \tag{9.127}$$

The original Lee-Pomeransky polynomial is a sum of 26 monomials. For the generalised Lee-Pomeransky polynomial G we therefore introduce 26 variables $x'_1, \dots, x'_{26}$. The generalised Lee-Pomeransky polynomial reads

$$\begin{aligned} G\left(z_1, \dots, z_7, x'_1, \dots, x'_{26}\right) = \\ & x'_1 z_1 z_5 + x'_2 z_1 z_6 + x'_3 z_1 z_7 + x'_4 z_2 z_5 + x'_5 z_2 z_6 + x'_6 z_2 z_7 + x'_7 z_3 z_5 + x'_8 z_3 z_6 + x'_9 z_3 z_7 + x'_{10} z_1 z_4 \\ & + x'_{11} z_2 z_4 + x'_{12} z_3 z_4 + x'_{13} z_4 z_5 + x'_{14} z_4 z_6 + x'_{15} z_4 z_7 + x'_{16} z_2 z_3 z_4 + x'_{17} z_2 z_3 z_5 + x'_{18} z_2 z_3 z_6 \\ & + x'_{19} z_2 z_3 z_7 + x'_{20} z_5 z_6 z_1 + x'_{21} z_5 z_6 z_2 + x'_{22} z_5 z_6 z_3 + x'_{23} z_5 z_6 z_4 + x'_{24} z_2 z_4 z_6 + x'_{25} z_3 z_4 z_5 \\ & + x'_{26} z_1 z_4 z_7. \end{aligned} \tag{9.128}$$

Thus we go from the case of one kinematic variable x to a $\mathcal{A}$ hypergeometric function in 26 variables. At the end we are interested in the limit, where the variables x'_j go either to 1 or x.

9.3 The Bernstein-Sato Polynomial

Given a polynomial $V(x)$ in n variables $x_1, \dots, x_n$, the Bernstein-Sato theorem [235, 236] states that there is an identity of the form

$$P(x, \partial) [V(x)]^{\nu+1} = B [V(x)]^{\nu}, \tag{9.129}$$

where $P(x, \partial)$ is a polynomial of x and $\partial = (\partial_1, \dots, \partial_n)$ with $\partial_j = \partial/\partial_j$. B and all coefficients of P are polynomials of ν and of the coefficients of $V(x)$. B is called the **Bernstein-Sato polynomial**. This can be generalised to several polynomials $V_1(x), \dots, V_r(x)$ in n variables $x_1, \dots, x_n$ [237]:

$$P(x, \partial) \prod_{j=1}^{r} [V_j(x)]^{\nu_j+1} = B \prod_{j=1}^{r} [V_j(x)]^{\nu_j}, \tag{9.130}$$

where as above $P(x, \partial)$ is a polynomial of x and $\partial = (\partial_1, \dots, \partial_n)$. B and all coefficients of P are polynomials of the ν_j's and of the coefficients of the $V_j(x)$'s.

As an example consider

$$V(x) = x^T A x + 2 w^T X + c, \tag{9.131}$$

where A is an invertible $(n \times n)$-matrix and w a n-vector. Then

$$P(x, \partial) [V(x)]^{\nu+1} = B [V(x)]^{\nu}, \tag{9.132}$$

with

$$P(x, \partial) = 1 - \frac{1}{2(\nu+1)} \left(x + w^T A^{-1}\right) \partial, \; B = c - w^T A^{-1} w. \tag{9.133}$$

The Bernstein-Sato polynomial has been applied in [237, 238] to one-loop integrals in the Feynman parameter representation: For one-loop integrals we have

$$I = \frac{e^{\varepsilon \gamma_E} \Gamma\left(\nu - \frac{D}{2}\right)}{\prod\limits_{j=1}^{n_{\mathrm{int}}} \Gamma(\nu_j)} \int\limits_{a_j \geq 0} d^{n_{\mathrm{int}}} a \; \delta\left(1 - \sum_{j=1}^{n_{\mathrm{int}}} a_j\right) \left(\prod_{j=1}^{n_{\mathrm{int}}} a_j^{\nu_j - 1}\right) [\mathcal{F}(a)]^{\frac{D}{2} - \nu}, \tag{9.134}$$

and the second graph polynomial $\mathcal{F}(a)$ is quadratic in the Feynman parameters. Using

$$[\mathcal{F}(a)]^{\frac{D}{2}-\nu} = \frac{1}{B} P\,(x, \partial)\,[\mathcal{F}(a)]^{1+\frac{D}{2}-\nu} \tag{9.135}$$

will raise the exponent of the second graph polynomial. Using partial integration for the differential operator $P(x, \partial)$ will either produce simpler integrals on the boundary of the Feynman parameter integration domain or integrals with modified exponents ν_j. This can be repeated, until the integer part of the exponent of the second graph polynomial is non-negative. At this point, all poles in the dimensional regularisation parameter ε originate from the various $1/B$-prefactors. The remaining integrals are finite and can be performed numerically.

Methods to compute $P(x, \partial)$ and B for a general single polynomial $V(x)$ can be found in the book by Saito, Sturmfels and Takayama [224].

Chapter 10
Sector Decomposition

Let us consider the Laurent expansion in the dimensional regularisation parameter ε of a Feynman integral:

$$I = \sum_{j=j_{\min}}^{\infty} \varepsilon^j \, I^{(j)}. \tag{10.1}$$

For precision calculations we are interested in the first few terms of this Laurent expansion. The coefficients $I^{(j)}$ are independent of ε and we may ask if there is a way to compute them numerically. In this section we will discuss the method of sector decomposition, which allows us to compute the coefficients $I^{(j)}$ by Monte Carlo integration. The essential step will be to manipulate the original integrand in such a way, that all $I^{(j)}$'s are given by integrable integrands. We will start from the Feynman parameter representation of Eq. (2.170)

$$I = \frac{e^{l\varepsilon\gamma_E}\Gamma\left(\nu - \frac{lD}{2}\right)}{\prod\limits_{j=1}^{n_{\text{int}}}\Gamma(\nu_j)} \int\limits_{a_j \geq 0} d^{n_{\text{int}}}a \; \delta\left(1 - \sum_{j=1}^{n_{\text{int}}} a_j\right) \left(\prod_{j=1}^{n_{\text{int}}} a_j^{\nu_j - 1}\right) \frac{\left[\mathcal{U}(a)\right]^{\nu - \frac{(l+1)D}{2}}}{\left[\mathcal{F}(a)\right]^{\nu - \frac{lD}{2}}}. \tag{10.2}$$

Depending on the exponents, potential singularities of the integrand come from the regions $a_j = 0$, $\mathcal{U} = 0$ or $\mathcal{F} = 0$. We already know that $\mathcal{U} \neq 0$ inside the integration, but there is the possibility that $\mathcal{U}$ vanishes on the boundary of the integration region. In order to keep the discussion simple we will assume in this chapter that all kinematic variables are in the Euclidean region. Then it follows that $\mathcal{F} \neq 0$ inside the integration, but $\mathcal{F}$ may vanish on the boundary of the integration region. It is possible to relax the condition that all kinematic variables should be in the Euclidean region by an appropriate deformation of the integration contour of the final integration.

S. Weinzierl, *Feynman Integrals*, UNITEXT for Physics,
https://doi.org/10.1007/978-3-030-99558-4_10

In mathematical terms, sector decomposition corresponds to a resolution of singularities by a sequence of blow-ups. We will discuss the relation between these topics.

Apart from the practical application of being able to compute numerically the coefficients of the Laurent expansion in ε, we may show that in the case where all kinematic variables are algebraic numbers in the Euclidean region the coefficients $I^{(j)}$ are numerical periods.

10.1 The Algorithm of Sector Decomposition

In this section we discuss the algorithm for iterated sector decomposition [239]. The starting point is an integral of the form

$$\int\limits_{x_j \geq 0} d^n x \; \delta(1 - \sum_{i=1}^{n} x_i) \left(\prod_{i=1}^{n} x_i^{\mu_i} \right) \prod_{j=1}^{r} \left[P_j(x) \right]^{\lambda_j}, \tag{10.3}$$

where $\mu_i = a_i + \varepsilon b_i$ and $\lambda_j = c_j + \varepsilon d_j$. The integration is over the standard simplex. The a's, b's, c's and d's are integers. The P's are polynomials in the variables $x_1, ..., x_n$. The polynomials are required to be non-zero inside the integration region, but may vanish on the boundaries of the integration region.

The Feynman parameter integral in Eq. (10.2) is—apart from a trivial prefactor—of the form as in Eq. (10.3). Equation (10.3) is slightly more general: We allow more than two polynomials, the polynomials are not required to be homogeneous and the exponents μ_i are not required to be integers, but are allowed to be of the form $\mu_i = a_i + \varepsilon b_i$, with $a_i, b_i \in \mathbb{Z}$.

The algorithm of sector decomposition consists of the following six steps:

Step 1: Convert all polynomials to homogeneous polynomials. This is easily done as follows: Due to the presence of the Dirac delta distribution in Eq. (10.3) we have

$$1 = x_1 + x_2 + \cdots + x_n. \tag{10.4}$$

For each polynomial P_j we determine the highest degree h_j and multiply all terms of lower degree by an appropriate power of $x_1 + x_2 + \cdots + x_n$. As an example consider $n = 2$ and $P = x_1 + x_1 x_2^2$. The homogenisation of P is

$$x_1 \left(x_1 + x_2 \right)^2 + x_1 x_2^2. \tag{10.5}$$

After step 1 all polynomials P_j are homogeneous polynomials of degree h_j.

Step 2: Decompose the integral into n primary sectors. This is done as follows: We write

$$\int\limits_{x_j \geq 0} d^n x = \sum_{l=1}^{n} \int\limits_{x_j \geq 0} d^n x \prod_{i=1, i \neq l}^{n} \theta(x_l \geq x_i). \tag{10.6}$$

The sum over l corresponds to the sum over the n primary sectors. In the l-th primary sector we make the substitution

$$x_j = x_l x_j' \quad \text{for } j \neq l. \tag{10.7}$$

As after step 1 each polynomial P_j is homogeneous of degree h_j we arrive at

$$\int\limits_{x_j \geq 0} d^n x\, \delta(1 - \sum_{i=1}^{n} x_i) \left(\prod_{i=1, i \neq l}^{n} \theta(x_l \geq x_i) \right) \left(\prod_{i=1}^{n} x_i^{a_i + \varepsilon b_i} \right) \prod_{j=1}^{r} \left[P_j(x) \right]^{c_j + \varepsilon d_j} = \tag{10.8}$$
$$\int\limits_0^1 \left(\prod_{i=1, i \neq l}^{n} dx_i\, x_i^{a_i + \varepsilon b_i} \right) \left(1 + \sum_{j=1, j \neq l}^{n} x_j \right)^c \prod_{j=1}^{r} \left[P_j(x_1, ..., x_{j-1}, 1, x_{j+1}, ..., x_n) \right]^{c_j + \varepsilon d_j},$$

where

$$c = -n - \sum_{i=1}^{n} (a_i + \varepsilon b_i) - \sum_{j=1}^{r} h_j \left(c_j + \varepsilon d_j \right). \tag{10.9}$$

Each primary sector is now a $(n-1)$-dimensional integral over the unit hyper-cube. Note that in the general case this decomposition introduces an additional polynomial factor

$$\left(1 + \sum_{j=1, j \neq l}^{n} x_j \right)^c. \tag{10.10}$$

Exercise 88 *Show that for a Feynman integral as in Eq. (10.2) we have in any primary sector $c = 0$ and therefore the additional factor is absent.*

The underlying reason for the statement that for any Feynman integral we always have $c = 0$ is the fact that the Feynman integral is a projective integral. In Eq. (10.3) we consider more general integrals. In particular we do not require that the integral in Eq. (10.3) descends to an integral on projective space. Therefore we have in general $c \neq 0$. In any case, this factor is just an additional polynomial. After an adjustment $(n-1) \to n$ and possibly $(r+1) \to r$ we therefore deal with integrals of the form

$$\int\limits_0^1 d^n x \prod_{i=1}^{n} x_i^{a_i + \epsilon b_i} \prod_{j=1}^{r} \left[P_j(x) \right]^{c_j + \epsilon d_j}. \tag{10.11}$$

This is now an integral over the unit hypercube. The polynomials P_j do not vanish inside the integration region. They may vanish on intersections of the boundary of the integration region with coordinate subspaces.

Step 3: Decompose the sectors iteratively into sub-sectors until each of the polynomials is of the form

$$P = x_1^{m_1} ... x_n^{m_n} \left(c + P'(x)\right), \tag{10.12}$$

where $c \neq 0$ and $P'(x)$ is a polynomial in the variables x_j without a constant term. In this case the monomial prefactor $x_1^{m_1} ... x_n^{m_n}$ can be factored out and the remainder contains a non-zero constant term. To convert P into the form (10.12) one chooses a subset $S = \{\alpha_1, ..., \alpha_k\} \subseteq \{1, ... n\}$ according to a strategy discussed in the next section. One decomposes the k-dimensional hypercube into k sub-sectors according to

$$\int_0^1 d^n x = \sum_{l=1}^{k} \int_0^1 d^n x \prod_{i=1, i \neq l}^{k} \theta\left(x_{\alpha_l} \geq x_{\alpha_i}\right). \tag{10.13}$$

In the l-th sub-sector one makes for each element of S the substitution

$$\begin{aligned} x_{\alpha_i} &= x'_{\alpha_l} x'_{\alpha_i} \quad \text{for } i \neq l, \\ x_{\alpha_i} &= x'_{\alpha_i} \quad\quad \text{for } i = l. \end{aligned} \tag{10.14}$$

This procedure is iterated, until all polynomials are of the form (10.12). At the end all polynomials contain a constant term. Each sub-sector integral is of the form as in Eq. (10.11), where every P_j is now different from zero in the whole integration domain. Hence the singular behaviour of the integral depends on the a_i and b_i, the a_i being integers.

Figure 10.1 illustrates this for the simple example $S = \{1, 2\}$. Equation (10.13) gives the decomposition into the two sectors $x_1 > x_2$ and $x_2 > x_1$. Equation (10.14) transforms the triangles into squares. This transformation is one-to-one for all points except the origin. The origin is replaced by the line $x_1 = 0$ in the first sector and by the line $x_2 = 0$ in the second sector. In mathematics this is known as a blow-up.

Step 4: The singular behaviour of the integral depends now only on the factor

$$\prod_{i=1}^{n} x_i^{a_i + \epsilon b_i}. \tag{10.15}$$

For every x_j with $a_j < 0$ we perform a Taylor expansion around $x_j = 0$ in order to extract the possible ϵ-poles. In the variable x_j we write

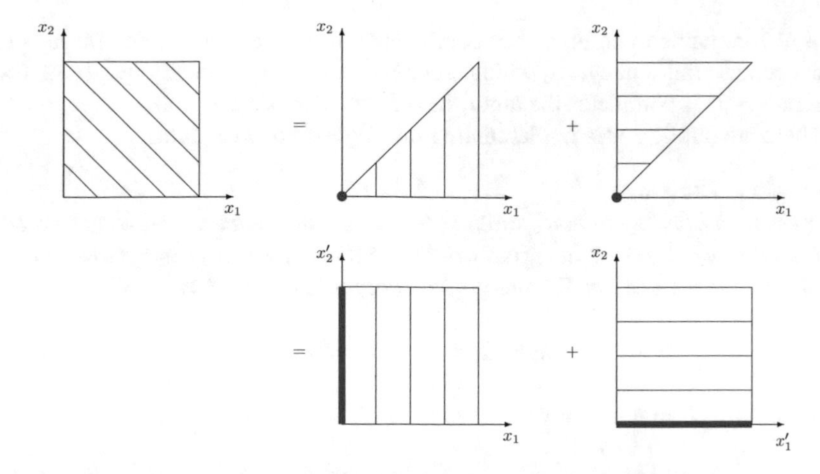

Fig. 10.1 Illustration of sector decomposition and blow-up for a simple example

$$\int_0^1 dx_j \, x_j^{a_j+b_j\varepsilon} \mathcal{I}(x_j) = \int_0^1 dx_j \, x_j^{a_j+b_j\varepsilon} \left(\sum_{p=0}^{|a_j|-1} \frac{x_j^p}{p!} \mathcal{I}^{(p)} + \mathcal{I}^{(R)}(x_j) \right) \quad (10.16)$$

where we defined $\mathcal{I}^{(p)} = \left(\frac{\partial}{\partial x_j}\right)^p \mathcal{I}(x_j)\Big|_{x_j=0}$. The remainder term

$$\mathcal{I}^{(R)}(x_j) = \mathcal{I}(x_j) - \sum_{p=0}^{|a_j|-1} \frac{x_j^p}{p!} \mathcal{I}^{(p)} \quad (10.17)$$

does not lead to ε-poles in the x_j-integration. The integration in the pole part can be carried out analytically:

$$\int_0^1 dx_j \, x_j^{a_j+b_j\varepsilon} \frac{x_j^p}{p!} \mathcal{I}^{(p)} = \frac{1}{a_j + b_j\varepsilon + p + 1} \frac{\mathcal{I}^{(p)}}{p!}. \quad (10.18)$$

This procedure is repeated for all variables x_j for which $a_j < 0$.

Step 5: All remaining integrals are now by construction finite. We can now expand all expressions in a Laurent series in ε

$$\sum_{i=A}^{B} C_i \epsilon^i + O\left(\epsilon^B\right) \quad (10.19)$$

and truncate to the desired order.

Step 6: It remains to compute the coefficients of the Laurent series. These coefficients contain finite integrals, which can be evaluated numerically by Monte Carlo integration. This completes the algorithm of sector decomposition.

There are public codes implementing this algorithm [240–246].

Digression. Blow-ups

Let's define a blow-up in mathematical terms: We start with the blow-up of a coordinate subspace. By choosing appropriate coordinates we may always arrange to be in this situation. Consider $\mathbb{C}^n$ *and the submanifold* Z *defined by*

$$x_1 = x_2 = ... = x_k = 0. \tag{10.20}$$

We have $\dim_{\mathbb{C}} Z = n - k$ *and*

$$Z = \left\{(0, ..., 0, x_{k+1}, ..., x_n) \mid x_j \in \mathbb{C},\ k+1 \le j \le n\right\}. \tag{10.21}$$

Let P *be the subset of*

$$\mathbb{C}^n \times \mathbb{CP}^{k-1}, \tag{10.22}$$

which satisfies the equations

$$x_i y_j - x_j y_i = 0, \qquad i, j \in \{1, ..., k\}, \tag{10.23}$$

where $[y_1 : y_2 : ... : y_k]$ *are homogeneous coordinates on* $\mathbb{CP}^{k-1}$. *In other words,*

$$P = \left\{(x, y) \in \mathbb{C}^n \times \mathbb{CP}^{k-1} | x_i y_j - x_j y_i = 0,\ 1 \le i, j \le k\right\}. \tag{10.24}$$

The blow-up map

$$\begin{aligned} \pi : P &\to \mathbb{C}^n \\ (x, y) &\to x, \end{aligned} \tag{10.25}$$

is an isomorphism away from Z*: To see this, let* $x \in \mathbb{C}^n \backslash Z$. *Then there is at least one* x_j *with* $x_j \neq 0$ *for* $1 \le j \le k$. *We then have*

$$y_i = \frac{x_i}{x_j} y_j, \qquad i \in \{1, ..., k\}. \tag{10.26}$$

Since $y \in \mathbb{CP}^{k-1}$ *it follows that* $y_j \neq 0$ *and we find*

$$y = [x_1 : x_2 : ... : x_k]. \tag{10.27}$$

On the other hand we have for $x \in Z$ that $x_j = 0$ for all $1 \leq j \leq k$ and the equations $x_i y_j - x_j y_i = 0$ are trivially satisfied for all $1 \leq i \leq k$. Therefore for $x \in Z$ any point $y \in \mathbb{CP}^{k-1}$ is allowed. We define

$$E = Z \times \mathbb{CP}^{k-1} \tag{10.28}$$

to be the **exceptional divisor**. *The restriction of π to E*

$$\pi|_E : E \to Z \tag{10.29}$$

gives a fibration with fibre $\mathbb{CP}^{k-1}$.

10.2 Hironaka's Polyhedra Game

In step 2 of the algorithm we choose at each iteration a subset $S = \{\alpha_1, \ldots, \alpha_k\} \subseteq \{1, \ldots n\}$, until we achieve the form of Eq. (10.12) for all polynomials P_j. Up to now we didn't specify how the subset S is chosen. We will now fill in the details.

Choosing the set S is a non-trivial issue. We have to ensure that we reach the form of Eq. (10.12) in a finite number of iterations. In particular, the iteration should not lead to an infinite loop.

To illustrate the problem, let us start with a strategy which does not work: Suppose we choose S as a minimal subset $S = \{\alpha_1, \ldots, \alpha_k\}$ such that at least one polynomial P_j vanishes for $x_{\alpha_1} = \cdots = x_{\alpha_k} = 0$. By a minimal set we mean a set which does not contain a proper subset having this property. If S contains only one element, $S = \{\alpha\}$, then the corresponding Feynman parameter factorises from P_j. A relative simple example shows, that this procedure may lead to an infinite loop: If one considers the polynomial

$$P(x_1, x_2, x_3) = x_1 x_3^2 + x_2^2 + x_2 x_3, \tag{10.30}$$

then the subset $S = \{1, 2\}$ is an allowed choice, as $P(0, 0, x_3) = 0$ and S is minimal. In the first sector the substitution (10.14) reads $x_1 = x_1'$, $x_2 = x_1' x_2'$, $x_3 = x_3'$. It yields

$$P(x_1, x_2, x_3) = x_1' x_3'^2 + x_1'^2 x_2'^2 + x_1' x_2' x_3' = x_1' \left(x_3'^2 + x_1' x_2'^2 + x_2' x_3' \right) = x_1' P\left(x_1', x_3', x_2'\right). \tag{10.31}$$

The original polynomial has been reproduced, which leads to an infinite recursion.

To avoid this situation we need a strategy for choosing S, for which we can show that the recursion always terminates. This is a highly non-trivial problem. It is closely related to the resolution of singularities of an algebraic variety over a field of characteristic zero [247]. The polyhedra game was introduced by Hironaka to illustrate the problem of resolution of singularities. The polyhedra game can be stated with little mathematics. Any solution to the polyhedra game will correspond to a strategy for choosing the subsets S within sector decomposition, which guarantee termination.

Hironaka's polyhedra game is played by two players, A and B. They are given a finite set M of points $m = (m_1, ..., m_n) \in \mathbb{N}_0^n$. We denote by $\Delta \subset \mathbb{R}_{\geq 0}^n$ the positive convex hull of the set M. It is given by the convex hull of the set

$$\bigcup_{m \in M} \left(m + \mathbb{R}_{\geq 0}^n \right). \tag{10.32}$$

The two players compete in the following game:

1. Player A chooses a non-empty subset $S \subseteq \{1, ..., n\}$.
2. Player B chooses one element i out of this subset S.

Then, according to the choices of the players, the components of all $(m_1, ..., m_n) \in M$ are replaced by new points $\left(m_1', ..., m_n'\right)$, given by:

$$\begin{aligned} m_j' &= m_j, \quad \text{if } j \neq i, \\ m_i' &= \sum_{j \in S} m_j - c, \end{aligned} \tag{10.33}$$

where for the moment we set $c = 1$. This defines the set M'. One then sets $M = M'$ and goes back to step 1. Player A wins the game if, after a finite number of moves, the polyhedron Δ is of the form

$$\Delta = m + \mathbb{R}_{\geq 0}^n, \tag{10.34}$$

i.e. generated by one point. If this never occurs, player B has won. The challenge of the polyhedra game is to show that player A always has a winning strategy, no matter how player B chooses his moves.

A simple illustration of Hironaka's polyhedra game in two dimensions is given in Fig. 10.2. Player A always chooses $S = \{1, 2\}$.

A winning strategy for Hironaka's polyhedra game translates directly into a strategy for choosing the sub-sectors within sector decomposition which guarantees termination. Without loss of generality we can assume that we have just one polynomial

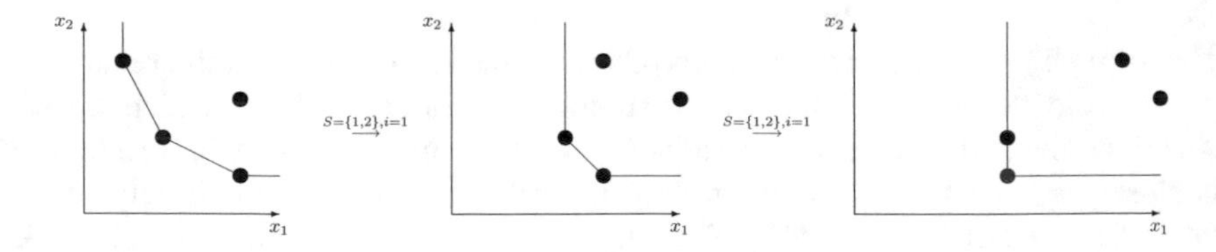

Fig. 10.2 Illustration of Hironaka's polyhedra game

P in Eq. (10.3). (If there are several polynomials, we obtain a single polynomial by multiplying them together. As only the zero-sets of the polynomials are relevant, the exponents λ_j can be neglected.) The polynomial P has the form

$$P = \sum_{i=1}^{p} c_i x_1^{m_1^{(i)}} x_2^{m_2^{(i)}} ... x_n^{m_n^{(i)}}. \tag{10.35}$$

The n-tuple $m^{(i)} = \left(m_1^{(i)}, ..., m_n^{(i)}\right)$ defines a point in $\mathbb{N}_0^n$ and $M = \left\{m^{(1)}, \ldots, m^{(p)}\right\}$ is the set of all such points. Substituting the parameters x_j according to equation (10.14) and factoring out a term x_i^c yields the same polynomial as replacing the powers m_j according to equation (10.33). In this sense, one iteration of the sector decomposition corresponds to one move in Hironaka's game. Reducing P to the form (10.12) is equivalent to achieving (10.34) in the polyhedra game. Finding a strategy which guarantees termination of the iterated sector decomposition corresponds to a winning strategy for player A in the polyhedra game. Note that we really need a strategy that guarantees player A's victory for every choice player B can take, because the sector decomposition has to be carried out in every appearing sector. In other words, we sample over all possible decisions of B.

There are winning strategies for Hironaka's polyhedra game [247–253]. Common to all strategies is a sequence of positive numbers associated to the polynomials. All strategies enforce this sequence to decrease with each step in the iteration with respect to lexicographical ordering. As the sequence cannot decrease forever, the algorithm is guaranteed to terminate. The actual construction of this sequence will differ for different strategies.

An an example we discuss here Spivakovsky's strategy [248], which was the first solution to Hironaka's polyhedra game. To state the strategy, we need a few auxiliary definitions: We define $\omega(\Delta) \in \mathbb{R}^n_{\geq 0}$ as the vector given by the minima of the individual coordinates of elements in Δ:

$$\omega_i = \min\{\nu_i \mid \nu \in \Delta\}, \quad i = 1, \ldots, n. \tag{10.36}$$

Furthermore we write $\tilde{\Delta} = \Delta - \omega(\Delta)$ and $\tilde{\nu}_i = \nu_i - \omega_i$. For a subset $\Gamma \subseteq \{1, ..., n\}$ we define

$$d_\Gamma(\Delta) = \min\left\{\sum_{j\in\Gamma} \nu_j \mid \nu \in \Delta\right\} \quad \text{and} \quad d(\Delta) = d_{\{1,\ldots,n\}}(\Delta). \tag{10.37}$$

We then define a sequence of sets

$$(I_0, \Delta_0, I_1, \Delta_1, ..., I_r, \Delta_r) \tag{10.38}$$

starting from

$$I_0 = \{1, \ldots, n\}, \quad \Delta_0 = \Delta. \tag{10.39}$$

For each Δ_k we define a set H_k by

$$H_k = \left\{ j \in I_k \mid \exists\, \nu \in \Delta_k \text{ such that } \sum_{i \in I_k} \nu_i = d(\Delta_k) \text{ and } \tilde{\nu}_j \neq 0 \right\}. \tag{10.40}$$

I_{k+1} is given by

$$I_{k+1} = I_k \backslash H_k. \tag{10.41}$$

In order to define Δ_{k+1} we first define for the two complementary subsets H_k and I_{k+1} of I_k the set

$$M_{H_k} = \left\{ \nu \in \mathbb{R}^{I_k}_{\geq 0} \mid \sum_{j \in H_k} \nu_j < 1 \right\} \tag{10.42}$$

and the projection

$$\begin{aligned} P_{H_k} &: M_{H_k} \longrightarrow \mathbb{R}^{I_{k+1}}_{\geq 0}, \\ P_{H_k}(\alpha, \beta) &= \frac{\alpha}{1 - |\beta|}, \quad \alpha \in \mathbb{R}^{I_{k+1}}_{\geq 0}, \quad \beta \in \mathbb{R}^{H_k}_{\geq 0}, \quad |\beta| = \sum_{j \in H_k} \beta_j. \end{aligned} \tag{10.43}$$

Then Δ_{k+1} is given by

$$\Delta_{k+1} = P_{H_k}\left(M_{H_k} \cap \left(\frac{\tilde{\Delta}_k}{d\left(\tilde{\Delta}_k\right)} \cup \Delta_k \right) \right), \tag{10.44}$$

where $\tilde{\Delta}_k = \Delta_k - \omega(\Delta_k)$. The sequence in Eq. (10.38) stops if either $d\left(\tilde{\Delta}_r\right) = 0$ or $\Delta_r = \emptyset$. Based on the sequence in Eq. (10.38) player A chooses now the set S as follows:

1. If $\Delta_r = \emptyset$, player A chooses $S = \{1, \ldots, n\} \backslash I_r$.
2. If $\Delta_r \neq \emptyset$, player A first chooses a minimal subset $\Gamma_r \subseteq I_r$, such that $\sum_{j \in \Gamma_r} \nu_j \geq 1$ for all $\nu \in \Delta_r$ and sets $S = (\{1, \ldots, n\} \backslash I_r) \cup \Gamma_r$.

To each stage of the game (i.e. to each Δ), we can associate a sequence of $2r + 2$ numbers

$$\delta(\Delta) = \left(d\left(\tilde{\Delta}\right), \#I_1, d\left(\tilde{\Delta}_1\right), \ldots, \#I_r, d\left(\tilde{\Delta}_r\right), d(\Delta_r) \right), \tag{10.45}$$

adopting the conventions $\tilde{\emptyset} = \emptyset$ and $d(\emptyset) = \infty$. The above strategy forces $\delta(\Delta)$ to decrease with each move with respect to lexicographical ordering. Further, it can be shown that $\delta(\Delta)$ cannot decrease forever. Hence player A is guaranteed to win. The proof is given in [248].

10.3 Numerical Periods

As a spin-off of the algorithm of sector decomposition we may prove a theorem related to the coefficients $I^{(j)}$ of the Laurent expansion in the dimensional regularisation parameter ε of a Feynman integral I.

In order to prepare the ground, we start with some sets of numbers: The natural numbers $\mathbb{N}$, the integer numbers $\mathbb{Z}$, the rational numbers $\mathbb{Q}$, the real numbers $\mathbb{R}$ and the complex numbers $\mathbb{C}$ are all well-known. More refined is already the set of algebraic numbers, denoted by $\overline{\mathbb{Q}}$. An algebraic number is a solution of a polynomial equation with rational coefficients:

$$x^n + a_{n-1}x^{n-1} + \cdots + a_0 = 0, \quad a_j \in \mathbb{Q}. \tag{10.46}$$

As all such solutions lie in $\mathbb{C}$, the set of algebraic numbers $\overline{\mathbb{Q}}$ is a sub-set of the complex numbers $\mathbb{C}$. Numbers which are not algebraic are called transcendental. The sets $\mathbb{N}$, $\mathbb{Z}$, $\mathbb{Q}$ and $\overline{\mathbb{Q}}$ are countable, whereas the sets $\mathbb{R}$, $\mathbb{C}$ and the set of transcendental numbers are uncountable.

We now introduce the set of numerical periods $\mathbb{P}$. The motivation originates in the theory of singly and doubly periodic functions $f(z)$ of a complex variable z: We know that the exponential function is a periodic function with period $(2\pi i)$:

$$\exp(z + 2\pi i) = \exp(z) \qquad \forall\, z \in \mathbb{C}. \tag{10.47}$$

In Chap. 13 we will discuss doubly periodic functions. A standard example for a doubly periodic function is Weierstrass's $\wp$-function. Let us denote the two periods of Weierstrass's $\wp$-function by ψ_1 and ψ_2:

$$\wp(z + \psi_1) \;=\; \wp(z + \psi_2) \;=\; \wp(z). \tag{10.48}$$

It is an observation that the period of the singly periodic function $\exp(z)$ and the periods ψ_1, ψ_2 of the doubly periodic function $\wp(z)$ can be expressed as integrals involving only algebraic functions: For the period of the exponential function we have

$$2\pi i = 2i \int\limits_{-1}^{1} \frac{dt}{\sqrt{1-t^2}}. \tag{10.49}$$

Weierstrass's $\wp$-function is associated to the elliptic curve $y^2 = 4x^3 - g_2 x - g_3$. Assume that the two constants g_2 and g_3 are algebraic numbers. The periods of Weierstrass's $\wp$-function can be written as

$$\psi_1 = 2 \int_{t_1}^{t_2} \frac{dt}{\sqrt{4t^3 - g_2 t - g_3}}, \quad \psi_2 = 2 \int_{t_3}^{t_2} \frac{dt}{\sqrt{4t^3 - g_2 t - g_3}}, \tag{10.50}$$

where t_1, t_2 and t_3 are the roots of the cubic equation $4t^3 - g_2 t - g_3 = 0$.

This observation motivated Kontsevich and Zagier [254] to define the set of numerical periods $\mathbb{P}$:

Numerical period:
A numerical period is a complex number whose real and imaginary parts are values of absolutely convergent integrals of rational functions with rational coefficients, over domains in $\mathbb{R}^n$ given by polynomial inequalities with rational coefficients.

We denote the set of numerical periods by $\mathbb{P}$. We may replace in the definition above any occurrence of "rational" with algebraic, this will not alter the set of numbers. The algebraic numbers are contained in the set of numerical periods: $\overline{\mathbb{Q}} \in \mathbb{P}$. In addition, $\mathbb{P}$ contains transcendental numbers, for example the transcendental number π

$$\pi = \iint\limits_{x^2+y^2 \le 1} dx\, dy, \tag{10.51}$$

or the transcendental number $\ln(2)$

$$\ln(2) = \int_1^2 \frac{dx}{x}. \tag{10.52}$$

On the other hand, it is conjectured that the basis of the natural logarithm e and Euler's constant γ_{E} are not periods. The number $(2\pi i)$ clearly is a period, but currently it is not known if the inverse $(2\pi i)^{-1}$ belongs to $\mathbb{P}$ or not. Although there are uncountably many numbers, which are not periods, only very recently an example for a number which is not a period has been found [255].

Periods are a countable set of numbers, lying between $\overline{\mathbb{Q}}$ and $\mathbb{C}$. The set of periods $\mathbb{P}$ is a $\overline{\mathbb{Q}}$-algebra. In particular the sum and the product of two periods are again periods.

Let us now turn to Feynman integrals:

Theorem 9 *Consider a Feynman integrals as in Eq. (10.2) with Laurent expansion as in Eq. (10.1). Assume that all kinematic invariants* $x_1, \dots, x_{N_B}$ *are in the Euclidean region and algebraic:*

$$x_j \geq 0, \;\; x_j \in \overline{\mathbb{Q}}, \qquad 1 \leq j \leq N_B. \tag{10.53}$$

Then the coefficients $I^{(j)}$ *of the Laurent expansion in* ε *are numerical periods:*

$$I^{(j)} \in \mathbb{P}. \tag{10.54}$$

The proof of this theorem follows from the algorithm of sector decomposition [256, 257]: If the kinematic variables x_j are in the Euclidean region and algebraic, the integral of each sector in the sector decomposition is a numerical period. As the sum of numerical periods is again a numerical period, the theorem follows.

10.4 Effective Periods and Abstract Periods

There is a more formal definition of periods as follows [254]: Let X be a smooth algebraic variety of dimension n defined over $\mathbb{Q}$ and $D \subset X$ a divisor with normal crossings. (A normal crossing divisor is a subvariety of dimension $n - 1$, which looks locally like a union of coordinate hyperplanes.) Further let ω be an algebraic differential form on X of degree n and Δ a singular n-chain on the complex manifold $X(\mathbb{C})$ with boundary on the divisor $D(\mathbb{C})$. We thus have a quadruple (X, D, ω, Δ). To each quadruple we can associate a complex number period(X, D, ω, Δ) called the period of the quadruple and given by the integral

$$\text{period}(X, D, \omega, \Delta) = \int\limits_{\Delta} \omega. \tag{10.55}$$

It is clear that the period of the quadruple is an element of $\mathbb{P}$, and that to any element $p \in \mathbb{P}$ one can find a quadruple, such that period$(X, D, \omega, \Delta) = p$. The period map is therefore surjective. The interesting question is whether the period map is also injective. As it stands above, the period map is certainly not injective for trivial reasons. For example, a simple change of variables can lead to a different quadruple, but does not change the period. One therefore considers equivalence classes of quadruples modulo relations induced by linearity in ω and Δ, changes of variables and Stokes' formula. The vector space over $\mathbb{Q}$ of the equivalence classes of quadruples (X, D, ω, Δ) is called the space of **effective periods** and denoted by $\mathcal{P}$. $\mathcal{P}$ is an algebra. It is conjectured that the period map from $\mathcal{P}$ to $\mathbb{P}$ is injective and therefore an isomorphism [254, 258, 259]. This would imply that all relations between numerical periods are due to linearity, change of variables and Stokes' formula. Let us summarise:

Effective period:
We consider quadruples (X, D, ω, Δ), where X is a smooth algebraic variety of dimension n defined over $\mathbb{Q}$, $D \subset X$ a divisor with normal crossings, ω an algebraic differential form on X of degree n and Δ a singular n-chain on the complex manifold $X(\mathbb{C})$ with boundary on the divisor $D(\mathbb{C})$.
Two quadruples (X, D, ω, Δ) and $(X', D', \omega', \Delta')$ are called equivalent, if

$$\int\limits_{\Delta} \omega = \int\limits_{\Delta'} \omega' \tag{10.56}$$

and this can be derived from linearity in ω and Δ, a change of variables and Stokes' formula.
The equivalence classes are called **effective periods** and the algebra of effective periods is denoted by $\mathcal{P}$. The period map

$$\begin{aligned} \text{period} : \mathcal{P} &\rightarrow \mathbb{P}, \\ (X, D, \omega, \Delta) &\rightarrow \int\limits_{\Delta} \omega \end{aligned} \tag{10.57}$$

maps every effective period to a numerical period.

In order to make the definition more concrete, we consider as an example the quadruple given by $X(\mathbb{C}) = \mathbb{C}\backslash\{0\}$, $D = \emptyset$, $\omega = dz/z$ and Δ the path along the unit circle in the counter-clockwise direction. We have

$$\text{period}\,(X, D, \omega, \Delta) = 2\pi i. \tag{10.58}$$

As a second example let us consider the quadruple $X(\mathbb{C}) = \mathbb{C}$, $D = \{1, 2\}$, $\omega = dz/z$ and Δ the path from 1 to 2 along the real line. We have

$$\text{period}\,(X, D, \omega, \Delta) = \ln(2)\,. \tag{10.59}$$

As in the case of numerical periods it is not known whether there is a quadruple in $\mathcal{P}$, whose period is $(2\pi i)^{-1}$. One therefore adjoins to $\mathcal{P}$ formally the inverse of the element whose period is $(2\pi i)$ and writes $\mathcal{P}[\frac{1}{2\pi i}]$ for the so obtained algebra. Elements of $\mathcal{P}[\frac{1}{2\pi i}]$ are called abstract periods.

Abstract period:
Adjoin formally to the algebra of effective periods an element I with

$$\text{period}\,(I) = \frac{1}{2\pi i}. \tag{10.60}$$

The enlarged algebra is called the algebra of abstract periods and denoted by

$$\mathcal{P}\left[\frac{1}{2\pi i}\right]. \tag{10.61}$$

Digression. Algebraic varieties over arbitrary fields
Let $\mathbb{F}$ be a field and $\overline{\mathbb{F}}$ its algebraic closure. Let $\mathbb{F}[t_1, ..., t_n]$ be the ring of polynomials over the field $\mathbb{F}$ in n variables $t_1, \ldots, t_n$. An element $f \in \mathbb{F}[t_1, ..., t_n]$ is a polynomial in $t_1, \ldots, t_n$ with coefficients from $\mathbb{F}$. Let $A \subset \mathbb{F}[t_1, ..., t_n]$ be a set of such polynomials. The corresponding affine algebraic set is given by

$$V\,(A) = \{\, x \in \overline{\mathbb{F}}^n \mid f(x) = 0 \;\; \forall f \in A \,\}. \tag{10.62}$$

Note that one takes the algebraic closure $\overline{\mathbb{F}}$ of $\mathbb{F}$.

Let's now specialise to the case where $\mathbb{F}$ is a subfield of $\mathbb{C}$. The most important example is given by $\mathbb{F} = \mathbb{Q}$ (the rational numbers). The algebraic closure is $\overline{\mathbb{F}} = \overline{\mathbb{Q}}$ (the algebraic numbers). Let $A \subset \mathbb{Q}[t_1, ..., t_n]$ and denote

$$X = \{\, x \in \overline{\mathbb{Q}}^n \mid f(x) = 0 \;\; \forall f \in A \,\}. \tag{10.63}$$

By $X\,(\mathbb{C})$ we understand the affine algebraic set

$$X\,(\mathbb{C}) = \{\, x \in \mathbb{C}^n \mid f(x) = 0 \;\; \forall f \in A \,\}. \tag{10.64}$$

To see the difference between X and $X\,(\mathbb{C})$ consider $A = \{t_1^2 + t_2^2 - 25\}$. The point

$$(x_1, x_2) = \left(\pi, \sqrt{25 - \pi^2}\right) \tag{10.65}$$

is a point of $X\,(\mathbb{C})$, but not of X (the coordinates of any point of X are algebraic numbers). We may equip $X\,(\mathbb{C})$ with the topology induced from the standard topology on $\mathbb{C}^n$. (This means in particular that we are not using the Zariski topology, which is otherwise widely used in algebraic geometry.) The set $X\,(\mathbb{C})$ together with this topology becomes then a topological space, usually denoted by X^{an} and called the **analytification** *of X.*

Chapter 11
Hopf Algebras, Coactions and Symbols

In this chapter we investigate more formal aspects of Feynman integrals. We first introduce Hopf algebras and discuss where they appear in particle physics. We then focus on multiple polylogarithms. There are three Hopf algebras (with different coproducts) associated with multiple polylogarithms. The first two are combinatorial in nature and stem from the shuffle algebra and the quasi-shuffle algebra, respectively. The third one is motivic.We introduce motivic multiple polylogarithms (sounds complicated, but in the end it boils down to the fact that we introduce objects, which have all the known relations of multiple polylogarithms and no other relations). We also introduce the de Rham multiple polylogarithms. The de Rham multiple polylogarithm form a Hopf algebra. This Hopf algebra coacts on the motivic multiple polylogarithms, turning the motivic multiple polylogarithms into a comodule. We then study the symbol and the iterated coaction of multiple polylogarithms. There is a practical application: The symbol (or the iterated coaction) can be used to simplify long expressions of multiple polylogarithms.

Multiple polylogarithms are in general multi-valued functions. In the last section of this chapter we will study a systematic method to associate a single-valued function to a multiple polylogarithm.

Textbooks and lecture notes on Hopf algebras can be found in [260–264].

11.1 Hopf Algebras

Let us start with a brief history of Hopf algebras: Hopf algebras were introduced in mathematics in 1941 to describe similar aspects of groups and algebras in a unified manner [265]. An article by Woronowicz in 1987 [266], which provided explicit examples of non-trivial (non-cocommutative) Hopf algebras, triggered the interest of the physics community. This led to applications of Hopf algebras in the field of integrable systems and quantum groups. In physics, Hopf algebras received a

S. Weinzierl, *Feynman Integrals*, UNITEXT for Physics,
https://doi.org/10.1007/978-3-030-99558-4_11

further boost in 1998, when Kreimer and Connes re-examined the renormalisation of quantum field theories and showed that the combinatorial aspects of renormalisation can be described by a Hopf algebra structure [267, 268]. Since then, Hopf algebras have appeared in several facets of physics.

Let us now consider the definition of a Hopf algebra. Let R be a commutative ring with unit 1. An unitial associative algebra over the ring R is an R-module together with an associative multiplication $\cdot$ and a unit e. In this chapter we will always assume that the algebra has a unit and that the multiplication is associative. In this chapter we simply write "algebra" whenever we mean a unitial associative algebra. In physics, the ring R will almost always be a field $\mathbb{F}$ (examples are the rational numbers $\mathbb{Q}$, the real numbers $\mathbb{R}$, or the complex number $\mathbb{C}$). In this case the R-module will actually be a $\mathbb{F}$-vector space. Note that the unit e can be viewed as a map from R to A and that the multiplication $\cdot$ can be viewed as a map from the tensor product $A \otimes A$ to A (e.g., one takes two elements from A, multiplies them, and obtains one element as the outcome):

$$\begin{array}{lrl} \text{Multiplication:} & \cdot \, : & A \otimes A \to A, \\ \text{Unit:} & e \, : & R \to A. \end{array} \tag{11.1}$$

Instead of multiplication and a unit, a coalgebra has the dual structures, obtained by reversing the arrows in Eq. (11.1): a comultiplication Δ and a counit $\bar{e}$. The **counit** $\bar{e}$ is a map from A to R, whereas the **comultiplication** Δ is a map from A to $A \otimes A$:

$$\begin{array}{lrl} \text{Comultiplication:} & \Delta \, : & A \to A \otimes A, \\ \text{Counit:} & \bar{e} \, : & A \to R. \end{array} \tag{11.2}$$

We will always assume that the comultiplication Δ is **coassociative**. But what does coassociativity mean? We can easily derive it from associativity as follows: For $a, b, c \in A$ associativity requires

$$(a \cdot b) \cdot c = a \cdot (b \cdot c) \,. \tag{11.3}$$

We can re-write condition (11.3) in the form of a commutative diagram:

$$\begin{array}{ccc} A \otimes A \otimes A & \xrightarrow{\text{id} \otimes \cdot} & A \otimes A \\ \Big\downarrow {\scriptstyle \cdot \otimes \text{id}} & & \Big\downarrow {\scriptstyle \cdot} \\ A \otimes A & \xrightarrow{\cdot} & A \end{array} \tag{11.4}$$

We obtain the condition for coassociativity by reversing all arrows and by exchanging multiplication with comultiplication. We thus obtain the following commutative diagram:

$$\begin{array}{ccc} A & \xrightarrow{\Delta} & A \otimes A \\ \downarrow{\scriptstyle \Delta} & & \downarrow{\scriptstyle \Delta \otimes \mathrm{id}} \\ A \otimes A & \xrightarrow{\mathrm{id} \otimes \Delta} & A \otimes A \otimes A \end{array} \tag{11.5}$$

The general form of the coproduct is

$$\Delta(a) = \sum_i a_i^{(1)} \otimes a_i^{(2)}, \tag{11.6}$$

where $a_i^{(1)}$ denotes an element of A appearing in the first slot of $A \otimes A$ and $a_i^{(2)}$ correspondingly denotes an element of A appearing in the second slot. **Sweedler's notation** [260] consists of omitting the dummy index i and the summation symbol:

$$\Delta(a) = a^{(1)} \otimes a^{(2)} \tag{11.7}$$

The sum is implicitly understood. This is similar to Einstein's summation convention, except that the dummy summation index i is also dropped. The superscripts $^{(1)}$ and $^{(2)}$ indicate that a sum is involved. Using Sweedler's notation, coassociativity is equivalent to

$$a^{(1)(1)} \otimes a^{(1)(2)} \otimes a^{(2)} = a^{(1)} \otimes a^{(2)(1)} \otimes a^{(2)(2)}. \tag{11.8}$$

As it is irrelevant whether we apply the second coproduct to the first or the second factor in the tensor product of $\Delta(a)$, we can simply write

$$\Delta^2(a) = a^{(1)} \otimes a^{(2)} \otimes a^{(3)}. \tag{11.9}$$

If the coproduct of an element $a \in A$ is of the form

$$\Delta(a) = a \otimes a, \tag{11.10}$$

then a is referred to as a **group-like element**. If the coproduct of a is of the form

$$\Delta(a) = a \otimes e + e \otimes a, \tag{11.11}$$

then a is referred to as a **primitive element**.

In an algebra we have for the unit 1 of the underlying ring R and the unit e of the algebra the relation

$$a \;=\; 1 \cdot a \;=\; e \cdot a \;=\; a \tag{11.12}$$

for any element $a \in A$ (together with the analogue relation $a = a \cdot 1 = a \cdot e = a$). In terms of commutative diagrams this is expressed as

$$\begin{array}{ccc} A \otimes A & = & A \otimes A \\ {\scriptstyle e\otimes\mathrm{id}}\uparrow & & \downarrow{\scriptstyle \cdot} \\ R \otimes A & \cong & A \end{array} \qquad \begin{array}{ccc} A \otimes A & = & A \otimes A \\ {\scriptstyle \mathrm{id}\otimes e}\uparrow & & \downarrow{\scriptstyle \cdot} \\ A \otimes R & \cong & A \end{array} \tag{11.13}$$

In a coalgebra we have the dual relations obtained from Eq. (11.13) by reversing all arrows and by exchanging multiplication with comultiplication as well as by exchanging the unit e with the counit $\bar{e}$:

$$\begin{array}{ccc} A \otimes A & = & A \otimes A \\ {\scriptstyle \bar{e}\otimes\mathrm{id}}\downarrow & & \uparrow{\scriptstyle \Delta} \\ R \otimes A & \cong & A \end{array} \qquad \begin{array}{ccc} A \otimes A & = & A \otimes A \\ {\scriptstyle \mathrm{id}\otimes \bar{e}}\downarrow & & \uparrow{\scriptstyle \Delta} \\ A \otimes R & \cong & A \end{array} \tag{11.14}$$

A **bi-algebra** is an algebra and a coalgebra at the same time, such that the two structures are compatible with each other. In terms of commutative diagrams, the compatibility condition between the product and the coproduct is expressed as

$$\begin{array}{ccccc} A \otimes A & \xrightarrow{\cdot} & A & \xrightarrow{\Delta} & A \otimes A \\ \downarrow{\scriptstyle \Delta\otimes\Delta} & & & & \uparrow{\scriptstyle \cdot\otimes\cdot} \\ A \otimes A \otimes A \otimes A & & \xrightarrow{\mathrm{id}\otimes\tau\otimes\mathrm{id}} & & A \otimes A \otimes A \otimes A \end{array} \tag{11.15}$$

where $\tau : A \otimes A \to A \otimes A$ is the map, which exchanges the entries in the two slots: $\tau(a \otimes b) = b \otimes a$. Using Sweedler's notation, the compatibility between the multiplication and comultiplication is expressed as

$$\Delta(a \cdot b) = \left(a^{(1)} \cdot b^{(1)}\right) \otimes \left(a^{(2)} \cdot b^{(2)}\right). \tag{11.16}$$

It is common practice to write the right-hand side of Eq. (11.16) as

$$\left(a^{(1)} \cdot b^{(1)}\right) \otimes \left(a^{(2)} \cdot b^{(2)}\right) = \Delta(a)\,\Delta(b). \tag{11.17}$$

In addition, there is a compatibility condition between the unit and the coproduct

$$\begin{array}{ccc} R \otimes R \cong R & \xrightarrow{e} & A \\ {\scriptstyle e\otimes e}\downarrow & & \downarrow{\scriptstyle \Delta} \\ A \otimes A & = & A \otimes A \end{array} \tag{11.18}$$

as well as a compatibility condition between the counit and the product, which is dual to Eq. (11.18):

$$\begin{array}{ccc} A & \xrightarrow{\bar{e}} & R \cong R \otimes R \\ \uparrow \cdot & & \uparrow \bar{e}\otimes\bar{e} \\ A \otimes A & = & A \otimes A \end{array} \tag{11.19}$$

The commutative diagrams in Eqs. (11.18) and (11.19) are equivalent to

$$\Delta e = e \otimes e, \quad \text{and} \quad \bar{e}\,(a \cdot b) = \bar{e}\,(a)\,\bar{e}\,(b)\,, \quad \text{respectively.} \tag{11.20}$$

An algebra A is **commutative** if for all $a, b \in A$ one has

$$a \cdot b = b \cdot a. \tag{11.21}$$

A coalgebra A is **cocommutative** if for all $a \in A$ one has

$$a^{(1)} \otimes a^{(2)} = a^{(2)} \otimes a^{(1)}. \tag{11.22}$$

With the help of the swap map τ we may express commutativity and cocommutativity equivalently as

$$\cdot\,\tau = \cdot, \qquad \text{and} \qquad \tau\Delta = \Delta, \qquad \text{respectively.} \tag{11.23}$$

A **Hopf algebra** is a bi-algebra with an additional map from A to A, known as the **antipode** S, which fulfils

$$\begin{array}{ccccc} A & \xrightarrow{\bar{e}} & R & \xrightarrow{e} & A \\ \downarrow \Delta & & & & \uparrow \cdot \\ A \otimes A & & \xrightarrow[S\otimes\text{id}]{\text{id}\otimes S} & & A \otimes A \end{array} \tag{11.24}$$

An equivalent formulation is

$$a^{(1)} \cdot S\left(a^{(2)}\right) \;=\; S\left(a^{(1)}\right) \cdot a^{(2)} \;=\; e \cdot \bar{e}(a). \tag{11.25}$$

If a bi-algebra has an antipode (satisfying the commutative diagram (11.24) or Eq. (11.25)), then the antipode is unique.

If a Hopf algebra A is either commutative or cocommutative, then

$$S^2 = \text{id}. \tag{11.26}$$

A bi-algebra A is **graded**, if it has a decomposition

$$A = \bigoplus_{n \geq 0} A_n, \tag{11.27}$$

with

$$A_n \cdot A_m \subseteq A_{n+m}, \quad \Delta(A_n) \subseteq \bigoplus_{k+l=n} A_k \otimes A_l. \tag{11.28}$$

Elements in A_n are said to have degree n. The bi-algebra is **graded connected**, if in addition one has

$$A_0 = R \cdot e. \tag{11.29}$$

It is useful to know that a graded connected bi-algebra is automatically a Hopf algebra [269]. In a graded Hopf algebra we denote by $\Delta_{i_1,\dots,i_k}(a)$ the projection of $\Delta^n(a)$ (where $n = i_1 + \cdots + i_k$) onto

$$A_{i_1} \otimes \cdots \otimes A_{i_k}, \tag{11.30}$$

i.e. we only keep those terms which have degree i_j in the the jth slot.

Let A be a graded connected Hopf algebra. We have

$$\text{Ker}(\bar{e}) = \bigoplus_{n \geq 1} A_n, \tag{11.31}$$

e.g $\bar{e}(a) = 0$ if $a \in A_n$ with $n \geq 1$. For the coproduct one often writes for $a \in A_n$ with $n \geq 1$

$$\Delta(a) = a \otimes e + e \otimes a + \tilde{\Delta}(a). \tag{11.32}$$

$\tilde{\Delta}$ is called the **reduced coproduct**. We have

$$\tilde{\Delta}(a) \in \bigoplus_{\substack{k+l=n \\ k,l>0}} A_k \otimes A_l \tag{11.33}$$

For a graded connected Hopf algebra A the antipode is recursively determined by $S(e) = e$ and

$$S(a) = -a - \cdot (S \otimes \text{id})\, \tilde{\Delta}(a) \tag{11.34}$$

for $a \in A_n$ with $n \geq 1$. In the second term on the right-hand side of Eq. (11.34) we first apply the reduced coproduct and apply the antipode to the first tensor slot, while the second is left as it is. By the definition of the reduced coproduct, the weight of the entry in the first tensor slot (as well as the weight in the second tensor slot) is less than the original weight of a, therefore the recursion terminates. At the end we multiply the two entries in the two tensor slots together, indicated by the little multiplication sign "$\cdot$" just in front of $(S \otimes \text{id})$.

Let us elaborate on the antipode. Let C be a coalgebra over the ring R and A an algebra over the ring R. Both are R-modules. Let us denote by $\text{Hom}(C, A)$ the set of linear maps from C to A, i.e. for $\varphi \in \text{Hom}(C, A)$ we have

$$\varphi(\lambda_1 a_1 + \lambda_2 a_2) = \lambda_1 \varphi(a_1) + \lambda_2 \varphi(a_2), \qquad \lambda_1, \lambda_2 \in R, \quad a_1, a_2 \in C. \quad (11.35)$$

As C is a coalgebra and A an algebra we may define a product in $\text{Hom}(C, A)$ as follows: For $\varphi_1, \varphi_2 \in \text{Hom}(C, A)$ we set

$$\varphi_1 * \varphi_2 = \cdot \, (\varphi_1 \otimes \varphi_2) \, \Delta. \quad (11.36)$$

This product is called the **convolution product**. Δ denotes the coproduct in C, the multiplication in A is denoted by "$\cdot$". The convolution product is associative.

Exercise 89 *Show that the convolution product is associative:*

$$(\varphi_1 * \varphi_2) * \varphi_3 = \varphi_1 * (\varphi_2 * \varphi_3). \quad (11.37)$$

The convolution product has a neutral element, given by $e\bar{e}$, where e denotes the unit in A and $\bar{e}$ the counit in C.

Exercise 90 *Show that* $1_{\text{Hom}} = e\bar{e} \in \text{Hom}(C, A)$ *is a neutral element for the convolution product, i.e.*

$$\varphi * 1_{\text{Hom}} = 1_{\text{Hom}} * \varphi = \varphi. \quad (11.38)$$

Let H be a Hopf algebra and let us specialise to the case $C = H$. In particular we now have a product in H. We therefore restrict our attention to algebra homomorphisms $\text{AlgHom}(H, A)$ from H to A which preserve the unit, e.g. in additon to Eq. (11.35) we require

$$\begin{aligned} \varphi(a_1 \cdot a_2) &= \varphi(a_1) \cdot \varphi(a_2), \qquad a_1, a_2 \in H, \\ \varphi(e_H) &= e_A, \end{aligned} \quad (11.39)$$

where e_H denotes the unit in H and e_A denotes the unit in A. As before $1_{\text{AlgHom}} = e_A \bar{e}_H \in \text{AlgHom}(H, A)$ is a neutral element for the convolution product. We may now ask, given $\varphi \in \text{AlgHom}(H, A)$ is there an inverse element $\varphi^{-1} \in \text{AlgHom}(H, A)$, such that

$$\varphi * \varphi^{-1} = \varphi^{-1} * \varphi = 1_{\text{AlgHom}}. \quad (11.40)$$

There is, and the inverse element is given by

$$\varphi^{-1} = \varphi S, \tag{11.41}$$

where S denotes the antipode in H.

Exercise 91 *Show that* $\varphi^{-1} = \varphi S$ *is an inverse element to* $\varphi \in \text{AlgHom}(H, A)$.

The Exercises 89–91 show that the unit-preserving algebra homomorphisms $\text{AlgHom}(H, A)$ form a group with the convolution product. If A is in addition commutative, we call a unit-preserving algebra homomorphism $\varphi \in \text{AlgHom}(H, A)$ a **character of the Hopf algebra H**.

Now let us specialise even further: We take $A = C = H$ and consider $\text{AlgHom}(H, H)$. The neutral element with respect to the convolution product is as before

$$1_{\text{AlgHom}} = e\bar{e} \in \text{AlgHom}(H, H), \tag{11.42}$$

where e denotes the unit in H and $\bar{e}$ denotes the counit in H. Let us now consider the identity map $\text{id} \in \text{AlgHom}(H, H)$,

$$\text{id}\,(a) = a, \qquad \forall\, a \in H. \tag{11.43}$$

Please don't confuse the maps 1_{AlgHom} and id, they are different. This is most easily seen for a graded connected Hopf algebra. Let $a \in A_n$ and $n \geq 1$. Then (see Eq. (11.31))

$$\begin{aligned} 1_{\text{AlgHom}}\,(a) = e\bar{e}\,(a) &= 0, \\ \text{id}\,(a) &= a. \end{aligned} \tag{11.44}$$

The inverse element of the identity map id with respect to the covolution product is given by the antipode

$$\text{id}^{-1} = S. \tag{11.45}$$

Let us now consider a few examples of Hopf algebras.

Example 11.1 The Group Algebra

Let $\mathbb{F}$ be a field and let G be a group. We denote by $\mathbb{F}[G]$ the vector space with basis G over the field $\mathbb{F}$. Then $\mathbb{F}[G]$ is an algebra with the multiplication given by the group multiplication. The counit, the coproduct, and the antipode are defined for the basis elements $g \in G$ as follows: The counit $\bar{e}$ is given by:

$$\bar{e}\,(g) = 1. \tag{11.46}$$

The coproduct Δ is given by:

$$\Delta\,(g) = g \otimes g. \tag{11.47}$$

Thus, the basis elements $g \in G$ are goup-like elements in $\mathbb{F}[G]$. The antipode S is given by:

$$S(g) = g^{-1}. \tag{11.48}$$

Having defined the counit, the coproduct, and the antipode for the basis elements $g \in G$, the corresponding definitions for arbitrary vectors in $\mathbb{F}[G]$ are obtained by linear extension. $\mathbb{F}[G]$ is a cocommutative Hopf algebra. $\mathbb{F}[G]$ is commutative if G is commutative.

Example 11.2 Lie Algebras

A Lie algebra $\mathfrak{g}$ is not necessarily associative nor does it have a unit. To overcome this obstacle one considers the **universal enveloping algebra** $U(\mathfrak{g})$, obtained from the tensor algebra $T(\mathfrak{g})$ by factoring out the ideal generated by

$$X \otimes Y - Y \otimes X - [X, Y], \tag{11.49}$$

with $X, Y \in \mathfrak{g}$. The universal enveloping algebra $U(\mathfrak{g})$ is a Hopf algebra. The counit $\bar{e}$ is given by:

$$\bar{e}(e) = 1, \quad \bar{e}(X) = 0. \tag{11.50}$$

The coproduct Δ is given by:

$$\Delta(e) = e \otimes e, \quad \Delta(X) = X \otimes e + e \otimes X. \tag{11.51}$$

Thus, the elements $X \in \mathfrak{g}$ are primitive elements in $U(\mathfrak{g})$. The antipode S is given by:

$$S(e) = e, \quad S(X) = -X. \tag{11.52}$$

$U(\mathfrak{g})$ is a non-commutative cocommutative Hopf algebra.

Example 11.3 Quantum SU(2)

The Lie algebra $su(2)$ is generated by three generators $H, X_\pm$ with

$$\left[H, X_\pm\right] = \pm 2X_\pm, \quad \left[X_+, X_-\right] = H. \tag{11.53}$$

To obtain the deformed universal enveloping algebra $U_q(su(2))$, the last relation is replaced with [262, 270]

$$\left[X_+, X_-\right] = \frac{q^H - q^{-H}}{q - q^{-1}}, \tag{11.54}$$

where q is the deformation parameter. The undeformed Lie algebra $su(2)$ is recovered in the limit $q \to 1$. The counit $\bar{e}$ is given by:

$$\bar{e}(e) = 1, \ \ \bar{e}(H) = \bar{e}(X_\pm) = 0. \tag{11.55}$$

The coproduct Δ is given by:

$$\begin{aligned} \Delta(H) &= H \otimes e + e \otimes H, \\ \Delta(X_\pm) &= X_\pm \otimes q^{H/2} + q^{-H/2} \otimes X_\pm. \end{aligned} \tag{11.56}$$

The antipode S is given by:

$$S(H) = -H, \ \ S(X_\pm) = -q^{\pm 1} X_\pm. \tag{11.57}$$

$U_q(su(2))$ is a non-commutative non-cocommutative Hopf algebra.

Example 11.4 Symmetric Algebras

Let V be a finite dimensional vector space with basis $\{v_i\}$. The symmetric algebra $\mathrm{Sym}(V)$ is the direct sum

$$\mathrm{Sym}(V) = \bigoplus_{n=0}^{\infty} \mathrm{Sym}^n(V), \tag{11.58}$$

where $\mathrm{Sym}^n(V)$ is spanned by elements of the form $v_{i_1} v_{i_2} \dots v_{i_n}$ with $i_1 \le i_2 \le \dots \le i_n$. The multiplication is defined by

$$\left(v_{i_1} v_{i_2} \dots v_{i_m}\right) \cdot \left(v_{i_{m+1}} v_{i_{m+2}} \dots v_{i_{m+n}}\right) = v_{i_{\sigma(1)}} v_{i_{\sigma(2)}} \dots v_{i_{\sigma(m+n)}}, \tag{11.59}$$

where σ is a permutation on $m+n$ elements such that $i_{\sigma(1)} \le i_{\sigma(2)} \le \dots \le i_{\sigma(m+n)}$. The counit $\bar{e}$ is given by:

$$\bar{e}(e) = 1, \qquad \bar{e}(v_1 v_2 \dots v_n) = 0. \tag{11.60}$$

The coproduct Δ is given for the basis elements v_i by:

$$\Delta(v_i) = v_i \otimes e + e \otimes v_i. \tag{11.61}$$

Using (11.16) one obtains for a general element of $\mathrm{Sym}(V)$

$$\begin{aligned} \Delta(v_1 v_2 \dots v_n) &= v_1 v_2 \dots v_n \otimes e + e \otimes v_1 v_2 \dots v_n \\ &+ \sum_{j=1}^{n-1} \sum_{\sigma} v_{\sigma(1)} \dots v_{\sigma(j)} \otimes v_{\sigma(j+1)} \dots v_{\sigma(n)}, \end{aligned} \tag{11.62}$$

where σ runs over all $(j, n-j)$-shuffles. A $(j, n-j)$-shuffle is a permutation σ of $(1, \dots, n)$ such that

$$\sigma(1) < \sigma(2) < \dots < \sigma(j) \text{ and } \sigma(j+1) < \sigma(j+2) < \dots < \sigma(n).$$

The antipode S is given by:

$$S(v_{i_1} v_{i_2} \dots v_{i_n}) = (-1)^n v_{i_1} v_{i_2} \dots v_{i_n}. \tag{11.63}$$

The symmetric algebra $\mathrm{Sym}(V)$ is a commutative cocommutative Hopf algebra.

Example 11.5 Shuffle Algebras

Recall the definition of a shuffle algebra from Sect. 8.3 (where we denoted the multiplication with the symbol "$\sqcup\!\sqcup$" instead of "$\cdot$"): Consider a set of letters A. The set A is known as the alphabet. A word is an ordered sequence of letters:

$$w = l_1 l_2 \dots l_k, \tag{11.64}$$

where $l_1, \dots, l_k \in A$. The word of length zero is denoted by e. The shuffle algebra $\mathcal{A}$ on the vector space spanned by words is defined by

$$(l_1 l_2 \dots l_k) \sqcup\!\sqcup (l_{k+1} \dots l_r) = \sum_{\text{shuffles } \sigma} l_{\sigma(1)} l_{\sigma(2)} \dots l_{\sigma(r)}, \tag{11.65}$$

where the sum runs over all permutations σ, which preserve the relative order of $1, 2, \dots, k$ and of $k+1, \dots, r$. The empty word e is the unit in this algebra:

$$e \sqcup\!\sqcup w = w \sqcup\!\sqcup e = w. \tag{11.66}$$

The shuffle algebra is a (non-cocommutative) Hopf algebra [197]. The counit $\bar{e}$ is given by:

$$\bar{e}(e) = 1, \quad \bar{e}(l_1 l_2 \dots l_n) = 0. \tag{11.67}$$

The coproduct Δ is given by:

$$\Delta(l_1 l_2 \dots l_k) = \sum_{j=0}^{k} \left(l_{j+1} \dots l_k\right) \otimes \left(l_1 \dots l_j\right). \tag{11.68}$$

This particular coproduct is also known as the **deconcatenation coproduct**. The antipode S is given by:

$$S(l_1 l_2 \dots l_k) = (-1)^k \, l_k l_{k-1} \dots l_2 l_1. \tag{11.69}$$

The shuffle multiplication is commutative, therefore the antipode satisfies

$$S^2 = \text{id}. \tag{11.70}$$

From Eq. (11.69) this is evident.

To summarise, the shuffle algebra $\mathcal{A}$ is a commutative non-cocommutative Hopf algebra.

Example 11.6 Quasi-Shuffle Algebras

In Sect. 8.4 we discussed quasi-shuffle algebras $\mathcal{A}_q$ with the quasi-shuffle product $\sqcup\!\sqcup_q$. They are similar to shuffle algebras. For a quasi-shuffle algebra we consider as for a shuffle algebra the vector space spanned by words, but now equipped with the quasi-shuffle product instead of the shuffle product. The quasi-shuffle product differs from the normal shuffle product only by terms of lower depth. Quasi-shuffle algebras are Hopf algebras [198].

Comultiplication and counit are defined as for the shuffle algebras. The counit $\bar{e}$ is given by:

$$\bar{e}\,(e) = 1, \quad \bar{e}\,(l_1 l_2 \dots l_n) = 0. \tag{11.71}$$

The coproduct Δ is given by:

$$\Delta\,(l_1 l_2 \dots l_k) = \sum_{j=0}^{k} \left(l_{j+1} \dots l_k\right) \otimes \left(l_1 \dots l_j\right). \tag{11.72}$$

The antipode S is recursively defined through

$$S\,(l_1 l_2 \dots l_k) = -l_1 l_2 \dots l_k - \sum_{j=1}^{k-1} S\left(l_{j+1} \dots l_k\right) \sqcup\!\sqcup_q \left(l_1 \dots l_j\right), \qquad S(e) = e. \tag{11.73}$$

The quasi-shuffle product is commutative, therefore the antipode satisfies

$$S^2 = \text{id}. \tag{11.74}$$

A quasi-shuffle algebra $\mathcal{A}_q$ is a commutative non-cocommutative Hopf algebra.

Example 11.7 Rooted Trees

A rooted tree is a tree where one vertex is marked as the root. It is common practice in mathematics to draw the root at the top. (Admittedly, this is a little bit counter-intuitive as a real tree in nature has its root below.) An individual rooted tree is shown in Fig. 11.1.

We consider the algebra generated by rooted trees. Elements of this algebra are sets of rooted trees, conventionally known as forests. The product of two forests is simply

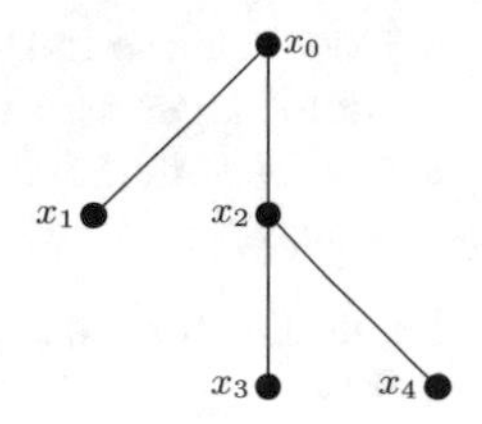

Fig. 11.1 Illustration of a rooted tree. The root is drawn at the top and is labeled x_0

the disjoint union of all trees from the two forests. The empty forest, consisting of no trees, will be denoted by e. Before we are able to define a coproduct, we first need the definition of an **admissible cut**. A single cut is a cut of an edge. An admissible cut of a rooted tree is any assignment of single cuts such that any path from any vertex of the tree to the root has at most one single cut. An admissible cut C maps a tree t to a monomial in trees $t_1 \cdot \ldots \cdot t_{n+1}$. Precisely one of these sub-trees t_j will contain the root of t. We denote this distinguished tree by $R^C(t)$, and the monomial consisting of the n other factors by $P^C(t)$. The counit $\bar{e}$ is given by:

$$\bar{e}(e) = 1, \qquad \bar{e}\,(t_1 \cdot \ldots \cdot t_k) = 0 \quad \text{for } k \geq 1. \tag{11.75}$$

The coproduct Δ is given by (t denotes a non-empty tree):

$$\begin{aligned} \Delta(e) &= e \otimes e, \\ \Delta(t) &= t \otimes e + e \otimes t + \sum_{\text{adm.cuts } C \text{of } t} P^C(t) \otimes R^C(t), \\ \Delta\,(t_1 \cdot \ldots \cdot t_k) &= \Delta\,(t_1) \; \ldots \; \Delta\,(t_k)\,. \end{aligned} \tag{11.76}$$

The antipode S is given by:

$$\begin{aligned} S(e) &= e, \\ S(t) &= -t - \sum_{\text{adm.cuts } C \text{ of } t} S\left(P^C(t)\right) \cdot R^C(t), \\ S\,(t_1 \cdot \ldots \cdot t_k) &= S\,(t_1) \cdot \ldots \cdot S\,(t_k)\,. \end{aligned} \tag{11.77}$$

The algebra of rooted trees is a commutative non-cocommutative Hopf algebra.

Exercise 92 *Which rooted trees are primitive elements in the Hopf algebra of rooted trees?*

It is possible to classify the examples discussed above into four groups according to whether they are commutative or cocommutative:

- Commutative and cocommutative: Examples are the group algebra of a commutative group or the symmetric algebras.
- Non-commutative and cocommutative: Examples are the group algebra of a non-commutative group or the universal enveloping algebra of a Lie algebra.

- Commutative and non-cocommutative: Examples are the shuffle algebra, the quasi-shuffle algebra or the algebra of rooted trees.
- Non-commutative and non-cocommutative: Examples are given by quantum groups.

Let us now turn to a few applications of Hopf algebras in perturbative quantum field theory.

11.1.1 Renormalisation

We start with revisiting the renormalisation of ultraviolet divergences in quantum field theory (see Sect. 4.2). A Feynman integral may have ultraviolet (or short-distance) singularities. These divergences are removed by renormalisation [271]. The combinatorics involved in the renormalisation are governed by a Hopf algebra [267, 268]. The relevant Hopf algebra is the Hopf algebra of decorated rooted trees. We discussed the Hopf algebra of rooted trees in Example 11.7. The relation between a Feynman integral and a rooted tree is as follows: A Feynman integral may have nested ultraviolet divergences, i.e. the associated Feynman graph may contain sub-graphs, which correspond to ultraviolet sub-integrals. The associated rooted tree of a Feynman graph (or of a Feynman integral) encodes the nested structure of the sub-divergences. This is best explained by an example. Figure 11.2 shows a three-loop two-point function. This Feynman integral has an overall ultraviolet divergence and two sub-divergences, corresponding to the two fermion self-energy corrections. We obtain the corresponding rooted tree by drawing boxes around all ultraviolet-divergent sub-graphs. The rooted tree is obtained from the nested structure of these boxes. Graphs with overlapping singularities correspond to a sum of rooted trees. This is illustrated for a two-loop example with an overlapping singularity in Fig. 11.3.

Given a Feynman graph G we may associate to G a rooted tree t as above. In addition, we may associate to each vertex of the rooted tree additional information: The sub-graph it corresponds to, as well as the momenta q_j, masses m_j and powers

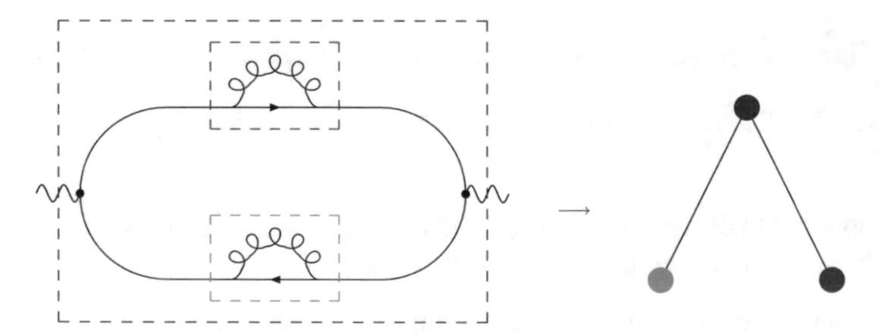

Fig. 11.2 A three-loop two-point function with an overall ultraviolet divergence and two sub-divergences. We find the corresponding rooted tree by first drawing boxes around all ultraviolet-divergent sub-graphs. The rooted tree is obtained from the nested structure of these boxes

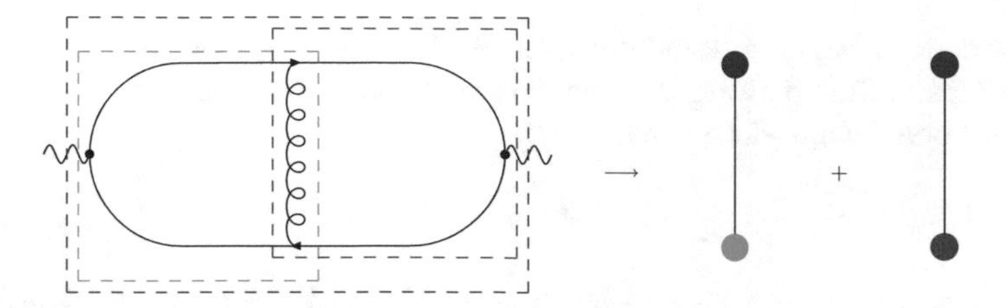

Fig. 11.3 Example with overlapping singularities. This graph corresponds to a sum of rooted trees

ν_j associated with the edges of the sub-graph. This ensures that no information is lost in passing from the Feynman graph G to the rooted tree t and we may recover G from t. A rooted tree with this additional information is called a **decorated rooted tree**. The decorations do not spoil the Hopf algebra structure. We denote the Hopf algebra of decorated rooted trees by H. Furthermore we denote by A the (commutative) algebra of Laurent series in the dimensional regularisation parameter ε. The Feynman integral $I(t)$ assigns any decorated rooted tree $t \in H$ a Laurent series. The map

$$I : H \to A \tag{11.78}$$

is a character of the Hopf algebra H with values in A.

We recall that the counit applied to any non-trivial rooted tree $t \neq e$ yields zero:

$$\bar{e}(t) = 0, \qquad t \neq e. \tag{11.79}$$

If we combine this with the unit e in A we have

$$e\bar{e}(t) = 0, \qquad t \neq e. \tag{11.80}$$

From the discussion of the convolution product we know that $e\bar{e} \in \text{AlgHom}(H, A)$ is the neutral element in $\text{AlgHom}(H, A)$ and that the inverse of $I : H \to A$ is given by $I^{-1} = IS$. Thus we may write Eq. (11.80) as

$$\left(I^{-1} * I\right)(t) = 0, \qquad t \neq e. \tag{11.81}$$

Equation (11.81) can also be writtend as

$$I^{-1}\left(t^{(1)}\right) \cdot I\left(t^{(2)}\right) = 0, \qquad t \neq e, \tag{11.82}$$

where we used Sweedler's notation. Equation (11.81) will be our starting point. However, rather than obtaining zero on the right-hand side, we are interested in a finite quantity. To keep the discussion simple, we only consider Feynman integrals which have ultraviolet but no infrared divergences. For example, this can be achieved by regulating all infrared divergences with a small non-zero mass.

In addition we introduce a map

$$R : A \to A \tag{11.83}$$

which does not alter the divergence structure and which satisfies the Rota-Baxter relation [272]:

$$R(a_1 a_2) + R(a_1) R(a_2) = R(a_1 R(a_2)) + R(R(a_1) a_2). \tag{11.84}$$

The map R defines a **renormalisation scheme**. An example is given by modified minimal subtraction scheme ($\overline{\text{MS}}$). With the conventions of this book, the $\overline{\text{MS}}$-scheme is defined by

$$R\left(\sum_{k=-L}^{\infty} c_k \varepsilon^k\right) = \sum_{k=-L}^{-1} c_k \varepsilon^k. \tag{11.85}$$

You may wonder, why there are no terms like $\ln(4\pi) - \gamma_{\text{E}}$ in Eq. (11.85): We eliminated these terms from the very start by an appropriate choice of the integration measure in Eq. (2.56). This motivates a posteriori the choice of the prefactors $e^{l\varepsilon\gamma_{\text{E}}}$ and $\pi^{-\frac{D}{2}}$ in Eq. (2.56).

Exercise 93 *Show that the map R in Eq. (11.85) fulfills the Rota-Baxter equation (11.84).*

The notation with the letter R for the map in Eq. (11.83) stems from **Bogoliubov's R-operation** [273]. One can now twist the map $I^{-1} = IS$ with R and define a new map I_R^{-1} recursively by

$$I_R^{-1}(t) = -R\left(I(t) + \sum_{\text{adm.cuts } C \text{ of } t} I_R^{-1}\left(P^C(t)\right) \cdot I\left(R^C(t)\right)\right). \tag{11.86}$$

From the multiplicativity constraint (11.84) it follows that

$$I_R^{-1}(t_1 t_2) = I_R^{-1}(t_1) I_R^{-1}(t_2). \tag{11.87}$$

If we replace I^{-1} by I_R^{-1} in (11.82) we no longer obtain zero on the right-hand side, but one may show that

$$I_R^{-1}\left(t^{(1)}\right) I\left(t^{(2)}\right) = \text{finite}, \qquad t \neq e. \tag{11.88}$$

This corresponds to the renormalised value of the Feynman integral. Equation (11.88) is equivalent to the forest formula [271]. It should be noted that R is not unique and different choices for R correspond to different renormalisation schemes. There is certainly more that could be said on the Hopf algebra of renormalisation and we refer the reader to the original literature [268, 274–280].

11.1.2 Wick's Theorem

Let us consider bosonic field operators, which we denote by $\phi_i = \phi(x_i)$. Wick's theorem relates the time-ordered product of n bosonic field operators to the normal product of these operators and contractions. As an example one has

$$\begin{aligned} T\left(\phi_1\phi_2\phi_3\phi_4\right) = \; &: \phi_1\phi_2\phi_3\phi_4 : + \left(\phi_1, \phi_2\right) : \phi_3\phi_4 : \\ &+ \left(\phi_1, \phi_3\right) : \phi_2\phi_4 : + \left(\phi_1, \phi_4\right) : \phi_2\phi_3 : + \left(\phi_2, \phi_3\right) : \phi_1\phi_4 : \\ &+ \left(\phi_2, \phi_4\right) : \phi_1\phi_3 : + \left(\phi_3, \phi_4\right) : \phi_1\phi_2 : + \left(\phi_1, \phi_2\right)\left(\phi_3, \phi_4\right) \\ &+ \left(\phi_1, \phi_3\right)\left(\phi_2, \phi_4\right) + \left(\phi_1, \phi_4\right)\left(\phi_2, \phi_3\right), \end{aligned} \tag{11.89}$$

where we used the notation

$$\left(\phi_i, \phi_j\right) = \left\langle 0 \left| T\left(\phi_i\phi_j\right) \right| 0 \right\rangle \tag{11.90}$$

to denote the contraction. One can use Wick's theorem to define the time-ordered product in terms of the normal product and the contraction. To establish the connection with Hopf algebras, let V be the vector space with basis $\{\phi_i\}$ and identify the normal product with the symmetric product introduced in Example 11.4, [281, 282]. This yields the symmetric algebra $S(V)$. The contraction defines a bilinear form $V \otimes V \to \mathbb{C}$. One extends this pairing to $S(V)$ by

$$\begin{aligned} (: N_1 N_2 :, M_1) &= \left(N_1, M_1^{(1)}\right)\left(N_2, M_1^{(2)}\right), \\ (N_1, : M_1 M_2 :) &= \left(N_1^{(1)}, M_1\right)\left(N_1^{(2)}, M_2\right). \end{aligned} \tag{11.91}$$

Here, N_1, N_2, M_1 and M_2 are arbitrary normal products of the ϕ_i. With the help of this pairing one defines a new product, called the circle product, as follows:

$$N \circ M = \left(N^{(1)}, M^{(1)}\right) \; : N^{(2)} M^{(2)} : \tag{11.92}$$

Again, N and M are normal products.

Figure 11.4 shows pictorially the definition of the circle product involving the coproduct, the pairing (..., ...) and the multiplication. It can be shown that the circle product is associative. Furthermore, one obtains that the circle product coincides with the time-ordered product. For example,

$$\phi_1 \circ \phi_2 \circ \phi_3 \circ \phi_4 = T\left(\phi_1\phi_2\phi_3\phi_4\right). \tag{11.93}$$

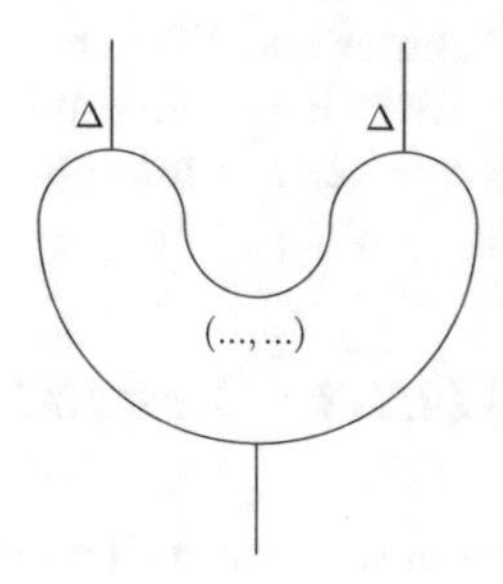

Fig. 11.4 The "sausage tangle": pictorial representation of the definition of the circle product

The reader is invited to verify the left-hand side of (11.93) with the help of the definitions (11.90)–(11.92).

11.1.3 Multiple Polylogarithms

Let us now turn to multiple polylogarithms. In Chap. 8 we introduced the shuffle algebra and the quasi-shuffle algebra related to the multiple polylogarithms. In the examples in Sect. 11.1 we saw that a shuffle algebra and a quasi-shuffle algebra are Hopf algebras. We therefore have two Hopf algebras associated with the multiple polylogarithms, one associated with the shuffle product and the integral representation, the other one with the quasi-shuffle product and the sum representation.

Let us start with the shuffle algebra. Consider an alphabet $A = \{z_1, z_2, \dots\}$ and denote by $\mathcal{A}$ the shuffle algebra of words in this alphabet. From Example 11.5 in Sect. 11.1 we know that $\mathcal{A}$ is a Hopf algebra. For fixed y we may view a multiple polylogarithm $G(z_1, \dots, z_r; y)$ as a map

$$\begin{aligned} G: \; & \mathcal{A} \to \mathbb{C}, \\ & w \to G(z_1, \dots, z_r; y), \quad \text{for } w = z_1 \dots z_r. \end{aligned} \tag{11.94}$$

This map is a character of the Hopf algebra $\mathcal{A}$. In particular it is an algebra homomorphism (see Eq. (8.54)) and we have

$$G(w_1 \sqcup\!\sqcup\, w_2) = G(w_1) \cdot G(w_2). \tag{11.95}$$

The empty word e is the unit in the shuffle algebra $\mathcal{A}$ and mapped to 1 in $\mathbb{C}$:

$$G(e) = G(;y) = 1. \tag{11.96}$$

Let us now see what the antipode gives us: We start (without any reference to Hopf algebras) with integration-by-parts identities for the multiple polylogarithms $G(z_1, \dots, z_r; y)$. The starting point is as follows:

$$
\begin{aligned}
G(z_1, \dots, z_k; y) &= \int\limits_0^y dt \left(\frac{\partial}{\partial t} G(z_1; t) \right) G(z_2, \dots, z_k; t) \\
&= G(z_1; y) G(z_2, \dots, z_k; y) - \int\limits_0^y dt \; G(z_1; t) g(z_2; t) G(z_3, \dots, z_k; t) \\
&= G(z_1; y) G(z_2, \dots, z_k; y) - \int\limits_0^y dt \left(\frac{\partial}{\partial t} G(z_2, z_1; t) \right) G(z_3, \dots, z_k; t).
\end{aligned}
\tag{11.97}
$$

Repeating this procedure one arrives at the following integration-by-parts identity:

$$
\begin{aligned}
& G(z_1, \dots, z_k; y) + (-1)^k G(z_k, \dots, z_1; y) \\
& \quad = G(z_1; y) G(z_2, \dots, z_k; y) - G(z_2, z_1; y) G(z_3, \dots, z_k; y) \\
& \quad \quad + \dots - (-1)^{k-1} G(z_{k-1}, \dots, z_1; y) G(z_k; y),
\end{aligned}
\tag{11.98}
$$

which relates the combination $G(z_1, \dots, z_k; y) + (-1)^k G(z_k, \dots, z_1; y)$ to G-functions of lower depth. This relation is useful in simplifying expressions. Equation (11.98) can also be derived in a different way. In the shuffle algebra $\mathcal{A}$ we have for any non-trivial element w the following relation involving the antipode:

$$
S\left(w^{(1)}\right) ⧢ w^{(2)} = 0. \tag{11.99}
$$

Here Sweedler's notation has been used. Composing Eq. (11.99) with the map $G : \mathcal{A} \to \mathbb{C}$ of Eq. (11.94) we obtain

$$
G\left(S\left(w^{(1)}\right) ⧢ w^{(2)}\right) = 0. \tag{11.100}
$$

Working out the relation (11.100) for the shuffle algebra of the functions $G(z_1, \dots, z_k; y)$, we recover (11.98).

Let us now turn to the quasi-shuffle algebra. We denote letters by $l_j = (m_j, z_j)$ and consider an alphabet $A = \{l_1, l_2, \dots\}$. We denote by $\mathcal{A}_q$ the quasi-shuffle algebra of words in this alphabet as in Sect. 8.4. From Example 11.6 in Sect. 11.1 we know that $\mathcal{A}_q$ is a Hopf algebra. We may view a multiple polylogarithm $\text{Li}_{m_1 \dots m_k}(x_1, \dots, x_k)$ as a map

$$
\begin{aligned}
\text{Li} : \; & \mathcal{A}_q \to \mathbb{C}, \\
& w \to \text{Li}_{m_1 \dots m_k}(x_1, \dots, x_k), \quad \text{for } w = l_1 \dots l_k \text{ and } l_j = (m_j, x_j).
\end{aligned}
\tag{11.101}
$$

We may be a little bit more general, fix an integer $n \in \mathbb{N}$ and consider Z-sums as in Sect. 9.1. For fixed n we consider the map

$$\begin{aligned} Z: \ & \mathcal{A}_q \rightarrow \mathbb{C}, \\ & w \rightarrow Z_{m_1 \dots m_k}(x_1, \dots, x_k; n), \quad \text{for } w = l_1 \dots l_k \text{ and } l_j = (m_j, x_j). \end{aligned} \tag{11.102}$$

The maps in Eqs. (11.101) and (11.102) are characters of the Hopf algebra $\mathcal{A}_q$. We may view Eq. (11.101) as the special case $n = \infty$ of Eq. (11.102). Equations (11.101) and (11.102) are algebra homomorphisms, therefore

$$\begin{aligned} \text{Li}\left(w_1 \sqcup\!\sqcup_q w_2\right) &= \text{Li}\left(w_1\right) \cdot \text{Li}\left(w_2\right), \\ Z\left(w_1 \sqcup\!\sqcup_q w_2\right) &= Z\left(w_1\right) \cdot Z\left(w_2\right). \end{aligned} \tag{11.103}$$

The empty word e is the unit in the quasi-shuffle algebra $\mathcal{A}_q$ and mapped to 1 in $\mathbb{C}$:

$$\begin{aligned} \text{Li}\left(e\right) = \text{Li}\left(\right) &= 1, \\ Z\left(e\right) = Z\left(;n\right) &= 1. \end{aligned} \tag{11.104}$$

We may now proceed and check if the antipode provides also a non-trivial relation for the quasi-shuffle algebra of Z-sums. This requires first some notation: A composition of a positive integer k is a sequence $I = (i_1, \dots, i_l)$ of positive integers such that $i_1 + \dots + i_l = k$. The set of all composition of k is denoted by $C(k)$. Compositions act on words $w = l_1 \dots l_k$ in $\mathcal{A}_q$ as

$$(i_1, \dots, i_l) \circ (l_1 l_2 \dots l_k) = l_1' l_2' \dots l_l', \tag{11.105}$$

with

$$l_1' = l_1 \circ \dots \circ l_{i_1}, \quad l_2' = l_{i_1+1} \circ \dots \circ l_{i_1+i_2}, \quad \dots \quad l_l' = l_{i_1+\dots+i_{l-1}+1} \circ \dots \circ l_{i_1+\dots+i_l}, \tag{11.106}$$

where $\circ$ in Eq. (11.106) denotes the operation defined in Eq. (8.65). Thus the first i_1 letters of the word are combined into one new letter l_1', the next i_2 letters are combined into the second new letter l_2', etc.. To give an example let $l_1 = (m_1, x_1)$, $l_2 = (m_2, x_2)$, $l_3 = (m_3, x_3)$, $w = l_1 l_2 l_3$ and $I = (2, 1)$. Then

$$\begin{aligned} & I \circ w = l_1' l_2', \\ & l_1' = (m_1 + m_2, x_1 \cdot x_2), \quad l_2' = (m_3, x_3). \end{aligned} \tag{11.107}$$

With this notation for compositions one obtains the following closed formula for the antipode in the quasi-shuffle algebra [198]:

$$S\left(l_1 l_2 \dots l_k\right) = (-1)^k \sum_{I \in C(k)} I \circ \left(l_k \dots l_2 l_1\right). \tag{11.108}$$

The analogue of Eq. (11.100) reads for $w \neq e$

$$Z\left(S\left(w^{(1)}\right) ⧢_q w^{(2)}\right) = 0. \tag{11.109}$$

Written more explicitly we have

$$Z\left(l_1, \dots, l_k\right) + (-1)^k Z\left(l_k, \dots, l_1\right)$$
$$= -\sum_{j=1}^{k-1} Z\left(S\left(l_{j+1} \dots l_k\right)\right) Z\left(l_1 \dots l_j\right) - (-1)^k \sum_{I \in C(k) \setminus (1,1,\dots,1)} Z\left(I \circ \left(l_k \dots l_2 l_1\right)\right). \tag{11.110}$$

Again, the combination $Z(n; m_1, \dots, m_k; x_1, \dots, x_k) + (-1)^k Z(n; m_k, \dots, m_1; x_k, \dots, x_1)$ reduces to Z-sums of lower depth, similar to the integration-by-parts identity in Eq. (11.98). We therefore obtained an "integration-by-parts" identity for objects, which don't have an integral representation. We first observed, that for the G-functions, which have an integral representation, the integration-by-parts identites are equal to the identities obtained from the antipode. After this abstraction towards an algebraic formulation, one can translate these relations to cases, which only have the appropriate algebra structure, but not necessarily a concrete integral representation. As an example we have

$$\begin{aligned}
Z(n; m_1, m_2, m_3; x_1, x_2, x_3) - Z(n; m_3, m_2, m_1; x_3, x_2, x_1)& \\
= Z(n; m_1; x_1) Z(n; m_2, m_3; x_2, x_3) - Z(n; m_2, m_1; x_2, x_1) Z(n; m_3; x_3)& \\
- Z(n; m_1 + m_2; x_1 x_2) Z(n; m_3; x_3) + Z(n; m_2 + m_3, m_1; x_2 x_3, x_1)& \\
+ Z(n; m_3, m_1 + m_2; x_3, x_1 x_2) + Z(n; m_1 + m_2 + m_3; x_1 x_2 x_3),&
\end{aligned} \tag{11.111}$$

which expresses the combination of the two Z-sums of depth 3 as Z-sums of lower depth. Taking $n = \infty$ in the equation above we obtain a relation among multiple polylogarithms:

$$\begin{aligned}
&\text{Li}_{m_1 m_2 m_3}(x_1, x_2, x_3) - \text{Li}_{m_3 m_2 m_1}(x_3, x_2, x_1) \\
&= \text{Li}_{m_1}(x_1) \text{Li}_{m_2 m_3}(x_2, x_3) - \text{Li}_{m_2 m_1}(x_2, x_1) \text{Li}_{m_3}(x_3) \\
&\quad - \text{Li}_{m_1+m_2}(x_1 x_2) \text{Li}_{m_3}(x_3) + \text{Li}_{(m_2+m_3) m_1}(x_2 x_3, x_1) + \text{Li}_{m_3 (m_1+m_2)}(x_3, x_1 x_2) \\
&\quad + \text{Li}_{m_1+m_2+m_3}(x_1 x_2 x_3).
\end{aligned} \tag{11.112}$$

The analog example for the shuffle algebra of the G-function reads:

$$G(z_1, z_2, z_3; y) - G(z_3, z_2, z_1; y) = G(z_1; y) G(z_2, z_3; y) - G(z_2, z_1; y) G(z_3; y). \tag{11.113}$$

Multiple polylogarithms obey both the quasi-shuffle algebra and the shuffle algebra. Therefore we have for multiple polylogarithms two relations, which are in general independent.

11.2 Coactions

In the previous section we saw that the shuffle algebra is a Hopf algebra and so is the quasi-shuffle algebra. When working with multiple polylogarithms we may either use the $G(z_1, \dots, z_r; y)$ notation and work with shuffle algebra or the $\text{Li}_{m_1 \dots m_k}(x_1, \dots, x_k)$ notation and work with the quasi-shuffle algebra. Whatever our choice is, the relations coming from the other algebra are not directly accessible. We would like to work with a structure, which contains all the relations we know and only those. As we may view G (for fixed y) as a map from the shuffle algebra A to $\mathbb{C}$, and Li as a map from the quasi-shuffle algebra A_q to $\mathbb{C}$, our first guess might be to look at the complex numbers $\mathbb{C}$. There we have all the relations we know about. But it is very hard to prove that there are no additional relations. (That is to say that one would need to prove that the period map is injective, this is currently a conjecture.) For this reason we construct a set of objects, called **motivic multiple polylogarithms**, which have exactly the relations we know about and only those. We denote the set of motivic multiple polylogarithms by $\mathcal{P}^{\mathfrak{m}}_{\text{MPL}}$. The set of motivic multiple polylogarithms is not quite a Hopf algebra, but it is a comodule. In this section we first introduce coactions and comodules and define then the motivic multiple polylogarithms. References for this section are [189, 206, 283, 284]. Applications towards Feynman integrals are considered in [285–287].

Let A be a unitial associative algebra over a ring R and M a left R-module. A linear map

$$\begin{aligned} \cdot : \;\; & A \otimes M \to M, \\ & (a, v) \to a \cdot v, \end{aligned} \tag{11.114}$$

with

$$\begin{aligned} e \cdot v &= v, \\ (a_1 \cdot a_2) \cdot v &= a_1 \cdot (a_2 \cdot v) \end{aligned} \tag{11.115}$$

defines a (left-) **action** of A on M (where e denotes the unit in A). This upgrades the left R-module M to a left A-module. Please note that we use the multiplication sign "$\cdot$" to denote the multiplication in the algebra (e.g. $a_1 \cdot a_2$) as well as for the action of A on M (e.g. $a \cdot v$).

Let C be a coalgebra. We always assume that C is coassociative and that C has a counit $\bar{e}$. We are now going to define a coaction and a comodule. This isn't too complicated, we just have to reverse the arrows of all maps. We start from a linear map

$$\begin{aligned} \Delta : \;\; & M \to C \otimes M, \\ & v \to \Delta(v). \end{aligned} \tag{11.116}$$

Again, please note that we use the symbol to denote on the one hand the coproduct in C (e.g. $\Delta(a)$ for $a \in C$) as well as the new map defined in Eq. (11.116) (e.g. $\Delta(v)$ for $v \in M$). This is unambiguous, as the argument determines what operation is meant. (It's like in C++ with operator overloading.) We will use Sweedler's notation to write

$$\Delta(v) = a^{(1)} \otimes v^{(2)}, \qquad a^{(1)} \in C, \quad v, v^{(2)} \in M. \tag{11.117}$$

Let's now work out the analogue relations of Eq. (11.115). For the map Δ defined in Eq. (11.116) we require

$$\begin{aligned} \cdot (\bar{e} \otimes \text{id}) \Delta(v) &= v, \\ (\Delta \otimes \text{id}) \Delta(v) &= (\text{id} \otimes \Delta) \Delta(v). \end{aligned} \tag{11.118}$$

Exercise 94 *Resolve the operator overloading: In Eq. (11.118) the symbols "$\cdot$", $\bar{e}$ and Δ appear in various places. Determine for each occurrence to which operation they correspond.*

A linear map as in Eqs. (11.116) and satisfying (11.118) defines a (left-) **coaction** of C on M. In this case we call M a (left-) **comodule**.

Let H be a Hopf algebra and M an algebra. We denote the unit in H by e_H and the unit in M by e_M. M is called a H-**comodule algebra** if M is a (left-) H-comodule and in addition

$$\begin{aligned} \Delta(e_M) &= e_H \otimes e_M, \\ \Delta(v_1 \cdot v_2) &= \Delta(v_1) \Delta(v_2), \qquad v_1, v_2 \in M. \end{aligned} \tag{11.119}$$

M is called a H-**module algebra** if M is a (left-) H-module and in addition

$$\begin{aligned} a \cdot e_M &= \bar{e}(a) \cdot e_M, \\ a \cdot (v_1 \cdot v_2) &= \left(a^{(1)} \cdot v_1\right) \cdot \left(a^{(2)} \cdot v_2\right), \quad a \in H, \quad v_1, v_2 \in M. \end{aligned} \tag{11.120}$$

Note that the definitions of a H-comodule algebra and of a H-module algebra are not dual to each other, as M is assumed to be in both cases an algebra. For H a Hopf algebra and M a coalgebra there are also the notions of a H-module coalgebra and of a H-comodule coalgebra. We will not need them, but their definitions are the duals of the definitions of a H-comodule algebra and of a H-module algebra.

Let us now discuss the application towards multiple polylogarithms. We first define for $z_1 \neq z_0$ and $z_r \neq z_{r+1}$

$$I(z_0; z_1, z_2, \dots, z_r; z_{r+1}) = \int\limits_{z_0}^{z_{r+1}} \frac{dt_r}{t_r - z_r} \int\limits_{z_0}^{t_r} \frac{dt_{r-1}}{t_{r-1} - z_{r-1}} \dots \int\limits_{z_0}^{t_2} \frac{dt_1}{t_1 - z_1}, \tag{11.121}$$

together with the convention

$$I(z_0; z_1) = 1. \tag{11.122}$$

The condition $z_1 \neq z_0$ ensures that there is no divergence at the lower integration boundary, the condition $z_r \neq z_{r+1}$ ensures that there is no divergence at the upper integration boundary. We then extend the definition to $z_1 = z_0$ and $z_r = z_{r+1}$ as follows: For $z_1 = z_0$ we use the shuffle product to isolate all divergences in powers of $I(z_0; z_0; z_2)$. We then set

$$I(z_0; z_0; z_2) = \ln(z_2 - z_0). \tag{11.123}$$

In a similar way we handle the case $z_r = z_{r+1}$: We use again the shuffle product and isolate all divergences in powers of $I(z_0; z_2; z_2)$. We then set

$$I(z_0; z_2; z_2) = -\ln(z_0 - z_2). \tag{11.124}$$

The two regularisation prescriptions in Eqs. (11.123) and (11.124) are compatible with the path decomposition formula: We have

$$I(z_0; z_1; z_2) = I(z_0; z_1; z_1) + I(z_1; z_1; z_2). \tag{11.125}$$

The definition in Eq. (11.121) is a slight generalisation of Eq. (8.7), allowing the starting point z_0 of the integration to be different from zero. We have

$$\begin{aligned} G(z_1, \dots, z_r; y) &= I(0; z_r, \dots, z_1; y), \qquad (z_1 \neq y,\ z_r \neq 0), \\ I(z_0; z_1, z_2, \dots, z_r; z_{r+1}) &= G(z_r - z_0, \dots, z_1 - z_0; z_{r+1} - z_0). \end{aligned} \tag{11.126}$$

Let us now consider formal objects $I^{\mathfrak{m}}(z_0; z_1, \dots, z_r; z_{r+1})$ (the $\mathfrak{m}$ stands for "motivic"). The set of all those objects (modulo an equivalence relation discussed below) will be denoted by $\mathcal{P}^{\mathfrak{m}}_{\mathrm{MPL}}$. We may think of the $I^{\mathfrak{m}}(z_0; z_1, \dots, z_r; z_{r+1})$'s in the same way we think about words $w = z_1 \dots z_r \in A$ in the shuffle algebra in the context of multiple polylogarithms. For fixed y and $z_j \in \mathbb{C}$ we have in the latter case an evaluation map (see Eq. (11.94))

$$\begin{aligned} G: \ & A \to \mathbb{C}, \\ & w \to G(z_1, \dots, z_r; y). \end{aligned} \tag{11.127}$$

In the same way we will assume that there is an evaluation map for $I^{\mathfrak{m}}(z_0; z_1, \dots, z_r; z_{r+1})$:

$$\begin{aligned} \text{period}: \ & \mathcal{P}^{\mathfrak{m}}_{\mathrm{MPL}} \to \mathbb{C}, \\ & I^{\mathfrak{m}}(z_0; z_1, \dots, z_r; z_{r+1}) \to I(z_0; z_1, \dots, z_r; z_{r+1}), \end{aligned} \tag{11.128}$$

sending the formal object to the concrete iterated integral of Eq. (11.121). The map in Eq. (11.128) will be called the **period map**.

Let's now assume that all z_j's are algebraic: $z_j \in \overline{\mathbb{Q}}$. In this case the period map takes values in $\mathbb{P}$, the set of numerical periods (see Sect. 10.3). We consider the $\mathbb{Q}$-algebra generated by the $I^{\mathrm{m}}(z_0; z_1, \dots, z_r; z_{r+1})$'s subject to the following relations:

1. In $\mathcal{P}^{\mathrm{m}}_{\mathrm{MPL}}$ we have any relation, which can be derived for the non-motivic multiple polylogarithms $I(z_0; z_1, \dots, z_r; z_{r+1})$ using linearity, a change of variables and Stokes' theorem.
2. Shuffle regularisation: An object $I^{\mathrm{m}}(z_0; z_1, \dots, z_r; z_{r+1})$ is said to have a trailing zero, if $z_1 = z_0$. It is said to have leading one, if $z_r = z_{r+1}$. Using the shuffle product, we isolate trailing zeros in powers of $I^{\mathrm{m}}(z_0; z_0; z_{r+1})$ and leading ones in powers of $I^{\mathrm{m}}(z_0; z_{r+1}; z_{r+1})$. We then set

$$\begin{aligned} I^{\mathrm{m}}\left(z_0; z_0; z_{r+1}\right) &= \ln^{\mathrm{m}}\left(z_{r+1} - z_0\right), \\ I^{\mathrm{m}}\left(z_0; z_{r+1}; z_{r+1}\right) &= -\ln^{\mathrm{m}}\left(z_0 - z_{r+1}\right). \end{aligned} \tag{11.129}$$

We call the objects $I^{\mathrm{m}}(z_0; z_1, \dots, z_r; z_{r+1})$ **motivic multiple polylogarithms** and we denote the algebra defined as above by $\mathcal{P}^{\mathrm{m}}_{\mathrm{MPL}}$. In point 1 we impose relations which can be obtained from linearity, a change of variables and Stokes' theorem. This is completely analogue to the definition of effective periods in Sect. 10.4. This includes shuffle relations, i.e. for identical start and end points we have

$$\begin{aligned} & I^{\mathrm{m}}(z_0; z_1, \dots, z_k; z_{r+1}) \cdot I^{\mathrm{m}}(z_0; z_{k+1}, \dots, z_r; z_{r+1}) \\ & = \sum_{\text{shuffles } \sigma} I^{\mathrm{m}}(z_0; z_{\sigma(1)}, \dots, z_{\sigma(r)}; z_{r+1}). \end{aligned} \tag{11.130}$$

It also includes path composition: For $y \in \overline{\mathbb{Q}}$ we have

$$I^{\mathrm{m}}(z_0; z_1, \dots, z_r; z_{r+1}) = \sum_{k=0}^{r} I^{\mathrm{m}}(z_0; z_1, \dots, z_k; y) \cdot I^{\mathrm{m}}(y; z_{k+1}, \dots, z_r; z_{r+1}), \tag{11.131}$$

as well as a relation for a vanishing integration cycle: For $r \geq 1$ we have

$$I^{\mathrm{m}}\left(z_0; z_1, \dots, z_r; z_0\right) = 0. \tag{11.132}$$

Point 2 implies that for example for $z \neq 0, 1$ we have

$$\begin{aligned} I^{\mathrm{m}}\left(0; 0, 0, z; 1\right) &= \frac{1}{2}\left[I^{\mathrm{m}}\left(0; 0; 1\right)\right]^2 I^{\mathrm{m}}\left(0; z; 1\right) \\ &\quad - I^{\mathrm{m}}\left(0; 0; 1\right) I^{\mathrm{m}}\left(0; z, 0; 1\right) + I^{\mathrm{m}}\left(0; z, 0, 0; 1\right) \\ &= \frac{1}{2}\left[\ln^{\mathrm{m}}\left(1\right)\right]^2 I^{\mathrm{m}}\left(0; z; 1\right) - \ln^{\mathrm{m}}\left(1\right) I^{\mathrm{m}}\left(0; z, 0; 1\right) \\ &\quad + I^{\mathrm{m}}\left(0; z, 0, 0; 1\right) \\ &= I^{\mathrm{m}}\left(0; z, 0, 0; 1\right). \end{aligned} \tag{11.133}$$

We now define the **de Rham multiple polylogarithms**. Roughly speaking, we may think of the de Rham multiple polylogarithms as the motivic multiple polylogarithms modulo $(2\pi i)$.

Let $r_3 = -\frac{1}{2} + \frac{i}{2}\sqrt{3}$ be the third root of unity. The algebra $\mathcal{P}^{\mathfrak{m}}_{\mathrm{MPL}}$ contains the element

$$I^{\mathfrak{m}}(1; 0; r_3) + I^{\mathfrak{m}}\left(r_3; 0; r_3^2\right) + I^{\mathfrak{m}}\left(r_3^2; 0; 1\right) \tag{11.134}$$

which we denote by $(2\pi i)^{\mathfrak{m}}$. The notation stands for "the motivic lift of $(2\pi i)$" (and in particular the super-script $\mathfrak{m}$ stands for "motivic", it does not denote an exponent). We have

$$\mathrm{period}\left((2\pi i)^{\mathfrak{m}}\right) = 2\pi i, \tag{11.135}$$

which explains the notation. We denote by $\mathcal{P}^{\mathfrak{dR}}_{\mathrm{MPL}}$ the algebra obtained by factoring out the ideal $\langle(2\pi i)^{\mathfrak{m}}\rangle$:

$$\mathcal{P}^{\mathfrak{dR}}_{\mathrm{MPL}} = \mathcal{P}^{\mathfrak{m}}_{\mathrm{MPL}} / \langle(2\pi i)^{\mathfrak{m}}\rangle. \tag{11.136}$$

The super-script $\mathfrak{dR}$ stands for "de Rham". Elements in $\mathcal{P}^{\mathfrak{dR}}_{\mathrm{MPL}}$ are denoted as

$$I^{\mathfrak{dR}}(z_0; z_1, \dots, z_r; z_{r+1}) \tag{11.137}$$

and called de Rham multiple polylogarithms. Note that for de Rham multiple polylogarithms there is no period map, as such a map would be ambiguous by terms proportional to $(2\pi i)$. $\mathcal{P}^{\mathfrak{dR}}_{\mathrm{MPL}}$ is a Hopf algebra [189, 288], the coproduct is given by

$$\Delta I^{\mathfrak{dR}}(z_0; z_1, z_2, \dots, z_r; z_{r+1}) = \sum_{k=0}^{r} \sum_{0=i_0<i_1<\dots<i_k<i_{k+1}=r+1}$$
$$\prod_{p=0}^{k} I^{\mathfrak{dR}}\left(z_{i_p}; z_{i_p+1}, z_{i_p+2}, \dots, z_{i_{p+1}-1}; z_{i_{p+1}}\right) \otimes I^{\mathfrak{dR}}\left(z_0; z_{i_1}, z_{i_2}, \dots, z_{i_k}; z_{r+1}\right). \tag{11.138}$$

As $\mathcal{P}^{\mathfrak{dR}}_{\mathrm{MPL}}$ is graded connected, the antipode is given by (see Eq. (11.34)):

$$\begin{aligned} S\left(I^{\mathfrak{dR}}(z_0; z_1)\right) &= I^{\mathfrak{dR}}(z_0; z_1), \\ S\left(I^{\mathfrak{dR}}(z_0; z_1, \dots, z_r; z_{r+1})\right) &= -I^{\mathfrak{dR}}(z_0; z_1, \dots, z_r; z_{r+1}) \\ &\quad - \cdot (S \otimes \mathrm{id})\,\tilde{\Delta}\left(I^{\mathfrak{dR}}(z_0; z_1, \dots, z_r; z_{r+1})\right). \end{aligned} \tag{11.139}$$

On the other hand, $\mathcal{P}^{\mathfrak{m}}_{\mathrm{MPL}}$ is not a Hopf algebra, it is just a $\mathcal{P}^{\mathfrak{dR}}_{\mathrm{MPL}}$-comodule. The coaction is given by

$$\Delta I^{\mathrm{m}}(z_0; z_1, z_2, \dots, z_r; z_{r+1}) = \sum_{k=0}^{r} \sum_{0=i_0<i_1<\dots<i_k<i_{k+1}=r+1}$$

$$\prod_{p=0}^{k} I^{\mathfrak{dR}}\left(z_{i_p}; z_{i_p+1}, z_{i_p+2}, \dots, z_{i_{p+1}-1}; z_{i_{p+1}}\right) \otimes I^{\mathrm{m}}\left(z_0; z_{i_1}, z_{i_2}, \dots, z_{i_k}; z_{r+1}\right). \quad (11.140)$$

This formula is very similar to Eq. (11.138), but note that the entry in the first slot belongs to $\mathcal{P}^{\mathfrak{dR}}_{\mathrm{MPL}}$, while the entry in the second slot belongs to $\mathcal{P}^{\mathrm{m}}_{\mathrm{MPL}}$. There is a graphical way to represent the formula for the coproduct/coaction. We may represent $I^{\mathfrak{dR}}(z_0; z_1, \dots, z_r; z_{r+1})$ and $I^{\mathrm{m}}(z_0; z_1, \dots, z_r; z_{r+1})$ as polygons drawn on a half-circle as shown in the left picture in Fig. 11.5. The points z_0 and z_{r+1} are drawn where the circle segment meets the line, the points $z_1, \dots, z_r$ are drawn in that order on the circle segment, such that z_1 is adjacent to z_0 (and z_r is adjacent to z_{r+1}). In order to obtain the coproduct or the coaction we consider all subsets of $\{i_1, \dots, i_k\} \in \{1, \dots, r\}$ (including the empty set and the full set). The entry of the second slot is defined by this subset and given by $I^{\mathfrak{dR}/\mathrm{m}}(z_0; z_{i_1}, z_{i_2}, \dots, z_{i_k}; z_{r+1})$. The entry in the first slot is the product of de Rham multiple polylogarithms corresponding to the smaller polygons, which have been omitted. This is shown in the right picture of Fig. 11.5 for one specific term obtained from the coaction on $I^{\mathrm{m}}(z_0; z_1, z_2, z_3, z_4; z_5)$. The full formula reads

$$\begin{aligned}
\Delta I^{\mathrm{m}}(z_0; z_1, z_2, z_3, z_4; z_5) &= 1 \otimes I^{\mathrm{m}}(z_0; z_1, z_2, z_3, z_4; z_5) + I^{\mathfrak{dR}}(z_0; z_1; z_2) \otimes I^{\mathrm{m}}(z_0; z_2, z_3, z_4; z_5) \\
&+ I^{\mathfrak{dR}}(z_1; z_2; z_3) \otimes I^{\mathrm{m}}(z_0; z_1, z_3, z_4; z_5) + I^{\mathfrak{dR}}(z_2; z_3; z_4) \otimes I^{\mathrm{m}}(z_0; z_1, z_2, z_4; z_5) \\
&+ I^{\mathfrak{dR}}(z_3; z_4; z_5) \otimes I^{\mathrm{m}}(z_0; z_1, z_2, z_3; z_5) + I^{\mathfrak{dR}}(z_0; z_1, z_2; z_3) \otimes I^{\mathrm{m}}(z_0; z_3, z_4; z_5) \\
&+ I^{\mathfrak{dR}}(z_1; z_2, z_3; z_4) \otimes I^{\mathrm{m}}(z_0; z_1, z_4; z_5) + I^{\mathfrak{dR}}(z_2; z_3, z_4; z_5) \otimes I^{\mathrm{m}}(z_0; z_1, z_2; z_5) \\
&+ I^{\mathfrak{dR}}(z_0; z_1; z_2) \cdot I^{\mathfrak{dR}}(z_2; z_3; z_4) \otimes I^{\mathrm{m}}(z_0; z_2, z_4; z_5) \\
&+ I^{\mathfrak{dR}}(z_0; z_1; z_2) \cdot I^{\mathfrak{dR}}(z_3; z_4; z_5) \otimes I^{\mathrm{m}}(z_0; z_2, z_3; z_5) \\
&+ I^{\mathfrak{dR}}(z_1; z_2; z_3) \cdot I^{\mathfrak{dR}}(z_3; z_4; z_5) \otimes I^{\mathrm{m}}(z_0; z_1, z_3; z_5) + I^{\mathfrak{dR}}(z_1; z_2, z_3, z_4; z_5) \otimes I^{\mathrm{m}}(z_0; z_1; z_5) \\
&+ I^{\mathfrak{dR}}(z_0; z_1; z_2) \cdot I^{\mathfrak{dR}}(z_2; z_3, z_4; z_5) \otimes I^{\mathrm{m}}(z_0; z_2; z_5) \\
&+ I^{\mathfrak{dR}}(z_0; z_1, z_2; z_3) \cdot I^{\mathfrak{dR}}(z_3; z_4; z_5) \otimes I^{\mathrm{m}}(z_0; z_3; z_5) \\
&+ I^{\mathfrak{dR}}(z_0; z_1, z_2, z_3; z_4) \otimes I^{\mathrm{m}}(z_0; z_4; z_5) + I^{\mathfrak{dR}}(z_0; z_1, z_2, z_3, z_4; z_5) \otimes 1.
\end{aligned} \quad (11.141)$$

The right picture of Fig. 11.5 represents one term in this expression (the term $I^{\mathfrak{dR}}(z_0; z_1; z_2) \cdot I^{\mathfrak{dR}}(z_3; z_4; z_5) \otimes I^{\mathrm{m}}(z_0; z_2, z_3; z_5)$).

Let us now see the reason why we introduced the motivic multiple polylogarithms and the de Rham multiple polylogarithms. We start with the coaction on $\mathrm{Li}^{\mathrm{m}}_n(x)$:

$$\Delta \mathrm{Li}^{\mathrm{m}}_n(x) = -\Delta I^{\mathrm{m}}(0; 1, \underbrace{0, \dots, 0}_{n-1}; x) \quad (11.142)$$

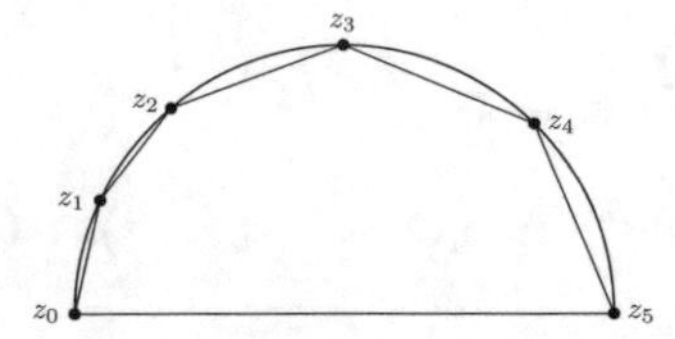

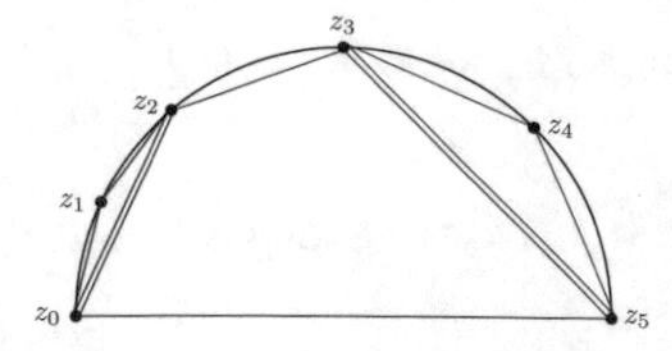

Fig. 11.5 Left figure: We may represent $I^{\mathrm{m}}(z_0; z_1, z_2, z_3, z_4; z_5)$ as a polygon on a half-circle. Right figure: The term $I^{\mathfrak{dR}}(z_0; z_1; z_2) \cdot I^{\mathfrak{dR}}(z_3; z_4; z_5) \otimes I^{\mathrm{m}}(z_0; z_2, z_3; z_5)$ appearing in the coaction. The polygon corresponding to $I^{\mathrm{m}}(z_0; z_2, z_3; z_5)$ is drawn in blue, the polygons corresponding to $I^{\mathfrak{dR}}(z_0; z_1; z_2)$ and $I^{\mathfrak{dR}}(z_3; z_4; z_5)$ are drawn in red

Since not all points are distinct, many terms in the coaction are zero due to Eqs. (11.129) and (11.132). We end up with

$$\Delta \mathrm{Li}_n^{\mathrm{m}}(x) = \mathrm{Li}_n^{\mathfrak{dR}}(x) \otimes 1 + \sum_{k=0}^{n-1} \frac{1}{k!} \left[\ln^{\mathfrak{dR}}(x)\right]^k \otimes \mathrm{Li}_{n-k}^{\mathrm{m}}(x), \qquad (11.143)$$

where

$$\ln^{\mathfrak{dR}}(x) = I^{\mathfrak{dR}}(1; 0; x). \qquad (11.144)$$

Now let us specialise to $x = 1$. Due to Eq. (11.132) we have

$$\ln^{\mathrm{m}}(1) = I^{\mathrm{m}}(1; 0; 1) = 0, \qquad (11.145)$$

and it follows that $\ln^{\mathfrak{dR}}(1) = 0$ as well. For $x = 1$ Eq. (11.143) reduces to

$$\Delta \zeta_n^{\mathrm{m}} = \zeta_n^{\mathfrak{dR}} \otimes 1 + 1 \otimes \zeta_n^{\mathrm{m}}. \qquad (11.146)$$

In Exercise 79 you were supposed to show that

$$\zeta_2^2 = \frac{5}{2} \zeta_4. \qquad (11.147)$$

This relation prohibits a coproduct for zeta values similar to Eq. (11.146). To see this, assume that H is a Hopf algebra, $\zeta_2, \zeta_4 \in H$ are non-zero elements with coproduct

$$\begin{aligned} \Delta(\zeta_2) &= \zeta_2 \otimes 1 + 1 \otimes \zeta_2, \\ \Delta(\zeta_4) &= \zeta_4 \otimes 1 + 1 \otimes \zeta_4, \end{aligned} \qquad (11.148)$$

and Eq. (11.147) holds in H. We consider $\Delta(\zeta_2^2)$. On the one hand we have

$$\Delta\left(\zeta_2^2\right) = \frac{5}{2}\Delta\left(\zeta_4\right) = \frac{5}{2}\left[\zeta_4 \otimes 1 + 1 \otimes \zeta_4\right] = \zeta_2^2 \otimes 1 + 1 \otimes \zeta_2^2, \quad (11.149)$$

on the other hand we obtain using the axiom of compatibility between multiplication and comultiplication in the Hopf algebra H

$$\begin{aligned}\Delta\left(\zeta_2^2\right) = \Delta\left(\zeta_2 \cdot \zeta_2\right) &= \Delta\left(\zeta_2\right) \cdot \Delta\left(\zeta_2\right) = \left[\zeta_2 \otimes 1 + 1 \otimes \zeta_2\right] \cdot \left[\zeta_2 \otimes 1 + 1 \otimes \zeta_2\right] \\ &= \zeta_2^2 \otimes 1 + 2\zeta_2 \otimes \zeta_2 + 1 \otimes \zeta_2^2. \end{aligned} \quad (11.150)$$

We assumed ζ_2 to be a non-zero element of H, hence $\zeta_2 \otimes \zeta_2 \neq 0$ and we have a contradiction.

Now let us return to the motivic zeta values and the de Rham zeta values: Let n be a positive even integer. From Eq. (8.90) we know that in this case the zeta value ζ_n is a rational number times a positive power of $(2\pi i)$. From the definition of the de Rham multiple polylogarithms $\mathcal{P}_{\mathrm{MPL}}^{\mathfrak{dR}}$ it follows that $\zeta_n^{\mathfrak{dR}}$ is equivalent to zero in $\mathcal{P}_{\mathrm{MPL}}^{\mathfrak{dR}}$. Thus for positive even integers the coaction on the zeta values reduces to

$$\Delta\left(\zeta_n^{\mathfrak{m}}\right) = 1 \otimes \zeta_n^{\mathfrak{m}}, \qquad n = 2, 4, 6, 8, \dots. \quad (11.151)$$

We further set

$$\Delta\left((2\pi i)^{\mathfrak{m}}\right) = 1 \otimes (2\pi i)^{\mathfrak{m}}. \quad (11.152)$$

Exercise 95 *Work out* $\Delta(\ln^{\mathfrak{m}}(x))$. *Note that* $\ln^{\mathfrak{m}}(x) = I^{\mathfrak{m}}(1; 0; x)$.

The coaction interacts with derivatives and discontinuities of multiple polylogarithms as follows: Let $I_n^{\mathfrak{m}} \in \mathcal{P}_{\mathrm{MPL}}^{\mathfrak{m}}$ be of weight n. Then

$$\Delta\left(\frac{\partial}{\partial z} I_n^{\mathfrak{m}}\right) = \left(\frac{\partial}{\partial z} \otimes \mathrm{id}\right) \Delta\left(I_n^{\mathfrak{m}}\right). \quad (11.153)$$

Let's verify this for the example $I_2^{\mathfrak{m}} = -I^{\mathfrak{m}}(0; 1, 0; x) = \mathrm{Li}_2^{\mathfrak{m}}(x)$. On the left-hand side we have

$$\Delta\left(\frac{\partial}{\partial x}\mathrm{Li}_2^{\mathfrak{m}}(x)\right) = \Delta\left(\frac{1}{x}\mathrm{Li}_1^{\mathfrak{m}}(x)\right) = \frac{1}{x}\left(\mathrm{Li}_1^{\mathfrak{dR}}(x) \otimes 1 + 1 \otimes \mathrm{Li}_1^{\mathfrak{m}}(x)\right). \quad (11.154)$$

On the right-hand side we have

$$\begin{aligned}\left(\frac{\partial}{\partial x} \otimes \mathrm{id}\right) \Delta\left(\mathrm{Li}_2^{\mathfrak{m}}(x)\right) &= \left(\frac{\partial}{\partial x} \otimes \mathrm{id}\right)\left(\mathrm{Li}_2^{\mathfrak{dR}}(x) \otimes 1 + \ln^{\mathfrak{dR}}(x) \otimes \mathrm{Li}_1^{\mathfrak{m}}(x) + 1 \otimes \mathrm{Li}_2^{\mathfrak{m}}(x)\right) \\ &= \left(\frac{\partial}{\partial x}\mathrm{Li}_2^{\mathfrak{dR}}(x)\right) \otimes 1 + \left(\frac{\partial}{\partial x}\ln^{\mathfrak{dR}}(x)\right) \otimes \mathrm{Li}_1^{\mathfrak{m}}(x) \\ &= \frac{1}{x}\left(\mathrm{Li}_1^{\mathfrak{dR}}(x) \otimes 1 + 1 \otimes \mathrm{Li}_1^{\mathfrak{m}}(x)\right). \end{aligned} \quad (11.155)$$

We also have

$$\frac{\partial}{\partial z} I_n^{\mathfrak{m}} = \cdot \left(\frac{\partial}{\partial z} \otimes 1 \right) \Delta_{1,n-1} \left(I_n^{\mathfrak{m}} \right). \tag{11.156}$$

As an alternative to Eq. (8.8) this formula can be used to calculate the derivative of a multiple polylogarithm. Suppose we would like to calculate

$$\frac{\partial}{\partial z} G\left(0, z; y\right). \tag{11.157}$$

From Eq. (8.8) we have

$$\frac{\partial}{\partial z} G\left(0, z; y\right) = G\left(z; y\right) \frac{\partial}{\partial z} \ln\left(\frac{y}{z}\right) + G\left(0; y\right) \frac{\partial}{\partial z} \ln\left(\frac{-z}{-z}\right) = -\frac{1}{z} G\left(z; y\right). \tag{11.158}$$

We have $G^{\mathfrak{m}}(0, z; y) = I^{\mathfrak{m}}(0; z, 0; y)$ and from Eq. (11.156) we obtain

$$\begin{aligned}
\frac{\partial}{\partial z} G^{\mathfrak{m}}\left(0, z; y\right) &= \cdot \left(\frac{\partial}{\partial z} \otimes 1 \right) \Delta_{1,1}\left(G^{\mathfrak{m}}\left(0, z; y\right) \right) \\
&= \cdot \left(\frac{\partial}{\partial z} \otimes 1 \right) \left(\ln^{\partial \mathfrak{R}}\left(\frac{y}{z}\right) \otimes G^{\mathfrak{m}}\left(z; y\right) \right) \\
&= \left(\frac{\partial}{\partial z} \ln^{\partial \mathfrak{R}}\left(\frac{y}{z}\right) \right) G^{\mathfrak{m}}\left(z; y\right) = -\frac{1}{z} G^{\mathfrak{m}}\left(z; y\right).
\end{aligned} \tag{11.159}$$

Exercise 96 *Consider*

$$I\left(0; x, x; y\right) = G\left(x, x; y\right) = G_{11}\left(1, 1; \frac{y}{x}\right) = \mathrm{Li}_{11}\left(\frac{y}{x}, 1\right) = H_{11}\left(\frac{y}{x}\right). \tag{11.160}$$

With the techniques of Chap. 8 it is not too difficult to show that the derivatives with respect to x and y are

$$\begin{aligned}
\frac{\partial}{\partial x} I\left(0; x, x; y\right) &= \frac{y}{x\left(x - y\right)} \ln\left(\frac{x - y}{x}\right), \\
\frac{\partial}{\partial y} I\left(0; x, x; y\right) &= \frac{1}{y - x} \ln\left(\frac{x - y}{x}\right).
\end{aligned} \tag{11.161}$$

Re-compute the derivatives using Eq. (11.156).

In Sect. 5.5.1 we defined the discontinuity of a function $f(z)$ across a branch cut as

$$\mathrm{Disc}_z\, f\left(z\right) = f\left(z + i\delta\right) - f\left(z - i\delta\right), \tag{11.162}$$

where $\delta > 0$ is infinitesimal. Let $I_n^{\mathrm{m}} \in \mathcal{P}_{\mathrm{MPL}}^{\mathrm{m}}$ be of weight n. Then

$$\Delta\left(\mathrm{Disc}_z I_n^{\mathrm{m}}\right) = \left(\mathrm{id} \otimes \mathrm{Disc}_z\right) \Delta\left(I_n^{\mathrm{m}}\right). \tag{11.163}$$

Let us also verify this with an example. The discontinuity of $\mathrm{Li}_2(z)$ across the branch cut $[1, \infty[$ is

$$\mathrm{Disc}_z \mathrm{Li}_2(z) = 2\pi i \ln(z). \tag{11.164}$$

The left-hand side of Eq. (11.163) gives

$$\begin{aligned}
\Delta\left(\mathrm{Disc}_z \mathrm{Li}_2^{\mathrm{m}}(z)\right) &= \Delta\left((2\pi i)^{\mathrm{m}} \ln^{\mathrm{m}}(z)\right) = \Delta\left((2\pi i)^{\mathrm{m}}\right) \cdot \Delta\left(\ln^{\mathrm{m}}(z)\right) \\
&= \left(1 \otimes (2\pi i)^{\mathrm{m}}\right) \cdot \left(\ln^{\mathfrak{dR}}(x) \otimes 1 + 1 \otimes \ln^{\mathrm{m}}(x)\right) \\
&= \ln^{\mathfrak{dR}}(x) \otimes (2\pi i)^{\mathrm{m}} + 1 \otimes (2\pi i)^{\mathrm{m}} \cdot \ln^{\mathrm{m}}(x)
\end{aligned} \tag{11.165}$$

The right-hand side of Eq. (11.163) gives

$$\begin{aligned}
\left(\mathrm{id} \otimes \mathrm{Disc}_z\right) \Delta\left(\mathrm{Li}_2^{\mathrm{m}}(z)\right) &= \left(\mathrm{id} \otimes \mathrm{Disc}_z\right)\left(\mathrm{Li}_2^{\mathfrak{dR}}(x) \otimes 1 + \ln^{\mathfrak{dR}}(x) \otimes \mathrm{Li}_1^{\mathrm{m}}(x) + 1 \otimes \mathrm{Li}_2^{\mathrm{m}}(x)\right) \\
&= \ln^{\mathfrak{dR}}(x) \otimes \mathrm{Disc}_z \mathrm{Li}_1^{\mathrm{m}}(x) + 1 \otimes \mathrm{Disc}_z \mathrm{Li}_2^{\mathrm{m}}(x) \\
&= \ln^{\mathfrak{dR}}(x) \otimes (2\pi i)^{\mathrm{m}} + 1 \otimes (2\pi i)^{\mathrm{m}} \cdot \ln^{\mathrm{m}}(x),
\end{aligned} \tag{11.166}$$

where we used $\mathrm{Disc}_z \mathrm{Li}_1(x) = 2\pi i$.

Let us summarise:

The **coaction commutes** with the operations of **differentiation** and **taking discontinuities** across branch cuts as follows: For $I^{\mathrm{m}} \in \mathcal{P}_{\mathrm{MPL}}^{\mathrm{m}}$ we have

$$\begin{aligned}
\Delta\left(\frac{\partial}{\partial z} I^{\mathrm{m}}\right) &= \left(\frac{\partial}{\partial z} \otimes \mathrm{id}\right) \Delta\left(I^{\mathrm{m}}\right), \\
\Delta\left(\mathrm{Disc}_z I^{\mathrm{m}}\right) &= \left(\mathrm{id} \otimes \mathrm{Disc}_z\right) \Delta\left(I^{\mathrm{m}}\right).
\end{aligned} \tag{11.167}$$

Please note that within the conventions of this book, the first entry of the tensor carries the information on the derivative, the last entry of the tensor carries the information on the discontinuities. This is a consequence of the definition of the coaction in Eq. (11.140). In this book we use the convention that $\mathcal{P}_{\mathrm{MPL}}^{\mathrm{m}}$ is a left $\mathcal{P}_{\mathrm{MPL}}^{\mathfrak{dR}}$-comodule.

Some authors use a different convention and consider $\mathcal{P}_{\mathrm{MPL}}^{\mathrm{m}}$ to be a right $\mathcal{P}_{\mathrm{MPL}}^{\mathfrak{dR}}$-comodule, in which case the roles of the tensor entries are exchanged.

11.3 Symbols

In this section we introduce symbols and the iterated coaction. Both operations forget information, but they are useful tools for simplifying expressions. If two expressions agree, their symbols and their iterated coactions must agree as well. However, the converse is in general not true.

The symbol is the coarser version, it forgets more. In particular, all constants are mapped to zero. The symbol is defined for transcendental functions, whose total differential is a linear combination of dlog-forms times transcendental functions of weight minus one. The transcendental functions appearing in the total differential are again requested to satisfy the same properties.

If a coaction is available we may use the iterated coaction. This is the finer version. The iterated coaction keeps the information on transcendental constants, algebraic constants are mapped to zero.

We start with the definition of the symbol for iterated integrals [289–292]. Let $\omega_1, \omega_2, \dots$ be a set of dlog-forms in the kinematic variables x. Thus

$$\omega_j = d \ln f_j, \tag{11.168}$$

where f_j is a function of x. In Sect. 6.3.3 we associated to any linear combination of iterated integrals of depth $\leq r$

$$I = \sum_{j=1}^{r} \sum_{i_1,\dots,i_j} c_{i_1 \dots i_j} I_\gamma \left(\omega_{i_1}, \dots, \omega_{i_j}; \lambda \right) \tag{11.169}$$

an element in the tensor algebra $T = \bigoplus\limits_{k=0}^{\infty} (\Omega(X))^{\otimes k}$

$$B = \sum_{j=1}^{r} \sum_{i_1,\dots,i_j} c_{i_1 \dots i_j} \left[\omega_{i_1} | \dots | \omega_{i_j} \right]. \tag{11.170}$$

Recall that the bar notation just denotes a tensor product:

$$[\omega_1 | \omega_2 | \dots | \omega_r] = \omega_1 \otimes \omega_2 \otimes \dots \otimes \omega_r. \tag{11.171}$$

As we are only considering dlog-forms, we may as well just denote the f_j's instead of the ω_j's:

$$S = \sum_{j=1}^{r} \sum_{i_1,\dots,i_j} c_{i_1 \dots i_j} \left(f_{i_1} \otimes \dots \otimes f_{i_j} \right). \tag{11.172}$$

S is called the **symbol** of I. B and S denote the information on the integrand of an iterated integral. The information on the integration path is not stored in the bar notation nor in the symbol. We alert the reader that in this section the letter S denotes the symbol, not an antipode.

For the linear combination I of iterated integrals in Eq. (11.169) we have the total differential

$$dI = \sum_{j=1}^{r} \sum_{i_1,\dots,i_j} c_{i_1\dots i_j} \omega_{i_1} I_\gamma \left(\omega_{i_2}, \dots, \omega_{i_j}; \lambda\right). \tag{11.173}$$

We extend the definition of the symbol from iterated integrals to functions, whose total differential can be written as

$$dF = \sum_i \left(d \ln f_i\right) \; F_i \tag{11.174}$$

with the requirement that the total differential of the function F_i can again be written in the same fashion. We then define the symbol recursively through

$$\begin{aligned} S\left(F\right) &= \sum_i f_i \otimes S\left(F_i\right), \\ S\left(\ln f\right) &= f. \end{aligned} \tag{11.175}$$

Due to Eq. (11.173) the definition in Eq. (11.172) for iterated integrals agrees with the definition of Eq. (11.175).

Please note the order in the tensor product: In the symbol $(f_{i_1} \otimes \dots \otimes f_{i_r})$ of an iterated integral $I_\gamma(\omega_{i_1}, \dots, \omega_{i_j}; \lambda)$ the first entry f_{i_1} corresponds to the outermost integration, while the last entry f_{i_r} corresponds to the innermost integration. This notation is consistent with the conventions used in this book: In writing $G(z_1, \dots, z_r; y)$ or $\text{Li}_{m_1 \dots m_k}(x_1, \dots, x_k)$, the variables z_1 and x_1 refer to the outermost integration and the outermost summation, respectively. The reader should be alerted that most literature on symbols uses the reversed notation. To make this clear, let's consider the classical polylogarithm

$$\text{Li}_n\left(x\right) = -G(\underbrace{0, \dots, 0}_{n-1}, 1; x) \;=\; -I_\gamma(\underbrace{\omega_0, \dots, \omega_0}_{n-1}, \omega_1), \tag{11.176}$$

where γ denotes an integration path from zero to x and $\omega_0 = d \ln(x)$ and $\omega_1 = d \ln(1-x)$. Thus $f_0 = x$ and $f_1 = 1 - x$. With the conventions of this book, the symbol of the classical polylogarithm is

$$S\left(\text{Li}_n\left(x\right)\right) = -(\underbrace{x \otimes \dots \otimes x}_{n-1} \otimes (1-x)). \tag{11.177}$$

As in the symbol the entries of the individual tensor slots denote arguments of dlog-forms we have the following rules:

$$f_1 \otimes \cdots \otimes (g_a g_b) \otimes \cdots \otimes f_r = (f_1 \otimes \cdots \otimes g_a \otimes \cdots \otimes f_r) + (f_1 \otimes \cdots \otimes g_b \otimes \cdots \otimes f_r),$$
$$f_1 \otimes \cdots \otimes (c f_j) \otimes \cdots \otimes f_r = f_1 \otimes \cdots \otimes f_j \otimes \cdots \otimes f_r, \tag{11.178}$$

where c is a constant (independent of x). Thus we have

$$S(\ln(2x)) = S(\ln(x)) = x. \tag{11.179}$$

Note that the minus sign on the right-hand side of Eq. (11.177) is outside the first tensor slot, it corresponds to $c_{i_1 \dots i_j}$ in Eq. (11.172).

Exercise 97 *Work out the symbols*

$$S(-\ln(x)) \text{ and } S(\ln(-x)). \tag{11.180}$$

For two iterated integrals $f = I_\gamma(\omega_1, \dots, \omega_k; \lambda)$ and $g = I_\gamma(\omega_{k+1}, \dots, \omega_r; \lambda)$ along the same path γ we have the shuffle product:

$$I_\gamma(\omega_1, \dots, \omega_k; \lambda) \cdot I_\gamma(\omega_{k+1}, \dots, \omega_r; \lambda) = \sum_{\text{shuffles } \sigma} I_\gamma(\omega_{\sigma(1)}, \omega_{\sigma(1)}, \dots, \omega_{\sigma(r)}; \lambda) \tag{11.181}$$

In Eq. (8.48) we showed this for the case of multiple polylogarithms. The proof carries over to iterated integrals. Taking the symbol on both sides of Eq. (11.181) we find that

$$S(f \cdot g) = S(f) ⧢ S(g), \tag{11.182}$$

where the shuffle product in the tensor algebra T is defined by

$$(f_1 \otimes f_2 \otimes \cdots \otimes f_k) ⧢ (f_{k+1} \otimes \cdots \otimes f_r) = \sum_{\text{shuffles } \sigma} f_{\sigma(1)} \otimes f_{\sigma(2)} \otimes \cdots \otimes f_{\sigma(r)}. \tag{11.183}$$

Let us now look at a simple example: We compute

$$S(\text{Li}_2(x) + \text{Li}_2(1-x)) = -(x \otimes (1-x) + (1-x) \otimes x). \tag{11.184}$$

We also have

$$S(\ln(x) \cdot \ln(1-x)) = S(\ln(x)) ⧢ S(\ln(1-x)) = (x) ⧢ (1-x)$$
$$= x \otimes (1-x) + (1-x) \otimes x. \tag{11.185}$$

Thus we obtain

$$S(\text{Li}_2(x) + \text{Li}_2(1-x) + \ln(x) \cdot \ln(1-x)) = 0. \tag{11.186}$$

This does not imply that

$$\mathrm{Li}_2(x) + \mathrm{Li}_2(1-x) + \ln(x) \cdot \ln(1-x) \tag{11.187}$$

is zero, but we know that the terms which we are missing are in the kernel of the symbol map. This could be a weight 2 constant, or a weight 1 constant times a logarithm of x. Evaluation the above expression at $x = 1$ we find that we should add $(-\zeta_2)$: Doing so, we already obtain the correct relation

$$\mathrm{Li}_2(x) + \mathrm{Li}_2(1-x) + \ln(x) \cdot \ln(1-x) - \zeta_2 = 0. \tag{11.188}$$

We may verify this relation by checking the relation at one point (say $x = 1$) and by showing that the derivative of the left-hand side equals zero. The derivative is of lower weight and repeating this procedure will prove the identity in a finite number of steps.

Transcendental constants like π or ζ_2 are in the kernel of the symbol map, and hence not seen at the level of the symbol. For multiple polylogarithms we also have a coaction. In order to get a handle on transcendental constants, we may use a finer variant of the symbol map, the iterated coaction. We start with $I_n^{\mathfrak{m}} \in \mathcal{P}_{\mathrm{MPL}}^{\mathfrak{m}}$ and assume that $I_n^{\mathfrak{m}}$ has homogeneous weight n. We then consider the $(n-1)$-fold iterated coproduct/coaction

$$\Delta^{n-1}\left(I_n^{\mathfrak{m}}\right) \tag{11.189}$$

Due to coassociativity and Eq. (11.118) it does not matter to which tensor slot the second and further coproducts/coactions are applied, the result will be the same. Equation (11.118) states that

$$(\Delta \otimes \mathrm{id})\, \Delta\left(I_n^{\mathfrak{m}}\right) = (\mathrm{id} \otimes \Delta)\, \Delta\left(I_n^{\mathfrak{m}}\right), \tag{11.190}$$

and this generalises to higher iterated coproducts/coactions. We may therefore simply write Δ^{n-1}, as we did in Eq. (11.189). We then look at $\Delta_{1,\dots,1}(I_n^{\mathfrak{m}})$ (with $1 + \dots + 1 = n$, i.e. the number 1 occurs n times). We call $\Delta_{1,\dots,1}(I_n^{\mathfrak{m}})$ the maximally iterated coaction, any further iteration would produce tensor slots of weight zero. In $\Delta_{1,\dots,1}(I_n^{\mathfrak{m}})$ the entries of all tensor slots are of weight one. We have for example

$$\Delta_{1,1}\left(\mathrm{Li}_2^{\mathfrak{m}}(x)\right) = \ln^{\partial\mathfrak{R}}(x) \otimes \mathrm{Li}_1^{\mathfrak{m}}(x). \tag{11.191}$$

Up to notation, this is identical to the result from the symbol map. However, the iterated coaction does not necessarily kill transcendental constants:

$$\Delta_{1,1}\left((2\pi i)^{\mathfrak{m}} \cdot \ln^{\mathfrak{m}}(x)\right) = \ln^{\partial\mathfrak{R}}(x) \otimes (2\pi i)^{\mathfrak{m}}. \tag{11.192}$$

Here the iterated coaction differs from the symbol map.

If we just look at the maximal iterated coaction we are not sensitive to transcendental constants of weight two or higher. For example

$$\Delta_{1,1,1}\left(\zeta_2^{\mathfrak{m}} \cdot \ln^{\mathfrak{m}}(x)\right) = 0, \tag{11.193}$$

since the coaction does not share out $\zeta_2^{\mathfrak{m}}$ into two weight one pieces. Hoever, this is easily fixed: There is actually no need to focus just on the maximally iterated coaction. Let $i_1 + \cdots + i_k = n$ with $i_j \in \mathbb{N}$. We may also look at

$$\Delta_{i_1,\dots,i_k}\left(I_n^{\mathfrak{m}}\right). \tag{11.194}$$

of the $(k-1)$-fold iterated coproduct/coaction $\Delta^{k-1}(I_n^{\mathfrak{m}})$. We have for example

$$\Delta_{1,2}\left(\zeta_2^{\mathfrak{m}} \cdot \ln^{\mathfrak{m}}(x)\right) = \ln^{\mathfrak{dR}}(x) \otimes \zeta_2^{\mathfrak{m}}. \tag{11.195}$$

The general idea is as follows: Suppose we know already relations for weight $< n$ and we would like to establish a new relation at weight n. Instead of dealing with a single expression of weight n, we use the iterated coaction $\Delta_{i_1,\dots,i_k}$. In each tensor slot the weight is lower than the original weight $(i_j < n)$ and we may use in a particular tensor slot relations which we already know. In summary, we may lower the weight of the objects which we would like to manipulate at the expense of raising the rank of the tensor.

Let's look at an example: We would like to relate the classical polylogarithm $\text{Li}_n(1/x)$ to $\text{Li}_n(x)$. We may derive the sought-after relation from the integral representation and the substitution $x' = 1/x$. Alternatively, we may derive the relation from the coaction. This derivation nicely illustrates how the coaction can be applied. This is an example taken from [293, 294]. Let $x \in \mathbb{R}_{>0}$ be a positive real number. We consider $x - i\delta \in \mathbb{C}$, where $i\delta$ denotes an infinitesimal small imaginary part. We therefore have

$$\ln(-x) = \ln(x) + i\pi. \tag{11.196}$$

At weight one we have

$$\begin{aligned}
\text{Li}_1\left(\frac{1}{x}\right) &= -\ln\left(1-\frac{1}{x}\right) = -\ln(1-x) + \ln(-x) \\
&= \text{Li}_1(x) + \ln(x) + i\pi.
\end{aligned} \tag{11.197}$$

At weight 2 we first consider the $\Delta_{1,1}$-part of the coaction

$$
\begin{aligned}
\Delta_{1,1}\left(\mathrm{Li}_2^{\mathfrak{m}}\left(\frac{1}{x}\right)\right) &= \ln^{\mathfrak{dR}}\left(\frac{1}{x}\right) \otimes \mathrm{Li}_1^{\mathfrak{m}}\left(\frac{1}{x}\right) \\
&= -\ln^{\mathfrak{dR}}(x) \otimes \left[\mathrm{Li}_1^{\mathfrak{m}}(x) + \ln^{\mathfrak{m}}(x) + (i\pi)^{\mathfrak{m}}\right] \\
&= -\ln^{\mathfrak{dR}}(x) \otimes \mathrm{Li}_1^{\mathfrak{m}}(x) - \ln^{\mathfrak{dR}}(x) \otimes \ln^{\mathfrak{m}}(x) - \ln^{\mathfrak{dR}}(x) \otimes (i\pi)^{\mathfrak{m}} \\
&= \Delta_{1,1}\left(-\mathrm{Li}_2^{\mathfrak{m}}(x) - \frac{1}{2}\left[\ln^{\mathfrak{m}}(x)\right]^2 - (i\pi)^{\mathfrak{m}} \ln^{\mathfrak{m}}(x)\right), \qquad (11.198)
\end{aligned}
$$

where we used in the second line the relation Eq. (11.197) for $\mathrm{Li}_1(1/x)$. The $\Delta_{1,1}$-part of the coaction will not detect all terms, in particular we will miss at weight 2 terms proportional to ζ_2. We make the ansatz

$$
\mathrm{Li}_2\left(\frac{1}{x}\right) = -\mathrm{Li}_2(x) - \frac{1}{2}\ln^2(x) - i\pi\ln(x) + c\zeta_2, \qquad (11.199)
$$

with some unknown rational coefficient c. Evaluating the equation at $x = 1$ we find $c = 2$. It is then easily verified (by taking derivatives and evaluating at special points) that

$$
\mathrm{Li}_2\left(\frac{1}{x}\right) = -\mathrm{Li}_2(x) - \frac{1}{2}\ln^2(x) - i\pi\ln(x) + 2\zeta_2 \qquad (11.200)
$$

is the correct relation.

Let us push this example further to weight three: We start with the maximal iterated coaction

$$
\begin{aligned}
\Delta_{1,1,1}\left(\mathrm{Li}_3^{\mathfrak{m}}\left(\frac{1}{x}\right)\right) &= \ln^{\mathfrak{dR}}\left(\frac{1}{x}\right) \otimes \ln^{\mathfrak{dR}}\left(\frac{1}{x}\right) \otimes \mathrm{Li}_1^{\mathfrak{m}}\left(\frac{1}{x}\right) \\
&= \ln^{\mathfrak{dR}}(x) \otimes \ln^{\mathfrak{dR}}(x) \otimes \left[\mathrm{Li}_1^{\mathfrak{m}}(x) + \ln^{\mathfrak{m}}(x) + (i\pi)^{\mathfrak{m}}\right] \\
&= \Delta_{1,1,1}\left(\mathrm{Li}_3^{\mathfrak{m}}(x) + \frac{1}{6}\left[\ln^{\mathfrak{m}}(x)\right]^3 + (i\pi)^{\mathfrak{m}} \frac{1}{2}\left[\ln^{\mathfrak{m}}(x)\right]^2\right). \qquad (11.201)
\end{aligned}
$$

This is not yet the final answer. In $\Delta_{1,1,1}$ we will not detect terms, which are proportional to ζ_2, ζ_3 or π^3. Let's first consider terms proportional to ζ_2. We may detect them in $\Delta_{1,2}$:

$$
\begin{aligned}
&\Delta_{1,2}\left(\mathrm{Li}_3^{\mathfrak{m}}\left(\frac{1}{x}\right) - \mathrm{Li}_3^{\mathfrak{m}}(x) - \frac{1}{6}\left[\ln^{\mathfrak{m}}(x)\right]^3 - (i\pi)^{\mathfrak{m}} \frac{1}{2}\left[\ln^{\mathfrak{m}}(x)\right]^2\right) \\
&= \ln^{\mathfrak{dR}}\left(\frac{1}{x}\right) \otimes \mathrm{Li}_2^{\mathfrak{m}}\left(\frac{1}{x}\right) \\
&= -\ln^{\mathfrak{dR}}(x) \otimes \mathrm{Li}_2^{\mathfrak{m}}(x) - \ln^{\mathfrak{dR}}(x) \otimes \frac{1}{2}\left[\ln^{\mathfrak{m}}(x)\right]^2 - \ln^{\mathfrak{dR}}(x) \otimes (i\pi)^{\mathfrak{m}} \ln^{\mathfrak{m}}(x) \\
&= -\ln^{\mathfrak{dR}}(x) \otimes \left[\mathrm{Li}_2^{\mathfrak{m}}\left(\frac{1}{x}\right) + \mathrm{Li}_2^{\mathfrak{m}}(x) + \frac{1}{2}\left[\ln^{\mathfrak{m}}(x)\right]^2 + (i\pi)^{\mathfrak{m}} \ln^{\mathfrak{m}}(x)\right]. \qquad (11.202)
\end{aligned}
$$

We may now use Eq. (11.200) and find

$$\Delta_{1,2}\left(\mathrm{Li}_3^{\mathfrak{m}}\left(\frac{1}{x}\right)-\mathrm{Li}_3^{\mathfrak{m}}(x)-\frac{1}{6}\left[\ln^{\mathfrak{m}}(x)\right]^3-(i\pi)^{\mathfrak{m}}\frac{1}{2}\left[\ln^{\mathfrak{m}}(x)\right]^2\right)$$
$$=-2\Delta_{1,2}\left(\zeta_2^{\mathfrak{m}}\ln^{\mathfrak{m}}(x)\right). \qquad (11.203)$$

Terms proportional to ζ_3 or π^3 cannot be detected from the coaction, as they do not share out. We make the ansatz

$$\mathrm{Li}_3\left(\frac{1}{x}\right)=\mathrm{Li}_3(x)+\frac{1}{6}\ln^3(x)+\frac{1}{2}i\pi\ln^2(x)-2\zeta_2\ln(x)+c_1\zeta_3+c_2 i\pi^3. \qquad (11.204)$$

Evaluating the expression at $x=1$ yields $c_1=c_2=0$ and we finally obtain

$$\mathrm{Li}_3\left(\frac{1}{x}\right)=\mathrm{Li}_3(x)+\frac{1}{6}\ln^3(x)+\frac{1}{2}i\pi\ln^2(x)-2\zeta_2\ln(x). \qquad (11.205)$$

This relation is then verified by taking derivatives and evaluating at special points.

We may continue in this way and systematically derive inversion relation for $\mathrm{Li}_n(1/x)$.

11.4 The Single-Valued Projection

Multiple polylogarithms are in general multi-valued functions. In relation to Feynman integrals this is what we want: The starting points of branch cuts of multiple polylogarithms are related to the thresholds of Feynman integrals. Nevertheless, we may ask if it is possible to define single-valued multiple polylogarithms. This is indeed possible and we will define single-valued multiple polylogarithms in this section. This will also shed some new light on the role of the de Rham multiple polylogarithms. Up to now we treated them as some formal objects, as we could not associate any numerical value to them. With the help of the single-valued multiple polylogarithms we may define an evaluation map for the de Rham multiple polylogarithms. References for this sections are [295–298].

In Sect. 11.2 we considered the algebras $\mathcal{P}^{\mathfrak{m}}_{\mathrm{MPL}}$ and $\mathcal{P}^{\mathfrak{dR}}_{\mathrm{MPL}}$. There is a projection

$$\pi^{\mathfrak{dR}}:\mathcal{P}^{\mathfrak{m}}_{\mathrm{MPL}}\rightarrow\mathcal{P}^{\mathfrak{dR}}_{\mathrm{MPL}} \qquad (11.206)$$

whose kernel is the ideal $\langle(2\pi i)^{\mathfrak{m}}\rangle$. We have for example

$$\pi^{\mathfrak{dR}}\left(\mathrm{Li}_2^{\mathfrak{m}}(x)+(i\pi)^{\mathfrak{m}}\ln^{\mathfrak{m}}(x)\right)=\mathrm{Li}_2^{\mathfrak{dR}}(x). \qquad (11.207)$$

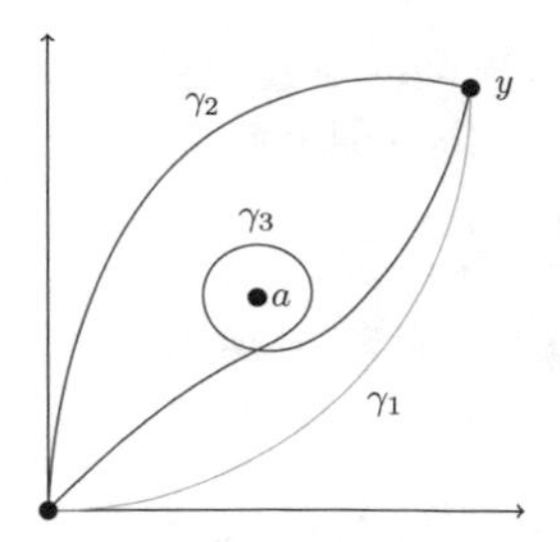

Fig. 11.6 Consider $I_{\gamma_i}(0; a; y)$. The three integration paths cannot be deformed continuously into each other without crossing the pole at a. The result for $I_{\gamma_i}(0; a; y)$ will depend on (the homotopy class of) the integration path. The difference is proportional to $(2\pi i)$, and therefore the de Rham logarithms $I^{\mathfrak{dR}}_{\gamma_i}(0; a; y)$ are equivalent. Phrased differently, the information on the integration path is lost in $\mathcal{P}^{\mathfrak{dR}}_{\text{MPL}}$

We started from the functions $I(z_0; z_1, \dots, z_r; z_{r+1})$ defined in Eq. (11.121). We implicitly assumed a standard integration path (say a straight line from z_0 to z_{r+1}, supplemented in the case of divergent integrals by a tangential base point prescription). We could have started from an extended definition $I_\gamma(z_0; z_1, \dots, z_r; z_{r+1})$, allowing arbitrary integration paths. $I_\gamma(z_0; z_1, \dots, z_r; z_{r+1})$ and $I^{\mathfrak{m}}_\gamma(z_0; z_1, \dots, z_r; z_{r+1})$ would then depend on the integration path. Let us now restrict our attention to linear combinations, which are homotopy functionals (see Fig. 11.6). For those linear combinations we may think about the de Rham version as multiple polylogarithms which have lost all information on the integration path. If we deform the integration path of a homotopy functional of ordinary multiple polylogarithm across a pole of an integrand, we should compensate by $(2\pi i)$ times the residue at the pole. However this equals zero in $\mathcal{P}^{\mathfrak{dR}}_{\text{MPL}}$ and hence this information is lost for the de Rham multiple polylogarithms. For this reason we did not define a period map for de Rham multiple polylogarithms, as a naive attempt would be ambiguous by terms of the form $(2\pi i) \times$ functions of weight $(n-1)$.

The essential ingredient for the definition of single-valued multiple polylogarithms is the map

$$
\begin{aligned}
\text{sv} : \ & \mathcal{P}^{\mathfrak{dR}}_{\text{MPL}} \to \mathcal{P}^{\mathfrak{m}}_{\text{MPL}} \\
& \text{sv}\left(I^{\mathfrak{dR}}\right) = \cdot \left(\text{id} \otimes F_\infty \Sigma\right) \Delta^{\mathfrak{m}} \left(I^{\mathfrak{dR}}\right).
\end{aligned}
\tag{11.208}
$$

Let us explain the ingredients. We recall that $\mathcal{P}^{\mathfrak{dR}}_{\text{MPL}}$ is a Hopf algebra, with coproduct $\Delta : \mathcal{P}^{\mathfrak{dR}}_{\text{MPL}} \to \mathcal{P}^{\mathfrak{dR}}_{\text{MPL}} \otimes \mathcal{P}^{\mathfrak{dR}}_{\text{MPL}}$ and antipode $S : \mathcal{P}^{\mathfrak{dR}}_{\text{MPL}} \to \mathcal{P}^{\mathfrak{dR}}_{\text{MPL}}$ defined in Eq. (11.138) and Eq. (11.139), respectively.

We now define maps

$$
\begin{aligned}
\Delta^{\mathfrak{m}} : \ & \mathcal{P}^{\mathfrak{dR}}_{\text{MPL}} \to \mathcal{P}^{\mathfrak{m}}_{\text{MPL}} \otimes \mathcal{P}^{\mathfrak{m}}_{\text{MPL}}, \\
S^{\mathfrak{m}} : \ & \mathcal{P}^{\mathfrak{m}}_{\text{MPL}} \to \mathcal{P}^{\mathfrak{m}}_{\text{MPL}}
\end{aligned}
\tag{11.209}
$$

by replacing the superscript $\mathfrak{dR}$ with $\mathfrak{m}$ on the right-hand side of Eq. (11.138) and everywhere in Eq. (11.139). We then define

$$\begin{aligned} \Sigma : \; & \mathcal{P}^{\mathfrak{m}}_{\mathrm{MPL}} \rightarrow \mathcal{P}^{\mathfrak{m}}_{\mathrm{MPL}} \\ & \Sigma \left(I^{\mathfrak{m}} \left(z_0; z_1, \dots, z_r; z_{r+1} \right) \right) = (-1)^r \, S^{\mathfrak{m}} \left(I^{\mathfrak{m}} \left(z_0; z_1, \dots, z_r; z_{r+1} \right) \right). \end{aligned} \tag{11.210}$$

The map F_∞ denotes the **real Frobenius**

$$F_\infty : \mathcal{P}^{\mathfrak{m}}_{\mathrm{MPL}} \rightarrow \mathcal{P}^{\mathfrak{m}}_{\mathrm{MPL}}. \tag{11.211}$$

We may think about F_∞ as complex conjugation. For us the important property will be

$$\mathrm{period} \left(F_\infty \left(I^{\mathfrak{m}} \right) \right) = \overline{\mathrm{period} \left(I^{\mathfrak{m}} \right)}, \tag{11.212}$$

where $\bar{z}$ denotes complex conjugation of z. The exact definition of the real Frobenius will be given in Chap. 14 in Eq. (14.175) and Eq. (14.184).

We may now combine the map sv with the period map of Eq. (11.128) and obtain a map

$$\begin{aligned} \mathrm{sv}^{\mathfrak{dR}} : \; & \mathcal{P}^{\mathfrak{dR}}_{\mathrm{MPL}} \rightarrow \mathbb{C}, \\ \mathrm{sv}^{\mathfrak{dR}} & = \mathrm{period} \circ \mathrm{sv}. \end{aligned} \tag{11.213}$$

This will assign a complex number to a de Rham multiple polylogarithm.

In a similar way we may combine the map $\pi^{\mathfrak{dR}}$ of Eq. (11.206) with the maps sv and period and obtain

$$\begin{aligned} \mathrm{sv}^{\mathfrak{m}} : \; & \mathcal{P}^{\mathfrak{m}}_{\mathrm{MPL}} \rightarrow \mathbb{C}, \\ \mathrm{sv}^{\mathfrak{m}} & = \mathrm{period} \circ \mathrm{sv} \circ \pi^{\mathfrak{dR}}. \end{aligned} \tag{11.214}$$

Equation (11.214) is called the **single-value projection**. It can be shown that Eqs. (11.213) and (11.214) define single-valued functions of x.

Let us consider a few examples: We have

$$\begin{aligned} \Delta \left(\ln^{\mathfrak{dR}} (x) \right) & = 1 \otimes \ln^{\mathfrak{dR}} (x) + \ln^{\mathfrak{dR}} (x) \otimes 1, \\ S \left(\ln^{\mathfrak{dR}} (x) \right) & = - \ln^{\mathfrak{dR}} (x), \end{aligned} \tag{11.215}$$

and therefore

$$\mathrm{sv}^{\mathfrak{m}} \left(\ln^{\mathfrak{m}} (x) \right) = \ln (\bar{x}) + \ln (x) = \ln \left(|x|^2 \right). \tag{11.216}$$

For the classical polylogarithms one has

$$\Delta\left(\mathrm{Li}_n^{\mathfrak{dR}}(x)\right) = \mathrm{Li}_n^{\mathfrak{dR}}(x) \otimes 1 + \sum_{k=0}^{n-1} \frac{1}{k!} \left[\ln^{\mathfrak{dR}}(x)\right]^k \otimes \mathrm{Li}_{n-k}^{\mathfrak{dR}}(x),$$

$$S\left(\mathrm{Li}_n^{\mathfrak{dR}}(x)\right) = -\sum_{k=0}^{n-1} \frac{1}{k!} \left[-\ln^{\mathfrak{dR}}(x)\right]^k \mathrm{Li}_{n-k}^{\mathfrak{dR}}(x). \tag{11.217}$$

We obtain

$$\mathrm{sv}^{\mathfrak{m}}\left(\mathrm{Li}_n^{\mathfrak{m}}(x)\right) = \mathrm{Li}_n(x) - (-1)^n \sum_{k=0}^{n-1} \frac{1}{k!} \left[-\ln\left(|x|^2\right)\right]^k \mathrm{Li}_{n-k}(\bar{x}), \tag{11.218}$$

e.g.

$$\mathrm{sv}^{\mathfrak{m}}\left(\mathrm{Li}_1^{\mathfrak{m}}(x)\right) = \mathrm{Li}_1(x) + \mathrm{Li}_1(\bar{x}),$$
$$\mathrm{sv}^{\mathfrak{m}}\left(\mathrm{Li}_2^{\mathfrak{m}}(x)\right) = \mathrm{Li}_2(x) - \mathrm{Li}_2(\bar{x}) + \ln\left(|x|^2\right)\mathrm{Li}_1(\bar{x}), \quad \text{etc..} \tag{11.219}$$

Exercise 98 *Fill in the details for the derivation of* $\mathrm{sv}^{\mathfrak{m}}(\mathrm{Li}_1^{\mathfrak{m}}(x))$ *and* $\mathrm{sv}^{\mathfrak{m}}(\mathrm{Li}_2^{\mathfrak{m}}(x))$.

Setting $x = -1$ in Eq. (11.216) shows that

$$\mathrm{sv}^{\mathfrak{m}}\left((i\pi)^{\mathfrak{m}}\right) = 0, \tag{11.220}$$

setting $x = 1$ in Eq. (11.218) yields

$$\mathrm{sv}^{\mathfrak{m}}\left(\zeta_n^{\mathfrak{m}}\right) = \mathrm{sv}^{\mathfrak{m}}\left(\mathrm{Li}_n^{\mathfrak{m}}(1)\right) = \begin{cases} 2\zeta_n, & n \text{ odd}, \\ 0, & n \text{ even}. \end{cases} \tag{11.221}$$

Equation (11.221) defines the so-called **single-valued zeta values**

$$\zeta_n^{\mathrm{sv}} = \mathrm{sv}^{\mathfrak{m}}\left(\zeta_n^{\mathfrak{m}}\right). \tag{11.222}$$

Although this may sound like an oxymoron (a zeta value is a number independent of x and therefore certainly single-valued as a function of x), it should be understood as follows: The ordinary zeta values are the values of the classical polylogarithms at $x = 1$. In exactly the same way we define the single-valued zeta values to be the values of the single-valued classical polylogarithms at $x = 1$.

11.5 Bootstrap

Let's look at a practical application of the symbol (or the iterated coaction): Suppose we expect that a certain Feynman integral can be written as a linear combination of multiple polylogarithms. Suppose further that we can figure out what the possible arguments of the multiple polylogarithms are. We can then write down an ansatz for the Feynman integral under consideration as a linear combination of multiple polylogarithms with the specific arguments and unknown coefficients. If the Feynman integral is of uniform weight, the unknown coefficients will be algebraic numbers. In the next step we determine the unknown coefficients. If the symbol of the Feynman integral is known, the symbol of the ansatz has to match the symbol of the Feynman integral. This information together boundary data (and possibly other information like information on discontinuities) can be sufficient to determine all coefficients. This is the **bootstrap approach**. It is an heuristic approach, as it depends on our original guess of the arguments of the multiple polylogarithms. However it is quite powerful. In particular it allows in special situations to bypass the need to rationalise square roots [299, 300]. We will illustrate this with an example below, taken from [301]. The bootstrap approach has also been applied to obtain results for scattering amplitudes to an impressive high loop-order, see for example [302–305].

Let's now look at an example: We consider the one-loop two-point function with equal internal masses introduced as Example 6.1 in Sect. 6.3.1. As master integrals we use (see Eqs. (6.236) and (6.277))

$$I_1' = -2\varepsilon I_{10}(2-2\varepsilon),$$
$$I_2' = -\varepsilon\sqrt{x(4+x)}I_{11}(2-2\varepsilon). \qquad (11.223)$$

The differential equation for $\vec{I}' = (I_1', I_2')^T$ is given by Eq. (6.237). With Eq. (6.239) we have

$$\left(d + A'\right)\vec{I}' = 0, \quad A' = \varepsilon\begin{pmatrix} 0 & 0 \\ 0 & 1 \end{pmatrix}\omega_1 - \varepsilon\begin{pmatrix} 0 & 0 \\ 1 & 0 \end{pmatrix}\omega_2, \qquad (11.224)$$
$$\omega_1 = d\ln(4+x), \qquad \omega_2 = d\ln\left(2+x+\sqrt{x(4+x)}\right).$$

The master integral I_1' is a tadpole integral and rather trivial:

$$I_1' = -2 - \zeta_2\varepsilon^2 + O\left(\varepsilon^3\right). \qquad (11.225)$$

In Sect. 6.4.4 we saw that the square root can be rationalised by the transformation

$$x = \frac{\left(1-x'\right)^2}{x'}, \quad x' = \frac{1}{2}\left(2+x-\sqrt{x(4+x)}\right), \qquad (11.226)$$

Under this transformation we have

$$\omega_1 = 2d\ln\left(x'+1\right) - d\ln\left(x'\right), \qquad \omega_2 = -d\ln\left(x'\right). \qquad (11.227)$$

The master integral I_2' vanishes at $x = 0$ (corresponding to $x' = 1$). With this boundary condition we may integrate the differential equation and obtain

$$I_2' = 2\varepsilon G\left(0; x'\right) + 2\varepsilon^2 \left[G\left(0,0;x'\right) - 2G\left(-1,0;x'\right) - \zeta_2\right] + O\left(\varepsilon^3\right). \tag{11.228}$$

We have

$$G\left(0; x'\right) = \ln\left(x'\right), \tag{11.229}$$
$$G\left(0,0; x'\right) = \frac{1}{2}\ln^2\left(x'\right), \qquad G\left(-1,0;x'\right) = \text{Li}_2\left(-x'\right) + \ln\left(x'\right)\ln\left(x'+1\right).$$

Suppose now that we don't know a rationalisation of the square root $\sqrt{x(4+x)}$. The symbol approach allows us to derive Eq. (11.228) without the need of rationalising the square root (so we forget Eqs. (11.226)–(11.228) for the moment). We will however assume that the result can be expressed in terms of multiple polylogarithms. We set

$$f_0 = 2, \quad f_1 = x+4, \quad f_2 = 2+x+r, \qquad r = \sqrt{x\left(4+x\right)}. \tag{11.230}$$

The set $\{f_0, f_1, f_2\}$ will be our alphabet. The subset $\{f_0, f_1\}$ is called the **rational part of the alphabet**, the subset $\{f_2\}$ is called the **algebraic part**. For an algebraic letter f we define the **conjugated letter** $\bar{f}$ with respect to the root r as the letter obtained by the substitution $r \to -r$. Thus

$$\bar{f}_2 = 2 + x - r. \tag{11.231}$$

The letters f_1 and f_2 can be directly read off from the differential equation (11.224). The inclusion of the letter $f_0 = 2$ seems a little bit artificial, after all 2 is a constant and we have

$$d \ln 2 = 0. \tag{11.232}$$

However, we require that $f_2 \bar{f}_2$ factorises over the rational part of the alphabet. We have

$$f_2 \bar{f}_2 = 4 = 2^2 = f_0^2. \tag{11.233}$$

From the differential equation (11.224) we may write down the symbol for I_2':

$$S\left(I_2'\right) = -2\varepsilon\left(f_2\right) + 2\varepsilon^2\left(f_1 \otimes f_2\right) + O\left(\varepsilon^3\right). \tag{11.234}$$

We now consider an ansatz in the form of a linear combination of multiple polylogarithms, such that the symbol of the ansatz matches the symbol of Eq. (11.234). At order ε^1 this is rather easy:

$$I^{(1)}_{\text{ansatz}} = -2\ln\left(f_2\right) \Rightarrow S\left(I^{(1)}_{\text{ansatz}}\right) = -2\left(f_2\right). \tag{11.235}$$

In the next step we check the total differential

$$d\left(I_2'^{(1)} - I_{\text{ansatz}}^{(1)}\right) = 0, \tag{11.236}$$

hence $I_2'^{(1)}$ and $I_{\text{ansatz}}^{(1)}$ can possibly only differ by a constant. From the boundary condition we obtain

$$I_2'^{(1)} = -2\left[\ln(f_2) - \ln(f_0)\right], \tag{11.237}$$

in agreement with Eq. (11.228). Note that

$$\frac{2}{2+x+r} = \frac{2+x-r}{2}. \tag{11.238}$$

The order ε^2 is more interesting: At weight two we expect dilogarithms and products of logarithms. Let us first consider the possible arguments of the dilogarithm: We start from a candidate argument of the form as a power product

$$y = f_0^{\alpha_0} f_1^{\alpha_1} f_2^{\alpha_2}, \qquad \alpha_j \in \mathbb{Q}. \tag{11.239}$$

Not every combination of $(\alpha_0, \alpha_1, \alpha_2)$ will be an allowed combination. To find the restrictions, we consider the symbol of the dilogarithm. We have

$$S\left(\text{Li}_2(y)\right) = -\left(y \otimes (1-y)\right). \tag{11.240}$$

In the first tensor slot the power product distributes according to the rules of Eq. (11.178):

$$(y) = \alpha_0 (f_0) + \alpha_1 (f_1) + \alpha_2 (f_2). \tag{11.241}$$

The second tensor slot is more problematic: We don't want any additional new dlog-forms. Therefore we require that $(1-y)$ is again a power product:

$$1 - y = f_0^{\beta_0} f_1^{\beta_1} f_2^{\beta_2}, \qquad \beta_j \in \mathbb{Q}. \tag{11.242}$$

Thus the allowed arguments y of Li_2 are such that for a given y in the form of Eq. (11.239) there exists $(\beta_0, \beta_1, \beta_2)$ such that Eq. (11.242) holds. Note that Eq. (11.242) is equivalent to

$$\ln(1-y) - \beta_0 \ln(f_0) - \beta_1 \ln(f_1) - \beta_2 \ln(f_2) = 0. \tag{11.243}$$

Given $\ln(1-y)$, $\ln(f_0)$, $\ln(f_1)$ and $\ln(f_2)$ we may use the PSLQ algorithm (discussed in Chap. 15) to check if $(\beta_0, \beta_1, \beta_2)$ exists.

For the case at hand one finds the allowed arguments

$$y \in \{y_1, y_2\}, \qquad y_1 = f_0^{-\frac{1}{2}} f_1^{-\frac{1}{2}} f_2^{\frac{1}{2}}, \qquad y_2 = f_0^{\frac{1}{2}} f_1^{-\frac{1}{2}} f_2^{-\frac{1}{2}}. \tag{11.244}$$

with symbols

$$\begin{aligned} S\left(\text{Li}_2\left(y_1\right)\right) = -\left(y_1 \otimes y_2\right) &= \frac{1}{4}\left(f_2 \otimes f_2 + f_2 \otimes f_1 - f_1 \otimes f_2 - f_1 \otimes f_1\right), \\ S\left(\text{Li}_2\left(y_2\right)\right) = -\left(y_2 \otimes y_1\right) &= \frac{1}{4}\left(f_2 \otimes f_2 - f_2 \otimes f_1 + f_1 \otimes f_2 - f_1 \otimes f_1\right). \end{aligned} \tag{11.245}$$

Let us now construct an ansatz, which matches the symbol. From

$$\begin{aligned} & S\left(-4\,\text{Li}_2\left(y_1\right)\right) = -f_2 \otimes f_2 - f_2 \otimes f_1 + f_1 \otimes f_2 + f_1 \otimes f_1, \\ & S\left(\ln\left(f_1\right) \ln\left(f_2\right)\right) = f_2 \otimes f_1 + f_1 \otimes f_2, \\ & S\left(-\frac{1}{2} \ln^2\left(f_1\right)\right) = -f_1 \otimes f_1, \\ & S\left(\frac{1}{2} \ln^2\left(f_2\right)\right) = f_2 \otimes f_2, \end{aligned} \tag{11.246}$$

it follows that for

$$I_{\text{ansatz}}^{(2)} = -4\,\text{Li}_2\left(y_1\right) + \ln\left(f_1\right) \ln\left(f_2\right) - \frac{1}{2} \ln^2\left(f_1\right) + \frac{1}{2} \ln^2\left(f_2\right) \tag{11.247}$$

we have

$$S\left(I_{\text{ansatz}}^{(2)}\right) = 2 f_1 \otimes f_2. \tag{11.248}$$

In the next step we check the derivative: From the differential equation we have

$$d I_2'^{(2)} = -\omega_1 I_2'^{(1)} = 2\left[\ln\left(f_2\right) - \ln\left(f_0\right)\right] d \ln\left(f_1\right). \tag{11.249}$$

However, the derivative of our ansatz is

$$d I_{\text{ansatz}}^{(2)} = 2\left[\ln\left(f_2\right) - \frac{1}{2} \ln\left(f_0\right)\right] d \ln\left(f_1\right) + \ln\left(f_0\right) d \ln\left(f_2\right). \tag{11.250}$$

This does not match:

$$d\left(I_2'^{(2)} - I_{\text{ansatz}}^{(2)}\right) = -\ln\left(f_0\right)\left[d \ln\left(f_1\right) + d \ln\left(f_2\right)\right]. \tag{11.251}$$

The difference is proportional to $\ln(f_0) = \ln(2)$. This comes to no surprise: A constant like $\ln(2)$ is in the kernel of the symbol map and terms proportional to $\ln(2)$ are not detected by the symbol. We can fix this issue as follows: We add to our ansatz a function, whose derivative is given by the right-hand side of Eq. (11.251). This is a problem of lower weight, as the sought-after function is proportional to the weight one constant $\ln(2)$. In our case it is rather trivial: We add

$$-\ln(f_0)\ln(f_1 f_2) \tag{11.252}$$

to our ansatz. Adding this term will not alter the symbol of our ansatz.

Finally, we match the boundary condition at $x = 0$ and we arrive at

$$\begin{aligned} I_2'^{(2)} &= -4\,\text{Li}_2(y_1) + \ln(f_1)\ln\left(\frac{f_2}{f_0}\right) - \frac{1}{2}\ln^2(f_1) + \frac{1}{2}\ln^2\left(\frac{f_2}{f_0}\right) + 2\zeta_2 \\ &= -4\,\text{Li}_2(y_1) + 2\ln^2(y_2) - \ln^2(f_1) + 2\zeta_2. \end{aligned} \tag{11.253}$$

The result in Eq. (11.253) does not look like our previous result in Eq. (11.228), but in the next exercise you are supposed to show that these two results are identical:

Exercise 99 *Show that Eqs. (11.228) and (11.253) agree in a neighbourhood of* $x = 0$.

In deriving the result of Eq. (11.253) we never had to rationalise the square root $r = \sqrt{x(4+x)}$. There are situations, where one can prove that a certain square root is not rationalisable [306], nevertheless the bootstrap approach is able to find a solution in terms of a linear combination of multiple polylogarithms [299]. This shows the power of the bootstrap approach.

Let us make a few more comments: The representation of the Feynman integral as a linear combination of multiple polylogarithms is not necessarily unique. This is already obvious from the two representations in Eqs. (11.228) and (11.253). Within the bootstrap approach we could as well have started with $4\,\text{Li}_2(y_2)$ instead of $(-4\,\text{Li}_2(y_1))$ and obtained yet another different representation. Using $y_1 + y_2 = 1$ and the relation (5.39) one may show that the so obtained representation is equivalent to the previous one.

One could argue that the result in Eq. (11.228) is simpler than the result in Eq. (11.253), as the former contains only a single square root

$$-x' = -\frac{1}{2}\left(2 + x - \sqrt{x(4+x)}\right), \tag{11.254}$$

whereas the latter contains a square root of a square root

$$y_1 = \sqrt{\frac{2 + x + \sqrt{x(4+x)}}{2(4+x)}}. \tag{11.255}$$

This is an artefact of the choice of our alphabet $\{f_0, f_1, f_2\}$. If we include another constant (-1) in our alphabet, we will find $(-x')$ as an allowed argument of the dilogarithm.

Exercise 100 *Let f_1, f_2, g_1, g_2 be algebraic functions of the kinematic variables x. Determine the symbols of*

$$\mathrm{Li}_{21}\left(f_1, f_2\right) \text{ and } G_{21}\left(g_1, g_2; 1\right). \tag{11.256}$$

Assume then $g_1 = 1/f_1$ and $g_2 = 1/(f_1 f_2)$. Show that in this case the two symbols agree.

From the two symbols deduce the constraints on the arguments f_1, f_2 of $\mathrm{Li}_{21}(f_1, f_2)$ and on the arguments g_1, g_2 of $G_{21}(g_1, g_2; 1)$.

From the previous exercise and Eq. (8.8) we may deduce the constraints on the arguments of a multiple polylogarithm in full generality:

Constraints on the arguments of a multiple polylogarithm:
Let

$$A = \left\{f_1, \ldots, f_{N_L}\right\} \tag{11.257}$$

be an alphabet. The f_j define dlog-forms $\omega_j = d \ln f_j$. Consider power products z_i of the form

$$z_i = \prod_{j=1}^{N_L} f_j^{\alpha_{ij}} \tag{11.258}$$

and set $Z = \{1, z_1, \ldots, z_k\}$. The symbol of the multiple polylogarithm $G(z_1, \ldots, z_k; 1)$ can be expressed in the alphabet A if any difference

$$z_i - z_j, \qquad z_i, z_j \in Z \tag{11.259}$$

can again be expressed as a power product in the form of Eq. (11.258).

The proof follows directly from Eq. (8.8).

In constructing an ansatz it is worth knowing that up to weight four all multiple polylogarithms can be expressed in terms of logarithms, $\mathrm{Li}_2(x_1)$, $\mathrm{Li}_3(x_1)$, $\mathrm{Li}_4(x_1)$ and $\mathrm{Li}_{22}(x_1, x_2)$ [307]. Thus up to weight four it is sufficient to consider only the functions

$$G_1\left(z_1; 1\right), \quad G_2\left(z_1; 1\right), \quad G_3\left(z_1; 1\right), \quad G_4\left(z_1; 1\right), \quad G_{22}\left(z_1, z_2; 1\right). \tag{11.260}$$

At a given weight, products of functions of lower weight may appear.

Chapter 12
Cluster Algebras

In order to motivate the content of this chapter let us look again at example 2 of Sect. 6.3.1: The one-loop four-point function with vanishing internal masses and one non-zero external mass. In this example we have two kinematic variables $x_1 = 2p_1 \cdot p_2/p_4^2$ and $x_2 = 2p_2 \cdot p_3/p_4^2$ (see Eq. (6.68)). By a suitable choice of master integrals we may transform the differential equation into an ε-form as we did in Eq. (6.78). We obtain an alphabet with five letters. The arguments of the dlog-forms are

$$x_1, \quad x_2, \quad x_1 - 1, \quad x_2 - 1, \quad x_1 + x_2 - 1. \tag{12.1}$$

We may now ask: Is there a relation between the initial kinematic variables x_1, x_2, the Feynman graph G and the alphabet in Eq. (12.1)?

This is where cluster algebras enter the game. Cluster algebras were introduced in mathematics by Fomin and Zelevinsky in 2001 [308, 309]. A cluster algebra is a commutative $\mathbb{Q}$-algebra generated by the so-called cluster variables. The cluster variables are grouped into overlapping subsets of fixed cardinality. These subsets are called clusters. Starting from an initial seed, the clusters are constructed recursively through mutations.

Introductory texts on cluster algebras are [310–314]. The relation between cluster algebras and scattering amplitudes in particle physics appeared for the first time in [315] in the context of $\mathcal{N} = 4$ supersymmetric Yang-Mills amplitudes. The relation of Feynman integrals to cluster integrals is an evolving field of research and in this chapter we merely touch the tip of an iceberg. A selection of current research literature on this subject is [316–324].

In Sect. 12.1 we introduce quivers and mutations. Cluster algebras arising from quivers and without coefficients are discussed in Sect. 12.2. These are the simplest ones. In Sect. 12.3 we introduce coefficients (or frozen vertices). Cluster algebras may also be defined in terms of matrices. In Sect. 12.4 we consider one further generalisation, going from anti-symmetric matrices to anti-symmetrisable matrices. In Sect. 12.5 we discuss the relation to Feynman integrals.

S. Weinzierl, *Feynman Integrals*, UNITEXT for Physics,
https://doi.org/10.1007/978-3-030-99558-4_12

12.1 Quivers and Mutations

A **quiver** is an oriented graph. A quiver therefore consists of a set of vertices V, a set of edges E and maps

$$\begin{aligned} \text{sink} &: E \to V, \\ \text{source} &: E \to V, \end{aligned} \tag{12.2}$$

assigning to each edge its sink and source, respectively. A quiver is called finite if both the sets V and E are finite sets. We recall that a self-loop is an edge e with

$$\text{sink}\,(e) = \text{source}\,(e)\,. \tag{12.3}$$

A two-cycle is a pair of two distinct edges e_1 and e_2 with

$$\text{sink}\,(e_1) \;=\; \text{source}\,(e_2) \text{ and } \text{sink}\,(e_2) \;=\; \text{source}\,(e_1)\,. \tag{12.4}$$

We will mainly deal with finite quivers without self-loops and two-cycles.

It is convenient to write $v_i \to v_j$, if there is an edge e with

$$\text{source}\,(e) \;=\; v_i \text{ and } \text{sink}\,(e) \;=\; v_j. \tag{12.5}$$

Let Q be a finite quiver without self-loops and two-cycles. We denote by $r = |V|$ the number of vertices of Q. We may associate an anti-symmetric $(r \times r)$-matrix B to Q as follows: The entry b_{ij} is given as the number of edges, which have v_i as source and v_j as sink minus the number of edges, which have v_i as sink and v_j as source. Since we exclude two-cycles at least one of these two numbers is zero. Furthermore, since we exclude self-loops and two-cycles, we may reconstruct uniquely the quiver Q from the matrix B. The matrix B is called the **exchange matrix** of Q.

Let Q be a finite quiver without self-loops and two-cycles and $v_k \in V$ a vertex of Q. The **mutation** of Q at the vertex v_k is a new quiver Q', obtained from Q as follows:

1. for each path $v_i \to v_k \to v_j$ add an edge $v_i \to v_j$,
2. reverse all arrows on the edges incident with v_k,
3. remove any two-cycles that may have formed.

Two quivers Q and Q' are called **mutation equivalent** if Q' can be obtained through a sequence of mutations from Q.

Under a mutation at the vertex v_k the matrix B transforms to a matrix B', whose entries are given by

$$b'_{ij} = \begin{cases} -b_{ij} & \text{if } i = k \text{ or } j = k, \\ b_{ij} + \text{sign}\,(b_{ik}) \cdot \max\left(0, b_{ik} b_{kj}\right), & \text{otherwise.} \end{cases} \tag{12.6}$$

We call two matrices B and B' mutation equivalent, if B' can be obtained through a sequence of mutations from B.

Exercise 101 *Show that the mutation of the matrix B at a fixed vertex v_k is an involution, i.e., mutating twice at the same vertex returns the original matrix B.*

Let us now associate a variable a_j to each vertex v_j. We define the mutation of these variables under a mutation at a vertex v_k as

$$a'_j = \begin{cases} a_j & j \neq k, \\ \frac{1}{a_k}\left(\prod\limits_{i \mid b_{ik}>0} a_i^{b_{ik}} + \prod\limits_{i \mid b_{ik}<0} a_i^{-b_{ik}} \right) & j = k. \end{cases} \tag{12.7}$$

We use the standard convention that an empty product equals one. The variables a_j are called the **cluster A-variables** or the **cluster A-coordinates**.

Let us set

$$x_j = \prod_i a_i^{b_{ij}} \tag{12.8}$$

Under a mutation at a vertex v_k the variables x_j transform as

$$x'_j = \begin{cases} x_j \left(1 + x_k^{-\operatorname{sign}(b_{kj})}\right)^{-b_{kj}} & j \neq k, \\ \frac{1}{x_k} & j = k. \end{cases} \tag{12.9}$$

The variables x_j are called the **cluster X-variables** or the **cluster X-coordinates** [325].

Exercise 102 *Derive the transformation in Eq. (12.9) from Eqs. (12.8), (12.7) and (12.6).*

Let's look at an example. We start with the quiver shown in the left picture of Fig. 12.1. The quiver has four vertices. The (4×4)-matrix B of the quiver Q reads

$$B = \begin{pmatrix} 0 & 1 & 0 & -1 \\ -1 & 0 & 1 & 0 \\ 0 & -1 & 0 & 1 \\ 1 & 0 & -1 & 0 \end{pmatrix}. \tag{12.10}$$

We consider the mutation at the vertex v_2. We obtain the mutated quiver Q' shown in the right picture of Fig. 12.1. In Q' the orientation of all edges incident to v_2 is reversed. As the original quiver Q contains the edges $v_1 \to v_2 \to v_3$, there is a new edge with the orientation $v_1 \to v_3$ (according to rule 1). The matrix B' associated to Q' reads

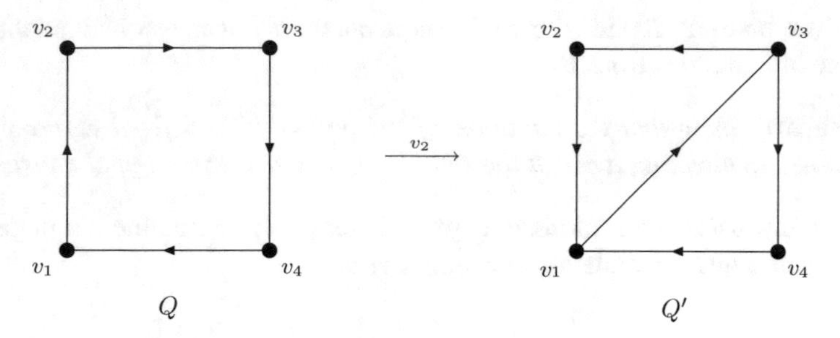

Fig. 12.1 The left picture shows the original quiver Q, the right picture the quiver Q' obtained through a mutation at the vertex v_2

$$B' = \begin{pmatrix} 0 & -1 & 1 & -1 \\ 1 & 0 & -1 & 0 \\ -1 & 1 & 0 & 1 \\ 1 & 0 & -1 & 0 \end{pmatrix}. \tag{12.11}$$

If we assign the variables (a_1, a_2, a_3, a_4) to the vertices (v_1, v_2, v_3, v_4) of the original quiver Q, the mutated variables (a'_1, a'_2, a'_3, a'_4) are

$$a'_1 = a_1, \quad a'_2 = \frac{a_1 + a_3}{a_2}, \quad a'_3 = a_3, \quad a'_4 = a_4. \tag{12.12}$$

The X-variables of the original quiver Q are given by

$$x_1 = \frac{a_4}{a_2}, \quad x_2 = \frac{a_1}{a_3}, \quad x_3 = \frac{a_2}{a_4}, \quad x_4 = \frac{a_3}{a_1}, \tag{12.13}$$

the X-variables of the mutated quiver Q' are given by

$$\begin{aligned} x'_1 &= x_1 (1 + x_2) &&= \frac{a'_2 a'_4}{a'_3} &&= \frac{(a_1 + a_3)\, a_4}{a_2 a_3}, \\ x'_2 &= \frac{1}{x_2} &&= \frac{a'_3}{a'_1} &&= \frac{a_3}{a_1}, \\ x'_3 &= \frac{x_2 x_3}{1 + x_2} &&= \frac{a'_1}{a'_2 a'_4} &&= \frac{a_1 a_2}{(a_1 + a_3)\, a_4}, \\ x'_4 &= x_4 &&= \frac{a'_3}{a'_1} &&= \frac{a_3}{a_1}. \end{aligned} \tag{12.14}$$

12.2 Cluster Algebras Without Coefficients

We start with the simplest cluster algebras: Cluster algebras obtained from a quiver and without coefficients.

A **seed** is pair (Q, a), where Q is a finite quiver without self-loops and two-cycles and $a = \{a_1, \dots, a_r\}$ a set of variables. The number r equals the number of vertices of the quiver: $r = |V|$. The set $a = \{a_1, \dots, a_r\}$ is a set of cluster A-variables.

We assume the variables $a_1, \dots, a_r$ to be independent, hence they generate the field of rational functions $\mathbb{Q}(a_1, \dots, a_r)$.

Let (Q', a') be a pair obtained through a sequence of mutations from the original seed (Q, a) (where a mutation transforms the a_j's according to Eq. (12.7)).

We call any a' so obtained (including the original a) the **clusters** with respect to Q. We call the union of all variables a'_j of all clusters the **cluster variables**. Finally, we define the **cluster algebra** A_Q to be the $\mathbb{Q}$-subalgebra of the field $\mathbb{Q}(a_1, \dots, a_r)$ generated by all the cluster variables. The cluster algebra is said to be of **finite type**, if the number of cluster variables is finite. Fomin and Zelevinsky [308, 309] have classified all cluster algebras of finite type. A cluster algebra A_Q generated by the seed (Q, a) is of finite type if Q is mutation equivalent to an orientation of a simply laced Dynkin diagram (i.e., a Dynkin diagram of type ADE). Dynkin diagrams are reviewed in Appendix D.

Let us look at an example.

Figure 12.2 shows the A_2-cluster algebra. We start with the seed shown on the left. The quiver has two vertices, and we may either mutate at the vertex v_1 or the vertex v_2. The mutations of the quiver are not too interesting: Independently of which vertex we choose, the mutation of the quiver has just the arrow reversed. The mutation of the cluster variables is more interesting: Performing mutations alternating at the vertices v_1 and v_2, we obtain the sequence shown in Fig. 12.2. We may easily verify that any other mutation reproduces a seed already shown in Fig. 12.2. There are only a finite number of possibilities to verify. The cluster variables are therefore

$$a_1, a_2, \frac{1+a_1}{a_2}, \frac{1+a_2}{a_1}, \frac{1+a_1+a_2}{a_1 a_2}. \tag{12.15}$$

This is a finite set and the A_2-cluster algebra is therefore of finite type. The A_2-cluster algebra is the algebra

$$\mathbb{Q}\left[a_1, a_2, \frac{1+a_1}{a_2}, \frac{1+a_2}{a_1}, \frac{1+a_1+a_2}{a_1 a_2}\right]. \tag{12.16}$$

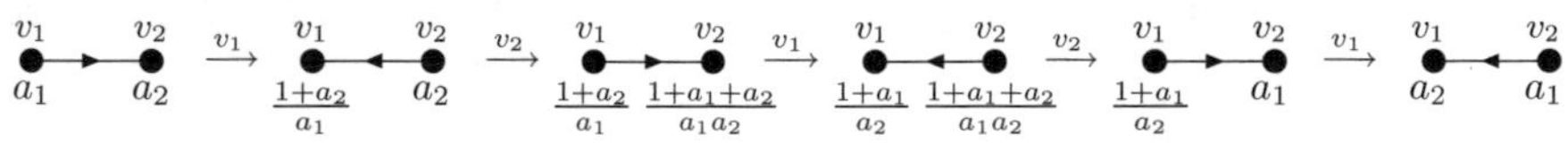

v_1 v_2: $a_1 \rightarrow a_2$ $\xrightarrow{v_1}$ $\frac{1+a_2}{a_1} \leftarrow a_2$ $\xrightarrow{v_2}$ $\frac{1+a_2}{a_1} \rightarrow \frac{1+a_1+a_2}{a_1 a_2}$ $\xrightarrow{v_1}$ $\frac{1+a_1}{a_2} \leftarrow \frac{1+a_1+a_2}{a_1 a_2}$ $\xrightarrow{v_2}$ $\frac{1+a_1}{a_2} \rightarrow a_1$ $\xrightarrow{v_1}$ $a_2 \leftarrow a_1$

Fig. 12.2 The A_2-cluster algebra

12.3 Cluster Algebras with Coefficients

In this section we introduce cluster algebras with coefficients.

Let $1 \leq r \leq s$ be integers. An **ice quiver** is a quiver with vertex set

$$V = \{v_1, \dots, v_r, v_{r+1}, \dots, v_s\} \tag{12.17}$$

such that there are no edges between vertices v_i and v_j if $i > r$ and $j > r$. The vertices $v_{r+1}, \dots, v_s$ are called **frozen vertices**. The **principal part** of Q is the subquiver consisting of the vertices $v_1, \dots, v_r$ and all the oriented edges between them. For an ice quiver we only allow mutations at vertices v_k with $k \leq r$. In addition, in a mutation no arrows are drawn between vertices v_i and v_j if $i > r$ and $j > r$.

To an ice quiver we associate a $(s \times r)$-matrix $\tilde{B}$ with entries b_{ij}. The entry b_{ij} (with $1 \leq i \leq s$ and $1 \leq j \leq r$) is given as before: The number of edges, which have v_i as source and v_j as sink minus the number of edges, which have v_i as sink and v_j as source. The matrix $\tilde{B}$ is called the **extended exchange matrix**. Under a mutation the entries b_{ij} transform as in Eq. (12.6).

The $(r \times r)$-submatrix B with entries b_{ij} with $1 \leq i, j \leq r$ is called – as before – the exchange matrix. B is also the exchange matrix of the principal part of Q.

In the initial seed we associate the variables

$$\{a_1, \dots, a_r, a_{r+1}, \dots, a_s\} \tag{12.18}$$

with the vertices $\{v_1, \dots, v_r, v_{r+1}, \dots, v_s\}$. We call $a_1, \dots, a_r$ the cluster variables, $a_1, \dots, a_r, a_{r+1}, \dots, a_s$ the **extended cluster variables** and $a_{r+1}, \dots, a_s$ the **coefficients**. Under a mutation the extended cluster variables transform as in Eq. (12.6). Note that the coefficients $a_{r+1}, \dots, a_s$ do not change under mutations.

For $1 \leq j \leq r$ we define the cluster X-variables as in Eq. (12.8). They transform under mutations as in Eq. (12.9). Note that there are no cluster X-variables with indices $r + 1, \dots, s$.

The cluster algebra of an ice quiver is the subalgebra of $\mathbb{Q}(a_1, \dots, a_r, a_{r+1}, \dots, a_s)$ generated by the extended cluster variables. The type of a cluster algebra with coefficients is the type of the cluster algebra generated by the principal part of the seed. Thus a cluster algebra with coefficients is of finite type, if its principal part is of finite type.

Figure 12.3 shows an example. It is standard practice to draw frozen vertices as boxes.

The left picture of Fig. 12.3 shows an ice quiver with two unfrozen vertices and four frozen vertices. A mutation at the vertex v_1 will produce the quiver Q' shown in the right picture of Fig. 12.3. The principal part of Q consists of two vertices and one edge shown in red in the left picture of Fig. 12.3. The cluster algebra generated by Q corresponds to a A_2-cluster algebra with coefficients.

Exercise 103 *Determine the cluster A-variables for the ice quiver Q' of Fig. 12.3 in terms of the cluster variables of the ice quiver Q.*

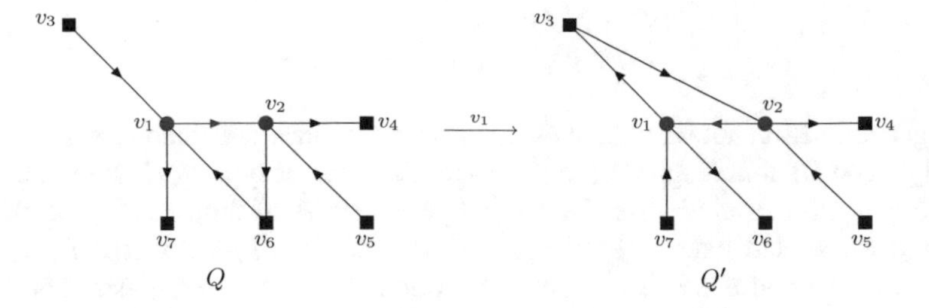

Fig. 12.3 The left picture shows an ice quiver Q with two unfrozen vertices (v_1, v_2) and four frozen vertices (v_3, v_4, v_5, v_6). The mutation at vertex v_1 gives the ice quiver Q', shown in the right picture. The principal parts of Q and Q' are drawn in red

Exercise 104 *Mutate the ice quiver Q' of Fig. 12.3 at the vertex v_2 to obtain an ice quiver Q''. Determine the cluster A-variables for the ice quiver Q'' in terms of the cluster variables of the ice quiver Q.*

12.4 Cluster Algebras from Anti-Symmetrisable Matrices

In order to arrive at a classification of all cluster algebras of finite type we need one more generalisation: Up to now we discussed cluster algebras in terms of quivers: Quivers without frozen vertices for cluster algebras without coefficients and ice quivers for cluster algebras with coefficients. Instead of quivers we could have used the exchange matrix B (for cluster algebras without coefficients) or the extended exchange matrix $\tilde{B}$ (for cluster algebras with coefficients). Up to now the exchange matrix B was always an anti-symmetric $(r \times r)$-matrix

$$B = -B^T. \tag{12.19}$$

We now relax this condition. We no longer require that B is anti-symmetric, but only that B is anti-symmetrisable.

A $(r \times r)$-matrix B is called **anti-symmetrisable** if there is a diagonal $(r \times r)$-matrix D with positive integers on the diagonal, such that $D \cdot B$ is anti-symmetric: $D \cdot B = -B^T D$. Let $D = \mathrm{diag}(d_1, \dots, d_r)$. An anti-symmetrisable matrix satisfies

$$d_i b_{ij} = -d_j b_{ji}. \tag{12.20}$$

Example: The matrix

$$B = \begin{pmatrix} 0 & -1 \\ 2 & 0 \end{pmatrix} \tag{12.21}$$

is not anti-symmetric. However, it is anti-symmetrisable:

$$\begin{pmatrix} 2 & 0 \\ 0 & 1 \end{pmatrix} \begin{pmatrix} 0 & -1 \\ 2 & 0 \end{pmatrix} = \begin{pmatrix} 0 & -2 \\ 2 & 0 \end{pmatrix}. \tag{12.22}$$

This generalisation allows us to include cluster algebras, which correspond to non-simply laced Dynkin diagrams (i.e., Dynkin diagrams of type B, C, F or G).

We now start from an extended exchange matrix $\tilde{B}$ of dimension $(s \times r)$, such that the associated exchange matrix B (i.e., the $(r \times r)$-submatrix B) is anti-symmetrisable and a set $\{a_1, \dots, a_s\}$ of extended cluster A-variables. The first r variables $\{a_1, \dots, a_r\}$ are the cluster A-variables, the remaining $(s - r)$ variables $a_{r+1}, \dots, a_s$ are the coefficients. Under a mutation the extended exchange matrix transforms as in Eq. (12.6). The cluster A-variables transform as in Eq. (12.7), the coefficients do not change. For $1 \le j \le r$ we may define cluster X-variables as in Eq. (12.8). They transform as in Eq. (12.9). The cluster algebra is the subalgebra of $\mathbb{Q}(a_1, \dots, a_r, a_{r+1}, \dots, a_s)$ generated by the extended cluster variables.

In order to state the theorem on the classification of cluster algebras of finite type, we need a little bit more terminology to establish the relation with Dynkin diagrams. A $(r \times r)$-matrix A with integer entries is called a **symmetrisable generalised Cartan matrix** if

1. all diagonal entries of A are equal to 2,
2. all off-diagonal entries of A are non-positive,
3. there exists a diagonal matrix D with positive diagonal entries such that the matrix $D \cdot A$ is symmetric.

A symmetrisable generalised Cartan matrix is called **positive**, if $D \cdot A$ is positive definite. This is equivalent to the positivity of all principal minors $|A[I]|$, where I denotes the rows and columns to be deleted. Now let us choose I as the subset of $(r - 2)$ elements, which leaves the i-th and j-th row and column undeleted. For a positive symmetrisable generalised Cartan matrix we then have

$$\det \begin{pmatrix} 2 & a_{ij} \\ a_{ji} & 2 \end{pmatrix} > 0, \tag{12.23}$$

or equivalently

$$a_{ij} a_{ji} \le 3. \tag{12.24}$$

A positive symmetrisable generalised Cartan matrix is called a **Cartan matrix of finite type**.

A symmetrisable generalised Cartan matrix is called **decomposable**, if by a simultaneous permutation of rows and columns it can be transformed to a block-diagonal matrix with at least two blocks. Otherwise it is called **indecomposable**. The indecomposable Cartan matrices of finite type can be classified into four families (A_n, B_n, C_n and D_n) and a finite number of exceptional cases (E_6, E_7, E_8, F_4 and G_2). We review this classification in Appendix D. The indecomposable Cartan matrices of finite type are in one-to-one correspondence with the respective Dynkin diagrams.

Given a Dynkin diagram with r vertices, we obtain the corresponding $r \times r$-Cartan matrix as follows: The entries of the Cartan matrix on the diagonal are $a_{ii} = 2$ (this is fixed by the definition of a symmetrisable generalised Cartan matrix). The off-diagonal entries a_{ij} are zero, unless the vertices i and j are connected in the Dynkin diagram. If they are connected in the Dynkin diagram, they may be connected by one line, two lines or three lines. In the last two cases there will be in addition an arrow in the Dynkin diagram. We have:

one line i ○—○ j $\quad a_{ij} = -1, \quad a_{ji} = -1,$

two lines i ○⇛○ j $\quad a_{ij} = -1, \quad a_{ji} = -2,$

three lines i ○⭆○ j $\quad a_{ij} = -1, \quad a_{ji} = -3,$

Let B be an anti-symmetrisable $(r \times r)$-matrix with integer entries. We associate to B a $(r \times r)$-symmetrisable generalised Cartan matrix $A(B)$ with entries a_{ij} as follows:

$$as_{ij} = \begin{cases} 2, \text{ if } & i = j, \\ -\left|b_{ij}\right|, \text{ if } & i \neq j. \end{cases} \tag{12.25}$$

We may now state the classification theorem for cluster algebras of finite type [309, 311]:

Classification of cluster algebras of finite type:

Theorem 10 *A cluster algebra is of finite type, if and only if it contains an exchange matrix B such that A(B) is a Cartan matrix of finite type.*

Please note that Dynkin diagrams and quivers are different objects. This is best seen by an example: The Dynkin diagram of type B_2 corresponds to the exchange matrices B

$$1 \text{○⇛○} 2 \;\Rightarrow B = \pm \begin{pmatrix} 0 & -1 \\ 2 & 0 \end{pmatrix}, \tag{12.26}$$

whereas the quiver Q, shown below, corresponds to the exchange matrix B'

$$1 \text{●⇉●} 2 \;\Rightarrow B' = \begin{pmatrix} 0 & 2 \\ -2 & 0 \end{pmatrix}. \tag{12.27}$$

The former generates a cluster algebra of finite type, the latter does not.

Exercise 105 *The B_2-cluster algebra: Determine the cluster variables from the initial seed*

$$B = \begin{pmatrix} 0 & -1 \\ 2 & 0 \end{pmatrix}, \quad a = (a_1, a_2). \tag{12.28}$$

12.5 The Relation of Cluster Algebras to Feynman Integrals

Let us now explore the relation of cluster algebras to Feynman integrals. In this section we denote the kinematic variables of a Feynman integrals by x, the cluster A-variables by a. In order to avoid a conflict of notation we will denote in this section cluster X-coordinates by $\tilde{x}$.

We define polylogarithmic cluster functions of weight w as follows [317]: Let us assume that the cluster A-variables a (or the cluster X-variables $\tilde{x}$) are functions of the kinematic variables x. Polylogarithmic cluster functions of weight 0 are constants, a polylogarithmic cluster functions $f^{(w)}$ of weight w has a differential of the form

$$df^{(w)} = \sum_j f_j^{(w-1)} d \ln a_j, \tag{12.29}$$

where the a_j's are cluster A-variables and the $f_j^{(w-1)}$'s are polylogarithmic cluster functions of weight $(w-1)$. A similar definition applies by substituting the cluster A-variables with cluster X-variables.

Let us now return to the one-loop box integral with one external mass and vanishing internal masses from the introductory remarks of this chapter. The four master integrals I_1', I_2', I_3' and I_4' (defined in Eq. (6.73)) can be expressed in terms of multiple polylogarithms with the five-letter alphabet

$$x_1, \quad x_2, \quad x_1 - 1, \quad x_2 - 1, \quad x_1 + x_2 - 1. \tag{12.30}$$

Furthermore, the term $I_i^{(j)\prime}$ appearing in the ε-expansion at order j

$$I_i' = \sum_{j=0}^{\infty} I_i^{(j)\prime} \cdot \varepsilon^j \tag{12.31}$$

is of uniform weight j. Setting

$$a_1 = -x_1, \; a_2 = -x_2 \tag{12.32}$$

shows that $I_i^{(j)\prime}$ is a polylogarithmic cluster functions of weight j for the A_2-cluster algebra (compare with Fig. 12.2). We set

$$a_1 = a_1, \quad a_2 = a_2, \quad a_3 = \frac{1+a_2}{a_1}, \quad a_4 = \frac{1+a_1+a_2}{a_1 a_2}, \quad a_5 = \frac{1+a_1}{a_2}. \tag{12.33}$$

Let us take the definition of ω_1-ω_5 from Eq. (6.77). We have

$$\begin{aligned}
\omega_1 &= d\ln(x_1) & & = d\ln(a_1), \\
\omega_2 &= d\ln(x_1 - 1) & = d\ln(1+a_1) & = d\ln(a_2) + d\ln(a_5), \\
\omega_3 &= d\ln(x_2) & & = d\ln(a_2), \\
\omega_4 &= d\ln(x_2 - 1) & = d\ln(1+a_2) & = d\ln(a_1) + d\ln(a_3), \\
\omega_5 &= d\ln(x_1 + x_2 - 1) & = d\ln(1+a_1+a_2) & = d\ln(a_1) + d\ln(a_2) + d\ln(a_4).
\end{aligned} \tag{12.34}$$

Let us now go to higher loops, keeping the same external kinematic. The planar and non-planar double box integrals are known [194, 326], as well as the planar triple box [327]. These can be expressed as multiple polylogarithms with a six-letter alphabet

$$x_1, \quad x_2, \quad x_1 - 1, \quad x_2 - 1, \quad x_1 + x_2 - 1, \quad x_1 + x_2. \tag{12.35}$$

The sixth letter $(x_1 + x_2)$ is not present in the one-loop case. We may again relate this alphabet to a cluster algebra. We need a cluster algebra with six cluster variables. The B_2-cluster algebra discussed in Exercise 105 has six cluster variables:

$$a_1, \; a_2, \; a_3 = \frac{1+a_2^2}{a_1}, \; a_4 = \frac{1+a_1+a_2^2}{a_1 a_2}, \; a_5 = \frac{1+2a_1+a_1^2+a_2^2}{a_1 a_2^2}, \; a_6 = \frac{1+a_1}{a_2}. \tag{12.36}$$

Setting [322]

$$x_1 = -\frac{a_2^2}{1+a_1}, \quad x_2 = -\frac{1+a_1+a_2^2}{a_1(1+a_1)} \tag{12.37}$$

allows us to express these Feynman integrals to all order in ε as polylogarithmic cluster functions. In detail we have

$$\begin{aligned}
\omega_1 &= d\ln(x_1) & &= d\ln(a_2) - d\ln(a_6), \\
\omega_2 &= d\ln(x_1 - 1) & &= d\ln(a_1) + d\ln(a_4) - d\ln(a_6), \\
\omega_3 &= d\ln(x_2) & &= d\ln(a_4) - d\ln(a_6), \\
\omega_4 &= d\ln(x_2 - 1) & &= d\ln(a_2) + d\ln(a_5) - d\ln(a_6), \\
\omega_5 &= d\ln(x_1 + x_2 - 1) & &= d\ln(a_2) + d\ln(a_4), \\
\omega_6 &= d\ln(x_1 + x_2) & &= d\ln(a_3).
\end{aligned} \tag{12.38}$$

Chapter 13
Elliptic Curves

Up to now we discussed mainly Feynman integrals, which can expressed in terms of multiple polylogarithms. Multiple polylogarithms are an important class of functions for Feynman integrals, but not every Feynman integral can be expressed in terms of multiple polylogarithms. We have already encountered one example: The two-loop sunrise integral with equal internal masses, discussed as example 4 in Sect. 6.3.1, is not expressible in terms of multiple polylogarithms.

This Feynman integral is related to an elliptic curve. We will see that by a suitable fibre transformation and by a suitable base transformation we may nevertheless transform the differential equation for this Feynman integral into the ε-form of Eq. (6.92). The solution is then again given as iterated integrals, in this specific case as iterated integrals of modular forms.

We call the Feynman integrals treated in this chapter "elliptic Feynman integrals". As a rough guide, elliptic Feynman integrals are the next-to-easiest Feynman integrals, with Feynman integrals evaluating to multiple polylogarithms being the easiest Feynman integrals. Of course, there are also more complicated Feynman integrals beyond these two categories [328–332].

In this chapter we will study elliptic functions, elliptic curves, modular transformations and the moduli space of a genus one curve with n marked points.

Textbooks on elliptic curves are Du Val [333] and Silverman [334], textbooks on modular forms are Stein [335], Miyake [336], Diamond and Shurman [337] and Cohen and Strömberg [338].

13.1 Algebraic Curves

We start with the definition of an **algebraic curve**. As ground field we take the complex numbers $\mathbb{C}$. An algebraic curve in $\mathbb{C}^2$ is defined by the zero set of a polynomial $P(x, y)$ in two variables x and y:

S. Weinzierl, *Feynman Integrals*, UNITEXT for Physics,
https://doi.org/10.1007/978-3-030-99558-4_13

$$P(x, y) = 0 \tag{13.1}$$

It is more common to consider algebraic curves not in the affine space $\mathbb{C}^2$, but in the projective space $\mathbb{CP}^2$. Let $[x : y : z]$ be homogeneous coordinates of $\mathbb{CP}^2$. An algebraic curve in $\mathbb{CP}^2$ is defined by the zero set of a homogeneous polynomial $P(x, y, z)$ in the three variables x, y and z:

$$P(x, y, z) = 0 \tag{13.2}$$

The requirement that $P(x, y, z)$ is a homogeneous polynomial is necessary to have a well-defined zero set on $\mathbb{CP}^2$.

We usually work in the chart $z = 1$. In this chart Eq. (13.2) reduces to

$$P(x, y, 1) = 0. \tag{13.3}$$

If d is the degree of the polynomial $P(x, y, z)$, the **arithmetic genus** of the algebraic curve is given by

$$g = \frac{1}{2}(d-1)(d-2). \tag{13.4}$$

For a smooth curve the arithmetic genus equals the **geometric genus**, therefore just using "genus" is unambiguous in the smooth case. Let's look at an example: The equation

$$y^2 z - x^3 - x z^2 = 0 \tag{13.5}$$

defines a smooth algebraic curve of genus 1.

Let us now turn to elliptic curves: An **elliptic curve** over $\mathbb{C}$ is a smooth algebraic curve in $\mathbb{CP}^2$ of genus one with one marked point. It is common practice to work in the chart $z = 1$ and to take as the marked point the "point at infinity". Equation (13.5) reads in the chart $z = 1$

$$y^2 - x^3 - x = 0, \tag{13.6}$$

The point at infinity, which is not contained in this chart, is given by $[x : y : z] = [0 : 1 : 0]$.

Over the complex numbers $\mathbb{C}$ any elliptic curve can be cast into the **Weierstrass normal form**. In the chart $z = 1$ the Weierstrass normal form reads

$$y^2 = 4x^3 - g_2 x - g_3. \tag{13.7}$$

A second important example is to define an elliptic curve by a quartic polynomial in the chart $z = 1$:

$$y^2 = (x - x_1)(x - x_2)(x - x_3)(x - x_4). \tag{13.8}$$

If all roots of the quartic polynomial on the right-hand side are distinct, this defines a smooth elliptic curve. (The attentive reader may ask, how this squares with the genus formula above. The answer is that the elliptic curve in $\mathbb{CP}^2$ is not given by the homogenisation $y^2z^2 = (x - x_1z)(x - x_2z)(x - x_3z)(x - x_4z)$. The latter curve is singular at infinity. However, there is a smooth elliptic curve, which in the chart $z = 1$ is isomorphic to the affine curve defined by Eq. (13.8).)

As one complex dimension corresponds to two real dimensions, we may consider a smooth algebraic curve (i.e. an object of complex dimension one) also as a real surface (i.e. an object of real dimension two). The latter objects are called **Riemann surfaces**, as the real surface inherits the structure of a complex manifold. We may therefore view an elliptic curve either as a complex one-dimensional smooth algebraic curve in $\mathbb{CP}^2$ with one marked point or as a real Riemann surface of genus one with one marked point. This is shown in Fig. 13.1. We get from the complex algebraic curve to the real Riemann surface as follows: Let's consider the curve defined by Eq. (13.8). We first note that for $x \notin \{x_1, x_2, x_3, x_4\}$ we have two possible values of y:

$$y = \pm\sqrt{(x - x_1)(x - x_2)(x - x_3)(x - x_4)}. \tag{13.9}$$

We denote by $[x_i, x_j]$ the line segment from x_i to x_j in the complex plane. We may define a single-valued square root for $x \in \mathbb{CP}^1\backslash\text{Cuts}$, where the cuts remove points between, say, x_1 and x_2 as well as between x_3 and x_4:

$$\text{Cuts} = [x_1, x_2] \cup [x_3, x_4] \tag{13.10}$$

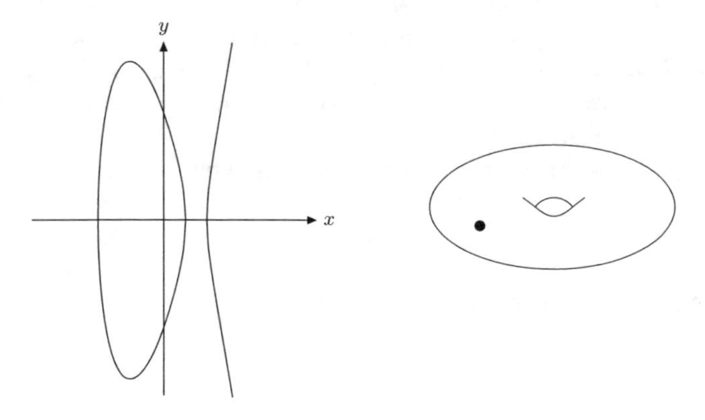

Fig. 13.1 The left picture shows the real part of an elliptic curve in the Weierstrass normal form $y^2 = 4x^3 - g_2x - g_3$. The marked point is at infinity. The right part shows a real Riemann surface of genus one with one marked point

For the other possible value of y we take a second copy of $\mathbb{CP}^1\backslash$Cuts. The two cuts on each copy can be deformed into circles. We then glue the two copies together along the circles originating from the cuts. This gives the torus shown in Fig. 13.1.

13.2 Elliptic Functions and Elliptic Curves

Let us now turn to periodic functions and periods. In Chap. 10 we already introduced the advanced concepts of numerical periods, effective periods and abstract periods. We didn't really discuss where the topic of periods originated from. We will now close this gap and review periodic functions of a single complex variable z.

We consider a non-constant meromorphic function f of a complex variable z. A **period** ψ of the function f is a constant such that for all z:

$$f(z+\psi) = f(z) \tag{13.11}$$

The set of all periods of f forms a lattice, which is either

- trivial (i.e. the lattice consists of $\psi = 0$ only),
- a **simple lattice**, generated by one period ψ: $\Lambda = \{n\psi \mid n \in \mathbb{Z}\}$,
- a **double lattice**, generated by two periods ψ_1, ψ_2 with $\mathrm{Im}(\psi_2/\psi_1) \neq 0$:

$$\Lambda = \{n_1\psi_1 + n_2\psi_2 \mid n_1, n_2 \in \mathbb{Z}\}. \tag{13.12}$$

It is common practice to order these two periods such that $\mathrm{Im}(\psi_2/\psi_1) > 0$.

There cannot be more possibilities: Assume that there is a third period ψ_3, which is not an element of the lattice Λ spanned by ψ_1 and ψ_2. In this case we may construct arbitrary small periods as linear combinations of ψ_1, ψ_2 and ψ_3 with integer coefficients. In the next step one shows that this implies that the derivative of $f(z)$ vanishes at any point z, hence $f(z)$ is a constant. This contradicts our assumption that f is a non-constant function.

An example for a singly periodic function is given by

$$\exp(z). \tag{13.13}$$

In this case the simple lattice is generated by $\psi = 2\pi i$.

Double periodic functions are called **elliptic functions**. An example for a doubly periodic function is given by **Weierstrass's $\wp$-function**. Let Λ be the lattice generated by ψ_1 and ψ_2. Then

$$\wp(z) = \frac{1}{z^2} + \sum_{\psi \in \Lambda \backslash \{0\}} \left(\frac{1}{(z+\psi)^2} - \frac{1}{\psi^2} \right). \tag{13.14}$$

$\wp(z)$ is periodic with periods ψ_1 and ψ_2. Weierstrass's $\wp$-function is an even function, i.e. $\wp(-z) = \wp(z)$.

Of particular interest are also the corresponding inverse functions. These are in general multivalued functions. In the case of the exponential function $x = \exp(z)$, the inverse function is given by

$$z = \ln(x). \tag{13.15}$$

The inverse function to Weierstrass's elliptic function $x = \wp(z)$ is an elliptic integral given by

$$z = \int\limits_{\infty}^{x} \frac{dt}{\sqrt{4t^3 - g_2 t - g_3}} \tag{13.16}$$

with

$$g_2 = 60 \sum_{\psi \in \Lambda \backslash \{0\}} \frac{1}{\psi^4}, \qquad g_3 = 140 \sum_{\psi \in \Lambda \backslash \{0\}} \frac{1}{\psi^6}. \tag{13.17}$$

Note that as $\wp(-z) = \wp(z)$,

$$z = -\int\limits_{\infty}^{x} \frac{dt}{\sqrt{4t^3 - g_2 t - g_3}} \tag{13.18}$$

is also an inverse function to $x = \wp(z)$. We may therefore choose any sign of the square root. In this book we use the convention as in Eq. (13.16) together with a branch cut of the square root along the negative real axis.

Elliptic integrals:

The standard elliptic integrals are classified as complete or incomplete elliptic integrals and as integrals of the first, second or third kind. The complete elliptic integrals are

$$\begin{aligned}
\text{First kind:} \quad K(x) &= \int_0^1 \frac{dt}{\sqrt{(1-t^2)(1-x^2t^2)}}, \\
\text{Second kind:} \quad E(x) &= \int_0^1 dt \frac{\sqrt{1-x^2t^2}}{\sqrt{1-t^2}}, \\
\text{Third kind:} \quad \Pi(v,x) &= \int_0^1 \frac{dt}{(1-vt^2)\sqrt{(1-t^2)(1-x^2t^2)}}.
\end{aligned} \tag{13.19}$$

The incomplete elliptic integrals are

$$\begin{aligned}
\text{First kind:} \quad F(z,x) &= \int_0^z \frac{dt}{\sqrt{(1-t^2)(1-x^2t^2)}}, \\
\text{Second kind:} \quad E(z,x) &= \int_0^z dt \frac{\sqrt{1-x^2t^2}}{\sqrt{1-t^2}}, \\
\text{Third kind:} \quad \Pi(v,z,x) &= \int_0^z \frac{dt}{(1-vt^2)\sqrt{(1-t^2)(1-x^2t^2)}}.
\end{aligned} \tag{13.20}$$

The complete elliptic integrals are a special case of the incomplete elliptic integrals and obtained from the incomplete elliptic integrals by setting the variable z to one.

The classification of elliptic integrals as integrals of the first, second or third kind follows the classification of **Abelian differentials**: An Abelian differential $f(z)dz$ is called Abelian differential of the first kind, if $f(z)$ is holomorphic. It is called an Abelian differential of the second kind, if $f(z)$ is meromorphic, but with all residues vanishing. It is called an Abelian differential of the third kind, if $f(z)$ is meromorphic with non-zero residues.

In $\mathbb{C}/\Lambda$ with coordinate z the differential dz is clearly an Abelian differential of the first kind.

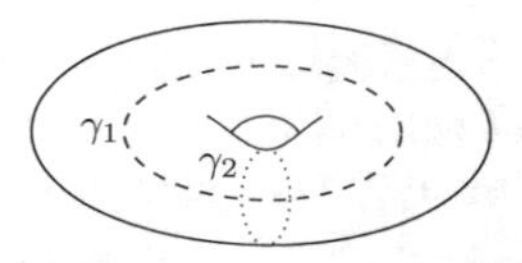

Fig. 13.2 A genus one Riemann surface, where the two independent cycles γ_1 and γ_2 are indicated

Exercise 106 *Consider the elliptic curve* $y^2 = 4x^3 - g_2 x - g_3$. *Show that*

$$dz = \frac{dx}{y}, \tag{13.21}$$

where $y = \sqrt{4x^3 - g_2 x - g_3}$. *This shows that* dx/y *is a holomorphic differential.*

So far we introduced elliptic curves and elliptic integrals. The link between the two is provided by the **periods of an elliptic curve**. An elliptic curve has one holomorphic differential (i.e. one Abelian differential of the first kind). If we view the elliptic curve as a genus one Riemann surface (i.e. a torus), we see that there are two independent cycles γ_1 and γ_2, as shown in Fig. 13.2. A period of an elliptic curve is the integral of the holomorphic differential along a cycle. As there are two independent cycles, there are two independent periods. Let's study this for an elliptic curve in the Legendre form

$$y^2 = x\,(x-1)\,(x-\lambda)\,, \tag{13.22}$$

where λ is a parameter not equal to 0, 1 or infinity. We may think of the elliptic curve as two copies of $\mathbb{CP}^1\backslash$Cuts, where the cuts are between 0 and λ as well as between 1 and ∞. dx/y is the holomorphic differential. Integrating between $x = 0$ and $x = \lambda$ will give a half-period, integrating from $x = \lambda$ to $x = 0$ on the other side of the cut gives another half-period. In order to obtain two periods, we choose two independent integration paths. (The integration between $x = 0$ and $x = \lambda$ will give a result proportional to the integration between $x = 1$ and $x = \infty$, therefore to obtain two independent periods, the two integrations should have one integration boundary in common and differ in the other integration boundary.) A possible choice for the two independent periods is

$$\psi_1 = 2\int\limits_0^\lambda \frac{dx}{y} = 4\,K\left(\sqrt{\lambda}\right), \quad \psi_2 = 2\int\limits_1^\lambda \frac{dx}{y} = 4i\,K\left(\sqrt{1-\lambda}\right). \tag{13.23}$$

Exercise 107 *Determine two independent periods for the elliptic curve defined by a quartic polynomial:*

$$y^2 = (x-x_1)\,(x-x_2)\,(x-x_3)\,(x-x_4)\,. \tag{13.24}$$

From Fig. 13.2 it is evident that the first homology group $H_1(E)$ of an elliptic curve E is isomorphic to $\mathbb{Z} \times \mathbb{Z}$ and generated by γ_1 and γ_2. The two periods are integrals of the holomorphic differential dx/y along γ_1 and γ_2, respectively. We have $\dim H_1(E) = 2$ and it follows that also the first cohomology group of an elliptic curve is two-dimensional. We already know that dx/y is an element of the first de Rham cohomology group

$$\frac{dx}{y} \in H^1_{\mathrm{dR}}(E). \tag{13.25}$$

For the elliptic curve of Eq. (13.22) we may take as a second generator

$$\frac{xdx}{y} \in H^1_{\mathrm{dR}}(E). \tag{13.26}$$

Integrating xdx/y over γ_1 and γ_2 defines the **quasi-periods** ϕ_1 and ϕ_2, respectively. We obtain

$$\phi_1 = 2\int_0^\lambda \frac{xdx}{y} = 4K\left(\sqrt{\lambda}\right) - 4E\left(\sqrt{\lambda}\right), \quad \phi_2 = 2\int_1^\lambda \frac{xdx}{y} = 4iE\left(\sqrt{1-\lambda}\right). \tag{13.27}$$

The **period matrix** is defined by

$$P = \begin{pmatrix} \psi_1 & \psi_2 \\ \phi_1 & \phi_2 \end{pmatrix} = \begin{pmatrix} 4K & 4iK' \\ 4K - 4E & 4iE' \end{pmatrix}, \tag{13.28}$$

where we used the abbreviations $K = K(\sqrt{\lambda})$, $E = E(\sqrt{\lambda})$, $K' = K(\sqrt{1-\lambda})$ and $E' = E(\sqrt{1-\lambda})$. The determinant of the period matrix is given by

$$\det P = 8\pi i. \tag{13.29}$$

This follows from the **Legendre relation**:

$$KE' + EK' - KK' = \frac{\pi}{2}. \tag{13.30}$$

The elliptic curve $y^2 = x(x-1)(x-\lambda)$ depends on a parameter λ, and so do the periods $\psi_1(\lambda)$ and $\psi_2(\lambda)$. We may now ask: How do the periods change, if we change λ? The variation is governed by a second-order differential equation: We have

$$\left[4\lambda(1-\lambda)\frac{d^2}{d\lambda^2} + 4(1-2\lambda)\frac{d}{d\lambda} - 1\right]\psi_j = 0, \qquad j = 1, 2. \tag{13.31}$$

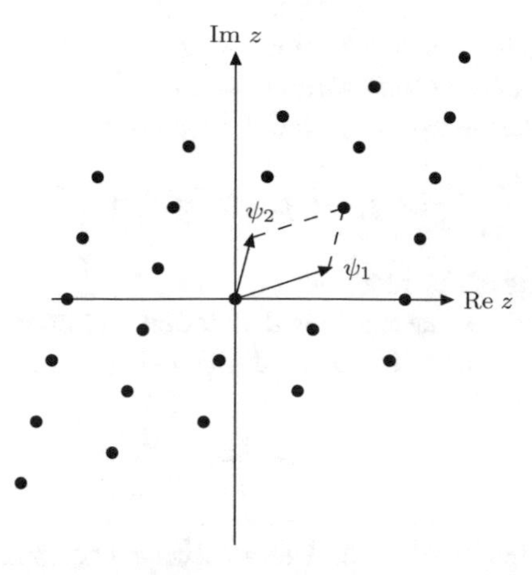

Fig. 13.3 $\mathbb{C}/\Lambda$, where Λ is a double lattice generated by ψ_1 and ψ_2. Points inside the fundamental parallelogram correspond to points on the elliptic curve. A point outside the fundamental parallelogram can always be shifted inside the fundamental parallelogram through the addition of some lattice vector

The differential operator

$$4\lambda\left(1-\lambda\right)\frac{d^2}{d\lambda^2}+4\left(1-2\lambda\right)\frac{d}{d\lambda}-1 \tag{13.32}$$

is called the **Picard-Fuchs operator** of the elliptic curve $y^2 = x(x-1)(x-\lambda)$.

There is a third possibility to represent an elliptic curve: We may also represent an elliptic curve as $\mathbb{C}/\Lambda$, where Λ is the double lattice generated by ψ_1 and ψ_2. This is shown in Fig. 13.3. Points, which differ by a lattice vector are considered to be equivalent. The different equivalence classes are represented by the points inside the fundamental parallelogram, as shown in Fig. 13.3. They correspond to points on the elliptic curve. Before we go into the details, let us first remark that this is not too surprising: If we start from the representation of an elliptic curve as a genus one Riemann surface and cut open this surface along the two cycles γ_1 and γ_2 shown in Fig. 13.2, we obtain a parallelogram.

Let's now fill in the technical detail: We would like to map a point on an elliptic curve, defined by a polynomial P, to a point in $\mathbb{C}/\Lambda$ and vice versa. For simplicity we assume that the elliptic curve is given in the Weierstrass normal form $y^2 - 4x^3 + g_2x + g_3 = 0$.

We start with the direction from the Weierstrass normal form to $\mathbb{C}/\Lambda$: Given a point (x, y) with $y^2 - 4x^3 + g_2x + g_3 = 0$ the corresponding point $z \in \mathbb{C}/\Lambda$ is given by the incomplete elliptic integral

$$z = \int\limits_{\infty}^{x} \frac{dt}{\sqrt{4t^3 - g_2t - g_3}}. \tag{13.33}$$

Let's now consider the reverse direction from $z \in \mathbb{C}/\Lambda$ to a point on the curve defined by the Weierstrass normal form. Given a point $z \in \mathbb{C}/\Lambda$ the corresponding point (x, y) on $y^2 - 4x^3 + g_2 x + g_3 = 0$ is given by

$$(x, y) = \left(\wp(z), \wp'(z)\right). \tag{13.34}$$

$\wp(z)$ denotes Weierstrass's $\wp$-function.

Let us now introduce some additional notation and conventions: It is common practise to normalise one period to one: $(\psi_2, \psi_1) \to (\tau, 1)$, where

$$\tau = \frac{\psi_2}{\psi_1}. \tag{13.35}$$

In addition one requires $\text{Im}(\tau) > 0$. This is always possible: If $\text{Im}(\tau) < 0$ simply exchange ψ_1 and ψ_2 and proceed as above. The possible values of τ lie therefore in the **complex upper half-plane**, defined by

$$\mathbb{H} = \{ \tau \in \mathbb{C} \mid \text{Im}(\tau) > 0 \}. \tag{13.36}$$

Let us now consider an elliptic curve as being given by $\mathbb{C}/\Lambda$, where Λ is a lattice. Two elliptic curves $E = \mathbb{C}/\Lambda$ and $E' = \mathbb{C}/\Lambda'$ are called **isomorphic**, if there is a complex number c such that

$$c\Lambda = \Lambda'. \tag{13.37}$$

For example, the two elliptic curves defined by the lattices with the periods (ψ_2, ψ_1) and $(\tau, 1) = (\psi_2/\psi_1, 1)$ are isomorphic. Two elliptic curves are **isogenic**, if there is a complex number c such that

$$c\Lambda \subset \Lambda', \tag{13.38}$$

i.e. $c\Lambda$ is a sub-lattice of Λ'.

13.2.1 Calculations with Elliptic Curves

Let us now turn to practicalities of doing calculations with elliptic curves. We consider the generic quartic case

$$E : v^2 - (u - u_1)(u - u_2)(u - u_3)(u - u_4) = 0. \tag{13.39}$$

where the roots u_j may depend on variables $x = (x_1, \dots, x_{N_B})$:

$$u_j = u_j(x), \qquad j \in \{1, 2, 3, 4\}. \tag{13.40}$$

In practical applications the x_j's will be the kinematic variables the Feynman integral depends on. In Eq. (13.39) we used the variables (u, v) instead of (x, y) to avoid a clash of notation with the kinematic variables. In this section we consider an elliptic curve together with a fixed choice of two independent periods ψ_1, ψ_2. In mathematical terms we are considering a **framed elliptic curve**. In Sect. 13.3 we will remove the framing and discuss arbitrary choices for the two periods.

But let us now proceed and define a standard choice for the two periods: We set

$$U_1 = (u_3 - u_2)(u_4 - u_1), \quad U_2 = (u_2 - u_1)(u_4 - u_3), \quad U_3 = (u_3 - u_1)(u_4 - u_2). \tag{13.41}$$

Note that we have

$$U_1 + U_2 = U_3. \tag{13.42}$$

We define the **modulus** k and the **complementary modulus** $\bar{k}$ of the elliptic curve E by

$$k^2 = \frac{U_1}{U_3}, \quad \bar{k}^2 = 1 - k^2 = \frac{U_2}{U_3}. \tag{13.43}$$

Note that there are six possibilities of defining k^2 (compare with Exercise 107). Our standard choice for the periods and quasi-periods is

$$\begin{aligned} \psi_1 &= \frac{4K(k)}{U_3^{\frac{1}{2}}}, \quad \psi_2 = \frac{4iK(\bar{k})}{U_3^{\frac{1}{2}}}, \\ \phi_1 &= \frac{4[K(k) - E(k)]}{U_3^{\frac{1}{2}}}, \quad \phi_2 = \frac{4iE(\bar{k})}{U_3^{\frac{1}{2}}}. \end{aligned} \tag{13.44}$$

This defines the framing of the elliptic curve. The Legendre relation for the periods and the quasi-periods reads

$$\psi_1\phi_2 - \psi_2\phi_1 = \frac{8\pi i}{U_3}. \tag{13.45}$$

As in Eq. (13.35) we define the modular parameter τ by

$$\tau = \frac{\psi_2}{\psi_1}. \tag{13.46}$$

In addition we define the **nome** q and the **nome squared** $\bar{q}$ by

$$q = \exp(i\pi\tau), \quad \bar{q} = \exp(2i\pi\tau). \tag{13.47}$$

(In the literature the letter q is either used for the nome or the nome squared. In this book we denote the nome by q and the nome squared by $\bar{q}$. Obviously we have $\bar{q} = q^2$. We will mainly use the nome squared $\bar{q}$.)

We assumed that the roots u_1, u_2, u_3, u_4 depend on the variables $x = (x_1, \dots, x_{N_B})$, hence also the periods and quasi-periods will depend on x. We would like to know how the periods and quasi-periods vary with x. The answer is provided by the following system of first-order differential equations

$$d \begin{pmatrix} \psi_i \\ \phi_i \end{pmatrix} = \begin{pmatrix} -\frac{1}{2} d \ln U_2 & \frac{1}{2} d \ln \frac{U_2}{U_1} \\ -\frac{1}{2} d \ln \frac{U_2}{U_3} & \frac{1}{2} d \ln \frac{U_2}{U_3^2} \end{pmatrix} \begin{pmatrix} \psi_i \\ \phi_i \end{pmatrix}, \qquad i \in \{1, 2\}, \tag{13.48}$$

where d denotes the differential with respect to the variables $x_1, \dots, x_{N_B}$, e.g.

$$df(x) = \sum_{j=1}^{N_B} \left(\frac{\partial f}{\partial x_j} \right) dx_j. \tag{13.49}$$

We further have

$$2\pi i \, d\tau = d \ln \bar{q} = \frac{2\pi i}{\psi_1^2} \frac{4\pi i}{U_3} d \ln \frac{U_2}{U_1}. \tag{13.50}$$

In general, our base space is N_B-dimensional (with coordinates $x_1, \dots, x_{N_B}$). Sometimes we want to restrict to a one-dimensional subspace. To this aim consider a path $\gamma : [0, 1] \to \mathbb{C}^{N_B}$ such that $x_i = x_i(\lambda)$, where the variable λ parametrises the path. For a path γ we may view the periods ψ_1 and ψ_2 as functions of the path variable λ. We may then write down a second-order Picard-Fuchs equation for the variation of the periods along the path γ (as we did in Eq. (13.31)):

$$\left[\frac{d^2}{d\lambda^2} + p_{1,\gamma} \frac{d}{d\lambda} + p_{0,\gamma} \right] \psi_i = 0, \qquad i \in \{1, 2\}. \tag{13.51}$$

The coefficients $p_{1,\gamma}$ and $p_{0,\gamma}$ are given by

$$\begin{aligned} p_{1,\gamma} &= \frac{d}{d\lambda} \ln U_3 - \frac{d}{d\lambda} \ln \left(\frac{d}{d\lambda} \ln \frac{U_2}{U_1} \right), \\ p_{0,\gamma} &= \frac{1}{2} \left(\frac{d}{d\lambda} \ln U_1 \right) \left(\frac{d}{d\lambda} \ln U_2 \right) - \frac{1}{2} \frac{\left(\frac{d}{d\lambda} U_1 \right) \left(\frac{d^2}{d\lambda^2} U_2 \right) - \left(\frac{d^2}{d\lambda^2} U_1 \right) \left(\frac{d}{d\lambda} U_2 \right)}{U_1 \left(\frac{d}{d\lambda} U_2 \right) - U_2 \left(\frac{d}{d\lambda} U_1 \right)} \\ &\quad + \frac{1}{4 U_3} \left[\frac{1}{U_1} \left(\frac{d}{d\lambda} U_1 \right)^2 + \frac{1}{U_2} \left(\frac{d}{d\lambda} U_2 \right)^2 \right]. \end{aligned} \tag{13.52}$$

This defines the Picard-Fuchs operator along the path γ:

$$L_\gamma = \frac{d^2}{d\lambda^2} + p_{1,\gamma}\frac{d}{d\lambda} + p_{0,\gamma}. \tag{13.53}$$

Note that Eqs. (13.51) and (13.53) follow from Eq. (13.48) by restricting to γ and by eliminating the quasi-periods ϕ_i.

The **Wronskian** of Eq. (13.51) is defined by

$$W_\gamma = \psi_1 \frac{d}{d\lambda}\psi_2 - \psi_2 \frac{d}{d\lambda}\psi_1, \tag{13.54}$$

and given by

$$W_\gamma = \frac{4\pi i}{U_3}\frac{d}{d\lambda}\ln\frac{U_2}{U_1}. \tag{13.55}$$

We further have

$$\begin{aligned} \frac{d}{d\lambda}W_\gamma &= -p_{1,\gamma}W_\gamma, \\ 2\pi i d\tau &= \frac{2\pi i\, W_\gamma}{\psi_1^2}d\lambda. \end{aligned} \tag{13.56}$$

Let us now consider a one-parameter family of elliptic curves, which we parametrise by a variable x. This occurs, if either we just have one kinematic variable x (i.e. $N_B = 1$) or if we restrict to a one-dimensional subspace as above. The discussion from above (Eqs. (13.51)–(13.56)) carries over, with λ replaced by x.

In this situation we may want to perform a base transformation as in Sect. 7.2 and change variables from x to τ (or from x to $\bar{q}$, the change from τ to $\bar{q}$ is rather trivial). Equations (13.46) and (13.47) gives us τ and $\bar{q}$ as a function of x. From Eq. (13.56) we have for the Jacobian of the transformation

$$\frac{d\tau}{dx} = \frac{W}{\psi_1^2}, \tag{13.57}$$

where W denotes the Wronskian, defined as in Eq. (13.54) and with λ replaced by x. However, what we really need is not τ or $\bar{q}$ as a function of x, but x as a function of τ or $\bar{q}$. In this context it is useful to know about the Jacobi theta functions and the Dedekind eta function.

Digression Jacobi theta functions and Dedekind eta function
Dedekind's eta function $\eta(\tau)$ is defined for $\tau \in \mathbb{H}$ by

$$\eta(\tau) = e^{\frac{i\pi\tau}{12}} \prod_{n=1}^{\infty}\left(1 - e^{2\pi i n\tau}\right). \tag{13.58}$$

With $\bar{q} = \exp(2\pi i \tau)$ *this becomes*

$$\eta(\tau) = \bar{q}^{\frac{1}{24}} \prod_{n=1}^{\infty} \left(1 - \bar{q}^n\right). \tag{13.59}$$

We have

$$\eta(\tau) = \sum_{n=-\infty}^{\infty} (-1)^n \bar{q}^{\frac{(6n-1)^2}{24}} = \bar{q}^{\frac{1}{24}} \left\{ 1 + \sum_{n=1}^{\infty} (-1)^n \left[\bar{q}^{\frac{1}{2}(3n-1)n} + \bar{q}^{\frac{1}{2}(3n+1)n} \right] \right\}. \tag{13.60}$$

Dedekind's eta function is related to the Jacobi theta function θ_2 *(defined below):*

$$\eta(\tau) = \frac{1}{\sqrt{3}} \theta_2\left(\frac{\pi}{6}, \bar{q}^{\frac{1}{6}}\right). \tag{13.61}$$

Under modular transformations we have

$$\eta(\tau + 1) = e^{\frac{2\pi i}{24}} \eta(\tau), \quad \eta\left(-\frac{1}{\tau}\right) = (-i\tau)^{\frac{1}{2}} \eta(\tau). \tag{13.62}$$

Dedekind's eta function is related to the **modular discriminant** $\Delta(\tau)$ *through*

$$\Delta(\tau) = (2\pi i)^{12} \eta(\tau)^{24}. \tag{13.63}$$

Let us now turn to the Jacobi theta functions. For historical reasons, they are defined through the nome $q = \exp(i\pi\tau)$ *and a variable* z*, which within the conventions of this book we later always will rescale as* $z \to \pi z$*. But for the moment, we follow the standard (historical) notation. The general theta function is defined for* $a, b \in \mathbb{R}$ *by*

$$\theta[a,b](z,q) = \theta[a,b](z|\tau) = \sum_{n=-\infty}^{\infty} q^{\left(n+\frac{1}{2}a\right)^2} e^{2i\left(n+\frac{1}{2}a\right)\left(z-\frac{1}{2}\pi b\right)}. \tag{13.64}$$

The theta function satisfies

$$4i \frac{\partial}{\partial \tau} \theta[a,b](z|\tau) = \pi \frac{\partial^2}{\partial z^2} \theta[a,b](z|\tau). \tag{13.65}$$

For $n \in \mathbb{Z}$ *we have*

$$\theta[a+2n,b](z|\tau) = \theta[a,b](z|\tau), \quad \theta[a,b+2n](z|\tau) = e^{-n\pi i a} \theta[a,b](z|\tau). \tag{13.66}$$

If $a, b \in \mathbb{Z}$ it is therefore sufficient to consider the four cases

$$\begin{aligned} \theta_1(z,q) &= \theta[1,1](z,q)\,, \quad \theta_2(z,q) = \theta[1,0](z,q)\,, \\ \theta_3(z,q) &= \theta[0,0](z,q)\,, \quad \theta_4(z,q) = \theta[0,1](z,q)\,. \end{aligned} \tag{13.67}$$

Explicitly the four theta functions are defined by

$$\begin{aligned} \theta_1(z,q) = \theta_1(z|\tau) &= -i\sum_{n=-\infty}^{\infty}(-1)^n q^{\left(n+\frac{1}{2}\right)^2} e^{i(2n+1)z} = 2\sum_{n=0}^{\infty}(-1)^n q^{\left(n+\frac{1}{2}\right)^2}\sin((2n+1)z)\,, \\ \theta_2(z,q) = \theta_2(z|\tau) &= \sum_{n=-\infty}^{\infty} q^{\left(n+\frac{1}{2}\right)^2} e^{i(2n+1)z} = 2\sum_{n=0}^{\infty} q^{\left(n+\frac{1}{2}\right)^2}\cos((2n+1)z)\,, \\ \theta_3(z,q) = \theta_3(z|\tau) &= \sum_{n=-\infty}^{\infty} q^{n^2} e^{2inz} = 1+2\sum_{n=1}^{\infty} q^{n^2}\cos(2nz)\,, \\ \theta_4(z,q) = \theta_4(z|\tau) &= \sum_{n=-\infty}^{\infty}(-1)^n q^{n^2} e^{2inz} = 1+2\sum_{n=1}^{\infty}(-1)^n q^{n^2}\cos(2nz)\,. \end{aligned} \tag{13.68}$$

The theta functions have a representation as infinite products:

$$\begin{aligned} \theta_1(z,q) &= 2q^{\frac{1}{4}}\sin z\prod_{n=1}^{\infty}\left(1-q^{2n}\right)\left(1-2q^{2n}\cos(2z)+q^{4n}\right), \\ \theta_2(z,q) &= 2q^{\frac{1}{4}}\cos z\prod_{n=1}^{\infty}\left(1-q^{2n}\right)\left(1+2q^{2n}\cos(2z)+q^{4n}\right), \\ \theta_3(z,q) &= \prod_{n=1}^{\infty}\left(1-q^{2n}\right)\left(1+2q^{2n-1}\cos(2z)+q^{4n-2}\right), \\ \theta_4(z,q) &= \prod_{n=1}^{\infty}\left(1-q^{2n}\right)\left(1-2q^{2n-1}\cos(2z)+q^{4n-2}\right). \end{aligned} \tag{13.69}$$

The functions θ_1 and θ_2 are periodic in z with period 2π, the functions θ_3 and θ_4 are periodic in z with period π.

$$\begin{aligned} \theta_1(z+2\pi,q) &= \theta_1(z,q)\,, \quad \theta_3(z+\pi,q) = \theta_3(z,q)\,, \\ \theta_2(z+2\pi,q) &= \theta_2(z,q)\,, \quad \theta_4(z+\pi,q) = \theta_4(z,q)\,. \end{aligned} \tag{13.70}$$

$\pi\tau$ is a quasi-period of the theta functions with periodicity factor $\pm\left(qe^{2iz}\right)^{-1}$:

$$\begin{aligned} \theta_1(z+\pi\tau,q) &= -\left(qe^{2iz}\right)^{-1}\theta_1(z,q)\,, \quad \theta_3(z+\pi\tau,q) = \left(qe^{2iz}\right)^{-1}\theta_3(z,q)\,, \\ \theta_2(z+\pi\tau,q) &= \left(qe^{2iz}\right)^{-1}\theta_2(z,q)\,, \quad \theta_4(z+\pi\tau,q) = -\left(qe^{2iz}\right)^{-1}\theta_4(z,q)\,. \end{aligned} \tag{13.71}$$

A prime denotes the derivative with respect to the first variable z

$$\theta_i'(z,q) = \frac{\partial}{\partial z}\theta_i(z,q), \qquad i \in \{1,2,3,4\}. \tag{13.72}$$

Useful relations are

$$\begin{aligned} \theta_1'(0,q) &= \theta_2(0,q)\,\theta_3(0,q)\,\theta_4(0,q), \\ \theta_3^4(0,q) &= \theta_2^4(0,q) + \theta_4^4(0,q). \end{aligned} \tag{13.73}$$

The theta functions can be used to express the modulus k, the complementary modulus k′, the complete elliptic integral of the first kind K and the complete elliptic integral of the second kind E as functions of the nome q:

$$\begin{aligned} k &= \frac{\theta_2^2(0,q)}{\theta_3^2(0,q)}, & k' &= \frac{\theta_4^2(0,q)}{\theta_3^2(0,q)}, \\ K &= \frac{\pi}{2}\theta_3^2(0,q), & E &= \frac{\pi}{2}\left(1 - \frac{\theta_4''(0,q)}{\theta_3^4(0,q)\,\theta_4(0,q)}\right)\theta_3^2(0,q). \end{aligned} \tag{13.74}$$

We have the following relations with Dedekind's eta function:

$$\theta_2(0,q) = 2\frac{\eta(2\tau)^2}{\eta(\tau)}, \qquad \theta_3(0,q) = \frac{\eta(\tau)^5}{\eta\left(\frac{\tau}{2}\right)^2\eta(2\tau)^2}, \qquad \theta_4(0,q) = \frac{\eta\left(\frac{\tau}{2}\right)^2}{\eta(\tau)}. \tag{13.75}$$

Exercise 108 *Express the modulus squared k^2 and the complementary modulus squared k'^2 as a quotient of eta functions.*

Equipped with the Jacobi theta functions and the Dedekind eta function we now return to our original problem, expressing x as a function of τ or $\bar{q}$. Let's look at the modulus squared k^2, defined in Eq. (13.43) for the elliptic curve E of Eq. (13.39). On the one hand we have from Eq. (13.43)

$$k^2 = \frac{U_1}{U_3} = \frac{(u_3 - u_2)(u_4 - u_1)}{(u_3 - u_1)(u_4 - u_2)}. \tag{13.76}$$

The right-hand side is a (known) function of x. On the other hand, we learned in Exercise 108 that

$$k^2 = 16\frac{\eta\left(\frac{\tau}{2}\right)^8\eta(2\tau)^{16}}{\eta(\tau)^{24}}. \tag{13.77}$$

Here, the right-hand side has an expansion in $\bar{q}^{\frac{1}{2}}$ (the square root originates from the argument $\tau/2$):

$$16\frac{\eta\left(\frac{\tau}{2}\right)^8 \eta\left(2\tau\right)^{16}}{\eta\left(\tau\right)^{24}} = 16\bar{q}^{\frac{1}{2}} - 128\bar{q} + 704\bar{q}^{\frac{3}{2}} - 3072\bar{q}^2 + 11488\bar{q}^{\frac{5}{2}} + O\left(\bar{q}^3\right). \tag{13.78}$$

Thus

$$\frac{U_1}{U_3} = 16\frac{\eta\left(\frac{\tau}{2}\right)^8 \eta\left(2\tau\right)^{16}}{\eta\left(\tau\right)^{24}}. \tag{13.79}$$

We may then solve Eq. (13.79) for x as a power series in $\bar{q}_N = \bar{q}^{\frac{1}{N}}$ for an appropriate N. It is usually possible with the help of computer algebra programs to obtain a large number of terms of this power series. In a second step we try to find a closed form for this power series. There are several heuristic methods how this can be done:

1. If we expect that the result should lie within a certain function space, we can start from an ansatz with unknown coefficients and determine the coefficients by comparing sufficient many terms in the power series.
2. If we expect that the result can be expressed as an eta quotient, we may use dedicated computer programs to find this eta quotient [339].
3. We may use the "On-Line Encyclopedia of Integer Sequences" [340] by typing the first few coefficients of the power series into the web interface.

Let us remark that the first method can be turned into a strict method by first proving that the result must lie within a finite-dimensional function space. Once this is established, we know that the result can be written as a finite linear combination of certain basis functions and it suffices to determine the coefficients by comparing sufficient many terms in the power series.

Let us now look at an example: We consider a family of elliptic curves

$$E: v^2 - u\left(u+4\right)\left[u^2 + 2\left(1+x\right)u + \left(1-x\right)^2\right] = 0, \tag{13.80}$$

depending on a parameter x. We denote the roots of the quartic polynomial in Eq. (13.80) by

$$u_1 = -4, \quad u_2 = -\left(1+\sqrt{x}\right)^2, \quad u_3 = -\left(1-\sqrt{x}\right)^2, \quad u_4 = 0. \tag{13.81}$$

We have

$$k^2 = \frac{U_1}{U_3} = \frac{16\sqrt{x}}{\left(1+\sqrt{x}\right)^3\left(3-\sqrt{x}\right)}, \tag{13.82}$$

and therefore

$$\frac{16\sqrt{x}}{\left(1+\sqrt{x}\right)^3\left(3-\sqrt{x}\right)} = 16\frac{\eta\left(\frac{\tau}{2}\right)^8 \eta\left(2\tau\right)^{16}}{\eta\left(\tau\right)^{24}}. \tag{13.83}$$

We first solve this equation for $\sqrt{x}$ as a power series in $\bar{q}_2 = \bar{q}^{\frac{1}{2}}$:

$$\sqrt{x} = 3\bar{q}_2 - 6\bar{q}_2^3 + 9\bar{q}_2^5 - 12\bar{q}_2^7 + 21\bar{q}_2^9 - 36\bar{q}_2^{11} + 51\bar{q}_2^{13} + O\left(\bar{q}_2^{15}\right). \quad (13.84)$$

Squaring the left-hand side and the right-hand side we obtain

$$x = 9\bar{q} - 36\bar{q}^2 + 90\bar{q}^3 - 180\bar{q}^4 + 351\bar{q}^5 - 684\bar{q}^6 + 1260\bar{q}^7 + O\left(\bar{q}^8\right). \quad (13.85)$$

In the last step we convert the power series to a closed form:

$$x = 9\frac{\eta(\tau)^4\,\eta(6\tau)^8}{\eta(3\tau)^4\,\eta(2\tau)^8}. \quad (13.86)$$

We therefore obtained an expression for x as a function of τ (or $\bar{q}$).

13.3 Modular Transformations and Modular Forms

In the previous section we considered a framed elliptic curve, i.e. an elliptic curve together with a fixed choice of periods ψ_1 and ψ_2, which generate the lattice Λ. In this section we investigate the implications of our freedom of choice for the periods ψ_1 and ψ_2. This will lead us to modular transformations.

We recall that we may represent an elliptic curve as $\mathbb{C}/\Lambda$, where Λ is a double lattice generated by ψ_1 and ψ_2. As only the lattice Λ matters, but not the specific generators, we may consider a different pair of periods (ψ_2', ψ_1'), which generate the same lattice Λ. An example is shown in Fig. 13.4: The generators τ and 1 generate the same lattice as the generators $\tau'$and 1.

Let's return to the general case and consider a change of basis from the pair of periods (ψ_2, ψ_1) to the pair of periods (ψ_2', ψ_1'). The new pair of periods (ψ_2', ψ_1') is again a pair of lattice vectors, so it can be written as

$$\begin{pmatrix} \psi_2' \\ \psi_1' \end{pmatrix} = \begin{pmatrix} a & b \\ c & d \end{pmatrix} \begin{pmatrix} \psi_2 \\ \psi_1 \end{pmatrix}, \quad (13.87)$$

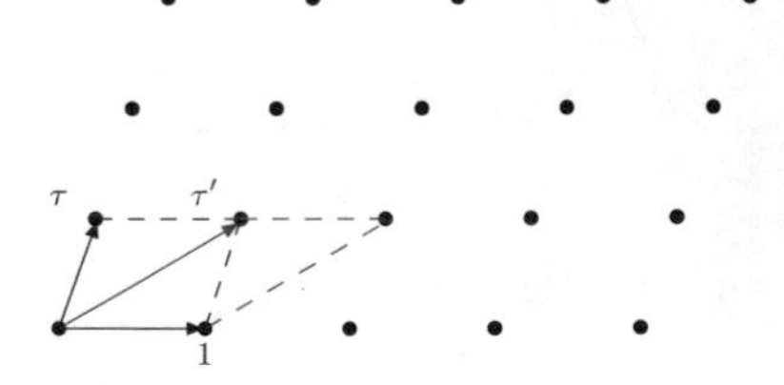

Fig. 13.4 The generators τ and 1 generate the same lattice as the generators $\tau'$and 1

with $a, b, c, d \in \mathbb{Z}$. The transformation should be invertible and (ψ_2, ψ_1) and (ψ_2', ψ_1') should generate the same lattice Λ. This implies

$$\begin{pmatrix} a & b \\ c & d \end{pmatrix} \in \mathrm{SL}_2(\mathbb{Z}). \tag{13.88}$$

The group $\mathrm{SL}_2(\mathbb{Z})$ is called the **modular group**. It is generated by the two matrices

$$T = \begin{pmatrix} 1 & 1 \\ 0 & 1 \end{pmatrix} \text{ and } S = \begin{pmatrix} 0 & -1 \\ 1 & 0 \end{pmatrix}. \tag{13.89}$$

In terms of τ and τ' we have

$$\tau' = \frac{a\tau + b}{c\tau + d}. \tag{13.90}$$

A transformation of the form as in Eq. (13.90) is called a **modular transformation**.

We may then look at functions $f(\tau)$, which transform under modular transformations in a particular way. This will lead us to modular forms. A meromorphic function $f : \mathbb{H} \to \mathbb{C}$ is a **modular form** of modular weight k for $\mathrm{SL}_2(\mathbb{Z})$ if

1. f transforms under modular transformations as

$$f\left(\frac{a\tau + b}{c\tau + d}\right) = (c\tau + d)^k \cdot f(\tau) \quad \text{for } \gamma = \begin{pmatrix} a & b \\ c & d \end{pmatrix} \in \mathrm{SL}_2(\mathbb{Z}), \tag{13.91}$$

2. f is holomorphic on $\mathbb{H}$,
3. f is holomorphic at $i\infty$.

The prefactor $(c\tau + d)^k$ in Eq. (13.91) is called **automorphic factor** and equals

$$(c\tau + d)^k = \left(\frac{\psi_1'}{\psi_1}\right)^k. \tag{13.92}$$

It is convenient to introduce the $|_k\gamma$ **operator**, defined by

$$(f|_k\gamma)(\tau) = (c\tau + d)^{-k} \cdot f(\gamma(\tau)). \tag{13.93}$$

Exercise 109 *Show that*

$$(f|_k\gamma_1)|_k\gamma_2 = f|_k(\gamma_1\gamma_2). \tag{13.94}$$

With the help of the $|_k\gamma$ operator we may rewrite Eq. (13.91) as

$$(f|_k\gamma) = f \quad \text{for } \gamma \in \mathrm{SL}_2(\mathbb{Z}). \tag{13.95}$$

A meromorphic function $f : \mathbb{H} \to \mathbb{C}$, which only satisfies Eq. (13.91) (or equivalently only Eq. (13.95)) is called **weakly modular** of weight k for $\mathrm{SL}_2(\mathbb{Z})$. It is clear that f is weakly modular of weight k for $\mathrm{SL}_2(\mathbb{Z})$ if Eq. (13.95) holds for the two generators of $\mathrm{SL}_2(\mathbb{Z})$:

$$f\left(T\left(\tau\right)\right) = f\left(\tau+1\right) = f\left(\tau\right) \quad \text{and} \quad f\left(S\left(\tau\right)\right) = f\left(\frac{-1}{\tau}\right) = \tau^k f\left(\tau\right). \tag{13.96}$$

From the periodicity $f(\tau + 1) = f(\tau)$ and the holomorphicity at the cusp $\tau = i\infty$ it follows that a modular form $f(\tau)$ of $\mathrm{SL}_2(\mathbb{Z})$ has a $\bar{q}$-expansion

$$f\left(\tau\right) = \sum_{n=0}^{\infty} a_n \bar{q}^n \qquad \text{with} \qquad \bar{q} = e^{2\pi i \tau}. \tag{13.97}$$

A modular form for $\mathrm{SL}_2(\mathbb{Z})$ is called a **cusp form** of $\mathrm{SL}_2(\mathbb{Z})$, if it vanishes at the cusp $\tau = i\infty$. This is the case if $a_0 = 0$ in the $\bar{q}$-expansion of $f(\tau)$. The set of modular forms of weight k for $\mathrm{SL}_2(\mathbb{Z})$ is denoted by $\mathcal{M}_k(\mathrm{SL}_2(\mathbb{Z}))$, the set of cusp forms of weight k for $\mathrm{SL}_2(\mathbb{Z})$ is denoted by $\mathcal{S}_k(\mathrm{SL}_2(\mathbb{Z}))$.

Apart from $\mathrm{SL}_2(\mathbb{Z})$ we may also look at congruence subgroups. The **standard congruence subgroups** are defined by

$$\begin{aligned}
\Gamma_0(N) &= \left\{ \begin{pmatrix} a & b \\ c & d \end{pmatrix} \in \mathrm{SL}_2(\mathbb{Z}) : c \equiv 0 \bmod N \right\}, \\
\Gamma_1(N) &= \left\{ \begin{pmatrix} a & b \\ c & d \end{pmatrix} \in \mathrm{SL}_2(\mathbb{Z}) : a, d \equiv 1 \bmod N, \;\; c \equiv 0 \bmod N \right\}, \\
\Gamma(N) &= \left\{ \begin{pmatrix} a & b \\ c & d \end{pmatrix} \in \mathrm{SL}_2(\mathbb{Z}) : a, d \equiv 1 \bmod N, \;\; b, c \equiv 0 \bmod N \right\}.
\end{aligned}$$

$\Gamma(N)$ is called the **principle congruence subgroup** of level N. The principle congruence subgroup $\Gamma(N)$ is a normal subgroup of $\mathrm{SL}_2(\mathbb{Z})$. In general, a subgroup Γ of $\mathrm{SL}_2(\mathbb{Z})$ is called a **congruence subgroup**, if there exists an N such that

$$\Gamma\left(N\right) \subseteq \Gamma. \tag{13.98}$$

The smallest such N is called the **level of the congruence subgroup**.

We may now define modular forms for a congruence subgroup Γ, by relaxing the transformation law in Eq. (13.91) to hold only for modular transformations from the subgroup Γ, plus holomorphicity on $\mathbb{H}$ and at the cusps. In detail: A meromorphic function $f : \mathbb{H} \to \mathbb{C}$ is a modular form of modular weight k for the congruence subgroup Γ if

1. f transforms as

$$\left(f|_k\gamma\right) = f \qquad \text{for } \gamma \in \Gamma, \tag{13.99}$$

2. f is holomorphic on $\mathbb{H}$,
3. $f|_k\gamma$ is holomorphic at $i\infty$ for all $\gamma \in \mathrm{SL}_2(\mathbb{Z})$.

Let Γ be a congruence subgroup of level N. Modular forms for Γ are invariant under $\tau' = \tau + N$, since

$$T_N = \begin{pmatrix} 1 & N \\ 0 & 1 \end{pmatrix} \in \Gamma. \tag{13.100}$$

In other words, they are periodic with period N: $f(\tau + N) = f(\tau)$. Depending on Γ, there might even be a smaller N' with $N'|N$ such that $T_{N'} \in \Gamma$. For example for $\Gamma_0(N)$ and $\Gamma_1(N)$ we have

$$\begin{pmatrix} 1 & 1 \\ 0 & 1 \end{pmatrix} \in \Gamma_0(N) \text{ and } \begin{pmatrix} 1 & 1 \\ 0 & 1 \end{pmatrix} \in \Gamma_1(N). \tag{13.101}$$

Let now N' be the smallest positive integer such that $T_{N'} \in \Gamma$. It follows that modular forms for Γ have a Fourier expansion in $\bar{q}_{N'} = \bar{q}^{\frac{1}{N'}}$:

$$f(\tau) = \sum_{n=0}^{\infty} a_n \bar{q}_{N'}^n. \tag{13.102}$$

We remark that Eq. (13.101) implies that modular forms for $\Gamma_0(N)$ and $\Gamma_1(N)$ have a Fourier expansion in $\bar{q}$:

$$f(\tau) = \sum_{n=0}^{\infty} a_n \bar{q}^n. \tag{13.103}$$

In the following we will use frequently the notation

$$\tau_N = \frac{\tau}{N}, \quad \bar{q}_N = e^{\frac{2\pi i \tau}{N}} \text{ and therefore } \bar{q}_N = \exp(2\pi i \tau_N) = \bar{q}^{\frac{1}{N}}. \tag{13.104}$$

A modular form $f(\tau)$ for Γ is called a **cusp form**, if $a_0 = 0$ in the Fourier expansion of $f|_k\gamma$ for all $\gamma \in \mathrm{SL}_2(\mathbb{Z})$.

For a congruence subgroup Γ of $\mathrm{SL}_2(\mathbb{Z})$ we denote by $\mathcal{M}_k(\Gamma)$ the space of modular forms of weight k, and by $\mathcal{S}_k(\Gamma)$ the space of cusp forms of weight k. The space $\mathcal{M}_k(\Gamma)$ is a finite dimensional $\mathbb{C}$-vector space. Furthermore, $\mathcal{M}_k(\Gamma)$ is the direct sum of two finite dimensional $\mathbb{C}$-vector spaces: the space of cusp forms $\mathcal{S}_k(\Gamma)$ and the Eisenstein subspace $\mathcal{E}_k(\Gamma)$.

From the inclusions

$$\Gamma(N) \subseteq \Gamma_1(N) \subseteq \Gamma_0(N) \subseteq \mathrm{SL}_2(\mathbb{Z}) \tag{13.105}$$

follow the inclusions

$$\mathcal{M}_k(\mathrm{SL}_2(\mathbb{Z})) \subseteq \mathcal{M}_k(\Gamma_0(N)) \subseteq \mathcal{M}_k(\Gamma_1(N)) \subseteq \mathcal{M}_k(\Gamma(N)). \tag{13.106}$$

For a given N, the space $\mathcal{M}_k(\Gamma(N))$ of modular forms of weight k for the principal congruence subgroup $\Gamma(N)$ is the largest one among the spaces listed in Eq. (13.106). By definition we have for $f \in \mathcal{M}_k(\Gamma(N))$ and $\gamma \in \Gamma(N)$

$$f|_k\gamma = f, \qquad \gamma \in \Gamma(N). \tag{13.107}$$

We may ask what happens if we transform by a $\gamma \in \mathrm{SL}_2(\mathbb{Z})$, which does not belong to the congruence subgroup $\Gamma(N)$. One may show that in this case we have

$$f|_k\gamma \in \mathcal{M}_k(\Gamma(N)), \qquad \gamma \in \mathrm{SL}_2(\mathbb{Z})\backslash\Gamma(N), \tag{13.108}$$

i.e. $f|_k\gamma$ is again a modular form of weight k for $\Gamma(N)$, although not necessarily identical to f. The proof relies on the fact that $\Gamma(N)$ is a normal subgroup of $\mathrm{SL}_2(\mathbb{Z})$. This is essential: If Γ is a non-normal congruence subgroup of $\mathrm{SL}_2(\mathbb{Z})$ one has in general $f|_k\gamma \notin \mathcal{M}_k(\Gamma)$.

Exercise 110 *Let* $f \in \mathcal{M}_k(\Gamma(N))$ *and* $\gamma \in \mathrm{SL}_2(\mathbb{Z})\backslash\Gamma(N)$. *Show that* $f|_k\gamma \in \mathcal{M}_k(\Gamma(N))$.

Let N' be a divisor of N. We have

$$\Gamma_0(N) \subseteq \Gamma_0\left(N'\right), \qquad \Gamma_1(N) \subseteq \Gamma_1\left(N'\right), \qquad \Gamma(N) \subseteq \Gamma\left(N'\right), \tag{13.109}$$

and therefore

$$\mathcal{M}_k\left(\Gamma_0\left(N'\right)\right) \subseteq \mathcal{M}_k\left(\Gamma_0(N)\right), \quad \mathcal{M}_k\left(\Gamma_1\left(N'\right)\right) \subseteq \mathcal{M}_k\left(\Gamma_1(N)\right), \quad \mathcal{M}_k\left(\Gamma\left(N'\right)\right) \subseteq \mathcal{M}_k\left(\Gamma(N)\right). \tag{13.110}$$

In other words, a modular form $f \in \mathcal{M}_k(\Gamma_0(N))$ is also a modular form for $\Gamma_0(K \cdot N)$ where $K \in \mathbb{N}$ (and similar for $\Gamma_1(N)$ and $\Gamma(N)$).

There is one more generalisation which we can do: We may consider for $\Gamma_0(N)$ modular forms with a character χ. In Appendix E we review in detail Dirichlet characters $\chi(n)$. In essence, a Dirichlet character of modulus N is a function $\chi : \mathbb{Z} \to \mathbb{C}$ satisfying

$$\begin{array}{lll} (i) & \chi(n) = \chi(n+N) & \forall\, n \in \mathbb{Z}, \\ (ii) & \chi(n) = 0 & \text{if } \gcd(n, N) > 1, \\ & \chi(n) \neq 0 & \text{if } \gcd(n, N) = 1, \\ (iii) & \chi(nm) = \chi(n)\chi(m) & \forall\, n, m \in \mathbb{Z}. \end{array} \tag{13.111}$$

Let N be a positive integer and let χ be a Dirichlet character modulo N. A meromorphic function $f : \mathbb{H} \to \mathbb{C}$ is a modular form of weight k for $\Gamma_0(N)$ with character χ if

1. f transforms as

$$f\left(\frac{a\tau+b}{c\tau+d}\right) = \chi(d)(c\tau+d)^k f(\tau) \ \ \text{for} \ \ \begin{pmatrix} a & b \\ c & d \end{pmatrix} \in \Gamma_0(N), \quad (13.112)$$

2. f is holomorphic on $\mathbb{H}$,
3. $f|_k\gamma$ is holomorphic at $i\infty$ for all $\gamma \in \mathrm{SL}_2(\mathbb{Z})$.

The space of modular forms of weight k and character χ for the congruence subgroup $\Gamma_0(N)$ is denoted by $\mathcal{M}_k(N,\chi)$, the associated space of cusp forms by $\mathcal{S}_k(N,\chi)$ and the Eisenstein subspace by $\mathcal{E}_k(N,\chi)$.

Introducing modular forms with characters is useful due to the following theorem:

Theorem 11 *The space $\mathcal{M}_k(\Gamma_1(N))$ is a direct sum of spaces of modular forms with characters:*

$$\mathcal{M}_k(\Gamma_1(N)) = \bigoplus_{\chi} \mathcal{M}_k(N,\chi), \quad (13.113)$$

where the sum runs over all Dirichlet characters modulo N. Similar decompositions hold for the space of cusp forms and the Eisenstein subspaces:

$$\mathcal{S}_k(\Gamma_1(N)) = \bigoplus_{\chi} \mathcal{S}_k(N,\chi), \ \ \mathcal{E}_k(\Gamma_1(N)) = \bigoplus_{\chi} \mathcal{E}_k(N,\chi). \quad (13.114)$$

The following exercise shows, why we only consider modular forms with characters for the congruence subgroup $\Gamma_0(N)$ (and subgroups thereof).

Exercise 111 *Let χ be a Dirichlet character with modulus N and $f \in \mathcal{M}_k(N,\chi)$. Let further $\gamma_1, \gamma_2 \in \Gamma_0(N)$ and set $\gamma_{12} = \gamma_1\gamma_2$. Show that*

$$f\left(\gamma_1\left(\gamma_2\left(\tau\right)\right)\right) = f\left(\gamma_{12}\left(\tau\right)\right). \quad (13.115)$$

We now introduce **iterated integrals of modular forms**. Let $f_1, \dots, f_n$ be modular forms (for $\mathrm{SL}_2(\mathbb{Z})$ or some congruence subgroup Γ of level N). For a modular form $f \in \mathcal{M}_k(\Gamma)$ of level N let N' be the smallest positive integer such that

$$\begin{pmatrix} 1 & N' \\ 0 & 1 \end{pmatrix} \in \Gamma. \quad (13.116)$$

We set

$$\omega^{\text{modular}}(f) = 2\pi i \ f(\tau) \frac{d\tau}{N'} \ = \ 2\pi i \ f(\tau) \, d\tau_{N'} \ = \ f(\tau) \frac{d\bar{q}_{N'}}{\bar{q}_{N'}}. \tag{13.117}$$

If the modular form $f(\tau)$ has the $\bar{q}_{N'}$-expansion

$$f(\tau) = \sum_{n=0}^{\infty} a_n \bar{q}_{N'}^n, \tag{13.118}$$

we have

$$\omega^{\text{modular}}(f) = \sum_{n=0}^{\infty} a_n \bar{q}_{N'}^{n-1} d\bar{q}_{N'}. \tag{13.119}$$

Let $\gamma : [a, b] \to \mathbb{H}$ be a path with $\gamma(a) = \tau_0$ and $\gamma(b) = \tau$. We set

$$I(f_1, \dots, f_n; \tau) = I_\gamma \left(\omega^{\text{modular}}(f_1), \dots, \omega^{\text{modular}}(f_n); b \right), \tag{13.120}$$

where the right-hand side refers to the general definition of an iterated integral given in Eq. (6.140). Explicitly

$$I(f_1, f_2, \dots, f_n; \tau) = \left(\frac{2\pi i}{N'} \right)^n \int_{\tau_0}^{\tau} d\tau_1 f_1(\tau_1) \int_{\tau_0}^{\tau_1} d\tau_2 f_2(\tau_2) \dots \int_{\tau_0}^{\tau_{n-1}} d\tau_n f_n(\tau_n). \tag{13.121}$$

As base point we usually take $\tau_0 = i\infty$. Please note that an integral over a modular form is in general not a modular form. This is not surprising if we consider the following analogy: An integral over a rational function is in general not a rational function.

We usually like iterated integrals appearing in solutions of Feynman integrals to have at worst simple poles. Let's study iterated integrals of modular forms. As modular forms are holomorphic in the complex upper half-plane, there are no poles there. So the only interesting points are the cusps. Let's consider as an example modular forms $f \in \mathcal{M}_k(\text{SL}_2(\mathbb{Z}))$, so the only cusp is at $\tau = i\infty$. By definition a modular form $f(\tau)$ is holomorphic at the cusp and has a $\bar{q}$-expansion

$$f(\tau) = a_0 + a_1 \bar{q} + a_2 \bar{q}^2 + \dots, \qquad \bar{q} = \exp(2\pi i \tau). \tag{13.122}$$

The transformation $\bar{q} = \exp(2\pi i \tau)$ transforms the point $\tau = i\infty$ to $\bar{q} = 0$ and we have

$$2\pi i \ f(\tau) d\tau = \frac{d\bar{q}}{\bar{q}} \left(a_0 + a_1 \bar{q} + a_2 \bar{q}^2 + \dots \right). \tag{13.123}$$

Thus a modular form non-vanishing at the cusp $\tau = i\infty$ has a simple pole at $\bar{q} = 0$.

13.3.1 Eisenstein Series

In applications towards Feynman integrals we will need the $\bar{q}_{N'}$-expansions of modular forms. A complete treatment is beyond the scope of this book and we refer the reader to textbooks on modular forms [335–338].

We limit ourselves to the aspects most relevant to Feynman integrals. In this section we will look at the $\bar{q}_{N'}$-expansions of modular forms spanning the Eisenstein subspace. We will study Eisenstein series for $\mathrm{SL}_2(\mathbb{Z})$, $\Gamma_1(N)$ and $\Gamma(N)$.

The case of the full modular group $\mathrm{SL}_2(\mathbb{Z})$ is rather simple and the main result is that any modular form for $\mathrm{SL}_2(\mathbb{Z})$ can be written as a polynomial in two Eisenstein series $e_4(\tau)$ and $e_6(\tau)$ of modular weight 4 and 6, respectively.

Eisenstein series for $\Gamma_1(N)$ appear in the simplest elliptic Feynman integrals. A basis for $\mathcal{E}_k(\Gamma_1(N))$ can be given explicitly.

For a modular form f of a congruence subgroup Γ of level N we also would like to know the transformation behaviour under $\gamma \in \mathrm{SL}_2(\mathbb{Z})\backslash\Gamma$. We first note that by the definition of a congruence subgroup we have $f \in \mathcal{M}_k(\Gamma(N))$. If in addition $f \in \mathcal{E}_k(\Gamma(N))$ we may answer this question explicitly.

13.3.1.1 Eisenstein Series for $\mathrm{SL}_2(\mathbb{Z})$

The z-dependent Eisenstein series $E_k(z, \tau)$ are defined by

$$E_k\left(z, \tau\right) = \sum_{(n_1,n_2)\in\mathbb{Z}^2}{}_{e} \frac{1}{\left(z + n_1 + n_2\tau\right)^k}. \tag{13.124}$$

The subscript e at the summation sign denotes the Eisenstein summation prescription defined by

$$\sum_{(n_1,n_2)\in\mathbb{Z}^2}{}_{e} f\left(z + n_1 + n_2\tau\right) = \lim_{N_2\to\infty} \sum_{n_2=-N_2}^{N_2} \left(\lim_{N_1\to\infty} \sum_{n_1=-N_1}^{N_1} f\left(z + n_1 + n_2\tau\right) \right). \tag{13.125}$$

The series in Eq. (13.124) is absolutely convergent for $k \geq 3$. For $k = 1$ and $k = 2$ the Eisenstein summation depends on the choice of generators. One further sets

$$e_k\left(\tau\right) = \sum_{(n_1,n_2)\in\mathbb{Z}^2\backslash(0,0)}{}_{e} \frac{1}{\left(n_1 + n_2\tau\right)^k}. \tag{13.126}$$

We have $e_k(\tau) = 0$ whenever k is odd. The $\bar{q}$-expansions of the first few Eisenstein series are

$$e_2(\tau) = 2(2\pi i)^2 \left[-\frac{1}{24} + \bar{q} + 3\bar{q}^2 + 4\bar{q}^3 + 7\bar{q}^4 + 6\bar{q}^5 + 12\bar{q}^6 \right] + O\left(\bar{q}^7\right), \tag{13.127}$$

$$e_4(\tau) = \frac{(2\pi i)^4}{3} \left[\frac{1}{240} + \bar{q} + 9\bar{q}^2 + 28\bar{q}^3 + 73\bar{q}^4 + 126\bar{q}^6 \right] + O\left(\bar{q}^7\right),$$

$$e_6(\tau) = \frac{(2\pi i)^6}{60} \left[-\frac{1}{504} + \bar{q} + 33\bar{q}^2 + 244\bar{q}^3 + 1057\bar{q}^4 + 3126\bar{q}^5 + 8052\bar{q}^6 \right] + O\left(\bar{q}^7\right).$$

For $k \geq 4$ the Eisenstein series $e_k(\tau)$ are modular forms of $\mathcal{M}_k(\mathrm{SL}_2(\mathbb{Z}))$. The space $\mathcal{M}_k(\mathrm{SL}_2(\mathbb{Z}))$ has a basis of the form

$$(e_4(\tau))^{\nu_4} (e_6(\tau))^{\nu_6}, \tag{13.128}$$

where ν_4 and ν_6 run over all non-negative integers with $4\nu_4 + 6\nu_6 = k$.

As an example, let us give the cusp form of modular weight 12 for $\mathrm{SL}_2(\mathbb{Z})$:

$$\Delta(\tau) = (2\pi i)^{12} \eta(\tau)^{24} = 10800 \left(20 (e_4(\tau))^3 - 49 (e_6(\tau))^2\right). \tag{13.129}$$

Note that $e_2(\tau)$ is not a modular form. Under modular transformations $e_2(\tau)$ transforms as

$$e_2\left(\frac{a\tau + b}{c\tau + d}\right) = (c\tau + d)^2 e_2(\tau) - 2\pi i c (c\tau + d). \tag{13.130}$$

Modularity is spoiled by the second term on the right-hand side.

Exercise 112 *Consider*

$$f(\tau) = e_2(\tau) - 2e_2(2\tau) \tag{13.131}$$

and work out the transformation properties under $\gamma \in \Gamma_0(2)$.

13.3.1.2 Eisenstein Series for $\Gamma_1(N)$

Let Γ be a congruence subgroup of $\mathrm{SL}_2(\mathbb{Z})$. By definition there exists an N, such that

$$\Gamma(N) \subseteq \Gamma. \tag{13.132}$$

This implies

$$\mathcal{M}_k(\Gamma) \subseteq \mathcal{M}_k(\Gamma(N)) \tag{13.133}$$

and this reduces in a first step the study of modular forms for an arbitrary congruence subgroup Γ to the study of modular forms of the principal congruence subgroup

$\Gamma(N)$. Now let $\eta(\tau) \in \mathcal{M}_k(\Gamma(N))$. Then [336]

$$\eta(N\tau) \in \mathcal{M}_k\left(\Gamma_1\left(N^2\right)\right), \tag{13.134}$$

which reduces in a second step the study of modular forms for an arbitrary congruence subgroup Γ to the study of modular forms of the congruence subgroup $\Gamma_1(N)$.

Let us therefore consider modular forms for the congruence subgroups $\Gamma_1(N)$, and here in particular the Eisenstein subspace $\mathcal{E}_k(\Gamma_1(N))$. Let us first note that

$$T_1 = \begin{pmatrix} 1 & 1 \\ 0 & 1 \end{pmatrix} \in \Gamma_1(N), \tag{13.135}$$

and therefore the modular forms $f \in \mathcal{M}_k(\Gamma_1(N))$ have an expansion in $\bar{q}$. From Eq. (13.114) we know that $\mathcal{E}_k(\Gamma_1(N))$ decomposes into Eisenstein spaces $\mathcal{E}_k(N, \chi)$.

$$\mathcal{E}_k(\Gamma_1(N)) = \bigoplus_{\chi} \mathcal{E}_k(N, \chi). \tag{13.136}$$

A basis for the Eisenstein subspace $\mathcal{E}_k(N, \chi)$ can be given explicitly. To this aim we first define **generalised Eisenstein series**. Let χ_a and χ_b be primitive Dirichlet characters with conductors d_a and d_b, respectively. We set

$$E_k(\tau, \chi_a, \chi_b) = a_0 + \sum_{n=1}^{\infty} \left(\sum_{d|n} \chi_a(n/d) \cdot \chi_b(d) \cdot d^{k-1} \right) \bar{q}^n, \tag{13.137}$$

The normalisation is such that the coefficient of $\bar{q}$ is one. The constant term a_0 is given by

$$a_0 = \begin{cases} -\frac{B_{k,\chi_b}}{2k}, & \text{if } d_a = 1, \\ 0, & \text{if } d_a > 1. \end{cases} \tag{13.138}$$

Note that the constant term a_0 depends on χ_a and χ_b. The **generalised Bernoulli numbers** B_{k,χ_b} are defined by

$$\sum_{n=1}^{d_b} \chi_b(n) \frac{x e^{nx}}{e^{d_b x} - 1} = \sum_{k=0}^{\infty} B_{k,\chi_b} \frac{x^k}{k!}. \tag{13.139}$$

Note that in the case of the trivial character $\chi_b = 1$, Eq. (13.139) reduces to

$$\frac{x e^x}{e^x - 1} = \sum_{k=0}^{\infty} B_{k,1} \frac{x^k}{k!}, \tag{13.140}$$

yielding $B_{1,1} = 1/2$. The ordinary Bernoulli numbers B_k are generated by $x/(e^x - 1)$ (i.e. without an extra factor e^x in the numerator) and yield $B_1 = -1/2$.

Let now χ_a, χ_b and k be such that

$$\chi_a(-1)\,\chi_b(-1) = (-1)^k \tag{13.141}$$

and if $k = 1$ one requires in addition

$$\chi_a(-1) \;=\; 1, \;\; \chi_b(-1) \;=\; -1. \tag{13.142}$$

Let K be a positive integer. We then set

$$E_{k,K}(\tau, \chi_a, \chi_b) = \begin{cases} E_k(K\tau, \chi_a, \chi_b), & (k, \chi_a, \chi_b) \neq (2, 1, 1),\; K \geq 1, \\ E_2(\tau, 1, 1) - K E_2(K\tau, 1, 1), & (k, \chi_a, \chi_b) = (2, 1, 1),\; K > 1. \end{cases} \tag{13.143}$$

Let N be an integer multiple of $(K \cdot d_a \cdot d_b)$. Then $E_{k,K}(\tau, \chi_a, \chi_b)$ is a modular form for $\Gamma_1(N)$ of modular weight k and level N:

$$E_{k,K}(\tau, \chi_a, \chi_b) \in \mathcal{E}_k(\Gamma_1(N)), \qquad (K \cdot d_a \cdot d_b) \mid N. \tag{13.144}$$

In more detail we have that for $(k, \chi_a, \chi_b) \neq (2, 1, 1)$

$$E_{k,K}(\tau, \chi_a, \chi_b) \in \mathcal{E}_k(N, \tilde{\chi}), \tag{13.145}$$

where $\tilde{\chi}$ is the Dirichlet character with modulus N induced by $\chi_a \chi_b$. Furthermore, for $(k, \chi_a, \chi_b) = (2, 1, 1)$ and $K > 1$ we have

$$E_{k,K}(\tau, \chi_a, \chi_b) \in \mathcal{E}_k(N, \tilde{1}) \;=\; \mathcal{E}_k(\Gamma_0(N)). \tag{13.146}$$

Theorem 12 *Let $(K \cdot d_a \cdot d_b)$ be a divisor of N. The $E_{k,K}(\tau, \chi_a, \chi_b)$ subject to the conditions outlined above (Eqs. (13.141)–(13.142)) form a basis of $\mathcal{E}_k(N, \tilde{\chi})$, where $\tilde{\chi}$ is the Dirichlet character with modulus N induced by $\chi_a \chi_b$.*

Of particular interest are characters which are obtained from the Kronecker symbol. These characters take the values $\{-1, 0, 1\}$. In general, the value of a Dirichlet character is a root of unity or zero. The restriction to Dirichlet characters obtained from the Kronecker symbol has the advantage that the $\bar{q}$-expansion of the Eisenstein series can be computed within the rational numbers.

Let a be an integer, which is either one or the discriminant of a quadratic field. In Appendix E.9 we give a criteria for a being the discriminant of a quadratic field. The Kronecker symbol, also defined in Appendix E.9, then defines a primitive Dirichlet character

$$\chi_a(n) = \left(\frac{a}{n}\right) \tag{13.147}$$

of conductor $|a|$. On the right-hand side of Eq. (13.147) we have the Kronecker symbol. Be alert that the right-hand side of Eq. (13.147) does not denote a fraction. Let a and b be integers, which are either one or the discriminant of a quadratic field.

$$\chi_a(n) = \left(\frac{a}{n}\right), \quad \chi_b(n) = \left(\frac{b}{n}\right) \tag{13.148}$$

we introduce the short-hand notations

$$\begin{aligned} E_{k,a,b}(\tau) &= E_k(\tau, \chi_a, \chi_b) \\ E_{k,a,b,K}(\tau) &= E_{k,K}(\tau, \chi_a, \chi_b). \end{aligned} \tag{13.149}$$

These Eisenstein series have a $\bar{q}$-expansion with rational coefficients.

Remark For k even and $k \geq 4$ the relation between the Eisenstein series $e_k(\tau)$ defined in Eq. (13.126) and the Eisenstein series with a trivial character is

$$e_k(\tau) = 2\frac{(2\pi i)^k}{(k-1)!} E_{k,1,1}(\tau). \tag{13.150}$$

13.3.1.3 Eisenstein Series for $\Gamma(N)$

Let us now turn to Eisenstein series for $\Gamma(N)$. For these Eisenstein series we may give the transformation law for any $\gamma \in \mathrm{SL}_2(\mathbb{Z})$ and not just $\gamma \in \Gamma(N)$. Let r, s be integers with $0 \leq r, s < N$. Following [341, 342] we set

$$h_{k,N,r,s}(\tau) = \sum_{n=0}^{\infty} a_n \bar{q}_N^n. \tag{13.151}$$

For $n \geq 1$ the coefficients are given by

$$a_n = \frac{1}{2N^k} \sum_{d|n} \sum_{c_1=0}^{N-1} d^{k-1} \left[e^{\frac{2\pi i}{N}\left(r\frac{n}{d}-(s-d)c_1\right)} + (-1)^k e^{-\frac{2\pi i}{N}\left(r\frac{n}{d}-(s+d)c_1\right)} \right]. \tag{13.152}$$

The constant term is given for $k \geq 2$ by

$$a_0 = -\frac{1}{2k} B_k\left(\frac{s}{N}\right), \tag{13.153}$$

where $B_k(x)$ is the k'th **Bernoulli polynomial** defined by

$$\frac{te^{xt}}{e^t - 1} = \sum_{k=0}^{\infty} \frac{B_k(x)}{k!} t^k. \tag{13.154}$$

For $k = 1$ the constant term is given by

$$a_0 = \begin{cases} \frac{1}{4} - \frac{s}{2N}, & s \neq 0, \\ 0, & (r,s) = (0,0), \\ \frac{i}{4} \cot\left(\frac{r}{N}\pi\right), & \text{otherwise.} \end{cases} \tag{13.155}$$

We have

$$h_{k,N,r,s}(\tau) = \frac{1}{2} \frac{(k-1)!}{(2\pi i)^k} \sum_{(n_1,n_2)\in\mathbb{Z}^2\setminus(0,0)}^{e} \frac{e^{\frac{2\pi i}{N}(n_1 s - n_2 r)}}{(n_1 + n_2\tau)^k}. \tag{13.156}$$

Exercise 113 *Show Eq. (13.156).*

The normalisation is compatible with the normalisation of the previous subsection. For example we have

$$h_{k,1,0,0}(\tau) = E_{k,1,1}(\tau). \tag{13.157}$$

With the exception of $(k, r, s) \neq (2, 0, 0)$ the $h_{k,N,r,s}(\tau)$ are Eisenstein series for $\Gamma(N)$:

$$h_{k,N,r,s}(\tau) \in \mathcal{E}_k(\Gamma(N)). \tag{13.158}$$

For $(k, N, r, s) = (2, 1, 0, 0)$ we have

$$h_{2,1,0,0}(\tau) = E_{2,1,1}(\tau) = \frac{1}{2(2\pi i)^2} e_2(\tau), \tag{13.159}$$

which is not a modular form.

The most important property of the Eisenstein series $h_{k,N,r,s}(\tau)$ is their transformation behaviour under the full modular group: The Eisenstein series $h_{k,N,r,s}(\tau)$ transform under modular transformations

$$\gamma = \begin{pmatrix} a & b \\ c & d \end{pmatrix} \in \mathrm{SL}_2(\mathbb{Z}) \tag{13.160}$$

of the full modular group $\mathrm{SL}_2(\mathbb{Z})$ as

$$h_{k,N,r,s}\left(\frac{a\tau + b}{c\tau + d}\right) = (c\tau + d)^k \, h_{k,N,(rd+sb) \bmod N,(rc+sa) \bmod N}(\tau), \tag{13.161}$$

or equivalently with the help of the $|_k\gamma$ operator

$$\left(h_{k,N,r,s}|_k\gamma\right)(\tau) = h_{k,N,(rd+sb) \bmod N,(rc+sa) \bmod N}(\tau). \tag{13.162}$$

Exercise 114 *Prove Eq. (13.162) for the case $k \geq 3$.*

13.3.2 The Modular Lambda Function and Klein's j-Invariant

In this paragraph we discuss the modular lambda function and Klein's j-invariant. The latter allows us to decide, if two elliptic curves are isomorphic. In Eq. (13.77) we have seen that

$$k^2 = 16\frac{\eta\left(\frac{\tau}{2}\right)^8 \eta(2\tau)^{16}}{\eta(\tau)^{24}}. \tag{13.163}$$

The right-hand side defines a function of τ, called the **modular lambda function**:

$$\lambda(\tau) = 16\frac{\eta\left(\frac{\tau}{2}\right)^8 \eta(2\tau)^{16}}{\eta(\tau)^{24}}. \tag{13.164}$$

The function $\lambda(\tau)$ is invariant under $\Gamma(2)$. The congruence subgroup $\Gamma(2)$ is generated by

$$\begin{pmatrix} 1 & 2 \\ 0 & 1 \end{pmatrix}, \quad \begin{pmatrix} 1 & 0 \\ 2 & 1 \end{pmatrix}, \quad \begin{pmatrix} -1 & 0 \\ 0 & -1 \end{pmatrix}. \tag{13.165}$$

Thus $\lambda(\tau)$ is invariant under

$$\tau' = \tau + 2, \text{ and } \tau' = \frac{\tau}{1+2\tau}. \tag{13.166}$$

Under the generators of $\mathrm{SL}_2(\mathbb{Z})$ the modular lambda function transforms as

$$\begin{aligned} \tau' &= \tau + 1: \ \lambda(\tau') = -\frac{\lambda(\tau)}{1-\lambda(\tau)}, \\ \tau' &= \frac{-1}{\tau}: \ \lambda(\tau') = 1 - \lambda(\tau). \end{aligned} \tag{13.167}$$

Klein's j-invariant is defined by

$$j(\tau) = \frac{1728 g_2(\tau)^3}{g_2(\tau)^3 - 27 g_3(\tau)^2}, \tag{13.168}$$

where

$$g_2(\tau) = 60e_4(\tau), \quad g_3(\tau) = 140e_6(\tau). \tag{13.169}$$

Note that the denominator in Eq. (13.168) is the modular discriminant

$$\Delta(\tau) = g_2(\tau)^3 - 27g_3(\tau)^2 = (2\pi i)^{12}\,\eta(\tau)^{24}. \tag{13.170}$$

Two elliptic curves are isomorphic if their j-invariants agree.

The relation of the modular lambda function to Klein's j-invariant is

$$j(\tau) = 256\frac{\left(1-\lambda+\lambda^2\right)^3}{\lambda^2(1-\lambda)^2}. \tag{13.171}$$

13.4 Moduli Spaces

For elliptic Feynman integrals, which only depend on one kinematic variable (e.g. $N_B = 1$) we now have all necessary tools: In essence, we perform in addition to a fibre transformation also a base transformation and change from the original kinematic variable x to the modular parameter τ. We then express the Feynman integrals as iterated integrals of modular forms. We will see in an example later on, how this is done in practice.

However, there are elliptic Feynman integrals, which depend on more than one kinematic variable (e.g. $N_B > 1$) and for those we need one more ingredient: We have to introduce the moduli space $\mathcal{M}_{1,n}$ of a smooth genus one curve with n marked points and we will see that an elliptic Feynman integral can be expressed as a linear combination of iterated integrals on a covering space of the moduli space $\mathcal{M}_{1,n}$ with integrands having only simple poles. This comes to no surprise, as a "simple" Feynman integral, which evaluates to multiple polylogarithms, can be expressed as a linear combination of iterated integrals on a covering space of the moduli space $\mathcal{M}_{0,n}$ of a genus zero curve with n marked points, again with integrands having only simple poles.

In Appendix F we give a detailed introduction into the moduli space $\mathcal{M}_{g,n}$ of a smooth algebraic curve of genus g with n marked points. Here in this section we briefly summarise the main points and proceed then to the relevant aspects for computing elliptic Feynman integrals.

Let's start with the short summary: Let X be a topological space. The configuration space of n ordered points in X is

$$\mathrm{Conf}_n(X) = \left\{ (x_1, \dots, x_n) \in X^n \middle| x_i \neq x_j \text{ for } i \neq j \right\}. \tag{13.172}$$

The non-trivial ingredient is the requirement that the points are distinct: $x_i \neq x_j$. Without this requirement we would simply look at X^n.

An example is the configuration space of n points on the Riemann sphere $\mathbb{CP}^1$:

$$\mathrm{Conf}_n\left(\mathbb{CP}^1\right) = \left\{ (z_1, \dots, z_n) \in \left(\mathbb{CP}^1\right)^n \middle| z_i \neq z_j \text{ for } i \neq j \right\}. \quad (13.173)$$

A Möbius transformation

$$z' = \frac{az+b}{cz+d} \quad (13.174)$$

transforms the Riemann sphere into itself. These transformations form a group PSL $(2, \mathbb{C})$. Usually we are not interested in configurations

$$(z_1, \dots, z_n) \in \mathrm{Conf}_n\left(\mathbb{CP}^1\right) \text{ and } (z'_1, \dots, z'_n) \in \mathrm{Conf}_n\left(\mathbb{CP}^1\right), \quad (13.175)$$

which differ only by a Möbius transformation:

$$z'_j = \frac{az_j+b}{cz_j+d}, \qquad j \in \{1, \dots, n\}. \quad (13.176)$$

This brings us to the definition of the moduli space of the Riemann sphere with n marked points:

$$\mathcal{M}_{0,n} = \mathrm{Conf}_n\left(\mathbb{CP}^1\right) / \mathrm{PSL}\,(2, \mathbb{C}). \quad (13.177)$$

We may use the freedom of Möbius transformations to fix three points (usually 0, 1 and ∞). Therefore

$$\begin{aligned} \dim\left(\mathrm{Conf}_n\left(\mathbb{CP}^1\right)\right) &= n, \\ \dim\left(\mathcal{M}_{0,n}\right) &= n - 3. \end{aligned} \quad (13.178)$$

Let's generalise this: We are interested in the situation, where the topological space X is a smooth algebraic curve C in $\mathbb{CP}^2$. This implies that there exists a homogeneous polynomial $P(z_1, z_2, z_3)$ such that

$$C = \left\{ [z_1 : z_2 : z_3] \in \mathbb{CP}^2 \middle| P(z_1, z_2, z_3) = 0 \right\}. \quad (13.179)$$

If d is the degree of the polynomial $P(z_1, z_2, z_3)$, the genus g of C is given by Eq. (13.4). Note that we may view C either as a complex curve (of complex dimension one) or as a real surface (of real dimension two). This is illustrated in Fig. 13.5.

Let's pause for a second and let us convince ourselves that this set-up is a generalisation of the previous case: The special case $C = \mathbb{CP}^1$ is obtained for example by the choice $P(z_1, z_2, z_3) = z_3$. The genus formula gives us that $\mathbb{CP}^1$ has genus zero.

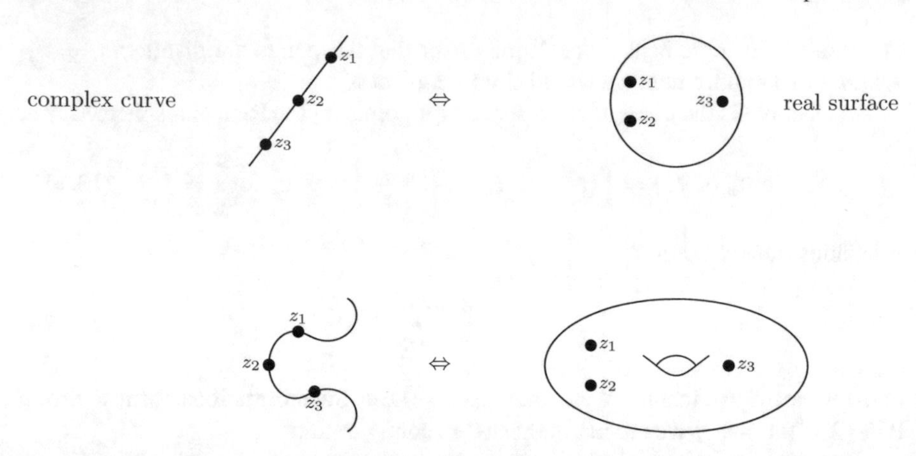

Fig. 13.5 The upper left figure shows a configuration of three marked points on a complex curve of genus zero, the upper right figure shows the corresponding configuration when the complex curve is viewed as a real Riemann surface. The lower figures show the analogous situation for a complex curve of genus one

Let us now consider a smooth curve C of genus g with n marked points. Two such curves $(C; z_1, \dots, z_n)$ and $(C'; z'_1, \dots, z'_n)$ are isomorphic if there is an isomorphism

$$\phi : C \to C' \qquad \text{such that} \;\; \phi(z_i) = z'_i. \tag{13.180}$$

The moduli space

$$\mathcal{M}_{g,n} \tag{13.181}$$

is the space of isomorphism classes of smooth curves of genus g with n marked points. For $g \geq 1$ the isomorphism classes do not only depend on the positions of the marked points, but also on the "shape" of the curve. For $g = 0$ there is only one "shape", the Riemann sphere. For $g = 1$ the shape of the torus is described by the modular parameter τ.

The dimension of $\mathcal{M}_{g,n}$ is

$$\dim\left(\mathcal{M}_{g,n}\right) = 3g + n - 3, \tag{13.182}$$

for $g = 0$ this formula agrees with the previous result in Eq. (13.178).

Let us now focus on the moduli spaces $\mathcal{M}_{0,n}$ and $\mathcal{M}_{1,n}$ and work out natural choices for coordinates on $\mathcal{M}_{0,n}$ and $\mathcal{M}_{1,n}$.

- We start with genus 0. We have $\dim \mathcal{M}_{0,n} = n - 3$. As mentioned above, the sphere has a unique shape. We may use Möbius transformations to fix three points, say $z_{n-2} = 1$, $z_{n-1} = \infty$, $z_n = 0$. This leaves

$$(z_1, \dots, z_{n-3}) \tag{13.183}$$

as coordinates on $\mathcal{M}_{0,n}$.
- We now turn to genus 1. From Eq. (13.182) we have $\dim \mathcal{M}_{1,n} = n$. We need one coordinate to describe the shape of the elliptic curve (or the shape of the torus or the shape of the parallelogram). We may take τ as defined in Eq. (13.35) for this. We may use translation transformations to fix one marked point, say $z_n = 0$. This gives

$$(\tau, z_1, \dots, z_{n-1}) \tag{13.184}$$

as coordinates on $\mathcal{M}_{1,n}$.

We then consider iterated integrals on $\mathcal{M}_{0,n}$ and $\mathcal{M}_{1,n}$. We recall from Sect. 6.3.3 that iterated integrals on a manifold M are defined by a set ω_1, ..., ω_k of differential 1-forms on M and a path $\gamma : [0, 1] \to M$ as

$$I_\gamma (\omega_1, \dots, \omega_k; \lambda) = \int\limits_0^{\lambda} d\lambda_1 f_1 (\lambda_1) \int\limits_0^{\lambda_1} d\lambda_2 f_2 (\lambda_2) \dots \int\limits_0^{\lambda_{k-1}} d\lambda_k f_k (\lambda_k), \tag{13.185}$$

where the pull-back of ω_j to the interval [0, 1] is denoted by

$$f_j (\lambda) \, d\lambda = \gamma^* \omega_j. \tag{13.186}$$

Let us briefly discuss iterated integrals on $\mathcal{M}_{0,n}$. We are interested in differential one-forms, which have only simple poles. Thus we consider

$$\omega = d \ln \left(z_i - z_j\right) = \frac{dz_i - dz_j}{z_i - z_j}. \tag{13.187}$$

Keeping $z_1, \dots, z_{n-4}$ fixed and integrating along $y = z_{n-3}$ leads to

$$\omega^{\mathrm{mpl}} = \frac{dy}{y - z_j}. \tag{13.188}$$

The iterated integrals constructed from these differential one-forms are the multiple polylogarithms:

$$G(z_1, \dots, z_k; y) = \int\limits_0^y \frac{dy_1}{y_1 - z_1} \int\limits_0^{y_1} \frac{dy_2}{y_2 - z_2} \dots \int\limits_0^{y_{k-1}} \frac{dy_k}{y_k - z_k}, \qquad z_k \neq 0, \tag{13.189}$$

discussed in detail in Sect. 8. As discussed in Sect. 8 we may relax the condition $z_k \neq 0$ and allow trailing zeros.

Let's now consider iterated integrals on $\mathcal{M}_{1,n}$. We recall that we may take $(\tau, z_1, \dots, z_{n-1})$ as coordinates on $\mathcal{M}_{1,n}$. We may decompose an arbitrary integration path into pieces along $d\tau$ (with $z_1 = \dots = z_{n-1} = \text{const}$) and pieces along the dz_j's (with $\tau = \text{const}$). Thus we obtain two classes of standardised iterated integrals: Iterated integrals on $\mathcal{M}_{1,n}$ with integration along $d\tau$ and iterated integrals on $\mathcal{M}_{1,n}$ with integration along the dz_j's.

In addition we have to specify the differential one-forms we want to integrate. The differential one-forms which we want to consider in the case of $\mathcal{M}_{1,n}$ are derived from the Kronecker function. The Kronecker function $F(x, y, \tau)$ is defined in terms of the first Jacobi theta function by

$$F(x, y, \tau) = \pi \theta_1'(0, q) \frac{\theta_1(\pi(x + y), q)}{\theta_1(\pi x, q) \theta_1(\pi y, q)}, \tag{13.190}$$

where $q = \exp(\pi i \tau)$ and the first Jacobi theta function $\theta_1(z, q)$ is defined in Eq. (13.68). θ_1' denotes the derivative with respect to the first argument. Please note that in order to make contact with the standard notation for the Jacobi theta functions we used here the nome $q = \exp(\pi i \tau)$ and not the nome squared $\bar{q} = q^2 = \exp(2\pi i \tau)$. The definition of the Kronecker function is cleaned up if we define

$$\bar{\theta}_1(z, \bar{q}) = \theta_1\left(\pi z, \bar{q}^{\frac{1}{2}}\right). \tag{13.191}$$

Then

$$F(x, y, \tau) = \bar{\theta}_1'(0, \bar{q}) \frac{\bar{\theta}_1(x + y, \bar{q})}{\bar{\theta}_1(x, \bar{q}) \bar{\theta}_1(y, \bar{q})}. \tag{13.192}$$

Digression Properties of the Kronecker function:

From the definition it is obvious that the Kronecker function is symmetric in the variables (x, y):

$$F(y, x, \tau) = F(x, y, \tau). \tag{13.193}$$

The (quasi-) periodicity properties are

$$F(x+1,y,\tau) = F(x,y,\tau), \quad F(x+\tau,y,\tau) = e^{-2\pi i y} F(x,y,\tau). \tag{13.194}$$

The function

$$\Omega(x,y,\tau) = \exp\left(2\pi i \frac{(y \text{Im}(x) + x \text{Im}(y))}{\text{Im}(\tau)}\right) F(x,y,\tau) \tag{13.195}$$

is symmetric in x and y and doubly periodic

$$\Omega(x+1,y,\tau) = \Omega(x+\tau,y,\tau) = \Omega(x,y,\tau), \tag{13.196}$$

but no longer meromorphic (that is to say that Ω also depends on $\bar{x}$, $\bar{y}$ and $\bar{\tau}$, the dependency on the anti-holomorphic variables enters through $2i\,\text{Im}(x) = x - \bar{x}$).

The Fay identity reads

$$\begin{aligned} & F(x_1,y_1,\tau)\, F(x_2,y_2,\tau) = \\ & \quad F(x_1,y_1+y_2,\tau)\, F(x_2-x_1,y_2,\tau) + F(x_2,y_1+y_2,\tau)\, F(x_1-x_2,y_1,\tau). \end{aligned} \tag{13.197}$$

We recall that the Kronecker function is symmetric in x and y. We are interested in the Laurent expansion in one of these variables. We define functions $g^{(k)}(z,\tau)$ through

$$F(z,\alpha,\tau) = \sum_{k=0}^{\infty} g^{(k)}(z,\tau)\, \alpha^{k-1}. \tag{13.198}$$

We are primarily interested in the coefficients $g^{(k)}(z,\tau)$ of the Kronecker function. Let us recall some of their properties [343–345].

Properties of the coefficients $g^{(k)}(z,\tau)$ of the Kronecker function:

1. The functions $g^{(k)}(z,\tau)$ have the symmetry

$$g^{(k)}(-z,\tau) = (-1)^k\, g^{(k)}(z,\tau). \tag{13.199}$$

2. When viewed as a function of z, the function $g^{(k)}(z,\tau)$ has only simple poles. More concretely, the function $g^{(1)}(z,\tau)$ has a simple pole with unit residue at every point of the lattice. For $k > 1$ the function $g^{(k)}(z,\tau)$ has a simple pole only at those lattice points that do not lie on the real axis.
3. The (quasi-) periodicity properties are

$$g^{(k)}(z+1,\tau) = g^{(k)}(z,\tau),$$
$$g^{(k)}(z+\tau,\tau) = \sum_{j=0}^{k} \frac{(-2\pi i)^j}{j!} g^{(k-j)}(z,\tau). \qquad (13.200)$$

We see that $g^{(k)}(z,\tau)$ is invariant under translations by 1, but not by τ. The translation invariance by τ is only spoiled by the terms with $j \geq 1$ in Eq. (13.200).

4. Under modular transformations the functions $g^{(k)}(z,\tau)$ transform as

$$g^{(k)}\left(\frac{z}{c\tau+d}, \frac{a\tau+b}{c\tau+d}\right) = (c\tau+d)^k \sum_{j=0}^{k} \frac{(2\pi i)^j}{j!} \left(\frac{cz}{c\tau+d}\right)^j g^{(k-j)}(z,\tau). \qquad (13.201)$$

Modular invariance is only spoiled by the terms with $j \geq 1$ in Eq. (13.201).

5. The $\bar{q}$-expansion of the $g^{(k)}(z,\tau)$ functions is given by (with $\bar{q} = \exp(2\pi i \tau)$ and $\bar{w} = \exp(2\pi i z)$)

$$g^{(0)}(z,\tau) = 1,$$
$$g^{(1)}(z,\tau) = -2\pi i \left[\frac{1+\bar{w}}{2(1-\bar{w})} + \overline{\mathrm{E}}_{0,0}(\bar{w};1;\bar{q})\right],$$
$$g^{(k)}(z,\tau) = -\frac{(2\pi i)^k}{(k-1)!}\left[-\frac{B_k}{k} + \overline{\mathrm{E}}_{0,1-k}(\bar{w};1;\bar{q})\right], \quad k>1, \qquad (13.202)$$

where B_k denotes the k-th Bernoulli number, defined in Eq. (7.86) and

$$\overline{\mathrm{E}}_{n;m}(\bar{u};\bar{v};\bar{q}) = \mathrm{ELi}_{n;m}(\bar{u};\bar{v};\bar{q}) - (-1)^{n+m}\mathrm{ELi}_{n;m}\left(\bar{u}^{-1};\bar{v}^{-1};\bar{q}\right),$$
$$\mathrm{ELi}_{n;m}(\bar{u};\bar{v};\bar{q}) = \sum_{j=1}^{\infty}\sum_{k=1}^{\infty} \frac{\bar{u}^j}{j^n}\frac{\bar{v}^k}{k^m}\bar{q}^{jk}. \qquad (13.203)$$

6. We may relate $g^{(k)}(z,K\tau)$ (with $K \in \mathbb{N}$) to functions with argument τ according to

$$g^{(k)}(z,K\tau) = \frac{1}{K}\sum_{l=0}^{K-1} g^{(k)}\left(\frac{z+l}{K},\tau\right). \qquad (13.204)$$

Having defined the functions $g^{(k)}(z, \tau)$, we may now state the differential one-forms which we would like to integrate on $\mathcal{M}_{1,n}$. To keep the discussion simple, we focus on $\mathcal{M}_{1,2}$ with coordinates (τ, z). (The general case $\mathcal{M}_{1,n}$ is only from a notational perspective more cumbersome.) We consider

$$\omega_k^{\mathrm{Kronecker}} = (2\pi i)^{2-k} \left[g^{(k-1)} \left(z - c_j, \tau\right) dz + (k-1)\, g^{(k)} \left(z - c_j, \tau\right) \frac{d\tau}{2\pi i} \right], \tag{13.205}$$

with c_j being a constant. The differential one-form $\omega_k^{\mathrm{Kronecker}}$ is closed

$$d\omega_k^{\mathrm{Kronecker}} = 0. \tag{13.206}$$

Let us first consider the integration along $d\tau$ (i.e. $z = \mathrm{const}$). Here, the part

$$\begin{aligned} \omega_k^{\mathrm{Kronecker},\tau} &= (2\pi i)^{2-k} (k-1)\, g^{(k)} \left(z - c_j, \tau\right) \frac{d\tau}{2\pi i} \\ &= \frac{(k-1)}{(2\pi i)^k} g^{(k)} \left(z - c_j, \tau\right) \frac{d\bar{q}}{\bar{q}} \end{aligned} \tag{13.207}$$

is relevant. This is supplemented by z-independent differential one-forms constructed from modular forms: Let $f_k(\tau) \in \mathcal{M}_k(\Gamma)$ be a modular form of weight k and level N for the congruence subgroup Γ. Let N' be the smallest positive integer such that $T_{N'} \in \Gamma$. We set as in Eq. (13.117)

$$\omega_k^{\mathrm{modular}} = (2\pi i)\, f_k(\tau) \frac{d\tau}{N'}. \tag{13.208}$$

Let us now assume for simplicity $N' = 1$ (otherwise we should use the variable $\bar{q}_{N'}$ instead of the variable $\bar{q}$ in the formulae below). Let ω_j with weight k_j be as in Eq. (13.207) or as in Eq. (13.208) and γ the path from $\tau = i\infty$ to τ, corresponding in $\bar{q}$-space to a path from $\bar{q} = 0$ to $\bar{q}$. We then consider in $\bar{q}$-space the iterated integrals

$$I_\gamma \left(\omega_1, \ldots, \omega_r; \bar{q}\right). \tag{13.209}$$

The integrands have no poles in $0 < |\bar{q}| < 1$. A simple pole at $\bar{q} = 0$ is possible and allowed. If ω_r has a simple pole at $\bar{q} = 0$ we say that the iterated integral has a trailing zero. We may split ω_r into a part proportional to $d\bar{q}/\bar{q}$ and a regular remainder. The singular part of a trailing zero can be treated in exactly the same way as we did in the case of multiple polylogarithms.

To summarise:

Iterated integrals along $d\tau$:

For the integration along $d\tau$ we consider in $\bar{q}$-space the iterated integrals

$$I_\gamma\left(\omega_1, \dots, \omega_r; \bar{q}\right), \tag{13.210}$$

where ω_j is of the form

$$\begin{aligned} \omega_{k_j}^{\text{Kronecker},\tau} &= \frac{\left(k_j-1\right)}{(2\pi i)^{k_j}} g^{(k_j)}\left(z-c_j,\tau\right)\frac{d\bar{q}}{\bar{q}} \\ \text{or}\, \omega_{k_j}^{\text{modular}} &= f_{k_j}\left(\tau\right)\frac{d\bar{q}}{\bar{q}}, \end{aligned} \tag{13.211}$$

with $f_{k_j}(\tau)$ being a modular form of weight k_j.

Let us now consider the integration along dz (i.e. $\tau = \text{const}$). For the integration along dz the part

$$\omega_k^{\text{Kronecker},z} = (2\pi i)^{2-k}\, g^{(k-1)}\left(z-c_j,\tau\right)dz \tag{13.212}$$

is relevant. The iterated integrals of the differential one-forms in Eq. (13.212) along a path γ from $z=0$ to z are the elliptic multiple polylogarithms $\widetilde{\Gamma}$, as defined in ref. [346]:

The elliptic multiple polylogarithms $\widetilde{\Gamma}$ (i.e. iterated integrals along dz):

$$\begin{aligned} &\widetilde{\Gamma}\left(\begin{smallmatrix} n_1 & \dots & n_r \\ c_1 & \dots & c_r \end{smallmatrix}; z; \tau\right) = \\ &\quad (2\pi i)^{n_1+\dots+n_r-r}\, I_\gamma\left(\omega_{n_1+1}^{\text{Kronecker},z}\left(c_1,\tau\right), \dots, \omega_{n_r+1}^{\text{Kronecker},z}\left(c_r,\tau\right); z\right). \end{aligned} \tag{13.213}$$

Let us stress that this is one possibility to define elliptic multiple polylogarithms. In the literature there exist various definitions of elliptic multiple polylogarithms due to the following problem: It is not possible that the differential one-forms ω entering the definition of elliptic multiple polylogarithms have at the same time the following three properties:

(i) ω is double-periodic,
(ii) ω is meromorphic,
(iii) ω has only simple poles.

We can only require two of these three properties. The definition of the $\tilde{\Gamma}$-functions selects meromorphicity and simple poles. Meromorphicity means that ω does not depend on the anti-holomorphic variables. The differential one-forms are not double-periodic. (This is spoiled by the quasi-periodicity of $g^{(k)}(z, \tau)$ with respect to $z \to z + \tau$.) However, this is what physics (i.e. the evaluation of Feynman integrals) dictates us to choose. The integrands are then either multi-valued functions on $\mathcal{M}_{1,n}$ or single-valued functions on a covering space, in the same way as $\ln(z)$ is a multi-valued function on $\mathbb{C}^\times$ or a single-valued function on a covering space of $\mathbb{C}^\times$. Of course, in mathematics one might also consider alternative definitions, which prioritise other properties. A definition of elliptic multiple polylogarithms, which implements properties (i) and (ii), but gives up property (iii) can be found in [347], a definition, which implements properties (i) and (iii), but gives up (ii) can be found in [344]. It is a little bit unfortunate that these different function are all named elliptic multiple polylogarithms. The reader is advised to carefully check what is meant by the name "elliptic multiple polylogarithm", this also concerns the definitions in [348, 349].

It is not unusual in Feynman integral calculations that we end up with an integration involving a square root of a quartic polynomial in combination with integrands known from multiple polylogarithms. There is a systematic way to convert these integrals to the elliptic multiple polylogarithms $\tilde{\Gamma}$ [345, 346]. The square root defines an elliptic curve

$$E : v^2 - (u - u_1)(u - u_2)(u - u_3)(u - u_4) = 0, \tag{13.214}$$

and we consider as integrands the field of rational functions of the elliptic curve E, i.e. rational functions in u and v subject to the relation $v^2 = (u - u_1)(u - u_2)(u - u_3)(u - u_4)$. Primitives, which are not in this function field originate from the integrands

$$\frac{du}{v}, \quad \frac{u\, du}{v}, \quad \frac{u^2\, du}{v}, \quad \frac{du}{u - c}, \quad \frac{du}{(u - c)\, v}, \tag{13.215}$$

where c is a constant. This list includes the integrands $du/(u - c)$ for the multiple polylogarithms. Note that in the quartic case $u\, du/v$ is a differential of the third kind, a differential of the second kind can be constructed from a linear combination of $u^2\, du/v$ and $u\, du/v$:

$$\left(u^2 - \frac{1}{2} s_1 u\right) \frac{du}{v}, \quad s_1 = u_1 + u_2 + u_3 + u_4. \tag{13.216}$$

In Sect. 13.2.1 we discussed the general quartic case and defined a pair of periods (see Eq. (13.44), we use the notation as in Sect. 13.2.1)

$$\psi_1 = 2 \int\limits_{u_2}^{u_3} \frac{du}{v} = \frac{4\,K(k)}{U_3^{\frac{1}{2}}}, \quad \psi_2 = 2 \int\limits_{u_4}^{u_3} \frac{du}{v} = \frac{4i\,K\left(\bar{k}\right)}{U_3^{\frac{1}{2}}}. \tag{13.217}$$

It is convenient to normalise one period to one, hence we define $\tau = \psi_2/\psi_1$ and we consider the lattice Λ generated by $(1, \tau)$. Abel's map

$$z = \frac{1}{\psi_1} \int\limits_{u_1}^{u} \frac{du}{v} \tag{13.218}$$

relates a point (u, v) on the elliptic curve to a point $z \in \mathbb{C}/\Lambda$. We have

$$\frac{2\pi i}{\psi_1} \frac{du}{v} = 2\pi i\; dz = \omega_1^{\text{Kronecker},z}. \tag{13.219}$$

Let us also discuss the case of modular weight $k = 2$. This includes the integrands for the multiple polylogarithms $du/(u - c)$. These integrands have a pole at $u = c$ and a pole at $u = \infty$. We define the images of $u = c$ and $u = \infty$ under Abel's map by

$$z_c = \frac{1}{\psi_1} \int\limits_{u_1}^{c} \frac{du}{v}, \quad z_\infty = \frac{1}{\psi_1} \int\limits_{u_1}^{\infty} \frac{du}{v}. \tag{13.220}$$

We then have

$$\frac{du}{u-c} = \left[g^{(1)}\left(z - z_c, \tau\right) + g^{(1)}\left(z + z_c, \tau\right) - g^{(1)}\left(z - z_\infty, \tau\right) - g^{(1)}\left(z + z_\infty, \tau\right) \right] dz. \tag{13.221}$$

The integrand $du/((u - c)v)$ translates to

$$\frac{v_c du}{(u-c)\,v} = \left[g^{(1)}\left(z - z_c, \tau\right) - g^{(1)}\left(z + z_c, \tau\right) + g^{(1)}\left(z_c - z_\infty, \tau\right) + g^{(1)}\left(z_c + z_\infty, \tau\right) \right] dz. \tag{13.222}$$

where $v_c^2 = (c - u_1)(c - u_2)(c - u_3)(c - u_4)$. For the integrand $u\, du/v$ one finds

$$\frac{(u - u_1)\; du}{v} = \left[g^{(1)}\left(z + z_\infty, \tau\right) - g^{(1)}\left(z - z_\infty, \tau\right) - 2 g^{(1)}\left(z_\infty, \tau\right) \right] dz. \tag{13.223}$$

The term $u_1 du/v$ can be translated with formula (13.219). It remains to treat the integrand $u^2\, du/v$ or equivalently the linear combination appearing in Eq. (13.216). This integrand has a double pole at $u = \infty$. We would like to have integrands with only simple poles. This can be enforced by introducing a primitive. To illustrate this with a simple example consider

$$\frac{dx}{x^2} = -d \left(\frac{1}{x} \right), \tag{13.224}$$

where $1/x$ has only a simple pole. In order to enforce simple poles, we therefore introduce

$$Z_4 = -\int_{u_1}^{u} \frac{du}{v} \left[u^2 - \frac{1}{2} s_1 u + \frac{1}{2} \left(u_1 u_2 + u_3 u_4 \right) - \frac{1}{2} U_3 \frac{\phi_1}{\psi_1} \right], \quad (13.225)$$

where ϕ_1 is the quasi-period defined in Eq. (13.44). The first two terms of the integrand are proportional to Eq. (13.216), the remainder of the integrand is proportional to the holomorphic one-form du/v. It can be shown that Z_4 has as a function of u only simple poles. One then finds

$$\frac{Z_4\, du}{v} = \left[g^{(1)}\left(z - z_\infty, \tau \right) + g^{(1)}\left(z + z_\infty, \tau \right) \right] dz. \quad (13.226)$$

However, enforcing simple poles with the introduction of Z_4 has a price: We now have to consider higher powers of Z_4 as well. This will give rise to an infinite tower of integrands, corresponding to modular weight $k > 2$. For the details we refer to [345, 346]. Equations (13.221), (13.222), (13.223) and (13.226) are the complete set of formulae at modular weight 2.

13.5 Elliptic Feynman Integrals

With the background in mathematics on elliptic curves, modular forms and moduli spaces we are now in a position to tackle the first Feynman integrals, which cannot be expressed in terms of multiple polylogarithms.

We do this in two steps: We start with elliptic Feynman integrals, which depend only on one kinematic variable x. The essential trick is to change from the variable x to a new variable, the modular parameter τ. We will find that the Feynman integrals can be expressed as iterated integrals on $\mathcal{M}_{1,1}$ (i.e. iterated integrals of modular forms).

In a second step we generalise to elliptic Feynman integrals which depend on several kinematic variables $x_1, x_2, \dots$. This will lead us to iterated integrals on $\mathcal{M}_{1,n}$.

13.5.1 Feynman Integrals Depending on One Kinematic Variable

In Sect. 6.3.1 we introduced already the two-loop sunrise integral with equal internal masses:

$$I_{\nu_1 \nu_2 \nu_3}\left(D, x\right) = e^{2\varepsilon\gamma_E} \left(m^2\right)^{\nu_{123}-D} \int \frac{d^D k_1}{i\pi^{\frac{D}{2}}} \frac{d^D k_2}{i\pi^{\frac{D}{2}}} \frac{1}{\left(-q_1^2+m^2\right)^{\nu_1} \left(-q_2^2+m^2\right)^{\nu_2} \left(-q_3^2+m^2\right)^{\nu_3}}, \quad (13.227)$$

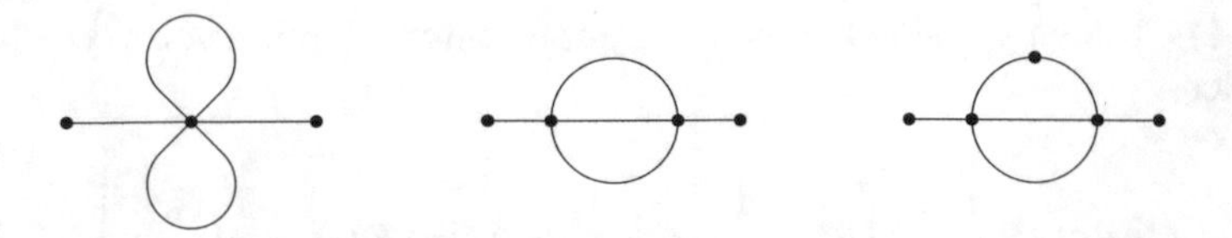

Fig. 13.6 The three master integrals for the family of the equal mass sunrise integral

with $x = -p^2/m^2$ and $q_1 = k_1, q_2 = k_2 - k_1, q_3 = -k_2 - p$. We have set $\mu^2 = m^2$. This is the simplest example of an elliptic Feynman integral. It has been studied intensively in the literature [350–358].

There are three master integrals and we start from the basis

$$\vec{I} = (I_{110}, I_{111}, I_{211})^T . \tag{13.228}$$

The three master integrals are shown in Fig. 13.6. The differential equation for this basis has been given in Eq. (6.87).

It is simpler to analyse this system in $D = 2 - 2\varepsilon$ dimensions. This is no restriction: With the help of the dimensional shift relations we may always relate integrals in $(2 - 2\varepsilon)$ dimensions to integrals in $(4 - 2\varepsilon)$ dimensions. We have for example

$$\begin{aligned}
I_{110}(2-2\varepsilon, x) &= I_{220}(4-2\varepsilon, x), \\
I_{111}(2-2\varepsilon, x) &= \frac{3}{(x+1)(x+9)} \left[(3+x) I_{220}(4-2\varepsilon, x) \right. \\
&\quad \left. + 2(1-2\varepsilon)(2-3\varepsilon) I_{111}(4-2\varepsilon, x) + 2(1-2\varepsilon)(3-x) I_{211}(4-2\varepsilon, x) \right], \\
I_{211}(2-2\varepsilon, x) &= \\
&\frac{1}{(x+1)^2(x+9)^2} \left\{ \left[3(x+1)(x+9) + \varepsilon \left(2x^3 + 34x^2 + 54x + 54 \right) \right] I_{220}(4-2\varepsilon, x) \right. \\
&\quad + 2(1-2\varepsilon)(2-3\varepsilon) \left[(x+1)(x+9) - 2\varepsilon (x-3)(x+3) \right] I_{111}(4-2\varepsilon, x) \\
&\quad \left. + 2(1-2\varepsilon) \left[3(x+1)(x+9) - \varepsilon \left(x^3 + 36x^2 + 45x - 54 \right) \right] I_{211}(4-2\varepsilon, x) \right\}.
\end{aligned} \tag{13.229}$$

The equal mass sunrise integral is the simplest Feynman integral related to an elliptic curve. The first question which we should address is how to obtain the elliptic curve associated to this integral. For the sunrise integral there are two possibilities, we may either obtain an elliptic curve from the Feynman graph polynomial or from the maximal cut. The sunrise integral has three propagators, hence we need three Feynman parameters, which we denote by a_1, a_2, a_3. The Feynman parameter representation for I_{111} reads

$$\begin{aligned}
I_{111}(2-2\varepsilon, x) &= e^{2\varepsilon\gamma_E} \Gamma(1+2\varepsilon) \int\limits_{a_j \geq 0} d^3a \, \delta\left(1 - \sum_{j=1}^{3} a_j\right) \frac{[\mathcal{U}(a)]^{3\varepsilon}}{[\mathcal{F}(a)]^{1+2\varepsilon}}, \\
\mathcal{U}(a) &= a_1 a_2 + a_2 a_3 + a_3 a_1, \\
\mathcal{F}(a) &= a_1 a_2 a_3 x + (a_1 + a_2 + a_3)(a_1 a_2 + a_2 a_3 + a_3 a_1).
\end{aligned} \tag{13.230}$$

The second graph polynomial defines an elliptic curve

$$E^{\mathrm{Feynman}} : a_1 a_2 a_3 x + (a_1 + a_2 + a_3)(a_1 a_2 + a_2 a_3 + a_3 a_1) = 0, \tag{13.231}$$

in $\mathbb{CP}^2$, with $[a_1 : a_2 : a_3]$ being the homogeneous coordinates of $\mathbb{CP}^2$. The elliptic curve varies with the kinematic variable x. In general, the Feynman parameter space can be viewed as $\mathbb{CP}^{n-1}$, with n being the number of propagators of the Feynman integral. It is clear that this approach does not generalise in a straightforward way to other elliptic Feynman integrals with more than three propagators. (For an elliptic curve we want the zero set of a single polynomial in $\mathbb{CP}^2$).

We therefore turn to the second method of obtaining the elliptic curve, which generalises easily: We study the maximal cut of the sunrise integral. Within the loop-by-loop approach

$$\mathrm{MaxCut}\ I_{111}(2-2\varepsilon, x) = \frac{(2\pi i)^3}{\pi^2} \int\limits_{\mathcal{C}_{\mathrm{MaxCut}}} \frac{dz}{\sqrt{z(z+4)\left[z^2 + 2(1-x)z + (1+x)^2\right]}} + O(\varepsilon), \tag{13.232}$$

Thus we obtain an elliptic curve as a quartic polynomial $P(u, v) = 0$:

$$E^{\mathrm{cut}} : v^2 - u(u+4)\left[u^2 + 2(1-x)u + (1+x)^2\right] = 0. \tag{13.233}$$

Also this elliptic curve varies with the kinematic variable x. Please note that these two elliptic curves E^{Feynman} and E^{cut} are not isomorphic, but only isogenic. For the sunrise integral we may work with either one of the two. In the following we will use E^{cut}.

Let us therefore consider E^{cut}, defined by the quartic polynomial in Eq. (13.233). We denote the roots by

$$u_1 = -4, \quad u_2 = -\left(1+\sqrt{-x}\right)^2, \quad u_3 = -\left(1-\sqrt{-x}\right)^2, \quad u_4 = 0. \tag{13.234}$$

We may then determine two independent periods ψ_1 and ψ_2 as in Sect. 13.2.1. Therefore, let ψ_1 and ψ_2 be defined by Eq. (13.44). We set $\tau = \psi_2/\psi_1$ and we denote the Wronskian by

$$W = \psi_1 \frac{d}{dx}\psi_2 - \psi_2 \frac{d}{dx}\psi_1 = -\frac{6\pi i}{x(x+1)(x+9)}. \tag{13.235}$$

We then perform a change of the basis of the master integrals from the pre-canonical basis $(I_{110}, I_{111}, I_{211})$ to

$$J_1 = 4\varepsilon^2 \, I_{110}\left(2-2\varepsilon, x\right),$$
$$J_2 = \varepsilon^2 \frac{\pi}{\psi_1} \, I_{111}\left(2-2\varepsilon, x\right),$$
$$J_3 = \frac{1}{\varepsilon} \frac{\psi_1^2}{2\pi i W} \frac{d}{dx} J_2 + \frac{\psi_1^2}{2\pi i W} \frac{\left(3x^2+10x-9\right)}{2x\left(x+1\right)\left(x+9\right)} J_2. \tag{13.236}$$

This transformation is not rational or algebraic in x, as can be seen from the prefactor $1/\psi_1$ in the definition of J_2. The period ψ_1 is a transcendental function of x.

The fibre transformation in Eq. (13.236) can be understood and motivated as follows: The definition of J_1 is straightforward and follows from Eq. (2.123). As far as J_2 is concerned, we first note that ψ_1 and ψ_2 are obtained by integrating the holomorphic one-form du/v of the elliptic curve along two cycles γ_1 and γ_2, respectively (see also Fig. 13.2). If we replace in Eq. (13.232) C_{MaxCut} by γ_1 we obtain

$$\frac{\left(2\pi i\right)^3}{\pi^2} \int\limits_{\gamma_1} \frac{dz}{\sqrt{z\left(z+4\right)\left[z^2+2\left(1-x\right)z+\left(1+x\right)^2\right]}} = -8i\pi^2 \frac{\psi_1}{\pi}. \tag{13.237}$$

J_2 is obtained by dividing I_{111} by ψ_1 and by adjusting powers of π and ε. Prefactors consisting of algebraic numbers (for example $(-8i)$) are not relevant.

Let us turn to J_3: It is well-known in mathematics, that the first cohomology group for a family of elliptic curves E_x, parametrised by x, is generated by the holomorphic one-form du/v and its x-derivative. This motivates an ansatz, consisting of J_2 and its τ-derivative:

$$J_3 = c_2\left(x\right) \frac{1}{2\pi i} \frac{d}{d\tau} J_2 + c_3\left(x\right) J_2, \tag{13.238}$$

with unknown functions $c_2(x)$ and $c_3(x)$. We determine $c_2(x)$ and $c_3(x)$ such that this ansatz transforms the differential equation into an ε-form. One finds

$$c_2\left(x\right) = \frac{1}{\varepsilon} \text{ and } c_3\left(x\right) = \frac{1}{24}\left(3x^2+10x-9\right)\frac{\psi_1^2}{\pi^2}, \tag{13.239}$$

and therefore

$$J_3 = \frac{1}{\varepsilon} \frac{1}{2\pi i} \frac{d}{d\tau} J_2 + \frac{1}{24}\left(3x^2+10x-9\right)\frac{\psi_1^2}{\pi^2} J_2. \tag{13.240}$$

This agrees with Eq. (13.236): From Eq. (13.56) it follows that

$$\frac{1}{2\pi i} \frac{d}{d\tau} = \frac{\psi_1^2}{2\pi i W} \frac{d}{dx}. \tag{13.241}$$

In the basis $J = (J_1, J_2, J_3)^T$ we have now

$$\left(\frac{d}{dx} + A_x \right) \vec{J} = 0 \tag{13.242}$$

with

$$A_x = \varepsilon \begin{pmatrix} 0 & 0 & 0 \\ 0 & \frac{(3x^2+10x-9)}{2x(x+1)(x+9)} & -\frac{2\pi i W}{\psi_1^2} \\ \frac{3i}{4} \frac{\psi_1}{W} \frac{1}{x(x+1)(x+9)} & -\frac{i}{288} \frac{W \psi_1^2}{\pi^3} (3-x)^4 & \frac{(3x^2+10x-9)}{2x(x+1)(x+9)} \end{pmatrix}. \tag{13.243}$$

Through the fibre transformation we have managed that the dimensional regularisation parameter ε is factored out, however the matrix A_x is not yet in a particular nice and suitable form. The situation is similar to Eq. (6.237), where a fibre transformation allowed us to factor out ε, but left us with a square root singularity. In Eq. (13.243) we have a transcendental function (i.e. ψ_1) appearing in the matrix A_x.

In order to put the matrix A into a nice form, we perform a base transformation and change the variable from x to τ. With the help of Eq. (13.241) we obtain

$$\left(\frac{1}{2\pi i} \frac{d}{d\tau} + A_\tau \right) \vec{J} = 0 \tag{13.244}$$

with

$$A_\tau = \varepsilon \begin{pmatrix} 0 & 0 & 0 \\ 0 & \eta_2 & \eta_0 \\ \eta_3 & \eta_4 & \eta_2 \end{pmatrix} \tag{13.245}$$

and

$$\begin{aligned} \eta_0 &= -1, \\ \eta_2 &= \frac{1}{24} \frac{\psi_1^2}{\pi^2} \left(3x^2 + 10x - 9\right), \\ \eta_3 &= -\frac{1}{96} \frac{\psi_1^3}{\pi^3} x\,(x+1)\,(x+9), \\ \eta_4 &= -\frac{1}{576} \frac{\psi_1^4}{\pi^4} (3-x)^4. \end{aligned} \tag{13.246}$$

It remains to express η_2, η_3 and η_4 as a function of τ. To this aim we first express x as a function of τ. Noting that upon the substitution $x \to (-x)$ the elliptic curve in Eq. (13.233) is identical to the one discussed in Eq. (13.80) it follows that

$$x = -9 \frac{\eta(\tau)^4 \, \eta(6\tau)^8}{\eta(3\tau)^4 \, \eta(2\tau)^8}. \tag{13.247}$$

We may then obtain the $\bar{q}$-expansions of η_2, η_3 and η_4. (For the complete elliptic integral K appearing in ψ_1 we use Eq. (13.74).) For example, for η_2 we obtain

$$\eta_2 = -\frac{1}{2} - 8\bar{q} - 4\bar{q}^2 - 44\bar{q}^3 + 4\bar{q}^4 - 48\bar{q}^5 - 40\bar{q}^6 + O\left(\bar{q}^7\right). \quad (13.248)$$

By comparing the $\bar{q}$-expansions one checks that η_0, η_2, η_3 and η_4 are modular forms of $\Gamma_1(6)$ of modular weight 0, 2, 3 and 4, respectively. In order to get an explicit expression we introduce a basis $\{b_1, b_2\}$ for the modular forms of modular weight 1 for the Eisenstein subspace $\mathcal{E}_1(\Gamma_1(6))$ as follows: We define two primitive Dirichlet characters χ_1 and $\chi_{(-3)}$ with conductors 1 and 3, respectively, through the Kronecker symbol

$$\chi_1 = \left(\frac{1}{n}\right), \quad \chi_{(-3)} = \left(\frac{-3}{n}\right). \quad (13.249)$$

Explicitly we have

$$\begin{aligned} \chi_1(n) &= 1, \qquad \forall n \in \mathbb{Z}, \\ \chi_{-3}(n) &= \begin{cases} 0, & n = 0 \mod 3, \\ 1, & n = 1 \mod 3, \\ -1, & n = 2 \mod 3, \end{cases} \end{aligned} \quad (13.250)$$

We then set

$$\begin{aligned} b_1 &= E_1\left(\tau; \chi_1, \chi_{(-3)}\right) = E_{1,1,-3,1}(\tau), \\ b_2 &= E_1\left(2\tau; \chi_1, \chi_{(-3)}\right) = E_{1,1,-3,2}(\tau). \end{aligned} \quad (13.251)$$

The generalised Eisenstein series $E_k(\tau, \chi_a, \chi_b)$ have been defined in Eq. (13.137), the generalised Eisenstein series $E_{k,a,b,K}(\tau)$ have been defined in Eq. (13.149). The integration kernels may be expressed as polynomials in b_1 and b_2:

$$\begin{aligned} \eta_2 &= -6\left(b_1^2 + 6b_1b_2 - 4b_2^2\right), \\ \eta_3 &= -9\sqrt{3}\left(b_1^3 - b_1^2b_2 - 4b_1b_2^2 + 4b_2^3\right), \\ \eta_4 &= -324b_1^4. \end{aligned} \quad (13.252)$$

Let us summarise: Through a fibre transformation and a base transformation we obtained the differential equation

$$(d + A)\,\vec{J} = 0 \quad (13.253)$$

with

$$A = 2\pi i \, \varepsilon \begin{pmatrix} 0 & 0 & 0 \\ 0 & \eta_2(\tau) & \eta_0(\tau) \\ \eta_3(\tau) & \eta_4(\tau) & \eta_2(\tau) \end{pmatrix} d\tau, \tag{13.254}$$

where $\eta_k(\tau)$ denotes a modular form of modular weight k for $\Gamma_1(6)$. The differential equation for the equal mass sunrise system is now in ε-form and the kinematic variable matches the standard coordinate on $\mathcal{M}_{1,1}$. With the additional information of a boundary value, the differential equation is now easily solved order by order in ε in terms of iterated integrals of modular forms. One finds for example

$$J_2 = \left[3\,\mathrm{Cl}_2\left(\frac{2\pi}{3}\right) + 4I\left(\eta_0, \eta_3; \tau\right) \right] \varepsilon^2 + O\left(\varepsilon^3\right). \tag{13.255}$$

The Clausen value $\mathrm{Cl}_2(2\pi/3)$ comes from the boundary value. The first few terms of the $\bar{q}$-expansion read

$$J_2^{(2)} = 3\,\mathrm{Cl}_2\left(\frac{2\pi}{3}\right)$$
$$-3\sqrt{3}\left[\bar{q} - \frac{5}{4}\bar{q}^2 + \bar{q}^3 - \frac{11}{16}\bar{q}^4 + \frac{24}{25}\bar{q}^5 - \frac{5}{4}\bar{q}^6 + \frac{50}{49}\bar{q}^7 - \frac{53}{64}\bar{q}^8 + \bar{q}^9\right] + O(\bar{q}^{10}). \tag{13.256}$$

13.5.1.1 Analytic Continuation

In the calculation above we solved the differential equation in a neighbourhood of $x = 0$ (corresponding to $\bar{q} = 0$) and the result is valid in a neighbourhood of $x = 0$. We are interested in extending this result to all $x \in \mathbb{R} - i\delta$, where the infinitesimal imaginary part originates from Feynman's $i\delta$-prescription. We have to ensure that the periods ψ_1 and ψ_2 vary smoothly, as x varies in $\mathbb{R} - i\delta$ [359].

Let's look at the details: The expressions for the modulus k and the complementary modulus k' of the elliptic curve E^{cut} are

$$k^2 = \frac{16\sqrt{-x}}{\left(1+\sqrt{-x}\right)^3\left(3-\sqrt{-x}\right)}, \quad k'^2 = \frac{\left(1-\sqrt{-x}\right)^3\left(3+\sqrt{-x}\right)}{\left(1+\sqrt{-x}\right)^3\left(3-\sqrt{-x}\right)}, \tag{13.257}$$

where Feynman's $i\delta$-prescription ($x \to x - i\delta$) is understood. We define the periods ψ_1 and ψ_2 for all $x \in \mathbb{R} - i\delta$ by

$$\begin{pmatrix} \psi_2 \\ \psi_1 \end{pmatrix} = \frac{4}{\left(1+\sqrt{-x}\right)^{\frac{3}{2}}\left(3-\sqrt{-x}\right)^{\frac{1}{2}}} \gamma \begin{pmatrix} iK\left(k'\right) \\ K\left(k\right) \end{pmatrix}, \tag{13.258}$$

where $K(x)$ denotes the complete elliptic integral of the first kind. The essential new ingredient is the 2×2-matrix γ given by

$$\gamma = \begin{cases} \begin{pmatrix} 1 & 0 \\ 2 & 1 \end{pmatrix}, & -\infty < x < -1. \\ \begin{pmatrix} 1 & 0 \\ 0 & 1 \end{pmatrix}, & -1 < x < -3 + 2\sqrt{3}, \\ \begin{pmatrix} 1 & 0 \\ 2 & 1 \end{pmatrix}, & -3 + 2\sqrt{3} < x < \infty, \end{cases} \tag{13.259}$$

The matrix γ ensures that the periods ψ_1 and ψ_2 vary smoothly as the kinematic variable x varies smoothly in $x \in \mathbb{R} - i\delta$ [359]. The complete elliptic integral $K(k)$ can be viewed as a function of k^2: We set $\tilde{K}(k^2) = K(k)$. The function $\tilde{K}(k^2)$ has a branch cut at $[1, \infty[$ in the complex k^2-plane. The matrix γ compensates for the discontinuity when we cross this branch cut. It is relatively easy to see that k^2 as a function of x crosses this branch cut at the point $x = -3 + 2\sqrt{3} \approx 0.46$, the corresponding value in the k^2-plane is $k^2 = 2$. The point $x = -1$ is a little bit more subtle. Let us parametrise a small path around $x = -1$ by

$$x(\phi) = -1 - \delta e^{i\phi}, \qquad \phi \in [0, \pi], \tag{13.260}$$

then

$$k^2 = 1 + \frac{1}{32}\delta^3 e^{3i\phi} + O\left(\delta^4\right), \tag{13.261}$$

and the path in k^2-space winds around the point $k^2 = -1$ by an angle 3π as the path in x-space winds around the point $x = 1$ by the angle π.

Equation (13.258) defines the periods ψ_1 and ψ_2 for all values $x \in \mathbb{R} - i\delta$. The periods take values in $\mathbb{C} \cup \{\infty\}$.

The original differential equation (Eq. (6.87)) has singular points at

$$x \in \{-9, -1, 0, \infty\}. \tag{13.262}$$

The point $x = -9$ is called the threshold, the point $x = -1$ is called the pseudo-threshold. The singular points of the differential equation are mapped in τ-space to

$$\tau(x = -9) = \frac{1}{3}, \quad \tau(x = -1) = 0, \quad \tau(x = 0) = i\infty, \quad \tau(x = \infty) = \frac{1}{2}. \tag{13.263}$$

The points

$$\left\{0, \frac{1}{3}, \frac{1}{2}, i\infty\right\} \tag{13.264}$$

in τ-space are the cusps of $\Gamma_1(6)$.

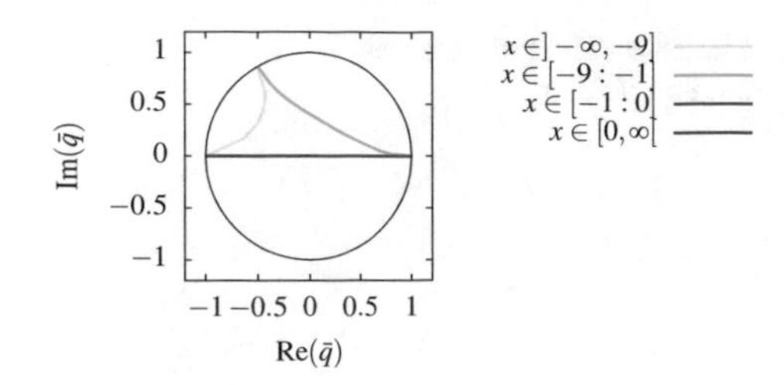

Fig. 13.7 The path in $\bar{q}$-space as x ranges over $\mathbb{R}$. We always have $|\bar{q}| \leq 1$ and $|\bar{q}| = 1$ only at $x \in \{-9, -1, \infty\}$

In Fig. 13.7 we plot the values of the variable $\bar{q}$ as x ranges over $\mathbb{R}$. We see that all values of $\bar{q}$ are inside the unit disc with the exception of the three points $x \in \{-9, -1, \infty\}$, where the corresponding $\bar{q}$-values are on the boundary of the unit disc. Once the periods ψ_1 and ψ_2 are defined as in Eq. (13.258), the $\bar{q}$-expansion in Eq. (13.255) gives the correct result for $x \in \mathbb{R}\backslash\{-9, -1, \infty\}$. Note that although the $\bar{q}$-expansion corresponds to an expansion around $x = 0$, it gives also the correct result for

$$x \ \in]-\infty, -9[\text{ and } x \ \in]-9, -1[. \tag{13.265}$$

These intervals correspond to the yellow and green segments in Fig. 13.7. The $\bar{q}$-expansion in Eq. (13.255) does not converge for $x \in \{-9, -1, \infty\}$ and in a neighbourhood of these points the convergence is slow. In the next subsection we will see how to improve the convergence in a neighbourhood of these points.

13.5.1.2 Modular Transformations

In this paragraph we address two questions: In the calculation of the equal mass two-loop sunrise integral we started with a specific choice of periods ψ_1 and ψ_2. Our first question is: What happens if we make a different choice ψ_1' and ψ_2'?

We have seen that we may express the equal mass two-loop sunrise integral as iterated integrals of modular forms. We may evaluate these integrals through their $\bar{q}$-expansion. The convergence of $\bar{q}$-series is fast, if $|\bar{q}| \ll 1$, however it is rather slow if $|\bar{q}| \lesssim 1$. The sunrise integral depends only on one kinematic variable, which we take as the modular parameter τ. By a modular transformation $\tau' = \gamma(\tau)$ with $\gamma \in \mathrm{SL}_2(\mathbb{Z})$ we may transform τ' into the fundamental domain $\mathcal{F}$ shown in Fig. 13.8 and defined by

$$\mathcal{F} = \left\{ \tau' \in \mathbb{H} \middle| |\tau'| > 1 \text{ and } -\frac{1}{2} < \mathrm{Re}\left(\tau'\right) \leq \frac{1}{2} \right\} \cup \left\{ \tau' \in \mathbb{H} \middle| |\tau'| = 1 \text{ and } 0 \leq \mathrm{Re}\left(\tau'\right) \leq \frac{1}{2} \right\}. \tag{13.266}$$

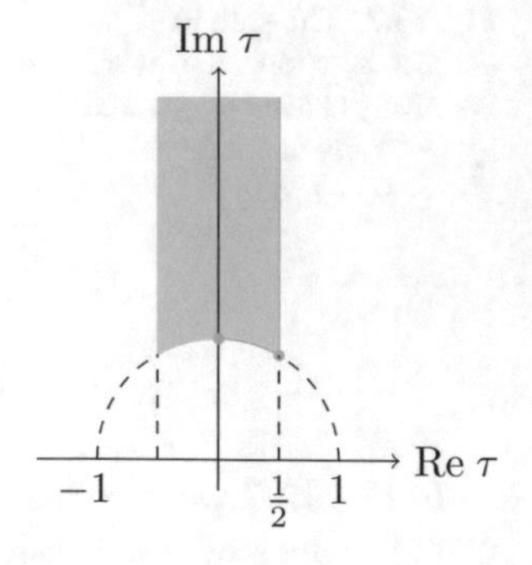

Fig. 13.8 The fundamental domain $\mathcal{F}$ for τ

Such a transformation achieves that

$$|\bar{q}'| \leq e^{-\pi\sqrt{3}} \approx 0.0043. \tag{13.267}$$

This is a small expansion parameter. Our second question is: How do elliptic Feynman integrals transform under modular transformations?

The two questions are of course related: A modular transformation corresponds exactly to the transformation (see Eq. (13.87))

$$\begin{pmatrix} \psi_2' \\ \psi_1' \end{pmatrix} = \gamma \begin{pmatrix} \psi_2 \\ \psi_1 \end{pmatrix}. \tag{13.268}$$

There are two complications we have to take into account: First of all, in order to transform τ into the fundamental domain $\mathcal{F}$ we need a $\gamma \in \mathrm{SL}_2(\mathbb{Z})$. Already in the simplest example of an elliptic Feynman integral we have seen that the modular forms appearing in the solution belong to a congruence subgroup Γ (e.g. $\Gamma_1(6)$ in the case of the sunrise integral). For $f \in \mathcal{M}_k(\Gamma)$ and $\gamma \in \mathrm{SL}_2(\mathbb{Z})$ we have in general

$$f|_k\gamma \notin \mathcal{M}_k(\Gamma). \tag{13.269}$$

The second complication is as follows: Suppose our Feynman integral is expressed as iterated integrals of modular forms for a congruence subgroup Γ and—in order to avoid the first complication—suppose that $\gamma \in \Gamma$. By changing variables from τ to $\tau' = \gamma(\tau)$ we leave the class of functions we started with and generate new integrands with additional appearances of $\ln(\bar{q})$. To see this in more detail, let's look at an example: Let f be a modular form of weight k for Γ. For simplicity we assume that f vanishes at the cusp $\tau = i\infty$. Let N' be the smallest positive integer such that $T_{N'} \in \Gamma$. The modular form has then the Fourier expansion

$$f = \sum_{n=1}^{\infty} a_n \bar{q}_{N'}^n. \tag{13.270}$$

Let $\gamma \in \Gamma$. We then have

$$I(f;\tau) = \frac{2\pi i}{N'} \int\limits_{i\infty}^{\tau} f(\tilde{\tau})\, d\tilde{\tau} = \sum_{n=1}^{\infty} \int\limits_{0}^{\bar{q}} a_n \tilde{q}_{N'}^{n} \frac{d\tilde{q}_{N'}}{\tilde{q}_{N'}} = \sum_{n=1}^{\infty} \frac{a_n}{n} \bar{q}_{N'}^{n}. \tag{13.271}$$

Let us now consider a coordinate transformation

$$\tau = \gamma^{-1}\left(\tau'\right) = \frac{a\tau' + b}{c\tau' + d}, \qquad \gamma^{-1} \in \Gamma. \tag{13.272}$$

It is simpler to consider the inverse transformation here. We have

$$\begin{aligned} I(f;\tau) &= \frac{2\pi i}{N'} \int\limits_{i\infty}^{\tau} f(\tilde{\tau})\, d\tilde{\tau} = \frac{2\pi i}{N'} \int\limits_{\gamma(i\infty)}^{\gamma(\tau)} f\left(\gamma^{-1}\left(\tilde{\tau}'\right)\right) \frac{d\tilde{\tau}'}{(c\tilde{\tau}' + d)^2} \\ &= \frac{2\pi i}{N'} \int\limits_{\gamma(i\infty)}^{\gamma(\tau)} \left(c\tilde{\tau}' + d\right)^{k-2} (f|_k\gamma^{-1})(\tilde{\tau}')\, d\tilde{\tau}'. \end{aligned} \tag{13.273}$$

As we only consider $\gamma \in \Gamma$, the expression $(f|_k\gamma^{-1})(\tilde{\tau}')$ is again a modular form for Γ, this is fine. However, we picked up a factor $(c\tilde{\tau}' + d)^{k-2}$. Only for the modular weight $k = 2$ this factor is absent. In general we leave the class of integrands constructed purely from modular forms and end up with integrands which contain in addition powers of the automorphic factor $(c\tau' + d)$.

The two complications are solved as follows [360]: For the first complication we note that if $f \in \mathcal{M}_k(\Gamma)$ is of level N we have $f \in \mathcal{M}_k(\Gamma(N))$. For any $\gamma \in \mathrm{SL}_2(\mathbb{Z})$ we then have (see Eq. (13.108))

$$f|_k\gamma \in \mathcal{M}_k\left(\Gamma\left(N\right)\right). \tag{13.274}$$

For the second complication we note that we may view the transformation $\tau' = \gamma(\tau)$ as a base transformation. In order to stay within the class of iterated integrals of modular forms we should accompany the base transformation by a fibre transformation.

Let us illustrate the two aspects with an example. We consider again the system of the equal mass sunrise integral with the master integrals $\vec{J} = (J_1, J_2, J_3)^T$ defined in Eq. (13.236) and the differential equation (see Eqs. (13.253)–(13.254))

$$(d + A)\, J = 0, \quad A = 2\pi i\, \varepsilon \begin{pmatrix} 0 & 0 & 0 \\ 0 & \eta_2(\tau) & \eta_0(\tau) \\ \eta_3(\tau) & \eta_4(\tau) & \eta_2(\tau) \end{pmatrix} d\tau. \tag{13.275}$$

For this particular example, the $\eta_k(\tau)$'s are modular forms of $\Gamma_1(6)$ and therefore also modular forms of $\Gamma(6)$. Let us now consider for

$$\gamma(\tau) = \frac{a\tau + b}{c\tau + d}, \qquad \gamma \in \mathrm{SL}_2(\mathbb{Z}) \tag{13.276}$$

the combined transformation

$$\begin{aligned} J' &= \begin{pmatrix} 1 & 0 & 0 \\ 0 & (c\tau + d)^{-1} & 0 \\ 0 & \frac{c}{2\pi i \varepsilon \eta_0} & (c\tau + d) \end{pmatrix} J, \\ \tau' &= \frac{a\tau + b}{c\tau + d}. \end{aligned} \tag{13.277}$$

Working out the transformed differential equation we obtain

$$\left(d + A'\right) J' = 0 \tag{13.278}$$

with

$$A' = 2\pi i\, \varepsilon \begin{pmatrix} 0 & 0 & 0 \\ 0 & (\eta_2|_2\gamma^{-1})(\tau') & (\eta_0|_0\gamma^{-1})(\tau') \\ (\eta_3|_3\gamma^{-1})(\tau') & (\eta_4|_4\gamma^{-1})(\tau') & (\eta_2|_2\gamma^{-1})(\tau') \end{pmatrix} d\tau'. \tag{13.279}$$

We have

$$\eta_k|_k\gamma^{-1} \in \mathcal{M}_k(\Gamma(6)) \tag{13.280}$$

and therefore we don't leave the space of modular forms with the combined transformation of Eq. (13.277). The transformed system may therefore again be solved for any $\gamma \in \mathrm{SL}_2(\mathbb{Z})$ in terms of iterated integrals of modular forms. In particular, we achieved that terms with additional automorphic factors do not occur.

The fact that we need to redefine the master integrals is not too surprising. Let's look at J_2. We originally defined J_2 by

$$J_2 = \varepsilon^2 \frac{\pi}{\psi_1} I_{111}(2 - 2\varepsilon, x), \tag{13.281}$$

i.e. we rescaled I_{111} (up to a constant) by $1/\psi_1$. This definition is tied to our initial choice of periods. Noting that the automorphic factor $(c\tau + d)$ is nothing than the ratio of two periods

$$c\tau + d = \frac{\psi_1'}{\psi_1}, \tag{13.282}$$

we find that J_2' is given by

$$J_2' = \varepsilon^2 \frac{\pi}{\psi_1'} I_{111}(2 - 2\varepsilon, x). \tag{13.283}$$

13.5.1.3 The Maximal Cut and the Period Matrix

We have seen that the basis J_1, J_2 and J_3 defined in Eq. (13.236) puts the differential equation for the system into an ε-form. In Sect. 7.1.7 we introduced a technique, which guesses a basis of master integrals from a set of master integrands φ_i and a set of master contours $\mathcal{C}_j$. The various master integrands integrated against the various master contours define a period matrix P with entries

$$P_{ij} = \langle \varphi_i | \mathcal{C}_j \rangle . \tag{13.284}$$

In the i-th row of this matrix we then look at the term of order $j_{\min}$ in the ε-expansion (with $j_{\min}$ defined as in Eq. (7.176)). This defines a matrix P^{leading} with entries

$$P_{ij}^{\text{leading}} = \text{coeff}\left(\langle \varphi_i | \mathcal{C}_j \rangle , \varepsilon^{j_{\min}} \right) \cdot \varepsilon^{j_{\min}} \tag{13.285}$$

Our strategy in Sect. 7.1.7 was to look for integrands φ_i such that all entries of P^{leading} are constants of weight zero. We already know that this is not a sufficient condition, as we may always multiply φ_i by an ε-dependent prefactor with leading term 1. We now show that it is neither a necessary condition. (But the condition is helpful in practice.)

We will focus on the maximal cut, hence only the integrals J_2 and J_3 are relevant. We work with the loop-by-loop Baikov representations, where we have four integration variables $z_1 - z_4$. Let $\mathcal{C}'$ be the integration domain selecting the maximal cut, i.e. a small counter-clockwise circle around $z_1 = 0$, a small counter-clockwise circle around $z_2 = 0$ and a small counter-clockwise circle around $z_3 = 0$. We set $z_4 = z$ in accordance with the notation used in Eq. (13.237). We denote by γ_1 and γ_2 the two cycles of the elliptic curve. They define the integration domain in the variable z. We define

$$\mathcal{C}_2 = \mathcal{C}' \cup \gamma_1, \quad \mathcal{C}_3 = \mathcal{C}' \cup \gamma_2. \tag{13.286}$$

We denote the integrands of J_2 and J_3 by φ_2 and φ_3, respectively. Let us look at the period matrix

$$P = \begin{pmatrix} \langle \varphi_2 | \mathcal{C}_2 \rangle & \langle \varphi_2 | \mathcal{C}_3 \rangle \\ \langle \varphi_3 | \mathcal{C}_2 \rangle & \langle \varphi_3 | \mathcal{C}_3 \rangle \end{pmatrix} \tag{13.287}$$

and the matrix P^{leading} defined as in Eq. (13.285). We compute the entries of P^{leading}. To this aim we first express J_2 and J_3 as a linear combination of I_{111} and I_{211}. One finds

$$J_2 = \varepsilon^2 \frac{\pi}{\psi_1} I_{111},$$

$$J_3 = \frac{\varepsilon^2}{24}\left(7x^2+50x+27\right)\frac{\psi_1}{\pi} I_{111} + \frac{\varepsilon}{12}(x+1)(x+9)\left(\frac{\psi_1}{\pi} + x\frac{d}{dx}\frac{\psi_1}{\pi}\right) I_{111}$$
$$-\frac{\varepsilon}{4}(x+1)(x+9)\frac{\psi_1}{\pi} I_{211}. \quad (13.288)$$

The derivative $d\psi_1/dx$ may be expressed with the help of Eq. (13.48) as a linear combination of ψ_1 and ϕ_1:

$$\frac{d}{dx}\psi_1 = -\frac{1}{2}\psi_1 \frac{d}{dx}\ln U_2 + \frac{1}{2}\phi_1 \frac{d}{dx}\ln\frac{U_2}{U_1}, \quad (13.289)$$
$$U_1 = 16\sqrt{-x}, \quad U_2 = \left(1-\sqrt{-x}\right)^3\left(3+\sqrt{-x}\right).$$

Let φ_{111} and φ_{211} denote the integrands of I_{111} and I_{211}, respectively. Replacing C_{MaxCut} by γ_1 or γ_2 in Eq. (13.232) we immediately have

$$\langle\varphi_{111}|C_2\rangle = -8i\pi\psi_1 + O(\varepsilon), \quad \langle\varphi_{111}|C_3\rangle = -8i\pi\psi_2 + O(\varepsilon), \quad (13.290)$$

For φ_{211} the following two integrals are helpful ($v = \sqrt{(u-u_1)(u-u_2)(u-u_3)(u-u_4)}$):

$$2\int\limits_{u_2}^{u_3} \frac{du}{(u-u_1)\,v} = \frac{1}{u_2-u_1}\psi_1 - \frac{u_4-u_2}{(u_2-u_1)(u_4-u_1)}\phi_1,$$
$$2\int\limits_{u_4}^{u_3} \frac{du}{(u-u_1)\,v} = \frac{1}{u_2-u_1}\psi_2 - \frac{u_4-u_2}{(u_2-u_1)(u_4-u_1)}\phi_2. \quad (13.291)$$

One finds

$$P^{\text{leading}} = 2i\begin{pmatrix} (2\pi i\varepsilon)^2 & (2\pi i\varepsilon)^2\,\tau \\ 0 & -(2\pi i\varepsilon) \end{pmatrix}. \quad (13.292)$$

We see that the second entry in the first row P_{12}^{leading} is not constant, this entry depends on the kinematic variable τ.

13.5.1.4 Further Examples

There are more Feynman integrals depending on one kinematic variable $x = -p^2/m^2$ and evaluating to iterated integrals of modular forms. Figure 13.9 shows some additional examples [350, 361–365]. In computing these integrals, we encounter two additional integration kernels

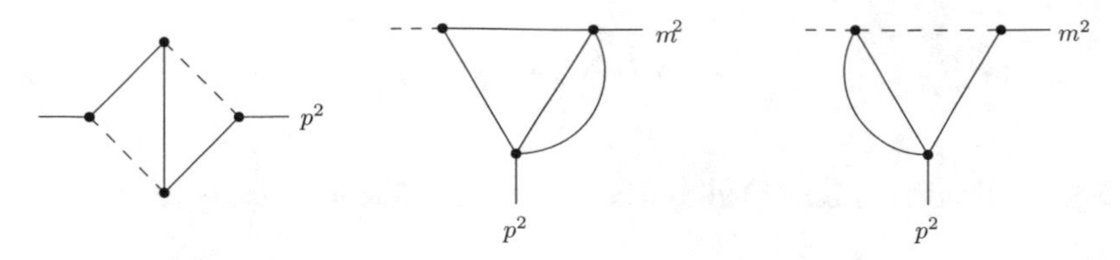

Fig. 13.9 Further examples of Feynman integrals evaluating to iterated integrals of modular forms. Internal solid lines correspond to a propagator with mass m^2, internal dashed lines to a massless propagator. External dashed lines indicate a light-like external momentum

$$\omega_0 = \frac{dx}{x}, \qquad \omega_1 = \frac{dx}{x+1}. \tag{13.293}$$

If we only would have those integration kernels in the differential equation, we would be able to express the result in terms of multiple polylogarithms. However, we have the ones appearing in Eq. (13.254) combined with the ones from Eq. (13.293). We therefore have to rewrite all integration kernels in terms of the variable τ. We may express ω_0 and ω_1 in terms of modular forms:

$$\omega_0 = 2\pi i \; \eta_{2,0} \; d\tau, \qquad \omega_1 = 2\pi i \; \eta_{2,1} \; d\tau. \tag{13.294}$$

The modular forms $\eta_{2,0}$ and $\eta_{2,1}$, both of modular weight 2, are given by

$$\begin{aligned}
\eta_{2,0} &= \frac{1}{2i\pi}\frac{\psi_1^2}{W}\frac{1}{x} &&= -12\left(b_1^2 - 4b_2^2\right), \\
\eta_{2,1} &= \frac{1}{2i\pi}\frac{\psi_1^2}{W}\frac{1}{x+1} &&= -18\left(b_1^2 + b_1 b_2 - 2b_2^2\right),
\end{aligned} \tag{13.295}$$

where b_1 and b_2 are the two Eisenstein series defined in Eq. (13.251).

13.5.2 Feynman Integrals Depending on Several Kinematic Variables

Let us now turn our attention to elliptic Feynman integrals, which depend on more than one kinematic variable (e.g. $N_B > 1$). In the case of just one kinematic variable ($N_B = 1$) we may think of the base space as a covering space of $\mathcal{M}_{1,1}$ with coordinate τ. In the case of more kinematic variables this generalises to a covering space of $\mathcal{M}_{1,n}$ with coordinates $(\tau, z_1, \dots, z_{n-1})$.

In Sect. 13.4 we introduced in Eq. (13.205) the differential one-form $\omega_k^{\text{Kronecker}}$. We can be a little bit more general than Eq. (13.205): Let $K \in \mathbb{N}$ and $L(z)$ a linear function of $z_1, \dots, z_{n-1}$:

$$L(z) = \sum_{j=1}^{n-1} \alpha_j z_j + \beta. \tag{13.296}$$

The generalisation of Eq. (13.205) which we would like to consider is

$$\omega_k(L(z), K\tau) = (2\pi i)^{2-k} \left[g^{(k-1)}(L(z), K\tau)\, dL(z) + K(k-1)\, g^{(k)}(L(z), K\tau) \frac{d\tau}{2\pi i} \right]. \tag{13.297}$$

The differential one-form $\omega_k(L(z), K\tau)$ is closed

$$d\omega_k(L(z), K\tau) = 0. \tag{13.298}$$

We may always reduce the case $K > 1$ to the case $K = 1$ with help of (compare with Eq. (13.204))

$$\omega_k(L(z), K\tau) = \sum_{l=0}^{K-1} \omega_k\left(\frac{L(z) + l}{K}, \tau \right). \tag{13.299}$$

It is therefore sufficient to focus on the case $K = 1$.

Let us study the case of elliptic Feynman integrals depending on several kinematic variables with a concrete example. We don't have to go very far, we may generalise the equal mass sunrise integral to the unequal mass sunrise integral. We now take the three masses squared m_1^2, m_2^2 and m_3^2 in the propagators to be pairwise distinct. We consider

$$I_{\nu_1\nu_2\nu_3}(D, x, y_1, y_2) = \\ e^{2\varepsilon\gamma_E} \left(m_3^2\right)^{\nu_{123}-D} \int \frac{d^D k_1}{i\pi^{\frac{D}{2}}} \frac{d^D k_2}{i\pi^{\frac{D}{2}}} \frac{1}{\left(-q_1^2 + m_1^2\right)^{\nu_1} \left(-q_2^2 + m_2^2\right)^{\nu_2} \left(-q_3^2 + m_3^2\right)^{\nu_3}}, \tag{13.300}$$

with $x = -p^2/m_3^2$, $y_1 = m_1^2/m_3^2$, $y_2 = m_2^2/m_3^2$. and as before $q_1 = k_1, q_2 = k_2 - k_1$, $q_3 = -k_2 - p$. We have set $\mu^2 = m_3^2$. Also this integral has been studied intensively in the literature [356, 366–374].

There are now seven master integrals and we may start from the basis

$$\vec{I} = (I_{110}, I_{101}, I_{011}, I_{111}, I_{211}, I_{121}, I_{112})^T. \tag{13.301}$$

In mathematical terms we are looking at a rank 7 vector bundle over $\mathcal{M}_{1,3}$.

Finding the elliptic curve proceeds exactly in the same way as discussed in the equal mass case. The second graph polynomial defines an elliptic curve

$$E^{\text{Feynman}} : a_1 a_2 a_3 x + (a_1 y_1 + a_2 y_2 + a_3)(a_1 a_2 + a_2 a_3 + a_3 a_1) = 0, \quad (13.302)$$

in $\mathbb{CP}^2$, with $[a_1 : a_2 : a_3]$ being the homogeneous coordinates of $\mathbb{CP}^2$.

Alternatively, the loop-by-loop approach for the maximal cut gives

$$E^{\text{cut}} : v^2 - \left[u^2 + 2(y_1 + y_2)u + (y_1 - y_2)^2\right]\left[u^2 + 2(1-x)u + (1+x)^2\right] = 0. \quad (13.303)$$

As in the equal mass case the two elliptic curves E^{Feynman} and E^{cut} are not isomorphic, but only isogenic. We may work with either of the two curves. In the following we will use E^{cut}.

In the next step we would like to change the kinematic variables from (x, y_1, y_2) to the standard coordinates (τ, z_1, z_2) on $\mathcal{M}_{1,3}$. This raises the question: How to express the new coordinates in terms of the old coordinates and vice versa? For τ the answer is straightforward: τ is again the ratio of the two periods

$$\tau = \frac{\psi_2}{\psi_1}, \quad (13.304)$$

and ψ_1 and ψ_2 are functions of x, y_1 and y_2, given by Eq. (13.44).

Also for z_1 and z_2 there is a simple geometric interpretation: In the Feynman parameter representation there are two geometric objects of interest: the domain of integration Δ (the simplex $a_1, a_2, a_3 \geq 0$, $a_1 + a_2 + a_3 \leq 1$) and the elliptic curve E^{Feynman} (the zero set of the second graph polynomial). The two objects E^{Feynman} and Δ intersect at three points, as shown in Fig. 13.10. The images of these three points in $\mathbb{C}/\Lambda^{\text{Feynman}}$ are

$$0, \; z_1^{\text{Feynman}}, \; z_2^{\text{Feynman}}, \quad (13.305)$$

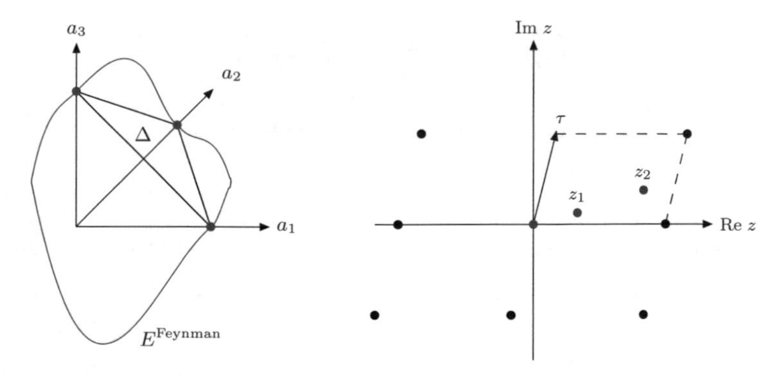

Fig. 13.10 E^{Feynman} and Δ intersect at three points, the images of these three points in $\mathbb{C}/\Lambda$ are $0, z_1, z_2$

where we used a translation transformation to fix one point at 0. The elliptic curves E^{Feynman} and E^{cut} are related by $\tau = \tau^{\text{cut}} = \frac{1}{2}\tau^{\text{Feynman}}$ and assuming that the points z_1^{Feynman} and z_2^{Feynman} are inside the fundamental parallelogram of E^{cut} we have

$$z_i = z_i^{\text{cut}} = z_i^{\text{Feynman}}, \quad i \in \{1, 2\}. \tag{13.306}$$

Working out the details we find

$$z_i = \frac{F(u_i, k)}{2K(k)}, \quad i \in \{1, 2\}, \tag{13.307}$$

where $K(x)$ denotes the complete elliptic integral of the first kind, $F(z, x)$ denotes the incomplete elliptic integral of the first kind and k denotes the modulus of the elliptic curve E^{cut} as defined by Eq. (13.43). The variables u_1 and u_2 are given by

$$\begin{aligned}
u_1 &= \frac{\sqrt{4y_1y_2 - 4x(y_1 + y_2) - (1 + x - y_1 - y_2)^2 + 8\sqrt{-xy_1y_2}}}{2\left(\sqrt{-xy_1} + \sqrt{y_2}\right)}, \\
u_2 &= \frac{\sqrt{4y_1y_2 - 4x(y_1 + y_2) - (1 + x - y_1 - y_2)^2 + 8\sqrt{-xy_1y_2}}}{2\left(\sqrt{-xy_2} + \sqrt{y_1}\right)}.
\end{aligned} \tag{13.308}$$

Equations (13.304) and (13.307) define the new coordinates (τ, z_1, z_2) as functions of the old coordinates (x, y_1, y_2). We also need the inverse relation, which gives us (x, y_1, y_2) as functions of (τ, z_1, z_2). One finds

$$\begin{aligned}
x &= \frac{(1-\kappa_1)(1-\kappa_2)\kappa_1\kappa_2\lambda^2}{(1-\lambda\kappa_1)(1-\lambda\kappa_2)}, \\
y_1 &= \frac{\kappa_1(1-\kappa_1)}{(1-\lambda\kappa_1)(\kappa_1-\kappa_2)^2(1-\kappa_1-\kappa_2+\lambda\kappa_1\kappa_2)^2} R, \\
y_2 &= \frac{\kappa_2(1-\kappa_2)}{(1-\lambda\kappa_2)(\kappa_1-\kappa_2)^2(1-\kappa_1-\kappa_2+\lambda\kappa_1\kappa_2)^2} R,
\end{aligned} \tag{13.309}$$

with

$$\begin{aligned}
R &= \left(1+\kappa_1^3\kappa_2^3\lambda^3\right)(\kappa_1+\kappa_2) - \left(1+\kappa_1^2\kappa_2^2\lambda^2\right)\left(\kappa_1^2+\kappa_2^2\right)(1+\lambda) - 8\left(1+\kappa_1^2\kappa_2^2\lambda^2\right)\kappa_1\kappa_2 \\
&+ (1+\lambda\kappa_1\kappa_2)(\kappa_1+\kappa_2)^3\lambda + 3(1+\lambda\kappa_1\kappa_2)\kappa_1\kappa_2(\kappa_1+\kappa_2)\lambda \\
&+ 8(1+\lambda\kappa_1\kappa_2)\kappa_1\kappa_2(\kappa_1+\kappa_2) + 4\kappa_1^2\kappa_2^2\lambda(1-\lambda) - 8\kappa_1\kappa_2(\kappa_1+\kappa_2)^2\lambda - 8\kappa_1^2\kappa_2^2 \\
&- 2\left(1-2\kappa_1+\lambda\kappa_1^2\right)\left(1-2\kappa_2+\lambda\kappa_2^2\right)\sqrt{\kappa_1\kappa_2(1-\kappa_1)(1-\kappa_2)(1-\lambda\kappa_1)(1-\lambda\kappa_2)}.
\end{aligned} \tag{13.310}$$

and

$$\lambda = \frac{\theta_2^4(0,q)}{\theta_3^4(0,q)},$$
$$\kappa_1 = \frac{\theta_3^2(0,q)\,\theta_1^2\left(\frac{\pi z_1}{2},q\right)}{\theta_2^2(0,q)\,\theta_4^2\left(\frac{\pi z_1}{2},q\right)},$$
$$\kappa_2 = \frac{\theta_3^2(0,q)\,\theta_1^2\left(\frac{\pi z_2}{2},q\right)}{\theta_2^2(0,q)\,\theta_4^2\left(\frac{\pi z_2}{2},q\right)}. \qquad (13.311)$$

Equations (13.309) and (13.311) together with $q = \exp(i\pi\tau)$ allow us to express (x, y_1, y_2) in terms of (τ, z_1, z_2).

The system of differential equations can again be transformed into an ε-form by a redefinition of the master integrals and a change of coordinates from (x, y_1, y_2) to (τ, z_1, z_2) [360, 374]. Let's look at the fibre transformation: We seek a matrix U relating the new master integrals $\vec{J}$ to the old master integrals $\vec{I}$

$$\vec{J} = U\vec{I}, \qquad (13.312)$$

such that in the transformed differential equation the dimensional regularisation parameter ε appears only as a prefactor. We construct U in two steps:

$$U = U_2 U_1, \qquad (13.313)$$

Let us set

$$\vec{J}_1 \;=\; U_1\vec{I},\;\; d\vec{J}_1 \;=\; \hat{A}_1\vec{J}_1. \qquad (13.314)$$

The entries of U_1 are constructed such that $\hat{A}_1$ is linear in ε and the ε^0-part is strictly lower triangular, i.e.

$$\hat{A}_1 = \hat{A}_1^{(0)} + \varepsilon\hat{A}_1^{(1)}, \qquad (13.315)$$

where $\hat{A}_1^{(0)}$ and $\hat{A}_1^{(1)}$ are independent of ε and $\hat{A}_1^{(0)}$ is strictly lower triangular. This can be done with a transformation, where the entries of U_1 are rational functions of ε, x, y_1, y_2, ψ_1 and $\partial_x\psi_1$ (i.e. compared to the full transformation U the entries of U_1 do not involve incomplete elliptic integrals. They only involve complete elliptic integrals related to ψ_1 and $\partial_x\psi_1$). The entries of U_1 are determined from an ansatz and with the help of the methods of Sect. 7.1. Explicitly, U_1 is given by setting $F_{54} = F_{64} = F_{74} = 0$ in the formula (13.322) below.

In a second step U_2 is constructed. U_2 eliminates the non-zero entries of $\hat{A}_1^{(0)}$. As $\hat{A}_1^{(0)}$ is strictly lower triangular, this can be done systematically by integration and will lead to incomplete elliptic integrals. In a final clean-up (and after the change of coordinates on the base manifold) we trade the incomplete elliptic integrals for $d\ln(y_1)/d\tau$ and $d\ln(y_2)/d\tau$.

In order to present the explicit formulae, we first introduce a few abbreviations: We introduce the monomial symmetric polynomials $M_{\lambda_1\lambda_2\lambda_3}(a_1, a_2, a_3)$ in the variables a_1, a_2 and a_3. These are defined by

$$M_{\lambda_1\lambda_2\lambda_3}(a_1, a_2, a_3) = \sum_{\sigma} (a_1)^{\sigma(\lambda_1)} (a_2)^{\sigma(\lambda_2)} (a_3)^{\sigma(\lambda_3)}, \qquad (13.316)$$

where the sum is over all distinct permutations σ of $(\lambda_1, \lambda_2, \lambda_3)$. A few examples are

$$\begin{aligned} M_{100}(a_1, a_2, a_3) &= a_1 + a_2 + a_3, \\ M_{111}(a_1, a_2, a_3) &= a_1 a_2 a_3, \\ M_{210}(a_1, a_2, a_3) &= a_1^2 a_2 + a_2^2 a_3 + a_3^2 a_1 + a_2^2 a_1 + a_3^2 a_2 + a_1^2 a_3. \end{aligned} \qquad (13.317)$$

As an abbreviation we then set

$$M_{\lambda_1\lambda_2\lambda_3} = M_{\lambda_1\lambda_2\lambda_3}(y_1, y_2, 1). \qquad (13.318)$$

As an example we have

$$M_{110} = M_{110}(y_1, y_2, 1) = y_1 y_2 + y_1 + y_2. \qquad (13.319)$$

In addition, we introduce the abbreviation

$$\Delta = 2M_{110} - M_{200} = 2y_1 y_2 + 2y_1 + 2y_2 - y_1^2 - y_2^2 - 1. \qquad (13.320)$$

We further set

$$W_x = \psi_1 \frac{d}{dx}\psi_2 - \psi_2 \frac{d}{dx}\psi_1. \qquad (13.321)$$

Let us now present the basis $\vec{J}$:

$$
\begin{aligned}
J_1 &= \varepsilon^2 I_{101}, \qquad\qquad (13.322)\\
J_2 &= \varepsilon^2 I_{011},\\
J_3 &= \varepsilon^2 I_{110},\\
J_4 &= \varepsilon^2 \frac{\pi}{\psi_1} I_{111},\\
J_5 &= \varepsilon \left[(y_1 + y_2 - 2) I_{111} + (3 - x - y_1 - 3y_2) y_1 I_{211} + (3 - x - 3y_1 - y_2) y_2 I_{121} + 2(1+x) I_{112} \right]\\
&+ \frac{2\varepsilon^2}{(3x^2 + 2M_{100}x + \Delta)} \left[7(y_1 + y_2 - 2) x^2 + 2\left(3y_1^2 + 3y_2^2 - 6 + y_1 + y_2 - 2y_1 y_2\right) x \right.\\
&\left. + (y_1 + y_2 - 2)\Delta \right] I_{111} + F_{54} J_4,\\
J_6 &= \varepsilon \left[(y_1 - y_2) I_{111} - (1 + x + y_1 - y_2) y_1 I_{211} + (1 + x - y_1 + y_2) y_2 I_{121} - 2(y_1 - y_2) I_{112} \right]\\
&+ \frac{2\varepsilon^2 (y_1 - y_2)}{(3x^2 + 2M_{100}x + \Delta)} \left[7x^2 + 2(3y_1 + 3y_2 - 1) x + \Delta \right] I_{111} + F_{64} J_4,\\
J_7 &= \frac{1}{\varepsilon} \frac{\psi_1^2}{2\pi i W_x} \frac{d}{dx} J_4 + \frac{\varepsilon^2}{8} \frac{1}{(3x^2 + 2M_{100}x + \Delta)^2} \left[9x^6 + 22M_{100}x^5 + (50M_{110} - M_{200}) x^4 \right.\\
&- (44M_{300} - 76M_{210} + 216M_{111}) x^3 - (41M_{400} - 84M_{310} + 86M_{220} + 52M_{211}) x^2\\
&\left. + 2\Delta (5M_{300} - 5M_{210} + 2M_{111}) x - \Delta^3 \right] \frac{\psi_1}{\pi} I_{111} - \frac{1}{8} F_{64} J_6 - \frac{1}{24} F_{54} J_5 + F_{74} J_4.
\end{aligned}
$$

The three functions F_{54}, F_{64}, F_{74}, appearing in the definition of J_5, J_6 and J_7 are given by

$$
\begin{aligned}
F_{54} &=\\
&\frac{6i}{(3x^2 + 2M_{100}x + \Delta)\psi_1} \left[(1 + x + y_1 - y_2) \frac{1}{y_1} \frac{dy_1}{d\tau} + (1 + x - y_1 + y_2) \frac{1}{y_2} \frac{dy_2}{d\tau} \right],\\
F_{64} &=\\
&\frac{2i}{(3x^2 + 2M_{100}x + \Delta)\psi_1} \left[(3y_1 + y_2 - 1 + 3x) \frac{1}{y_1} \frac{dy_1}{d\tau} - (y_1 + 3y_2 - 1 + 3x) \frac{1}{y_2} \frac{dy_2}{d\tau} \right],\\
F_{74} &=\\
&- \frac{1}{(3x^2 + 2M_{100}x + \Delta)^2 \psi_1^2} \left[\left(3y_1^2 + y_2^2 + 1 - 2y_2 + 6y_1 x + 3x^2\right) \left(\frac{1}{y_1} \frac{dy_1}{d\tau}\right)^2 \right.\\
&- \left(3y_1^2 + 3y_2^2 - 1 + 2y_1 y_2 - 2y_1 - 2y_2 + 6(y_1 + y_2 - 1) x + 3x^2\right) \left(\frac{1}{y_1} \frac{dy_1}{d\tau}\right) \left(\frac{1}{y_2} \frac{dy_2}{d\tau}\right)\\
&\left. + \left(y_1^2 + 3y_2^2 + 1 - 2y_1 + 6y_2 x + 3x^2\right) \left(\frac{1}{y_2} \frac{dy_2}{d\tau}\right)^2 \right]. \qquad (13.323)
\end{aligned}
$$

In this basis the differential equation reads

$$
(d + A) J = 0, \qquad (13.324)
$$

with

$$A = \varepsilon \begin{pmatrix} a_{11} & 0 & 0 & 0 & 0 & 0 & 0 \\ 0 & a_{22} & 0 & 0 & 0 & 0 & 0 \\ 0 & 0 & a_{33} & 0 & 0 & 0 & 0 \\ 0 & 0 & 0 & a_{44} & a_{45} & a_{46} & a_{47} \\ a_{51} & a_{52} & a_{53} & a_{54} & a_{55} & a_{56} & a_{57} \\ a_{61} & a_{62} & a_{63} & a_{64} & a_{65} & a_{66} & a_{67} \\ a_{71} & a_{72} & a_{73} & a_{74} & a_{75} & a_{76} & a_{77} \end{pmatrix}. \tag{13.325}$$

In order to present the entries of A in a compact form we introduce $z_3 = -z_1 - z_2$ and a constant $\beta = 1$. We define (for arbitrary β)

$$\begin{aligned} \Omega_k(z, \beta, \tau) &= \frac{1}{2}\omega_k(z, \tau) + \frac{1}{4}\omega_k(z - \beta, \tau) + \frac{1}{4}\omega_k(z + \beta, \tau) \\ &\quad -2(k-1)\left[\omega_k\left(\frac{z}{2}, \tau\right) + \frac{1}{2}\omega_k\left(\frac{z-\beta}{2}, \tau\right) + \frac{1}{2}\omega_k\left(\frac{z+\beta}{2}, \tau\right)\right], \end{aligned} \tag{13.326}$$

where $\omega_k(z, \tau)$ denotes the one-form defined in Eq. (13.297). For $\beta = 1$ we have

$$\Omega_k(z, 1, \tau) = \omega_k(z, \tau) - 2(k-1)\,\omega_k(z, 2\tau). \tag{13.327}$$

We will encounter $\Omega_2(z, \beta, \tau)$ and $\Omega_3(z, \beta, \tau)$. For the entries a_{ij} we also need two differential forms $\eta_2(\tau)$ and $\eta_4(\tau)$, which depend on τ, but not on the z_i's. These are defined by

$$\eta_2(\tau) = \left[e_2(\tau) - 2e_2(2\tau)\right]\frac{d\tau}{2\pi i}, \quad \eta_4(\tau) = \frac{1}{(2\pi i)^2}e_4(\tau)\frac{d\tau}{2\pi i}, \tag{13.328}$$

where $e_k(\tau)$ denotes the standard Eisenstein series, defined in Eq. (13.126). We have

$$e_2(\tau) - 2e_2(2\tau) \in \mathcal{M}_2(\Gamma_0(2)), \quad e_4(\tau) \in \mathcal{M}_4(\mathrm{SL}_2(\mathbb{Z})). \tag{13.329}$$

For the entries of A we have the following relations

$$\begin{aligned} a_{45} &= \frac{1}{24}a_{57}, & a_{46} &= \frac{1}{8}a_{67}, & a_{33} &= a_{11} + a_{22}, \\ a_{53} &= a_{11} + a_{22} - a_{51} - a_{52}, & a_{56} &= 3a_{65}, & a_{77} &= a_{44}, \\ a_{61} &= 2a_{11} - a_{51}, & a_{62} &= -2a_{11} + a_{51}, & a_{63} &= a_{11} - a_{22}, \\ a_{75} &= \frac{1}{24}a_{54}, & a_{76} &= \frac{1}{8}a_{64}, \end{aligned} \tag{13.330}$$

and the following symmetries

$$\begin{aligned} a_{22}(z_1, z_2, z_3) &= a_{11}(z_2, z_1, z_3), & a_{52}(z_1, z_2, z_3) &= a_{51}(z_2, z_1, z_3), \\ a_{72}(z_1, z_2, z_3) &= a_{71}(z_2, z_1, z_3), & a_{73}(z_1, z_2, z_3) &= a_{71}(z_1, z_3, z_2). \end{aligned} \tag{13.331}$$

Thus we need to specify only a few entries. We group them by modular weight.

Modular weight 0:

$$a_{4,7} = \omega_0 (z, \tau) \;=\; -2\pi i d\tau. \tag{13.332}$$

Modular weight 1:

$$\begin{aligned} a_{5,7} &= 6i \left[\omega_1 (z_1, \tau) + \omega_1 (z_2, \tau)\right], \\ a_{6,7} &= 2i \left[\omega_1 (z_1, \tau) - \omega_1 (z_2, \tau)\right]. \end{aligned} \tag{13.333}$$

Note that $\omega_1(z, \tau) = 2\pi i dz$ is independent of τ.

Modular weight 2:

$$\begin{aligned} a_{1,1} &= -2 \left[\Omega_2 (z_1, \beta, \tau) - \Omega_2 (z_3, \beta, \tau)\right], \\ a_{4,4} &= \omega_2 (z_1, \tau) + \omega_2 (z_2, \tau) + \omega_2 (z_3, \tau) - \Omega_2 (z_1, \beta, \tau) - \Omega_2 (z_2, \beta, \tau) + 3\Omega_2 (z_3, \beta, \tau) \\ &\quad -6\eta_2 (\tau), \\ a_{5,1} &= -2 \left[\Omega_2 (z_1, \beta, \tau) - \Omega_2 (z_2, \beta, \tau) - 2\Omega_2 (z_3, \beta, \tau)\right], \\ a_{5,5} &= -3\omega_2 (z_3, \tau) - \Omega_2 (z_1, \beta, \tau) - \Omega_2 (z_2, \beta, \tau) + 3\Omega_2 (z_3, \beta, \tau) - 6\eta_2 (\tau), \\ a_{6,5} &= -\omega_2 (z_1, \tau) + \omega_2 (z_2, \tau), \\ a_{6,6} &= -2\omega_2 (z_1, \tau) - 2\omega_2 (z_2, \tau) + \omega_2 (z_3, \tau) - \Omega_2 (z_1, \beta, \tau) - \Omega_2 (z_2, \beta, \tau) + 3\Omega_2 (z_3, \beta, \tau) \\ &\quad -6\eta_2 (\tau), \end{aligned} \tag{13.334}$$

Modular weight 3:

$$\begin{aligned} a_{5,4} &= 12i \left[\omega_3 (z_1, \tau) + \omega_3 (z_2, \tau) - 2\omega_3 (z_3, \tau)\right], \\ a_{6,4} &= 12i \left[\omega_3 (z_1, \tau) - \omega_3 (z_2, \tau)\right], \\ a_{7,1} &= i \left[\Omega_3 (z_1, \beta, \tau) - \Omega_3 (z_2, \beta, \tau) + \Omega_3 (z_3, \beta, \tau)\right]. \end{aligned} \tag{13.335}$$

Modular weight 4:

$$a_{7,4} = 12 \left[\omega_4 (z_1, \tau) + \omega_4 (z_2, \tau) + \omega_4 (z_3, \tau) - 6\eta_4 (\tau)\right]. \tag{13.336}$$

We have managed to transform the differential equation for the family of the unequal mass sunrise integral (e.g. an elliptic Feynman integral depending on three kinematic variables) into an ε-form. This differential equation can be solved in terms of the iterated integrals discussed in Sect. 13.4.

Let us close this section with a discussion of the behaviour of the system under a modular transformation

$$\gamma \;=\; \begin{pmatrix} a & b \\ c & d \end{pmatrix} \in \mathrm{SL}_2(\mathbb{Z}). \tag{13.337}$$

The coordinate transform as

$$z'_1 = \frac{z_1}{c\tau + d}, \qquad z'_2 = \frac{z_2}{c\tau + d}, \qquad \tau' = \frac{a\tau + b}{c\tau + d}. \tag{13.338}$$

The constant $\beta = 1$ transforms as

$$\beta' = \frac{\beta}{c\tau + d}, \tag{13.339}$$

so in general we will have $\beta' \neq 1$. We may view β as being a further marked point in a higher dimensional space $\mathcal{M}_{1,n'}$ with $n' > n$. We set again $z'_3 = -z'_1 - z'_2$. We also need to redefine the master integrals. We set

$$J' = U_\gamma \, J, \tag{13.340}$$

where U_γ is given by

$$U_\gamma = \begin{pmatrix} 1 & 0 & 0 & 0 & 0 & 0 & 0 \\ 0 & 1 & 0 & 0 & 0 & 0 & 0 \\ 0 & 0 & 1 & 0 & 0 & 0 & 0 \\ 0 & 0 & 0 & \frac{1}{c\tau+d} & 0 & 0 & 0 \\ 0 & 0 & 0 & \frac{6ic(z_1+z_2)}{c\tau+d} & 1 & 0 & 0 \\ 0 & 0 & 0 & \frac{2ic(z_1-z_2)}{c\tau+d} & 0 & 1 & 0 \\ 0 & 0 & 0 & -\frac{c}{2\pi i\varepsilon} + \frac{c^2\left(z_1^2+z_1z_2+z_2^2\right)}{c\tau+d} & -\frac{ic(z_1+z_2)}{4} & -\frac{ic(z_1-z_2)}{4} & c\tau + d \end{pmatrix}. \tag{13.341}$$

The transformation matrix U is not too difficult to construct, if one starts from the assumption that the first elliptic master integral (i.e. J_4) should be rescaled as

$$J'_4 = \frac{\omega_1}{\omega'_1} J_4 = \frac{1}{c\tau + d} J_4. \tag{13.342}$$

Under this combined transformation the differential equation for the transformed system reads then

$$\left(d + A'\right) J' = 0, \tag{13.343}$$

where A' is obtained (with one exception) from A by replacing all unprimed variables with primed variables. For example $a'_{7,1}$ is given by

$$a'_{7,1} = i \left[\Omega_3\left(z'_1, \beta', \tau'\right) - \Omega_3\left(z'_2, \beta', \tau'\right) + \Omega_3\left(z'_3, \beta', \tau'\right) \right]. \tag{13.344}$$

The only exception is $\eta_2(\tau)$. For $\gamma \in \Gamma_0(2)$ the differential one-form $\eta_2(\tau)$ transforms into $\eta_2(\tau')$. For a general $\gamma \in \mathrm{SL}_2(\mathbb{Z})$ let us set $b_2(\tau) = e_2(\tau) - 2e_2(2\tau)$. Then $\eta_2(\tau)$ is replaced by

$$(b_2|_2\gamma^{-1})(\tau') \frac{d\tau'}{2\pi i}. \tag{13.345}$$

$(b_2|_2\gamma^{-1})(\tau')$ is again a modular form for $\Gamma(2)$, but not necessarily identical to $b_2(\tau')$.

It remains to work out $(b_2|_2\gamma^{-1})(\tau')$. To this aim we first express $b_2(\tau)$ in terms of Eisenstein series for $\Gamma(2)$. We find

$$b_2(\tau) = 4(2\pi i)^2 h_{2,2,0,1}(\tau), \tag{13.346}$$

where the Eisenstein series $h_{k,N,r,s}(\tau)$ have been defined in Eq. (13.151). The transformation law for $b_2(\tau)$ follows then from the transformation law for $h_{k,N,r,s}(\tau)$ given in Eq. (13.161). We obtain

$$(b_2|_2\gamma^{-1})(\tau') = 4(2\pi i)^2 h_{2,2,b \bmod 2,d \bmod 2}\left(\tau'\right), \qquad \gamma^{-1} = \begin{pmatrix} d & -b \\ -c & a \end{pmatrix}. \tag{13.347}$$

Further examples of elliptic Feynman integrals can be found in [375–384].

Chapter 14
Motives and Mixed Hodge Structures

In most chapters of this book we followed the pattern to show how known facts in mathematics may help a physicist (to compute Feynman integrals). In this chapter we reverse the pattern. We would like to illustrate how known facts in physics (Feynman integrals we know how to calculate) may help a mathematician. This concerns the theory of motives. In part, such a theory is conjectural (hence the interest of mathematicians). Feynman integrals provide non-trivial examples such a theory should contain. In this chapter we introduce the mathematical language and the main ideas behind motives.

Motives were introduced by Alexander Grothendieck to unify cohomology theories. In order to see at least two different cohomology theories we discuss in Sect. 14.1 Betti cohomology and de Rham cohomology. Unification will require a more abstract language, and in Sect. 14.2 we introduce categories as the appropriate language. In Sect. 14.3 we try to give a glimpse on what a theory of motives should be about. As remarked earlier, this is still partly a conjectural theory and Sects. 14.2 and 14.3 mainly serve to acquaint the reader with the main ideas and the language of this field.

We can be more concrete if we look at the Hodge realisation of a motive (also called an H-motive). Here we focus on the interplay between Betti cohomology and de Rham cohomology. The theory is based on Hodge structures, which we discuss in Sect. 14.4. Finally, in Sect. 14.5 we discuss examples from Feynman integrals. We will also see how this formalism offers us an alternative way to compute the differential equation for a Feynman integral.

Readers interested in the topics of this chapter would almost certainly consult additional text books. Introductory texts for categories are the review article by Deligne and Milne [385] and the books by Mac Lane [386], Leinster [387] and Etingof, Gelaki, Nikshych and Ostrik [388]. More information on motives can be found in the book by André [389]. Hodge structures are treated in the books by Voisin [390] and Peters and Steenbrink [391]. The proceedings of a summer school on Hodge structures [392] give also very useful information. Review articles related to the content of this chapter are [284, 393, 394].

S. Weinzierl, *Feynman Integrals*, UNITEXT for Physics,
https://doi.org/10.1007/978-3-030-99558-4_14

14.1 Cohomology

In this section we review Betti cohomology and de Rham cohomology. We also discuss relative Betti cohomology and relative de Rham cohomology, as well as algebraic de Rham cohomology.

14.1.1 Betti Cohomology

Betti cohomology (also known as **singular cohomology**) is based on a triangulation of a topological space by simplices. We already encountered simplices in Sect. 9.2.1. Let $a_0, a_1, \dots, a_n$ be $(n+1)$ linear independent vectors in $\mathbb{R}^d$ (with $d \geq n+1$). The n-simplex $\sigma_n = \langle a_0, a_1, \dots, a_n \rangle$ is the polytope

$$\sigma_n = \langle a_0, a_1, \dots, a_n \rangle = \left\{ \alpha_0 a_0 + \alpha_1 a_1 + \dots + \alpha_n a_n | \alpha_j \geq 0, \sum_{j=0}^{n} \alpha_j = 1 \right\}. \quad (14.1)$$

For example a 0-simplex is a point, a 1-simplex is a line interval, a 2-simplex is a triangle and a 3-simplex is a tetrahedron. The a_i are called the vertices of the simplex σ_n. Faces and facets (we recall that a facet is a face of codimension one) of simplices are defined as for polytopes (see Sect. 9.2.1). They correspond to k-dimensional simplices defined by a subset of $(k+1)$ vertices of σ_n. We denote the i-th facet of the simplex $\sigma_n = \langle a_0, a_1, \dots, a_n \rangle$ by

$$\sigma_{n-1}^i = \langle a_0, \dots, \hat{a}_i, \dots, a_n \rangle, \quad (14.2)$$

where the hat denotes that the corresponding vertex is omitted. The **incident number** $[\sigma_n : \sigma_{n-1}^i]$ is defined as

$$[\sigma_n : \sigma_{n-1}^i] = \begin{cases} +1 \text{ if } (i, 0, \dots, i-1, i+1, \dots, n) \text{ is an even permutation of } (0, 1, \dots, n), \\ -1 \text{ if } (i, 0, \dots, i-1, i+1, \dots, n) \text{ is an odd permutation of } (0, 1, \dots, n). \end{cases} \quad (14.3)$$

A **simplicial complex** B is a set of a finite number of simplexes in $\mathbb{R}^d$ satisfying the following two properties:

1. an arbitrary face of a simplex σ of B belongs to B,
2. if σ and σ' are two simplexes of B, the intersection $\sigma \cap \sigma'$ is either empty or a face of σ and σ'.

The union of all simplices of B defines a subset of $\mathbb{R}^d$, which we denote by $|B|$. Let X be a topological space. If there exists a simplicial complex B and a homeomorphism $t : |B| \to X$ then X is said to be triangulable and the pair (B, t) is called a **triangulation** of X.

For a simplex σ_n we define the boundary of the simplex by

$$\partial \sigma_n = \sum_{i=0}^{n} [\sigma_n : \sigma_{n-1}^i] \sigma_{n-1}^i. \tag{14.4}$$

In other words, the boundary of a simplex is a linear combination of its facets with the coefficients being given by the incident numbers.

Let B be a simplicial complex. An **integral k-chain** of B is a function mapping the set of k-simplexes $\{\sigma_{k,1}, \sigma_{k,2}, \dots\}$ into the integers $\mathbb{Z}$, such that $\sigma_{k,j} \to n_j$. An integral k-chain is denoted as

$$c_k = \sum_j n_j \, \sigma_{k,j}. \tag{14.5}$$

The integral k-chains form a group with respect to addition. We denote this group by $C_k(B)$. The **boundary of a k-chain** is given by

$$\partial c_k = \sum_j \sum_{i=0}^{k} n_j \, [\sigma_{k,j} : \sigma_{k-1,j}^i] \sigma_{k-1,j}^i, \tag{14.6}$$

where $\sigma_{k-1,j}^i$ denotes the i-th facet of the k-dimensional simplex $\sigma_{k,j}$.

A **k-cycle** z_k is defined by

$$\partial z_k = 0. \tag{14.7}$$

A k-chain b_k is called a **boundary** if there exists a $(k+1)$-chain c_{k+1} such that

$$\partial c_{k+1} = b_k. \tag{14.8}$$

Due to the alternating signs originating from the incidence numbers we have $\partial\partial c_k = 0$ for any chain, so each boundary $b_k = \partial c_{k+1}$ is also a cycle (since $\partial b_k = \partial\partial c_{k+1} = 0$). The sets of all k-chains, k-cycles and k-boundaries form Abelian groups, denoted by $C_k(B)$, $Z_k(B)$ and $B_k(B)$, respectively. The boundary operator $\partial : C_k(B) \to C_{k-1}(B)$, which maps k-chains into $(k-1)$-chains, is also a group homomorphism. The group of boundaries $B_k(B)$ forms a subgroup of the group of cycles $Z_k(B)$. Since all groups are Abelian, the factor group

$$H_k(B) = \frac{Z_k(B)}{B_k(B)} \tag{14.9}$$

is well defined and is called the **k-th homology group** $H_k(B)$ of the complex B.

A sequence of Abelian groups $\dots, C_0, C_1, C_2, \dots$ together with homomorphisms $\partial_k : C_k \to C_{k-1}$ such that $\partial_{k-1} \circ \partial_k = 0$ in an example of a **chain complex**. A chain complex is denoted as

$$\dots \xrightarrow{\partial_3} C_2 \xrightarrow{\partial_2} C_1 \xrightarrow{\partial_1} C_0 \xrightarrow{\partial_0} \dots . \tag{14.10}$$

The chain groups $C_k(B)$ form a chain complex.

The cohomology groups are defined as follows: Given an Abelian group K and the k-chain groups $C_k(B)$ of a simplicial complex B we consider the k-**cochain group**

$$C^k(B, K) = \mathrm{Hom}(C_k(B), K). \tag{14.11}$$

The group K is called the **coefficient group**. A typical choice would be $K = \mathbb{Z}$. We define the **coboundary operator** $d : C^k(B, K) \to C^{k+1}(B, K)$ by

$$\left(df^k\right)(c_{k+1}) = f^k\left(\partial c_{k+1}\right) \tag{14.12}$$

for all $f^k \in C^k(B, K)$ and $c_{k+1} \in C_{k+1}(B)$. **Cocycles** and **coboundaries** are then defined in the usual way: A k-cochain f^k is called a k-cocylce if $df^k = 0$, it is called a k-coboundary if there is a $(k-1)$-cochain g^{k-1} such that $f^k = dg^{k-1}$. The k-cohomology group $H^k(B, K)$ is defined as

$$H^k(B, K) = \frac{Z^k(B, K)}{B^k(B, K)}, \tag{14.13}$$

where $Z^k(B, K)$ is the group of k-cocycles and $B^k(B, K)$ is the group of k-coboundaries.

Let us now define Betti homology/cohomology for a triangulable topological space. For a triangulable topological space X one first chooses a triangulation (B, t) and defines the Betti homology/cohomology (or singular homology/cohomology) by Eq. (14.9) and Eq. (14.13), respectively. One can show that this is independent of the chosen triangulation. We denote the k-th cohomology group by

$$H^k_B(X) \tag{14.14}$$

The subscript B stands for "Betti", not the simplicial complex B. As we mentioned above, $H^k_B(X)$ is independent of the chosen triangulation (B, t). If we want to emphasise the coefficient group K, we write $H^k_B(X, K)$.

We may also define **relative homology groups** and **relative cohomology groups**: Whereas a k-cycle of a complex B is a chain with no boundary at all, we can relax this condition by requiring that the boundary lies only within some specified subcomplex A. If A is a subcomplex of B, we define the relative chain group as

$$C_k(B, A) = \frac{C_k(B)}{C_k(A)}. \tag{14.15}$$

The relative cycle group $Z_k(B, A)$ and the relative boundary group $B_k(B, A)$ are defined in a similar way: A chain $z_k \in C_k(B)$ defines a cycle $[z_k] \in C_k(B, A)$ if

$$\partial z_k \in A. \tag{14.16}$$

A chain $b_k \in C_k(B)$ defines a boundary $[b_k] \in C_k(B, A)$ if there is a chain $c_{k+1} \in C_{k+1}(B)$ with

$$b_k - \partial c_{k+1} \in A. \tag{14.17}$$

It can be shown that $B_k(B, A)$ is a subgroup of $Z_k(B, A)$ and thus the relative homology group

$$H_k(B, A) = \frac{Z_k(B, A)}{B_k(B, A)} \tag{14.18}$$

is well defined.

Exercise 115 *Show that a relative boundary is a relative cycle.*

The relative cohomology groups are obtained from the relative homology groups in complete analogy to the way the cohomology groups are obtained from the homology groups, starting with the relative k-cochain group

$$C^k(B, A, K) = \mathrm{Hom}(C_k(B, A), K). \tag{14.19}$$

14.1.2 De Rham Cohomology

If X is a differentiable manifold, we may consider de Rham cohomology (as we did in Sect. 6.3.2): The k-th **de Rham cohomology group** $H^k_{\mathrm{dR}}(X)$ is the set of equivalence classes of closed k-forms modulo exact k-forms. The group law is the addition of k-forms.

Let us now look at the relation between de Rham cohomology and Betti cohomology: Let X be a differentiable manifold. To this manifold we can on the one hand associate the de Rham cohomology $H^k_{\mathrm{dR}}(X)$, as well as the Betti cohomology $H^k_{\mathrm{B}}(X)$. There is an isomorphism between the de Rham and Betti cohomology:

$$\begin{aligned} \mathrm{comparison} :\ & H^k_{\mathrm{dR}}(X) \otimes \mathbb{C} \to H^k_{\mathrm{B}}(X) \otimes \mathbb{C}, \\ & \omega \to \left(\gamma \to \int\limits_\gamma \omega \right). \end{aligned} \tag{14.20}$$

If we fix a basis for $H^k_{\mathrm{dR}}(X)$ and $H^k_{\mathrm{B}}(X)$ the isomorphism is given explicitly as follows: Let $n = \dim H^k_{\mathrm{dR}}(X) = \dim H^k_{\mathrm{B}}(X)$. Let us denote by $\omega_1, \dots, \omega_n$ a basis for $H^k_{\mathrm{dR}}(X)$ and by $\gamma_1, \dots, \gamma_n$ a basis for the Betti homology $H^{\mathrm{B}}_k(X)$. A basis for the Betti cohomology $H^k_{\mathrm{B}}(X)$ is then given by the duals $\gamma_1^*, \dots, \gamma_n^*$ and satisfies $\gamma_i^*(\gamma_j) = \delta_{ij}$. We then have

$$\omega_i = \sum_j p_{ij} \gamma_j^*, \quad p_{ij} = \int\limits_{\gamma_j} \omega_i. \tag{14.21}$$

The coefficients p_{ij} are called **periods** and the $n \times n$-matrix P with entries p_{ij} is called the **period matrix**.

If Y is a closed submanifold of X, we may consider **relative de Rham cohomology**. We first define $\Omega^k(X, Y)$ to be the space of differential k-forms on X, whose restriction to Y is zero. The relative de Rham cohomology group $H^k_{\mathrm{dR}}(X, Y)$ is then given by the equivalence classes of the closed forms in $\Omega^k(X, Y)$ modulo the exact ones.

Let's look at an example: We take $X = \mathbb{C}^* = \mathbb{C}\backslash\{0\}$ and $Y = \{1, 2\}$. For the non-relative cohomology we have

$$\dim H^1_{\mathrm{B}}(X) \;=\; 1, \;\; \dim H^1_{\mathrm{dR}}(X) \;=\; 1. \tag{14.22}$$

A basis for $H^{\mathrm{B}}_1(X)$ is given by an anti-clockwise circle γ_1 around $z = 0$, a basis for $H^1_{\mathrm{dR}}(X)$ is given by $\omega_1 = dz/z$. For the relative cohomology we have

$$\dim H^1_{\mathrm{B}}(X, Y) \;=\; 2, \;\; \dim H^1_{\mathrm{dR}}(X, Y) \;=\; 2. \tag{14.23}$$

A basis for $H^{\mathrm{B}}_1(X, Y)$ is now given by γ_1 as above and the line segment γ_2 from $z = 1$ to $z = 2$. The boundary of this line segments are the points $z = 1$ and $z = 2$, which are in Y. A basis for the relative de Rham cohomology $H^1_{\mathrm{dR}}(X, Y)$ is given by $\omega_1 = dz/z$ and $\omega_2 = dz$. Note that in the relative case dz is no longer an exact form, as the zero form (i.e. the function) $f(z) = z$ does not belong to $\Omega^0(X, Y)$. All functions from $\Omega^0(X, Y)$ are required to vanish on Y and $f(z)$ does not (for example $f(1) = 1 \neq 0$). As a side-remark, let us note that the restriction of a k-form to a $(k-1)$-dimensional submanifold always vanishes, there are simply not enough linear independent tangential vectors to contract into the k-form. Thus $\omega_1, \omega_2 \in \Omega^1(X, Y)$. There is again an isomorphism between the de Rham and Betti cohomology:

$$H^k_{\mathrm{dR}}(X, Y) \otimes \mathbb{C} \to H^k_{\mathrm{B}}(X, Y) \otimes \mathbb{C}. \tag{14.24}$$

The period matrix is given by

$$P = \begin{pmatrix} 2\pi i & \ln 2 \\ 0 & 1 \end{pmatrix}. \tag{14.25}$$

14.1.3 Algebraic de Rham Cohomology

On a differential manifold M we have the complex of differential forms $\Omega^\bullet(M)$. Let $U \subset M$ be an open subset. On U we may write any differential k-form as

$$\omega = \sum_{i_1 < \cdots < i_k} f_{i_1 \ldots i_k}(x)\, dx^{i_1} \wedge \cdots \wedge dx^{i_k}, \qquad f_{i_1 \ldots i_k} \in C^\infty(U, \mathbb{C}). \quad (14.26)$$

Let us stress that there is no point in singling out polynomial or rational functions: A polynomial or rational function in one coordinate system will in general not be a polynomial or rational function in another coordinate system.

However, the situation changes if we consider algebraic varieties. Algebraic varieties are defined by polynomial equations and we may single out the variables defining the variety. For an algebraic variety X we may consider regular functions on X, these are functions which we may write locally on an open set U as a rational function such that the denominator polynomial is nowhere vanishing on U. One denotes the regular functions on X by $\mathcal{O}(X)$. An **algebraic form** on X is given by

$$\omega = \sum_{i_1 < \cdots < i_k} f_{i_1 \ldots i_k}(x)\, dx^{i_1} \wedge \cdots \wedge dx^{i_k}, \qquad f_{i_1 \ldots i_k} \in \mathcal{O}(X). \quad (14.27)$$

We denote the complex of algebraic forms on X by $\Omega^\bullet_{\mathrm{alg}}(X)$. Differentiation is defined in the usual way.

If $\mathbb{F}$ is a subfield of $\mathbb{C}$ and X is defined over $\mathbb{F}$, we denote by X^{an} the analytification of X (see Sect. 10.4). This means that the coordinates of the points in X^{an} may be complex (and not just in $\overline{\mathbb{F}}$) and X^{an} is equipped with the standard topology (the one which is Hausdorff).

Algebraic de Rham cohomology is the cohomology obtained by restricting ourselves to the algebraic forms of Eq. (14.27). We denote algebraic de Rham cohomology groups by

$$H^k_{\mathrm{alg\ dR}}(X). \quad (14.28)$$

Any cohomology class in $H^k_{\mathrm{alg\ dR}}(X)$ defines a cohomology class in $H^k_{\mathrm{d}R}(X^{\mathrm{an}})$ and we trivially have

$$H^k_{\mathrm{alg\ dR}}(X) \subseteq H^k_{\mathrm{dR}}(X^{\mathrm{an}}). \quad (14.29)$$

Non-trivial is the following theorem [258].

Theorem 13

$$H^k_{\mathrm{alg\ dR}}(X) \otimes \mathbb{C} = H^k_{\mathrm{dR}}(X^{\mathrm{an}}). \quad (14.30)$$

This theorem states that every cohomology class of $H^k_{dR}(X^{an})$ has a representative as an algebraic cohomology class. It is therefore sufficient to consider just $H^k_{\text{alg dR}}(X)$. This allows us to work just with rational functions.

Let's look at an example. We consider

$$xy - 1 \in \mathbb{Q}[x, y] \,. \tag{14.31}$$

This defines a variety over $\mathbb{Q}$:

$$X = \left\{ (x, y) \in \overline{\mathbb{Q}}^2 \mid xy - 1 = 0 \right\}. \tag{14.32}$$

For x we may choose any algebraic number not equal to zero, y is then given by $y = 1/x$. Hence X is isomorphic to

$$X \cong \overline{\mathbb{Q}} \backslash \{0\}. \tag{14.33}$$

The analytification of X is given by

$$X^{an} \cong \mathbb{C} \backslash \{0\} \;=\; \mathbb{C}^*. \tag{14.34}$$

We have

$$\begin{aligned}
\Omega^0_{\text{alg}}(X) &= \mathbb{Q}[x, y] \,/\, \langle xy - 1 \rangle \;=\; \mathbb{Q}\left[x, \frac{1}{x}\right], \\
\Omega^1_{\text{alg}}(X) &= \mathbb{Q}\left[x, \frac{1}{x}\right] \cdot dx, \\
\Omega^2_{\text{alg}}(X) &= 0.
\end{aligned} \tag{14.35}$$

An algebraic one-form is a sum of terms $x^n \cdot dx$ with $n \in \mathbb{Z}$. The only term which is not exact is the one with $n = -1$. Hence, $H^1_{\text{alg dR}}(X)$ is generated by dx/x.

There is a generalisation of Theorem 13 to the relative case:

Theorem 14

$$H^k_{\text{alg dR}}(X, Y) \otimes \mathbb{C} = H^k_{\text{dR}}(X^{an}, Y^{an}) \,. \tag{14.36}$$

As Y is allowed to intersect itself, the definition of $H^k_{\text{alg dR}}(X, Y)$ is more involved. We illustrate here the construction of the relative algebraic de Rham cohomology groups for the case where X is a smooth affine algebraic variety and Y a simple normal crossing divisor. For the general case, where X is a smooth variety and Y a closed subvariety we refer to the book by Huber and Müller-Stach [395].

A codimension 1 closed subvariety $Y \subset X$ is called a **normal crossing divisor**, if for every point $x \in Y$ there is an open neighbourhood $U \subseteq X$ of x with local coordinates $x_1, \dots, x_n$ such that Y is locally given by

$$x_1 \cdot x_2 \cdot \dots \cdot x_k = 0, \qquad \text{for some } 1 \leq k \leq n. \tag{14.37}$$

Y is called a **simple normal crossing divisor** if in addition the irreducible components of Y are smooth. In other words, Y looks locally like a union of coordinate hyperplanes.

Let us assume that Y is a union of r irreducible components Y_i:

$$Y = Y_1 \cup Y_2 \cup \dots \cup Y_r. \tag{14.38}$$

Let $I \subseteq \{1, \dots, r\}$. We set

$$Y_I = \bigcap_{i \in I} Y_i, \quad Y^p = \begin{cases} X, & p = 0, \\ \bigsqcup_{|I|=p} Y_I, & p \geq 1. \end{cases} \tag{14.39}$$

Given a subset $I = \{i_0, \dots, i_p\} \subseteq \{1, \dots, r\}$ with $(p + 1)$ elements, we define I_l to be the subset with p elements, obtained from I by removing the l-th element:

$$I_l = \left\{ i_0, \dots, \hat{i}_l, \dots, i_p \right\}. \tag{14.40}$$

As Y_{I_l} is the intersection of p irreducible components and Y_I is the intersection of Y_{I_l} with Y_{i_l} we have

$$Y_I \subset Y_{I_l}. \tag{14.41}$$

With these definitions we may now look at the double complex $K^{p,q} = \Omega^q_{\text{alg}}(Y^p)$. In the double complex $K^{\bullet,\bullet}$ we have two differentials: The first one is defined by

$$\begin{aligned} d_{\text{vertical}} \,:\, K^{p,q} &\to K^{p,q+1}, \\ d_{\text{vertical}} &= (-1)^p \, d, \end{aligned} \tag{14.42}$$

where d is the ordinary exterior derivative. The alternating sign is required to turn $K^{p,q}$ into a complex. The second one is given by

$$\begin{aligned} d_{\text{horizontal}} \,:\, K^{p,q} &\to K^{p+1,q}, \\ d_{\text{horizontal}} &= \bigoplus_{|I|=p+1} \bigoplus_{l=0}^{p} (-1)^l \, r_{I_l I}, \end{aligned} \tag{14.43}$$

and

$$r_{I_l I} : \Omega^q_{\text{alg}} \left(Y_{I_l} \right) \to \Omega^q_{\text{alg}} \left(Y_I \right) \tag{14.44}$$

is the restriction map. In the next step one considers the total complex

$$\Omega^n_{\text{alg}}(X, Y) = \bigoplus_{p+q=n} K^{p,q} \tag{14.45}$$

together with the differential

$$d_{\text{total}} = d_{\text{vertical}} + d_{\text{horizontal}}. \tag{14.46}$$

The relative algebraic de Rham cohomology group $H^k_{\text{alg dR}}(X, Y)$ is then the k-th cohomology group of $\Omega^k_{\text{alg}}(X, Y)$ with respect to d_{total}.

We may visualise the double complex $K^{p,q}$ as follows:

$$\begin{array}{ccccccc}
\dots & & \dots & & \dots & & \\
d_{\text{vertical}}\uparrow & & d_{\text{vertical}}\uparrow & & d_{\text{vertical}}\uparrow & & \\
\Omega^2_{\text{alg}}(X) & \xrightarrow{d_{\text{horizontal}}} & \bigoplus\limits_i \Omega^2_{\text{alg}}(Y_i) & \xrightarrow{d_{\text{horizontal}}} & \bigoplus\limits_{i<j} \Omega^2_{\text{alg}}\left(Y_i \cap Y_j\right) & \xrightarrow{d_{\text{horizontal}}} & \dots \\
d_{\text{vertical}}\uparrow & & d_{\text{vertical}}\uparrow & & d_{\text{vertical}}\uparrow & & \\
\Omega^1_{\text{alg}}(X) & \xrightarrow{d_{\text{horizontal}}} & \bigoplus\limits_i \Omega^1_{\text{alg}}(Y_i) & \xrightarrow{d_{\text{horizontal}}} & \bigoplus\limits_{i<j} \Omega^1_{\text{alg}}\left(Y_i \cap Y_j\right) & \xrightarrow{d_{\text{horizontal}}} & \dots \\
d_{\text{vertical}}\uparrow & & d_{\text{vertical}}\uparrow & & d_{\text{vertical}}\uparrow & & \\
\Omega^0_{\text{alg}}(X) & \xrightarrow{d_{\text{horizontal}}} & \bigoplus\limits_i \Omega^0_{\text{alg}}(Y_i) & \xrightarrow{d_{\text{horizontal}}} & \bigoplus\limits_{i<j} \Omega^0_{\text{alg}}\left(Y_i \cap Y_j\right) & \xrightarrow{d_{\text{horizontal}}} & \dots
\end{array} \tag{14.47}$$

The total complex $\Omega^\bullet_{\text{alg}}(X, Y)$ is obtained by summing up the diagonals. For example

$$\Omega^2_{\text{alg}}(X, Y) = \Omega^2_{\text{alg}}(X) \oplus \left(\bigoplus_i \Omega^1_{\text{alg}}(Y_i) \right) \oplus \left(\bigoplus_{i<j} \Omega^0_{\text{alg}}\left(Y_i \cap Y_j\right) \right). \tag{14.48}$$

The construction above works the same way if we drop the restriction to algebraic forms. It is instructive to see how this definition reduces to the one we had previously in Sect. 14.1.2. To this aim we consider the case where Y has only one irreducible component without self-intersections. In this case

$$\Omega^k(X, Y) = \Omega^k(X) \oplus \Omega^{k-1}(Y). \tag{14.49}$$

We may therefore write an element in $\Omega^k(X, Y)$ as a pair (φ, ξ) with $\varphi \in \Omega^k(X)$ and $\xi \in \Omega^{k-1}(Y)$. The differential d_{total} works out as

$$d_{\text{total}}(\varphi, \xi) = \left(d\varphi, \iota^*\varphi - d\xi\right), \tag{14.50}$$

where $\iota : Y \to X$ denotes the inclusion. In Sect. 14.1.2 we defined the relative de Rham cohomology as equivalence classes of closed differential forms, which vanish on Y. Let us denote a representative of such a class by ω. We would like to show that the two definitions are equivalent: Any $\omega \in H^k_{\mathrm{dR}}(X, Y)$ (according to the definition in Sect. 14.1.2) defines

$$(\varphi, \xi) = (\omega, 0) . \tag{14.51}$$

ω is closed and vanishes on Y, hence $\iota^* \omega = 0$. We therefore have

$$d_{\mathrm{total}} (\omega, 0) = \left(d\omega, \iota^* \omega \right) \;=\; (0, 0) . \tag{14.52}$$

On the other hand, any $(\varphi, \xi) \in H^k_{\mathrm{dR}}(X, Y)$ (according to the definition of this subsection) defines

$$\omega = \varphi - d \left(h \pi^* \xi \right) . \tag{14.53}$$

This requires some explanation. We start from a tubular neighbourhood T of Y in X and denote by $\pi : T \to Y$ the projection. Furthermore h is a bump function. Let T_1 and T_2 be two further tubular neighbourhoods of Y in X with $T_1 \subset T_2 \subset T$. The bump function h is equal to 1 on T_1 and equal to zero on $X \backslash T_2$. One may show that the cohomology class defined by ω is independent of the choices made for T and h. If (φ, ξ) is closed, we have $d\varphi = 0$ and $\iota^* \varphi = d\xi$. It follows that ω is closed ($d\omega = d\varphi = 0$) and vanishes when restricted to Y:

$$\iota^* \omega = \iota^* \varphi - \iota^* d \left(h \pi^* \xi \right) \;=\; d\xi - d\xi \;=\; 0. \tag{14.54}$$

Therefore ω defines a cohomology class according to the definition of Sect. 14.1.2.

14.2 Categories

In mathematics it is quite common to consider for example vector spaces together with linear maps between vector spaces (i.e. vector space homomorphisms) or groups together with group homomorphisms. There are proofs which work the same way for vector spaces as they do for groups. It is therefore useful to introduce another layer of abstraction, where the common properties are treated in a unified way. This brings us to category theory.

Let us expand on the examples of vector spaces and groups: We may view the vector spaces (or groups) as objects and we denote morphisms between two objects X and Y (with X as source and Y as target) by an arrow $X \stackrel{\alpha}{\to} Y$.

A **category** consists of

- a class of **objects** denoted by $\text{Obj}(\mathcal{C})$,
- for every pair $X, Y \in \text{Obj}(\mathcal{C})$ a class of **morphisms** $X \xrightarrow{\alpha} Y$ from X to Y denoted by $\text{Hom}_{\mathcal{C}}(X, Y)$ or simply $\text{Hom}(X, Y)$ if no confusion arises,
- for every ordered triple of objects X, Y, Z a map from $\text{Hom}(X, Y) \times \text{Hom}(Y, Z)$ to $\text{Hom}(X, Z)$ called **composition**, the composition of $\alpha \in \text{Hom}(X, Y)$ with $\beta \in \text{Hom}(Y, Z)$ is denoted by $\beta \circ \alpha$,
- for $\alpha \in \text{Hom}(W, X)$, $\beta \in \text{Hom}(X, Y)$ and $\gamma \in \text{Hom}(Y, Z)$ we have the **associativity** law $\gamma \circ (\beta \circ \alpha) = (\gamma \circ \beta) \circ \alpha$,
- an **identity morphism** $\text{id} : X \to X$ for every object X, such that $\alpha \circ \text{id}_X = \alpha$ and $\text{id}_Y \circ \alpha = \alpha$ for any $\alpha \in \text{Hom}(X, Y)$.

A morphism $\alpha \in \text{Hom}(X, Y)$ is called an **isomorphism** if there exists a morphism $\alpha^{-1} \in \text{Hom}(Y, X)$ such that

$$\alpha^{-1} \circ \alpha = \text{id}_X \text{ and } \alpha \circ \alpha^{-1} = \text{id}_Y. \tag{14.55}$$

Examples of categories are the category of sets, denoted by **Set**, where the objects are sets and the morphisms are maps between sets, the category of groups, denoted by **Grp**, where the objects are groups and the morphisms are group homomorphisms and the category of finite-dimensional $\mathbb{F}$-vector spaces, denoted $\textbf{Vect}_{\mathbb{F}}$, where the objects are finite-dimensional vector space over the field $\mathbb{F}$ and the morphisms are linear maps between vector spaces. A less standard example of a category is a quiver, where every vertex has a self-loop attached to it and for any two oriented edges $v_i \to v_j$ and $v_j \to v_k$ there is an oriented edge $v_i \to v_k$, subject to the associativity law. The objects are the vertices and the morphisms are the oriented edges.

In the definition of a category we wrote "class of objects" and "class of morphisms". This wording avoids Russell's "set of all sets"-contradiction. A category $\mathcal{C}$ is called a **small category** if the class of all morphisms of $\mathcal{C}$ is a set. (This implies automatically that $\text{Obj}(\mathcal{C})$ is a set too, as objects are in one-to-one correspondence with the identity maps.) A category $\mathcal{C}$ is called a **locally small category** if for each $X, Y \in \text{Obj}(\mathcal{C})$ the class $\text{Hom}(X, Y)$ is a set.

Given a category $\mathcal{C}$ the **dual category** $\mathcal{C}^*$ is given by

$$\begin{aligned} \text{Obj}\left(\mathcal{C}^*\right) &= \text{Obj}\left(\mathcal{C}\right), \\ \text{Hom}_{\mathcal{C}^*}(X, Y) &= \text{Hom}_{\mathcal{C}}(Y, X). \end{aligned} \tag{14.56}$$

In other words, the dual category $\mathcal{C}^*$ is obtained from the category $\mathcal{C}$ by reversing all arrows.

Maps between categories which preserve composition and identities are called functors. In detail:

A **covariant functor** $T : \mathcal{C} \to \mathcal{D}$ from a category $\mathcal{C}$ to a category $\mathcal{D}$ consists of

- a map $T : \mathrm{Obj}(\mathcal{C}) \to \mathrm{Obj}(\mathcal{D})$, and
- maps $T = T_{XY} : \mathrm{Hom}(X, Y) \to \mathrm{Hom}(TX, TY)$, which preserve composition and identities, i.e.

$$\begin{aligned} T(\beta \circ \alpha) &= (T\beta) \circ (T\alpha) && \text{for all morphisms } X \xrightarrow{\alpha} Y \xrightarrow{\beta} Z \text{ in } \mathcal{C}, \\ T(\mathrm{id}_X) &= \mathrm{id}_{TX} && \text{for all } X \in \mathrm{Obj}(\mathcal{C}). \end{aligned} \tag{14.57}$$

If no confusion arises we will simply write T for T_{XY} (as we did already above).

A functor is called **faithful**, if for any $X, Y \in \mathrm{Obj}(\mathcal{C})$ the map

$$T_{XY} : \mathrm{Hom}_{\mathcal{C}}(X, Y) \to \mathrm{Hom}_{\mathcal{D}}(TX, TY) \tag{14.58}$$

is injective, the functor is called **full**, if the map is surjective and the functor is called **fully faithful**, if the map is bijective.

There is also the notion of a contravariant functor: A contravariant functor $T : \mathcal{C} \to \mathcal{D}$ is a covariant functor $\mathcal{C} \to \mathcal{D}^*$ (or equivalently a covariant functor $\mathcal{C}^* \to \mathcal{D}$). Spelled out as in Eq. (14.57) we have

A **contravariant functor** $T : \mathcal{C} \to \mathcal{D}$ from a category $\mathcal{C}$ to a category $\mathcal{D}$ consists of

- a map $T : \mathrm{Obj}(\mathcal{C}) \to \mathrm{Obj}(\mathcal{D})$, and
- maps $T = T_{XY} : \mathrm{Hom}(X, Y) \to \mathrm{Hom}(TY, TX)$, which preserve composition and identities, i.e.

$$\begin{aligned} T(\beta \circ \alpha) &= (T\alpha) \circ (T\beta) && \text{for all morphisms } X \xrightarrow{\alpha} Y \xrightarrow{\beta} Z \text{ in } \mathcal{C}, \\ T(\mathrm{id}_X) &= \mathrm{id}_{TX} && \text{for all } X \in \mathrm{Obj}(\mathcal{C}). \end{aligned} \tag{14.59}$$

A contravariant functor reverses all arrows.

A **natural transformation** $\Phi : T_1 \to T_2$ between two functors $T_1 : \mathcal{C}_1 \to \mathcal{C}_2$ and $T_2 : \mathcal{C}_1 \to \mathcal{C}_2$ (between the same categories) is a map that assigns to each object $X \in \mathrm{Obj}(\mathcal{C}_1)$ a morphism $\Phi_X \in \mathrm{Hom}_{\mathcal{C}_2}(T_1(X), T_2(X))$ such that for any morphism $\alpha \in \mathrm{Hom}_{\mathcal{C}_1}(X, Y)$

$$\Phi_Y \circ T_1(\alpha) = T_2(\alpha) \circ \Phi_X. \tag{14.60}$$

In terms of a commutative diagram:

$$\begin{array}{ccc} T_1(X) & \xrightarrow{\Phi_X} & T_2(X) \\ {\scriptstyle T_1(\alpha)}\downarrow & & \downarrow{\scriptstyle T_2(\alpha)} \\ T_1(Y) & \xrightarrow{\Phi_Y} & T_2(Y) \end{array} \tag{14.61}$$

A natural transformation Φ is called a **natural equivalence of functors** if each map Φ_X is an isomorphism. Φ_X is called a **functorial isomorphism**.

For a category C we denote by $I_C : C \to C$ the identity functor, which assigns each object and morphism to itself. Two categories C_1 and C_2 are called **equivalent categories**, if there exists two functors $T_{21} : C_1 \to C_2$ and $T_{12} : C_2 \to C_1$ such that $T_{12} \circ T_{21} : C_1 \to C_1$ and $I_{C_1} : C_1 \to C_1$ are natural equivalent as well as $T_{21} \circ T_{12} : C_2 \to C_2$ and $I_{C_2} : C_2 \to C_2$.

We are interested in categories, which have additional structures. We therefore introduce various specialisations: Monoidal categories, tensor categories, Abelian categories and finally Tannakian categories. It is the last one (Tannakian categories) which is most relevant to us.

14.2.1 Monoidal Categories

A monoidal category C is a category equipped with a functor from $C \times C$ into C, denoted by $\otimes$, a unit object $\mathbf{1}$ and an isomorphism $\iota : \mathbf{1} \otimes \mathbf{1} \to \mathbf{1}$, subject to the following constraints:

For any three objects $X, Y, Z \in \mathrm{Obj}(C)$ there is a functorial isomorphism

$$\Phi_{X,Y,Z} : X \otimes (Y \otimes Z) \to (X \otimes Y) \otimes Z \tag{14.62}$$

satisfying the pentagon identity:

$$(\Phi_{X,Y,Z} \otimes \mathrm{id}) \circ \Phi_{X,Y\otimes Z,W} \circ (\mathrm{id} \otimes \Phi_{Y,Z,W}) = \Phi_{X\otimes Y,Z,W} \circ \Phi_{X,Y,Z\otimes W}. \tag{14.63}$$

$\Phi_{X,Y,Z}$ is also called an **associativity constraint**. The pentagon identity reads in terms of a commutative diagram

$$\begin{array}{ccccc} X \otimes (Y \otimes (Z \otimes W)) & \xrightarrow{\Phi_{X,Y,Z\otimes W}} & (X \otimes Y) \otimes (Z \otimes W) & \xrightarrow{\Phi_{X\otimes Y,Z,W}} & ((X \otimes Y) \otimes Z) \otimes W \\ \downarrow{\scriptstyle \mathrm{id}_X \otimes \Phi_{Y,Z,W}} & & & & \uparrow{\scriptstyle \Phi_{X,Y,Z} \otimes \mathrm{id}_W} \\ X \otimes ((Y \otimes Z) \otimes W) & & \xrightarrow{\Phi_{X,Y\otimes Z,W}} & & (X \otimes (Y \otimes Z)) \otimes W \end{array} \tag{14.64}$$

For the unit object $\mathbf{1}$ one requires that there are functorial isomorphisms

$$l_X \ : \ \mathbf{1} \otimes X \ \to \ X \text{ and } r_X \ : \ X \otimes \mathbf{1} \ \to \ X, \tag{14.65}$$

satisfying the triangle diagram

$$\begin{array}{ccc} X \otimes (\mathbf{1} \otimes Y) & \xrightarrow{\Phi_{X,\mathbf{1},Y}} & (X \otimes \mathbf{1}) \otimes Y \\ & {}_{\mathrm{id}_X \otimes l_Y} \searrow \quad \swarrow {}_{r_X \otimes \mathrm{id}_Y} & \\ & X \otimes Y & \end{array} \tag{14.66}$$

Instead of Eq. (14.66) we may require that the functors $L_{\mathbf{1}} : \mathcal{C} \to \mathcal{C}$ and $R_{\mathbf{1}} : \mathcal{C} \to \mathcal{C}$ acting on the objects of $\mathcal{C}$ as

$$\begin{aligned} L_{\mathbf{1}} : X &\to \mathbf{1} \otimes X, \\ R_{\mathbf{1}} : X &\to X \otimes \mathbf{1} \end{aligned} \tag{14.67}$$

are autoequivalences of $\mathcal{C}$ [388]. We denote a monoidal category by $(\mathcal{C}, \otimes, \Phi, \mathbf{1}, \iota)$. The functorial isomorphisms l_X and r_X are related to this data as

$$\begin{aligned} L_{\mathbf{1}}\,(l_X) &= (\iota \otimes \mathrm{id}_X) \circ \phi_{\mathbf{1},\mathbf{1},X}, \\ R_{\mathbf{1}}\,(r_X) &= (\mathrm{id}_X \otimes \iota) \circ \phi^{-1}_{X,\mathbf{1},\mathbf{1}}. \end{aligned} \tag{14.68}$$

One of the simplest examples of a monoidal category is $\mathbf{Vect}_{\mathbb{F}}$, the category of finite-dimensional vector spaces over the field $\mathbb{F}$. The unit object $\mathbf{1}$ in this category is the one-dimensional vector space isomorphic to $\mathbb{F}$.

A monoidal category $\mathcal{C}$ is called (left) **rigid** (or has left duals), if for each object X there is an object X^* and morphisms $\mathrm{ev}_X : X^* \otimes X \to \mathbf{1}$, $\mathrm{coev}_X : \mathbf{1} \to X \otimes X^*$ such that

$$\begin{aligned} X &\xrightarrow{\mathrm{coev}_X} (X \otimes X^*) \otimes X \xrightarrow{\Phi^{-1}_{X,X^*,X}} X \otimes (X^* \otimes X) \xrightarrow{\mathrm{ev}_X} X, \\ X^* &\xrightarrow{\mathrm{coev}_X} X^* \otimes (X \otimes X^*) \xrightarrow{\Phi_{X^*,X,X^*}} (X^* \otimes X) \otimes X \xrightarrow{\mathrm{ev}_X} X^* \end{aligned} \tag{14.69}$$

compose to id_X and id_{X^*}, respectively.

14.2.2 Tensor Categories

A tensor category is a monoidal category such that the two functors $\mathcal{C} \times \mathcal{C} \to \mathcal{C}$ given by $(X, Y) \to X \otimes Y$ and $(X, Y) \to Y \otimes X$ are naturally equivalent, i.e. there exists a functorial isomorphism

$$\Psi_{X,Y} : X \otimes Y \to Y \otimes X. \tag{14.70}$$

The functional isomorphism $\Psi_{X,Y}$ is required to satisfy

$$\Psi_{Y,X} \circ \Psi_{X,Y} = \mathrm{id}_{X \otimes Y} \tag{14.71}$$

and the hexagon identity

$$(\Psi_{X,Z} \otimes \mathrm{id}) \circ \Phi_{X,Z,Y} \circ (\mathrm{id} \otimes \Psi_{Y,Z}) = \Phi_{Z,X,Y} \circ \Psi_{X \otimes Y,Z} \circ \Phi_{X,Y,Z}. \tag{14.72}$$

$\Psi_{X,Y}$ is also called a **commutativity constraint**. The hexagon identity ensures that the associativity constraint and the commutativity constraint are compatible. In terms of a commutative diagram we have

$$\begin{array}{ccccc} X \otimes (Y \otimes Z) & \xrightarrow{\Phi_{X,Y,Z}} & (X \otimes Y) \otimes Z & \xrightarrow{\Psi_{X \otimes Y,Z}} & Z \otimes (X \otimes Y) \\ \big\downarrow \mathrm{id}_X \otimes \Psi_{Y,Z} & & & & \big\downarrow \Phi_{Z,X,Y} \\ X \otimes (Z \otimes Y) & \xrightarrow{\Phi_{X,Z,Y}} & (X \otimes Z) \otimes Y & \xrightarrow{\Psi_{X,Z} \otimes \mathrm{id}_Y} & (Z \otimes X) \otimes Y \end{array} \tag{14.73}$$

Let $\mathcal{C}$ and $\mathcal{C}'$ be two tensor categories. A **tensor functor** is a pair (T, c) consisting of a functor $T : \mathcal{C} \to \mathcal{C}'$ and functorial isomorphisms $c_{X,Y} : T(X) \otimes T(Y) \to T(X \otimes Y)$ satisfying

1. for all $X, Y, Z \in \mathrm{Obj}(\mathcal{C})$ the following diagram commutes:

$$\begin{array}{ccccc} T(X) \otimes (T(Y) \otimes T(Z)) & \xrightarrow{\mathrm{id}_{T(X)} \otimes c_{Y,Z}} & T(X) \otimes T(Y \otimes Z) & \xrightarrow{c_{X,Y \otimes Z}} & T(X \otimes (Y \otimes Z)) \\ \big\downarrow \Phi'_{T(X),T(Y),T(Z)} & & & & \big\downarrow T(\Phi_{X,Y,Z}) \\ (T(X) \otimes T(Y)) \otimes T(Z) & \xrightarrow{c_{X,Y} \otimes \mathrm{id}_{T(Z)}} & T(X \otimes Y) \otimes T(Z) & \xrightarrow{c_{X \otimes Y,Z}} & T((X \otimes Y) \otimes Z) \end{array} \tag{14.74}$$

2. for all $X, Y \in \mathrm{Obj}(\mathcal{C})$ the following diagram commutes:

$$\begin{array}{ccc} T(X) \otimes T(Y) & \xrightarrow{c_{X,Y}} & T(X \otimes Y) \\ \big\downarrow \Psi'_{T(X),T(Y)} & & \big\downarrow T(\Psi_{X,Y}) \\ T(Y) \otimes T(X) & \xrightarrow{c_{Y,X}} & T(Y \otimes X) \end{array} \tag{14.75}$$

3. we have $\mathbf{1}' = T(\mathbf{1})$ and $\iota' = T(\iota)$ up to a unique isomorphism.

14.2.3 Abelian Categories

An **additive category** $\mathcal{C}$ is a locally small category where

1. every set $\text{Hom}_{\mathcal{C}}(X, Y)$ is an Abelian group (written additively) such that for $\alpha_1, \alpha_2 \in \text{Hom}(X, Y)$ and $\beta_1, \beta_2 \in \text{Hom}(Y, Z)$ we have

$$(\beta_1 + \beta_2) \circ (\alpha_1 + \alpha_2) = \beta_1 \circ \alpha_1 + \beta_1 \circ \alpha_2 + \beta_2 \circ \alpha_1 + \beta_2 \circ \alpha_2. \qquad (14.76)$$

2. There exists a zero object $\mathbf{0} \in \text{Obj}(\mathcal{C})$ such that $\text{Hom}_{\mathcal{C}}(\mathbf{0}, \mathbf{0}) = 0$.
3. For all objects $X_1, X_2 \in \text{Obj}(\mathcal{C})$ there exists an object $Y \in \text{Obj}(\mathcal{C})$ and morphisms $i_1 \in \text{Hom}(X_1, Y)$, $i_2 \in \text{Hom}(X_2, Y)$, $p_1 \in \text{Hom}(Y, X_1)$ and $p_2 \in \text{Hom}(Y, X_2)$ such that

$$p_1 \circ i_1 = \text{id}_{X_1}, \qquad p_2 \circ i_2 = \text{id}_{X_2}, \qquad i_1 \circ p_1 + i_2 \circ p_2 = \text{id}_Y. \qquad (14.77)$$

It can be shown that the object Y is unique up to a unique isomorphism. One denotes this element as $Y = X_1 \oplus X_2$ and calls it the **direct sum** of X_1 and X_2. This defines a bifunctor $\oplus : \mathcal{C} \times \mathcal{C} \to \mathcal{C}$.

One of the simplest examples of an additive category is again $\mathbf{Vect}_{\mathbb{F}}$, the category of finite-dimensional vector spaces over the field $\mathbb{F}$. The zero object $\mathbf{0}$ in this category is the zero-dimensional vector space consisting only of the zero vector.

Let $\mathbb{F}$ be a field. An additive category $\mathcal{C}$ is said to be **$\mathbb{F}$-linear** if every set $\text{Hom}_{\mathcal{C}}(X, Y)$ is a $\mathbb{F}$-vector space, such that the composition of morphisms is $\mathbb{F}$-linear.

Let us now turn to the definition of an Abelian category: An Abelian category is an additive category, where kernels and cokernels exist. In order to define kernels and cokernels in an additive category, we have to do some gymnastics:

Let $\mathcal{C}$ be an additive category and $\alpha \in \text{Hom}_{\mathcal{C}}(X, Y)$ a morphism. Suppose there exists an object $K \in \text{Obj}(\mathcal{C})$ and a morphism $k \in \text{Hom}_{\mathcal{C}}(K, X)$ such that $\alpha \circ k = 0$ and if $k' \in \text{Hom}_{\mathcal{C}}(K', X)$ is such that $\alpha \circ k' = 0$ then there exists a unique $l \in \text{Hom}_{\mathcal{C}}(K', K)$ such that $kl = k'$. The pair (K, k) is called the **kernel** of α and denoted $\text{Ker}(\alpha)$.

The cokernel is defined in a similar way: Suppose there exists an object $C \in \text{Obj}(\mathcal{C})$ and a morphism $c \in \text{Hom}_{\mathcal{C}}(Y, C)$ such that $c \circ \alpha = 0$ and if $c' \in \text{Hom}_{\mathcal{C}}(Y, C')$ is such that $c' \circ \alpha = 0$ then there exists a unique $l \in \text{Hom}_{\mathcal{C}}(C, C')$ such that $lc = c'$. The pair (C, c) is called the **cokernel** of α and denoted $\text{Coker}(\alpha)$.

An **Abelian category** $\mathcal{C}$ is an additive category, where for every $\alpha \in \text{Hom}_{\mathcal{C}}(X, Y)$ there exists a sequence

$$K \xrightarrow{k} X \xrightarrow{i} I \xrightarrow{j} Y \xrightarrow{c} C \qquad (14.78)$$

with $\alpha = j \circ i$ and

$$\text{Ker}\,(\alpha) = (K, k)\,, \qquad \text{Coker}\,(\alpha) = (C, c)\,,$$
$$\text{Ker}\,(c) = (I, j)\,, \qquad \text{Coker}\,(k) = (I, i)\,. \tag{14.79}$$

A functor is called **exact**, if it preserves short exact sequences, e.g.

$$\mathbf{0} \longrightarrow X \stackrel{\alpha}{\longrightarrow} Y \stackrel{\beta}{\longrightarrow} Z \longrightarrow \mathbf{0} \tag{14.80}$$

implies

$$\mathbf{0} \longrightarrow T(X) \stackrel{T(\alpha)}{\longrightarrow} T(Y) \stackrel{T(\beta)}{\longrightarrow} T(Z) \longrightarrow \mathbf{0}. \tag{14.81}$$

14.2.4 Tannakian Categories

We now have all ingredients to define Tannakian categories. As before, we denote by $\mathbb{F}$ a field and by $\mathbf{Vect}_{\mathbb{F}}$ the category of finite-dimensional $\mathbb{F}$-vector spaces.

Let R be a ring. We denote by $\mathbf{Mod}_R$ the category of finitely generated R-modules and by $\mathbf{Proj}_R$ the category of finitely generated projective R-modules.

Let $\mathcal{C}$ be a category. For $X \in \text{Obj}(\mathcal{C})$ we denote by $\text{End}(X) = \text{Hom}(X, X)$ the endomorphisms of X, and in particular $\text{End}(\mathbf{1}) = \text{Hom}(\mathbf{1}, \mathbf{1})$.

A **neutral Tannakian category** $\mathcal{C}$ over $\mathbb{F}$ is a rigid Abelian tensor category with $\text{End}(\mathbf{1}) = \mathbb{F}$ and a $\mathbb{F}$-linear exact faithful tensor functor

$$\omega : \mathcal{C} \rightarrow \mathbf{Vect}_{\mathbb{F}}, \tag{14.82}$$

called the **fibre functor**. We say that the fibre functor ω takes values in $\mathbb{F}$.

We may generalise the definition of a neutral Tannakian category towards a Tannakian category as follows: Let R be a non-zero $\mathbb{F}$-algebra. We define a fibre functor on $\mathcal{C}$ which takes values in R as a $\mathbb{F}$-linear exact faithful tensor functor $\eta : \mathcal{C} \rightarrow \mathbf{Mod}_R$ that takes values in the subcategory $\mathbf{Proj}_R$ of $\mathbf{Mod}_R$.

With these preparations we finally arrive at the definition of a Tannakian category:

A **Tannakian category** $\mathcal{C}$ over $\mathbb{F}$ is a rigid Abelian tensor category with $\text{End}(\mathbf{1}) = \mathbb{F}$ and a fibre functor with values in the $\mathbb{F}$-algebra R.

Let's look at an example: Let V be a finite-dimensional vector space and $\text{GL}(V)$ the group of automorphisms of V. For example $V = \mathbb{C}^n$ and $\text{GL}(V) = \text{GL}(n, \mathbb{C})$. Let G be a group. A representation of G is a homomorphism $\rho : G \rightarrow \text{GL}(V)$. Since ρ is a homomorphism we have

$$\rho\,(g_1 g_2) = \rho\,(g_1)\,\rho\,(g_2)\,. \tag{14.83}$$

By abuse of notation we denote by ρ also the map $\rho : G \times V \to V$, $v \to \rho(g)v$. Let's now fix the ground field to be $\mathbb{C}$. The finite-dimensional representations of G form a category, which we denote by $\mathbf{Rep}_{\mathbb{C}}(G)$. The objects in this category are pairs $(\rho, V) \in \mathrm{Obj}(\mathbf{Rep}_{\mathbb{C}}(G))$. The morhpisms in this category are as follows: Let (ρ_1, V_1) and (ρ_2, V_2) be two objects in $\mathbf{Rep}_{\mathbb{C}}(G)$. The class of morphisms $\mathrm{Hom}_{\mathbf{Rep}_{\mathbb{C}}(G)}((\rho_1, V_1), (\rho_2, V_2))$ consists of maps $\alpha : V_1 \to V_2$ such that for all $g \in G$ the following diagram commutes:

$$\begin{array}{ccc} G \times V_1 & \xrightarrow{\rho_1} & V_1 \\ \Big\downarrow \mathrm{id}_G \times \alpha & & \Big\downarrow \alpha \\ G \times V_2 & \xrightarrow{\rho_2} & V_2 \end{array} \tag{14.84}$$

$\mathbf{Rep}_{\mathbb{C}}(G)$ is a (neutral) Tannakian category. The fibre functor is given by the forgetful functor, which associates to any object $(\rho, V) \in \mathrm{Obj}(\mathbf{Rep}_{\mathbb{C}}(G))$ the object $V \in \mathrm{Obj}(\mathbf{Vect}_{\mathbb{C}})$.

A second example is given by the category of mixed Hodge structures **MHS**, discussed in Sect. 14.4.

14.3 Motives

Motives are a conjectured framework to unify different cohomology theories. For an algebraic variety X there is more than one cohomology theory. Most relevant to us are de Rham cohomology or Betti cohomology. Other cohomology theories are for example l-adic cohomology or crystalline cohomology. Before outlining the main ideas behind motives, we have to introduce correspondences, which will play the role of morphisms in the category of motives.

Digression Correspondences and adequate equivalence relations

Let X be an algebraic variety. The group of cycles $Z(X)$ is the free Abelian group generated by the set of subvarieties of X. We write a cycle as a formal linear combination

$$Z = \sum_j n_j Y_j, \tag{14.85}$$

where $n_j \in \mathbb{Z}$ and Y_j a subvariety of X.

Let X and Y be two algebraic varieties. A **correspondence** *between X and Y is a cycle of $Z(X \times Y)$.*

Two cycles Z_1 and Z_2 are called **rational equivalent**, *if there is a cycle W on $\mathbb{P}^1 \times X$ (i.e. a correspondence between $\mathbb{P}^1$ and X) and $t_1, t_2 \in \mathbb{P}^1$ such that*

$$Z_1 - Z_2 = W \cap (\{t_1\} \times X) - W \cap (\{t_2\} \times X)\,. \tag{14.86}$$

We write

$$Z_1 \sim_{\text{rat}} Z_2. \tag{14.87}$$

The group

$$Z(X)/\sim_{\text{rat}} \tag{14.88}$$

is called the **Chow group** *of* X.

Rational equivalence is the statement that we may interpolate between the cycles Z_1 *and* Z_2 *with a parameter* t*, being the coordinate of a curve of genus zero (i.e.* $\mathbb{P}^1$*). This can be generalised to curves of higher genus: Two cycles* Z_1 *and* Z_2 *are called* **algebraic equivalent***, if there is an irreducible curve* C *and a cycle* W *on* $C \times X$ *and* $t_1, t_2 \in C$ *such that*

$$Z_1 - Z_2 = W \cap (\{t_1\} \times X) - W \cap (\{t_2\} \times X). \tag{14.89}$$

We write

$$Z_1 \sim_{\text{alg}} Z_2. \tag{14.90}$$

Clearly, rational equivalence implies algebraic equivalence.

Two cycles Z_1 *and* Z_2 *are called* **numerical equivalent***, if* $\deg(Z_1 \cap W) = \deg(Z_2 \cap W)$ *for any cycle with* $\dim W = \operatorname{codim} Z$*. We write*

$$Z_1 \sim_{\text{num}} Z_2. \tag{14.91}$$

$\deg(Z_1 \cap Z_2)$ *is the intersection number of* Z_1 *and* Z_2*. If* $\dim Z_1 = \operatorname{codim} Z_2$ *and if the intersection is a set of points (counted with multiplicities)*

$$Z_1 \cap Z_2 = \sum_j n_j P_j, \qquad P_j \in X \tag{14.92}$$

one has

$$\deg(Z_1 \cap Z_2) = \sum_j n_j. \tag{14.93}$$

We have the implications

$$\sim_{\text{rat}} \;\Rightarrow\; \sim_{\text{alg}} \;\Rightarrow\; \sim_{\text{num}}, \tag{14.94}$$

hence rational equivalence is the finest equivalence relation and numerical equivalence is the coarsest equivalence relation.

Let us now give a short summary on the conjectured theory of motives: Let us denote by $\mathbf{Var}_{\mathbb{Q}}$, the category of algebraic varieties defined over $\mathbb{Q}$ and by $\mathbf{SmProj}_{\mathbb{Q}}$ the subcategory of smooth projective varieties over $\mathbb{Q}$. It is conjectured that there exists a Tannakian category of mixed motives **MixMot** and a functor

$$h : \mathbf{Var}_{\mathbb{Q}} \to \mathbf{MixMot}, \tag{14.95}$$

such that the cohomologies H_{dR} and H_B factor through h, e.g. there exist commutative diagrams

$$\begin{array}{ccc} \mathbf{Var}_{\mathbb{Q}} & \xrightarrow{h} & \mathbf{MixMot} \\ & \searrow^{H_{\mathrm{dR}}} & \downarrow \eta_{\mathrm{dR}} \\ & & \mathbf{Vect}_{\mathbb{Q}} \end{array}, \qquad \begin{array}{ccc} \mathbf{Var}_{\mathbb{Q}} & \xrightarrow{h} & \mathbf{MixMot} \\ & \searrow^{H_B} & \downarrow \eta_B \\ & & \mathbf{Vect}_{\mathbb{Q}} \end{array}. \tag{14.96}$$

η_{dR} and η_B are fibre functors in **MixMot**. If we just consider the subcategory of smooth projective varieties $\mathbf{SmProj}_{\mathbb{Q}}$ one expects

$$h : \mathbf{SmProj}_{\mathbb{Q}} \to \mathbf{PureMot}, \tag{14.97}$$

where **PureMot** denotes the subcategory of pure motives.

Morphisms in **MixMot** are given by correspondences.

One further expects that motives extend to the relative setting, i.e. to any pair (X, Y) with X a smooth algebraic variety and Y a closed subvariety there is a motive $h(X, Y)$.

Motives are often studied through their Hodge realisation. The Hodge realisation is a functor

$$\mathbf{MixMot} \to \mathbf{MHS}, \tag{14.98}$$

where **MHS** denotes the category of mixed Hodge structures (again a Tannakian category). Restricted to the category of pure motives we have

$$\mathbf{PureMot} \to \mathbf{HS}, \tag{14.99}$$

where **HS** denotes the category of (pure) Hodge structures.

With this short interlude on motives we now leave the field of conjectural mathematics and return to solid grounds. In the next section we introduce Hodge structures, followed by examples from Feynman integrals.

14.4 Hodge Structures

Hodge structures have their origin in the study of compact Kähler manifolds [396]. Let M be a complex manifold with complex structure J. A Riemannian metric g on M is called Hermitian, if it is compatible with the complex structure J, in other words for vector fields X, Y on M we have

$$g\left(JX, JY\right) = g\left(X, Y\right). \tag{14.100}$$

For a Hermitian manifold one defines an associated differential two-form by

$$K\left(X, Y\right) = g\left(JX, Y\right). \tag{14.101}$$

K is called the Kähler form. A Hermitian manifold is called a **Kähler manifold**, if the two-form K is closed:

$$dK = 0. \tag{14.102}$$

Examples of compact Kähler manifolds are provided by compact Riemann surfaces. Riemann surfaces are complex manifolds of complex dimension one and the Kähler form of any Hermitian metric is necessarily closed. A second example is given by the complex projective space $\mathbb{P}^n(\mathbb{C})$. As a third example we mention complex submanifolds of Kähler manifolds. These submanifolds are again Kähler.

On a compact Kähler manifold we have the following decomposition of the cohomology groups

$$H^k\left(X\right) \otimes \mathbb{C} = \bigoplus_{p+q=k} H^{p,q}(X), \qquad \overline{H^{p,q}(X)} = H^{q,p}(X). \tag{14.103}$$

For a fixed k this provides an example of a pure Hodge structure of weight k.

14.4.1 Pure Hodge Structures

Let us now define pure Hodge structures. Let $V_{\mathbb{Z}}$ be a $\mathbb{Z}$-module of finite rank and $V_{\mathbb{C}} = V_{\mathbb{Z}} \otimes_{\mathbb{Z}} \mathbb{C}$ its complexification.

A **pure Hodge structure** of weight k on the $\mathbb{Z}$-module $V_{\mathbb{Z}}$ is a direct sum decomposition

$$V_{\mathbb{C}} = \bigoplus_{p+q=k} V^{p,q} \quad \text{with} \quad \overline{V^{p,q}} = V^{q,p}. \tag{14.104}$$

If one replaces $\mathbb{Z}$ by $\mathbb{Q}$ or $\mathbb{R}$, one speaks about a rational or real Hodge structure, respectively. The bar in $\overline{V^{q,p}}$ denotes complex conjugation with respect to the real structure $V_\mathbb{C} = V_\mathbb{R} \otimes_\mathbb{R} \mathbb{C}$ (if we start from $V_\mathbb{Z}$ we have $V_\mathbb{R} = V_\mathbb{Z} \otimes_\mathbb{Z} \mathbb{R}$).

The numbers

$$h^{p,q}(V) = \dim V^{p,q} \tag{14.105}$$

are called the **Hodge numbers**.

There is a second definition of a pure Hodge structure, which is more adapted for generalisations. The second definition is based on a **Hodge filtration**: Let $F^\bullet V_\mathbb{C}$ be a finite decreasing filtration:

$$V_\mathbb{C} \supseteq \cdots \supseteq F^{p-1} V_\mathbb{C} \supseteq F^p V_\mathbb{C} \supseteq F^{p+1} V_\mathbb{C} \supseteq \cdots \supseteq (0) \tag{14.106}$$

such that

$$V_\mathbb{C} = F^p V_\mathbb{C} \oplus \overline{F^{k-p+1} V_\mathbb{C}}. \tag{14.107}$$

Then V carries a pure Hodge structure of weight k. These two definitions are equivalent: Given the Hodge decomposition, we can define the corresponding Hodge filtration by

$$F^p V_\mathbb{C} = \bigoplus_{j \geq p} V^{j,k-j}. \tag{14.108}$$

Conversely, given a Hodge filtration we obtain the Hodge decomposition by

$$V^{p,q} = F^p V_\mathbb{C} \cap \overline{F^q V_\mathbb{C}}. \tag{14.109}$$

Hodge structures behave under the operations of direct sums, tensor products and duality as follows:

- Direct sum: If V and W are Hodge structures of weight k, then also $V \oplus W$ is a Hodge structure of weight k.
- Tensor product: If V is a Hodge structure of weight k, and W is a Hodge structure of weight l, then the tensor product $V \otimes W$ is a Hodge structure of weight $(k \cdot l)$.
- Duality: If V is a Hodge structure of weight k, then $\mathrm{Hom}\,(V, \mathbb{Z})$ is a Hodge structure of weight $(-k)$.

Let look at a few examples:

Example 1: Let (e_1, e_2) be a basis of $\mathbb{R}^2$ and consider $V_\mathbb{Z} = \mathbb{Z}e_1 \oplus \mathbb{Z}e_2$. We can define a Hodge structure of weight 1 with the decomposition

$$V_\mathbb{C} = V^{1,0} \oplus V^{0,1}, \tag{14.110}$$

by setting

$$V^{1,0} = \mathbb{C}\,(e_1 - ie_2)\,, \quad V^{0,1} = \mathbb{C}\,(e_1 + ie_2)\,. \tag{14.111}$$

Note that the definition $V^{1,0} = \mathbb{C}e_1$, $V^{0,1} = \mathbb{C}e_2$ would not work: Since $\overline{e_1} = e_1$ and $\overline{e_2} = e_2$, we have $\overline{V^{1,0}} = \mathbb{C}e_1 \neq V^{0,1} = \mathbb{C}e_2$.

Example 2: The **Tate Hodge structure** $\mathbb{Z}(1)$ is the Hodge structure with underlying $\mathbb{Z}$-module given by $V_{\mathbb{Z}} = \mathbb{Z}(1) = 2\pi i\ \mathbb{Z}$. One sets

$$V_{\mathbb{C}} = \mathbb{Z}(1) \otimes_{\mathbb{Z}} \mathbb{C} = \mathbb{C}. \tag{14.112}$$

For the Hodge decomposition one sets

$$V_{\mathbb{C}} = V^{-1,-1}, \tag{14.113}$$

hence $V^{p,q} = 0$ for $(p, q) \neq (-1, -1)$. The Tate Hodge structure $\mathbb{Z}(1)$ is a pure Hodge structure of weight -2.

One further defines

$$\mathbb{Z}(m) = \mathbb{Z}^{\otimes m}. \tag{14.114}$$

We therefore have

$$\mathbb{Z}(m) = (2\pi i)^m\ \mathbb{Z}. \tag{14.115}$$

$\mathbb{Z}(m)$ is a pure Hodge structure of weight $(-2m)$ with the decomposition

$$\mathbb{Z}(m) \otimes \mathbb{C} = V^{-m,-m}. \tag{14.116}$$

Given a Hodge structure on $V_{\mathbb{Z}}$ of weight k, one defines the **Tate twist** $V(m)$ as the Hodge structure of weight $k - 2m$ with underlying $\mathbb{Z}$-module

$$V(m)_{\mathbb{Z}} = (2\pi i)^m \otimes V_{\mathbb{Z}} \tag{14.117}$$

and Hodge decomposition

$$V(m)_{\mathbb{C}} = \bigoplus_{p+q=k-2m} V(m)^{p,q} \qquad \text{with} \quad V(m)^{p,q} = V^{p+m,q+m}. \tag{14.118}$$

A **polarisation** of a pure Hodge structure V of weight k is a non-degenerate bilinear form

$$Q : V_{\mathbb{Z}} \otimes V_{\mathbb{Z}} \to \mathbb{Z}, \tag{14.119}$$

with $Q(v, w) = (-1)^k Q(w, v)$. For the complex extension $Q : V_{\mathbb{C}} \otimes V_{\mathbb{C}} \to \mathbb{C}$ one requires for $v \in V^{p,q}$ and $w \in V^{p',q'}$

$$Q(v, w) = 0 \text{ for } (p', q') \neq (k - p, k - q) \tag{14.120}$$

and

$$i^{p-q} Q(v, \bar{v}) > 0 \tag{14.121}$$

for $v \neq 0$.

As an example let us consider polarised Hodge structures of dimension 2 and weight 1: Let $V_{\mathbb{Z}}$ be generated by e_1, e_2. For $v, w \in V_{\mathbb{Z}}$ we write $v = v_1 e_1 + v_2 e_2$, $w = w_1 e_1 + w_2 e_2$. Let us assume that the bilinear form defining the polarisation $Q : V_{\mathbb{Z}} \otimes V_{\mathbb{Z}} \to \mathbb{Z}$ is given by

$$Q(v, w) = (w_2, w_1) \begin{pmatrix} 0 & 1 \\ -1 & 0 \end{pmatrix} \begin{pmatrix} v_2 \\ v_1 \end{pmatrix} \tag{14.122}$$

We assume that $V_{\mathbb{Z}}$ is a Hodge structure of weight 1, hence

$$V_{\mathbb{C}} = V^{1,0} \otimes V^{0,1} \tag{14.123}$$

and $h^{1,0} = h^{0,1} = 1$. Let $\psi = \psi_1 e_1 + \psi_2 e_2$ with $\psi_1, \psi_2 \in \mathbb{C}$ be a basis of $V^{1,0}$. From

$$i Q\left(\psi, \bar{\psi}\right) > 0 \tag{14.124}$$

it follows that

$$i\left(\bar{\psi}_2 \psi_1 - \bar{\psi}_1 \psi_2\right) > 0. \tag{14.125}$$

This implies in particular $\psi_1 \neq 0$. We may therefore rescale the generator of $V^{1,0}$ by $1/\psi_1$. We then have

$$V^{1,0} = \langle e_1 + \tau e_2 \rangle, \qquad \tau = \frac{\psi_2}{\psi_1} \tag{14.126}$$

and

$$i Q\left(e_1 + \tau e_2, e_1 + \bar{\tau} e_2\right) = -i\left(\tau - \bar{\tau}\right) = 2\text{Im}\tau > 0. \tag{14.127}$$

This shows that all Hodge structures of dimension 2 and weight 1 with the polarisation form as in Eq. (14.122) are parametrised by $\tau \in \mathbb{H}$. $V^{0,1}$ is generated by

$$V^{0,1} = \overline{V^{1,0}} = \langle e_1 + \bar{\tau} e_2 \rangle. \tag{14.128}$$

It is instructive to discuss this concretely for elliptic curves: Let

$$E : y^2 = 4x(x-1)(x-\lambda) \tag{14.129}$$

be an elliptic curve. We denote by γ_1 and γ_2 two independent cycles. As they are independent, they form a basis of the first Betti homology group $H_1^{\mathrm{B}}(E)$. We then consider Betti cohomology. Let $\gamma_1^*, \gamma_2^* \in H_{\mathrm{B}}^1(E)$ be the dual basis, i.e. the basis which satisfies

$$\left\langle \gamma_i^*, \gamma_i \right\rangle = \delta_{ij}. \tag{14.130}$$

γ_1^* and γ_2^* correspond to e_1 and e_2 in the discussion above. From Sect. 13.2 we know that a basis of $H_{\mathrm{dR}}^1(E)$ is given by

$$\omega_1 = \frac{dx}{y}, \quad \omega_2 = \frac{x\,dx}{y}. \tag{14.131}$$

We denote the period matrix by

$$P = \begin{pmatrix} \langle \omega_1, \gamma_1 \rangle & \langle \omega_1, \gamma_2 \rangle \\ \langle \omega_2, \gamma_1 \rangle & \langle \omega_2, \gamma_2 \rangle \end{pmatrix} = \begin{pmatrix} \psi_1 & \psi_2 \\ \phi_1 & \phi_2 \end{pmatrix}. \tag{14.132}$$

From the Legendre relation we have

$$\det P = 2\pi i. \tag{14.133}$$

We recall that instead of (x, y) we may use a complex coordinate z through

$$z = \int\limits_{\infty}^{x} \frac{dt}{\sqrt{4t\,(t-1)\,(t-\lambda)}}, \quad (x, y) = \left(\wp\,(z)\,,\wp'\,(z)\right). \tag{14.134}$$

In terms of the complex coordinate z, the one-form ω_1 is given by $\omega_1 = dz$. On the other hand, we may express ω_1 as a linear combination of γ_1^* and γ_2^*: We make the ansatz $\omega_1 = c_1\gamma_1^* + c_2\gamma_2^*$, contract with γ_1 and γ_2 and find

$$\omega_1 = \psi_1\gamma_1^* + \psi_2\gamma_2^*. \tag{14.135}$$

This corresponds to $\psi = \psi_1 e_1 + \psi_2 e_2$ just before Eq. (14.124). Thus we have

$$H_{\mathrm{B}}^1(E)_{\mathbb{C}} = H^{1,0} \otimes H^{0,1}, \tag{14.136}$$

with $H^{1,0}$ being generated by $\omega_1 = dz$ and $H^{0,1}$ being generated by $\overline{\omega}_1 = d\bar{z}$.

Exercise 116 *We now have two bases of $H_{\mathrm{dR}}^1(E)$: on the one hand (ω_1, ω_2), on the other hand $(dz, d\bar{z})$. We already know $\omega_1 = dz$. Work out the full relation between the two bases.*

14.4.2 Mixed Hodge Structures

Pure Hodge structures are relevant for smooth projective algebraic varieties, these are necessarily compact. If one gives up the requirement of smoothness or compactness one is lead to a generalisation called mixed Hodge structure [397–399].

A **mixed Hodge structure** is given by a $\mathbb{Z}$-module $V_{\mathbb{Z}}$ of finite rank, a finite increasing filtration on $V_{\mathbb{Q}} = V_{\mathbb{Z}} \otimes_{\mathbb{Z}} \mathbb{Q}$, called the **weight filtration**:

$$(0) \subseteq \cdots \subseteq W_{k-1} V_{\mathbb{Q}} \subseteq W_k V_{\mathbb{Q}} \subseteq W_{k+1} V_{\mathbb{Q}} \subseteq \cdots \subseteq V_{\mathbb{Q}}, \quad (14.137)$$

and a finite decreasing filtration on $V_{\mathbb{C}} = V_{\mathbb{Z}} \otimes_{\mathbb{Z}} \mathbb{C}$, called the **Hodge filtration**:

$$V_{\mathbb{C}} \supseteq \cdots \supseteq F^{p-1} V_{\mathbb{C}} \supseteq F^p V_{\mathbb{C}} \supseteq F^{p+1} V_{\mathbb{C}} \supseteq \cdots \supseteq (0), \quad (14.138)$$

such that $F^\bullet$ induces a pure Hodge structure of weight k on

$$\mathrm{Gr}_k^W V_{\mathbb{Q}} = W_k V_{\mathbb{Q}} / W_{k-1} V_{\mathbb{Q}}. \quad (14.139)$$

A mixed Hodge structure is called a **mixed Tate Hodge structure** if

$$h^{p,q} = 0 \qquad \text{for } p \neq q. \quad (14.140)$$

Example 1: Let us fix two independent vectors e_0 and e_{-1} and a complex number x. We set

$$V_{\mathbb{Q}} = \langle e_0 + \ln x \cdot e_{-1}, 2\pi i e_{-1} \rangle . \quad (14.141)$$

On $V_{\mathbb{Q}}$ we define the weight filtration by

$$\begin{aligned} W_0 V_{\mathbb{Q}} &= V_{\mathbb{Q}} = \langle e_0 + \ln x \cdot e_{-1}, 2\pi i e_{-1} \rangle , \\ W_{-1} V_{\mathbb{Q}} &= W_{-2} V_{\mathbb{Q}} = \langle 2\pi i e_{-1} \rangle , \\ W_{-3} V_{\mathbb{Q}} &= 0. \end{aligned} \quad (14.142)$$

On $V_{\mathbb{C}}$ we define the Hodge filtration by

$$F^1 V_{\mathbb{C}} = 0, \quad F^0 V_{\mathbb{C}} = \langle e_0 \rangle , \quad F^{-1} V_{\mathbb{C}} = \langle e_0, e_{-1} \rangle . \quad (14.143)$$

We have

$$
\begin{aligned}
&\mathrm{Gr}_0^W V_{\mathbb{C}} = \langle e_0 + \ln x \cdot e_{-1}, 2\pi i e_{-1} \rangle \,/\, \langle 2\pi i e_{-1} \rangle = \langle e_0, e_{-1} \rangle \,/\, \langle e_{-1} \rangle \cong \langle e_0 \rangle \,, \\
&\mathrm{Gr}_{-1}^W V_{\mathbb{C}} = 0, \\
&\mathrm{Gr}_{-2}^W V_{\mathbb{C}} = \langle 2\pi i e_{-1} \rangle \,.
\end{aligned}
\tag{14.144}
$$

In the decomposition

$$
\mathrm{Gr}_0^W V_{\mathbb{C}} = \bigoplus_p V^{p,-p} \tag{14.145}
$$

one easily finds that $V^{p,-p} = 0$ for $p \geq 1$, since $F^1 V_{\mathbb{C}} = 0$. From $V^{-p,p} = \overline{V^{p,-p}}$ it follows then that also $V^{-p,p} = 0$ for $p \geq 1$. Therefore

$$
\mathrm{Gr}_0^W V_{\mathbb{C}} = V^{0,0} \cong \langle e_0 \rangle \,. \tag{14.146}
$$

In a similar way one finds

$$
\mathrm{Gr}_{-2}^W V_{\mathbb{C}} = V^{-1,-1} = \langle 2\pi i e_{-1} \rangle \,. \tag{14.147}
$$

Therefore the Hodge structure of $\mathrm{Gr}_0^W V_{\mathbb{Q}}$ is isomorph to $\mathbb{Q}(0)$, and the Hodge structure of $\mathrm{Gr}_{-2}^W V_{\mathbb{Q}}$ is isomorph to $\mathbb{Q}(1)$. Thus, $V_{\mathbb{Q}}$ defines a mixed Tate Hodge structure with $V^{0,0}$ and $V^{-1,-1}$ non-zero.

Note that $V_{\mathbb{Q}}$ is spanned by the columns of

$$
P = \begin{pmatrix} 1 & 0 \\ \ln x & 2\pi i \end{pmatrix}. \tag{14.148}
$$

With

$$
C_0 = \begin{pmatrix} 0 & 0 \\ 1 & 0 \end{pmatrix} \tag{14.149}
$$

we have

$$
\frac{d}{dx} P \;=\; \frac{C_0}{x} P, \;\; \mathcal{M}_0 P \;=\; P \exp\left(C_0\right), \tag{14.150}
$$

where $\mathcal{M}_0$ denotes the monodromy operator around $x = 0$ (see Eq. (8.95)).

Example 2: In a similar spirit let us consider the $(n+1) \times (n+1)$-matrix

$$
P = \begin{pmatrix}
1 & 0 & 0 & \cdots & 0 \\
-\mathrm{Li}_1(x) & 2\pi i & 0 & \cdots & 0 \\
-\mathrm{Li}_2(x) & 2\pi i \ln(x) & (2\pi i)^2 & \cdots & 0 \\
\cdots & \cdots & \cdots & \cdots & \cdots \\
-\mathrm{Li}_n(x) & 2\pi i \frac{\ln^{n-1}(x)}{(n-1)!} & (2\pi i)^2 \frac{\ln^{n-2}(x)}{(n-2)!} & \cdots & (2\pi i)^n
\end{pmatrix} \tag{14.151}
$$

For later use we note that with

$$C_0 = \begin{pmatrix} 0\,0 \cdots\cdots 0 \\ 0\,0 \cdots\cdots 0 \\ 0\,1\ \ddots \quad 0 \\ \vdots\ \vdots\ \ddots\ \ddots\ \vdots \\ 0\,0 \cdots\ 1\ 0 \end{pmatrix}, \quad C_1 = \begin{pmatrix} 0\,0\,0 \cdots 0 \\ 1\,0\,0 \cdots 0 \\ 0\,0\,0 \cdots 0 \\ \vdots\ \vdots\ \vdots \quad \vdots \\ 0\,0\,0 \cdots 0 \end{pmatrix} \tag{14.152}$$

we have

$$\frac{d}{dx} P = \left(\frac{C_0}{x} + \frac{C_1}{x-1} \right) P, \quad \mathcal{M}_0 P = P \exp(C_0), \quad \mathcal{M}_1 P = P \exp(C_1). \tag{14.153}$$

$\mathcal{M}_0$ and $\mathcal{M}_1$ denote the monodromy operators around $x = 0$ and $x = 1$, respectively.
We introduce independent vectors $e_0, e_{-1}, \ldots, e_{-n}$ and set

$$\begin{aligned} v_0 &= e_0 - \mathrm{Li}_1(x)e_{-1} - \mathrm{Li}_2(x)e_{-2} - \ldots \qquad - \mathrm{Li}_n(x)e_{-n}, \\ v_1 &= \qquad 2\pi i e_{-1} + 2\pi i \ln(x) e_{-2} + \cdots \quad + 2\pi i \frac{\ln^{n-1}(x)}{(n-1)!} e_{-n}, \\ v_2 &= \qquad\qquad (2\pi i)^2 e_{-2} + \cdots + (2\pi i)^2 \frac{\ln^{n-2}(x)}{(n-2)!} e_{-n}, \\ &\ldots \\ v_n &= \qquad\qquad\qquad (2\pi i)^n e_{-n}. \end{aligned} \tag{14.154}$$

We then consider

$$V_{\mathbb{Q}} = \langle v_0, v_1, \ldots, v_n \rangle. \tag{14.155}$$

$V_{\mathbb{Q}}$ is a mixed Hodge structure with the weight filtration

$$\begin{aligned} W_0 V_{\mathbb{Q}} &= \langle v_0, v_1, v_2, \ldots, v_n \rangle, \\ W_{-1} V_{\mathbb{Q}} = W_{-2} V_{\mathbb{Q}} &= \langle v_1, v_2, \ldots, v_n \rangle, \\ W_{-3} V_{\mathbb{Q}} = W_{-4} V_{\mathbb{Q}} &= \langle v_2, \ldots, v_n \rangle, \\ \ldots \qquad \ldots \quad &\qquad \ldots \\ W_{-2n+1} V_{\mathbb{Q}} = W_{-2n} V_{\mathbb{Q}} &= \langle v_n \rangle, \\ W_{-2n-1} V_{\mathbb{Q}} &= 0, \end{aligned} \tag{14.156}$$

and the Hodge filtration

$$\begin{aligned} F^1 V_{\mathbb{C}} &= 0, \\ F^0 V_{\mathbb{C}} &= \langle e_0 \rangle, \\ F^{-1} V_{\mathbb{C}} &= \langle e_0, e_{-1} \rangle, \\ &\dots \\ F^{-n} V_{\mathbb{C}} &= \langle e_0, e_{-1}, \dots, e_{-n} \rangle \;=\; V_{\mathbb{C}}. \end{aligned} \tag{14.157}$$

$V_{\mathbb{Q}}$ is a mixed Tate Hodge structure.

Exercise 117 *Work out all $V^{p,q}$ and show that $V_{\mathbb{Q}}$ is mixed Tate.*

14.4.3 Variations of Hodge Structures

We are in particular interested in families of (mixed) Hodge structures, parametrised by a manifold B (the B stands for "base"). We assume that for every point $x \in B$ we have a mixed Hodge structure. This will lead us to a variation of mixed Hodge structures [400, 401]. In detail:

A **variation of mixed Hodge structure** on the manifold B consists of

- a local system $\mathcal{L}_{\mathbb{Z}}$ of $\mathbb{Z}$-modules of finite rank,
- a finite increasing filtration $\mathcal{W}$ of $\mathcal{L}_{\mathbb{Q}} = \mathcal{L}_{\mathbb{Z}} \otimes \mathbb{Q}$ by sublocal systems of rational vector spaces,
- a finite decreasing filtration $\mathcal{F}$ of $\mathcal{L}_{\mathcal{O}_B} = \mathcal{L}_{\mathbb{Z}} \otimes \mathcal{O}_B$, satisfying Griffiths' transversality condition:

$$\nabla \left(\mathcal{F}^p \right) \subset \mathcal{F}^{p-1} \otimes \Omega^1_B. \tag{14.158}$$

- the filtrations $\mathcal{W}$ and $\mathcal{F}$ define a mixed Hodge structure on each fibre $(\mathcal{L}_{\mathcal{O}_B}(x), \mathcal{W}(x), \mathcal{F}(x))$ of the bundle $\mathcal{L}_{\mathcal{O}_B}(x)$ at point x.

Here, $\mathcal{O}_B$ denotes the sheaf of holomorphic functions on B, and Ω^1_B denotes the sheaf of differential one-forms on B.

Let's continue with example 2 from the previous section: $V_{\mathbb{Q}}$ is generated by $\langle v_0, v_1, \dots, v_n \rangle$, where the vectors v_j are given by the columns of the matrix P in Eq. (14.151). From Eq. (14.153) we deduce

$$\left[d - C_0 \, d \ln(x) - C_1 \, d \ln(x-1) \right] v_j = 0, \tag{14.159}$$

with the $(n+1) \times (n+1)$-matrices C_0 and C_1 defined in Eq. (14.152). The connection ∇ is therefore given by

$$\nabla = d - C_0 \, d \ln(x) - C_1 \, d \ln(x-1) \tag{14.160}$$

and we write

$$\nabla v_j = 0. \tag{14.161}$$

In physics jargon we call ∇ a covariant derivative and say that the v_j's are covariantly constant (or parallel transported with respect to ∇). In the mathematical language one says that the v_j's define a locally constant sheaf.

Let's consider $U = \mathbb{C}\backslash(]-\infty, 0] \cup [1, \infty[)$. In this region P is single-valued and from Sect. 14.4.2 we know already that for any $x \in U$ the columns of P define a mixed Tate Hodge structure. It remains to verify Griffiths' transversality condition. We need to determine ∇e_{-j} for $j \in \{0, \dots, n\}$. We find

$$\begin{aligned}
\nabla e_0 &= -\frac{dx}{x-1} e_{-1}, \\
\nabla e_{-j} &= -\frac{dx}{x} e_{-j-1}, \qquad 1 \le j < n, \\
\nabla e_{-n} &= 0.
\end{aligned} \tag{14.162}$$

With Eq. (14.162) it follows that

$$\nabla \left(\mathcal{F}^p\right) \subset \mathcal{F}^{p-1} \otimes \Omega^1_U. \tag{14.163}$$

Exercise 118 *Derive Eq. (14.162).*

14.4.4 Mixed Hodge Structures on Cohomology Groups

Let X be a complex algebraic variety and Y a (possibly empty) closed subvariety. It has been shown by Deligne [398, 399] that the relative cohomology group $H^k(X, Y)$ carries a mixed Hodge structure. If Y is empty, this reduces to the (non-relative) cohomology $H^k(X)$.

Let us look at the details: To simplify life we assume that X is smooth and Y a simple normal crossing divisor. If Y is not a simple normal crossing divisor, one uses first the resolution of singularities (see Chap. 10) to achieve this condition.

For the mixed Hodge structure we have to define the weight filtration and the Hodge filtration. This is best explained with differential forms and de Rham cohomology. Due to Theorem 14 we may replace de Rham cohomology with algebraic de Rham cohomology. But let us stick in this section with de Rham cohomology. We denote by Ω^p_X the sheaf of holomorphic differential forms of degree p on X and by $\Omega^p_X(\log Y)$ the sheaf of meromorphic differential forms of degree p on X with at most logarithmic poles along Y. A meromorphic differential form ω has at most logarithmic poles along Y if both ω and $d\omega$ have at most a simple pole along Y.

This implies that in the coordinate system of Eq. (14.37) a differential one-form $\omega \in \Omega^1_X$ can locally be written as

$$\omega = \sum_{j=1}^{n} f_j(x)\, dx_j, \tag{14.164}$$

with f_j holomorphic, while a differential one-form $\omega \in \Omega^1_X(\log Y)$ can locally be written as

$$\omega = \sum_{j=1}^{k} f_j(x) \frac{dx_j}{x_j} + \sum_{j=k+1}^{n} f_j(x)\, dx_j, \tag{14.165}$$

again with f_j holomorphic. The differential p-forms in Ω^p_X and $\Omega^p_X(\log Y)$ are then obtained from the p-fold wedge product of forms in Ω^1_X and $\Omega^1_X(\log Y)$, respectively.

Let's now consider $\Omega^\bullet_X(\log Y)$. There is a trivial filtration:

$$F^p \Omega^\bullet_X(\log Y) = \bigoplus_{r \geq p} \Omega^p_X(\log Y). \tag{14.166}$$

This filtration induces the Hodge filtration on the cohomology.

For the weight filtration we count the number of dlog-forms. One sets

$$W_m \Omega^p_X(\log Y) = \begin{cases} 0 & \text{for } m < 0, \\ \Omega^{p-m}_X \wedge \Omega^m_X(\log Y) & \text{for } 0 \leq m \leq p, \\ \Omega^p_X(\log Y) & \text{for } p < m. \end{cases} \tag{14.167}$$

This filtration induces the weight filtration on the cohomology. Without going into the details let us note that in the k-th cohomology group differential forms with m dlog-one-forms will contribute at weight $m + k$.

Let us illustrate the weight filtration with an example. We take $X = \mathbb{C}^3$ and

$$Y = \{x_1 = 0\} \cup \{x_2 = 0\} \cup \{x_3 = 1\}. \tag{14.168}$$

We consider the three-forms

$$\begin{aligned} \omega_1 &= -\frac{dx_1}{x_1} \wedge \frac{dx_2}{x_2} \wedge \frac{dx_3}{1 - x_3}, \\ \omega_2 &= -\frac{dx_1}{x_1} \wedge dx_2 \wedge \frac{dx_3}{1 - x_3}. \end{aligned} \tag{14.169}$$

As both are three-forms, we have

$$\omega_1, \omega_2 \in F^3 \Omega^\bullet_X(\log Y). \tag{14.170}$$

In a local coordinate system around $(x_1, x_2, x_3) = (0, 0, 1)$ the three-form ω_1 has three dlog forms, while ω_2 has only two. Hence

$$\omega_1 \in W_3\Omega^3_X(\log Y), \quad \omega_2 \in W_2\Omega^3_X(\log Y). \tag{14.171}$$

The counting of the dlog forms as weight is illustrated as follows: Consider the integration domain $\gamma : 0 \le x_3 \le x_2 \le x_1 \le 1$. We then have

$$\int\limits_{\gamma} \omega_1 = \zeta_3,$$
$$\int\limits_{\gamma} \omega_2 = -\zeta_2 + 2. \tag{14.172}$$

Thus $\langle \omega_1 | \gamma \rangle$ is pure of (polylogarithmic) weight 3, while $\langle \omega_2 | \gamma \rangle$ is mixed with highest (polylogarithmic) weight 2.

The attentive reader might have noticed that we already discussed weight filtrations of mixed Hodge structures in two examples. From Eqs. (14.142) and (14.156) we would expect that the weight of ζ_n within Hodge theory is $(-2n)$. This deserves some explanation. Let's look at ζ_3: We start from the three-form ω_1, hence we are interested in the relative cohomology group $H^3(X, Y)$. The form ω_1 is a wedge product of three dlog-one-forms. We therefore have $k = 3$ and $m = 3$. The comment after Eq. (14.167) implies that ω_1 contributes at weight $m + k = 6$ in cohomology. Instead of looking at cohomology we may also look at homology, which is just the dual. Under dualisation the weight of the weight filtration changes sign, in our example $6 \to -6$, in agreement with the examples of Eqs. (14.142) and (14.156).

14.4.5 Motivic Periods

We may now bring the various pieces together and define motivic periods. We remind the reader that we discussed effective periods in Chap. 10 and that we already discussed motivic periods in Chap. 11. The definition which we provide now is a slight generalisation of the definition of effective periods discussed in Sect. 10.4. The motivic periods discussed in Chap. 11 correspond to the subset of motivic mixed Tate periods.

Let X be a smooth variety defined by polynomials with coefficients in $\mathbb{Q}$. Let Y be a closed subvariety. We denote by X^{an} and Y^{an} the analytifications. Let $\omega \in H^k_{\mathrm{alg\ dR}}(X, Y)$ be a class of the k-th relative algebraic de Rham cohomology and let $\gamma \in H_k^{\mathrm{B}}(X^{\mathrm{an}}, Y^{\mathrm{an}}, \mathbb{Q})$ be a class in the k-th relative Betti homology.

From the previous section we know that we have mixed Hodge structure on $H^k(X, Y)$. That is to say that $H^k_{\mathrm{alg\ dR}}(X, Y)$ is equipped with a Hodge filtration $F^\bullet$

and a weight filtration $W_\bullet$, that the Betti cohomology $H^k_{\mathrm{B}}(X^{\mathrm{an}}, Y^{\mathrm{an}}, \mathbb{Q})$ is equipped with a weight filtration $W_\bullet$ and there is a comparison isomorphism

$$\text{comparison} : H^k_{\mathrm{alg\ dR}}(X, Y) \otimes \mathbb{C} \to H^k_{\mathrm{B}}(X^{\mathrm{an}}, Y^{\mathrm{an}}, \mathbb{Q}) \otimes \mathbb{C} \tag{14.173}$$

compatible with the weight filtration.

The triple

$$M^k(X, Y) = \left[H^k_{\mathrm{alg\ dR}}(X, Y), H^k_{\mathrm{B}}(X^{\mathrm{an}}, Y^{\mathrm{an}}, \mathbb{Q}), \text{comparison} \right]^{\mathrm{m}} \tag{14.174}$$

is called an **H-motive** (or "motive" for short, the long official name is "Hodge realisation of the motive").

The **real Frobenius** is a linear involution

$$F_\infty : H^k_{\mathrm{B}}(X^{\mathrm{an}}, Y^{\mathrm{an}}, \mathbb{Q}) \to H^k_{\mathrm{B}}(X^{\mathrm{an}}, Y^{\mathrm{an}}, \mathbb{Q}) \tag{14.175}$$

defined as follows: We denote by $\mathrm{conj}_{\mathrm{alg\ dR}}$ the $\mathbb{C}$-antilinear involution on $H^k_{\mathrm{alg\ dR}}(X, Y) \otimes \mathbb{C}$ given by

$$\mathrm{conj}_{\mathrm{alg\ dR}}(\omega \otimes z) = \omega \otimes \bar{z} \tag{14.176}$$

and by $\mathrm{conj}_{\mathrm{B}}$ the analogous $\mathbb{C}$-antilinear involution on $H^k_{\mathrm{B}}(X^{\mathrm{an}}, Y^{\mathrm{an}}) \otimes \mathbb{C}$. Then F_∞ is defined such that the following diagram commutes:

$$\begin{array}{ccc} H^k_{\mathrm{alg\ dR}}(X, Y) \otimes \mathbb{C} & \xrightarrow{\text{comparison}} & H^k_{\mathrm{B}}(X^{\mathrm{an}}, Y^{\mathrm{an}}) \otimes \mathbb{C} \\ \mathrm{conj}_{\mathrm{alg\ dR}} \big\downarrow & & \big\downarrow (F_\infty \otimes \mathrm{id})\, \mathrm{conj}_{\mathrm{B}} \\ H^k_{\mathrm{alg\ dR}}(X, Y) \otimes \mathbb{C} & \xrightarrow{\text{comparison}} & H^k_{\mathrm{B}}(X^{\mathrm{an}}, Y^{\mathrm{an}}) \otimes \mathbb{C} \end{array} \tag{14.177}$$

The following exercise illustrates the action of the real Frobenius:

Exercise 119 *Consider $X = \mathbb{C} \backslash \{0\}$ and $Y = \emptyset$. Take $\omega = dx/x$ as a basis of $H^1_{\mathrm{alg\ dR}}(X)$ and let γ be a small counter-clockwise circle around $x = 0$. γ is a basis of $H_1^{\mathrm{B}}(X)$. Denote by γ^* the dual basis of $H^1_{\mathrm{B}}(X)$. Work out*

$$F_\infty \left(\gamma^* \right). \tag{14.178}$$

Let us then consider the $\mathbb{Q}$-vector space of equivalence classes of triples

$$\left[M^k(X, Y), \omega, \gamma \right]^{\mathrm{m}} \tag{14.179}$$

modulo the relations induced by linearity in ω and γ, changes of variables and Stokes' formula. Linearity states that for $c_1, c_2 \in \mathbb{Q}$

$$\left[M^k(X,Y), c_1\omega_1 + c_2\omega_2, \gamma\right]^{\mathrm{m}} = c_1\left[M^k(X,Y), \omega_1, \gamma\right]^{\mathrm{m}} + c_2\left[M^k(X,Y), \omega_2, \gamma\right]^{\mathrm{m}},$$
$$\left[M^k(X,Y), \omega, c_1\gamma_1 + c_2\gamma_2\right]^{\mathrm{m}} = c_1\left[M^k(X,Y), \omega, \gamma_1\right]^{\mathrm{m}} + c_2\left[M^k(X,Y), \omega, \gamma_2\right]^{\mathrm{m}}. \quad (14.180)$$

Let $f : X \to X'$ be a regular map and $Y' = f(Y)$. A change of variables implies

$$\left[M^k(X,Y), f^*\omega', \gamma\right]^{\mathrm{m}} = \left[M^k(X',Y'), \omega', f_*\gamma\right]^{\mathrm{m}}. \quad (14.181)$$

For Stokes' theorem consider $X \supset Y \supset Z$ and denote the connecting morphism by $d : H^{k-1}(Y,Z) \to H^k(X,Y)$. Then

$$\left[M^k(X,Y), d\omega, \gamma\right]^{\mathrm{m}} = \left[M^{k-1}(Y,Z), \omega, \partial\gamma\right]^{\mathrm{m}}. \quad (14.182)$$

The equivalence classes with respect to these relations are called **effective motivic periods** and denoted as $\mathcal{P}^{\mathrm{m}}$. To each effective motivic period we can associate a numerical period. In other words, there is a map

$$\begin{aligned} \mathrm{period} : \mathcal{P}^{\mathrm{m}} &\to \mathbb{P}, \\ \left[M^k(X,Y), \omega, \gamma\right]^{\mathrm{m}} &\to \int_\gamma \omega. \end{aligned} \quad (14.183)$$

Every motivic period comes with a mixed Hodge structure (the one discussed in Sect. 14.4.4) and in particular a weight filtration. If we forget about the extra information on the mixed Hodge structure we have an effective period as introduced in Chap. 10, which can be specified by the quadruple (X, Y, ω, γ).

The real Frobenius $F_\infty : H^k_{\mathrm{B}}(X^{\mathrm{an}}, Y^{\mathrm{an}}, \mathbb{Q}) \to H^k_{\mathrm{B}}(X^{\mathrm{an}}, Y^{\mathrm{an}}, \mathbb{Q})$ of Eq. (14.175) induces a map

$$F_\infty : \mathcal{P}^{\mathrm{m}} \to \mathcal{P}^{\mathrm{m}} \quad (14.184)$$

through

$$\left[M^k(X,Y), \omega, \gamma\right]^{\mathrm{m}} \to \left[M^k(X,Y), \omega, F_\infty(\gamma)\right]^{\mathrm{m}}. \quad (14.185)$$

By abuse of notation we also denote this map by F_∞. The latter map we already encountered in Eq. (11.211).

14.4.6 The Motivic Galois Group

In order not to raise false expectations let us state from the beginning that although despite the title of this section we will be dealing with the motivic Galois group only indirectly through its dual, which is a Hopf algebra. But that is o.k., as this is the way we would like to apply it anyway. Of course, it is possible to construct the group from its dual.

We start by introducing algebraic groups. An **affine algebraic group** G is a group defined by polynomial equations such that the group law is polynomial.

To give an example, consider $\mathrm{SL}_2(\overline{\mathbb{Q}})$, the set of (2×2)-matrices with entries from $\overline{\mathbb{Q}}$ and determinant 1:

$$g = \begin{pmatrix} z_{11} & z_{12} \\ z_{21} & z_{22} \end{pmatrix}, \qquad z_{11}, z_{12}, z_{21}, z_{22} \in \overline{\mathbb{Q}}, \qquad \det g = 1. \tag{14.186}$$

Alternatively, we may view $\mathrm{SL}_2(\overline{\mathbb{Q}})$ as an affine algebraic variety:

$$\mathrm{SL}_2(\overline{\mathbb{Q}}) = \left\{ (z_{11}, z_{12}, z_{21}, z_{22}) \in \overline{\mathbb{Q}}^4 \mid z_{11}z_{22} - z_{12}z_{21} - 1 = 0 \right\} \tag{14.187}$$

Exercise 120 *Let $\mathbb{F}$ be a sub-field of $\mathbb{C}$. Show that $\mathrm{GL}_n(\mathbb{F})$ (the group of $(n \times n)$-matrices with entries from $\mathbb{F}$ and non-zero determinant) can be defined by a polynomial equation.*

Let us now consider functions on the group, i.e. maps

$$G \to \mathbb{F}, \tag{14.188}$$

where we take $\mathbb{F}$ to be a sub-field of $\mathbb{C}$. If we specialise to matrix groups, like $\mathrm{GL}_n(\mathbb{F})$, the entry of the i-th row and j-th column provides a function on the group:

$$\begin{aligned} a_{ij} : \; & \mathrm{GL}_n(\mathbb{F}) \to \mathbb{F}, \\ & g \to z_{ij}. \end{aligned} \tag{14.189}$$

A polynomial function on the group is a function on the group, which is a polynomial in the a_{ij}'s. We denote by H the set of polynomial functions on the group. H is a Hopf algebra. The coproduct comes from the multiplication in the group. Let $g_1, g_2 \in \mathrm{GL}_n(\mathbb{F})$ with entries $z_{ij}^{(1)}$ and $z_{ij}^{(2)}$, respectively, and consider

$$\begin{aligned} a_{ij}(g_1 \cdot g_2) &= \sum_{k=1}^{n} z_{ik}^{(1)} z_{kj}^{(2)} = \sum_{k=1}^{n} a_{ik}(g_1)\, a_{kj}(g_2) \\ &= \cdot \left(\sum_{k=1} a_{ik} \otimes a_{kj} \right) (g_1 \otimes g_2). \end{aligned} \tag{14.190}$$

This gives us the coproduct

$$\begin{aligned} \Delta : \; H &\rightarrow H \otimes H, \\ a_{ij} &\rightarrow \sum_{k=1}^{n} a_{ik} \otimes a_{kj}. \end{aligned} \tag{14.191}$$

It can be verified that H is indeed a Hopf algebra.

Let us summarise: Given a group $G = \mathrm{GL}_n(\mathbb{F})$ we obtain a Hopf algebra H by considering the polynomial functions on G. This also goes in the reverse direction: One may show that one can construct the group G from the Hopf algebra H.

A group may act on a vector space. Also in this case we may consider the dual picture. For concreteness, we take again $G = \mathrm{GL}_n(\mathbb{F})$ and consider $V = \mathbb{F}^n$. Let

$$v = \begin{pmatrix} x_1 \\ \vdots \\ x_n \end{pmatrix} \in V. \tag{14.192}$$

We consider functions on V. We start from the coordinate functions

$$\begin{aligned} b_i : \mathbb{F}^n &\rightarrow \mathbb{F}, \\ v &\rightarrow x_i. \end{aligned} \tag{14.193}$$

Let M denote the set of polynomial functions on V (i.e. functions which are given as polynomials in the b_i's). The action of G on V induces a coaction of H on M: Consider

$$\begin{aligned} b_i \left(g \cdot v\right) = \sum_{k=1}^{n} z_{ik} x_k \; &= \sum_{k=1}^{n} a_{ik} \left(g\right) b_k \left(v\right) \\ &= \cdot \left(\sum_{k=1}^{n} a_{ik} \otimes b_k \right) \left(g \otimes v\right). \end{aligned} \tag{14.194}$$

This gives us the coaction (which we also denote by Δ)

$$\begin{aligned} \Delta : \; M &\rightarrow H \otimes M, \\ b_i &\rightarrow \sum_{k=1}^{n} a_{ik} \otimes b_k. \end{aligned} \tag{14.195}$$

Let's now apply these ideas to motivic periods: We first define motivic de Rham periods. These are triples of the form

$$\left[M^k \left(X, Y\right), \omega, \omega^* \right]^{\mathfrak{dR}}, \tag{14.196}$$

where $\omega \in H^k_{\text{alg dR}}(X, Y)$ and $\omega^* \in (H^k_{\text{alg dR}}(X, Y))^*$ (the dual). As usual, we consider these triples modulo the relations of linearity Eq. (14.180), change of variables Eq. (14.181) and Stokes Eq. (14.182). We call the set of equivalence classes **motivic de Rham periods** and denote this set by $\mathcal{P}^{\mathfrak{dR}}$.

Let us denote by $\omega_1, \dots, \omega_r \in H^k_{\text{alg dR}}(X, Y)$ a basis of $H^k_{\text{alg dR}}(X, Y)$ and by $\omega_1^*, \dots, \omega_r^* \in (H^k_{\text{alg dR}}(X, Y))^*$ the dual basis of $(H^k_{\text{alg dR}}(X, Y))^*$. On $\mathcal{P}^{\mathfrak{dR}}$ we have a coproduct

$$\Delta : \mathcal{P}^{\mathfrak{dR}} \to \mathcal{P}^{\mathfrak{dR}} \otimes \mathcal{P}^{\mathfrak{dR}} \tag{14.197}$$

which is given by

$$\Delta \left[M^k(X, Y), \omega_i, \omega_k^* \right]^{\mathfrak{dR}} = \sum_{l=1}^{r} \left[M^k(X, Y), \omega_i, \omega_l^* \right]^{\mathfrak{dR}} \otimes \left[M^k(X, Y), \omega_l, \omega_k^* \right]^{\mathfrak{dR}}. \tag{14.198}$$

$\mathcal{P}^{\mathfrak{dR}}$ is a Hopf algebra. The **motivic Galois group** $G^{\mathfrak{dR}}$ is the dual of the Hopf algebra $\mathcal{P}^{\mathfrak{dR}}$. (This is the definition of the motivic Galois group. As mentioned in the introduction of this section, it is an indirect definition.) $\mathcal{P}^{\mathfrak{dR}}$ coacts on $\mathcal{P}^{\text{m}}$, this gives us the motivic coaction:

Motivic coaction:
We have a coaction of the motivic de Rham periods $\mathcal{P}^{\mathfrak{dR}}$ on the effective motivic periods $\mathcal{P}^{\text{m}}$

$$\Delta : \mathcal{P}^{\text{m}} \to \mathcal{P}^{\mathfrak{dR}} \otimes \mathcal{P}^{\text{m}} \tag{14.199}$$

given by

$$\Delta \left[M^k(X, Y), \omega, \gamma \right]^{\text{m}} = \sum_{l=1}^{r} \left[M^k(X, Y), \omega, \omega_l^* \right]^{\mathfrak{dR}} \otimes \left[M^k(X, Y), \omega_l, \gamma \right]^{\text{m}}. \tag{14.200}$$

Eq. (14.200) is called the **motivic coaction**.

Let's look at an example. We elaborate on the example from Sect. 14.1.2. We take $X = \mathbb{C}^* = \mathbb{C}\backslash\{0\}$ and $Y = \{1, x\}$ (in Sect. 14.1.2 we considered the special case $x = 2$). A basis for $H_1^{\text{B}}(X, Y)$ is given by an anti-clockwise circle γ_1 around $z = 0$ and the line segment γ_2 from $z = 1$ to $z = x$. A basis for $H^1_{\text{dR}}(X, Y)$ is given by $\omega_1 = dz/z$ and $\omega_2 = dz/(x - 1)$. The period matrix is given by

$$P = \begin{pmatrix} 2\pi i & \ln x \\ 0 & 1 \end{pmatrix}. \tag{14.201}$$

We have chosen the normalisation of ω_2 such that

$$\int_{\gamma_2} \omega_2 = 1. \tag{14.202}$$

We set

$$\ln^{\mathfrak{m}}(x) = \left[M^1(X,Y), \omega_1, \gamma_2\right]^{\mathfrak{m}}. \tag{14.203}$$

The motivic coaction gives us

$$\Delta\left(\ln^{\mathfrak{m}}(x)\right) = \left[M^1(X,Y), \omega_1, \omega_1^*\right]^{\mathfrak{dR}} \otimes \left[M^1(X,Y), \omega_1, \gamma_2\right]^{\mathfrak{m}} + \left[M^1(X,Y), \omega_1, \omega_2^*\right]^{\mathfrak{dR}} \otimes \left[M^1(X,Y), \omega_2, \gamma_2\right]^{\mathfrak{m}}. \tag{14.204}$$

Let's look at the individual expressions: The period map of Eq. (14.183) sends $[M^1(X,Y), \omega_2, \gamma_2]^{\mathfrak{m}}$ to 1. This can be read off from Eq. (14.202). Assuming that the period map is injective, we set

$$\left[M^1(X,Y), \omega_2, \gamma_2\right]^{\mathfrak{m}} = 1^{\mathfrak{m}}. \tag{14.205}$$

$[M^1(X,Y), \omega_1, \omega_1^*]^{\mathfrak{dR}}$ is an element of $\mathcal{P}^{\mathfrak{dR}}$. The motivic de Rham periods form a Hopf algebra and it can be shown that the counit maps $[M^1(X,Y), \omega_1, \omega_1^*]^{\mathfrak{dR}}$ to 1. We denote

$$\left[M^1(X,Y), \omega_1, \omega_1^*\right]^{\mathfrak{dR}} = 1^{\mathfrak{dR}}. \tag{14.206}$$

If we finally define

$$\ln^{\mathfrak{dR}}(x) = \left[M^1(X,Y), \omega_1, \omega_2^*\right]^{\mathfrak{dR}} \tag{14.207}$$

we arrive at

$$\Delta\left(\ln^{\mathfrak{m}}(x)\right) = 1^{\mathfrak{dR}} \otimes \ln^{\mathfrak{m}}(x) + \ln^{\mathfrak{dR}}(x) \otimes 1^{\mathfrak{m}}, \tag{14.208}$$

in agreement with Eq. (J.527). This provides the link with the coaction defined for multiple polylogarithms in Sect. 11.2.

14.5 Examples from Feynman Integrals

Feynman integrals provide ample examples for motivic periods. Throughout this section we assume that all kinematic variables satisfy

$$x_j \geq 0, \qquad \text{(Euclidean region)} \tag{14.209}$$

and

$$x_j \in \mathbb{Q}. \tag{14.210}$$

Condition (14.209) ensures that in the Feynman parametrisation the singularities of the integrand are at worst on the boundary of the integration domain, but not inside. Condition (14.210) ensures that the variety where the integrand is singular is defined over $\mathbb{Q}$.

14.5.1 Feynman Motives Depending only on One Graph Polynomial

We recall that with these assumptions we may use the algorithm of sector decomposition (see Chap. 10) and express any term $I^{(j)}$ in the ε-expansion of any Feynman integral I

$$I = \sum_{j=j_{\min}}^{\infty} \varepsilon^j I^{(j)} \tag{14.211}$$

as an absolute convergent integral. This will also ensure that the singularities have normal crossings. From Theorem 9 we know that each $I^{(j)}$ gives a numerical period. Denoting by η the integrand (a k-form), γ the integration domain (a relative k-cycle with boundary contained in B), P the space containing γ and Y the variety where η is singular we obtain a motive

$$M^k \left(P \backslash Y, B \backslash \left(B \cap Y\right)\right) \tag{14.212}$$

and the motivic period

$$\left[M^k \left(P \backslash Y, B \backslash \left(B \cap Y\right)\right), \eta, \gamma\right]^{\mathrm{m}}. \tag{14.213}$$

Let us now specialise to two cases, where Y is determined by a single graph polynomial. (The case where Y is determined by both graph polynomials is only from a notational perspective more cumbersome.) We start from the Feynman parameter representation Eq. (2.198) and set from the beginning $\nu_1 = \dots = \nu_{n_{\mathrm{int}}} = 1$:

$$I = e^{l\varepsilon\gamma_{\mathrm{E}}} \Gamma\left(n_{\mathrm{int}} - \frac{lD}{2}\right) \int\limits_{\Delta} \frac{\mathcal{U}^{n_{\mathrm{int}} - \frac{(l+1)D}{2}}}{\mathcal{F}^{n_{\mathrm{int}} - \frac{lD}{2}}} \omega. \tag{14.214}$$

The differential $(n_{\mathrm{int}} - 1)$-form ω is defined in Eq. (2.196). Let us further assume that the integration over the Feynman parameters is well-defined without regularisation.

Fig. 14.1 The 4-loop wheel with four spokes graph (left) and the 6-loop zigzag graph (right)

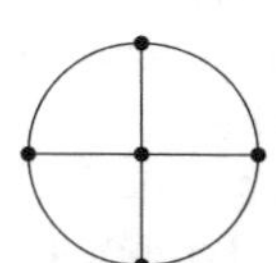

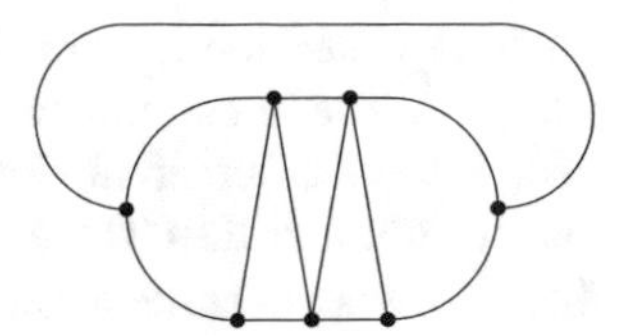

This implies that the Feynman integral is either finite or has at most an overall ultraviolet singularity, which manifests itself in the prefactor $\Gamma(n_{\text{int}} - lD/2)$. Let us further assume that either the exponent of the $\mathcal{F}$-polynomial is zero or the exponent of the $\mathcal{U}$-polynomial is zero. In the former case we end up with (apart from prefactors)

$$\int\limits_{\Delta} \frac{\omega}{\mathcal{U}^{\frac{D}{2}}}, \tag{14.215}$$

in the latter case with

$$\int\limits_{\Delta} \frac{\omega}{\mathcal{F}^{\frac{D}{2}}}. \tag{14.216}$$

In both cases the integrand depends only on one graph polynomial. For D an even integer, the integrand is a rational function.

Let's see if there are interesting examples matching our assumptions: Setting $D = 2$ in Eq. (14.215) yields $n_{\text{int}} = l$ (since the exponent of the $\mathcal{F}$-polynomial has to vanish). This only allows for a product of tadpole integrals and is not so interesting. However, $D = 4$ leads to $n_{\text{int}} = 2l$ and there are interesting graphs. There is the family of the wheel with l spokes graphs and the family of zigzag graphs [402–404]. Two examples are shown in Fig. 14.1.

In the case of Eq. (14.216) we may set $D = 2$. We obtain then $n_{\text{int}} = (l + 1)$. This gives the family of banana graphs shown in Fig. 14.2 [368, 405]. For non-zero internal masses these integrals are finite (in $D = 2$ space-time dimensions).

Thus the examples which we are going to consider are

$$\int\limits_{\Delta} \frac{\omega}{\mathcal{U}^2} \;\; \text{(wheel with} l \text{spokes graphs, zigzag graphs)}, \qquad \int\limits_{\Delta} \frac{\omega}{\mathcal{F}} \;\; \text{(banana graphs)}. \tag{14.217}$$

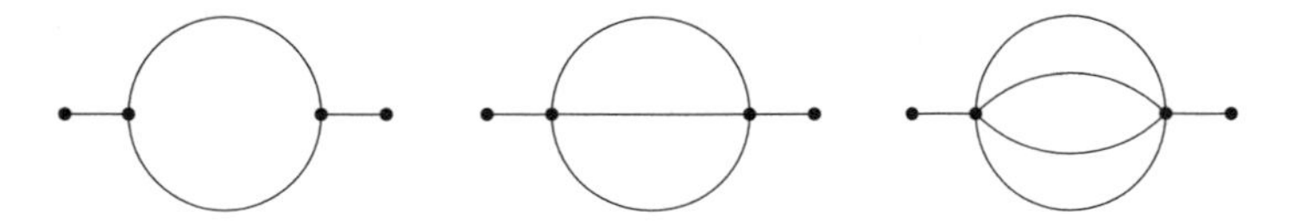

Fig. 14.2 The first three members of the family of banana graphs

In both cases the integrand is a rational differential form on $\mathbb{CP}^{n_{\text{int}}-1}$ and $\Delta = \mathbb{RP}^{n_{\text{int}}-1}_{\geq 0}$. The integration region has a boundary, which we denote by $\partial\Delta$. Despite the fact that we assumed the integrals are finite, we cannot conclude that the integrand has no singularities: There might be (and will be) integrable singularities. This will happen, whenever the relevant graph polynomial vanishes. Let us set $\mathcal{X} = \mathcal{U}$ for the wheel with l spokes graphs and the zigzag graphs and $\mathcal{X} = \mathcal{F}$ for the banana graphs. We define

$$X = \left\{ a \in \mathbb{CP}^{n_{\text{int}}-1} \mid \mathcal{X}(a) = 0 \right\}. \tag{14.218}$$

X is an algebraic variety, defined as the zero set of the graph polynomial $\mathcal{X}$. In the integration, we encounter a singularity wherever X and Δ intersect. It can be shown for the first graph polynomial $\mathcal{U}$ that

$$X \cap \Delta \subset \partial\Delta, \tag{14.219}$$

E.g. intersections happens only on the boundary $\partial\Delta$ of the integration region Δ. This is the case because $\mathcal{U}$ is a sum of monomials with all coefficients equal to 1. Equation (14.219) also holds true for the second graph polynomial $\mathcal{F}$, if we restrict ourselves for the kinematics to the Euclidean region (e.g. all kinematic variables $x_j \geq 0$). In this case $\mathcal{F}$ is a sum of monomials with all coefficients positive. This is the reason why we imposed assumption Eq. (14.209).

If $X \cap \partial\Delta$ is non-empty, we blow-up $\mathbb{CP}^{n_{\text{int}}-1}$ in this region. Let us denote the blow-up by P. We further denote the strict transform of X by Y and we denote the total transform of the set $\{x_1 \cdot x_2 \cdot \dots \cdot x_n = 0\}$ by B.

We view the integrals of Eq. (14.217) as periods of a mixed Hodge structures, obtained from two geometric objects: The algebraic variety X and the domain of integration Δ. As the domain of integration Δ has a boundary we have to consider relative cohomology. In order to avoid (integrable) singularities of the integrand we consider the blow-up. This leads us to the mixed Hodge structure given by the relative cohomology group [51, 402, 406–409]

$$H^{n_{\text{int}}-1}\left(P \backslash Y, B \backslash (B \cap Y)\right). \tag{14.220}$$

The Feynman integral is then a period of this cohomology class. The **motive associated to the Feynman integral** is given by

$$M^{n_{\text{int}}-1}\left(P \backslash Y, B \backslash (B \cap Y)\right) = \tag{14.221}$$
$$\left[H^{n_{\text{int}}-1}_{\text{alg dR}}\left(P \backslash Y, B \backslash (B \cap Y)\right), H^{n_{\text{int}}-1}_{\text{B}}\left((P \backslash Y)^{\text{an}}, (B \backslash (B \cap Y))^{\text{an}}, \mathbb{Q}\right), \text{comparison} \right]^{\mathfrak{m}}.$$

14.5.2 The Sunrise Motive

Let us now look at a concrete example: The two-loop sunrise integral with unequal masses. We encountered this Feynman integral already in Eq. (13.300). With the notation as in Eq. (13.300) we consider

$$I_{111}\left(2, x, y_1, y_2\right) = \int \frac{d^2k_1}{i\pi} \frac{d^2k_2}{i\pi} \frac{m_3^2}{\left(-q_1^2+m_1^2\right)\left(-q_2^2+m_2^2\right)\left(-q_3^2+m_3^2\right)}. \tag{14.222}$$

In the Feynman parameter representation we have

$$I_{111}\left(2, x, y_1, y_2\right) = \int\limits_{\Delta} \frac{\omega}{\mathcal{F}}, \tag{14.223}$$

with

$$\begin{aligned} \mathcal{F} &= a_1 a_2 a_3 x + \left(a_1 y_1 + a_2 y_2 + a_3\right)\left(a_1 a_2 + a_2 a_3 + a_3 a_1\right), \\ \omega &= a_1 da_2 \wedge da_3 - a_2 da_1 \wedge da_3 + a_3 da_1 \wedge da_2, \\ \Delta &= \mathbb{RP}^2_{\geq 0}. \end{aligned} \tag{14.224}$$

We define the variety where $\mathcal{F}$ vanishes:

$$X = \left\{ \left[a_1 : a_2 : a_3\right] \in \mathbb{CP}^2 \mid \mathcal{F}(a) = 0 \right\}. \tag{14.225}$$

X and Δ intersect in the three points

$$[1:0:0], \quad [0:1:0], \quad [0:0:1]. \tag{14.226}$$

Let P be the blow-up of $\mathbb{CP}^2$ in these three points. The exceptional divisors in these three points are denoted by E_1, E_2 and E_3, respectively. We denote the strict transform of X by Y and we denote by B the total transform of $a_1 \cdot a_2 \cdot a_3 = 0$. The mixed Hodge structure (or motive) associated to the Feynman integral $I_{111}(2, x, y_1, y_2)$ is then

$$H^2\left(P \backslash Y, B \backslash B \cap Y\right). \tag{14.227}$$

The Feynman integral is a period of this motive.

What can be said about this motive? One can show that there is a short exact sequence of mixed Hodge structures [368]

$$0 \longrightarrow \mathbb{Z}(-1) \longrightarrow H^2\left(P \backslash Y, B \backslash B \cap Y\right) \longrightarrow H^2\left(P \backslash Y\right) \longrightarrow 0, \tag{14.228}$$

and for $H^2(P\backslash Y)$ we have the short exact sequence

$$0 \longrightarrow \mathbb{Z}E_1 \oplus \mathbb{Z}E_2 \oplus \mathbb{Z}E_3 \longrightarrow H^2(P\backslash Y) \xrightarrow{\text{res}} H^1(Y) \longrightarrow 0, \quad (14.229)$$

This sequence is split as a sequence of mixed Hodge structures via

$$\begin{array}{ccc} H^2(P\backslash Y) & \xrightarrow{\text{res}} & H^1(Y) \\ \pi^* \uparrow & & \downarrow \cong \\ H^2(\mathbb{CP}^2\backslash X) & \xrightarrow[\cong]{\text{res}} & H^1(X). \end{array} \quad (14.230)$$

Digression Exact sequences
We start from a category where kernels and cokernels are defined. Typical examples are the category of groups, the category of vector spaces or the category of modules. We denote the objects by O_i and the morphisms by f_j. Consider a sequence of morphisms

$$O_0 \xrightarrow{f_1} O_1 \xrightarrow{f_2} O_2 \xrightarrow{f_3} \ldots \xrightarrow{f_n} O_n. \quad (14.231)$$

The sequence is said to be **exact**, *if*

$$\text{im}(f_i) = \ker(f_{i+1}). \quad (14.232)$$

An exact sequence of the form

$$0 \longrightarrow O_1 \xrightarrow{f} O_2 \quad (14.233)$$

states that f is injective, i.e. f is a monomorphism. The image of 0 under the first map is 0, which by the assumption of exactness is the kernel of f. Hence the kernel of f is trivial and f is injective.

By a similar reasoning, the exact sequence

$$O_1 \xrightarrow{f} O_2 \longrightarrow 0 \quad (14.234)$$

expresses that f is surjective, i.e. f is an epimorphism: The kernel of the rightmost morphism is O_2. Since the sequence is supposed to be exact, this equals the image of f, hence f is surjective.

A **short exact sequence** *is an exact sequence of the form*

$$0 \longrightarrow O_1 \xrightarrow{f} O_2 \xrightarrow{g} O_3 \longrightarrow 0. \quad (14.235)$$

This implies that f is injective and g is surjective. A short exact sequence is said to be **split**, *if there is a morphism $h : O_3 \to O_2$ such that $g \circ h$ is the identity map on O_3.*

In order to distinguish a general exact sequence from the special case of a short exact sequence, the term **long exact sequence** *is also used for the former.*

Now let us discuss how these considerations can be turned into a practical tool. Suppose we would like to compute the differential equation of $I_{111}(2, x, y_1, y_2)$ with respect to x. Of course, we already know one possibility how to do this: We could start with integration-by-parts identities in D dimensions as in Sect. 6.1, derive a coupled system of first-order differential equations as in Sect. 6.3 and convert the coupled system of first-order differential equations to a single higher order differential equation as in Sect. 7.1.3. Doing so, we would expect an inhomogeneous fourth order differential equation for $I_{111}(D, x, y_1, y_2)$. (We expect a fourth order differential equation because there are four master integrals in the top sector. We expect an inhomogeneous differential equation because there are sub-topologies.) For generic D we would indeed obtain an inhomogeneous fourth order differential equation. However we are interested in the finite integral $I_{111}(2, x, y_1, y_2)$ in two space-time dimensions. This raises the question if we can carry out the calculation directly in $D = 2$ space-time dimensions without the need of introducing an additional symbolic variable D. Secondly, we will see shortly that in two space-time dimensions there are only two independent master integrals in the top sector (instead of four master integrals for generic D). Thus we would like to avoid to work in intermediate stages with four master integrals in the top sector, if only two master integrals are required.

These questions are particularly important for cutting-edge calculations, where additional variables or additional master integrals can push the required computing resources (memory and/or CPU time) beyond the available resources. Therefore we would like to calculate only these parts which are strictly necessary. Motivic methods allow us to do this [368].

Let's look at the details. We would like to derive the differential equation for $I_{111}(2, x, y_1, y_2)$ with respect to the kinematic variable x. We treat the two additional kinematic variables y_1 and y_2 as (fixed) parameters. We seek

$$L \, I_{111}(2, x, y_1, y_2) = Q(x), \tag{14.236}$$

where L is a differential operator

$$L = \sum_{j=0}^{r} P_j(x) \frac{d^j}{dx^j} \tag{14.237}$$

and $Q(x)$ the inhomogeneous term. The order r of the differential operator L is a priori unknown.

Let's focus first on the differential operator L. We start from the algebraic variety X defined by the second Symanzik polynomial $\mathcal{F}$:

$$a_1 a_2 a_3 x + (a_1 y_1 + a_2 y_2 + a_3)(a_1 a_2 + a_2 a_3 + a_3 a_1) = 0. \quad (14.238)$$

This defines for generic values of the parameters x, y_1 and y_2 an elliptic curve. The elliptic curve varies smoothly with the parameters x, y_1 and y_2. By a birational change of coordinates this equation can brought into the Weierstrass normal form (instead of the common x, y, z we use u, v, w as coordinates in $\mathbb{CP}^2$)

$$v^2 w - u^3 - f_2(x) u w^2 - f_3(x) w^3 = 0. \quad (14.239)$$

f_2 and f_3 are functions of the kinematic variables x, y_1 and y_2. As we are mainly interested in the dependence on x, we suppress in the notation the dependence on y_1 and y_2. In the chart $w = 1$ the above equation reduces to

$$v^2 - u^3 - f_2(x) u - f_3(x) = 0. \quad (14.240)$$

In these coordinates $H^1(X)$ is generated by

$$\eta = \frac{du}{v} \text{ and } \eta' = \frac{d}{dx}\eta. \quad (14.241)$$

Since $H^1(X)$ is two-dimensional it follows that $\eta'' = \frac{d^2}{dx^2}\eta$ must be a linear combination of η and η'. In other words we must have a relation of the form

$$\eta'' + R_1(x)\eta' + R_0(x)\eta = 0. \quad (14.242)$$

It is convenient to bring this equation onto a common denominator. Doing so and carrying out the derivatives with respect to x we have

$$\begin{aligned}
\eta &= \left(u^3 + f_2 u + f_3\right)^2 \frac{du}{v^5}, \\
\eta' &= -\frac{1}{2}\left(f_2' u + f_3'\right)\left(u^3 + f_2 u + f_3\right)\frac{du}{v^5}, \\
\eta'' &= \left[-\frac{1}{2}\left(f_2'' u + f_3''\right)\left(u^3 + f_2 u + f_3\right) + \frac{3}{4}\left(f_2' u + f_3'\right)^2\right]\frac{du}{v^5}.
\end{aligned} \quad (14.243)$$

The numerator of Eq. (14.242) is then a polynomial of degree 6 in the single variable u. Since we work in $H^1(X)$, we can simplify the expression by adding an exact form

$$d\left(\frac{u^n}{v^3}\right) = u^{n-1}\left[\left(n - \frac{9}{2}\right)u^3 + \left(n - \frac{3}{2}\right) f_2 u + n f_3\right]\frac{du}{v^5}. \quad (14.244)$$

This allows us to reduce the numerator polynomial from degree six to a linear polynomial. The two coefficients of this linear polynomial have to vanish, on account of Eq. (14.242). We obtain therefore two equations for the two unknowns R_1 and R_0. Solving for R_1 and R_0 we find

$$R_1(x) = \frac{P_1(x)}{P_2(x)}, \quad R_0(x) = \frac{P_0(x)}{P_2(x)}, \tag{14.245}$$

with

$$\begin{aligned}
P_2(x) &= -x\left[3x^2 + 2M_{100}x - M_{200} + 2M_{110}\right]\left[x^4 + 4M_{100}x^3 + 2(3M_{200} + 2M_{110})x^2\right. \\
&\quad \left. +4(M_{300} - M_{210} + 10M_{111})x + (M_{200} - 2M_{110})^2\right], \\
P_1(x) &= -9x^6 - 32M_{100}x^5 - (37M_{200} + 70M_{110})x^4 - (8M_{300} + 56M_{210} \\
&\quad +144M_{111})x^3 + (13M_{400} - 36M_{310} + 46M_{220} - 124M_{211})x^2 \\
&\quad -(-8M_{500} + 24M_{410} - 16M_{320} - 96M_{311} + 144M_{221})x \\
&\quad +(M_{600} - 6M_{510} + 15M_{420} - 20M_{330} + 18M_{411} - 12M_{321} - 6M_{222}), \\
P_0(x) &= -3x^5 - 7M_{100}x^4 - (2M_{200} + 16M_{110})x^3 + (6M_{300} - 14M_{210})x^2 \\
&\quad +(5M_{400} - 8M_{310} + 6M_{220} - 8M_{211})x + (M_{500} - 3M_{410} + 2M_{320} \\
&\quad +8M_{311} - 10M_{221}).
\end{aligned} \tag{14.246}$$

Here we used the same notation as in Sect. 13.5.2:

$$M_{\lambda_1\lambda_2\lambda_3} = M_{\lambda_1\lambda_2\lambda_3}(y_1, y_2, 1) \tag{14.247}$$

and $M_{\lambda_1\lambda_2\lambda_3}(a_1, a_2, a_3)$ is defined in Eq. (13.316).

We therefore obtain the Picard-Fuchs operator for $\eta \in H^1(X)$ as

$$L = P_2(x)\frac{d^2}{dx^2} + P_1(x)\frac{d}{dx} + P_0(x), \tag{14.248}$$

with P_2, P_1 and P_0 defined in Eq. (14.246). This is also the Picard-Fuchs operator of the Feynman form

$$\varphi = \frac{\omega}{\mathcal{F}} \in H^2(P\backslash Y) \tag{14.249}$$

due to the splitting of the sequence in Eq. (14.229) and the flatness of the system $\mathbb{Z}E_1 \oplus \mathbb{Z}E_2 \oplus \mathbb{Z}E_3$. So for any cycle $\gamma \in H_2(P\backslash Y)$ we have

$$L\left(\int_\gamma \varphi\right) = 0. \tag{14.250}$$

Let us now turn to the inhomogeneous part $Q(x)$. The integration domain for the Feynman integral I_{111} is not a cycle $\gamma \in H_2(P\backslash Y)$, but the relative cycle $\Delta \in H_2(P\backslash Y, B\backslash B \cap Y)$. From the short exact sequence in Eq. (14.228) we deduce that $L\varphi$ is exact, i.e. there is a one-form β such that

$$L\,\varphi = d\beta. \tag{14.251}$$

We make the ansatz [410]

$$\beta = \frac{1}{\mathcal{F}^2} \left[(a_2 q_3 - a_3 q_2)\, da_1 + (a_3 q_1 - a_1 q_3)\, da_2 + (a_1 q_2 - a_2 q_1)\, da_3 \right], \quad (14.252)$$

where q_1, q_2 and q_3 are polynomials of degree 4 in the variables a_1, a_2 and a_3. The most general form is

$$\begin{aligned} q_i = & \; c^{(i)}_{400} a_1^4 + c^{(i)}_{040} a_2^4 + c^{(i)}_{004} a_3^4 + c^{(i)}_{310} a_1^3 a_2 + c^{(i)}_{301} a_1^3 a_3 + c^{(i)}_{130} a_1 a_2^3 \\ & + c^{(i)}_{103} a_1 a_3^3 + c^{(i)}_{031} a_2^3 a_3 + c^{(i)}_{013} a_2 a_3^3 + c^{(i)}_{211} a_1^2 a_2 a_3 + c^{(i)}_{121} a_1 a_2^2 a_3 \\ & + c^{(i)}_{112} a_1 a_2 a_3^2 + c^{(i)}_{220} a_1^2 a_2^2 + c^{(i)}_{202} a_1^2 a_3^2 + c^{(i)}_{022} a_2^2 a_3^2. \end{aligned} \quad (14.253)$$

We would like β to be finite on the boundary $\partial\sigma$. This implies

$$c^{(1)}_{040} = c^{(1)}_{004} = c^{(2)}_{400} = c^{(2)}_{004} = c^{(3)}_{400} = c^{(3)}_{040} = 0. \quad (14.254)$$

The remaining 39 coefficients $c^{(i)}_{jkl}$ are found by solving the linear system of equations obtained from inserting the ansatz into Eq. (14.251). The solution of this linear system is not unique, corresponding to the fact that β can be changed by a closed one-form. The solutions for the coefficients $c^{(i)}_{jkl}$ are rather lengthy and not listed here. In the next step we integrate β along the boundary $\partial\Delta$ to get $Q(x)$:

$$Q(x) = \int\limits_{\partial\Delta} \beta. \quad (14.255)$$

Note that the integration is in the blow-up P of $\mathbb{P}^2$. This is most easily done as follows: We start from the integration domain $\mathbb{RP}^2_{\geq 0}$, which we take as the union of three squares

$$\Delta_{12,3} \cup \Delta_{23,1} \cup \Delta_{31,2} \quad (14.256)$$

with

$$\Delta_{ij,k} = \left\{ [a_1 : a_2 : a_3] \;|\; 0 \leq a_i, a_j \leq 1, \; a_k = 1 \right\}. \quad (14.257)$$

Each square contains one point, which needs to be blown-up. In the square $\Delta_{ij,k}$ this is the point $a_i = a_j = 0$, $a_k = 1$. The blow-up can be done as described in Sect. 10.1. The blow-up of each square can be covered with two charts. Thus we get six charts in total.

The integration contour $\partial\Delta$ is sketched in Fig. 14.3. Performing the integration we obtain

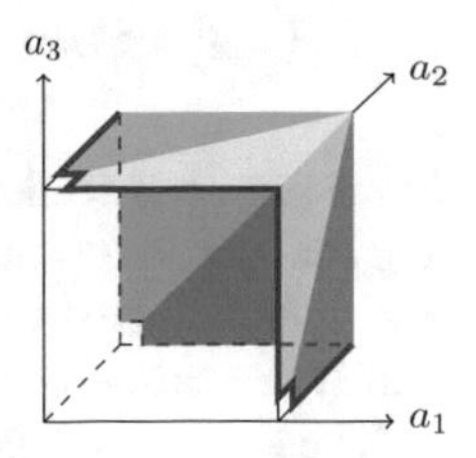

Fig. 14.3 Sketch of the integration domain Δ covered by six charts. Each chart is drawn in a different colour. The boundary $\partial\Delta$ is shown in red

$$\begin{aligned}Q(x) = &-18x^4 - 24M_{100}x^3 + (4M_{200} - 40M_{110})x^2 - (-8M_{300} + 8M_{210} + 48M_{111})x \\ &+(-2M_{400} + 8M_{310} - 12M_{220} - 8M_{211}) \\ &+2c(x, y_1, y_2, 1)\ln y_1 + 2c(x, y_2, 1, y_1)\ln y_2 \end{aligned} \tag{14.258}$$

and

$$\begin{aligned}c\ \ (x, y_1, y_2, y_3) = &\\ &(2y_1 - y_2 - y_3)x^3 + \left(6y_1^2 - 3y_2^2 - 3y_3^2 - 7y_1y_2 - 7y_1y_3 + 14y_2y_3\right)x^2 \\ &-\left(-6y_1^3 + 3y_2^3 + 3y_3^3 + 11y_1^2y_2 + 11y_1^2y_3 - 8y_1y_2^2 - 8y_1y_3^2 - 3y_2^2y_3 - 3y_2y_3^2\right)x \\ &+\left(2y_1^4 - y_2^4 - y_3^4 - 5y_1^3y_2 - 5y_1^3y_3 + y_1y_2^3 + y_1y_3^3 + 4y_2^3y_3 + 4y_2y_3^3\right. \\ &\left.+3y_1^2y_2^2 + 3y_1^2y_3^2 - 6y_2^2y_3^2 + 2y_1^2y_2y_3 - y_1y_2^2y_3 - y_1y_2y_3^2\right). \end{aligned} \tag{14.259}$$

The coefficients $c(x, y_i, y_j, y_k)$ of the logarithms of the masses vanish for equal masses.

Putting everything together the sought-after differential equation reads

$$\left[P_2(x)\frac{d^2}{dx^2} + P_1(x)\frac{d}{dx} + P_0(x)\right] I_{111}(2, x, y_1, y_2) = Q(x). \tag{14.260}$$

This is an inhomogeneous second-order differential equation for the sunrise integral with unequal masses in two space-time dimensions. As it is second-order, we have two master integrals in the top sector in two space-time dimensions. We recall that for generic D we have four master integrals in the top sector. The two additional master integrals can be chosen such that they vanish in the limit $D \to 2$ [370].

Exercise 121 *Consider the one-loop two-point function with equal internal masses*

$$I_{11}(2, x) = \int \frac{d^2k}{i\pi} \frac{m^2}{\left(-q_1^2 + m_1^2\right)\left(-q_2^2 + m_2^2\right)}, \qquad x = -\frac{p^2}{m^2}. \tag{14.261}$$

Derive with the methods of this section the differential equation for $I_{11}(2, x)$ with respect to the kinematic variable x.

14.5.3 Banana Motives

In the previous section we considered the one-loop bubble integral (in Exercise 121) and the two-loop sunrise integral, both in two space-time dimensions. These are the first two members of the family of banana graphs. The first three members of this family are shown in Fig. 14.2.

The l-loop banana integral has $(l+1)$ propagators and is given in D space-time dimensions by

$$I_{\nu_1 \dots \nu_{l+1}}(D) = e^{l\varepsilon\gamma_E} \left(\mu^2\right)^{\nu - \frac{lD}{2}} \int \prod_{a=1}^{l} \frac{d^D k_a}{i\pi^{\frac{D}{2}}} \delta^D\left(p - \sum_{b=1}^{l+1} k_b\right) \left(\prod_{c=1}^{l+1} \frac{1}{\left(-k_c^2 + m_c^2\right)^{\nu_j}}\right). \quad (14.262)$$

We introduce

$$x = \frac{-p^2}{\mu^2}, \quad y_1 = \frac{m_1^2}{\mu^2}, \quad y_2 = \frac{m_2^2}{\mu^2}, \quad \dots, \quad y_{l+1} = \frac{m_{l+1}^2}{\mu^2}. \quad (14.263)$$

As usual we may set one variable to one. The l-loop banana integral depends therefore in the generic case (i.e. all internal masses pairwise distinct and $(-p^2)$ non-zero and not equal to any internal mass squared) on $N_B = l+1$ kinematic variables. Of particular interest is also the equal mass case $m_1 = \dots = m_{l+1} = m$, in which case the l-loop banana integral depends on one kinematic variable, which we may take as $x = -p^2/m^2$.

The Feynman parameter representation of the l-loop banana integral is given by

$$I_{\nu_1 \dots \nu_{l+1}}(D) = \frac{e^{l\varepsilon\gamma_E}\Gamma\left(\nu - \frac{lD}{2}\right)}{\prod\limits_{j=1}^{l+1}\Gamma(\nu_j)} \int\limits_{\Delta} \omega \left(\prod_{j=1}^{l+1} a_j^{\nu_j - 1}\right) \frac{\mathcal{U}^{\nu - \frac{(l+1)D}{2}}}{\mathcal{F}^{\nu - \frac{lD}{2}}}, \quad (14.264)$$

with $\Delta = \mathbb{RP}^l_{\ge 0}$ and

$$\omega = \sum_{j=1}^{l+1} (-1)^{j-1}\, a_j\, da_1 \wedge \dots \wedge \widehat{da_j} \wedge \dots \wedge da_n. \quad (14.265)$$

The hat indicates that the corresponding term is omitted. The graph polynomials are given by

$$\mathcal{U} = \left(\prod_{i=1}^{l+1} a_i\right) \cdot \left(\sum_{j=1}^{l+1} \frac{1}{a_j}\right), \quad \mathcal{F} = x\left(\prod_{i=1}^{l+1} a_i\right) + \left(\sum_{i=1}^{l+1} a_i y_i\right) \mathcal{U}. \quad (14.266)$$

At one, two and three loops we have

$$\begin{aligned}
l = 1: \quad & \mathcal{F} = a_1 a_2 x + (a_1 + a_2)(a_1 y_1 + a_2 y_2), \\
l = 2: \quad & \mathcal{F} = a_1 a_2 a_3 x + (a_1 a_2 + a_1 a_3 + a_2 a_3)(a_1 y_1 + a_2 y_2 + a_3 y_3), \\
l = 3: \quad & \mathcal{F} = a_1 a_2 a_3 a_4 x \\
& \quad + (a_1 a_2 a_3 + a_1 a_2 a_4 + a_1 a_3 a_4 + a_2 a_3 a_4)(a_1 y_1 + a_2 y_2 + a_3 y_3 + a_4 y_4).
\end{aligned} \tag{14.267}$$

Any sub-topology of the l-loop banana integral is a product of l one-loop tadpole integrals. If all masses are distinct and $(-p^2)$ and D generic, we have for the l-loop banana family

$$N_{\text{master}} = 2^{l+1} - 1 \tag{14.268}$$

master integrals. We have $(l + 1)$ sub-topologies and there is one master integral for each sub-topology. These may be taken as

$$I_{011...11}, \; I_{101...11}, \; \dots, \; I_{111...10}. \tag{14.269}$$

The top sector has

$$2^{l+1} - l - 2 \tag{14.270}$$

master integrals. A basis for the top sector is given by

$$\nu_j \in \{1, 2\}, \; l + 1 \leq \nu \leq 2l, \tag{14.271}$$

where as usual $\nu = \nu_1 + \dots + \nu_{l+1}$. At one, two and three loops a basis for the top sector is given by

$$\begin{aligned}
l = 1: \quad & I_{11}, \\
l = 2: \quad & I_{111}, \; I_{211}, \; I_{121}, \; I_{112}, \\
l = 3: \quad & I_{1111}, \; I_{2111}, \; I_{1211}, \; I_{1121}, \; I_{1112}, \; I_{2211}, \; I_{2121}, \; I_{2112}, \; I_{1221}, \; I_{1212}, \; I_{1122}.
\end{aligned} \tag{14.272}$$

In the equal mass case the number of master integrals is

$$N_{\text{master}} = l + 1. \tag{14.273}$$

There is only one sub-sector with one master integral. The top sector has l master integrals in the equal mass case, which can be taken as

$$I_{11...1}, \; I_{21...1}, \; I_{31...1}, \; \dots, \; I_{(l-1)1...1}, \; I_{l1...1}. \tag{14.274}$$

Let us now study the l-loop banana integral in $D = 2$ space-time dimensions. We go back to the unequal mass case. As already observed in the two-loop case, the master integrals may be ramified for special values of D. For $D = 2$ one has in the unequal mass case in the top sector only

$$2^{l+1} - \binom{l+2}{\lfloor \frac{l+2}{2} \rfloor} \qquad (14.275)$$

master integrals. For example, there are 1, 2, 6 master integrals in the top sector for 1, 2, 3 loops, respectively.

We are in particular interested in the case $\nu_1 = \cdots = \nu_{l+1} = 1$ and $D = 2$. Equation (14.264) simplifies in this case to

$$I_{1\dots1}(2) = \int\limits_{\Delta} \frac{\omega}{\mathcal{F}}. \qquad (14.276)$$

As a sideremark we note that there exists a one-dimensional integral representation for $I_{1\dots1}(2)$ involving Bessel functions [411, 412]:

$$I_{1\dots1}(2) = 2^l \int\limits_0^\infty dt\ t\ J_0\left(t\sqrt{x}\right) \prod_{i=1}^{l+1} K_0\left(t\sqrt{y_i}\right). \qquad (14.277)$$

However, our main interest here is the geometry underlying the l-loop banana integral. The geometry is determined by the variety where $\mathcal{F}$ vanishes:

$$X = \left\{ \left[a_1 : a_2 : \cdots : a_{l+1}\right] \in \mathbb{CP}^l \mid \mathcal{F}(a) = 0 \right\}. \qquad (14.278)$$

The second graph polynomial is a homogeneous polynomial of degree $(l+1)$. For generic kinematic variables the hypersurface $X \in \mathbb{CP}^l$ is smooth and defines by Theorem 15 below a Calabi-Yau $(l-1)$-fold. In particular we have at two-loops an elliptic curve and at three-loops a K3 surface.

The motive associated to the Feynman integral $I_{1\dots1}(2)$ is obtained along the lines of Sect. 14.5.1. We denote the blow-up of $\mathbb{CP}^l$ by P, the strict transform of X by Y and the total transform of the set $\{x_1 \cdot x_2 \cdot \ldots \cdot x_n = 0\}$ by B. The motive associated to the Feynman integral $I_{1\dots1}(2)$ is then

$$M^l\left(P \backslash Y, B \backslash \left(B \cap Y\right)\right). \qquad (14.279)$$

The banana integrals are the simplest example, where higher-dimensional algebraic varieties (i.e. Calabi-Yau $(l-1)$-folds) enter the computation of Feynman integrals. The banana integrals have been studied in [331, 332, 368, 405, 413–418], other Feynman integrals related to Calabi-Yau manifolds have been studied in [328–330].

Digression Calabi-Yau manifolds

A Calabi-Yau manifold of complex dimension n (or a **Calabi-Yau n-fold** *for short) is a compact Kähler manifold M of complex dimension n, satisfying one of the following equivalent conditions:*

1. *The first Chern class of M vanishes (over $\mathbb{R}$).*

2. *M has a Kähler metric with vanishing Ricci curvature.*
3. *M has a Kähler metric with local holonomy* $\mathrm{Hol}(p) \subseteq \mathrm{SU}(n)$ *(where* $p \in M$ *denotes a point).*
4. *A positive power of the canonical bundle of M is trivial.*
5. *M has a finite cover that is a product of a torus and a simply connected manifold with trivial canonical bundle.*

If M is simply connected, the following conditions are equivalent to the ones above:

6. *M has a holomorphic n-form that vanishes nowhere.*
7. *The canonical bundle of M is trivial.*

Note that the conditions (1)–(5) are in general weaker than the conditions (6)–(7). The exact definition of Calabi-Yau manifolds varies slightly in the literature. Some authors define Calabi-Yau manifolds by conditions (6)–(7), such a definition excludes for example Enriques surfaces as Calabi-Yau manifolds. Other authors are even more restrictive and require

8. *M has a Kähler metric with local holonomy* $\mathrm{Hol}(p) = \mathrm{SU}(n)$.

Defining a Calabi-Yau manifold by condition (8) excludes for example Abelian surfaces as Calabi-Yau manifolds. On the other hand, condition (8) ensures that the Hodge numbers $h^{j,0}$ *vanish for* $0 < j < n$. *It can be shown that condition (8) implies condition (6) [419]. Condition (6) is equivalent to condition (7) and condition (7) clearly implies condition (4). Therefore requiring condition (8) defines a Calabi-Yau manifold according to (1)–(5).*

A simply connected Calabi-Yau 2-fold is called a **K3 surface** *(after Kummer, Kähler and Kodaira). The requirement of simple connectedness in the definition of a K3 surface excludes complex tori.*

In Table 14.1 we show the Hodge diamonds for a K3 surface, an Abelian surface and an Enriques surface. These Hodge diamonds illustrate that condition (7) ensures

$$h^{n,0} = 1, \; h^{j,0} = 0, \qquad 0 < j < n. \tag{14.280}$$

Table 14.1 The Hodge diamonds for a K3 surface, an Abelian surface and an Enriques surface

$$\begin{array}{ccccc} & & 1 & & \\ & 0 & & 0 & \\ 1 & & 20 & & 1 \\ & 0 & & 0 & \\ & & 1 & & \end{array} \qquad \begin{array}{ccccc} & & 1 & & \\ & 2 & & 2 & \\ 1 & & 4 & & 1 \\ & 2 & & 2 & \\ & & 1 & & \end{array} \qquad \begin{array}{ccccc} & & 1 & & \\ & 0 & & 0 & \\ 0 & & 10 & & 0 \\ & 0 & & 0 & \\ & & 1 & & \end{array}$$

K3 surface — Abelian surface — Enriques surface

Examples of Calabi-Yau manifolds can be obtained from the following theorem:

Theorem 15 *Let X be the hypersurface defined by a homogeneous polynomial P of degree $d = n + 2$ in $\mathbb{CP}^{n+1}$. If X is smooth, than X is a Calabi-Yau n-fold.*

In the definitions above we used some technical terms, which we now explain:

We start from a Kähler manifold M. First of all, a Kähler manifold is a special case of a Riemannian manifold of real dimension (2n), so M comes with a Riemannian metric g and the associated Levi-Civita connection ∇. M is also a complex manifold, so there is a complex structure J. The manifold M is further a Hermitian manifolds, this means that the complex structure J is compatible with the metric g:

$$g\,(JX, JY) = g\,(X, Y) \tag{14.281}$$

for $X, Y \in T_pM$. Finally, the fact that M is Kähler implies that the complex structure satisfies

$$\nabla J = 0. \tag{14.282}$$

From the Levi-Civita connection one defines the Riemann curvature tensor R and the Ricci tensor Ric. *The* **Ricci form** *is then defined by*

$$\rho\,(X, Y) = \mathrm{Ric}\,(JX, Y)\,. \tag{14.283}$$

The Ricci form of a Kähler manifold is a real closed $(1, 1)$-form and can be written locally as

$$\rho = -i\partial\bar{\partial}\ln\det g. \tag{14.284}$$

As ρ is closed, it defines a class $[\rho] \in H^{1,1}(M, \mathbb{R}) \subset H^2(M, \mathbb{R})$. The **first Chern class** *of M is given by*

$$c_1 = \frac{1}{2\pi}\,[\rho]\,. \tag{14.285}$$

The connection ∇ naturally defines a transformation group at each tangent space T_pM as follows: Let $p \in M$ be a point and consider the set of closed loops at p, $\{\gamma(t)|0 \leq t \leq 1, \gamma(0) = \gamma(1) = p\}$. Take a vector $X \in T_pM$ and parallel transport X along the curve γ. After a trip along γ we end up with a new vector $X_\gamma \in T_pM$. Thus the loop γ and the connection ∇ induces a linear transformation

$$P_\gamma : T_pM \to T_pM. \tag{14.286}$$

The set of these transformations is denoted $\mathrm{Hol}(p)$ *and is called the* **holonomy group** *at p. The holonomy of a Riemannian manifold of real dimension $(2n)$ is contained in* $\mathrm{O}(2n)$*. If the manifold is orientable this becomes* $\mathrm{SO}(2n)$*. If M is flat, the holonomy* *group consists only of the identity element. If M is Kähler, the holonomy group is contained in* $\mathrm{U}(n)$ *and M is Calabi-Yau if the holonomy group is contained in* $\mathrm{SU}(n)$.

The **canonical bundle** *of a smooth algebraic variety X of (complex) dimension n is the line bundle of holomorphic n-forms*

$$K_X = \bigwedge^n \Omega_X^1. \tag{14.287}$$

The k-th tensor product K_X^k of the canonical bundle is again a line bundle. The **plurigenus** *of X is the dimension of the vector space of global sections of K_X^k:*

$$P_k = \dim H^0\left(X, K_X^k\right). \tag{14.288}$$

The **Kodaira dimension** *κ of X is defined to be $(-\infty)$, if all plurigenera P_k are zero for $k > 0$, otherwise it is defined as the minimum such that P_k/k^κ is bounded. In other words, the Kodaira dimension κ gives the rate of growth of the plurigenera:*

$$P_k = O\left(k^\kappa\right). \tag{14.289}$$

Let's look at an example in (complex) dimension 1*: The smooth algebraic curves are classified by their genus g. One has*

$$\begin{array}{llll} g = 0: & \kappa = -\infty, & P_k = 0, & k > 0, \\ g = 1: & \kappa = 0, & P_k = 1, & k \geq 0, \\ g \geq 2: & \kappa = 1, & P_k = (2k-1)(g-1), & k \geq 2. \end{array} \tag{14.290}$$

An Abelian surface is an Abelian variety of (complex) dimension two. An **Abelian variety** *is a projective algebraic variety that is also an algebraic group. The group law is necessarily commutative. It can be shown that the Abelian varieties are exactly those complex tori, which can be embedded into projective space. To give an example: An elliptic curve is an Abelian variety of (complex) dimension one.*

An **Enriques surface** *is a smooth projective minimal algebraic surface of Kodaira dimension* 0 *with Betti numbers $b_1 = 0$ and $b_2 = 10$* [420].

Chapter 15
Numerics

At the end of the day of an analytic calculation of Feynman integrals we would like to get a number. This requires methods for the numerical evaluation of all functions appearing in the final result for a Feynman integral. In this chapter we discuss methods how this can be done for multiple polylogarithms and the elliptic generalisations discussed in Chap. 13. All these methods can be pushed to obtain numerical results with a precision of hundred or thousand digits. For phenomenological applications in physics this is certainly overkill, but there is one application, where high-precision numerics is extremely useful: The PSLQ algorithm allows us to find relations among dependent transcendental constants. This algorithm is often used in the context of Feynman integral calculations to find a simple form for the boundary constants. We discuss this algorithm in Sect. 15.4.

15.1 The Dilogarithm

As a warm-up example let us start with the numerical evaluation of the dilogarithm [97]: The dilogarithm is defined by

$$\text{Li}_2(x) = -\int\limits_0^x \frac{dt}{t} \ln(1-t), \tag{15.1}$$

and has a branch cut along the positive real axis, starting at the point $x = 1$. For $|x| \leq 1$ one has the convergent power series expansion

$$\text{Li}_2(x) = \sum_{n=1}^{\infty} \frac{x^n}{n^2}. \tag{15.2}$$

S. Weinzierl, *Feynman Integrals*, UNITEXT for Physics,
https://doi.org/10.1007/978-3-030-99558-4_15

The first step for a numerical evaluation consists in mapping an arbitrary (complex) argument x into the region, where the power series in Eq. (15.2) converges. This can be done with the help of the reflection identity (see Eq. (5.39))

$$\text{Li}_2(x) = -\text{Li}_2\left(\frac{1}{x}\right) - \frac{\pi^2}{6} - \frac{1}{2}\left(\ln(-x)\right)^2, \tag{15.3}$$

which is used to map the argument x, lying outside the unit circle into the unit circle. The function $\ln(-x)$ appearing on the right-hand side of Eq. (15.3) is considered to be "simpler", e.g., it is assumed that a numerical evaluation routine for this function is known. In addition we can shift the argument into the range $-1 \leq \text{Re}(x) \leq 1/2$ with the help of

$$\text{Li}_2(x) = -\text{Li}_2(1-x) + \frac{\pi^2}{6} - \ln(x)\ln(1-x). \tag{15.4}$$

Although one can now attempt a brute force evaluation of the power series in Eq. (15.2), it is more efficient to rewrite the dilogarithm as a series involving the Bernoulli numbers B_j (defined in Eq. (7.85)):

$$\text{Li}_2(x) = \sum_{j=0}^{\infty} \frac{B_j}{(j+1)!} z^{j+1}, \qquad z = -\ln(1-x). \tag{15.5}$$

Therefore the numerical evaluation of the dilogarithm consists in using Eqs. (15.3) and (15.4) to map any argument x into the unit circle with the additional condition $\text{Re}(x) \leq 1/2$. One then uses the series expansion in terms of Bernoulli numbers Eq. (15.5).

Exercise 122 *Derive Eq. (15.5).*

15.2 Multiple Polylogarithm

Let us now consider multiple polylogarithms. We used several notations for them (see Chap. 8): A long notation related to the integral representation

$$G\left(z_1, z_2 \dots, z_r; y\right), \tag{15.6}$$

where the z_j's are allowed to be zero, a short notation related to the integral representation

$$G_{m_1 \dots m_k}(z_1, \dots, z_k; y), \tag{15.7}$$

where all z_j's are assumed to be non-zero and a representation related to the sum representation

$$\text{Li}_{m_1 \dots m_k}(x_1, \dots, x_k). \tag{15.8}$$

These notations are related by Eqs. (8.13), (8.19) and (8.20). We call k the depth of the multiple polylogarithm.

We would like to evaluate numerically the multiple polylogarithms in Eq. (15.6) for arbitrary complex arguments. Let us first note that with the help of the shuffle product we may always remove trailing zeros (as discussed in Sect. 8.3). If $z_r \neq 0$ we may use the scaling relation Eq. (8.10) to scale y to 1 (or a positive real number). Let us therefore assume without loss of generality that y is a positive real number. Our integration path is then the line segment from 0 to y along the positive real axis. The z'_j are assumed not to lie on this line segment (but they are allowed to be infinitesimal close to this line segment):

$$z_j \in \mathbb{C} \backslash [0, y]. \tag{15.9}$$

As we already removed trailing zeros, we may use the notation as in Eq. (15.7) or Eq. (15.8). The principal ideas for the algorithm for the numerical evaluation of multiple polylogarithms are very similar to the example of the dilogarithm discussed above. We first use the integral representation to transform all arguments into a region, where the sum representation converges. Truncating the sum representation to an appropriate order provides a numerical evaluation. In addition, there are methods which can be used to accelerate the convergence for the series representation of the multiple polylogarithms.

Let's look at the details: In most physical applications, the z_j's appearing in the integral representation will be real numbers. To distinguish if the integration contour runs above or below a cut, we define the abbreviations $z_\pm$, meaning that a small positive, respectively negative imaginary part is to be added to the value of the variable:

$$z_+ = z + i\delta, \quad z_- = z - i\delta, \qquad \delta > 0. \tag{15.10}$$

The sum representation $\text{Li}_{m_1,\dots,m_k}(x_1, \dots, x_k)$ is convergent, if

$$\left|x_1 x_2 \dots x_j\right| \leq 1 \text{ for all } j \in \{1, \dots, k\} \text{ and } (m_1, x_1) \neq (1, 1). \tag{15.11}$$

Therefore the function $G_{m_1,\dots,m_k}(z_1, \dots, z_k; y)$ has a convergent series representation if

$$|y| \leq \left|z_j\right| \quad \text{for all } j, \tag{15.12}$$

e.g., no element in the set $\{|z_1|, \dots, |z_k|, |y|\}$ is smaller than $|y|$ and in addition if $m_1 = 1$ we have $y/z_1 \neq 1$.

15.2.1 Transformation into the Region Where the Sum Representation Converges

If Eq. (15.12) is not satisfied, we first transform into the domain, where the sum representation is convergent. This transformation is based on the integral representation. We start from the function

$$G_{m_1,\dots,m_k}\left(z_1, \dots, z_{j-1}, s, z_{j+1}, \dots, z_k; y\right), \tag{15.13}$$

with the assumption that $|s|$ is the smallest element in the set $\{|z_1|, ..., |z_{j-1}|, |s|, |z_{j+1}|, ..., |z_k|, |y|\}$. The algorithm goes by induction and introduces the more general structure

$$\int\limits_0^{y_1} \frac{ds_1}{s_1 - b_1} \dots \int\limits_0^{s_{r-1}} \frac{ds_r}{s_r - b_r} G(a_1, \dots, s_r, \dots, a_w; y_2), \tag{15.14}$$

where $|y_1|$ is the smallest element in the set $\{|y_1|, |b_1'|, \dots, |b_r'|, |a_1'|, \dots, |a_w'|, |y_2|\}$. The prime indicates that only the non-zero elements of a_i and b_j are considered. If the integrals over s_1 to s_r are absent, we recover the original G-function in Eq. (15.13). Since we can always remove trailing zeroes with the help of the algorithm in Sect. 8.3, we can assume that $a_w \neq 0$. We first consider the case where the G-function is of depth one, e.g.,

$$\int\limits_0^{y_1} \frac{ds_1}{s_1 - b_1} \dots \int\limits_0^{s_{r-1}} \frac{ds_r}{s_r - b_r} G(\underbrace{0, \dots, 0}_{m-1}, s_r; y_2) = \int\limits_0^{y_1} \frac{ds_1}{s_1 - b_1} \dots \int\limits_0^{s_{r-1}} \frac{ds_r}{s_r - b_r} G_m(s_r; y_2), \tag{15.15}$$

and show that we can relate the function $G_m(s_r; y_2)$ to $G_m\,(y_2; s_r)$, powers of $\ln(s_r)$ and functions, which do not depend on s_r. For $m = 1$ we have

$$G_1\left(s_{r\,\pm}; y_2\right) = G_1\left(y_{2\,\mp}; s_r\right) - G(0; s_r) + \ln\left(-\,y_{2\,\mp}\right). \tag{15.16}$$

For $m \geq 2$ one can use the transformation $1/y$ and one obtains:

$$G_m\left(s_{r\,\pm}; y_2\right) = -\zeta_m + \int\limits_0^{y_2} \frac{dt}{t} G_{m-1}\left(t_\pm; y_2\right) - \int\limits_0^{s_r} \frac{dt}{t} G_{m-1}\left(t_\pm; y_2\right). \tag{15.17}$$

One sees that the first and second term in Eq. (15.17) yield functions independent of s_r. The third term has a reduced weight and we may therefore use recursion. This completes the discussion for $G_m\,(s_r; y_2)$. We now turn to the general case with a G-

function of depth greater than one in Eq. (15.14). Here we first consider the sub-case, that s_r appears in the last place in the parameter list and $(m-1)$ zeroes precede s_r, e.g.,

$$\int_0^{y_1} \frac{ds_1}{s_1 - b_1} \cdots \int_0^{s_{r-1}} \frac{ds_r}{s_r - b_r} G(a_1, \dots, a_k, \underbrace{0, \dots, 0}_{m-1}, s_r; y_2). \tag{15.18}$$

Since we assumed that the G-function has a depth greater than one, we have $a_k \neq 0$. Here we use the shuffle relation to relate this case to the case where s_r does not appear in the last place:

$$G(a_1, \dots, a_k, \underbrace{0, \dots, 0}_{m-1}, s_r; y_2) =$$
$$G(a_1, \dots, a_k; y_2) G(\underbrace{0, \dots, 0}_{m-1}, s_r; y_2) - \sum_{\text{shuffles}'} G(\alpha_1, \dots, \alpha_{k+m}; y_2), \tag{15.19}$$

where the sum runs over all shuffles of $(a_1, \dots, a_k)$ with $(0, \dots, 0, s_r)$ and the prime indicates that $(\alpha_1, \dots, \alpha_{k+m}) = (a_1, \dots, a_k, 0, \dots, 0, s_r)$ is to be excluded from this sum. In the first term on the right-hand side of Eq. (15.19) the factor $G(a_1, \dots, a_k; y_2)$ is independent of s_r , whereas the second factor $G(0, \dots, 0, s_r; y_2)$ is of depth one and can be treated with the methods discussed above. The terms corresponding to the sum over the shuffles in Eq. (15.19) have either s_r not appearing in the last place in the parameter list or a reduced number of zeroes preceding s_r. In the last case we may use recursion to remove s_r from the last place in the parameter list. It remains to discuss the case, where the G-function has depth greater than one and s_r does not appear in the last place in the parameter list, e.g.,

$$\int_0^{y_1} \frac{ds_1}{s_1 - b_1} \cdots \int_0^{s_{r-1}} \frac{ds_r}{s_r - b_r} G(a_1, \dots, a_{i-1}, s_r, a_{i+1}, \dots, a_w; y_2), \tag{15.20}$$

with $a_w \neq 0$. Obviously, we have

$$G(a_1, \dots, a_{i-1}, s_r, a_{i+1}, \dots, a_w; y_2) = G(a_1, \dots, a_{i-1}, 0, a_{i+1}, \dots, a_w; y_2)$$
$$+ \int_0^{s_r} ds_{r+1} \frac{\partial}{\partial s_{r+1}} G(a_1, \dots, a_{i-1}, s_{r+1}, a_{i+1}, \dots, a_w; y_2). \tag{15.21}$$

The first term $G(a_1, \dots, a_{i-1}, 0, a_{i+1}, \dots, a_w; y_2)$ does no longer depend on s_r and has a reduced depth. For the second term we first write out the integral representation of the G-function. We then use

$$\frac{\partial}{\partial s}\frac{1}{t-s} = -\frac{\partial}{\partial t}\frac{1}{t-s}, \tag{15.22}$$

followed by partial integration in t and finally partial fraction decomposition according to

$$\frac{1}{(t-\alpha)(t-s)} = \frac{1}{s-\alpha}\left(\frac{1}{t-s} - \frac{1}{t-\alpha}\right). \tag{15.23}$$

If s_r is not in the first place of the parameter list, we obtain

$$\begin{aligned}
&\int_0^{s_r} ds_{r+1} \frac{\partial}{\partial s_{r+1}} G\left(a_1, \dots, a_{i-1}, s_{r+1}, a_{i+1}, \dots, a_w; y_2\right) \\
&= -\int_0^{s_r} \frac{ds_{r+1}}{s_{r+1}-a_{i-1}} G\left(a_1, \dots, a_{i-2}, s_{r+1}, a_{i+1}, \dots, a_w; y_2\right) \\
&\quad + \int_0^{s_r} \frac{ds_{r+1}}{s_{r+1}-a_{i-1}} G\left(a_1, \dots, a_{i-2}, a_{i-1}, a_{i+1}, \dots, a_w; y_2\right) \\
&\quad + \int_0^{s_r} \frac{ds_{r+1}}{s_{r+1}-a_{i+1}} G\left(a_1, \dots, a_{i-1}, s_{r+1}, a_{i+2}, \dots, a_w; y_2\right) \\
&\quad - \int_0^{s_r} \frac{ds_{r+1}}{s_{r+1}-a_{i+1}} G\left(a_1, \dots, a_{i-1}, a_{i+1}, a_{i+2}, \dots, a_w; y_2\right).
\end{aligned} \tag{15.24}$$

Each G-function has a weight reduced by one unit and we may use recursion. If s_r appears in the first place we have the following special case:

$$\begin{aligned}
&\int_0^{s_r} ds_{r+1} \frac{\partial}{\partial s_{r+1}} G\left(s_{r+1}, a_{i+1}, \dots, a_w; y_2\right) = \int_0^{s_r} \frac{ds_{r+1}}{s_{r+1}-y_2} G\left(a_{i+1}, \dots, a_w; y_2\right) \\
&\quad + \int_0^{s_r} \frac{ds_{r+1}}{s_{r+1}-a_{i+1}} G\left(s_{r+1}, a_{i+2}, \dots, a_w; y_2\right) - \int_0^{s_r} \frac{ds_{r+1}}{s_{r+1}-a_{i+1}} G\left(a_{i+1}, a_{i+2}, \dots, a_w; y_2\right).
\end{aligned} \tag{15.25}$$

There is however a subtlety: If α_{i-1} or α_{i+1} are zero, the algorithm generates terms of the form

$$\int_0^y \frac{ds}{s} F(s) - \int_0^y \frac{ds}{s} F(0). \tag{15.26}$$

Although the sum of these two terms is finite, individual pieces diverge at $s = 0$. We regularise the individual contributions with a lower cut-off λ:

$$\int\limits_{\lambda}^{y} \frac{ds}{s} F(s) - \int\limits_{\lambda}^{y} \frac{ds}{s} F(0). \tag{15.27}$$

In individual contributions we therefore obtain at the end of the day powers of $\ln \lambda$ from integrals of the form

$$\int\limits_{\lambda}^{y} \frac{ds_1}{s_1} \int\limits_{\lambda}^{s_1} \frac{ds_2}{s_2} = \frac{1}{2} \ln^2 y - \ln y \ln \lambda + \frac{1}{2} \ln^2 \lambda. \tag{15.28}$$

In the final result, all powers of $\ln \lambda$ cancel, and we are left with G-functions with trailing zeros. These are then converted by standard algorithms to G-functions without trailing zeros. The G-functions without trailing zeros can then be evaluated numerically by their power series expansion.

In addition, the algorithms may introduce in intermediate steps G-functions with leading ones, e.g., $G(1, \dots, z_k; 1)$. These functions are divergent, but the divergence can be factorised and expressed in terms of the basic divergence $G(1; 1)$. The algorithm is very similar to the one for the extraction of trailing zeroes. In the end all divergences cancel.

15.2.2 Series Acceleration

The G-function $G_{m_1,\dots,m_k}(z_1, \dots, z_k; y)$ has a convergent sum representation if the conditions in Eq. (15.12) are met. This does not necessarily imply, that the convergence is sufficiently fast, such that the power series expansion can be used in a straightforward way. In particular, if z_1 is close to y the convergence is rather poor. In this paragraph we consider methods to improve the convergence. The main tool will be the Hölder convolution.

The multiple polylogarithms satisfy the **Hölder convolution** [190]. For $z_1 \neq 1$ and $z_r \neq 0$ this identity reads

$$\begin{aligned} & G\left(z_1, \dots, z_r; 1\right) \\ & = \sum_{j=0}^{r} (-1)^j \, G\left(1 - z_j, 1 - z_{j-1}, \dots, 1 - z_1; 1 - \frac{1}{p}\right) G\left(z_{j+1}, \dots, z_r; \frac{1}{p}\right). \end{aligned} \tag{15.29}$$

The Hölder convolution can be used to improve the rate of convergence for the series representation of multiple polylogarithms.

Let us see how this is done: We consider $G_{m_1,\dots,m_k}(z_1, \dots, z_k; y)$ and assume that the conditions of Eq. (15.12) are met (i.e., the multiple polylogarithm has a convergent sum representation). By assumption we have $z_k \neq 0$, and therefore we can normalise y to one. We are therefore considering $G_{m_1,\dots,m_k}(z_1, \dots, z_k; 1)$. Convergence implies then, that we have $|z_j| \geq 1$ and $(z_1, m_1) \neq (1, 1)$. If some z_j is close to the unit circle, say,

$$1 \leq |z_j| \leq 2, \tag{15.30}$$

we use the Hölder convolution Eq. (15.29) with $p = 2$ to rewrite the G-functions as

$$\begin{aligned} G(z_1, \dots, z_r; 1) &= G(2z_1, \dots, 2z_r; 1) + (-1)^r G\left(2(1-z_r), 2(1-z_{r-1}), \dots, 2(1-z_1); 1\right) \\ &+ \sum_{j=1}^{r-1} (-1)^j G\left(2(1-z_j), 2(1-z_{j-1}), \dots, 2(1-z_1); 1\right) G\left(2z_{j+1}, \dots, 2z_r; 1\right). \end{aligned} \tag{15.31}$$

Here, we normalised the right-hand side to one and explicitly wrote the first and last term of the sum. We observe, that the first term $G(2z_1, \dots, 2z_r; 1)$ has all arguments outside $|2z_j| \geq 2$. This term has therefore a better convergence. Let us now turn to the second term in Eq. (15.31). If some z_j lies within $|z_j - 1| < 1/2$, the Hölder convolution transforms the arguments out of the region of convergence. In this case, we repeat the steps above, e.g., transformation into the region of convergence, followed by a Hölder convolution, if necessary. While this is a rather simple recipe to implement into a computer program, it is rather tricky to proof that this procedure does not lead to an infinite recursion, and besides that, does indeed lead to an improvement in the convergence. For the proof we have to understand how the algorithms for the transformation into the region of convergence act on the arguments of a G-function with length r. In particular we have to understand how in the result the G-functions of length r are related to the original G-function. Products of G-functions of lower length are "simpler" and not relevant for the argument here. We observe, that this algorithm for the G-function $G(z_1, \dots, z_r; y)$ substitutes y by the element with the smallest non-zero modulus from the set $\{|z_1|, \dots, |z_r|, |y|\}$, permutes the remaining elements into an order, which is of no relevance here and possibly substitutes some non-zero elements by zero. The essential point is, that it does not introduce any non-trivial new arguments (e.g., new non-zero arguments). The details can be found in [421].

15.2.3 Series Expansion

With the preparations of the previous paragraphs we may now assume that we have a multiple polylogarithm $G_{m_1,\dots,m_k}(z_1, \dots, z_k; y)$, which has a sufficient fast converging sum representation. With the help of Eq. (8.20) we switch to the Li-notation

$$\mathrm{Li}_{m_1 \dots m_k}(x_1, \dots, x_k) = \sum_{n_1 > n_2 > \dots > n_k > 0}^{\infty} \frac{x_1^{n_1}}{{n_1}^{m_1}} \dots \frac{x_k^{n_k}}{{n_k}^{m_k}}. \tag{15.32}$$

Let us write

$$\mathrm{Li}_{m_1 \dots m_k}(x_1, \dots, x_k) = \sum_{n_1=1}^{\infty} d_{n_1}, \quad d_{n_1} = \frac{x_1^{n_1}}{n_1^{m_1}} \sum_{n_2=1}^{n_1-1} \frac{x_2^{n_2}}{n_2^{m_2}} \dots \sum_{n_k=1}^{n_{k-1}-1} \frac{x_k^{n_k}}{n_k^{m_k}}. \tag{15.33}$$

We may approximate $\mathrm{Li}_{m_1 \dots m_k}(x_1, \dots, x_k)$ by

$$I^{\mathrm{approx}}(N) = \sum_{n_1=1}^{N} d_{n_1} \tag{15.34}$$

for some $N \in \mathbb{N}$. This is a finite sum and can be evaluated on a computer. Choosing N large enough, such that the neglected terms contribute below the numerical precision gives the numerical evaluation of the iterated integral.

In more detail, let us define for two numbers a and b an equivalence relation. We say $a \sim b$, if they have exactly the same floating-point representation within a given numerical precision. A reasonable truncation criteria is as follows: We truncate the iterated integral at N if

$$I^{\mathrm{approx}}(N) \sim I^{\mathrm{approx}}(N-1) \text{ and } d_N \neq 0. \tag{15.35}$$

This gives reliable results in most cases.

15.2.4 Examples

The algorithms for the numerical evaluation of multiple polylogarithm are implemented in the computer algebra program GiNaC [421]. GiNaC is a C++ library and allows symbolic calculations as well as numerical calculations with arbitrary precision within C++. Alternatively, GiNaC offers also a small interactive shell called ginsh.

Let us consider as a first example

$$\mathrm{Li}_{31}\left(\frac{1}{2}, \frac{3}{4}\right). \tag{15.36}$$

This multiple polylogarithm is evaluated numerically in ginsh as follows:

```
> Li({3,1},{0.5,0.75});
0.029809219570239646653
```

We may change the number of digits:

```
> Digits=30;
30
> Li({2,2,1},{3.0,2.0,0.2});
0.0298092195702396466595180002639066394709
```

We may also use the $G(z_1, z_2 \dots, z_r; y)$-notation: We have

$$\mathrm{Li}_{31}\left(\frac{1}{2}, \frac{3}{4}\right) = G_{31}\left(6, 8; 3\right) = G\left(0, 0, 6, 8; 3\right). \tag{15.37}$$

The G-function is evaluated as follows:

```
> Digits=30;
30
> G({0,0,6.0,8.0},3.0);
0.0298092195702396466595180002639066394709
```

Let us also consider

$$G\left(\frac{1}{2} \pm i\delta; 1\right) = \pm i\pi, \tag{15.38}$$

where δ is an infinitesimal small positive number. Here we have to specify a small imaginary part, which indicates, if the pole at $\frac{1}{2}$ lies above or below the integration path. In `ginsh` we may include the signs of the small imaginary parts of the z_j's as an optional second list:

```
> G({0.5},{1},1.0);
3.14159265358979323856*I
> G({0.5},{-1},1.0);
-3.14159265358979323856*I
```

The default choice is a small positive imaginary part for the z_j's:

```
> G({0.5},1.0);
3.14159265358979323856*I
```

Implementations, which work with floating-point data types are `handyG` [422] and `FastGPL` [423]. These programs offer only a fixed precision, but are significantly faster and therefore better suited to be used in situations where multiple polylogarithms need to be evaluated several million times (like in Monte Carlo integrations).

Furthermore, there are dedicated implementations for the subclass of harmonic polylogarithms [217, 424, 425].

15.3 Iterated Integrals in the Elliptic Case

Let us now turn to the numerical evaluation of iterated integrals related to elliptic Feynman integrals. We introduced these integrals in Sect. 13.4. Let us recall that these are iterated integrals on a covering space of the moduli space $\mathcal{M}_{1,n}$. Standard coordinates on $\mathcal{M}_{1,n}$ are $(\tau, z_1, \dots, z_{n-1})$ (see Sect. 13.4) and we may decompose an arbitrary integration path into pieces along $d\tau$ (with $z_1 = \dots = z_{n-1} = \text{const}$) and pieces along the dz_j's (with $\tau = \text{const}$). By choosing appropriate boundary values it is sufficient to limit ourselves to iterated integrals with integration along $d\tau$. Thus we consider in this section iterated integrals of the form as in Eq. (13.210).

Comparing the numerical evaluation of these iterated integrals to the numerical evaluation of multiple polylogarithms, there are several similarities, but also two fundamental differences. We recall that the essential steps for the evaluation of multiple polylogarithms were (i) removal of trailing zeros, (ii) transformation into a region, where the series representation converges, (iii) series acceleration and (iv) evaluation of the truncated series.

The first difference of the numerical evaluation of iterated integrals in $d\tau$ with the numerical evaluation of multiple polylogarithms is actually good news: There are no poles in τ-space along the integration path. There might be poles at the starting point of the integration path ("trailing zeros") or at the endpoint, but not in between. This implies that the iterated integrals in τ-space always have a convergent series representation except for a few points. These few points correspond in $\bar{q}$-space to an integration up to $|\bar{q}| = 1$. In physical terms, this corresponds to a threshold or (more generally) to a singularity of the differential equation. Thus, for iterated integrals in $d\tau$ we do not need to consider point (ii) from the list above (i.e., no need for a transformation into a region, where the series representation converges). Let us emphasize, that this is not true for the iterated integrals in dz, e.g., the elliptic multiple polylogarithms $\widetilde{\Gamma}$ defined in Eq. (13.213): As in the case of ordinary multiple polylogarithms we may integrate in the case of elliptic multiple polylogarithms $\widetilde{\Gamma}$ past poles. However, as already mentioned above, there is no need for the functions $\widetilde{\Gamma}$. With appropriate boundary values we may always integrate along $d\tau$.

On the other hand, the second difference is not so pleasant: For endpoints of the integration path in $\bar{q}$-space close to $|\bar{q}| \lesssim 1$ the series expansion of the iterated integral converges rather slowly and we would like to apply methods to accelerate the series convergence. A modular transformation is the natural candidate. By a modular transformation $\tau' = \gamma(\tau)$ with $\gamma \in \mathrm{SL}_2(\mathbb{Z})$ we may transform τ' into the fundamental domain

$$\mathcal{F} = \left\{ \tau' \in \mathbb{H} \middle| |\tau'| > 1 \text{ and } -\frac{1}{2} < \mathrm{Re}\left(\tau'\right) \leq \frac{1}{2} \right\} \cup \left\{ \tau' \in \mathbb{H} \middle| |\tau'| = 1 \text{ and } 0 \leq \mathrm{Re}\left(\tau'\right) \leq \frac{1}{2} \right\} \tag{15.39}$$

and achieve that

$$\left|\bar{q}'\right| \leq e^{-\pi\sqrt{3}} \approx 0.0043. \tag{15.40}$$

This is a small expansion parameter. So far, so good. However, as discussed in Sects. 13.5.1 and 13.5.2 individual iterated integrals of the form as in Eq. (13.210) do in general not transform nicely under modular transformations. In general, we will leave through a modular transformation the space of iterated integrals of the form as in Eq. (13.210). We will stay inside the space of iterated integrals of the form as in Eq. (13.210) if we perform simultaneously a fibre transformation and change the basis of our master integrals. (We have seen explicit examples in Eqs. (13.276) and (13.339).) Unfortunately, this implies that the acceleration techniques are tied to the specific family of Feynman integrals under consideration and that we cannot implement acceleration techniques based on modular transformations into a black-box algorithm for iterated integrals of the form as in Eq. (13.210).

What can be done, is the following: Assuming that the iterated integral under consideration has a sufficiently fast converging series expansion, we may implement points (i) (removal of trailing zeros) and (iv) (evaluation of the truncated series) into a black-box algorithm. On the positive side, this can be done with a generality which exceeds the specific forms of Eq. (13.211). We will now discuss this in more detail.

Setup

Let M be a one-dimensional complex manifold with coordinate x and let ω_1, ..., ω_r be differential 1-forms on M. Let $\lambda_0 \in \mathbb{R}_{>0}$ and denote by U the domain $U = \{x \in \mathbb{C} | |x| \leq \lambda_0\}$. Let us assume that all ω_j are holomorphic in $U \backslash \{0\}$ and have at most a simple pole at $x = 0$. In other words

$$\omega_j = f_j(x)\, dx = \sum_{n=0}^{\infty} c_{j,n}\, x^{n-1} dx, \qquad c_{j,n} \in \mathbb{C}. \tag{15.41}$$

We say that ω_j has a **trailing zero**, if $c_{j,0} \neq 0$. We denote by

$$L_0 = d \ln(x) = \frac{dx}{x} \tag{15.42}$$

the logarithmic form with $c_0 = 1$ and $c_n = 0$ for $n > 0$.

We set

$$I(\underbrace{L_0, \dots, L_0}_{r}; x_0) = \frac{1}{r!} \ln^r(x_0) \tag{15.43}$$

and define recursively

$$I(\omega_1, \omega_2, \dots, \omega_r; x_0) = \int_0^{x_0} dx_1 f_1(x_1)\, I(\omega_2, \dots, \omega_r; x_1). \tag{15.44}$$

We say that the iterated integral $I(\omega_1, \dots, \omega_r; x_0)$ has a **trailing zero**, if ω_r has a trailing zero. If ω_r has a trailing zero, we may always write

$$\omega_r = c_{r,0} L_0 + \omega_r^{\text{reg}}, \tag{15.45}$$

with

$$\omega_r^{\text{reg}} = \sum_{n=1}^{\infty} c_{j,n} \, x^{n-1} dx \tag{15.46}$$

having no trailing zero. In the case where $I(\omega_1, \omega_2, \ldots, \omega_r; x_0)$ has no trailing zero, the definition in Eq. (15.44) agrees with the previous definition of iterated integrals in Eq. (6.140). Furthermore, we do not need to specify the path: As all ω_j's are holomorphic in U and $\dim M = 1$, the iterated integral is path-independent.

Shuffle product and trailing zeros
Iterated integrals always come with a shuffle product (compare with Sect. 8.3):

$$I(\omega_1, \ldots, \omega_k; x_0) \cdot I(\omega_{k+1}, \ldots, \omega_r; x_0) = \sum_{\text{shuffles } \sigma} I(\omega_{\sigma(1)}, \ldots, \omega_{\sigma(r)}; x_0), \tag{15.47}$$

where the sum runs over all shuffles σ of $(1, \ldots, k)$ with $(k+1, \ldots, r)$. The proof of this formula is identical to the proof of the shuffle product formula for multiple polylogarithms given in Sect. 8.3. We may use the shuffle product and Eq. (15.45) to remove trailing zeros, for example if $c_{1,0} = 0$ and $c_{2,0} = 1$ we have

$$\begin{aligned} I(\omega_1, \omega_2; x_0) &= I(\omega_1, L_0; x_0) + I(\omega_1, \omega_2^{\text{reg}}; x_0) \\ &= I(L_0; x_0) I(\omega_1; x_0) - I(L_0, \omega_1; x_0) + I(\omega_1, \omega_2^{\text{reg}}; x_0). \end{aligned} \tag{15.48}$$

This isolates all trailing zeros in integrals of the form (15.43), for which we may use the explicit formula in Eq. (15.43). It is therefore sufficient to focus on iterated integrals with no trailing zeros. For

$$I(\omega_1, \ldots, \omega_r; x_0) \tag{15.49}$$

this means $c_{r,0} = 0$. Please note that $c_{k,0} \neq 0$ is allowed for $k < r$ and in particular that the form L_0 is allowed in positions $k < r$.

For integrals with no trailing zeros we introduce the notation

$$\begin{aligned} & I_{m_1,\ldots,m_r}(\omega_1, \ldots, \omega_r; x_0) \\ &= I(\underbrace{L_0, \ldots, L_0}_{m_1-1}, \omega_1, \ldots, \omega_{r-1}, \underbrace{L_0, \ldots, L_0}_{m_r-1}, \omega_r; x_0), \end{aligned} \tag{15.50}$$

where we assumed that $\omega_k \neq L_0$ and $(m_k - 1)$ L_0's precede ω_k. This notation resembles the notation of multiple polylogarithms. The motivation for this notation is as follows: The iterated integrals $I_{m_1,\ldots,m_r}(\omega_1, \ldots, \omega_r; x_0)$ have just a r-fold series expansion, and not a $(m_1 + \cdots + m_r)$-fold one.

Series expansion

With the same assumptions as in the previous subsection (all ω_j are holomorphic in $U \backslash \{0\}$ and have at most a simple pole at $x = 0$) an iterated integral with no trailing zero has a convergent series expansion in U:

$$I_{m_1,\dots,m_r}(\omega_1, \dots, \omega_r; x_0) = \sum_{i_1=1}^{\infty} \sum_{i_2=1}^{i_1} \dots \sum_{i_r=1}^{i_{r-1}} x_0^{i_1} \frac{c_{1,i_1-i_2} \dots c_{r-1,i_{r-1}-i_r} c_{r,i_r}}{i_1^{m_1} i_2^{m_2} \cdot \dots \cdot i_r^{m_r}}, \quad (15.51)$$

where the Laurent expansion around $x = 0$ of the differential one-forms ω_j is given by Eq. (15.41). Equation (15.51) can be used for the numerical evaluation of the iterated integral: We truncate the outer sum over at $i_1 = N$. Let us write Eq. (15.51) as

$$I_{m_1,\dots,m_r}(\omega_1, \dots, \omega_r; x_0) = \sum_{i_1=1}^{\infty} d_{i_1},$$

$$d_{i_1} = x_0^{i_1} \sum_{i_2=1}^{i_1} \dots \sum_{i_r=1}^{i_{r-1}} \frac{c_{1,i_1-i_2} \dots c_{r-1,i_{r-1}-i_r} c_{r,i_r}}{i_1^{m_1} i_2^{m_2} \cdot \dots \cdot i_r^{m_r}}. \quad (15.52)$$

This gives a numerical approximation $I^{\text{approx}}(N)$ of the iterated integral

$$I^{\text{approx}}(N) = \sum_{i_1=1}^{N} d_{i_1}. \quad (15.53)$$

As truncation criteria we may again use Eq. (15.35).

Iterated integrals along $d\tau$

The discussion of the previous paragraphs applies to the iterated integrals along $d\tau$, introduced in Eq. (13.210). For the integration along $d\tau$ we consider in $\bar{q}$-space the iterated integrals

$$I_\gamma(\omega_1, \dots, \omega_r; \bar{q}), \quad (15.54)$$

where ω_j is of the form

$$\omega_{k_j}^{\text{Kronecker},\tau} = \frac{(k_j - 1)}{(2\pi i)^{k_j}} g^{(k_j)}\left(z - c_j, \tau\right) \frac{d\bar{q}}{\bar{q}} \quad \text{or} \quad \omega_{k_j}^{\text{modular}} = f_{k_j}(\tau) \frac{d\bar{q}}{\bar{q}}, \quad (15.55)$$

with $f_{k_j}(\tau)$ being a modular form of weight k_j. The $\bar{q}$-expansion of $\omega_{k_j}^{\text{Kronecker},\tau}$ is given in Eq. (13.202), the $\bar{q}$-expansion of Eisenstein series has been discussed in Sect. 13.3.1. Note that if a modular form is non-vanishing at the cusp $\tau = i\infty$, then it has a simple pole at $\bar{q} = 0$ in $\bar{q}$-space (and no further poles inside the unit disk $|\bar{q}| < 1$). The simple pole at $\bar{q} = 0$ comes from the Jacobian of the transformation from τ to $\bar{q}$:

$$2\pi i d\tau = \frac{d\bar{q}}{\bar{q}}. \tag{15.56}$$

Example

We may use GiNaC to evaluate numerically iterated integrals of the form as in Eq. (13.210) [426]. Let us see how this works in a full example. We consider the equal mass sunrise integral in two space-time dimensions:

$$I_{111}(2,x) = \frac{m^2}{\pi^2} \int d^2k_1 \int d^2k_2 \int d^2k_3 \frac{\delta^2(p-k_1-k_2-k_3)}{\left(k_1^2-m^2\right)\left(k_1^2-m^2\right)\left(k_1^2-m^2\right)} \tag{15.57}$$

with $x = -p^2/m^2$. Feynman's $i\delta$-prescription translates into an infinitesimal small negative imaginary part of x. In Sect. 13.5.1 we worked out this integral and found

$$I_{111}(2,x) = \frac{\psi_1}{\pi}\left[3\,\mathrm{Cl}_2\left(\frac{2\pi}{3}\right) + 4I(\eta_0, \eta_3; \tau)\right]. \tag{15.58}$$

This involves the iterated integral

$$I(\eta_0, \eta_3; \tau), \tag{15.59}$$

with

$$\begin{aligned} \eta_0 &= -1, \\ \eta_3 &= -9\sqrt{3}\left(b_1^3 - b_1^2 b_2 - 4b_1 b_2^2 + 4b_2^3\right) \end{aligned} \tag{15.60}$$

and

$$\begin{aligned} b_1 &= E_1\left(\tau; \chi_1, \chi_{(-3)}\right) &= E_{1,1,-3,1}(\tau), \\ b_2 &= E_1\left(2\tau; \chi_1, \chi_{(-3)}\right) &= E_{1,1,-3,2}(\tau). \end{aligned} \tag{15.61}$$

The following C++ code computes the Feynman integral $I_{111}(2,x)$ for $x \in \mathbb{R}\backslash\{-9,-1,0\}$:

```
#include <ginac/ginac.h>

int main()
{
  using namespace std;
  using namespace GiNaC;

  Digits = 30;

  // input x = -p^2/m^2, x real and not equal to {-9,-1,0}
  numeric x = numeric(-1,100);
```

```
  numeric sqrt_3  = sqrt(numeric(3));
  numeric sqrt_mx = sqrt(-x);
  numeric k2      = 16*sqrt_mx/pow(1+sqrt_mx,numeric(3))/(3-sqrt_mx);

  ex  pre = 4*pow(1+sqrt_mx,numeric(-3,2))*pow(3-sqrt_mx,numeric(-1,2));
  if (x < -9) pre = -pre;
  ex psi1 = pre*EllipticK(sqrt(k2));
  ex psi2 = pre*I*EllipticK(sqrt(1-k2));
  if ((x < -1) || (x > -3+2*sqrt_3)) psi1 += 2*psi2;
  if ((x > -9) && (x < -1)) psi1 += 2*psi2;
  ex tau  = psi2/psi1;
  ex qbar = exp(2*Pi*I*tau);

  ex L0   = basic_log_kernel();
  ex b1   = Eisenstein_kernel(1, 6, 1, -3, 1);
  ex b2   = Eisenstein_kernel(1, 6, 1, -3, 2);
  ex eta3 = modular_form_kernel(3, -9*sqrt_3*(pow(b1,3)-pow(b1,2)*b2
              -4*b1*pow(b2,2)+4*pow(b2,3)));

  ex Cl2  = numeric(1,2)/I*(Li(2,exp(2*Pi*I/3))-Li(2,exp(-2*Pi*I/3)));
  ex I111 = psi1/Pi*(3*Cl2-4*iterated_integral(lst{L0,eta3},qbar));

  cout << "I111 = " << I111.evalf() << endl;

  return 0;
}
```

Let us explain the code: The input is given in the line

```
  numeric x = numeric(-1,100);
```

One may change this value to any other value except $x \notin \{-9, -1, 0\}$. The program computes then the two periods ψ_1 and ψ_2, the modular parameter τ and the variable $\bar{q}$. The line

```
  if ((x < -1) || (x > -3+2*sqrt_3)) psi1 += 2*psi2;
```

corresponds to Eq. (13.258).

There is a convention how mathematical software should evaluate a function on a branch cut: Implementations shall map a cut so the function is continuous as the cut is approached coming around the finite endpoint of the cut in a counter clockwise direction [207]. `GiNaC` follows this convention. In physics, Feynman's $i\delta$-prescription dictates how a function should be evaluated on a branch cut. The lines

```
  if (x < -9) pre = -pre;
  if ((x < -9) && (x < -1)) psi1 += 2*psi2;
```

correct for a mismatch between the standard convention for mathematical software and Feynman's $i\delta$-prescription.

We then define the modular forms. `basic_log_kernel()` represents

$$2\pi i \, d\tau = \frac{d\bar{q}}{\bar{q}}. \tag{15.62}$$

The Eisenstein series $E_{k,a,b,K}(\tau)$ for $\Gamma_1(N)$ are defined by

```
Eisenstein_kernel(k, N, a, b, K);
```

Finally,

```
iterated_integral(lst{L0,eta3},qbar);
```

defines the iterated integral $I(1, \eta_3; \tau) = -I(\eta_0, \eta_3; \tau)$.

Running the code for

$$x = -0.01 \tag{15.63}$$

yields

```
I111 = 2.3450544099124155711465801399777317976
```

15.4 The PSLQ Algorithm

The possibility to evaluate numerically transcendental functions to high precision allows us to simplify boundary constants. Suppose the boundary value of a Feynman integral at a certain kinematic point is given by a linear combination of harmonic polylogarithms at $x = 1$:

$$G(z_1, \ldots, z_r; 1), \quad z_j \in \{-1, 0, 1\}. \tag{15.64}$$

These are just transcendental numbers. Excluding trailing zeros and leading ones we have at weight $r \geq 2$

$$4 \cdot 3^{r-2} \tag{15.65}$$

transcendental numbers. However, they are not linearly independent.

The PSLQ algorithm [427–430] may be used to find relations among a set of transcendental numbers. The input to the PSLQ algorithm are high-precision numerical values for the transcendental numbers. The PSLQ algorithm then finds integer coefficients, such that the linear combination with these integer coefficients is close to zero within the numerical precision. This does not provide a strict mathematical proof that the linear combination is indeed zero. However, we may increase the numerical precision and if the coefficients stay constant we may be confident that the relation is correct.

Let us start with an example: We consider

$$G(-1, 0, -1, -1; 1) \tag{15.66}$$

and we ask if there is a relation between $G(-1, 0, -1, -1; 1)$ and the simpler constants

$$\text{Li}_4\left(\frac{1}{2}\right), \quad \zeta_4, \quad \zeta_3 \ln(2), \quad \zeta_2 \ln^2(2), \quad \ln^4(2). \tag{15.67}$$

We evaluate numerically all quantities to 50 digits:

$$\begin{aligned}
G(-1,0,-1,-1;1) &\approx 0.032893195194356041326359565602868932538\,7, \\
\text{Li}_4\left(\frac{1}{2}\right) &\approx 0.51747906167389938633075816189886294561\,8, \\
\zeta_4 &\approx 1.0823232337111381915160036965411679027\,76, \\
\zeta_3 \ln(2) &\approx 0.83320235329769199344576252966156010389\,4, \\
\zeta_2 \ln^2(2) &\approx 0.79031353011395460877291733568064410420\,4, \\
\ln^4(2) &\approx 0.23083509858308345188749771776781277151\,7.
\end{aligned} \tag{15.68}$$

The PSLQ algorithm gives then

$$8G(-1,0,-1,-1;1) - 24\text{Li}_4\left(\frac{1}{2}\right) + 24\zeta_4 - 22\zeta_3 \ln(2) + 6\zeta_2 \ln^2(2) - \ln^4(2) = 0$$

and therefore

$$G(-1,0,-1,-1;1) = 3\text{Li}_4\left(\frac{1}{2}\right) - 3\zeta_4 + \frac{11}{4}\zeta_3 \ln(2) - \frac{3}{4}\zeta_2 \ln^2(2) + \frac{1}{8}\ln^4(2). \tag{15.69}$$

We may repeat the calculation with a higher number of digits. The empirical relation will stay the same. This gives us confidence that the relation is correct.

On the other hand, if no relation exists, the PSLQ algorithm will tell us that no relation with integer coefficients smaller than a certain bound exists. The bound depends on the numerical precision.

The input to the PSLQ algorithm is a vector $x = (x_1, \ldots, x_n) \in \mathbb{R}^n$ of real numbers, given as floating-point numbers to a certain precision. An integer relation is given by a vector $m = (m_1, \ldots, m_n) \in \mathbb{Z}^n$ of integer numbers such that

$$m_1 x_1 + m_2 x_2 + \cdots + m_n x_n = 0. \tag{15.70}$$

The name of the PSLQ algorithm derives from the **partial sums**

$$s_k = \sqrt{\sum_{j=k}^{n} x_j^2} \tag{15.71}$$

and the **LQ-decomposition of matrices**: Any $(n \times m)$-matrix M may be written as

$$M = L \cdot Q, \tag{15.72}$$

where L is a lower trapezoidal $(n \times m)$-matrix and Q is an orthogonal $(m \times m)$-matrix (i.e., $Q^{-1} = Q^T$). A $(n \times m)$-matrix L is called a **lower trapezoidal matrix**, if

$$L_{ij} = 0 \qquad \text{for} \quad i < j. \tag{15.73}$$

The PSLQ algorithm is not too complicated to state. For the proof why the algorithm works we refer to the literature [428, 429]. In order to state the algorithm we denote for a real number x by $[x]$ the rounded value to the nearest integer. The PSQL algorithm depends on two parameters γ and δ. The first parameter γ is chosen as

$$\gamma \geq \sqrt{\frac{4}{3}} \tag{15.74}$$

and determines the weighting of the diagonal elements of a matrix H in the first iteration step below. The second parameter δ defines the detection threshold for an integer relation. As a rule of thumb, if we expect an integer relation between n input number $x_1, \dots, x_n$ with integer coefficients of maximum size d digits, we should work with a precision of $(n \cdot d)$ digits (i.e., the input numbers x_j have to be given with this precision, and all internal arithmetic has to be carried out with this precision). In order to tolerate numerical rounding errors, we set the detection threshold a few orders of magnitude greater than 10^{-nd}, e.g.,

$$\delta = 10^{-nd+o}, \tag{15.75}$$

where o is a small positive integer. Let us now state the algorithm:

Algorithm 3 *The PSLQ algorithm*

The algorithm is divided into an initialisation phase and an iteration phase.

Initialisation:

1. *Initialise two integer $(n \times n)$-matrices A and B by*

$$A = \mathbf{1}_{n\times n}, \ B = \mathbf{1}_{n\times n}. \tag{15.76}$$

2. *Initialise s_k by Eq. (15.71) and set*

$$y_k = \frac{x_k}{s_1}, \qquad 1 \leq k \leq n. \tag{15.77}$$

3. *Initialise a lower trapezoidal* $n \times (n-1)$*-matrix* H *by*

$$H_{ij} = \begin{cases} -\frac{x_i x_j}{s_j s_{j+1}}, \; i > j, \\ \frac{s_{j+1}}{s_j}, \; i = j, \\ 0, \; i < j. \end{cases} \tag{15.78}$$

4. *Reduce* H*:*

```
for i = 2 to n do
    for j = i − 1 to 1 step −1 do
        t ← [H_ij / H_jj]
        y_j ← y_j + t y_i
        for k = 1 to j do
            H_ik ← H_ik − t H_jk
        end for
        for k = 1 to n do
            A_ik ← A_ik − t A_jk
            B_kj ← B_kj − t B_ki
        end for
    end for
end for
```

Iteration:

1. *Select* l *such that* $\gamma^i |H_{ii}|$ *is maximal for* $i = l$*.*
2. *Exchange the entries of* y *indexed* l *and* $(l+1)$*, the corresponding rows of* A *and* H*, and the corresponding columns of* B*.*
3. *Remove corner:*

```
if l ≤ n − 2 then
    t_0 ← sqrt(H_ll^2 + H_l(l+1)^2)
    t_1 ← H_ll / t_0
    t_2 ← H_l(l+1) / t_0
    for i = l to n do
        t_3 ← H_il
        t_4 ← H_i(l+1)
        H_il ← t_1 t_3 + t_2 t_4
        H_i(l+1) ← −t_2 t_3 + t_1 t_4
    end for
end if
```

4. *Reduce* H*:*

```
for i = l + 1 to n do
    for j = min(i − 1, l + 1) to 1 step −1 do
        t ← [H_ij / H_jj]
```

$y_j \leftarrow y_j + t y_i$
for $k = 1$ *to* j ***do***
$H_{ik} \leftarrow H_{ik} - t H_{jk}$
end for
for $k = 1$ *to* n ***do***
$A_{ik} \leftarrow A_{ik} - t A_{jk}$
$B_{kj} \leftarrow B_{kj} - t B_{ki}$
end for
end for
end for

5. *Termination test: If the largest entry of* A *exceeds the numerical precision, then no relation exists where the Euclidean norm of the vector* m *is less than* $1/\max_j |H_{jj}|$. *If the smallest entry of the vector* y *is less than the detection threshold* δ, *return the corresponding column of* B. *Otherwise go back to step 1 of the iteration.*

The PSLQ algorithm is implemented in many commercial computer algebra systems.

Chapter 16
Final Project

In the last chapter of this book, let's do a final project. In this way we are going to review many techniques introduced in the previous chapters.

16.1 A Two-Loop Penguin Integral

We are going to compute the Feynman integral of the penguin graph shown in Fig. 16.1. We are going to neglect all light quark masses ($m_s = m_b = 0$) and only keep the heavy masses m_W, m_H, m_t non-zero. The integral corresponding to the graph shown in Fig. 16.1 has six loop propagators (and one tree-like propagator: the gluon propagator drawn by a curly line at the bottom). We notice that two of the six loop propagators are identical: These are the W-boson propagators making up the left and the right shoulder of the penguin. The momenta flowing through these lines is the same, as is the internal mass (i.e., m_W). Thus we actually only need to consider a two-loop integral with five loop propagators, where one propagator is raised to the power two. The corresponding diagram (a distorted penguin) is shown in Fig. 16.2. In Fig. 16.2 we also indicated the external momenta. As usual our convention is to take all momenta as outgoing, therefore momentum conservation reads

$$p_1 + p_2 + p_3 + p_4 = 0. \tag{16.1}$$

The external particles are assumed to be on the mass-shell, and since we assumed $m_b = m_s = 0$ we have

$$p_1^2 = m_b^2 = 0, \; p_2^2 = p_3^2 = p_4^2 = m_s^2 = 0. \tag{16.2}$$

It is clear from the diagram that the scalar Feynman integrals will only depend on $p_1, p_2, (p_3 + p_4)$, but not on p_3 nor p_4 individually. Thus the only non-vanishing Lorentz invariant is

S. Weinzierl, *Feynman Integrals*, UNITEXT for Physics,
https://doi.org/10.1007/978-3-030-99558-4_16

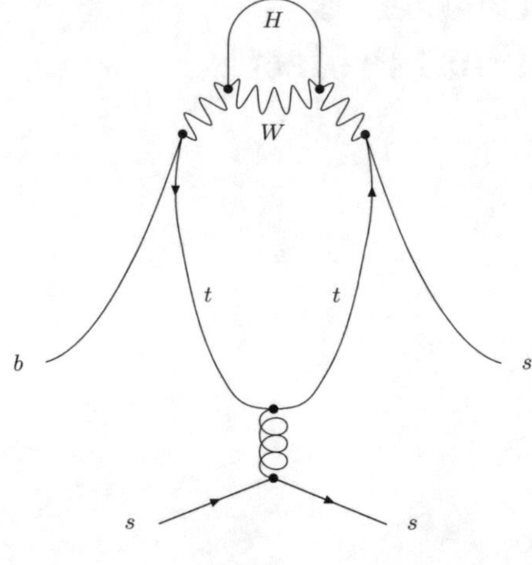

Fig. 16.1 A two-loop penguin diagram

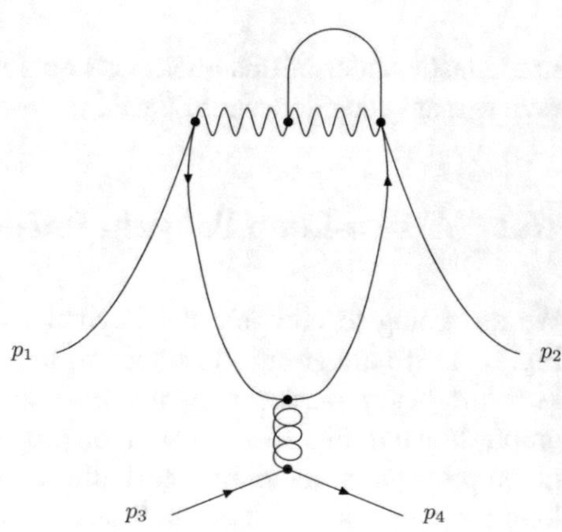

Fig. 16.2 The distorted penguin: A two-loop diagram with five loop propagators

$$s = (p_1 + p_2)^2 \;=\; (p_3 + p_4)^2 \,. \tag{16.3}$$

Therefore our kinematic variables are

$$x_1 = \frac{-s}{\mu^2}, \qquad x_2 = \frac{m_W^2}{\mu^2}, \qquad x_3 = \frac{m_H^2}{\mu^2}, \qquad x_4 = \frac{m_t^2}{\mu^2}. \tag{16.4}$$

In our calculation we may set (temporarily) μ to any value we want, and in particular the choice $\mu = m_t$ sets the last variable equal to one: $x_4 = 1$. As μ enters only as a trivial prefactor the definition of Feynman integrals, the μ dependence can be restored at the end of the calculation. We therefore have a problem with three kinematic variables x_1, x_2, x_3 and $N_B = 3$ (see the discussion in Sect. 2.5.1 and in Sect. 6.4.2).

Fig. 16.3 A two-loop three-point function. A green line denotes a propagator with mass m_t, a red line denotes a propagator with mass m_W, a blue line denotes a propagator with mass m_H. A thick black external line indicates that $p^2 = s$, a thin black external line indicates that $p^2 = 0$

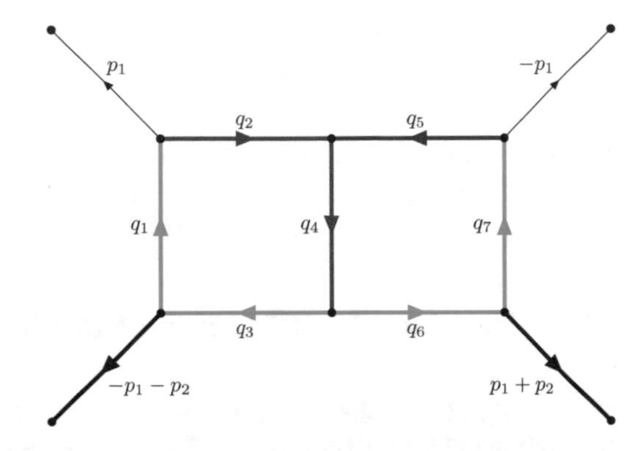

Fig. 16.4 The auxiliary graph $\tilde{G}$ with seven propagators. The arrow indicate the momentum flow. The colour coding is as in Fig. 16.3

We recall from Sect. 4.3 that all tensor integrals can be reduced to scalar integrals. Therefore we only need to consider the relevant scalar integrals. In Fig. 16.3 we draw the relevant Feynman graph in a more standard way: We are interested in a two-loop three-point function with five propagators as shown in Fig. 16.3.

We are going to use integration-by-parts identities. We have two linear independent external momenta (p_1 and p_2) and two independent loop momenta (which we label k_1 and k_2). Therefore we have (see Sect. 2.5.5)

$$N_V = \frac{1}{2} l\,(l+1) + el \;=\; 7 \tag{16.5}$$

linear independent scalar products involving the loop momenta. We therefore consider an auxiliary graph with seven loop propagators, such that any scalar product involving the loop momenta can be expressed as a linear combination of the propagators and a constant and vice versa. This is to say that we seek an auxiliary graph such that Eq. (2.225) holds. This is not too complicated, the double-box graph $\tilde{G}$ shown in Fig. 16.4 will do the job. Note that this double-box graph depends only on two linear independent external momenta.

We set $\mu = m_t$ and we consider the family of Feynman integrals

$$\begin{aligned} & I_{\nu_1\nu_2\nu_3\nu_4\nu_5\nu_6\nu_7}\left(D, x_1, x_2, x_3\right) \\ &= e^{2\varepsilon\gamma_{\mathrm{E}}} \left(m_t^2\right)^{\nu-D} \int \frac{d^D k_1}{i\pi^{\frac{D}{2}}} \frac{d^D k_2}{i\pi^{\frac{D}{2}}} \prod_{j=1}^{7} \frac{1}{\left(-q_j^2 + m_j^2\right)^{\nu_j}}, \end{aligned} \tag{16.6}$$

with

$$
\begin{aligned}
q_1 &= k_1 + p_1 + p_2, & m_1 &= m_t, \\
q_2 &= k_1 + p_2, & m_2 &= m_W, \\
q_3 &= k_1, & m_3 &= m_t, \\
q_4 &= k_1 + k_2, & m_4 &= m_W, \\
q_5 &= k_2 - p_2, & m_5 &= m_H, \\
q_6 &= k_2, & m_6 &= m_t, \\
q_7 &= k_2 - p_1 - p_2, & m_7 &= m_t.
\end{aligned} \tag{16.7}
$$

We are interested in the integrals with $\nu_6 \leq 0$, $\nu_7 \leq 0$, these correspond to the topology shown in Fig. 16.3.

16.2 Deriving the Differential Equation

We first determine the two graph polynomials $\mathcal{U}$ and $\mathcal{F}$ of the graph $\tilde{G}$ with the help of the methods from Chap. 3. The first graph polynomial can actually be copied directly from Eq. (2.164), for the second graph polynomial we have to take into account that our kinematic configuration is different. We obtain

$$
\begin{aligned}
\mathcal{U} &= (a_1 + a_2 + a_3)(a_5 + a_6 + a_7) + a_4 (a_1 + a_2 + a_3 + a_5 + a_6 + a_7), \\
\mathcal{F} &= x_1 [a_1 a_3 (a_4 + a_5 + a_6 + a_7) + a_6 a_7 (a_1 + a_2 + a_3 + a_4) + a_1 a_4 a_6 + a_3 a_4 a_7] \\
&\quad + [a_1 + a_3 + a_6 + a_7 + x_2 (a_2 + a_4) + x_3 a_5] \,\mathcal{U}.
\end{aligned} \tag{16.8}
$$

In the next step we generate the integration-by-parts identities. We do this with the help of a computer program, like `Fire` [113, 114], `Reduze` [115, 116] or `Kira` [117, 118]. We may modify the set-up from Exercise 44. For example, if we are using `Kira` the file `integralfamilies.yaml` should now read

```
integralfamilies:
  - name: "doublebox"
    loop_momenta: [k1, k2]
    top_level_sectors: [31]
    propagators:
      - [ "k1+p1+p2", "mt2" ]
      - [ "k1+p2", "mW2" ]
      - [ "k1", "mt2" ]
      - [ "k1+k2", "mW2" ]
      - [ "k2-p2", "mH2" ]
      - [ "k2", "mt2" ]
      - [ "k2-p1-p2", "mt2" ]
```

and the file `kinematics.yaml` should read

```
kinematics :
  incoming_momenta: []
  outgoing_momenta: [p1, p2, p3]
  momentum_conservation: [p3,-p1-p2]
  kinematic_invariants:
    - [s,  2]
    - [mW2,2]
    - [mH2,2]
    - [mt2,2]
  scalarproduct_rules:
    - [[p1,p1],  0]
    - [[p2,p2],  0]
    - [[p1+p2,p1+p2],  s]
  symbol_to_replace_by_one: mt2
```

Running an integration-by-parts reduction program we also obtain a list of master integrals. This list is not unique and will depend on the chosen ordering criteria for the Laporta algorithm (see Sect. 6.1). A possible basis of master integrals is tabulated in Table 16.1. We have 15 master integrals, hence $N_{\text{master}} = 15$. There are 12 sectors. The sectors (or master topologies) are shown in Fig. 16.5. Eleven sectors have only one master integral per sector, while one sector (with sector id $N_{\text{id}} = 29$) has four master integrals.

Table 16.1 Overview of the set of master integrals. The first column denotes the number of propagators, the second column labels consecutively the sectors or topologies, the third column gives the sector id N_{id} (defined in Eq. (6.16)), the fourth column lists the master integrals in the basis $\vec{I}$, the fifth column the corresponding ones in the basis $\vec{J}$. The last column denotes the kinematic dependence

Number of propagators	Block	Sector	Master integrals basis $\vec{I}$	Master integrals basis $\vec{J}$	Kinematic dependence
2	1	9	$I_{1001000}$	J_1	x_2
	2	10	$I_{0101000}$	J_2	x_2
	3	17	$I_{1000100}$	J_3	x_3
	4	18	$I_{0100100}$	J_4	x_2, x_3
3	5	13	$I_{1011000}$	J_5	x_1, x_2
	6	21	$I_{1010100}$	J_6	x_1, x_3
	7	25	$I_{1001100}$	J_7	x_2, x_3
	8	26	$I_{0101100}$	J_8	x_2, x_3
4	9	15	$I_{1111000}$	J_9	x_1, x_2
	10	23	$I_{1110100}$	J_{10}	x_1, x_2, x_3
	11	29	$I_{1011100}$, $I_{1(-1)11100}$, $I_{10111(-1)0}$, $I_{1(-2)11100}$	$J_{11}, J_{12}, J_{13}, J_{14}$	x_1, x_2, x_3
5	12	31	$I_{1111100}$	J_{15}	x_1, x_2, x_3

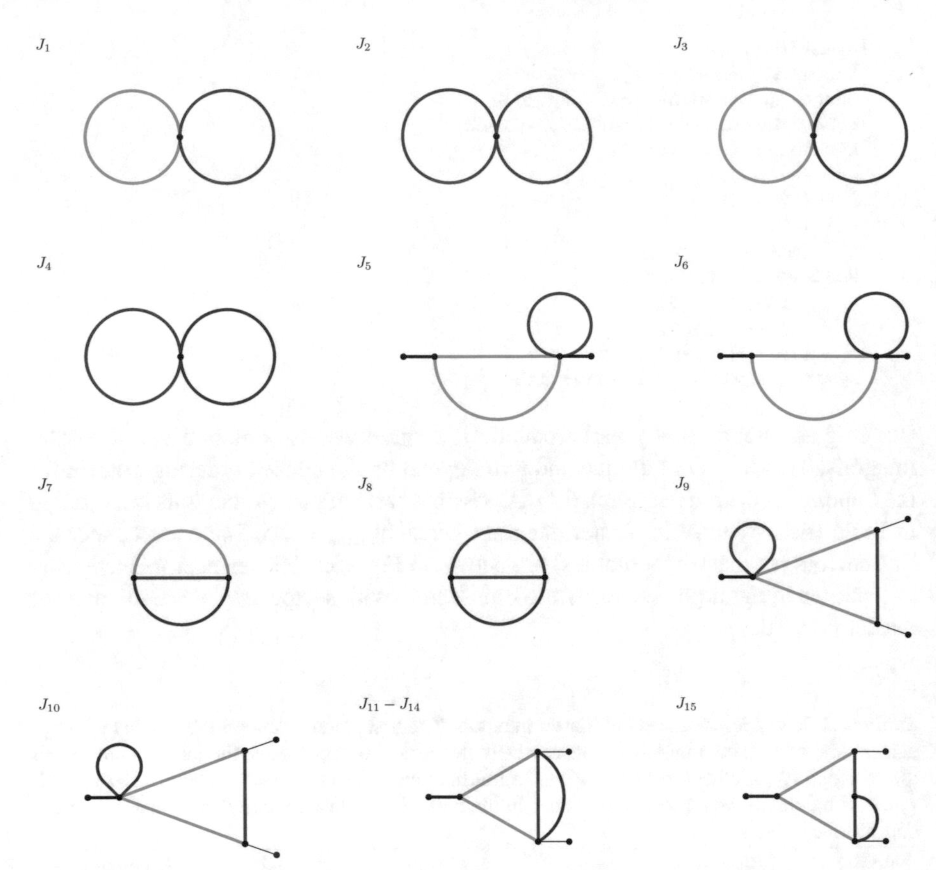

Fig. 16.5 The master topologies

We set

$$\vec{I} = (I_{1001000}, I_{0101000}, I_{1000100}, I_{0100100}, I_{1011000}, I_{1010100}, I_{1001100}, I_{0101100}, \\ I_{1111000}, I_{1110100}, I_{1011100}, I_{1(-1)11100}, I_{10111(-1)0}, I_{1(-2)11100}, I_{1111100})^T. \quad (16.9)$$

In the next step one derives the differential equation for $\vec{I}$, as explained in Sect. 6.3. For the derivatives with respect to x_1, x_2, x_3 we may use Eq. (6.46). As we are considering a two-loop integral we get from $\mathcal{F}'_{x_j}(\mathbf{1}^+, \dots, \mathbf{n}_{\mathrm{int}}{}^+)$ three raising operators and we therefore need integration-by-parts reduction identities for up to three dots. We also need the dimensional shift relations, which involves $\mathcal{U}(\mathbf{1}^+, \dots, \mathbf{n}_{\mathrm{int}}{}^+)$. As for a two-loop graph $\mathcal{U}$ is homogeneous of degree 2, this leads to two raising operators and requires integration-by-parts reduction identities for up to two dots. These are a subset of the ones required for up to three dots.

For this particular example the full differential equation can actually be derived with reduction identities for only one dot. This saves quite some computer memory and CPU time. The trick is as follows: We do not set $\mu = m_t$ in the beginning and first compute the derivatives with respect to x_2, x_3, x_4. These are the derivatives with respect to the internal masses and we may use Eq. (6.50). This involves only one raising operator (and no dimensional shift), therefore reduction identities for one dot are sufficient. The derivative with respect to x_1 is then obtained from the scaling relation Eq. (6.51). Having obtained the derivatives with respect to x_1, x_2, x_3 we may set $\mu = m_t$.

In this way we obtain the differential equation

$$(d + A)\,\vec{I} = 0, \tag{16.10}$$

where

$$d = \sum_{j=1}^{3} dx_j \frac{\partial}{\partial x_j}, \quad A = \sum_{j=1}^{3} dx_j \, A_{x_j}, \tag{16.11}$$

and the A_{x_j} are (15×15)-matrices, whose entries are rational functions of x_1, x_2, x_3 and ε. These matrices are not in a form to be printed here, but they may be computed in a straightforward way and stored on a computer. The matrices $A_{x_1}, A_{x_2}, A_{x_3}$ have to satisfy the integrability condition of Eq. (6.57). Spelled out in components we must have

$$\begin{aligned}
\partial_{x_1} A_{x_2} - \partial_{x_2} A_{x_1} + \left[A_{x_1}, A_{x_2}\right] &= 0, \\
\partial_{x_1} A_{x_3} - \partial_{x_3} A_{x_1} + \left[A_{x_1}, A_{x_3}\right] &= 0, \\
\partial_{x_2} A_{x_3} - \partial_{x_3} A_{x_2} + \left[A_{x_2}, A_{x_3}\right] &= 0,
\end{aligned} \tag{16.12}$$

where $[A, B] = A \cdot B - B \cdot A$ denotes the commutator of the two matrices A and B. It is highly recommended to check these relations at this stage.

16.3 Fibre Transformation

Let us set $D = 4 - 2\varepsilon$. In the next step we perform a fibre transformation, e.g., we redefine the master integrals

$$\vec{J} = U(\varepsilon, x)\,\vec{I}. \tag{16.13}$$

We seek a transformation U such that in the transformed differential equation

$$\left(d + A'\right)\vec{J} = 0, \qquad A' = U A U^{-1} + U d U^{-1} \tag{16.14}$$

the dimensional regularisation parameter ε appears only as a prefactor of A'. To this aim we define a new basis of master integrals

$$\vec{J} = (J_1, J_2, \dots, J_{15})^T \tag{16.15}$$

such that the J_i's are of uniform weight. Expressing the new J_i's as a linear combination of the old basis $\vec{I} = (I_1, \dots, I_{15})^T$ defines the matrix U:

$$J_i = \sum_{j=1}^{15} U_{ij} I_j. \tag{16.16}$$

We may use the methods of Chap. 7 to construct the J_i's. However, it is usually the case that the first few J_i's may already be obtained from known examples. This is also the case here. The first four master integrals are each products of two one-loop tadpole integrals. The tadpole integral was the first Feynman integral we calculated and from Eq. (2.125) we know that $\varepsilon\ T_1(2-2\varepsilon)$ is of uniform weight. We have set $D = 4 - 2\varepsilon$, hence we may write

$$\varepsilon\ T_1\left(2-2\varepsilon\right) = \varepsilon\ \mathbf{D}^- T_1\left(4-2\varepsilon\right). \tag{16.17}$$

From the dimensional shift relation Eq. (2.126) we have $T_1(2-2\varepsilon) = T_2(4-2\varepsilon)$ and therefore

$$\varepsilon\ T_1\left(2-2\varepsilon\right) = \varepsilon\ \mathbf{D}^- T_1\left(4-2\varepsilon\right) \ = \ \varepsilon\ T_2\left(4-2\varepsilon\right). \tag{16.18}$$

Thus the first four master integrals of uniform weight are

$$\begin{aligned}
J_1 &= \varepsilon^2\ \mathbf{D}^- I_{1001000} \ = \ \varepsilon^2\ I_{2002000}, \\
J_2 &= \varepsilon^2\ \mathbf{D}^- I_{0101000} \ = \ \varepsilon^2\ I_{0202000}, \\
J_3 &= \varepsilon^2\ \mathbf{D}^- I_{1000100} \ = \ \varepsilon^2\ I_{2000200}, \\
J_4 &= \varepsilon^2\ \mathbf{D}^- I_{0100100} \ = \ \varepsilon^2\ I_{0200200}.
\end{aligned} \tag{16.19}$$

The next two sectors (sectors 13 and 21) are again products of one-loop integrals, in this case the product of a tadpole integral and a bubble integral. The bubble integral is the one discussed as example 1 in Sect. 6.3.1. In Eq. (6.277) we have given the corresponding master integral of uniform weight. This involves the root

$$r_1 = \sqrt{x_1\left(4+x_1\right)} \ = \ \frac{1}{m_t^2}\sqrt{-s\left(4m_t^2 - s\right)}. \tag{16.20}$$

Thus

$$\begin{aligned}
J_5 &= -\varepsilon^2 r_1\ \mathbf{D}^- I_{1011000} = 2\varepsilon^2 \frac{r_1}{4+x_1}\left[\left(1-\varepsilon\right) I_{1002000} + \left(1-2\varepsilon\right) I_{1012000}\right], \\
J_6 &= -\varepsilon^2 r_1\ \mathbf{D}^- I_{1010100} = 2\varepsilon^2 \frac{r_1}{4+x_1}\left[\left(1-\varepsilon\right) I_{1000200} + \left(1-2\varepsilon\right) I_{1010200}\right].
\end{aligned} \tag{16.21}$$

The sectors 25 and 26 are genuine two-loop topologies. In order to find master integrals of uniform weight we look at maximal cuts and constant leading singularities (see Sect. 7.1.7). For the sector 25 the maximal cut in $D = 2$ space-time dimensions is given by

$$\text{MaxCut } I_{1001100}(2) = \frac{(2\pi i)^3 m_t^2}{\pi^2} \int\limits_{\mathcal{C}_{\text{MaxCut}}} dz_2 \frac{1}{\left(z_2 + m_t^2 - m_W^2\right)\sqrt{4m_H^2 m_W^2 - \left(z_2 + m_H^2\right)^2}}$$
$$= \int\limits_{\mathcal{C}_{\text{MaxCut}}} \varphi. \quad (16.22)$$

The last equation defines the integrand φ. We now replace the integration domain $\mathcal{C}_{\text{MaxCut}}$ by a simpler integration domain $\mathcal{C}$, given by a small anti-clockwise circle around $z_2 = m_W^2 - m_t^2$. This gives

$$\langle \varphi | \mathcal{C} \rangle = \frac{16\pi^2}{\sqrt{-\lambda\left(x_2, x_3, 1\right)}}, \quad (16.23)$$

where $\lambda(x, y, z)$ denotes the Källén function

$$\lambda\left(x, y, z\right) = x^2 + y^2 + z^2 - 2xy - 2yz - 2zx. \quad (16.24)$$

The analysis for sector 26 is similar and gives instead of $\lambda(x_2, x_3, 1)$ the expression $\lambda(x_2, x_3, x_2)$. We introduce two new square roots

$$r_2 = \sqrt{-\lambda\left(x_2, x_3, 1\right)} = \sqrt{2x_2x_3 + 2x_2 + 2x_3 - x_2^2 - x_3^2 - 1} = \frac{1}{m_t^2}\sqrt{-\lambda\left(m_W^2, m_H^2, m_t^2\right)},$$
$$r_3 = \sqrt{-\lambda\left(x_2, x_3, x_2\right)} = \sqrt{4x_2x_3 - x_3^2} = \frac{1}{m_t^2}\sqrt{m_H^2\left(4m_W^2 - m_H^2\right)}. \quad (16.25)$$

Our tentative guess for J_7 and J_8 is

$$J_7 = \frac{1}{2}\varepsilon^2 r_2 \, \mathbf{D}^- I_{1001100},$$
$$J_8 = \frac{1}{2}\varepsilon^2 r_3 \, \mathbf{D}^- I_{0101100}. \quad (16.26)$$

This guess is obtained (apart from irrelevant rational prefactors) by dividing I_7 and I_8 by the appropriate period $\langle \varphi | \mathcal{C} \rangle$ and by replacing π by ε^{-1}. There may be additional terms proportional to sub-topologies. In these two examples we verify a posteriori that there are no additional terms and J_7 and J_8 define master integrals of uniform weight.

The sectors 15 and 23 are again products of one-loop integrals. We need the one-loop triangle of uniform weight, which again can be obtained from the maximal cut. Proceeding as above we obtain

$$J_9 = \varepsilon^3 x_1 \; I_{1112000},$$
$$J_{10} = \varepsilon^3 x_1 \; I_{1110200}. \tag{16.27}$$

We now come to sector 29. This is the most challenging sector, as it has four master integrals. We started from an ISP-basis, where we chose

$$I_{1011100}, I_{1(-1)11100}, I_{10111(-1)0}, I_{1(-2)11100} \tag{16.28}$$

as a basis for this sector. We could have chosen a dot-basis, in which case

$$I_{1011100}, I_{1011200}, I_{1012100}, I_{1021100} \tag{16.29}$$

would be an appropriate basis. Our strategy is to put first the (4×4)-diagonal block into an ε-form and to treat the off-diagonal blocks (corresponding to sub-topologies) in a second stage. For the diagonal block we may work on the maximal cut. For the maximal cut we use the loop-by-loop approach (see Sect. 2.5.5), this yields a one-fold integral representation in the Baikov variable z_2 for the maximal cut. We look at the maximal cuts for various sets of indices ν_j and various values of the space-time dimension. In these integral representations we recognise in the denominators a few recurring expressions:

$$\begin{aligned}
P_1 &= z_2 - m_W^2, \\
P_2 &= \left(z_2 + m_t^2 - m_W^2\right)^2 - s\left(z_2 - m_W^2\right), \\
R &= \sqrt{4 m_W^2 m_H^2 - \left(z_2 + m_H^2\right)^2}.
\end{aligned} \tag{16.30}$$

P_1 is a linear polynomial in z_2, P_2 is a quadratic polynomial in z_2 and R is the square root of a quadratic polynomial in z_2. In terms of maximal cuts we have for example

$$\begin{aligned}
\mathrm{MaxCut}\; I_{1012100}(4) &= 4\pi^2 i \int\limits_{\mathcal{C}_{\mathrm{MaxCut}}} \frac{m_t^2\left(m_H^2 - z_2\right) dz_2}{s P_1 R}, \\
\mathrm{MaxCut}\; I_{1011200}(4) &= 4\pi^2 i \int\limits_{\mathcal{C}_{\mathrm{MaxCut}}} \frac{m_t^2\left(2m_W^2 - m_H^2 - z_2\right) dz_2}{s P_1 R}, \\
\frac{1}{\varepsilon}\mathrm{MaxCut}\; I_{1011100}(2) &= -8\pi^2 i \int\limits_{\mathcal{C}_{\mathrm{MaxCut}}} \frac{m_t^4 dz_2}{P_2 R},
\end{aligned} \tag{16.31}$$

where the z_2-dependent expressions in the denominator are P_1, P_2 and R. We have four master integrals for this sector, hence we look for four master contours $\mathcal{C}_1, \dots, \mathcal{C}_4$. The four master contours have to be independent. We recall that on a Riemann sphere with 5 punctures we may define four independent contours as small

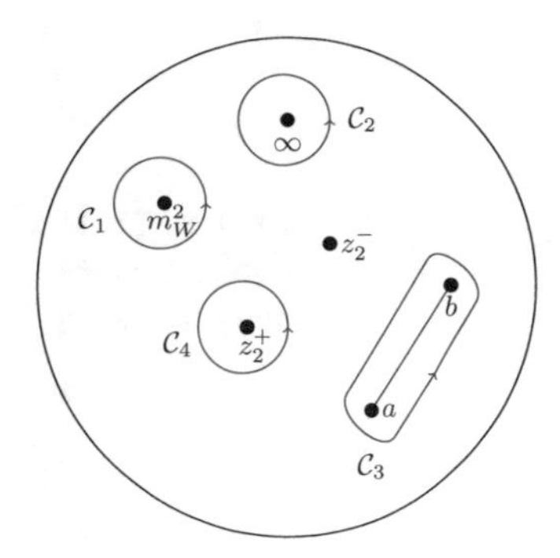

Fig. 16.6 The integration contours $C_1, \dots, C_4$ on the Riemann sphere. Note that a small counter clockwise circle around z_2^- is equivalent to $(-C_1 - \cdots - C_4)$

circles around four of the five punctures. The small circle around the fifth puncture is equivalent to minus the sum of the first four contours.

We may also count a square root $\sqrt{(z-a)(z-b)}$ as a deformed puncture: Consider for simplicity the square root $f = \sqrt{x}\sqrt{4+x}$ with the usual branch cut of the square root along the negative real axis. The function f is single valued for $x \in \mathbb{C}\backslash[-4, 0]$. The essential point is that f is single valued in a neighbourhood of the negative real axis for $x < -4$, the two sign ambiguities cancel each other. Thus we may define $\sqrt{(z-a)(z-b)}$ as a single-valued function on $\mathbb{C}\backslash[a, b]$, where $[a, b]$ denotes a slit between a and b. The situation is shown in Fig. 16.6. Let us now investigate which integrands give constant residues. We start with the square root $\sqrt{(z-a)(z-b)}$. Let C be an anticlockwise closed contour around the slit $[a, b]$. Then

$$\int\limits_C \frac{dz}{\sqrt{(z-a)(z-b)}} = 2\pi i. \tag{16.32}$$

For the combination of the square root $\sqrt{(z-a)(z-b)}$ with a simple pole $(z-c)$ in the denominator one finds that the integrand

$$\frac{1}{2\pi i} \frac{\sqrt{(c-a)\,(c-b)}}{(z-c)\,\sqrt{(z-a)\,(z-b)}} \tag{16.33}$$

gives ± 1, when integrated around a small circle at $z = c$ or along a cycle around the slit $[a, b]$.

Let us now turn to quadratic polynomials in the denominator. We consider the quadratic polynomial

$$z^2 - 2cz + c^2 - r^2 = (z-c-r)\,(z-c+r)\,, \tag{16.34}$$

which we may factorise at the expense of introducing the z-independent square root r. We consider integrals of the form

$$\int_C \frac{(az+b)\,dz}{(z-c-r)\,(z-c+r)} \tag{16.35}$$

where $C = n_+ C_+ + n_- C_-$ is a $\mathbb{Z}$-linear combination of small anticlockwise circles around the two poles $z = c + r$ (contour C_+) and $z = c - r$ (contour C_-). For $C = C_+ + C_-$ the square root r will not show up in the final result:

$$\int_{C_+ + C_-} \frac{(az+b)\,dz}{(z-c-r)\,(z-c+r)} = 2\pi i\; a. \tag{16.36}$$

However, on a punctured Riemann sphere this contour is equivalent to minus the sum of all other contours and hence not independent of those. We need the integral around one pole (or the difference between them). We have

$$\int_{C_+ - C_-} \frac{(az+b)\,dz}{(z-c-r)\,(z-c+r)} = 2\pi i\, \frac{ac+b}{r}. \tag{16.37}$$

Dividing the integrand on the left-hand side by the right-hand side we obtain an integrand with unit residue.

Let us now return to sector 29 of our example. We denote the two roots of P_2 by

$$z_2^{\pm} = \frac{1}{2}\left[2m_W^2 - 2m_t^2 + s \pm \sqrt{-s\left(4m_t^2 - s\right)}\right]. \tag{16.38}$$

We will need the value of R at $z_2 = z_2^{\pm}$. We introduce

$$r_4 = \sqrt{-\lambda\,(x_2, x_3, 1) - \frac{x_1^2}{2} - x_1\,(2 - x_2 - x_3) + r_1\left(1 + \frac{x_1}{2} - x_2 - x_3\right)},$$
$$r_5 = \sqrt{-\lambda\,(x_2, x_3, 1) - \frac{x_1^2}{2} - x_1\,(2 - x_2 - x_3) - r_1\left(1 + \frac{x_1}{2} - x_2 - x_3\right)}. \tag{16.39}$$

Then $R(z_2^+) = m_t^2 r_4$ and $R(z_2^-) = m_t^2 r_5$. With these preparations it is now clear from Eqs. (16.30) and (16.31) that a suitable set of master contours is

$$\begin{aligned}
&C_1 : \text{small anticlockwise circle around } z_2 = m_W^2. \\
&C_2 : \text{small anticlockwise circle around } z_2 = \infty. \\
&C_3 : \text{anticlockwise cycle around the slit } [-m_H^2 - 2m_H m_W, -m_H^2 + 2m_H m_W]. \\
&C_4 : \text{small anticlockwise circle around } z_2 = z_2^+.
\end{aligned} \tag{16.40}$$

The integration contours are sketched in Fig. 16.6. Inspecting Eq. (16.31) we define (note that kinematic independent algebraic prefactors don't matter here)

$$\begin{aligned}
\varphi_1 &= i\pi^2\varepsilon^3 \frac{(m_H^2 - z_2)}{P_1 R} dz_2, \\
\varphi_2 &= i\pi^2\varepsilon^3 \frac{(2m_W^2 - m_H^2 - z_2)}{P_1 R} dz_2, \\
\varphi_3 &= \pi^2\varepsilon^3 \frac{m_t^2 r_4 \left(z_2 - z_2^-\right)}{P_2 R} dz_2, \\
\varphi_4 &= \pi^2\varepsilon^3 \frac{m_t^2 r_5 \left(z_2 - z_2^+\right)}{P_2 R} dz_2.
\end{aligned} \tag{16.41}$$

Note that the singularities of φ_i are all as in Eqs. (16.32) and (16.33). In particular we have $P_2 = (z_2 - z_2^+)(z_2 - z_2^-)$. Hence the factor $(z_2 - z_2^-)$ in the numerator in the definition of φ_3 cancels the same factor of P_2 in the denominator, leaving just a single factor $(z_2 - z_2^+)$ in the denominator.

We verify that $\langle \varphi_i | C_j \rangle$ is a constant of weight zero and that the so defined (4×4)-matrix is invertible. Indeed, we have

$$\langle \varphi_i | C_j \rangle = 2i\pi^3\varepsilon^3 \begin{pmatrix} 1 & 1 & -2 & 0 \\ -1 & 1 & 0 & 0 \\ 0 & 0 & -1 & 1 \\ 0 & 0 & -1 & 0 \end{pmatrix}. \tag{16.42}$$

As the matrix is invertible, $\varphi_1, \dots, \varphi_4$ form a basis. We now look for Feynman integrals, whose maximal cut on the sector 29 gives the differential forms of Eq. (16.41) up to irrelevant kinematic independent algebraic prefactors (that is to say we do not care about prefactors like 2, i or $\sqrt{3}$). We do however care about kinematic dependent prefactors (like x_1). Comparing Eq. (16.41) with Eq. (16.31) we see that φ_1 matches MaxCut $I_{1012100}(4)$ and φ_2 matches MaxCut $I_{1011200}(4)$. The last two cases (φ_3 and φ_4) are only slightly more complicated: We first see that the denominator of φ_3 and φ_4 matches the one of MaxCut $I_{1011100}(2)$. A factor z_2 in the numerator of φ_3 or φ_4 corresponds to an irreducible scalar product in the Feynman integral and translates to $\nu_2 = -1$. We therefore deduce

$$\begin{aligned}
J_{11} &= \varepsilon^3 x_1 I_{1012100}, \\
J_{12} &= \varepsilon^3 x_1 I_{1011200}, \\
J_{13} &= \varepsilon^2 r_4 \left[\mathbf{D}^- I_{1(-1)11100} + \left(1 + \frac{1}{2}x_1 - x_2 + \frac{1}{2}r_1\right) \mathbf{D}^- I_{1011100} \right], \\
J_{14} &= \varepsilon^2 r_5 \left[\mathbf{D}^- I_{1(-1)11100} + \left(1 + \frac{1}{2}x_1 - x_2 - \frac{1}{2}r_1\right) \mathbf{D}^- I_{1011100} \right].
\end{aligned} \tag{16.43}$$

Note that

$$z_2^\pm = -m_t^2 \left(1 + \frac{1}{2}x_1 - x_2 \mp \frac{1}{2}r_1\right). \tag{16.44}$$

As we work on the maximal cut, there might be corrections corresponding to sub-topologies. Also in this case we are lucky and we verify (by inspecting A') that no correction terms are required.

It remains to treat the last sector. The sector 31 has again only one master integral, and we first consider the maximal cut. The steps are similar to what we did for J_7 and J_8. From the maximal cut we obtain the tentative guess

$$J_{15}^{\text{tentative}} = \varepsilon^2 r_3 \left[(1-x_2)^2 - x_1 x_2 \right] \mathbf{D}^- I_{1111100}. \tag{16.45}$$

Again, there might be corrections due to sub-topologies (which are not detected at the maximal cut). In this case there are actually corrections. We could use the methods of Sect. 7.1 to systematically obtain them. However, with an educated guess we might reach our goal faster: From the differential equation for $J_{15}^{\text{tentative}}$ we see that we need correction terms proportional to the master integrals of sector 29. In Eq. (16.45) we recognise $(1-x_2)^2 - x_1 x_2$ as the constant part of the polynomial P_2:

$$P_2|_{z_2=0} = m_t^4 \left[(1-x_2)^2 - x_1 x_2 \right]. \tag{16.46}$$

Terms proportional to z_2 cancel the second propagator of $I_{1111100}$ and correspond to Feynman integrals from sector 29. Therefore we take for the educated guess the terms proportional to z_2 and z_2^2 of P_2 into account:

$$J_{15} = \varepsilon^2 r_3 \left\{ \left[(1-x_2)^2 - x_1 x_2 \right] \mathbf{D}^- I_{1111100} + (2 + x_1 - 2x_2) \mathbf{D}^- I_{1011100} + \mathbf{D}^- I_{1(-1)11100} \right\}. \tag{16.47}$$

Plugging this ansatz into the differential equation we verify that this guess factorises ε out. We obtain

$$\left(d + A'\right) \vec{J} = 0 \tag{16.48}$$

with

$$A' = \varepsilon \sum_{j=1}^{25} C_j \omega_j, \qquad \omega_j = d \ln f_j. \tag{16.49}$$

The f_j are algebraic functions of the kinematic variables. Some of them are polynomials in the kinematic variables. These define the rational part of the alphabet:

$$\begin{aligned}
& f_1 = x_1, && f_2 = x_1 + 4, && f_3 = x_2, \\
& f_4 = x_2 - 1, && f_5 = x_3, && f_6 = (1-x_2)^2 - x_1 x_2, \\
& f_7 = 4x_2 - x_3, && f_8 = -\lambda(x_2, x_3, 1), && f_9 = \left[x_1 (x_2 + x_3) - \lambda(x_2, x_3, 1) \right]^2 - 4x_1 (4 + x_1) x_2 x_3.
\end{aligned} \tag{16.50}$$

The algebraic part of the alphabet can be taken as

$$
\begin{aligned}
f_{10} &= 2 + x_1 - r_1, & f_{11} &= 2 + x_1 - 2x_2 - r_1, \\
f_{12} &= -\lambda\left(x_2, x_3, 1\right) - x_1 - \left(1 + \frac{x_1}{2} - x_2 - x_3\right)\left(x_1 - r_1\right), & f_{13} &= 1 + x_2 - x_3 - ir_2, \\
f_{14} &= 1 - x_2 + x_3 - ir_2, & f_{15} &= x_3 - ir_3, \\
f_{16} &= 1 - x_2 + ir_2 + ir_3, & f_{17} &= 1 + \frac{x_1}{2} + x_2 - x_3 - \frac{r_1}{2} - ir_4, \\
f_{18} &= 1 + \frac{x_1}{2} - x_2 + x_3 - \frac{r_1}{2} - ir_4, & f_{19} &= 1 + \frac{x_1}{2} + x_2 - x_3 + \frac{r_1}{2} - ir_5, \\
f_{20} &= 1 + \frac{x_1}{2} - x_2 + x_3 + \frac{r_1}{2} - ir_5, & f_{21} &= \left(r_1 - ir_4 - ir_5\right)\left(r_1 + ir_4 + ir_5\right), \\
f_{22} &= x_1 - r_1 - 2ir_2 - 2ir_4, & f_{23} &= 1 + \frac{x_1}{2} - x_2 - \frac{r_1}{2} + ir_3 - ir_4, \\
f_{24} &= x_1 + r_1 - 2ir_2 - 2ir_5, & f_{25} &= 1 + \frac{x_1}{2} - x_2 + \frac{r_1}{2} + ir_3 - ir_5.
\end{aligned}
\tag{16.51}
$$

Note that the choice of the ω_j's and of the f_j's is not unique. For example, we may always transform the ω_j's by a $\mathrm{GL}_{N_L}(\mathbb{Q})$-transformation.

Note further that all roots are included in our alphabet:

$$
r_1^2 = f_1 f_2, \quad r_2^2 = f_8, \quad r_3^2 = f_5 f_7, \quad r_4^2 = f_{12}, \quad r_5^2 = \frac{f_9}{f_{12}}. \tag{16.52}
$$

Rewriting the ω_j's as dlog-forms is not entirely straightforward. The exercises 123 (for the polynomial case) and 125 (for the algebraic case) show how to do this.

Exercise 123 *Rewrite the differential one-form*

$$
\omega = -\frac{\left(1 - x_2\right)^2 dx_1}{x_1\left[\left(1 - x_2\right)^2 - x_1 x_2\right]} - \frac{x_1\left(1 + x_2\right) dx_2}{\left(1 - x_2\right)\left[\left(1 - x_2\right)^2 - x_1 x_2\right]} \tag{16.53}
$$

as a dlog-form.

The C_j are 15×15 matrices, whose entries are of the form $\mathbb{Q} + i\mathbb{Q}$. They are usually sparse matrices. As an example we have

$$
C_1 = \begin{pmatrix}
0 & 0 & 0 & 0 & 0 & 0 & 0 & 0 & 0 & 0 & 0 & 0 & 0 & 0 & 0 \\
0 & 0 & 0 & 0 & 0 & 0 & 0 & 0 & 0 & 0 & 0 & 0 & 0 & 0 & 0 \\
0 & 0 & 0 & 0 & 0 & 0 & 0 & 0 & 0 & 0 & 0 & 0 & 0 & 0 & 0 \\
0 & 0 & 0 & 0 & 0 & 0 & 0 & 0 & 0 & 0 & 0 & 0 & 0 & 0 & 0 \\
0 & 0 & 0 & 0 & 0 & 0 & 0 & 0 & 0 & 0 & 0 & 0 & 0 & 0 & 0 \\
0 & 0 & 0 & 0 & 0 & 0 & 0 & 0 & 0 & 0 & 0 & 0 & 0 & 0 & 0 \\
0 & 0 & 0 & 0 & 0 & 0 & 0 & 0 & 0 & 0 & 0 & 0 & 0 & 0 & 0 \\
0 & 0 & 0 & 0 & 0 & 0 & 0 & 0 & 0 & 0 & 0 & 0 & 0 & 0 & 0 \\
0 & 0 & 0 & 0 & 0 & 0 & 0 & 0 & 1 & 0 & 0 & 0 & 0 & 0 & 0 \\
0 & 0 & 0 & 0 & 0 & 0 & 0 & 0 & 0 & 1 & 0 & 0 & 0 & 0 & 0 \\
0 & 0 & 0 & 0 & 0 & 0 & 0 & 0 & 0 & 0 & 1 & 0 & 0 & 0 & 0 \\
0 & 0 & 0 & 0 & 0 & 0 & 0 & 0 & 0 & 0 & 0 & 1 & 0 & 0 & 0 \\
0 & 0 & 0 & 0 & 0 & 0 & -2 & 0 & 0 & 0 & 0 & 0 & \frac{1}{2} & \frac{1}{2} & 0 \\
0 & 0 & 0 & 0 & 0 & 0 & -2 & 0 & 0 & 0 & 0 & 0 & \frac{1}{2} & \frac{1}{2} & 0 \\
0 & 0 & 0 & 0 & 0 & 0 & 0 & -2 & 0 & 0 & 0 & 0 & 0 & 0 & 1
\end{pmatrix}. \tag{16.54}
$$

The matrix C_1 accompanies the differential one-form $\omega_1 = dx_1/x_1$. From the lists in Eqs. (16.50) and (16.51) one may check that we have chosen the remaining ω_j's such that they do not have a pole on $x_1 = 0$. For the family of Feynman integrals under consideration we do not expect for the master integrals J_1–J_{15} any logarithmic singularities in the limit $x_1 \to 0$. This translates into the requirement that there should be no trailing zeros with respect to the integration in x_1. From Eq. (16.54) we see that trailing zeros in the x_1-integration are absent if

$$\begin{aligned} 0 = \lim_{x_1 \to 0} J_9 &= \lim_{x_1 \to 0} J_{10} = \lim_{x_1 \to 0} J_{11} = \lim_{x_1 \to 0} J_{12} \\ &= \lim_{x_1 \to 0} \left(J_{13} + J_{14} - 4J_7\right) = \lim_{x_1 \to 0} \left(J_{15} - 2J_8\right). \end{aligned} \tag{16.55}$$

With our choice of the ω_j's as in Eqs. (16.50) and (16.51) only ω_1 has a pole along $x_1 = 0$. We may always choose the remaining ω_j's such that they do not have a pole along $x_1 = 0$. The following exercise shows that this is not entirely trivial:

Exercise 124 *Let*

$$\begin{aligned} \tilde{f}_1 &= \lambda\left(x_2, x_3, 1\right) + 8x_3 - x_1\left(x_2 - x_3\right) - r_4 r_5, \\ \tilde{f}_2 &= \lambda\left(x_2, x_3, 1\right) + 8x_2 + x_1\left(x_2 - x_3\right) - r_4 r_5, \\ \tilde{f}_3 &= \lambda\left(x_2, x_3, 1\right) - x_1\left(x_2 - x_3\right) + r_4 r_5, \\ \tilde{f}_4 &= \lambda\left(x_2, x_3, 1\right) + x_1\left(x_2 - x_3\right) + r_4 r_5, \end{aligned} \tag{16.56}$$

Show that

$$\tilde{\omega} = d \ln\left(\frac{\tilde{f}_1 \tilde{f}_2}{\tilde{f}_3 \tilde{f}_4}\right) \tag{16.57}$$

has a pole along $x_1 = 0$, *while*

$$\omega = d \ln\left(x_1^2 \frac{\tilde{f}_1 \tilde{f}_2}{\tilde{f}_3 \tilde{f}_4}\right) \tag{16.58}$$

does not.

16.4 Base Transformation

The differential equation Eq. (16.48) for the Feynman integrals involve dlog-forms with algebraic arguments (square roots). A sufficient but not necessary criteria to express the result in terms of multiple polylogarithms is the possibility to rationalise simultaneously all occurring square roots. Unfortunately, we are not so lucky here: With the methods of Sect. 7.2 we may rationalise four out the five square roots r_1–r_5.

One square root (either r_4 or r_5) remains unrationalised. Nevertheless it is instructive to see how four of the square roots are rationalised.

The root r_1 occurs frequently in Feynman integrals and we encountered this root already before in Sects. 6.4.3 and 6.4.4. The transformation (see Eq. (6.249))

$$x_1 = \frac{\left(1 - x_1'\right)^2}{x_1'} \tag{16.59}$$

rationalises the root r_1:

$$r_1 = \frac{1 - x_1'^2}{x_1'}. \tag{16.60}$$

We note that in terms of the new variable x_1' the roots r_4 and r_5 are given by

$$r_4 = \sqrt{-\lambda\left(x_2, x_3, x_1'\right)}, \quad r_5 = \sqrt{-\lambda\left(x_2, x_3, \frac{1}{x_1'}\right)}. \tag{16.61}$$

The roots r_2 and r_3 depend only on x_2 and x_3. With the help of the methods from Sect. 7.2 one finds that these roots are rationalised simultaneously for example by

$$x_2 = \frac{\left(1 + x_2'^2\right)\left(2 - x_3'\right)}{x_3'\left(3 - 2x_3' - x_2'^2\right)}, \quad x_3 = \frac{4\left(2 - x_3'\right)}{x_3'\left(3 - 2x_3' - x_2'^2\right)}. \tag{16.62}$$

We have

$$r_2 = \frac{2i\left(3 - 4x_3' - x_2'^2 + x_3'^2\right)}{x_3'\left(3 - 2x_3' - x_2'^2\right)}, \quad r_3 = \frac{4x_2'\left(2 - x_3'\right)}{x_3'\left(3 - 2x_3' - x_2'^2\right)}. \tag{16.63}$$

This leaves us with the roots r_4 and r_5. From Eq. (16.61) it is clear that the argument of the square root r_4 or r_5 is quadratic in x_1'. It is therefore straightforward to rationalise in addition either r_4 or r_5 (but not both). For example, r_4 is rationalised by

$$x_1' = \frac{\left[x_4'^2 - (x_2 - x_3)^2\right]}{2\left(x_4' - x_2 - x_3\right)}, \quad r_4 = i\frac{\lambda\left(x_2, x_3, x_4'\right)}{2\left(x_4' - x_2 - x_3\right)}, \tag{16.64}$$

while r_5 is rationalised by

$$x_1' = \frac{2\left(x_5' - x_2 - x_3\right)}{\left[x_5'^2 - (x_2 - x_3)^2\right]}, \quad r_5 = i\frac{\lambda\left(x_2, x_3, x_5'\right)}{2\left(x_5' - x_2 - x_3\right)}. \tag{16.65}$$

The variables (x_2', x_3', x_4') rationalise simultaneously the roots $\{r_1, r_2, r_3, r_4\}$, the variables (x_2', x_3', x_5') rationalise simultaneously the roots $\{r_1, r_2, r_3, r_5\}$.

Rationalisations are also helpful to convert the ω_j's to dlog-forms, as the following exercise shows:

Exercise 125 *Rewrite the differential one-form*

$$\omega = -\frac{(2 - 2x_2 - x_1 x_2)\, r_1 dx_1}{x_1\,(4 + x_1)\left[(1 - x_2)^2 - x_1 x_2\right]} - \frac{r_1 dx_2}{\left[(1 - x_2)^2 - x_1 x_2\right]} \tag{16.66}$$

where $r_1 = \sqrt{x_1(4 + x_1)}$ *as a dlog-form.*

16.5 Boundary Values

In order to solve the differential equation we need boundary values for the master integrals $\vec{J}$. A convenient boundary point is the point

$$x_1 = 0, \quad x_2 = 1, \quad x_3 = 1. \tag{16.67}$$

At this point we have

$$r_1 = 0, \; r_2 = r_3 = r_4 = r_5 = \sqrt{3}. \tag{16.68}$$

At the boundary point, the four tadpole integrals J_1–J_4 are all equal and given by

$$J_1 = J_2 = J_3 = J_4 = 1 + \zeta_2 \varepsilon^2 - \frac{2}{3}\zeta_3 \varepsilon^3 + \frac{7}{4}\zeta_4 \varepsilon^4 + O\left(\varepsilon^5\right). \tag{16.69}$$

The master integrals J_5 and J_6 vanish at the boundary point (due to the prefactor r_1):

$$J_5 = J_6 = 0. \tag{16.70}$$

The master integrals J_7 and J_8 are equal at the boundary point. The value is given by the value of the equal mass sunrise integral at zero external momentum squared (times a trivial prefactor). Let us introduce the following linear combination of harmonic polylogarithms

$$\overline{H}_{m_1 \ldots m_k}(x) = H_{m_1 \ldots m_k}(x) - H_{m_1 \ldots m_k}\left(x^{-1}\right). \tag{16.71}$$

Then [354]

$$J_7 = J_8 = \frac{3}{2i}\left\{\varepsilon^2 \overline{H}_2\left(e^{\frac{2\pi i}{3}}\right) + \varepsilon^3\left[-2\overline{H}_{2,1}\left(e^{\frac{2\pi i}{3}}\right) - \overline{H}_3\left(e^{\frac{2\pi i}{3}}\right) - \ln(3)\,\overline{H}_2\left(e^{\frac{2\pi i}{3}}\right)\right]\right.$$
$$+\varepsilon^4\left[4\overline{H}_{2,1,1}\left(e^{\frac{2\pi i}{3}}\right) - 2\overline{H}_{3,1}\left(e^{\frac{2\pi i}{3}}\right) + \overline{H}_4\left(e^{\frac{2\pi i}{3}}\right) + \frac{2}{9}\pi^2\overline{H}_2\left(e^{\frac{2\pi i}{3}}\right)\right.$$
$$\left.\left.+\ln(3)\left[2\overline{H}_{2,1}\left(e^{\frac{2\pi i}{3}}\right) + \overline{H}_3\left(e^{\frac{2\pi i}{3}}\right)\right] + \frac{1}{2}\ln^2(3)\,\overline{H}_2\left(e^{\frac{2\pi i}{3}}\right)\right]\right\}$$
$$+O\left(\varepsilon^5\right). \tag{16.72}$$

The master integrals J_9 and J_{10} vanish again at the boundary point and so do the master integrals J_{11} and J_{12}:

$$\begin{aligned} J_9 &= J_{10} = 0, \\ J_{11} &= J_{12} = 0. \end{aligned} \tag{16.73}$$

From the definition of the master integrals J_{13}, J_{14} and J_{15} it follows that they are equal at the boundary point:

$$J_{13} = J_{14} = J_{15} = \varepsilon^2\sqrt{3}\,\mathbf{D}^- I_{1(-1)11100}. \tag{16.74}$$

The integral $I_{1(-1)11100}$ reduces at the boundary point to the equal mass sunrise integral at zero external momentum squared and we find

$$J_{13} = J_{14} = J_{15} = 2J_7, \tag{16.75}$$

with J_7 given by Eq. (16.72).

Note that boundary information on the master integrals $J_9 - J_{15}$ can already be extracted from the matrix C_1 in Eq. (16.54). In fact, this matrix does not only give information on the boundary point $(x_1, x_2, x_3) = (0, 1, 1)$, but on the complete hyperplane $x_1 = 0$. Assuming that the master integrals do not have any logarithmic singularities at $x_1 = 0$ and using the fact that J_{13} and J_{14} are equal for $x_1 = 0$ (this follows from the definition of J_{13} and J_{14}), it follows that in the hyperplane $x_1 = 0$ we have (compare with Eq. (16.55))

$$J_9 = J_{10} = J_{11} = J_{12} = 0, \quad J_{13} = J_{14} = 2J_7, \quad J_{15} = 2J_8. \tag{16.76}$$

16.6 Integrating the Differential Equation

With the differential equation and the boundary values we have everything at hand to solve for the master integrals $\vec{J}$ in terms of iterated integrals. Let $\gamma : [0, 1] \to \mathbb{C}^3$ be a path from the boundary point $\gamma(0) = (0, 1, 1)$ to the point of interest $\gamma(1) =$

(x_1, x_2, x_3). We integrate the differential equation order by order in ε as described in Sect. 6.3.3. Thus we may express any master integral as a linear combination of iterated integrals along the path γ. For example

$$J_1 = 1 - I_\gamma(\omega_3)\,\varepsilon + \left[I_\gamma(\omega_3, \omega_3) + \zeta_2\right]\varepsilon^2 + O\left(\varepsilon^3\right). \tag{16.77}$$

It remains to express these iterated integrals in terms of more commonly used functions.

For the master integrals J_1–J_{10} this is straightforward: The roots r_4 and r_5 do not enter these master integrals. The master integrals J_1–J_{10} depend only on the roots r_1, r_2 and r_3. These roots can be rationalised simultaneously and we may express the master integrals J_1–J_{10} to any order in ε in terms of multiple polylogarithms. A trivial example is

$$J_1 = 1 - \ln(x_2)\,\varepsilon + \left[\frac{1}{2}\ln^2(x_2) + \zeta_2\right]\varepsilon^2 + O\left(\varepsilon^3\right). \tag{16.78}$$

The situation is more complicated for the master integrals J_{11}–J_{15}. These involve all five roots

$$r_1 = \sqrt{x_1(4+x_1)}, \qquad r_2 = \sqrt{-\lambda(x_2, x_3, 1)}, \qquad r_3 = \sqrt{-\lambda(x_2, x_3, x_2)},$$
$$r_4 = \sqrt{-\lambda\left(x_2, x_3, x_1'\right)}, \quad r_5 = \sqrt{-\lambda\left(x_2, x_3, \frac{1}{x_1'}\right)}, \quad x_1' = \frac{1}{2}(2 + x_1 - r_1). \tag{16.79}$$

For $x_1 = 0$ (corresponding to $x_1' = 1$) we have

$$r_1 = 0, \;\; r_4 = r_5 = r_2. \tag{16.80}$$

This means that on the hyperplane $x_1 = 0$ we only deal with two square roots r_2 and r_3, which can be rationalised simultaneously. Thus for the special kinematic configuration $x_1 = 0$ we may express all master integrals in terms of multiple polylogarithms. For a generic kinematic configuration with $x_1 \neq 0$ we are left with an (iterated) integration along the x_1-direction. From Sect. 16.4 we know that we may rationalise four of the five square roots. We may treat the variable x_1 for the variable x_4'. In the variables (x_2', x_3', x_4') only the square root r_5 remains unrationalised. In these variables r_5 is the square root of a quartic polynomial in x_4'. Thus we are in the situation that we have an integration in a single variable (x_4') involving a single square root of a quartic polynomial.

With the methods of Sect. 13.4 we may transform these iterated integrals to elliptic multiple polylogarithms. This does not exclude the possibility that the (first few terms of the ε-expansion of the) master integrals can be expressed in terms of simpler functions (i.e., multiple polylogarithms).

We may search for a representation in terms of multiple polylogarithms with the help of the bootstrap approach described in Sect. 11.5.

Exercise 126 *Let*

$$g = \frac{2 + x_1 - r_1}{2 + x_1 + r_1}. \tag{16.81}$$

Express g and $(1 - g)$ as a power product in the letters of the alphabet defined by Eqs. (16.50), (16.51) and the constant $f_0 = 2$.

Exercise 127 *The master integral J_{15} starts at order $O(\varepsilon^2)$. The weight two term of J_{15} is given in terms of iterated integrals by*

$$J_{15}^{(2)} = 2i I_\gamma \left(2\omega_{15} - \omega_3 - \omega_5, \omega_5 - \omega_3; 1\right) + 2J_7^{(2)}(0, 1, 1), \tag{16.82}$$

where $J_7^{(2)}(0, 1, 1)$ denotes the boundary value of Eq. (16.72):

$$J_7^{(2)}(0, 1, 1) = \frac{3}{2i} \overline{H}_2\left(e^{\frac{2\pi i}{3}}\right) = \frac{3}{2i}\left[\text{Li}_2\left(e^{\frac{2\pi i}{3}}\right) - \text{Li}_2\left(e^{-\frac{2\pi i}{3}}\right)\right]. \tag{16.83}$$

Express $J_{15}^{(2)}$ in terms of multiple polylogarithms.

16.7 Final Result

Let's now return to our original problem: The scalar Feynman integral corresponding to the penguin diagram in Fig. 16.1 is $I_{1211100}$. As a pedagogical example we work out the first non-vanishing term in the ε-expansion of this Feynman integral. This illustrates the general procedure. (Of course, if we are only interested in this particular coefficient, there are simpler ways to obtain the result.)

Using integration-by-parts identities we may express the integral $I_{1211100}$ as a linear combination of the master integrals $\vec{I}$. By using

$$\vec{I} = U^{-1} \vec{J} \tag{16.84}$$

we express $I_{1211100}$ as a linear combination of the master integrals $\vec{J}$. The integral $I_{1211100}(4 - 2\varepsilon)$ has a Laurent expansion in the dimensional regularisation parameter starting with ε^{-1}:

$$I_{1211100}\left(4 - 2\varepsilon, x_1, x_2, x_3\right) = \sum_{j=-1}^{\infty} \varepsilon^j \, I_{1211100}^{(j)}\left(x_1, x_2, x_3\right). \tag{16.85}$$

Let's focus on the first non-vanishing coefficient $I^{(-1)}_{1211100}$. For the master integrals we have a similar expansion in ε

$$J_i\left(4-2\varepsilon, x_1, x_2, x_3\right) = \sum_{j=0}^{\infty} \varepsilon^j\, J_i^{(j)}\left(x_1, x_2, x_3\right). \tag{16.86}$$

In terms of the master integrals $\vec{J}$ we obtain for $I^{(-1)}_{1211100}$

$$\begin{aligned}
I^{(-1)}_{1211100} = \frac{1}{\left[(1-x_2)^2 - x_1 x_2\right]} & \left\{ \frac{(1+x_2)(2x_2-x_3)}{2x_2(1-x_2)} \left(J_2^{(1)} - J_1^{(1)}\right) \right. \\
& \left. + \frac{(1+x_2)\,x_3}{2x_2(1-x_2)} \left(J_4^{(1)} - J_3^{(1)}\right) + \frac{(2x_2-x_3)\,r_1}{4x_1x_2} J_5^{(1)} + \frac{x_3 r_1}{4x_1x_2} J_6^{(1)} \right\} \\
& + \frac{(x_2-x_3)}{2x_1x_2^2} \left(J_9^{(2)} - J_{10}^{(2)}\right) + \frac{(3x_2-x_3)\,r_3}{2x_1x_2^2\,(4x_2-x_3)} \left(J_{15}^{(2)} - 2J_8^{(2)}\right)
\end{aligned} \tag{16.87}$$

We then substitute the results for $\vec{J}$ and obtain

$$I^{(-1)}_{1211100} = \frac{1}{\left[(1-x_2)^2 - x_1x_2\right]} \left\{ \sqrt{\frac{4+x_1}{x_1}}\, I_\gamma\left(\omega_{10}\right) - \frac{1+x_2}{1-x_2} I_\gamma\left(\omega_3\right) \right\}. \tag{16.88}$$

With

$$\omega_3 = d\ln\left(x_2\right), \quad \omega_{10} = d\ln\left(2+x_1-r_1\right) = \frac{1}{2} d\ln\left(\frac{2+x_1-r_1}{2+x_1+r_1}\right) \tag{16.89}$$

we finally obtain

$$\begin{aligned}
& I_{1211100}\left(4-2\varepsilon, x_1, x_2, x_3\right) = \\
& \frac{1}{\varepsilon} \frac{1}{\left[(1-x_2)^2 - x_1x_2\right]} \left\{ \frac{1}{2} \sqrt{\frac{4+x_1}{x_1}} \ln\left(\frac{2+x_1-r_1}{2+x_1+r_1}\right) - \frac{1+x_2}{1-x_2} \ln\left(x_2\right) \right\} + \mathcal{O}\left(\varepsilon^0\right).
\end{aligned} \tag{16.90}$$

Note that $I^{(-1)}_{1211100}$ is finite in the $(x_1 \to 0)$-limit and in the $(x_2 \to 1)$-limit. Note further that Eq. (16.90) does not contain any weight two terms. Although we might expect from Eq. (16.87) terms of weight two, they cancel out in this order. Finally note that $I^{(-1)}_{1211100}$ is independent of x_3. The $1/\varepsilon$-term $I^{(-1)}_{1211100}$ originates from the ultraviolet divergence of the sub-graph formed by propagators 4 and 5. The ultraviolet divergence is independent of the masses propagating in the sub-graph. As x_3 (or m_H^2) enters only this sub-graph, the result for $I^{(-1)}_{1211100}$ is independent of x_3. Essentially, $I^{(-1)}_{1211100}$ is given by the pole term of the sub-graph times a one-loop three-point function obtained by contracting the sub-graph to a point.

Let's plug in some numbers: With

$$(x_1, x_2, x_3) = \left(\frac{5^2}{173^2}, \frac{80^2}{173^2}, \frac{125^2}{173^2} \right) \tag{16.91}$$

we obtain

$$I^{(-1)}_{1211100} \left(\frac{25}{29929}, \frac{6400}{29929}, \frac{15625}{29929} \right) \approx 0.61751141179432938382213384. \tag{16.92}$$

It is always recommended to perform an independent cross check. Sector decomposition offers the possibility to check a Feynman integral at a specific kinematic point. The following C++ code uses the program `sector_decomposition` [240]:

```
#include <iostream>
#include <stdexcept>
#include <vector>
#include <ginac/ginac.h>
#include "sector_decomposition/sector_decomposition.h"

int main()
{
  using namespace sector_decomposition;
  using namespace GiNaC;

  symbol eps("eps");

  int n            =  5;
  int loops        =  2;
  int order        = -1;
  int D_int_over_2 =  2;

  std::vector<ex> nu = {1,2,1,1,1};

  ex x1  = numeric(25,29929);
  ex x2  = numeric(6400,29929);
  ex x3  = numeric(15625,29929);

  // --------------------------------------------------------------

  int verbose_level = 0;

  CHOICE_STRATEGY = STRATEGY_C;

  monte_carlo_parameters mc_parameters = monte_carlo_parameters( 5, 15,
  100000, 1000000 );

  // --------------------------------------------------------------

  symbol a1("a1"), a2("a2"), a3("a3"), a4("a4"), a5("a5");
  std::vector<ex> parameters = { a1, a2, a3, a4, a5 };
```

```
  ex U = a1*a4+a1*a5+a4*a3+a5*a3+a2*a4+a5*a4+a5*a2;
  ex F = a1*a3*(a4+a5)*x1 + U*( (a2+a4)*x2 + a5*x3 + (a1+a3) );

  std::vector<ex> poly_list = {U,F};

  std::vector<exponent> nu_minus_1(n);
  for (int i1=0; i1<n; i1++) nu_minus_1[i1] = exponent(nu[i1]-1,0);

  std::vector<exponent> c(poly_list.size());
  c[0] = exponent( n-(loops+1)*D_int_over_2, loops+1 );
  c[1] = exponent( -n+loops*D_int_over_2, -loops );
  for (int k=0; k<n; k++)
  {
    c[0].sum_up(nu_minus_1[k]);
    c[1].subtract_off(nu_minus_1[k]);
  }

  integrand my_integrand = integrand(nu_minus_1, poly_list, c);

  // ---------------------------------------------------------------

  integration_data global_data(parameters, eps, order);

  monte_carlo_result res =
      do_sector_decomposition(global_data, my_integrand, mc_parameters,
      verbose_level);

  std::cout << "Order " << pow(eps,order) << ": " << res.get_mean()
        << " +/- " << res.get_error() << std::endl;

  return 0;
}
```

Running this program will print out

```
Order eps^(-1): 0.617517 +/- 9.57571e-06
```

in agreement with Eq. (16.92).

Appendix A
Spinors

In this appendix we summarise properties of spinors in four space-time dimensions. Although we use dimensional regularisation throughout this book, four-dimensional formulae can be used for the external kinematic and within some variants of dimensional regularisation like the FDH-scheme. Note that some formulae (like the Schouten identity) are specific to four space-time dimensions.

A.1 The Dirac Equation

The Lagrange density for a Dirac field in four space-time dimensions depends on four-component spinors $\psi_\alpha(x)$ ($\alpha = 1, 2, 3, 4$) and $\bar{\psi}_\alpha(x) = \left(\psi^\dagger(x)\gamma^0\right)_\alpha$:

$$\mathcal{L}(\psi, \bar{\psi}, \partial_\mu \psi) = i\bar{\psi}(x)\gamma^\mu \partial_\mu \psi(x) - m\bar{\psi}(x)\psi(x) \tag{A.1}$$

Here, the (4×4)-Dirac matrices satisfy the anti-commutation rules

$$\{\gamma^\mu, \gamma^\nu\} = 2g^{\mu\nu}\mathbf{1}, \;\; \{\gamma^\mu, \gamma_5\} = 0, \;\; \gamma_5 = i\gamma^0\gamma^1\gamma^2\gamma^3 = \frac{i}{24}\varepsilon_{\mu\nu\rho\sigma}\gamma^\mu\gamma^\nu\gamma^\rho\gamma^\sigma. \tag{A.2}$$

The Dirac equations read

$$\left(i\gamma^\mu\partial_\mu - m\right)\psi(x) \;=\; 0, \quad \bar{\psi}(x)\left(i\gamma^\mu \overleftarrow{\partial}_\mu + m\right) \;=\; 0. \tag{A.3}$$

It is useful to have an explicit representation of the Dirac matrices. There are several widely used representations. A particular useful one is the **Weyl representation** of the Dirac matrices:

$$\gamma^\mu = \begin{pmatrix} \mathbf{0} & \sigma^\mu \\ \bar{\sigma}^\mu & \mathbf{0} \end{pmatrix}, \;\; \gamma_5 = i\gamma^0\gamma^1\gamma^2\gamma^3 = \begin{pmatrix} \mathbf{1} & \mathbf{0} \\ \mathbf{0} & -\mathbf{1} \end{pmatrix} \tag{A.4}$$

S. Weinzierl, *Feynman Integrals*, UNITEXT for Physics,
https://doi.org/10.1007/978-3-030-99558-4

Here, **0** denotes the (2×2)-zero matrix and **1** denotes the (2×2)-unit matrix. The 4-dimensional σ^μ-matrices are defined by

$$\sigma^\mu_{A\dot{B}} = (\mathbf{1}, -\vec{\sigma}), \quad \bar{\sigma}^{\mu\dot{A}B} = (\mathbf{1}, \vec{\sigma}). \tag{A.5}$$

and $\vec{\sigma} = (\sigma_x, \sigma_y, \sigma_z)$ are the standard Pauli matrices:

$$\sigma_x = \begin{pmatrix} 0 & 1 \\ 1 & 0 \end{pmatrix}, \quad \sigma_y = \begin{pmatrix} 0 & -i \\ i & 0 \end{pmatrix}, \quad \sigma_z = \begin{pmatrix} 1 & 0 \\ 0 & -1 \end{pmatrix}. \tag{A.6}$$

The σ^μ-matrices satisfy the Fierz identities

$$\sigma^\mu_{A\dot{A}} \bar{\sigma}^{\dot{B}B}_\mu = 2\delta_A^{\ B} \delta_{\dot{A}}^{\ \dot{B}}, \quad \sigma^\mu_{A\dot{A}} \sigma_{\mu B\dot{B}} = 2\varepsilon_{AB}\varepsilon_{\dot{A}\dot{B}}, \quad \bar{\sigma}^{\mu\dot{A}A} \bar{\sigma}^{\dot{B}B}_\mu = 2\varepsilon^{\dot{A}\dot{B}}\varepsilon^{AB}. \tag{A.7}$$

Let us now look for plane wave solutions of the Dirac equation. We make the ansatz

$$\psi(x) = \begin{cases} u(p)e^{-ipx}, \ p^0 > 0, \ p^2 = m^2, \text{ incoming fermion}, \\ v(p)e^{+ipx}, \ p^0 > 0, \ p^2 = m^2, \text{ outgoing anti-fermion}. \end{cases} \tag{A.8}$$

$u(p)$ describes incoming particles, $v(p)$ describes outgoing anti-particles. Similar,

$$\bar{\psi}(x) = \begin{cases} \bar{u}(p)e^{+ipx}, \ p^0 > 0, \ p^2 = m^2, \text{ outgoing fermion}, \\ \bar{v}(p)e^{-ipx}, \ p^0 > 0, \ p^2 = m^2, \text{ incoming anti-fermion}, \end{cases} \tag{A.9}$$

where

$$\bar{u}(p) = u^\dagger(p)\gamma^0, \qquad \bar{v}(p) = v^\dagger(p)\gamma^0. \tag{A.10}$$

$\bar{u}(p)$ describes outgoing particles, $\bar{v}(p)$ describes incoming anti-particles. Then

$$\begin{aligned} (\not{p} - m)\, u(p) &= 0, & (\not{p} + m)\, v(p) &= 0, \\ \bar{u}(p)\, (\not{p} - m) &= 0, & \bar{v}(p)\, (\not{p} + m) &= 0, \end{aligned} \tag{A.11}$$

There are two solutions for $u(p)$ (and the other spinors $\bar{u}(p)$, $v(p)$, $\bar{v}(p)$). We will label the various solutions with $\lambda \in \{+, -\}$ and we will use the notation $\bar{\lambda} = -\lambda$. The degeneracy is related to the additional spin degree of freedom. We require that the two solutions satisfy the orthogonality relations

$$\begin{aligned} \bar{u}(p, \bar{\lambda})u(p, \lambda) &= 2m\delta_{\bar{\lambda}\lambda}, \\ \bar{v}(p, \bar{\lambda})v(p, \lambda) &= -2m\delta_{\bar{\lambda}\lambda}, \\ \bar{u}(p, \bar{\lambda})v(p, \lambda) &= \bar{v}(\bar{\lambda})u(\lambda) \ = \ 0, \end{aligned} \tag{A.12}$$

and the completeness relations

$$\sum_{\lambda} u(p,\lambda)\bar{u}(p,\lambda) \;=\; \not{p} + m, \quad \sum_{\lambda} v(p,\lambda)\bar{v}(p,\lambda) \;=\; \not{p} - m. \quad \text{(A.13)}$$

A.2 Massless Spinors in the Weyl Representation

Let us now try to find explicit solutions for the spinors $u(p)$, $v(p)$, $\bar{u}(p)$ and $\bar{v}(p)$. The simplest case is the one of a massless fermion:

$$m = 0. \quad \text{(A.14)}$$

In this case the Dirac equation for the u- and the v-spinors are identical and it is sufficient to consider

$$\not{p}u(p) = 0, \qquad \bar{u}(p)\not{p} = 0. \quad \text{(A.15)}$$

In the Weyl representation $\not{p}$ is given by

$$\not{p} = \begin{pmatrix} \mathbf{0} & p_\mu \sigma^\mu \\ p_\mu \bar{\sigma}^\mu & \mathbf{0} \end{pmatrix}, \quad \text{(A.16)}$$

therefore the 4×4-matrix equation for $u(p)$ (or $\bar{u}(p)$) decouples into two 2×2-matrix equations. We introduce the following notation: Four-component Dirac spinors are constructed out of two **Weyl spinors** as follows:

$$u(p) = \begin{pmatrix} |p+\rangle \\ |p-\rangle \end{pmatrix} = \begin{pmatrix} |p\rangle \\ |p] \end{pmatrix} = \begin{pmatrix} p_A \\ p^{\dot{B}} \end{pmatrix} = \begin{pmatrix} u_+(p) \\ u_-(p) \end{pmatrix}. \quad \text{(A.17)}$$

Bra-spinors are given by

$$\bar{u}(p) = (\, \langle p-|,\ \langle p+|\,) = (\, \langle p|,\ [p|\,) = \left(p^A,\ p_{\dot{B}} \right) = (\, \bar{u}_-(p),\ \bar{u}_+(p)\,). \quad \text{(A.18)}$$

In the literature there exists various notations for Weyl spinors. Equations (A.17) and (A.18) show four of them and the way how to translate from one notation to another notation. By a slight abuse of notation we will in the following not distinguish between a two-component Weyl spinor and a Dirac spinor, where either the upper two components or the lower two components are zero. If we define the chiral projection operators

$$P_+ \;=\; \frac{1}{2}(1+\gamma_5) = \begin{pmatrix} \mathbf{1} & \mathbf{0} \\ \mathbf{0} & \mathbf{0} \end{pmatrix}, \quad P_- \;=\; \frac{1}{2}(1-\gamma_5) = \begin{pmatrix} \mathbf{0} & \mathbf{0} \\ \mathbf{0} & \mathbf{1} \end{pmatrix}, \quad \text{(A.19)}$$

then (with the slight abuse of notation mentioned above)

$$u_\pm(p) = P_\pm u(p), \quad \bar{u}_\pm(p) = \bar{u}(p) P_\mp. \tag{A.20}$$

The two solutions of the Dirac equation

$$\not{p} u(p, \lambda) = 0 \tag{A.21}$$

are then

$$u(p, +) = u_+(p), \quad u(p, -) = u_-(p). \tag{A.22}$$

We now have to solve

$$\begin{aligned} p_\mu \bar{\sigma}^\mu \left| p+ \right\rangle &= 0, \quad & p_\mu \sigma^\mu \left| p- \right\rangle &= 0, \\ \left\langle p+ \right| p_\mu \bar{\sigma}^\mu &= 0, \quad & \left\langle p- \right| p_\mu \sigma^\mu &= 0. \end{aligned} \tag{A.23}$$

It it convenient to express the four-vector $p^\mu = (p^0, p^1, p^2, p^3)$ in terms of light-cone coordinates:

$$p^+ = \frac{1}{\sqrt{2}}\left(p^0 + p^3\right), \quad p^- = \frac{1}{\sqrt{2}}\left(p^0 - p^3\right), \quad p^\perp = \frac{1}{\sqrt{2}}\left(p^1 + ip^2\right), \quad p^{\perp *} = \frac{1}{\sqrt{2}}\left(p^1 - ip^2\right).$$

Note that $p^{\perp *}$ does not involve a complex conjugation of p^1 or p^2. For null-vectors one has

$$p^{\perp *} p^\perp = p^+ p^-. \tag{A.24}$$

Then the equation for the ket-spinors becomes

$$\begin{pmatrix} p^- & -p^{\perp *} \\ -p^\perp & p^+ \end{pmatrix} \left| p+ \right\rangle = 0, \quad \begin{pmatrix} p^+ & p^{\perp *} \\ p^\perp & p^- \end{pmatrix} \left| p- \right\rangle = 0, \tag{A.25}$$

and similar equations can be written down for the bra-spinors. This is a problem of linear algebra. Solutions for ket-spinors are

$$\left| p+ \right\rangle = p_A = c_1 \begin{pmatrix} p^{\perp *} \\ p^- \end{pmatrix}, \quad \left| p- \right\rangle = p^{\dot{A}} = c_2 \begin{pmatrix} p^- \\ -p^\perp \end{pmatrix}, \tag{A.26}$$

with some yet unspecified multiplicative constants c_1 and c_2. Solutions for bra-spinors are

$$\left\langle p+ \right| = p_{\dot{A}} = c_3 \left(p^\perp, p^-\right), \quad \left\langle p- \right| = p^A = c_4 \left(p^-, -p^{\perp *}\right), \tag{A.27}$$

with some further constants c_3 and c_4. Let us now introduce the 2-dimensional anti-symmetric tensor:

$$\varepsilon_{AB} = \begin{pmatrix} 0 & 1 \\ -1 & 0 \end{pmatrix}, \qquad \varepsilon_{BA} = -\varepsilon_{AB} \tag{A.28}$$

Furthermore we set

$$\varepsilon^{AB} = \varepsilon^{\dot{A}\dot{B}} = \varepsilon_{AB} = \varepsilon_{\dot{A}\dot{B}}. \tag{A.29}$$

Note that these definitions imply

$$\varepsilon^{AC}\varepsilon_{BC} = \delta^{A}_{B}, \qquad \varepsilon^{\dot{A}\dot{C}}\varepsilon_{\dot{B}\dot{C}} = \delta^{\dot{A}}_{\dot{B}}. \tag{A.30}$$

We would like to have the following relations for raising and lowering a spinor index A or $\dot{B}$:

$$\begin{aligned} p^{A} &= \varepsilon^{AB} p_{B}, \qquad p^{\dot{A}} = \varepsilon^{\dot{A}\dot{B}} p_{\dot{B}}, \\ p_{\dot{B}} &= p^{\dot{A}} \varepsilon_{\dot{A}\dot{B}}, \qquad p_{B} = p^{A} \varepsilon_{AB}. \end{aligned} \tag{A.31}$$

Note that raising an index is done by left-multiplication, whereas lowering is performed by right-multiplication. Postulating these relations implies

$$c_1 = c_4, \qquad c_2 = c_3. \tag{A.32}$$

In addition we normalise the spinors according to

$$\langle p \pm | \gamma^{\mu} | p \pm \rangle = 2 p^{\mu}. \tag{A.33}$$

This implies

$$c_1 c_3 = \frac{\sqrt{2}}{p^-}, \qquad c_2 c_4 = \frac{\sqrt{2}}{p^-}. \tag{A.34}$$

Equations (A.31) and (A.33) determine the spinors only up to a scaling

$$p_A \to \lambda p_A, \qquad p_{\dot{A}} \to \frac{1}{\lambda} p_{\dot{A}}. \tag{A.35}$$

This scaling freedom is referred to as **little group scaling**. Keeping the scaling freedom, we define the spinors as

$$\begin{aligned} |p+\rangle = p_A &= \frac{\lambda_p 2^{\frac{1}{4}}}{\sqrt{p^-}} \begin{pmatrix} p^{\perp *} \\ p^- \end{pmatrix}, \qquad |p-\rangle = p^{\dot{A}} = \frac{2^{\frac{1}{4}}}{\lambda_p \sqrt{p^-}} \begin{pmatrix} p^- \\ -p^{\perp} \end{pmatrix}, \\ \langle p+| = p_{\dot{A}} &= \frac{2^{\frac{1}{4}}}{\lambda_p \sqrt{p^-}} \left(p^{\perp}, p^- \right), \qquad \langle p-| = p^{A} = \frac{\lambda_p 2^{\frac{1}{4}}}{\sqrt{p^-}} \left(p^-, -p^{\perp *} \right). \end{aligned} \tag{A.36}$$

Popular choices for λ_p are

$$\begin{aligned}\lambda_p &= 1 : \text{symmetric},\\ \lambda_p &= 2^{\frac{1}{4}}\sqrt{p^-} : p_A \text{ linear in } p^\mu,\\ \lambda_p &= \frac{1}{2^{\frac{1}{4}}\sqrt{p^-}} : p_{\dot{A}} \text{ linear in } p^\mu.\end{aligned} \tag{A.37}$$

Note that all formulae in this Sect. A.2 work not only for real momenta p^μ but also for complex momenta p^μ. This will be useful later on, where we encounter situations with complex momenta. However there is one exception: The relations $p_A^\dagger = p_{\dot{A}}$ and $p^{A\dagger} = p^{\dot{A}}$ (or equivalently $\bar{u}(p) = u(p)^\dagger\gamma^0$) encountered in previous sub-sections are valid only for real momenta $p^\mu = (p^0, p^1, p^2, p^3)$. If on the other hand the components (p^0, p^1, p^2, p^3) are complex, these relations will in general not hold. In the latter case p_A and $p_{\dot{A}}$ are considered to be independent quantities. The reason, why the relations $p_A^\dagger = p_{\dot{A}}$ and $p^{A\dagger} = p^{\dot{A}}$ do not hold in the complex case lies in the definition of $p^{\perp *}$: We defined $p^{\perp *}$ as $p^{\perp *} = (p^1 - ip^2)/\sqrt{2}$, and not as $(p^1 - i(p^2)^*)/\sqrt{2}$. With the former definition $p^{\perp *}$ is a holomorphic function of p^1 and p^2. There are applications where holomorphicity is more important than nice properties under hermitian conjugation.

A.3 Spinor Products

Let us now make the symmetric choice $\lambda_p = 1$. Spinor products are defined by

$$\begin{aligned}\langle pq\rangle &= \langle p - |q+\rangle = p^A q_A = \frac{\sqrt{2}}{\sqrt{p^-}\sqrt{q^-}}\left(p^- q^{\perp *} - q^- p^{\perp *}\right),\\ [qp] &= \langle q + |p-\rangle = q_{\dot{A}} p^{\dot{A}} = \frac{\sqrt{2}}{\sqrt{p^-}\sqrt{q^-}}\left(p^- q^{\perp} - q^- p^{\perp}\right),\end{aligned} \tag{A.38}$$

where the last expression in each line used the choice $\lambda_p = \lambda_q = 1$. We have

$$\langle pq\rangle\,[qp] = 2pq. \tag{A.39}$$

If p^μ and q^μ are real we have

$$[qp] = \langle pq\rangle^* \;\text{sign}(p^0)\,\text{sign}(q^0). \tag{A.40}$$

The spinor products are anti-symmetric

$$\langle qp\rangle = -\langle pq\rangle, \quad [pq] = -[qp]. \tag{A.41}$$

From the Schouten identity for the 2-dimensional antisymmetric tensor

$$\varepsilon_{AB}\varepsilon_{CD} + \varepsilon_{BC}\varepsilon_{AD} + \varepsilon_{CA}\varepsilon_{BD} = 0. \tag{A.42}$$

one derives

$$\begin{aligned} \langle p_1 p_2 \rangle \langle p_3 p_4 \rangle + \langle p_2 p_3 \rangle \langle p_1 p_4 \rangle + \langle p_3 p_1 \rangle \langle p_2 p_4 \rangle &= 0, \\ [p_1 p_2][p_3 p_4] + [p_2 p_3][p_1 p_4] + [p_3 p_1][p_2 p_4] &= 0. \end{aligned} \tag{A.43}$$

The Fierz identity reads

$$\langle p_1 + |\gamma_\mu| p_2 + \rangle \langle p_3 - |\gamma^\mu| p_4 - \rangle = 2[p_1 p_4] \langle p_3 p_2 \rangle. \tag{A.44}$$

Note that with our slight abuse of notation we identify a two-component Weyl spinor with a Dirac spinor, where the other two components are zero. Therefore

$$\langle p_1 + |\gamma_\mu| p_2 + \rangle = \langle p_1 + |\bar{\sigma}_\mu| p_2 + \rangle, \ \langle p_3 - |\gamma^\mu| p_4 - \rangle = \langle p_3 - |\sigma^\mu| p_4 - \rangle. \tag{A.45}$$

We further have the reflection identities

$$\begin{aligned} \langle p \pm |\gamma^{\mu_1} \dots \gamma^{\mu_{2n+1}}| q \pm \rangle &= \langle q \mp |\gamma^{\mu_{2n+1}} \dots \gamma^{\mu_1}| p \mp \rangle, \\ \langle p \pm |\gamma^{\mu_1} \dots \gamma^{\mu_{2n}}| q \mp \rangle &= -\langle q \pm |\gamma^{\mu_{2n}} \dots \gamma^{\mu_1}| p \mp \rangle. \end{aligned} \tag{A.46}$$

A.4 Massive Spinors

As in the massless case, a massive spinor satisfying the Dirac equation has a two-fold degeneracy. We will label the two different eigenvectors by "+" and "−". Let p be a massive four-vector with $p^2 = m^2$, and let q be an arbitrary light-like four-vector. With the help of q we can construct a light-like vector $p^\flat$ associated to p:

$$p^\flat = p - \frac{p^2}{2 p \cdot q} q. \tag{A.47}$$

We define [431–433]

$$\begin{aligned} u(p,+) &= \frac{1}{\langle p^\flat + | q - \rangle} (\not{p} + m) \, |q-\rangle, \ \ v(p,-) = \frac{1}{\langle p^\flat + | q - \rangle} (\not{p} - m) \, |q-\rangle, \\ u(p,-) &= \frac{1}{\langle p^\flat - | q + \rangle} (\not{p} + m) \, |q+\rangle, \ \ v(p,+) = \frac{1}{\langle p^\flat - | q + \rangle} (\not{p} - m) \, |q+\rangle. \end{aligned} \tag{A.48}$$

For the conjugate spinors we have

$$\bar{u}(p,+) = \frac{1}{\langle q - | p^{\flat} + \rangle} \langle q - | (\not{p} + m), \; \bar{v}(p,-) = \frac{1}{\langle q - | p^{\flat} + \rangle} \langle q - | (\not{p} - m),$$
$$\bar{u}(p,-) = \frac{1}{\langle q + | p^{\flat} - \rangle} \langle q + | (\not{p} + m), \; \bar{v}(p,+) = \frac{1}{\langle q + | p^{\flat} - \rangle} \langle q + | (\not{p} - m). \tag{A.49}$$

These spinors satisfy the Dirac equations of Eq. (A.11), the orthogonality relations of Eq. (A.12) and the completeness relations of Eq. (A.13). We further have

$$\bar{u}(p,\bar{\lambda})\gamma^{\mu}u(p,\lambda) = 2p^{\mu}\delta_{\bar{\lambda}\lambda}, \quad \bar{v}(p,\bar{\lambda})\gamma^{\mu}v(p,\lambda) = 2p^{\mu}\delta_{\bar{\lambda}\lambda}. \tag{A.50}$$

In the massless limit the definition reduces to

$$u(p,+) = v(p,-) = |p+\rangle, \quad \bar{u}(p,+) = \bar{v}(p,-) = \langle p+|,$$
$$u(p,-) = v(p,+) = |p-\rangle, \quad \bar{u}(p,-) = \bar{v}(p,+) = \langle p-|, \tag{A.51}$$

and the spinors are in the massless limit independent of the reference spinors $|q+\rangle$ and $\langle q+|$.

Appendix B
Scalar One-Loop Integrals

In this appendix we list basic scalar one-loop integrals for massless theories in $D = 4 - 2\varepsilon$ dimensions as an expansion in the dimensional regularisation parameter ε up to and including the $O(\varepsilon^0)$ term. The basic scalar integrals consist of the scalar two-point, the scalar three-point and the scalar four-point functions. The scalar one-point function vanishes in dimensional regularisation. Since we restrict ourselves to massless quantum field theories, all internal propagators are massless and we only have to distinguish the momentum configurations of the external momenta. We call an external momentum p "massive" if $p^2 \neq 0$. All scalar integrals have been known for a long time in the literature. Classical papers on scalar integrals are [97, 434]. Scalar integrals within dimensional regularization are treated in [90, 435]. Useful information on the three-mass triangle can be found in [436–438]. The scalar boxes have been recalculated in [439, 440]. The compilation given here is based on [103].

The basic scalar integrals with internal masses constitute a longer list. These integrals can be found in the literature [441, 442].

B.1 Massless Scalar One-Loop Integrals

We recall the notation for massless one-loop integrals from Eq. (5.14):

$$I_{n_{\mathrm{ext}}} = e^{\varepsilon \gamma_{\mathrm{E}}} \mu^{2\varepsilon} \int \frac{d^D k}{i \pi^{\frac{D}{2}}} \frac{1}{(-k^2)(-(k-p_1)^2)...(-(k-p_1-...p_{n_{\mathrm{ext}}-1})^2)}. \quad \text{(B.1)}$$

The Two-Point Function

The scalar two-point function is given by

S. Weinzierl, *Feynman Integrals*, UNITEXT for Physics,
https://doi.org/10.1007/978-3-030-99558-4

$$I_2(p_1^2, \mu^2) = \frac{1}{\varepsilon} + 2 - \ln\left(\frac{-p_1^2}{\mu^2}\right) + \mathcal{O}(\varepsilon). \tag{B.2}$$

Three-Point Functions

For the three-point functions we have three different cases: One external mass, two external masses and three external masses. The one-mass scalar triangle with $p_1^2 \neq 0$, $p_2^2 = p_3^2 = 0$ is given by

$$I_3^{1m}(p_1^2, \mu^2) = \frac{1}{\varepsilon^2 p_1^2} - \frac{1}{\varepsilon p_1^2} \ln\left(\frac{-p_1^2}{\mu^2}\right) + \frac{1}{2p_1^2} \ln^2\left(\frac{-p_1^2}{\mu^2}\right) - \frac{1}{2p_1^2}\zeta_2 + \mathcal{O}(\varepsilon). \tag{B.3}$$

The two-mass scalar triangle with $p_1^2 \neq 0$, $p_2^2 \neq 0$ and $p_3^2 = 0$ is given by

$$\begin{aligned} I_3^{2m}(p_1^2, p_2^2, \mu^2) = & \frac{1}{\varepsilon} \frac{1}{(p_1^2 - p_2^2)} \left[-\ln\left(\frac{-p_1^2}{\mu^2}\right) + \ln\left(\frac{-p_2^2}{\mu^2}\right) \right] \\ & + \frac{1}{2(p_1^2 - p_2^2)} \left[\ln^2\left(\frac{-p_1^2}{\mu^2}\right) - \ln^2\left(\frac{-p_2^2}{\mu^2}\right) \right] + \mathcal{O}(\varepsilon). \end{aligned} \tag{B.4}$$

The three-mass scalar triangle with $p_1^2 \neq 0$, $p_2^2 \neq 0$ and $p_3^2 \neq 0$: This integral is finite and we have

$$I_3^{3m}\left(p_1^2, p_2^2, p_3^2, \mu^2\right) = -\int\limits_0^1 d^3\alpha \frac{\delta\left(1 - \alpha_1 - \alpha_2 - \alpha_3\right)}{-\alpha_1\alpha_2 p_1^2 - \alpha_2\alpha_3 p_2^2 - \alpha_3\alpha_1 p_3^2} + \mathcal{O}(\varepsilon). \tag{B.5}$$

With the notation

$$\begin{aligned} & \delta_1 = p_1^2 - p_2^2 - p_3^2, \quad \delta_2 = p_2^2 - p_3^2 - p_1^2, \quad \delta_3 = p_3^2 - p_1^2 - p_2^2, \\ & \Delta_3 = \left(p_1^2\right)^2 + \left(p_2^2\right)^2 + \left(p_3^2\right)^2 - 2p_1^2 p_2^2 - 2p_2^2 p_3^2 - 2p_3^2 p_1^2, \end{aligned} \tag{B.6}$$

the three-mass triangle I_3^{3m} is expressed in the region $p_1^2, p_2^2, p_3^2 < 0$ and $\Delta_3 < 0$ by

$$\begin{aligned} I_3^{3m} = & -\frac{2}{\sqrt{-\Delta_3}} \\ & \times \left[\mathrm{Cl}_2\left(2\arctan\left(\frac{\sqrt{-\Delta_3}}{\delta_1}\right)\right) + \mathrm{Cl}_2\left(2\arctan\left(\frac{\sqrt{-\Delta_3}}{\delta_2}\right)\right) + \mathrm{Cl}_2\left(2\arctan\left(\frac{\sqrt{-\Delta_3}}{\delta_3}\right)\right) \right] \\ & + \mathcal{O}(\varepsilon). \end{aligned} \tag{B.7}$$

The Clausen function $\mathrm{Cl}_2(\theta)$ is defined by

$$\mathrm{Cl}_2(\theta) = \frac{1}{2i}\left[\mathrm{Li}_n\left(e^{i\theta}\right) - \mathrm{Li}_n\left(e^{-i\theta}\right)\right] \tag{B.8}$$

and discussed in more detail in Chap. 8.

In the region $p_1^2, p_2^2, p_3^2 < 0$ and $\Delta_3 > 0$ as well as in the region $p_1^2, p_3^2 < 0$, $p_2^2 > 0$ (for which Δ_3 is always positive) the integral I_3^{3m} is given by

$$I_3^{3m} = \frac{1}{\sqrt{\Delta_3}}\mathrm{Re}\left[2\left(\mathrm{Li}_2(-\rho x) + \mathrm{Li}_2(-\rho y)\right) + \ln(\rho x)\ln(\rho y) + \ln\left(\frac{y}{x}\right)\ln\left(\frac{1+\rho x}{1+\rho y}\right) + \frac{\pi^2}{3}\right]$$
$$+ \frac{i\pi\theta(p_2^2)}{\sqrt{\Delta_3}}\ln\left(\frac{\left(\delta_1+\sqrt{\Delta_3}\right)\left(\delta_3+\sqrt{\Delta_3}\right)}{\left(\delta_1-\sqrt{\Delta_3}\right)\left(\delta_3-\sqrt{\Delta_3}\right)}\right) + \mathcal{O}(\varepsilon), \tag{B.9}$$

where

$$x = \frac{p_1^2}{p_3^2}, \quad y = \frac{p_2^2}{p_3^2}, \quad \rho = \frac{2p_3^2}{\delta_3 + \sqrt{\Delta_3}}. \tag{B.10}$$

The step function $\theta(x)$ is defined as $\theta(x) = 1$ for $x > 0$ and $\theta(x) = 0$ otherwise.

Four-Point Functions

For the four-point function we use the invariants

$$s = (p_1 + p_2)^2, \qquad t = (p_2 + p_3)^2 \tag{B.11}$$

together with the external masses $m_i^2 = p_i^2$.
The zero-mass box ($m_1^2 = m_2^2 = m_3^2 = m_4^2 = 0$):

$$I_4^{0m}\left(s,t,\mu^2\right) = \frac{4}{\varepsilon^2 st} - \frac{2}{\varepsilon st}\left[\ln\left(\frac{-s}{\mu^2}\right) + \ln\left(\frac{-t}{\mu^2}\right)\right]$$
$$+ \frac{1}{st}\left[\ln^2\left(\frac{-s}{\mu^2}\right) + \ln^2\left(\frac{-t}{\mu^2}\right) - \ln^2\left(\frac{-s}{-t}\right) - 8\zeta_2\right] + \mathcal{O}(\varepsilon). \tag{B.12}$$

The one-mass box ($m_1^2 = m_2^2 = m_3^2 = 0$):

$$I_4^{1m}\left(s,t,m_4^2,\mu^2\right) = \frac{2}{\varepsilon^2 st} - \frac{2}{\varepsilon st}\left[\ln\left(\frac{-s}{\mu^2}\right) + \ln\left(\frac{-t}{\mu^2}\right) - \ln\left(\frac{-m_4^2}{\mu^2}\right)\right] + \frac{1}{st}\left[\ln^2\left(\frac{-s}{\mu^2}\right)\right.$$
$$+ \ln^2\left(\frac{-t}{\mu^2}\right) - \ln^2\left(\frac{-m_4^2}{\mu^2}\right) - \ln^2\left(\frac{-s}{-t}\right) - 2\,\mathrm{Li}_2\left(1 - \frac{(-m_4^2)}{(-s)}\right) - 2\,\mathrm{Li}_2\left(1 - \frac{(-m_4^2)}{(-t)}\right)$$
$$\left. - 3\zeta_2\right] + \mathcal{O}(\varepsilon). \tag{B.13}$$

The easy two-mass box ($m_1^2 = m_3^2 = 0$):

$$
\begin{aligned}
& I_4^{2me}\left(s,t,m_2^2,m_4^2,\mu^2\right) = \\
& \quad -\frac{2}{\varepsilon\left(st-m_2^2m_4^2\right)}\left[\ln\left(\frac{-s}{\mu^2}\right)+\ln\left(\frac{-t}{\mu^2}\right)-\ln\left(\frac{-m_2^2}{\mu^2}\right)-\ln\left(\frac{-m_4^2}{\mu^2}\right)\right] \\
& \quad +\frac{1}{st-m_2^2m_4^2}\left[\ln^2\left(\frac{-s}{\mu^2}\right)+\ln^2\left(\frac{-t}{\mu^2}\right)-\ln^2\left(\frac{-m_2^2}{\mu^2}\right)-\ln^2\left(\frac{-m_4^2}{\mu^2}\right)-\ln^2\left(\frac{-s}{-t}\right)\right. \\
& \quad -2\,\mathrm{Li}_2\left(1-\frac{(-m_2^2)}{(-s)}\right)-2\,\mathrm{Li}_2\left(1-\frac{(-m_2^2)}{(-t)}\right)-2\,\mathrm{Li}_2\left(1-\frac{(-m_4^2)}{(-s)}\right) \\
& \quad \left.-2\,\mathrm{Li}_2\left(1-\frac{(-m_4^2)}{(-t)}\right)+2\,\mathrm{Li}_2\left(1-\frac{(-m_2^2)}{(-s)}\frac{(-m_4^2)}{(-t)}\right)\right]+\mathcal{O}(\varepsilon).
\end{aligned}
\tag{B.14}
$$

The hard two-mass box ($m_1^2 = m_2^2 = 0$):

$$
\begin{aligned}
I_4^{2mh}\left(s,t,m_3^2,m_4^2,\mu^2\right) &= \frac{1}{\varepsilon^2 st}-\frac{1}{\varepsilon st}\left[\ln\left(\frac{-s}{\mu^2}\right)+2\ln\left(\frac{-t}{\mu^2}\right)-\ln\left(\frac{-m_3^2}{\mu^2}\right)-\ln\left(\frac{-m_4^2}{\mu^2}\right)\right] \\
& +\frac{1}{st}\left[\frac{3}{2}\ln^2\left(\frac{-s}{\mu^2}\right)+\ln^2\left(\frac{-t}{\mu^2}\right)-\frac{1}{2}\ln^2\left(\frac{-m_3^2}{\mu^2}\right)-\frac{1}{2}\ln^2\left(\frac{-m_4^2}{\mu^2}\right)-\ln^2\left(\frac{-s}{-t}\right)\right. \\
& -\ln\left(\frac{-s}{\mu^2}\right)\ln\left(\frac{-m_3^2}{\mu^2}\right)-\ln\left(\frac{-s}{\mu^2}\right)\ln\left(\frac{-m_4^2}{\mu^2}\right)+\ln\left(\frac{-m_3^2}{\mu^2}\right)\ln\left(\frac{-m_4^2}{\mu^2}\right) \\
& \left.-2\,\mathrm{Li}_2\left(1-\frac{(-m_3^2)}{(-t)}\right)-2\,\mathrm{Li}_2\left(1-\frac{(-m_4^2)}{(-t)}\right)-\frac{1}{2}\zeta_2\right]+\mathcal{O}(\varepsilon).
\end{aligned}
\tag{B.15}
$$

The three-mass box ($m_1^2 = 0$):

$$
\begin{aligned}
& I_4^{3m}\left(s,t,m_2^2,m_3^2,m_4^2,\mu^2\right) = \\
& \quad -\frac{1}{\varepsilon\left(st-m_2^2m_4^2\right)}\left[\ln\left(\frac{-s}{\mu^2}\right)+\ln\left(\frac{-t}{\mu^2}\right)-\ln\left(\frac{-m_2^2}{\mu^2}\right)-\ln\left(\frac{-m_4^2}{\mu^2}\right)\right] \\
& \quad +\frac{1}{st-m_2^2m_4^2}\left[\frac{3}{2}\ln^2\left(\frac{-s}{\mu^2}\right)+\frac{3}{2}\ln^2\left(\frac{-t}{\mu^2}\right)-\frac{1}{2}\ln^2\left(\frac{-m_2^2}{\mu^2}\right)-\frac{1}{2}\ln^2\left(\frac{-m_4^2}{\mu^2}\right)-\ln^2\left(\frac{-s}{-t}\right)\right. \\
& \quad -\ln\left(\frac{-s}{\mu^2}\right)\ln\left(\frac{-m_3^2}{\mu^2}\right)-\ln\left(\frac{-s}{\mu^2}\right)\ln\left(\frac{-m_4^2}{\mu^2}\right)+\ln\left(\frac{-m_3^2}{\mu^2}\right)\ln\left(\frac{-m_4^2}{\mu^2}\right) \\
& \quad -\ln\left(\frac{-t}{\mu^2}\right)\ln\left(\frac{-m_2^2}{\mu^2}\right)-\ln\left(\frac{-t}{\mu^2}\right)\ln\left(\frac{-m_3^2}{\mu^2}\right)+\ln\left(\frac{-m_2^2}{\mu^2}\right)\ln\left(\frac{-m_3^2}{\mu^2}\right) \\
& \quad \left.-2\,\mathrm{Li}_2\left(1-\frac{(-m_2^2)}{(-s)}\right)-2\,\mathrm{Li}_2\left(1-\frac{(-m_4^2)}{(-t)}\right)+2\,\mathrm{Li}_2\left(1-\frac{(-m_2^2)}{(-s)}\frac{(-m_4^2)}{(-t)}\right)\right]+\mathcal{O}(\varepsilon).
\end{aligned}
\tag{B.16}
$$

The four-mass box:

$$I_4^{4m}\left(s,t,m_1^2,m_2^2,m_3^2,m_4^2,\mu^2\right) = I_3^{3m}(st, m_1^2 m_3^2, m_2^2 m_4^2, \mu^2) + K(s,t,m_1^2,m_3^2,m_2^2,m_4^2), \tag{B.17}$$

where

$$K(s_1,t_1,s_2,t_2,s_3,t_3) = -\frac{2\pi i}{\lambda}\sum_{i=1}^{3}\theta(-s_i)\theta(-t_i) \\ \times\left[\ln\left(\sum_{j\neq i} s_j t_j - (s_i t_i - \lambda)(1+i\delta)\right) - \ln\left(\sum_{j\neq i} s_j t_j - (s_i t_i + \lambda)(1+i\delta)\right)\right], \tag{B.18}$$

and

$$\lambda = \sqrt{(s_1t_1)^2 + (s_2t_2)^2 + (s_3t_3)^2 - 2s_1t_1s_2t_2 - 2s_2t_2s_3t_3 - 2s_3t_3s_1t_1}. \tag{B.19}$$

$\delta > 0$ is an infinitesimal quantity.

B.2 Analytic Continuation

In one-loop integrals the functions

$$\ln\left(\frac{-s}{-t}\right),\quad \text{Li}_2\left(1 - \frac{(-s)}{(-t)}\right) \tag{B.20}$$

and generalizations thereof occur. The analytic continuation is defined by giving all quantities a small imaginary part, e.g.

$$s \to s + i\delta, \tag{B.21}$$

with $\delta > 0$ being an infinitesimal quantity. Explicitly, the imaginary parts of the logarithm and the dilogarithm are given by

$$\ln\left(\frac{-s}{-t}\right) = \ln\left(\left|\frac{s}{t}\right|\right) - i\pi\left[\theta(s) - \theta(t)\right], \\ \text{Li}_2\left(1 - \frac{(-s)}{(-t)}\right) = \text{ReLi}_2\left(1 - \frac{s}{t}\right) - i\theta\left(-\frac{s}{t}\right)\ln\left(1 - \frac{s}{t}\right)\text{Im}\ln\left(\frac{-s}{-t}\right). \tag{B.22}$$

This generalizes as follows:

$$\ln\left(\frac{(-s_1)}{(-t_1)}\frac{(-s_2)}{(-t_2)}\right) = \ln\left(\left|\frac{s_1 s_2}{t_1 t_2}\right|\right) - i\pi\left[\theta(s_1) + \theta(s_2) - \theta(t_1) - \theta(t_2)\right],$$

$$\mathrm{Li}_2\left(1 - \frac{(-s_1)}{(-t_1)}\frac{(-s_2)}{(-t_2)}\right) = \mathrm{Re}\mathrm{Li}_2\left(1 - \frac{s_1 s_2}{t_1 t_2}\right) - i\ln\left(1 - \frac{(-s_1)}{(-t_1)}\frac{(-s_2)}{(-t_2)}\right)\mathrm{Im}\ln\left(\frac{(-s_1)}{(-t_1)}\frac{(-s_2)}{(-t_2)}\right),$$

where

$$\ln\left(1 - \frac{(-s_1)}{(-t_1)}\frac{(-s_2)}{(-t_2)}\right) = \ln\left|1 - \frac{s_1 s_2}{t_1 t_2}\right| - \frac{1}{2}i\pi\left[\theta(s_1) + \theta(s_2) - \theta(t_1) - \theta(t_2)\right]\theta\left(\frac{s_1 s_2}{t_1 t_2} - 1\right).$$

Appendix C
Transcendental Functions

In this appendix we summarise definitions and properties of a few transcendental functions. We start with hypergeometric functions in one variable in Sect. C.1. Appell functions are generalisations to two variables and discussed in Sect. C.2. Lauricella functions are particular generalisations to n variables and briefly discussed in Sect. C.3. The general case with n variables is known as Horn functions. These are introduced in Sect. C.4.

In defining these functions, the Pochhammer symbol occurs frequently. The Pochhammer symbol $(a)_n$ is defined by

$$(a)_n = \frac{\Gamma(a+n)}{\Gamma(a)}. \tag{C.1}$$

C.1 Hypergeometric Functions

The generalised hypergeometric function ${}_AF_B$ (or hypergeometric function for short) is defined by

$${}_AF_B(a_1, \dots, a_A; b_1, \dots, b_B; x) = \sum_{n=0}^{\infty} \frac{(a_1)_n \dots (a_A)_n}{(b_1)_n \dots (b_B)_n} \frac{x^n}{n!}. \tag{C.2}$$

We are mainly interested in the case ${}_{A+1}F_A$. The case ${}_1F_0$ is trivial

$${}_1F_0(a; ; x) = (1-x)^{-a}. \tag{C.3}$$

The first non-trivial case in the family ${}_{A+1}F_A$ is the function

$${}_2F_1(a_1, a_2; b_1; x) = \sum_{n=0}^{\infty} \frac{(a_1)_n (a_2)_n}{(b_1)_n} \frac{x^n}{n!}. \tag{C.4}$$

S. Weinzierl, *Feynman Integrals*, UNITEXT for Physics,
https://doi.org/10.1007/978-3-030-99558-4

This function is sometimes referred to as "the" hypergeometric function or the Gauß hypergeometric function. In this book we will call any function of the form as in Eq. (C.2) a hypergeometric function. From the definition it is easy to verify that

$$
\begin{aligned}
{}_{A+1}F_{B+1}(a_1, \ldots, a_A, a_{A+1}; b_1, \ldots, b_B, b_{B+1}; x) &= \frac{\Gamma(b_{B+1})}{\Gamma(a_{A+1})\Gamma(b_{B+1} - a_{A+1})} \\
&\times \int\limits_0^1 dt \; t^{a_{A+1}-1}(1-t)^{b_{B+1}-a_{A+1}-1} \, {}_AF_B(a_1, \ldots, a_A; b_1, \ldots, b_B; tx). \qquad \text{(C.5)}
\end{aligned}
$$

This allows us to deduce a A-fold integral representation for ${}_{A+1}F_A$. In particular

$$
{}_2F_1(a_1, a_2; b_1; x) = \frac{\Gamma(b_1)}{\Gamma(a_2)\Gamma(b_1 - a_2)} \int\limits_0^1 dt \; t^{a_2-1}(1-t)^{b_1-a_2-1}(1-tx)^{-a_1}, \qquad \text{(C.6)}
$$

$$
\begin{aligned}
{}_3F_2(a, b_1, b_2; c_1, c_2; x) &= \frac{\Gamma(c_1)}{\Gamma(b_1)\Gamma(c_1 - b_1)} \frac{\Gamma(c_2)}{\Gamma(b_2)\Gamma(c_2 - b_2)} \\
&\quad \int\limits_0^1 du \int\limits_0^1 dv \; u^{b_1-1}(1-u)^{c_1-b_1-1} v^{b_2-1}(1-v)^{c_2-b_2-1}(1-uvx)^{-a}.
\end{aligned}
$$

C.2 Appell Functions

In this section we discuss the four Appell functions and the Kampé de Fériet function [443, 444]. They are generalisations of the hypergeometric function from one variable x to two variables x_1 and x_2.

The Appell function of the first kind is defined by

$$
F_1(a, b_1, b_2; c; x_1, x_2) = \sum_{m_1=0}^{\infty} \sum_{m_2=0}^{\infty} \frac{(a)_{m_1+m_2}(b_1)_{m_1}(b_2)_{m_2}}{(c)_{m_1+m_2}} \frac{x_1^{m_1}}{m_1!} \frac{x_2^{m_2}}{m_2!}. \qquad \text{(C.7)}
$$

The Appell function of the second kind is defined by

$$
F_2(a, b_1, b_2; c_1, c_2; x_1, x_2) = \sum_{m_1=0}^{\infty} \sum_{m_2=0}^{\infty} \frac{(a)_{m_1+m_2}(b_1)_{m_1}(b_2)_{m_2}}{(c_1)_{m_1}(c_2)_{m_2}} \frac{x_1^{m_1}}{m_1!} \frac{x_2^{m_2}}{m_2!}. \qquad \text{(C.8)}
$$

The Appell function of the third kind is defined by

$$
F_3(a_1, a_2, b_1, b_2; c; x_1, x_2) = \sum_{m_1=0}^{\infty} \sum_{m_2=0}^{\infty} \frac{(a_1)_{m_1}(a_2)_{m_2}(b_1)_{m_1}(b_2)_{m_2}}{(c)_{m_1+m_2}} \frac{x_1^{m_1}}{m_1!} \frac{x_2^{m_2}}{m_2!}. \qquad \text{(C.9)}
$$

The Appell function of the fourth kind is defined by

$$F_4(a,b;c_1,c_2;x_1,x_2) = \sum_{m_1=0}^{\infty}\sum_{m_2=0}^{\infty} \frac{(a)_{m_1+m_2}(b)_{m_1+m_2}}{(c_1)_{m_1}(c_2)_{m_2}} \frac{x_1^{m_1}}{m_1!}\frac{x_2^{m_2}}{m_2!}. \tag{C.10}$$

The generalized Kampé de Fériet function S_1 is defined by

$$S_1(a_1,a_2,b_1;c,c_1;x_1,x_2) = \sum_{m_1=0}^{\infty}\sum_{m_2=0}^{\infty} \frac{(a_1)_{m_1+m_2}(a_2)_{m_1+m_2}(b_1)_{m_1}}{(c)_{m_1+m_2}(c_1)_{m_1}} \frac{x_1^{m_1}}{m_1!}\frac{x_2^{m_2}}{m_2!}. \tag{C.11}$$

We have the following integral representations:

$$F_1(a,b_1,b_2;c;x_1,x_2) = \frac{\Gamma(c)}{\Gamma(a)\Gamma(c-a)} \int_0^1 dy\, y^{a-1}(1-y)^{c-a-1}(1-x_1 y)^{-b_1}(1-x_2 y)^{-b_2}$$

$$= \frac{\Gamma(c)}{\Gamma(b_1)\Gamma(b_2)\Gamma(c-b_1-b_2)} \int d^3y\, \delta(1-\sum_{j=1}^{3} y_j)\, y_1^{b_1-1} y_2^{b_2-1} y_3^{c-b_1-b_2-1} (1-x_1y_1-x_2y_2)^{-a},$$

$$F_2(a,b_1,b_2;c_1,c_2;x_1,x_2) = \frac{\Gamma(c_1)}{\Gamma(b_1)\Gamma(c_1-b_1)}\frac{\Gamma(c_2)}{\Gamma(b_2)\Gamma(c_2-b_2)}$$

$$\times \int_0^1 du \int_0^1 dv\, u^{b_1-1}(1-u)^{c_1-b_1-1} v^{b_2-1}(1-v)^{c_2-b_2-1}(1-ux_1-vx_2)^{-a},$$

$$F_3(a_1,a_2,b_1,b_2;c;x_1,x_2) = \frac{\Gamma(c)}{\Gamma(b_1)\Gamma(b_2)\Gamma(c-b_1-b_2)}$$

$$\times \int d^3y\, \delta(1-\sum_{j=1}^{3} y_j)\, y_1^{b_1-1} y_2^{b_2-1} y_3^{c-b_1-b_2-1} (1-x_1y_1)^{-a_1}(1-x_2y_2)^{-a_2},$$

$$F_4(a,b;c_1,c_2;x_1(1-x_2),x_2(1-x_1)) = \frac{\Gamma(c_1)}{\Gamma(a)\Gamma(c_1-a)}\frac{\Gamma(c_2)}{\Gamma(b)\Gamma(c_2-b)}$$

$$\times \int_0^1 du \int_0^1 dv\, u^{a-1}(1-u)^{c_1-a-1} v^{b-1}(1-v)^{c_2-b-1}(1-ux_1)^{a-c_1-c_2+1}(1-vx_2)^{b-c_1-c_2+1}$$

$$\cdot (1-ux_1-vx_2)^{c_1+c_2-a-b-1},$$

$$S_1(a_1,a_2,b_1;c,c_1;x_1,x_2) = \frac{\Gamma(c)}{\Gamma(a_1)\Gamma(c-a_1)}\frac{\Gamma(c_1)}{\Gamma(b_1)\Gamma(c_1-b_1)}$$

$$\int_0^1 du \int_0^1 dv\, u^{a_1-1}(1-u)^{c-a_1-1} v^{b_1-1}(1-v)^{c_1-b_1-1}(1-uvx_1-ux_2)^{-a_2}. \tag{C.12}$$

Note that in the equation above the arguments of the fourth Appell function are $x_1(1-x_2)$ and $x_2(1-x_1)$.

C.3 Lauricella Functions

The Lauricella functions are generalisations to n variables $x_1, \ldots, x_n$. The four Lauricella functions are defined by [445]

$$\begin{aligned}
&F_A(a, b_1, \ldots, b_n; c_1, \ldots, c_n; x_1, \ldots, x_n) = \\
&\quad \sum_{m_1=0}^{\infty} \cdots \sum_{m_n=0}^{\infty} \frac{(a)_{m_1+\cdots+m_n} (b_1)_{m_1} \ldots (b_n)_{m_n}}{(c_1)_{m_1} \ldots (c_n)_{m_n}} \frac{x_1^{m_1}}{m_1!} \cdots \frac{x_n^{m_n}}{m_n!}, \\
&F_B(a_1, \ldots, a_n, b_1, \ldots, b_n; c; x_1, \ldots, x_n) = \\
&\quad \sum_{m_1=0}^{\infty} \cdots \sum_{m_n=0}^{\infty} \frac{(a_1)_{m_1} \ldots (a_n)_{m_n} (b_1)_{m_1} \ldots (b_n)_{m_n}}{(c)_{m_1+\cdots+m_n}} \frac{x_1^{m_1}}{m_1!} \cdots \frac{x_n^{m_n}}{m_n!}, \\
&F_C(a, b; c_1, \ldots, c_n; x_1, \ldots, x_n) = \\
&\quad \sum_{m_1=0}^{\infty} \cdots \sum_{m_n=0}^{\infty} \frac{(a)_{m_1+\cdots+m_n} (b)_{m_1+\cdots+m_n}}{(c_1)_{m_1} \ldots (c_n)_{m_n}} \frac{x_1^{m_1}}{m_1!} \cdots \frac{x_n^{m_n}}{m_n!}, \\
&F_D(a, b_1, \ldots, b_n; c; x_1, \ldots, x_n) = \\
&\quad \sum_{m_1=0}^{\infty} \cdots \sum_{m_n=0}^{\infty} \frac{(a)_{m_1+\cdots+m_n} (b_1)_{m_1} \ldots (b_n)_{m_n}}{(c)_{m_1+\cdots+m_n}} \frac{x_1^{m_1}}{m_1!} \cdots \frac{x_n^{m_n}}{m_n!}.
\end{aligned} \tag{C.13}$$

C.4 Horn Functions

In his original publication, Jakob Horn extended the list of hypergeometric functions in two variables (examples are the four Appell functions and the Kampé de Fériet function from Sect. C.2) to 34 functions [446]. There is no point in listing them all here.

In the modern literature the name "Horn-type hypergeometric function" is used for a function of the following type: With the multi-index notation

$$\mathbf{x} = (x_1, \ldots, x_n), \quad \mathbf{i} = (i_1, \ldots, i_n), \quad \mathbf{x}^{\mathbf{i}} = x_1^{i_1} \cdot \ldots \cdot x_n^{i_n} \tag{C.14}$$

a Horn-type hypergeometric function is defined by

$$\begin{aligned}
H &= \sum_{\mathbf{i} \in \mathbb{N}_0^n} C_{\mathbf{i}} \mathbf{x}^{\mathbf{i}}, \\
C_{\mathbf{i}} &= \frac{\prod_{j=1}^{p} \Gamma\left(\sum_{k=1}^{n} A_{jk} i_k + u_j\right)}{\prod_{j=1}^{q} \Gamma\left(\sum_{k=1}^{n} B_{jk} i_k + v_j\right)},
\end{aligned} \tag{C.15}$$

with $A_{jk}, B_{jk} \in \mathbb{Z}$ and $u_j, v_j \in \mathbb{C}$. Let us denote

$$\mathbf{i} + \mathbf{e}_j = \left(i_1, \ldots, i_{j-1}, i_j + 1, i_{j+1}, \ldots, i_n\right). \tag{C.16}$$

It follows from $\Gamma(z+1) = z\Gamma(z)$ that the ratio $C_{\mathbf{i}+\mathbf{e}_j}/C_{\mathbf{i}}$ is a rational function in $i_1, \ldots, i_n$:

$$\frac{C_{\mathbf{i}+\mathbf{e}_j}}{C_{\mathbf{i}}} = \frac{P_j(\mathbf{i})}{Q_j(\mathbf{i})}, \tag{C.17}$$

with $P_j(\mathbf{i})$ and $Q_j(\mathbf{i})$ being polynomials in $(i_1, \ldots, i_n)$. Let us denote the Euler operators by

$$\boldsymbol{\theta} = (\theta_1, \ldots, \theta_n)^T = \left(x_1 \frac{\partial}{\partial x_1}, \ldots, x_n \frac{\partial}{\partial x_n}\right)^T. \tag{C.18}$$

The function $H(\mathbf{x})$ satisfies the differential equation

$$\left(1 + \theta_j\right)\left[Q_j(\theta)\frac{1}{x_j} - P_j(\boldsymbol{\theta})\right] H(\mathbf{x}) = 0, \quad j = 1, \ldots, n. \tag{C.19}$$

Exercise 128 *Prove Eq. (C.19).*

As $\sum_{k=1}^{n} A_{jk} i_k$ is an integer we may use the reflection identity Eq. (2.112) to write

$$\Gamma\left(\sum_{k=1}^{n} A_{jk} i_k + u_j\right) = (-1)^{\sum_{k=1}^{n} A_{jk} i_k} \frac{\Gamma(u_j)\Gamma(1-u_j)}{\Gamma\left(-\sum_{k=1}^{n} A_{jk} i_k + 1 - u_j\right)}. \tag{C.20}$$

This converts the series representation of the Horn functions of Eq. (C.15) into a form similar to the Γ-series defined in Eq. (9.99).

Note that the Horn functions define a rather large class of functions: If we consider the system of differential equations they satisfy, this system may not be holonomic [447]. This implies that the space of local solutions of the system of differential equations may be infinite dimensional.

The relation of Feynman integrals with Horn-type hypergeometric function is discussed in [448–450].

Appendix D
Lie Groups and Lie Algebras

In this appendix we give a short introduction to the theory of Lie groups and Lie algebras. Lie groups figure prominently as symmetry groups in particle physics, and we assume that most readers have already come across Lie groups and Lie algebras. The main point of this appendix is the classification of the simple Lie algebras.

Lie groups and Lie algebras are treated in many textbooks, examples are the books by Helgason [451], Fulton and Harris [452], Bourbaki [453] and Weyl [454].

D.1 Definitions

We start with the definitions:

Lie group:
A Lie group G is a group which is also an analytic manifold such that the mapping $(a, b) \rightarrow ab^{-1}$ of the product manifold $G \times G$ into G is analytic.

Lie algebra:
A Lie algebra $\mathfrak{g}$ over a field $\mathbb{F}$ is a vector space together with a bilinear mapping $[\cdot, \cdot] : \mathfrak{g} \times \mathfrak{g} \rightarrow \mathfrak{g}$, $(X, Y) \rightarrow [X, Y]$ such that for $X, Y, Z \in \mathfrak{g}$:

$$\begin{aligned} [X, X] &= 0, \\ [[X, Y], Z] + [[Y, Z], X] + [[Z, X], Y] &= 0. \end{aligned} \tag{D.1}$$

Exercise 129 *Show that* $[X, X] = 0$ *implies the anti-symmetry of the Lie bracket* $[X, Y] = -[Y, X]$. *Show further that also the converse is true, provided* char $\mathbb{F} \neq 2$. *Explain, why the argument does not work for* char $\mathbb{F} = 2$.

S. Weinzierl, *Feynman Integrals*, UNITEXT for Physics,
https://doi.org/10.1007/978-3-030-99558-4

We are mainly interested in the case where the ground field are the real numbers $\mathbb{R}$ or the complex numbers $\mathbb{C}$.

Let $\mathfrak{g}$ be a Lie algebra and $X^1, \dots, X^n$ a basis of $\mathfrak{g}$ as a vector space. $[X^a, X^b]$ is again in $\mathfrak{g}$ and can be expressed as a linear combination of the basis vectors X^k:

$$\left[X^a, X^b\right] = \sum_{c=1}^{n} c^{abc} X^c. \tag{D.2}$$

The coefficients c^{abc} are called the **structure constants** of the Lie algebra. For matrix algebras the X^a's are anti-hermitian matrices.

The notation above is mainly used in the mathematical literature. In physics a slightly different convention is often used (which corresponds for matrix algebras to having hermitian matrices as a basis): Denote by $T^1, \dots, T^n$ a basis of $\mathfrak{g}$ as a (complex) vector space and write

$$\left[T^a, T^b\right] = i \sum_{c=1}^{n} f^{abc} T^c. \tag{D.3}$$

We can get from one convention to the other one by letting

$$T^a = i X^a. \tag{D.4}$$

In this case we have

$$f^{abc} = c^{abc}. \tag{D.5}$$

The standard normalisation of the generators is

$$\mathrm{Tr}\left(T^a T^b\right) = \frac{1}{2}\delta^{ab}. \tag{D.6}$$

The relation between Lie groups and Lie algebras is as follows: Let G be a Lie group. It is therefore a manifold and a group. Let n be the dimension of G as a manifold. Choose a local coordinate system with coordinates $(\theta_1, \dots, \theta_n)$, such that the identity element e is given by

$$e = g(0, \dots, 0). \tag{D.7}$$

In a neighbourhood of e we may write

$$\begin{aligned} g(0, \dots, \theta_a, \dots, 0) &= g(0, \dots, 0, \dots, 0) + \theta_a X^a + O(\theta^2) \\ &= g(0, \dots, 0, \dots, 0) - i\theta_a T^a + O(\theta^2), \end{aligned} \tag{D.8}$$

where

$$X^a = \lim_{\theta_a \to 0} \frac{g(0, \dots, \theta_a, \dots, 0) - g(0, \dots, 0, \dots, 0)}{\theta_a},$$
$$T^a = i \lim_{\theta_a \to 0} \frac{g(0, \dots, \theta_a, \dots, 0) - g(0, \dots, 0, \dots, 0)}{\theta_a}. \quad \text{(D.9)}$$

The T^a's (and the X^a's) are called the **generators** of the Lie group G.

Theorem 16 *The generators T^a of a Lie group are a basis of a Lie algebra $\mathfrak{g}$, in particular the commutators of the generators are linear combinations of the generators:*

$$\left[T^a, T^b\right] = i \sum_{c=1}^{n} f^{abc} T^c. \quad \text{(D.10)}$$

$\mathfrak{g}$ is called the Lie algebra of the Lie group G.

We will often use Einstein's summation convention and write Eq. (D.10) as

$$\left[T^a, T^b\right] = i f^{abc} T^c. \quad \text{(D.11)}$$

We have seen that given a Lie group G we obtain its Lie algebra from Eq. (D.9). We may now ask if the converse is also possible: Given a Lie algebra $\mathfrak{g}$, can we reconstruct the Lie group G? The answer is that this can almost be done. Note that a Lie group need not be connected. Given a Lie algebra we have information about the connected component in which the identity lies. The exponential map takes us from the Lie algebra into the group, more precisely into the connected component in which the identity lies. In a neighbourhood of the identity we have

$$g(\theta_1, \dots, \theta_n) = \exp\left(-i \sum_{a=1}^{n} \theta_a T^a\right). \quad \text{(D.12)}$$

A few examples of Lie groups are:

1. $\mathrm{SU}(n, \mathbb{C})$: The group of special unitary $(n \times n)$-matrices defined through

$$U U^\dagger = \mathbf{1} \text{ and } \det(U) = 1. \quad \text{(D.13)}$$

   The group $\mathrm{SU}(n, \mathbb{C})$ has $n^2 - 1$ real parameters.
2. $\mathrm{SO}(n, \mathbb{R})$: The group of special orthogonal $(n \times n)$-matrices defined through

$$R R^T = \mathbf{1} \text{ and } \det(R) = 1. \quad \text{(D.14)}$$

   The group $\mathrm{SO}(n, \mathbb{R})$ has $n(n-1)/2$ real parameters.
3. $\mathrm{Sp}(n, \mathbb{R})$: The symplectic group is the group of $(2n \times 2n)$-matrices satisfying

$$M^T \begin{pmatrix} \mathbf{0} & \mathbf{1} \\ -\mathbf{1} & \mathbf{0} \end{pmatrix} M = \begin{pmatrix} \mathbf{0} & \mathbf{1} \\ -\mathbf{1} & \mathbf{0} \end{pmatrix}, \tag{D.15}$$

where $\mathbf{0}$ denotes the $(n \times n)$-zero matrix and $\mathbf{1}$ denotes the $(n \times n)$-identity matrix. The group $\mathrm{Sp}(n, \mathbb{R})$ has $(2n+1)n$ real parameters. The group $\mathrm{Sp}(n, \mathbb{R})$ can also be defined as the transformation group of a real $(2n)$-dimensional vector space with coordinates $(x_1, \dots, x_n, x_{n+1}, \dots, x_{2n})$, which preserves the inner product

$$\sum_{j=1}^{n} \left(x_j y_{j+n} - x_{j+n} y_j \right). \tag{D.16}$$

The corresponding Lie algebras are denotes $\mathfrak{su}(n)$, $\mathfrak{so}(n)$ and $\mathfrak{sp}(n)$, respectively.

Let us now focus on Lie algebras. Let $\mathfrak{g}$ be the Lie algebra of a Lie group G.

Definition 6 The **rank of a Lie algebra** is the number of simultaneously diagonalisable generators.

Definition 7 A **Casimir operator** is an operator, which commutes with all the generators of the group.

Theorem 17 *The number of independent Casimir operators is equal to the rank of the Lie algebra.*

Definition 8 A Lie algebra is called **simple** if it is non-Abelian and has no non-trivial ideals. ($\mathfrak{g}$ and $\{0\}$ are the two trivial ideals every Lie algebra has.)

Definition 9 A Lie algebra is called **semi-simple** if it has no non-trivial Abelian ideals.

Theorem 18 *A Lie algebra $\mathfrak{g}$ is semi-simple if and only if*

$$\det(g) \neq 0, \tag{D.17}$$

where

$$g^{ab} = f^{acd} f^{bcd}. \tag{D.18}$$

Definition 10 A Lie algebra is called **reductive** if it is the sum of a semi-simple and an Abelian Lie algebra.

A simple Lie algebra is also semi-simple and a semi-simple Lie algebra is also reductive. Let us look at a few examples: The Lie algebras

$$\mathfrak{su}(n), \quad \mathfrak{so}(n), \quad \mathfrak{sp}(n) \tag{D.19}$$

are simple. Semi-simple Lie algebras are sums of simple Lie algebras, for example

$$\mathfrak{su}(n_1) \oplus \mathfrak{su}(n_2). \tag{D.20}$$

Reductive Lie algebras may have in addition an Abelian part, for example

$$\mathfrak{u}(1) \oplus \mathfrak{su}(2) \oplus \mathfrak{su}(3). \tag{D.21}$$

The Abelian Lie algebras are rather trivial, they only have one-dimensional irreducible representations. Therefore the classification of all reductive Lie algebras essentially boils down to the classification of all simple Lie algebras.

D.2 The Cartan Basis and Root Systems

We now take the complex numbers $\mathbb{C}$ as the ground field. Consider $A, X \in \mathfrak{g}$ with $X \neq 0$ and assume that

$$[A, X] = \rho X, \qquad \rho \in \mathbb{C}. \tag{D.22}$$

ρ is called a **root** of the Lie algebra $\mathfrak{g}$. We write

$$A = \sum_{a=1}^{n} c_a T^a, \quad X = \sum_{a=1}^{n} x_a T^a. \tag{D.23}$$

For a non-trivial solution $X \neq 0$ of Eq. (D.22) we must have

$$\det\left(c_a i f^{abc} - \rho \delta^{bc}\right) = 0. \tag{D.24}$$

Equation (D.24) is called the **secular equation**. In general the secular equation will give a n-th order polynomial in ρ. Solving for ρ one obtains n roots. One root may occur more than once. The degree of degeneracy is called the **multiplicity of the root**.

Exercise 130 *Derive Eq. (D.24) from Eq. (D.22).*

Theorem 19 *If A is chosen such that the secular equation has the maximum number of distinct roots, then only the root $\rho = 0$ is degenerate. Further if r is the multiplicity of that root, there exist r linearly independent generators H_i, which mutually commute*

$$\left[H_i, H_j\right] = 0, \qquad i, j \in \{1, \dots, r\}. \tag{D.25}$$

The multiplicity r of the root $\rho = 0$ equals the rank of the Lie algebra.

The generators $H_1, \dots, H_r$ generate an Abelian sub-algebra of $\mathfrak{g}$. This sub-algebra is called the **Cartan sub-algebra** of $\mathfrak{g}$.

It is a standard convention to use Latin indices $i \in \{1, \ldots, r\}$ to denote the r mutually commuting generators H_i and greek indices and the letter E to denote the remaining $(n-r)$ generators E_α.

Theorem 20 *For any semi-simple Lie algebra, non-zero roots occur in pairs of opposite sign and are denoted E_α and $E_{-\alpha}$ with $\alpha \in \{1, \ldots, \frac{1}{2}(n-r)\}$.*

We thus have the **Cartan standard form** or **Cartan basis**:

$$\begin{aligned} \left[H_i, H_j\right] &= 0, \\ \left[H_i, E_\alpha\right] &= \rho\,(\alpha, i)\, E_\alpha. \end{aligned} \tag{D.26}$$

We write $\alpha_i = \rho(\alpha, i)$. With this notation the last equation may be written as

$$\left[H_i, E_\alpha\right] = \alpha_i E_\alpha. \tag{D.27}$$

The standard normalisation for the Cartan basis is

$$\sum_{\alpha=1}^{\frac{1}{2}(n-r)} \alpha_i \alpha_j = \delta_{ij}. \tag{D.28}$$

For a fixed generator E_α we collect the r numbers $\alpha_1, \ldots, \alpha_r$ into one vector

$$\vec{\alpha} = \left(\alpha_1, \ldots, \alpha_r\right)^T. \tag{D.29}$$

The vector $\vec{\alpha}$ is called the **root vector** of E_α. The set of root vectors $\vec{\alpha}$ for all E_α is called the **root system** of $\mathfrak{g}$.

Exercise 131 *Consider the Lie algebra $\mathfrak{su}(2)$: Start from the generators*

$$I^1 = \frac{1}{2}\begin{pmatrix} 0 & 1 \\ 1 & 0 \end{pmatrix}, \quad I^2 = \frac{1}{2}\begin{pmatrix} 0 & -i \\ i & 0 \end{pmatrix}, \quad I^3 = \frac{1}{2}\begin{pmatrix} 1 & 0 \\ 0 & -1 \end{pmatrix}. \tag{D.30}$$

These generators are proportional to the Pauli matrices and normalised as

$$\mathrm{Tr}\left(I^a I^b\right) = \frac{1}{2}\delta^{ab}. \tag{D.31}$$

The commutators are given by

$$\left[I^a, I^b\right] = i\varepsilon^{abc} I^c, \tag{D.32}$$

where ε^{abc} denotes the totally antisymmetric tensor. Start from $A = I^3$. Determine for this choice the roots, the Cartan standard form and the root vectors.

Let us look at an example: The Cartan basis for $\mathfrak{su}(3)$ is

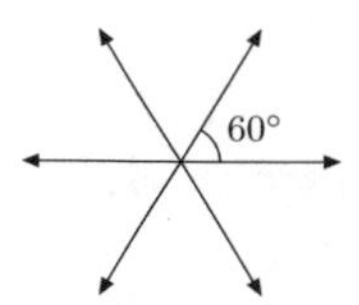

Fig. D.1 The root system of $\mathfrak{su}(3)$ (or equivalently of A_2)

$$
\begin{aligned}
H_1 &= \frac{1}{\sqrt{6}}\begin{pmatrix} 1 & 0 & 0 \\ 0 & -1 & 0 \\ 0 & 0 & 0 \end{pmatrix}, \quad H_2 = \frac{1}{3\sqrt{2}}\begin{pmatrix} 1 & 0 & 0 \\ 0 & 1 & 0 \\ 0 & 0 & -2 \end{pmatrix}, \\
E_1 &= \frac{1}{\sqrt{3}}\begin{pmatrix} 0 & 1 & 0 \\ 0 & 0 & 0 \\ 0 & 0 & 0 \end{pmatrix}, \quad E_2 = \frac{1}{\sqrt{3}}\begin{pmatrix} 0 & 0 & 1 \\ 0 & 0 & 0 \\ 0 & 0 & 0 \end{pmatrix}, \quad E_3 = \frac{1}{\sqrt{3}}\begin{pmatrix} 0 & 0 & 0 \\ 0 & 0 & 1 \\ 0 & 0 & 0 \end{pmatrix}, \\
E_{-1} &= \frac{1}{\sqrt{3}}\begin{pmatrix} 0 & 0 & 0 \\ 1 & 0 & 0 \\ 0 & 0 & 0 \end{pmatrix}, \quad E_{-2} = \frac{1}{\sqrt{3}}\begin{pmatrix} 0 & 0 & 0 \\ 0 & 0 & 0 \\ 1 & 0 & 0 \end{pmatrix}, \quad E_{-3} = \frac{1}{\sqrt{3}}\begin{pmatrix} 0 & 0 & 0 \\ 0 & 0 & 0 \\ 0 & 1 & 0 \end{pmatrix}.
\end{aligned}
\tag{D.33}
$$

The roots of E_1, E_2 and E_3 are

$$
\begin{aligned}
[H_1, E_1] &= \frac{1}{3}\sqrt{6}E_1, & [H_2, E_1] &= 0, \\
[H_1, E_2] &= \frac{1}{6}\sqrt{6}E_2, & [H_2, E_2] &= \tfrac{1}{2}\sqrt{2}E_2 \\
[H_1, E_3] &= -\frac{1}{6}\sqrt{6}E_3, & [H_2, E_3] &= \tfrac{1}{2}\sqrt{2}E_3,
\end{aligned}
\tag{D.34}
$$

and similar for E_{-1}, E_{-2} and E_{-3}. For the root vectors we obtain

$$
\begin{aligned}
\vec{\alpha}(E_1) &= \begin{pmatrix} \frac{1}{3}\sqrt{6} \\ 0 \end{pmatrix}, & \vec{\alpha}(E_2) &= \begin{pmatrix} \frac{1}{6}\sqrt{6} \\ \frac{1}{2}\sqrt{2} \end{pmatrix}, & \vec{\alpha}(E_3) &= \begin{pmatrix} -\frac{1}{6}\sqrt{6} \\ \frac{1}{2}\sqrt{2} \end{pmatrix}, \\
\vec{\alpha}(E_{-1}) &= \begin{pmatrix} -\frac{1}{3}\sqrt{6} \\ 0 \end{pmatrix}, & \vec{\alpha}(E_{-2}) &= \begin{pmatrix} -\frac{1}{6}\sqrt{6} \\ -\frac{1}{2}\sqrt{2} \end{pmatrix}, & \vec{\alpha}(E_{-3}) &= \begin{pmatrix} \frac{1}{6}\sqrt{6} \\ -\frac{1}{2}\sqrt{2} \end{pmatrix}.
\end{aligned}
\tag{D.35}
$$

Figure D.1 shows the root system for $\mathfrak{su}(3)$.

There are a few theorems on root vectors:

Theorem 21 *If $\vec{\alpha}$ is a root vector, so is $-\vec{\alpha}$, (since roots always occur in pairs of opposite sign).*

Theorem 22 *If $\vec{\alpha}$ and $\vec{\beta}$ are root vectors then*

$$
\frac{2\vec{\alpha}\cdot\vec{\beta}}{|\vec{\alpha}|^2} \text{ and } \frac{2\vec{\alpha}\cdot\vec{\beta}}{|\vec{\beta}|^2}
\tag{D.36}
$$

are integers.

Theorem 23 *If $\vec{\alpha}$ and $\vec{\beta}$ are root vectors so is*

$$\vec{\gamma} = \vec{\beta} - \frac{2\vec{\alpha} \cdot \vec{\beta}}{|\vec{\alpha}|^2}\vec{\alpha} \tag{D.37}$$

Let us now investigate the implications of these theorems. We start with Theorem 22. Denote the two integers by p and q, i.e.

$$\frac{2\vec{\alpha} \cdot \vec{\beta}}{|\vec{\alpha}|^2} = p, \quad \frac{2\vec{\alpha} \cdot \vec{\beta}}{|\vec{\beta}|^2} = q. \tag{D.38}$$

Then

$$\frac{\left(\vec{\alpha} \cdot \vec{\beta}\right)^2}{|\vec{\alpha}|^2|\vec{\beta}|^2} = \frac{pq}{4} = \cos^2\theta \leq 1, \tag{D.39}$$

where θ denotes the angle between the root vectors $\vec{\alpha}$ and $\vec{\beta}$. Therefore

$$pq \leq 4. \tag{D.40}$$

As p and q are integers, this puts strong constraints on the angle between $\vec{\alpha}$ and $\vec{\beta}$ and the ratio of their lengths. We have

$$\cos^2\theta \in \left\{0, \frac{1}{4}, \frac{1}{2}, \frac{3}{4}, 1\right\}. \tag{D.41}$$

This restricts the angle between two root vectors to

$$0°, 30°, 45°, 60°, 90°, 120°, 135°, 150°, 180°. \tag{D.42}$$

This is a finite list. Let's go through all possibilities:

- Case $\cos^2\theta = 1$: This implies $\cos\theta = \pm 1$ and the angle is either $\theta = 0°$ or $\theta = 180°$. We further have $|\vec{\alpha}| = |\vec{\beta}|$ and therefore either $\vec{\alpha} = \vec{\beta}$ or $\vec{\alpha} = -\vec{\beta}$.
- Case $\cos^2\theta = \frac{3}{4}$: This implies $\cos\theta = \pm\frac{1}{2}\sqrt{3}$ and the angle is either $\theta = 30°$ or $\theta = 150°$. We have $pq = 3$ and therefore either $p = 1, q = 3$ or $p = 3, q = 1$. Let us first discuss $p = 1, q = 3$. This means

$$\frac{2\vec{\alpha} \cdot \vec{\beta}}{|\vec{\alpha}|^2} = 1, \quad \frac{2\vec{\alpha} \cdot \vec{\beta}}{|\vec{\beta}|^2} = 3. \tag{D.43}$$

Therefore

$$\frac{|\vec{\alpha}|^2}{|\vec{\beta}|^2} = 3. \tag{D.44}$$

The case $p = 3, q = 1$ is similar and in summary we obtain

$$\frac{|\vec{\alpha}|^2}{|\vec{\beta}|^2} \in \left\{ \frac{1}{3}, 3 \right\}. \tag{D.45}$$

- Case $\cos^2\theta = \frac{1}{2}$: This implies $\cos\theta = \pm\frac{1}{2}\sqrt{2}$ and the angle is either $\theta = 45°$ or $\theta = 135°$. We have $pq = 2$ and either $p = 1, q = 2$ or $p = 2, q = 1$. It follows

$$\frac{|\vec{\alpha}|^2}{|\vec{\beta}|^2} \in \left\{ \frac{1}{2}, 2 \right\}. \tag{D.46}$$

- Case $\cos^2\theta = \frac{1}{4}$: This implies $\cos\theta = \pm\frac{1}{2}$ and the angle is either $\theta = 60°$ or $\theta = 120°$. We have $pq = 1$ and hence $p = 1, q = 1$. It follows

$$\frac{|\vec{\alpha}|^2}{|\vec{\beta}|^2} = 1. \tag{D.47}$$

- Case $\cos^2\theta = 0$: This implies $\cos\theta = 0$ and the angle is $\theta = 90°$. In this case we have $p = 0$ and $q = 0$. This leaves the ratio $|\vec{\alpha}|^2/|\vec{\beta}|^2$ undetermined.

Let us now explore Theorem 23. We have already seen that if $\vec{\alpha}$ and $\vec{\beta}$ are root vectors so is

$$\vec{\gamma} = \vec{\beta} - \frac{2\vec{\alpha}\cdot\vec{\beta}}{|\vec{\alpha}|^2}\vec{\alpha}. \tag{D.48}$$

Let us now put this a little bit more formally. For any root vector α we define a mapping W_α from the set of root vectors to the set of root vectors by

$$W_\alpha(\beta) = \vec{\beta} - \frac{2\vec{\alpha}\cdot\vec{\beta}}{|\vec{\alpha}|^2}\vec{\alpha}. \tag{D.49}$$

W_α can be described as the reflection by the plane Ω_α perpendicular to α. It is clear that this mapping is an involution: After two reflections one obtains the original root vector again. The set of all these mappings W_α generates a group, which is called the **Weyl group**.

Since W_α maps a root vector to another root vector, we have the following corollary:

Corollary 24 *The set of root vectors is invariant under the Weyl group.*

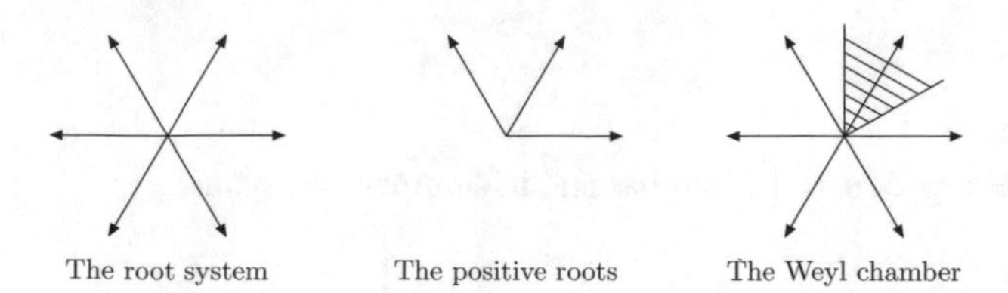

Fig. D.2 The root system, the positive roots and the Weyl chamber for $\mathfrak{su}(3)$ (or equivalently A_2)

For root vectors we define an **ordering** as follows: $\vec{\alpha}$ is said to be higher than $\vec{\alpha}'$ if the rth component of $(\vec{\alpha} - \vec{\alpha}')$ is positive (if zero look at the $(r-1)$th component etc.). If $\vec{\alpha}$ is higher than $\vec{\alpha}'$ we write $\vec{\alpha} > \vec{\alpha}'$.

Definition 11 A root vector $\vec{\alpha}$ is called **positive**, if $\vec{\alpha} > \vec{0}$.

Therefore the set of non-zero root vectors R decomposes into

$$R = R^+ \cup R^-, \tag{D.50}$$

where R^+ denotes the positive roots and R^- denotes the negative roots.

Definition 12 The (closed) **Weyl chamber** relative to a given ordering is the set of points $\vec{x}$ in the r-dimensional space of root vectors, such that

$$2\frac{\vec{x} \cdot \vec{\alpha}}{|\vec{\alpha}|^2} \geq 0 \quad \forall\, \vec{\alpha} \in R^+. \tag{D.51}$$

An example of a root system, the positive roots and the Weyl chamber is shown for the Lie algebra $\mathfrak{su}(3)$ in Fig. D.2.

Let us summarise: The root system R of a Lie algebra $\mathfrak{g}$ has the following properties:

1. R is a finite set.
2. If $\vec{\alpha} \in R$, then also $-\vec{\alpha} \in R$.
3. For any $\vec{\alpha} \in R$ the reflection W_α maps R to itself.
4. If $\vec{\alpha}$ and $\vec{\beta}$ are root vectors then $2\vec{\alpha} \cdot \vec{\beta}/|\alpha|^2$ is an integer.

This puts strong constraints on the geometry of a root system. Before we embark on the general classification, let us investigate the possible root systems of rank 1 and 2. For rank 1 the root vectors are one-dimensional and the only possibility is the one shown in Fig. D.3. This is the root system of $\mathfrak{su}(2)$.

For rank 2 we first note that due to property (3) the angle between two roots must be the same for any pair of adjacent roots. It will turn out that any of the four angles 90°, 60°, 45° and 30° can occur. Once this angle is specified, the relative lengths of the roots are fixed except for the case of right angles.

Let us start with the case $\theta = 90°$. Up to rescaling the root system is the one shown in Fig. D.4. This corresponds to $\mathfrak{su}(2) \oplus \mathfrak{su}(2)$. This Lie algebra is semi-simple, but not simple.

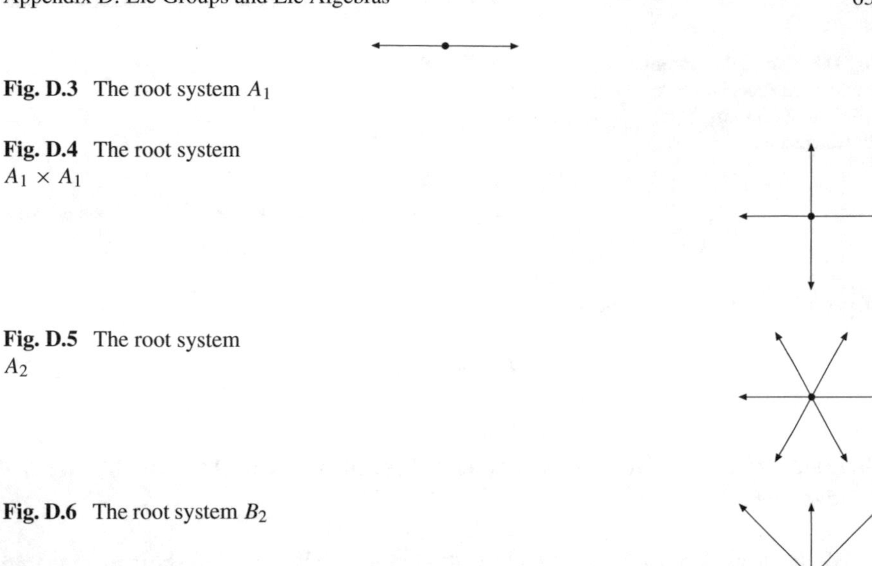

Fig. D.3 The root system A_1

Fig. D.4 The root system $A_1 \times A_1$

Fig. D.5 The root system A_2

Fig. D.6 The root system B_2

For the angle $\theta = 60°$ we have the root system shown in Fig. D.5. This is the root system of $\mathfrak{su}(3)$.

For the angle $\theta = 45°$ we have the root system shown in Fig. D.6. This is the root system of $\mathfrak{so}(5)$.

Finally, for $\theta = 30°$ we have the root system shown in Fig. D.7. This is the root system of the exceptional Lie group G_2.

D.3 Dynkin Diagrams

Let us try to reduce further the data of a root system. We already learned that with the help of an ordering we can divide the non-zero root vectors into a disjoint union

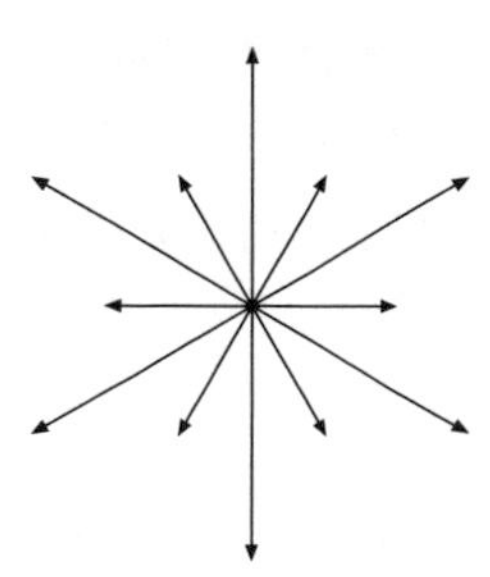

Fig. D.7 The root system G_2

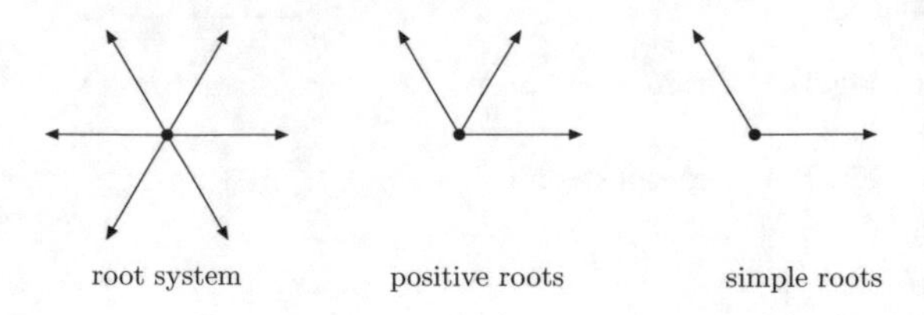

Fig. D.8 The root system, the positive roots and the simple roots for $\mathfrak{su}(3)$ (or equivalently A_2)

of positive and negative roots:

$$R = R^+ \cup R^-. \tag{D.52}$$

Definition 13 A positive root vector is called **simple** if it is not the sum of two other positive roots.

We illustrate this for the Lie algebra $\mathfrak{su}(3)$ in Fig. D.8. The angle between the two simple roots is $\theta = 120°$.

The **Dynkin diagram** of the root system is constructed by drawing one vertex ∘ for each simple root and joining two vertices by a number of lines depending on the angle θ between the two roots:

no lines ○ ○ if $\theta = 90°$

one line ○—○ if $\theta = 120°$

two lines ○⇒○ if $\theta = 135°$

three lines ○⇛○ if $\theta = 150°$

When there is one line, the roots have the same length. If two roots are connected by two or three lines, an arrow is drawn pointing from the longer to the shorter root. Example: The Dynkin diagram of $\mathfrak{su}(3)$ is

○—○

We have the following theorems:

Theorem 25 *Two complex semi-simple Lie algebras are isomorphic if and only if they have the same Dynkin diagram.*

Theorem 26 *A complex semi-simple Lie algebra is simple if and only if its Dynkin diagram is connected.*

Theorem 27 *Classification of simple Lie algebras: The connected Dynkin diagrams can be grouped into four families (A_n, B_n, C_n, D_n) and a set of five exceptional Dynkin diagrams. The four families are*

- *A_n for $n \geq 1$*

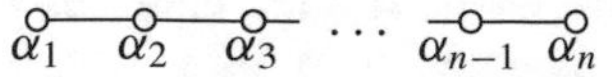

$$\alpha_1 \quad \alpha_2 \quad \alpha_3 \quad \cdots \quad \alpha_{n-1} \quad \alpha_n$$

- *B_n for $n \geq 2$*

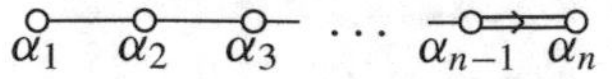

$$\alpha_1 \quad \alpha_2 \quad \alpha_3 \quad \cdots \quad \alpha_{n-1} \quad \alpha_n$$

- *C_n for $n \geq 3$*

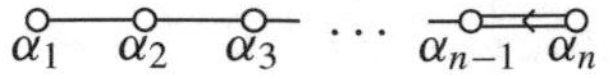

$$\alpha_1 \quad \alpha_2 \quad \alpha_3 \quad \cdots \quad \alpha_{n-1} \quad \alpha_n$$

- *D_n for $n \geq 4$*

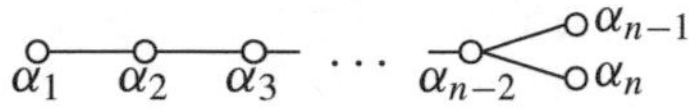

$$\alpha_1 \quad \alpha_2 \quad \alpha_3 \quad \cdots \quad \alpha_{n-2} \quad \alpha_{n-1} \quad \alpha_n$$

The exceptional Dynkin diagrams are

- E_6

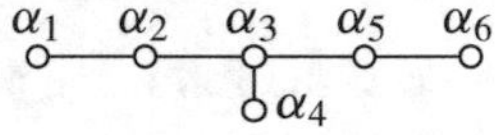

$$\alpha_1 \quad \alpha_2 \quad \alpha_3 \quad \alpha_5 \quad \alpha_6 \quad \alpha_4$$

- E_7

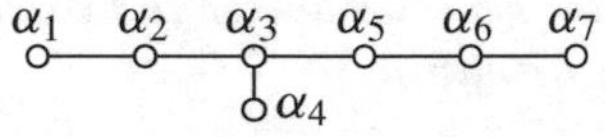

$$\alpha_1 \quad \alpha_2 \quad \alpha_3 \quad \alpha_5 \quad \alpha_6 \quad \alpha_7 \quad \alpha_4$$

- E_8

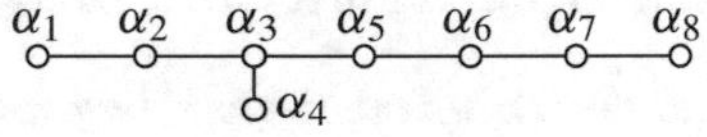

$$\alpha_1 \quad \alpha_2 \quad \alpha_3 \quad \alpha_5 \quad \alpha_6 \quad \alpha_7 \quad \alpha_8 \quad \alpha_4$$

- F_4

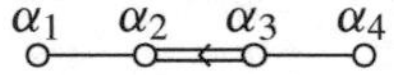

$$\alpha_1 \quad \alpha_2 \quad \alpha_3 \quad \alpha_4$$

- G_2

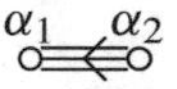

$$\alpha_1 \quad \alpha_2$$

Up to now we considered Lie algebras over the complex numbers $\mathbb{C}$. We are also interested in real Lie groups, whose Lie algebra is a real Lie algebra. Starting from a real simple Lie algebra $\mathfrak{g}$ we consider the complexification $\mathfrak{g}^{\mathbb{C}}$ of $\mathfrak{g}$. The latter is classified by Theorem 27. Thus a classification of all real simple Lie algebras amounts to a classification of all real forms of the complex simple Lie algebras $\mathfrak{g}^{\mathbb{C}}$ from Theorem 27.

The types of the Lie algebras of the classical real compact simple Lie groups are

$$\begin{aligned} \mathrm{SU}\,(n+1) &: A_n, \\ \mathrm{SO}\,(2n+1) &: B_n, \\ \mathrm{Sp}\,(n) &: C_n, \\ \mathrm{SO}\,(2n) &: D_n. \end{aligned} \tag{D.53}$$

In order to prove Theorem 27 we have to show that the only possible connected Dynkin diagrams are the ones mentioned in Theorem 27 and that for every connected Dynkin diagrams from the list of Theorem 27 there is a simple Lie algebra corresponding to it.

For the first part of the proof it is sufficient to consider only the angles between the simple roots, the relative lengths do not enter the proof. We may therefore drop the arrows in the Dynkin diagrams. Such diagrams, without the arrows to indicate the relative lengths, are called **Coxeter diagrams**. Let us now consider a Coxeter diagram with n vertices. Two vertices are connected by either 0, 1, 2 or 3 lines. We call a Coxeter diagram **admissible** if there are n independent unit vectors $\vec{e}_1, \dots, \vec{e}_n$ in an Euclidean space with the angle θ between $\vec{e}_i$ and $\vec{e}_j$ as follows:

no lines	○ ○	if $\theta = 90°$
one line	○—○	if $\theta = 120°$
two lines	○═○	if $\theta = 135°$
three lines	○≡○	if $\theta = 150°$

We prove Theorem 27 with the help of the following proposition:

Proposition 28 *The only connected admissible Coxeter graphs are the ones listed in Theorem 27 (without the arrows).*

To prove this proposition, we will first prove the following four lemmata:

Lemma 29 *Any sub-diagram of an admissible diagram, obtained by removing some vertices and all lines attached to them, will also be admissible.*

Proof Suppose we have an admissible diagram with n vertices. By definition there are n vectors $\vec{e}_j$, such that the angle between a pair of vectors is in the set

$$\{90°, 120°, 135°, 150°\} \tag{D.54}$$

Removing some of the vectors $\vec{e}_j$ does not change the angles between the remaining ones. Therefore any sub-diagram of an admissible diagram is again admissible.

Lemma 30 *There are at most* $(n-1)$ *pairs of vertices that are connected by lines. The diagram has no loops.*

Proof We have

$$2\vec{e}_i \cdot \vec{e}_j \in \{0, -1, -\sqrt{2}, -\sqrt{3}\} \tag{D.55}$$

Therefore if $\vec{e}_i$ and $\vec{e}_j$ are connected we have $\theta > 90°$ and

$$2\vec{e}_i \cdot \vec{e}_j \leq -1. \tag{D.56}$$

Now

$$0 < \left(\sum_i \vec{e}_i\right) \cdot \left(\sum_j \vec{e}_i\right) = n + 2\sum_{i<j} \vec{e}_i \cdot \vec{e}_j < n - \# \text{ connected pairs.} \tag{D.57}$$

Therefore

$$\# \text{ connected pairs} < n. \tag{D.58}$$

Connecting n vertices with $(n-1)$ connections (of either 1, 2 or 3 lines) implies that there are no loops.

Lemma 31 *No vertex has more than three lines attached to it.*

Proof We first note that

$$\left(2\vec{e}_i \cdot \vec{e}_j\right)^2 = \# \text{ number of lines between } \vec{e}_i \text{ and } \vec{e}_j. \tag{D.59}$$

Consider the vertex $\vec{e}_1$ and let $\vec{e}_2, \ldots \vec{e}_j$, be the vertices connected to $\vec{e}_1$. We want to show

$$\sum_{i=2}^{j} (2\vec{e}_1 \cdot \vec{e}_i)^2 < 4. \tag{D.60}$$

Since there are no loops, no pair of $\vec{e}_2, \ldots, \vec{e}_j$ is connected. Therefore $\vec{e}_2$, ..., $\vec{e}_j$ are perpendicular unit vectors. Further, by assumption $\vec{e}_1, \vec{e}_2, \ldots, \vec{e}_j$ are linearly independent vectors. Therefore $\vec{e}_1$ is not in the span of $\vec{e}_2, \ldots, \vec{e}_j$. It follows

$$1 = (\vec{e}_1 \cdot \vec{e}_1)^2 > \sum_{i=2}^{j} (\vec{e}_1 \cdot \vec{e}_i)^2 \tag{D.61}$$

and therefore

$$\sum_{i=2}^{j} (\vec{e}_1 \cdot \vec{e}_i)^2 < 1 \text{ and } \sum_{i=2}^{j} (2\vec{e}_1 \cdot \vec{e}_i)^2 < 4. \tag{D.62}$$

□

Lemma 32 *In an admissible diagram, any chain of vertices connected to each other by one line, with none but the ends of the chain connected to any other vertices, can be collapsed to one vertex, and the resulting diagram remains admissible.*

Proof Let us consider a chain of r vertices:

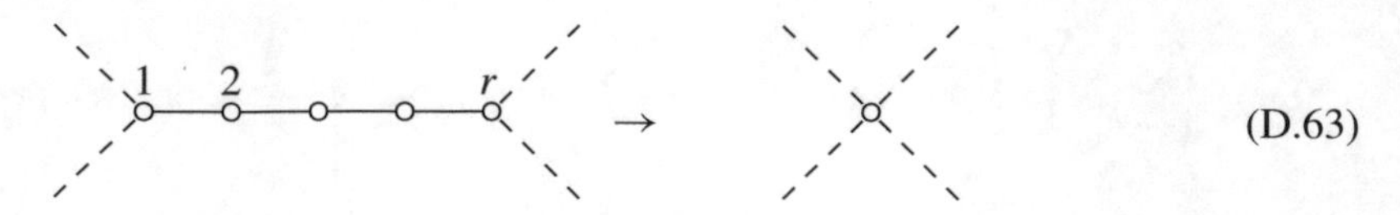

(D.63)

If $\vec{e}_1$, ..., $\vec{e}_r$ are the unit vectors corresponding to the chain of vertices as indicated above, then

$$\vec{e}' = \vec{e}_1 + \cdots + \vec{e}_r \tag{D.64}$$

is a unit vector since

$$\begin{aligned} \vec{e}' \cdot \vec{e}' &= (\vec{e}_1 + \cdots + \vec{e}_r)^2 = r + 2\vec{e}_1 \cdot \vec{e}_2 + 2\vec{e}_2 \cdot \vec{e}_3 + \cdots + 2\vec{e}_{r-1} \cdot \vec{e}_r \\ &= r - (r-1) = 1. \end{aligned} \tag{D.65}$$

Furthermore, $\vec{e}'$ satisfies the same conditions with respect to the other vectors since $\vec{e}' \cdot \vec{e}_j$ is either $\vec{e}_1 \cdot \vec{e}_j$ or $\vec{e}_r \cdot \vec{e}_j$. □

With the help of these lemmata we can now prove Proposition 28:

Proof From Lemma 31 it follows that the only connected diagram with a triple line is G_2.

Furthermore we cannot have a diagram with two double lines, otherwise we would have a sub-diagram, which we could contract as

(D.66)

contradicting again Lemma 31. By the same reasoning we cannot have a diagram with a double line and a vertex with three single lines attached to it:

(D.67)

Again this contradicts Lemma 31.
To finish the case with double lines, we rule out the diagram

1 2 3 4 5

(D.68)

Consider the vectors

$$\vec{v} = \vec{e}_1 + 2\vec{e}_2, \qquad \vec{w} = 3\vec{e}_3 + 2\vec{e}_4 + \vec{e}_5. \tag{D.69}$$

We find

$$(\vec{v} \cdot \vec{w})^2 = 18, \quad |\vec{v}|^2 = 3, \quad |\vec{w}|^2 = 6. \tag{D.70}$$

This violates the Cauchy-Schwarz inequality

$$(\vec{v} \cdot \vec{w})^2 < |\vec{v}|^2 \cdot |\vec{w}|^2 . \tag{D.71}$$

By a similar reasoning one rules out the following (sub-) graphs with single lines:

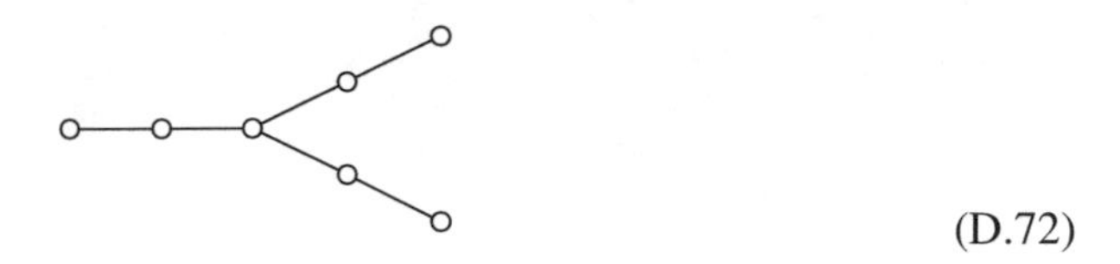

(D.72)

(D.73)

(D.74)

These sub-diagrams rules out all graphs not in the list of Theorem 27. To finish the proof of the proposition it remains to show that all graphs in the list are admissible. This is equivalent to show that for each Dynkin diagram in the list there exists a corresponding Lie algebra. (The simple root vectors of such a Lie algebra will then have automatically the corresponding angles of the Coxeter diagram.)

To prove the existence it is sufficient to give for each Dynkin diagram an example of a Lie algebra corresponding to it. For the four families A_n, B_n, C_n and D_n we have already seen that they correspond to the Lie algebras of $\mathfrak{su}(n+1)$, $\mathfrak{so}(2n+1)$, $\mathfrak{sp}(n)$ and $\mathfrak{so}(2n)$ In addition one can write down explicit matrix representations for the Lie algebras corresponding to the five exceptional groups E_6, E_7, E_8, F_4 and G_2. $\square$

Appendix E
Dirichlet Characters

E.1 Definition

Let N be a positive integer. We denote by $\mathbb{Z}_N^*$ the set of invertible elements in $\mathbb{Z}_N$. These are all elements $a \in \mathbb{Z}_N$ with $\gcd(a, N) = 1$. The set $\mathbb{Z}_N^*$ is an Abelian group with respect to multiplication. We further denote $\mathbb{C}^* = \mathbb{C}\backslash\{0\}$. A **Dirichlet character** modulo N is a function

$$\chi : \mathbb{Z}_N^* \to \mathbb{C}^* \tag{E.1}$$

that is a homomorphism of groups, i.e.

$$\chi(nm) = \chi(n)\chi(m) \qquad \text{for all } \; n, m \in \mathbb{Z}_N^*. \tag{E.2}$$

We may extend χ to a function $\chi : \mathbb{Z}_N \to \mathbb{C}$ by setting $\chi(n) = 0$ if $\gcd(n, N) > 1$ and then further extend to a function $\chi : \mathbb{Z} \to \mathbb{C}$ by setting $\chi(n) = \chi(n \mod N)$. By abuse of notation we denote both extensions again by χ. This function satisfies

$$\begin{array}{lll} (i) & \chi(n) = \chi(n+N) & \forall\, n \in \mathbb{Z}, \\ (ii) & \chi(n) = 0 & \text{if } \; \gcd(n, N) > 1, \\ & \chi(n) \neq 0 & \text{if } \; \gcd(n, N) = 1, \\ (iii) & \chi(nm) = \chi(n)\chi(m) & \forall\, n, m \in \mathbb{Z}. \end{array} \tag{E.3}$$

Property (ii) and (iii) imply for any Dirichlet character χ:

$$\chi(1) = 1. \tag{E.4}$$

Let's look at a few examples of Dirichlet characters with modulus $N \in \{1, 2, 3, 4, 5, 6\}$.

Modulus 1: For $N = 1$ there is only the trivial character

S. Weinzierl, *Feynman Integrals*, UNITEXT for Physics,
https://doi.org/10.1007/978-3-030-99558-4

n	0
$\chi_{1,1}(n)$	1

Modulus 2: There is one character modulo 2

n	0	1
$\chi_{2,1}(n)$	0	1

Modulus 3: There are two characters modulo 3

n	0	1	2
$\chi_{3,1}(n)$	0	1	1
$\chi_{3,2}(n)$	0	1	-1

Modulus 4: There are two characters modulo 4

n	0	1	2	3
$\chi_{4,1}(n)$	0	1	0	1
$\chi_{4,2}(n)$	0	1	0	-1

Modulus 5: There are four characters modulo 5

n	0	1	2	3	4
$\chi_{5,1}(n)$	0	1	1	1	1
$\chi_{5,2}(n)$	0	1	i	$-i$	-1
$\chi_{5,3}(n)$	0	1	-1	-1	1
$\chi_{5,4}(n)$	0	1	$-i$	i	-1

Modulus 6: There are two characters modulo 6

n	0	1	2	3	4	5
$\chi_{6,1}(n)$	0	1	0	0	0	1
$\chi_{6,2}(n)$	0	1	0	0	0	-1

We denote by $\chi_{N,1}$ the trivial character modulo N (with $\chi_{N,1}(n) = 1$ if $\gcd(n, N) = 1$ and $\chi_{N,1}(n) = 0$ otherwise). If no confusion with the modulus arises, we simply write χ_1 instead of $\chi_{N,1}$. The trivial character modulo 1 is denoted by $1 = \chi_{1,1}$ (we have $\chi_{1,1}(n) = 1$ for all n, hence the notation).

The **conductor** of χ is the smallest positive divisor $d|N$ such that there is a character χ' modulo d with

$$\chi(n) = \chi'(n) \qquad \forall\, n \in \mathbb{Z} \text{ with } \gcd(n, N) = 1. \tag{E.5}$$

A Dirichlet character is called **primitive**, if its modulus equals its conductor.

To give an example, the Dirichlet character $\chi_{4,1}$ of modulus 4 from the examples above has conductor 1, since

$$\chi_{4,1}(1) \;=\; \chi_{1,1}(1), \qquad \chi_{4,1}(3) \;=\; \chi_{1,1}(3). \tag{E.6}$$

On the other hand, $\chi_{4,2}$ has conductor 4 and is therefore a primitive Dirichlet character.

If χ is a Dirichlet character modulo N and M a positive integer, χ induces a Dirichlet character $\tilde{\chi}$ with modulus $(M \cdot N)$ by setting

$$\tilde{\chi}(n) = \begin{cases} \chi(n), & \text{if } \gcd(n, M \cdot N) = 1, \\ 0, & \text{if } \gcd(n, M \cdot N) \neq 1. \end{cases} \tag{E.7}$$

We call $\tilde{\chi}$ the **induced character** of modulus $(M \cdot N)$ induced by χ. In the examples above $\chi_{N,1}$ (the trivial character modulo N) is the induced character of modulus N induced by $\chi_{1,1}$ (the trivial character modulo 1).

In the other direction we may associate to a Dirichlet character χ with modulus N and conductor d a primitive Dirichlet character $\bar{\chi}$ with modulus d as follows: We first note if $\gcd(n, d) = 1$ there exists an integer n' such that $\gcd(n', N) = 1$ and $n' \equiv n \mod d$. We set

$$\bar{\chi}(n) = \begin{cases} \chi(n'), & \text{if } \gcd(n, d) = 1, \\ 0, & \text{if } \gcd(n, d) \neq 1. \end{cases} \tag{E.8}$$

$\bar{\chi}$ is called the **primitive character associated with** χ.

E.2 The Kronecker Symbol

In this section we introduce the Kronecker symbol

$$\left(\frac{a}{n}\right). \tag{E.9}$$

(There is an overloading of the name "Kronecker symbol": In this section we mean the symbol as in Eq. (E.9), not δ_{ij}. The two symbols are not related.) The Kronecker symbol defines a Dirichlet character, which takes values $\{-1, 0, 1\}$. In addition, we may give a criteria under which condition this Dirichlet character is primitive.

Let a be an integer and n a non-zero integer with prime factorisation $n = u p_1^{\alpha_1} p_2^{\alpha_2} \cdots p_k^{\alpha_k}$, where $u \in \{1, -1\}$ is a unit and the p_j's are prime numbers. The **Kronecker symbol** is defined by

$$\left(\frac{a}{n}\right) = \left(\frac{a}{u}\right) \left(\frac{a}{p_1}\right)^{\alpha_1} \left(\frac{a}{p_2}\right)^{\alpha_2} \cdots \left(\frac{a}{p_k}\right)^{\alpha_k}. \tag{E.10}$$

The individual factors are defined as follows: For a unit u we define

$$\left(\frac{a}{u}\right) = \begin{cases} 1, \; u = 1, \\ 1, \; u = -1, \; a \geq 0, \\ -1, \; u = -1, \; a < 0. \end{cases} \tag{E.11}$$

For $p = 2$ we define

$$\left(\frac{a}{2}\right) = \begin{cases} 1, \; a \equiv \pm 1 \mod 8, \\ -1, \; a \equiv \pm 3 \mod 8, \\ 0, \; a \text{ even}. \end{cases} \tag{E.12}$$

For an odd prime p we have

$$\left(\frac{a}{p}\right) = a^{\frac{p-1}{2}} \mod p = \begin{cases} 1, \; a \equiv b^2 \mod p, \\ -1, \; a \not\equiv b^2 \mod p, \\ 0, \; a \equiv 0 \mod p. \end{cases} \tag{E.13}$$

We further set

$$\left(\frac{a}{0}\right) = \begin{cases} 1, \; a = \pm 1 \\ 0, \text{ otherwise}. \end{cases} \tag{E.14}$$

For any non-zero integer a the mapping

$$n \to \left(\frac{a}{n}\right) \tag{E.15}$$

is a Dirichlet character, which we denote by χ_a:

$$\chi_a(n) = \left(\frac{a}{n}\right). \tag{E.16}$$

If a is the discriminant of a quadratic field, then it is a primitive Dirichlet character with conductor $|a|$. One may give a condition for a being the discriminant of a quadratic field [336]. We first set for p being a prime number, -1 or -2

$$p^* = \begin{cases} p, \text{ if } & p \equiv 1 \mod 4, \\ -p, \text{ if } & p \equiv -1 \mod 4 \text{ and } p \neq -1, \\ -4, \text{ if } & p = -1, \\ 8, \text{ if } & p = 2, \\ -8, \text{ if } & p = -2. \end{cases} \tag{E.17}$$

Then an integer a is the discriminant of a quadratic field if and only if a is a product of distinct p^*'s.

Including the trivial character (for which $a = 1$) the possible values for a with smallest absolute value are

$$1, -3, -4, 5, -7, 8, -8, -11, 12, \ldots \tag{E.18}$$

Appendix F
The Moduli Space $\mathcal{M}_{g,n}$

In this appendix we discuss the moduli space of a smooth algebraic curve of genus g with n marked points. This moduli space is denoted by $\mathcal{M}_{g,n}$.

We start in Sect. F.1 from configuration spaces of n points in a topological space X and mod out configurations which are isomorphic. This gives us the moduli space. We are in particular interested in the case, where the space X is a smooth complex algebraic curve of genus g. These complex curves, and their equivalence to real Riemann surfaces are discussed in Sect. F.2.

With these preparations, we specialise in Sect. F.3 to the main topic of this appendix: The moduli space $\mathcal{M}_{g,n}$ of a smooth algebraic curve of genus g with n marked points. This moduli space is non-compact, and one is interested in its compactification $\overline{\mathcal{M}}_{g,n}$. The compactification includes configurations, where points and/or the algebraic curve degenerates.

The cases of genus zero and genus one are the most important ones for applications towards Feynman integrals. We discuss the moduli space $\mathcal{M}_{0,n}$ in Sect. F.4 and the moduli space $\mathcal{M}_{1,n}$ in Sect. F.5.

F.1 Configuration Spaces

Let X be a topological space. The **configuration space** of n ordered points in X is

$$\mathrm{Conf}_n(X) = \left\{ (x_1, \dots, x_n) \in X^n \middle| x_i \neq x_j \text{ for } i \neq j \right\}. \tag{F.1}$$

Please note that we require that the points are distinct: $x_i \neq x_j$. As a simple example consider the configuration space of 2 ordered points in $\mathbb{R}$:

$$\mathrm{Conf}_2(\mathbb{R}) = \left\{ (x_1, x_2) \in \mathbb{R}^2 \middle| x_1 \neq x_2 \right\}. \tag{F.2}$$

S. Weinzierl, *Feynman Integrals*, UNITEXT for Physics,
https://doi.org/10.1007/978-3-030-99558-4

$\mathrm{Conf}_2(\mathbb{R})$ is the plane $\mathbb{R}^2$ with the diagonal $x_1 = x_2$ removed. It is a two-dimensional space.

As a second example consider the configuration space of 2 ordered points in the complex projective space $\mathbb{CP}^1$ (i.e. the Riemann sphere):

$$\mathrm{Conf}_2\left(\mathbb{CP}^1\right) = \left\{ (z_1, z_2) \in \left(\mathbb{CP}^1\right)^2 \middle| z_1 \neq z_2 \right\}. \tag{F.3}$$

This is again a two-dimensional space. A Möbius transformation

$$z' = \frac{az + b}{cz + d} \tag{F.4}$$

transforms the Riemann sphere into itself. These transformations form the group $\mathrm{PSL}(2, \mathbb{C})$. Usually we are not interested in configurations

$$(z_1, \dots, z_n) \in \mathrm{Conf}_n\left(\mathbb{CP}^1\right) \text{ and } (z'_1, \dots, z'_n) \in \mathrm{Conf}_n\left(\mathbb{CP}^1\right), \tag{F.5}$$

which differ only by a Möbius transformation. This brings us to the definition of the **moduli space** of the Riemann sphere with n marked points:

$$\mathcal{M}_{0,n} = \mathrm{Conf}_n\left(\mathbb{CP}^1\right) / \mathrm{PSL}(2, \mathbb{C}). \tag{F.6}$$

We may use the freedom of Möbius transformations to fix three points (usually 0, 1 and ∞). Therefore

$$\begin{aligned} \dim\left(\mathrm{Conf}_n\left(\mathbb{CP}^1\right)\right) &= n, \\ \dim\left(\mathcal{M}_{0,n}\right) &= n - 3. \end{aligned} \tag{F.7}$$

F.2 Complex Algebraic Curves and Riemann Surfaces

We are mainly interested in the situation, where the topological space X is a Riemann surface C. Let us start with a compact, connected and smooth Riemann surface C.

On the one hand, we may view C as a two-dimensional real surface (hence Riemann surface) with a complex structure.

On the other hand we may view C as an algebraic curve (i.e. of complex dimension one) in $\mathbb{CP}^2$: There exists a homogeneous polynomial $P(z_1, z_2, z_3)$ such that

$$C : \left\{ [z_1 : z_2 : z_3] \in \mathbb{CP}^2 \middle| P(z_1, z_2, z_3) = 0 \right\} \tag{F.8}$$

If d is the degree of the polynomial $P(z_1, z_2, z_3)$, the **arithmetic genus** of C is given by

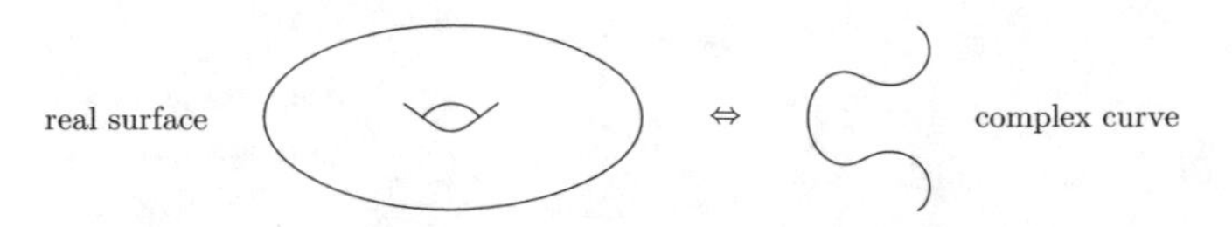

Fig. F.1 We may view C either as a Riemann surface (a two-dimensional real surface) or as a complex algebraic curve (of complex dimension one, corresponding to two real dimensions)

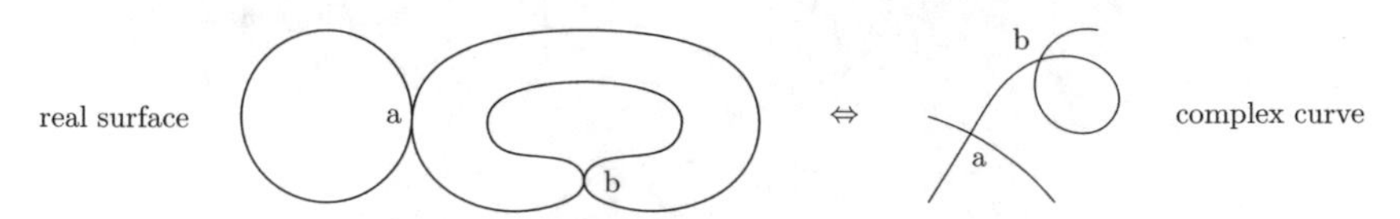

Fig. F.2 A nodal curve, shown as a real surface (left) or as a complex curve (right)

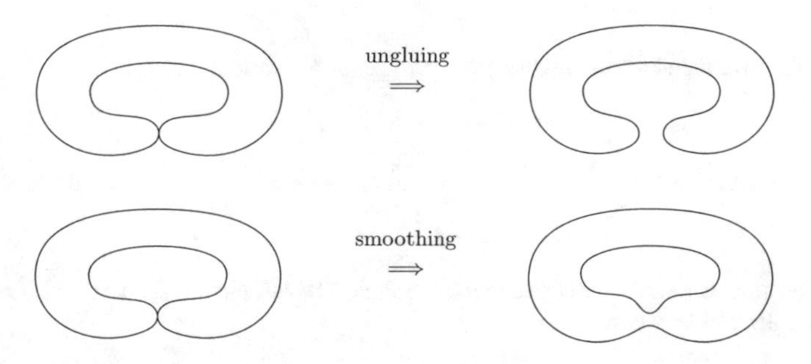

Fig. F.3 On a node we may perform the operations of ungluing (top) and smoothing (bottom)

$$g = \frac{1}{2}(d-1)(d-2). \tag{F.9}$$

Example:

$$y^2 z - x^3 - x z^2 = 0 \tag{F.10}$$

is a smooth curve of genus 1. The fact that we may view C either as a Riemann surface (of real dimension two) or as a complex algebraic curve is illustrated in Fig. F.1.

The requirement that the curve is smooth is a little bit too restrictive and we consider the generalisation towards nodal curves. A **node** of a curve is a singularity isomorphic to

$$xy = 0 \quad \text{in} \quad \mathbb{C}^2. \tag{F.11}$$

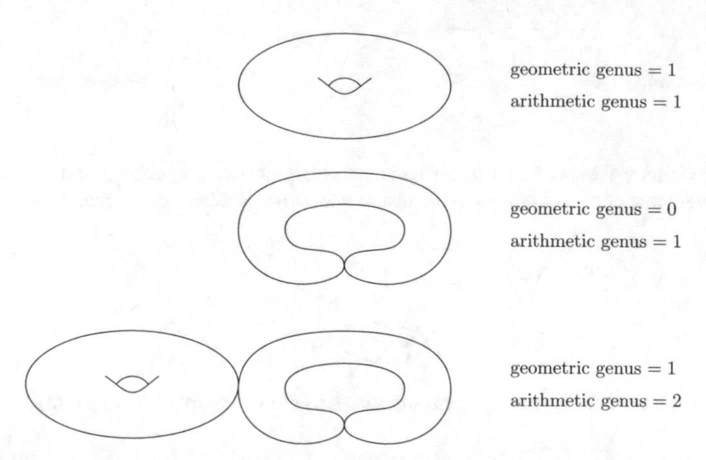

Fig. F.4 The arithmetic genus and the geometric genus for various examples

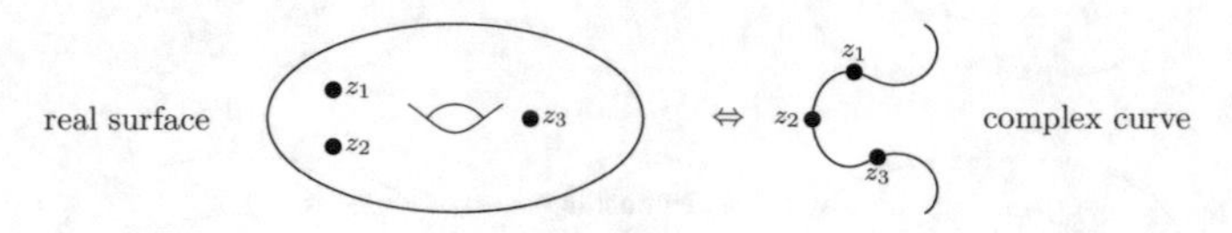

Fig. F.5 A (smooth) curve with three marked points. The left picture shows the real surface, the right picture the complex curve

A **nodal curve** is a compact, connected curve which is smooth except for a finite number of points, which are nodes. Figure F.2 shows an example.

There are two operations, which we may perform on a node: We may **unglue** a node or we may **smoothen** a node. These two operations are illustrated in Fig. F.3.

For a nodal curve we have to distinguish the geometric genus and the arithmetic genus. The **geometric genus** of an irreducible nodal curve is its genus once all of the nodes are unglued, and the geometric genus of a (reducible) nodal curve is the sum of the geometric genera of the irreducible components. The **arithmetic genus** of a nodal curve is the genus of the curve obtained by smoothing. Figure F.4 shows a few examples.

If the curve C has s nodes and k irreducible components the relation between the arithmetic genus g_{arithm} and the geometric genus g_{geom} is

$$g_{\text{arithm}} = g_{\text{geom}} + 1 + s - k. \tag{F.12}$$

Let us now consider nodal curves with n marked points. An example with three marked points is shown in Fig. F.5. To a nodal curve we may associate a **dual graph** as follows:

- The irreducible components of the curve C are drawn as vertices, labelled with their geometric genera.
- The nodes are drawn as edges.

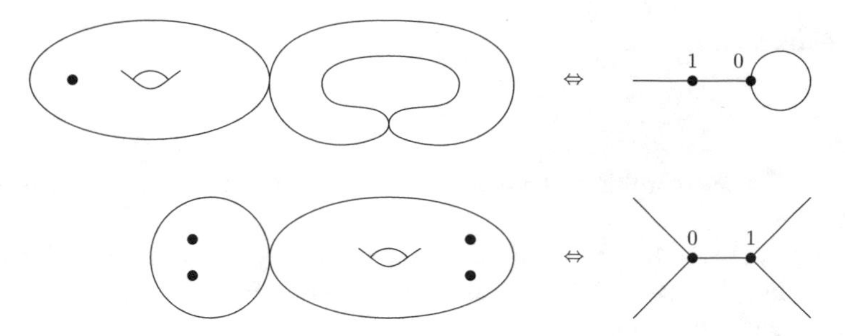

Fig. F.6 The correspondence between a nodal curve and its dual graph

- The marked points are drawn as half-edges.

Figure F.6 shows a few examples. We call a nodal curve with marked points a **stable curve**, if in the dual graph

- each genus 0 vertex has valence ≥ 3,
- each genus 1 vertex has valence ≥ 1.

This definition implies that smooth curves of genus g with n marked points are not stable if

$$(g, n) \in \{(0, 0), (0, 1), (0, 2), (1, 0)\}. \tag{F.13}$$

The smooth stable curves with n marked points are the ones with

$$\chi = 2 - 2g - n < 0. \tag{F.14}$$

χ is the **Euler characteristic** of the smooth curve with n marked points.

Let us remark that for a smooth curve the arithmetic genus equals the geometric genus, therefore just using "genus" is unambiguous in the smooth case.

F.3 The Moduli Space of a Curve of Genus g with n Marked Points

Let us now consider a smooth curve C of genus g with n marked points. Two such curves $(C; z_1, \dots, z_n)$ and $(C'; z'_1, \dots, z'_n)$ are **isomorphic** if there is an isomorphism

$$\phi : C \to C' \qquad \text{such that } \phi(z_i) = z'_i. \tag{F.15}$$

The **moduli space**

$$\mathcal{M}_{g,n} \tag{F.16}$$

is the space of isomorphism classes of smooth curves of genus g with n marked points.

For $g \geq 1$ the isomorphism classes do not only depend on the positions of the marked points, but also on the "shape" of the curve. For $g = 0$ there is only one "shape", the Riemann sphere.

The dimension of $\mathcal{M}_{g,n}$ is

$$\dim\left(\mathcal{M}_{g,n}\right) = 3g + n - 3. \tag{F.17}$$

The Euler characteristic of $\mathcal{M}_{g,n}$ is

$$\chi\left(\mathcal{M}_{g,n}\right) = (-1)^n \frac{(2g+n-3)!}{2g\,(2g-2)!} B_{2g}, \tag{F.18}$$

where B_j are the Bernoulli numbers.
The moduli space is not compact. We are interested in a specific compactification, the **Deligne-Mumford-Knudsen compactification** [455–458], which we denote by $\overline{\mathcal{M}}_{g,n}$. The space $\overline{\mathcal{M}}_{g,n}$ is the moduli space of stable nodal curves of arithmetic genus g with n marked points. $\mathcal{M}_{g,n}$ is an open subset in $\overline{\mathcal{M}}_{g,n}$. A generic element of $\overline{\mathcal{M}}_{g,n}$ is drawn in the dual graph picture as

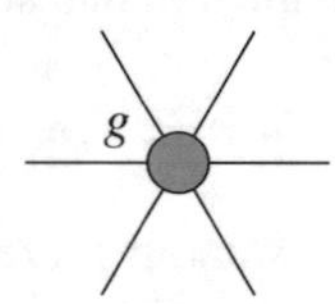

If the curve is smooth, the grey blob corresponds to a single genus g vertex. If the curve is non-smooth, the grey blob represents the appropriate sub-graph.
Stratification: Let Γ be the dual graph of a stable curve. We denote by $\mathcal{M}_\Gamma$ the subset of $\overline{\mathcal{M}}_{g,n}$ of stable curves with dual graph Γ. The various $\mathcal{M}_\Gamma$'s give a stratification of $\overline{\mathcal{M}}_{g,n}$. The dense open set $\mathcal{M}_{g,n}$ is one stratum, all other strata are called **boundary strata**. The closure of the codimension 1 strata are called **boundary divisors**. The divisors meet transversely along smaller strata. Let us denote the number of internal edges of Γ by n_e. Then

$$\dim\left(\mathcal{M}_\Gamma\right) = 3g + n - 3 - n_e. \tag{F.19}$$

Let us look at a few examples:

$$\Gamma_1 = \qquad \Rightarrow \qquad \dim\left(\mathcal{M}_{\Gamma_1}\right) = 0$$

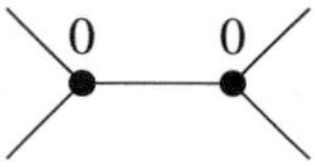

$$\Gamma_2 = \qquad \Rightarrow \qquad \dim\left(\mathcal{M}_{\Gamma_2}\right) = 0$$

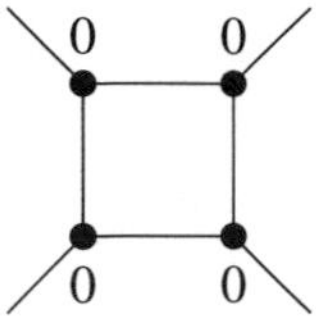

$$\Gamma_3 = \qquad \Rightarrow \qquad \dim\left(\mathcal{M}_{\Gamma_3}\right) = 1$$

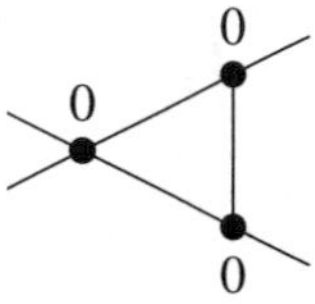

$$\Gamma_4 = \qquad \Rightarrow \qquad \dim\left(\mathcal{M}_{\Gamma_4}\right) = 2$$

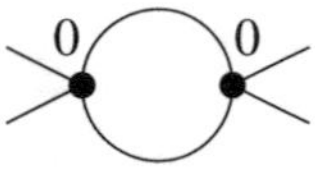

Of course we may also consider dual graphs with vertices of higher genus:

$$\Gamma_5 = \qquad \Rightarrow \qquad \dim\left(\mathcal{M}_{\Gamma_5}\right) = 3$$

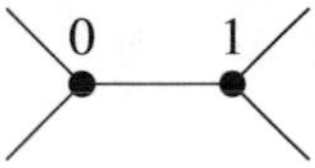

The **forgetful morphism**: Consider a stable nodal curve with n marked points (and assume $n > 0$, $(g, n) \neq (0, 3), (1, 1)$). We may forget the n-th point. This gives a nodal curve with $(n-1)$ marked points. This curve may not be stable. One stabilises the curve by contracting all components to a point, which correspond to genus 0 vertices of valency 2. This gives the forgetful morphism

$$\overline{\mathcal{M}}_{g,n} \to \overline{\mathcal{M}}_{g,n-1}. \tag{F.20}$$

Example:

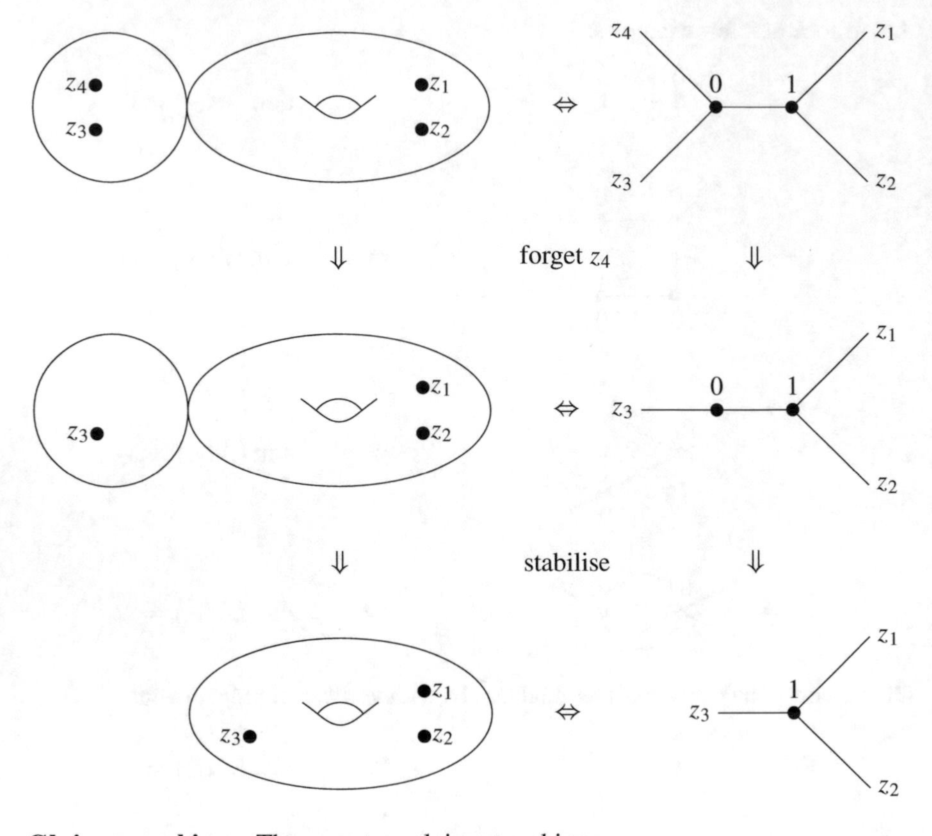

Gluing morphisms: There are two gluing morphisms

$$\begin{aligned} \overline{\mathcal{M}}_{g_1,n_1+1} \times \overline{\mathcal{M}}_{g_2,n_2+1} &\to \overline{\mathcal{M}}_{g_1+g_2,n_1+n_2}, \\ \overline{\mathcal{M}}_{g,n+2} &\to \overline{\mathcal{M}}_{g+1,n}. \end{aligned} \tag{F.21}$$

We may represent them graphically as

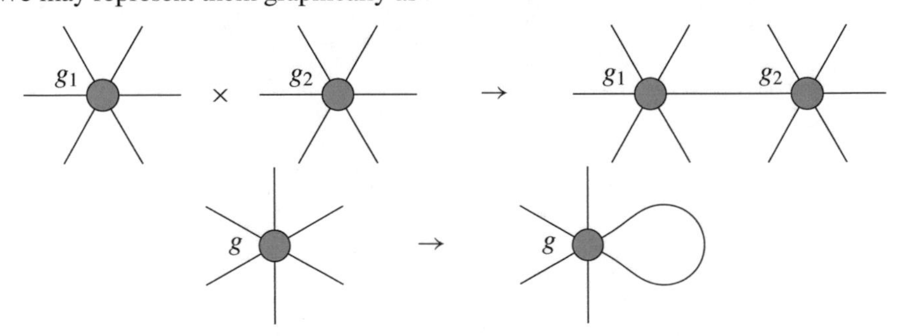

Note that the images of these morphisms are necessarily non-smooth.
Remark: We are not restricted to forget the last marked point in the forgetful morphism, nor are we restricted to glue the last two points together in the gluing mor-

phisms. We therefore have various forgetful morphisms and gluing morphisms, which we may index by the point, which we forget, or by the two points which are glued together, respectively.

F.4 The Genus Zero Case

Let us now consider a Riemann surface of genus 0, i.e. the Riemann sphere. Such a surface is isomorphic to $\mathbb{CP}^1$ and the group of Möbius transformations PSL (2, $\mathbb{C}$)

$$z \to \frac{az + b}{cz + d} \tag{F.22}$$

acts as an automorphism. We mark on the Riemann sphere n points. The moduli space is denoted by $\mathcal{M}_{0,n}$.

$$\mathcal{M}_{0,n} = \left\{ (z_1, \ldots, z_n) \mid z_i \in \mathbb{CP}^1, z_i \neq z_j \right\} / \mathrm{PSL}\,(2, \mathbb{C})\,. \tag{F.23}$$

This is an affine variety of dimension

$$\dim \mathcal{M}_{0,n} = n - 3. \tag{F.24}$$

We may use the freedom of PSL(2, $\mathbb{C}$)-transformations to fix three points. The standard choice will be $z_1 = 0$, $z_{n-1} = 1$ and $z_n = \infty$. Thus

$$\mathcal{M}_{0,n} = \left\{ (z_2, \ldots, z_{n-2}) \in \mathbb{C}^{n-3} \;:\; z_i \neq z_j,\; z_i \neq 0,\; z_i \neq 1 \right\}. \tag{F.25}$$

The variables $z_2, \ldots, z_{n-2}$ are called **simplicial coordinates**. We denote the **set of real points** by $\mathcal{M}_{0,n}(\mathbb{R})$:

$$\mathcal{M}_{0,n}(\mathbb{R}) = \left\{ (z_2, \ldots, z_{n-2}) \in \mathbb{R}^{n-3} \;:\; z_i \neq z_j,\; z_i \neq 0,\; z_i \neq 1 \right\}. \tag{F.26}$$

Let us look at a few examples.

1. For $n = 3$ we have $\dim \mathcal{M}_{0,3} = 0$ and $\mathcal{M}_{0,3}$ consists of a single point. We may use PSL (2, $\mathbb{C}$)-invariance to take

$$(z_1, z_2, z_3) = (0, 1, \infty) \tag{F.27}$$

   as a representative of this point.
2. For $n = 4$ we have $\dim \mathcal{M}_{0,4} = 1$ and elements of $\mathcal{M}_{0,4}$ can be represented by

$$(z_1, z_2, z_3, z_4) = (0, z, 1, \infty) \tag{F.28}$$

with

$$z \in \mathbb{CP}^1 \setminus \{0, 1, \infty\}\,. \tag{F.29}$$

Thus

$$\mathcal{M}_{0,4} \;\simeq\; \mathbb{CP}^1 \setminus \{0, 1, \infty\} \;\simeq\; \mathbb{C} \setminus \{0, 1\}\,. \tag{F.30}$$

3. For $n = 5$ we have dim $\mathcal{M}_{0,5} = 2$ and $\mathcal{M}_{0,5}$ is isomorphic to

$$\mathcal{M}_{0,5} \;\simeq\; \{(z_1, z_2) \mid z_i \in \mathbb{C},\; z_i \neq 0, 1,\;\; z_1 \neq z_2\}\,. \tag{F.31}$$

Thus $\mathcal{M}_{0,5}$ is isomorphic to the complement of the five lines $z_1 = 0$, $z_1 = 1$, $z_2 = 0$, $z_2 = 1$ and $z_1 = z_2$ in $\mathbb{C}^2$.

In Fig. (F.7) we sketch the moduli space $\mathcal{M}_{0,5}(\mathbb{R})$. In Fig. F.7 we indicated in red a region X of $\mathcal{M}_{0,5}(\mathbb{R})$. This region is bounded by $z_2 = 0$, $z_3 = 1$ and $z_2 = z_3$. In general there will be points, where the boundaries do not cross normally. For the region X in the example above this occurs for $(z_2, z_3) = (0, 0)$ and $(z_2, z_3) = (1, 1)$. We denote by $\overline{\mathcal{M}}_{0,n}$ the blow-up of $\mathcal{M}_{0,n}$ in all those points, such that in $\overline{\mathcal{M}}_{0,n}$ all boundaries cross normally. In this way the region X of our example transforms from a triangle in $\mathcal{M}_{0,5}(\mathbb{R})$ into a pentagon in $\overline{\mathcal{M}}_{0,5}(\mathbb{R})$.

For a set of points $\{z_i, z_j, z_k, z_l\} \subseteq \{z_1, z_2, \ldots, z_n\}$ we define a **cross-ratio** as follows

$$[i, j|k, l] = \frac{(z_i - z_k)\left(z_j - z_l\right)}{(z_i - z_l)\left(z_j - z_k\right)}. \tag{F.32}$$

The cross-ratios are invariant under Möbius transformations PSL $(2, \mathbb{C})$. We have the relations

$$\begin{aligned}
[i, j|k, l] &= [i, j|l, k]^{-1}\,, \\
[i, j|k, l] &= [j, i|l, k]^{-1}\,, \\
[i, j|k, l] &= [k, l|i, j]\,, \\
[i, j|k, l] &= 1 - [i, k|j, l]\,.
\end{aligned} \tag{F.33}$$

F.4.1 The Deligne-Mumford-Knudsen Compactification

Let us now review a systematic way to construct $\overline{\mathcal{M}}_{0,n}$. There is a smooth compactification

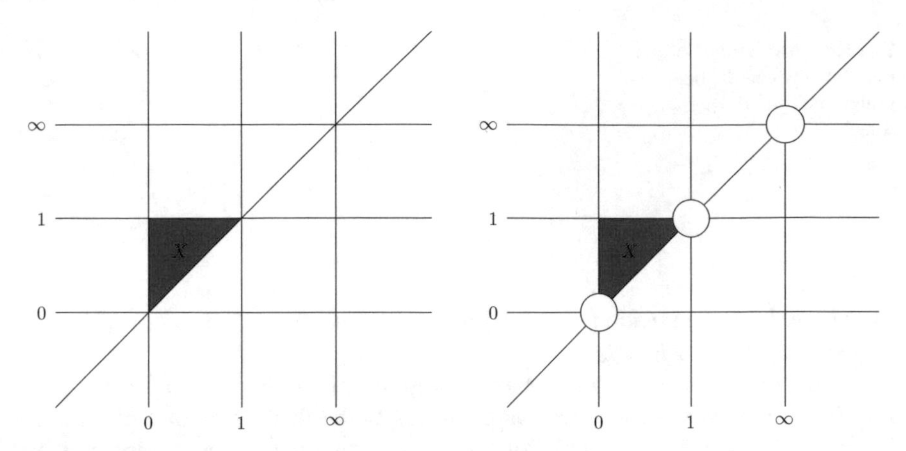

Fig. F.7 The moduli space $\mathcal{M}_{0,5}(\mathbb{R})$ (left). The region X is bounded by $z_2 = 0$, $z_3 = 1$ and $z_2 = z_3$. The right figure shows $\overline{\mathcal{M}}_{0,5}(\mathbb{R})$, obtained from $\mathcal{M}_{0,5}(\mathbb{R})$ by blowing up the points $(z_2, z_3) = (0, 0)$, $(z_2, z_3) = (1, 1)$ and $(z_2, z_3) = (\infty, \infty)$

$$\mathcal{M}_{0,n} \subset \overline{\mathcal{M}}_{0,n}, \tag{F.34}$$

known as the Deligne-Mumford-Knudsen compactification [455–458], such that $\overline{\mathcal{M}}_{0,n} \backslash \mathcal{M}_{0,n}$ is a smooth normal crossing divisor. In order to describe $\overline{\mathcal{M}}_{0,n}$ we follow Ref. [459].

Let $\pi = (\pi_1, \dots, \pi_n)$ be a permutation of $(1, \dots, n)$. A **cyclic order** is defined as a permutation modulo cyclic permutations $(\pi_1, \pi_2, \dots, \pi_n) \rightarrow (\pi_2, \dots, \pi_n, \pi_1)$. We may represent a cyclic order by an n-gon, where the edges of the n-gon are indexed clockwise by $\pi_1, \pi_2, \dots, \pi_n$. A **dihedral structure** is defined as a permutation modulo cyclic permutations and reflection $(\pi_1, \pi_2, \dots, \pi_n) \rightarrow (\pi_n, \dots, \pi_2, \pi_1)$. We may represent a dihedral structure by an n-gon, where the edges of the n-gon are indexed either clockwise or anti-clockwise by $\pi_1, \pi_2, \dots, \pi_n$.

The construction of $\overline{\mathcal{M}}_{0,n}$ proceeds through intermediate spaces $\mathcal{M}^{\pi}_{0,n}$, labelled by a dihedral structure π, such that

$$\mathcal{M}_{0,n} \subset \mathcal{M}^{\pi}_{0,n} \subset \overline{\mathcal{M}}_{0,n}. \tag{F.35}$$

Let $z = (z_1, \dots, z_n)$ denote the (ordered) set of the n marked points on the curve. In the following we will use the notation

$$\mathcal{M}_{0,z} \tag{F.36}$$

for $\mathcal{M}_{0,n}$. This notation allows us to distinguish $\mathcal{M}_{0,z'}$ from $\mathcal{M}_{0,z''}$ if z' and z'' are two non-identical subsets of z with k elements each (i.e. $z' \neq z''$ but $|z'| = |z''| = k$). Let π denote a permutation of $(1, \dots, n)$, which defines a dihedral structure. We may draw a **regular** n**-gon**, where the edges are labelled by $z_{\pi_1}, z_{\pi_2}, \dots, z_{\pi_n}$ in this order.

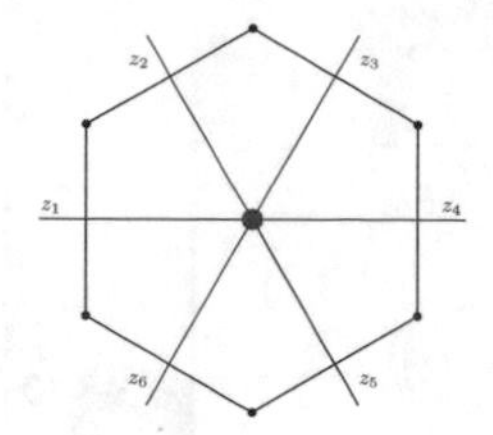

Fig. F.8 An example of the relation between the dual graph (red) and the n-gon (black)

In order to keep the notation simple let us assume that $\pi = (1, 2, \dots, n)$. Then the edges are labelled by $z_1, z_2, \dots, z_n$.

We may think of the n-gon as the "dual graph of the dual graph", as shown in Fig. F.8. The words "of the dual graph" refers to the dual graph of a nodal curve introduced in Sect. F.2, the words "dual graph of" refer to the construction of a dual graph from a planar graph as discussed in Sect. 3.4. In this example one starts from a Riemann sphere with six marked points z_1-z_6. The dual graph is given by a genus-0 vertex with six external legs. Choosing a dihedral structure defines in particular a cyclic order and we draw the dual graph with this cyclic order in the plane. We then construct the dual graph of the dual graph, this gives the hexagon. Please note that the term "dual graph" is used with two different meanings: Once we mean the graph dual to a Riemann sphere with marked points, the second meaning refers to the dual of a graph drawn in a plane.

A **chord** of the polygon connects two non-adjacent vertices and may be specified by giving the two edges preceding the two vertices in the clockwise orientation. Thus (i, j) denotes the chord from the vertex between edge z_i and z_{i+1} to the vertex between the edge z_j and z_{j+1}. There are

$$\frac{1}{2} n \left(n-3\right) \tag{F.37}$$

chords for a regular n-gon. We denote by $\chi(z, \pi)$ the set of all chords of the n-gon defined by the set z and the dihedral structure π. Each chord defines a cross-ratio as follows (for the dihedral structure $\pi = (1, 2, \dots, n)$):

$$u_{i,j} = [i, i+1 | j+1, j] \;=\; \frac{\left(z_i - z_{j+1}\right)\left(z_{i+1} - z_j\right)}{\left(z_i - z_j\right)\left(z_{i+1} - z_{j+1}\right)} \tag{F.38}$$

For an arbitrary dihedral structure π we set

$$u^{\pi}_{i,j} = \left[\pi_i, \pi_{i+1} | \pi_{j+1}, \pi_j\right] \;=\; \frac{\left(z_{\pi_i} - z_{\pi_{j+1}}\right)\left(z_{\pi_{i+1}} - z_{\pi_j}\right)}{\left(z_{\pi_i} - z_{\pi_j}\right)\left(z_{\pi_{i+1}} - z_{\pi_{j+1}}\right)}. \tag{F.39}$$

As already mentioned, the cross-ratio is invariant under PSL$(2, \mathbb{C})$-transformations. Each cross-ratio defines a function

$$\mathcal{M}_{0,z} \to \mathbb{CP}^1 \backslash \{0, 1, \infty\}, \tag{F.40}$$

or equivalently

$$\mathcal{M}_{0,z} \to \mathbb{C} \backslash \{0, 1\}. \tag{F.41}$$

If a cross-ratio takes a value from the set $\{0, 1, \infty\}$, one can show that two points of the z_i's coincide (which contradicts the assumption that all marked points are distinct). The set of all cross-ratios for a given dihedral structure π defines an embedding

$$\mathcal{M}_{0,z} \to \mathbb{C}^{n(n-3)/2}. \tag{F.42}$$

We may now consider the Zariski closure of the image of this embedding and take the Zariski closure as a chart of the dihedral extension $\mathcal{M}^{\pi}_{0,z}$. This defines the dihedral extension $\mathcal{M}^{\pi}_{0,z}$. Since the chart and the dihedral extension $\mathcal{M}^{\pi}_{0,z}$ are homeomorphic, one usually does not distinguish between the two. The Deligne-Mumford-Knudsen compactification is obtained by gluing these charts together:

$$\overline{\mathcal{M}}_{0,z} = \bigcup_{\pi} \mathcal{M}^{\pi}_{0,z}, \tag{F.43}$$

where π ranges over the $(n-1)!/2$ inequivalent dihedral structures.

Example 1: The simplest case is $n = 4$ and from $(z_1, z_2, z_3, z_4) = (0, z, 1, \infty)$ we concluded that

$$\mathcal{M}_{0,4} \simeq \mathbb{CP}^1 \backslash \{0, 1, \infty\} \simeq \mathbb{C} \backslash \{0, 1\}. \tag{F.44}$$

For $n = 4$ there are three inequivalent dihedral structures, which we may take as

$$\pi_1 = (1, 2, 3, 4), \quad \pi_2 = (2, 3, 1, 4), \quad \pi_3 = (3, 1, 2, 4). \tag{F.45}$$

For each dihedral structure there are two chords. Let's start with π_1. We have

$$u^{\pi_1}_{1,3} = [1, 2|4, 3] = 1 - z, \quad u^{\pi_1}_{2,4} = [2, 3|1, 4] = z. \tag{F.46}$$

The embedding is given by

$$\begin{aligned} \mathcal{M}_{0,4} = \mathbb{C} \backslash \{0, 1\} &\to \mathbb{C}^2, \\ z &\to \begin{pmatrix} 1 - z \\ z \end{pmatrix}. \end{aligned} \tag{F.47}$$

Taking the closure of the image, we add the points

$$\begin{pmatrix} 1 \\ 0 \end{pmatrix}, \quad \begin{pmatrix} 0 \\ 1 \end{pmatrix}. \tag{F.48}$$

Therefore

$$\mathcal{M}_{0,4}^{\pi_1} = \mathbb{C} \;=\; \mathbb{CP}^1 \backslash \{\infty\}. \tag{F.49}$$

The point at infinity has not been added. In order to get the point at infinity we look at the other dihedral extensions. We have

$$u_{1,3}^{\pi_2} \;=\; [2,3|4,1] \;=\; \frac{1}{z}, \quad u_{2,4}^{\pi_2} \;=\; [3,1|2,4] \;=\; 1 - \frac{1}{z}, \tag{F.50}$$

and therefore

$$\mathcal{M}_{0,4}^{\pi_2} = \mathbb{CP}^1 \backslash \{0\}. \tag{F.51}$$

Furthermore

$$u_{1,3}^{\pi_3} = [3,1|4,2] = 1 - \frac{1}{1-z}, \quad u_{2,4}^{\pi_3} = [1,2|3,4] = \frac{1}{1-z}, \tag{F.52}$$

and hence

$$\mathcal{M}_{0,4}^{\pi_3} = \mathbb{CP}^1 \backslash \{1\}. \tag{F.53}$$

In total we find

$$\overline{\mathcal{M}}_{0,4} \;=\; \mathbb{CP}^1, \quad \overline{\mathcal{M}}_{0,4} \backslash \mathcal{M}_{0,4} \;=\; \{0, 1, \infty\}. \tag{F.54}$$

We have constructed three charts for $\overline{\mathcal{M}}_{0,4}$, indexed by the dihedral structures π_1, π_2 and π_3. A single chart does not cover $\overline{\mathcal{M}}_{0,4}$, we need at least two of them. The boundary divisor $\overline{\mathcal{M}}_{0,4} \backslash \mathcal{M}_{0,4}$ consists of three points. This is a particular simple example, where blow-ups do not yet enter.

Example 2: In order to see how blow-ups enter the game, we discuss the next more complicated example given by $n = 5$. With $(z_1, z_2, z_3, z_4, z_5) = (0, z_2, z_3, 1, \infty)$ we have

$$\mathcal{M}_{0,5} \;\simeq\; \{(z_2, z_3) \;|\; z_i \in \mathbb{C}, \; z_i \neq 0, 1, \;\; z_2 \neq z_3\}. \tag{F.55}$$

There are now 12 inequivalent dihedral structures. Let us take $\pi = (1, 2, 3, 4, 5)$. For each dihedral structure we have 5 chords. We have

$$u_{1,3} = [1,2|4,3] \;=\; \frac{z_3 - z_2}{(1 - z_2)\, z_3},$$

$$\begin{aligned} u_{1,4} = [1, 2|5, 4] &= 1 - z_2, \\ u_{2,4} = [2, 3|5, 4] &= \frac{1 - z_3}{1 - z_2}, \\ u_{2,5} = [2, 3|1, 5] &= \frac{z_2}{z_3}, \\ u_{3,5} = [3, 4|1, 5] &= z_3. \end{aligned} \tag{F.56}$$

The five cross-ratios define the embedding

$$\mathcal{M}_{0,5} \to \mathbb{C}^5, \tag{F.57}$$

the image of this embedding is contained in a plane. The image does not include five lines, given by the intersection of the plane with one of the hyperplanes defined by

$$u_{1,3} = 0, \quad u_{1,4} = 0, \quad u_{2,4} = 0, \quad u_{2,5} = 0, \quad u_{3,5} = 0. \tag{F.58}$$

One checks, that the intersection of the plane with one of the hyperplanes defined by

$$u_{1,3} = 1, \quad u_{1,4} = 1, \quad u_{2,4} = 1, \quad u_{2,5} = 1, \quad u_{3,5} = 1 \tag{F.59}$$

does not give new lines, for example $u_{3,5} = 1$ is equivalent to $u_{2,4} = 0$. For the Zariski closure we add these five lines back. Let us now understand how we get from the red triangle X in the left picture of Fig. F.7 to the pentagon in the right picture. It is clear that we have away from critical values the correspondence

$$\begin{aligned} u_{1,3} &= 0 \Rightarrow z_2 = z_3, \\ u_{2,5} &= 0 \Rightarrow z_2 = 0, \\ u_{2,4} &= 0 \Rightarrow z_3 = 1. \end{aligned} \tag{F.60}$$

This gives us three edges of the pentagon. Let us now see what happens, if we blow-up the point $(z_2, z_3) = (0, 0)$ by $\mathbb{CP}^1$. For $\mathbb{CP}^1$ we need two charts. In the first chart $(z_2 \neq 0)$ we use coordinates (z_2, z_3'), where z_3' is related to the old coordinates by

$$z_3 = z_2 z_3' \Leftrightarrow z_3' = \frac{z_3}{z_2}. \tag{F.61}$$

In the second chart $(z_3 \neq 0)$ we use coordinates (z_2', z_3), where z_2' is related to the old coordinates by

$$z_2 = z_3 z_2' \Leftrightarrow z_2' = \frac{z_2}{z_3}. \tag{F.62}$$

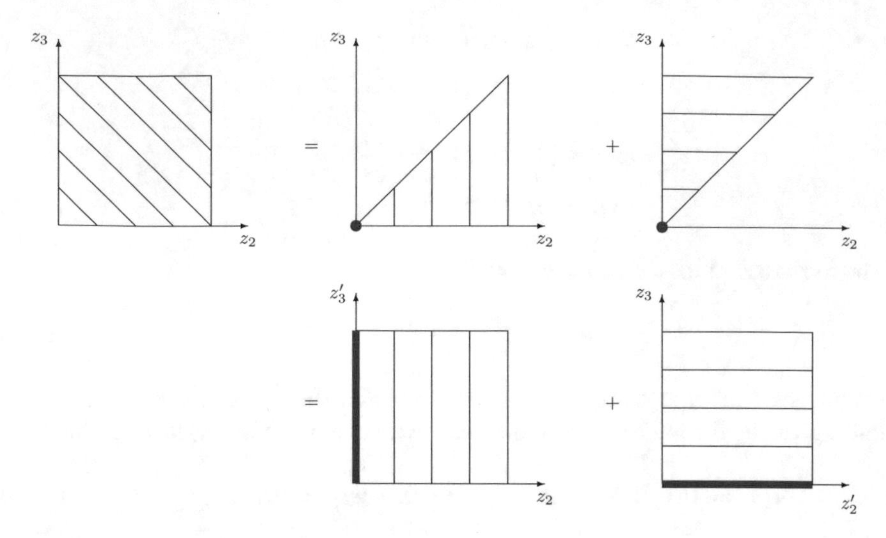

Fig. F.9 The blow-up of the point $(z_2, z_3) = (0, 0)$

Graphically this is shown in Fig. F.9. Note that the procedure for the blow-up of this point is completely analogous to the discussion in Chap. 10. The condition $u_{3,5} = 0$ gives in the second chart the line $z_3 = 0$, i.e. the blow-up of the point $(0, 0)$.

A similar argumentation holds for the blow-up of the point $(z_2, z_3) = (1, 1)$ and the condition $u_{1,4} = 0$.

F.4.2 The Associahedron

Let us discuss the dihedral extension $\mathcal{M}^{\pi}_{0,z}$ in more detail. We recall that the construction of $\mathcal{M}^{\pi}_{0,z}$ requires the specification of a dihedral structure π (i.e. a permutation up to cyclic permutations and reflection). We will need a few properties of the dihedral extension $\mathcal{M}^{\pi}_{0,z}$ [459]:

1. The complement $\mathcal{M}^{\pi}_{0,z} \backslash \mathcal{M}_{0,z}$ is a normal crossing divisor, whose irreducible components are

$$D_{ij} = \left\{ u_{i,j} = 0 \right\}, \tag{F.63}$$

indexed by the chords $(i, j) \in \chi(z, \pi)$.

2. Each divisor is again a product of spaces of the same type: Let us consider a chord (i, j). This chord decomposes the original polygon (z, π) into two smaller polygons, as shown in Fig. F.10. We denote the new edge by z_e. The set of edges for the two smaller polygons are $z' \cup \{z_e\}$ and $z'' \cup \{z_e\}$, where $z = z' \cup z''$ and $z' \cap z'' = \emptyset$. The two smaller polygons inherit their dihedral structures π' and π''

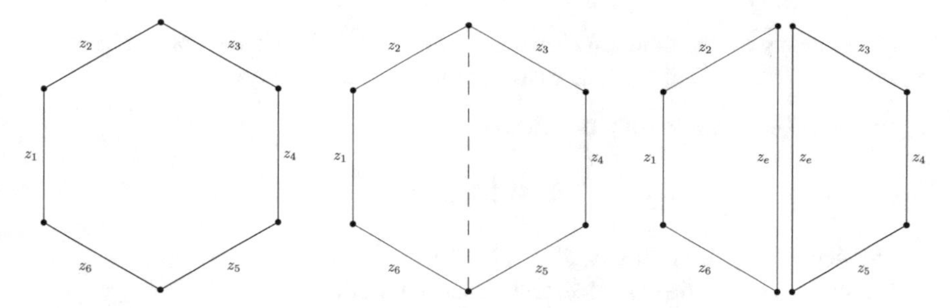

Fig. F.10 A hexagon, where the edges are labelled by the cyclic ordered variables $(z_1, z_2, \ldots, z_6)$ (left picture). The middle picture shows the chord $(2, 5)$. Right picture: A chord divides the hexagon into two lower n-gons, in this case two quadrangles

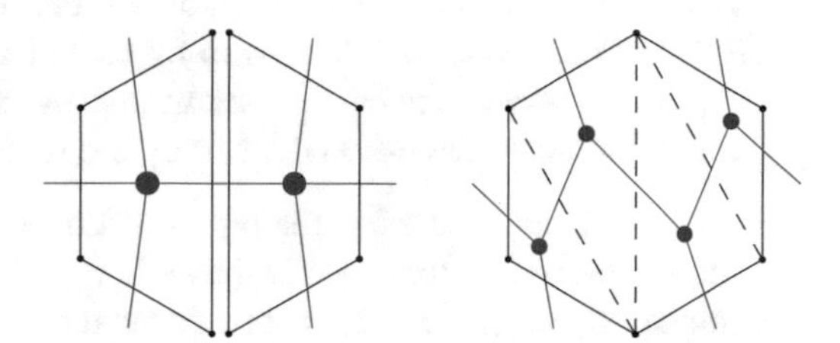

Fig. F.11 The limit $u_{i,j} \to 0$ leads to a factorisation of the dual graph (left). A complete triangulation of the n-gon leads to a dual graph with trivalent vertices only (right)

from π and the chord (i, j). We have

$$D_{ij} \cong \mathcal{M}_{0,z'\cup\{z_e\}}^{\pi'} \times \mathcal{M}_{0,z''\cup\{z_e\}}^{\pi''}. \tag{F.64}$$

This factorisation translates to the dual graphs, as shown in Fig. F.11. Iteration of this procedure corresponds to a triangulation of the n-gon or equivalently to a dual tree graph with three-valent vertices only.

Let us now consider the space of real points. For a given set z and dihedral structure π we set

$$X_{0,z}^{\pi} = \left\{u_{i,j} > 0 \;:\; (i, j) \in \chi(z, \pi)\right\} \tag{F.65}$$

and

$$\overline{X}_{0,z}^{\pi} = \left\{u_{i,j} \geq 0 \;:\; (i, j) \in \chi(z, \pi)\right\}. \tag{F.66}$$

One has

$$\mathcal{M}_{0,n}(\mathbb{R}) = \bigsqcup_{\pi} X_{0,z}^{\pi}, \tag{F.67}$$

where π ranges again over the $(n-1)!/2$ inequivalent dihedral structures.

For a given set z and dihedral structure π the cell $\overline{X}^{\pi}_{0,z}$ is called a Stasheff polytope or associahedron [460–463]. The associahedron has the properties

1. Its facets (i.e. codimension one faces)
$$F_{ij} = \left\{ u_{i,j} = 0 \right\}, \tag{F.68}$$
are indexed by the chords $(i, j) \in \chi(z, \pi)$.
2. From Eq. (F.64) it follows that each facet is a product
$$F_{ij} = \overline{X}^{\pi'}_{0,z'\cup\{z_e\}} \times \overline{X}^{\pi''}_{0,z''\cup\{z_e\}}. \tag{F.69}$$
3. Two facets F_{ij} and F_{kl} meet if and only if the chords (i, j) and (k, l) do not cross.
4. Faces of codimension k are given by sets of k non-crossing chords. In particular, the set of vertices of $\overline{X}^{\pi}_{0,z}$ are in one-to-one correspondence with the set of triangulations of the n-gon defined by the set z and the dihedral structure π.

Properties (1) and (2) are the analogues of Eqs. (F.63) and (F.64), respectively.

The associahedron for $n = 5$ is shown in Fig. F.12. Let us now have a closer look at coordinates on $\mathcal{M}^{\pi}_{0,z}$. We already introduced the simplicial coordinates $(z_2, \dots, z_{n-2})$ in Eq. (F.25). Let us fix a dihedral structure π. Without loss of generality we may take the cyclic order to be $(1, 2, \dots, n)$. Let us consider a chord from $\chi(z, \pi)$. Due to cyclic invariance we may limit ourselves to chords of the form (i, n). With the gauge choice $z_1 = 0$, $z_{n-1} = 1$ and $z_n = \infty$ we have

$$u_{2,n} = \frac{z_2}{z_3}, \quad \dots \quad u_{(n-3),n} = \frac{z_{n-3}}{z_{n-2}}, \quad u_{(n-2),n} = z_{n-2}, \tag{F.70}$$

and hence

$$z_i = \prod_{j=i}^{n-2} u_{j,n}, \qquad i \in \{2, \dots, n-2\}. \tag{F.71}$$

Thus we may use as coordinates on $\mathcal{M}^{\pi}_{0,z}$ instead of the $(n-3)$ coordinates $(z_2, \dots, z_{n-2})$ the $(n-3)$ cross-ratios $(u_{2,n}, \dots, u_{n-2,n})$. For $n = 6$ this is illustrated in Fig. F.13.

We have

$$d^{n-3}z = \left(\prod_{j=3}^{n-2} u_{j,n}^{j-2} \right) d^{n-3}u. \tag{F.72}$$

Let us further set

$$x_j = u_{j,n}^{-1}, \qquad 2 \le j \le n-2. \tag{F.73}$$

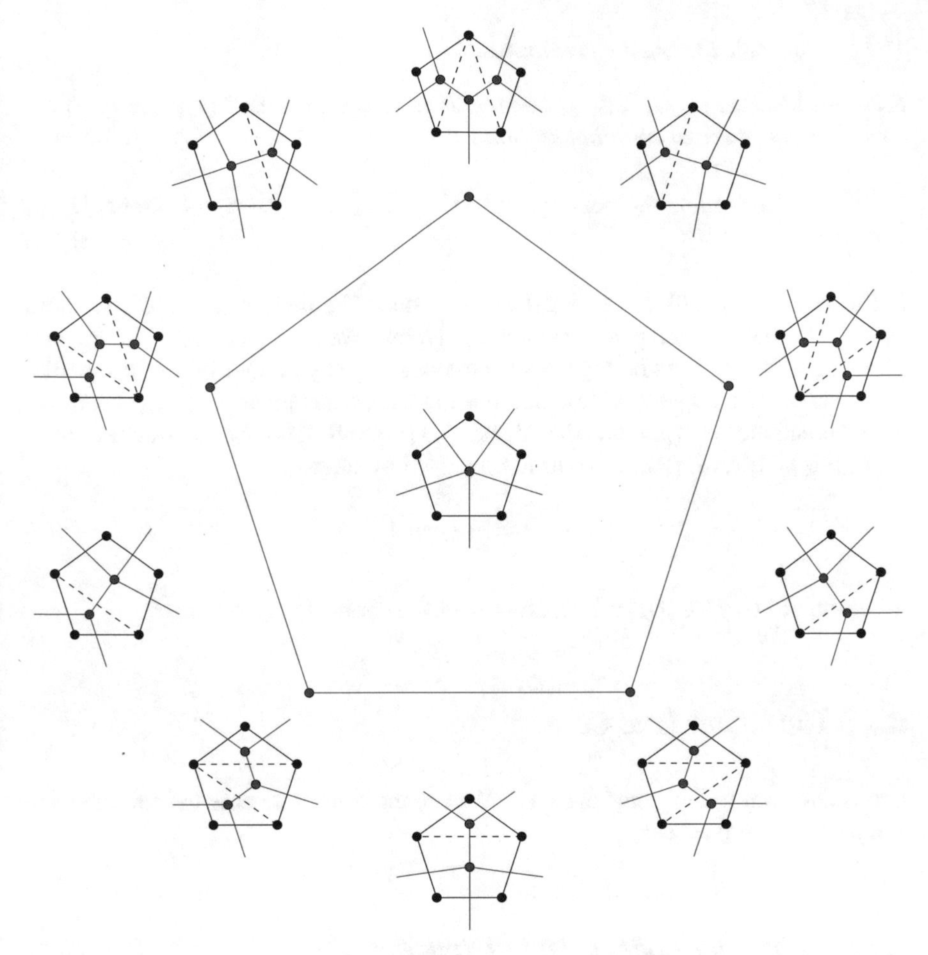

Fig. F.12 The associahedron for $n = 5$ (blue). For the Riemann sphere with n marked points the associahedron is a $(n - 3)$-dimensional object. The codimension k faces are either indexed by an n-gon with k non-crossing chords (black) or by dual graphs with k internal edges (red)

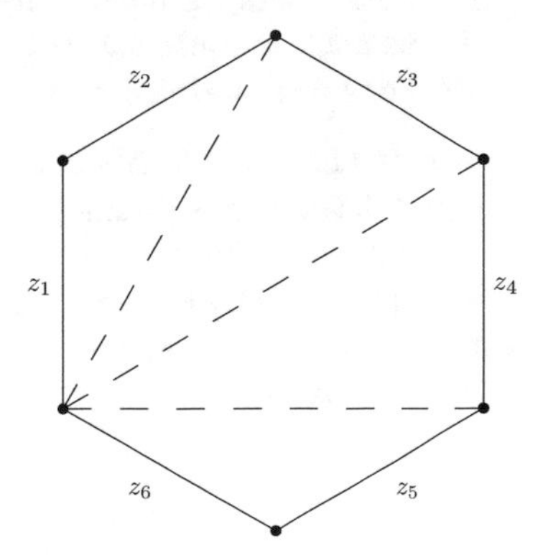

Fig. F.13 The three chords (2, 6), (3, 6) and (4, 6) define cross-ratios $u_{2,6}$, $u_{3,6}$ and $u_{4,6}$, which may be used as coordinates on $\mathcal{M}^{\pi}_{0,6}$ (for $\pi = (1, 2, 3, 4, 5, 6)$)

The x_j's are called **cubical coordinates**.

Exercise 132 *Let* $(z_2, \dots, z_{n-2})$ *be simplicial coordinates and* $(x_2, \dots, x_{n-2})$ *the corresponding cubical coordinates. Show that*

$$\text{Li}_{m_{n-2}\dots m_3 m_2}(x_{n-2}, \dots, x_3, x_2) = (-1)^{n-3} G_{m_{n-2}\dots m_3 m_2}(z_{n-2}, \dots, z_3, z_2; 1) \tag{F.74}$$

Let us now fix $i_0 \in \{2, \dots, n-2\}$. We will study the limit $u_{i_0,n} \to 0$. The chord (i_0, n) splits the polygon into two smaller polygons. We set $z' = (z_1, z_2, \dots, z_{i_0})$ and $z'' = (z_{i_0+1}, \dots, z_n)$. As before we label the new edge by z_e. One of the two smaller polygons has the edges $z' \cup \{z_e\}$ and the dihedral structure $\pi' = (1, 2, \dots, i_0, e)$, the other smaller polygon has the edges $z'' \cup \{z_e\}$ and the dihedral structure $\pi'' = (e, i_0 + 1, i_0 + 2, \dots, n)$. In the limit $u_{i_0,n} \to 0$ we have

$$\lim_{u_{i_0,n} \to 0} u_{i,j} = 1 \tag{F.75}$$

for any chord $(i, j) \in \chi(z, \pi)$ which crosses the chord $(i_0, n) \in \chi(z, \pi)$.

F.5 The Genus One Case

Let us now turn to the genus one case. We use the genus one case to introduce fine and coarse moduli spaces.

F.5.1 Fine and Coarse Moduli Spaces

A moduli space is a space (or a scheme or a stack), whose points represent isomorphism classes of algebro-geometric objects.

Let us now introduce the concepts of a fine moduli space and of a coarse moduli space. For a **fine moduli space** $\mathcal{M}$ we require that

- there is a universal family of objects $C \to \mathcal{M}$, such that the fibre over $z \in \mathcal{M}$ is the object the point z is parametrising.
  Example: If we consider the moduli space $\mathcal{M}_g$ of genus g curves, the fibre over z would be given by the corresponding curve C.
- for any family of objects, parametrised by some base B, say $C_B \to B$, we require that there is a map

$$f : B \to \mathcal{M}, \tag{F.76}$$

  and C_B is isomorphic to $f^* C$.

$$\begin{array}{ccc} C_B & & C \\ \downarrow & & \downarrow \\ B & \xrightarrow{f} & \mathcal{M} \end{array} \tag{F.77}$$

For a **coarse moduli space** M we require that

- for any family of objects, parametrised by some base B, say $C_B \to B$, we require that there is a map

$$f : B \to M, \tag{F.78}$$

which sends the fibres of B to their isomorphism classes.

A coarse moduli space does not necessarily carry any family of appropriate objects, let alone a universal one. In other words, a fine moduli space includes both a base space $\mathcal{M}$ and an universal family $C \to \mathcal{M}$, while a coarse moduli space only has the base space M.

F.5.2 Framed Elliptic Curves

We recall that an elliptic curve is a cubic curve in $\mathbb{CP}^2$ with a marked (rational) point. Equivalently, we may represent an elliptic curve as the quotient of $\mathbb{C}$ by a lattice Λ:

$$\mathbb{C}/\Lambda. \tag{F.79}$$

The origin of $\mathbb{C}$ corresponds to the marked point and we denote the curve together with the marked point by $(\mathbb{C}/\Lambda, 0)$.

A **framed elliptic curve** is an elliptic curve E together with an ordered basis γ_1, γ_2 of $H_1(E, \mathbb{Z})$ such that the intersection number $\gamma_1 \cdot \gamma_2 = 1$. A framing of a lattice Γ in $\mathbb{C}$ is an ordered basis ψ_1, ψ_2 such that

$$\text{Im}\left(\frac{\psi_2}{\psi_1}\right) > 0. \tag{F.80}$$

We may think of a framed elliptic curve as $\mathbb{C}/\Lambda$ together with the choice of an ordered basis ψ_1, ψ_2 satisfying Eq. (F.80). Two elliptic curves are isomorphic if there is a $c \in \mathbb{C}^*$ such that

$$\Lambda' = c\Lambda. \tag{F.81}$$

Let Λ be generated by (ψ_1, ψ_2). We may therefore rescale the lattice such that $\psi_1 = 1$ and $\text{Im}(\psi_2) > 0$. We label the basis vectors $(1, \tau)$. The same lattice is generated if we perform a SL $(2, \mathbb{Z})$-transformation. Thus we are tempted to consider as a set

$$M_{1,1} \cong \mathbb{H}/\mathrm{SL}\,(2,\mathbb{Z})\,. \tag{F.82}$$

We will later see that this is just a coarse moduli space, not a fine one. In order to get a fine moduli space we have to consider the **orbifold**

$$\mathcal{M}_{1,1} \cong \mathbb{H} \,/\!\!/\, \mathrm{SL}\,(2,\mathbb{Z})\,. \tag{F.83}$$

We will define orbifolds in a second. But let us first discuss why the set $M_{1,1}$ is just a coarse moduli space. All points of $M_{1,1}$ have a non-trivial stabiliser group. The stabiliser group (or isotropy group or little group) of the point τ is

$$\{\gamma \in \mathrm{SL}\,(2,\mathbb{Z})\,|\gamma\,(\tau) = \tau\}\,. \tag{F.84}$$

There is an isomorphism between the stabilizer group at the point τ and the automorphism group of the corresponding elliptic curve:

$$\mathrm{Aut}\,(\mathbb{C}/\Lambda_\tau, 0) \cong \{\gamma \in \mathrm{SL}\,(2,\mathbb{Z})\,|\gamma\,(\tau) = \tau\}\,. \tag{F.85}$$

Each point in $\mathbb{H}$ is invariant under

$$\gamma = \begin{pmatrix} -1 & 0 \\ 0 & -1 \end{pmatrix}. \tag{F.86}$$

In addition, the point $\tau = i$ is invariant under S, while the point $\tau = r_3 = \exp(2\pi i/3)$ is invariant under $U = ST$, where S and T are the usual generators of SL (2, $\mathbb{Z}$):

$$T = \begin{pmatrix} 1 & 1 \\ 0 & 1 \end{pmatrix}, \quad S = \begin{pmatrix} 0 & -1 \\ 1 & 0 \end{pmatrix}, \tag{F.87}$$

and hence

$$U = \begin{pmatrix} 0 & -1 \\ 1 & 1 \end{pmatrix}. \tag{F.88}$$

The element S is of order 4, the element U is of order 6. Furthermore, $S^2 = U^3 = -I$. Thus

$$\mathrm{SL}\,(2,\mathbb{Z}) = \left\langle S, T | S^2 = (ST)^3\,, S^4 = I \right\rangle. \tag{F.89}$$

We therefore have

$$\mathrm{Aut}\,(\mathbb{C}/\Lambda_\tau, 0) \cong \mathbb{Z}_2, \qquad \tau \neq i, r_3, \tag{F.90}$$

and

$$\text{Aut}\left(\mathbb{C}/\Lambda_i, 0\right) \cong \mathbb{Z}_4, \quad \text{Aut}\left(\mathbb{C}/\Lambda_{r_3}, 0\right) \cong \mathbb{Z}_6. \tag{F.91}$$

F.5.3 The Universal Family of Framed Elliptic Curves

We first discuss framed elliptic curves, i.e. elliptic curves together with a fixed choice of an ordered basis ψ_1, ψ_2. We may construct a **universal family of framed elliptic curves**: Let us consider $\mathbb{C} \times \mathbb{H}$ with an $\mathbb{Z}^2$-action given by

$$(n_2, n_1) : (z, \tau) \rightarrow (z + n_2\tau + n_1, \tau). \tag{F.92}$$

The $\mathbb{Z}^2$-action corresponds to the translation of z by a lattice vector. We set

$$\mathcal{C}_{\mathbb{H}} = (\mathbb{C} \times \mathbb{H}) / \mathbb{Z}^2. \tag{F.93}$$

There is a projection

$$\begin{aligned} \pi : \mathcal{C}_{\mathbb{H}} &\rightarrow \mathbb{H}, \\ (z, \tau) &\rightarrow \tau \end{aligned} \tag{F.94}$$

such that the fibre over τ is $\mathbb{C}/\Lambda_\tau$.

A family of elliptic curves is **framed**, if it has a locally constant framing. This means that the cycles $\gamma_1, \gamma_2 \in H_1(E, \mathbb{Z})$ defining the framing vary smoothly. The family $\mathcal{C}_{\mathbb{H}} \rightarrow \mathbb{H}$ is framed. Let $\mathcal{C}_B \rightarrow B$ be a family of framed elliptic curves. We have a function

$$\begin{aligned} f : B &\rightarrow \mathbb{H}, \\ t &\rightarrow \frac{\int\limits_{\gamma_2(t)} \omega_t}{\int\limits_{\gamma_1(t)} \omega_t}, \end{aligned} \tag{F.95}$$

where ω_t is any non-zero holomorphic differential one-form on $\mathcal{C}_t$. This mapping is called the **period mapping**. The period mapping is holomorphic. Furthermore, $\mathcal{C}_B$ is isomorphic to $f^*\mathcal{C}_{\mathbb{H}}$. Thus $\mathbb{H}$ is a fine moduli space for families of framed elliptic curves.

Let us now try to remove the framing. In other words, we not only allow for z translations by lattice vectors, but also allow a change of basis of the lattice vectors by a modular transformation. We consider again $\mathbb{C} \times \mathbb{H}$, but now with the action of the semi-direct product $\text{SL}(2, \mathbb{Z}) \ltimes \mathbb{Z}^2$. We denote elements of this group by

$$(\gamma, \vec{n}), \quad \gamma \in \text{SL}(2, \mathbb{Z}), \quad \vec{n} = (n_2, n_1) \in \mathbb{Z}^2. \tag{F.96}$$

The group composition is given by

$$(\gamma_1, \vec{n}_1)(\gamma_2, \vec{n}_2), = (\gamma_1\gamma_2, \vec{n}_1\gamma_2 + \vec{n}_2). \tag{F.97}$$

The group $\mathrm{SL}(2, \mathbb{Z}) \ltimes \mathbb{Z}^2$ acts on $\mathbb{C} \times \mathbb{H}$ as

$$z' = \frac{z + n_2\tau + n_1}{c\tau + d}, \qquad \tau' = \frac{a\tau + b}{c\tau + d}. \tag{F.98}$$

In order to grasp the point of what follows, the following simple exercise is helpful:

Exercise 133 *Let*

$$\gamma = \begin{pmatrix} -1 & 0 \\ 0 & -1 \end{pmatrix}, \qquad \vec{n} = (0, 0). \tag{F.99}$$

Work out z' and τ'.

Let us now consider

$$\begin{aligned} C_{M_{1,1}} &= (\mathbb{C} \times \mathbb{H}) / \left(\mathrm{SL}(2, \mathbb{Z}) \ltimes \mathbb{Z}^2\right), \\ M_{1,1} &= \mathbb{H}/\mathrm{SL}(2, \mathbb{Z}). \end{aligned} \tag{F.100}$$

There is a projection $\pi : C_{M_{1,1}} \to M_{1,1}$ given by $(z, \tau) \to \tau$, but $C_{M_{1,1}} \to M_{1,1}$ is not a universal elliptic curve. To see this, let us consider the fibre above $\tau \in M_{1,1}$. We have

$$\pi^{-1}(\tau) = (\mathbb{C}/\Lambda_\tau, 0) / \mathrm{Aut}(\mathbb{C}/\Lambda_\tau, 0). \tag{F.101}$$

For a generic value of τ we have $\mathrm{Aut}(\mathbb{C}/\Lambda_\tau, 0) \cong \mathbb{Z}_2$ and $\mathbb{Z}_2$ acts on z (see Exercise 133) by

$$z' = -z. \tag{F.102}$$

This additional symmetry makes $\pi^{-1}(\tau)$ isomorphic to the Riemann sphere $\mathbb{CP}^1$. In order to see this, consider first the parallelogram spanned by 1 and τ. The additional $\mathbb{Z}_2$-symmetry identifies the points

$$z \text{ and } 1 + \tau - z. \tag{F.103}$$

Thus, we have to consider only half of the points, i.e. for example just the triangle as shown in the figure below.

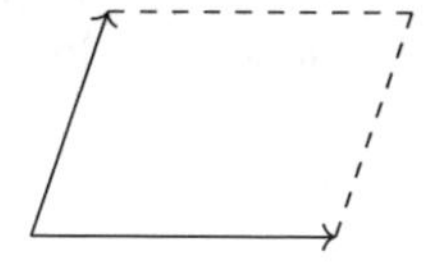

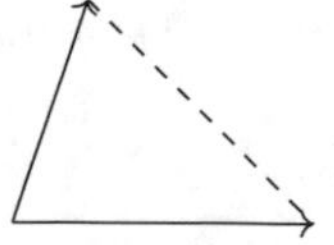

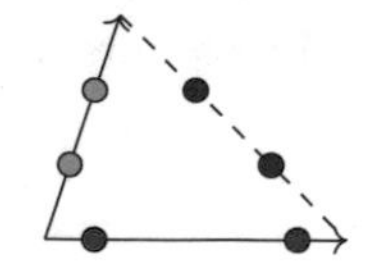

Along the edges of the triangle we then identify points which are symmetric around the mid-point, as shown by the points of the same colour in the figure above. This gives a sphere.
In particular, no fibre of $C_{M_{1,1}}$ is an elliptic curve. Thus, $C_{M_{1,1}} \rightarrow M_{1,1}$ is not a universal elliptic curve and $M_{1,1}$ is just a coarse moduli space.

F.5.4 Orbifolds

In order to get a fine moduli space, we have to introduce orbifolds. We start with the definition of an **orbifold chart** (also called **uniformising system**) [464, 465]. Let U_i be a non-empty connected topological space. An orbifold chart of dimension n for U_i is a quadruple $(V_i, \Gamma_i, \rho_i, \phi_i)$, where

- V_i is a connected and simply connected open subset of $\mathbb{R}^n$,
- Γ_i is a **finite group**,
- $\rho_i : \Gamma_i \rightarrow \text{Aut}\,(V_i)$ is a (**not necessarily injective**) homomorphism from Γ_i to the group of smooth automorphisms of V_i. We set

$$\text{Ker}\,(\Gamma_i) \;=\; \text{Ker}\,(\rho_i) \;\subseteq\; \Gamma_i, \qquad \Gamma_i^{\text{red}} \;=\; \rho_i\,(\Gamma_i) \;\subseteq\; \text{Aut}\,(V_i)\,. \quad \text{(F.104)}$$

- $\phi_i : V_i \rightarrow U_i$ is a continuous and surjective map from V_i to U_i invariant under Γ_i, which defines a homeomorphism

$$V_i / \Gamma_i^{\text{red}} \rightarrow U_i. \qquad \text{(F.105)}$$

Note that we do not require that Γ_i acts effectively on V_i.
The chart $(V_i, \Gamma_i, \rho_i, \phi_i)$ is called **linear**, if the Γ_i-action on $\mathbb{R}^n$ is linear.
Next we define **embeddings** [466]: Let us consider $U_i \subset U_j$, this defines an inclusion $\iota : U_i \rightarrow U_j$. Let $(V_i, \Gamma_i, \rho_i, \phi_i)$ be a chart for U_i and let $(V_j, \Gamma_j, \rho_j, \phi_j)$ be a chart of U_j. An embedding is given by a pair (ψ_{ji}, ρ_{ji}), where

- $\rho_{ji} : \Gamma_i \rightarrow \Gamma_j$ is an injective group homomorphism, which induces a group isomorphism $\rho_{ji} : \text{Ker}\,(\Gamma_i) \rightarrow \text{Ker}\left(\Gamma_j\right)$,
- $\psi_{ji} : V_i \rightarrow V_j$ is a homeomorphism, called **gluing map**, of V_i onto an open subset of V_j,
- the gluing map is compatible with the chart:

$$\begin{array}{ccc} V_i & \xrightarrow{\psi_{ji}} & V_j \\ {\scriptstyle \phi_i}\downarrow & & \downarrow{\scriptstyle \phi_j} \\ U_i & \xrightarrow{\iota} & U_j \end{array} \qquad \phi_j \circ \psi_{ji} \;=\; \iota \circ \phi_i. \qquad \text{(F.106)}$$

- the gluing map is equivariant:

$$\begin{array}{ccc} V_i & \xrightarrow{\psi_{ji}} & V_j \\ \rho_i(\gamma_i)\downarrow & & \downarrow\rho_j(\rho_{ji}(\gamma_i)) \\ V_i & \xrightarrow{\psi_{ji}} & V_j \end{array} \qquad \psi_{ji}\left(\rho_i\left(\gamma_i\right)\cdot x\right) = \rho_j\left(\rho_{ji}\left(\gamma_i\right)\right)\cdot\psi_{ji}\left(x\right) \tag{F.107}$$

In order to simplify the notation we drop the maps ρ_i and ρ_j (which are understood implicitly), thus

$$\psi_{ji}\left(\gamma_i \cdot x\right) = \rho_{ji}\left(\gamma_i\right)\cdot\psi_{ji}\left(x\right). \tag{F.108}$$

- the gluing map ψ_{ji} is unique up to a right action of Γ_i and a left action of Γ_j.

An **orbifold atlas** on X is a family of orbifold charts, which cover X and are compatible in the following sense: Given a chart $(V_i, \Gamma_i, \rho_i, \phi_i)$ of U_i and a chart $(V_j, \Gamma_j, \rho_j, \phi_j)$ of U_j, and given any $x \in U_i \cap U_j$, there exists an open neighbourhood $U_k \subset U_i \cap U_j$ and a chart $(V_k, \Gamma_k, \rho_k, \phi_k)$ of U_k, such there are embeddings

$$\begin{aligned} (V_k, \Gamma_k, \rho_k, \phi_k) &\to (V_i, \Gamma_i, \rho_i, \phi_i), \\ (V_k, \Gamma_k, \rho_k, \phi_k) &\to \left(V_j, \Gamma_j, \rho_j, \phi_j\right). \end{aligned} \tag{F.109}$$

Two such atlases are said to be equivalent, if they have a common refinement.
An **orbifold** O is the space X together with an equivalence class of orbifold atlases. The space X is called the **underlying space**. An orbifold contains more information than just its underlying space X.
Let M be a manifold and Γ a finite group acting properly on M. This defines an orbifold, which we denote by $M \mathbin{/\!\!/} \Gamma$ with underlying space M/Γ.
If $x \in X$ and $v \in \phi^{-1}(x)$ is a point in the inverse image of x in some local chart, then the stabiliser group at v is independent of the chart. We call this group the **local group** at x and denote it by Γ_x. The **singular points** of an orbifold are the points $x \in X$ with a non-trivial local group Γ_x.
Example 1: The circle S^1, defined by

$$x^2 + y^2 = 1, \tag{F.110}$$

together with the group $\Gamma : y \to -y$. Then $S^1 \mathbin{/\!\!/} \Gamma$ is an orbifold. The singular points are $(-1, 0)$ and $(1, 0)$.
Example 2: The complex upper-half plane $\mathbb{H}$ with the group $\mathrm{PSL}(2, \mathbb{Z})$. Then

$$\mathbb{H} \mathbin{/\!\!/} \mathrm{PSL}\left(2, \mathbb{Z}\right) \tag{F.111}$$

is an orbifold. The singular points are $\tau = i$ and $\tau = r_3$.
Example 3: The complex upper-half plane $\mathbb{H}$ with the group $\mathrm{SL}(2, \mathbb{Z})$. Then

$$\mathbb{H} \mathbin{/\!/} \mathrm{SL}\,(2, \mathbb{Z}) \tag{F.112}$$

is an orbifold. All points of $\mathbb{H}/\mathrm{SL}\,(2, \mathbb{Z})$ are singular points of the orbifold, since

$$\begin{pmatrix} -1 & 0 \\ 0 & -1 \end{pmatrix} \tag{F.113}$$

acts trivially.

Let O_1 and O_2 be two orbifolds. A **smooth map** between O_1 and O_2 is defined as follows: There is a continuous map

$$f : X_1 \;\to\; X_2 \tag{F.114}$$

between the underlying spaces such that for any point $x_1 \in X_1$ there is a chart $(V_1, \Gamma_1, \rho_1, \phi_1)$ around x_1 and a chart $(V_2, \Gamma_2, \rho_2, \phi_2)$ around $f(x_1)$ together with a group homomorphism $\rho : \Gamma_1 \to \Gamma_2$ with the properties that f maps $\phi_1(V_1)$ into $\phi_2(V_2)$ and can be lifted to a smooth map

$$\tilde{f} : V_1 \;\to\; V_2 \tag{F.115}$$

such that

$$\begin{aligned} \phi_2 \circ \tilde{f} &= f \circ \phi_1, \\ \tilde{f}\,(\gamma_1 \cdot x) &= \rho\,(\gamma_1) \cdot \tilde{f}\,(x)\,. \end{aligned} \tag{F.116}$$

In terms of commutative diagrams:

$$\begin{array}{ccc} V_1 & \xrightarrow{\tilde{f}} & V_2 \\ {\scriptstyle \phi_1}\big\downarrow & & \big\downarrow{\scriptstyle \phi_2} \\ U_1 & \xrightarrow{f} & U_2 \end{array} \qquad\qquad \begin{array}{ccc} V_1 & \xrightarrow{\tilde{f}} & V_2 \\ {\scriptstyle \gamma_1}\big\downarrow & & \big\downarrow{\scriptstyle \rho(\gamma_1)} \\ V_1 & \xrightarrow{\tilde{f}} & V_2 \end{array} \tag{F.117}$$

Let us discuss an example: We consider $O_1 = (\mathbb{C} \times \mathbb{C}) \mathbin{/\!/} (C_2 \times C_2)$ and $O_2 = \mathbb{C} \mathbin{/\!/} C_2$, where $C_2 = \{1, -1\}$ denotes the multiplicative group with two elements, together with the group actions

$$\begin{aligned} O_1 : (\lambda_1, \lambda_2) \cdot (z_1, z_2) &= (\lambda_1 z_1, \lambda_2 z_2)\,, \\ O_2 : \lambda \cdot z &= \lambda z. \end{aligned} \tag{F.118}$$

An orbifold map $O_1 \to O_2$ is defined by

$$\begin{aligned} f : (\mathbb{C} \times \mathbb{C}) \,/\, (C_2 \times C_2) &\to \mathbb{C} \mathbin{/\!/} C_2, \\ (z_1, z_2) &\to z_1, \end{aligned} \tag{F.119}$$

and

$$\begin{aligned}\rho : C_2 \times C_2 &\rightarrow C_2, \\ (\lambda_1, \lambda_2) &\rightarrow \lambda_1. \end{aligned} \tag{F.120}$$

This is a projection and the fibre above the point $[\pm z_1]$ is given by the second factor $O_3 = \mathbb{C} \mathbin{/\!/} C_2$.

Now let us discuss a slight modification of this example: We consider the case that some groups act trivially:

$$\begin{aligned}\tilde{O}_1 : (\lambda_1, \lambda_2) \cdot (z_1, z_2) &= (z_1, \lambda_2 z_2), \\ \tilde{O}_2 : \lambda \cdot z &= z. \end{aligned} \tag{F.121}$$

Note that $\tilde{O}_1$ is not identical to O_1 and $\tilde{O}_2$ is not identical to O_2, as the group actions are different. We consider an orbifold map $\tilde{O}_1 \rightarrow \tilde{O}_2$ with f and ρ as in Eq. (F.119) and Eq. (F.120), respectively. This is again a projection, where the fibre above the point $[z_1]$ is given by the second factor $\tilde{O}_3 = \mathbb{C} \mathbin{/\!/} C_2$. Note that λ_1 acts trivially on $\tilde{O}_1$ and (through ρ) trivially on $\tilde{O}_2$, but does not act on $\tilde{O}_3$.

F.5.5 The Universal Family of Elliptic Curves

With the preparation on orbifolds we are now in a position to present the universal family of elliptic curves (without any framing). We set

$$\begin{aligned}\mathcal{C} &= (\mathbb{C} \times \mathbb{H}) \mathbin{/\!/} \left(\mathrm{SL}(2, \mathbb{Z}) \ltimes \mathbb{Z}^2\right), \\ \mathcal{M}_{1,1} &= \mathbb{H} \mathbin{/\!/} \mathrm{SL}(2, \mathbb{Z}). \end{aligned} \tag{F.122}$$

The projection $\mathbb{C} \times \mathbb{H} \rightarrow \mathbb{H}$, given by $(z, \tau) \rightarrow \tau$, induces an orbifold morphism

$$\mathcal{C} \rightarrow \mathcal{M}_{1,1}. \tag{F.123}$$

In more detail, this orbifold morphism is given by

$$\begin{aligned}f : (\mathbb{C} \times \mathbb{H}) / \left(\mathrm{SL}(2, \mathbb{Z}) \ltimes \mathbb{Z}^2\right) &\rightarrow \mathbb{H}/\mathrm{SL}(2, \mathbb{Z}), \\ (z, \tau) &\rightarrow \tau, \end{aligned} \tag{F.124}$$

and

$$\begin{aligned}\rho : \mathrm{SL}(2, \mathbb{Z}) \ltimes \mathbb{Z}^2 &\rightarrow \mathrm{SL}(2, \mathbb{Z}), \\ (\gamma, \vec{n}) &\rightarrow \gamma. \end{aligned} \tag{F.125}$$

The fibre above τ is $\mathbb{C} \mathbin{/\!/} \mathbb{Z}^2$, i.e. an elliptic curve, and $\mathcal{C} \rightarrow \mathcal{M}_{1,1}$ is a fine moduli space of smooth genus one curves with one marked point.

F.5.6 Compactification of $\mathcal{M}_{1,1}$

The moduli space $\mathcal{M}_{1,1}$ parametrises equivalence classes of smooth genus one curves with one marked point, i.e. smooth elliptic curves. For the compactification $\overline{\mathcal{M}}_{1,1}$ we add one point, corresponding to a nodal genus one curve with one marked point (the marked point does not coincide with the node). Technically this is done as follows: We denote by $\mathbb{D}$ the (open) unit disc

$$\mathbb{D} = \{ \bar{q} \in \mathbb{C} \mid |\bar{q}| < 1 \} \tag{F.126}$$

and by $\mathbb{D}^*$ the punctured unit disc

$$\mathbb{D}^* = \mathbb{D} \backslash \{0\}. \tag{F.127}$$

We construct $\overline{\mathcal{M}}_{1,1}$ with the help of two charts. The first chart

$$\mathcal{M}_{1,1} = \mathbb{H} \mathbin{/\!/} \mathrm{SL}(2, \mathbb{Z}) \tag{F.128}$$

covers $\mathcal{M}_{1,1}$ and has been discussed above. The second chart is given by

$$\mathbb{D} \mathbin{/\!/} C_2, \tag{F.129}$$

where C_2 acts trivially on $\mathbb{D}$ and is the relict of the modular transformation

$$\begin{pmatrix} -1 & 0 \\ 0 & -1 \end{pmatrix} \tag{F.130}$$

The mapping between the coordinate τ on $\mathcal{M}_{1,1}$ and $\bar{q}$ on $\mathbb{D} \mathbin{/\!/} C_2$ is given by

$$\bar{q} = \exp(2\pi i \tau). \tag{F.131}$$

The two charts overlap on $\mathbb{D}^* \mathbin{/\!/} C_2$.

We call an action of a group Γ on a space X **virtually free**, if Γ has a finite index subgroup Γ', that acts freely.

Let Γ be a discrete group which acts virtually free and properly discontinuous on X. Let Γ' be the finite index normal subgroup, which acts freely. The **orbifold Euler characteristic** is defined by

$$\chi\left(X /\!\!/ \Gamma\right) = \frac{1}{[\Gamma : \Gamma']}\chi\left(X/\Gamma'\right). \tag{F.132}$$

For the two moduli spaces $\mathcal{M}_{1,1}$ and $\overline{\mathcal{M}}_{1,1}$ one finds the orbifold Euler characteristics

$$\chi\left(\mathcal{M}_{1,1}\right) = -\frac{1}{12}, \quad \chi\left(\overline{\mathcal{M}}_{1,1}\right) = \frac{5}{12}. \tag{F.133}$$

Appendix G
Algebraic Geometry

In the main part of this book we made no reference to sheaves or schemes, although they are fundamental concepts in algebraic geometry. As mathematicians always aim to state theorems as general as possible, this inevitably involves generalisations and abstraction, leading to sheaves and schemes. It is therefore no surprise that one will encounter these terms in the mathematical literature quite frequently. While physicists are certainly familiar with concrete examples of sheaves and schemes, the abstract language makes it sometimes difficult to read mathematical literature. In this appendix we give a short introduction to sheaves and schemes. This appendix may serve as a survival kit for reading the mathematical literature.

References for sheaves and schemes are the books of Hartshorne [467], Eisenbud and Harris [468] and Holme [469].

G.1 Topology

We start with the definition of a topology:

Let X be a set and $\mathcal{T}$ a collection of subsets of X. $\mathcal{T}$ is called a **topology** of X if

1. $\emptyset \in \mathcal{T}$ and $X \in \mathcal{T}$,
2. $U_1, U_2 \in \mathcal{T} \Rightarrow U_1 \cap U_2 \in \mathcal{T}$,
3. $U_\alpha \in \mathcal{T}$ for all $\alpha \in I \Rightarrow \bigcup\limits_{\alpha \in I} U_\alpha \in \mathcal{T}$.

The pair $(X, \mathcal{T})$ is called a **topological space**. A subset $U \subseteq X$ is called **open** if $U \in \mathcal{T}$. A subset $A \subseteq X$ is called **closed** if the complement $X \backslash A$ is open. For closed sets we have

S. Weinzierl, *Feynman Integrals*, UNITEXT for Physics,
https://doi.org/10.1007/978-3-030-99558-4

1. $\emptyset$ and X are closed sets,
2. if A and B are closed, then $A \cup B$ is closed,
3. if A_α is closed, where $\alpha \in I$, then $\bigcap_{\alpha \in I} A_\alpha$ is closed.

The closure of an open set U is denoted by $\overline{U}$. We defined a topological space by its open sets. Alternatively, we may define a topological space by its closed sets and the requirement that the closed sets satisfy items 1-3 for closed sets.

A map between topological spaces is called **continuous** if the pre-image of any open set is again open. A bijective map which is continuous in both directions is called a **homeomorphism**.

A topological space is called **Hausdorff** if for any two distinct points $x_1, x_2 \in X$ there exist open sets $U_1, U_2 \in \mathcal{T}$ with

$$p_1 \in U_1, \quad p_2 \in U_2, \quad U_1 \cap U_2 = \emptyset. \tag{G.1}$$

For a subset $X' \subseteq X$ we define the **induced topology** by

$$\mathcal{T}' = \left\{ U' | U' = U \cap X', U \in \mathcal{T} \right\}. \tag{G.2}$$

Note that U' is open in X only if X' is open in X.

A topological space X is called **irreducible**, if it cannot be expressed as the union of two proper closed subsets, otherwise the topological space is called reducible.

Let $Y \subseteq X$. Y is called an **irreducible component** of X if Y is irreducible (within the relative topology $\mathcal{T}'$) and if Y is maximal (i.e. $Y \subseteq Y'$ and Y' irreducible implies $Y = Y'$). Note that if Y is an irreducible component of X, then $Y = \overline{Y}$.

Let X be a topological space. The **dimension** of X is defined to be the supremum of all integers n such that there exists a chain

$$Y_0 \subset Y_1 \subset \cdots \subset Y_n \tag{G.3}$$

of distinct irreducible subsets of X.

Theorem 33 *A topological space X is the union of irreducible components Y_α:*

$$X = \bigcup_\alpha Y_\alpha \tag{G.4}$$

A topological space X is called a **Noetherian space**, if each ascending chain of open sets $U_1 \subseteq U_2 \subseteq \dots$ becomes stationary, i.e. there exists a $k \in \mathbb{N}$ with $U_j = U_k$ for all $j \geq k$. This is equivalent to the requirement that each descending chain of closed sets $A_1 \supseteq A_2 \supseteq \dots$ becomes stationary. A third equivalent formulation is, that each non-empty set of open sets contains a maximal element.

Theorem 34 *Let X be a Noetherian topological space. Then X has only a finite number of irreducible components $Y_1, Y_2, \dots, Y_r$:*

$$X = Y_1 \cup Y_2 \cup \cdots \cup Y_r. \tag{G.5}$$

If we require $Y_i \not\subseteq Y_j$, this decomposition is unique up to ordering.

G.2 Rings

Rings play an essential part in the the theory of schemes. In this section we recall the most important facts and definitions.

A set $(R, +, \cdot)$ is called a **ring** if

(R1): $(R, +)$ is an **Abelian group**,
(R2): the operation $\cdot$ is **associative**: $a \cdot (b \cdot c) = (a \cdot b) \cdot c$,
(R3): the operation $\cdot$ is **distributive** with respect to the operation $+$:

$$\begin{aligned} a \cdot (b + c) &= (a \cdot b) + (a \cdot c), \\ (a + b) \cdot c &= (a \cdot c) + (a \cdot b). \end{aligned} \tag{G.6}$$

The **trivial ring** (or zero ring) is the ring consisting of one element 0 with $0 + 0 = 0$ and $0 \cdot 0 = 0$. The trivial ring is denoted by $\{0\}$.

If there is a neutral element 1 for the operation $\cdot$, the ring is called a **ring with** 1. Other names for a ring with 1 are unital ring, ring with unity or ring with identity.

A ring $(R, +, \cdot)$ is called **commutative** if the operation $\cdot$ is commutative :

$$a \cdot b = b \cdot a. \tag{G.7}$$

Let R be a ring with 1. An element $a \in R$ is called **invertible** or **unit** in R, if a has a left-inverse and a right-inverse with respect to multiplication. We denote the set of invertible elements by R^*. R^* is a group with respect to multiplication.

An element $a \in R$, $a \neq 0$ is called **zero-divisor**, if there is a $b \in R$, $b \neq 0$ such that $ab = 0$ or $ba = 0$.

A non-trivial commutative ring with 1 and with no zero-divisors is called an **integral domain**.

Of particular importance are ideals of a ring:

A sub-group I of $(R, +)$ is called an **ideal** (or two-sided ideal), if

(I1): $r \cdot a \in I$ for all $r \in R$ and $a \in I$ (or short $RI \subseteq I$),
(I2): $a \cdot r \in I$ for all $r \in R$ and $a \in I$ (or short $IR \subseteq I$).

If a sub-group I of $(R, +)$ satisfies only $RI \subseteq I$, we call I a left-ideal. Similar, if a sub-group I of $(R, +)$ satisfies only $IR \subseteq I$, we call I a right-ideal.

Every ring R has the ideals $\{0\}$ and R. We call a non-trivial ring **simple**, if these are the only ideals. An ideal I is called a **proper ideal**, if $I \neq R$.

An ideal I is called a **principal ideal**, if it is generated by one element:

$$I = \langle a \rangle . \tag{G.8}$$

R is called a **principal ideal ring** if every ideal I of R is a principal ideal. If R is an integral domain and every ideal I of R is a principal ideal, then R is called a **principal ideal domain**.

Let R be a commutative ring with 1. An ideal P is called a **prime ideal**, if P is a proper ideal and

$$a \cdot b \in P \Rightarrow a \in P \text{ or } b \in P. \tag{G.9}$$

An ideal P in a commutative ring R with 1 is prime if and only if R/P is an integral domain.

An ideal M is called a **maximal ideal**, if M is a proper ideal and for any other ideal I with

$$M \subseteq I \subseteq R \tag{G.10}$$

it follows that $I = M$ or $I = R$. Every ring with 1 contains a maximal ideal. An ideal M in a commutative ring R with 1 is maximal if and only if R/M is a field. Every maximal ideal M in a commutative ring R with 1 is a prime ideal.

The **height** of a prime ideal P is the supremum of all integers n such that there exists a chain

$$P_0 \subset P_1 \subset \cdots \subset P_n = P \tag{G.11}$$

of distinct prime ideals. The dimension or **Krull dimension** of the ring R is defined to be the supremum of the heights of all prime ideals.

Let R be a commutative ring with 1. R is called a **local ring** if R contains exactly one maximal ideal. In order to check if a ring is local, the following theorem is useful:

Theorem 35 *Let R be a commutative ring with* 1*. R is a local ring if and only if the set of non-invertible elements $N = R\backslash R^*$ is an ideal in R. If N is an ideal in R then N is a maximal ideal and the only maximal ideal in R.*

The concept of **localisation** is of central importance for the theory of schemes: Let R be a commutative ring with 1 and S a subset of R closed under multiplication together with the conditions $1 \in S$ and $0 \notin S$. One defines the quotient ring

$$R_S = \left\{ \left[\frac{r}{s}\right] \mid r \in R, s \in S \; : \; \frac{r_1}{s_1} \sim \frac{r_2}{s_2} \Leftrightarrow \exists\, s \in S \text{ such that } s\,(s_2 r_1 - s_1 r_2) = 0 \right\} \tag{G.12}$$

The ring R is a subring of R_S via the identification

$$r = \left[\frac{r}{1}\right]. \tag{G.13}$$

If P is a prime ideal of R, we may consider $S_P = R\backslash P$. By definition of a prime ideal this set is closed under multiplication. The quotient ring R_{S_P} is a local ring. We denote the maximal ideal by P_{S_P}. One often simplifies the notation and writes

$$R_P = R_{S_P}, \quad P = P_P = P_{S_P}. \tag{G.14}$$

Exercise 134 *Let R be a commutative ring with* 1 *and P a prime ideal. Set $S_P = R\backslash P$. Show that S_P is closed under multiplication.*

Exercise 135 *Consider the commutative ring $\mathbb{Z}$ and the prime ideal $P = \langle 5 \rangle$. Define $S_P = \mathbb{Z}\backslash\langle 5 \rangle$. Describe the quotient ring R_{S_P} and its maximal ideal P_{S_P}.*

Finally, let us introduce Noetherian rings and Artinian rings:

A ring R is called a **Noetherian ring** if one of the following conditions is satisfied:

1. Every ascending sequence $I_1 \subseteq I_2 \subseteq \dots$ of ideals I_j of R becomes stationary, i.e. there exists a $k \in \mathbb{N}$ with $I_j = I_k$ for all $j \geq k$.
2. Every non-empty set of ideals of R, partially ordered by inclusion, has a maximal element with respect to set inclusion.
3. Every ideal I of R is finitely generated.

A ring R is calleda **Artinian ring** if every descending sequence $I_1 \supseteq I_2 \supseteq \dots$ of ideals I_j of R becomes stationary, i.e. there exists a $k \in \mathbb{N}$ with $I_j = I_k$ for all $j \geq k$.

G.3 Algebraic Varieties

Affine Algebraic Varieties

Let $\mathbb{A}$ be an algebraically closed field. The set of all n-tuples of elements of $\mathbb{A}$ defines the affine n-space over $\mathbb{A}$, which we denote by $\mathbb{A}^n$. An element $x \in \mathbb{A}^n$ will be called a point, and if $x = (x_1, \ldots, x_n)$ with $x_i \in \mathbb{A}$, then the x_i will be called the coordinates of x.

Let $T = (t_1, \ldots, t_n)$ be an n-tuple of independent variables, $\mathbb{A}[T] = \mathbb{A}[t_1, \ldots, t_n]$ the ring of polynomials over the field $\mathbb{A}$ and let $A \subset \mathbb{A}[T]$. An **affine algebraic set** is given by

$$V(A) = \{\, x \in \mathbb{A}^n \mid f(x) = 0 \;\; \forall f \in A \,\}. \tag{G.15}$$

Note that $V(A) = V(\mathcal{A})$, where $\mathcal{A} = \langle A \rangle$ is the ideal generated by A in $\mathbb{A}[T]$. Note further that $\mathbb{A}[T]$ is a Noetherian ring, therefore every ideal is generated by a finite set of polynomials. If $\mathcal{A} = \langle f \rangle$ is a principal ideal, e.g. generated by one element f, then $V(f)$ is called a **hyperplane** in $\mathbb{A}^n$. V is antiton, i.e.

$$\mathcal{A} \subseteq \mathcal{B} \Rightarrow V(\mathcal{B}) \subseteq V(\mathcal{A}). \tag{G.16}$$

The empty set $\emptyset$ and the whole space $\mathbb{A}^n$ are algebraic sets:

$$\emptyset = V(\langle 1 \rangle), \quad \mathbb{A}^n = V(\{0\}). \tag{G.17}$$

We further have

$$\begin{aligned} V(\mathcal{A}) \cup V(\mathcal{B}) &= V(\mathcal{A}\mathcal{B}) = V(\mathcal{A} \cap \mathcal{B}), \\ \bigcap_{i \in I} V(\mathcal{A}_i) &= V(\sum_{i \in I} \mathcal{A}_i). \end{aligned} \tag{G.18}$$

Thus the algebraic sets can be viewed as closed sets. The open sets are then given as the complements of the closed sets. This defines the **Zariski topology**.

> An **affine algebraic variety** is an irreducible closed subset of $\mathbb{A}^n$ (within the induced topology). An open subset of an affine algebraic variety is called a **quasi-affine algebraic variety**.

Note that the terminology differs in the literature: Some authors use the term "affine algebraic variety" for an affine algebraic set and indicate explicitly if they refer to an irreducible set.

Let's look at an example: We take $\mathbb{A} = \mathbb{C}$ and consider the affine line $\mathbb{A}^1 = \mathbb{C}^1$. The proper irreducible closed subsets of $\mathbb{C}^1$ are sets $\{x\}$ consisting of a single point x. The non-trivial open sets are the complements: $\mathbb{C}^1 \backslash \{x\}$. Please note that the affine line $\mathbb{C}^1$ together with the Zariski topology is not a Hausdorff space: It is impossible to find open sets U_1 and U_2 with $x_1 \in U_1$, $x_2 \in U_2$ and $U_1 \cap U_2 = \emptyset$.

The **annihilation ideal** of a set $Y \subseteq \mathbb{A}^n$ is defined by

$$I(Y) = \{ f \in \mathbb{A}[T] \mid f(x) = 0 \ \forall x \in Y \}. \tag{G.19}$$

We have

$$I(\emptyset) = \mathbb{A}[T], \quad I(\mathbb{A}^n) = \{0\}, \tag{G.20}$$

and

$$Y_1 \subseteq Y_2 \Rightarrow I(Y_2) \subseteq I(Y_1). \tag{G.21}$$

For any two subsets $Y_1, Y_2 \subseteq \mathbb{A}^n$ we have

$$I(Y_1 \cup Y_2) = I(Y_1) \cap I(Y_2). \tag{G.22}$$

We have for any ideal $\mathcal{A}$

$$I(V(\mathcal{A})) = \text{Rad}\mathcal{A}, \tag{G.23}$$

where Rad$\mathcal{A}$ denotes the **radical** of $\mathcal{A}$:

$$\text{Rad}\mathcal{A} = \left\{ f \in \mathbb{A}[T] \mid f^r \in \mathcal{A} \text{ for some } r \in \mathbb{N} \right\}. \tag{G.24}$$

We further have for any set $Y \subseteq \mathbb{A}^n$

$$V(I(Y)) = \overline{Y}. \tag{G.25}$$

There is a one-to-one inclusion-reversing correspondence between algebraic sets in $\mathbb{A}^n$ and radical ideals (i.e. ideals which are equal to their own radical) in $\mathbb{A}[T]$, given by

$$Y \to I(Y), \quad \mathcal{A} \to V(\mathcal{A}). \tag{G.26}$$

Furthermore, an algebraic set is irreducible if and only if its ideal is a prime ideal.

Let Y be an affine algebraic set. The **affine coordinate ring** of Y is defined by

$$\mathbb{A}[T]/I(Y). \tag{G.27}$$

If Y is an affine algebraic set, then the dimension of Y is equal to the dimension of its affine coordinate ring.

Projective algebraic varieties

Let's now consider the projective case. We denote by $P^n(\mathbb{A})$ (or $\mathbb{P}^n$ for short) the n-dimensional projective space over the field $\mathbb{A}$. Points $x \in \mathbb{P}^n$ will be denoted by $x = [x_0 : x_1 : \cdots : x_n]$. We now consider **homogeneous** polynomials $f \in \mathbb{A}[t_0, \dots, t_n]$. We denote by $\mathbb{A}^h[T]$ the set of homogeneous polynomials in $\mathbb{A}[t_0, \dots, t_n]$. Let $A \subseteq \mathbb{A}^h[T]$. A **projective algebraic set** is given by

$$V(A) = \{ x \in \mathbb{P}^n \mid f(x) = 0 \ \ \forall f \in A \}. \tag{G.28}$$

A **projective algebraic variety** is an irreducible projective algebraic set. An open subset of a projective algebraic variety is a **quasi-projective variety**.

As in the affine case the algebraic sets are the closed sets within the Zariski topology. The **annihilation ideal** of a set $Y \subseteq \mathbb{P}^n$ is defined by

$$I(Y) = \{ f \in \mathbb{A}^h[T] \mid f(x) = 0 \ \forall x \in Y \}. \tag{G.29}$$

Note that all f's are homogeneous polynomials.

Let Y be a projective algebraic set. The **homogeneous coordinate ring** of Y is defined by

$$\mathbb{A}[T]/I(Y). \tag{G.30}$$

If Y is a projective algebraic set, then the dimension of Y is equal to the dimension of its homogeneous coordinate ring minus one.

Every affine variety is also a quasi-projective variety. Furthermore the complement of an algebraic set in an affine variety is a quasi-projective variety. In the following we will take "**variety**" to mean either an affine, a quasi-affine, a projective or a quasi-projective variety.

Regular functions

Let X be a **quasi-affine** algebraic variety in $\mathbb{A}^n$. We consider functions

$$f : X \to \mathbb{A}. \tag{G.31}$$

A function $f : X \to \mathbb{A}$ is **regular** at a point x if there is an open neighbourhood U with $x \in U \subseteq X$, and polynomials $p, q \in \mathbb{A}[T]$, such that q is nowhere zero on U, and

$$f = \frac{p}{q} \tag{G.32}$$

on U. The function f is regular on X if it is regular at every point of X.

Let X be now a **quasi-projective** algebraic variety in $\mathbb{P}^n$. A function $f : X \to \mathbb{A}$ is **regular** at a point x if there is an open neighbourhood U with $x \in U \subseteq X$, and homogeneous polynomials $p, q \in \mathbb{A}[t_0, t_1, \dots, t_n]$ of the same degree, such that q is nowhere zero on U, and

$$f = \frac{p}{q} \tag{G.33}$$

on U. The function f is regular on X if it is regular at every point of X.

Let X be a variety. We denote by $\mathcal{O}(X)$ the **ring of all regular functions** on X.

Let x be a point of X. We define the **local ring** $\mathcal{O}_x$ of x on X to be the ring of germs of regular functions on X near x. In other words, an element of $\mathcal{O}_x$ is a pair (U, f), where U is an open subset of X containing x, and f is a regular function on U, and we identify two such pairs (U_1, f_1) and (U_2, f_2) if $f_1 = f_2$ on $U_1 \cap U_2$. $\mathcal{O}_x$ is a local ring: its maximal ideal M is the set of germs of regular functions which vanish at x. The residue field $\mathcal{O}_x/M$ is isomorphic to $\mathbb{A}$.

The **function field** $\mathbb{A}(X)$ of X is defined as follows: an element of $\mathbb{A}(X)$ is an equivalence class of pairs (U, f), where U is a non-empty open subset of X, f is a regular function on U, and where we identify two pairs (U_1, f_1) and (U_2, f_2) if $f_1 = f_2$ on $U_1 \cap U_2$. The elements of $\mathbb{A}(X)$ are called **rational functions** on X.

G.4 Sheaves

The law of physics are often given locally, i.e. as differential equations. Prominent examples are the equations of motion for a particle or – in the context of this book – the differential equations for a family of Feynman integrals. We almost never state it explicitly, but the first step is usually to study these systems in an open neighbourhood of a point of interest. Thus we have open sets U (with a time coordinate t in the case of equation of motions and with kinematic coordinates x in the case of Feynman integrals) and we are interested in local sections with values in some space (position space in the case of a particle or a vector space with dimension equal to the number of master integrals in the case of Feynman integrals), giving us the trajectory of a particle as a function of t or the values of the Feynman integrals as a function of x. Of course, the result should not change if we restrict to a slightly smaller open set $U' \subset U$. Furthermore, if we have two overlapping open sets U_1 and U_1 the results on the intersection $U_1 \cap U_2$ should be compatible.

Sheaves formalise this concept. We may think of a sheaf as data attached to open sets of a topological space, compatible with restriction to smaller open set and compatible with intersections of open sets.

Let's now consider the definition: We start with the definition of a **presheaf**. Let X be a topological space, and let $\mathcal{C}$ be a category. Usually $\mathcal{C}$ is taken to be the category of sets, the category of groups, the category of Abelian groups or the category of commutative rings. A presheaf $\mathcal{F}$ on X assigns to each open set $U \subset X$ an object

$\mathcal{F}(U) \in \mathcal{C}$, called the **sections** of $\mathcal{F}$ over U, and to each inclusion of open sets $U \subset V$ a morphism $r_{V,U} : \mathcal{F}(V) \to \mathcal{F}(U)$, called the **restriction map**, satisfying:

- For every open set $U \in X$, the restriction morphism $r_{U,U} : \mathcal{F}(U) \to \mathcal{F}(U)$ is the identity.
- For any triple $U \subset V \subset W$ of open sets,

$$r_{W,U} = r_{V,U} \circ r_{W,V}. \tag{G.34}$$

By virtue of this relation, we may write $\sigma|_U$ for $r_{V,U}(\sigma)$ without loss of information.

A **sheaf** is a presheaf, which in addition satisfies the following two conditions:

- **Locality**: Let (U_i) be an open covering of an open set U, and $\sigma, \tau \in \mathcal{F}(U)$, such that $\sigma|_{U_i} = \tau|_{U_i}$ for each subset U_i of the covering, then

$$\sigma = \tau. \tag{G.35}$$

This means that a section over U is determined by all its restrictions to subsets U_i of U.
- **Gluing**: Let (U_i) be an open covering of an open set U. If for each i a section $\sigma_i \in \mathcal{F}(U_i)$ is given with the property that for any pair i, j we have

$$\sigma_i|_{U_i \cap U_j} = \sigma_j\big|_{U_i \cap U_j}, \tag{G.36}$$

then there exists a section $\sigma \in \mathcal{F}(U)$ such that

$$\sigma|_{U_i} = \sigma_i \tag{G.37}$$

for all i. This allows the passage from local data to global data: A section ρ on U may be assembled from the local data on the subsets U_i of U.

Let $\mathcal{F}$ be a sheaf on X and $x \in X$ a point. We define the **stalk** $\mathcal{F}_x$ at x to be the direct limit of $\mathcal{F}(U)$ for all open sets U containing x

$$\mathcal{F}_x = \varinjlim_{U \ni x} \mathcal{F}(U). \tag{G.38}$$

The direct limit is taken over all open subsets of X containing the point x with the restriction map. An element $\sigma \in \mathcal{F}_x$ is called a **germ**.

Let's look at a few examples of sheaves which occur frequently. Example of sheaves, which take values in the category of Abelian groups, are:

$\mathcal{O}(U)$	holomorphic functions on U with addition,
$\mathcal{O}^*(U)$	holomorphic functions on U which are nowhere zero with multiplication,
$\mathcal{M}(U)$	meromorphic functions on U with addition,
$\mathcal{M}^*(U)$	meromorphic functions on U without the zero function with multiplication,
$\Lambda^p(U)$	p-forms on U with addition,
$\Omega^p(U)$	holomorphic p-forms on U with addition,
$\mathbb{Z}(U)$	locally constants $\mathbb{Z}$-valued functions on U with addition.

Example of sheaves, which take values in the category of commutative rings with 1, are:

$\mathcal{O}(U)$	holomorphic functions on U with addition and multiplication,
$\mathbb{Z}(U)$	locally constants $\mathbb{Z}$-valued functions on U with addition and multiplication.

Example of sheaves, which take values in the category of rings with 1, are:

$\Lambda^\bullet(U)$	differential forms on U with addition and the wedge product,
$\Omega^\bullet(U)$	holomorphic differential forms on U with addition and the wedge product.

The notation $\mathcal{F}_U$ instead of $\mathcal{F}(U)$ is also used frequently.

G.5 The Spectrum of a Ring

Let R be a commutative ring with 1. The spectrum $\text{Spec}(R)$ of R is a pair $(S, \mathcal{O})$, where S is a set with a topology defined on it (hence a topological space) and $\mathcal{O}$ a sheaf on S. The sheaf $\mathcal{O}$ is called the **structure sheaf**. By abuse of notation, the set S is also often denoted as $\text{Spec}(R)$. It should be clear from the context if the set S or the pair $(S, \mathcal{O})$ is meant.

The set S is given as the set of all prime ideals of R:

$$S = \text{Spec}\,(R) = \{\, P \mid P \text{ prime ideal of } R \,\}. \tag{G.39}$$

Example:

$$\text{Spec}\,(\mathbb{Z}) = \{(0)\,, (2)\,, (3)\,, (5)\,, (7)\,, (11)\,, \dots\}. \tag{G.40}$$

Let I be an ideal of R. We set

$$V\,(I) = \{\, P \in \text{Spec}\,(R) \mid I \subseteq P \,\}. \tag{G.41}$$

We have $V(R) = \emptyset$, $V(\{0\}) = \text{Spec}\,(R)$ and

$$V(I_1 \cdot I_2) = V(I_1) \cup V(I_2),$$
$$V\left(\sum_\alpha I_\alpha\right) = \bigcap_\alpha V(I_\alpha). \tag{G.42}$$

This defines a topology on S, where the $V(I)$ are the closed sets.

Now let P be a prime ideal of R and denote by R_P the localisation of R at P. We now define the structure sheaf O. Let $U \subseteq \mathrm{Spec}(R)$ be an open set. The sheaf $O(U)$ consists of functions

$$\sigma : U \to \bigsqcup_{P \in U} R_P, \tag{G.43}$$

where $\bigsqcup$ denotes the disjoint union, such that

1. $$\sigma(P) \in R_P \tag{G.44}$$
2. σ is locally a quotient of elements from R: This means that for all $P \in U$ there exists a V with $P \in V$ and $V \subseteq U$ as well as $a, b \in R$ such that for all $Q \in V$ we have $b \notin Q$ and
$$\sigma(Q) = \frac{a}{b} \tag{G.45}$$
in R_Q.

Let's look at an example: The structure sheaf of $\mathrm{Spec}(\mathbb{Z})$ has $\mathbb{Q}$ as stalk in the point (0). For a prime number p the stalk at (p) is $\mathbb{Z}$ localised at (p) (see Exercise 135).

G.6 Schemes

A pair (X, O_X) of a topological space X and a sheaf of commutative rings with 1 on X is called a **ringed space**. The sheaf O_X is called the **structure sheaf** of the space. If all the stalks of the structure sheaf are local rings, the pair (X, O_X) is called a **locally ringed space**.

Examples:

1. An arbitrary topological space X can be considered a locally ringed space by taking O_X to be the sheaf of real-valued (or complex-valued) continuous functions on open subsets of X. The stalk at a point x can be thought of as the set of all germs of continuous functions at x; this is a local ring with maximal ideal consisting of those germs whose value at x is 0.

Remark: There may exist continuous functions over open subsets of X that are not the restriction of any continuous function over X.

2. If X is a differentiable manifold, we may take the sheaf of differentiable functions. If X is a complex manifold, we may take the sheaf of holomorphic functions. Both of these give rise to locally ringed spaces.
3. If X is an algebraic variety with the Zariski topology, we can define a locally ringed space by taking $O_X(U)$ to be the ring of rational mappings defined on the open set U that do not become infinite on U.

We now have all ingredients to define a scheme. We do this in two steps: We first define an affine scheme and then a (general) scheme. You may think about the relation between an affine scheme and a scheme as being similar to the relation between a coordinate patch and a manifold, the latter being described by a collection of coordinate patches.

An **affine scheme** is a locally ringed space (X, O_X) which is isomorphic (as a locally ringed space) to the spectrum of some ring R.

A **scheme** is a locally ringed space (X, O_X) such that every point $x \in X$ has an open neighbourhood U such that $(U, O_X|_U)$ is an affine scheme.

Appendix H
Algorithms for Polynomial Rings

In this appendix we review a few basic algorithms related to polynomial rings. We discuss algorithms for computing a Gröbner basis, a Nullstellensatz certificate, an annihilator and the syzygy module.

H.1 Computing a Gröbner Basis

One of the essential tools is the computation of a Gröbner basis. Most computer algebra systems offer implementations to do this. Here we review the basics. A standard reference is the book by Adams and Loustaunau [146].

Let $\mathbb{F}$ be a field and $\mathbb{F}[x_1, \dots, x_n]$ the ring of polynomials in n variables $x_1, \dots, x_n$ with coefficients from the field $\mathbb{F}$. We fix a term order, which we denote by $<$.

Consider now $f, q_1, \dots, q_r \in \mathbb{F}[x_1, \dots, x_n]$. Using long division (almost as in primary school) we may write

$$f = \sum_{j=1}^{r} h_j q_j + r, \tag{H.1}$$

with $h_j q_j \leq f$, $r \leq f$ and no term in r is divisible by any leading term $\mathrm{lt}(q_j)$. r is called a remainder for f with respect to $\{q_1, \dots, q_r\}$. The division algorithm proceeds as follows: We start with one polynomial from the set $q_1, \dots, q_r$, say q_1 and check if $\mathrm{lt}(q_1)$ divides $\mathrm{lt}(f)$. If this is the case, we reduce f, if this is not the case, we try the next polynomial q_2. If none of the leading terms $\mathrm{lt}(q_j)$ divides $\mathrm{lt}(f)$, we move $\mathrm{lt}(f)$ from f to the remainder and continue with the next term of f. Note that the result of the division algorithm is not necessarily unique. The result may depend on the order in which we try the polynomials $q_1, \dots, q_r$.

For two polynomials $f_i, f_j \in \mathbb{F}[x_1, \dots, x_n]$, the S-polynomial is defined by

S. Weinzierl, *Feynman Integrals*, UNITEXT for Physics,
https://doi.org/10.1007/978-3-030-99558-4

$$S\left(f_i, f_j\right) = \frac{l_{ij}}{\mathrm{lt}\left(f_i\right)} f_i - \frac{l_{ij}}{\mathrm{lt}\left(f_j\right)} f_j, \qquad l_{ij} = \mathrm{lcm}\left(\mathrm{lt}\left(f_i\right), \mathrm{lt}\left(f_j\right)\right). \quad \text{(H.2)}$$

Consider now the ideal I generated by $q_1, \ldots, q_r$:

$$I = \langle q_1, \ldots, q_r \rangle . \quad \text{(H.3)}$$

A basic algorithm for the computation of a Gröbner basis G of the ideal I is as follows: One starts from $G = \{q_1, \ldots, q_r\}$ and one computes for a pair $f_i, f_j \in G$ the S-polynomial $S(f_i, f_j)$ and reduces $S(f_i, f_j)$ with the multivariate division algorithm relative to G. If the remainder is non-zero, it is added to G. This process is iterated until for any pair $f_i, f_j \in G$ the S-polynomial $S(f_i, f_j)$ reduces to zero relative to G.

A Gröbner basis G for I is not necessarily unique. There are two places were choices are made: The order of the polynomials in the multivariate division algorithm already discussed above and the order in which pairs $f_i, f_j \in G$ are selected.

A Gröbner basis G is called a **reduced Gröbner basis**, if for any $f_i \in G$ the coefficient of $\mathrm{lt}(f_i)$ is one and no term of f_i is divisible by $\mathrm{lt}(f_j)$ for any $j \neq i$. For a given term order, the reduced Gröbner basis is unique.

H.2 Computing a Nullstellensatz Certificate

Let $\mathbb{F}$ be a field and $\mathbb{F}[x_1, \ldots, x_n]$ the ring of polynomials in n variables $x_1, \ldots, x_n$ with coefficients from the field $\mathbb{F}$. If a set of r polynomials $q_1, \ldots, q_r \in \mathbb{F}[x_1, \ldots, x_n]$ have a common zero, Hilbert's Nullstellensatz guarantees that there exist r polynomials $h_1, \ldots, h_r \in \mathbb{F}[x_1, \ldots, x_n]$ such that

$$\sum_{j=1}^{r} h_j q_j = 1. \quad \text{(H.4)}$$

We may compute the h_j's as follows: Consider the ideal

$$I = \langle q_1, \ldots, q_r \rangle . \quad \text{(H.5)}$$

Eq. (H.4) states that $1 \in I$. Hence, a Gröbner basis G for the ideal I is given by

$$G = \{1\} . \quad \text{(H.6)}$$

Thus, we may just compute a Gröbner basis for the ideal I together with the transformation matrix, which expresses any element of the Gröbner basis as a linear combination of the input polynomials $q_1, \ldots, q_r$. We already know that the Gröbner basis will be $\{1\}$, and we are only interested in the transformation matrix, which expresses the generator 1 as a linear combination of the polynomials $q_1, \ldots, q_r$. The coefficients are the sought-after polynomials $h_1, \ldots, h_r$.

H.3 Computing an Annihilator

Let $\mathbb{F}$ be a field and $\mathbb{F}[x_1, \ldots, x_n]$ the ring of polynomials in n variables $x_1, \ldots, x_n$ with coefficients from the field $\mathbb{F}$. A set of r polynomials $q_1, \ldots, q_r \in \mathbb{F}[x_1, \ldots, x_n]$ is said to be algebraically dependent, if there is a non-zero polynomial $a(y_1, \ldots, y_r) \in \mathbb{F}[y_1, \ldots, y_r]$ in r variables $y_1, \ldots, y_r$ with coefficients in $\mathbb{F}$ such that

$$a\,(q_1, \ldots, q_r) = 0. \tag{H.7}$$

The polynomial $a(y_1, \ldots, y_r)$ is called an **annihilating polynomial** of $q_1, \ldots, q_r$. In order to compute an annihilating polynomial we start from the ideal

$$I = \langle y_1 - q_1, \ldots, y_r - q_r \rangle \ \in \ \mathbb{F}[x_1, \ldots, x_n, y_1, \ldots, y_r]. \tag{H.8}$$

Let G be a Gröbner basis of I with respect to the lexicographic order

$$x_1 \ > \ \ldots \ > \ x_n \ > \ y_1 \ > \ \ldots \ > \ y_r. \tag{H.9}$$

Set

$$G_Y = G \cap \mathbb{F}[y_1, \ldots, y_r]. \tag{H.10}$$

G_Y is a Gröbner basis for the ideal $I \cap \mathbb{F}[y_1, \ldots, y_r]$ and any $a \in G_Y$ is an annihilating polynomial.

H.4 Computing the Syzygy Module

Let $\mathbb{F}$ be a field and $\mathbb{F}[x_1, \ldots, x_n]$ the ring of polynomials in n variables $x_1, \ldots, x_n$ with coefficients from the field $\mathbb{F}$. Consider a set of r polynomials $q_1, \ldots, q_r \in \mathbb{F}[x_1, \ldots, x_n]$ and the ideal

$$I = \langle q_1, \ldots, q_r \rangle\,. \tag{H.11}$$

A syzygy is a relation

$$\sum_{j=1}^{r} h_j q_j = 0, \tag{H.12}$$

with $h_1, \dots, h_r \in \mathbb{F}[x_1, \dots, x_n]$. In comparison with Eq. (H.4) note that Eq. (H.4) has a one on the right-hand side, while Eq. (H.12) has a zero on the right-hand side.

The difference with an annihilating polynomial as in Eq. (H.7) is as follows: A syzygy is linear in the q_j's, while an annihilating polynomial is allowed to be polynomial in the q_j's. On the other hand, the coefficients h_j of a syzygy are allowed to be in $\mathbb{F}[x_1, \dots, x_n]$, while the coefficients of an annihilating polynomial are in $\mathbb{F}$.

The set of syzygies form a module over $\mathbb{F}[x_1, \dots, x_n]$. We may compute a basis of this module as follows: We first compute a Gröebner basis for the ideal I (as outlined in Sect. H.1). Let therefore

$$G = \{f_1, f_2, \dots, f_s\} \tag{H.13}$$

be a Gröbner basis for I. For each pair (f_i, f_j) with $1 \le i, j \le s$ we consider the S-polynomial $S(f_i, f_j)$. As G is a Gröbner basis, multivariate division with remainder reduces $S(f_i, f_j)$ to zero relative to G. That is to say, we find by multivariate division polynomials $g_{ijk} \in \mathbb{F}[x_1, \dots, x_n]$ such that

$$S\left(f_i, f_j\right) = \sum_{k=1}^{s} g_{ijk} f_k. \tag{H.14}$$

We set

$$r_{ij} = \frac{l_{ij}}{\mathrm{lt}\,(f_i)} y_i - \frac{l_{ij}}{\mathrm{lt}\,\left(f_j\right)} y_j - \sum_{k=1}^{s} g_{ijk} y_k, \qquad l_{ij} = \mathrm{lcm}\left(\mathrm{lt}\,(f_i)\,, \mathrm{lt}\,\left(f_j\right)\right). \tag{H.15}$$

The syzygy module is generated by all relations r_{ij} with $1 \le i < j \le s$. Substituting f_k for y_k gives a syzygy relation of the form as in Eq. (H.12). More refined algorithms to compute the syzygy module are given in [470].

Syzygies are of interest in extending the unitarity-based methods discussed in Sect. 5.5 from one-loop to higher loops [471–475].

Appendix I
Finite Fields Methods

Integration-by-parts identities and the reduction to master integrals are at the core of many Feynman integral computations. On the positive side we note that this only involves linear algebra and rational functions in the kinematic variables x and the dimension of space-time D. However, the simplification of the rational functions (i.e. cancelling common factors in the numerator and in the denominator) is actually the bottle-neck. Finite-field methods can be used to improve the performance. In this appendix we first discuss Euclid's algorithm for the greatest common divisor and gradually turn to finite field methods.

I.1 The Greatest Common Divisor and the Euclidean Algorithm

It is often required to simplify rational functions by cancelling common factors in the numerator and denominator. As an example let us consider

$$\frac{(x+y)^2(x-y)^3}{(x+y)(x^2-y^2)} = (x-y)^2. \tag{I.1}$$

One factor of $(x+y)$ is trivially removed, the remaining factors are cancelled once we noticed that $(x^2-y^2) = (x+y)(x-y)$. For the implementation in a computer algebra system this is however not the way to proceed. The factorization of the numerator and the denominator into irreducible polynomials is a very expensive calculation and actually not required. To cancel the common factors in the numerator and in the denominator it is sufficient to calculate the **greatest common divisor (gcd)** of the two expressions. The efficient implementation of an algorithm for the calculation of the greatest common divisor is essential for many other algorithms. Like in the example above, most gcd calculations are done in polynomial rings. It is therefore useful to recall first some basic definitions from ring theory:

S. Weinzierl, *Feynman Integrals*, UNITEXT for Physics,
https://doi.org/10.1007/978-3-030-99558-4

A **commutative ring** $(R, +, \cdot)$ is a set R with two operations $+$ and $\cdot$, such that $(R, +)$ is an Abelian group and $\cdot$ is associative, distributive and commutative. In addition we always assume that there is a unit element for the multiplication. An example for a commutative ring would be $\mathbb{Z}_8$, i.e. the set of integers modulo 8. In this ring one has for example $3 + 7 = 2$ and $2 \cdot 4 = 0$. From the last equation one sees that it is possible to obtain zero by multiplying two non-zero elements.

An **integral domain** D is a commutative ring with the additional requirement

$$a \cdot b = 0 \Rightarrow a = 0 \;\text{ or }\; b = 0 \quad \text{(no zero divisors).} \tag{I.2}$$

Sometimes an integral domain D is defined by requiring

$$a \cdot b = a \cdot c \;\text{ and }\; a \neq 0 \Rightarrow b = c \quad \text{(cancellation law).} \tag{I.3}$$

It can be shown that these two requirements are equivalent. An example for an integral domain would be the subset of the complex numbers defined by

$$S = \left\{ a + bi\sqrt{5} \;\middle|\; a, b \in \mathbb{Z} \right\} \tag{I.4}$$

An element $u \in D$ is called **unit** or **invertible** if u has a multiplicative inverse in D. The only units in the example Eq. (I.4) are 1 and (-1). We further say that a divides b if there is an element $x \in D$ such that $b = ax$. In that case one writes $a|b$. Two elements $a, b \in D$ are called **associates** if a divides b and b divides a. In the integral domain S defined in Eq. (I.4) the elements 1 and (-1) are associates.

We can now define the **greatest common divisor**: An element $c \in D$ is called the greatest common divisor of a and b if $c|a$ and $c|b$ and c is a multiple of every other element which divides both a and b. Closely related to the greatest common divisor is the **least common multiple (lcm)** of two elements a and b: d is called least common multiple of a and b if $a|d$ and $b|d$ and d is a divisor of every other element which is a multiple of both a and b. Since gcd and lcm are related by

$$\text{lcm}(a, b) = \frac{ab}{\text{gcd}(a, b)} \tag{I.5}$$

it is sufficient to focus on an algorithm for the calculation of the greatest common divisor.

An element $p \in D\backslash\{0\}$ is called **prime** if from $p|ab$ it follows that $p|a$ or $p|b$. An element $p \in D\backslash\{0\}$ is called **irreducible** if p is not a unit and whenever $p = ab$ either a or b is a unit. In an integral domain, any prime element is automatically also an irreducible element. However, the reverse is in general not true. This requires some additional properties in the ring.

Let us now turn to these additional properties: An integral domain D is called a **unique factorization domain** if for all $a \in D\backslash\{0\}$, either a is a unit or else a can be expressed as a finite product of irreducible elements such that this factorization into

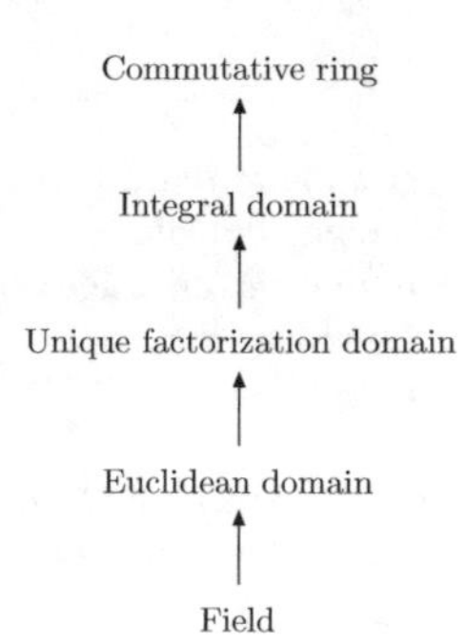

Fig. I.1 Hierarchy of domains. Arrows $A \to B$ indicate that A is a specialisation of B (the notation is borrowed from derived classes in C++)

irreducible elements is unique up to associates and reordering. It can be shown that in an unique factorization domain the notions of irreducible element and prime element are equivalent. In a unique factorization domain the greatest common divisor exists and is unique (up to associates and reordering). The integral domain S in Eq. (I.4) is not a unique factorization domain, since for example

$$21 = 3 \cdot 7 = \left(1 - 2i\sqrt{5}\right)\left(1 + 2i\sqrt{5}\right) \tag{I.6}$$

are two factorizations into irreducible elements. An example for a unique factorization domains is the polynomial ring $\mathbb{Z}[x]$ in one variable with integer coefficients.

An **Euclidean domain** is an integral domain D with a valuation map $v : D\backslash\{0\} \to \mathbb{N}_0$ into the non-negative integer numbers, such that $v(ab) \geq v(a)$ for all $a, b \in D\backslash\{0\}$, and for all $a, b \in D$ with $b \neq 0$, there exist elements $q, r \in D$ such that

$$a = bq + r, \tag{I.7}$$

where either $r = 0$ or $v(r) < v(b)$. This means that in an Euclidean domain division with remainder is possible. An example for an Euclidean domain is given by the integer numbers $\mathbb{Z}$.

Finally, a **field** is a commutative ring in which every non-zero element has a multiplicative inverse, e.g. $R\backslash\{0\}$ is an Abelian group. Any field is automatically an Euclidean domain. Examples for fields are given by the rational numbers $\mathbb{Q}$, the real numbers $\mathbb{R}$, the complex numbers $\mathbb{C}$ or $\mathbb{Z}_p$, the integers modulo p with p a prime number. $\mathbb{Z}_p$ is a **finite field**, it has p elements $0, 1, \dots, (p-1)$.

Figure I.1 summarises the relationships between the various domains. Of particular importance are polynomial rings in one or several variables. Figure I.2 summarises the structure of these domains. Note that a multivariate polynomial ring $R[x_1, \dots, x_n]$ can always be viewed as an univariate polynomial ring in one variable x_n with coefficients in the ring $R[x_1, \dots, x_{n-1}]$.

The algorithm for the calculation of the gcd in an Euclidean domain dates back to Euclid [476]. It is based on the fact that if $a = bq + r$, then

R	R[x]	$R[x_1, x_2, \ldots, x_n]$
commutative ring	commutative ring	commutative ring
integral domain	integral domain	integral domain
unique factorization domain	unique factorization domain	unique factorization domain
Euclidean domain	unique factorization domain	unique factorization domain
field	Euclidean domain	unique factorization domain

Fig. I.2 Structure of polynomial rings in one variable and several variables depending on the underlying coefficient ring R

$$\gcd(a, b) = \gcd(b, r). \tag{I.8}$$

This is easily seen as follows: Let $c = \gcd(a, b)$ and $d = \gcd(b, r)$. Since $r = a - bq$ we see that c divides r, therefore it also divides d. On the other hand d divides $a = bq + r$ and therefore it also divides c. We now have $c|d$ and $d|c$ and therefore c and d are associates.
It is clear that for $r = 0$, e.g. $a = bq$ we have $\gcd(a, b) = b$. Let us denote the remainder as $r = \text{rem}(a, b)$. We can now define a sequence $r_0 = a$, $r_1 = b$ and $r_i = \text{rem}(r_{i-2}, r_{i-1})$ for $i \geq 2$. Then there is a finite index k such that $r_{k+1} = 0$ (since the valuation map applied to the remainders is a strictly decreasing function). We have

$$\gcd(a, b) = \gcd(r_0, r_1) = \gcd(r_1, r_2) = \cdots = \gcd(r_{k-1}, r_k) = r_k. \tag{I.9}$$

This is the Euclidean algorithm. We briefly mention that as a side product one can find elements s, t such that

$$sa + tb = \gcd(a, b). \tag{I.10}$$

This is called the **extended Euclidean algorithm**. For the extended Euclidean algorithm one defines three sequences r_i, s_i and t_i, starting from

$$\begin{aligned} r_0 &= a, & s_0 &= 1, & t_0 &= 0, \\ r_1 &= b, & s_1 &= 0, & t_1 &= 1, \end{aligned} \tag{I.11}$$

and updates these sequences with $q_i = \lfloor \frac{r_{i-2}}{r_{i-1}} \rfloor$ as

$$r_i = \text{rem}(r_{i-2}, r_{i-1}) = r_{i-2} - q_i r_{i-1}, \quad s_i = s_{i-2} - q_i s_{i-1}, \quad t_i = t_{i-2} - q_i t_{i-1}. \tag{I.12}$$

As before, the algorithm stops whenever $r_{k+1} = 0$. Then $\gcd(a, b) = r_k$, $s = s_k$ and $t = t_k$. This allows the solution of the **Diophantine equation**

$$sa + tb = c, \tag{I.13}$$

for s and t whenever $\gcd(a, b)$ divides c.

We are primarily interested in gcd computations in polynomial rings. However, polynomial rings are usually only unique factorization domains, but not Euclidean domains, e.g. division with remainder is in general not possible. As an example consider the polynomials $a(x) = x^2 + 2x + 3$ and $b(x) = 5x + 7$ in $\mathbb{Z}[x]$. It is not possible to write $a(x)$ in the form $a(x) = b(x)q(x) + r(x)$, where the polynomials $q(x)$ and $r(x)$ have integer coefficients. However in $\mathbb{Q}[x]$ we have

$$x^2 + 2x + 3 = (5x + 7)\left(\frac{1}{5}x + \frac{3}{25}\right) + \frac{54}{25} \tag{I.14}$$

and we see that the obstruction arises from the leading coefficient of $b(x)$. It is therefore appropriate to introduce a **pseudo-division with remainder**. Let $D[x]$ be a polynomial ring over a unique factorization domain D. For $a(x) = a_n x^n + \cdots + a_0$, $b(x) = b_m x^m + \cdots + b_0$ with $n \geq m$ and $b(x) \neq 0$ there exists $q(x), r(x) \in D[x]$ such that

$$b_m^{n-m+1} a(x) = b(x)q(x) + r(x) \tag{I.15}$$

with $\deg(r(x)) < \deg(b(x))$. This pseudo-division property is sufficient to extend the Euclidean algorithm to polynomial rings over unique factorization domains.

Unfortunately, the Euclidean algorithm as well as the extended algorithm with pseudo-division have a severe drawback: Intermediate expressions can become quite long. This can be seen in the following example, where we would like to calculate the gcd of the polynomials

$$\begin{aligned} a(x) &= x^8 + x^6 - 3x^4 - 3x^3 + 8x^2 + 2x - 5, \\ b(x) &= 3x^6 + 5x^4 - 4x^2 - 9x + 21, \end{aligned} \tag{I.16}$$

in $\mathbb{Z}[x]$. Calculating the pseudo-remainder sequence $r_i(x)$ we obtain

$$\begin{aligned} r_2(x) &= -15x^4 + 3x^2 - 9, \\ r_3(x) &= 15795x^2 + 30375x - 59535, \\ r_4(x) &= 1254542875143750x - 1654608338437500, \\ r_5(x) &= 12593338795500743100931141992187500. \end{aligned} \tag{I.17}$$

This implies that $a(x)$ and $b(x)$ are relatively prime, but the numbers which occur in the calculation are large. An analysis of the problem shows, that the large numbers can be avoided if each polynomial is split into a content part and a primitive part. The **content of a polynomial** is the gcd of all it's coefficients. For example we have

$$15795x^2 + 30375x - 59535 = 1215\left(13x^2 + 25x + 49\right) \tag{I.18}$$

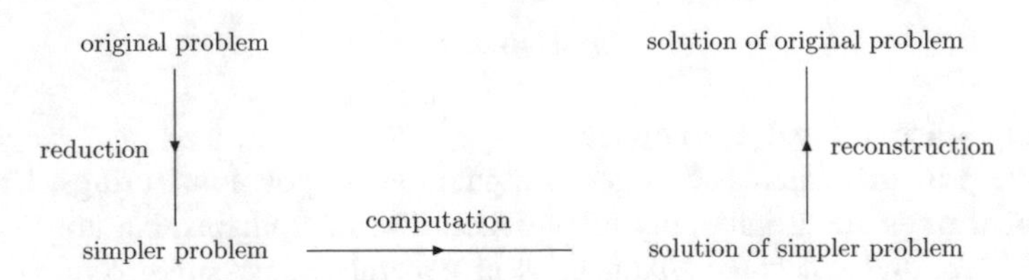

Fig. I.3 The modular approach: Starting from the original problem, one first tries to find a related simpler problem. The solution of the simpler problem is used to reconstruct a solution of the original problem

and 1215 is the content and $13x^2 + 25x + 49$ the **primitive part**. Taking out the content of a polynomial in each step requires a gcd calculation in the coefficient domain and avoids large intermediate expressions in the example above. However the extra cost for the gcd calculation in the coefficient domain is prohibitive for multivariate polynomials. The art of gcd calculations consists in finding an algorithm which keeps intermediate expressions at reasonable size and which at the same time does not involve too much computational overhead. An acceptable algorithm is given by the **subresultant algorithm** [477, 478]: Similar to the methods discussed above, one calculates a polynomial remainder sequence $r_0(x), r_1(x), \dots, r_k(x)$. This sequence is obtained through $r_0(x) = a(x)$, $r_1(x) = b(x)$ and

$$c_i^{\delta_i+1} r_{i-1}(x) = q_i(x) r_i(x) + d_i r_{i+1}(x), \tag{I.19}$$

where c_i is the leading coefficient of $r_i(x)$, $\delta_i = \deg(r_{i-1}(x)) - \deg(r_i(x))$ and $d_1 = (-1)^{\delta_1+1}$, $d_i = -c_{i-1}\psi_i^{\delta_i}$ for $2 \le i \le k$. The ψ_i are defined by $\psi_1 = -1$ and

$$\psi_i = (-c_{i-1})^{\delta_{i-1}} \psi_{i-1}^{1-\delta_{i-1}}. \tag{I.20}$$

Then the primitive part of the last non-vanishing remainder equals the primitive part of the greatest common divisor $\gcd(a(x), b(x))$.

I.1.1 Heuristic Methods

In order to further speed up the calculation of polynomial gcds one may resort to heuristic algorithms [479]. In general a heuristic algorithm maps a problem to a simpler problem, solves the simpler problem and tries to reconstruct the solution of the original problem from the solution of the simpler problem. This is illustrated in Fig. I.3. This approach is at the core of finite field methods.

Let's see how this works with a concrete example: For the calculation of polynomial gcds one evaluates the polynomials at a specific point and one considers the gcd of the results in the coefficient domain. Since gcd calculations in the coefficient domain are cheaper, this can lead to a sizeable speed-up, if both the evaluation of the polynomial and the reconstruction of the polynomial gcd can be done at reasonable cost. Let us consider the polynomials

$$\begin{aligned} a(x) &= 6x^4 + 21x^3 + 35x^2 + 27x + 7, \\ b(x) &= 12x^4 - 3x^3 - 17x^2 - 45x + 21. \end{aligned} \tag{I.21}$$

Evaluating these polynomials at the point $\xi = 100$ yields $a(100) = 621352707$ and $b(100) = 1196825521$. The gcd of theses two numbers is

$$c = \gcd(621352707, 1196825521) = 30607. \tag{I.22}$$

To reconstruct the polynomial gcd one writes c in the ξ-adic representation

$$c = c_0 + c_1\xi + \cdots + c_n\xi^n, \quad -\frac{\xi}{2} < c_i \le \frac{\xi}{2}. \tag{I.23}$$

Then the candidate for the polynomial gcd is

$$g(x) = c_0 + c_1 x + \cdots + c_n x^n. \tag{I.24}$$

In our example we have

$$30607 = 7 + 6 \cdot 100 + 3 \cdot 100^2 \tag{I.25}$$

and the candidate for the polynomial gcd is $g(x) = 3x^2 + 6x + 7$. A theorem guarantees now if ξ is chosen such that

$$\xi > 1 + 2 \min \left(||a(x)||_\infty, ||b(x)||_\infty \right), \tag{I.26}$$

then $g(x)$ is the greatest common divisor of $a(x)$ and $b(x)$ if and only if $g(x)$ divides $a(x)$ and $b(x)$. This can easily be checked by a trial division. In the example above, $g(x) = 3x^2 + 6x + 7$ divides both $a(x)$ and $b(x)$ and is therefore the gcd of the two polynomials.

Note that there is no guarantee that the heuristic algorithm will succeed in finding the gcd. But if it does, this algorithm is usually faster than the subresultant algorithm discussed previously. Therefore, a strategy for a computer algebra system could be to try first a few times the heuristic algorithm with various evaluation points and to fall back onto the subresultant algorithm, if the greatest common divisor has not been found by the heuristic algorithm.

I.2 The Chinese Remainder Theorem

An important theorem is the Chinese remainder theorem, which allows us to use several (cheap) calculations in the rings $\mathbb{Z}_{n_1}, \mathbb{Z}_{n_2}, \dots, \mathbb{Z}_{n_k}$ to obtain the corresponding result in the larger ring $\mathbb{Z}_{n_1 \cdot \dots \cdot n_k}$, provided no pair (n_i, n_j) has a common factor greater than one. This can be advantageous, as the computational cost for the calculations in the rings $\mathbb{Z}_{n_1}, \dots, \mathbb{Z}_{n_k}$ plus the computational cost for the reconstruction can be significantly lower than the computational cost for a direct calculation in the ring $\mathbb{Z}_{n_1 \cdot \dots \cdot n_k}$.

Let's see how this is done: Two integers $n_1, n_2 \in \mathbb{Z}$ are called **coprime**, if $\gcd(n_1, n_2) = 1$. Let $n_1, \dots, n_k$ be a set of natural numbers, which are pairwise coprime. Set

$$n = n_1 \cdot n_2 \cdot \dots \cdot n_k. \tag{I.27}$$

Let $N \in \mathbb{Z}$ be an (unknown) integer. Assume that we know all remainders of N mod n_j:

$$r_j = N \bmod n_j, \qquad 1 \le j \le k. \tag{I.28}$$

We have $0 \le r_j < n_j$. Given the remainders $r_1, \dots, r_k$ the Chinese remainder theorem allows us to compute the remainder $r = N \bmod n$: We first set

$$\tilde{n}_j = \prod_{\substack{i=1 \\ i \neq j}}^{k} n_i = \frac{n}{n_j}. \tag{I.29}$$

The integers n_j and $\tilde{n}_j$ are coprime $\gcd(n_j, \tilde{n}_j) = 1$, hence there exist integers s_j and t_j such that

$$s_j \tilde{n}_j + t_j n_j = 1. \tag{I.30}$$

s_j and t_j can be obtained with the extended Euclidean algorithm. Then

$$r = \sum_{j=1}^{k} r_j s_j \tilde{n}_j \bmod n. \tag{I.31}$$

In mathematical terms, the Chinese remainder theorem says that there is a ring homomorphism

$$\mathbb{Z}_{n_1} \times \mathbb{Z}_{n_2} \times \dots \times \mathbb{Z}_{n_k} \cong \mathbb{Z}_n. \tag{I.32}$$

In one direction, this isomorphism is given by Eq. (I.31). In the other direction, the isomorphism is trivially given by

$$r_j = r \bmod n_j. \tag{I.33}$$

Exercise 136 *Let r be defined by Eq. (I.31). Determine*

$$r \bmod n_i. \tag{I.34}$$

I.3 Black Box Reconstruction

Suppose we are given a routine, which returns the numerical value of a rational function f in several variables $x = (x_1, \dots, x_n)$ for a choice of the input variables x. The black box reconstruction problem asks, if it is possible to find the analytic form of the rational function f by evaluating numerically the function f for sufficiently many distinct input values x.

A typical application in the context of Feynman integrals would be the following [480–482]: Integration-by-parts reduction expresses any Feynman integral as a linear combination of master integrals. The coefficients of this linear combination are rational functions in the number of space-time dimensions D and the kinematic variables x. It is much faster to run the integration-by-parts reduction algorithm for specific values of the kinematic variables x (and D). Doing this several times we may reconstruct the coefficients.

The numerical evaluation is usually done with finite field arithmetic. This is exact (i.e. avoids rounding errors) and limits the size of the numbers at intermediate stages (a finite field contains only finitely many numbers). On the other side, there is some loss of information, as in a finite field there are relations (like $1 + 1 = 0$ in $\mathbb{Z}_2$), which do not hold in a field of characteristic 0.

A **finite field** is a field with finitely many elements. It can be shown that the number of elements must always be a power of a prime number: A finite field has p^k elements, where p is a prime number and $k \in \mathbb{N}$. For us it is sufficient to focus on the finite fields $\mathbb{F}_p = \mathbb{Z}_p$, where p is a prime number. The field $\mathbb{F}_p$ has p elements

$$\{0, 1, \dots, p-1\}. \tag{I.35}$$

Addition, subtraction and multiplication are done modulo p. For example

$$2 \cdot 3 = 1 \quad \text{in} \quad \mathbb{Z}_5. \tag{I.36}$$

Division is only slightly more complicated: First of all, a divided by b is nothing than the multiplication of a with the inverse of b. As in characteristic zero, we assume that $b \neq 0$. As p is a prime number and $1 \leq b < p$ we have $\gcd(b, p) = 1$. From the extended Euclidean algorithm we find s and t such that

$$s \cdot b + t \cdot p = 1. \tag{I.37}$$

s is the inverse of b in $\mathbb{Z}_p$, since

$$s \cdot b = 1 \bmod p. \tag{I.38}$$

Let us remark that b^{-1} exists in $\mathbb{Z}_n$ (with n not necessarily prime) if b and n are coprime.

Let $\frac{p}{q} \in \mathbb{Q}$ be a rational number and assume that we know the image

$$c = \frac{p}{q} \bmod n \qquad \text{with} \quad \gcd(q, n) \;=\; 1. \tag{I.39}$$

In general, there is no inverse mapping from $\mathbb{Z}_n$ to $\mathbb{Q}$. ($\mathbb{Z}_n$ has finitely many elements, while $\mathbb{Q}$ has countable many elements.) We may use Wang's algorithm [483] to obtain a guess for (p, q): We use the extended Euclidean algorithm for $a = c$ and $b = n$ and monitor the sequence (r_i, s_i). We run the extended Euclidean algorithm until $r_i^2 \le n/2$. If in addition $s_i^2 \le n/2$ and $\gcd(r_i, s_i) = 1$ one returns $(p, q) = (r_i, s_i)$. If on the other hand $s_i^2 > n/2$ or $\gcd(r_i, s_i) = 1$, the reconstruction failed.

Let us look at an example: We consider the image of $\frac{2}{3} \in \mathbb{Q}$ in $\mathbb{Z}_{37}$. From the extended Euclidean algorithm we obtain

$$25 \cdot 3 - 2 \cdot 37 = 1, \tag{I.40}$$

and hence $3^{-1} = 25$ in $\mathbb{Z}_{37}$. Thus the image of $\frac{2}{3} \in \mathbb{Q}$ in $\mathbb{Z}_{37}$ is

$$(2 \cdot 25) \bmod 37 = 50 \bmod 37 \;=\; 13. \tag{I.41}$$

Let us now consider the other direction: Assume we know $c = 13 \in \mathbb{Z}_{37}$. Can we find a rational number $\frac{p}{q} \in \mathbb{Q}$ such that the image of $\frac{p}{q}$ in $\mathbb{Z}_{37}$ is $c = 13$? We use Wang's algorithm from above. We first note that

$$\left\lfloor \sqrt{\frac{37}{2}} \right\rfloor = 4. \tag{I.42}$$

From the extended Euclidean algorithm with $a = 13$ and $b = 37$ we obtain

$$\begin{aligned}
r_0 &= 13, & s_0 &= 1, & t_0 &= 0, \\
r_1 &= 37, & s_1 &= 0, & t_1 &= 1, \\
r_2 &= 13, & s_2 &= 1, & t_2 &= 0, \\
r_3 &= 11, & s_3 &= -2, & t_3 &= 1, \\
r_4 &= 2, & s_4 &= 3, & t_4 &= -1, \\
r_5 &= 1, & s_5 &= -17, & t_5 &= 6, \\
r_6 &= 0, & s_6 &= 37, & t_6 &= -13.
\end{aligned} \tag{I.43}$$

We stop with r_4, as

$$r_4 \leq \left\lfloor \sqrt{\frac{37}{2}} \right\rfloor . \tag{I.44}$$

In addition we have $s_4 = 3 \leq \lfloor \sqrt{\frac{37}{2}} \rfloor$ and $\gcd(r_4, s_4) = \gcd(2, 3) = 1$. Hence Wang's algorithm returns $(p, q) = (2, 3)$, which is the correct result.

I.3.1 Univariate Polynomials

Let us start with the reconstruction of a polynomial $f(x)$ in one variable x. A degree d polynomial is uniquely specified by the values y_i at $(d + 1)$ points x_i:

$$y_i = f(x_i), \qquad 0 \leq i \leq d. \tag{I.45}$$

There is only a small problem: We do not know the degree of the polynomial a priori. In this situation, a representation in terms of Newton polynomials is convenient. We write

$$\begin{aligned} f(x) &= \sum_{j=0}^{d} a_j \prod_{i=0}^{j-1} (x - x_i) \qquad\qquad (I.46) \\ &= a_0 + (x - x_0)\,(a_1 + (x - x_1)\,(a_2 + (x - x_2)\,(\cdots + (x - x_{d-1})\,a_d))) . \end{aligned}$$

The coefficients a_i are computed recursively as

$$\begin{aligned} a_0 &= y_0, \\ a_1 &= \frac{y_1 - a_0}{x_1 - x_0}, \\ a_2 &= \left(\frac{y_2 - a_0}{x_2 - x_0} - a_1 \right) \frac{1}{x_2 - x_1}, \\ &\dots \\ a_d &= \left(\left(\frac{y_d - a_0}{x_d - x_0} - a_1 \right) \frac{1}{x_d - x_1} - \dots - a_{d-1} \right) \frac{1}{x_d - x_{d-1}} . \end{aligned} \tag{I.47}$$

This representation has the advantage that additional evaluation points will not change the values of the already computed coefficients a_j. If f is of degree d and if we probe more than $(d + 1)$ points, the coefficients a_j with $j > d$ will be zero. This can be used as a termination criteria: We fix a positive integer η and terminate the algorithm if

$$a_{d+1} = a_{d+2} = \dots = a_{d+\eta} = 0. \tag{I.48}$$

This is a heuristic algorithm: It may fail if for example f is of degree $(d + \eta + 1)$ and we accidentally choose the points x_i such that $a_{d+1} = \cdots = a_{d+\eta} = 0$. However, it can be shown that the probability that we obtain the correct polynomial in $\mathbb{Z}_p$ is no less than [484, 485]

$$1 - (d+1)\left(\frac{d}{p}\right)^{\eta}. \tag{I.49}$$

I.3.2 Univariate Rational Functions

For the reconstruction of a rational function f in one variable x we may use Thiele's formula [486], which is based on a continued fraction

$$\begin{aligned} f(x) &= a_0 + \cfrac{x - x_0}{a_1 + \cfrac{x - x_1}{a_2 + \cfrac{x - x_2}{\cdots + \cfrac{x - x_{d-1}}{a_d}}}} \\ &= a_0 + (x - x_0)\left(a_1 + (x - x_1)\left(a_2 + (x - x_2)\left(\cdots + \frac{(x - x_{d-1})}{a_d}\right)^{-1}\right)^{-1}\right)^{-1}. \end{aligned} \tag{I.50}$$

The coefficients a_i are computed recursively as

$$\begin{aligned} a_0 &= y_0, \\ a_1 &= \frac{x_1 - x_0}{y_1 - a_0}, \\ a_2 &= \left(\frac{x_2 - x_0}{y_2 - a_0} - a_1\right)^{-1} (x_2 - x_1), \\ &\dots \\ a_d &= \left(\left(\frac{x_d - x_0}{y_d - a_0} - a_1\right)^{-1} (x_d - x_1) - \cdots - a_{d-1}\right)^{-1} (x_d - x_{d-1}). \end{aligned} \tag{I.51}$$

As in the case of Newton interpolation, the coefficient a_j depends only on the evaluations $0, 1, \dots, j$ and will not change if we add additional evaluations.

It should be noted that this algorithm may lead to spurious singularities (for example if $y_1 = a_0$) in which case the algorithm fails. In this case one may re-try with different values of the x_j's.

I.3.3 Multivariate Polynomials

Let us now turn to multivariate polynomials. Let f be a polynomial in the variables $x = (x_1, \dots, x_n)$. We denote the values of the i-th variable where we probe the function f by

$$x_{i,0}, x_{i,1}, x_{i,2}, \dots . \tag{I.52}$$

We may view a multivariate polynomial $f(x_1, \dots, x_n)$ as a univariate polynomial in the variable x_n with coefficients in the ring $\mathbb{F}[x_1, \dots, x_{n-1}]$. Thus we may reconstruct f recursively: We start with Newton interpolation in the variable x_n:

$$f(x_1, \dots, x_n) = \sum_{j=0}^{d} a_j(x_1, \dots, x_{n-1}) \prod_{i=0}^{j-1} \left(x_n - x_{n,i}\right) \tag{I.53}$$

The coefficients $a_j(x_1, \dots, x_{n-1})$ are polynomials in $(n-1)$ variables, i.e. one variable less. Given a numerical black box routine for f and choices $x_{n,0}, x_{n,1}, \dots$ for the last variable x_n, we immediately have a numerical black box routine for $a_0(x_1, \dots, x_{n-1})$ by choosing $x_n = x_{n,0}$. We therefore first reconstruct $a_0(x_1, \dots, x_{n-1})$. Once $a_0(x_1, \dots, x_{n-1})$ is known, we obtain a numerical black box routine for $a_1(x_1, \dots, x_{n-1})$ by using

$$a_1(x_1, \dots, x_{n-1}) = \frac{f(x_1, \dots, x_n) - a_0(x_1, \dots, x_{n-1})}{x_n - x_{n,0}}. \tag{I.54}$$

By using Eq. (I.47) we may continue this process to reconstruct $a_2(x_1, \dots, x_{n-1})$, $a_3(x_1, \dots, x_{n-1})$, etc.. Each $a_j(x_1, \dots, x_{n-1})$ is a polynomial in $(n-1)$ variables.

In order to reconstruct $a_j(x_1, \dots, x_{n-1})$ we repeat the ansatz and view $a_j(x_1, \dots, x_{n-1})$ as an univariate polynomial in x_{n-1} with coefficients in $\mathbb{F}[x_1, \dots, x_{n-2}]$. This recursion terminates with the reconstruction of univariate polynomials in the variable x_1 with coefficients in the field $\mathbb{F}$.

For sparse multivariate polynomials the algorithm above can be improved. A multivariate polynomial in n variables and total degree d has only a finite number of coefficients. A (multivariate) polynomial is called **sparse**, if most of its coefficients are zero. The contrary of a sparse polynomial is a **dense polynomial**, where most of its coefficients are non-zero. We may always convert a polynomial from Newton's representation as in Eqs. (I.46) and (I.53) to the standard monomial representation

$$f(x) = \sum_{\alpha} c_\alpha x^\alpha, \tag{I.55}$$

where we use the multi-index notation $x^\alpha = x_1^{\alpha_1} \cdot \dots \cdot x_n^{\alpha_n}$. The main idea of Zippel's algorithm [487, 488] for sparse polynomials is the following: Suppose we are in the recursive algorithm above at a stage where we reconstruct a polynomial in the variables $x_1, \dots, x_k$ (with $k < n$). Assume further that the coefficient $c_{\alpha_1 \dots \alpha_k}(x_{k+1, j_{k+1}}, \dots, x_{n, j_n})$ of the monomial

$$c_{\alpha_1 \dots \alpha_k} x_1^{\alpha_1} \cdot \ldots \cdot x_k^{\alpha_k} \tag{I.56}$$

turns out to be zero for the chosen numerical values of the remaining variables $x_{k+1}, \dots, x_n$. Zippel's algorithm assumes that this remains true in $\mathbb{F}[x_{k+1}, \dots, x_n]$, e.g. the final polynomial will not contain a term of the form as in Eq. (I.56) with a coefficient $c_{\alpha_1 \dots \alpha_k} \in \mathbb{F}[x_{k+1}, \dots, x_n]$. This can lead to a significant speed-up. Let us stress that this is a guess (which may or may not be true). In order to minimise the risk of accidental zeros, one chooses the numerical values $x_{k+1, j_{k+1}}, \dots, x_{n, j_n}$ with care. A typical choice is given by random values for the variables $x_{i,0}$ and the values

$$x_{i,j} = \left(x_{i,0}\right)^{j+1}, \tag{I.57}$$

for the remaining variables.

I.3.4 Multivariate Rational Functions

We finally turn to the reconstruction of multivariate rational functions. We may write a multivariate rational function as

$$f(x) = \frac{p(x)}{q(x)}, \tag{I.58}$$

where $p(x)$ and $q(x)$ are multivariate polynomials in n variables $x_1, \dots, x_n$. In order to make this representation unique, one may require that the coefficient of the smallest monomial of $q(x)$ with respect to a chosen term order equals one. If $q(x)$ contains a constant term, this implies that the constant term equals one.

Let us now discuss the main ideas of the algorithm of Cuyt and Lee [489] for the reconstruction of a multivariate rational function. We focus on the ideas how this can be done in principle. For the details how this is implemented efficiently we refer to the literature [480, 482, 489].

We introduce an auxiliary variable t and we consider the function $g(t, x)$ defined by

$$g(t, x) = f(tx_1, \dots, tx_n). \tag{I.59}$$

We view g as a function of t, depending in addition on parameters $x_1, \dots, x_n$. Clearly

$$f(x) = g(1, x). \tag{I.60}$$

Let us first assume that the denominator polynomial $q(x)$ of the rational function $f(x)$ contains a constant term. We choose the normalisation where this constant term equals one. Then $g(t, x)$ can be written as

$$g(t,x) = \frac{\sum_{r=0}^{d_p} p_r(x)\, t^r}{1 + \sum_{r'=1}^{d_q} q_{r'}(x)\, t^{r'}}, \tag{I.61}$$

where the p_r and q_r are homogeneous polynomials of degree r in the variables $x = (x_1, \dots, x_n)$.

For chosen numerical values $x_{1,j_1}, \dots, x_{n,j_n}$ for the variables $x_1, \dots, x_n$ we may use Thiele's algorithm for the reconstruction of the univariate rational function g in t. This reconstruction will provide

1. a verification (or falsification) of our assumption that the denominator contains a constant term,
2. the degrees d_p and d_q of the numerator polynomial and of the denominator polynomial, respectively,
3. numerical black box routines for the polynomials p_r and q_r (given as the coefficient of t^r in the numerator and denominator, respectively).

With the numerical black box routines for the polynomials p_r and q_r at hand, we may use any algorithm for the reconstruction of multivariate polynomials (for example the algorithm discussed above). As we know that p_r and q_r are homogeneous of degree r, one usually uses a dedicated algorithm for homogeneous polynomials.

It may happen that our initial assumption that the denominator polynomial $q(x)$ contains a constant term is not justified. In this case one considers first a modified function

$$\tilde{f}(x_1, \dots, x_n) = f(x_1 + s_1, \dots, x_n + s_n) \tag{I.62}$$

obtained by a random shift $(s_1, \dots, s_n)$. One reconstructs $\tilde{f}$ and obtains $f(x_1, \dots, x_n) = \tilde{f}(x_1 - s_1, \dots, x_n - s_n)$. The denominator polynomials $\tilde{q}$ of the rational function $\tilde{f}$ will in general have a constant term (if accidentally this is not the case, one may try a different shift). This completes the algorithm for the reconstruction of a multivariate rational function. However, the last step comes with a caveat: If the numerator polynomial p and the denominator polynomial q of the original rational function f are sparse polynomials, a random shift will in general lead to dense polynomials $\tilde{p}$ and $\tilde{q}$ of the shifted rational function $\tilde{f}$: For example, x^5 is a sparse polynomial in one variable x, while

$$(x+1)^5 = x^5 + 5x^4 + 10x^3 + 10x^2 + 5x + 1 \tag{I.63}$$

is a dense polynomial.

Appendix J
Solutions to the Exercises

Exercise 1 *Consider a connected graph G with the notation as above. Show that momentum conservation at each vertex of valency* > 1 *implies momentum conservation of the external momenta:*

$$\sum_{e_j \in E^{\text{in}}} q_j = \sum_{e_j \in E^{\text{out}}} q_j. \tag{J.1}$$

If we choose an orientation such that all external edges have a vertex of valency 1 *as sink (e.g.* $E^{\text{in}} = \emptyset$*) this translates to*

$$\sum_{j=1}^{n_{\text{ext}}} p_j = 0. \tag{J.2}$$

Solution: *We proof the claim by induction on the number of internal edges* n_{int}*. For* $n_{\text{int}} = 0$ *our graph looks like*

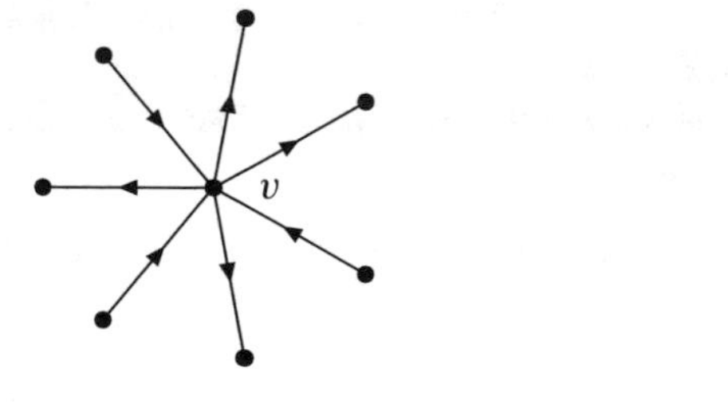

(J.3)

and has exactly one vertex v *of valency* > 1*. Momentum conservation at this vertex reads*

S. Weinzierl, *Feynman Integrals*, UNITEXT for Physics,
https://doi.org/10.1007/978-3-030-99558-4

$$\sum_{e_j \in E^{\text{in}}} q_j = \sum_{e_j \in E^{\text{out}}} q_j. \tag{J.4}$$

and corresponds exactly to the claim.

Let us now assume that the claim is correct for $(n_{\text{int}} - 1)$*. Consider now a graph with* n_{int} *internal edges. Pick one internal edge* e_i*. Denote by* v_a *its source and by* v_b *its sink. Let us write down momentum conservation at* v_a *and* v_b*:*

$$\begin{aligned} v_a : \quad q_j + \sum_{e_r \in E^{\text{source}}(v_a) \backslash \{e_j\}} q_r &= \sum_{e_r \in E^{\text{sink}}(v_a)} q_r, \\ v_b : \quad \sum_{e_r \in E^{\text{source}}(v_b)} q_r &= q_j + \sum_{e_r \in E^{\text{sink}}(v_b) \backslash \{e_j\}} q_r. \end{aligned} \tag{J.5}$$

We may eliminate q_i *from these two equations. We obtain a single equation*

$$v : \sum_{e_r \in E^{\text{source}}(v_a) \backslash \{e_j\}} q_r + \sum_{e_r \in E^{\text{source}}(v_b)} q_r = \sum_{e_r \in E^{\text{sink}}(v_a)} q_r + \sum_{e_r \in E^{\text{sink}}(v_b) \backslash \{e_j\}} q_r. \tag{J.6}$$

The momentum q_j *appears at no other vertex. As far as momentum conservation is concerned, we may replace the graph G with a new graph* $\tilde{G}$*, where the edge* e_j *has been contracted (e.g. the edge* e_j *is removed and the vertices* v_a *and* v_b *are merged to a new vertex* v*. Pictorially we have*

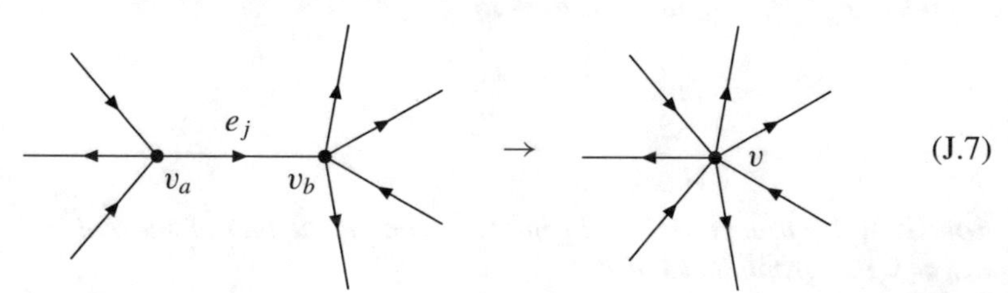

(J.7)

The new graph $\tilde{G}$ *has* $(n_{\text{int}} - 1)$ *internal edges and we may use the induction hypothesis. As G and* $\tilde{G}$ *only differ by the contraction of an internal edge, the claim holds for the graph G as well.*

If we choose an orientation such that all external edges have a vertex of valency 1 *as sink we have*

$$E^{\text{in}} = \emptyset, \quad E^{5out} = \left\{ e_{n_{\text{int}}+1}, \ldots, e_{n_{\text{int}}+n_{\text{ext}}} \right\} \tag{J.8}$$

and

$$0 = \sum_{e_j \in E^{\text{in}}} q_j = \sum_{e_j \in E^{\text{out}}} q_j = \sum_{j=1}^{n_{\text{ext}}} p_j. \tag{J.9}$$

Exercise 2 *Consider the one-loop graph shown in Fig. 2.4. Write down the equations expressing momentum conservation at each vertex of valency* > 1*. Use* p_1, p_2, p_3

as independent external momenta and $k_1 = q_4$ *as the independent loop momentum. Express all other momenta as linear combinations of these.*

Solution: *Momentum conservation at the four vertices* $v_1, \dots, v_4$ *reads*

$$\begin{aligned} v_1 &: \; p_1 + q_1 = q_4, \\ v_2 &: \; p_2 + q_2 = q_1, \\ v_3 &: \; p_3 + q_3 = q_2, \\ v_4 &: \; p_4 + q_4 = q_3. \end{aligned} \tag{J.10}$$

With $k_1 = q_4$ *we express* q_1, q_2, q_3 *and* p_4 *in terms of* k_1, p_1, p_2 *and* p_3 *as*

$$\begin{aligned} q_1 &= k_1 - p_1, \\ q_2 &= k_1 - p_1 - p_2, \\ q_3 &= k_1 - p_1 - p_2 - p_3, \\ p_4 &= -p_1 - p_2 - p_3. \end{aligned} \tag{J.11}$$

Exercise 3 *Prove*

$$T_\nu(D) = \nu T_{\nu+1}(D+2). \tag{J.12}$$

Solution: *The tadpole integral is given by Eq. (2.123). The left-hand side of our equation equals*

$$T_\nu(D) = \frac{e^{\varepsilon\gamma_E}\Gamma\left(\nu - \frac{D}{2}\right)}{\Gamma(\nu)} \left(\frac{m^2}{\mu^2}\right)^{\frac{D}{2}-\nu}. \tag{J.13}$$

Let's work out the right-hand side:

$$\begin{aligned} \nu T_{\nu+1}(D+2) &= \nu \frac{e^{\varepsilon\gamma_E}\Gamma\left((\nu+1) - \frac{D+2}{2}\right)}{\Gamma(\nu+1)} \left(\frac{m^2}{\mu^2}\right)^{\frac{D+2}{2}-(\nu+1)} \\ &= \frac{e^{\varepsilon\gamma_E}\Gamma\left(\nu - \frac{D}{2}\right)}{\Gamma(\nu)} \left(\frac{m^2}{\mu^2}\right)^{\frac{D}{2}-\nu}, \end{aligned} \tag{J.14}$$

where we used $\Gamma(\nu+1) = \nu\Gamma(\nu)$.

Exercise 4 *Derive Eq. (2.132).*

Solution: *We repeat the steps from the calculation of the tadpole integral. Starting from*

$$\tilde{T} = e^{\varepsilon\gamma_E} \left(\mu^2\right)^{\nu-\frac{D}{2}-a} \int \frac{d^D k}{i\pi^{\frac{D}{2}}} \frac{\left(-k^2\right)^a}{\left(-Uk^2 + F\right)^\nu} \tag{J.15}$$

we perform a Wick rotation and obtain

$$\tilde{T} = e^{\varepsilon \gamma_E} \left(\mu^2\right)^{\nu - \frac{D}{2} - a} \int \frac{d^D K}{\pi^{\frac{D}{2}}} \frac{\left(K^2\right)^a}{\left(U K^2 + F\right)^{\nu}}. \tag{J.16}$$

We then introduce spherical coordinates and integrate over the angles. This yields

$$\tilde{T} = \frac{e^{\varepsilon \gamma_E} \left(\mu^2\right)^{\nu - \frac{D}{2} - a}}{\Gamma\left(\frac{D}{2}\right)} \int\limits_0^\infty dK^2 \frac{\left(K^2\right)^{\frac{D}{2}+a-1}}{\left(U K^2 + F\right)^{\nu}}. \tag{J.17}$$

We then substitute $t = U K^2 / F$. We obtain

$$\tilde{T} = \frac{e^{\varepsilon \gamma_E} \left(\mu^2\right)^{\nu - \frac{D}{2} - a}}{\Gamma\left(\frac{D}{2}\right)} \frac{U^{-\frac{D}{2}-a}}{F^{\nu - \frac{D}{2} - a}} \int\limits_0^\infty dt \frac{t^{\frac{D}{2}+a-1}}{(t+1)^{\nu}}. \tag{J.18}$$

The remaining integral is again Euler's beta integral and we finally obtain

$$\tilde{T} = e^{\varepsilon \gamma_E} \left(\mu^2\right)^{\nu - \frac{D}{2} - a} \frac{\Gamma\left(\frac{D}{2} + a\right)}{\Gamma\left(\frac{D}{2}\right)} \frac{\Gamma\left(\nu - \frac{D}{2} - a\right)}{\Gamma(\nu)} \frac{U^{-\frac{D}{2}-a}}{F^{\nu - \frac{D}{2} - a}}. \tag{J.19}$$

Exercise 5 *Derive Eq. (2.136).*

Hint: Split the D-dimensional integration into a D_{int}-dimensional part and a (-2ε)-dimensional part. Equation (2.136) can be derived by just considering the (-2ε)-dimensional part.

Solution: $f(k_{(D_{\text{int}})}, k^2_{(-2\varepsilon)})$ *may depend arbitrarily on k^0, k^1, ..., $k^{D_{\text{int}}-1}$, but the dependence on $k^{D_{\text{int}}}$, $k^{D_{\text{int}}+1}$, ..., k^{D-1} is only through $k^2_{(-2\varepsilon)}$. We split the integration into a D_{int}-dimensional part and a (-2ε)-dimensional part. We write*

$$\frac{d^D k}{i \pi^{\frac{D}{2}}} = \frac{d^{D_{\text{int}}} k}{i \pi^{\frac{D_{\text{int}}}{2}}} \frac{d^{(-2\varepsilon)} k}{i \pi^{-\varepsilon}}. \tag{J.20}$$

From Eqs. (2.131) and (2.132) we have

$$\int \frac{d^{(-2\varepsilon)} k}{i \pi^{-\varepsilon}} \left(-k^2_{(-2\varepsilon)}\right)^r f\left(k_{(D_{\text{int}})}, k^2_{(-2\varepsilon)}\right) = \frac{\Gamma(r-\varepsilon)}{\Gamma(-\varepsilon)} \int \frac{d^{(-2\varepsilon+2r)} k}{i \pi^{-\varepsilon+r}} f\left(k_{(D_{\text{int}})}, k^2_{(-2\varepsilon)}\right). \tag{J.21}$$

Integrating then also over the D_{int}-dimensional part gives

$$\int \frac{d^D k}{i \pi^{\frac{D}{2}}} \left(-k^2_{(-2\varepsilon)}\right)^r f\left(k_{(D_{\text{int}})}, k^2_{(-2\varepsilon)}\right) = \frac{\Gamma(r-\varepsilon)}{\Gamma(-\varepsilon)} \int \frac{d^{D+2r} k}{i \pi^{\frac{D+2r}{2}}} f\left(k_{(D_{\text{int}})}, k^2_{(-2\varepsilon)}\right). \tag{J.22}$$

Exercise 6 *Derive Eq. (2.137).*

Hint: Consider the mass dimension of the integral to prove the statement for $D/2 + a \neq 0$ and the normalisation of the integral measure in Eq. (2.77) to prove the statement for $D/2 + a = 0$.

Solution: *We start with the case $D/2 + a \neq 0$. The integral*

$$\int \frac{d^D k}{i\pi^{\frac{D}{2}}} \left(-k^2\right)^a \tag{J.23}$$

has mass dimension $(D + 2a)$, so it must be proportional to some scale μ raised to the power of the mass dimension:

$$\left(\mu^2\right)^{\frac{D}{2}+a}. \tag{J.24}$$

However, the integral is a scaleless integral. Thus there is no such scale available, hence the integral must be zero.

Let us now turn to the case $D/2 + a = 0$. Let us consider $a \in \mathbb{N}$. The space-time dimension is then necessarily even and negative: $D = -2a$. For this reason, the use of Eq. (2.137) is sometimes called the **negative dimension approach**. *After Wick rotation we have to show*

$$\int \frac{d^D K}{\pi^{\frac{D}{2}}} \left(K^2\right)^a = (-1)^{\frac{D}{2}}\, \Gamma\left(1 - \frac{D}{2}\right), \quad \textit{for} \quad \frac{D}{2} + a = 0 \quad \textit{and} \quad a \in \mathbb{N}. \tag{J.25}$$

The normalisation of the integral measure reads

$$\int \frac{d^D K}{\pi^{\frac{D}{2}}} \exp\left(-K^2\right) = 1. \tag{J.26}$$

We expand the left-hand side

$$\int \frac{d^D K}{\pi^{\frac{D}{2}}} \exp\left(-K^2\right) = \sum_{a=0}^{\infty} \frac{(-1)^a}{a!} \int \frac{d^D K}{\pi^{\frac{D}{2}}} \left(K^2\right)^a. \tag{J.27}$$

We already know that integrals with $D/2 + a \neq 0$ vanish, thus only the term $a = -D/2$ survives:

$$\int \frac{d^D K}{\pi^{\frac{D}{2}}} \exp\left(-K^2\right) = \frac{(-1)^{-\frac{D}{2}}}{\left(-\frac{D}{2}\right)!} \int \frac{d^D K}{\pi^{\frac{D}{2}}} \left(K^2\right)^{-\frac{D}{2}}. \tag{J.28}$$

This should be equal to 1, hence

$$\int \frac{d^D K}{\pi^{\frac{D}{2}}} \left(K^2\right)^{-\frac{D}{2}} = (-1)^{\frac{D}{2}}\, \Gamma\left(1 - \frac{D}{2}\right). \tag{J.29}$$

This proves Eq. (J.25). Analytic continuation in D (or the parameter a) on the variety $D/2 + a = 0$ completes the proof.

Exercise 7 *Consider again the one-loop box graph shown in Fig.2.4. Assume first that all internal masses are non-zero and pairwise distinct and that the external momenta are as generic as possible. How many kinematic variables are there? Secondly, assume that all internal masses are zero and that the external momenta satisfy $p_1^2 = p_2^2 = p_3^2 = p_4^2 = 0$. How many kinematic variables are there now?*

Solution: *We start with the case where all internal masses are non-zero and pairwise distinct and the external momenta are as generic as possible. The external momenta still satisfy momentum conservation. If we take all momenta outgoing, momentum conservation reads*

$$p_1 + p_2 + p_3 + p_4 = 0. \tag{J.30}$$

Momentum conservation allows us to eliminate p_4. Thus we have

$$\frac{-p_1^2}{\mu^2}, \quad \frac{-p_1 \cdot p_2}{\mu^2}, \quad \frac{-p_1 \cdot p_3}{\mu^2}, \quad \frac{-p_2^2}{\mu^2}, \quad \frac{-p_2 \cdot p_3}{\mu^2}, \quad \frac{-p_3^2}{\mu^2}, \quad \frac{m_1^2}{\mu^2}, \quad \frac{m_2^2}{\mu^2}, \quad \frac{m_3^2}{\mu^2}, \quad \frac{m_4^2}{\mu^2} \tag{J.31}$$

as kinematic variables, where $m_1, \dots, m_4$ denote the internal masses. Due to the scaling relation in Eq. (2.144) we may set one kinematic variable to one (say the last one m_4^2/μ^2). This gives us $N_B = 9$ for the most general one-loop box graph. Quite often one uses instead of the scalar products $p_1 \cdot p_2$, $p_2 \cdot p_3$, $p_1 \cdot p_3$ the **Mandelstam variables**

$$s = (p_1 + p_2)^2, \quad t = (p_2 + p_3)^2, \quad u = (p_1 + p_3)^2. \tag{J.32}$$

These satisfy the relation (inherited from momentum conservation)

$$s + t + u = p_1^2 + p_2^2 + p_3^2 + p_4^2. \tag{J.33}$$

Thus we may either eliminate p_4^2 (as we did above) or u (another popular choice). Let us now discuss the second part, where we assume that all internal masses are zero and that the external momenta satisfy $p_1^2 = p_2^2 = p_3^2 = p_4^2 = 0$. The relations $p_1^2 = p_2^2 = p_3^2 = m_1^2 = m_2^2 = m_3^2 = m_4^2 = 0$ leave only

$$\frac{-p_1 \cdot p_2}{\mu^2}, \quad \frac{-p_1 \cdot p_3}{\mu^2}, \quad \frac{-p_2 \cdot p_3}{\mu^2} \tag{J.34}$$

from the list above. However, we haven't used $p_4^2 = 0$ yet, which allows to eliminate another kinematic variable. In the massless case we have $s = 2p_1 \cdot p_2$, $t = 2p_2 \cdot p_3$ and $u = 2p_1 \cdot p_3$ and the Mandelstam relation simplifies to

$$s + t + u = 0. \tag{J.35}$$

Thus we may trade $p_4^2 = 0$ to eliminate u (e.g. $p_1 \cdot p_3/\mu^2$). This brings us down to two kinematic variables, which can be taken as $-2p_1 \cdot p_2/\mu^2$ and $-2p_2 \cdot p_3/\mu^2$. As above we may set one kinematic variable to one, giving us $N_B = 1$ in the massless case. It is quite common to use as kinematic variable

$$x = \frac{s}{t} \tag{J.36}$$

in this case.

Exercise 8 *Determine with the method above the graph polynomials $\mathcal{U}$ and $\mathcal{F}$ for the graph shown in Fig. 2.6 for the case where all internal masses are zero.*

Solution: *Let us first note that the graph shown in Fig. 2.6 may equally well be drawn as in Fig. J.1. As independent loop momenta we take*

$$k_1 = q_1, \quad k_2 = q_5. \tag{J.37}$$

Then

$$q_2 = k_1 + p_1, \quad q_3 = -k_1 - k_2, \quad q_4 = -k_1 - k_2 + p_2, \quad q_6 = k_2 + p_3, \tag{J.38}$$

and $p_3 = -p_1 - p_2$. We work out

$$\sum_{j=1}^{6} \alpha_j \left(-q_j^2\right) = -(\alpha_1 + \alpha_2 + \alpha_3 + \alpha_4) k_1^2 - 2(\alpha_3 + \alpha_4) k_1 \cdot k_2 - (\alpha_3 + \alpha_4 + \alpha_5 + \alpha_6) k_2^2$$
$$+ 2(\alpha_4 p_2 - \alpha_2 p_1) \cdot k_1 + 2(\alpha_4 p_2 - \alpha_6 p_3) \cdot k_2 - \alpha_2 p_1^2 - \alpha_4 p_2^2 - \alpha_6 p_3^2. \tag{J.39}$$

In comparing with Eq. (2.156) we find

$$M = \begin{pmatrix} \alpha_1 + \alpha_2 + \alpha_3 + \alpha_4 & \alpha_3 + \alpha_4 \\ \alpha_3 + \alpha_4 & \alpha_3 + \alpha_4 + \alpha_5 + \alpha_6 \end{pmatrix},$$
$$v = \begin{pmatrix} \alpha_4 p_2 - \alpha_2 p_1 \\ \alpha_4 p_2 - \alpha_6 p_3 \end{pmatrix},$$
$$J = \alpha_2 (-p_1)^2 + \alpha_4 (-p_2)^2 + \alpha_6 (-p_3)^2. \tag{J.40}$$

From momentum conservation we have $(p_1 + p_2)^2 = p_3^2$ and hence

$$2p_1 \cdot p_2 = p_3^2 - p_1^2 - p_2^2,$$
$$2p_2 \cdot p_3 = p_1^2 - p_2^2 - p_3^2,$$

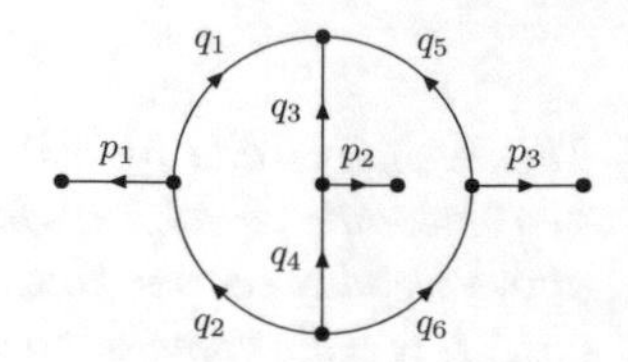

Fig. J.1 The two-loop non-planar vertex graph of Fig. 2.6 drawn in an alternative way

$$2p_3 \cdot p_1 = p_2^2 - p_3^2 - p_1^2. \tag{J.41}$$

Using this and Eq. (2.157) we finally obtain

$$\begin{aligned}
\mathcal{U} &= (\alpha_1 + \alpha_2)(\alpha_3 + \alpha_4) + (\alpha_1 + \alpha_2)(\alpha_5 + \alpha_6) + (\alpha_3 + \alpha_4)(\alpha_5 + \alpha_6), \\
\mathcal{F} &= [\alpha_1\alpha_4\alpha_6 + \alpha_2\alpha_3\alpha_5 + \alpha_1\alpha_2(\alpha_3 + \alpha_4 + \alpha_5\alpha_6)]\left(\frac{-p_1^2}{\mu^2}\right) \\
&\quad + [\alpha_3\alpha_2\alpha_6 + \alpha_4\alpha_1\alpha_5 + \alpha_3\alpha_4(\alpha_1 + \alpha_2 + \alpha_5\alpha_6)]\left(\frac{-p_2^2}{\mu^2}\right) \\
&\quad + [\alpha_5\alpha_2\alpha_4 + \alpha_6\alpha_1\alpha_3 + \alpha_5\alpha_6(\alpha_1 + \alpha_2 + \alpha_3\alpha_4)]\left(\frac{-p_3^2}{\mu^2}\right).
\end{aligned} \tag{J.42}$$

Exercise 9 *Prove Eq. (2.171).*

Solution: *The attentive reader might have noticed that we (implicitly) already gave a proof of Eq. (2.171) when we we derived the Feynman parameter representation from the Schwinger parameter representation. For clarity, let's distil the proof here: For $A_j > 0$ and $\mathrm{Re}(\nu_j) > 0$ we have*

$$\frac{1}{A_j^{\nu_j}} = \frac{1}{\Gamma(\nu_j)} \int\limits_0^\infty d\alpha_j\, \alpha_j^{\nu_j - 1}\, e^{-\alpha A_j}, \tag{J.43}$$

and therefore

$$\prod_{j=1}^n \frac{1}{A_j^{\nu_j}} = \frac{1}{\prod\limits_{j=1}^n \Gamma(\nu_j)} \int\limits_{\alpha_j \geq 0} d^n\alpha \left(\prod_{j=1}^n \alpha_j^{\nu_j - 1}\right) \exp\left(-\sum_{j=1}^n \alpha_j A_j\right). \tag{J.44}$$

Clearly, $\sum\limits_{j=1}^n \alpha_j \geq 0$. We insert

$$1 = \int\limits_0^\infty dt\, \delta\left(t - \sum_{j=1}^n \alpha_j\right) \tag{J.45}$$

and change variables according to $a_j = \alpha_j / t$. This gives

$$\prod_{j=1}^{n} \frac{1}{A_j^{\nu_j}} = \frac{1}{\prod\limits_{j=1}^{n} \Gamma(\nu_j)} \int\limits_{a_j \geq 0} d^n a \; \delta\left(1 - \sum_{j=1}^{n} a_j\right) \left(\prod_{j=1}^{n} a_j^{\nu_j - 1}\right) \int\limits_0^{\infty} dt \; t^{\nu-1} e^{-t \sum\limits_{j=1}^{n} a_j A_j}. \tag{J.46}$$

A further change of variable $t \rightarrow t / (\sum\limits_{j=1}^{n} a_j A_j)$ yields

$$\prod_{j=1}^{n} \frac{1}{A_j^{\nu_j}} = \frac{1}{\prod\limits_{j=1}^{n} \Gamma(\nu_j)} \int\limits_{a_j \geq 0} d^n a \; \delta\left(1 - \sum_{j=1}^{n} a_j\right) \left(\prod_{j=1}^{n} a_j^{\nu_j - 1}\right) \frac{1}{\left(\sum\limits_{j=1}^{n} a_j A_j\right)^{\nu}} \int\limits_0^{\infty} dt \; t^{\nu-1} e^{-t}. \tag{J.47}$$

With

$$\int\limits_0^{\infty} dt \; t^{\nu-1} e^{-t} = \Gamma(\nu) \tag{J.48}$$

we arrive at

$$\prod_{j=1}^{n} \frac{1}{A_j^{\nu_j}} = \frac{\Gamma(\nu)}{\prod\limits_{j=1}^{n} \Gamma(\nu_j)} \int\limits_{a_j \geq 0} d^n a \; \delta\left(1 - \sum_{j=1}^{n} a_j\right) \left(\prod_{j=1}^{n} a_j^{\nu_j - 1}\right) \frac{1}{\left(\sum\limits_{j=1}^{n} a_j A_j\right)^{\nu}}. \tag{J.49}$$

Exercise 10 *Calculate with the help of the Feynman parameter representation the one-loop triangle integral*

$$I_{\nu_1 \nu_2 \nu_3} = e^{\varepsilon \gamma_E} \left(\mu^2\right)^{\nu - \frac{D}{2}} \int \frac{d^D k}{i \pi^{\frac{D}{2}}} \frac{1}{\left(-q_1^2\right)^{\nu_1} \left(-q_2^2\right)^{\nu_2} \left(-q_3^2\right)^{\nu_3}}, \tag{J.50}$$

shown in Fig. 2.8 for the case where all internal masses are zero ($m_1 = m_2 = m_3 = 0$) and for the kinematic configuration $p_1^2 = p_2^2 = 0$, $p_3^2 \neq 0$.

Solution: *The second graph polynomial is given by*

$$\mathcal{F} = a_1 a_3 \left(\frac{-p_3^2}{\mu^2}\right) \tag{J.51}$$

and the Feynman parameter representation reads

$$I_{\nu_1\nu_2\nu_3} = \frac{e^{\varepsilon\gamma_E}\Gamma\left(\nu - \frac{D}{2}\right)}{\Gamma(\nu_1)\Gamma(\nu_2)\Gamma(\nu_3)} \int\limits_{a_j \geq 0} d^3a\, \delta(1 - a_1 - a_2 - a_3)\, \frac{a_1^{\nu_1-1} a_2^{\nu_2-1} a_3^{\nu_3-1}}{\left[\mathcal{F}(a)\right]^{\nu - \frac{D}{2}}} \tag{J.52}$$

$$= \frac{e^{\varepsilon\gamma_E}\Gamma\left(\nu - \frac{D}{2}\right)}{\Gamma(\nu_1)\Gamma(\nu_2)\Gamma(\nu_3)} \left(\frac{-p_3^2}{\mu^2}\right)^{\frac{D}{2}-\nu} \int\limits_{a_j \geq 0} d^3a\, \delta(1 - a_1 - a_2 - a_3)\, a_1^{\frac{D}{2}-\nu_{23}-1} a_2^{\nu_2-1} a_3^{\frac{D}{2}-\nu_{12}-1}.$$

The Feynman parameter integral is a generalisation of Euler's beta function: For $n \in \mathbb{N}$ we have

$$\int\limits_{a_j \geq 0} d^n a\, \delta\left(1 - \sum_{j=1}^{n} a_j\right) \left(\prod_{j=1}^{n} a_j^{\nu_j - 1}\right) = \frac{\prod\limits_{j=1}^{n} \Gamma(\nu_j)}{\Gamma(\nu_1 + \cdots + \nu_n)} \tag{J.53}$$

and therefore

$$I_{\nu_1\nu_2\nu_3} = \frac{e^{\varepsilon\gamma_E}\Gamma\left(\nu - \frac{D}{2}\right)\Gamma\left(\frac{D}{2} - \nu_{12}\right)\Gamma\left(\frac{D}{2} - \nu_{23}\right)}{\Gamma(\nu_1)\Gamma(\nu_3)\Gamma(D - \nu)} \left(\frac{-p_3^2}{\mu^2}\right)^{\frac{D}{2}-\nu}. \tag{J.54}$$

Exercise 11 *Consider again the one-loop box graph in Fig. 2.4, this time for the kinematic configuration*

$$p_2^2 = p_4^2 = 0, \quad m_1 = m_2 = m_3 = m_4 = 0. \tag{J.55}$$

Write down the Feynman parameter representation as in Eq. (2.170). Obtain a second integral representation by first combining propagators 1 and 2 with a pair of Feynman parameters, then combining propagators 3 and 4 with a second pair of Feynman parameters and finally the two results with a third pair of Feynman parameters.

Solution: *Let us start with the standard (democratic) Feynman parameter representation. The graph polynomials are*

$$\mathcal{U} = a_1 + a_2 + a_3 + a_4,$$
$$\mathcal{F} = a_2 a_4 \left(\frac{-s}{\mu^2}\right) + a_1 a_3 \left(\frac{-t}{\mu^2}\right) + a_1 a_4 \left(\frac{-p_1^2}{\mu^2}\right) + a_2 a_3 \left(\frac{-p_3^2}{\mu^2}\right). \tag{J.56}$$

The democratic Feynman parameter representation is then

$$I = \frac{e^{\varepsilon\gamma_E}\Gamma\left(\nu - \frac{D}{2}\right)}{\prod\limits_{j=1}^{4} \Gamma(\nu_j)} \int\limits_{a_j \geq 0} d^4a\, \delta\left(1 - \sum_{j=1}^{4} a_j\right) \left(\prod_{j=1}^{4} a_j^{\nu_j - 1}\right) \frac{1}{\mathcal{F}^{\nu - \frac{D}{2}}}. \tag{J.57}$$

Note that we may ignore the $\mathcal{U}$-polynomial due to the delta distribution.

Let us now follow a hierarchical approach: It will be convenient to use the following notation: $\bar{a} = 1 - a$, $\bar{b} = 1 - b$, $\bar{c} = 1 - c$ *and* $\nu_{i_1 \dots i_k} = \nu_{i_1} + \dots + \nu_{i_k}$. *We first combine propagators* 1 *and* 2

$$\frac{1}{\left(-q_1^2\right)^{\nu_1}\left(-q_2^2\right)^{\nu_1}} = \frac{\Gamma\left(\nu_{12}\right)}{\Gamma\left(\nu_1\right)\Gamma\left(\nu_2\right)} \int\limits_0^1 da\, a^{\nu_1 - 1} \bar{a}^{\nu_2 - 1} \frac{1}{\left(-a q_1^2 - \bar{a} q_2^2\right)^{\nu_{12}}}, \tag{J.58}$$

then propagators 3 *and* 4

$$\frac{1}{\left(-q_3^2\right)^{\nu_1}\left(-q_4^2\right)^{\nu_1}} = \frac{\Gamma\left(\nu_{34}\right)}{\Gamma\left(\nu_3\right)\Gamma\left(\nu_4\right)} \int\limits_0^1 db\, b^{\nu_3 - 1} \bar{b}^{\nu_4 - 1} \frac{1}{\left(-b q_3^2 - \bar{b} q_4^2\right)^{\nu_{34}}}, \tag{J.59}$$

and finally the two intermediate results:

$$\frac{1}{\left(-a q_1^2 - \bar{a} q_2^2\right)^{\nu_{12}}\left(-b q_3^2 - \bar{b} q_4^2\right)^{\nu_{34}}} = \frac{\Gamma\left(\nu_{1234}\right)}{\Gamma\left(\nu_{12}\right)\Gamma\left(\nu_{34}\right)} \int\limits_0^1 dc\, \frac{c^{\nu_{12}-1}\bar{c}^{\nu_{34}-1}}{\left(-ac q_1^2 - \bar{a}c q_2^2 - b\bar{c} q_3^2 - \bar{b}\bar{c} q_4^2\right)^{\nu_{1234}}}. \tag{J.60}$$

Let's work out the denominator:

$$\begin{aligned} & - ac q_1^2 - \bar{a} c q_2^2 - b \bar{c} q_3^2 - \bar{b}\bar{c} q_4^2 = \\ & \quad - \left(k - ac p_1 - \bar{a} c \left(p_1 + p_2\right) - b \bar{c}\left(p_1 + p_2 + p_3\right)\right)^2 + c\bar{c}\left[\bar{a}\bar{b}\left(-s\right) + ab\left(-t\right) + a\bar{b}\left(-p_1^2\right) + \bar{a}b\left(-p_3^2\right)\right]. \end{aligned} \tag{J.61}$$

We see that the c-dependence in the second term factors out. We may now use Eq. (2.133) and obtain

$$\begin{aligned} I = & \frac{e^{\varepsilon \gamma_E}\Gamma\left(\nu - \frac{D}{2}\right)}{\prod\limits_{j=1}^4 \Gamma(\nu_j)} \int\limits_0^1 da\, a^{\nu_1 - 1}\bar{a}^{\nu_2 - 1} \int\limits_0^1 db\, b^{\nu_3 - 1}\bar{b}^{\nu_4 - 1} \frac{1}{\left[\bar{a}\bar{b}\left(-s\right) + ab\left(-t\right) + a\bar{b}\left(-p_1^2\right) + \bar{a}b\left(-p_3^2\right)\right]^{\nu - \frac{D}{2}}} \\ & \times \int\limits_0^1 dc\, c^{\frac{D}{2} - \nu_{34} - 1}\bar{c}^{\frac{D}{2} - \nu_{12} - 1}. \end{aligned} \tag{J.62}$$

The integral over c is trivial and gives Euler's beta function. We obtain

$$\begin{aligned} I = & \\ & \frac{e^{\varepsilon\gamma_E}\Gamma\left(\nu - \frac{D}{2}\right)\Gamma\left(\frac{D}{2} - \nu_{12}\right)\Gamma\left(\frac{D}{2} - \nu_{34}\right)}{\Gamma\left(D - \nu\right)\prod\limits_{j=1}^4 \Gamma(\nu_j)} \int\limits_0^1 da \int\limits_0^1 db \frac{a^{\nu_1 - 1}\bar{a}^{\nu_2 - 1} b^{\nu_3 - 1}\bar{b}^{\nu_4 - 1}}{\left[\bar{a}\bar{b}\left(-s\right) + ab\left(-t\right) + a\bar{b}\left(-p_1^2\right) + \bar{a}b\left(-p_3^2\right)\right]^{\nu - \frac{D}{2}}}. \end{aligned} \tag{J.63}$$

This leaves two non-trivial integrations, compared to three non-trivial integrations within the standard Feynman parameter representations.

Exercise 12 *Prove Eq. (2.195).*

Solution: *We have to prove*

$$\sum_{j=1}^{n_{\mathrm{int}}} a_j \frac{\partial}{\partial a_j} f(a) = -n_{\mathrm{int}} f(a). \tag{J.64}$$

$f(a)$ is a product of three factors, each factor is a homogeneous polynomial raised to some power. The derivatives act by the product rules. It is therefore sufficient to prove

$$\sum_{j=1}^{n_{\mathrm{int}}} a_j \frac{\partial}{\partial a_j} (p(a))^{\nu} = \nu h p(a) \tag{J.65}$$

for a homogeneous polynomial $p(a)$ of degree h. We may simplify the task further by noting that it is sufficient to prove

$$\sum_{j=1}^{n_{\mathrm{int}}} a_j \frac{\partial}{\partial a_j} p(a) = h p(a). \tag{J.66}$$

Let us now write

$$p(a) = \sum_i c_i \prod_{j=1}^{n_{\mathrm{int}}} a_j^{\nu_{ij}} \tag{J.67}$$

We assumed that $p(a)$ is homogeneous of degree h, therefore we have for all i

$$\sum_{j=1}^{n_{\mathrm{int}}} \nu_{ij} = h. \tag{J.68}$$

It is easy to show that for each term we have

$$\sum_{j=1}^{n_{\mathrm{int}}} a_j \frac{\partial}{\partial a_j} \left(c_i \prod_{j=1}^{n_{\mathrm{int}}} a_j^{\nu_{ij}} \right) = \left(\sum_{j=1}^{n_{\mathrm{int}}} \nu_{ij} \right) \left(c_i \prod_{j=1}^{n_{\mathrm{int}}} a_j^{\nu_{ij}} \right) = h \left(c_i \prod_{j=1}^{n_{\mathrm{int}}} a_j^{\nu_{ij}} \right), \tag{J.69}$$

which completes the proof.

Exercise 13 *Show explicitly that Eq. (2.198) is equivalent to Eq. (2.170).*

Solution: *Let us denote by*

$$\tilde{\Delta} = \left\{ \left(a_1, \dots, a_{n_{\text{int}}-1}\right) \in \mathbb{R}^{n_{\text{int}}-1} \middle| \sum_{j=1}^{n_{\text{int}}-1} a_j \le 1, a_j \ge 0 \right\}. \tag{J.70}$$

$\tilde{\Delta}$ *is a coordinate patch for the standard simplex. Let us further agree that in this exercise we always define*

$$a_{n_{\text{int}}} = 1 - \sum_{j=1}^{n_{\text{int}}-1} a_j. \tag{J.71}$$

The Feynman parameter representation from Eq. (2.170) is then

$$I = \frac{e^{l\varepsilon\gamma_{\text{E}}}\Gamma\left(\nu - \frac{lD}{2}\right)}{\prod\limits_{j=1}^{n_{\text{int}}}\Gamma(\nu_j)} \int\limits_{\tilde{\Delta}} d^{n_{\text{int}}-1}a \left(\prod_{j=1}^{n_{\text{int}}} a_j^{\nu_j-1}\right) \frac{[\mathcal{U}(a)]^{\nu-\frac{(l+1)D}{2}}}{[\mathcal{F}(a)]^{\nu-\frac{lD}{2}}}. \tag{J.72}$$

In other words, we have used the Dirac delta distribution to integrate out $a_{n_{\text{int}}}$.

Let us now consider Eq. (2.198). We work out ω *for our coordinate chart: From Eq. (J.71) we have*

$$da_{n_{\text{int}}} = -\sum_{j=1}^{n_{\text{int}}-1} da_j. \tag{J.73}$$

and hence

$$\begin{aligned}
\omega &= \sum_{j=1}^{n_{\text{int}}} (-1)^{n_{\text{int}}-j}\, a_j\, da_1 \wedge \dots \wedge \widehat{da_j} \wedge \dots \wedge da_{n_{\text{int}}} \\
&= (-1)^{n_{\text{int}}-1} a_1 da_2 \wedge \dots \wedge da_{n_{\text{int}}-1} \wedge (-da_1) + (-1)^{n_{\text{int}}-2} a_2 da_1 \wedge da_3 \wedge \dots \wedge da_{n_{\text{int}}-1} \wedge (-da_2) + \dots \\
&\quad + (-1)^1 a_{n_{\text{int}}-1} da_1 \wedge \dots \wedge da_{n_{\text{int}}-2} \wedge \left(-da_{n_{\text{int}}-1}\right) + \left(1 - \sum_{j=1}^{n_{\text{int}}-1} a_j\right) da_1 \wedge \dots \wedge da_{n_{\text{int}}-2} \wedge da_{n_{\text{int}}-1} \\
&= da_1 \wedge \dots \wedge da_{n_{\text{int}}-1} = d^{n_{\text{int}}-1}a.
\end{aligned} \tag{J.74}$$

Exercise 14 *Prove Eq. (2.200).*

Solution: *Let us write with Einstein's summation convention*

$$\omega = \frac{1}{(n_{\text{int}}-1)!}\omega_{i_2 \dots i_{n_{\text{int}}}} da_{i_2} \wedge \dots \wedge da_{i_{n_{\text{int}}}}, \text{ with } \omega_{i_2 \dots i_{n_{\text{int}}}} = (-1)^{n_{\text{int}}-1} \varepsilon_{i_1 i_2 \dots i_{n_{\text{int}}}} a_{i_1}, \tag{J.75}$$

where $\varepsilon_{i_1 i_2 \dots i_{n_{\mathrm{int}}}}$ denotes the totally antisymmetric tensor with $\varepsilon_{12\dots n_{\mathrm{int}}} = 1$. For

$$X = \lambda_i e_i \tag{J.76}$$

the interior product is given by

$$\iota_X \omega = \frac{1}{(n_{\mathrm{int}} - 2)!} \lambda_{i_2} \omega_{i_2 i_3 \dots i_{n_{\mathrm{int}}}} da_{i_3} \wedge \dots \wedge da_{i_{n_{\mathrm{int}}}}. \tag{J.77}$$

Integrating along the radial direction we have $a_i = \lambda_i t$, where t is the curve parameter. We then have

$$\lambda_{i_2} \omega_{i_2 i_3 \dots i_{n_{\mathrm{int}}}} = (-1)^{n_{\mathrm{int}}-1} a_{i_1} \lambda_{i_2} \varepsilon_{i_1 i_2 i_3 \dots i_{n_{\mathrm{int}}}} = (-1)^{n_{\mathrm{int}}-1} \lambda_{i_1} \lambda_{i_2} \varepsilon_{i_1 i_2 i_3 \dots i_{n_{\mathrm{int}}}} t. \tag{J.78}$$

This vanishes, as $\lambda_{i_1} \lambda_{i_2}$ is symmetric under the exchange of i_1 and i_2, while $\varepsilon_{i_1 i_2 i_3 \dots i_{n_{\mathrm{int}}}}$ is antisymmetric.

Exercise 15 *Prove Eq. (2.201).*

Solution: *From Leibniz's rule we have*

$$d(f\omega) = (df) \wedge \omega + f(d\omega). \tag{J.79}$$

We work out $d\omega$ first:

$$\begin{aligned} d\omega &= \sum_{j=1}^{n_{\mathrm{int}}} (-1)^{n_{\mathrm{int}}-j} da_j \wedge da_1 \wedge \dots \wedge \widehat{da_j} \wedge \dots \wedge da_{n_{\mathrm{int}}} \\ &= n_{\mathrm{int}} (-1)^{n_{\mathrm{int}}-1} da_1 \wedge \dots da_{n_{\mathrm{int}}}. \end{aligned} \tag{J.80}$$

df is given by

$$df = \sum_{j=1}^{n_{\mathrm{int}}} \frac{\partial f}{\partial a_j} da_j. \tag{J.81}$$

In the wedge product with ω only a single sum survives:

$$\begin{aligned} (df) \wedge \omega &= \left(\sum_{j_1=1}^{n_{\mathrm{int}}} \frac{\partial f}{\partial a_{j_1}} da_{j_1} \right) \wedge \left(\sum_{j_2=1}^{n_{\mathrm{int}}} (-1)^{n_{\mathrm{int}}-j_2} a_{j_2} \wedge da_1 \wedge \dots \wedge \widehat{da_{j_2}} \wedge \dots \wedge da_{n_{\mathrm{int}}} \right) \\ &= \sum_{j=1}^{n_{\mathrm{int}}} (-1)^{n_{\mathrm{int}}-j} a_j \frac{\partial f}{\partial a_j} da_j \wedge da_1 \wedge \dots \wedge \widehat{da_j} \wedge \dots \wedge da_{n_{\mathrm{int}}} \\ &= (-1)^{n_{\mathrm{int}}-1} \left(\sum_{j=1}^{n_{\mathrm{int}}} a_j \frac{\partial f}{\partial a_j} \right) da_1 \wedge \dots \wedge da_{n_{\mathrm{int}}}. \end{aligned} \tag{J.82}$$

With Eq. (2.195) we have

$$(df) \wedge \omega = -n_{\text{int}} (-1)^{n_{\text{int}}-1} f da_1 \wedge \cdots \wedge da_{n_{\text{int}}}. \tag{J.83}$$

Exercise 16 *An alternative proof of the Cheng-Wu theorem: Prove the Cheng-Wu theorem directly from the Schwinger parameter representation by inserting*

$$1 = \int\limits_{-\infty}^{\infty} dt \; \delta \left(t - \sum_{j \in S} \alpha_j \right) = \int\limits_{0}^{\infty} dt \; \delta \left(t - \sum_{j \in S} \alpha_j \right), \tag{J.84}$$

where in the last step we used again the fact that the sum of the Schwinger parameters is non-negative.

Solution: *We start from the Schwinger parameter representation*

$$I = \frac{e^{l \varepsilon \gamma_{\text{E}}}}{\prod\limits_{j=1}^{n_{\text{int}}} \Gamma(\nu_j)} \int\limits_{\alpha_j \geq 0} d^{n_{\text{int}}} \alpha \left(\prod_{j=1}^{n_{\text{int}}} \alpha_j^{\nu_j - 1} \right) [\mathcal{U}(\alpha)]^{-\frac{D}{2}} \exp\left(-\frac{\mathcal{F}(\alpha)}{\mathcal{U}(\alpha)} \right) \tag{J.85}$$

and insert

$$1 = \int\limits_{0}^{\infty} dt \; \delta \left(t - \sum_{j \in S} \alpha_j \right). \tag{J.86}$$

We then change variables as $a_j = \alpha_j / t$ (for all $j \in \{1, \dots, n_{\text{int}}\}$) and obtain:

$$I = \frac{e^{l \varepsilon \gamma_{\text{E}}}}{\prod\limits_{j=1}^{n_{\text{int}}} \Gamma(\nu_j)} \int\limits_{a_j \geq 0} d^{n_{\text{int}}} a \; \delta \left(1 - \sum_{j \in S} a_j \right) \left(\prod_{j=1}^{n_{\text{int}}} a_j^{\nu_j - 1} \right) [\mathcal{U}(a)]^{-\frac{D}{2}} \int\limits_{0}^{\infty} dt \; t^{\nu - \frac{lD}{2} - 1} \exp\left(-t \frac{\mathcal{F}(a)}{\mathcal{U}(a)} \right). \tag{J.87}$$

The remaining steps are as in the derivation of the Feynman parameter representation from the Schwinger parameter representation and yield the result

$$I = \frac{e^{l \varepsilon \gamma_{\text{E}}} \Gamma\left(\nu - \frac{lD}{2}\right)}{\prod\limits_{j=1}^{n_{\text{int}}} \Gamma(\nu_j)} \int\limits_{a_j \geq 0} d^{n_{\text{int}}} a \; \delta \left(1 - \sum_{j \in S} a_j \right) \left(\prod_{j=1}^{n_{\text{int}}} a_j^{\nu_j - 1} \right) \frac{[\mathcal{U}(a)]^{\nu - \frac{(l+1)D}{2}}}{[\mathcal{F}(a)]^{\nu - \frac{lD}{2}}}. \tag{J.88}$$

Exercise 17 *Prove Eq. (2.261).*

Solution: *We have to show*

$$\det G^{\text{eucl}} (K, P_1, \dots, P_e) = K_{\perp}^2 \det G^{\text{eucl}} (P_1, \dots, P_e). \tag{J.89}$$

We start with the following lemma:

$$\det G^{\text{eucl}}\left(K+P_j, P_1, \dots, P_e\right) = \det G^{\text{eucl}}\left(K, P_1, \dots, P_e\right), \qquad 1 \le j \le e. \tag{J.90}$$

The $(e+1)$ *vectors* $K, P_1, \dots, P_e$ *span at most a vector space of dimension* $(e+1)$. *Let* V *be a vector space of dimension* $(e+1)$ *containing these vectors and define (with respect to a basis of* V*)*

$$J\left(K, P_1, \dots, P_e\right) = \begin{pmatrix} K^0 & K^1 & \dots & K^e \\ P_1^0 & P_1^1 & \dots & P_1^e \\ & & \dots & \\ P_e^0 & P_e^1 & \dots & P_e^e \end{pmatrix},$$

where we labelled the coordinates with respect to the basis of V *from* 0 *to* e. *It is clear that*

$$\det J\left(K+P_j, P_1, \dots, P_e\right) = \det J\left(K, P_1, \dots, P_e\right), \tag{J.91}$$

since we may always add a linear dependent row inside a determinant. We have

$$\det G^{\text{eucl}}\left(K, P_1, \dots, P_e\right) = \det\left(J\left(K, P_1, \dots, P_e\right) \cdot J\left(K, P_1, \dots, P_e\right)^T\right) \tag{J.92}$$

and the lemma $\det G^{\text{eucl}}(K+P_j, P_1, \dots, P_e) = \det G^{\text{eucl}}(K, P_1, \dots, P_e)$ *follows.*

Thus we have to show

$$\det G^{\text{eucl}}\left(K_\perp, P_1, \dots, P_e\right) = K_\perp^2 \det G^{\text{eucl}}\left(P_1, \dots, P_e\right). \tag{J.93}$$

We are free to choose an appropriate basis of V. *We choose a basis* $V_0, V_1, \dots V_e$ *such that*

$$\langle K_\perp \rangle = \langle V_0 \rangle, \quad \langle P_1, \dots P_e \rangle = \langle V_1, \dots, V_e \rangle. \tag{J.94}$$

Then

$$J\left(K_\perp, P_1, \dots, P_e\right) = \begin{pmatrix} K_\perp^0 & 0 & \dots & 0 \\ 0 & P_1^1 & \dots & P_1^e \\ & & \dots & \\ 0 & P_e^1 & \dots & P_e^e \end{pmatrix}$$

and the claim follows.

Exercise 18 *Perform the integration in Eq. (2.269).*

Solution: *We have to compute*

$$T_\nu(D,x) = \frac{e^{\gamma_E \varepsilon} \left(\mu^2\right)^{\nu-\frac{D}{2}}}{\Gamma\left(\frac{D}{2}\right)} \int\limits_{m^2}^{\infty} dz_1 \left[z_1 - m^2\right]^{\frac{D}{2}-1} \frac{1}{z_1^\nu}. \tag{J.95}$$

We first set $s = (z_1 - m^2)/\mu^2$*:*

$$T_\nu(D,x) = \frac{e^{\gamma_E \varepsilon}}{\Gamma\left(\frac{D}{2}\right)} \int\limits_0^{\infty} ds\, s^{\frac{D}{2}-1} \frac{1}{(s+x)^\nu}. \tag{J.96}$$

A further substitution $s = tx$ *gives*

$$T_\nu(D,x) = \frac{e^{\gamma_E \varepsilon}}{\Gamma\left(\frac{D}{2}\right)} x^{\frac{D}{2}-\nu} \int\limits_0^{\infty} dt\, t^{\frac{D}{2}-1} \frac{1}{(t+1)^\nu}. \tag{J.97}$$

The integral gives Euler's beta function:

$$\int\limits_0^{\infty} dt\, t^{\frac{D}{2}-1} \frac{1}{(t+1)^\nu} = \frac{\Gamma\left(\frac{D}{2}\right)\Gamma\left(\nu-\frac{D}{2}\right)}{\Gamma(\nu)} \tag{J.98}$$

and we obtain

$$T_\nu(D,x) = e^{\gamma_E \varepsilon} \frac{\Gamma\left(\nu-\frac{D}{2}\right)}{\Gamma(\nu)} x^{\frac{D}{2}-\nu} \tag{J.99}$$

in agreement with Eq. (2.123).

Exercise 19 *Derive the Baikov representation of the graph shown in Fig. 2.16 within the democratic approach and within the loop-by-loop approach. Assume that all internal masses are non-zero and equal.*

Solution: *We start with the democratic approach. We have*

$$e = \dim\langle p, -p\rangle = 1, \tag{J.100}$$

and thus

$$N_V = \frac{1}{2} l (l+1) + el = 5. \tag{J.101}$$

For the democratic approach we need a graph $\tilde{G}$ *with* $n_{\mathrm{int}} = 5$ *internal propagators, which allows us to express any scalar products involving the loop momenta in terms of inverse propagators and terms independent of the loop momenta. The graph* $\tilde{G}$ *shown in Fig. J.2 has this property. This graph is known as the kite graph. We denote*

$$q_1 = k_1, \quad q_2 = k_2 - k_1, \quad q_3 = -k_2 - p, \quad q_4 = -k_1 - p, \quad q_5 = k_2. \qquad \text{(J.102)}$$

Propagators four and five are auxiliary propagators. We may associate any mass to them. The simplest choice is that propagators four and five are massless propagators. The momentum representation of the Feynman integral for the kite graph is

$$I_{\nu_1\nu_2\nu_3\nu_4\nu_5} = e^{2\varepsilon\gamma_{\mathrm{E}}} \left(\mu^2\right)^{\nu-D} \int \frac{d^D k_1}{i\pi^{\frac{D}{2}}} \frac{d^D k_2}{i\pi^{\frac{D}{2}}} \frac{1}{\left(-q_1^2+m^2\right)^{\nu_j} \left(-q_2^2+m^2\right)^{\nu_j} \left(-q_3^2+m^2\right)^{\nu_j} \left(-q_4^2\right)^{\nu_j} \left(-q_5^2\right)^{\nu_j}}. \qquad \text{(J.103)}$$

The original sunrise integral $S_{\nu_1\nu_2\nu_3}$ is simply

$$S_{\nu_1\nu_2\nu_3} = I_{\nu_1\nu_2\nu_3 00}. \qquad \text{(J.104)}$$

The detour through the kite integral is necessary to satisfy the condition that we may express any internal inverse propagator as a linear combination of the linear independent scalar products involving the loop momenta and terms independent of the loop momenta. The democratic Baikov representation of the kite integral reads

$$I_{\nu_1\nu_2\nu_3\nu_4\nu_5} = \frac{e^{2\varepsilon\gamma_{\mathrm{E}}} \left(\mu^2\right)^{\nu-D} \left(-p^2\right)^{1-\frac{D}{2}}}{8\pi^{\frac{3}{2}} \Gamma\left(\frac{D-1}{2}\right) \Gamma\left(\frac{D-2}{2}\right)} \int\limits_C d^5 z \, \left[\mathcal{B}(z)\right]^{\frac{D}{2}-2} \prod_{s=1}^{5} z_s^{-\nu_s}, \qquad \text{(J.105)}$$

Here we used

$$\det C = 8, \quad \det G(p) = -p^2. \qquad \text{(J.106)}$$

The Baikov polynomial reads

$$\begin{aligned}
\mathcal{B}(z) &= \det G(k_1, k_2, p) \\
&= \frac{1}{4} \{(z_1 z_3 - z_4 z_5)(z_4 + z_5 - z_1 - z_3) + z_2 (z_1 - z_4)(z_3 - z_5) \\
&\quad + [z_1 (z_1 - z_2 - z_4) + z_3 (z_3 - z_2 - z_5) + 3 z_1 z_3 + z_2 (z_4 + z_5) - 3 z_5 z_4] m^2 \\
&\quad + [z_2 (z_2 - z_1 - z_3 - z_4 - z_5) - (z_1 - z_5)(z_3 - z_4)] p^2 - z_2 \left(p^2\right)^2 + 2 (z_1 + z_3) m^2 p^2 \\
&\quad + (z_2 - 2 z_1 - 2 z_3) \left(m^2\right)^2 + m^2 \left(m^2 - p^2\right)^2 \}.
\end{aligned} \qquad \text{(J.107)}$$

Let us now consider the loop-by-loop approach. We start with the loop formed by the internal edges e_1 and e_2 in the sunrise graph. The external momenta with respect to this loop are k_2 and $-k_2$. Thus we need for the first loop only the auxiliary edge e_5, but not e_4. Re-writing the measure gives us

$$\frac{d^D k_1}{i\pi^{\frac{D}{2}}} = \frac{1}{2\sqrt{\pi}\Gamma\left(\frac{D-1}{2}\right)} \left[\det G(k_2)\right]^{1-\frac{D}{2}} \left[\det G(k_1, k_2)\right]^{\frac{D-3}{2}} dz_1 dz_2. \qquad \text{(J.108)}$$

Fig. J.2 The two-loop kite diagram

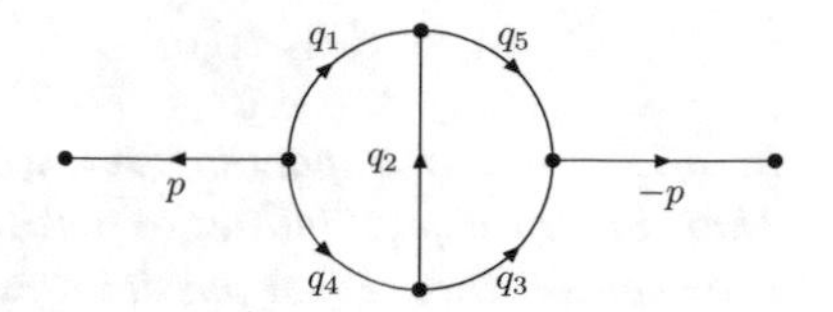

Having done the first loop, we turn to the second loop. We still have the edge e_3. In addition, we introduced the auxiliary edge e_5 in the previous step. Thus we deal again with a one-loop two-point function. The loop momentum is k_2, the external momenta are p and $-p$. Re-writing the second measure in terms of Baikov variables gives

$$\frac{d^D k_2}{i\pi^{\frac{D}{2}}} = \frac{1}{2\sqrt{\pi}\Gamma\left(\frac{D-1}{2}\right)} \left[\det G\left(p\right)\right]^{1-\frac{D}{2}} \left[\det G\left(k_2, p\right)\right]^{\frac{D-3}{2}} dz_3 dz_5. \quad \text{(J.109)}$$

Putting everything together we arrive at the Baikov representation within the loop-by-loop approach:

$$S_{\nu_1\nu_2\nu_3} = \quad \text{(J.110)}$$
$$\frac{e^{2\varepsilon\gamma_E}\left(\mu^2\right)^{\nu-D}\left(-p^2\right)^{1-\frac{D}{2}}}{4\pi\left[\Gamma\left(\frac{D-1}{2}\right)\right]^2} \int\limits_C dz_1 dz_2 dz_3 dz_5 \left[\det G\left(k_2, p\right)\right]^{\frac{D-3}{2}} \left[\det G\left(k_1, k_2\right)\right]^{\frac{D-3}{2}} z_5^{1-\frac{D}{2}} \prod_{s=1}^{3} z_s^{-\nu_s}.$$

Here, we already used

$$\det G\left(p\right) = -p^2, \quad \det G\left(k_2\right) = -k_2^2 = z_5. \quad \text{(J.111)}$$

The remaining Gram determinants, expressed in terms of the Baikov variables, read

$$\begin{aligned}\det G\left(k_2, p\right) &= \frac{1}{4}\left[-\left(m^2 - p^2 - z_3\right)^2 + 2\left(z_3 - p^2 - m^2\right) z_5 - z_5^2\right],\\ \det G\left(k_1, k_2\right) &= \frac{1}{4}\left[-\left(z_1 - z_2\right)^2 + 2\left(z_1 + z_2 - 2m^2\right) z_5 - z_5^2\right].\end{aligned} \quad \text{(J.112)}$$

It is worth noting that within the loop-by-loop approach we only have four integration variables (z_1, z_2, z_3, z_5), compared to five integration variables in the democratic approach.

Exercise 20 *Re-compute the first graph polynomial $\mathcal{U}$ for the graph shown in Fig. 2.6 from the set of spanning trees.*

Solution: *The graph shown in Fig. 2.6 can alternatively be drawn as shown in Fig. J.3. We have to find all spanning trees for this graph. The graph has three chains, which we may take as*

$$C_1 = \{e_1, e_2\}, \qquad C_2 = \{e_3, e_4\}, \qquad C_3 = \{e_5, e_6\}. \tag{J.113}$$

In order to obtain a spanning tree, we have to delete from two chains one edge each. There are three possibilities to pick two chains out of three. For any choice of these two chains there are four possibilities to delete one edge from each chain. Thus there are in total $3 \cdot 4 = 12$ possibilities. This is the number of spanning trees. The first graph polynomial $\mathcal{U}$ contains for each spanning tree a monomial corresponding to the edges, which have been deleted to obtain the spanning tree. Thus

$$\mathcal{U} = (\alpha_1 + \alpha_2)(\alpha_3 + \alpha_4) + (\alpha_1 + \alpha_2)(\alpha_5 + \alpha_6) + (\alpha_3 + \alpha_4)(\alpha_5 + \alpha_6), \tag{J.114}$$

in agreement with Eq. (J.42).

Exercise 21 *Re-compute the first graph polynomial $\mathcal{U}$ for the graph shown in Fig. 2.6 from the Laplacian of the graph.*

Solution: *Let us label the internal vertices as shown in Fig. J.3. The Laplacian for this graph is given by*

$$L_{\text{int}} = \begin{pmatrix} a_1 + a_2 & 0 & 0 & -a_1 & -a_2 \\ 0 & a_3 + a_4 & 0 & -a_3 & -a_4 \\ 0 & 0 & a_5 + a_6 & -a_5 & -a_6 \\ -a_1 & -a_3 & -a_5 & a_1 + a_3 + a_5 & 0 \\ -a_2 & -a_4 & -a_6 & 0 & a_2 + a_4 + a_6 \end{pmatrix}. \tag{J.115}$$

We obtain the Kirchhoff polynomial by deleting row and column j and taking the determinant afterwards. Here, j can be any number $j \in \{1, 2, 3, 4, 5\}$. Let's take $j = 5$:

$$\mathcal{K}_{\text{int}}(a_1, a_2, a_3, a_4, a_5) = \det\ L_{\text{int}}[5] = \begin{vmatrix} a_1 + a_2 & 0 & 0 & -a_1 \\ 0 & a_3 + a_4 & 0 & -a_3 \\ 0 & 0 & a_5 + a_6 & -a_5 \\ -a_1 & -a_3 & -a_5 & a_1 + a_3 + a_5 \end{vmatrix}. \tag{J.116}$$

We then obtain the first graph polynomial from Eq. (3.16):

$$\begin{aligned} \mathcal{U}(a_1, a_2, a_3, a_4, a_5) &= a_1 a_2 a_3 a_4 a_5\ \mathcal{K}_{\text{int}}\left(\frac{1}{a_1}, \frac{1}{a_2}, \frac{1}{a_3}, \frac{1}{a_4}, \frac{1}{a_5}\right) \\ &= (\alpha_1 + \alpha_2)(\alpha_3 + \alpha_4) + (\alpha_1 + \alpha_2)(\alpha_5 + \alpha_6) + (\alpha_3 + \alpha_4)(\alpha_5 + \alpha_6). \end{aligned} \tag{J.117}$$

Again, we find agreement with Eq. (J.42).

Exercise 22 *Consider a massless theory. Show that in this case the Lee-Pomeransky polynomial $\mathcal{G}$ satisfies for any regular edge e_k the recursion*

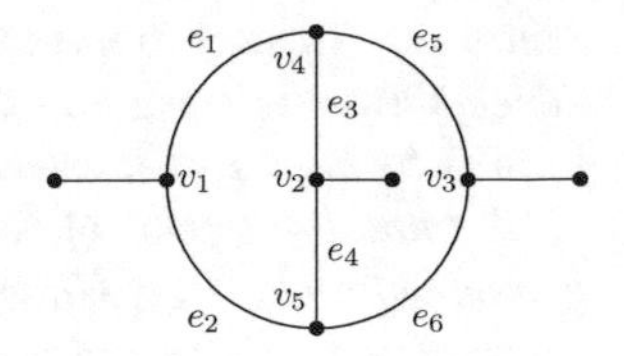

Fig. J.3 The labelling of the internal vertices for the two-loop non-planar vertex graph of Fig. 2.6

$$\mathcal{G}(G) = \mathcal{G}(G/e_k) + a_k \mathcal{G}(G - e_k). \tag{J.118}$$

Solution: *In a massless theory we have*

$$\mathcal{F} = \mathcal{F}_0. \tag{J.119}$$

From Eq. (3.53) we have for any regular edge e_k the recursion

$$\begin{aligned} \mathcal{U}(G) &= \mathcal{U}(G/e_k) + a_k \mathcal{U}(G - e_k), \\ \mathcal{F}_0(G) &= \mathcal{F}_0(G/e_k) + a_k \mathcal{F}_0(G - e_k). \end{aligned} \tag{J.120}$$

The Lee-Pomeransky polynomial $\mathcal{G}$ is given by

$$\mathcal{G} = \mathcal{U} + \mathcal{F} = \mathcal{U} + \mathcal{F}_0, \tag{J.121}$$

and hence the claim follows for any regular edge e_k:

$$\mathcal{G}(G) = \mathcal{G}(G/e_k) + a_k \mathcal{G}(G - e_k). \tag{J.122}$$

Exercise 23 *Let G be a graph with n_{int} edges and n_{ext} edges and set $n = n_{\text{int}} + n_{\text{ext}}$. Label the edges as*

$$\begin{aligned} &\text{internal edges} : \{e_1, e_2, \dots, e_{n_{\text{int}}}\}, \\ &\text{external edges} : \{e_{n_{\text{int}}+1}, e_{n_{\text{int}}+2}, \dots, e_{n_{\text{int}}+n_{\text{ext}}}\}. \end{aligned} \tag{J.123}$$

Let G_{int} be the internal graph of G. Define $\mathcal{U}$, $\mathcal{K}$ and $\mathcal{K}_{\text{int}}$ as before. Define $\tilde{\mathcal{U}}$ by

$$\tilde{\mathcal{U}}(a_1, \dots, a_n) = a_1 \dots a_n \; \mathcal{K}\left(\frac{1}{a_1}, \dots, \frac{1}{a_n}\right). \tag{J.124}$$

Show

$$\begin{aligned} \tilde{\mathcal{U}}(a_1, \dots, a_n) &= \mathcal{U}\left(a_1, \dots, a_{n_{\text{int}}}\right), \\ \mathcal{K}(a_1, \dots, a_n) &= a_{n_{\text{int}}+1} \dots a_n \mathcal{K}\left(a_1, \dots, a_{n_{\text{int}}}\right). \end{aligned} \tag{J.125}$$

Solution: *The key to the solution is to realise that there is a one-to-one correspondence between the spanning trees of G and G_{int}. The internal graph G_{int} is obtained from G by deleting all external vertices and edges. Now consider a spanning tree of G. A spanning tree T of G is obtained from G by deleting l edges, such that T is connected and a tree. However, we cannot delete an external edge: If we delete an external edge, the resulting graph is disconnected. Thus, any spanning tree of G can be mapped onto a spanning tree of G_{int} and vice versa. From Eq. (3.14)*

$$\mathcal{K}(a_1, \dots, a_n) = \sum_{T \in \mathcal{T}_1} \prod_{e_j \in T} a_j \tag{J.126}$$

and $\mathcal{K}_{\text{int}}(G) = \mathcal{K}(G_{\text{int}})$ it follows that

$$\mathcal{K}(a_1, \dots, a_n) = a_{n_{\text{int}}+1} \dots a_n \mathcal{K}\left(a_1, \dots, a_{n_{\text{int}}}\right). \tag{J.127}$$

The second equation

$$\tilde{\mathcal{U}}(a_1, \dots, a_n) = \mathcal{U}\left(a_1, \dots, a_{n_{\text{int}}}\right) \tag{J.128}$$

follows then from Eq. (3.16) and the definition of $\tilde{\mathcal{U}}$.

Exercise 24 *Determine the number of loops for K_5 and $K_{3,3}$.*

Solution: *The loop number is given by (see Eq. (2.15))*

$$l = n - r + k, \tag{J.129}$$

where n denotes the number of edges, r denotes the number of vertices and k denotes the number of connected components. Both K_5 and $K_{3,3}$ are connected, hence $k = 1$ in both cases. It remains to count for each graph the number of edges and vertices. The graph K_5 has 10 edges and 5 vertices. We therefore find

$$l_{K_5} = 10 - 5 + 1 = 6. \tag{J.130}$$

The graph $K_{3,3}$ has 9 edges and 6 vertices. We therefore find

$$l_{K_{3,3}} = 9 - 6 + 1 = 4. \tag{J.131}$$

Exercise 25 *Consider the two graphs G_1 and G_2 shown in Fig. 3.19, which differ by a self-loop. For each of the two graphs, give the Kirchhoff polynomial $\mathcal{K}$ and the first graph polynomial $\mathcal{U}$. Show that the cycle matroids are not isomorphic.*

Solution: *We may work out the Kirchhoff polynomial and the first graph polynomial from the spanning trees (or alternatively from the Laplacian). The result is:*

$$\mathcal{K}(G_1) = (a_1 + a_2)(a_3 + a_4) + a_3 a_4,$$

$$\mathcal{K}(G_2) = (a_1 + a_2)(a_3 + a_4) + a_3 a_4,$$
$$\mathcal{U}(G_1) = (a_1 + a_2)(a_3 + a_4) + a_1 a_2,$$
$$\mathcal{U}(G_2) = [(a_1 + a_2)(a_3 + a_4) + a_1 a_2] a_5. \tag{J.132}$$

The two graphs G_1 and G_2 have the same set of spanning trees. Hence the Kirchhoff polynomials are equal $\mathcal{K}(G_1) = \mathcal{K}(G_2)$. On the other hand, in the first graph polynomial $\mathcal{U}$ the deleted edges enter. In the graph G_2 we have to delete 3 edges to obtain a tree graph, whereas in the graph G_1 we only have to deleted two edges. Hence the first graph polynomials differ: $\mathcal{U}(G_1) \neq \mathcal{U}(G_2)$.

Now let's look at the cycle matroids: It is clear that we cannot transform G_2 by a sequence of vertex identifications, vertex cleavings and twistings into G_1. We may detach the self-loop formed by e_5 from the rest of the graph by vertex cleaving, but we cannot delete the self-loop. From Whitney's theorem it follows that the cycle matroids are not isomorphic.

It is instructive to go back to the definition of a matroid: A matroid is specified by a ground set E and the set of independent sets I. We get the ground set E and the set of independent sets I from the incidence matrix:

$$B_{\text{incidence}}(G_1) = \begin{pmatrix} 1 & 1 & 1 & 0 \\ 0 & 0 & 1 & 1 \\ 1 & 1 & 0 & 1 \end{pmatrix}, \quad B_{\text{incidence}}(G_2) = \begin{pmatrix} 1 & 1 & 1 & 0 & 0 \\ 0 & 0 & 1 & 1 & 0 \\ 1 & 1 & 0 & 1 & 0 \end{pmatrix}. \tag{J.133}$$

The column j in the incidence matrix corresponds to the edge e_j. We have

$$E(G_1) = \{e_1, e_2, e_3, e_4\},$$
$$I(G_1) = \{\emptyset, \{e_1\}, \{e_2\}, \{e_3\}, \{e_4\},$$
$$\{e_1, e_3\}, \{e_1, e_4\}, \{e_2, e_3\}, \{e_2, e_4\}, \{e_3, e_4\}\},$$
$$E(G_2) = \{e_1, e_2, e_3, e_4, e_5\},$$
$$I(G_2) = \{\emptyset, \{e_1\}, \{e_2\}, \{e_3\}, \{e_4\},$$
$$\{e_1, e_3\}, \{e_1, e_4\}, \{e_2, e_3\}, \{e_2, e_4\}, \{e_3, e_4\}\}. \tag{J.134}$$

Although

$$I(G_1) = I(G_2), \tag{J.135}$$

the two cycle matroids are not isomorphic, as there is no bijection from $E(G_1)$ to $E(G_2)$. $E(G_2)$ has one element more than $E(G_1)$.

Exercise 26 *Consider first the two graphs G_1 and G_2 shown in Fig. 3.20, both with three external legs. Assume that all internal masses vanish. Show that*

$$\mathcal{U}(G_1) = \mathcal{U}(G_2), \quad \mathcal{F}(G_1) = \mathcal{F}(G_2). \tag{J.136}$$

Consider then the graphs G_3 and G_4 with four external legs. Show that

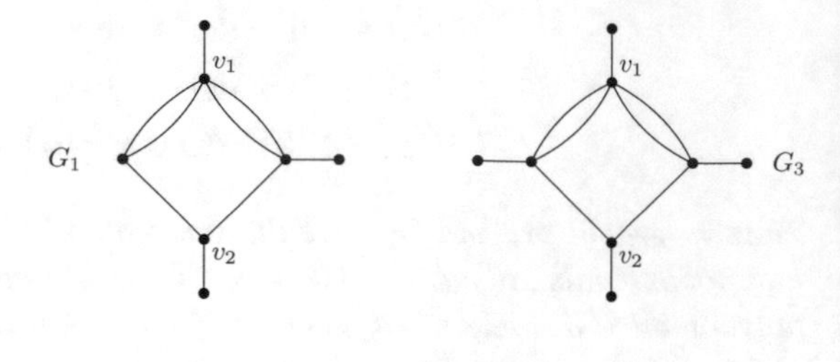

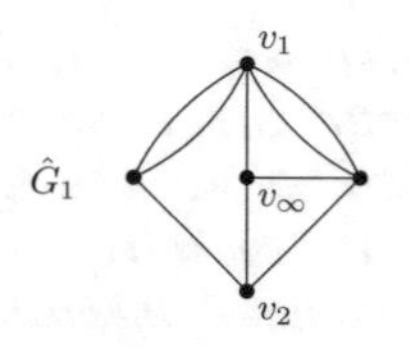

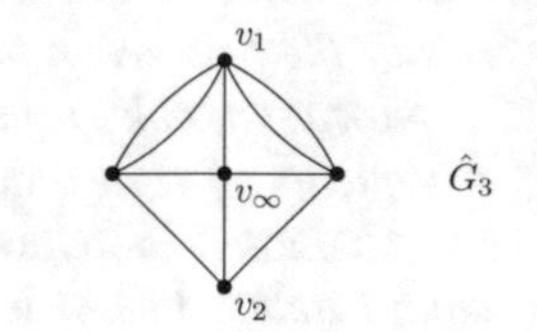

Fig. J.4 The upper part shows the graphs G_1 and G_3 with two labelled vertices. The lower part shows the graphs $\hat{G}_1$ and $\hat{G}_3$

$$\mathcal{U}(G_3) = \mathcal{U}(G_4), \tag{J.137}$$

but

$$\mathcal{F}(G_3) \neq \mathcal{F}(G_4). \tag{J.138}$$

Solution: *Let us first show*

$$\mathcal{U}(G_1) \;=\; \mathcal{U}(G_2), \quad \mathcal{U}(G_3) \;=\; \mathcal{U}(G_4). \tag{J.139}$$

Consider Fig. J.4. We obtain G_2 from G_1 by twisting at the vertices v_1 and v_2 of G_1. Similarly, we obtain G_4 from G_3 by twisting at the vertices v_1 and v_2 of G_3. Hence $\mathcal{U}(G_1) = \mathcal{U}(G_2)$ and $\mathcal{U}(G_3) = \mathcal{U}(G_4)$.

For the second graph polynomial $\mathcal{F}$ we have to consider the graphs $\hat{G}_1$, $\hat{G}_2$, $\hat{G}_3$ and $\hat{G}_4$. The graph $\hat{G}_1$ is shown in the lower left part of Fig. J.4. By twisting at the vertices v_1 and v_2 of $\hat{G}_1$ we obtain $\hat{G}_2$ and the relation $\mathcal{F}(G_1) = \mathcal{F}(G_2)$ follows. The graph $\hat{G}_3$ is shown in the lower right part of Fig. J.4. For the twisting operation we have to split the graph into two disjoint pieces by cleaving at exactly two vertices. In order to obtain $\hat{G}_4$ from $\hat{G}_3$ we would have to cleave at the three vertices v_1, v_2 and v_∞. But the twisting operation requires cleaving at exactly two vertices. Hence we cannot obtain $\hat{G}_4$ from $\hat{G}_3$ by the operations of vertex identifications, vertex cleavings and twisting. A short calculation shows that for generic external momenta $\mathcal{F}(G_3) \neq \mathcal{F}(G_4)$.

Exercise 27 *Derive Eq. (4.38) from Eqs. (4.36) and (4.37).*

Solution: *With*

$$P^{\mu\nu\, ab}(x) = \partial_\rho \partial^\rho g^{\mu\nu} \delta^{ab} - \left(1 - \frac{1}{\xi}\right) \partial^\mu \partial^\nu \delta^{ab} \tag{J.140}$$

and

$$\left(P^{-1}\right)^{ab}_{\mu\nu}(x) = \int \frac{d^D q}{(2\pi)^D} e^{-iq\cdot x} \left(\tilde{P}^{-1}\right)^{ab}_{\mu\nu}(q). \tag{J.141}$$

we have

$$P^{\mu\sigma\, ac}(x)\left(P^{-1}\right)^{cb}_{\sigma\nu}(x-y) = \int \frac{d^D q}{(2\pi)^D} e^{-iq\cdot(x-y)} q^2 \left[-g^{\mu\sigma} + \left(1-\frac{1}{\xi}\right)\frac{q^\mu q^\sigma}{q^2}\right]\delta^{ac}\left(\tilde{P}^{-1}\right)^{cb}_{\sigma\nu}(q).$$

This should be equal to

$$g^\mu{}_\nu \delta^{ab}\delta^D(x-y) = \int \frac{d^D q}{(2\pi)^D} e^{-iq\cdot(x-y)} g^\mu{}_\nu \delta^{ab}. \tag{J.142}$$

We have for

$$M^{\mu\nu} = -g^{\mu\nu} + \left(1-\frac{1}{\xi}\right)\frac{q^\mu q^\nu}{q^2} \text{ and } N_{\mu\nu} = -g_{\mu\nu} + (1-\xi)\frac{q_\mu q_\nu}{q^2} \tag{J.143}$$

the relation

$$M^{\mu\sigma} N_{\sigma\nu} = g^\mu{}_\nu. \tag{J.144}$$

Therefore

$$\mu, a \quad \text{(gluon propagator)} \quad \nu, b \quad = \frac{i}{q^2}\left(-g_{\mu\nu} + (1-\xi)\frac{q_\mu q_\nu}{q^2}\right)\delta^{ab}. \tag{J.145}$$

Exercise 28 *Compute the four-gluon amplitude $\mathcal{A}_4^{(0)}$ from the four diagrams shown in Fig. 4.1. Assume that all momenta are outgoing. Derive the Mandelstam relation*

$$s + t + u = 0. \tag{J.146}$$

Solution: *Let us start with the Mandelstam relation. Using momentum conservation and the on-shell relations we have*

$$\begin{aligned} 0 = p_4^2 \;&=\; (p_1+p_2+p_3)^2 \;=\; 2p_1\cdot p_2 + 2p_2\cdot p_3 + 2p_1\cdot p_3 \\ = (p_1+p_2)^2 &+ (p_2+p_3)^2 + (p_1+p_3)^2 \;=\; s+t+u. \end{aligned} \tag{J.147}$$

We then turn to the computation of the amplitude. Let us first examine the colour factors. The first diagrams has a colour factor $C_s = (if^{a_1a_2b})(if^{ba_3a_4})$. The second diagram we may equally well draw with legs 1 and 4 exchanged. The colour factor is then given by $C_t = (if^{a_2a_3b})(if^{ba_1a_4})$. (If we read off the colour factor directly from diagram 2, we find $(if^{a_2a_3b})(if^{ba_4a_1}) = -(if^{a_2a_3b})(if^{ba_1a_4})$. The minus sign cancels *with another minus sign from the kinematic part.) The third diagram has the colour factor $C_u = (if^{a_3 a_1 b})(if^{b a_2 a_4})$. The fourth diagram with the four-gluon vertex gives three terms, one contributing to each colour structure. We may therefore write the amplitude as*

$$\mathcal{A}_4^{(0)} = ig^2 \left[\frac{(if^{a_1 a_2 b})(if^{b a_3 a_4})\, N_s}{s} + \frac{(if^{a_2 a_3 b})(if^{b a_1 a_4})\, N_t}{t} + \frac{(if^{a_3 a_1 b})(if^{b a_2 a_4})\, N_u}{u} \right]. \tag{J.148}$$

N_s is given by

$$N_s = \{[g^{\mu_1\mu_2}(p_1^\nu - p_2^\nu) + g^{\mu_2\nu}(p_2^{\mu_1} - p_{34}^{\mu_1}) + g^{\nu\mu_1}(p_{34}^{\mu_2} - p_1^{\mu_2})]\, g_{\nu\rho}\, [g^{\mu_3\mu_4}(p_3^\rho - p_4^\rho) + g^{\mu_4\rho}(p_4^{\mu_3} - p_{12}^{\mu_3}) \\ + g^{\rho\mu_3}(p_{12}^{\mu_4} - p_3^{\mu_4})] + 2p_1\cdot p_2\,(g^{\mu_1\mu_3}g^{\mu_2\mu_4} - g^{\mu_2\mu_3}g^{\mu_1\mu_4})\}\, \varepsilon_1^{\mu_1}\varepsilon_2^{\mu_2}\varepsilon_3^{\mu_3}\varepsilon_4^{\mu_4}, \tag{J.149}$$

where we used the notation $p_{ij} = p_i + p_j$. Using momentum conservation and $p_j \cdot \varepsilon_j = 0$ we may simplify this expression to

$$N_s = \{[g^{\mu_1\mu_2}(p_1^\nu - p_2^\nu) + 2g^{\mu_2\nu}p_2^{\mu_1} - 2g^{\nu\mu_1}p_1^{\mu_2}]\, g_{\nu\rho}\, [g^{\mu_3\mu_4}(p_3^\rho - p_4^\rho) + 2g^{\mu_4\rho}p_4^{\mu_3} - 2g^{\rho\mu_3}p_3^{\mu_4}] \\ + 2p_1\cdot p_2\,(g^{\mu_1\mu_3}g^{\mu_2\mu_4} - g^{\mu_2\mu_3}g^{\mu_1\mu_4})\}\, \varepsilon_1^{\mu_1}\varepsilon_2^{\mu_2}\varepsilon_3^{\mu_3}\varepsilon_4^{\mu_4}. \tag{J.150}$$

The contraction of indices for long expressions is best done with the help of a computer algebra program. Here is a short FORM *program, which performs the contractions in Eq. (J.150):*

```
* Example program for FORM

V p1,p2,p3,p4, e1,e2,e3,e4;
I mu1,mu2,mu3,mu4,nu,rho;

L  Ns = ((d_(mu1,mu2)*(p1(nu)-p2(nu)) + 2*d_(mu2,nu)*p2(mu1) - 2*d_(nu,mu1)*p1(mu2))*
          d_(nu,rho)*
          (d_(mu3,mu4)*(p3(rho)-p4(rho)) + 2*d_(mu4,rho)*p4(mu3) - 2*d_(rho,mu3)*p3(mu4))
          +2*p1(nu)*p2(nu)*(d_(mu1,mu3)*d_(mu2,mu4)-d_(mu2,mu3)*d_(mu1,mu4)))*
        e1(mu1)*e2(mu2)*e3(mu3)*e4(mu4);

print;

.end
```

The same can be done in C++, using the GiNaC *library:*

```
// Example in C++ with GiNaC

#include <iostream>
#include <string>
#include <sstream>
#include <ginac/ginac.h>
using namespace std;
using namespace GiNaC;

string itos(int arg)
{
  ostringstream buffer;
  buffer << arg;
  return buffer.str();
}
```

```
int main()
{
  varidx mu1(symbol("mu1"),4), mu2(symbol("mu2"),4),
         mu3(symbol("mu3"),4), mu4(symbol("mu4"),4),
         nu(symbol("nu"),4), rho(symbol("rho"),4);

  symbol p1("p1"), p2("p2"), p3("p3"), p4("p4"),
         e1("e1"), e2("e2"), e3("e3"), e4("e4");

  vector<ex> p_vec = { p1, p2, p3, p4 };
  vector<ex> e_vec = { e1, e2, e3, e4 };

  scalar_products sp;
  for (int i=0; i<4; i++)
    {
      for (int j=i+1; j<4; j++)
        {
          sp.add(p_vec[i],p_vec[j],symbol( string("p")+itos(i+1)+string("p")+itos(j+1) ));
          sp.add(e_vec[i],e_vec[j],symbol( string("e")+itos(i+1)+string("e")+itos(j+1) ));
          sp.add(p_vec[i],e_vec[j],symbol( string("p")+itos(i+1)+string("e")+itos(j+1) ));
          sp.add(e_vec[i],p_vec[j],symbol( string("p")+itos(j+1)+string("e")+itos(i+1) ));
        }
    }

  ex Ns = ((lorentz_g(mu1,mu2)*(indexed(p1,nu)-indexed(p2,nu))
        + 2*lorentz_g(mu2,nu)*indexed(p2,mu1) - 2*lorentz_g(nu,mu1)*indexed(p1,mu2))
       * lorentz_g(nu.toggle_variance(),rho.toggle_variance())
       * (lorentz_g(mu3,mu4)*(indexed(p3,rho)-indexed(p4,rho))
          + 2*lorentz_g(mu4,rho)*indexed(p4,mu3) - 2*lorentz_g(rho,mu3)*indexed(p3,mu4))
           + 2*indexed(p1,nu)*indexed(p2,nu.toggle_variance())
       *(lorentz_g(mu1,mu3)*lorentz_g(mu2,mu4) - lorentz_g(mu2,mu3)*lorentz_g(mu1,mu4)))
    *indexed(e1,mu1.toggle_variance())*indexed(e2,mu2.toggle_variance())
    *indexed(e3,mu3.toggle_variance())*indexed(e4,mu4.toggle_variance());

  Ns = Ns.expand();
  Ns = Ns.simplify_indexed(sp);

  cout << Ns << endl;
}
```

After a few additional simplifications one finds

$$\begin{aligned} N_s = {} & 4\,(p_1 \cdot \varepsilon_2)\,(p_3 \cdot \varepsilon_4)\,(\varepsilon_1 \cdot \varepsilon_3) - 4\,(p_1 \cdot \varepsilon_2)\,(p_4 \cdot \varepsilon_3)\,(\varepsilon_1 \cdot \varepsilon_4) + 4\,(p_2 \cdot \varepsilon_1)\,(p_4 \cdot \varepsilon_3)\,(\varepsilon_2 \cdot \varepsilon_4) \\ & - 4\,(p_2 \cdot \varepsilon_1)\,(p_3 \cdot \varepsilon_4)\,(\varepsilon_2 \cdot \varepsilon_3) + 4\left[(p_1 \cdot \varepsilon_3)\,(p_2 \cdot \varepsilon_4) - (p_1 \cdot \varepsilon_4)\,(p_2 \cdot \varepsilon_3)\right](\varepsilon_1 \cdot \varepsilon_2) \\ & + 4\left[(p_3 \cdot \varepsilon_1)\,(p_4 \cdot \varepsilon_2) - (p_3 \cdot \varepsilon_2)\,(p_4 \cdot \varepsilon_1)\right](\varepsilon_3 \cdot \varepsilon_4) + 2\,(p_1 \cdot p_2)\,(\varepsilon_1 \cdot \varepsilon_3)\,(\varepsilon_2 \cdot \varepsilon_4) \\ & - 2\,(p_1 \cdot p_2)\,(\varepsilon_1 \cdot \varepsilon_4)\,(\varepsilon_2 \cdot \varepsilon_3) - 2\,(p_2 \cdot p_3 - p_1 \cdot p_3)\,(\varepsilon_1 \cdot \varepsilon_2)\,(\varepsilon_3 \cdot \varepsilon_4)\,. \end{aligned} \tag{J.151}$$

The numerator N_t is obtained from the numerator N_s by the substitution $(1, 2, 3) \rightarrow (2, 3, 1)$, *the numerator N_u is obtained from the numerator N_s by the substitution* $(1, 2, 3) \rightarrow (3, 1, 2)$.

Let us add the following remark: The colour factors satisfy (obviously) the Jacobi identity

$$C_s + C_t + C_u = 0. \tag{J.152}$$

It is an easy exercise to check, that the numerators N_s, N_t and N_u, as determined above, satisfy the Jacobi-like identity

$$N_s + N_t + N_u = 0. \tag{J.153}$$

Exercise 29 *Let $n \in \mathbb{N}$. Show that the action of $(\mathbf{j}^+)^n$ on the integrand of the Schwinger parameter representation is given by*

$$\left(\mathbf{j}^+\right)^n I_{\nu_1 \dots \nu_j \dots \nu_{n_{\mathrm{int}}}}(D) = \frac{e^{l \varepsilon \gamma_{\mathrm{E}}}}{\prod\limits_{k=1}^{n_{\mathrm{int}}} \Gamma(\nu_k)} \int\limits_{\alpha_k \geq 0} d^{n_{\mathrm{int}}} \alpha \left(\prod_{k=1}^{n_{\mathrm{int}}} \alpha_k^{\nu_k - 1} \right) \frac{\alpha_j^n}{\mathcal{U}^{\frac{D}{2}}} e^{-\frac{\mathcal{F}}{\mathcal{U}}}. \tag{J.154}$$

Solution: *By definition the operator $(\mathbf{j}^+)^n$ acts on $I_{\nu_1 \dots \nu_j \dots \nu_{n_{\mathrm{int}}}}(D)$ as*

$$\begin{aligned} \left(\mathbf{j}^+\right)^n I_{\nu_1 \dots \nu_j \dots \nu_{n_{\mathrm{int}}}}(D) &= \nu_j \left(\nu_j + 1\right) \cdot \ldots \left(\nu_j + n - 1\right) \cdot I_{\nu_1 \dots (\nu_j + n) \dots \nu_{n_{\mathrm{int}}}}(D) \\ &= \frac{\Gamma\left(\nu_j + n\right)}{\Gamma\left(\nu_j\right)} I_{\nu_1 \dots (\nu_j + n) \dots \nu_{n_{\mathrm{int}}}}(D). \end{aligned} \tag{J.155}$$

For $I_{\nu_1 \dots (\nu_j + n) \dots \nu_{n_{\mathrm{int}}}}(D)$ we use the Schwinger parameter representation

$$I_{\nu_1 \dots (\nu_j + n) \dots \nu_{n_{\mathrm{int}}}}(D) = \frac{e^{l \varepsilon \gamma_{\mathrm{E}}}}{\Gamma(\nu_j + n) \prod\limits_{\substack{k=1 \\ k \neq j}}^{n_{\mathrm{int}}} \Gamma(\nu_k)} \int\limits_{\alpha_k \geq 0} d^{n_{\mathrm{int}}} \alpha \left(\prod_{k=1}^{n_{\mathrm{int}}} \alpha_k^{\nu_k - 1} \right) \frac{\alpha_j^n}{\mathcal{U}^{\frac{D}{2}}} e^{-\frac{\mathcal{F}}{\mathcal{U}}}. \tag{J.156}$$

Thus

$$\begin{aligned} \left(\mathbf{j}^+\right)^n I_{\nu_1 \dots \nu_j \dots \nu_{n_{\mathrm{int}}}}(D) &= \frac{\Gamma\left(\nu_j + n\right)}{\Gamma\left(\nu_j\right)} I_{\nu_1 \dots (\nu_j + n) \dots \nu_{n_{\mathrm{int}}}}(D) \\ &= \frac{e^{l \varepsilon \gamma_{\mathrm{E}}}}{\prod\limits_{k=1}^{n_{\mathrm{int}}} \Gamma(\nu_k)} \int\limits_{\alpha_k \geq 0} d^{n_{\mathrm{int}}} \alpha \left(\prod_{k=1}^{n_{\mathrm{int}}} \alpha_k^{\nu_k - 1} \right) \frac{\alpha_j^n}{\mathcal{U}^{\frac{D}{2}}} e^{-\frac{\mathcal{F}}{\mathcal{U}}}, \end{aligned} \tag{J.157}$$

as claimed.

Exercise 30 *Work out the corresponding formula for*

$$\int \frac{d^D k}{i \pi^{D/2}} k^{\mu_1} k^{\mu_2} k^{\mu_3} k^{\mu_4} k^{\mu_5} k^{\mu_6} f(k^2). \tag{J.158}$$

Solution: *The result must be proportional to a symmetric tensor $T^{\mu_1 \mu_2 \mu_3 \mu_4 \mu_5 \mu_6}$ build from the metric tensor. Let's first construct this tensor. It has 15 terms. This can be seen as follows: Start from index μ_1: There are five possibilities how this index can be paired with another index μ_j into $g^{\mu_1 \mu_j}$: Any choice $j \in \{2, 3, 4, 5, 6\}$ is allowed.*

The remaining four indices must form a symmetric tensor of rank 4, such a rank 4 tensor has three terms (compare with Eq. (4.104)). Thus

$$\begin{aligned} T^{\mu_1\mu_2\mu_3\mu_4\mu_5\mu_6} = & \; g^{\mu_1\mu_2} g^{\mu_3\mu_4} g^{\mu_5\mu_6} + g^{\mu_1\mu_2} g^{\mu_3\mu_5} g^{\mu_4\mu_6} + g^{\mu_1\mu_2} g^{\mu_3\mu_6} g^{\mu_4\mu_5} \\ & + g^{\mu_1\mu_3} g^{\mu_2\mu_4} g^{\mu_5\mu_6} + g^{\mu_1\mu_3} g^{\mu_2\mu_5} g^{\mu_4\mu_6} + g^{\mu_1\mu_3} g^{\mu_2\mu_6} g^{\mu_4\mu_5} \\ & + g^{\mu_1\mu_4} g^{\mu_3\mu_2} g^{\mu_5\mu_6} + g^{\mu_1\mu_4} g^{\mu_3\mu_5} g^{\mu_2\mu_6} + g^{\mu_1\mu_4} g^{\mu_3\mu_6} g^{\mu_2\mu_5} \\ & + g^{\mu_1\mu_5} g^{\mu_3\mu_4} g^{\mu_2\mu_6} + g^{\mu_1\mu_5} g^{\mu_3\mu_2} g^{\mu_4\mu_6} + g^{\mu_1\mu_5} g^{\mu_3\mu_6} g^{\mu_4\mu_2} \\ & + g^{\mu_1\mu_6} g^{\mu_3\mu_4} g^{\mu_5\mu_2} + g^{\mu_1\mu_6} g^{\mu_3\mu_5} g^{\mu_4\mu_2} + g^{\mu_1\mu_6} g^{\mu_3\mu_2} g^{\mu_4\mu_5}. \end{aligned} \tag{J.159}$$

Thus we have the ansatz

$$\int \frac{d^D k}{i\pi^{D/2}} k^{\mu_1} k^{\mu_2} k^{\mu_3} k^{\mu_4} k^{\mu_5} k^{\mu_6} f(k^2) = T^{\mu_1\mu_2\mu_3\mu_4\mu_5\mu_6} \int \frac{d^D k}{i\pi^{D/2}} g(k^2) f(k^2) \tag{J.160}$$

for some unknown function $g(k^2)$. We contract both sides with $g_{\mu_1\mu_2} g_{\mu_3\mu_4} g_{\mu_5\mu_6}$. On the left-hand side we obtain

$$g_{\mu_1\mu_2} g_{\mu_3\mu_4} g_{\mu_5\mu_6} k^{\mu_1} k^{\mu_2} k^{\mu_3} k^{\mu_4} k^{\mu_5} k^{\mu_6} = \left(k^2\right)^3, \tag{J.161}$$

on the right-hand side we obtain

$$g_{\mu_1\mu_2} g_{\mu_3\mu_4} g_{\mu_5\mu_6} T^{\mu_1\mu_2\mu_3\mu_4\mu_5\mu_6} = D\,(D+2)\,(D+4), \tag{J.162}$$

hence

$$\int \frac{d^D k}{i\pi^{D/2}} k^{\mu_1} k^{\mu_2} k^{\mu_3} k^{\mu_4} k^{\mu_5} k^{\mu_6} f(k^2) = -\frac{T^{\mu_1\mu_2\mu_3\mu_4\mu_5\mu_6}}{D\,(D+2)\,(D+4)} \int \frac{d^D k}{i\pi^{D/2}} \left(-k^2\right)^3 f(k^2). \tag{J.163}$$

Exercise 31 *Prove Eqs. (4.124)–(4.126).*

Solution: *We start with*

$$\mathrm{Tr}\left(\gamma^{\mu}_{(4)} \gamma^{\nu}_{(4)}\right) = 4 g^{\mu\nu}_{(4)}. \tag{J.164}$$

From the cyclic property of trace and the anti-commutation relation of Eq. (4.119) we have

$$2\mathrm{Tr}\left(\gamma^{\mu}_{(4)} \gamma^{\nu}_{(4)}\right) = \mathrm{Tr}\left(\gamma^{\mu}_{(4)} \gamma^{\nu}_{(4)}\right) + \mathrm{Tr}\left(\gamma^{\nu}_{(4)} \gamma^{\mu}_{(4)}\right) = 2 g^{\mu\nu}_{(4)} \mathrm{Tr}\,\mathbf{1} = 8 g^{\mu\nu}_{(4)}. \tag{J.165}$$

In order to prove

$$\mathrm{Tr}\left(\gamma_{(4)}^{\mu_1}\gamma_{(4)}^{\mu_2}\cdots\gamma_{(4)}^{\mu_{2n}}\right)=\sum_{j=2}^{2n}(-1)^j\,g_{(4)}^{\mu_1\mu_j}\mathrm{Tr}\left(\gamma_{(4)}^{\mu_2}\cdots\gamma_{(4)}^{\mu_{j-1}}\gamma_{(4)}^{\mu_{j+1}}\cdots\gamma_{(4)}^{\mu_{2n}}\right) \tag{J.166}$$

we anti-commute the first Dirac matrix $\gamma_{(4)}^{\mu_1}$ from the first place to the last place, using the anti-commutation relation of Eq. (4.119). This yields

$$\mathrm{Tr}\left(\gamma_{(4)}^{\mu_1}\gamma_{(4)}^{\mu_2}\cdots\gamma_{(4)}^{\mu_{2n}}\right)=2\sum_{j=2}^{2n}(-1)^j\,g_{(4)}^{\mu_1\mu_j}\mathrm{Tr}\left(\gamma_{(4)}^{\mu_2}\cdots\gamma_{(4)}^{\mu_{j-1}}\gamma_{(4)}^{\mu_{j+1}}\cdots\gamma_{(4)}^{\mu_{2n}}\right)+\mathrm{Tr}\left(\gamma_{(4)}^{\mu_2}\cdots\gamma_{(4)}^{\mu_{2n}}\gamma_{(4)}^{\mu_1}\right). \tag{J.167}$$

From the cyclicity of the trace we have

$$\mathrm{Tr}\left(\gamma_{(4)}^{\mu_2}\cdots\gamma_{(4)}^{\mu_{2n}}\gamma_{(4)}^{\mu_1}\right)=\mathrm{Tr}\left(\gamma_{(4)}^{\mu_1}\gamma_{(4)}^{\mu_2}\cdots\gamma_{(4)}^{\mu_{2n}}\right) \tag{J.168}$$

and the result follows. In order to show that the trace of an odd number of Dirac matrices vanishes

$$\mathrm{Tr}\left(\gamma_{(4)}^{\mu_1}\gamma_{(4)}^{\mu_2}\cdots\gamma_{(4)}^{\mu_{2n-1}}\right)=0 \tag{J.169}$$

we use $\gamma_5^2=\mathbf{1}$, the anti-commutation relations of γ_5 (we anti-commute one γ_5 from the second place to the last place) and the cyclicity of the trace:

$$\begin{aligned}\mathrm{Tr}\left(\gamma_{(4)}^{\mu_1}\gamma_{(4)}^{\mu_2}\cdots\gamma_{(4)}^{\mu_{2n-1}}\right)&=\mathrm{Tr}\left(\gamma_5\gamma_5\gamma_{(4)}^{\mu_1}\gamma_{(4)}^{\mu_2}\cdots\gamma_{(4)}^{\mu_{2n-1}}\right)=-\mathrm{Tr}\left(\gamma_5\gamma_{(4)}^{\mu_1}\gamma_{(4)}^{\mu_2}\cdots\gamma_{(4)}^{\mu_{2n-1}}\gamma_5\right)\\&=-\mathrm{Tr}\left(\gamma_5\gamma_5\gamma_{(4)}^{\mu_1}\gamma_{(4)}^{\mu_2}\cdots\gamma_{(4)}^{\mu_{2n-1}}\right)=-\mathrm{Tr}\left(\gamma_{(4)}^{\mu_1}\gamma_{(4)}^{\mu_2}\cdots\gamma_{(4)}^{\mu_{2n-1}}\right).\end{aligned} \tag{J.170}$$

Thus

$$2\mathrm{Tr}\left(\gamma_{(4)}^{\mu_1}\gamma_{(4)}^{\mu_2}\cdots\gamma_{(4)}^{\mu_{2n-1}}\right)=0. \tag{J.171}$$

For traces involving γ_5

$$\begin{aligned}\mathrm{Tr}\left(\gamma_5\right)&=0,\\\mathrm{Tr}\left(\gamma_{(4)}^{\mu_1}\gamma_{(4)}^{\mu_2}\gamma_5\right)&=0,\\\mathrm{Tr}\left(\gamma_{(4)}^{\mu_1}\gamma_{(4)}^{\mu_2}\gamma_{(4)}^{\mu_3}\gamma_{(4)}^{\mu_4}\gamma_5\right)&=4i\,\varepsilon^{\mu_1\mu_2\mu_3\mu_4}\end{aligned} \tag{J.172}$$

we insert the definition of γ_5

$$\gamma_5=\frac{i}{24}\varepsilon_{\nu_1\nu_2\nu_3\nu_4}\gamma_{(4)}^{\nu_1}\gamma_{(4)}^{\nu_2}\gamma_{(4)}^{\nu_3}\gamma_{(4)}^{\nu_4}. \tag{J.173}$$

We then have to evaluate traces with an even number of Dirac matrices, which we can do with the rule proven above. In the first two cases we have to evaluate a trace over four, respectively six Dirac matrices. Each term of the result necessarily contains a factor $g^{\nu_i \nu_j}$, which vanishes when contracted into $\varepsilon_{\nu_1 \nu_2 \nu_3 \nu_4}$. In the third case we have to evaluate a trace over eight Dirac matrices. Here, terms of the form

$$g^{\mu_1 \nu_{\pi(1)}} g^{\mu_2 \nu_{\pi(2)}} g^{\mu_3 \nu_{\pi(3)}} g^{\mu_4 \nu_{\pi(4)}} \tag{J.174}$$

survive, where π is a permutation of $(1, 2, 3, 4)$. We may either work out the result by brute force, or – more elegantly – first establish that the final result must be proportional to $\varepsilon^{\mu_1 \mu_2 \mu_3 \mu_4}$, as any symmetric part vanishes:

$$\begin{aligned}
\mathrm{Tr}\left(\gamma_{(4)}^{\mu_2}\gamma_{(4)}^{\mu_1}\gamma_{(4)}^{\mu_3}\gamma_{(4)}^{\mu_4}\gamma_5\right) &= -\mathrm{Tr}\left(\gamma_{(4)}^{\mu_1}\gamma_{(4)}^{\mu_2}\gamma_{(4)}^{\mu_3}\gamma_{(4)}^{\mu_4}\gamma_5\right) + 2g_{(4)}^{\mu_1\mu_2}\mathrm{Tr}\left(\gamma_{(4)}^{\mu_3}\gamma_{(4)}^{\mu_4}\gamma_5\right) \\
&= -\mathrm{Tr}\left(\gamma_{(4)}^{\mu_1}\gamma_{(4)}^{\mu_2}\gamma_{(4)}^{\mu_3}\gamma_{(4)}^{\mu_4}\gamma_5\right).
\end{aligned} \tag{J.175}$$

Thus

$$\mathrm{Tr}\left(\gamma_{(4)}^{\mu_1}\gamma_{(4)}^{\mu_2}\gamma_{(4)}^{\mu_3}\gamma_{(4)}^{\mu_4}\gamma_5\right) = c\varepsilon^{\mu_1\mu_2\mu_3\mu_4} \tag{J.176}$$

for some constant c. We then obtain the constant by contracting with $i\varepsilon_{\mu_1\mu_2\mu_3\mu_4}/24$. On the left-hand side we find

$$\frac{i}{24}\varepsilon_{\mu_1\mu_2\mu_3\mu_4}\mathrm{Tr}\left(\gamma_{(4)}^{\mu_1}\gamma_{(4)}^{\mu_2}\gamma_{(4)}^{\mu_3}\gamma_{(4)}^{\mu_4}\gamma_5\right) = \mathrm{Tr}\left(\gamma_5\gamma_5\right) = \mathrm{Tr}\left(\mathbf{1}\right) = 4. \tag{J.177}$$

On the right-hand side we obtain (using $\varepsilon_{\mu_1\mu_2\mu_3\mu_4}\varepsilon^{\mu_1\mu_2\mu_3\mu_4} = -24$)

$$\frac{i}{24}\varepsilon_{\mu_1\mu_2\mu_3\mu_4} \cdot c\varepsilon^{\mu_1\mu_2\mu_3\mu_4} = -ic, \tag{J.178}$$

and hence $c = 4i$.

Exercise 32 *Show that with the definitions and conventions as above the rules for the traces of Dirac matrices carry over to D dimensions. In detail, show:*

1. *Traces of an even number of Dirac matrices are evaluated with the rules*

$$\begin{aligned}
\mathrm{Tr}\left(\gamma_{(D)}^{\mu}\gamma_{(D)}^{\nu}\right) &= 4g_{(D)}^{\mu\nu}, \\
\mathrm{Tr}\left(\gamma_{(D)}^{\mu_1}\gamma_{(D)}^{\mu_2}\cdots\gamma_{(D)}^{\mu_{2n}}\right) &= \sum_{j=2}^{2n}(-1)^j\, g_{(D)}^{\mu_1\mu_j}\mathrm{Tr}\left(\gamma_{(D)}^{\mu_2}\cdots\gamma_{(D)}^{\mu_{j-1}}\gamma_{(D)}^{\mu_{j+1}}\cdots\gamma_{(D)}^{\mu_{2n}}\right).
\end{aligned} \tag{J.179}$$

2. *Traces of an odd number of Dirac matrices vanish:*

$$\mathrm{Tr}\left(\gamma_{(D)}^{\mu_1}\gamma_{(D)}^{\mu_2}\cdots\gamma_{(D)}^{\mu_{2n-1}}\right) = 0 \tag{J.180}$$

3. *For traces involving* γ_5 *we have*

$$\begin{aligned}
\mathrm{Tr}\left(\gamma_5\right) &= 0,\\
\mathrm{Tr}\left(\gamma_{(D)}^{\mu}\gamma_{(D)}^{\nu}\gamma_5\right) &= 0,\\
\mathrm{Tr}\left(\gamma_{(D)}^{\mu}\gamma_{(D)}^{\nu}\gamma_{(D)}^{\rho}\gamma_{(D)}^{\sigma}\gamma_5\right) &= \begin{cases} 4i\varepsilon^{\mu\nu\rho\sigma}, & \mu,\nu,\rho,\sigma\in\{0,1,2,3\},\\ 0, & \text{otherwise}.\end{cases}
\end{aligned} \tag{J.181}$$

Solution: *The proof of point 1 follows exactly the proof in four space-time dimensions. Note that we assume the normalisation as in Eq. (4.128)*

$$\mathrm{Tr}\left(\mathbf{1}\right) = 4, \tag{J.182}$$

otherwise there would be small modifications.

In order to prove that the trace of an odd number of Dirac matrices vanishes, we proceed by induction. For $n = 1$ *we have*

$$\begin{aligned}
D\mathrm{Tr}\left(\gamma_{(D)}^{\mu}\right) &= \mathrm{Tr}\left(\gamma_{\nu}^{(D)}\gamma_{(D)}^{\nu}\gamma_{(D)}^{\mu}\right) \; = \; 2g_{(D)}^{\mu\nu}\mathrm{Tr}\left(\gamma_{\nu}^{(D)}\right) - \mathrm{Tr}\left(\gamma_{\nu}^{(D)}\gamma_{(D)}^{\mu}\gamma_{(D)}^{\nu}\right)\\
&= (2-D)\,\mathrm{Tr}\left(\gamma_{(D)}^{\mu}\right)
\end{aligned} \tag{J.183}$$

and hence

$$2\,(D-1)\,\mathrm{Tr}\left(\gamma_{(D)}^{\mu}\right) = 0. \tag{J.184}$$

As this has to hold for any D, we conclude

$$\mathrm{Tr}\left(\gamma_{(D)}^{\mu}\right) = 0. \tag{J.185}$$

Let us now assume that a trace over $(2n-3)$ *Dirac matrices vanishes. We consider*

$$\begin{aligned}
&D\mathrm{Tr}\left(\gamma_{(D)}^{\mu_1}\gamma_{(D)}^{\mu_2}\cdots\gamma_{(D)}^{\mu_{2n-1}}\right) = \mathrm{Tr}\left(\gamma_{\nu}^{(D)}\gamma_{(D)}^{\nu}\gamma_{(D)}^{\mu_1}\gamma_{(D)}^{\mu_2}\cdots\gamma_{(D)}^{\mu_{2n-1}}\right)\\
&= 2\sum_{j=1}^{2n-1}(-1)^{j-1}\,\mathrm{Tr}\left(\gamma_{(D)}^{\mu_j}\gamma_{(D)}^{\mu_1}\cdots\gamma_{(D)}^{\mu_{j-1}}\gamma_{(D)}^{\mu_{j+1}}\cdots\gamma_{(D)}^{\mu_{2n-1}}\right) - D\mathrm{Tr}\left(\gamma_{(D)}^{\mu_1}\gamma_{(D)}^{\mu_2}\cdots\gamma_{(D)}^{\mu_{2n-1}}\right)\\
&= \left[2\,(2n-1) - D\right]\mathrm{Tr}\left(\gamma_{(D)}^{\mu_1}\gamma_{(D)}^{\mu_2}\cdots\gamma_{(D)}^{\mu_{2n-1}}\right),
\end{aligned} \tag{J.186}$$

and hence

$$2\,(D-2n+1)\,\mathrm{Tr}\left(\gamma_{(D)}^{\mu_1}\gamma_{(D)}^{\mu_2}\cdots\gamma_{(D)}^{\mu_{2n-1}}\right) = 0, \tag{J.187}$$

from which the claim follows.

The proof of point 3 is to a large extent identical to the proof for four space-time dimensions. It remains to show that

$$\text{Tr}\left(\gamma^{\mu}_{(D)}\gamma^{\nu}_{(D)}\gamma^{\rho}_{(D)}\gamma^{\sigma}_{(D)}\gamma_5\right) = 0, \tag{J.188}$$

if at least one index is not an element of $\{0, 1, 2, 3\}$. *Assume* $\mu \notin \{0, 1, 2, 3\}$. *We insert the definition of* γ_5

$$\gamma_5 = i\gamma^{0}_{(D)}\gamma^{1}_{(D)}\gamma^{2}_{(D)}\gamma^{3}_{(D)} \tag{J.189}$$

and evaluate a trace over eight Dirac matrices. Each term which does nor vanish for other reasons, will contain a factor

$$g^{\mu\tau}, \tag{J.190}$$

with $\mu \notin \{0, 1, 2, 3\}$ *and* $\tau \in \{0, 1, 2, 3\}$. *As the metric tensor is diagonal, any non-diagonal element vanishes.*

Exercise 33 *Derive Eq. (4.147).*

Solution: *In order to show*

$$\begin{aligned}
\text{Tr}\, \not{q}_1\gamma_\beta\not{q}_0\gamma_\alpha\not{q}_2\not{k}_{(-2\varepsilon)}\gamma_5 &= k^2_{(-2\varepsilon)} \cdot 4i\varepsilon_{\alpha\lambda\beta\kappa}p_1^\lambda p_2^\kappa + \dots, \\
\text{Tr}\, \not{q}_2\gamma_\alpha\not{q}_0\gamma_\beta\not{q}_1\not{k}_{(-2\varepsilon)}\gamma_5 &= k^2_{(-2\varepsilon)} \cdot 4i\varepsilon_{\alpha\lambda\beta\kappa}p_1^\lambda p_2^\kappa + \dots,
\end{aligned} \tag{J.191}$$

we decompose q_0, q_1 *and* q_2 *into a four-dimension part and a* (-2ε)*-dimensional part:*

$$q_0 = k_{(4)} + k_{(-2\varepsilon)}, \qquad q_1 = k_{(4)} + k_{(-2\varepsilon)} - p_2, \qquad q_2 = k_{(4)} + k_{(-2\varepsilon)} + p_1. \tag{J.192}$$

The external momenta p_1 *and* p_2 *are four-dimensional. Inside the traces we already have one* $\not{k}_{(-2\varepsilon)}$, *we need to pick up from the substitution of* q_0, q_1, q_2 *exactly one other* (-2ε)*-dimensional part. In all other cases the traces vanish. Thus*

$$\begin{aligned}
\text{Tr}\, \not{q}_1\gamma_\beta\not{q}_0\gamma_\alpha\not{q}_2\not{k}_{(-2\varepsilon)}\gamma_5 &= \text{Tr}\, \not{k}_{(-2\varepsilon)}\gamma_\beta\not{k}_{(4)}\gamma_\alpha\left(\not{k}_{(4)} + \not{p}_1\right)\not{k}_{(-2\varepsilon)}\gamma_5 \\
&+ \text{Tr}\,\left(\not{k}_{(4)} - \not{p}_2\right)\gamma_\beta\not{k}_{(-2\varepsilon)}\gamma_\alpha\left(\not{k}_{(4)} + \not{p}_1\right)\not{k}_{(-2\varepsilon)}\gamma_5 + \text{Tr}\,\left(\not{k}_{(4)} - \not{p}_2\right)\gamma_\beta\not{k}_{(4)}\gamma_\alpha\not{k}_{(-2\varepsilon)}\not{k}_{(-2\varepsilon)}\gamma_5.
\end{aligned} \tag{J.193}$$

We permute the $\not{k}_{(-2\varepsilon)}$ *next to each other and use Eq. (4.140):*

$$\begin{aligned}
&\text{Tr}\, \not{q}_1\gamma_\beta\not{q}_0\gamma_\alpha\not{q}_2\not{k}_{(-2\varepsilon)}\gamma_5 = \\
&\left(k^2_{(-2\varepsilon)}\right) \cdot 4i\varepsilon_{\alpha\lambda\beta\kappa}\left[k^\kappa_{(4)}\left(k^\lambda_{(4)} + p_1^\lambda\right) - \left(k^\kappa_{(4)} - p_2^\kappa\right)\left(k^\lambda_{(4)} + p_1^\lambda\right) + \left(k^\kappa_{(4)} - p_2^\kappa\right)k^\lambda_{(4)}\right].
\end{aligned} \tag{J.194}$$

Here we used the fact, that traces like

$$\text{Tr}\ \left(\not{k}_{(4)} - \not{p}_2\right) \gamma_\beta \left(\not{k}_{(4)} + \not{p}_1\right) \not{k}_{(-2\varepsilon)} \gamma_5 = 0 \tag{J.195}$$

vanish. In Eq. (J.194) terms quadratic in $k_{(4)}$ vanish due to the contraction with the totally anti-symmetric tensor, terms linear in $k_{(4)}$ vanish after integration. Thus

$$\text{Tr}\ \not{q}_1 \gamma_\beta \not{q}_0 \gamma_\alpha \not{q}_2 \not{k}_{(-2\varepsilon)} \gamma_5 = \left(k^2_{(-2\varepsilon)}\right) \cdot 4i\,\varepsilon_{\alpha\lambda\beta\kappa}\, p_1^\lambda p_2^\kappa + \dots. \tag{J.196}$$

The derivation of

$$\text{Tr}\ \not{q}_2 \gamma_\alpha \not{q}_0 \gamma_\beta \not{q}_1 \not{k}_{(-2\varepsilon)} \gamma_5 = k^2_{(-2\varepsilon)} \cdot 4i\,\varepsilon_{\alpha\lambda\beta\kappa}\, p_1^\lambda p_2^\kappa + \dots \tag{J.197}$$

follows along the same lines.

Exercise 34 *Show*

$$\int \frac{d^D k}{(2\pi)^D i} \frac{k^2_{(-2\varepsilon)}}{k_0^2 k_1^2 k_2^2} = -\frac{1}{2} \frac{1}{(4\pi)^2} + O(\varepsilon). \tag{J.198}$$

Solution: *From Eq. (2.136) it follows that*

$$\int \frac{d^D k}{(2\pi)^D i} \frac{k^2_{(-2\varepsilon)}}{k_0^2 k_1^2 k_2^2} = 4\pi\varepsilon \int \frac{d^{D+2} k}{(2\pi)^{D+2} i} \frac{1}{k_0^2 k_1^2 k_2^2}. \tag{J.199}$$

The Feynman parameter representation of the integral in $D + 2 = 6 - 2\varepsilon$ space-time dimensions reads

$$\int \frac{d^{D+2} k}{(2\pi)^{D+2} i} \frac{1}{k_0^2 k_1^2 k_2^2} = -\frac{\Gamma(\varepsilon)}{(4\pi)^{3-\varepsilon}} \left(\mu^2\right)^{-\varepsilon} \int\limits_{a_j \ge 0} d^3 a\ \delta\left(1 - a_0 - a_1 - a_2\right) \frac{1}{[\mathcal{F}(a)]^\varepsilon} \tag{J.200}$$

with

$$\mathcal{F} = a_0 a_2 \left(\frac{-p_1^2}{\mu^2}\right) + a_0 a_1 \left(\frac{-p_2^2}{\mu^2}\right) + a_1 a_2 \left(\frac{-\left(p_1 + p_2\right)^2}{\mu^2}\right). \tag{J.201}$$

We only need the pole part. The prefactor $\Gamma(\varepsilon)$ delivers a pole in ε. The Feynman parameter integral is finite and we may therefore set $\varepsilon = 0$ in the exponent of $\mathcal{F}$. Then the Feynman parameter integral becomes trivial:

$$\int\limits_{a_j \ge 0} d^3 a\ \delta\left(1 - a_0 - a_1 - a_2\right) \frac{1}{\left[\mathcal{F}(a)\right]^\varepsilon} = \int\limits_{a_j \ge 0} d^3 a\ \delta\left(1 - a_0 - a_1 - a_2\right) + O(\varepsilon) = \frac{1}{2} + O(\varepsilon). \tag{J.202}$$

Thus

$$\int \frac{d^{D+2}k}{(2\pi)^{D+2} i} \frac{1}{k_0^2 k_1^2 k_2^2} = -\frac{1}{2\varepsilon} \frac{1}{(4\pi)^3} + O\left(\varepsilon^0\right) \tag{J.203}$$

and hence we obtain in $D = 4 - 2\varepsilon$ space-time dimensions

$$\int \frac{d^D k}{(2\pi)^D i} \frac{k_{(-2\varepsilon)}^2}{k_0^2 k_1^2 k_2^2} = -\frac{1}{2} \frac{1}{(4\pi)^2} + O(\varepsilon). \tag{J.204}$$

Exercise 35 *Reduce the tensor integral*

$$A^{\mu_1\mu_2\mu_3\mu_4}(m) = e^{\varepsilon\gamma_E} \mu^{2\varepsilon} \int \frac{d^D k}{i\pi^{D/2}} \frac{k^{\mu_1} k^{\mu_2} k^{\mu_3} k^{\mu_4}}{(-k^2 + m^2)} \tag{J.205}$$

to $A_0(m)$.

Solution: *The integral does not depend on any external momenta, so it must be proportional to a symmetric tensor build from $g^{\mu\nu}$:*

$$A^{\mu_1\mu_2\mu_3\mu_4}(m) = \left(g^{\mu_1\mu_2} g^{\mu_3\mu_4} + g^{\mu_1\mu_3} g^{\mu_2\mu_4} + g^{\mu_1\mu_4} g^{\mu_2\mu_3}\right) A_4(m). \tag{J.206}$$

Contraction with $g_{\mu_1\mu_2} g_{\mu_3\mu_4}$ yields

$$A_4(m) = \frac{1}{D(D+2)} e^{\varepsilon\gamma_E} \mu^{2\varepsilon} \int \frac{d^D k}{i\pi^{D/2}} \frac{\left(k^2\right)^2}{(-k^2 + m^2)}. \tag{J.207}$$

We further have

$$\begin{aligned} \int \frac{d^D k}{i\pi^{D/2}} \frac{\left(k^2\right)^2}{(-k^2 + m^2)} &= \left(m^2\right)^2 \int \frac{d^D k}{i\pi^{D/2}} \frac{1}{(-k^2 + m^2)} - m^2 \int \frac{d^D k}{i\pi^{D/2}} + \int \frac{d^D k}{i\pi^{D/2}} (-k^2) \\ &= \left(m^2\right)^2 \int \frac{d^D k}{i\pi^{D/2}} \frac{1}{(-k^2 + m^2)}, \end{aligned} \tag{J.208}$$

since the scaleless integrals vanish for $D \neq 0, -2$ (see Eq. (2.137)). Thus

$$A^{\mu_1\mu_2\mu_3\mu_4}(m) = \frac{\left(m^2\right)^2}{D(D+2)} \left(g^{\mu_1\mu_2} g^{\mu_3\mu_4} + g^{\mu_1\mu_3} g^{\mu_2\mu_4} + g^{\mu_1\mu_4} g^{\mu_2\mu_3}\right) A_0(m). \tag{J.209}$$

Exercise 36 *Reduce*

$$g_{\mu_1\mu_2} g_{\mu_3\mu_4} C^{\mu_1\mu_2\mu_3\mu_4}(p_1, p_2, 0, 0, 0) = -g_{\mu_1\mu_2} g_{\mu_3\mu_4} e^{\varepsilon\gamma_E} \mu^{2\varepsilon} \int \frac{d^D k}{i\pi^{D/2}} \frac{k^{\mu_1} k^{\mu_2} k^{\mu_3} k^{\mu_4}}{k^2 (k - p_1)^2 (k - p_1 - p_2)^2} \tag{J.210}$$

to scalar integrals.

Solution: *We have*

$$g_{\mu_1\mu_2} g_{\mu_3\mu_4} C^{\mu_1\mu_2\mu_3\mu_4}(p_1, p_2, 0, 0, 0) = -e^{\varepsilon\gamma_E}\mu^{2\varepsilon} \int \frac{d^D k}{i\pi^{D/2}} \frac{(k^2)^2}{k^2 (k-p_1)^2 (k-p_1-p_2)^2}. \tag{J.211}$$

There are two powers of k^2 in the numerator. One factor of k^2 in the numerator cancels the factor of k^2 in the denominator:

$$g_{\mu_1\mu_2} g_{\mu_3\mu_4} C^{\mu_1\mu_2\mu_3\mu_4}(p_1, p_2, 0, 0, 0) = -e^{\varepsilon\gamma_E}\mu^{2\varepsilon} \int \frac{d^D k}{i\pi^{D/2}} \frac{k^2}{(k-p_1)^2 (k-p_1-p_2)^2}. \tag{J.212}$$

However, there is no factor k^2 left in the denominator to cancel the one in the numerator. The purpose of this exercise is to show how to handle this case. We realise that the integral in Eq. (J.212) is no longer a three-point function, but just a two-point function. With $k' = k - p_1$ we have

$$\begin{aligned} g_{\mu_1\mu_2} g_{\mu_3\mu_4} C^{\mu_1\mu_2\mu_3\mu_4}(p_1, p_2, 0, 0, 0) &= -e^{\varepsilon\gamma_E}\mu^{2\varepsilon} \int \frac{d^D k'}{i\pi^{D/2}} \frac{(k'+p_1)^2}{k'^2 (k'-p_2)^2} \\ &= -g_{\mu_1\mu_2} B^{\mu_1\mu_2}(p_2, 0, 0) - 2p_{1,\mu_1} B^{\mu_1}(p_2, 0, 0) - p_1^2 B_0(p_2, 0, 0). \end{aligned} \tag{J.213}$$

With

$$g_{\mu_1\mu_2} B^{\mu_1\mu_2}(p_2, 0, 0) = 0 \tag{J.214}$$

and

$$B^{\mu_1}(p_2, 0, 0) = \frac{1}{2} p_2^{\mu_1} B_0(p_2, 0, 0) \tag{J.215}$$

we finally obtain

$$g_{\mu_1\mu_2} g_{\mu_3\mu_4} C^{\mu_1\mu_2\mu_3\mu_4}(p_1, p_2, 0, 0, 0) = -p_1 \cdot (p_1 + p_2) B_0(p_2, 0, 0).$$

Exercise 37 *Determine the constants α_1 and α_2 in Eq. (5.45) from the requirement that l_1 and l_2 are light-like, i.e. $l_1^2 = l_2^2 = 0$. Distinguish the cases*

(i) *p_i and p_j are light-like.*
(ii) *p_i is light-like, p_j is not.*
(iii) *both p_i and p_j are not light-like.*

Solution: *We start with the case (i): If $p_i^2 = p_j^2 = 0$ there is not much to be done: We set $\alpha_1 = \alpha_2 = 0$ and*

$$l_1 = p_i, \quad l_2 = p_j \tag{J.216}$$

is the desired solution. Let now consider the case (ii): We assume $p_i^2 = 0$ and $p_j^2 \neq 0$. (The case $p_i^2 \neq 0$, $p_j^2 = 0$ is similar and obtained by $p_i \leftrightarrow p_j$.) With

$$\alpha_1 = 0, \quad \alpha_2 = \frac{p_j^2}{2 p_i p_j} \tag{J.217}$$

we have

$$l_1 = p_i, \quad l_2 = -\alpha_2 p_i + p_j. \tag{J.218}$$

We verify that l_2 is light-like:

$$l_2^2 = p_j^2 - 2\alpha_2 p_i \cdot p_j = 0. \tag{J.219}$$

Let us now turn to case (iii): We assume that both p_i and p_j are not light-like, i.e. $p_i^2 \neq 0$ and $p_j^2 \neq 0$. For $2 p_i p_j > 0$ we set

$$\alpha_1 = \frac{2 p_i p_j - \sqrt{\Delta}}{2 p_j^2}, \quad \alpha_2 = \frac{2 p_i p_j - \sqrt{\Delta}}{2 p_i^2}. \tag{J.220}$$

For $2 p_i p_j < 0$ we set

$$\alpha_1 = \frac{2 p_i p_j + \sqrt{\Delta}}{2 p_j^2}, \quad \alpha_2 = \frac{2 p_i p_j + \sqrt{\Delta}}{2 p_i^2}. \tag{J.221}$$

Here,

$$\Delta = \left(2 p_i p_j\right)^2 - 4 p_i^2 p_j^2. \tag{J.222}$$

The signs are chosen in such away that the light-like limit $p_i^2 \to 0$ (or $p_j^2 \to 0$) is approached smoothly. Note that l_1, l_2 are real for $\Delta > 0$. For $\Delta < 0$, l_1 and l_2 acquire imaginary parts.

Exercise 38 *The method above does not apply to a tensor two-point function, as there is only one linear independent external momentum. However, the tensor two-point functions is easily reduced with standard methods to the scalar two-point function. In this exercise you are asked to work this out for the massless tensor two-point function. The most general massless tensor two-point function is given by*

$$I_2^{\mu_1 \dots \mu_r, s} = e^{\varepsilon \gamma_E} \mu^{2\varepsilon} \int \frac{d^D k}{i \pi^{\frac{D}{2}}} \left(-k_{(-2\varepsilon)}^2\right)^s \frac{k^{\mu_1} \dots k^{\mu_r}}{k^2 (k-p)^2}. \tag{J.223}$$

Reduce this tensor integral to a scalar integral.

Solution: *We first use Feynman parametrisation and obtain*

$$I_2^{\mu_1 \dots \mu_r, s} = e^{\varepsilon \gamma_E} \mu^{2\varepsilon} \int\limits_0^1 da \int \frac{d^D k}{i\pi^{\frac{D}{2}}} \left(-k_{(-2\varepsilon)}^2\right)^s (k+ap)^{\mu_1} \dots (k+ap)^{\mu_r} \left[-k^2 + a(1-a)\left(-p^2\right)\right]^{-2}.$$

Expanding $(k+ap)^{\mu_1} \dots (k+ap)^{\mu_r}$ *yields terms of the form*

$$a^{r-2t} k^{\mu_{\sigma(1)}} \dots k^{\mu_{\sigma(2t)}} p^{\mu_{\sigma(2t+1)}} \dots p^{\mu_{\sigma(r)}}. \quad (J.224)$$

Note that terms with an odd number of k^μ*'s vanish after integration (see Eq. (4.103)). For terms with an even number of* k^μ*'s let us recall Eq. (4.104) and let us generalise the formulae given there to arbitrary even tensor rank. We find*

$$\int \frac{d^D k}{i\pi^{\frac{D}{2}}} k^{\mu_1} \dots k^{\mu_{2w}} f(k^2) = 2^{-w} \frac{\Gamma\left(\frac{D}{2}\right)}{\Gamma\left(\frac{D}{2}+w\right)} \left(g^{\mu_1\mu_2} \dots g^{\mu_{2w-1}\mu_{2w}} + \textit{permutations}\right) \int \frac{d^D k}{i\pi^{\frac{D}{2}}} \left(k^2\right)^w f(k^2).$$

The fully symmetric tensor structure

$$S^{\mu_1 \dots \mu_{2w}} = g^{\mu_1\mu_2} \dots g^{\mu_{2w-1}\mu_{2w}} + \textit{permutations} \quad (J.225)$$

has $(2w-1)!! = (2w-1)(2w-3)\dots 1$ *terms. We obtain in the absence of powers of* $k_{(-2\varepsilon)}^2$

$$\begin{aligned} & e^{\varepsilon \gamma_E} \mu^{2\varepsilon} \int\limits_0^1 da\, a^{r-2t} \int \frac{d^D k}{i\pi^{\frac{D}{2}}} k^{\mu_1} \dots k^{\mu_{2t}} \left[-k^2 + a(1-a)\left(-p^2\right)\right]^{-2} = \\ & = \left(-\frac{p^2}{2}\right)^t S^{\mu_1 \dots \mu_{2t}} \frac{\Gamma(1+r-t-\varepsilon)\Gamma(2-2\varepsilon)}{\Gamma(1-\varepsilon)\Gamma(2+r-2\varepsilon)} I_2 \\ & = \left(-\frac{p^2}{2}\right)^t S^{\mu_1 \dots \mu_{2t}} \frac{(r-t)!}{(r+1)!} \left\{1 + \varepsilon\left[2S_1(r+1) - S_1(r-t) - 2\right] + \mathcal{O}(\varepsilon^2)\right\} I_2, \end{aligned} \quad (J.226)$$

where $S_1(n)$ *is a harmonic sum*

$$S_1(n) = \sum_{j=1}^{n} \frac{1}{j}, \quad (J.227)$$

and I_2 *is the scalar two-point function:*

$$I_2 = e^{\varepsilon \gamma_E} \left(\frac{-p^2}{\mu^2}\right)^{-2\varepsilon} \frac{\Gamma(-\varepsilon)\Gamma(1-\varepsilon)^2}{\Gamma(2-2\varepsilon)} = \frac{1}{\varepsilon} + 2 - \ln\left(\frac{-p^2}{\mu^2}\right) + \mathcal{O}(\varepsilon). \quad (J.228)$$

Since I_2 starts at $1/\varepsilon$ we can neglect $\mathcal{O}(\varepsilon^2)$ terms in Eq. (J.226). If powers of $k^2_{(-2\varepsilon)}$ are present, we obtain if all indices are contracted into four-dimensional quantities

$$e^{\varepsilon\gamma_E}\mu^{2\varepsilon}\int\limits_0^1 da\, a^{r-2t}\int\frac{d^Dk}{i\pi^{\frac{D}{2}}}\left(-k^2_{(-2\varepsilon)}\right)^s k^{\mu_1}\dots k^{\mu_{2t}}\left[-k^2+a(1-a)\left(-p^2\right)\right]^{-2} =$$

$$= -\varepsilon\left(p^2\right)^s\left(-\frac{p^2}{2}\right)^t S^{\mu_1\dots\mu_{2t}}\frac{(s-1)!(r+s-t)!}{(r+2s+1)!}I_2+\mathcal{O}(\varepsilon)$$

$$+\text{ terms, which vanish when contracted into 4-dimensional quantities.} \qquad \text{(J.229)}$$

Exercise 39 *Consider a one-loop four-point function with external momenta p_1, p_2, p_3, p_4 and $p_1^2 = p_2^2 = p_3^2 = p_4^2 = 0$. The external momenta satisfy momentum conservation $p_1 + p_2 + p_3 + p_4 = 0$. For $j \in \{1, 2, 3, 4\}$ set $q_j = k - p_j^{\text{sum}}$. Solve the equations for the quadruple cut*

$$q_1^2 = q_2^2 = q_3^2 = q_4^2 = 0. \qquad \text{(J.230)}$$

Hint: Start from an ansatz

$$k_\mu = c\left\langle a-\left|\gamma_\mu\right|b-\right\rangle, \qquad \text{(J.231)}$$

with $c \in \mathbb{C}$ and a, b light-like.

Solution: *We have $q_4 = k - p_1 - p_2 - p_3 - p_4 = k$ due to momentum conservation. The ansatz of Eq. (J.231) automatically satisfies the equation $q_4^2 = 0$:*

$$q_4^2 \;=\; k^2 \;=\; c^2\left\langle a-\left|\gamma_\mu\right|b-\right\rangle\left\langle a-\left|\gamma^\mu\right|b-\right\rangle \;=\; 2c^2\left\langle a-|a+\right\rangle\left[b+|b-\right] \;=\; 0, \qquad \text{(J.232)}$$

as $\langle a-|a+\rangle = 0$ and $[b+|b-] = 0$. With $k^2 = 0$ (and $p_1^2 = p_2^2 = p_3^2 = p_4^2 = 0$) the other three equations which we have to satisfy reduce to

$$\begin{aligned} 2k\cdot p_1 &= 0,\\ 2k\cdot p_2 &= 2p_1\cdot p_2,\\ 2k\cdot p_4 &= 0 \end{aligned} \qquad \text{(J.233)}$$

Here we used $p_4 = -p_1 - p_2 - p_3$. Let's consider the first equation $2k\cdot p_1 = 0$. If we choose $a = p_1$ (or $b = p_1$) this equation is trivially satisfied:

$$c\left\langle p_1-\left|\gamma_\mu\right|b-\right\rangle p_1^\mu \;=\; c\left\langle p_1-|p_1+\right\rangle\left[p_1+|b-\right] \;=\; 0. \qquad \text{(J.234)}$$

If we choose $a = p_1$ we may then use the freedom to choose b to satisfy the third equation $2k\cdot p_4 = 0$. This will lead to the choice $b = p_4$. We now have one remaining *parameter c left in our ansatz*

$$k_\mu = c \left\langle p_1 - |\gamma_\mu| p_4 - \right\rangle, \tag{J.235}$$

which we use to satisfy the second equation $2k \cdot p_2 = 2p_1 \cdot p_2$. *This yields*

$$c = \frac{[p_2 p_1]}{[p_2 p_4]}. \tag{J.236}$$

Of course, we could have chosen as well $a = p_4$ *in the beginning. This will give us the second solution. In summary we obtain the two solutions*

$$k_\mu^+ = \frac{[p_2 p_1]}{[p_2 p_4]} \left\langle p_1 - |\gamma_\mu| p_4 - \right\rangle, \quad k_\mu^- = \frac{\langle p_1 p_2 \rangle}{\langle p_4 p_2 \rangle} \left\langle p_4 - |\gamma_\mu| p_1 - \right\rangle. \tag{J.237}$$

Exercise 40 *Consider the one-loop eight-point amplitude in massless* ϕ^4 *theory. Verify Eq. (5.72) for the box coefficient.*

Solution: *We consider the box coefficient, where the external momenta are distributed as in Fig. J.5. We first perform the calculation with standard Feynman rules. There is only one Feynman diagram contributing to the coefficient* $c_{0246}^{(0)}$. *This is the diagram shown in Fig. J.5. The contribution of this Feynman diagram to the one-loop amplitude is given by (including a factor* $(4\pi)^{-\varepsilon} e^{\varepsilon \gamma_E}$ *corresponding to the* $\overline{\text{MS}}$*-scheme)*

$$i \mathcal{A}_8^{(1)} \Big|_{\text{Box}} = (i\lambda)^4 (4\pi)^{-\varepsilon} e^{\varepsilon \gamma_E} \mu^{2\varepsilon} \int \frac{d^D k}{(2\pi)^D} \frac{i}{q_0^2} \frac{i}{q_2^2} \frac{i}{q_4^2} \frac{i}{q_6^2} = \frac{i\lambda^4}{(4\pi)^2} I_4^{(0246)}. \tag{J.238}$$

The coefficient of the box integral is therefore given by

$$c_{0246}^{(0)} = \frac{\lambda^4}{(4\pi)^2}. \tag{J.239}$$

Let us now consider Eq. (5.72). There are four tree amplitudes. We have $I_1 = \{1, 2\}$, $I_2 = \{3, 4\}$, $I_3 = \{5, 6\}$ *and* $I_4 = \{7, 8\}$. *In this simple example each tree amplitude equals*

$$i \mathcal{A}_{|I_1|+2}^{(0)} = i \mathcal{A}_{|I_2|+2}^{(0)} = i \mathcal{A}_{|I_3|+2}^{(0)} = i \mathcal{A}_{|I_4|+2}^{(0)} = i\lambda^4, \tag{J.240}$$

and is independent of the external momenta. Thus we do not need to know the concrete solutions of the on-shell conditions

$$q_0^2 = q_2^2 = q_4^2 = q_6^2 = 0, \tag{J.241}$$

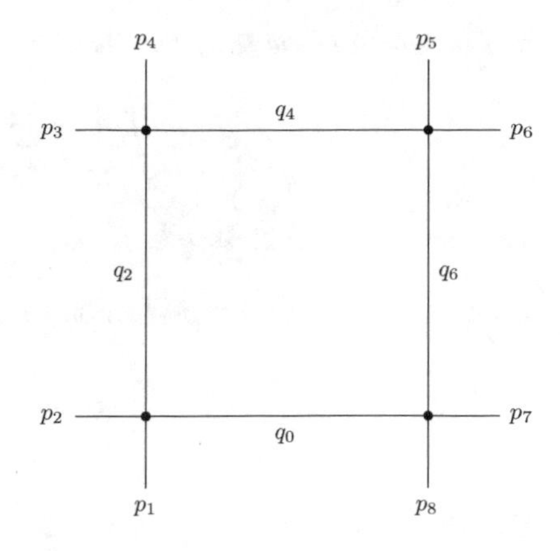

Fig. J.5 A diagram in ϕ^4 theory

it suffices to know that there are two solutions. We consider a scalar theory, hence there are no spins. Eq. (5.72) gives us then

$$c^{(0)}_{0246} = \frac{1}{2}\frac{1}{(4\pi)^2}\sum_{\sigma=\pm} \mathcal{A}^{(0)}_{|I_1|+2}\mathcal{A}^{(0)}_{|I_2|+2}\mathcal{A}^{(0)}_{|I_3|+2}\mathcal{A}^{(0)}_{|I_4|+2} = \frac{\lambda^4}{(4\pi)^2}, \tag{J.242}$$

which agrees with our previous calculation.

Exercise 41 *Show that Eq. (6.2) holds for $q = k$.*

Hint: Consider the scaling relation Eq. (2.76).

Solution: *The scaling relation reads*

$$\int \frac{d^D k}{i\pi^{\frac{D}{2}}} f(k) = (1+\lambda)^D \int \frac{d^D k}{i\pi^{\frac{D}{2}}} f(k+\lambda k). \tag{J.243}$$

The right-hand side has to be independent of λ. This implies in particular that the $O(\lambda)$-term has to vanish. There are now two contributions to the $O(\lambda)$-term. On the one hand we have

$$(1+\lambda)^D = 1 + D\lambda + O\left(\lambda^2\right), \tag{J.244}$$

on the other hand we have

$$f(k+\lambda k) = f(k) + \lambda k^\mu \frac{\partial}{\partial k^\mu} f(k) + O\left(\lambda^2\right). \tag{J.245}$$

Note that

$$k^\mu \frac{\partial}{\partial k^\mu} f(k) = \frac{\partial}{\partial k^\mu}\left[k^\mu \cdot f(k)\right] - D f(k) \tag{J.246}$$

and in summary the $O(\lambda)$-term is given by

$$\int \frac{d^D k}{i\pi^{\frac{D}{2}}} \frac{\partial}{\partial k^\mu} \left[k^\mu \cdot f(k) \right]. \tag{J.247}$$

This proves Eq. (6.2) for $q = k$.

Exercise 42 *Repeat the derivation with $q_{\mathrm{IBP}} = k$ and show*

$$(D - \nu_1 - 2\nu_2) I_{\nu_1\nu_2} - \nu_1 I_{(\nu_1+1)(\nu_2-1)} + \nu_1 (2+x) I_{(\nu_1+1)\nu_2} + 2\nu_2 I_{\nu_1(\nu_2+1)} = 0. \tag{J.248}$$

Solution: *We have*

$$\begin{aligned}
0 &= e^{\varepsilon\gamma_E} \left(m^2\right)^{\nu_{12}-\frac{D}{2}} \int \frac{d^D k}{i\pi^{\frac{D}{2}}} \frac{\partial}{\partial k^\mu} \frac{k^\mu}{\left(-q_1^2+m^2\right)^{\nu_1} \left(-q_2^2+m^2\right)^{\nu_2}} \\
&= e^{\varepsilon\gamma_E} \left(m^2\right)^{\nu_{12}-\frac{D}{2}} \int \frac{d^D k}{i\pi^{\frac{D}{2}}} \left[\frac{\nu_1 \left(q_1^2+q_2^2-p^2\right)}{\left(-q_1^2+m^2\right)^{\nu_1+1} \left(-q_2^2+m^2\right)^{\nu_2}} + \frac{2\nu_2 q_2^2}{\left(-q_1^2+m^2\right)^{\nu_1} \left(-q_2^2+m^2\right)^{\nu_2+1}} \right. \\
&\quad \left. + \frac{D}{\left(-q_1^2+m^2\right)^{\nu_1} \left(-q_2^2+m^2\right)^{\nu_2}} \right] \\
&= \nu_1 \left[-I_{\nu_1\nu_2} - I_{(\nu_1+1)(\nu_2-1)} + (2+x) I_{(\nu_1+1)\nu_2} \right] + 2\nu_2 \left[-I_{\nu_1\nu_2} + I_{\nu_1(\nu_2+1)} \right] + D I_{\nu_1\nu_2}.
\end{aligned} \tag{J.249}$$

Thus

$$(D - \nu_1 - 2\nu_2) I_{\nu_1\nu_2} - \nu_1 I_{(\nu_1+1)(\nu_2-1)} + \nu_1 (2+x) I_{(\nu_1+1)\nu_2} + 2\nu_2 I_{\nu_1(\nu_2+1)} = 0. \tag{J.250}$$

Exercise 43 *Derive the integration-by-parts identity for the integral*

$$I_{0\nu_2} = e^{\varepsilon\gamma_E} \left(m^2\right)^{\nu_2-\frac{D}{2}} \int \frac{d^D k}{i\pi^{\frac{D}{2}}} \frac{1}{\left(-k^2+m^2\right)^{\nu_2}}. \tag{J.251}$$

Verify the identity with the explicit result from Eq. (2.123).

Solution: *The tadpole integral does not depend on any external momentum, hence the only integration-by-parts identity is*

$$\begin{aligned}
0 &= e^{\varepsilon\gamma_E} \left(m^2\right)^{\nu_2-\frac{D}{2}} \int \frac{d^D k}{i\pi^{\frac{D}{2}}} \frac{\partial}{\partial k^\mu} \frac{k^\mu}{\left(-k^2+m^2\right)^{\nu_2}} \\
&= e^{\varepsilon\gamma_E} \left(m^2\right)^{\nu_2-\frac{D}{2}} \int \frac{d^D k}{i\pi^{\frac{D}{2}}} \left[\frac{2\nu_2 k^2}{\left(-k^2+m^2\right)^{\nu_2+1}} + \frac{D}{\left(-k^2+m^2\right)^{\nu_2}} \right] \\
&= (D - 2\nu_2) I_{0\nu_2} + 2\nu_2 I_{0(\nu_2+1)}.
\end{aligned} \tag{J.252}$$

From Eq. (2.123) we have

$$I_{0\nu_2} = \frac{e^{\varepsilon\gamma_E}\Gamma\left(\nu_2 - \frac{D}{2}\right)}{\Gamma\left(\nu_2\right)}. \tag{J.253}$$

Substituting this into the integration-by-parts identity and using $\Gamma(x+1) = x\Gamma(x)$ *we obtain*

$$\begin{aligned}
\left(D - 2\nu_2\right) I_{0\nu_2} + 2\nu_2 I_{0(\nu_2+1)} &= \left(D - 2\nu_2\right) \frac{e^{\varepsilon\gamma_E}\Gamma\left(\nu_2 - \frac{D}{2}\right)}{\Gamma\left(\nu_2\right)} + 2\nu_2 \frac{e^{\varepsilon\gamma_E}\Gamma\left(\nu_2 + 1 - \frac{D}{2}\right)}{\Gamma\left(\nu_2 + 1\right)} \\
&= \left(D - 2\nu_2\right) \frac{e^{\varepsilon\gamma_E}\Gamma\left(\nu_2 - \frac{D}{2}\right)}{\Gamma\left(\nu_2\right)} + 2\left(\nu_2 - \frac{D}{2}\right) \frac{e^{\varepsilon\gamma_E}\Gamma\left(\nu_2 - \frac{D}{2}\right)}{\Gamma\left(\nu_2\right)} \\
&= 0.
\end{aligned} \tag{J.254}$$

Exercise 44 *Consider the double-box graph G shown in Fig. 2.3 and the auxiliary graph* $\tilde{G}$ *with nine propagators shown in Fig. 2.11. This exercise is about the family of Feynman integrals*

$$I_{\nu_1\nu_2\nu_3\nu_4\nu_5\nu_6\nu_7\nu_8\nu_9} \tag{J.255}$$

with $\nu_8, \nu_9 \leq 0$. *Use the notation of the momenta as in Fig. 2.11. Assume that all external momenta are light-like (*$p_1^2 = p_2^2 = p_3^2 = p_4^2 = 0$*) and that all internal propagators are massless. Use one of the public available computer programs* `Kira`, `Reduze` *or* `Fire` *to reduce the Feynman integral*

$$I_{1111111(-1)(-1)} \tag{J.256}$$

to master integrals. For the choice of master integrals you may use the default ordering criteria of the chosen computer program.

Solution: *We are considering the integral*

$$I_{\nu_1\nu_2\nu_3\nu_4\nu_5\nu_6\nu_7\nu_8\nu_9} = e^{2\varepsilon\gamma_E}\left(\mu^2\right)^{\nu-D} \int \frac{d^Dk_1}{i\pi^{\frac{D}{2}}} \frac{d^Dk_2}{i\pi^{\frac{D}{2}}} \prod_{j=1}^{9} \frac{1}{\left(-q_j^2\right)^{\nu_j}} \tag{J.257}$$

with

$$\begin{aligned}
q_1 &= k_1 - p_1, & q_2 &= k_1 - p_1 - p_2, & q_3 &= k_1, \\
q_4 &= k_1 + k_2, & q_5 &= k_2 + p_1 + p_2, & q_6 &= k_2, \\
q_7 &= k_2 + p_1 + p_2 + p_3, & q_8 &= k_1 - p_1 - p_3, & q_9 &= k_2 + p_1 + p_3.
\end{aligned} \tag{J.258}$$

The aim of this exercise is to get acquainted with one of the integration-by-parts reduction programs `Kira`, `Reduze` *or* `Fire`. *We show how each of the three programs can be applied to the problem at hand. The actual syntax may differ for different versions of the same program. The solutions shown below refer to* Kira *version 2.0,* Reduze *version 2.4 and* Fire *version 6.4.2. In addition one should consult the manuals of these programs. Please note that* Kira *and* Reduze *use the convention*

$$\frac{1}{q_j^2 - m_j^2} \quad \textit{instead of} \quad \frac{1}{-q_j^2 + m_j^2} \tag{J.259}$$

for propagators. This implies a minus sign for every integral where ν is odd between the Kira*/*Reduze *notation and the notation used in this book. The CPU timings refer to a standard laptop with* 2.6 GHz.

We start with Kira*. We prepare the following files*

```
job.yaml
myreduction.in
config/integralfamilies.yaml
config/kinematics.yaml
```

The files job.yaml *and* myreduction.in *reside in a directory. This directory also contains a subdirectory* config*. The* config*-subdirectory contains the files* integralfamilies.yaml *and* kinematics.yaml.

The file job.yaml *is the main file and specifies what should be done:*

```
jobs:
  - reduce_sectors:
      reduce:
        - {topologies: [doublebox], sectors: [127], r: 8, s: 2}
      select_integrals:
        select_mandatory_recursively:
          - {topologies: [doublebox], sectors: [127], r: 8, s: 2}
  - kira2form:
      target:
      - [doublebox,myreductions.in]
      reconstruct_mass: true
```

The file myreduction.in *contains the integrals, which should be reduced. In our case it only contains a single line*

```
doublebox[1,1,1,1,1,1,1,-1,-1]
```

We need to give the information on the family of Feynman integrals we are interested in. This is done in the file integralfamilies.yaml*:*

```
integralfamilies:
  - name: "doublebox"
    loop_momenta: [k1, k2]
    top_level_sectors: [127]
    propagators:
      - [ "k1-p1", 0 ]
      - [ "k1-p1-p2", 0 ]
      - [ "k1", 0 ]
      - [ "k1+k2", 0 ]
      - [ "k2+p1+p2", 0 ]
      - [ "k2", 0 ]
      - [ "k2+p1+p2+p3", 0 ]
      - [ "k1-p1-p3", 0 ]
      - [ "k2+p1+p3", 0 ]
```

Finally, we need to specify the kinematics. This is done in the file kinematics.yaml*:*

```
kinematics :
  incoming_momenta: []
  outgoing_momenta: [p1, p2, p3, p4]
  momentum_conservation: [p4,-p1-p2-p3]
  kinematic_invariants:
    - [s,  2]
    - [t,  2]
  scalarproduct_rules:
    - [[p1,p1],  0]
    - [[p2,p2],  0]
    - [[p3,p3],  0]
    - [[p1+p2,p1+p2],  s]
    - [[p2+p3,p2+p3],  t]
    - [[p1+p3,p1+p3],  -s-t]
  symbol_to_replace_by_one: s
```

With these preparations we may now run `Kira` *with the command*

```
kira job.yaml
```

This will produce a file `kira_myreductions.in.inc` *in the directory* `results/doublebox` *with the content*

```
id doublebox(1,1,1,1,1,1,1,-1,-1) =
 + doublebox(1,1,1,1,1,1,1,-1,0)*((-4*t-3*s)*den(2))
 + doublebox(1,1,1,1,1,1,1,0,0)*(-t^2+(-s)*t)
 + doublebox(1,0,1,1,1,0,1,0,0)*((-18*t^2+(-27*s)*t-9*s^2)*den((2*s)*t))
 + doublebox(1,1,1,1,0,0,1,0,0)*(((12*d-36)*t+(9*d-27)*s)*den((d-4)*t))
 + doublebox(1,0,0,1,0,0,1,0,0)*(((162*d^3-1458*d^2+4356*d-4320)*t+(99*d^3-891*d^2+2662*d
 -2640)*s)*den(((2*d^3-24*d^2+96*d-128)*s)*t^2))
 + doublebox(1,0,0,1,1,1,0,0,0)*(((12*d^2-76*d+120)*t^2+((66*d^2-418*d+660)*s)*t+(27*d^2
 -171*d+270)*s^2)*den(((2*d^2-16*d+32)*s^2)*t))
 + doublebox(0,1,1,0,1,1,0,0,0)*(((4*d^2-16*d+12)*t+(4*d^2-16*d+12)*s)*den((d^2-8*d+16)
 *s^2))
 + doublebox(0,0,1,1,1,0,0,0,0)*(((-144*d^2+864*d-1280)*t^2+((54*d^3-630*d^2+2316*d-2720)
 *s)*t+(81*d^3-729*d^2+2178*d-2160)*s^2)*den(((2*d^3-24*d^2+96*d-128)*s^3)*t))
;
```

This gives the desired reduction in `FORM` *notation.* `Kira` *uses by default a ISP-basis. There are eight master integrals. The total run time on a standard laptop is about* 30 s.

Let us now turn to `Reduze`. *Using* `Reduze` *we prepare the following files:*

```
job.yaml
myreduction.in
config/integralfamilies.yaml
config/kinematics.yaml
config/global.yaml
```

The syntax of `Reduze` *is very similar to the syntax of* `Kira`. *The file* `job.yaml` *reads now*

```
jobs:
  - setup_sector_mappings: {}
  - reduce_sectors:
      conditional: true
      sector_selection:
        select_recursively: [ [doublebox, 127] ]
      identities:
        ibp:
          - { r: [t, 8], s: [0, 2] }
        lorentz:
          - { r: [t, 8], s: [0, 2] }
        sector_symmetries:
          - { r: [t, 8], s: [0, 2] }
  - select_reductions:
```

```
      input_file: "myreductions.in"
      output_file: "myreductions.tmp.1"
  - reduce_files:
      equation_files:
          - "myreductions.tmp.1"
      output_file: "myreductions.tmp.2"
  - export:
      input_file: "myreductions.tmp.2"
      output_file: "myreductions.sol"
      output_format: "maple"
```

The file `myreduction.in` *contains again a list of the integrals to be reduced, for the case at hand it is given by*

```
{
INT["doublebox",{1,1,1,1,1,1,1,-1,-1}]
}
```

The file `integralfamilies.yaml` *specifies the family of Feynman integrals under consideration:*

```
integralfamilies:
  - name: "doublebox"
    loop_momenta: [k1, k2]
    propagators:
      - [ "k1-p1", 0 ]
      - [ "k1-p1-p2", 0 ]
      - [ "k1", 0 ]
      - [ "k1+k2", 0 ]
      - [ "k2+p1+p2", 0 ]
      - [ "k2", 0 ]
      - [ "k2+p1+p2+p3", 0 ]
      - [ "k1-p1-p3", 0 ]
      - [ "k2+p1+p3", 0 ]
    permutation_symmetries: []
```

The file `kinematics.yaml` *is identical to the corresponding file for* `Kira`*:*

```
kinematics :
  incoming_momenta: []
  outgoing_momenta: [p1,p2,p3,p4]
  momentum_conservation: [p4,-p1-p2-p3]
  kinematic_invariants:
    - [s,  2]
    - [t,  2]
  scalarproduct_rules:
    - [[p1,p1],  0]
    - [[p2,p2],  0]
    - [[p3,p3],  0]
    - [[p1+p2,p1+p2],  s]
    - [[p2+p3,p2+p3],  t]
    - [[p1+p3,p1+p3],  -s-t]
  symbol_to_replace_by_one: s
```

In addition, there is a file `global.yaml` *containing*

```
global_symbols:
  space_time_dimension: d

paths:
  fermat: /usr/local/fermat/ferl64/fer64
```

The last line gives the absolute path to the `Fermat`*-executable and should be modified accordingly. Running*

```
 reduze job.yaml
```

will produce a file `myreductions.sol`

```
myreductions := [

INT("doublebox",7,127,7,2,[1,1,1,1,1,1,1,-1,-1]) =
  INT("doublebox",7,127,7,1,[1,1,1,1,1,1,1,-1,0]) *
      (-2*t-3/2*s) +
  INT("doublebox",7,127,7,0,[1,1,1,1,1,1,1,0,0]) *
      (-t^2-t*s) +
  INT("doublebox",5,93,5,0,[1,0,1,1,1,0,1,0,0]) *
      (-9/2*(2*t^2+3*t*s+s^2)*t^(-1)*s^(-1)) +
  INT("doublebox",5,79,5,0,[1,1,1,1,0,0,1,0,0]) *
      (-3*(4*t-t*d)^(-1)*(3*d*s-12*t+4*t*d-9*s)) +
  INT("doublebox",4,57,4,0,[1,0,0,1,1,1,0,0,0]) *
      (1/2*(12*t^2*d^2+66*t*d^2*s+120*t^2-171*d*s^2-76*t^2*d+660*t*s+27*d^2*s^2+270*s^2
      -418*t*d*s)*(16*t-8*t*d+t*d^2)^(-1)*s^(-2)) +
  INT("doublebox",4,54,4,0,[0,1,1,0,1,1,0,0,0]) *
      (-4*(16+d^2-8*d)^(-1)*(4*d*s-3*t+4*t*d-d^2*s-3*s-t*d^2)*s^(-2)) +
  INT("doubleboxx123",3,28,3,0,[0,0,1,1,1,0,0,0,0]) *
      (-1/2*(2662*d*s-4320*t+4356*t*d+99*d^3*s-891*d^2*s+162*t*d^3-2640*s-1458*t*d^2)
      *(12*t^2*d^2+64*t^2-t^2*d^3-48*t^2*d)^(-1)*s^(-1)) +
  INT("doublebox",3,28,3,0,[0,0,1,1,1,0,0,0,0]) *
      (1/2*(144*t^2*d^2+630*t*d^2*s+1280*t^2-54*t*d^3*s-2178*d*s^2-864*t^2*d+2720*t*s
      +729*d^2*s^2+2160*s^2-81*d^3*s^2-2316*t*d*s)*(64*t-48*t*d-t*d^3+12*t*d^2)^(-1)
      *s^(-3))
];
```

This gives the reduction in `Maple` *format.* `Reduze` *uses by default a ISP-basis. The running time on a standard laptop is about* 1270 s.

Let us now turn to `Fire`*. In order to get the same number of master integrals we use it in combination with* `Litered` *[490, 491]. The program* `Litered` *provides symmetry relations to* `Fire` *and ensures that we end up in the example under consideration with eight master integrals as above. We prepare the following files:*

```
prepare1.m
prepare2.m
prepare3.m
readout.m
work/data.m
work/myreductions.m
work/doublebox.config
```

The file `data.m` *defines the family of Feynman integrals*

```
Internal = {k1, k2};
External = {p1, p2, p3};
Propagators = { -(k1-p1)^2, -(k1-p1-p2)^2, -k1^2, -(k1+k2)^2, -(k2+p1+p2)^2, -k2^2,
-(k2+p1+p2+p3)^2, -(k1-p1-p3)^2, -(k2+p1+p3)^2 };
Replacements = { p1^2 -> 0, p2^2 -> 0, p3^2 -> 0, p1 p2 -> s/2, p2 p3 -> t/2,
p1 p3 -> (-s-t)/2 };
```

The file `myreductions.m` *contains the list of Feynman integrals, which we would like to reduce. For the case at hand*

```
{
 {1,{1,1,1,1,1,1,1,-1,-1}}
}
```

`Fire` *runs partly within* `Mathematica` *and partly in* `C++`*. First, three preparation steps are done within* `Mathematica`*, specified by the files* `prepare1.m`*,* `prepare2.m` *and* `prepare3.m`*. The file* `prepare1.m` *reads*

```
Get["FIRE6.m"];
Get["work/data.m"];
PrepareIBP[];
Prepare[AutoDetectRestrictions -> True];
SaveStart["work/doublebox"];
```

Issuing in Mathematica *the command*

```
Get["prepare1.m"];
```

will generate the file

```
work/doublebox.start
```

The file prepare2.m *reads*

```
SetDirectory["extra/LiteRed/Setup/"];
Get["LiteRed.m"];
SetDirectory["../../../"];
Get["FIRE6.m"];
Get["work/data.m"];
CreateNewBasis[doublebox, Directory -> "work/litered"];
GenerateIBP[doublebox];
AnalyzeSectors[doublebox, {__,0,0}];
FindSymmetries[doublebox, EMs->True];
DiskSave[doublebox];
```

Issuing in Mathematica *the command*

```
Get["prepare2.m"];
```

will generate a subdirectory

```
work/litered
```

containing several files. The file prepare3.m *reads*

```
Get["FIRE6.m"];
LoadStart["work/doublebox"];
TransformRules["work/litered", "work/doublebox.lbases", 1];
SaveSBases["work/doublebox"];
```

Issuing in Mathematica *the command*

```
Get["prepare3.m"];
```

will generate the two files

```
work/doublebox.lbases
work/doublebox.sbases
```

After these preparation step the C++ *program can be called. We need a configuration file* doublebox.config *containing*

```
#threads          1
#variables        d, s, t
#start
#folder           work/
#problem          1 doublebox.sbases
#lbases           doublebox.lbases
#integrals        myreductions.m
#output           myreductions.tables
```

Running

```
bin/FIRE6 -c work/doublebox
```

will generate the file

```
work/myreductions.tables
```

The readout is again done within Mathematica. *The file* readout.m *reads*

```
Get["FIRE6.m"];
LoadStart["work/doublebox", 1];
Burn[];
LoadTables["work/myreductions.tables"];
res = F[1, {1,1,1,1,1,1,1,-1,-1}];
Save["work/myreductions.out",res];
```

Issuing in `Mathematica` *the command*

```
Get["readout.m"];
```

will generate the file `myreductions.out` *with*

```
res = ((-10 + 3*d)*(-8 + 3*d)*(-1350*s^2 + 990*d*s^2 - 234*d^2*s^2 +
        18*d^3*s^2 - 238*s*t - 56*d*s*t + 65*d^2*s*t - 9*d^3*s*t + 1616*t^2 -
        1448*d*t^2 + 404*d^2*t^2 - 36*d^3*t^2)*
      G[1, {0, 0, 1, 1, 1, 0, 0, 0, 0}])/(2*(-5 + d)^2*(-4 + d)^3*s^3*t) -
    (2*(-3 + d)*(46*s - 33*d*s + 5*d^2*s + 58*t - 40*d*t + 6*d^2*t)*
      G[1, {0, 1, 1, 0, 1, 1, 0, 0, 0}])/((-5 + d)*(-4 + d)^2*s^2) -
    ((-3 + d)*(-10 + 3*d)*(-5760*s^2 + 2772*d*s^2 - 414*d^2*s^2 +
        18*d^3*s^2 - 21744*s*t + 12352*d*s*t - 2326*d^2*s*t + 145*d^3*s*t -
        13152*t^2 + 8368*d*t^2 - 1768*d^2*t^2 + 124*d^3*t^2)*
      G[1, {0, 1, 1, 1, 0, 0, 1, 0, 0}])/(8*(-6 + d)*(-5 + d)^2*(-4 + d)^2*
      s^2*t) + (5*(-3 + d)*(-10 + 3*d)*(-8 + 3*d)*(2*s + 3*t)*
      G[1, {1, 0, 0, 1, 0, 0, 1, 0, 0}])/((-4 + d)^3*s*t^2) +
    (3*(-3 + d)*(64 - 18*d + d^2)*(3*s + 4*t)*
      G[1, {1, 0, 0, 1, 1, 1, 1, 0, 0}])/(2*(-6 + d)*(-5 + d)*(-4 + d)*t) -
    (3*(s + t)*(3*s + 5*t)*G[1, {1, 0, 1, 1, 1, 0, 1, 0, 0}])/(s*t) +
    ((-42*s^2 + 9*d*s^2 - 64*s*t + 14*d*s*t - 16*t^2 + 4*d*t^2)*
      G[1, {1, 1, 1, 1, 1, 1, 1, 0, 0}])/(4*(-4 + d)) -
    ((-6 + d)*s*t*(3*s + 4*t)*G[1, {1, 1, 1, 1, 1, 1, 2, 0, 0}])/
     (4*(-5 + d)*(-4 + d))
```

`Fire` *uses by default a dot-basis. The running time for the individual parts sum up to about* 30 s *on a standard laptop.*

Exercise 45 *Consider the example of the one-loop two-point function with equal internal masses, discussed below Eq. (6.6). Let*

$$\vec{I} = \begin{pmatrix} I_{10}\left(D, x\right) \\ I_{11}\left(D, x\right) \end{pmatrix} \tag{J.260}$$

be a basis in D space-time dimensions and

$$\vec{I}' = \begin{pmatrix} I_{10}\left(D+2, x\right) \\ I_{11}\left(D+2, x\right) \end{pmatrix} \tag{J.261}$$

be a basis in $(D+2)$ space-time dimensions. Work out the 2×2-matrices S and S^{-1}.

Solution: *For the one-loop two-point function the graph polynomial $\mathcal{U}$ is given by*

$$\mathcal{U}\left(\alpha_1, \alpha_2\right) = \alpha_1 + \alpha_2. \tag{J.262}$$

Hence

$$I_{10}\left(D, x\right) = \mathcal{U}\left(\mathbf{1}^+, \mathbf{2}^+\right) I_{10}\left(D, x\right) \;=\; \left(\mathbf{1}^+ + \mathbf{2}^+\right) I_{10}\left(D+2, x\right)$$

$$= I_{20}\,(D+2, x)\,. \tag{J.263}$$

Note that

$$\mathbf{2}^{+} I_{10}\,(D+2, x) = 0 \cdot I_{11}\,(D+2, x) \;=\; 0, \tag{J.264}$$

which shows that an index, which is zero, cannot be raised. For I_{10} we could also have used alternatively the graph polynomial for the one-loop one-point function, which is simpler and given by $\mathcal{U}(\alpha_1) = \alpha_1$, yielding the same result.

For the second master integral we have

$$\begin{aligned} I_{11}\,(D, x) &= \mathcal{U}\big(\mathbf{1}^{+}, \mathbf{2}^{+}\big)\, I_{11}\,(D, x) \;=\; \big(\mathbf{1}^{+} + \mathbf{2}^{+}\big)\, I_{11}\,(D+2, x) \\ &= I_{21}\,(D+2, x) + I_{12}\,(D+2, x) \;=\; 2 I_{21}\,(D+2, x)\,. \end{aligned} \tag{J.265}$$

Here we used the fact that for the equal-mass two-point function we have the symmetry $I_{\nu_1\nu_2} = I_{\nu_2\nu_1}$.

In the next step we reduce each of the integrals in $I_{20}(D+2, x)$ and $I_{21}(D+2, x)$ to a linear combination of the master integrals $I_{10}(D+2, x)$ and $I_{11}(D+2, x)$. From Eq. (6.14) we obtain

$$I_{20}\,(D+2, x) = \left(1 - \frac{(D+2)}{2}\right) I_{10}\,(D+2, x) \;=\; -\frac{D}{2} I_{10}\,(D+2, x)\,. \tag{J.266}$$

Please note that we use Eq. (6.14) with D substituted by $(D+2)$. From Eq. (6.11) we obtain

$$\begin{aligned} & x\,(4+x)\, I_{21}\,(D+2, x) = \\ &\quad = [3 - (D+2)]\, x I_{11}\,(D+2, x) + (2+x)\, I_{20}\,(D+2, x) - 2 I_{02}\,(D+2, x) \\ &\quad = -\,(D-1)\, x I_{11}\,(D+2, x) + x I_{20}\,(D+2, x) \\ &\quad = -\,(D-1)\, x I_{11}\,(D+2, x) - \frac{D}{2} x I_{10}\,(D+2, x)\,. \end{aligned} \tag{J.267}$$

Putting everything together we have

$$\begin{pmatrix} I_{10}\,(D, x) \\ I_{11}\,(D, x) \end{pmatrix} = \begin{pmatrix} -\frac{D}{2} & 0 \\ -\frac{D}{4+x} & -\frac{2(D-1)}{4+x} \end{pmatrix} \begin{pmatrix} I_{10}\,(D+2, x) \\ I_{11}\,(D+2, x) \end{pmatrix}, \tag{J.268}$$

and therefore

$$S \;=\; \begin{pmatrix} -\frac{D}{2} & 0 \\ -\frac{D}{4+x} & -\frac{2(D-1)}{4+x} \end{pmatrix}, \quad S^{-1} \;=\; \begin{pmatrix} -\frac{2}{D} & 0 \\ \frac{1}{D-1} & -\frac{4+x}{2(D-1)} \end{pmatrix}. \tag{J.269}$$

Exercise 46 *Show that*

$$\sum_{j=1}^{N_B+1} x_j \frac{\partial}{\partial x_j} I_{\nu_1 \dots \nu_{n_{\text{int}}}} = \left(\frac{lD}{2} - \nu \right) \cdot I_{\nu_1 \dots \nu_{n_{\text{int}}}}. \tag{J.270}$$

Hint: Consider the Feynman parameter representation.

Solution: *We apply the differential operator to the Feynman parameter representation*

$$I_{\nu_1 \dots \nu_{n_{\text{int}}}} = \frac{e^{l \varepsilon \gamma_E} \Gamma\left(\nu - \frac{lD}{2}\right)}{\prod\limits_{k=1}^{n_{\text{int}}} \Gamma(\nu_k)} \int\limits_{a_k \geq 0} d^{n_{\text{int}}} a \; \delta\left(1 - \sum_{k=1}^{n_{\text{int}}} a_k\right) \left(\prod_{k=1}^{n_{\text{int}}} a_k^{\nu_k - 1}\right) \frac{[\mathcal{U}(a)]^{\nu - \frac{(l+1)D}{2}}}{[\mathcal{F}(a)]^{\nu - \frac{lD}{2}}}. \tag{J.271}$$

We obtain

$$\begin{aligned}
& \sum_{j=1}^{N_B+1} x_j \frac{\partial}{\partial x_j} I_{\nu_1 \dots \nu_{n_{\text{int}}}} = \\
& = \left(\frac{lD}{2} - \nu \right) \sum_{j=1}^{N_B+1} \frac{e^{l \varepsilon \gamma_E} \Gamma\left(\nu - \frac{lD}{2}\right)}{\prod\limits_{k=1}^{n_{\text{int}}} \Gamma(\nu_k)} \int\limits_{a_k \geq 0} d^{n_{\text{int}}} a \; \delta\left(1 - \sum_{k=1}^{n_{\text{int}}} a_k\right) \left(\prod_{k=1}^{n_{\text{int}}} a_k^{\nu_k - 1}\right) \frac{[\mathcal{U}(a)]^{\nu - \frac{(l+1)D}{2}} x_j \cdot \mathcal{F}'_{x_j}(a)}{\left[\mathcal{F}(a)\right]^{\nu - \frac{lD}{2} + 1}} \\
& = \left(\frac{lD}{2} - \nu \right) \frac{e^{l \varepsilon \gamma_E} \Gamma\left(\nu - \frac{lD}{2}\right)}{\prod\limits_{k=1}^{n_{\text{int}}} \Gamma(\nu_k)} \int\limits_{a_k \geq 0} d^{n_{\text{int}}} a \; \delta\left(1 - \sum_{k=1}^{n_{\text{int}}} a_k\right) \left(\prod_{k=1}^{n_{\text{int}}} a_k^{\nu_k - 1}\right) \frac{[\mathcal{U}(a)]^{\nu - \frac{(l+1)D}{2}}}{\left[\mathcal{F}(a)\right]^{\nu - \frac{lD}{2}}} \\
& = \left(\frac{lD}{2} - \nu \right) I_{\nu_1 \dots \nu_{n_{\text{int}}}}.
\end{aligned} \tag{J.272}$$

Exercise 47 *The steps from Eq. (6.64) to Eq. (6.65) can still be carried out by hand. Fill in the missing details.*

Solution: *From Eq. (6.64) we have*

$$\begin{aligned}
\frac{\partial}{\partial x} I_{10}(D, x) &= 0, \\
\frac{\partial}{\partial x} I_{11}(D, x) &= -I_{22}(D+2, x).
\end{aligned} \tag{J.273}$$

We have to express $I_{22}(D+2, x)$ as a linear combination of $I_{10}(D, x)$ and $I_{11}(D, x)$. We first express $I_{22}(D+2, x)$ as a linear combination of $I_{10}(D+2, x)$ and $I_{11}(D+2, x)$. Using Eq. (6.11) with $(D+2)$ we obtain

$$I_{22}(D+2, x) = \frac{4 + (4-D)x}{x(4+x)} I_{21}(D+2, x) - \frac{4}{x(4+x)} I_{30}(D+2, x). \tag{J.274}$$

Here we used the symmetry $I_{\nu_2\nu_1}(D+2,x) = I_{\nu_1\nu_2}(D+2,x)$. Using again Eq. (6.11), this time for $I_{21}(D+2,x)$ we obtain

$$I_{21}(D+2,x) = \frac{1-D}{4+x} I_{11}(D+2,x) + \frac{1}{4+x} I_{20}(D+2,x). \quad \text{(J.275)}$$

From Eq. (6.14) we have

$$\begin{aligned} I_{20}(D+2,x) &= -\frac{D}{2} I_{10}(D+2,x), \\ I_{30}(D+2,x) &= \frac{D(D-2)}{8} I_{10}(D+2,x), \end{aligned} \quad \text{(J.276)}$$

and hence

$$I_{22}(D+2,x) = \frac{(1-D)[4+(4-D)x]}{x(4+x)^2} I_{11}(D+2,x) - \frac{D[2(D-1)+x]}{x(4+x)^2} I_{10}(D+2,x). \quad \text{(J.277)}$$

From Exercise 45 we have

$$\begin{aligned} I_{10}(D+2,x) &= -\frac{2}{D} I_{10}(D,x), \\ I_{11}(D+2,x) &= \frac{1}{D-1} I_{10}(D,x) - \frac{4+x}{2(D-1)} I_{11}(D,x), \end{aligned} \quad \text{(J.278)}$$

and therefore

$$I_{22}(D+2,x) = \frac{D-2}{x(4+x)} I_{10}(D,x) + \frac{4+(4-D)x}{2x(4+x)} I_{11}(D,x). \quad \text{(J.279)}$$

Exercise 48 *This example depends on two kinematic variables x_1 and x_2, hence the integrability condition is non-trivial. Check explicitly the integrability condition*

$$dA + A \wedge A = 0. \quad \text{(J.280)}$$

Solution: *With $A = A_{x_1} dx_1 + A_{x_2} dx_2$ we have*

$$\begin{aligned} dA &= \left(\partial_{x_1} A_{x_2} - \partial_{x_2} A_{x_1}\right) dx_1 \wedge dx_2, \\ A \wedge A &= \left(A_{x_1} A_{x_2} - A_{x_2} A_{x_1}\right) dx_1 \wedge dx_2 = \left[A_{x_1}, A_{x_2}\right] dx_1 \wedge dx_2. \end{aligned} \quad \text{(J.281)}$$

The integrability condition translates to

$$\partial_{x_1} A_{x_2} - \partial_{x_2} A_{x_1} + \left[A_{x_1}, A_{x_2}\right] = 0. \quad \text{(J.282)}$$

With the explicit expressions for A_{x_1} and A_{x_2} given in Eq. (6.71) one verifies that this equation holds. We have

$$dA = -A \wedge A \tag{J.283}$$
$$= \begin{pmatrix} 0 & 0 & 0 & 0 \\ 0 & 0 & 0 & 0 \\ 0 & 0 & 0 & 0 \\ \frac{2(D-3)(x_1-x_2)}{x_1(1-x_1)x_2(1-x_2)(1-x_1-x_2)} & \frac{2(D-3)(1-2x_1)}{x_1^2(1-x_1)x_2(1-x_1-x_2)} & -\frac{2(D-3)(1-2x_2)}{x_1x_2^2(1-x_2)(1-x_1-x_2)} & 0 \end{pmatrix} dx_1 \wedge dx_2.$$

Exercise 49 *Let* $X = \mathbb{C}^2$ *and*

$$Y = \{x \in X | x_1 + x_2 = 0\}. \tag{J.284}$$

Compute

$$\mathrm{Res}_Y \left(\frac{x_1 x_2^2 dx_1 \wedge dx_2}{x_1 + x_2} \right). \tag{J.285}$$

Solution: *We have*

$$\frac{x_1 x_2^2 dx_1 \wedge dx_2}{x_1 + x_2} = \frac{d(x_1 + x_2)}{x_1 + x_2} \wedge \left(x_1 x_2^2 dx_2 \right), \tag{J.286}$$

hence

$$\mathrm{Res}_Y \left(\frac{x_1 x_2^2 dx_1 \wedge dx_2}{x_1 + x_2} \right) = x_1 x_2^2 dx_2 \big|_Y = -x_2^3 dx_2 \big|_Y. \tag{J.287}$$

Note that alternatively we could have used

$$\frac{x_1 x_2^2 dx_1 \wedge dx_2}{x_1 + x_2} = \frac{d(x_1 + x_2)}{x_1 + x_2} \wedge \left(-x_1 x_2^2 dx_1 \right) \tag{J.288}$$

and

$$\mathrm{Res}_Y \left(\frac{x_1 x_2^2 dx_1 \wedge dx_2}{x_1 + x_2} \right) = -x_1 x_2^2 dx_1 \big|_Y = -x_1^3 dx_1 \big|_Y. \tag{J.289}$$

On Y *we have* $x_2 = -x_1$ *and* $dx_2 = -dx_1$. *Therefore the two forms agree on* Y.

Exercise 50 *Let*

$$\omega = 3dx_1 + (5 + x_1)dx_2 + x_3 dx_3 \tag{J.290}$$

and

$$\gamma : [0, 1] \to \mathbb{C}^3, \qquad \gamma(\lambda) = \begin{pmatrix} \lambda \\ \lambda^2 \\ 1 + \lambda \end{pmatrix}. \tag{J.291}$$

Compute

$$\int_{\gamma} \omega. \tag{J.292}$$

Solution: *We have*

$$\begin{aligned}
\int_{\gamma} \omega &= \int_{0}^{1} d\lambda \left[3\frac{d}{d\lambda}\lambda + (5+\lambda)\frac{d}{d\lambda}\lambda^2 + (1+\lambda)\frac{d}{d\lambda}(1+\lambda) \right] \\
&= \int_{0}^{1} d\lambda \; [3 + 2\lambda(5+\lambda) + (1+\lambda)] = \int_{0}^{1} d\lambda \; \left[4 + 11\lambda + 2\lambda^2\right] \\
&= 4 + \frac{11}{2} + \frac{2}{3} = \frac{61}{6}.
\end{aligned} \tag{J.293}$$

Exercise 51 *Prove Eq. (6.143) for the case $r = 2$, i.e. show*

$$I_{\gamma_2 \circ \gamma_1}(\omega_1, \omega_2; \lambda) = I_{\gamma_1}(\omega_1, \omega_2; \lambda) + I_{\gamma_2}(\omega_1; \lambda) \, I_{\gamma_1}(\omega_2; \lambda) + I_{\gamma_2}(\omega_1, \omega_2; \lambda). \tag{J.294}$$

Solution: *Without loss of generality we take $a = 0$ and $b = \lambda = 1$, i.e. we consider*

$$\begin{aligned}
\gamma_1 &: [0, 1] \to X, \\
\gamma_2 &: [0, 1] \to X, \\
\gamma_2 \circ \gamma_1 &: [0, 1] \to X.
\end{aligned} \tag{J.295}$$

We decompose the full integration region $0 \le \lambda_1 \le 1$, $0 \le \lambda_2 \le \lambda_1$ of $I_{\gamma_2 \circ \gamma_1}(\omega_1, \omega_2; 1)$ into three regions as shown in Fig. J.6. The integration over the region $0 \le \lambda_1 \le 1/2$, $0 \le \lambda_2 \le \lambda_1$ (red region) gives $I_{\gamma_1}(\omega_1, \omega_2; 1)$, the integration over the region $1/2 \le \lambda_1 \le 1$, $0 \le \lambda_2 \le 1/2$ (green region) gives $I_{\gamma_2}(\omega_1; 1) I_{\gamma_1}(\omega_2; 1)$ and the integration over the region $1/2 \le \lambda_1 \le 1$, $1/2 \le \lambda_2 \le \lambda_1$ (blue region) gives $I_{\gamma_2}(\omega_1, \omega_2; 1)$.

Exercise 52 *Prove Eq. (6.145).*

Solution: *Without loss of generality we take $[a, b] = [0, 1]$. For the path $\gamma : [0, 1] \to X$ and a differential one-form ω_j we write for the pull-back*

$$f_j(\lambda) \, d\lambda = \gamma^* \omega_j. \tag{J.296}$$

For the reversed path $\gamma^{-1} : [0, 1] \to X$ we write for the pull-back of ω_j

$$f_j^{\text{rev}}(\lambda') \, d\lambda' = \left(\gamma^{-1}\right)^* \omega_j. \tag{J.297}$$

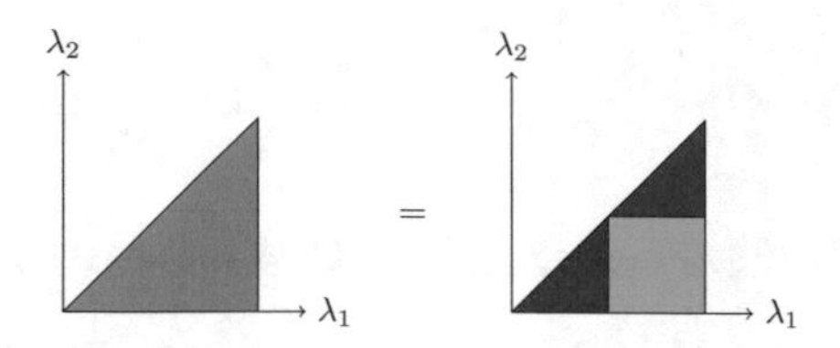

Fig. J.6 The iterated integral of depth 2 for the combined path $\gamma_2 \circ \gamma_1$: The full integration region (shown in grey on the left) is decomposed into three regions (shown in red, green and blue on the right)

We have

$$f_j^{\mathrm{rev}}\left(\lambda'\right) = -f_j\left(1-\lambda'\right). \tag{J.298}$$

We have to consider

$$I_{\gamma^{-1}}\left(\omega_1, \dots, \omega_r; 1\right) = \int\limits_0^1 d\lambda_1' f_1^{\mathrm{rev}}\left(\lambda_1'\right) \dots \int\limits_0^{\lambda_{r-1}'} d\lambda_r' f_r^{\mathrm{rev}}\left(\lambda_r'\right). \tag{J.299}$$

With $\lambda_j = 1 - \lambda_j'$ *we have for the integrand*

$$\begin{aligned} f_1^{\mathrm{rev}}\left(\lambda_1'\right) \dots f_r^{\mathrm{rev}}\left(\lambda_r'\right) &= (-1)^r f_1\left(1-\lambda_1'\right) \dots f_r\left(1-\lambda_r'\right) \\ &= (-1)^r f_1\left(\lambda_1\right) \dots f_r\left(\lambda_r\right). \end{aligned} \tag{J.300}$$

The integration domain

$$0 \le \lambda_r' \le \dots \le \lambda_1' \le 1 \tag{J.301}$$

transforms into

$$0 \le \lambda_1 \le \dots \le \lambda_r \le 1. \tag{J.302}$$

Thus

$$\begin{aligned} I_{\gamma^{-1}}\left(\omega_1, \dots, \omega_r; 1\right) &= (-1)^r \int\limits_0^1 d\lambda_r f_r\left(\lambda_r\right) \dots \int\limits_0^{\lambda_1} d\lambda_1 f_1\left(\lambda_1\right) \\ &= (-1)^r I_\gamma\left(\omega_r, \dots, \omega_1; 1\right). \end{aligned} \tag{J.303}$$

Exercise 53 *Show that* I_4' *is given at the kinematic point* $(x_1, x_2) = (1, 1)$ *by*

$$I_4' = e^{\varepsilon\gamma_E} \frac{\Gamma(1+\varepsilon)\Gamma(1-\varepsilon)^2}{\Gamma(1-2\varepsilon)} \left(1 - \sum_{k=2}^{\infty} \zeta_k \varepsilon^k\right). \quad (J.304)$$

Hint: Use the trick from Exercise 11 and the Mellin-Barnes technique.

Solution: *From Exercise 11 we know that we may write I_4' with generic x_1 and x_2 as*

$$I_4' = \frac{\varepsilon^2 e^{\varepsilon\gamma_E}\Gamma(2+\varepsilon)\Gamma(-\varepsilon)^2}{2\Gamma(-2\varepsilon)} x_1 x_2 \int_0^1 da \int_0^1 db \left[abx_1 + \bar{a}\bar{b}x_2 + \bar{a}b\right]^{-2-\varepsilon}, \quad (J.305)$$

with $\bar{a} = 1-a$ and $\bar{b} = 1-b$. Specialising to $x_1 = x_2 = 1$ we have

$$I_4' = \frac{\varepsilon^2 e^{\varepsilon\gamma_E}\Gamma(2+\varepsilon)\Gamma(-\varepsilon)^2}{2\Gamma(-2\varepsilon)} \int_0^1 da \int_0^1 db \, [1-a+ab]^{-2-\varepsilon}. \quad (J.306)$$

We now use the Mellin-Barnes technique and split $1-a+ab$ into $(1-a)$ and (ab):

$$I_4' = \frac{\varepsilon^2 e^{\varepsilon\gamma_E}\Gamma(-\varepsilon)^2}{2\Gamma(-2\varepsilon)} \frac{1}{2\pi i} \int d\sigma \, \Gamma(-\sigma)\Gamma(\sigma+2+\varepsilon) \int_0^1 da \int_0^1 db \, a^\sigma b^\sigma (1-a)^{-\sigma-2-\varepsilon}. \quad (J.307)$$

The integrals over a and b are now easily done, yielding

$$I_4' = \frac{\varepsilon^2 e^{\varepsilon\gamma_E}\Gamma(-\varepsilon)}{2\Gamma(-2\varepsilon)} \frac{1}{2\pi i} \int d\sigma \, \frac{\Gamma(-\sigma)\Gamma(-\sigma-1-\varepsilon)\Gamma(\sigma+1)^2\Gamma(\sigma+2+\varepsilon)}{\Gamma(\sigma+2)}. \quad (J.308)$$

We close the contour to the right and sum up the residues from $\Gamma(-\sigma)$ and $\Gamma(-\sigma-1-\varepsilon)$. This yields

$$I_4' = \frac{\varepsilon^2 e^{\varepsilon\gamma_E}\Gamma(-\varepsilon)}{2\Gamma(-2\varepsilon)} \left\{ \sum_{n=0}^{\infty} \frac{(-1)^n}{n!} \frac{\Gamma(-n-1-\varepsilon)\Gamma(n+1)^2\Gamma(n+2+\varepsilon)}{\Gamma(n+2)} + \sum_{n=0}^{\infty} \frac{(-1)^n}{n!} \frac{\Gamma(-n+1+\varepsilon)\Gamma(n-\varepsilon)^2\Gamma(n+1)}{\Gamma(n+1-\varepsilon)} \right\}. \quad (J.309)$$

With the help of

$$(-1)^n \Gamma(-n-1-\varepsilon)\Gamma(n+2+\varepsilon) = \Gamma(-1-\varepsilon)\Gamma(2+\varepsilon),$$

$$(-1)^n \Gamma(-n+1+\varepsilon)\Gamma(n-\varepsilon) = \Gamma(1+\varepsilon)\Gamma(-\varepsilon), \tag{J.310}$$

and

$$\Gamma(-1-\varepsilon)\Gamma(2+\varepsilon), = -\Gamma(1+\varepsilon)\Gamma(-\varepsilon) \tag{J.311}$$

this simplifies to

$$I_4' = \frac{e^{\varepsilon\gamma_E}\Gamma(1-\varepsilon)^2\Gamma(1+\varepsilon)}{\Gamma(1-2\varepsilon)}\left[1-\varepsilon\sum_{n=1}^{\infty}\left(\frac{1}{n}-\frac{1}{n-\varepsilon}\right)\right]. \tag{J.312}$$

Finally, expanding the geometric series

$$\frac{1}{n-\varepsilon} = \frac{1}{n}\sum_{k=0}^{\infty}\left(\frac{\varepsilon}{n}\right)^k \tag{J.313}$$

we obtain

$$\begin{aligned} I_4' &= \frac{e^{\varepsilon\gamma_E}\Gamma(1-\varepsilon)^2\Gamma(1+\varepsilon)}{\Gamma(1-2\varepsilon)}\left[1-\sum_{k=2}^{\infty}\varepsilon^k\sum_{n=1}^{\infty}\frac{1}{n^k}\right] \\ &= \frac{e^{\varepsilon\gamma_E}\Gamma(1-\varepsilon)^2\Gamma(1+\varepsilon)}{\Gamma(1-2\varepsilon)}\left[1-\sum_{k=2}^{\infty}\zeta_k\varepsilon^k\right]. \end{aligned} \tag{J.314}$$

Exercise 54 *Show the equivalence of the $O(\varepsilon^2)$-term of I_4' between Eqs. (6.177) and (6.178).*

Solution: *We have to show that the expressions*

$$G(0,0;x_1) + G(0,0;x_2) - G(1,0;x_1) - G(1,0;x_2) + G(0;x_1)G(0;x_2) + \frac{1}{2}\zeta_2 \tag{J.315}$$

and

$$-\text{Li}_2(x_1) - \text{Li}_2(x_2) + \frac{1}{2}\ln^2(x_1) + \frac{1}{2}\ln^2(x_2) + \ln(x_1)\ln(x_2) - \ln(x_1)\ln(1-x_1) - \ln(x_2)\ln(1-x_2) + \frac{1}{2}\zeta_2 \tag{J.316}$$

agree. We have

$$\begin{aligned} G(0;x) &= \ln(x), \\ G(0,0;x) &= \frac{1}{2}\ln^2(x) \end{aligned} \tag{J.317}$$

and

$$G(1,0;x) = \int_0^x dt \frac{\ln t}{t-1} = \int_1^{1-x} dt \frac{\ln(1-t)}{t} = \int_0^{1-x} dt \frac{\ln(1-t)}{t} - \int_0^1 dt \frac{\ln(1-t)}{t}$$
$$= \zeta_2 - \text{Li}_2(1-x) \qquad \text{(J.318)}$$

From Eq. (5.39) we have

$$\text{Li}_2(x) = -\text{Li}_2(1-x) + \frac{1}{6}\pi^2 - \ln(x)\ln(1-x), \qquad \text{(J.319)}$$

and therefore

$$G(1,0;x) = \text{Li}_2(x) + \ln(x)\ln(1-x). \qquad \text{(J.320)}$$

Plugging all expressions into Eq. (J.315) shows the equivalence of Eq. (J.315) with Eq. (J.316).

Exercise 55 *Derive Eq. (6.192) from Eq. (6.190).*

Solution: *Let $\gamma : [a,b] \to M$ be a curve in M with $\gamma(0) = x$. A tangent vector at x is given by*

$$X = \left. \frac{d}{dt}\gamma(t) \right|_{t=0} \qquad \text{(J.321)}$$

In order to keep the notation simple we will suppress maps between a manifold and an appropriate coordinate chart. We have to show

$$A_2(X) = \left(U A_1 U^{-1} + U dU^{-1} \right)(X), \qquad \text{(J.322)}$$

where A_1 and A_2 are defined by

$$A_1 = \sigma_1^* \omega, \quad A_2 = \sigma_2^* \omega, \qquad \text{(J.323)}$$

and the sections σ_1 and σ_2 are related by

$$\sigma_2 = \sigma_1 U^{-1}. \qquad \text{(J.324)}$$

Let us choose a local trivialisation (x, g) of $P(M, G)$ and work out $\sigma_{2*}X$. With $U_0 = U(\gamma(0))$ we have

$$\sigma_{2*}X = \sigma_{2*}\left(\left.\frac{d}{dt}\gamma(t)\right|_{t=0} \right) = \left.\frac{d}{dt}\left(\gamma(t), \sigma_2(\gamma(t))\right)\right|_{t=0} = \left.\frac{d}{dt}\left(\gamma(t), \sigma_1(\gamma(t))\, U(\gamma(t))^{-1}\right)\right|_{t=0}$$
$$= \left(X, \left.\frac{d}{dt}\sigma_1(\gamma(t))\right|_{t=0} U_0^{-1} + \sigma_1(\gamma(0)) \left.\frac{d}{dt} U(\gamma(t))^{-1}\right|_{t=0} \right)$$

$$= R_{U_0^{-1}*}(\sigma_{1*}X) + \left(X, \sigma_2(\gamma(0)) U_0 \frac{d}{dt} U(\gamma(t))^{-1} \Big|_{t=0} \right). \tag{J.325}$$

We then have, using (CF1) and (CF2),

$$\begin{aligned} A_2(X) = \omega(\sigma_{2*}X) &= \omega_{(x,\sigma_2(\gamma(0)))}\left(R_{U_0^{-1}*}(\sigma_{1*}X) \right) + \omega_{(x,\sigma_2(\gamma(0)))}\left(U_0 \frac{d}{dt} U(\gamma(t))^{-1} \Big|_{t=0} \right) \\ &= U_0 \left(\omega_{(x,\sigma_1(\gamma(0)))}(\sigma_{1*}X) \right) U_0^{-1} + \left(U_0 dU^{-1} \right)(X) \\ &= \left(U A_1 U^{-1} + U dU^{-1} \right)(X). \end{aligned} \tag{J.326}$$

Exercise 56 *Work out the maximal cut of the double box integral $I_{111111100}$ shown in Fig. 6.5. Use the notation as in example 2 in Sect. 6.3.1. To work out the maximal cut it is simpler to use the loop-by-loop approach as discussed in Sect. 2.5.5.*

Solution: *We label the internal edges as in Fig. 2.11. We first consider the loop with edges e_1, e_2, e_3 and e_4 and then the second loop with edges e_5, e_6, e_7 and an edge with momentum $k_2 + p_1$. The latter edge is introduced by integrating out the first loop with loop momentum k_1. The loop-by-loop approach has the advantage that we only need eight Baikov variables z_1-z_7 and z_9, the variable z_8 is absent. The Baikov representation reads*

$$\begin{aligned} I_{111111100} &= \frac{e^{2\varepsilon\gamma_E}\left(\mu^2\right)^{7-D}}{64\pi^3\Gamma\left(\frac{D-3}{2}\right)^2} \left[G(p_1,p_2,p_3)\right]^{\frac{4-D}{2}} \\ &\int\limits_C d^8z \left[G(p_1,p_2,k_2)\right]^{\frac{4-D}{2}} \left[G(k_1,p_1,p_2,k_2)\right]^{\frac{D-5}{2}} \left[G(k_2,p_1,p_2,p_3)\right]^{\frac{D-5}{2}} \frac{1}{z_1 z_2 z_3 z_4 z_5 z_6 z_7}. \end{aligned} \tag{J.327}$$

We have

$$\begin{aligned} G(p_1,p_2,p_3) &= \frac{1}{4} s t (s+t), \\ G(p_1,p_2,k_2)\big|_{z_1=z_2=z_3=z_4=z_5=z_6=z_7=0} &= \frac{1}{4} s (s+t-z_9)(t-z_9), \\ G(k_1,p_1,p_2,k_2)\big|_{z_1=z_2=z_3=z_4=z_5=z_6=z_7=0} &= \frac{1}{16} s^2 (t-z_9)^2, \\ G(k_2,p_1,p_2,p_3)\big|_{z_1=z_2=z_3=z_4=z_5=z_6=z_7=0} &= \frac{1}{16} s^2 z_9^2. \end{aligned} \tag{J.328}$$

Thus

$$\begin{aligned} &\text{MaxCut}\; I_{111111100} = \\ &(2\pi i)^7 \frac{e^{2\varepsilon\gamma_E}\left(\mu^2\right)^{7-D} 2^{6-2D}}{\pi^3\Gamma\left(\frac{D-3}{2}\right)^2} s^{D-6} t^{\frac{4-D}{2}} (s+t)^{\frac{4-D}{2}} \int\limits_{C'} dz_9\, z_9^{D-5} (t-z_9)^{\frac{D-6}{2}} (s+t-z_9)^{\frac{4-D}{2}}. \end{aligned} \tag{J.329}$$

The integration region C' is obtained from the conditions

$$\left. \frac{G(k_2, p_1, p_2, p_3)}{G(p_1, p_2, p_3)} \right|_{z_1=z_2=z_3=z_4=z_5=z_6=z_7=0} = \frac{1}{4} \frac{s z_9^2}{t(s+t)} > 0,$$
$$\left. \frac{G(k_1, p_1, p_2, k_2)}{G(p_1, p_2, k_2)} \right|_{z_1=z_2=z_3=z_4=z_5=z_6=z_7=0} = \frac{1}{4} \frac{s(t-z_9)}{(s+t-z_9)} > 0. \quad \text{(J.330)}$$

Let's assume $t > 0$ and $s < -t$. Then

$$C' =]-\infty, s+t] \cup [t, \infty[, \quad \text{(J.331)}$$

and

$$\int\limits_{-\infty}^{s+t} dz_9 \, z_9^{D-5} (t-z_9)^{\frac{D-6}{2}} (s+t-z_9)^{\frac{4-D}{2}} =$$
$$\frac{\Gamma\left(\frac{6-D}{2}\right)\Gamma(5-D)}{\Gamma\left(\frac{16-3D}{2}\right)} (s+t)^{D-5} {}_2F_1\left(5-D, \frac{6-D}{2}, \frac{16-3D}{2}; \frac{t}{s+t}\right),$$
$$\int\limits_{t}^{\infty} dz_9 \, z_9^{D-5} (t-z_9)^{\frac{D-6}{2}} (s+t-z_9)^{\frac{4-D}{2}} =$$
$$-\frac{\Gamma\left(\frac{D-4}{2}\right)\Gamma\left(\frac{5-D}{2}\right)}{\sqrt{\pi}} 2^{4-D} t^{D-5} {}_2F_1\left(5-D, \frac{D-4}{2}, \frac{6-D}{2}; \frac{s+t}{t}\right). \quad \text{(J.332)}$$

Combining the results we obtain

$$\text{MaxCut } I_{111111100}(4-2\varepsilon) = (2\pi i)^7 \frac{(\mu^2)^3}{4\pi^4 s^2 t \varepsilon} + O\left(\varepsilon^0\right). \quad \text{(J.333)}$$

Exercise 57 *Show that the Landau equations imply $\mathcal{F} = 0$.*

Solution: *The graph polynomial $\mathcal{F}$ is homogeneous of degree $(l+1)$ in the Feynman parameters a_j. Hence*

$$\sum_{j=1}^{n_{\text{int}}} a_j \frac{\partial}{\partial a_j} \mathcal{F} = (l+1)\, \mathcal{F}. \quad \text{(J.334)}$$

The Landau equations imply that the left-hand side vanishes: We either have $a_j = 0$ (note that $\partial \mathcal{F}/\partial a_j$ is again a polynomial in the Feynman parameters) or $\partial \mathcal{F}/\partial a_j = 0$. Since $(l+1) \neq 0$ it follows that $\mathcal{F} = 0$.

Exercise 58 *Work out the Landau discriminant for the double box graph discussed in Exercise 44.*

Solution: *The graph polynomial $\mathcal{F}$ reads*

$$\mathcal{F} = [a_2 a_3 (a_4 + a_5 + a_6 + a_7) + a_5 a_6 (a_1 + a_2 + a_3 + a_4) + a_2 a_4 a_6 + a_3 a_4 a_5] x + a_1 a_4 a_7. \tag{J.335}$$

The Landau equations for the leading Landau singularity read

$$\begin{aligned}
a_4 a_7 + a_5 a_6 x &= 0, \\
[a_3 (a_4 + a_5 + a_6 + a_7) + (a_4 + a_5) a_6] x &= 0, \\
[a_2 (a_4 + a_5 + a_6 + a_7) + (a_4 + a_6) a_5] x &= 0, \\
a_1 a_7 + (a_2 + a_5)(a_3 + a_6) x &= 0, \\
[a_6 (a_1 + a_2 + a_3 + a_4) + (a_2 + a_4) a_3] x &= 0, \\
[a_5 (a_1 + a_2 + a_3 + a_4) + (a_3 + a_4) a_2] x &= 0, \\
a_1 a_4 + a_2 a_3 x &= 0.
\end{aligned} \tag{J.336}$$

We then solve these equation for $(a_1, \dots, a_7, x)$. For the leading Landau singularity we are interested in solutions with $a_j \neq 0$. We find that only for $x = -1$ we have such a solution and hence

$$D_{\text{Landau}} = \{-1\}. \tag{J.337}$$

Exercise 59 *Show that for φ as in Eq. (6.300) and ω as in Eq. (6.309) the differential n-form φ is closed with respect to ∇_ω:*

$$\nabla_\omega \varphi = 0. \tag{J.338}$$

Solution: *Let $1 \leq j \leq n$. We have*

$$\nabla_\omega = \sum_{j=1}^{n} \left(\frac{\partial}{\partial z_j} + \omega_j \right) dz_j + \sum_{j=1}^{n} \left(\frac{\partial}{\partial \bar{z}_j} \right) d\bar{z}_j. \tag{J.339}$$

Hence

$$\begin{aligned}
\left(\frac{\partial}{\partial z_j} \frac{q}{p_1^{n_1} \dots p_m^{n_m}} + \omega_j \right) dz_j \wedge dz_n \wedge \dots \wedge dz_1 &= 0, \\
\left(\frac{\partial}{\partial \bar{z}_j} \frac{q}{p_1^{n_1} \dots p_m^{n_m}} \right) d\bar{z}_j \wedge dz_n \wedge \dots \wedge dz_1 &= 0.
\end{aligned} \tag{J.340}$$

In the first line, the wedge product contains $dz_j \wedge dz_j$, while in the second line the derivative in the bracket vanishes.

Exercise 60 *Proof Eq. (6.353) for the special case $n_1 = n_2 = n_3 = n_4 = 0$.*

Solution: *We have to show that for*

$$\omega = \gamma_1 \frac{dz}{z} - \gamma_2 \frac{dz}{1-z} \tag{J.341}$$

and

$$\varphi_L = \frac{dz}{z(1-z)}, \quad \varphi_R = \frac{dz}{z(1-z)}, \tag{J.342}$$

we have

$$\langle \varphi_L | \varphi_R \rangle_\omega = \frac{(\gamma_1 + \gamma_2)}{\gamma_1 \gamma_2} = \frac{1}{\gamma_1} + \frac{1}{\gamma_2}. \tag{J.343}$$

For the case at hand, $\psi_{L,1}$ *is given by*

$$\psi_{L,1} = \frac{1}{\gamma_1} + O(z), \tag{J.344}$$

and $\psi_{L,2}$ *is given by*

$$\psi_{L,1} = -\frac{1}{\gamma_2} + O(z-1). \tag{J.345}$$

We further have

$$\begin{aligned} \mathrm{Res}_{D_1}\left(\psi_{L,1}\varphi_R\right) &= \mathrm{Res}_{\{0\}}\left(\frac{dz}{\gamma_1 z(1-z)}\right) = \frac{1}{\gamma_1}, \\ \mathrm{Res}_{D_2}\left(\psi_{L,2}\varphi_R\right) &= \mathrm{Res}_{\{1\}}\left(-\frac{dz}{\gamma_2 z(1-z)}\right) = \frac{1}{\gamma_2}. \end{aligned} \tag{J.346}$$

Therefore

$$\langle \varphi_L | \varphi_R \rangle_\omega = \frac{1}{\gamma_1} + \frac{1}{\gamma_2}. \tag{J.347}$$

Exercise 61 *Consider the monomials*

$$p_1 = x_1^2 x_2 x_3, \quad p_2 = x_1 x_2^3. \tag{J.348}$$

Order the two monomials with respect to the degree lexicographic order and the degree reverse lexicographic order (assuming $x_1 > x_2 > x_3$*).*

Solution: *Both polynomials have total degree four. Let us write*

$$\begin{aligned} p_1 &= x_1^2 x_2 x_3 = x_1^{m_1} x_2^{m_2} x_3^{m_3} \Rightarrow (m_1, m_2, m_3) = (2, 1, 1)\,, \\ p_2 &= x_1 x_2^3 = x_1^{m_1'} x_2^{m_2'} x_3^{m_3'} \Rightarrow \left(m_1', m_2', m_3'\right) = (1, 3, 0)\,. \end{aligned} \tag{J.349}$$

We have

$$\left(m_1 - m_1', m_2 - m_2', m_3 - m_3'\right) = (1, -2, 1)\,. \tag{J.350}$$

For the degree lexicographic order we have

$$p_1 >_{\text{deglex}} p_2, \tag{J.351}$$

as $m_1 - m_1' = 1 > 0$*, while for the degree reverse lexicographic order we have*

$$p_1 <_{\text{degrevlex}} p_2, \tag{J.352}$$

as $m_3' - m_3 = -1 < 0$.

Exercise 62 *Assume* $C_{\lambda+1} = C_{\lambda-1} = 0$*. Show that this implies*

$$\dim H_{\lambda+1}(M) = 0, \qquad \dim H_\lambda(M) = C_\lambda, \qquad \dim H_{\lambda-1}(M) = 0. \tag{J.353}$$

Solution: *Let us denote* $b_k = \dim H_k(M)$*. We write down the Morse inequalities for* $(\lambda+1)$*,* λ*,* $(\lambda-1)$ *and* $(\lambda-2)$*, using* $C_{\lambda+1} = C_{\lambda-1} = 0$*. Multiplying in addition the first and third equation by* (-1) *we obtain*

$$\begin{aligned} \sum_{k=0}^{\lambda+1} (-1)^{\lambda-k}\, b_k &\geq C_\lambda + \sum_{k=0}^{\lambda-2} (-1)^{\lambda-k}\, C_k, \\ \sum_{k=0}^{\lambda} (-1)^{\lambda-k}\, b_k &\leq C_\lambda + \sum_{k=0}^{\lambda-2} (-1)^{\lambda-k}\, C_k, \\ \sum_{k=0}^{\lambda-1} (-1)^{\lambda-k}\, b_k &\geq \sum_{k=0}^{\lambda-2} (-1)^{\lambda-k}\, C_k, \\ \sum_{k=0}^{\lambda-2} (-1)^{\lambda-k}\, b_k &\leq \sum_{k=0}^{\lambda-2} (-1)^{\lambda-k}\, C_k. \end{aligned} \tag{J.354}$$

From the first two inequalities we obtain

$$\sum_{k=0}^{\lambda} (-1)^{\lambda-k}\, b_k \leq \sum_{k=0}^{\lambda+1} (-1)^{\lambda-k}\, b_k \tag{J.355}$$

and therefore

$$0 \le -b_{\lambda+1}. \tag{J.356}$$

As $b_{\lambda+1}$ cannot be negative, it follows that $b_{\lambda+1} = \dim H_{\lambda+1}(M) = 0$. In a similar way we obtain from the third and fourth equation $b_{\lambda-1} = \dim H_{\lambda-1}(M) = 0$.

Having shown $b_{\lambda+1} = b_{\lambda-1} = 0$, the first two equations simplify to

$$\begin{aligned} b_\lambda + \sum_{k=0}^{\lambda-2} (-1)^{\lambda-k}\, b_k &\ge C_\lambda + \sum_{k=0}^{\lambda-2} (-1)^{\lambda-k}\, C_k, \\ b_\lambda + \sum_{k=0}^{\lambda-2} (-1)^{\lambda-k}\, b_k &\le C_\lambda + \sum_{k=0}^{\lambda-2} (-1)^{\lambda-k}\, C_k, \end{aligned} \tag{J.357}$$

and this implies

$$b_\lambda + \sum_{k=0}^{\lambda-2} (-1)^{\lambda-k}\, b_k = C_\lambda + \sum_{k=0}^{\lambda-2} (-1)^{\lambda-k}\, C_k, \tag{J.358}$$

In a similar way one obtains from the third and the fourth equation

$$\sum_{k=0}^{\lambda-2} (-1)^{\lambda-k}\, b_k = \sum_{k=0}^{\lambda-2} (-1)^{\lambda-k}\, C_k. \tag{J.359}$$

Hence $b_\lambda = C_\lambda = \dim H_\lambda(M)$.

Exercise 63 *Derive Eq. (6.417) from Eq. (6.415).*

Solution: *The proof is similar to the previous exercise. Let us denote $b_k = \dim H_k(M)$. With $\dim M = n$ we have trivially $b_{n+1} = \dim H_{n+1}(M) = 0$ and $C_{n+1} = 0$. We write down the Morse inequalities for $(n+1)$ and n, using $b_{n+1} = C_{n+1} = 0$. Multiplying in addition the first equation by (-1) we obtain*

$$\begin{aligned} \sum_{k=0}^{n} (-1)^{n-k}\, b_k &\ge \sum_{k=0}^{n} (-1)^{n-k}\, C_k, \\ \sum_{k=0}^{n} (-1)^{n-k}\, b_k &\le \sum_{k=0}^{n} (-1)^{n-k}\, C_k. \end{aligned} \tag{J.360}$$

Hence

$$\sum_{k=0}^{n} (-1)^{n-k}\, b_k = \sum_{k=0}^{n} (-1)^{n-k}\, C_k. \tag{J.361}$$

and

$$\chi(M) = \sum_{k=0}^{n} (-1)^k b_k = \sum_{k=0}^{n} (-1)^k C_k. \tag{J.362}$$

Exercise 64 *Let $N_{\text{master}} = 1$, $N_B = 2$ and*

$$A = d \ln \left(\frac{x_1}{x_1 + x_2} \right). \tag{J.363}$$

Show that B_λ, defined as in Eq. (7.29), equals zero.

Solution: *We have*

$$A = A_{x_1} dx_1 + A_{x_2} dx_2 = \left(\frac{1}{x_1} - \frac{1}{x_1 + x_2} \right) dx_1 - \frac{1}{x_1 + x_2} dx_2. \tag{J.364}$$

For $\alpha = [\alpha_1 : 1]$ we have

$$B_\lambda = \alpha_1 \left(\frac{1}{\alpha_1 \lambda} - \frac{1}{\alpha_1 \lambda + \lambda} \right) - \frac{1}{\alpha_1 \lambda + \lambda} = 0. \tag{J.365}$$

Exercise 65 *Prove the two relations in Eq. (7.59).*

Solution: *We start with*

$$\frac{d^k}{dx^k} = x^{-k} \prod_{j=0}^{k-1} (\theta - j). \tag{J.366}$$

We prove this relation by induction. For $k = 1$ the right-hand side equals

$$x^{-1} \prod_{j=0}^{0} (\theta - j) = x^{-1} \theta = \frac{d}{dx}. \tag{J.367}$$

Let us now assume that the relation is correct for $(k - 1)$. We have

$$\frac{d^k}{dx^k} = \frac{d}{dx} \frac{d^{k-1}}{dx^{k-1}} = \frac{1}{x} \theta \left[x^{-k+1} \prod_{j=0}^{k-2} (\theta - j) \right]. \tag{J.368}$$

We further have the operator relation

$$\theta x^{-k+1} = x^{-k+1} (\theta - k + 1) \tag{J.369}$$

and therefore

$$\frac{d^k}{dx^k} = x^{-k}\left(\theta - k + 1\right)\prod_{j=0}^{k-2}\left(\theta - j\right) \;=\; x^{-k}\prod_{j=0}^{k-1}\left(\theta - j\right). \tag{J.370}$$

Let us now look at

$$\theta^k = \sum_{j=1}^{k} S\left(k, j\right) x^j \frac{d^j}{dx^j}. \tag{J.371}$$

For $k = 1$ *the right-hand side equals*

$$\sum_{j=1}^{1} S\left(1, j\right) x^j \frac{d^j}{dx^j} = S\left(1, 1\right) x \frac{d}{dx} \;=\; \theta, \tag{J.372}$$

where we used $S(1, 1) = 1$. *Let us now assume that the relation is correct for* $(k - 1)$. *We have*

$$\begin{aligned}
\theta^k = \theta\theta^{k-1} \;&=\; x\frac{d}{dx}\left[\sum_{j=1}^{k-1} S\left(k-1, j\right) x^j \frac{d^j}{dx^j}\right] \\
&= \sum_{j=1}^{k-1} S\left(k-1, j\right)\left[j x^j \frac{d^j}{dx^j} + x^{j+1}\frac{d^{j+1}}{dx^{j+1}}\right] \\
&= \sum_{j=1}^{k} \left[j S\left(k-1, j\right) + S\left(k-1, j-1\right)\right] x^j \frac{d^j}{dx^j}.
\end{aligned} \tag{J.373}$$

In the last line we used $S(k-1, 0) = S(k-1, k) = 0$. *The Stirling numbers of the second kind satisfy the recurrence relation*

$$S\left(k, j\right) = j S\left(k-1, j\right) + S\left(k-1, j-1\right). \tag{J.374}$$

With the help of this relation the claim follows:

$$\theta^k = \sum_{j=1}^{k} S\left(k, j\right) x^j \frac{d^j}{dx^j}. \tag{J.375}$$

Exercise 66 *Rewrite*

$$L_2 = x\left(x+1\right)\left(x+9\right)\frac{d^2}{dx^2} + \left(3x^2 + 20x + 9\right)\frac{d}{dx} + x + 3 \tag{J.376}$$

in Euler operators. (This is the differential operator of Eq. (7.55) multiplied with $x(x+1)(x+9)$*).*

Solution: *With*

$$\frac{d^2}{dx^2} = \frac{1}{x^2}\theta(\theta-1), \quad \frac{d}{dx} = \frac{1}{x}\theta \tag{J.377}$$

we first obtain

$$L_2 = \frac{(x+1)(x+9)}{x}\theta^2 + 2(x+5)\theta + x + 3. \tag{J.378}$$

Multiplication with x gives

$$\tilde{L}_2 = (x+1)(x+9)\theta^2 + 2x(x+5)\theta + x(x+3). \tag{J.379}$$

Exercise 67 *Consider*

$$\tilde{L} = (\theta-\alpha)^\lambda. \tag{J.380}$$

Show that the solution space is spanned by

$$x^\alpha,\ x^\alpha \ln(x),\ \dots,\ \frac{x^\alpha \ln^{\lambda-1}(x)}{(\lambda-1)!}. \tag{J.381}$$

Solution: *Set*

$$f_j(x) = \frac{1}{j!}x^\alpha \ln^j(x). \tag{J.382}$$

We have

$$(\theta-\alpha) f_0(x) = \left(x\frac{d}{dx} - \alpha\right) x^\alpha = 0. \tag{J.383}$$

For $j > 0$ we have

$$(\theta-\alpha) f_j(x) = f_{j-1}(x), \tag{J.384}$$

hence all functions in Eq. (J.381) are annihilated by $(\theta-\alpha)^\lambda$.

Exercise 68 *Consider the differential operators*

$$\begin{aligned}\tilde{L}_a &= (\theta-1)(\theta-x),\\ \tilde{L}_b &= (\theta-x)(\theta-1).\end{aligned} \tag{J.385}$$

Construct for both operators two independent solutions around $x_0 = 0$.

Solution: *The indicial equation reads in both cases*

$$(\alpha - 1)\,\alpha = 0, \tag{J.386}$$

hence the indicials are in both cases $\alpha_1 = 0$ and $\alpha_2 = 1$.

Let's first consider the case $\tilde{L}_a$. We have

$$\tilde{L}_a = (\theta - 1)\,(\theta - x) \;=\; x^2 \frac{d^2}{dx^2} - x^2 \frac{d}{dx} \tag{J.387}$$

and it is clear that $f_{a,1}(x) = 1$ is a solution. $f_{a,1}$ corresponds to the indicial $\alpha_1 = 0$. We may view $f_{a,1}$ as a power series which terminates after the first term. We construct the second solution (corresponding to the indicial $\alpha_2 = 1$) from the ansatz

$$x + \sum_{j=2}^{\infty} c_{2,j} x^j. \tag{J.388}$$

One finds

$$f_{a,2}\,(x) = \sum_{j=1}^{\infty} \frac{x^j}{j!} \;=\; e^x - 1. \tag{J.389}$$

Let us now look at $\tilde{L}_b$. We have

$$\tilde{L}_a = (\theta - x)\,(\theta - 1) \;=\; x^2 \frac{d^2}{dx^2} - x^2 \frac{d}{dx} + x. \tag{J.390}$$

One solution is again trivial: One easily checks that

$$f_{b,2}\,(x) = x \tag{J.391}$$

is a solution. The purpose of this exercise is to show how the solution $f_{b,1}(x)$, which starts at order x^0, is extended to higher orders in x. We start from the ansatz

$$f_{b,1}\,(x) = 1 + \sum_{j=1}^{\infty} \left[c_{1,j,0} \ln(x) + c_{1,j,1}\right] x^j. \tag{J.392}$$

Inserting this ansatz into the differential equation one finds

$$\begin{aligned} 0 = x + \sum_{j=1}^{\infty} \Big\{ & \left[j\,(j-1)\,c_{1,j,1} + (2j-1)\,c_{1,j,0}\right] x^j - \left[(j-1)\,c_{1,j,1} + c_{1,j,0}\right] x^{j+1} \\ & + j\,(j-1)\,c_{1,j,0} x^j \ln(x) - (j-1)\,c_{1,j,0} x^{j+1} \ln(x) \Big\}. \end{aligned} \tag{J.393}$$

The coefficients of x^j and $x^j \ln(x)$ have to vanish separately. From the term x^1 and the logarithmic terms we obtain

$$c_{1,1,0} = -1, \; c_{1,j,0} = 0 \quad \textit{for } j \geq 2. \tag{J.394}$$

The differential equation does not constrain $c_{1,1,1}$, as this term corresponds to the second independent solution $f_{b,2}(x)$. Therefore we may set $c_{1,1,1} = 0$. For the higher terms one finds $c_{1,2,1} = -1/2$ and the recursion formula

$$c_{1,j,1} = \frac{(j-2)}{j\,(j-1)} c_{1,j-1,1}, \qquad j \geq 3. \tag{J.395}$$

Thus

$$f_{b,1}(x) = 1 - x \ln(x) - \frac{1}{2}x^2 - \frac{1}{12}x^3 - \frac{1}{72}x^4 - \frac{1}{480}x^5 + O\left(x^6\right). \tag{J.396}$$

Exercise 69 *Show the equivalence of Eq. (7.80) with Eq. (7.77).*

Solution: *We have to show*

$$\begin{aligned}\mathcal{P}\exp\left(-\int\limits_0^x dx_1 A(x_1)\right) &= 1 - \int\limits_0^x dx_1 A(x_1) + \int\limits_0^x dx_1 A(x_1) \int\limits_0^{x_1} dx_2 A(x_2) \\ &- \int\limits_0^x dx_1 A(x_1) \int\limits_0^{x_1} dx_2 A(x_2) \int\limits_0^{x_2} dx_3 A(x_3) + \dots\end{aligned} \tag{J.397}$$

We start with the left-hand side and expand the exponential function:

$$\mathcal{P}\exp\left(-\int\limits_0^x dx_1 A(x_1)\right) = \mathcal{P}\sum_{n=0}^{\infty} \frac{1}{n!}\left[-\int\limits_0^x dx_1 A(x_1)\right]^n. \tag{J.398}$$

The n-th term in this sum has n integrations. Let's label the integration variables $x_1, \dots, x_n$. The integration region is an n-dimensional cube

$$[0, x]^n. \tag{J.399}$$

We divide the n-dimensional cube into $n!$ simplices defined by

$$x \geq x_{\sigma_1} > s_{\sigma_2} > \dots > x_{\sigma_n} \geq 0. \tag{J.400}$$

Each simplex is uniquely specified by a permutation $\sigma \in S_n$. For each simplex, the path ordering operator gives

$$\mathcal{P}(A(x_1)A(x_2)\dots A(x_n)) = A\left(x_{\sigma_1}\right)A\left(x_{\sigma_2}\right)\dots A\left(x_{\sigma_n}\right). \tag{J.401}$$

By a relabelling of the integration variables we see that each of the $n!$ simplices gives exactly the same contribution, cancelling the $1/n!$ factor in Eq. (J.398). Thus we obtain

$$\mathcal{P}\exp\left(-\int_0^x dx_1 A(x_1)\right) = \sum_{n=0}^{\infty}(-1)^n \int_0^x dx_1 A(x_1)\int_0^{x_1} dx_2 A(x_2)\dots\int_0^{x_{n-1}} dx_n A(x_n). \tag{J.402}$$

Exercise 70 *Prove Eq. (7.93).*

Solution: *We have to show*

$$\frac{d}{dx}\left(e^{\Omega[D_x](x)}\,e^{\Omega[N'_x](x)}\right) = -\left(D_x(x)+N_x(x)\right)e^{\Omega[D_x](x)}\,e^{\Omega[N'_x](x)}. \tag{J.403}$$

In general, the Magnus series $\Omega[A_x](x)$ satisfies

$$\frac{d}{dx}\left(e^{\Omega[A_x](x)}\right) = -A_x(x)\,e^{\Omega[A_x](x)}. \tag{J.404}$$

Therefore

$$\frac{d}{dx}\left(e^{\Omega[D_x](x)}\,e^{\Omega[N'_x](x)}\right) = -D_x(x)\,e^{\Omega[D_x](x)}\,e^{\Omega[N'_x](x)} - e^{\Omega[D_x](x)}\,N'_x(x)\,e^{\Omega[N'_x](x)}. \tag{J.405}$$

We have

$$\begin{aligned} -e^{\Omega[D_x](x)}\,N'_x(x)\,e^{\Omega[N'_x](x)} &= -e^{\Omega[D_x](x)}e^{-\Omega[D_x](x)}N_x(x)\,e^{\Omega[D_x](x)}e^{\Omega[N'_x](x)} \\ &= -N_x(x)\,e^{\Omega[D_x](x)}e^{\Omega[N'_x](x)} \end{aligned} \tag{J.406}$$

and therefore

$$-D_x(x)\,e^{\Omega[D_x](x)}\,e^{\Omega[N'_x](x)} - e^{\Omega[D_x](x)}\,N'_x(x)\,e^{\Omega[N'_x](x)} = -\left(D_x(x)+N_x(x)\right)e^{\Omega[D_x](x)}\,e^{\Omega[N'_x](x)}. \tag{J.407}$$

Exercise 71 *Show that the transformation in Eq. (7.95) transforms the differential Eq. (7.96) into Eq. (7.97).*

Solution: *We have to show that*

$$\vec{I}' = U\vec{I}, \qquad U = e^{-\Omega[A_x^{(0)}](x)} \tag{J.408}$$

transforms the differential equation

$$\left(\frac{d}{dx} + A_x^{(0)} + \varepsilon A_x^{(1)}\right) \vec{I} = 0 \tag{J.409}$$

into

$$\left(\frac{d}{dx} + A'_x\right) \vec{I}' = 0, \tag{J.410}$$

with A'_x given by

$$A'_x = \varepsilon U A_x^{(1)} U^{-1}. \tag{J.411}$$

A'_x is given by

$$A'_x = U A_x U^{-1} + U \frac{d}{dx} U^{-1}. \tag{J.412}$$

We have $U^{-1} = e^{\Omega[A_x^{(0)}](x)}$ and

$$\frac{d}{dx} U^{-1} = -A_x^{(0)} e^{\Omega[A_x^{(0)}](x)} = -A_x^{(0)} U^{-1}. \tag{J.413}$$

Thus

$$A'_x = U\left(A_x^{(0)} + \varepsilon A_x^{(1)}\right) U^{-1} - U A_x^{(0)} U^{-1} = \varepsilon U A_x^{(1)} U^{-1}. \tag{J.414}$$

Exercise 72 *Assume that A_x is in Fuchsian form (i.e. of the form as in Eq. (7.106)). Show that the matrix residue at $x = \infty$ is given by*

$$M_{\infty,1}(\varepsilon) = -\sum_{x_i \in S'} M_{x_i,1}(\varepsilon). \tag{J.415}$$

Solution: *We map the point $x = \infty$ to $x' = 0$ with the transformation $x' = 1/x$. We have*

$$dx = -\frac{dx'}{x'^2} \tag{J.416}$$

and with $x'_i = 1/x_i$

$$A_x dx = \sum_{x_i \in S'} M_{x_i,1}(\varepsilon) \frac{1}{(x - x_i)} dx = -\sum_{x_i \in S'} M_{x_i,1}(\varepsilon) \frac{x'_i}{x'\left(x'_i - x'\right)} dx'$$

$$= - \sum_{x_i \in S'} M_{x_i,1}(\varepsilon) \frac{1}{x'} dx' + \dots, \tag{J.417}$$

where the dots stand for terms regular at $x' = 0$. Thus

$$M_{\infty,1}(\varepsilon) = - \sum_{x_i \in S'} M_{x_i,1}(\varepsilon). \tag{J.418}$$

Exercise 73 *Show that P and $\mathbf{1} - P$ are projectors, i.e.*

$$P^2 = P, \quad (\mathbf{1} - P)^2 = \mathbf{1} - P. \tag{J.419}$$

Show further

$$\left[(\mathbf{1} - P) + \frac{x - x_2}{x - x_1} P \right] \left[(\mathbf{1} - P) + \frac{x - x_1}{x - x_2} P \right] = \mathbf{1}. \tag{J.420}$$

Solution: *We have*

$$P^2 = \frac{\vec{v}_{R,x_1} \vec{v}^T_{L,x_2}}{\left(\vec{v}^T_{L,x_2} \cdot \vec{v}_{R,x_1}\right)} \cdot \frac{\vec{v}_{R,x_1} \vec{v}^T_{L,x_2}}{\left(\vec{v}^T_{L,x_2} \cdot \vec{v}_{R,x_1}\right)} = \frac{\vec{v}_{R,x_1} \left(\vec{v}^T_{L,x_2} \cdot \vec{v}_{R,x_1}\right) \vec{v}^T_{L,x_2}}{\left(\vec{v}^T_{L,x_2} \cdot \vec{v}_{R,x_1}\right)^2} = \frac{\vec{v}_{R,x_1} \vec{v}^T_{L,x_2}}{\left(\vec{v}^T_{L,x_2} \cdot \vec{v}_{R,x_1}\right)} = P. \tag{J.421}$$

Further

$$(\mathbf{1} - P)^2 = \mathbf{1} - 2P + P^2 = \mathbf{1} - 2P + P = \mathbf{1} - P. \tag{J.422}$$

With

$$P(\mathbf{1} - P) = (\mathbf{1} - P) P = 0 \tag{J.423}$$

we also have

$$\begin{aligned} &\left[(\mathbf{1} - P) + \frac{x - x_2}{x - x_1} P \right] \left[(\mathbf{1} - P) + \frac{x - x_1}{x - x_2} P \right] = \\ &= (\mathbf{1} - P)^2 + \frac{x - x_2}{x - x_1} P (\mathbf{1} - P) + \frac{x - x_1}{x - x_2} (\mathbf{1} - P) P + P^2 = (\mathbf{1} - P) + P = \mathbf{1}. \end{aligned} \tag{J.424}$$

Exercise 74 *Consider the square roots*

$$r_1 = \sqrt{x(4 + x)} \quad \text{and} \quad r_2 = \sqrt{x(36 + x)}. \tag{J.425}$$

Find a transformation, which simultaneously rationalises r_1 and r_2.

Solution: *We rationalise the two roots sequentially. We start with the root r_1. We already know that the transformation*

$$x = \frac{(1-x')^2}{x'} \qquad \text{(J.426)}$$

rationalises r_1. The root r_2 expressed in terms of x' reads

$$r_2 = \frac{1-x'}{x'}\sqrt{1+34x'+x'^2} \qquad \text{(J.427)}$$

We have

$$1+34x'+x'^2 = \left(x'+17-12\sqrt{2}\right)\left(x'+17+12\sqrt{2}\right) \qquad \text{(J.428)}$$

and Eq. (7.229) gives us

$$x' = -\frac{6\sqrt{2}}{x''}\left(1+\frac{17}{12}\sqrt{2}x''+x''^2\right). \qquad \text{(J.429)}$$

The variable x'' simultaneously rationalises r_1 and r_2:

$$r_1 = 12\sqrt{2}\frac{\left(x''+\sqrt{2}\right)\left(2x''+\sqrt{2}\right)\left(3x''^2+4\sqrt{2}x''+3\right)}{x''\left(3x''+2\sqrt{2}\right)\left(4x''+3\sqrt{2}\right)},$$
$$r_2 = 36\sqrt{2}\frac{\left(x''+\sqrt{2}\right)\left(2x''+\sqrt{2}\right)\left(1-x''^2\right)}{x''\left(3x''+2\sqrt{2}\right)\left(4x''+3\sqrt{2}\right)}. \qquad \text{(J.430)}$$

The appearance of the root $\sqrt{2}$ is unaesthetic, but not a principal problem. It stems from the fact that in using Eq. (7.229) we have implicitly chosen the non-rational point $(r, x') = (0, -17-12\sqrt{2})$ on the hypersurface

$$f(r, x') = r^2 - x'^2 - 34x' - 1. \qquad \text{(J.431)}$$

This hypersurface has rational points, for example $(r, x') = (6, 1)$. Repeating the exercise one finds for example that the transformation

$$x = \frac{(\tilde{x}^2-9)^2}{\tilde{x}(\tilde{x}+1)(\tilde{x}+9)} \qquad \text{(J.432)}$$

rationalises both r_1 and r_2

$$r_1 = \frac{\left(\tilde{x}^2 - 9\right)\left(\tilde{x}^2 + 2\tilde{x} + 9\right)}{\tilde{x}\left(\tilde{x}+1\right)\left(\tilde{x}+9\right)}, \quad r_2 = \frac{\left(\tilde{x}^2 - 9\right)\left(\tilde{x}^2 + 18\tilde{x} + 9\right)}{\tilde{x}\left(\tilde{x}+1\right)\left(\tilde{x}+9\right)}, \tag{J.433}$$

and only contains rational coefficients.

Exercise 75 *Prove Eq. (8.19).*

Solution: *We have to show*

$$\text{Li}_{m_1 \dots m_k}(x_1, \dots, x_k) = (-1)^k G_{m_1 \dots m_k}\left(\frac{1}{x_1}, \frac{1}{x_1 x_2}, \dots, \frac{1}{x_1 \dots x_k}; 1\right), \tag{J.434}$$

where we may assume that

$$\left|x_1 x_2 \dots x_j\right| \leq 1 \quad \textit{for all } j \in \{1, \dots, k\} \textit{ and } (m_1, x_1) \neq (1, 1). \tag{J.435}$$

Set $r = m_1 + \dots + m_k$. The integral representation has depth r. Let us introduce some notation to facilitate the proof: We set

$$b_j = \frac{1}{x_1 x_2 \dots x_j}. \tag{J.436}$$

and introduce the following notation for iterated integrals

$$\int_0^y \frac{dt}{t - z_1} \circ \dots \circ \frac{dt}{t - z_r} = \int_0^y \frac{dt_1}{t_1 - z_1} \int_0^{t_1} \frac{dt_2}{t_2 - z_2} \dots \int_0^{t_{r-1}} \frac{dt_r}{t_r - z_r}, \tag{J.437}$$

together with the short hand notation

$$\int_0^y \left(\frac{dt}{t}\circ\right)^m \frac{dt}{t - z} = \int_0^y \underbrace{\frac{dt}{t} \circ \dots \frac{dt}{t}}_{m \text{ times}} \circ \frac{dt}{t - z}. \tag{J.438}$$

The integral representation $G_{m_1 \dots m_k}(b_1, \dots, b_k; 1)$ reads then

$$(-1)^k G_{m_1 \dots m_k}(b_1, \dots, b_k; 1) = (-1)^k \int_0^1 \left(\frac{dt}{t}\circ\right)^{m_1 - 1} \frac{dt}{t - b_1} \circ \dots \circ \left(\frac{dt}{t}\circ\right)^{m_k - 1} \frac{dt}{t - b_k}. \tag{J.439}$$

For all integration variables we have $|t_j| \leq 1$ (with $j \in \{1, \dots, r\}$). We have k terms of the form $1/(t_j - b_j)$. As $|t_j / b_j| = |x_1 x_2 \dots x_j t_j| \leq 1$ we expand the geometric series (the case $x_1 x_2 \dots x_j t_j = 1$ is handled by first replacing the outermost integration limit 1 by $y < 1$ and taking the limit $y \to 1$ in the end. With $y < 1$ we have

$|t_j| < 1$, which is sufficient for the convergence of the geometric series. The limit $y \to 1$ may be taken for $(m_1, x_1) \neq (1, 1)$):

$$\frac{1}{t_j - b_j} = -\frac{1}{b_j} \sum_{i=0}^{\infty} \left(\frac{t_j}{b_j} \right)^i = - \sum_{i=1}^{\infty} \frac{t_j^{i-1}}{b_j^i}. \tag{J.440}$$

Integrating term-by-term gives

$$\begin{aligned}(-1)^k G_{m_1 \dots m_k}(b_1, \dots, b_k; 1) &= \sum_{i_1=1}^{\infty} \dots \sum_{i_k=1}^{\infty} \frac{1}{(i_1 + \dots + i_k)^{m_1}} \frac{1}{b_1^{i_1}} \cdots \frac{1}{(i_{k-1} + i_k)^{m_{k-1}}} \frac{1}{b_{k-1}^{i_{k-1}}} \frac{1}{i_k^{m_k}} \frac{1}{b_k^{i_k}} \\ &= \sum_{i_1=1}^{\infty} \dots \sum_{i_k=1}^{\infty} \frac{x_1^{i_1}}{(i_1 + \dots + i_k)^{m_1}} \cdots \frac{(x_1 \dots x_{k-1})^{i_{k-1}}}{(i_{k-1} + i_k)^{m_{k-1}}} \frac{(x_1 \dots x_k)^{i_k}}{i_k^{m_k}} \\ &= \sum_{i_1=1}^{\infty} \dots \sum_{i_k=1}^{\infty} \frac{x_1^{i_1+\dots+i_k}}{(i_1 + \dots + i_k)^{m_1}} \cdots \frac{x_{k-1}^{i_{k-1}+i_k}}{(i_{k-1} + i_k)^{m_{k-1}}} \frac{x_k^{i_k}}{i_k^{m_k}}. \end{aligned} \tag{J.441}$$

Changing the summation indices according to $n_1 = i_1 + \dots + i_k$, $n_2 = i_2 + \dots + i_k$, ..., $n_{k-1} = i_{k-1} + i_k$ and $n_k = i_k$ yields

$$\begin{aligned}(-1)^k G_{m_1 \dots m_k}(b_1, \dots, b_k; 1) &= \sum_{n_1=1}^{\infty} \sum_{n_2=1}^{n_1-1} \dots \sum_{n_{k-1}=1}^{n_{k-2}-1} \sum_{n_k=1}^{n_{k-1}-1} \frac{x_1^{n_1}}{n_1^{m_1}} \frac{x_2^{n_2}}{n_2^{m_2}} \cdots \frac{x_{k-1}^{n_{k-1}}}{n_{k-1}^{m_{k-1}}} \frac{x_k^{n_k}}{n_k^{m_k}} \\ &= \mathrm{Li}_{m_1 \dots m_k}(x_1, \dots, x_k). \end{aligned} \tag{J.442}$$

Exercise 76 *Consider the alphabet $A = \{l_1, l_2\}$ with $l_1 < l_2$. Write down all Lyndon words of depth ≤ 3.*

Solution: *At depth 1 we have the words l_1 and l_2. Both of them are Lyndon words. At depth 2 we have to consider the words $l_1 l_1$, $l_1 l_2$, $l_2 l_1$ and $l_2 l_2$. Out of these only*

$$l_1 l_2 \tag{J.443}$$

is a Lyndon word. For example $w = l_1 l_1$ may be written as $w = uv$, $u = l_1$, $v = l_1$ and $v < w$. At depth 3 we have to consider the words $l_1 l_1 l_1$, $l_1 l_1 l_2$, $l_1 l_2 l_1$, $l_1 l_2 l_2$, $l_2 l_1 l_1$, $l_2 l_1 l_2$, $l_2 l_2 l_1$ and $l_2 l_2 l_2$. Out of these the Lyndon words are

$$l_1 l_1 l_2, \quad l_1 l_2 l_2. \tag{J.444}$$

Exercise 77 *Express the product*

$$G_2(z; y) \cdot G_3(z; y) \tag{J.445}$$

as a linear combination of multiple polylogarithms.

Solution: *We have*

$$G_2(z;y) = G(0,z;y), \quad G_3(z;y) = G(0,0,z;y). \tag{J.446}$$

We first work out the shuffle product for $G(z_1,z_2;y) \cdot G(z_3,z_4,z_5;y)$ and set $z_1 = z_3 = z_4 = 0$ and $z_2 = z_5 = z$ in the end. We have

$$\begin{aligned} & G(z_1,z_2;y) \cdot G(z_3,z_4,z_5;y) = \\ & \quad G(z_1,z_2,z_3,z_4,z_5;y) + G(z_1,z_3,z_2,z_4,z_5;y) + G(z_1,z_3,z_4,z_2,z_5;y) + G(z_1,z_3,z_4,z_5,z_2;y) \\ & \quad + G(z_3,z_1,z_2,z_4,z_5;y) + G(z_3,z_1,z_4,z_2,z_5;y) + G(z_3,z_1,z_4,z_5,z_2;y) + G(z_3,z_4,z_1,z_2,z_5;y) \\ & \quad + G(z_3,z_4,z_1,z_5,z_2;y) + G(z_3,z_4,z_5,z_1,z_2;y). \end{aligned} \tag{J.447}$$

Setting $z_1 = z_3 = z_4 = 0$ and $z_2 = z_5 = z$ yields

$$G(0,z;y) \cdot G(0,0,z;y) = G(0,z,0,0,z;y) + 3G(0,0,z,0,z;y) + 6G(0,0,0,z,z;y), \tag{J.448}$$

or

$$G_2(z;y) \cdot G_3(z;y) = G_{23}(z,z;y) + 3G_{32}(z,z;y) + 6G_{41}(z,z;y). \tag{J.449}$$

Exercise 78 *Work out the quasi-shuffle product*

$$\mathrm{Li}_{m_1 m_2}(x_1,x_2) \cdot \mathrm{Li}_{m_3 m_4}(x_3,x_4). \tag{J.450}$$

Solution: *We obtain*

$$\begin{aligned} & \mathrm{Li}_{m_1 m_2}(x_1,x_2) \cdot \mathrm{Li}_{m_3 m_4}(x_3,x_4). = \\ & \quad \mathrm{Li}_{m_1 m_2 m_3 m_4}(x_1,x_2,x_3,x_4) + \mathrm{Li}_{m_1 m_3 m_2 m_4}(x_1,x_3,x_2,x_4) + \mathrm{Li}_{m_1 m_3 m_4 m_2}(x_1,x_3,x_4,x_2) \\ & \quad + \mathrm{Li}_{m_3 m_1 m_2 m_4}(x_3,x_1,x_2,x_4) + \mathrm{Li}_{m_3 m_1 m_4 m_2}(x_3,x_1,x_4,x_2) + \mathrm{Li}_{m_3 m_4 m_1 m_2}(x_3,x_4,x_1,x_2) \\ & \quad + \mathrm{Li}_{(m_1+m_3) m_2 m_4}(x_1 \cdot x_3,x_2,x_4) + \mathrm{Li}_{(m_1+m_3) m_4 m_2}(x_1 \cdot x_3,x_4,x_2) + \mathrm{Li}_{m_1 (m_2+m_3) m_4}(x_1,x_2 \cdot x_3,x_4) \\ & \quad + \mathrm{Li}_{m_3 (m_1+m_4) m_2}(x_3,x_1 \cdot x_4,x_2) + \mathrm{Li}_{m_1 m_3 (m_2+m_4)}(x_1,x_3,x_2 \cdot x_4) + \mathrm{Li}_{m_3 m_1 (m_2+m_4)}(x_3,x_1,x_2 \cdot x_4) \\ & \quad + \mathrm{Li}_{(m_1+m_3)(m_2+m_4)}(x_1 \cdot x_3,x_2 \cdot x_4). \end{aligned} \tag{J.451}$$

Exercise 79 *Use the (regularised) double-shuffle relations to show*

$$\zeta_2^2 = \frac{5}{2}\zeta_4. \tag{J.452}$$

Solution: *As in Eq. (8.80) we set*

$$L = -\ln\lambda = \mathrm{Li}_1(1-\lambda) = -G(1;1-\lambda). \tag{J.453}$$

In the quasi-shuffle algebra we have

$$L \cdot \zeta_3 = L \cdot \mathrm{Li}_3(1) = \mathrm{Li}_{13}(1-\lambda,1) + \mathrm{Li}_{31}(1,1-\lambda) + \mathrm{Li}_4(1-\lambda). \tag{J.454}$$

In the shuffle algebra we consider

$$\begin{aligned} -L \cdot G(0,0,1;1-\lambda) &= G(1;1-\lambda) \cdot G(0,0,1;1-\lambda) \\ &= G(1,0,0,1;1-\lambda) + G(0,1,0,1;1-\lambda) + 2G(0,0,1,1;1-\lambda). \end{aligned} \tag{J.455}$$

In the Li*-notation this is equivalent to*

$$L \cdot \mathrm{Li}_3(1-\lambda) = \mathrm{Li}_{13}(1-\lambda, 1) + \mathrm{Li}_{22}(1-\lambda, 1) + 2\mathrm{Li}_{31}(1-\lambda, 1). \tag{J.456}$$

We have $\mathrm{Li}_3(1-\lambda) - \zeta_3 = O(\lambda)$ *and subtracting the quasi-shuffle relation from the shuffle relation*

$$\zeta_{22} + \zeta_{31} - \zeta_4 = 0 \tag{J.457}$$

follows. From Eq. (8.76) we know already that $\zeta_{31} = \frac{1}{4}\zeta_4$, *and therefore*

$$\zeta_{22} = \frac{3}{4}\zeta_4. \tag{J.458}$$

Substituting this result into Eq. (8.78) the sought-after relation

$$\zeta_2^2 = \frac{5}{2}\zeta_4. \tag{J.459}$$

follows.

Exercise 80 *Let*

$$f_0(x) = \frac{1}{r!}\ln^r(x), \quad f_1(x) = \frac{(-1)^r}{r!}\ln^r(1-x). \tag{J.460}$$

Determine

$$\mathcal{M}_0 f_0(x), \quad \mathcal{M}_0 f_1(x), \quad \mathcal{M}_1 f_0(x), \quad \mathcal{M}_1 f_1(x). \tag{J.461}$$

Solution: *Let us first discuss the case* $r = 1$*:*

$$h_0(x) = \ln(x), \quad h_1(x) = -\ln(1-x). \tag{J.462}$$

We already know that (recall that $\ln(x)$ *is regular at* $x = 1$ *and* $\ln(1-x)$ *is regular at* $x = 0$*)*

$$\begin{aligned} \mathcal{M}_0 h_0(x) &= h_0(x) + 2\pi i, & \mathcal{M}_0 h_1(x) &= h_1(x), \\ \mathcal{M}_1 h_0(x) &= h_0(x), & \mathcal{M}_1 h_1(x) &= h_1(x) - 2\pi i. \end{aligned} \tag{J.463}$$

We then have

$$
\begin{aligned}
\mathcal{M}_0 f_0(x) &= \frac{1}{r!}\left[\mathcal{M}_0 h_0(x)\right]^r &=& \frac{1}{r!}\left[\ln(x) + 2\pi i\right]^r, \\
\mathcal{M}_0 f_1(x) &= \frac{1}{r!}\left[\mathcal{M}_0 h_1(x)\right]^r &=& f_1(x), \\
\mathcal{M}_1 f_0(x) &= \frac{1}{r!}\left[\mathcal{M}_1 h_0(x)\right]^r &=& f_0(x), \\
\mathcal{M}_1 f_1(x) &= \frac{1}{r!}\left[\mathcal{M}_1 h_1(x)\right]^r &=& \frac{1}{r!}\left[-\ln(1-x) - 2\pi i\right]^r.
\end{aligned}
\tag{J.464}
$$

Exercise 81 *Compute the monodromy of $G(1,1;y)$ around $y=1$.*

Solution: *The simple solution is as follows: As $G(1,1;y) = \frac{1}{2}\ln^2(1-y)$ it follows from Exercise 80 that*

$$
\begin{aligned}
\mathcal{M}_1 G(1,1;y) &= \frac{1}{2}\left[-\ln(1-y) - 2\pi i\right]^2 \\
&= G(1,1;y) + 2\pi i G(1;y) + \frac{1}{2}(2\pi i)^2.
\end{aligned}
\tag{J.465}
$$

The purpose of this exercise is to compute the monodromy with the help of Eq. (8.110). Of course, we should obtain the same result. With $G(1;y) = \ln(1-y)$ and $\mathcal{M}_1 G(1;y) = G(1;y) + 2\pi i$ we obtain from Eqs. (8.110) and (8.112)

$$
\begin{aligned}
\mathcal{M}_1 G(1,1;y) &= G(1,1;y) + \lim_{\varepsilon\to 0}\left\{\oint \frac{dy'}{y'-1} G(1;y') + \int\limits_{1+\varepsilon}^{y} \frac{dy'}{y'-1}\left[\mathcal{M}_z G(1;y') - G(1;y')\right]\right\} \\
&= G(1,1;y) + 2\pi i \lim_{\varepsilon\to 0}\left\{\int\limits_0^1 dt G\left(1; 1+\varepsilon e^{2\pi i t}\right) + \int\limits_{1+\varepsilon}^{y} \frac{dy'}{y'-1}\right\}.
\end{aligned}
\tag{J.466}
$$

For the first integral we have

$$
\int\limits_0^1 dt G\left(1; 1+\varepsilon e^{2\pi i t}\right) = \int\limits_0^1 dt \ln\left(-\varepsilon e^{2\pi i t}\right) = \ln(-\varepsilon) + 2\pi i \int\limits_0^1 t\, dt = \ln(-\varepsilon) + \frac{1}{2}(2\pi i),
\tag{J.467}
$$

the second integral gives

$$
\int\limits_{1+\varepsilon}^{y} \frac{dy'}{y'-1} = \ln(1-y) - \ln(-\varepsilon).
\tag{J.468}
$$

In total we obtain

$$
\mathcal{M}_1 G(1,1;y) = G(1,1;y) + 2\pi i G(1;y) + \frac{1}{2}(2\pi i)^2,
\tag{J.469}
$$

in agreement with our previous result.

Exercise 82 *Prove Eq. (8.38) from Chap. 8.*

Solution: *We may write the sum representation of* $\mathrm{Li}_{m_1\dots 0\dots m_k}(x_1,\dots,x_i,\dots,x_k)$ *as*

$$\mathrm{Li}_{m_1\dots 0\dots m_k}(x_1,\dots,x_i,\dots,x_k) = \sum_{n_1=1}^{\infty} \frac{x_1^{n_1}}{n_1^{m_1}} \cdots \sum_{n_{i-1}=1}^{n_{i-2}-1} \frac{x_{i-1}^{n_{i-1}}}{n_{i-1}^{m_{i-1}}} Z_{0,m_{i+1},\dots,m_k}(x_i,x_{i+1},\dots,x_k;n_{i-1}-1). \tag{J.470}$$

Consider now

$$Z_0(x_i;n_{i-1}-1)Z_{m_{i+1},\dots,m_k}(x_{i+1},\dots,x_k;n_{i-1}-1). \tag{J.471}$$

On the one hand we have

$$Z_0(x_i;n_{i-1}-1) = \frac{x_i}{1-x_i} - \frac{x_i^{n_{i-1}}}{1-x_i} = \mathrm{Li}_0(x_i) - \mathrm{Li}_0(x_i)x_i^{n_{i-1}-1}, \tag{J.472}$$

on the other hand we may use the quasi-shuffle product for the Z-sums:

$$\begin{aligned} Z_0(x_i;n_{i-1}-1)Z_{m_{i+1},\dots,m_k}(x_{i+1},\dots,x_k;n_{i-1}-1) = \\ \sum_{j=i}^{k} Z_{m_{i+1}\dots m_j 0 m_{j+1}\dots m_k}(x_{i+1},\dots,x_j,x_i,x_{j+1},\dots,x_k;n_{i-1}-1) \\ + \sum_{j=i+1}^{k} Z_{m_{i+1}\dots m_j\dots m_k}(x_{i+1},\dots,x_i\cdot x_j,\dots,x_k;n_{i-1}-1) \end{aligned} \tag{J.473}$$

Combining these two equations and noting that

$$\frac{1}{x_i}\mathrm{Li}_0(x_i) = 1 + \mathrm{Li}_0(x_i) \tag{J.474}$$

proves Eq. (8.38).

Exercise 83 *Consider* I_{111} *from Eq. (9.1) with* $\mu^2 = -p_3^2$ *and* $x = p_1^2/p_3^2$ *in* $D = 4 - 2\varepsilon$ *space-time dimensions:*

$$I_{111} = e^{\varepsilon\gamma_{\mathrm{E}}} \frac{\Gamma(-\varepsilon)\Gamma(1-\varepsilon)}{\Gamma(1-2\varepsilon)} \sum_{n=0}^{\infty} \frac{\Gamma(n+1+\varepsilon)}{\Gamma(n+2)} (1-x)^n . \tag{J.475}$$

Expand the sum in ε *and give the first two terms of the* ε*-expansion for the full expression.*

Solution: *With the substitution* $n \to n+1$ *we have*

$$I_{111} = e^{\varepsilon \gamma_E} \frac{\Gamma(-\varepsilon)\Gamma(1-\varepsilon)}{\Gamma(1-2\varepsilon)} \sum_{n=1}^{\infty} \frac{\Gamma(n+\varepsilon)}{\Gamma(n+1)} (1-x)^{n-1}. \tag{J.476}$$

Expanding $\Gamma(n+\varepsilon)$ according to Eq. (9.11) one obtains:

$$I_{111} = e^{\varepsilon \gamma_E} \frac{\Gamma(-\varepsilon)\Gamma(1-\varepsilon)\Gamma(1+\varepsilon)}{\Gamma(1-2\varepsilon)} \frac{1}{1-x} \sum_{n=1}^{\infty} \varepsilon^{n-1} H_{\underbrace{1\dots1}_{n}}(1-x). \tag{J.477}$$

In this special case all harmonic polylogarithms can be expressed in terms of powers of the standard logarithm:

$$H_{\underbrace{1\dots1}_{n}}(1-x) = \frac{(-1)^n}{n!} (\ln x)^n. \tag{J.478}$$

Therefore

$$I_{111} = -e^{\varepsilon \gamma_E} \frac{\Gamma(1-\varepsilon)\Gamma(1-\varepsilon)\Gamma(1+\varepsilon)}{\varepsilon^2 \Gamma(1-2\varepsilon)} \frac{1}{1-x} \sum_{n=1}^{\infty} \varepsilon^n \frac{(-1)^n}{n!} (\ln x)^n. \tag{J.479}$$

The expansion of the prefactor is

$$e^{\varepsilon \gamma_E} \frac{\Gamma(1-\varepsilon)\Gamma(1-\varepsilon)\Gamma(1+\varepsilon)}{\varepsilon^2 \Gamma(1-2\varepsilon)} = \frac{1}{\varepsilon^2} \left(1 - \frac{1}{2}\zeta_2 \varepsilon^2\right) + O(\varepsilon). \tag{J.480}$$

Thus

$$I_{111} = \frac{1}{1-x} \left[\frac{1}{\varepsilon} \ln x - \frac{1}{2} (\ln x)^2 \right] + O(\varepsilon). \tag{J.481}$$

The result for this example is particular simple and one recovers from Eq. (J.479) the well-known all-order result

$$I_{111} = e^{\varepsilon \gamma_E} \frac{\Gamma(1-\varepsilon)^2 \Gamma(1+\varepsilon)}{\Gamma(1-2\varepsilon)} \frac{1}{\varepsilon^2} \frac{1-x^{-\varepsilon}}{1-x}, \tag{J.482}$$

which (for this simple example) can also be obtained by direct integration. If we expand this result in ε we recover Eq. (J.481).

Exercise 84 *Consider the k-dimensional standard simplex in $\mathbb{R}^{k+1}$. This is the polytope with vertices given by the $(k+1)$ standard unit vectors $e_j \in \mathbb{R}^{k+1}$. Show that the standard simplex has Euclidean volume $1/k!$ and therefore the normalised volume 1.*

Solution: *The Euclidean volume of the standard k-dimensional simplex Δ is given by the integral*

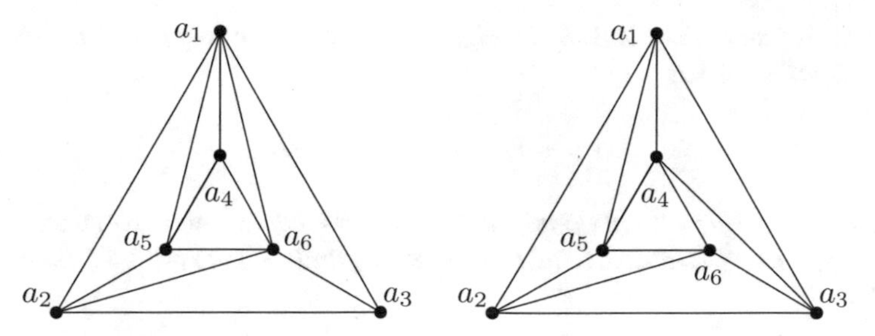

Fig. J.7 The labelling of the points for the regular triangulation of the polytope P (left) and the non-regular triangulation of the polytope P (right)

$$\text{vol}\,(\Delta) = \int\limits_{\alpha_j \geq 0} d^{k+1}\alpha \; \delta\left(1 - \sum_{j=1}^{k+1} \alpha_j\right). \tag{J.483}$$

From Eq. (2.278) we have

$$\text{vol}\,(\Delta) = \frac{1}{\Gamma(k+1)} = \frac{1}{k!}, \tag{J.484}$$

and hence

$$\text{vol}_0\,(\Delta) = k! \; \text{vol}\,(\Delta) \; = \; 1. \tag{J.485}$$

Exercise 85 *Show that the left picture of Fig. 9.3 defines a regular triangulation.*

Solution: *We label the points as in Fig. J.7. In order to show that the triangulation is regular, we have to give a height vector. The height vector*

$$h = (1, 2, 3, 0, 0, 0)^T \tag{J.486}$$

does the job. This is most easily seen by drawing $(\tilde{a}_1, \tilde{a}_2, \tilde{a}_3, \tilde{a}_4, \tilde{a}_5, \tilde{a}_6) \in \mathbb{R}^4$. *As* $(a_1, a_2, a_3, a_4, a_5, a_6) \in \mathbb{R}^3$ *lie in a plane with normal vector* $n = (1, 1, 1)$, *it is sufficient to draw the points* $(\tilde{a}_1, \tilde{a}_2, \tilde{a}_3, \tilde{a}_4, \tilde{a}_5, \tilde{a}_6)$ *in* $\mathbb{R}^3$, *with two coordinates for the plane and one coordinate the height. One easily sees that the lift is convex. The lift is convex for* $h_4 = h_5 = h_6 = 0$ *and*

$$0 \; < \; h_1 \; < \; h_2 \; < \; h_3. \tag{J.487}$$

The chosen height vector fulfils this condition.

Exercise 86 *Show that the right picture of Fig. 9.3 defines a non-regular triangulation.*

Solution: *We label the points as in Fig. J.7. Let us look at the points* a_1, a_2, a_4 *and* a_5. *These points satisfy*

$$a_1 - a_2 - 4a_4 + 4a_5 = 0. \tag{J.488}$$

This is easily verified by plugging in the defining coordinates. Now consider the simplex σ_{125}, *i.e. the triangle with vertices* a_1, a_2 *and* a_5. *Let* $r_{125} \in \mathbb{R}^3$ *be the vector such that*

$$\begin{aligned} r_{125} \cdot a_j &= h_j \qquad j \in \{1, 2, 5\}, \\ r_{125} \cdot a_j &< h_j \qquad j \in \{3, 4, 6\}. \end{aligned} \tag{J.489}$$

Contracting Eq. (J.488) with r_{125} *we obtain*

$$0 = h_1 - h_2 - 4r_{125} \cdot r_4 + 4h_5 \;>\; h_1 - h_2 - 4h_4 + 4h_5 \tag{J.490}$$

and hence

$$h_1 - h_2 - 4h_4 + 4h_5 < 0. \tag{J.491}$$

Repeating the argumentation for the points a_2, a_3, a_5, a_6 *and the triangle* σ_{236} *we obtain*

$$h_2 - h_3 - 4h_5 + 4h_6 < 0. \tag{J.492}$$

Repeating once more the argumentation for the points a_1, a_3, a_4, a_6 *and the triangle* σ_{134} *we obtain*

$$h_3 - h_1 - 4h_6 + 4h_4 < 0. \tag{J.493}$$

Adding up Eqs. (J.491), (J.492) and (J.493) we obtain

$$0 < 0, \tag{J.494}$$

this is a contradiction, hence a height vector does not exist and the triangulation is non-regular.

Exercise 87 *In this exercise we are going to prove Theorem 8. Let* $G(z, x') = G(z_1, \dots, z_{n_{\mathrm{int}}}, x'_1, \dots, x'_n)$ *be a generalised Lee-Pomeransky polynomial such that the associated* $(n_{\mathrm{int}} + 1) \times n$*-matrix* $\mathcal{A}$ *satisfies Eq. (9.90). Consider the integral*

$$I = C \int\limits_{z_j \geq 0} d^{n_{\mathrm{int}}} z \left(\prod_{j=1}^{n_{\mathrm{int}}} z_j^{\nu_j - 1} \right) \left[G\left(z, x'\right) \right]^{-\frac{D}{2}}. \tag{J.495}$$

Show that I satisfies the differential equations in Eqs. (9.92) and (9.93) with $c = (-D/2, -\nu_1, \dots, -\nu_{n_{\mathrm{int}}})^T$.

Solution: *The entries a_{ij} of $\mathcal{A}$ define $G(z, x')$:*

$$G\left(z, x'\right) = \sum_{j=1}^{n} x'_j \, z_1^{a_{1j}} \dots z_{n_{\mathrm{int}}}^{a_{n_{\mathrm{int}} j}}. \tag{J.496}$$

We have

$$\frac{\partial}{\partial x'_j} G\left(z, x'\right) = z_1^{a_{1j}} \dots z_{n_{\mathrm{int}}}^{a_{n_{\mathrm{int}} j}}. \tag{J.497}$$

We have to verify the set of equations in Eqs. (9.92) and (9.93). These equations are equivalent to Eqs. (9.96) and (9.98). We start with Eq. (9.96). Let $u, v \in \mathbb{N}_0^n$ with $\mathcal{A}u = \mathcal{A}v$ and set

$$|u| = \sum_{j=1}^{n} u_j, \quad |v| = \sum_{j=1}^{n} v_j. \tag{J.498}$$

Then

$$\begin{aligned}
\partial^u I &= \frac{\Gamma\left(1-\frac{D}{2}\right)}{\Gamma\left(1-\frac{D}{2}-|u|\right)} C \int\limits_{z_j \geq 0} d^{n_{\mathrm{int}}} z \, \left(\prod_{j=1}^{n_{\mathrm{int}}} z_j^{\nu_j + a_{jk} u_k - 1} \right) \left[G\left(z, x'\right) \right]^{-\frac{D}{2} - |u|}, \\
\partial^v I &= \frac{\Gamma\left(1-\frac{D}{2}\right)}{\Gamma\left(1-\frac{D}{2}-|v|\right)} C \int\limits_{z_j \geq 0} d^{n_{\mathrm{int}}} z \, \left(\prod_{j=1}^{n_{\mathrm{int}}} z_j^{\nu_j + a_{jk} v_k - 1} \right) \left[G\left(z, x'\right) \right]^{-\frac{D}{2} - |v|}.
\end{aligned} \tag{J.499}$$

From $\mathcal{A}u = \mathcal{A}v$ we have $a_{jk}u_k = a_{jk}v_k$ (a sum over k is implied). The first row of $\mathcal{A}$ is $(1, \dots, 1)$. This implies $|u| = |v|$. Hence

$$\partial^u I = \partial^v I. \tag{J.500}$$

Let us now turn to Eq. (9.98). We start with the first row of $\mathcal{A}$ with the entries $a_{01} = \dots = a_{0n} = 1$: We have

$$\sum_{j=1}^{n} x'_j \frac{\partial}{\partial x'_j} I = -\frac{D}{2} C \int\limits_{z_j \geq 0} d^{n_{\mathrm{int}}} z \, \left(\prod_{j=1}^{n_{\mathrm{int}}} z_j^{\nu_j - 1} \right) \left[G\left(z, x'\right) \right]^{-\frac{D}{2}} = -\frac{D}{2} I. \tag{J.501}$$

We then consider the other rows ($1 \le i \le n_{\text{int}}$): We first have

$$\sum_{j=1}^{n} a_{ij} x'_j \frac{\partial}{\partial x'_j} G\left(z, x'\right) = \sum_{j=1}^{n} a_{ij} x'_j \, z_1^{a_{1j}} \dots z_{n_{\text{int}}}^{a_{n_{\text{int}} j}} = z_i \frac{\partial}{\partial z_i} G\left(z, x'\right). \qquad \text{(J.502)}$$

Hence

$$\begin{aligned} \sum_{j=1}^{n} a_{ij} x'_j \frac{\partial}{\partial x'_j} I &= C \int\limits_{z_j \ge 0} d^{n_{\text{int}}} z \left(\prod_{j=1}^{n_{\text{int}}} z_j^{\nu_j - 1} \right) z_i \frac{\partial}{\partial z_i} \left[G\left(z, x'\right) \right]^{-\frac{D}{2}} \\ &= -\nu_i C \int\limits_{z_j \ge 0} d^{n_{\text{int}}} z \left(\prod_{j=1}^{n_{\text{int}}} z_j^{\nu_j - 1} \right) \left[G\left(z, x'\right) \right]^{-\frac{D}{2}} \\ &= -\nu_i I, \end{aligned} \qquad \text{(J.503)}$$

where we used partial integration. Therefore we have with $c = (-D/2, -\nu_1, \dots, -\nu_{n_{\text{int}}})^T$

$$\left(\mathcal{A} \boldsymbol{\theta} - c \right) I = 0. \qquad \text{(J.504)}$$

Exercise 88 *Show that for a Feynman integral as in Eq. (10.2) we have in any primary sector $c = 0$ and therefore the additional factor is absent.*

Solution: *For the Feynman integral in the Feynman parameter representation of Eq. (10.2) we have*

$$a_i + \varepsilon b_i = \nu_i - 1, \qquad 1 \le i \le n_{\text{int}} \qquad \text{(J.505)}$$

and $r = 2$. We take $P_1 = \mathcal{U}$ and $P_2 = \mathcal{F}$. We have

$$c_1 + \varepsilon d_1 = \nu - \frac{(l+1)\, D}{2}, \qquad c_2 + \varepsilon d_2 = \frac{l D}{2} - \nu. \qquad \text{(J.506)}$$

We further know that $\mathcal{U}$ is homogeneous of degree $h_1 = l$ and $\mathcal{F}$ is homogeneous of degree $h_2 = (l+1)$. Thus

$$\begin{aligned} c &= -n_{\text{int}} - \sum_{i=1}^{n_{\text{int}}} \left(a_i + \varepsilon b_i \right) - \sum_{j=1}^{2} h_j \left(c_j + \varepsilon d_j \right) \\ &= -n_{\text{int}} - \left(\nu - n_{\text{int}} \right) - l \left(\nu - \frac{(l+1)\, D}{2} \right) - (l+1) \left(\frac{l D}{2} - \nu \right) \\ &= 0. \end{aligned} \qquad \text{(J.507)}$$

Exercise 89 *Show that the convolution product is associative:*

$$(\varphi_1 * \varphi_2) * \varphi_3 = \varphi_1 * (\varphi_2 * \varphi_3)\,. \tag{J.508}$$

Solution: *Let $a \in C$. We have*

$$\begin{aligned}((\varphi_1 * \varphi_2) * \varphi_3)(a) &= \cdot\,((\varphi_1 * \varphi_2) \otimes \varphi_3)\,\Delta\,(a) \;=\; \cdot\,(\cdot\,(\varphi_1 \otimes \varphi_2)\,\Delta \otimes \varphi_3)\,\Delta\,(a)\\ &= \cdot\,(\cdot \otimes \mathrm{id})\,(\varphi_1 \otimes \varphi_2 \otimes \varphi_3)\,(\Delta \otimes \mathrm{id})\,\Delta\,(a)\\ &= \cdot\,(\mathrm{id} \otimes \cdot)\,(\varphi_1 \otimes \varphi_2 \otimes \varphi_3)\,(\mathrm{id} \otimes \Delta)\,\Delta\,(a)\\ &= \cdot\,(\varphi_1 \otimes \cdot\,(\varphi_2 \otimes \varphi_3)\,\Delta)\,\Delta\,(a) \;=\; \cdot\,(\varphi_1 \otimes (\varphi_2 * \varphi_3))\,\Delta\,(a)\\ &= (\varphi_1 * (\varphi_2 * \varphi_3))\,(a)\,.\end{aligned} \tag{J.509}$$

Exercise 90 *Show that $1_{\mathrm{Hom}} = e\bar{e} \in \mathrm{Hom}(C, A)$ is a neutral element for the convolution product, i.e.*

$$\varphi * 1_{\mathrm{Hom}} \;=\; 1_{\mathrm{Hom}} * \varphi = \varphi. \tag{J.510}$$

Solution: *Let $a \in C$ and write $\Delta(a) = a^{(1)} \otimes a^{(2)}$, using Sweedler's notation. We have*

$$(\varphi * 1_{\mathrm{Hom}})\,(a) = \cdot\;(\varphi \otimes e\bar{e})\,\Delta\,(a) \;=\; \cdot\;(\varphi \otimes e\bar{e})\left(a^{(1)} \otimes a^{(2)}\right). \tag{J.511}$$

Using the axiom of the counit in Eq. (11.14) this equals

$$(\varphi * 1_{\mathrm{Hom}})\,(a) = \cdot\;(\varphi \otimes e)\,(a \otimes 1)\,, \tag{J.512}$$

where 1 denotes the unit in R. We obtain

$$(\varphi * 1_{\mathrm{Hom}})\,(a) = \cdot\;(\varphi\,(a) \otimes e\,(1)) \;=\; \varphi\,(a)\,. \tag{J.513}$$

The proof for $1_{\mathrm{Hom}} * \varphi = \varphi$ is similar.

Exercise 91 *Show that $\varphi^{-1} = \varphi S$ is an inverse element to $\varphi \in \mathrm{AlgHom}(H, A)$.*

Solution: *We have to show that*

$$\varphi * \varphi^{-1} \;=\; \varphi^{-1} * \varphi = 1_{\mathrm{AlgHom}}. \tag{J.514}$$

We have

$$\begin{aligned}\left(\varphi * \varphi^{-1}\right)(a) &= \cdot\;(\varphi \otimes (\varphi S))\,\Delta\,(a) \;=\; \cdot\;(\varphi \otimes \varphi)\,(\mathrm{id} \otimes S)\,\Delta\,(a)\\ &= \varphi \cdot (\mathrm{id} \otimes S)\,\Delta\,(a)\,.\end{aligned} \tag{J.515}$$

In the last step we used the fact that φ is an algebra homomorphism. Using the axiom of the antipode Eq. (11.24) we have

$$\left(\varphi * \varphi^{-1}\right)(a) = \varphi e_H \bar{e}\,(a) \;=\; e_A \bar{e}\,(a) \;=\; 1_{\mathrm{AlgHom}}\,(a)\,, \tag{J.516}$$

since $\varphi(e_H) = e_A$. The proof of $\varphi^{-1} * \varphi = 1_{\text{AlgHom}}$ is similar.

Exercise 92 *Which rooted trees are primitive elements in the Hopf algebra of rooted trees?*

Solution: *For a primitive element t_{prim} the coproduct has to be*

$$\Delta\left(t_{\text{prim}}\right) = t_{\text{prim}} \otimes e + e \otimes t_{\text{prim}}. \tag{J.517}$$

For a rooted tree t, the coproduct is in general

$$\Delta(t) = t \otimes e + e \otimes t + \sum_{\text{adm.cuts } C \text{of } t} P^C(t) \otimes R^C(t), \tag{J.518}$$

thus for a primitive rooted tree the sum over all admissible cuts has to be absent. This is the case if the rooted tree consists only of a single vertex, the root.

Exercise 93 *Show that the map R in Eq. (11.85) fulfils the Rota-Baxter equation (11.84).*

Solution: *Let*

$$a_1 = \sum_{j=-L_1}^{\infty} b_j \varepsilon^j \in A, \quad a_2 = \sum_{k=-L_2}^{\infty} c_k \varepsilon^k \in A. \tag{J.519}$$

Then

$$\begin{aligned} & R\left(a_1 a_2\right) + R\left(a_1\right) R\left(a_2\right) - R\left(a_1 R\left(a_2\right)\right) - R\left(R\left(a_1\right) a_2\right) = \\ & = \sum_{j+k<0} b_j c_k \varepsilon^{j+k} + \sum_{j<0,k<0} b_j c_k \varepsilon^{j+k} - \sum_{j+k<0,k<0} b_j c_k \varepsilon^{j+k} - \sum_{j+k<0,j<0} b_j c_k \varepsilon^{j+k} \\ & = 0. \end{aligned} \tag{J.520}$$

Exercise 94 *Resolve the operator overloading: In Eq. (11.118) the symbols "$\cdot$", $\bar{e}$ and Δ appear in various places. Determine for each occurrence to which operation they correspond.*

Solution: *We start with the equation*

$$\cdot \left(\bar{e} \otimes \text{id}\right) \Delta\left(v\right) = v. \tag{J.521}$$

We read the left-hand side right-to-left. $\Delta(v)$ with $v \in M$ corresponds to the map $\Delta : M \to C \otimes M$ of Eq. (11.116). Let us write

$$\Delta\left(v\right) = a^{(1)} \otimes v^{(2)}, \qquad a^{(1)} \in C, \qquad v, v^{(2)} \in M. \tag{J.522}$$

Then

$$(\bar{e} \otimes \mathrm{id})\, \Delta(v) = \bar{e}\left(a^{(1)}\right) \otimes v^{(2)}, \tag{J.523}$$

and $\bar{e} : C \to R$ denotes the counit in C (there is no other counit). Finally, as $\bar{e}(a) \in R$, the final multiplication "$\cdot$" is the scalar multiplication of the R-module: $\cdot : R \times M \to M$.

Let us now turn to

$$(\Delta \otimes \mathrm{id})\, \Delta(v) = (\mathrm{id} \otimes \Delta)\, \Delta(v)\,. \tag{J.524}$$

Again, we read both sides right-to-left. The first operation $\Delta(v)$ is the operation $\Delta : M \to C \otimes M$ of Eq. (11.116). With Eq. (J.522) our original equation becomes

$$\Delta\left(a^{(1)}\right) \otimes v^{(2)} = a^{(1)} \otimes \Delta\left(v^{(2)}\right). \tag{J.525}$$

On the left-hand side $\Delta(a^{(1)})$ refers to the comultiplication in C, i.e. $\Delta : C \to C \otimes C$. On the right-hand side $\Delta(v^{(2)})$ refers again to the operation $\Delta : M \to C \otimes M$ of Eq. (11.116).

Exercise 95 *Work out $\Delta(\ln^{\mathfrak{m}}(x))$. Note that $\ln^{\mathfrak{m}}(x) = I^{\mathfrak{m}}(1; 0; x)$.*

Solution: *From Eq. (11.140) we have*

$$\Delta\left(I^{\mathfrak{m}}(1; 0; x)\right) = I^{\mathfrak{dR}}(1; 0; x) \otimes 1 + 1 \otimes I^{\mathfrak{m}}(1; 0; x), \tag{J.526}$$

and therefore

$$\Delta\left(\ln^{\mathfrak{m}}(x)\right) = \ln^{\mathfrak{dR}}(x) \otimes 1 + 1 \otimes \ln^{\mathfrak{m}}(x). \tag{J.527}$$

Exercise 96 *Consider*

$$I(0; x, x; y) = G(x, x; y) = G_{11}\left(1, 1; \frac{y}{x}\right) = \mathrm{Li}_{11}\left(\frac{y}{x}, 1\right) = H_{11}\left(\frac{y}{x}\right). \tag{J.528}$$

With the techniques of Chap. 8 it is not too difficult to show that the derivatives with respect to x and y are

$$\begin{aligned} \frac{\partial}{\partial x} I(0; x, x; y) &= \frac{y}{x(x-y)} \ln\left(\frac{x-y}{x}\right), \\ \frac{\partial}{\partial y} I(0; x, x; y) &= \frac{1}{y-x} \ln\left(\frac{x-y}{x}\right). \end{aligned} \tag{J.529}$$

Re-compute the derivatives using Eq. (11.156).

Solution: *We first compute the coaction*

$$\Delta\left(I^{\mathfrak{m}}(0;x,x;y)\right) = 1\otimes I^{\mathfrak{m}}(0;x,x;y) + \left(I^{\partial\mathfrak{R}}(0;x;x) + I^{\partial\mathfrak{R}}(x;x;y)\right)\otimes I^{\mathfrak{m}}(0;x;y) + I^{\partial\mathfrak{R}}(0;x,x;y)\otimes 1, \tag{J.530}$$

and therefore

$$\Delta_{1,1}\left(I^{\mathfrak{m}}(0;x,x;y)\right) = \left(I^{\partial\mathfrak{R}}(0;x;x) + I^{\partial\mathfrak{R}}(x;x;y)\right)\otimes I^{\mathfrak{m}}(0;x;y) = \left(-\ln^{\partial\mathfrak{R}}(-x) + \ln^{\partial\mathfrak{R}}(y-x)\right)\otimes \ln^{\mathfrak{m}}\left(\frac{x-y}{x}\right). \tag{J.531}$$

We then have

$$\frac{\partial}{\partial x} I^{\mathfrak{m}}(0;x,x;y) = \cdot\left(\frac{\partial}{\partial x}\otimes 1\right)\Delta_{1,1}\left(I^{\mathfrak{m}}(0;x,x;y)\right) = \cdot\left(\frac{\partial}{\partial x}\otimes 1\right)\left[\left(-\ln^{\partial\mathfrak{R}}(-x) + \ln^{\partial\mathfrak{R}}(y-x)\right)\otimes \ln^{\mathfrak{m}}\left(\frac{x-y}{x}\right)\right] = \cdot\left[\left(-\frac{1}{x} - \frac{1}{y-x}\right)\otimes \ln^{\mathfrak{m}}\left(\frac{x-y}{x}\right)\right] = \frac{y}{x(x-y)}\ln^{\mathfrak{m}}\left(\frac{x-y}{x}\right). \tag{J.532}$$

In a similar way we obtain

$$\frac{\partial}{\partial y} I^{\mathfrak{m}}(0;x,x;y) = \cdot\left(\frac{\partial}{\partial y}\otimes 1\right)\Delta_{1,1}\left(I^{\mathfrak{m}}(0;x,x;y)\right) = \cdot\left(\frac{\partial}{\partial x}\otimes 1\right)\left[\left(-\ln^{\partial\mathfrak{R}}(-x) + \ln^{\partial\mathfrak{R}}(y-x)\right)\otimes \ln^{\mathfrak{m}}\left(\frac{x-y}{x}\right)\right] = \frac{1}{y-x}\ln^{\mathfrak{m}}\left(\frac{x-y}{x}\right). \tag{J.533}$$

Exercise 97 *Work out the symbols*

$$\mathcal{S}(-\ln(x)) \text{ and } \mathcal{S}(\ln(-x)). \tag{J.534}$$

Solution: *We start with $\mathcal{S}(-\ln(x))$. We have*

$$\mathcal{S}(-\ln(x)) = -\mathcal{S}(\ln(x)) = -(x). \tag{J.535}$$

On the other hand, we have for $\mathcal{S}(\ln(-x))$

$$\mathcal{S}(\ln(-x)) = \mathcal{S}(\ln(x)) = (x). \tag{J.536}$$

Exercise 98 *Fill in the details for the derivation of* $\mathrm{sv}^{\mathfrak{m}}(\mathrm{Li}_1^{\mathfrak{m}}(x))$ *and* $\mathrm{sv}^{\mathfrak{m}}(\mathrm{Li}_2^{\mathfrak{m}}(x))$.

Solution: *We start with* $\mathrm{Li}_1^{\mathfrak{m}}(x)$. *From Eq. (11.138) we have*

$$\Delta\left(\mathrm{Li}_1^{\mathfrak{dR}}(x)\right) = \mathrm{Li}_1^{\mathfrak{dR}}(x) \otimes 1 + 1 \otimes \mathrm{Li}_1^{\mathfrak{dR}}(x). \tag{J.537}$$

Eq. (11.139) gives us

$$S\left(\mathrm{Li}_1^{\mathfrak{dR}}(x)\right) = -\mathrm{Li}_1^{\mathfrak{dR}}(x). \tag{J.538}$$

We have

$$\Delta^{\mathfrak{m}}\left(\mathrm{Li}_1^{\mathfrak{dR}}(x)\right) = \mathrm{Li}_1^{\mathfrak{m}}(x) \otimes 1 + 1 \otimes \mathrm{Li}_1^{\mathfrak{m}}(x), \tag{J.539}$$

and

$$\begin{aligned} \Sigma\left(1\right) &= 1, \\ \Sigma\left(\mathrm{Li}_1^{\mathfrak{m}}(x)\right) &= \mathrm{Li}_1^{\mathfrak{m}}(x). \end{aligned} \tag{J.540}$$

Thus

$$\mathrm{sv}^{\mathfrak{m}}\left(\mathrm{Li}_1^{\mathfrak{m}}(x)\right) = \mathrm{period}\left(\mathrm{Li}_1^{\mathfrak{m}}(x)\right) + \mathrm{period}\left(F_\infty \mathrm{Li}_1^{\mathfrak{m}}(x)\right) \;=\; \mathrm{Li}_1\left(x\right) + \mathrm{Li}_1\left(\bar{x}\right). \tag{J.541}$$

Let us now turn to $\mathrm{Li}_2^{\mathfrak{m}}(x)$. *From Eq. (11.138) we have*

$$\Delta\left(\mathrm{Li}_2^{\mathfrak{dR}}(x)\right) = \mathrm{Li}_2^{\mathfrak{dR}}(x) \otimes 1 + 1 \otimes \mathrm{Li}_2^{\mathfrak{dR}}(x) + \ln^{\mathfrak{dR}}(x) \otimes \mathrm{Li}_1^{\mathfrak{dR}}(x). \tag{J.542}$$

Eq. (11.139) gives us

$$S\left(\mathrm{Li}_2^{\mathfrak{dR}}(x)\right) = -\mathrm{Li}_2^{\mathfrak{dR}}(x) + \ln^{\mathfrak{dR}}(x)\mathrm{Li}_1^{\mathfrak{dR}}(x). \tag{J.543}$$

We work out

$$\begin{aligned} &\left(\mathrm{id} \otimes F_\infty \Sigma\right) \Delta^{\mathfrak{m}}\left(\mathrm{Li}_2^{\mathfrak{dR}}(x)\right) = \\ &\quad = \left(\mathrm{id} \otimes F_\infty \Sigma\right)\left[\mathrm{Li}_2^{\mathfrak{m}}(x) \otimes 1 + 1 \otimes \mathrm{Li}_2^{\mathfrak{m}}(x) + \ln^{\mathfrak{m}}(x) \otimes \mathrm{Li}_1^{\mathfrak{m}}(x)\right] \\ &\quad = \mathrm{Li}_2^{\mathfrak{m}}(x) \otimes 1 + 1 \otimes F_\infty\left(-\mathrm{Li}_2^{\mathfrak{m}}(x) + \ln^{\mathfrak{m}}(x)\mathrm{Li}_1^{\mathfrak{m}}(x)\right) + \ln^{\mathfrak{m}}(x) \otimes \left(F_\infty \mathrm{Li}_1^{\mathfrak{m}}(x)\right), \end{aligned} \tag{J.544}$$

and therefore

$$\begin{aligned} \mathrm{sv}^{\mathfrak{m}}\left(\mathrm{Li}_2^{\mathfrak{m}}(x)\right) &= \mathrm{Li}_2\left(x\right) - \mathrm{Li}_2\left(\bar{x}\right) + \ln\left(\bar{x}\right)\mathrm{Li}_1\left(\bar{x}\right) + \ln\left(x\right) \cdot \mathrm{Li}_1\left(\bar{x}\right) \\ &= \mathrm{Li}_2\left(x\right) - \mathrm{Li}_2\left(\bar{x}\right) + \ln\left(|x|^2\right) \cdot \mathrm{Li}_1\left(\bar{x}\right). \end{aligned} \tag{J.545}$$

Exercise 99 *Show that Eqs. (11.228) and (11.253) agree in a neighbourhood of $x = 0$.*

Solution: *We have to show that*

$$\begin{aligned} {I'_{2,a}}^{(2)}(x) &= 2\left[G\left(0, 0; x'\right) - 2G\left(-1, 0; x'\right) - \zeta_2\right], \\ {I'_{2,b}}^{(2)}(x) &= -4\,\mathrm{Li}_2\left(y_1\right) + 2\ln^2\left(y_2\right) - \ln^2\left(f_1\right) + 2\zeta_2 \end{aligned} \tag{J.546}$$

agree in a neighbourhood of $x = 0$. ${I'_{2,a}}^{(2)}(x)$ and ${I'_{2,b}}^{(2)}(x)$ are functions of a single variable x. Two functions are identical, if their derivatives are identical and the two functions have the same value at a single point.

We first check the value at $x = 0$ (corresponding to $x' = 1$ and $y_1 = y_2 = 1/2$):

$${I'_{2,a}}^{(2)}(0) = 0, \quad {I'_{2,b}}^{(2)}(0) = 0. \tag{J.547}$$

In the second step we check the derivatives. We set $r = \sqrt{x(4+x)}$. Carrying out the derivatives and collecting terms we first obtain

$$\begin{aligned} \frac{d}{dx}{I'_{2,a}}^{(2)}(x) &= 2\frac{x-r}{r\left(4+x-r\right)}\ln\left(\frac{2+x-r}{2}\right), \\ \frac{d}{dx}{I'_{2,b}}^{(2)}(x) &= -\frac{2}{r}\ln\left(\frac{y_2}{1-y_1}\right) - \frac{2}{4+x}\ln\left(\left(1-y_1\right)y_2\left(4+x\right)\right). \end{aligned} \tag{J.548}$$

We then simplify these expressions. We first make the denominator of $d{I'_{2,a}}^{(2)}/dx$ rational. We multiply the numerator and the denominator with $(4 + x + r)$. Noting that

$$\begin{aligned} \left(4+x-r\right)\left(4+x+r\right) &= 4\left(4+x\right), \\ \left(x-r\right)\left(4+x+r\right) &= -4r \end{aligned} \tag{J.549}$$

we obtain

$$\frac{d}{dx}{I'_{2,a}}^{(2)}(x) = -\frac{2}{(4+x)}\ln\left(\frac{2+x-r}{2}\right). \tag{J.550}$$

In order to simplify $d{I'_{2,b}}^{(2)}/dx$ we first notice that

$$y_2 = 1 - y_1. \tag{J.551}$$

Thus

$$\frac{d}{dx}{I'_{2,b}}^{(2)}(x) = -\frac{2}{4+x}\ln\left(y_2^2\left(4+x\right)\right) = -\frac{2}{4+x}\ln\left(\frac{2}{2+x+r}\right) = -\frac{2}{4+x}\ln\left(\frac{2+x-r}{2}\right). \tag{J.552}$$

Exercise 100 *Let f_1, f_2, g_1, g_2 be algebraic functions of the kinematic variables x. Determine the symbols of*

$$\text{Li}_{21}(f_1, f_2) \text{ and } G_{21}(g_1, g_2; 1). \tag{J.553}$$

Assume then $g_1 = 1/f_1$ and $g_2 = 1/(f_1 f_2)$. Show that in this case the two symbols agree.

From the two symbols deduce the constraints on the arguments f_1, f_2 of $\text{Li}_{21}(f_1, f_2)$ and on the arguments g_1, g_2 of $G_{21}(g_1, g_2; 1)$.

Solution: *We start with $G_{21}(g_1, g_2; 1)$. According to Eq. (8.8) we have*

$$\begin{aligned} dG_{21}(g_1, g_2; 1) &= dG(0, g_1, g_2; 1) = -G(g_1, g_2; 1)\, d\ln(g_1) + G(0, g_2; 1)\, d\ln\left(\frac{g_1}{g_2 - g_1}\right) \\ &\quad + G(0, g_1; 1)\, d\ln\left(\frac{g_2 - g_1}{g_2}\right), \end{aligned} \tag{J.554}$$

and therefore

$$\begin{aligned} &S(G_{21}(g_1, g_2; 1)) = \\ &- (g_1 \otimes S(G(g_1, g_2; 1))) - \left(\frac{(g_2 - g_1)}{g_1} \otimes S(G(0, g_2; 1))\right) + \left(\frac{(g_2 - g_1)}{g_2} \otimes S(G(0, g_1; 1))\right). \end{aligned} \tag{J.555}$$

The symbols of the functions of weight two are

$$\begin{aligned} S(G(g_1, g_2; 1)) &= \left(\frac{(1 - g_1)}{(g_2 - g_1)} \otimes \frac{(1 - g_2)}{g_2}\right) + \left(\frac{(g_2 - g_1)}{g_2} \otimes \frac{(1 - g_1)}{g_1}\right), \\ S(G(0, g_2; 1)) &= -\left(g_2 \otimes \frac{(1 - g_2)}{g_2}\right), \\ S(G(0, g_1; 1)) &= -\left(g_1 \otimes \frac{(1 - g_1)}{g_1}\right). \end{aligned} \tag{J.556}$$

Putting everything together we obtain

$$\begin{aligned} S(G_{21}(g_1, g_2; 1)) &= \left(g_1 \otimes \frac{(g_2 - g_1)}{(1 - g_1)} \otimes \frac{(1 - g_2)}{g_2}\right) + \left(\frac{(g_2 - g_1)}{g_1} \otimes g_2 \otimes \frac{(1 - g_2)}{g_2}\right) \\ &\quad - \left(g_1 \otimes \frac{(g_2 - g_1)}{g_2} \otimes \frac{(1 - g_1)}{g_1}\right) - \left(\frac{(g_2 - g_1)}{g_2} \otimes g_1 \otimes \frac{(1 - g_1)}{g_1}\right) \end{aligned} \tag{J.557}$$

Let us now turn to $\text{Li}_{21}(f_1, f_2)$. According to Eq. (8.35) we have

$$d\text{Li}_{21}(f_1, f_2) = \text{Li}_{11}(f_1, f_2)\, d\ln(f_1) + \text{Li}_{20}(f_1, f_2)\, d\ln(f_2). \tag{J.558}$$

We may rewrite Li_{20} with the help of Eq. (8.38) as

$$\text{Li}_{20}\left(f_1, f_2\right) = \text{Li}_0\left(f_2\right)\text{Li}_2\left(f_1\right) - \text{Li}_2\left(f_1 f_2\right) - \text{Li}_0\left(f_2\right)\text{Li}_2\left(f_1 f_2\right). \quad (J.559)$$

This generates terms with Li_0. *We haven't defined the symbol of a weight zero function. Let's work out the prescription: We consider* $\text{Li}_1(x)$. *We know its symbol from Eq. (11.175):*

$$S\left(\text{Li}_1(x)\right) = S\left(-\ln(1-x)\right) = -(1-x). \quad (J.560)$$

On the other hand, Eq. (8.35) gives us

$$d\text{Li}_1(x) = \text{Li}_0(x)\, d\ln(x). \quad (J.561)$$

Since

$$\text{Li}_0(x) = \frac{x}{1-x} \quad (J.562)$$

we have

$$\text{Li}_0(x)\, d\ln(x) = \frac{x}{1-x} \cdot \frac{dx}{x} = \frac{dx}{1-x} = -d\ln(1-x). \quad (J.563)$$

Hence we have

$$\begin{aligned} d\text{Li}_1(x) &= -d\ln(1-x), \\ S\left(\text{Li}_1(x)\right) &= -(1-x). \end{aligned} \quad (J.564)$$

Thus we combine any Li_0 *function with the accompanying dlog-form. This gives*

$$\begin{aligned} S\left(\text{Li}_{21}\left(f_1, f_2\right)\right) = &\left(f_1 \otimes S\left(\text{Li}_{11}\left(f_1, f_2\right)\right)\right) - \left(f_2 \otimes S\left(\text{Li}_2\left(f_1 f_2\right)\right)\right) \\ &+ \left(\left(1-f_2\right) \otimes S\left(\text{Li}_2\left(f_1 f_2\right) - \text{Li}_2\left(f_1\right)\right)\right). \end{aligned} \quad (J.565)$$

With

$$S\left(\text{Li}_{11}\left(f_1, f_2\right)\right) = \left(\left(1-f_2\right) \otimes \left(1-f_1\right)\right) + \left(\frac{\left(1-f_1\right) f_2}{\left(1-f_2\right)} \otimes \left(1 - f_1 f_2\right)\right) \quad (J.566)$$

we arrive at

$$\begin{aligned} S\left(\text{Li}_{21}\left(f_1, f_2\right)\right) = &\left(f_1 \otimes \left(1-f_2\right) \otimes \left(1-f_1\right)\right) + \left(\left(1-f_2\right) \otimes f_1 \otimes \left(1-f_1\right)\right) \\ &+ \left(f_1 \otimes \frac{\left(1-f_1\right) f_2}{\left(1-f_2\right)} \otimes \left(1-f_1 f_2\right)\right) + \left(\frac{f_2}{1-f_2} \otimes f_1 f_2 \otimes \left(1-f_1 f_2\right)\right). \end{aligned} \quad (J.567)$$

Now let us substitute $g_1 = 1/f_1$ and $g_2 = 1/(f_1 f_2)$ in Eq. (J.557):

$$S\left(G_{21}\left(\frac{1}{f_1}, \frac{1}{f_1 f_2}; 1\right)\right) = \left(f_1 \otimes \frac{(1-f_1) f_2}{(1-f_2)} \otimes (1 - f_1 f_2)\right) - \left(\frac{1-f_2}{f_2} \otimes f_1 f_2 \otimes (1 - f_1 f_2)\right) + (f_1 \otimes (1-f_2) \otimes (1-f_1)) + ((1-f_2) \otimes f_1 \otimes (1-f_1)). \quad \text{(J.568)}$$

This agrees with Eq. (J.567).

Let's assume that f_1 and f_2 are power products of the letters of the alphabet. From the symbol of $\text{Li}_{21}(f_1, f_2)$ we deduce that then

$$1 - f_1, \quad 1 - f_2, \quad 1 - f_1 f_2 \quad \text{(J.569)}$$

should also be power products of the letters of the alphabet.

Assuming that g_1 and g_2 are power products of the letters of the alphabet, we deduce from the symbol of $G_{21}(g_1, g_2; 1)$ that

$$1 - g_1, \quad 1 - g_2, \quad g_2 - g_1 \quad \text{(J.570)}$$

should also be power products of the letters of the alphabet.

Exercise 101 *Show that the mutation of the matrix B at a fixed vertex v_k is an involution, i.e. mutating twice at the same vertex returns the original matrix B.*

Solution: *We mutate the matrix B at the vertex v_k twice. We denote the matrix after the first mutation by B', the one after the second mutation by B''. For $i = k$ or $j = k$ we have*

$$b''_{ij} = -b'_{ij} = b_{ij}. \quad \text{(J.571)}$$

For $i \neq k$ and $j \neq k$ we have

$$\begin{aligned} b''_{ij} &= b'_{ij} + \text{sign}\left(b'_{ik}\right) \cdot \max\left(0, b'_{ik} b'_{kj}\right) \\ &= b_{ij} + \text{sign}\left(b_{ik}\right) \cdot \max\left(0, b_{ik} b_{kj}\right) + \text{sign}\left(-b_{ik}\right) \cdot \max\left(0, b_{ik} b_{kj}\right) = b_{ij}. \end{aligned} \quad \text{(J.572)}$$

Exercise 102 *Derive the transformation in Eq. (12.9) from Eqs. (12.8), (12.7) and (12.6).*

Solution: *We start with the case $j = k$: We have*

$$x'_k = \prod_i \left(a'_i\right)^{b'_{ik}} = \prod_i \left(a_i\right)^{-b_{ik}} = \frac{1}{x_k}. \quad \text{(J.573)}$$

The case $j \neq k$ requires more work:

$$
\begin{aligned}
x'_j &= \prod_i (a'_i)^{b'_{ij}} = (a'_k)^{b'_{kj}} \prod_{i \neq k} (a'_i)^{b'_{ij}} \\
&= a_k^{b_{kj}} \left(\prod_{i \mid b_{ik}>0} a_i^{b_{ik}} + \prod_{i \mid b_{ik}<0} a_i^{-b_{ik}} \right)^{-b_{kj}} \prod_{i \neq k} a_i^{b_{ij}+\operatorname{sign}(b_{ik})\cdot\max(0,b_{ik}b_{kj})} \\
&= x_j \left(\prod_{i \mid b_{ik}>0} a_i^{b_{ik}} + \prod_{i \mid b_{ik}<0} a_i^{-b_{ik}} \right)^{-b_{kj}} \prod_{i \neq k} a_i^{\operatorname{sign}(b_{ij})\cdot\max(0,b_{ij}b_{kj})} \\
&= x_j \left(\prod_i a_i^{b_{ik}} + 1 \right)^{-b_{kj}} \left(\prod_{i \mid b_{ik}<0} a_i^{b_{ik}b_{kj}} \right) \left(\prod_{i \neq k} a_i^{\operatorname{sign}(b_{ik})\cdot\max(0,b_{ik}b_{kj})} \right) \\
&= x_j \left(1 + x_k\right)^{-b_{kj}} \left(\prod_{i \mid b_{ik}<0} a_i^{b_{ik}b_{kj}} \right) \left(\prod_{i \neq k} a_i^{\operatorname{sign}(b_{ik})\cdot\max(0,b_{ik}b_{kj})} \right). \qquad \text{(J.574)}
\end{aligned}
$$

We now distinguish the cases $b_{kj} > 0$ and $b_{kj} < 0$. For $b_{kj} > 0$ we have

$$
\left(\prod_{i \mid b_{ik}<0} a_i^{b_{ik}b_{kj}} \right) \left(\prod_{i \neq k} a_i^{\operatorname{sign}(b_{ik})\cdot\max(0,b_{ik}b_{kj})} \right) = \prod_i a_i^{b_{ik}b_{kj}} = x_k^{b_{kj}}, \qquad \text{(J.575)}
$$

while for $b_{kj} < 0$ we have

$$
\left(\prod_{i \mid b_{ik}<0} a_i^{b_{ik}b_{kj}} \right) \left(\prod_{i \neq k} a_i^{\operatorname{sign}(b_{ik})\cdot\max(0,b_{ik}b_{kj})} \right) = 1. \qquad \text{(J.576)}
$$

We therefore obtain

$$
x'_j = \begin{cases} x_j \left(\frac{x_k}{1+x_k} \right)^{b_{kj}} = x_j \left(1 + \frac{1}{x_k} \right)^{-b_{kj}}, & b_{kj} > 0, \\ x_j \left(1 + x_k\right)^{-b_{kj}}, & b_{kj} < 0. \end{cases} \qquad \text{(J.577)}
$$

Combining the two cases into one formula we find

$$
x'_j = x_j \left(1 + x_k^{-\operatorname{sign}(b_{kj})} \right)^{-b_{kj}}. \qquad \text{(J.578)}
$$

Note that we never used the anti-symmetry of b_{ij}.

Exercise 103 *Determine the cluster A-variables for the ice quiver Q' of Fig. 12.3 in terms of the cluster variables of the ice quiver Q.*

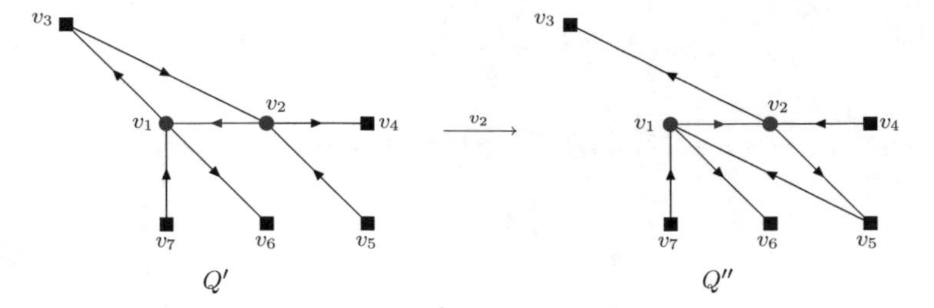

Fig. J.8 The mutation of the ice quiver Q' at the vertex v_2 yields the ice quiver Q''

Solution: *We have*

$$a'_1 = \frac{a_3a_6 + a_2a_7}{a_1},$$
$$a'_2 = a_2. \tag{J.579}$$

The frozen A-variables are not changed.

Exercise 104 *Mutate the ice quiver Q' of Fig. 12.3 at the vertex v_2 to obtain an ice quiver Q''. Determine the cluster A-variables for the ice quiver Q'' in terms of the cluster variables of the ice quiver Q.*

Solution: *The mutated quiver Q'' is shown in Fig. J.8. For the cluster A-variables we have*

$$\begin{aligned} a''_1 = a'_1 &= \frac{a_3a_6 + a_2a_7}{a_1}, \\ a''_2 = \frac{a'_3a'_5 + a'_1a'_4}{a'_2} &= \frac{a_1a_3a_5 + a_2a_4a_7 + a_3a_4a_6}{a_1a_2} \end{aligned} \tag{J.580}$$

Exercise 105 *The B_2-cluster algebra: Determine the cluster variables from the initial seed*

$$B = \begin{pmatrix} 0 & -1 \\ 2 & 0 \end{pmatrix}, \quad a = (a_1, a_2). \tag{J.581}$$

Solution: *We start from the seed (B, a). As two mutations on the same vertex will give us back the original seed, we alternate the vertices where we perform mutations. The exchange matrix B changes sign under each mutation. We start with a mutation at v_1. This yields*

$$(a'_1, a'_2) = \left(\frac{1 + a_2^2}{a_1}, a_2\right). \tag{J.582}$$

We set $a_3 = (1 + a_2^2)/a_1$. We then mutate at vertex v_2. This yields

$$\left(a_1'', a_2''\right) = \left(\frac{1+a_2^2}{a_1}, \frac{1+a_1+a_2^2}{a_1 a_2}\right). \tag{J.583}$$

We set $a_4 = (1 + a_1 + a_2^2)/(a_1 a_2)$. *Continuing in this way we obtain*

$$a_{n+1} = \begin{cases} \frac{1+a_n^2}{a_{n-1}}, & \text{if } n \text{ is even}, \\ \frac{1+a_n}{a_{n-1}}, & \text{if } n \text{ is odd}. \end{cases} \tag{J.584}$$

This will give a sequence with period 6*:*

$$a_{n+6} = a_n. \tag{J.585}$$

The first six terms are the cluster variables:

$$a_1, \quad a_2, \quad \frac{1+a_2^2}{a_1}, \quad \frac{1+a_1+a_2^2}{a_1 a_2}, \quad \frac{1+2a_1+a_1^2+a_2^2}{a_1 a_2^2}, \quad \frac{1+a_1}{a_2}. \tag{J.586}$$

Exercise 106 *Consider the elliptic curve* $y^2 = 4x^3 - g_2 x - g_3$ *. Show that*

$$dz = \frac{dx}{y}, \tag{J.587}$$

where $y = \sqrt{4x^3 - g_2 x - g_3}$. *This shows that* dx/y *is a holomorphic differential.*

Solution: *The variables* z *and* x *are related by Eq. (13.16):*

$$z = \int\limits_{\infty}^{x} \frac{dt}{\sqrt{4t^3 - g_2 t - g_3}}. \tag{J.588}$$

We therefore have

$$dz = \frac{dx}{\sqrt{4x^3 - g_2 x - g_3}} = \frac{dx}{y}. \tag{J.589}$$

Exercise 107 *Determine two independent periods for the elliptic curve defined by a quartic polynomial:*

$$y^2 = (x - x_1)(x - x_2)(x - x_3)(x - x_4). \tag{J.590}$$

Solution: *We would like to express* y *as the square root of the right hand side of Eq. (J.590). We denote by* $[x_i, x_j]$ *the line segment from* x_i *to* x_j *in the complex plane. We may express* y *as a single-valued and continuous function on* $\mathbb{C} \backslash ([x_l, x_i] \cup [x_j, x_k])$ *through*

$$y = \pm (x_i - x_l)\left(x_k - x_j\right)\sqrt{\frac{x - x_l}{x_i - x_l}}\sqrt{\frac{x - x_i}{x_i - x_l}}\sqrt{\frac{x - x_j}{x_k - x_j}}\sqrt{\frac{x - x_k}{x_k - x_j}}. \quad \text{(J.591)}$$

For a given choice (i, j, k, l) of branch cuts $[x_l, x_i]$ and $[x_j, x_k]$ the transformation

$$T(x) = \frac{(x_k - x_l)}{(x_k - x_i)}\frac{(x - x_i)}{(x - x_l)} \quad \text{(J.592)}$$

maps the points x_i, x_k, x_l to $0, 1, \infty$, respectively. The point x_j is then mapped to

$$\lambda = \frac{(x_k - x_l)}{(x_k - x_i)}\frac{\left(x_j - x_i\right)}{\left(x_j - x_l\right)}. \quad \text{(J.593)}$$

We denote the cross-ratio by

$$[j, k|i, l] = \frac{(x_k - x_l)}{(x_k - x_i)}\frac{\left(x_j - x_i\right)}{\left(x_j - x_l\right)}. \quad \text{(J.594)}$$

The cross-ratios satisfy

$$\begin{aligned} [i, j|k, l] &= [k, l|i, j], \\ [i, j|k, l] &= [j, i|l, k], \\ [i, j|k, l] &= [i, j|l, k]^{-1}, \\ [i, j|k, l] + [i, k|j, l] &= 1. \end{aligned} \quad \text{(J.595)}$$

Let δ_1 and δ_2 be two independent cycles as shown in Fig. J.9. The cycles δ_1 and δ_2 have intersection number $+1$. We define the periods by

$$\begin{aligned} \psi_1 = 2\int\limits_{x_i}^{x_j} \frac{dx}{y} &= \frac{4}{\sqrt{\left(x_j - x_l\right)(x_k - x_i)}} K\left(\sqrt{\frac{\left(x_j - x_i\right)(x_k - x_l)}{\left(x_j - x_l\right)(x_k - x_i)}}\right), \\ \psi_2 = 2\int\limits_{x_k}^{x_j} \frac{dx}{y} &= \frac{4i}{\sqrt{\left(x_j - x_l\right)(x_k - x_i)}} K\left(\sqrt{\frac{(x_i - x_l)\left(x_k - x_j\right)}{\left(x_j - x_l\right)(x_k - x_i)}}\right). \end{aligned} \quad \text{(J.596)}$$

For ψ_2 the square root y is evaluated in the complex x-plane below the cut. Let us now discuss the possibilities of choosing x_i, x_j, x_k, x_l: Due to the symmetries $[i, j|k, l] = [k, l|i, j]$ and $[i, j|k, l] = [j, i|l, k]$ we may fix $x_l = x_4$. This leaves six possibilities for λ. These are

$$(i, j, k, l) = (2, 3, 1, 4): \ [3, 1|2, 4] = \lambda = \frac{(x_1 - x_4)}{(x_1 - x_2)}\frac{(x_3 - x_2)}{(x_3 - x_4)},$$

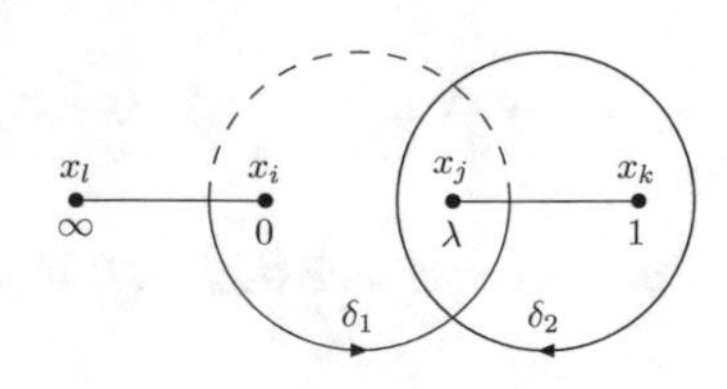

Fig. J.9 Branch cuts and cycles for the computation of the periods of an elliptic curve

$$
\begin{aligned}
(i,j,k,l) = (2,1,3,4): \ [1,3|2,4] &= \frac{1}{\lambda}, \\
(i,j,k,l) = (3,1,2,4): \ [1,2|3,4] &= \frac{\lambda-1}{\lambda}, \\
(i,j,k,l) = (3,2,1,4): \ [2,1|3,4] &= \frac{\lambda}{\lambda-1}, \\
(i,j,k,l) = (1,2,3,4): \ [2,3|1,4] &= \frac{1}{1-\lambda}, \\
(i,j,k,l) = (1,3,2,4): \ [3,2|1,4] &= 1-\lambda.
\end{aligned} \tag{J.597}
$$

It is worth noting that we have three possibilities for $\lambda(1-\lambda)$. These are

$$
\begin{aligned}
[3,1|2,4] \cdot [3,2|1,4] &= \lambda\,(1-\lambda)\,, \\
[1,2|3,4] \cdot [1,3|2,4] &= -\frac{(1-\lambda)}{\lambda^2}, \\
[2,3|1,4] \cdot [2,1|3,4] &= -\frac{\lambda}{(1-\lambda)^2}.
\end{aligned} \tag{J.598}
$$

Exercise 108 *Express the modulus squared k^2 and the complementary modulus squared k'^2 as a quotient of eta functions.*

Solution: *From Eqs. (13.74) and (13.75) we have*

$$
k^2 = \frac{\theta_2^4\,(0,q)}{\theta_3^4\,(0,q)} = 16\frac{\eta\left(\frac{\tau}{2}\right)^8 \eta\,(2\tau)^{16}}{\eta\,(\tau)^{24}} \tag{J.599}
$$

and

$$
k'^2 = \frac{\theta_4^4\,(0,q)}{\theta_3^4\,(0,q)} = \frac{\eta\left(\frac{\tau}{2}\right)^{16} \eta\,(2\tau)^{8}}{\eta\,(\tau)^{24}}. \tag{J.600}
$$

Exercise 109 *Show that*

$$
(f\,|_k\gamma_1)\,|_k\gamma_2 = f\,|_k\,(\gamma_1\gamma_2)\,. \tag{J.601}
$$

Solution: *Let*

$$
\gamma_1 = \begin{pmatrix} a_1 & b_1 \\ c_1 & d_1 \end{pmatrix}, \quad \gamma_2 = \begin{pmatrix} a_2 & b_2 \\ c_2 & d_2 \end{pmatrix}, \quad \gamma_{12} = \begin{pmatrix} a_{12} & b_{12} \\ c_{12} & d_{12} \end{pmatrix} \tag{J.602}
$$

with $\gamma_{12} = \gamma_1 \cdot \gamma_2$. From matrix multiplication we have

$$c_{12} = c_1 a_2 + d_1 c_2, \quad d_{12} = c_1 b_2 + d_1 d_2. \tag{J.603}$$

On the one hand we have

$$\begin{aligned}(f|_k\gamma_1)|_k\gamma_2 &= \left((c_1\tau + d_1)^{-k} f(\gamma_1(\tau))\right)|_k\gamma_2 \\ &= (c_2\tau + d_2)^{-k}(c_1\gamma_2(\tau) + d_1)^{-k} f(\gamma_1(\gamma_2(\tau))) \\ &= ((c_1a_2 + d_1c_2)\tau + c_1b_2 + d_1d_2)^{-k} f(\gamma_1(\gamma_2(\tau))) \\ &= (c_{12}\tau + d_{12})^{-k} f(\gamma_{12}(\tau)).\end{aligned} \tag{J.604}$$

On the other hand we have

$$f|_k(\gamma_1\gamma_2) = f|_k\gamma_{12} = (c_{12}\tau + d_{12})^{-k} \cdot f(\gamma_{12}(\tau)). \tag{J.605}$$

Exercise 110 *Let $f \in \mathcal{M}_k(\Gamma(N))$ and $\gamma \in \mathrm{SL}_2(\mathbb{Z})\backslash\Gamma(N)$. Show that $f|_k\gamma \in \mathcal{M}_k(\Gamma(N))$.*

Solution: *Let $\gamma_1 \in \Gamma(N)$. We have to show*

$$(f|_k\gamma)|_k\gamma_1 = f|_k\gamma. \tag{J.606}$$

As $\Gamma(N)$ is a normal subgroup of $\mathrm{SL}_2(\mathbb{Z})$ there exist for $\gamma \in \mathrm{SL}_2(\mathbb{Z})$ and $\gamma_1 \in \Gamma(N)$ a $\gamma_2 \in \Gamma(N)$ such that

$$\gamma\gamma_1 = \gamma_2\gamma. \tag{J.607}$$

As $\gamma_2 \in \Gamma(N)$ we have

$$f|_k\gamma_2 = f. \tag{J.608}$$

We have with $(f|_k\gamma_1)|_k\gamma_2 = f|_k(\gamma_1\gamma_2)$

$$(f|_k\gamma)|_k\gamma_1 = (f|_k(\gamma\gamma_1)) = (f|_k(\gamma_2\gamma)) = (f|_k\gamma_2)|_k\gamma = f|_k\gamma. \tag{J.609}$$

Exercise 111 *Let χ be a Dirichlet character with modulus N and $f \in \mathcal{M}_k(N, \chi)$. Let further $\gamma_1, \gamma_2 \in \Gamma_0(N)$ and set $\gamma_{12} = \gamma_1\gamma_2$. Show that*

$$f(\gamma_1(\gamma_2(\tau))) = f(\gamma_{12}(\tau)). \tag{J.610}$$

Solution: *Let*

$$\gamma_1 = \begin{pmatrix} a_1 & b_1 \\ c_1 & d_1 \end{pmatrix}, \quad \gamma_2 = \begin{pmatrix} a_2 & b_2 \\ c_2 & d_2 \end{pmatrix}, \quad \gamma_{12} = \begin{pmatrix} a_{12} & b_{12} \\ c_{12} & d_{12} \end{pmatrix}. \tag{J.611}$$

Since $\gamma_{12} = \gamma_1 \cdot \gamma_2$ we have

$$c_{12} = c_1 a_2 + d_1 c_2, \quad d_{12} = c_1 b_2 + d_1 d_2. \tag{J.612}$$

Let us set $\tau' = \gamma_2(\tau)$. We have

$$f\left(\gamma_1\left(\gamma_2\left(\tau\right)\right)\right) = \chi(d_1)\left(c_1\tau' + d_1\right)^k f\left(\tau'\right) = \chi(d_1)\chi(d_2)\left(c_1\tau' + d_1\right)^k \left(c_2\tau + d_2\right)^k f\left(\tau\right)$$
$$= \chi(d_1)\chi(d_2)\left(c_{12}\tau + d_{12}\right)^k f\left(\tau\right). \tag{J.613}$$

On the other hand

$$f\left(\gamma_{12}\left(\tau\right)\right) = \chi\left(d_{12}\right)\left(c_{12}\tau + d_{12}\right)^k f\left(\tau\right). \tag{J.614}$$

Thus we see that $f(\gamma_{12}(\tau)) = f(\gamma_1(\gamma_2(\tau)))$ requires $\chi(d_{12}) = \chi(d_1)\chi(d_2)$. Since $\gamma_1 \in \Gamma_0(N)$ we have $c_1 = 0 \mod N$ and therefore

$$\chi\left(d_{12}\right) = \chi\left(c_1 b_2 + d_1 d_2\right) = \chi\left(d_1 d_2\right) = \chi\left(d_1\right)\chi\left(d_2\right). \tag{J.615}$$

Exercise 112 *Consider*

$$f\left(\tau\right) = e_2\left(\tau\right) - 2e_2\left(2\tau\right) \tag{J.616}$$

and work out the transformation properties under $\gamma \in \Gamma_0(2)$.

Solution: *Let*

$$\gamma = \begin{pmatrix} a & b \\ c & d \end{pmatrix} = \Gamma_0(2). \tag{J.617}$$

This implies that c is even and hence

$$\begin{pmatrix} a & 2b \\ \frac{c}{2} & d \end{pmatrix} \in \mathrm{SL}_2(\mathbb{Z}). \tag{J.618}$$

We have

$$\tau' = \frac{a\tau + b}{c\tau + d} \text{ and } 2\tau' = \frac{a\left(2\tau\right) + 2b}{\frac{c}{2}\left(2\tau\right) + d}. \tag{J.619}$$

Hence

$$f\left(\tau'\right) = e_2\left(\tau'\right) - 2e_2\left(2\tau'\right)$$
$$= \left(c\tau + d\right)^2 e_2\left(\tau\right) - 2\pi i c\left(c\tau + d\right) - 2\left[\left(\frac{c}{2}\left(2\tau\right) + d\right)^2 e_2\left(2\tau\right) - 2\pi i \frac{c}{2}\left(\frac{c}{2}\left(2\tau\right) + d\right)\right]$$
$$= \left(c\tau + d\right)^2 \left[e_2\left(\tau\right) - 2e_2\left(2\tau\right)\right] = \left(c\tau + d\right)^2 f\left(\tau\right). \tag{J.620}$$

This shows that $f(\tau)$ is a modular form of weight 2 for $\Gamma_0(2)$.

Exercise 113 *Show Eq. (13.156).*

Solution: *Let us introduce the short-hand notation*

$$\sum_{n_1 \in \mathbb{Z}} f(n_1) = \lim_{N_1 \to \infty} \sum_{n_1=-N_1}^{N_1} f(n_1). \tag{J.621}$$

Then we may write Eisenstein's summation prescription as

$$\begin{aligned}\sum_{(n_1,n_2)\in\mathbb{Z}^2}{}_e\, f(z+n_1+n_2\tau) &= \lim_{N_2\to\infty} \sum_{n_2=-N_2}^{N_2} \left(\lim_{N_1\to\infty} \sum_{n_1=-N_1}^{N_1} f(z+n_1+n_2\tau) \right) \\ &= \sum_{n_2\in\mathbb{Z}} \left(\sum_{n_1\in\mathbb{Z}} f(z+n_1+n_2\tau) \right). \end{aligned} \tag{J.622}$$

Let's now turn to the problem at hand. We split the outer n_2-summation into $n_2 = 0$ and $n_2 \neq 0$ and obtain

$$\begin{aligned}\sum_{(n_1,n_2)\in\mathbb{Z}^2\setminus(0,0)}{}_e\, \frac{e^{\frac{2\pi i}{N}(n_1 s - n_2 r)}}{(n_1+n_2\tau)^k} &= \mathrm{Li}_k\left(e^{2\pi i \frac{s}{N}}\right) + (-1)^k\, \mathrm{Li}_k\left(e^{-2\pi i \frac{s}{N}}\right) \\ &+ \sum_{n_2=1}^{\infty} \sum_{n_1\in\mathbb{Z}} \frac{e^{\frac{2\pi i}{N}(n_1 s - n_2 r)} + (-1)^k e^{-\frac{2\pi i}{N}(n_1 s - n_2 r)}}{(n_1+n_2\tau)^k}. \end{aligned} \tag{J.623}$$

The essential trick is the following identity

$$\sum_{n_1\in\mathbb{Z}} \frac{1}{n_1+\tau} = \pi \cot(\pi\tau) = -2\pi i \left[\frac{1}{2} + \sum_{n=1}^{\infty} \bar{q}^n \right], \qquad \bar{q} = e^{2\pi i \tau}. \tag{J.624}$$

Taking $(k-1)$-times the derivative with respect to τ gives for $k \geq 2$

$$\sum_{n_1\in\mathbb{Z}} \frac{1}{(n_1+\tau)^k} = \frac{(-2\pi i)^k}{(k-1)!} \sum_{n=1}^{\infty} n^{k-1} \bar{q}^n. \tag{J.625}$$

Let us first consider the case $k \geq 2$. We apply Eq. (J.625) to

$$\sum_{n_1\in\mathbb{Z}} \frac{\left(e^{2\pi i \frac{s}{N}}\right)^{n_1}}{(n_1+n_2\tau)^k}. \tag{J.626}$$

For $k \geq 2$ the sum is absolutely convergent and we may reorder the terms. For $N \in \mathbb{N}$, $s \in \mathbb{N}_0$ the numerator is periodic with period N. We therefore have

$$\sum_{n_1\in\mathbb{Z}} \frac{\left(e^{2\pi i \frac{s}{N}}\right)^{n_1}}{(n_1+n_2\tau)^k} = \sum_{c_1=0}^{N-1}\sum_{n_1'\in\mathbb{Z}} \frac{\left(e^{2\pi i \frac{s}{N}}\right)^{n_1'N+c_1}}{\left(n_1'N+c_1+n_2\tau\right)^k}$$
$$= \frac{1}{N^k}\sum_{c_1=0}^{N-1} e^{2\pi i \frac{sc_1}{N}} \sum_{n_1'\in\mathbb{Z}} \frac{1}{\left(n_1'+\frac{c_1}{N}+\frac{n_2\tau}{N}\right)^k}$$
$$= \frac{(-2\pi i)^k}{(k-1)!N^k}\sum_{c_1=0}^{N-1}\sum_{d=1}^{\infty} d^{k-1} e^{2\pi i\left(\frac{sc_1}{N}+\frac{c_1 d}{N}+\frac{n_2 d}{N}\tau\right)}. \tag{J.627}$$

Thus

$$\frac{1}{2}\frac{(k-1)!}{(2\pi i)^k}\sum\nolimits^{e}_{(n_1,n_2)\in\mathbb{Z}^2\setminus(0,0)} \frac{e^{\frac{2\pi i}{N}(n_1 s - n_2 r)}}{(n_1+n_2\tau)^k} = \frac{1}{2}\frac{(k-1)!}{(2\pi i)^k}\left(\mathrm{Li}_k\left(e^{2\pi i\frac{s}{N}}\right)+(-1)^k\,\mathrm{Li}_k\left(e^{-2\pi i\frac{s}{N}}\right)\right)$$
$$+\frac{1}{2}\frac{(-1)^k}{N^k}\sum_{n_2=1}^{\infty}\sum_{c_1=0}^{N-1}\sum_{d=1}^{\infty} d^{k-1}\left[e^{2\pi i\left(\frac{sc_1}{N}-\frac{n_2 r}{N}+\frac{c_1 d}{N}+\frac{n_2 d}{N}\tau\right)}+(-1)^k e^{2\pi i\left(-\frac{sc_1}{N}+\frac{n_2 r}{N}+\frac{c_1 d}{N}+\frac{n_2 d}{N}\tau\right)}\right]. \tag{J.628}$$

The first term on the right-hand side yields with Eq. (8.33)

$$a_0 = \frac{1}{2}\frac{(k-1)!}{(2\pi i)^k}\left(\mathrm{Li}_k\left(e^{2\pi i\frac{s}{N}}\right)+(-1)^k\,\mathrm{Li}_k\left(e^{-2\pi i\frac{s}{N}}\right)\right) = -\frac{1}{2k}B_k\left(\frac{s}{N}\right). \tag{J.629}$$

The second term on the right-hand side of Eq. (J.628) we may rearrange as follows:

$$\frac{1}{2}\frac{(-1)^k}{N^k}\sum_{n_2=1}^{\infty}\sum_{c_1=0}^{N-1}\sum_{d=1}^{\infty} d^{k-1}\left[e^{2\pi i\left(\frac{sc_1}{N}-\frac{n_2 r}{N}+\frac{c_1 d}{N}+\frac{n_2 d}{N}\tau\right)}+(-1)^k e^{2\pi i\left(-\frac{sc_1}{N}+\frac{n_2 r}{N}+\frac{c_1 d}{N}+\frac{n_2 d}{N}\tau\right)}\right]$$
$$= \frac{(-1)^k}{2N^k}\sum_{n=1}^{\infty}\sum_{c_1=0}^{N-1}\sum_{d|n} d^{k-1}\left[e^{2\pi i\left(\frac{sc_1}{N}-\frac{nr}{dN}+\frac{c_1 d}{N}\right)}+(-1)^k e^{2\pi i\left(-\frac{sc_1}{N}+\frac{nr}{dN}+\frac{c_1 d}{N}\right)}\right]\bar{q}_N^n$$
$$= \frac{1}{2N^k}\sum_{n=1}^{\infty}\sum_{c_1=0}^{N-1}\sum_{d|n} d^{k-1}\left[e^{\frac{2\pi i}{N}\left(r\frac{n}{d}-(s-d)c_1\right)}+(-1)^k e^{-\frac{2\pi i}{N}\left(r\frac{n}{d}-(s+d)c_1\right)}\right]\bar{q}_N^n. \tag{J.630}$$

For $k=1$ we have to be more careful about absolute convergence and reordering of terms. The easiest approach is to consider Eq. (J.627) for $k=2$ and to integrate in $\tau'=n_2\tau$. This yields

$$\sum_{n_1\in\mathbb{Z}} \frac{\left(e^{2\pi i\frac{s}{N}}\right)^{n_1}}{n_1+n_2\tau} = \tilde{C} - \frac{2\pi i}{N}\sum_{c_1=0}^{N-1}\sum_{d=1}^{\infty} e^{2\pi i\left(\frac{sc_1}{N}+\frac{c_1 d}{N}+\frac{n_2 d}{N}\tau\right)}, \tag{J.631}$$

with some unknown constant $\tilde{C}$. We then repeat the steps as in Eqs. (J.628) and (J.630) and obtain

$$\begin{aligned}
&\frac{1}{2}\frac{1}{2\pi i}\sum_{(n_1,n_2)\in\mathbb{Z}^2\setminus(0,0)}{}_{e}\frac{e^{\frac{2\pi i}{N}(n_1 s-n_2 r)}}{n_1+n_2\tau}\\
&= a_0+\frac{1}{2N}\sum_{n=1}^{\infty}\sum_{c_1=0}^{N-1}\sum_{d|n}\left[e^{\frac{2\pi i}{N}\left(r\frac{n}{d}-(s-d)c_1\right)}-e^{-\frac{2\pi i}{N}\left(r\frac{n}{d}-(s+d)c_1\right)}\right]\bar{q}_N^{n}, \qquad \text{(J.632)}
\end{aligned}$$

with another unknown constant a_0. We determine a_0 by evaluating both sides at $\tau = i\infty$. On the right-hand side only a_0 survives. On the left-hand side we consider the three cases (i) $s = r = 0 \bmod N$, (ii) $s = 0 \bmod N$, $r \neq 0 \bmod N$ and (iii) $s \neq 0 \bmod N$. We start with case (i): We have

$$\lim_{\tau\to i\infty}\frac{1}{2}\frac{1}{2\pi i}\sum_{(n_1,n_2)\in\mathbb{Z}^2\setminus(0,0)}{}_{e}\frac{1}{n_1+n_2\tau} = \lim_{\tau\to i\infty}\frac{1}{2}\frac{1}{2\pi i}e_1(\tau) = 0, \qquad \text{(J.633)}$$

and therefore $a_0 = 0$. In the case (ii) we have

$$\begin{aligned}
&\lim_{\tau\to i\infty}\frac{1}{2}\frac{1}{2\pi i}\sum_{(n_1,n_2)\in\mathbb{Z}^2\setminus(0,0)}{}_{e}\frac{e^{-2\pi i\frac{r}{N}n_2}}{n_1+n_2\tau}\\
&= \lim_{\tau\to i\infty}\frac{1}{2}\frac{1}{2\pi i}\sum_{n_2=1}^{\infty}\left[e^{-2\pi i\frac{r}{N}n_2}\sum_{n_1\in\mathbb{Z}}\frac{1}{n_1+n_2\tau}+e^{2\pi i\frac{r}{N}n_2}\sum_{n_1\in\mathbb{Z}}\frac{1}{n_1-n_2\tau}\right]\\
&= \lim_{\tau\to i\infty}\frac{1}{4i}\sum_{n_2=1}^{\infty}\left[e^{-2\pi i\frac{r}{N}n_2}-e^{2\pi i\frac{r}{N}n_2}\right]\cot(\pi\tau n_2)\\
&= -\frac{1}{4}\sum_{n_2=1}^{\infty}\left[e^{-2\pi i\frac{r}{N}n_2}-e^{2\pi i\frac{r}{N}n_2}\right] = \frac{i}{2}\sum_{n_2=1}^{\infty}\sin\left(2\pi\frac{r}{N}n_2\right)\\
&= \frac{i}{4}\cot\left(\pi\frac{r}{N}\right), \qquad \text{(J.634)}
\end{aligned}$$

and therefore $a_0 = \frac{i}{4}\cot(\frac{r}{N}\pi)$. In the case (iii) one first shows that

$$\lim_{\tau\to i\infty}\sum_{n_1\in\mathbb{Z}}\frac{e^{2\pi i\frac{s}{N}n_1}}{n_1+n_2\tau} = 0. \qquad \text{(J.635)}$$

Then

$$\lim_{\tau\to i\infty} \frac{1}{2}\frac{1}{2\pi i} \sum_{(n_1,n_2)\in\mathbb{Z}^2\backslash(0,0)}{}_e \frac{e^{\frac{2\pi i}{N}(n_1 s - n_2 r)}}{n_1+n_2\tau} = \frac{1}{4\pi i}\sum_{n_1=1}^{\infty} \frac{e^{2\pi i \frac{s}{N} n_1} - e^{-2\pi i \frac{s}{N} n_1}}{n_1}$$
$$= \frac{1}{4\pi i}\left[\mathrm{Li}_1\left(e^{2\pi i \frac{s}{N}}\right) - \mathrm{Li}_1\left(e^{-2\pi i \frac{s}{N}}\right)\right]$$
$$= \frac{1}{2\pi}\mathrm{Gl}_1\left(2\pi\frac{s}{N}\right) = \frac{1}{4} - \frac{s}{2N}, \quad (J.636)$$

and therefore $a_0 = \frac{1}{4} - \frac{s}{2N}$.

Exercise 114 *Prove Eq. (13.162) for the case* $k \geq 3$.

Solution: *Let*

$$\tau' = \frac{a\tau+b}{c\tau+d}. \quad (J.637)$$

We set

$$\gamma = \begin{pmatrix} a & b \\ c & d \end{pmatrix}, \quad \gamma^{-1} = \begin{pmatrix} d & -b \\ -c & a \end{pmatrix}. \quad (J.638)$$

In this exercise we only consider the case $k \geq 3$. *In this case the sums are absolutely convergent and we may drop the Eisenstein summation prescription. We consider*

$$2\frac{(2\pi i)^k}{(k-1)!}\left(h_{k,N,r,s}|_k\gamma\right)(\tau) = (c\tau+d)^{-k} \sum_{(n_1,n_2)\in\mathbb{Z}^2\backslash(0,0)} \frac{e^{\frac{2\pi i}{N}(n_1 s - n_2 r)}}{(n_1+n_2\tau')^k}. \quad (J.639)$$

We have

$$n_1 + n_2\tau' = \frac{1}{c\tau+d}(n_2,n_1)\begin{pmatrix} a & b \\ c & d \end{pmatrix}\begin{pmatrix} \tau \\ 1 \end{pmatrix}. \quad (J.640)$$

Thus

$$(c\tau+d)^{-k} \sum_{(n_1,n_2)\in\mathbb{Z}^2\backslash(0,0)} \frac{e^{\frac{2\pi i}{N}(n_1 s - n_2 r)}}{(n_1+n_2\tau')^k} = \sum_{(n_1,n_2)\in\mathbb{Z}^2\backslash(0,0)} \frac{e^{\frac{2\pi i}{N}(n_1 s - n_2 r)}}{\left[(n_2,n_1)\begin{pmatrix} a & b \\ c & d \end{pmatrix}\begin{pmatrix} \tau \\ 1 \end{pmatrix}\right]^k}. \quad (J.641)$$

For $k \geq 3$ *the sum is absolutely convergent and we may sum over the individual terms in a different order. We set*

$$\left(n_2', n_1'\right) = (n_2, n_1)\begin{pmatrix} a & b \\ c & d \end{pmatrix}, \quad (n_2, n_1) = \left(n_2', n_1'\right)\begin{pmatrix} d & -b \\ -c & a \end{pmatrix} \quad \text{(J.642)}$$

and sum over (n_1', n_2'). *Then*

$$(c\tau + d)^{-k} \sum_{(n_1,n_2)\in\mathbb{Z}^2\setminus(0,0)} \frac{e^{\frac{2\pi i}{N}(n_1 s - n_2 r)}}{(n_1 + n_2\tau')^k} = \sum_{(n_1',n_2')\in\mathbb{Z}^2\setminus(0,0)} \frac{e^{\frac{2\pi i}{N}\left[n_1'(as+cr) - n_2'(bs+dr)\right]}}{\left(n_1' + n_2'\tau\right)^k}. \quad \text{(J.643)}$$

Thus we have shown for $k \geq 3$

$$\left(h_{k,N,r,s}|_k\gamma\right)(\tau) = h_{k,N,(rd+sb) \bmod N,(rc+sa) \bmod N}(\tau). \quad \text{(J.644)}$$

Exercise 115 *Show that a relative boundary is a relative cycle.*

Solution: *Elements in the relative chain group* $C_k(B, A)$ *are equivalence classes in* $C_k(B)$. *If* $c_k \in C_k(B)$ *we denote the corresponding equivalence class by* $[c_k] = c_k + A$.

Now, let us consider a boundary $[b_k] \in B_k(B, A)$. *By definition, there exists a* $c_{k+1} \in C_{k+1}(B, A)$ *such that*

$$b_k - \partial c_{k+1} \in A. \quad \text{(J.645)}$$

We have to show that $[b_k]$ *is a relative cycle, i.e.* $\partial b_k \in A$. *As* A *is a subcomplex, we have for any* $a_k \in A$ *that* $\partial a_k \in A$. *From Eq. (J.645) we have then*

$$\partial\left(b_k - \partial c_{k+1}\right) = \partial b_k - \partial\partial c_{k+1} \in A. \quad \text{(J.646)}$$

Since $\partial\partial c_{k+1} = 0$ *we have* $\partial b_k \in A$.

Exercise 116 *We now have two bases of* $H^1_{\mathrm{dR}}(E)$: *on the one hand* (ω_1, ω_2), *on the other hand* $(dz, d\bar{z})$. *We already know* $\omega_1 = dz$. *Work out the full relation between the two bases.*

Solution: *We first relate* (ω_1, ω_2) *to* (γ_1^*, γ_2^*): *We make the ansatz* $\omega_i = c_{i,1}\gamma_1^* + c_{i,2}\gamma_2^*$. *Integrating over* γ_j *gives*

$$\left\langle \omega_i, \gamma_j \right\rangle = c_{i,1}\left\langle \gamma_1^*, \gamma_j \right\rangle + c_{i,2}\left\langle \gamma_2^*, \gamma_j \right\rangle. \quad \text{(J.647)}$$

Using Eq. (14.130) we find

$$\begin{pmatrix} \omega_1 \\ \omega_2 \end{pmatrix} = \begin{pmatrix} \psi_1 & \psi_2 \\ \phi_1 & \phi_2 \end{pmatrix}\begin{pmatrix} \gamma_1^* \\ \gamma_2^* \end{pmatrix}. \quad \text{(J.648)}$$

As $dz = \omega_1$ *we have*

$$\begin{pmatrix} dz \\ d\bar{z} \end{pmatrix} = \begin{pmatrix} \psi_1 & \psi_2 \\ \overline{\psi}_1 & \overline{\psi}_2 \end{pmatrix} \begin{pmatrix} \gamma_1^* \\ \gamma_2^* \end{pmatrix}. \tag{J.649}$$

The inverse of the period matrix is

$$\begin{pmatrix} \psi_1 & \psi_2 \\ \phi_1 & \phi_2 \end{pmatrix}^{-1} = \frac{1}{2\pi i} \begin{pmatrix} \phi_2 & -\psi_2 \\ -\phi_1 & \psi_1 \end{pmatrix} \tag{J.650}$$

and we obtain

$$\begin{pmatrix} dz \\ d\bar{z} \end{pmatrix} = \begin{pmatrix} 1 & 0 \\ \frac{1}{2\pi i}\left(\overline{\psi}_1\phi_2 - \overline{\psi}_2\phi_1\right) & -\frac{1}{2\pi i}\left(\overline{\psi}_1\psi_2 - \overline{\psi}_2\psi_1\right) \end{pmatrix} \begin{pmatrix} \omega_1 \\ \omega_2 \end{pmatrix}. \tag{J.651}$$

Exercise 117 *Work out all $V^{p,q}$ and show that $V_{\mathbb{Q}}$ is mixed Tate.*

Solution: *The four conditions*

$$W_0 V_{\mathbb{Q}} = V_{\mathbb{Q}}, \qquad W_{-2n-1} V_{\mathbb{Q}} = 0, \qquad F^1 V_{\mathbb{C}} = 0, \qquad \overline{F^1 V_{\mathbb{C}}} = 0 \tag{J.652}$$

allow only a finite number of $V^{p,q}$'s to be non-zero, namely the ones with

$$p \le 0, \qquad q \le 0, \qquad p + q \ge -2n. \tag{J.653}$$

For $n = 2$ these are shown in Fig. J.10. We have over $\mathbb{C}$

$$\begin{aligned}
\mathrm{Gr}_0^W V_{\mathbb{C}} &= W_0 V_{\mathbb{C}} / W_{-1} V_{\mathbb{C}} = \langle v_0, v_1, v_2, \dots, v_n \rangle / \langle v_1, v_2, \dots, v_n \rangle = \langle e_0, v_1, v_2, \dots, v_n \rangle / \langle v_1, v_2, \dots, v_n \rangle \\
&\cong \langle e_0 \rangle, \\
\mathrm{Gr}_{-1}^W V_{\mathbb{C}} &= W_{-1} V_{\mathbb{C}} / W_{-2} V_{\mathbb{C}} = \langle v_1, v_2, \dots, v_n \rangle / \langle v_1, v_2, \dots, v_n \rangle \\
&\cong 0, \\
\mathrm{Gr}_{-2}^W V_{\mathbb{C}} &= W_{-2} V_{\mathbb{C}} / W_{-3} V_{\mathbb{C}} = \langle v_1, v_2, \dots, v_n \rangle / \langle v_2, \dots, v_n \rangle = \langle 2\pi i e_1, v_2, \dots, v_n \rangle / \langle v_2, \dots, v_n \rangle \\
&\cong \langle 2\pi i e_1 \rangle,
\end{aligned} \tag{J.654}$$

and more generally for $0 \le j \le n$

$$\mathrm{Gr}_{-2j}^W V_{\mathbb{C}} \cong \left\langle (2\pi i)^j \, e_j \right\rangle, \quad \mathrm{Gr}_{-2j-1}^W V_{\mathbb{C}} \cong 0. \tag{J.655}$$

Let's look at the even weights $(-2j)$: As the e_j's are independent we have

$$(2\pi i)^j e_j \notin F^{j-1} V_{\mathbb{C}} = \langle e_0, e_{-1}, \dots, e_{-j+1} \rangle \tag{J.656}$$

and therefore $V^{p,q} = 0$ for $p > q$ and $p + q = -2j$. From $V^{q,p} = \overline{V^{p,q}}$ it follows then that also $V^{p,q} = 0$ for $p < q$ and $p + q = -2j$. This shows that at weight $(-2j)$ only $V^{-j,-j}$ can be non-zero. This proves that $V_{\mathbb{Q}}$ is mixed Tate and we have

$$V^{-j,-j} \cong \left\langle (2\pi i)^j \, e_j \right\rangle. \tag{J.657}$$

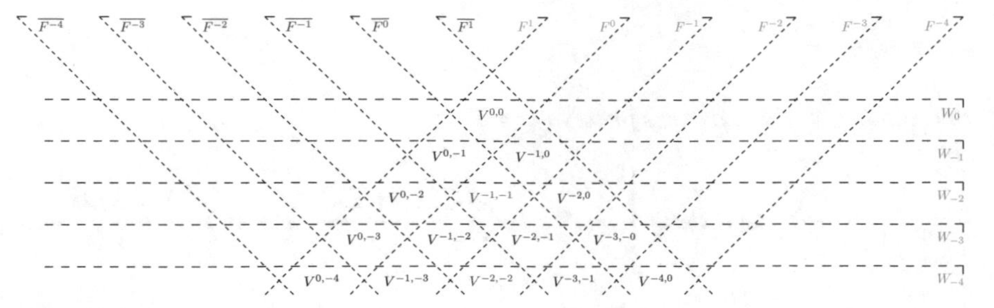

Fig. J.10 For $n = 2$ the four conditions $W_0 V_{\mathbb{Q}} = V_{\mathbb{Q}}$, $W_{-2n-1} V_{\mathbb{Q}} = 0$, $F^1 V_{\mathbb{C}} = 0$ and $\overline{F^1 V_{\mathbb{C}}} = 0$ allow only the shown $V^{p,q}$ (shown in red and black) to be non-zero. In the text we then show that only the $V^{p,q}$ shown in red are non-zero

Exercise 118 *Derive Eq. (14.162).*

Solution: *The starting point is $\nabla v_j = 0$ for $0 \leq j \leq n$ together with the relation of the v_j's to the e_{-j}'s given in Eq. (14.154). We prove Eq. (14.162) recursively, starting with e_{-n}. As $v_n = (2\pi i)^n e_{-n}$, the equation $\nabla v_n = 0$ implies*

$$\nabla e_{-n} = 0. \tag{J.658}$$

We further have

$$v_{n-1} = (2\pi i)^{n-1} \left[e_{-n+1} + \ln(x) e_{-n} \right] \tag{J.659}$$

and $\nabla v_{n-1} = 0$ implies

$$0 = \nabla \left[e_{-n+1} + \ln(x) e_{-n} \right] = \nabla e_{-n+1} + \frac{dx}{x} e_{-n}, \tag{J.660}$$

and therefore

$$\nabla e_{-n+1} = -\frac{dx}{x} e_{-n}. \tag{J.661}$$

Let now assume that

$$\nabla e_{-j} = -\frac{dx}{x} e_{-j-1} \tag{J.662}$$

for $j \in \{n, n-1, \dots, k+1\}$ and $k > 0$. We show that Eq. (J.662) holds also for $j = k$: From $\nabla v_k = 0$ we have

$$0 = \nabla \left[\sum_{j=k}^{n} \frac{\ln^{j-k}(x)}{(j-k)!} e_{-j} \right] = \nabla e_{-k} + \sum_{j=k+1}^{n} \frac{\ln^{j-k-1}(x) dx}{(j-k-1)! x} e_{-j} - \sum_{j=k+1}^{n-1} \frac{\ln^{j-k}(x) dx}{(j-k)! x} e_{-j-1}$$

$$= \nabla e_{-k} + \frac{dx}{x} e_{-j}. \tag{J.663}$$

In a similar way one derives from $\nabla v_0 = 0$

$$0 = \nabla \left[e_0 - \sum_{j=1}^{n} \text{Li}_j(x) e_{-j} \right] = \nabla e_0 - \sum_{j=1}^{n} \frac{\text{Li}_{j-1}(x)dx}{x} e_{-j} + \sum_{j=1}^{n-1} \frac{\text{Li}_j(x)dx}{x} e_{-j-1}$$
$$= \nabla e_0 - \frac{\text{Li}_0(x)dx}{x} e_{-1} = \nabla e_0 + \frac{dx}{x-1} e_{-1}. \tag{J.664}$$

Exercise 119 *Consider* $X = \mathbb{C}\backslash\{0\}$ *and* $Y = \emptyset$. *Take* $\omega = dx/x$ *as a basis of* $H^1_{\text{alg dR}}(X)$ *and let* γ *be a small counter-clockwise circle around* $x = 0$. γ *is a basis of* $H^{\text{B}}_1(X)$. *Denote by* γ^* *the dual basis of* $H^1_{\text{B}}(X)$. *Work out*

$$F_\infty\left(\gamma^*\right). \tag{J.665}$$

Solution: *Let's first work out the comparison isomorphism*

$$\text{comparison} : H^1_{\text{alg dR}}(X) \otimes \mathbb{C} \to H^1_{\text{B}}(X, \mathbb{Q}) \otimes \mathbb{C} \tag{J.666}$$

The period is

$$p = \int\limits_\gamma \omega = 2\pi i. \tag{J.667}$$

We then have with Eq. (14.21)

$$\text{comparison}(\omega) = 2\pi i \gamma^*. \tag{J.668}$$

Let us now consider $\omega \otimes 1 \in H^1_{\text{alg dR}}(X) \otimes \mathbb{C}$. *On the one hand we have*

$$\begin{aligned} \text{comparison}\left(\text{conj}_{\text{alg dR}}(\omega \otimes 1)\right) &= \text{comparison}(\omega \otimes 1) \\ &= \gamma^* \otimes (2\pi i). \end{aligned} \tag{J.669}$$

On the other hand we have

$$\begin{aligned} (F_\infty \otimes \text{id})\ \text{conj}_{\text{B}}(\text{comparison}(\omega \otimes 1)) &= (F_\infty \otimes \text{id})\ \text{conj}_{\text{B}}\left(\gamma^* \otimes (2\pi i)\right) \\ &= (F_\infty \otimes \text{id})\ \left(\gamma^* \otimes (-2\pi i)\right) \\ &= -F_\infty\left(\gamma^*\right) \otimes (2\pi i). \end{aligned} \tag{J.670}$$

This should be equal to $\gamma^* \otimes (2\pi i)$ *and therefore*

$$F_\infty\left(\gamma^*\right) = -\gamma^*. \tag{J.671}$$

The dual of $(-\gamma^*)$ *is* $(-\gamma)$*, a small circle in the clockwise direction around* $x = 0$.

Exercise 120 *Let* $\mathbb{F}$ *be a sub-field of* $\mathbb{C}$. *Show that* $\mathrm{GL}_n(\mathbb{F})$ *(the group of* $(n \times n)$*-matrices with entries from* $\mathbb{F}$ *and non-zero determinant) can be defined by a polynomial equation.*

Solution: *An element* $g \in \mathrm{GL}_n(\mathbb{F})$ *is given by* n^2 *elements* $z_{ij} \in \mathbb{F}$

$$g = \begin{pmatrix} z_{11} & z_{12} & \dots & z_{1n} \\ \vdots & \vdots & & \vdots \\ z_{n1} & z_{n2} & \dots & z_{nn} \end{pmatrix}, \tag{J.672}$$

such that

$$\det g \neq 0. \tag{J.673}$$

The determinant $\det g$ *is a polynomial in the* z_{ij}*'s. In order to convert this inequality to a polynomial equation, we introduce another variable* $z_0 \in \mathbb{F}$. $\mathrm{GL}_n(\mathbb{F})$ *is then isomorph to the set of points in* $\mathbb{F}^{n^2+1}$ *satisfying the polynomial equation*

$$z_0 \det g - 1 = 0. \tag{J.674}$$

Exercise 121 *Consider the one-loop two-point function with equal internal masses*

$$I_{11}(2, x) = \int \frac{d^2k}{i\pi} \frac{m^2}{\left(-q_1^2 + m_1^2\right)\left(-q_2^2 + m_2^2\right)}, \qquad x = -\frac{p^2}{m^2}. \tag{J.675}$$

Derive with the methods of this section the differential equation for $I_{11}(2, x)$ *with respect to the kinematic variable* x.

Solution: *The Feynman parametrisation reads*

$$I_{11}(2, x) = \int\limits_{\Delta} \frac{\omega}{\mathcal{F}}, \tag{J.676}$$

with

$$\mathcal{F} = a_1 a_2 x + (a_1 + a_2)^2, \quad \omega = a_1 da_2 - a_2 da_1, \quad \Delta = \mathbb{RP}^1_{\geq 0}. \tag{J.677}$$

We set

$$\varphi = \frac{\omega}{\mathcal{F}}. \tag{J.678}$$

We first look for a differential equation of the form

$$L^{(r)}\varphi = d\beta, \tag{J.679}$$

where

$$L^{(r)} = \sum_{j=0}^{r} R_j(x) \frac{d^j}{dx^j}, \qquad R_r(x) = 1. \tag{J.680}$$

is a Picard-Fuchs operator of order r. For the example of the one-loop two-point function β is a 0*-form, depending on the Feynman parameters a_i. The differential d is with respect to the Feynman parameters a_i. For β we make the following ansatz*

$$\beta = \frac{1}{\mathcal{F}^r}\left(a_1 q_2 - a_2 q_1\right), \tag{J.681}$$

where the q_i are homogeneous polynomials of degree $(2r-1)$ *in the variables a_i. The ansatz is based on the fact, that the singularities of the integrand φ are given by powers of the graph polynomial $\mathcal{F}$. Acting with $L^{(r)}$ on the integrand φ will only increase the power of $\mathcal{F}$ in the denominator by r, but will not introduce singularities on new algebraic varieties. Each polynomial q_i contains only a finite number of monomials in the Feynman parameters with a priori unknown coefficients. We therefore take these coefficients and the variables R_0, ..., R_{r-1} as the set of our unknown variables. Plugging the ansatz (J.681) in Eq. (J.679) gives a linear system of equations for the unknown variables. This system may or may not have a solution. In order to find the differential equation of minimal order we start at $r = 1$ and try to solve the linear system of equations. If no solution is found, we increase r by one and repeat the exercise, until a solution is found. This is then the solution of minimal order r.*

For the case of the one-loop two-point function we obtain a solution for $r = 1$. Thus we obtain the Picard-Fuchs operator

$$L^{(1)} = \frac{d}{dx} + \frac{x+2}{x(x+4)} \tag{J.682}$$

and a possible solution for β is given by

$$\beta = \frac{1}{\mathcal{F}} \frac{1}{x(x+4)} \left[(x+2)a_1 a_2 + 2a_2^2\right]. \tag{J.683}$$

The boundary of Δ is given by the two points $[1:0]$ *and* $[0:1]$*. The integration of the inhomogeneous term yields*

$$\int\limits_{\partial\Delta} \beta = \frac{2}{x(x+4)}. \tag{J.684}$$

Putting everything together, we obtain the differential equation

$$\left[x(x+4)\frac{d}{dx} + x + 2 \right] I_{11}(2, x) = 2. \tag{J.685}$$

This agrees with Eq. (6.65) if we use the result for the tadpole integral

$$I_{10}(2 - 2\varepsilon, x) = \frac{1}{\varepsilon} + O(\varepsilon^0). \tag{J.686}$$

Exercise 122 *Derive Eq. (15.5).*

Solution: *We start from the integral representation, substitute $t = 1 - e^{-\tilde{z}}$, expand and integrate term-by-term:*

$$\text{Li}_2(x) = -\int\limits_0^x \frac{dt}{t} \ln(1-t) = \int\limits_0^z d\tilde{z} \frac{\tilde{z}}{e^{\tilde{z}} - 1} = \int\limits_0^z d\tilde{z} \sum_{j=0}^{\infty} \frac{B_j}{j!} \tilde{z}^j = \sum_{j=0}^{\infty} \frac{B_j}{(j+1)!} z^{j+1}. \tag{J.687}$$

Exercise 123 *Rewrite the differential one-form*

$$\omega = -\frac{(1-x_2)^2 \, dx_1}{x_1 \left[(1-x_2)^2 - x_1 x_2 \right]} - \frac{x_1 (1+x_2) \, dx_2}{(1-x_2) \left[(1-x_2)^2 - x_1 x_2 \right]} \tag{J.688}$$

as a dlog-form.

Solution: *We first look at the polynomials which appear in the denominator of ω: There are three distinct polynomials*

$$p_1 = x_1, \qquad p_2 = x_2 - 1, \qquad p_3 = (1-x_2)^2 - x_1 x_2. \tag{J.689}$$

We then make the ansatz

$$\omega^{\text{ansatz}} = d \ln \left(p_1^{n_1} p_2^{n_2} p_3^{n_3} \right), \qquad n_1, n_2, n_3 \in \mathbb{Z}. \tag{J.690}$$

Let's now study $\omega - \omega^{\text{ansatz}}$. We partial fraction the terms proportional to dx_1 with respect to x_1, and the terms proportional to dx_2 with respect to x_2. This yields

$$\omega - \omega^{\text{ansatz}} = \left[-\frac{1+n_1}{x_1} - \frac{x_2(1-n_3)}{(1-x_2)^2 - x_1 x_2} \right] dx_1 + \left[-\frac{2+n_2}{x_2 - 1} - \frac{(1-n_3)(2+x_1-2x_2)}{(1-x_2)^2 - x_1 x_2} \right] dx_2. \tag{J.691}$$

This gives the linear system of equations

$$\begin{aligned} 1 + n_1 &= 0, \\ 2 + n_2 &= 0, \\ 1 - n_3 &= 0, \end{aligned} \tag{J.692}$$

whose unique solution is $(n_1, n_2, n_3) = (-1, -2, 1)$. *Hence*

$$\omega = d \ln p_3 - d \ln p_1 - 2d \ln p_2. \tag{J.693}$$

Exercise 124 *Let*

$$\begin{aligned}
\tilde{f}_1 &= \lambda(x_2, x_3, 1) + 8x_3 - x_1(x_2 - x_3) - r_4 r_5, \\
\tilde{f}_2 &= \lambda(x_2, x_3, 1) + 8x_2 + x_1(x_2 - x_3) - r_4 r_5, \\
\tilde{f}_3 &= \lambda(x_2, x_3, 1) - x_1(x_2 - x_3) + r_4 r_5, \\
\tilde{f}_4 &= \lambda(x_2, x_3, 1) + x_1(x_2 - x_3) + r_4 r_5,
\end{aligned} \tag{J.694}$$

Show that

$$\tilde{\omega} = d \ln \left(\frac{\tilde{f}_1 \tilde{f}_2}{\tilde{f}_3 \tilde{f}_4} \right) \tag{J.695}$$

has a pole along $x_1 = 0$, *while*

$$\omega = d \ln \left(x_1^2 \frac{\tilde{f}_1 \tilde{f}_2}{\tilde{f}_3 \tilde{f}_4} \right) \tag{J.696}$$

does not.

Solution: *There are several ways to show this: We may compute the residue of* $\tilde{\omega}$ *and* ω *along* $Y = \{x_1 = 0\}$. *One finds*

$$\text{Res}_Y(\tilde{\omega}) = -2, \quad \text{Res}_Y(\omega) = 0. \tag{J.697}$$

Alternatively we may compute the limits $x_1 \to 0$ *of* $\tilde{f}_1$-$\tilde{f}_4$*:*

$$\begin{aligned}
\tilde{f}_1 \tilde{f}_2 &= 4\left[1 - (x_2 - x_3)^2\right]^2 + O(x_1), \\
\tilde{f}_3 \tilde{f}_4 &= 4x_1^2 x_2 x_3 \left[\frac{1 - (x_2 - x_3)^2}{\lambda(x_2, x_3, 1)}\right]^2 + O\left(x_1^3\right).
\end{aligned} \tag{J.698}$$

The simple pole of $\tilde{\omega}$ *along* $x_1 = 0$ *originates from* $\tilde{f}_3 \tilde{f}_4$, *which vanishes as* x_1^2 *in the limit* $x_1 \to 0$.

Exercise 125 *Rewrite the differential one-form*

$$\omega = -\frac{(2 - 2x_2 - x_1 x_2) r_1 dx_1}{x_1(4 + x_1)\left[(1 - x_2)^2 - x_1 x_2\right]} - \frac{r_1 dx_2}{\left[(1 - x_2)^2 - x_1 x_2\right]} \tag{J.699}$$

where $r_1 = \sqrt{x_1(4 + x_1)}$ *as a dlog-form.*

Solution: *We use the rationalisation*

$$x_1 = \frac{\left(1 - x_1'\right)^2}{x_1'}, \quad x_1' = \frac{1}{2}\left(2 + x_1 - r_1\right). \tag{J.700}$$

We have

$$dx_1 = -\frac{\left(1 - x_1'^2\right)}{x_1'^2} dx_1' \tag{J.701}$$

and

$$\omega = \left[\frac{1}{x_1'} + \frac{1}{x_1' - x_2} + \frac{x_2}{1 - x_1' x_2}\right] dx_1' + \left[\frac{1}{x_2 - x_1'} + \frac{x_1'}{1 - x_1' x_2}\right] dx_2. \tag{J.702}$$

We are now back to the rational case and we find

$$\omega = d \ln \left(\frac{x_1' \left(x_1' - x_2\right)}{\left(1 - x_1' x_2\right)}\right). \tag{J.703}$$

We substitute back from x_1' to x_1 and obtain

$$\omega = d \ln \left(\frac{2 + x_1 - 2x_2 - r_1}{2 + x_1 - 2x_2 + r_1}\right) = 2d \ln\left(2 + x_1 - 2x_2 - r_1\right) - d \ln\left(\left(1 - x_2\right)^2 - x_1 x_2\right). \tag{J.704}$$

Exercise 126 *Let*

$$g = \frac{2 + x_1 - r_1}{2 + x_1 + r_1}. \tag{J.705}$$

Express g and $(1 - g)$ as a power product in the letters of the alphabet defined by Eq. (16.50), Eq. (16.51) and the constant $f_0 = 2$.

Solution: *We start with g: We multiply the numerator and the denominator with $(2 + x_1 + r_1)$:*

$$g = \frac{2 + x_1 - r_1}{2 + x_1 + r_1} = \frac{\left(2 + x_1 - r_1\right)^2}{\left(2 + x_1 + r_1\right)\left(2 + x_1 - r_1\right)}. \tag{J.706}$$

The third binomial formula gives

$$\left(2 + x_1 + r_1\right)\left(2 + x_1 - r_1\right) = \left(2 + x_1\right)^2 - r_1^2 = \left(4 + 4x_1 + x_1^2\right) - x_1\left(4 + x_1\right) = 4. \tag{J.707}$$

Hence

$$g = \frac{(2 + x_1 - r_1)^2}{4} = f_0^{-2} f_{10}. \tag{J.708}$$

Let us now turn to $(1 - g)$. *We have*

$$1 - g = 1 - \frac{2 + x_1 - r_1}{2 + x_1 + r_1} = \frac{2r_1}{2 + x_1 + r_1} = \frac{2r_1 (2 + x_1 - r_1)}{4} \tag{J.709}$$

From Eq. (16.52) we have $r_1^2 = f_1 f_2$ *and therefore*

$$1 - g = f_0^{-1} f_1 f_2 f_{10}. \tag{J.710}$$

Thus g *is an allowed argument of* Li_n.

Exercise 127 *The master integral* J_{15} *starts at order* $O(\varepsilon^2)$. *The weight two term of* J_{15} *is given in terms of iterated integrals by*

$$J_{15}^{(2)} = 2i\, I_\gamma\left(2\omega_{15} - \omega_3 - \omega_5, \omega_5 - \omega_3; 1\right) + 2J_7^{(2)}(0, 1, 1), \tag{J.711}$$

where $J_7^{(2)}(0, 1, 1)$ *denotes the boundary value of Eq. (16.72):*

$$J_7^{(2)}(0, 1, 1) = \frac{3}{2i} \overline{H}_2\left(e^{\frac{2\pi i}{3}}\right) = \frac{3}{2i}\left[\mathrm{Li}_2\left(e^{\frac{2\pi i}{3}}\right) - \mathrm{Li}_2\left(e^{-\frac{2\pi i}{3}}\right)\right]. \tag{J.712}$$

Express $J_{15}^{(2)}$ *in terms of multiple polylogarithms.*

Solution: *We have to convert the iterated integral* $I_\gamma(2\omega_{15} - \omega_3 - \omega_5, \omega_5 - \omega_3; 1)$ *to multiple polylogarithms. We have*

$$2\omega_{15} - \omega_3 - \omega_5 = d \ln\left(\frac{f_{15}^2}{f_3 f_5}\right), \quad \omega_5 - \omega_3 = d \ln\left(\frac{f_5}{f_3}\right) \tag{J.713}$$

and

$$\frac{f_{15}^2}{f_3 f_5} = -2 \frac{(2x_2 - x_3 + i r_3)}{x_2}, \quad \frac{f_5}{f_3} = \frac{x_3}{x_2}. \tag{J.714}$$

This involves only the square root r_3, *which we may rationalise. With the rationalisation of Eq. (16.62) we obtain*

$$\begin{aligned} 2\omega_{15} - \omega_3 - \omega_5 &= d \ln\left(x_2' + i\right) - d \ln\left(x_2' - i\right), \\ \omega_5 - \omega_3 &= -d \ln\left(x_2' + i\right) - d \ln\left(x_2' - i\right). \end{aligned} \tag{J.715}$$

We are in the lucky situation that after the change of variables $(x_2, x_3) \rightarrow (x_2', x_3')$ *the integrand depends only on* x_2' *but not on* x_3'. *The inverse transformation is given by*

$$x_2' = \sqrt{\frac{4x_2 - x_3}{x_3}}. \tag{J.716}$$

The boundary point $(x_2, x_3) = (1, 1)$ *corresponds to* $x_2' = \sqrt{3}$. *Thus we have to integrate in* x_2'*-space from* $\sqrt{3}$ *to the final value* x_2'. *We obtain*

$$I_\gamma\left(2\omega_{15} - \omega_3 - \omega_5, \omega_5 - \omega_3; 1\right) = \int\limits_{\sqrt{3}}^{x_2'} dt_1 \left(\frac{1}{t_1 - i} - \frac{1}{t_1 + i}\right) \int\limits_{\sqrt{3}}^{t_1} dt_2 \left(\frac{1}{t_2 - i} + \frac{1}{t_2 + i}\right). \tag{J.717}$$

This is not quite yet the standard definition of multiple polylogarithms, the lower integration boundary equals $\sqrt{3}$, *not* 0. *However this is easily adjusted, for example*

$$\int\limits_{\sqrt{3}}^{t_1} \frac{dt_2}{t_2 - i} = \int\limits_{0}^{t_1} \frac{dt_2}{t_2 - i} - \int\limits_{0}^{\sqrt{3}} \frac{dt_2}{t_2 - i} \tag{J.718}$$

and we obtain

$$\begin{aligned} I_\gamma & \left(2\omega_{15} - \omega_3 - \omega_5, \omega_5 - \omega_3; 1\right) = \\ & G\left(i, i; x_2'\right) - G\left(-i, i; x_2'\right) + G\left(i, -i; x_2'\right) - G\left(-i, -i; x_2'\right) \\ & - \left[G\left(i; \sqrt{3}\right) + G\left(-i; \sqrt{3}\right)\right]\left[G\left(i; x_2'\right) - G\left(-i; x_2'\right)\right] \\ & + G\left(i, i; \sqrt{3}\right) + G\left(-i, i; \sqrt{3}\right) - G\left(i, -i; \sqrt{3}\right) - G\left(-i, -i; \sqrt{3}\right). \end{aligned} \tag{J.719}$$

We could stop here, as we managed to express everything in terms of multiple polylogarithms. However it is instructive to consider also the bootstrap approach. As a benefit, this will allow us to simplify the expression above. The symbol of $I_\gamma(2\omega_{15} - \omega_3 - \omega_5, \omega_5 - \omega_3; 1)$ *is*

$$S\left(I_\gamma\left(2\omega_{15} - \omega_3 - \omega_5, \omega_5 - \omega_3; 1\right)\right) = \frac{f_{15}^2}{f_3 f_5} \otimes \frac{f_5}{f_3}. \tag{J.720}$$

We expect $1 \pm i x_2'$ *to be possible arguments of the* Li_n*-functions. Let us check if they are allowed. We extend the alphabet by the constants* $f_{-1} = -1$ *and* $f_0 = 2$. *We have*

$$ix_2' = f_{-1}^{\frac{1}{2}} f_7^{\frac{1}{2}} f_5^{-\frac{1}{2}}, \quad 1 - ix_2' = f_5^{-1} f_{15}, \quad 1 + ix_2' = f_0^2 f_3 f_{15}^{-1}. \tag{J.721}$$

In a similar way one checks that also $(1+ix_2')/(1-ix_2')$ *and* $(1-ix_2')/(1+ix_2')$ *are admissible arguments of the* Li_n*-functions. We calculate a few symbols:*

$$\begin{aligned} S\left[2\text{Li}_2\left(1-ix_2'\right) - 2\text{Li}_2\left(1+ix_2'\right)\right] &= \left(\frac{f_{15}}{f_3 f_5} \otimes \frac{f_5}{f_7}\right), \\ S\left[\text{Li}_2\left(\frac{1+ix_2'}{1-ix_2'}\right) - \text{Li}_2\left(\frac{1-ix_2'}{1+ix_2'}\right)\right] &= \left(\frac{f_{15}}{f_3 f_5} \otimes \frac{f_7}{f_3}\right). \end{aligned} \tag{J.722}$$

We see that the sum of the two terms matches the symbol. We then check the derivative with respect to x_2'*. This requires us to add the term*

$$i\pi \ln\left(x_2'^2 + 1\right) \tag{J.723}$$

to our ansatz. Finally we check the value at a specific point. This will instruct us that we should add the constant

$$iC_7^{(2)} \tag{J.724}$$

to our ansatz for $I_\gamma(2\omega_{15} - \omega_3 - \omega_5, \omega_5 - \omega_3; 1)$*. Putting everything together we arrive at*

$$J_{15}^{(2)} = 2i\left[2\text{Li}_2\left(1-ix_2'\right) - 2\text{Li}_2\left(1+ix_2'\right) + \text{Li}_2\left(\frac{1+ix_2'}{1-ix_2'}\right) - \text{Li}_2\left(\frac{1-ix_2'}{1+ix_2'}\right) + i\pi\ln\left(x_2'^2+1\right)\right]. \tag{J.725}$$

Exercise 128 *Prove Eq. (C.19).*

Solution: $P_j(\boldsymbol{\theta})$ *acts on a monomial* $\mathbf{x}^{\mathbf{i}} = x_1^{i_1} \cdot \ldots \cdot x_n^{i_n}$ *as*

$$P_j(\boldsymbol{\theta})\mathbf{x}^{\mathbf{i}} = P_j(\mathbf{i})\mathbf{x}^{\mathbf{i}}. \tag{J.726}$$

Therefore

$$\left(1+\theta_j\right) P_j(\boldsymbol{\theta}) H\left(\mathbf{x}\right) = \sum_{\mathbf{i}\in\mathbb{N}_0^n} C_{\mathbf{i}} \left(1+\theta_j\right) P_j(\boldsymbol{\theta})\mathbf{x}^{\mathbf{i}} = \sum_{\mathbf{i}\in\mathbb{N}_0^n} C_{\mathbf{i}} \left(1+i_j\right) P_j(\mathbf{i})\mathbf{x}^{\mathbf{i}}. \tag{J.727}$$

Let's now study

$$\left(1+\theta_j\right) Q_j(\boldsymbol{\theta}) \frac{1}{x_j} H\left(\mathbf{x}\right) = \sum_{\mathbf{i}\in\mathbb{N}_0^n} C_{\mathbf{i}} \left(1+\theta_j\right) Q_j(\boldsymbol{\theta})\mathbf{x}^{\mathbf{i}-e_j} = \sum_{\mathbf{i}\in\mathbb{N}_0^n} C_{\mathbf{i}}\, i_j\, Q_j(\mathbf{i}-e_j)\mathbf{x}^{\mathbf{i}-e_j}. \tag{J.728}$$

Due to the factor i_j *all terms with* $i_j = 0$ *vanish and the sum over* i_j *starts at* $i_j = 1$. *The substitution* $i_j \to i_j - 1$ *transforms the summation range back to* $\mathbb{N}_0$. *We thus have*

$$\left(1+\theta_j\right) Q_j(\boldsymbol{\theta}) \frac{1}{x_j} H\left(\mathbf{x}\right) = \sum_{\mathbf{i}\in\mathbb{N}_0^n, i_j\geq 1} C_{\mathbf{i}}\, i_j\, Q_j(\mathbf{i}-e_j)\mathbf{x}^{\mathbf{i}-e_j} = \sum_{\mathbf{i}\in\mathbb{N}_0^n} C_{\mathbf{i}+e_j}\, \left(1+i_j\right)\, Q_j(\mathbf{i})\mathbf{x}^{\mathbf{i}}. \tag{J.729}$$

Hence

$$\left(1+\theta_j\right)\left[Q_j(\boldsymbol{\theta})\frac{1}{x_j} - P_j(\boldsymbol{\theta})\right] H\left(\mathbf{x}\right) = \sum_{\mathbf{i}\in\mathbb{N}_0^n} \left(1+i_j\right)\left[C_{\mathbf{i}+e_j}\, Q_j(\mathbf{i}) - C_{\mathbf{i}}\, P_j(\boldsymbol{\theta})\right]\mathbf{x}^{\mathbf{i}}. \tag{J.730}$$

From Eq. (C.17) we have

$$C_{\mathbf{i}+e_j}\, Q_j(\mathbf{i}) - C_{\mathbf{i}}\, P_j(\boldsymbol{\theta}) = 0 \tag{J.731}$$

and therefore

$$\left(1+\theta_j\right)\left[Q_j(\boldsymbol{\theta})\frac{1}{x_j} - P_j(\boldsymbol{\theta})\right] H\left(\mathbf{x}\right) = 0. \tag{J.732}$$

Exercise 129 *Show that* $[X, X] = 0$ *implies the anti-symmetry of the Lie bracket* $[X, Y] = -[Y, X]$. *Show further that also the converse is true, provided* $\mathrm{char}\ \mathbb{F} \neq 2$. *Explain, why the argument does not work for* $\mathrm{char}\ \mathbb{F} = 2$.

Solution:

$$0 = [X+Y, X+Y] = [X, Y] + [Y, X] \tag{J.733}$$

and therefore $[X, Y] = -[Y, X]$. *Now let us consider the other direction. Assuming* $[X, Y] = -[Y, X]$ *we have for* $X = Y$ *the relation* $[X, X] = -[X, X]$ *or equivalently*

$$2\,[X, X] = 0. \tag{J.734}$$

For $\mathrm{char}\ \mathbb{F} \neq 2$ *it follows that* $[X, X] = 0$. *For* $\mathrm{char}\ \mathbb{F} = 2$ *we have* $2 = 0 \mod 2$ *and Eq. (J.734) does not give any constraint on* $[X, X]$.

Exercise 130 *Derive Eq. (D.24) from Eq. (D.22).*

Solution: *Substituting*

$$A = \sum_{a=1}^{n} c_a T^a, \quad X = \sum_{a=1}^{n} x_a T^a \tag{J.735}$$

into Eq. (D.22) we obtain

$$i c_a x_b f^{abc} T^c = \rho x_c T^c. \tag{J.736}$$

This is equivalent to

$$\left(c_a x_b i f^{abc} - \rho x_c\right) = 0 \tag{J.737}$$

and

$$\left(c_a i f^{abc} - \rho \delta^{bc}\right) x_b = 0. \tag{J.738}$$

Exercise 131 *Consider the Lie algebra* $\mathfrak{su}(2)$*: Start from the generators*

$$I^1 = \frac{1}{2}\begin{pmatrix} 0 & 1 \\ 1 & 0 \end{pmatrix}, \quad I^2 = \frac{1}{2}\begin{pmatrix} 0 & -i \\ i & 0 \end{pmatrix}, \quad I^3 = \frac{1}{2}\begin{pmatrix} 1 & 0 \\ 0 & -1 \end{pmatrix}. \tag{J.739}$$

These generators are proportional to the Pauli matrices and normalised as

$$\mathrm{Tr}\left(I^a I^b\right) = \frac{1}{2}\delta^{ab}. \tag{J.740}$$

The commutators are given by

$$\left[I^a, I^b\right] = i\varepsilon^{abc} I^c, \tag{J.741}$$

where ε^{abc} *denotes the totally antisymmetric tensor. Start from* $A = I^3$*. Determine for this choice the roots, the Cartan standard form and the root vectors.*

Solution: *A root satisfies the equation*

$$\left[I^3, X\right] = \rho X. \tag{J.742}$$

The secular equation reads

$$\det\left(i\varepsilon^{3bc} - \rho\delta^{bc}\right) = 0. \tag{J.743}$$

Working this out yields

$$\begin{vmatrix} -\rho & i & 0 \\ -i & -\rho & 0 \\ 0 & 0 & -\rho \end{vmatrix} = 0,$$

$$-\rho^3 + \rho = 0,$$
$$\rho\left(\rho^2 - 1\right) = 0. \tag{J.744}$$

Therefore the roots are $0, \pm 1$. *We have*

$$\begin{array}{llll} \rho = 0: & \left[I^3, X\right] = 0 & \Rightarrow X & = I^3 = H_1, \\ \rho = 1: & \left[I^3, X\right] = X & \Rightarrow X & = \frac{1}{\sqrt{2}}\left(I^1 + iI^2\right) = E_1, \\ \rho = -1: & \left[I^3, X\right] = -X & \Rightarrow X & = \frac{1}{\sqrt{2}}\left(I^1 - iI^2\right) = E_{-1}. \end{array} \tag{J.745}$$

Thus we obtain the Cartan standard form of $\mathfrak{su}(2)$*:*

$$H_1 = \frac{1}{2}\begin{pmatrix} 1 & 0 \\ 0 & -1 \end{pmatrix}, \quad E_1 = \frac{1}{\sqrt{2}}\begin{pmatrix} 0 & 1 \\ 0 & 0 \end{pmatrix}, \quad E_{-1} = \frac{1}{\sqrt{2}}\begin{pmatrix} 0 & 0 \\ 1 & 0 \end{pmatrix}. \tag{J.746}$$

The roots are

$$[H, E_1] = E_1, \quad \left[H, E_{-1}\right] = -E_1. \tag{J.747}$$

The Lie algebra $\mathfrak{su}(2)$ *has rank* 1 *(there is one generator denoted by* H*) and the root vectors are one-dimensional. They are given by*

$$\vec{\alpha}\left(E_1\right) = (1), \quad \vec{\alpha}\left(E_{-1}\right) = (-1). \tag{J.748}$$

Exercise 132 *Let* $(z_2, \dots, z_{n-2})$ *be simplicial coordinates and* $(x_2, \dots, x_{n-2})$ *the corresponding cubical coordinates. Show that*

$$\text{Li}_{m_{n-2}\dots m_3 m_2}\left(x_{n-2}, \dots, x_3, x_2\right) = (-1)^{n-3}\, G_{m_{n-2}\dots m_3 m_2}\left(z_{n-2}, \dots, z_3, z_2; 1\right) \tag{J.749}$$

Solution: *Let's first work out the relation between the simplicial coordinates* z_j *and the cubical coordinates* x_j*. From Eqs. (F.71) and (F.73) we have*

$$z_i = \frac{1}{\prod\limits_{j=i}^{n-2} x_j}, \qquad i \in \{2, \dots, n-2\}. \tag{J.750}$$

Hence, we have to show

$$\text{Li}_{m_{n-2}\dots m_3 m_2}\left(x_{n-2}, \dots, x_3, x_2\right) = (-1)^{n-3}\, G_{m_{n-2}\dots m_3 m_2}\left(\frac{1}{x_{n-2}}, \dots, \frac{1}{x_3 \dots x_{n-2}}, \frac{1}{x_2 \dots x_{n-2}}; 1\right) \tag{J.751}$$

However, this follows immediately from Eq. (8.19).

Exercise 133 *Let*

$$\gamma = \begin{pmatrix} -1 & 0 \\ 0 & -1 \end{pmatrix}, \qquad \vec{n} = (0, 0). \tag{J.752}$$

Work out z' and τ'.

Solution: *Let start with τ': With $a = -1, b = 0, c = 0, d = -1$ we have*

$$\tau' = \frac{a\tau + b}{c\tau + d} = \frac{-\tau + 0}{-1} = \tau. \tag{J.753}$$

Let's now work out z'. We have in addition $n_1 = n_2 = 0$ and therefore

$$z' = \frac{z + n_2\tau + n_1}{c\tau + d} = \frac{z}{-1} = -z. \tag{J.754}$$

Thus we see that the variable z changes sign.

Exercise 134 *Let R be a commutative ring with 1 and P a prime ideal. Set $S_P = R\backslash P$. Show that S_P is closed under multiplication.*

Solution: *Let $a \in S_P$ and $b \in S_P$. Assume that $a \cdot b \notin S_P$. This means*

$$a \cdot b \in P. \tag{J.755}$$

However, P is assumed to be a prime ideal. This implies if $a \cdot b \in P$ then either $a \in P$ or $b \in P$. This contradicts our assumption that $a \notin P$ and $b \notin P$. Hence $a \cdot b \in S_P$ and we have shown that S_P is closed under multiplication.

Exercise 135 *Consider the commutative ring $\mathbb{Z}$ and the prime ideal $P = \langle 5 \rangle$. Define $S_P = \mathbb{Z}\backslash\langle 5 \rangle$. Describe the quotient ring R_{S_P} and its maximal ideal P_{S_P}.*

Solution: *The prime ideal $P = \langle 5 \rangle$ consists of all integer numbers (zero included), which are divisible by 5. The set S_P is then the set of all integer numbers, which are not divisible by 5. We have $0 \notin S_P$. The ring $\mathbb{Z}$ is an integral domain, hence the condition*

$$s\left(s_2 r_1 - s_1 r_2\right) = 0, \qquad s, s_1, s_2 \in S_P, \quad r_1, r_2 \in \mathbb{Z}, \tag{J.756}$$

implies that either $s = 0$ or $s_2 r_1 - s_1 r_2 = 0$. As $0 \notin S$ it follows that $s_2 r_1 - s_1 r_2 = 0$. Hence R_{S_P} is the set

$$R_{S_P} = \left\{ \frac{p}{q} \mid p \in \mathbb{Z},\ q \in \mathbb{N},\ \gcd(p, q) = 1,\ 5 \nmid q \right\} \subset \mathbb{Q}. \tag{J.757}$$

In plain text: R_{S_P} is the set of rational numbers p/q where q is not divisible by 5 (and in order to have a unique representative we require $\gcd(p, q) = 1$ and $q > 0$).

Now let's look at the non-invertible elements in the ring R_{S_P}: These are all elements, where p in the numerator is divisible by 5: If we would invert these elements in $\mathbb{Q}$, we would get a prime factor 5 in the denominator, however these inverses are not in R_{S_P}. Let's define

$$P_{S_P} = \left\{ \frac{p}{q} \mid p \in \mathbb{Z},\ q \in \mathbb{N},\ \gcd(p,q) = 1,\ 5|p,\ 5 \nmid q \right\}. \tag{J.758}$$

One easily shows that P_{S_P} is an ideal. From Theorem 35 we conclude that P_{S_P} is the only maximal ideal in R_{S_P}. It is easy to see that P_{S_P} is generated by $5 = \frac{5}{1}$ in R_{S_P}:

$$P_{S_P} = \langle 5 \rangle . \tag{J.759}$$

This motivates the abuse of notation $P = P_{S_P}$. Note that

$$\begin{aligned} \langle 5 \rangle \subset \mathbb{Z} : \langle 5 \rangle &= \{ 5r \mid r \in \mathbb{Z} \}, \\ \langle 5 \rangle \subset R_{S_P} : \langle 5 \rangle &= \left\{ 5r \mid r \in R_{S_P} \right\}. \end{aligned} \tag{J.760}$$

Exercise 136 *Let r be defined by Eq. (I.31). Determine*

$$r \bmod n_i . \tag{J.761}$$

Solution: *We have with $s_i \tilde{n}_i + t_i n_i = 1$*

$$\begin{aligned} r \bmod n_i &= \left(\sum_{j=1}^{k} r_j s_j \tilde{n}_j \bmod n \right) \bmod n_i = \sum_{j=1}^{k} r_j s_j \tilde{n}_j \bmod n_i \\ &= r_i s_i \tilde{n}_i \bmod n_i = r_i \left(1 - t_i n_i\right) \bmod n_i = r_i \bmod n_i \\ &= r_i . \end{aligned} \tag{J.762}$$

References

1. Hwa, R.C., Teplitz, V.L.: Homology and Feynman Integrals. Benjamin, W. A (1966)
2. Nakanishi, N.: Graph Theory and Feynman Integrals. Gordon and Breach (1971)
3. Todorov, I.T.: Analytic Properties of Feynman Diagrams in Quantum Field Theory. Pergamon Press (1971)
4. Smirnov, V.A.: Evaluating Feynman integrals. Springer Tracts Mod. Phys. **211**, 1 (2004)
5. Smirnov, V.A.: Feynman Integral Calculus. Springer, Berlin (2006)
6. Smirnov, V.A.: Analytic tools for Feynman integrals. Springer Tracts Mod. Phys. **250**, 1 (2012)
7. Bogner, C., Borowka, S., Hahn, T., Heinrich, G., Jones, S.P., Kerner, M., et al.: Loopedia, a database for loop integrals. Comput. Phys. Commun. **225**, 1 (2018). arXiv:1709.01266
8. Bogner, C., Weinzierl, S.: Feynman graph polynomials. Int. J. Mod. Phys. A **25**, 2585 (2010). arXiv:1002.3458
9. Weinzierl, S.: Computer algebra in particle physics (2002). arXiv:hep-ph/0209234
10. Weinzierl, S.: Algebraic algorithms in perturbative calculations. In: Les Houches School of Physics: Frontiers in Number Theory, Physics and Geometry, pp. 737–757 (2007). arXiv:hep-th/0305260
11. Weinzierl, S.: Hopf algebra structures in particle physics. Eur. Phys. J. C **33**, S871 (2004). arXiv:hep-th/0310124
12. Weinzierl, S.: The art of computing loop integrals. Fields Inst. Commun. **50**, 345 (2006). arXiv:hep-ph/0604068
13. Weinzierl, S.: Introduction to Feynman integrals. In: 6th Summer School on Geometric and Topological Methods for Quantum Field Theory, pp. 144–187 (2013). arXiv:1005.1855
14. Weinzierl, S.: Feynman graphs. In: LHCPhenoNet School: Integration, Summation and Special Functions in Quantum Field Theory, pp. 381–406 (2013). arXiv:1301.6918
15. Weinzierl, S.: Hopf algebras and Dyson-Schwinger equations. Front. Phys. **11**, 111206 (2016). arXiv:1506.09119
16. Weinzierl, S.: Iterated integrals related to Feynman integrals associated to elliptic curves. In: Antidifferentiation and The Calculation of Feynman Amplitudes, vol. 12 (2020). arXiv:2012.08429
17. Vermaseren, J.A.M.: Axodraw. Comput. Phys. Commun. **83**, 45 (1994)
18. Tantau, T., Menke, H., Pohlmann, J., Purton, D., Hoftich, M., Tzanev, K., et al.: TikZ & PGF. https://github.com/pgf-tikz/pgf (2021)
19. Brun, R., Rademakers, F.: Root–an object oriented data analysis framework. Nucl. Inst. Meth. Phys. Res. **A389**, 81 (1997)
20. Wick, G.: Properties of Bethe-Salpeter wave functions. Phys. Rev. **96**, 1124 (1954)
21. 't Hooft, G., Veltman, M.J.G.: Regularization and renormalization of gauge fields. Nucl. Phys. **B44**, 189 (1972)

S. Weinzierl, *Feynman Integrals*, UNITEXT for Physics,
https://doi.org/10.1007/978-3-030-99558-4

22. Bollini, C.G., Giambiagi, J.J.: Dimensional renormalization: the number of dimensions as a regularizing parameter. Nuovo Cim. B **12**, 20 (1972)
23. Cicuta, G.M., Montaldi, E.: Analytic renormalization via continuous space dimension. Nuovo Cim. Lett. **4**, 329 (1972)
24. Wilson, K.G.: Quantum field theory models in less than four-dimensions. Phys. Rev. D **7**, 2911 (1973)
25. Collins, J.: Renormalization. Cambridge University Press (1984)
26. Weinzierl, S.: Equivariant dimensional regularization. arXiv:hep-ph/9903380
27. Kinoshita, T.: Mass singularities of Feynman amplitudes. J. Math. Phys. **3**, 650 (1962)
28. Lee, T.D., Nauenberg, M.: Degenerate systems and mass singularities. Phys. Rev. **133**, B1549 (1964)
29. Cheng, H., Wu, T.T.: Expanding Protons: Scattering at High Energies. MIT Press, Cambridge, Massachusetts (1987)
30. Lee, R.N., Pomeransky, A.A.: Critical points and number of master integrals. JHEP **11**, 165 (2013). arXiv:1308.6676
31. Baikov, P.A.: Explicit solutions of the multi-loop integral recurrence relations and its application. Nucl. Instrum. Meth. **A389**, 347 (1997). arXiv:hep-ph/9611449
32. Frellesvig, H., Papadopoulos, C.G.: Cuts of Feynman integrals in Baikov representation. JHEP **04**, 083 (2017). arXiv:1701.07356
33. Mellin, H.: Abriß einer einheitlichen Theorie der Gamma- und der hypergeometrischen Funktionen. Math. Ann. **68**, 305–337 (1910)
34. Barnes, E.: A transformation of generalised hypergeometric series. Quart. J. Math. **41**, 136–140 (1910)
35. Tutte, W.T.: Graph Theory. Encyclopedia of Mathematics and Its Applications, vol. 21. Addison-Wesley (1984)
36. Chaiken, S.: A combinatorial proof of the all minors matrix tree theorem. SIAM J. Alg. Disc. Meth. **3**, 319 (1982)
37. Chen, W.K.: Applied Graph Theory. North Holland, Graphs and Electrical Networks (1982)
38. Moon, J.: Some determinant expansions and the matrix-tree theorem. Discrete Math. **124**, 163 (1994)
39. Godsil, C., Royle, G.: Algebraic Graph Theory. Springer (2001)
40. Sokal, A.D.: The multivariate Tutte polynomial (alias Potts model) for graphs and matroids. In: Webb, B.S. (ed.) Surveys in Combinatorics. Cambridge University Press (2005). arXiv:math/0503607
41. Tutte, W.T.: A ring in graph theory. Proc. Cambridge Phil. Soc. **43**, 26 (1947)
42. Tutte, W.T.: A contribution to the theory of chromatic polynomials. Can. J. Math. **6**, 80 (1954)
43. Tutte, W.T.: On dichromatic polynomials. J. Combin. Theory **2**, 301 (1967)
44. Ellis-Monaghan, J., Merino, C.: Graph polynomials and their applications I: the Tutte polynomial. In: Dehmer, M. (ed.) Structural Analysis of Complex Networks. Birkhäuser (2010). arXiv:0803.3079
45. Ellis-Monaghan, J., Merino, C.: Graph polynomials and their applications II: interrelations and interpretations. In: Dehmer, M. (ed.) Structural Analysis of Complex Networks. Birkhäuser (2010). arXiv:0806.4699
46. Krajewski, T., Rivasseau, V., Tanasa, A., Wang, Z.: Topological graph polynomials and quantum field theory, Part I: Heat kernel theories. J. Noncommut. Geom. **4**, 29 (2010). arXiv:0811.0186
47. Dodgson, C.L.: Condensation of determinants. Proc. Roy. Soc. London **15**, 150 (1866)
48. Zeilberger, D.: Dodgson's determinant-evaluation rule proved by two-timing men and women. Electron. J. Combin. **4**, 2 (1997)
49. Stembridge, J.: Counting points on varieties over finite fields related to a conjecture of Kontsevich. Ann. Combin. **2**, 365 (1998)
50. Brown, F.: The massless higher-loop two-point function. Commun. Math. Phys. **287**, 925 (2008). arXiv:0804.1660
51. Brown, F.: On the periods of some Feynman integrals. arXiv:0910.0114

52. Brown, F., Yeats, K.: Spanning forest polynomials and the transcendental weight of Feynman graphs. Commun. Math. Phys. **301**, 357 (2011). arXiv:0910.5429
53. Kuratowski, C.: Sur le probleme de courbes gauches en topologie. Fund. Math. **15**, 271 (1930)
54. Wagner, K.: Über die Eigenschaft der ebenen Komplexe. Math. Ann. **114**, 570 (1937)
55. Diestel, R.: Graph Theory. Springer (2005)
56. Oxley, J.: Matroid Theory. Oxford University Press (2006)
57. Oxley, J.: What is a matroid? Cubo **5**, 179 (2003)
58. Whitney, H.: 2-isomorphic graphs. Am. J. Math. **55**, 245 (1933)
59. Oxley, J.: Graphs and series-parallel networks. In: White, N. (ed.) Theory of Matroids (1986)
60. Truemper, K.: On Whitney's 2-isomorphism theorem for graphs. J. Graph Theory **4**, 43 (1980)
61. Peskin, M.E., Schroeder, D.V.: An Introduction to Quantum Field Theory. Perseus Books (1995)
62. Böhm, M., Denner, A., Joos, H.: Gauge Theories of the Strong and Electroweak Interaction. Teubner, B.G. (2001)
63. Srednicki, M.: Quantum Field Theory. Cambridge University Press (2007)
64. Schwartz, M.D.: Quantum Field Theory and the Standard Model. Cambridge University Press (2014)
65. Weinzierl, S.: Tales of 1001 Gluons. Phys. Rep. **676**, 1 (2017). arXiv:1610.05318
66. Vermaseren, J.A.M.: New features of FORM. arXiv:math-ph/0010025
67. Bauer, C., Frink, A., Kreckel, R.: Introduction to the GiNaC framework for symbolic computation within the C++ programming language. J. Symb. Comput. **33**, 1 (2002). arXiv:cs.sc/0004015
68. Giele, W.T., Glover, E.W.N.: Higher order corrections to jet cross-sections in e+ e- annihilation. Phys. Rev. D **46**, 1980 (1992)
69. Giele, W.T., Glover, E.W.N., Kosower, D.A.: Higher order corrections to jet cross-sections in hadron colliders. Nucl. Phys. B **403**, 633 (1993). arXiv:hep-ph/9302225
70. Keller, S., Laenen, E.: Next-to-leading order cross sections for tagged reactions. Phys. Rev. D **59**, 114004 (1999). arXiv:hep-ph/9812415
71. Frixione, S., Kunszt, Z., Signer, A.: Three jet cross-sections to next-to-leading order. Nucl. Phys. B **467**, 399 (1996). arXiv:hep-ph/9512328
72. Catani, S., Seymour, M.H.: A general algorithm for calculating jet cross-sections in NLO QCD. Nucl. Phys. B **485**, 291 (1997). arXiv:hep-ph/9605323
73. Dittmaier, S.: A general approach to photon radiation off fermions. Nucl. Phys. B **565**, 69 (2000). arXiv:hep-ph/9904440
74. Phaf, L., Weinzierl, S.: Dipole formalism with heavy fermions. JHEP **04**, 006 (2001). arXiv:hep-ph/0102207
75. Catani, S., Dittmaier, S., Seymour, M.H., Trocsanyi, Z.: The dipole formalism for next-to-leading order QCD calculations with massive partons. Nucl. Phys. B **627**, 189 (2002). arXiv:hep-ph/0201036
76. Tarasov, O.V.: Connection between Feynman integrals having different values of the space-time dimension. Phys. Rev. D **54**, 6479 (1996). arXiv:hep-th/9606018
77. Tarasov, O.V.: Generalized recurrence relations for two-loop propagator integrals with arbitrary masses. Nucl. Phys. B **502**, 455 (1997). arXiv:hep-ph/9703319
78. Breitenlohner, P., Maison, D.: Dimensional renormalization and the action principle. Commun. Math. Phys. **52**, 11 (1977)
79. Siegel, W.: Supersymmetric dimensional regularization via dimensional reduction. Phys. Lett. B **84**, 193 (1979)
80. Siegel, W.: Inconsistency of supersymmetric dimensional regularization. Phys. Lett. B **94**, 37 (1980)
81. Gnendiger, C., et al.: To d, or not to d: recent developments and comparisons of regularization schemes. Eur. Phys. J. **C77**, 471 (2017). arXiv:1705.01827
82. Bern, Z., Kosower, D.A.: The computation of loop amplitudes in gauge theories. Nucl. Phys. B **379**, 451 (1992)

83. Bern, Z., De Freitas, A., Dixon, L., Wong, H.L.: Supersymmetric regularization, two-loop QCD amplitudes and coupling shifts. Phys. Rev. D **66**, 085002 (2002). arXiv:hep-ph/0202271
84. Kunszt, Z., Signer, A., Trocsanyi, Z.: One loop helicity amplitudes for all $2 \to 2$ processes in QCD and N = 1 supersymmetric Yang-Mills theory. Nucl. Phys. B **411**, 397 (1994). arXiv:hep-ph/9305239
85. Signer, A.: Helicity method for next-to-leading order corrections in QCD. Ph.D. Thesis, Diss. ETH Nr. 11143
86. Catani, S., Seymour, M.H., Trocsanyi, Z.: Regularization scheme independence and unitarity in QCD cross sections. Phys. Rev. D **55**, 6819 (1997). arXiv:hep-ph/9610553
87. Passarino, G., Veltman, M.J.G.: One-loop corrections for e+ e- annihilation into mu+ mu- in the Weinberg model. Nucl. Phys. B **160**, 151 (1979)
88. Melrose, D.B.: Reduction of Feynman diagrams. Nuovo Cim. **40**, 181 (1965)
89. van Neerven, W.L., Vermaseren, J.A.M.: Large loop integrals. Phys. Lett. B **137**, 241 (1984)
90. Bern, Z., Dixon, L.J., Kosower, D.A.: Dimensionally regulated pentagon integrals. Nucl. Phys. **B412**, 751 (1994). arXiv:hep-ph/9306240
91. Binoth, T., Guillet, J.P., Heinrich, G.: Reduction formalism for dimensionally regulated one-loop N-point integrals. Nucl. Phys. B **572**, 361 (2000). arXiv:hep-ph/9911342
92. Fleischer, J., Jegerlehner, F., Tarasov, O.V.: Algebraic reduction of one-loop Feynman graph amplitudes. Nucl. Phys. B **566**, 423 (2000). arXiv:hep-ph/9907327
93. Denner, A., Dittmaier, S.: Reduction of one-loop tensor 5-point integrals. Nucl. Phys. B **658**, 175 (2003). arXiv:hep-ph/0212259
94. Duplancic, G., Nizic, B.: Reduction method for dimensionally regulated one-loop N-point Feynman integrals. Eur. Phys. J. C **35**, 105 (2004). arXiv:hep-ph/0303184
95. Binoth, T., Guillet, J.P., Heinrich, G., Pilon, E., Schubert, C.: An algebraic/numerical formalism for one-loop multi-leg amplitudes. JHEP **10**, 015 (2005). arXiv:hep-ph/0504267
96. Giele, W.T., Glover, E.W.N.: A calculational formalism for one-loop integrals. JHEP **04**, 029 (2004). arXiv:hep-ph/0402152
97. ’t Hooft, G., Veltman, M.J.G.: Scalar one loop integrals. Nucl. Phys. **B153**, 365 (1979)
98. Pittau, R.: A simple method for multi-leg loop calculations. Comput. Phys. Commun. **104**, 23 (1997). arXiv:hep-ph/9607309
99. Pittau, R.: A simple method for multi-leg loop calculations. II: A general algorithm. Comput. Phys. Commun. **111**, 48 (1998). arXiv:hep-ph/9712418
100. Weinzierl, S.: Reduction of multi-leg loop integrals. Phys. Lett. B **450**, 234 (1999). arXiv:hep-ph/9811365
101. del Aguila, F., Pittau, R.: Recursive numerical calculus of one-loop tensor integrals. JHEP **07**, 017 (2004). arXiv:hep-ph/0404120
102. Pittau, R.: Formulae for a numerical computation of one-loop tensor integrals. In: International Conference on Linear Colliders (LCWS 04), 6, 2004. arXiv:hep-ph/0406105
103. van Hameren, A., Vollinga, J., Weinzierl, S.: Automated computation of one-loop integrals in massless theories. Eur. Phys. J. C **41**, 361 (2005). arXiv:hep-ph/0502165
104. Bern, Z., Dixon, L., Dunbar, D.C., Kosower, D.A.: One loop n point gauge theory amplitudes, unitarity and collinear limits. Nucl. Phys. B **425**, 217 (1994). arXiv:hep-ph/9403226
105. Bern, Z., Dixon, L., Dunbar, D.C., Kosower, D.A.: Fusing gauge theory tree amplitudes into loop amplitudes. Nucl. Phys. B **435**, 59 (1995). arXiv:hep-ph/9409265
106. Cutkosky, R.: Singularities and discontinuities of Feynman amplitudes. J. Math. Phys. **1**, 429 (1960)
107. Britto, R., Cachazo, F., Feng, B.: Generalized unitarity and one-loop amplitudes in N = 4 super-Yang-Mills. Nucl. Phys. B **725**, 275 (2005). arXiv:hep-th/0412103
108. Ossola, G., Papadopoulos, C.G., Pittau, R.: Reducing full one-loop amplitudes to scalar integrals at the integrand level. Nucl. Phys. B **763**, 147 (2007). arXiv:hep-ph/0609007
109. Ossola, G., Papadopoulos, C.G., Pittau, R.: On the rational terms of the one-loop amplitudes. JHEP **05**, 004 (2008). arXiv:0802.1876
110. Tkachov, F.V.: A theorem on analytical calculability of four loop renormalization group functions. Phys. Lett. B **100**, 65 (1981)

111. Chetyrkin, K.G., Tkachov, F.V.: Integration by parts: the algorithm to calculate beta functions in 4 loops. Nucl. Phys. B **192**, 159 (1981)
112. Laporta, S.: High-precision calculation of multi-loop Feynman integrals by difference equations. Int. J. Mod. Phys. A **15**, 5087 (2000). arXiv:hep-ph/0102033
113. Smirnov, A.: Algorithm FIRE–Feynman Integral REduction. JHEP **10**, 107 (2008). arXiv:0807.3243
114. Smirnov, A., Chuharev, F.: FIRE6: Feynman Integral REduction with Modular Arithmetic. arXiv:1901.07808
115. Studerus, C.: Reduze-Feynman integral reduction in C++. Comput. Phys. Commun. **181**, 1293 (2010). arXiv:0912.2546
116. von Manteuffel, A., Studerus, C.: Reduze 2–Distributed Feynman Integral Reduction. arXiv:1201.4330
117. Maierhöfer, P., Usovitsch, J., Uwer, P.: Kira—a Feynman integral reduction program. Comput. Phys. Commun. **230**, 99 (2018). arXiv:1705.05610
118. Klappert, J., Lange, F., Maierhöfer, P., Usovitsch, J.: Integral reduction with Kira 2.0 and finite field methods. Comput. Phys. Commun. **266**, 108024 (2021). arXiv:2008.06494
119. Kotikov, A.V.: Differential equations method: new technique for massive Feynman diagrams calculation. Phys. Lett. B **254**, 158 (1991)
120. Kotikov, A.V.: Differential equation method: the calculation of N point Feynman diagrams. Phys. Lett. B **267**, 123 (1991)
121. Remiddi, E.: Differential equations for Feynman graph amplitudes. Nuovo Cim. A **110**, 1435 (1997). arXiv:hep-th/9711188
122. Gehrmann, T., Remiddi, E.: Differential equations for two-loop four-point functions. Nucl. Phys. B **580**, 485 (2000). arXiv:hep-ph/9912329
123. Henn, J.M.: Multiloop integrals in dimensional regularization made simple. Phys. Rev. Lett. **110**, 251601 (2013). arXiv:1304.1806
124. Leray, J.: Le calcul différentiel et intégral sur une variété analytique complexe. (problème de cauchy. iii.) Bull. Soc. Math. France **87**, 81 (1959)
125. Griffiths, P., Harris, J.: Principles of Algebraic Geometry. Wiley, New York (1994)
126. Chen, K.-T.: Iterated path integrals. Bull. Amer. Math. Soc. **83**, 831 (1977)
127. Nakahara, M.: Geometry, Topology and Physics. Institute of Physics Publishing (2003)
128. Isham, C.J.: Modern Differential Geometry for Physicists. World Scientific (1999)
129. Primo, A., Tancredi, L.: On the maximal cut of Feynman integrals and the solution of their differential equations. Nucl. Phys. **B916**, 94 (2017). arXiv:1610.08397
130. Bosma, J., Sogaard, M., Zhang, Y.: Maximal cuts in arbitrary dimension. JHEP **08**, 051 (2017). arXiv:1704.04255
131. Harley, M., Moriello, F., Schabinger, R.M.: Baikov-Lee representations of cut Feynman integrals. JHEP **06**, 049 (2017). arXiv:1705.03478
132. Eden, R.J., Landshoff, P.V., Olive, D.I., Polkinghorne, J.C.: The Analytic S-Matrix. Cambridge University Press (1966)
133. Mizera, S., Telen, S.: Landau Discriminants. arXiv:2109.08036
134. Yoshida, M.: Hypergeometric Functions, My Love. Vieweg (1997)
135. Aomoto, K., Kita, M.: Theory of Hypergeometric Functions. Springer (2011)
136. Mastrolia, P., Mizera, S.: Feynman integrals and intersection theory. JHEP **02**, 139 (2019). arXiv:1810.03818
137. Frellesvig, H., Gasparotto, F., Mandal, M.K., Mastrolia, P., Mattiazzi, L., Mizera, S.: Vector space of Feynman integrals and multivariate intersection numbers. Phys. Rev. Lett. **123**, 201602 (2019). arXiv:1907.02000
138. Frellesvig, H., Gasparotto, F., Laporta, S., Mandal, M.K., Mastrolia, P., Mattiazzi, L., et al.: Decomposition of Feynman integrals by multivariate intersection numbers. JHEP **03**, 027 (2021). arXiv:2008.04823
139. Mizera, S.: Status of intersection theory and Feynman integrals. PoS **MA2019**, 016 (2019). arXiv:2002.10476

140. Cacciatori, S.L., Conti, M., Trevisan, S.: Co-homology of differential forms and Feynman diagrams. Universe **7** (2021). arXiv:2107.14721
141. Cho, K., Matsumoto, K.: Intersection theory for twisted cohomologies and twisted Riemann's period relations. I. Nagoya Math. J. **139**, 67 (1995)
142. Mizera, S.: Scattering amplitudes from intersection theory. Phys. Rev. Lett. **120**, 141602 (2018). arXiv:1711.00469
143. Aomoto, K.: On vanishing of cohomology attached to certain many valued meromorphic functions. J. Math. Soc. Jpn. **27**, 248 (1975)
144. Kita, M., Noumi, M.: On the structure of cohomology groups attached to the integral of certain many-valued analytic functions. Proc. Jpn. Acad. Ser. A Math. Sci. **58**, 97 (1982)
145. Cho, K.: A generalization of Kita and Noumi's vanishing theorems of cohomology groups of local system. Nagoya Math. J. **147**, 63 (1997)
146. Adams, W.W., Loustaunau, P.: An Introduction to Gröbner Bases. American Mathematical Society (1994)
147. Buchberger, B.: An algorithm for finding a basis for the residue class ring of a zero-dimensional polynomial ideal. Ph.D. Thesis (in German), University of Innsbruck, Math. Inst. (1965)
148. Milnor, J.: Morse Theory. Princeton University Press (1963)
149. Mizera, S.: Aspects of scattering amplitudes and moduli space localization. Ph.D. Thesis, Perimeter Inst. Theor. Phys. (2019). arXiv:1906.02099
150. Weinzierl, S.: On the computation of intersection numbers for twisted cocycles. J. Math. Phys. **62**, 072301 (2021). arXiv:2002.01930
151. Cattani, E., Dickenstein, A.: Introduction to residues and resultants. In: Bronstein, M. et al. (eds.) Solving Polynomial Equations, Algorithms and Computation in Mathematics, vol. 14. Springer (2005)
152. Søgaard, M., Zhang, Y.: Scattering equations and global duality of residues. Phys. Rev. **D93**, 105009 (2016). arXiv:1509.08897
153. Frellesvig, H., Gasparotto, F., Laporta, S., Mandal, M.K., Mastrolia, P., Mattiazzi, L., et al.: Decomposition of Feynman integrals on the maximal cut by intersection numbers. JHEP **05**, 153 (2019). arXiv:1901.11510
154. Mizera, S., Pokraka, A.: From infinity to four dimensions: higher residue pairings and Feynman integrals. JHEP **02**, 159 (2020). arXiv:1910.11852
155. Chen, J., Jiang, X., Xu, X., Yang, L.L.: Constructing canonical Feynman integrals with intersection theory. Phys. Lett. B **814**, 136085 (2021). arXiv:2008.03045
156. Caron-Huot, S., Pokraka, A.: Duals of Feynman integrals. Part I. Differential equations. JHEP **12**, 045 (2021). arXiv:2104.06898
157. Caron-Huot, S., Pokraka, A.: Duals of Feynman Integrals, vol. 2. Generalized Unitarity. arXiv:2112.00055
158. Gehrmann, T., von Manteuffel, A., Tancredi, L., Weihs, E.: The two-loop master integrals for $q\overline{q} \to VV$. JHEP **06**, 032 (2014). arXiv:1404.4853
159. Adams, L., Chaubey, E., Weinzierl, S.: Simplifying differential equations for multi-scale Feynman integrals beyond multiple polylogarithms. Phys. Rev. Lett. **118**, 141602 (2017). arXiv:1702.04279
160. Müller-Stach, S., Weinzierl, S., Zayadeh, R.: Picard-Fuchs equations for Feynman integrals. Commun. Math. Phys. **326**, 237 (2014). arXiv:1212.4389
161. Tancredi, L.: Integration by parts identities in integer numbers of dimensions. A criterion for decoupling systems of differential equations. Nucl. Phys. **B901**, 282 (2015). arXiv:1509.03330
162. Ince, E.L.: Ordinary Differential Equations. Dover Publications, New York (1944)
163. Magnus, W.: On the exponential solution of differential equations for a linear operator. Commun. Pure Appl. Math. **7**, 649 (1954)
164. Argeri, M., Di Vita, S., Mastrolia, P., Mirabella, E., Schlenk, J., Schubert, U., et al.: Magnus and Dyson series for master integrals. JHEP **03**, 082 (2014). arXiv:1401.2979
165. Moser, J.: The order of a singularity in Fuchs' theory. Mathematische Zeitschrift **1**, 379 (1959)

166. Lee, R.N.: Reducing differential equations for multiloop master integrals. JHEP **04**, 108 (2015). arXiv:1411.0911
167. Lee, R.N., Pomeransky, A.A.: Normalized Fuchsian form on Riemann sphere and differential equations for multiloop integrals. arXiv:1707.07856
168. Prausa, M.: Epsilon: a tool to find a canonical basis of master integrals. Comput. Phys. Commun. **219**, 361 (2017). arXiv:1701.00725
169. Gituliar, O., Magerya, V.: Fuchsia: a tool for reducing differential equations for Feynman master integrals to epsilon form. Comput. Phys. Commun. **219**, 329 (2017). arXiv:1701.04269
170. Lee, R.N.: Libra: a package for transformation of differential systems for multiloop integrals. Comput. Phys. Commun. **267**, 108058 (2021). arXiv:2012.00279
171. Barkatou, M.A., Pflügel, E.: Computing super-irreducible forms of systems of linear differential equations via Moser-reduction: a new approach. In: Proceedings of the 2007 International Symposium on Symbolic and Algebraic Computation, ISSAC '07, New York, NY, USA, pp. 1–8. Association for Computing Machinery (2007)
172. Barkatou, M.A., Pflügel, E.: On the Moser- and super-reduction algorithms of systems of linear differential equations and their complexity. J. Symb. Comput. **44**, 1017 (2009)
173. Leinartas, E.K.: Factorization of rational functions of several variables into partial fractions. Izv. Vyssh. Uchebn. Zaved. Mat. **22**, 47 (1978)
174. Meyer, C.: Transforming differential equations of multi-loop Feynman integrals into canonical form. JHEP **04**, 006 (2017). arXiv:1611.01087
175. Meyer, C.: Algorithmic transformation of multi-loop master integrals to a canonical basis with CANONICA. Comput. Phys. Commun. **222**, 295 (2018). arXiv:1705.06252
176. Heller, M., von Manteuffel, A.: MultivariateApart: generalized partial fractions. Comput. Phys. Commun. **271**, 108174 (2022). arXiv:2101.08283
177. Cachazo, F.: Sharpening The Leading Singularity. arXiv:0803.1988
178. Arkani-Hamed, N., Bourjaily, J.L., Cachazo, F., Trnka, J.: Local integrals for planar scattering amplitudes. JHEP **06**, 125 (2012). arXiv:1012.6032
179. Ifrah, G.: The Universal History of Numbers: From Prehistory to the Invention of the Computer. Wiley (1999)
180. Besier, M., Van Straten, D., Weinzierl, S.: Rationalizing roots: an algorithmic approach. Commun. Num. Theor. Phys. **13**, 253 (2019). arXiv:1809.10983
181. Besier, M., Wasser, P., Weinzierl, S.: RationalizeRoots: software package for the rationalization of square roots. Comput. Phys. Commun. **253**, 107197 (2020). arXiv:1910.13251
182. Besier, M., Festi, D.: Rationalizability of square roots. J. Symb. Comput. **106**, 48 (2021). arXiv:2006.07121
183. Kummer, E.: Über die transcendenten, welche aus wiederholten integrationen rationaler Formeln entstehen. J. Reine Angew. Math. **21**, 74 (1840)
184. Kummer, E.: Über die transcendenten, welche aus wiederholten integrationen rationaler Formeln entstehen. (Fortsetzung). J. Reine Angew. Math. **21**, 193 (1840)
185. Kummer, E.: Über die transcendenten, welche aus wiederholten integrationen rationaler Formeln entstehen. (Fortsetzung). J. Reine Angew. Math. **21**, 328 (1840)
186. Poincaré, H.: Sur les groupes des équations linéaires. Acta Math. **4**, 201–312 (1884)
187. Lappo-Danilevsky, J.A.: Mémoires sur la théorie des systémes des équations différentielles linéaires, vol. III. Chelsea Publishing Company, New York (1953)
188. Goncharov, A.B.: Multiple polylogarithms, cyclotomy and modular complexes. Math. Res. Lett. **5**, 497 (1998)
189. Goncharov, A.B.: Multiple polylogarithms and mixed Tate motives. arXiv:math.AG/0103059
190. Borwein, J.M., Bradley, D.M., Broadhurst, D.J., Lisonek, P.: Special values of multiple polylogarithms. Trans. Amer. Math. Soc. **353**(3), 907 (2001). arXiv:math.CA/9910045
191. Waldschmidt, M.: Multiple polylogarithms: an introduction. In: Agarwal, A.K., Berndt, B.C., Krattenthaler, C.F., Mullen, G.L., Ramachandra, K., Waldschmidt, M. (eds.) Conference on Number Theory and Discrete Mathematics in Honour of Srinivasa Ramanujan, Chandigarh, India. Hindustan Book Agency (2000). https://hal.archives-ouvertes.fr/hal-00416166

192. Nielsen, N.: Der Eulersche Dilogarithmus und seine Verallgemeinerungen. Nova Acta Leopoldina (Halle) **90**, 123 (1909)
193. Remiddi, E., Vermaseren, J.A.M.: Harmonic polylogarithms. Int. J. Mod. Phys. **A15**, 725 (2000). arXiv:hep-ph/9905237
194. Gehrmann, T., Remiddi, E.: Two-loop master integrals for gamma* –> 3jets: the planar topologies. Nucl. Phys. B **601**, 248 (2001). arXiv:hep-ph/0008287
195. Ablinger, J., Blümlein, J., Schneider, C.: Harmonic sums and polylogarithms generated by cyclotomic polynomials. J. Math. Phys. **52**, 102301 (2011). arXiv:1105.6063
196. Ecalle, J.: Ari/gari, la dimorphie et l'arithmétique des multizêtas: un premier bilan. J. Théorie Nombres Bordeaux **15**, 411 (2003)
197. Reutenauer, C.: Free Lie Algebras. Clarendon Press, Oxford (1993)
198. Hoffman, M.E.: Quasi-shuffle products. J. Algebraic Combin. **11**, 49 (2000). arXiv:math.QA/9907173
199. Guo, L., Keigher, W.: Baxter algebras and shuffle products. Adv. Math. **150**, 117 (2000). arXiv:math.RA/0407155
200. Minh, H.M., Petitot, M.: Lyndon words, polylogarithms and the riemann ζ function. Discrete Math. **217**, 273 (2000)
201. Blümlein, J., Broadhurst, D.J., Vermaseren, J.A.M.: The multiple Zeta value data mine. Comput. Phys. Commun. **181**, 582 (2010). arXiv:0907.2557
202. Hain, R.M.: Classical polylogarithms. Proc. Sympos. Pure Math. **55**, 3 (1994). arXiv:alg-geom/9202022
203. Knizhnik, V.G., Zamolodchikov, A.B.: Current algebra and Wess-Zumino model in two-dimensions. Nucl. Phys. B **247**, 83 (1984)
204. Drinfeld, V.G.: On quasitriangular quasi-Hopf algebras and on a group that is closely connected with Gal($\overline{\mathbb{Q}}/\mathbb{Q}$). Leningrad Math. J. **2**, 829 (1991)
205. Moch, S., Vermaseren, J.A.M.: Deep inelastic structure functions at two loops. Nucl. Phys. B **573**, 853 (2000). arXiv:hep-ph/9912355
206. Brown, F.: Mixed Tate motives over F. Ann. Math. **175**(2), 949 (2012). arXiv:1102.1312
207. The British Standards Institution. The C Standard, Wiley (2003)
208. Panzer, E.: Algorithms for the symbolic integration of hyperlogarithms with applications to Feynman integrals. Comput. Phys. Commun. **188**, 148 (2014). arXiv:1403.3385]
209. Hidding, M., Moriello, F.: All orders structure and efficient computation of linearly reducible elliptic Feynman integrals. JHEP **01**, 169 (2019). [arXiv: 1712.04441]
210. Bourjaily, J.L., McLeod, A.J., von Hippel, M., Wilhelm, M.: Rationalizing loop integration. JHEP **08**, 184 (2018). arXiv:1805.10281
211. Bourjaily, J.L., He, Y.-H., McLeod, A.J., Spradlin, M., Vergu, C., Volk, M., et al.: Direct integration for multi-leg amplitudes: tips, tricks, and when they fail. In: Antidifferentiation and the Calculation of Feynman Amplitudes, vol. 3 (2021). arXiv:2103.15423
212. Bierenbaum, I., Weinzierl, S.: The massless two-loop two-point function. Eur. Phys. J. C **32**, 67 (2003). arXiv:hep-ph/0308311
213. Moch, S., Uwer, P., Weinzierl, S.: Nested sums, expansion of transcendental functions and multi-scale multi-loop integrals. J. Math. Phys. **43**, 3363 (2002). arXiv:hep-ph/0110083
214. Vermaseren, J.A.M.: Harmonic sums, Mellin transforms and integrals. Int. J. Mod. Phys. A **14**, 2037 (1999). arXiv:hep-ph/9806280
215. Weinzierl, S.: Symbolic expansion of transcendental functions. Comput. Phys. Commun. **145**, 357 (2002). arXiv:math-ph/0201011
216. Moch, S., Uwer, P.: Xsummer: transcendental functions and symbolic summation in form. Comput. Phys. Commun. **174**, 759 (2006). arXiv:math-ph/0508008
217. Maitre, D.: HPL, a mathematica implementation of the harmonic polylogarithms. Comput. Phys. Commun. **174**, 222 (2006). arXiv:hep-ph/0507152
218. Huber, T., Maitre, D.: HypExp, a mathematica package for expanding hypergeometric functions around integer-valued parameters. Comput. Phys. Commun. **175**, 122 (2006). arXiv:hep-ph/0507094

219. Weinzierl, S.: Expansion around half-integer values, binomial sums and inverse binomial sums. J. Math. Phys. **45**, 2656 (2004). arXiv:hep-ph/0402131
220. Ziegler, G.: Lectures on Polytopes. Springer, New York (2007)
221. Loera, J.D., Rambau, J., Santos, F.: Triangulations. Springer, Berlin (2010)
222. Björk, J.-E.: Rings of Differential Operators. North-Holland, Amsterdam (1979)
223. Coutinho, S.: A Primer of Algebraic D-modules. Cambridge University Press (1995)
224. Saito, N.T.M., Sturmfels, B.: Gröbner Deformations of Hypergeometric Differential Equations. Springer, Berlin (2000)
225. Stienstra, J.: GKZ hypergeometric structures. In: Instanbul 2005: CIMPA Summer School on Arithmetic and Geometry Around Hypergeometric Functions, vol. 11 (2005). arXiv:math/0511351
226. Cattani, E.: Three lectures on hypergeometric functions (2006)
227. Gelfand, A.Z.I., Kapranov, M.: Discriminants, Resultants and Multidimensional Determinants. Birkhäuser, Boston (1994)
228. Gelfand, I.M., Zelevinskii, A.V., Kapranov, M.M.: Hypergeometric functions and toral manifolds. Funct. Anal. Appl. **23**, 94 (1989)
229. Gelfand, I.M., Zelevinskii, A.V., Kapranov, M.M.: A correction to the paper "Hypergeometric functions and toric varieties". Funct. Anal. Appl. **27**, 295 (1993)
230. Gelfand, I., Kapranov, M., Zelevinsky, A.: Generalized Euler integrals and A-hypergeometric functions. Adv. Math. **84**, 255 (1990)
231. Nasrollahpoursamami, E.: Periods of Feynman Diagrams and GKZ D-Modules. arXiv:1605.04970
232. Vanhove, P.: Feynman integrals, toric geometry and mirror symmetry. In: KMPB Conference: Elliptic Integrals, Elliptic Functions and Modular Forms in Quantum Field Theory, pp. 415–458 (2019). arXiv:1807.11466
233. de la Cruz, L.: Feynman integrals as A-hypergeometric functions. JHEP **12**, 123 (2019). arXiv:1907.00507
234. Klausen, R.P.: Hypergeometric series representations of Feynman integrals by GKZ hypergeometric systems. JHEP **04**, 121 (2020). arXiv:1910.08651
235. Bernshtein, I.N.: Modules over a ring of differential operators. Study of the fundamental solutions of equations with constant coefficients. Funct. Anal. Appl. **5**, 89–101 (1971)
236. Sato, M., Shintani, T.: On Zeta functions associated with prehomogeneous vector spaces. Proc. Natl. Acad. Sci. **69**, 1081 (1972)
237. Tkachov, F.V.: Algebraic algorithms for multiloop calculations. The first 15 years. What's next?. Nucl. Instrum. Meth. A **389**, 309 (1997). arXiv:hep-ph/9609429
238. Passarino, G.: An approach toward the numerical evaluation of multiloop Feynman diagrams. Nucl. Phys. B **619**, 257 (2001). arXiv:hep-ph/0108252
239. Binoth, T., Heinrich, G.: An automatized algorithm to compute infrared divergent multi-loop integrals. Nucl. Phys. B **585**, 741 (2000). arXiv:hep-ph/0004013
240. Bogner, C., Weinzierl, S.: Resolution of singularities for multi-loop integrals. Comput. Phys. Commun. **178**, 596 (2008). arXiv:0709.4092
241. Smirnov, A.V., Tentyukov, M.N.: Feynman integral evaluation by a sector decomposition approach (FIESTA). Comput. Phys. Commun. **180**, 735 (2009). arXiv:0807.4129
242. Smirnov, A.V., Smirnov, V.A., Tentyukov, M.: FIESTA 2: parallelizeable multiloop numerical calculations. Comput. Phys. Commun. **182**, 790 (2011). arXiv:0912.0158
243. Carter, J., Heinrich, G.: SecDec: a general program for sector decomposition. Comput. Phys. Commun. **182**, 1566 (2011). arXiv:1011.5493
244. Borowka, S., Carter, J., Heinrich, G.: Numerical evaluation of multi-loop integrals for arbitrary kinematics with SecDec 2.0. Comput. Phys. Commun. **184**, 396 (2013). arXiv:1204.4152
245. Borowka, S., Heinrich, G., Jones, S.P., Kerner, M., Schlenk, J., Zirke, T.: SecDec-3.0: numerical evaluation of multi-scale integrals beyond one loop. Comput. Phys. Commun. **196**, 470 (2015). arXiv:1502.06595
246. Borowka, S., Heinrich, G., Jahn, S., Jones, S.P., Kerner, M., Schlenk, J., et al.: pySecDec: a toolbox for the numerical evaluation of multi-scale integrals. Comput. Phys. Commun. **222**, 313 (2018). arXiv:1703.09692

247. Hironaka, H.: Resolution of singularities of an algebraic variety over a field of characteristic zero. Ann. Math. **79**, 109 (1964)
248. Spivakovsky, M.: A solution to Hironaka's polyhedra game. Progr. Math. **36**, 419 (1983)
249. Encinas, S., Hauser, H.: Strong resolution of singularities in characteristic zero. Comment. Math. Helv. **77**, 821 (2002)
250. Hauser, H.: The Hironaka theorem on resolution of singularities (or: a proof we always wanted to understand). Bull. Amer. Math. Soc. **40**, 323 (2003)
251. Zeillinger, D.: A short solution to Hironaka's polyhedra game. Enseign. Math. **52**, 143 (2006)
252. Smirnov, A.V., Smirnov, V.A.: Hepp and Speer sectors within modern strategies of sector decomposition. JHEP **05**, 004 (2009). arXiv:0812.4700
253. Kaneko, T., Ueda, T.: A geometric method of sector decomposition. Comput. Phys. Commun. **181**, 1352 (2010). arXiv:0908.2897
254. Kontsevich, M., Zagier, D.: Periods. In: Engquist, B., Schmid, W. (eds.) Mathematics Unlimited - 2001 and Beyond, 771 (2001)
255. Yoshinaga, M.: Periods and elementary real numbers. arXiv:0805.0349
256. Belkale, P., Brosnan, P.: Periods and Igusa local zeta functions. Int. Math. Res. Not. **49**, 2655 (2003)
257. Bogner, C., Weinzierl, S.: Periods and Feynman integrals. J. Math. Phys. **50**, 042302 (2009). arXiv:0711.4863
258. Grothendieck, A.: On the de Rham cohomology of algebraic varieties. Publ. Math. Inst. Hautes Études Sci. **29**, 95 (1966)
259. Ayoub, J.: Une version relative de la conjecture des périodes de Kontsevich-Zagier. Ann. Math. **181**, 905 (2011). arXiv:K-theory/1010
260. Sweedler, M.: Hopf Algebras. Benjamin, New York (1969)
261. Kassel, C.: Quantum Groups. Springer, New York (1995)
262. Majid, S.: Quasitriangular Hopf algebras and Yang-Baxter equations. Int. J. Mod. Phys. A **5**, 1 (1990)
263. Manchon, D.: Hopf algebras, from basics to applications to renormalization. In: 5th Mathematical Meeting of Glanon: Algebra, Geometry and Applications to Physics (2001). arXiv:math/0408405
264. Frabetti, A.: Renormalization Hopf algebras and combinatorial groups. In: Geometric and Topological Methods for Quantum Field Theory Proceedings of the 2007 Villa de Leyva Summer School, pp. 159–219. Cambridge University Press (2010). arXiv:0805.4385
265. Hopf, H.: Über die Topologie der Gruppen-Manigfaltigkeiten und ihren Verallgemeinerungen. Ann. Math. **42**, 22 (1941)
266. Woronowicz, S.L.: Compact matrix pseudogroups. Commun. Math. Phys. **111**, 613 (1987)
267. Kreimer, D.: On the Hopf algebra structure of perturbative quantum field theories. Adv. Theor. Math. Phys. **2**, 303 (1998). arXiv:q-alg/9707029
268. Connes, A., Kreimer, D.: Hopf algebras, renormalization and noncommutative geometry. Commun. Math. Phys. **199**, 203 (1998). arXiv:hep-th/9808042
269. Ehrenborg, R.: On posets and Hopf algebras. Adv. Math. **119**, 1 (1996)
270. Schupp, P.: Quantum groups, noncommutative differential geometry and applications. Ph.D. Thesis, UC, Berkeley (1993). arXiv:hep-th/9312075
271. Zimmermann, W.: Convergence of Bogoliubov's method of renormalization in momentum space. Commun. Math. Phys. **15**, 208 (1969)
272. Ebrahimi-Fard, K., Guo, L.: Rota-Baxter algebras in renormalization of perturbative quantum field theory. Fields Inst. Commun. **50**, 47 (2007). arXiv:hep-th/0604116
273. Bogoliubov, N.N., Parasiuk, O.S.: Über die Multiplikation der Kausalfunktionen in der Quantentheorie der Felder. Acta Math. **97**, 227 (1957)
274. Kreimer, D.: On overlapping divergences. Commun. Math. Phys. **204**, 669 (1999). arXiv:hep-th/9810022
275. Krajewski, T., Wulkenhaar, R.: On Kreimer's Hopf algebra structure of Feynman graphs. Eur. Phys. J. C **7**, 697 (1999). arXiv:hep-th/9805098

276. Connes, A., Kreimer, D.: Renormalization in quantum field theory and the Riemann-Hilbert problem. 1. The Hopf algebra structure of graphs and the main theorem. Commun. Math. Phys. **210**, 249 (2000). arXiv:hep-th/9912092
277. Connes, A., Kreimer, D.: Renormalization in quantum field theory and the Riemann-Hilbert problem. 2. The beta function, diffeomorphisms and the renormalization group. Commun. Math. Phys. **216**, 215 (2001). arXiv:hep-th/0003188
278. van Suijlekom, W.D.: Renormalization of gauge fields: a Hopf algebra approach. Commun. Math. Phys. **276**, 773 (2007). arXiv:hep-th/0610137
279. Ebrahimi-Fard, K., Patras, F.: Exponential renormalization. Ann. Henri Poincaré **11**, 943 (2010). arXiv:1003.1679
280. Ebrahimi-Fard, K., Patras, F.: Exponential renormalisation. II. Bogoliubov's R-operation and momentum subtraction schemes. J. Math. Phys. **53**, 083505 (2012). arXiv:1104.3415
281. Fauser, B.: On the Hopf algebraic origin of Wick normal-ordering. J. Phys. A **34**, 105 (2001). arXiv:hep-th/0007032
282. Brouder, C.: A quantum field algebra. arXiv:math-ph/0201033
283. Brown, F.: On the decomposition of motivic multiple zeta values. Adv. Stud. Pure Math. **63**, 31 (2012). arXiv:1102.1310
284. Brown, F.: Notes on motivic periods. Commun. Num. Theor. Phys. **11**, 557 (2017). arXiv:1512.06410
285. Abreu, S., Britto, R., Duhr, C., Gardi, E.: Cuts from residues: the one-loop case. JHEP **06**, 114 (2017). arXiv:1702.03163
286. Abreu, S., Britto, R., Duhr, C., Gardi, E.: Algebraic structure of cut Feynman integrals and the diagrammatic coaction. Phys. Rev. Lett. **119**, 051601 (2017). arXiv:1703.05064
287. Abreu, S., Britto, R., Duhr, C., Gardi, E., Matthew, J.: The diagrammatic coaction beyond one loop. JHEP **10**, 131 (2021). arXiv:2106.01280
288. Goncharov, A.B.: Galois symmetries of fundamental groupoids and noncommutative geometry. Duke Math. J. **128**, 209 (2005). arXiv:math.AG/0208144
289. Goncharov, A.B., Spradlin, M., Vergu, C., Volovich, A.: Classical polylogarithms for amplitudes and Wilson loops. Phys. Rev. Lett. **105**, 151605 (2010). arXiv:1006.5703
290. Spradlin, M., Volovich, A.: Symbols of one-loop integrals from mixed Tate motives. JHEP **11**, 084 (2011). arXiv:1105.2024
291. Duhr, C., Gangl, H., Rhodes, J.R.: From polygons and symbols to polylogarithmic functions. JHEP **10**, 075 (2012). arXiv:1110.0458
292. Duhr, C.: Hopf algebras, coproducts and symbols: an application to Higgs boson amplitudes. JHEP **08**, 043 (2012). arXiv:1203.0454
293. Duhr, C.: Mathematical aspects of scattering amplitudes. In: Theoretical Advanced Study Institute in Elementary Particle Physics: Journeys Through the Precision Frontier: Amplitudes for Colliders, pp. 419–476 (2015). arXiv:1411.7538
294. Duhr, C.: Scattering amplitudes, Feynman integrals and multiple polylogarithms. Contemp. Math. **648**, 109 (2015)
295. Brown, F.: Polylogarithmes multiples uniformes en une variable. C. R. Acad. Sci. Paris **338**, 527 (2004)
296. Brown, F.: Single-valued motivic periods and multiple zeta values. SIGMA **2**, e25 (2014). arXiv:1309.5309
297. Schnetz, O.: Graphical functions and single-valued multiple polylogarithms. Commun. Num. Theor. Phys. **08**, 589 (2014). arXiv:1302.6445
298. Charlton, S., Duhr, C., Gangl, H.: Clean single-valued polylogarithms. SIGMA **17**, 107 (2021). arXiv:2104.04344
299. Heller, M., von Manteuffel, A., Schabinger, R.M.: Multiple polylogarithms with algebraic arguments and the two-loop EW-QCD Drell-Yan master integrals. Phys. Rev. D **102**, 016025 (2020). arXiv:1907.00491
300. Heller, M.: Planar two-loop integrals for $\mu\, e$ scattering in QED with finite lepton masses. arXiv:2105.08046

301. Heller, M.: Radiative corrections to Compton processes on the proton and to the Drell-Yan process. Ph.D. Thesis, University of Mainz
302. Caron-Huot, S., Dixon, L.J., McLeod, A., von Hippel, M.: Bootstrapping a five-loop amplitude using Steinmann relations. Phys. Rev. Lett. **117**, 241601 (2016). arXiv:1609.00669
303. Dixon, L.J., Drummond, J., Harrington, T., McLeod, A.J., Papathanasiou, G., Spradlin, M.: Heptagons from the Steinmann cluster bootstrap. JHEP **02**, 137 (2017). arXiv:1612.08976
304. Caron-Huot, S., Dixon, L.J., Dulat, F., von Hippel, M., McLeod, A.J., Papathanasiou, G.: Six-Gluon amplitudes in planar $\mathcal{N} = 4$ super-Yang-Mills theory at six and seven loops. JHEP **08**, 016 (2019). arXiv:1903.10890
305. Caron-Huot, S., Dixon, L.J., Dulat, F., Von Hippel, M., McLeod, A.J., Papathanasiou, G.: The Cosmic Galois group and extended Steinmann relations for planar $\mathcal{N} = 4$ SYM amplitudes. JHEP **09**, 061 (2019). arXiv:1906.07116
306. Besier, M., Festi, D., Harrison, M., Naskręcki, B.: Arithmetic and geometry of a K3 surface emerging from virtual corrections to Drell–Yan scattering. Commun. Num. Theor. Phys. **14**, 863 (2020). arXiv:1908.01079
307. Frellesvig, H., Tommasini, D., Wever, C.: On the reduction of generalized polylogarithms to Li_n and $\text{Li}_{2,2}$ and on the evaluation thereof. JHEP **03**, 189 (2016). arXiv:1601.02649
308. Fomin, S., Zelevinsky, A.: Cluster algebras I: Foundations. J. Amer. Math. Soc. **15**, 497 (2002). arXiv:math/0104151
309. Fomin, S., Zelevinsky, A.: Cluster algebras II: Finite type classification. Invent. Math. **154**, 63 (2003). arXiv:math/0208229
310. Fomin, S., Williams, L., Zelevinsky, A.: Introduction to cluster algebras. Chapters 1–3. arXiv:1608.05735
311. Fomin, S., Williams, L., Zelevinsky, A.: Introduction to cluster algebras. Chapters 4–5. arXiv:1707.07190
312. Fomin, S., Williams, L., Zelevinsky, A.: Introduction to cluster algebras. Chapter 6. arXiv:2008.09189
313. Fomin, S., Williams, L., Zelevinsky, A.: Introduction to cluster algebras. Chapter 7. arXiv:2106.02160
314. Keller, B.: Cluster algebras, quiver representations and triangulated categories. In: Holm, T., Jørgensen, P., Rouquier, R. (eds.) Triangulated Categories. London Mathematical Society Lecture Note Series, pp. 76–160. Cambridge University Press (2010). arXiv:0807.1960
315. Golden, J., Goncharov, A.B., Spradlin, M., Vergu, C., Volovich, A.: Motivic amplitudes and cluster coordinates. JHEP **01**, 091 (2014). arXiv:1305.1617
316. Drummond, J.M., Papathanasiou, G., Spradlin, M.: A symbol of uniqueness: the cluster bootstrap for the 3-loop MHV heptagon. JHEP **03**, 072 (2015). arXiv:1412.3763
317. Parker, D., Scherlis, A., Spradlin, M., Volovich, A.: Hedgehog bases for A_n cluster polylogarithms and an application to six-point amplitudes. JHEP **11**, 136 (2015). arXiv:1507.01950
318. Harrington, T., Spradlin, M.: Cluster functions and scattering amplitudes for six and seven points. JHEP **07**, 016 (2017). arXiv:1512.07910
319. Mago, J., Schreiber, A., Spradlin, M., Volovich, A.: Symbol alphabets from plabic graphs. JHEP **10**, 128 (2020). arXiv:2007.00646
320. Mago, J., Schreiber, A., Spradlin, M., Srikant, A.Y., Volovich, A.: Symbol alphabets from plabic graphs II: rational letters. JHEP **04**, 056 (2021). arXiv:2012.15812
321. Mago, J., Schreiber, A., Spradlin, M., Yelleshpur Srikant, A., Volovich, A.: Symbol alphabets from plabic graphs III: n = 9. JHEP **09**, 002 (2021). arXiv:2106.01406
322. Chicherin, D., Henn, J.M., Papathanasiou, G.: Cluster algebras for Feynman integrals. Phys. Rev. Lett. **126**, 091603 (2021). arXiv:2012.12285
323. He, S., Li, Z., Yang, Q.: Notes on cluster algebras and some all-loop Feynman integrals. JHEP **06**, 119 (2021). arXiv:2103.02796
324. He, S., Li, Z., Yang, Q.: Kinematics, cluster algebras and Feynman integrals. arXiv:2112.11842
325. Fock, V., Goncharov, A.: Cluster ensembles, quantization and the dilogarithm. Ann. Sci. Ecole Normale Superieure **42**, 865 (2003). arXiv:math/0311245

326. Gehrmann, T., Remiddi, E.: Two-loop master integrals for gamma* –> 3jets: the non-planar topologies. Nucl. Phys. B **601**, 287 (2001). arXiv:hep-ph/0101124
327. Di Vita, S., Mastrolia, P., Schubert, U., Yundin, V.: Three-loop master integrals for ladder-box diagrams with one massive leg. JHEP **09**, 148 (2014). arXiv:1408.3107
328. Brown, F., Schnetz, O.: A K3 in phi4. Duke Math. J. **161**, 1817 (2012). arXiv:1006.4064
329. Bourjaily, J.L., McLeod, A.J., von Hippel, M., Wilhelm, M.: Bounded collection of Feynman integral Calabi-Yau geometries. Phys. Rev. Lett. **122**, 031601 (2019). arXiv:1810.07689
330. Bourjaily, J.L., McLeod, A.J., Vergu, C., Volk, M., Von Hippel, M., Wilhelm, M.: Embedding Feynman integral (Calabi-Yau) geometries in weighted projective space. JHEP **01**, 078 (2020). arXiv:1910.01534
331. Klemm, A., Nega, C., Safari, R.: The l-loop Banana amplitude from GKZ systems and relative Calabi-Yau periods. JHEP **04**, 088 (2020). arXiv:1912.06201
332. Bönisch, K., Fischbach, F., Klemm, A., Nega, C., Safari, R.: Analytic structure of all loop banana integrals. JHEP **05**, 066 (2021). arXiv:2008.10574
333. Du Val, P.: Elliptic Functions and Elliptic Curves. London Mathematical Society Lecture Note Series. Cambridge University Press (1973)
334. Silverman, J.: The Arithmetic of Elliptic Curves. Springer (1986)
335. Stein, W.A.: Modular Forms, a Computational Approach. American Mathematical Society (2007)
336. Miyake, T.: Modular Forms. Springer (1989)
337. Diamond, F., Shurman, J.: A First Course in Modular Forms. Springer (2005)
338. Cohen, H., Strömberg, F.: Modular Forms: A Classical Approach. American Mathematical Society (2017)
339. Garvan, F.: A q-product tutorial for a q-series MAPLE package. In: Foata, D., Han, G.-N.: The Andrews Festschrift, vol. 111 (2001). arXiv:math/9812092
340. OEIS Foundation: On-Line Encyclopedia of Integer Sequences. https://www.oeis.org
341. Broedel, J., Duhr, C., Dulat, F., Penante, B., Tancredi, L.: Elliptic symbol calculus: from elliptic polylogarithms to iterated integrals of Eisenstein series. JHEP **08**, 014 (2018). arXiv:1803.10256
342. Duhr, C., Tancredi, L.: Algorithms and tools for iterated Eisenstein integrals. JHEP **02**, 105 (2020). arXiv:1912.00077
343. Zagier, D.: Periods of modular forms and Jacobi theta functions. Invent. Math. **104**, 449 (1991)
344. Brown, F., Levin, A.: Multiple elliptic polylogarithms. arXiv:1110.6917
345. Broedel, J., Duhr, C., Dulat, F., Penante, B., Tancredi, L.: Elliptic Feynman integrals and pure functions. JHEP **01**, 023 (2019). arXiv:1809.10698
346. Broedel, J., Duhr, C., Dulat, F., Tancredi, L.: Elliptic polylogarithms and iterated integrals on elliptic curves. Part I: general formalism. JHEP **05**, 093 (2018). arXiv:1712.07089
347. Levin, A., Racinet, G.: Towards multiple elliptic polylogarithms. arXiv:math/0703237
348. Passarino, G.: Elliptic polylogarithms and basic hypergeometric functions. Eur. Phys. J. C **77**, 77 (2017). arXiv:1610.06207
349. Remiddi, E., Tancredi, L.: An elliptic generalization of multiple polylogarithms. Nucl. Phys. **B925**, 212 (2017). arXiv:1709.03622
350. Sabry, A.: Fourth order spectral functions for the electron propagator. Nucl. Phys. **33**, 401 (1962)
351. Broadhurst, D.J., Fleischer, J., Tarasov, O.: Two loop two point functions with masses: asymptotic expansions and Taylor series, in any dimension. Z. Phys. C **60**, 287 (1993). arXiv:hep-ph/9304303
352. Laporta, S., Remiddi, E.: Analytic treatment of the two loop equal mass sunrise graph. Nucl. Phys. B **704**, 349 (2005). arXiv:hep-ph/0406160
353. Bloch, S., Vanhove, P.: The elliptic dilogarithm for the sunset graph. J. Numb. Theor. **148**, 328 (2015). arXiv:1309.5865
354. Adams, L., Bogner, C., Weinzierl, S.: The iterated structure of the all-order result for the two-loop sunrise integral. J. Math. Phys. **57**, 032304 (2016). arXiv:1512.05630

355. Adams, L., Weinzierl, S.: Feynman integrals and iterated integrals of modular forms. Commun. Num. Theor. Phys. **12**, 193 (2018). arXiv:1704.08895
356. Broedel, J., Duhr, C., Dulat, F., Tancredi, L.: Elliptic polylogarithms and iterated integrals on elliptic curves II: an application to the sunrise integral. Phys. Rev. **D97**, 116009 (2018). arXiv:1712.07095
357. Adams, L., Weinzierl, S.: The ε-form of the differential equations for Feynman integrals in the elliptic case. Phys. Lett. **B781**, 270 (2018). arXiv:1802.05020
358. Hönemann, I., Tempest, K., Weinzierl, S.: Electron self-energy in QED at two loops revisited. Phys. Rev. **D98**, 113008 (2018). arXiv:1811.09308
359. Bogner, C., Schweitzer, A., Weinzierl, S.: Analytic continuation and numerical evaluation of the kite integral and the equal mass sunrise integral. Nucl. Phys. **B922**, 528 (2017). arXiv:1705.08952
360. Weinzierl, S.: Modular transformations of elliptic Feynman integrals. Nucl. Phys. B **964**, 115309 (2021). arXiv:2011.07311
361. Remiddi, E., Tancredi, L.: Differential equations and dispersion relations for Feynman amplitudes. The two-loop massive sunrise and the kite integral. Nucl. Phys. **B907**, 400 (2016). arXiv:1602.01481
362. Adams, L., Bogner, C., Schweitzer, A., Weinzierl, S.: The kite integral to all orders in terms of elliptic polylogarithms. J. Math. Phys. **57**, 122302 (2016). arXiv:1607.01571
363. Adams, L., Chaubey, E., Weinzierl, S.: Planar double box integral for top pair production with a closed top loop to all orders in the dimensional regularization parameter. Phys. Rev. Lett. **121**, 142001 (2018). arXiv:1804.11144
364. Adams, L., Chaubey, E., Weinzierl, S.: Analytic results for the planar double box integral relevant to top-pair production with a closed top loop. JHEP **10**, 206 (2018). arXiv:1806.04981
365. Bezuglov, M.A., Onishchenko, A.I., Veretin, O.L.: Massive kite diagrams with elliptics. Nucl. Phys. B **963**, 115302 (2021). arXiv:2011.13337
366. Berends, F.A., Buza, M., Böhm, M., Scharf, R.: Closed expressions for specific massive multiloop selfenergy integrals. Z. Phys. C **63**, 227 (1994)
367. Caffo, M., Czyz, H., Laporta, S., Remiddi, E.: The master differential equations for the 2-loop sunrise selfmass amplitudes. Nuovo Cim. A **111**, 365 (1998). arXiv:hep-th/9805118
368. Müller-Stach, S., Weinzierl, S., Zayadeh, R.: A second-order differential equation for the two-loop sunrise graph with arbitrary masses. Commun. Num. Theor. Phys. **6**, 203 (2012). arXiv:1112.4360
369. Adams, L., Bogner, C., Weinzierl, S.: The two-loop sunrise graph with arbitrary masses. J. Math. Phys. **54**, 052303 (2013). arXiv:1302.7004
370. Remiddi, E., Tancredi, L.: Schouten identities for Feynman graph amplitudes; the master integrals for the two-loop massive sunrise graph. Nucl. Phys. B **880**, 343 (2014). arXiv:1311.3342
371. Adams, L., Bogner, C., Weinzierl, S.: The two-loop sunrise graph in two space-time dimensions with arbitrary masses in terms of elliptic dilogarithms. J. Math. Phys. **55**, 102301 (2014). arXiv:1405.5640
372. Adams, L., Bogner, C., Weinzierl, S.: The two-loop sunrise integral around four space-time dimensions and generalisations of the Clausen and Glaisher functions towards the elliptic case. J. Math. Phys. **56**, 072303 (2015). arXiv:1504.03255
373. Bloch, S., Kerr, M., Vanhove, P.: Local mirror symmetry and the sunset Feynman integral. Adv. Theor. Math. Phys. **21**, 1373 (2017). arXiv:1601.08181
374. Bogner, C., Müller-Stach, S., Weinzierl, S.: The unequal mass sunrise integral expressed through iterated integrals on $\overline{\mathcal{M}}_{1,3}$. Nucl. Phys. B **954**, 114991 (2020). arXiv:1907.01251
375. Søgaard, M., Zhang, Y.: Elliptic functions and maximal unitarity. Phys. Rev. D **91**, 081701 (2015). arXiv:1412.5577
376. Bonciani, R., Del Duca, V., Frellesvig, H., Henn, J.M., Moriello, F., Smirnov, V.A.: Two-loop planar master integrals for Higgs $\rightarrow$ 3 partons with full heavy-quark mass dependence. JHEP **12**, 096 (2016). arXiv:1609.06685
377. von Manteuffel, A., Tancredi, L.: A non-planar two-loop three-point function beyond multiple polylogarithms. JHEP **06**, 127 (2017). arXiv:1701.05905

378. Ablinger, J., Blümlein, J., De Freitas, A., van Hoeij, M., Imamoglu, E., Raab, C.G., et al.: Iterated elliptic and hypergeometric integrals for Feynman diagrams. J. Math. Phys. **59**, 062305 (2018). arXiv:1706.01299
379. Bourjaily, J.L., McLeod, A.J., Spradlin, M., von Hippel, M., Wilhelm, M.: The elliptic double-box integral: massless amplitudes beyond polylogarithms. Phys. Rev. Lett. **120**, 121603 (2018). arXiv:1712.02785
380. Broedel, J., Duhr, C., Dulat, F., Penante, B., Tancredi, L.: Elliptic polylogarithms and Feynman parameter integrals. JHEP **05**, 120 (2019). arXiv:1902.09971
381. Abreu, S., Becchetti, M., Duhr, C., Marzucca, R.: Three-loop contributions to the ρ parameter and iterated integrals of modular forms. JHEP **02**, 050 (2020). arXiv:1912.02747
382. Kniehl, B.A., Kotikov, A.V., Onishchenko, A.I., Veretin, O.L.: Two-loop diagrams in non-relativistic QCD with elliptics. Nucl. Phys. **B948**, 114780 (2019). arXiv:1907.04638
383. Campert, L.G.J., Moriello, F., Kotikov, A.: Sunrise integrals with two internal masses and pseudo-threshold kinematics in terms of elliptic polylogarithms. JHEP **09**, 072 (2021). arXiv:2011.01904
384. Kristensson, A., Wilhelm, M., Zhang, C.: Elliptic double box and symbology beyond polylogarithms. Phys. Rev. Lett. **127**, 251603 (2021). arXiv:2106.14902
385. Deligne, P., Milne, J.S.: Tannakian categories. In: Hodge Cycles, Motives, and Shimura Varieties, pp. 101–228. Springer, Berlin, Heidelberg (1982)
386. MacLane, S.: Categories for the Working Mathematician. Springer-Verlag, New York (1971)
387. Leinster, T.: Basic Category Theory. Cambridge Studies in Advanced Mathematics. Cambridge University Press (2014). arXiv:1612.09375
388. Etingof, P., Gelaki, S., Nikshych, D., Ostrik, V.: Tensor Categories. American Mathematical Society (2015)
389. André, Y.: Une Introduction aux Motifs (Motifs Purs, Motifs Mixtes, Périodes). Société Mathématique de France (2004)
390. Voisin, C.: Théorie de Hodge et géométrie algébrique complexe. Société Mathématique de France (2002)
391. Peters, C., Steenbrink, J.: Mixed Hodge Structures. Springer, Berlin (2008)
392. Cattani, E., Zein, F.E., Griffiths, P.A., Tráng, L.D.: Hodge Theory. Princeton University Press (2014)
393. Brown, F.: Feynman amplitudes, coaction principle, and cosmic Galois group. Commun. Num. Theor. Phys. **11**, 453 (2017). arXiv:1512.06409
394. Rella, C.: An introduction to motivic Feynman integrals. SIGMA **17**, 032 (2021). arXiv:2009.00426
395. Huber, A., Müller-Stach, S.: Periods and Nori Motives. Springer, Berlin (2017)
396. Ballmann, W.: Lectures on Kähler Manifolds. European Mathematical Society (2006)
397. Deligne, P.: Théorie de Hodge I, p. 425. Actes du congrès international des mathématiciens, Nice (1970)
398. Deligne, P.: Théorie de Hodge II. Publ. Math. Inst. Hautes Études Sci. **40**, 5 (1971)
399. Deligne, P.: Théorie de Hodge III. Publ. Math. Inst. Hautes Études Sci. **44**, 5 (1974)
400. Griffiths, P.: Periods of integrals on algebraic manifolds I. Am. J. Math. **90**, 568 (1968)
401. Griffiths, P.: Periods of integrals on algebraic manifolds II. Am. J. Math. **90**, 808 (1968)
402. Bloch, S., Esnault, H., Kreimer, D.: On motives associated to graph polynomials. Commun. Math. Phys. **267**, 181 (2006). arXiv:math.AG/0510011
403. Doryn, D.: Cohomology of graph hypersurfaces associated to certain Feynman graphs. Commun. Num. Theor. Phys. **4**, 365 (2010)
404. Brown, F., Schnetz, O.: Single-valued multiple polylogarithms and a proof of the zig-zag conjecture. J. Number Theory **148**, 478 (2015). arXiv:1208.1890
405. Aluffi, P., Marcolli, M.: Feynman motives of banana graphs. Commun. Num. Theor. Phys. **3**, 1 (2009). arXiv:0807.1690
406. Bloch, S., Kreimer, D.: Mixed Hodge structures and renormalization in physics. Commun. Num. Theor. Phys. **2**, 637 (2008). arXiv:0804.4399

407. Connes, A., Marcolli, M.: Quantum fields and motives. J. Geom. Phys. **56**, 55 (2006). arXiv:hep-th/0504085
408. Aluffi, P., Marcolli, M.: Parametric Feynman integrals and determinant hypersurfaces. Adv. Theor. Math. Phys. **14**, 911 (2010). arXiv:0901.2107
409. Schnetz, O.: Quantum periods: a census of ϕ^4-transcendentals. Commun. Num. Theor. Phys. **4**, 1 (2010). arXiv:0801.2856
410. Griffiths, P.A.: On the periods of certain rational integrals: I. Ann. Math. **90**, 460 (1969)
411. Groote, S., Körner, J.G., Pivovarov, A.A.: On the evaluation of a certain class of Feynman diagrams in x-space: sunrise-type topologies at any loop order. Ann. Phys. **322**, 2374 (2007). arXiv:hep-ph/0506286
412. Groote, S., Körner, J., Pivovarov, A.: A numerical test of differential equations for one- and two-loop sunrise diagrams using configuration space techniques. Eur. Phys. J. C **72**, 2085 (2012). arXiv:1204.0694
413. Bloch, S., Kerr, M., Vanhove, P.: A Feynman integral via higher normal functions. Compos. Math. **151**, 2329 (2015). arXiv:1406.2664
414. Vanhove, P.: The physics and the mixed Hodge structure of Feynman integrals. Proc. Symp. Pure Math. **88**, 161 (2014). arXiv:1401.6438
415. Primo, A., Tancredi, L.: Maximal cuts and differential equations for Feynman integrals. An application to the three-loop massive banana graph. Nucl. Phys. **B921**, 316 (2017). arXiv:1704.05465
416. Broedel, J., Duhr, C., Dulat, F., Marzucca, R., Penante, B., Tancredi, L.: An analytic solution for the equal-mass banana graph. JHEP **09**, 112 (2019). arXiv:1907.03787
417. Bönisch, K., Duhr, C., Fischbach, F., Klemm, A., Nega, C.: Feynman Integrals in Dimensional Regularization and Extensions of Calabi-Yau Motives. arXiv:2108.05310
418. Broedel, J., Duhr, C., Matthes, N.: Meromorphic modular forms and the three-loop equal-mass banana integral. arXiv:2109.15251
419. Mark Gross, D.H., Joyce, D.: Calabi-Yau Manifolds and Related Geometries. Springer, Berlin (2003)
420. Dolgachev, I.V.: A brief introduction to Enriques surfaces. Adv. Stud. Pure Math. **69**, 1 (2016). arXiv:1412.7744
421. Vollinga, J., Weinzierl, S.: Numerical evaluation of multiple polylogarithms. Comput. Phys. Commun. **167**, 177 (2005). arXiv:hep-ph/0410259
422. Naterop, L., Signer, A., Ulrich, Y.: handyG —Rapid numerical evaluation of generalised polylogarithms in Fortran. Comput. Phys. Commun. **253**, 107165 (2020). arXiv:1909.01656
423. Wang, Y., Yang, L.L., Zhou, B.: FastGPL: a C++ library for fast evaluation of generalized polylogarithms. arXiv:2112.04122
424. Gehrmann, T., Remiddi, E.: Numerical evaluation of harmonic polylogarithms. Comput. Phys. Commun. **141**, 296 (2001). arXiv:hep-ph/0107173
425. Ablinger, J., Blümlein, J., Round, M., Schneider, C.: Numerical implementation of harmonic polylogarithms to weight w = 8. Comput. Phys. Commun. **240**, 189 (2019). arXiv:1809.07084
426. Walden, M., Weinzierl, S.: Numerical evaluation of iterated integrals related to elliptic Feynman integrals. Comput. Phys. Commun. **265**, 108020 (2021). arXiv:2010.05271
427. Ferguson, H.R.P., Forcade, R.W.: Generalization of the Euclidean algorithm for real numbers to all dimensions higher than two. Bull. Amer. Math. Soc. (N.S.) **1**, 912 (1979)
428. Ferguson, H.R.P., Bailey, D.H.: A polynomial time, numerically stable integer relation algorithm. RNR Technical Report RNR-91-032 (1992)
429. Ferguson, H., Bailey, D., Arno, S.: Analysis of PSLQ, an integer relation finding algorithm. Math. Comput. **68**, 351 (1999)
430. Bailey, D.H., Broadhurst, D.J.: Parallel integer relation detection: techniques and applications. Math. Comput. **70**, 1719 (2001). arXiv:math/9905048
431. Kleiss, R., Stirling, W.J.: Cross-sections for the production of an arbitrary number of photons in electron - positron annihilation. Phys. Lett. B **179**, 159 (1986)
432. Schwinn, C., Weinzierl, S.: Scalar diagrammatic rules for born amplitudes in QCD. JHEP **05**, 006 (2005). arXiv:hep-th/0503015

433. Rodrigo, G.: Multigluonic scattering amplitudes of heavy quarks. JHEP **09**, 079 (2005). arXiv:hep-ph/0508138
434. Denner, A., Nierste, U., Scharf, R.: A compact expression for the scalar one loop four point function. Nucl. Phys. B **367**, 637 (1991)
435. Bern, Z., Dixon, L.J., Kosower, D.A.: Dimensionally regulated one loop integrals. Phys. Lett. B **302**, 299 (1993). arXiv:hep-ph/9212308
436. Lu, H.J., Perez, C.A.: Massless one loop scalar three point integral and associated Clausen, Glaisher and L functions. SLAC-PUB-5809 (1992)
437. Ussyukina, N.I., Davydychev, A.I.: An approach to the evaluation of three and four point ladder diagrams. Phys. Lett. B **298**, 363 (1993)
438. Bern, Z., Dixon, L., Kosower, D.A., Weinzierl, S.: One-loop amplitudes for e+ e- –> anti-q q anti-q q. Nucl. Phys. B **489**, 3 (1997). arXiv:hep-ph/9610370
439. Duplancic, G., Nizic, B.: Dimensionally regulated one-loop box scalar integrals with massless internal lines. Eur. Phys. J. C **20**, 357 (2001). arXiv:hep-ph/0006249
440. Duplancic, G., Nizic, B.: IR finite one-loop box scalar integral with massless internal lines. Eur. Phys. J. C **24**, 385 (2002). arXiv:hep-ph/0201306
441. Ellis, R.K., Zanderighi, G.: Scalar one-loop integrals for QCD. JHEP **02**, 002 (2008). arXiv:0712.1851
442. Denner, A., Dittmaier, S.: Scalar one-loop 4-point integrals. Nucl. Phys. B **844**, 199 (2011). arXiv:1005.2076
443. Appell, P.: Sur les séries hypergéométriques de deux variables et sur dés équations différentielles linéaires aux dérivés partielles. C. R. Acad. Sci. Paris **90**, 296 (1880)
444. Appell, P., de Fériet, J.K.: Fonctions Hypergéométriques et Hypersphériques. Polynomes D'Hermite, Gauthiers-Villars, Paris (1926)
445. Lauricella, G.: Sulle funzioni ipergeometriche a piu variabili. Rend. Circ. Matem. **7**, 111–158 (1893)
446. Horn, J.: Hypergeometrische Funktionen zweier Veränderlichen. Math. Ann. **105**, 381–407 (1931)
447. Dickenstein, A., Matusevich, L., Sadykov, T.: Bivariate hypergeometric D-modules. Adv. Math. **196**, 78 (2005). arXiv:math/0310003
448. Bytev, V.V., Kalmykov, M.Y., Moch, S.-O.: HYPERgeometric functions DIfferential REduction (HYPERDIRE): MATHEMATICA based packages for differential reduction of generalized hypergeometric functions: F_D and F_S Horn-type hypergeometric functions of three variables. Comput. Phys. Commun. **185**, 3041 (2014). arXiv:1312.5777
449. Bytev, V., Kniehl, B., Moch, S.: Derivatives of Horn-type hypergeometric functions with respect to their parameters. arXiv:1712.07579
450. Kalmykov, M., Bytev, V., Kniehl, B.A., Moch, S.-O., Ward, B.F.L., Yost, S.A.: Hypergeometric functions and Feynman diagrams. In: Antidifferentiation and the Calculation of Feynman Amplitudes, vol. 12 (2020). arXiv:2012.14492
451. Helgason, S.: Differential Geometry, Lie Groups and Symmetric Spaces. American Mathematical Society (2001)
452. Fulton, W., Harris, J.: Representation Theory - A First Course. Springer, New York (2004)
453. Bourbaki, N.: Groupes et algèbres de Lie. Hermann (1972)
454. Weyl, H.: The Classical Groups. Princeton University Press (1946)
455. Deligne, P., Mumford, D.: The irreducibility of the space of curves of given genus. Publ. Math. Inst. Hautes Études Sci. **36**, 75 (1969)
456. Knudsen, F., Mumford, D.: The projectivity of the moduli space of stable curves I: preliminaries on "det" and "Div". Math. Scand. **39**, 19 (1976)
457. Knudsen, F.: The projectivity of the moduli space of stable curves II: The stacks $M_{g,n}$. Math. Scand. **52**, 161 (1983)
458. Knudsen, F.: The projectivity of the moduli space of stable curves III: The line bundles on $M_{g,n}$, and a proof of the projectivity of $\overline{M}_{g,n}$ in characteristic 0. Math. Scand. **52**, 200 (1983)
459. Brown, F.: Multiple zeta values and periods of moduli spaces $\overline{\mathcal{M}}_{0,n}$. C. R. Acad. Sci. Paris **342**, 949 (2006)

460. Stasheff, J.D.: Homotopy associativity of H-spaces I. Trans. Am. Math. Soc. **108**, 275 (1963)
461. Stasheff, J.D.: Homotopy associativity of H-spaces II. Trans. Am. Math. Soc. **108**, 293 (1963)
462. Devadoss, S.: Tessellations of Moduli spaces and the Mosaic operad. Contemp. Math. **239**, 91–114 (1999). arXiv:math/9807010
463. Devadoss, S.: Combinatorial equivalence of real moduli spaces. Notices AMS **51**, 620 (2004). arXiv:math-ph/0405011
464. Chen, W., Ruan, Y.: Orbifold Gromov-Witten theory. Contemp. Math. **310**, 25–85 (2002). arXiv:math/0103156
465. Pronk, D., Scull, L., Tommasini, M.: Atlases for Ineffective Orbifolds. arXiv:1606.04439
466. Henriques, A., Metzler, D.: Presentations of noneffective orbifolds. Trans. Amer. Math. Soc. **356**(6), 2481 (2004). arXiv:math/0302182
467. Hartshorne, R.: Algebraic Geometry. Springer, New York (1977)
468. Eisenbud, D., Harris, J.: The Geometry of Schemes. Springer, New York (2000)
469. Holme, A.: A Royal Road to Algebraic Geometry. Springer, Berlin (2012)
470. Eröcal, B., Motsak, O., Schreyer, F.-O., Steenpaß, A.: Refined algorithms to compute syzygies. J. Symb. Comput. **74**, 308 (2016). arXiv:1502.01654
471. Gluza, J., Kajda, K., Kosower, D.A.: Towards a basis for planar two-loop integrals. Phys. Rev. D **83**, 045012 (2011). arXiv:1009.0472
472. Schabinger, R.M.: A new algorithm for the generation of unitarity-compatible integration by parts relations. JHEP **01**, 077 (2012). arXiv:1111.4220
473. Ita, H.: Two-loop integrand decomposition into master integrals and surface terms. Phys. Rev. **D94**, 116015 (2016). arXiv:1510.05626
474. Larsen, K.J., Zhang, Y.: Integration-by-parts reductions from unitarity cuts and algebraic geometry. Phys. Rev. **D93**, 041701 (2016). arXiv:1511.01071
475. Zhang, Y.: Lecture Notes on Multi-loop Integral Reduction and Applied Algebraic Geometry. arXiv:1612.02249
476. Euclid, Elements, c. 300 B.C
477. Collins, G.E.: Subresultants and reduced polynomial remainder sequences. J. ACM **14**, 128 (1967)
478. Brown, W.S.: On Euclid's algorithm and the computation of polynomial greatest divisors. J. ACM **18**, 476 (1971)
479. Char, B.W., Geddes, K.O., Gonnet, G.H.: GCDHEU heuristic polynomial GCD algorithm based on integer GCD computation. J. Symb. Comp. **9**, 31 (1989)
480. Peraro, T.: Scattering amplitudes over finite fields and multivariate functional reconstruction. JHEP **12**, 030 (2016). arXiv:1608.01902
481. Peraro, T.: FiniteFlow: multivariate functional reconstruction using finite fields and dataflow graphs. JHEP **07**, 031 (2019). arXiv:1905.08019
482. Klappert, J., Lange, F.: Reconstructing rational functions with FireFly. Comput. Phys. Commun. **247**, 106951 (2020). arXiv:1904.00009
483. Wang, P.S.: A p-adic algorithm for univariate partial fractions. In: Proceedings of the Fourth ACM Symposium on Symbolic and Algebraic Computation, SYMSAC '81, New York, NY, USA, pp. 212–217. Association for Computing Machinery (1981)
484. Kaltofen, E., Villard, G. (eds.) Proceedings of the 2001 International Symposium on Symbolic and Algebraic Computation, ISSAC 2001, ORCCA & University of Western Ontario, London, Ontario, Canada, July 22–25, 2001. ACM (2001)
485. Kaltofen, E., Lee, W.: Early termination in sparse interpolation algorithms. J. Symb. Comput. **36**, 365 (2003)
486. Abramowitz, M., Stegun, I.A.: Handbook of Mathematical Functions with Formulas, Graphs, and Mathematical Tables, Dover, New York, ninth Dover printing, tenth GPO printing ed. (1964)
487. Zippel, R.: Probabilistic algorithms for sparse polynomials. In: Ng, E.W. (ed.) Symbolic and Algebraic Computation, pp. 216–226. Springer, Berlin, Heidelberg (1979)
488. Zippel, R.: Interpolating polynomials from their values. J. Symb. Comput. **9**, 375–403 (1990)

489. Cuyt, A., Lee, W.-S.: Symbolic-numeric sparse interpolation of multivariate rational functions. ACM Commun. Comput. Algebra **43**, 79–80 (2010)
490. Lee, R.N.: Presenting LiteRed: a tool for the Loop InTEgrals REDuction. arXiv:1212.2685
491. Lee, R.N.: LiteRed 1.4: a powerful tool for reduction of multiloop integrals. J. Phys. Conf. Ser. **523**, 012059 (2014). arXiv:1310.1145

Index

S. Weinzierl, *Feynman Integrals*, UNITEXT for Physics,
https://doi.org/10.1007/978-3-030-99558-4

D